A Sourcebook for the Biological Sciences

SECOND EDITION

TEACHING SCIENCE SERIES

A Book of Methods, Paul F. Brandwein, Fletcher G. Watson, and Paul E. Blackwood

A Sourcebook for Elementary Science, Elizabeth B. Hone, Alexander Joseph, and Edward Victor; under the General Editorship of Paul F. Brandwein

A Sourcebook for the Biological Sciences, Second Edition, Evelyn Morholt, Paul F. Brandwein, and Alexander Joseph

A Sourcebook for the Physical Sciences, Alexander Joseph, Paul F. Brandwein, Evelyn Morholt, Harvey Pollack, and Joseph F. Castka

A Sourcebook for the Biological Sciences

SECOND EDITION

Evelyn Morholt

Chairman, Department of Biological Sciences,
Fort Hamilton High School, Brooklyn

Paul F. Brandwein

Director, Pinchot Institute for Conservation Studies;
Adjunct Professor,
University of Pittsburgh;
Consultant to elementary and secondary schools

Alexander Joseph

Chairman, Division of Science and Mathematics,
College of Police Science,
The City University of New York;
Director, NSF In-Service Institute for Secondary
    School Teachers of Physics and Related Sciences,
The City College,
The City University of New York

Harcourt, Brace & World, Inc.

New York / Chicago / San Francisco / Atlanta

COVER PHOTO: *Hydrodictyon,* a fresh-water green alga (courtesy of General Biological Supply House, Inc., Chicago).

SECTION-OPENING PHOTOS: p. 1, vegetative and reproductive polyps in colony of *Obelia* (courtesy of General Biological Supply House, Inc., Chicago); p. 19, epidermis of *Sedum* (courtesy of General Biological Supply House, Inc., Chicago); p. 577, stained skeleton of *Rana pipiens,* embedded in Bio-Plastic (courtesy of Ward's Natural Science Establishment, Inc., Rochester, N.Y.); p. 709, biology laboratory (Harbrace photo).

ISBN: 0-15-582850-9

Library of Congress Catalog Card Number: 66-13536

Printed in the United States of America

Preface to the Second Edition

Since the first edition of this book was published, many new materials have been introduced into the teaching of biology, such as, for example, the excellent materials of the Biological Sciences Curriculum Study. To meet the changing demands of the field, we have revised *A Sourcebook for the Biological Sciences,* incorporating new materials and adding scope to established demonstrations and activities.

This edition is a substantial revision; little of the first edition has been omitted, and approximately 270 pages and 340 illustrations have been added, including photographs and drawings of many organisms rarely shown in high school biology textbooks. In particular, we have expanded the sections that develop concepts in photosynthesis and respiration, classic genetics and heredity in microorganisms, growth, embryology and differentiation, gross anatomy, biochemistry of cells, and ecological interrelationships.

As in the first edition, we have included in this book more techniques, procedures, demonstrations, projects, investigations, and suggestions for readings than any teacher can use in one year. The result, we hope, is a truly flexible source from which teachers may select techniques for creating a learning environment that is appropriate for their students and the specific teaching situation.

We express our sincere gratitude to Irving Reich, who read the galley proof of this edition and made many valuable suggestions.

EVELYN MORHOLT
PAUL F. BRANDWEIN
ALEXANDER JOSEPH

Preface to the First Edition:
Caution to the Reader

The teaching procedures—of all manner and description—offered in this book have been tested over more years than the authors care to remember in actual teaching of high school students, of different levels of ability, in general science, biology, health science, and field biology. The topics to which these procedures and techniques apply were selected on the basis of a study of fifty-eight courses in general science, biological science, and health given in typical communities (cities, towns, and county organizations) in the United States.

A good number of these procedures were also tested in a course titled "Laboratory Techniques in the Teaching of Biology" given as an in-service course in the New York City schools by Dr. Morris Rabinowitz, George I. Schwartz, and Paul Brandwein. A mimeographed text of the techniques developed in the course was also put together in 1940–42 under the title of the course.

Certainly there are in this book more techniques, procedures, demonstrations, projects, experiments, and suggestions than any teacher can use in any one year. These procedures are typical of the kinds of things that teachers do, or want to do, as they teach biological science—in the variety of forms this takes. Since the method of teaching is and must remain a personal invention, we leave to each teacher the selection of those techniques which are useful in a specific teaching situation.

We have included techniques and procedures not only for the "average" class, if there is such a thing, but for the variety of individual students who make up classes, no matter how they have been grouped. There are demonstrations involving visual effects, those requiring manipulation of various materials—from simple clay models to advanced histological techniques—those requiring reading, observing, thinking: all the many processes which make up science. There are suggestions for short- and long-range projects. It is our hope that there are here demonstrations and techniques for every student, no matter what his beginning interest or ability in science may be.

This volume is one of a series; following the Table of Contents are outlines of the three companion volumes, *A Book of Methods*, by P. F. Brandwein, F. G. Watson, and P. E. Blackwood, 1958; *A Sourcebook for Elementary Science*, by E. B. Hone, A. Joseph, and E. Victor, 1962; and *A Sourcebook for the Physical Sciences*, by A. Joseph, P. F. Brandwein, E. Morholt, H. Pollack, and J. F. Castka, 1961. The series is published by Harcourt, Brace & World.

We take this opportunity to express our appreciation to the many teachers from whom we have drawn inspiration; we particularly offer thanks to Herbert Drapkin and Leon Rintel for their careful and encouraging critiques of the book in galley proof.

We should like nothing better than to correspond with teachers who find certain of these techniques difficult to perform, or difficult to apply, or wanting in any manner. Certainly if a better technique is available, or if one we have described has been improved, we should consider it a privilege to include the contribution in a revised edition with appropriate credit. The opportunity we seek is that of being of service to teachers.

EVELYN MORHOLT
PAUL F. BRANDWEIN
ALEXANDER JOSEPH

Contents

Chapter **8**   **Continuity of the Organism**   341

Chapter **9**   **Growth and Differentiation**   439

A Book of Methods

by Paul F. Brandwein, Fletcher G. Watson, and Paul E. Blackwood

CONTENTS

A Sourcebook for Elementary Science

by Elizabeth B. Hone, Alexander Joseph, and Edward Victor
under the General Editorship of Paul F. Brandwein

CONTENTS

A Sourcebook for the Physical Sciences

by Alexander Joseph, Paul F. Brandwein, Evelyn Morholt, Harvey Pollack, and Joseph F. Castka

CONTENTS

Section *ONE*

Introduction

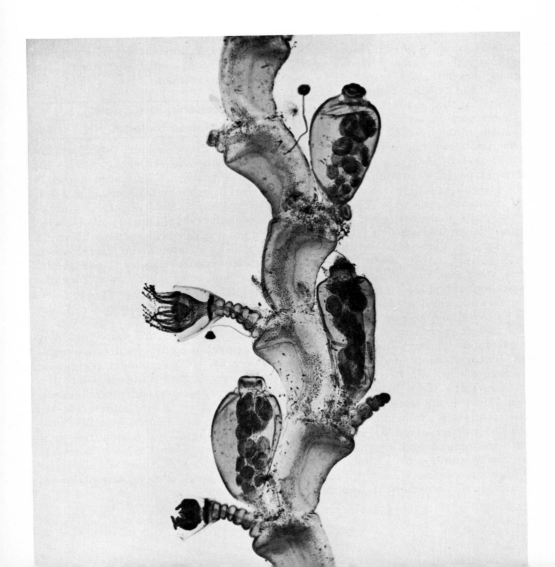

Patterns in the Use of
Laboratory Techniques and Procedures

The scientist explores the material universe, seeking orderly explanations of the objects and events he observes; he demands that these explanations be testable. Congruently, the science teacher creates situations in which students may explore the material universe, seeking orderly explanations of the objects and events they observe; the teacher demands that his students test their explanations.

In essence, the science teacher creates situations through which his students discover relationships among objects and events; that is, they uncover conceptual relationships. In uncovering concepts, students critically investigate arrays of objects and events and seek their common attributes; often these are what Bronowski terms "hidden likenesses." In finding hidden likenesses students uncover the unity in apparently unlike objects or events. Thus, yeast and man have their unity in respiration; rabbit and turtle share unity in embryonic development.

The techniques and procedures in this book include a variety of activities and investigations through which students may uncover conceptual relationships, and each teacher will have his own ways for using them. He will use them selectively: He will use some to create situations central to individual laboratory work; he will use others for group work, field trips, projects and investigations, his own demonstrations, or any of a number of other approaches. The central purpose of each activity, hopefully, is to encourage the student to investigate, to inquire, to furnish the energy for his own learning, to do, to act, to participate, to invent.

The purpose of this introductory chapter is to indicate how some teachers have used the procedures in this book to build their personal inventions—their own teaching techniques.

The sourcebook: in support of instruction in biology

Clearly, the technique, procedure, device—whatever one chooses to call it—fits within a pattern, a *curriculum pattern*. A teacher, or group of teachers, plans a unit of work. After completing their study of the unit, the students are to have mastered a large idea. To help students learn this large idea or concept, the teacher selects methods and materials from those available to him, adapting them to meet the total philosophy and practice—both administrative and pedagogical—of his school.

To illustrate how this sourcebook may be used within a curriculum of biological science, we have correlated portions of some currently used textbooks (which are, after all, curriculums) with procedures and techniques given in this sourcebook. We have selected five chapters from five "textbook-curriculums" to demonstrate how this book may be used *to support instruction. The techniques are never used in isolation, but are part and parcel of the teaching development within a unit.*

EXAMPLE 1   A unit of instruction in *High School Biology* (BSCS Green Version, Rand McNally, 1963): Chapter 3, "Communities and Ecosystems"

Textbook subjects	Sourcebook subjects (with page numbers)
The ecological approach	Adaptations, 528, 533–42; evolution, Chapter 10; field work, nature trails around school, 545, 547, 530, Table 11-2; indoor field trips, 548; laboratory block on *Chlorella*, 190–92, 542–44; reproduction and heredity, Chapter 8; school grounds, 520–24, 547
The biotic community	Classification of animals, 510–12, 550–52; classification of plants, 549–50; growth, Chapter 9; producers and consumers, Chapters 3, 4; use of materials, Chapter 6; web of life, Chapter 11
Community structure	
Kinds of interspecies relationships	Acidity of soils, 522; adaptations, 500–09, 520, 528–31; competition among paramecia, 449; competition between *Lemna minor* and *L. gibba*, 450; food chains, 532–44; modes of nutrition, 161; overproduction among living things, 498–500
Evaluating relationships	Antagonism between *Bacillus subtilis* and *Sarcina subflava*, 450; bacteria of decay, 521; consumers, Chapter 4; interrelationships in a fast-moving stream, 532; interrelationships in a pond, 528–31; nitrogen-fixing bacteria, 538; organisms in soil, 521; oxygen–carbon dioxide cycles, Chapters 3, 6; parasitism, 534; producers, Chapter 3; saprophytism, 536; soil bacteria, 521; symbiosis in lichens, 537; symbiosis in termites, 537; water cycle, 527
Succession	Adaptations, 533–42, 528; collecting animals, Chapters 11, 12; collecting plants, Chapters 11, 13; culturing animals, Chapter 12; culturing plants, Chapter 13; evolution, Chapter 10; reproduction and heredity, Chapter 8; succession in a microaquarium, 528–40, 529; succession in ponds, 530
The ecosystem	
Effects of organisms on the abiotic environment	Investigations, 190, 193, 493, 494; relations of organisms to their environment, 95, 222, 231, 234, 265, 268, 272, 277, 282, 284, 286, 451, 458, 460, 462, 494, 521–27, 542
Niches	Investigations, 190, 493; kinds of niches, 530, 535; readings, 193, 195, 263, 292
Continuity of the ecosystem	Chemical factors affecting the ecosystem, 520; growth in animals, 462; interrelations among living things, 528; laboratory block on *Chlorella*, 188; mineral nutrition in plants, 458–62; physical factors affecting ecosystems, 520

EXAMPLE 2   A unit of instruction in *Biological Science: An Inquiry into Life* (BSCS Yellow Version, Harcourt, Brace & World, 1963): Chapter 3, "Basic Structure"

Textbook subjects	Sourcebook subjects (with page numbers)
Early knowledge of plants and animals	Spontaneous generation (Redi, Spallanzani, Pasteur), 341–43
Scientific knowledge	Continuity of the organism, Chapter 8; evolution of living things and chemical evolution, 493–94; readings about cells, 97
The discovery of cells	
Early work with the microscope	Leeuwenhoek's letters, 97
Hooke's *Micrographia*	Bibliographies on history of cells, 152; general bibliography, Appendix A
The significance of Hooke's observations	Bibliography and readings, Chapter 2, Appendix A
Scientific societies and publications	Bibliography, Appendix A; Leeuwenhoek's letters, 97
The cell theory	
Accumulation of the evidence	Dissections, Chapter 1; microscopic examination of several kinds of cells and tissues, Chapter 2
The interpretation by Schwann and Schleiden	Bibliography (especially histories), Appendix A; microscopic examinations, Chapter 2
The origin of cells	Cells and compound microscope, Chapter 2; cells and electron microscope, 97; evolution of photosynthesizing systems, 159–62
Animal cells	Collecting and culturing, Chapter 12; dissections, Chapter 1; evolution, Chapter 10; growth, Chapter 9; microscopic examination of multicellular tissues, Chapter 2; microscopic examination of protists, Chapter 2; reproduction and heredity, Chapter 8; web of life, Chapter 11
Plant cells	Collecting and culturing, Chapter 13; dissections, Chapter 1; evolution, Chapter 10; growth, Chapter 9; microscopic examination of tissues and of protists, Chapter 2; physiology, Chapter 3; reproduction and heredity, Chapter 8; web of life, Chapter 11
Sizes of cells	Measuring, 91–94
Concluding remarks	Bibliography, Appendix A; chromatography, 137; contents of cells, 134–40; enzymes of cells, 143–48

EXAMPLE 3   A unit of instruction in *Biological Science: Molecules to Man* (BSCS Blue Version, Houghton Mifflin, 1963): Chapter 19, "Photosynthetic Systems"

Textbook subjects	Sourcebook subjects (with page numbers)
Evolution of photosynthetic systems	
The evolutionary pathway	Evolution of photosynthesizing systems, 159; plants as producers, Chapter 3; reading concerning chemical evolution, 494; reading concerning the first living things, 493
The final product of photosynthetic evolution: the multicellular green plant	Autotrophs and heterotrophs, 493; photosynthesis as process, Chapter 3; web of life, Chapter 11
Problems of the multicellular green plant	Adaptations, Chapters 2, 3, 10, 11; ecological relations, Chapter 11; growth, Chapter 9; mineral requirements, Chapter 4; reproduction and heredity, Chapter 8; transport of materials, Chapter 5; tropisms and coordination, Chapter 7
The structure of multicellular plants in relation to photosynthesis	
The entry of carbon dioxide into multicellular plants	Distribution of materials, Chapter 5; evidences of photosynthesis, Chapter 3; microscopic examination of leaves and stomates, Chapters 2, 3
The pathway of oxygen in multicellular plants	Cell respiration, Chapter 2; use of materials, Chapter 6
Photosynthesis occurs in the chloroplasts	Bibliography, Appendix A; chromatograms of pigments of chloroplasts, 137; composition of chloroplasts, 182; evidences of photosynthesis, Chapter 3; historic developments, 159; readings, 97; transport of materials, Chapter 5
Chlorophyll absorbs light energy	Absorption spectrum, 166; action spectrum, 165; effect of wavelength on rate of photosynthesis, 164; readings, 193; testing role of chlorophyll and light in photosynthesis, 182
The entry and transport of materials in the multicellular plant	Anatomy of stems, roots, leaves, Chapter 1; diffusion, 204; lifting power, 238; microscopic examination of plant tissues, Chapters 2, 3; movement of water in plants, 231; oxidation, Chapter 2; transpiration stream, 232; use of materials, Chapter 6
Factors which affect the rate of photosynthesis in multicellular plants	
Photosynthesis and the growth rate of the multicellular plant	Factors affecting growth, Chapter 9; growth as evidence of photosynthesis, Chapter 3; use of materials, Chapter 6
Carbon dioxide concentration	Absorption of carbon dioxide in photosynthesis, 166; investigations, 190; use of indicators, 169, 272

The effect of light intensity on photosynthesis	Role of light in photosynthesis, 162–66
The effect of temperature on photosynthesis	Photosynthesis as process, 155; preludes to inquiry—a laboratory block, 190
Mineral nutrition and its effect on photosynthesis	Acidity and alkalinity, 167; formation of chlorophyll, 182; mineral requirements of plants, 222
The effect of water concentration on photosynthesis	Movement of water in plants, Chapter 5

EXAMPLE 4 A unit of instruction in *The World of Living Things* (P. F. Brandwein et al., Harcourt, Brace & World, 1964): Chapter 4, "Nonvascular Plants"

Textbook subjects	Sourcebook subjects (with page numbers)
The smallest plants: bacteria	
How bacteria reproduce	Conditions required for growth, 449–51; culturing, 630–35; reproduction, 343; stained slides, 119
Useful bacteria	Bacteria in yoghurt, 104; competition between bacteria, 450; soil bacteria and nitrogen-fixing bacteria, Chapter 11; staining, 119, 124
Growing bacteria	Culturing, 630; growth of bacteria, 441–53; sterilizing media and plating of bacteria, Chapter 13; studies of growth of populations, 447
Simple green plants: algae	
Different kinds of algae	Culturing and isolating, Chapter 13; laboratory blocks on *Chlorella*, 190, 542; microscopic examination, 95–97, 99, 122; photosynthesis, 167; requirements for growth, 439–49
Plants without chlorophyll: fungi	
(Parasites and saprophytes)	Examination under the microscope, Chapter 2; modes of nutrition, 533; parasites, 534; requirements for growth, Chapter 13; saprophytes, 536
Lichens	Collecting and culturing, 642; microscopic examination, 95–99; symbiosis, 537
Flat-bodied plants: mosses and their relatives	
(Liverworts and mosses)	
Kinds of mosses	Collecting and culturing, Chapters 11, 13; reproduction, Chapter 8

EXAMPLE 5   A unit of instruction in *Design for Life* (R. F. Trump and D. L. Fagle, Holt, Rinehart and Winston, 1963): Chapter 13, "Coordination—Chemical Control"

Textbook subjects	Sourcebook subjects (with page numbers)
Two evolving systems	
(Hormones in animals)	Coordination through hormones, 323
(Hormones in plants)	Growth-promoting substances, 328; phytohormones, 326; other plant hormones, 330
Coordination in plants	
The discovery of auxins	Experiments of Boysen-Jensen, Darwin, and Went, 326; investigations using indoleacetic acid and other auxins on shoots, buds, stems, roots, cuttings, 326–30
How auxins work (phototropism and geotropism)	Light and the growth of shoots, 297; roots and growth, 327
Other auxin effects	α-naphthaleneacetic acid and β-indolebutyric acid, 328
Tracing the effects of auxins	Investigations, 327–30; polarity, Chapter 9
The gibberellins	Gibberellins, 330–32
Other chemical regulators	Florigens and kinetin, 330; readings, 330
Regulators in the interest of man	Growth retardants, Phosfon, Chapter 9; herbicides, 2,4-D, 328; phytochrome and photoperiodism, 333
The animal hormones	Dissections, Chapter 1; role of hormones, Chapter 7
Kinds of hormone functions	
Hormones and metabolism	Acetylcholine on frog's heart, 326; adrenalin and heartbeat of *Daphnia,* 245; adrenalin and heartbeat of a frog, 244; adrenalin and circulation in webbed foot of a frog, 247; thyroxin on tadpoles, 324; vasocontractor effect on arterioles, 248, 324
Interaction of hormones	Induction of ovulation in frogs, 360; organizer in frog's egg, 485; references to investigations, 324, 488
Insects and the juvenile hormone	References to readings and suggestions for investigations, Chapter 7
Hormones and color (chromatophores)	Melanophores of fish, 110
How hormones work	Effect of thyroxin (or iodine) on growth of tadpoles, 324–26; induction of ovulation in frogs by injection of pituitary glands, 360
Regulating the regulators	Photoperiodism, 332; readings and suggestions for investigation, Chapters 5, 7, 9

Some techniques and procedures within the daily lesson

The techniques and procedures given in this sourcebook may be used in a variety of patterns in the classroom, the laboratory, and the field. The pattern that each teacher uses will depend on the special circumstances and on his personal preference. Obviously, there is no single best way to create the learning situation.

To indicate the variety of situations that can be developed, we give below some procedures that other teachers have used. Throughout this sourcebook, we offer many other examples at several levels of sophistication. Teachers and students will undoubtedly devise many modifications. (Refer to the "Capsule Lessons" at the ends of Chapters 1 to 13.)

Planning a self-demonstration

Many teachers provide students with directions to guide them in a learning situation. For example, a teacher may give students directions on the following activities: typing blood, dissecting a frog, tracing the path of blood through a fresh calf's heart, using the microscope and preparing a slide, and measuring the acidity of a solution. An example of one such direction follows.

WHY MUST STARCH BE DIGESTED? You may discover the answer to this question in the following way. Add 1 g of cornstarch or arrowroot starch to 1 liter of water. Heat this gently to form a thin starch paste. Now test a sample of this paste for the presence of glucose by boiling a bit of it in Benedict's solution in a Pyrex test tube. If it turns green or reddish-orange, glucose is present. If no color change takes place, you may use the mixture.

Fill a test tube half full of starch paste. Get a goldbeater's membrane from your teacher, wet it, and fasten it tightly over the mouth of the test tube with a rubber band. Then invert the test tube in a beaker of water as in Fig. 0-1. The starch paste in the tube represents food in the intestine. Let the membrane over the mouth of the tube stand for the lining of the intestine.

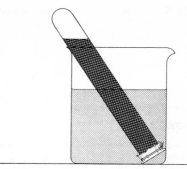

Fig. 0-1 Can starch pass through a membrane? Test tube of starch paste covered with a membrane inverted in a beaker of water.

The question is: Can starch pass through the membrane? Prepare two duplicates of Fig. 0-1 but substitute these materials for the starch paste: in one, dilute corn syrup, molasses, or glucose solution; in the other test tube, starch paste and some of your saliva. Cover each tube with a membrane as before; stand each in an individual beaker of water.

Within a half-hour you may test the water in the first of the three beakers with iodine. Did starch go through the membrane? In a Pyrex test tube heat to boiling, with Benedict's solution, a sample of water from the second beaker. Did simple sugar go through the membrane in the second tube? Now test the water in the third beaker for glucose (simple sugar). From where did the glucose come when the tube originally contained only starch paste and saliva? In the light of these results, explain why starch must be digested.

Using a specific exercise in planning controls

Students learn to use this method of scientists by "doing."

IS CARBON DIOXIDE USED BY GREEN PLANTS SUCH AS ELODEA? Do water plants use carbon dioxide? Since carbon dioxide is a colorless gas, we shall need a harmless indicator, such as brom thymol blue, to keep track of the amount of carbon dioxide in water. Fill a large beaker with brom thymol blue solution and blow into it through a straw. What happens? The carbon dioxide in your breath caused this change in color from blue to yellow. What color should result if the carbon dioxide is removed from the solution?

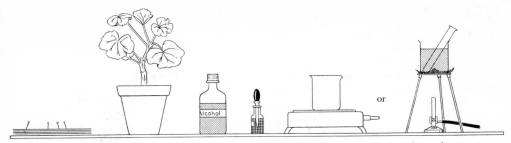

Fig. 0-2 Materials for demonstrating that green leaves need light to make starch: carbon paper or aluminum foil, pins, geranium, alcohol, dilute iodine, and beaker on electric hot plate or test tube in water bath over a burner.

Fill four test tubes with this brom thymol yellow (rich in carbon dioxide). Add sprigs of elodea plants to two of these tubes. Wrap one tube containing elodea in carbon paper; also wrap another tube without the plant in carbon paper. Set all the tubes in bright sunlight (or electric light). Within an hour examine the tubes for any change in color. In which tube (or tubes) has the brom thymol yellow changed to blue? Explain this change in color. Did the plant in the dark use carbon dioxide? Explain your answer. Compare your results with those of your classmates. Why should you make this comparison? In summary, do green plants use carbon dioxide in light? What two common tactics of scientists have you used here?

Would the results have been similar if green algae such as *Chlorella* had been used? Suppose an additional quantity of carbon dioxide were added to the water. What predictions could have been made if we had used our knowledge that carbon dioxide is involved in photosynthesis by green plants? What simple investigations could you have devised to test your predictions?

Other techniques are suggested in Chapters 3, 6, and 11. Students will also want to explore some of the "Preludes to Inquiry" (suggestions for investigations) described in Chapters 3 and 11.

Raising questions in planning an experiment

Sometimes a teacher may modify a technique described as a demonstration, and turn it to another purpose, making it a simple problem for students to work out independently or in small groups.

IS LIGHT NEEDED BY GREEN LEAVES TO MAKE STARCH? Could you plan an experiment to answer this question? Let us assume that you were given the materials in Fig. 0-2; write an explanation of your plan of action and show it to your teacher before you start work. Have you provided for a control? How could you find out whether leaves made starch? If leaves were covered, would they make starch? Compare your results with those of other students in class. In this way, you will be able to draw conclusions from many investigations, not just one. What is your conclusion?

Other possibilities for this sort of procedure come to mind. Do paramecia react to changes in their environment? Does thyroxin affect the growth of frog tadpoles? What kinds of chemicals affect the rate of heartbeat of *Daphnia*? What is the smallest part of a planarian that can regenerate a whole worm? Do variations in temperature affect the development of frogs' eggs? Of pond snails' eggs?

Beginning a simple investigation

This kind of activity may be used as a demonstration performed by the teacher or as an illustrative investigation conducted by students.

WHAT PIGMENTS ARE IN CHLOROPHYLL? Use the apparatus shown in Fig. 0-3 to determine what pigments are in chlorophyll. Use either frozen spinach leaves or freshly collected thin leaves (as described on pp. 183–86).

Students may also want to compare the amounts of different chlorophylls and carotenes in several plants: early leaves and fully grown leaves of maple, linden, or elm. Or they may wish to compare chromatograms of shoots of seedlings grown in solutions deficient in certain minerals with seedlings grown in enriched mineral solutions.

DO ALGAE ABSORB CARBON DIOXIDE IN LIGHT? Prepare test tubes of algae, such as *Chlorella*, grown in solutions to which indicators such as phenolphthalein and an alkali have been added (see pp. 166–68). If twice as much light is available, is carbon dioxide absorbed twice as fast?

DOES THE PRESENCE OF OXYGEN OFFSET THE PHOTOSENSITIVITY OF BLEPHARISMA? Read about the sensitivity to light of the pink protozoan *Blepharisma*. Also learn about the role of oxygen in protecting organisms against radiation. How might you plan your own investigation?

Making models

Students may use many kinds of models besides commercial models or charts. In this sourcebook we suggest the use of test tubes for villi, Florence flasks for alveoli, electric wire for chromosomes, and so forth.

WHAT IS THE APPEARANCE OF CHROMOSOMES AT DIFFERENT STAGES IN MITOSIS? Using Fig. 0-4 as a guide, *each* student can prepare this animated model of the separation of a chromosome and its replica along spindle fibers in mitosis. Use a length of insulated, double wire with a rubber band fastened by string. Assume that each wire "chromosome" has duplicated itself so that a pair of "chromatids" is represented.

From this step, the transition must be made to more abstract, more sophisticated levels. The chromosome is composed of many DNA molecules. How are these arranged? How do they become duplicated? What makes up a gene? Is the chromosome a physiological unit? Refer to ways to model DNA molecules (p. 428) and to preparation of squash slides of chromosomes (p. 116).

In essence, a model (the simpler the better) may reduce confusion among stu-

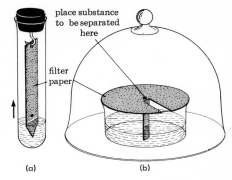

Fig. 0-3 Two techniques for paper chromatography: (a) strip filter paper method; (b) circular filter paper method.

dents as questions arise in the context of a lesson. Or the model may be used as a confrontation, challenging students' concepts of the way the world works. Further, the same model might well be used in the first grade or in graduate school. For example, a hard-boiled egg can be used as a model for a learning situation in the first grade: Children will know that a hen's egg will give rise to a chicken, not a pigeon. At the graduate level, the same hen's egg may serve to motivate questions of genetic continuity, and the "how" of chemical differentiation: How is the DNA of the individual chromosomes translated into functioning cells, tissues, and organs that have a biochemical individuality that makes one organism different from another organism?

Reporting back to the class

A student might make and show to the class a short film showing the progressive results over several months of an investigation he has made of seeds exposed to X-rays or treated with hormones or other chemicals. Or students might prepare and show 2 × 2 in. color slides of an examination of a quadrant of a forest community or of changes in kinds of vegetation in a drying pond or small lake. Or a filmloop might be useful; or overlays may be prepared to show patterns of growth of algae, eggs of frogs, or seeds exposed to

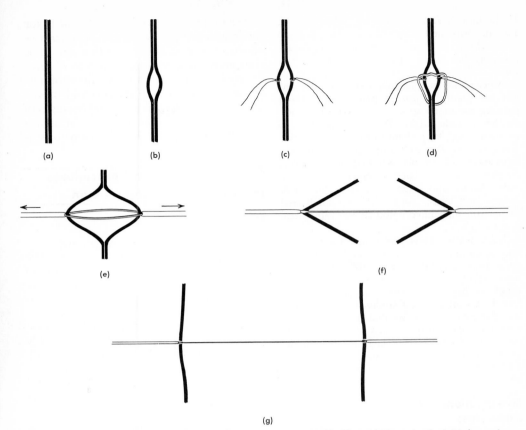

Fig. 0-4 Construction of wire model of chromosomes and spindle fibers during mitosis: (a) take a piece of insulated electrical cord; (b) separate the middle portion; (c) tie a piece of string to each piece of cord; (d) slip a rubber band around the cord, through the strings; (e) and (f) pull the strings apart slowly until (g) the wire splits in two, much as chromosomes are drawn out along spindle fibers.

different conditions. Collections of pictures may be shown by means of an opaque projector.

From some of the papers in *Scientific American* and from histories of science, students might easily prepare short talks describing historic experiments in our "breakthroughs" in animal and plant physiology, early studies in genetics, or tactics and strategies in unraveling photosynthesis.

Planning a field trip

Throughout this book, specific procedures are suggested for undertaking many out-of-doors studies in biology. In Chapter 11, "Web of Life: Ecological Patterns," many activities that take students out-of-doors are described. Chapters 12 and 13 offer methods for keeping organisms alive in the laboratory or classroom.

HOW TO COLLECT PLANTS. If a group of students, or your class, should plan a field trip, you'll need this equipment: collecting nets, jars with covers, pen knives, pads of paper and pencils for taking notes, comfortable shoes, and a first-aid kit. Plan where you intend to go so that you will know what to look for. Take with you some books as guides to identification of plants, for example, H. S. Conrad, *How to Know the Mosses and Liverworts,* ed. by H. E. Jaques (Wm. C. Brown, 1956); H. Durand, *Field Book of Common Ferns,* rev. ed. (Putnam's, 1949); and

H. E. Jaques, *How to Know the Trees* (Wm. C. Brown, 1946) and *Plant Families: How to Know Them* (Wm. C. Brown, 1948). Your library probably has many other references you can use.

Water plants may be collected with a dip net; put the plants in jars of pond water. Avoid packing too many specimens in one container, for they will die quickly. If you find mushrooms or other fungi, keep them in dry jars. You may wrap patches of mosses in newspaper. Transport ferns and small seed plants in folds of newspaper. As soon as you get back to school, transfer the water plants into large open jars. Put fungi, mosses, and ferns, along with any small seed plants you may have found in the same area, into terraria.

Twigs, leaves, seeds, and fruits of seed plants may be mounted on cardboard to show their life cycle. Be sure to label the plants clearly. If you have use of bulletin boards and hall cases, arrange to display the specimens you have collected. A committee of students might change the display regularly so that your classmates have the opportunity to learn a wide variety of plants. If you collect regularly throughout the year, you can display all the stages of some seed plants—winter buds, flowering stages, and seeds.

Investigations for the entire class

The following investigation may be conducted by groups outside of class time. You will find other investigations for groups suggested throughout this book.

WORKING WITH OTHER MEMBERS OF YOUR CLASS. Find the height and weight of each member of your class. Make two tables, one showing the number of students of each height (disregard fractions of an inch) and one showing the number of students in weight groups (76 to 80 lb, 81 to 85 lb, and so on).

Then prepare a bar graph to show the variations in height in your class. Put the units of height on the Y-axis, or vertical line, and the number of students on the X-axis, or horizontal line.

Now in your notebook prepare a graph similar to the one just described to show variations of weight in your class. Decide what units should be along the bottom line and what units along the side.

Refer especially to Chapters 9 and 11 for investigations that require the cooperative efforts of several students if they are to be completed in a reasonable length of time. Many investigations on the effects of hormones on plants and animals and on the embryonic development of early stages of frogs' eggs need large numbers of organisms that can be tended by teams of students.

Some techniques and procedures in extending the daily lesson

Homework

Sometimes students raise questions, or the teacher finds it desirable to extend the lesson into the home. Several kinds of procedures have been found useful. Usually students do not have the apparatus at home to carry on actual experimentation or self-demonstration. Such procedures for which apparatus is readily available—for example, collecting—will, of course, be used. However, there are some paper and pencil procedures and some simple techniques in maintaining cultures which stress various aspects of investigative procedures in science. You will find other types of procedures like these throughout the text.

You may want to have students prepare overlays for projection in class as they trace some trait in their "family tree," or give their own interpretation after reading accounts of Spemann's experiments on differentiation of tissue, or describe the effect of auxins in causing a positive phototropism in the stems of plants.

STUDYING A TRAIT IN YOUR FAMILY. Try to trace the occurrence of some trait in your family as a biology student diagramed a pedigree of a trait inherited in his family (Fig. 0-5). In this family sugar in the urine (but not in the blood) was frequent. Look at the large number of relatives included in this pedigree (the males are represented by squares, the females by circles). How many generations are shown? Which sex seems to be affected in this family? From this one family history, could you draw reliable conclusions as to the way the trait is inherited?

Select a definite trait such as tasting PTC paper, freckles, hair color, attached or free ear lobes, eye defects, long eyelashes, and so on. Develop a key as in Fig. 0-5 to show squares (males) and circles (females); shaded symbols represent those who have the trait, unshaded symbols those lacking the trait. (You may use a half-shaded symbol for a trait which seems to exist to a lesser degree than usual.)

GROWING INDIVIDUAL CULTURES OF PROTISTS. You can try to add a bit of mashed, hard-boiled egg yolk to pond water to develop a culture of protists. Cursory studies at home can be followed in a week's time by individual investigations with a microscope in the classroom. In class you will have the opportunity of exchanging with other students samples of your teeming cultures for extensive studies of protists.

NEW PLANTS WITHOUT SEEDS. You can prepare several examples of vegetative propagation at home. Try to grow a new plant from a sweet potato, a garlic bulblet, a slip of begonia, a tuber such as a white potato, a carrot, or runners from an aquarium plant or a snake plant. (Other examples are given on pp. 474–80.)

DOES SALIVA CHANGE STARCH? If you have your parents' permission, try this at home. Mix a bit of starch (arrowroot or cornstarch) in a cup of cold water; add to boiling water and stir for a few minutes. While the starch paste is cooling, prepare the rest of the materials for your experiment. Add a drop of iodine from the medicine chest to half a glass of water. Now with a teaspoon, transfer an amount of starch paste about the size of a drop to a flat dish. Add a drop of the diluted iodine. What causes this light blue color? Now add a large drop of your own saliva to the blue drop. Why does the blue color disappear?

Inventory-type projects

Many biology activities do not require special apparatus. How many kinds of insects can be found in one vacant lot? What are the names of the trees around the school? How many different birds have you seen in a week of observation? What is the best time to observe birds? Collect a 1-ft square of soil, and sift it into a basin or onto a white sheet of paper: How many kinds of organisms can you identify?

SEEDLINGS ON THE LAWN. Examine a number of trees in fruit, such as maples and elms. Estimate the number of seeds a tree produces. After a month or so return to this region and look for seedlings. How many do you find growing on the lawns? Why are there so few seedlings when there was an abundance of seeds? What factors tend to check the possibility of growth of all the seeds? What is survival of the fittest?

Investigations in the field

Other activities, such as a study of succession in a culture of pond water or succession of organisms in a temporary pool, or a field trip to a museum or fish hatchery, are examples of investigations that students might undertake in their study of biology. Also refer to Chapters 1, 11, 12, and 13.

EVIDENCES OF EROSION. On a trip around the school grounds or nearby look for evidences of bare slopes where soil has been washed away. Can you find a region cut through by erosion so that you see the roots of trees and vegetation?

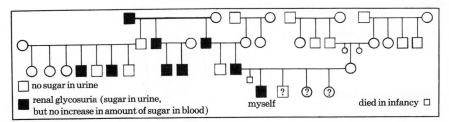

Fig. 0-5 Family pedigree of a student who suffered from renal glycosuria, excess of sugar in the urine. (Drawing by Marion A. Cox. From *Experiences in Biology*, by Evelyn Morholt and Ella Thea Smith, copyright, 1954, by Harcourt, Brace & World, Inc., and reproduced with permission of the publisher and the artist.)

How do roots help to slow down erosion of soil? Perhaps you can see the top layer of dark soil rich in organic matter which supports vegetation.

Look for evidence of layers of soil. What are the characteristics of the layers under the top-soil layer? What constructive measures might be put into practice to correct the kinds of erosion you find?

Some techniques and procedures for long-range investigations

Clearly we have not exhausted even the most common patterns involving the use of laboratory techniques and procedures as employed by teachers. But we cannot ignore patterns that involve long-range investigations. The examples below are included here to show the nature of an investigation which takes 2 to 3 years and which is "original" in the high school meaning of the term. Note that students began with a simple technique (which helped resolve a relatively simple problem), then evolved the problem into a full-scale investigation.

The serious student who already has a deep commitment to work in biology will find leads to original investigations in any one of the four volumes of *Research Problems in Biology: Investigations for Students,* prepared under the direction of the Biological Sciences Curriculum Study (BSCS) (Anchor Books, Doubleday, 1963, 1965).

The teacher will find some aid to identification of potential biologists, some practices currently in use over the country, and a bibliography on the nature of "giftedness" in P. F. Brandwein *et al., Teaching High School Biology: A Guide to Working with Potential Biologists* (BSCS Bull. 2, AIBS, 1962).

A long-range investigation: concerning planaria

There are hundreds of possibilities for long-range investigations by students. By way of example, we reproduce below one such investigation devised by a biologist, Sister M. C. Lockett. It is reprinted from

Research Problems in Biology: Investigations for Students, Series 2.

THE AMINO ACID CONTENT
AND REGENERATIVE PROPERTIES
OF PLANARIA DUGESIA DOROTOCEPHALA

Background

The phenomenon of regeneration is both perplexing and important to biologists. Regeneration, the restoration of lost or amputated parts, is a capacity manifested to a high degree among invertebrates and many cold-blooded vertebrates. It diminishes as we ascend the evolutionary scale, to wound-healing processes in the warm-blooded animals. Planarians, in particular, have been the subject of extensive research. Certain general facts ascertained from this research apply to all the lower animals. In the first place, any piece of such animals usually retains the same polarity it had while in the whole animal. That is, the regenerated head grows out of the cut piece that faced the anterior end in the whole animal, and the regenerated tail grows out of the cut end that faced the posterior end. Another generalization drawn from these experiments is that the capacity for generation is greatest near the anterior end and decreases toward the posterior end.

Problem

If the freshwater planarian, *Dugesia dorotocephala,* is cut into four pieces (see accompanying

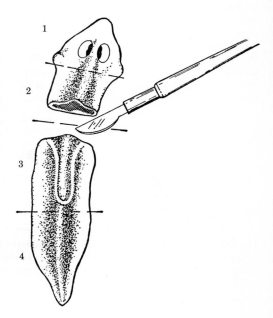

figure), Section 3 will produce a large number of abnormal regenerates, some with reduced heads and some completely headless. Sections 2 and 4 produce regenerates with normal heads. A primary purpose of this project is to investigate the possibility of a relationship between amino acid content and the regenerative properties of different sections. Since it has been observed that Section 3 produces abnormalities in head development, while Section 2 and Section 4 produce regenerates with normal heads—Section 2, however, regenerating faster than Section 4—it is our problem to determine:

1. Whether the abnormality in head development of Section 3 may be due to a deficiency of a particular amino acid in this section.

2. Whether there is a greater concentration of a particular amino acid in Section 2 which gives this section the capacity to regenerate faster than Section 4.

Suggested approach

1. Obtain planaria, *Dugesia dorotocephala,* and keep in enameled (glass) containers filled with dechlorinated tap water. Feed beef liver approximately twice a week; keep in temperature ranging from 16° to 20° C.

2. After three or four weeks of acclimatization to the laboratory, starve the worms for seven to ten days in order to maintain uniformity among the experimental animals.

3. Cut the worms into four sections as shown in previous figure and keep sections in dechlorinated water; observe the regeneration of each section—the time it takes and the number of normal head regenerates in each section. At the same time, section worms for the determination of the amino acid content.

4. Determine the amino acid content by paper chromatography.

(a) Hydrolysis

(1) Place a sample consisting of similar sections of ten planarias in 4 ml of Krebs–Ringer Buffer.

(2) Add 40 mg of pancreatin to each flask.

(3) Adjust pH to 8.5 with 0.1 N NaOH (about six drops) and cover with toluene.

(4) Hydrolyse at 37° C for approximately 24 hours.

(b) Extraction

(1) Acidify hydrolyzate to pH with 5 N HCl (approximately 4 drops). Do not remove toluene.

(2) Add 6 ml Butanol saturated with 0.1 N HCl, shake, centrifuge and decant with eyedropper after 15 minutes.

(3) Add 3 ml Butanol and shake. Centrifuge for 15 minutes and decant. (Add top layer to above decantate.)

(4) Add 3 ml Butanol and shake. Centrifuge for 15 minutes and decant.

(5) Centrifuge the three combined portions and decant with eyedropper to water layer.

(6) Take down to dryness under reduced pressure.

(7) Add 0.3 ml pure Butanol.

(8) Put extract on chromatographic paper.

(c) Preparing strips for separation of the amino acids. (The procedure listed below should be run on solutions of pure amino acids simultaneously with the experimental solutions.)

(1) Place 10 or 25 lambda of solution on chromatographic paper 8 cm from bottom or as suggested in Lederer (p. 129).

(2) Place two strips of each sample in each of the two solvents: Butanol-Dioxane-Ammonia (BDN) and Butanol-Acetic Acid-Water (BH). See Lederer (p. 308).

(3) Keep cabs in room of constant temperature.

(4) Remove strips from cabs when the solvent front is near the top (time depends upon the length of the strip and the height of the cab).

(5) Hang to dry. When dry, spray with ninhydrin.

(6) Determine R_f values of the experimental strips as well as those of the standard solutions of amino acids. *See* Lederer (p. 138) for determination of R_f value and page 139 for quantitative determination.

Preparation of Solutions

Butanol-Dioxane-Ammonia

    800 mls normal Butanol

    200 mls Dioxane (freshly distilled)

    1000 mls 2 N NH$_4$OH—132 mls concentrated to a liter. Shake in separatory funnel. Use upper layer—usually takes overnight for clear separation between layers.

Butanol–Acetic Acid–Water

    780 mls normal Butanol

    50 mls Glacial Acetic Acid

    170 mls distilled water

      Shake

      Will form one phase

Krebs–Ringer Buffer

    27.8   g NaCl

    16.89   g Na$_2$HPO$_4$

    3717.5   ml H$_2$O

    120.0   ml 1.15% KCl

    30.0   ml 3.82% MgSO$_4$ · 7H$_2$O

    0.986 ml conc HCl

Try using single amino acids on Section 3 to see whether the added nutrient affects the regenerative properties of these sections. Are there chemical factors contained in the other sections but lacking in Section 3? Prepare homogenates and extracts of each of the other sections and see how they affect the regeneration of Section 3.

How do physical agents such as ultraviolet light affect regenerative properties of planaria? How do chemical agents such as nitrogen mustard, colchicine, maleic acid hydrazide and other growth-stunting substances affect the regenerative properties of planaria?

References

GENERAL

1. Block, R. J., E. L. Durrum, and G. Zweig. 1958. A manual of paper chromatography and paper electrophoresis. 2d ed. Academic Press, New York.
2. Child, C. M. 1941. Patterns and problems of development. University of Chicago Press, Chicago.
3. Hyman, L. H. 1951. The invertebrates. Vol. 2. McGraw-Hill Book Co., New York.
4. Lederer, E., and M. Lederer. 1957. Chromatography. 2d ed. D. Van Nostrand Co., Princeton, N.J.
5. Needham, A. E. 1952. Regeneration and wound-healing. John Wiley & Sons, New York.
6. Simmonds, S., and J. S. Fruton. 1958. General biochemistry. 2d ed. John Wiley & Sons, New York.
7. Thompson, D. W. 1952. On growth and form. 2 vols. 2d ed. Cambridge University Press, New York.

SPECIFIC

8. Brøndsted, Agnes, and H. V. Brøndsted. 1953. The acceleration of regeneration in starved planarians by ribonucleic acid. J. Embryol. Exptl. Morphol. 1(1):49–54.
9. Flichinger, Reed A. 1959. A gradient of protein synthesis in planaria and reversal of axial polarity of regenerates. Growth 23(3):251–271.
10. Hammett, F. S. 1943. The role of the amino acids and nucleic acid components in developmental growth. Growth 7:331–399.
11. Henderson, R. F., and R. E. Eakin. 1959. Alteration of regeneration in planaria treated with lipoic acid. J. Exptl. Zool. 141(1):175–190.
12. Jenkins, M. M. 1958. The effects of thiourea and some related compounds on regeneration in planarians. Biol. Bull. 116(1):106–114.
13. Morrill, John B., Jr. 1958. A study of amino acids and proteins in two species of Tubularia and their relationship to regeneration. Dissertation Abstrs. 18(5):1906–1907. Florida State University, Tallahassee.
14. Navarra, J. G., and T. Gerne. 1962. Paper chromatography. Duquesne Sci. Counselor 25(3):66–69. (This article has very practical suggestions for separation of amino acids by paper chromatography.)
15. Pedersen, K. J. 1958. Morphogenetic activities during planarian regeneration as influenced by triethylene melanine. J. Embryol. Exptl. Morphol. 6(2):308–334.[1]

Other long-range investigations: Westinghouse projects

The titles of some of the winning papers in recent Westinghouse Science Talent Search examinations give some idea of the level of sophistication of the work of some seniors in our high schools.

"Amount of Light Required for Setting the 'Biological Clock' of Hamsters," John F. Emmel, 17, Dorsey H. S., Los Angeles, Calif.
"Conditions of Biosynthesis of Betacyanin Pigments in *Beta vulgaris*," Jonathan S. Fruchter, 17, Stephen H. Austin H. S., Austin, Tex.
"Effects of Glucose on Ice Formation in *Elodea canadensis*," Colin A. Connery, 17, Jamaica H. S., Jamaica, N.Y.
"Effects of Puffball Extracts on Bacteriophage," Jo L. Birkhead, 17, Harding H. S., Oklahoma City, Okla.
"Ethnic and Sex Variance in PTC Allelic Frequency Among Roosevelt High School Students," Bernice Wai Sheong Chang, 17, Roosevelt H.S., Honolulu, Haw.
"Function of Thyroid Hormones," Joseph W.

[1] Sister M. C. Lockett, "The Amino Acid Content and Regenerative Properties of Planaria *Dugesia dorotocephala*," from the book *Research Problems in Biology: Investigations for Students,* Series Two (1963). Copyright © 1963 by American Institute of Biological Sciences. Reprinted by permission of Doubleday & Company, Inc.

Bell Jr., 17, Robert E. Lee H.S., Springfield, Va.

"Genetic Influences and Other Studies on Erythropoietic Stimulating Factor in the Laboratory Mouse," Robert S. Zucker, 16, Central H.S., Philadelphia, Pa.

"An Original Key to Nineteen Genera of Western Atlantic Myctophid Fishes," Jane C. Johnson, 17, Coral Gables Senior H.S., Coral Gables, Fla.

"Qualitative Analysis of Activities of Bacteria and Fungi in the Soil," Katherine M. Gates, 17, Elyria Catholic H.S., Elyria, O.

"Tissue Culture of Functioning Single Neurons," Christopher G. Cherniak, 16, Melbourne H.S., Melbourne, Fla.

The following is an excerpt from the science report written by one of the winners in the Westinghouse Science Talent Search: Arthur Maurice Shapiro, 17, Central H.S., Philadelphia, Pa.

GENETICS AND VARIATION
OF COLIAS BUTTERFLIES

This report included: Environment-influenced Variation, Ground Color, Albinism, Semi-albinism, Ventral Median Flush, Submarginal Spots, Borders, Discocellular Spot of Hindwing, Discocellular Spot of Forewing, General Analysis and Summary.

Introductory remarks

The genus Colias has always presented a thorny problem to taxonomists and geneticists alike. In many cases it has been very difficult to define the limits of the species which are, as it were, in a fluid evolutionary state. Such has been the case with the representatives of the genus in eastern North America, which have been generally regarded as two species, "C. eurytheme Boisduval" and "C. philodice Latreille," both of which, however, intergrade completely. They are subject, moreover, to extensive individual variation, and to the regular production of white ("alba") females in addition to the normally colored (orange or yellow) ones. In experimental breeding conducted during the summers of 1961 and 1962, the writer obtained considerable evidence clarifying the status of the more important of these variations, from both an environmental and a genetic standpoint. The experiments and their results are herein fully described, with the objective of a re-evaluation of speciation in this particular section of Colias . . . .

Environment-influenced variation

. . . Collecting of wild Colias at a number of localities by the writer over a period of nearly seven years, which involved the examination of many thousands of specimens, has produced certain data which shed important light on the actual importance of environmental (climatic) influences in the variation of the (orange or yellow) ground color . . . .

The reduction of the orange ground-color to discal patches ("ariadne") in *eurytheme* subjects was most pronounced in those individuals which had been refrigerated immediately after pupation, and only slight reduction was noted in the other set. The yellow ground-color of *C. philodice* was, as expected, not affected.

The "control" group of butterflies from the same broods as the chilled subjects did not show any of the cold-influenced modifications described above . . . .

The incidence of "ariadne" is at all times higher than that of "ochracea," even when cold weather cannot be a factor. Furthermore, summer "ariadne" invariably lack the modifications of black pigment noted as resulting from exposure to cold. The experiments with chilling also resulted in no production of "ochracea." These phenomena have been explained by studies of "hybridization" in Colias, which are reported upon in the following section . . . .

Ground color and "hybridization"

. . . It has been suggested by various authors that hybridization accounts for at least some of the "intermediate" forms between *Colias philodice* and *eurytheme*. We will proceed for the remainder of this section on the assumption that the two are in fact separate species, although the primary reason for this designation is the difference in ground-color . . . .

Wild individuals of both species (but no white females) were selected for their "typical" qualities relative to either species and were hand-paired intraspecifically. There were two reasons for this: (1) to insure to the maximum that the stock used for hybridization was free of any hybrid ancestry and (2) to serve as controls against which the vigor of the possible hybrids might be measured. Selected individuals of the brood reared from these matings were hand-paired interspecifically. Twelve crosses were obtained, and five of these produced fertile ova. The fertility data of all broods are listed in Table I [not reprinted here]. Broods from each cross were reared separately under conditions described below. Matings of less than 45 minutes duration are invariably sterile, and are designated "insufficient duration" on the Tables. In addition, other pairings have failed to produce fertile ova, probably as a result of improper formation of a spermatophore . . . .

Secondary crosses

An attempt to hand-pair two of the hybrids in 1961 failed. In 1962, seven more broods of hybrids were raised (corroborating in all respects the results of the 1961 experiments, by the way) and from these four such matings were made, one of which was successful. The fertility rates again were depressed but not unduly low. The larvae were highly variable, as in the F_1 hybrids . . . .

The results of the various types of crosses suggest a certain set of genetic circumstances to be involved, with the specific characters of the two parental forms acting as regular Mendelian units. The "graduated blending" observed in the backcrosses, and particularly the segregarion observed in the F_2 hybrid, indicate (1) that "orange" ground-color is incompletely dominant over "yellow" and (2) that more than one pair of genes governs the ground-color, the effect being additive. For reasons to be discussed shortly, we will assume two pairs of genes to be involved, the "orange" genes 00′ being dominant over the "yellow" genes 00′. That these are autosomal is shown by the fact that no reciprocal differences were observed . . .

Summary

In summary, it may be stated that all of the apparent, superficial characters in the Colias population at Philadelphia, Pa., have been investigated from a genetic standpoint, and, to some extent, with reference to temperature-induced modifications. The genetics of variation in immature stages was not studied, but observations were noted casually in the case of "hybridization . . . ."[2]

A word in closing

Science is not "chalk-talk"; it is *experience in search of meaning*. In this chapter we have tried to indicate how certain techniques can be fitted into patterns that are characteristic of the science teacher's approach to teaching biology. As we have said, the entire sourcebook is given over to a description of activities and investigations that encompass a vast range of experience which serves this search for meaning.

[2] Arthur Maurice Shapiro, "Genetics and Variation of Colias Butterflies," from *How You Can Search for Science Talent,* Science Clubs of America, Washington, D.C., 1963. Reprinted with permission.

Section *TWO*

Techniques in the laboratory and classroom

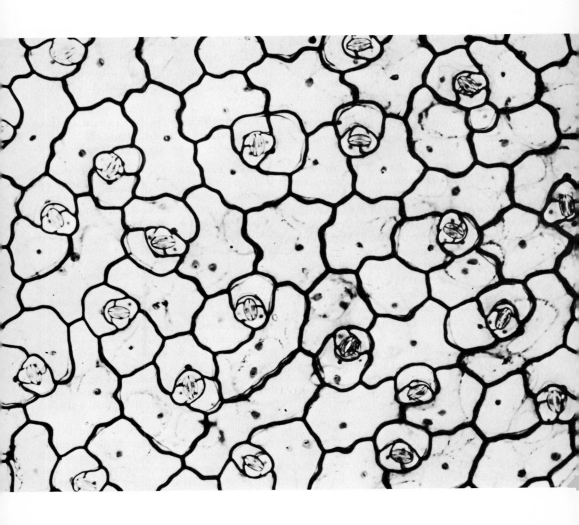

Chapter 1

Anatomy of
Representative Organisms

One of the functions, pleasures, and privileges of the biology laboratory is that it permits students to observe the elegant diversity of structure among organisms. Such observations lead to an understanding of the complementarity of structure and function as well as the beauty and economy in the patterning of different organisms. When students know one or more organisms well, the organism per se becomes a focus for developing many broad concepts and conceptual schemes related to growth, photosynthesis, energy transfers of various kinds, differentiation, heredity, or homeostasis.

The substance of Chapter 1 may serve as a guide for the dissection of some representative fair-sized animals, both invertebrate and vertebrate, as well as for a study of the anatomy of plants. Yet it must be remembered that the descriptions are only guides to an introductory study; there are a number of references, particularly laboratory manuals, that give detailed guides to dissection of specific plants and animals. (Representative laboratory manuals are listed in the bibliography at the end of this chapter.)

Nevertheless, a major function of dissection is to allow students to see for themselves.

Chapter 2, on the other hand, is devoted to a description of the microscopic anatomy of animals and plants—both unicellular microorganisms, or protists, and tissue cells of multicellular organisms. Several of the smaller members of the Crustacea, such as *Daphnia,* the fairy shrimp, *Cyclops,* or some of the transparent worms, need to be studied under the microscope. The preparation of these organisms for study is described in Chapter 2.

Before students begin individual investigation of any organism, it would be well to familiarize them with the following principles adopted by the Committee on Animal Care of the National Academy of Sciences.

GUIDING PRINCIPLES IN THE USE OF ANIMALS
BY SECONDARY SCHOOL STUDENTS
AND SCIENCE CLUB MEMBERS

1. The basic aim of scientific studies that involve animals is to achieve an understanding of life, and to advance our knowledge of the processes of life. Such studies lead to a respect for life.

2. A qualified adult supervisor must assume primary responsibility for the purposes and conditions of any experiment that involves living animals.

3. No experiment should be undertaken that involves anesthetic drugs, surgical procedures, pathogenic organisms, toxicological products, carcinogens, or radiation unless a biologist, physician, dentist, or veterinarian trained in the experimental procedure involved assumes direct responsibility for the proper conduct of the experiment.

4. Any experiment must be performed with the animal under appropriate anesthesia if the pain involved is greater than that attending anesthetization.

5. The comfort of the animal used in any study shall be a prime concern of the student investigator. Gentle handling, proper feeding, and provision of appropriate sanitary quarters shall be strictly observed at all times. Any experiment

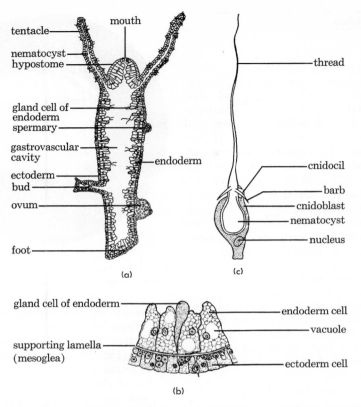

Fig. 1-1 *Hydra*: (a) sagittal section; (b) detail of cross section of the body wall; (c) a nematocyst. (Courtesy of General Biological Supply House, Inc., Chicago.)

in nutritional deficiency may proceed only to the point where symptoms of the deficiency appear. Appropriate measures shall then be taken to correct the deficiency, or the animal killed by humane methods.

6. All animals used must be lawfully acquired in accordance with state and local laws.

7. Experimental animals should not be carried over school vacation periods unless adequate housing is provided and a qualified caretaker is assigned specific duties of care and feeding.

Adopted by the Committee on Animal Care July 12, 1958[1]

In the following descriptions of plants and animals, we begin with macroscopic organisms, such as representatives of the

[1] Reproduced by permission of Committee on Animal Care, National Academy of Sciences–National Research Council, Washington, D.C.

phylum Coelenterata. We have omitted a description of the anatomy of sponges (Porifera), since these are not generally studied in high school biology classes. Also, algae and protozoa are presented later, in Chapter 2, since the preparation of material for study under the microscope is described there. Methods of cultivating algae and protozoa, as well as many of the other forms under discussion, are offered in Chapters 12 and 13.

Where possible, both fresh-water and marine forms have been presented as the representative or "classic" forms that demonstrate the certain special adaptations of organisms as they evolved from invertebrates to vertebrates. The phyla of plants show the development of special adaptations for life on land—specialized means of reproduction and of support.

Coelenterates

Hydra

This fresh-water coelenterate, which looks like a white thread about ¼ in. long, is representative of the hydroid body with many specialized cells. Living specimens may be found attached to submerged leaves in ponds or may be purchased from supply houses. Laboratory uses and culture methods are given in Chapter 12.

Examine an undisturbed specimen in a small dish with a hand lens or under a binocular microscope (Fig. 1-1). Note the sac-like hydroid body composed of two layers of cells, ectoderm and endoderm, separated by a mesoglea. Observe the active tentacles surrounding the hypostome, the only opening of the gastrovascular cavity. Hydras may be gray (for example, *Hydra vulgaris*); brown (for example, *H. fusca*); or, if they possess symbiotic green algae (*Chlorella*, sometimes called zoochlorellae), green (for example, *H. viridis*). Try to find a specimen in locomotion (Fig. 1-2).

Add a few small *Tubifex* worms or some

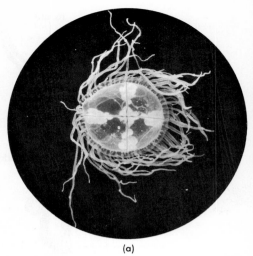

(a)

Fig. 1-3 *Gonionemus:* (a) photograph of the medusa stage.

Daphnia to the water. Observe, if possible, the discharge of nematocysts, or stinging cells, that are characteristic of coelenterates (Fig. 1-1c). Watch how the paralyzed organisms are carried to the mouth region of the hydra during ingestion.

Look for buds on well-fed specimens. Both temporary spermaries and ovaries may be found on one individual (Fig. 1-1a). The ovaries, which are larger than the spermaries, are nearer the basal end of the animals. Also look for the asexual buds, which are miniature hydras.

Living specimens may be used in experiments in regeneration (Chapter 9) or in studies of the effect of carbon dioxide on reproduction (see *Sci. Am.,* April 1959).

To study the specialized cells of hydras, prepared stained slides are needed.

Some marine coelenterates

Gonionemus (Fig. 1-3), an example of a marine hydrozoan showing the typical features of the medusoid form, is available in some regions. *Physalia,* the Portuguese man-of-war, is a highly specialized colony of organisms (Fig. 1-4). Students may explore life functions of *Aurelia* (Fig. 1-5), *Obelia* (Fig. 1-6), or *Metridium* (Fig. 1-7).

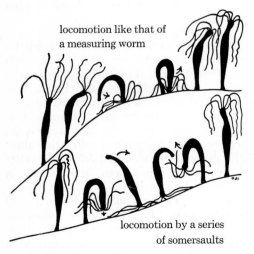

locomotion like that of a measuring worm

locomotion by a series of somersaults

Fig. 1-2 *Hydra,* showing its two methods of locomotion. (Reprinted with permission of the publisher from *College Zoology* by R. W. Hegner and K. A. Stiles, 7th ed., © The Macmillan Co., New York, 1959.)

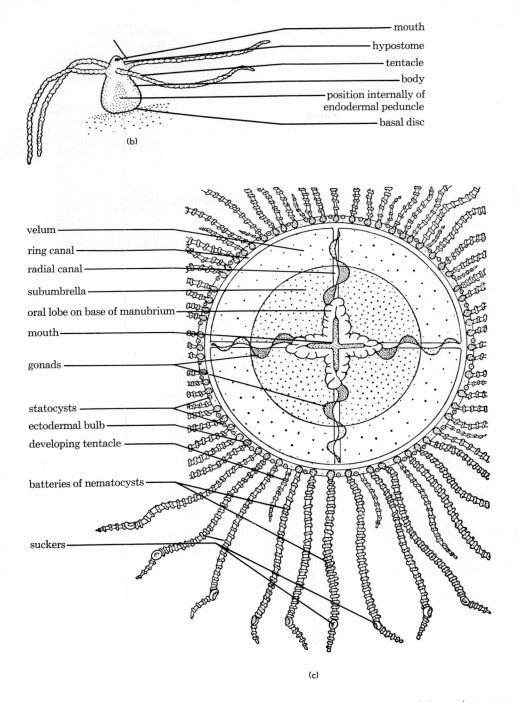

mouth
hypostome
tentacle
body
position internally of endodermal peduncle
basal disc

(b)

velum
ring canal
radial canal
subumbrella
oral lobe on base of manubrium
mouth

gonads

statocysts
ectodermal bulb
developing tentacle

batteries of nematocysts

suckers

(c)

Fig. 1-3 (continued) *Gonionemus:* (b) diagram of the hydroid stage; (c) diagram of the medusa stage as seen from the oral side. (a, courtesy of General Biological Supply House, Inc., Chicago; b, c, from W. S. Bullough, *Practical Invertebrate Anatomy*, 2nd ed., Macmillan, London, 1958, reprinted 1962.)

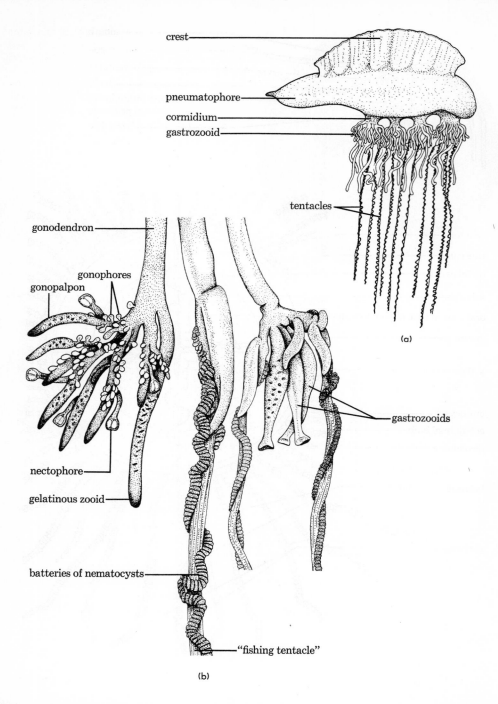

crest

pneumatophore

cormidium

gastrozooid

tentacles

(a)

gonodendron

gonophores

gonopalpon

gastrozooids

nectophore

gelatinous zooid

batteries of nematocysts

"fishing tentacle"

(b)

Fig. 1-4 *Physalia,* the Portuguese man-of-war: (a) a colony; (b) a portion of a colony: individuals exhibiting specialization of structure and function. (From D. E. Beck and L. F. Braithwaite, *Invertebrate Zoology: Laboratory Workbook,* 2nd ed., 1962; courtesy of Burgess Publishing Co., Minneapolis.)

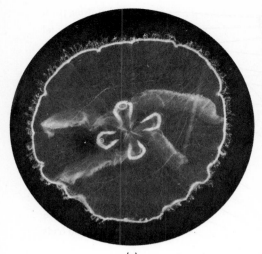

(a)

Fig. 1-5 *Aurelia:* (a) photograph of the entire organism; (b) details of anatomy as seen from above (in the top half of the drawing) and an oral, or subumbrella, view (in the bottom half). (a, courtesy of General Biological Supply House, Inc., Chicago; b, from W. S. Bullough, *Practical Invertebrate Anatomy*, 2nd ed., Macmillan, London, 1958, reprinted 1962.)

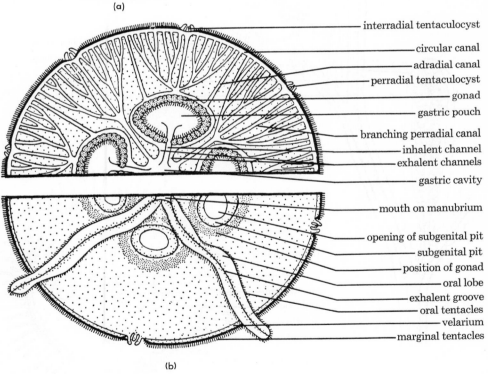

interradial tentaculocyst
circular canal
adradial canal
perradial tentaculocyst
gonad
gastric pouch
branching perradial canal
inhalent channel
exhalent channels
gastric cavity
mouth on manubrium
opening of subgenital pit
subgenital pit
position of gonad
oral lobe
exhalent groove
oral tentacles
velarium
marginal tentacles

(b)

Platyhelminths

Planarians

These free-living flatworms are chiefly aquatic; in contrast to a vast number of flatworms, such as tapeworms and flukes, they are not parasitic. If possible, living planarians should be studied by students. These may either be collected (pp. 592, 482) or purchased from supply houses

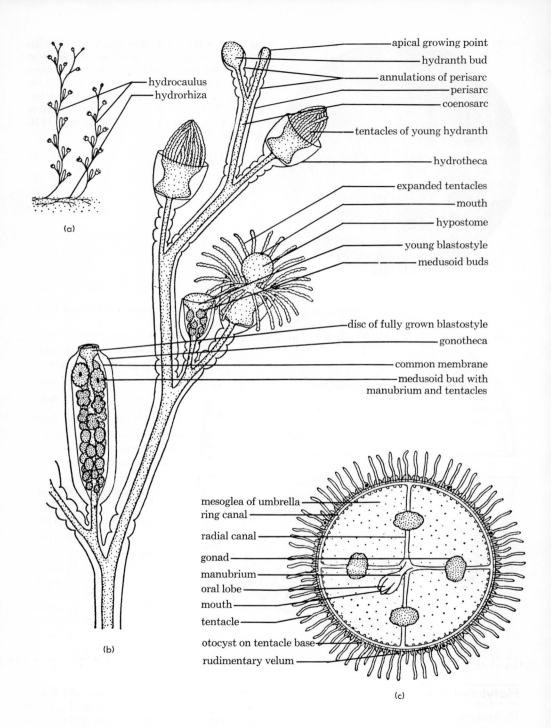

apical growing point
hydranth bud
annulations of perisarc
perisarc
coenosarc
tentacles of young hydranth
hydrotheca
expanded tentacles
mouth
hypostome
young blastostyle
medusoid buds
disc of fully grown blastostyle
gonotheca
common membrane
medusoid bud with
manubrium and tentacles

hydrocaulus
hydrorhiza

(a)

(b)

mesoglea of umbrella
ring canal
radial canal
gonad
manubrium
oral lobe
mouth
tentacle
otocyst on tentacle base
rudimentary velum

(c)

Fig. 1-6 *Obelia:* (a) a colony; (b) part of the hydrocaulus; (c) the medusa stage. (From W. S. Bullough, *Practical Invertebrate Anatomy*, 2nd ed., Macmillan, London, 1958, reprinted 1962.)

(Appendix C); they may readily be maintained in the laboratory (Chapter 12).

For study, place a few planarians in small vials. Notice how they glide smoothly, with the anterior region slightly raised. Cilia on the ventral surface produce this smooth motion. Planarians also crawl by means of strong body-muscle contractions. Introduce into the vials some brine shrimps (*Artemia*) or small worms such as *Tubifex* (Fig. 1-22) or *Enchytraeus* (Fig. 1-21); with a hand lens or under the binocular microscope examine the action of the hose-like pharynx as it is quickly extruded to bring the mouth near the source of food. When not actively ingesting food, the pharynx is withdrawn into a pharyngeal pouch in the middle of the body (Fig. 1-8a).

Students may notice the pigmentation, the darkly pigmented eyespots that enable planarians to react to light stimulus. Are planarians oriented positively or negatively toward light? Also note the two ear-like regions that are tactile receptors.

Use a prepared slide of a planarian such as *Dugesia* or *Dendrocoelum* to trace the organ systems. The branches of the digestive tract are blind ends in the sense that food materials enter and leave only through the mouth region. Enzymes pour into the branched regions, and food is digested. With the aid of Figs. 1-8 and 1-9 students can trace the excretory canals, which run along the length of the body and open into two excretory pores on the dorsal surface just back of the eyespots. Diffusion of gases is accomplished through the surface and coelomic fluid. Both testes and ovaries are present in these hermaphrodites. A genital pore is located near the posterior pointed end. Also look for two masses of ganglia in the head end, which continue as two strands of nerve fibers posteriorly along the body.

While planarians are representative of the flatworms as a group, they are also useful in laboratory studies of behavior per se and in regeneration experiments. Both areas of investigation have been combined in recent studies on training and memory

Fig. 1-7 *Metridium marginatum*, the sea anemone. (Courtesy of Carolina Biological Supply Co.)

and RNA.[2] Briefly, planarians that have been trained to go through a maze may be bisected, and when regenerated they learn the maze faster than untrained flatworms. If conditioned planarians are chopped up and fed to untrained worms, the untrained worms, now possessing the RNA of the chopped-up trained worms, learn the maze faster than those fed on the usual prosaic diet. (Also refer to Chapter 7.)

Some students may also want to stain planarian worms and prepare permanent slides. W. S. Bullough[3] gives instructions.

Nematodes and other aschelminths

Roundworms, the largest class of aschelminths, are numerous and widespread. Free-living forms, fresh-water and terrestrial, are often microscopic, or may be up to 1 mm long (and some marine forms are as large as 5 cm). No respiratory or circulatory system is present in round-

[2] J. McConnell, V. Jacobson, and D. Kimble, "The Effects of Regeneration upon Retention of a Conditioned Response in the Planarian," *J. Comp. Physiol. Psychol.* 52:1–5, 1959.
[3] W. S. Bullough, *Practical Invertebrate Anatomy*, 2nd ed., Macmillan, London, 1958, pp. 76–77.

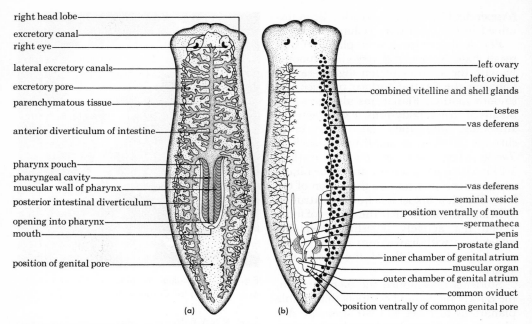

right head lobe

excretory canal
right eye

lateral excretory canals

excretory pore

parenchymatous tissue

anterior diverticulum of intestine

pharynx pouch
pharyngeal cavity
muscular wall of pharynx
posterior intestinal diverticulum

opening into pharynx
mouth

position of genital pore

left ovary
left oviduct
combined vitelline and shell glands

testes
vas deferens

vas deferens
seminal vesicle
position ventrally of mouth
spermatheca
penis
prostate gland
inner chamber of genital atrium
muscular organ
outer chamber of genital atrium
common oviduct
position ventrally of common genital pore

(a)   (b)

Fig. 1-8 *Dendrocoelum,* a planarian worm: (a) ventral view, showing the alimentary canal and excretory system; (b) dorsal view, showing the hermaphrodite reproductive system. (For clarity, only the male reproductive system is diagramed on the right side; and on the left side, the male reproductive system has been omitted to show the female reproductive system.) (After W. S. Bullough, *Practical Invertebrate Anatomy,* 2nd ed., Macmillan, London, 1958.)

worms. In club work, or as part of a study of representative invertebrates, students can learn the cycle of parasite-host relationship by examining prepared slides of the hookworm (Fig. 1-10), *Trichinella* larvae encysted in muscle (Fig. 1-11), or the microfilarian worm *Wuchereria bancrofti* (Fig. 1-12).

Free-living vinegar eels (Fig. 1-13) are rare examples of roundworms that are not parasitic. These are available in undistilled vinegar, in which the vinegar eels feed upon the fungi in the medium, or as prepared slides. With the aid of illustrations, students can identify the major organ systems.

A larger roundworm, *Ascaris,* may be dissected. *Ascaris megalocephala* is parasitic in the intestine of the horse, *A. suillae* in the intestine of the pig, and *A. lumbricoides* in the human intestine.

The largest number of invertebrates with a false coelom, or pseudocoel, are

grouped in Phylum Aschelminthes. These include nematodes, gastrotrichs, and rotifers. On the other hand, the majority of invertebrates have a true coelom; these include echinoderms, annelids, arthropods, and mollusks. Those interested in the several theories of the evolution of the coelom may want to refer to *Smaller Coelomate Groups,* Vol. V of L. Hyman's *The Invertebrates.*[4]

[4] L. Hyman, *The Invertebrates,* 5 vols., McGraw-Hill, New York, 1940–59.

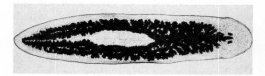

Fig. 1-9 A planarian, whole mount, for examination under the microscope; the digestive system is stained. (Courtesy of Carolina Biological Supply Co.)

Ascaris

The female *Ascaris* is larger than the male. Female *Ascaris lumbricoides* may be some 8 to 16 in. long and ¼ in. wide. Prepared whole specimens of *Ascaris* may be pinned down in a dissection pan containing water. Observe the external cuticle layer; notice that the worms are more cylindrical than round. Look for the terminal mouth with three lips; and on the ventral surface, just behind the mouth region, find the excretory pore (Fig. 1-14). In the male, the postanal region, or tail, is sharply curved down, while the contour of the female's tail is a straight tube. Be sure to locate the four distinguishing longitudinal lines which serve as identifying structures in dissection: the dorsal and ventral lines and the two brownish, thicker lateral lines.

After examining the external appearance (using Fig. 1-14 as a guide), cut through the dorsal wall anteriorly to the tip, pinning back the body wall along the way. Look for the longitudinal muscles underlying the epidermis; there are no circular muscles. Trace the organs of the digestive system: the muscular esophagus; the thin-walled intestine, which is somewhat obscured by the coils of the gonads; the short rectum, which opens to the surface at the anus. The *Ascaris* has no digestive glands (Fig. 1-15), and its food is digested material from the host.

If the specimen is a female, locate the genital opening on the ventral side, just posterior to the head region. Trace the reproductive system from the vagina to the uteri; these continue into oviducts and back to the ovaries. The ovaries are two thin, coiled sacs, which produce thousands of eggs.

In the male, there is a single coiled testis (Fig. 1-15), which extends into a vas deferens leading into a wider seminal vesicle. Trace the genital opening into the cloaca in the male.

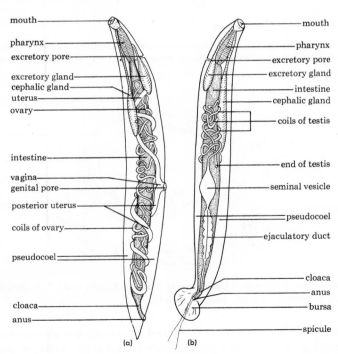

Fig. 1-10 *Necator*, the hookworm: (a) female; (b) male. (After D. E. Beck and L. F. Braithwaite, *Invertebrate Zoology: Laboratory Workbook*, 2nd ed., Burgess, 1962.)

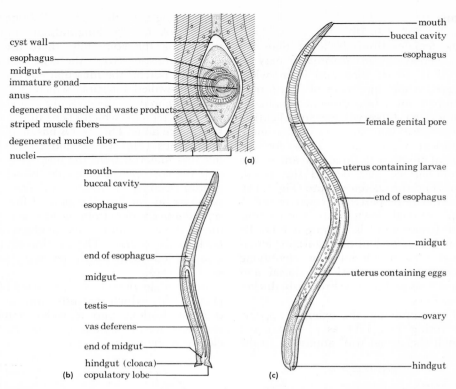

Fig. 1-11 *Trichinella*, a roundworm: (a) coiled larva encysted in striated muscle; (b) adult male; (c) adult female. (After W. S. Bullough, *Practical Invertebrate Anatomy*, 2nd ed., Macmillan, London, 1958.)

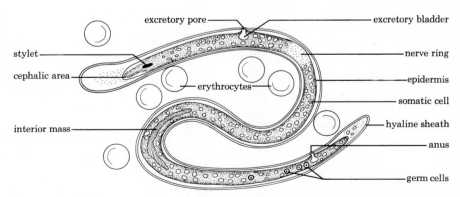

Fig. 1-12 *Wuchereria bancrofti*, a microfilarian worm. (After D. E. Beck and L. F. Braithwaite, *Invertebrate Zoology: Laboratory Workbook*, 2nd ed., Burgess, 1962.)

Examine the lateral lines and find the excretory system (Fig. 1-15), which opens ventrally into the aperture close to the anterior end of the worm. Try the difficult task of tracing the elements of the nervous system. Look for a ring of ganglia forming a branching mass around the pharynx. From this ring come a dozen nerves, eight leading to the anterior end, and four to the posterior region of the body. Two of the posterior nerves are on the dorsal and ventral lines and can be easily found.

Distinct respiratory or circulatory systems are absent in these parasitic worms. Exchanges of gases occur through the surface and open circulation of fluids throughout the body.

Rotifers

Although they bear a superficial resemblance to protozoa, these wheel animalcules are multicellular. Rotifers are grouped as a class of Phylum Aschelminthes, with other curious forms such as gastrotrichs (see Fig. 1-17) and the widely distributed nematodes (p. 27). There is a splendid chapter on rotifers, giving illustrations of many species of genera, in ad-

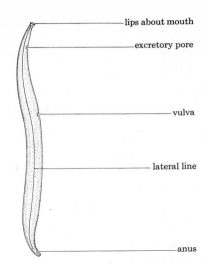

Fig. 1-14 *Ascaris*, external view. (After R. W. Hegner and K. A. Stiles, *College Zoology*, 7th ed., Macmillan, New York, 1959.)

dition to keys to identification, in *Fresh-Water Biology*.[5]

The most commonly found genera are *Hydatina* and *Philodina* (Figs. 1-16, 2-10a). These are often found with *Euglena* or *Chlamydomonas* in green pond water, or in old cultures of protozoa. While rotifers resemble one-celled forms, they have a well-developed digestive tube, nervous system, gonads, and paired nephridial tubes (Fig. 1-16).

Under the microscope, locate a quiet form, usually found feeding in a mass of

[5] H. B. Ward and G. C. Whipple, *Fresh-Water Biology*, 2nd ed., edited by W. T. Edmondson, Wiley, New York, 1959.

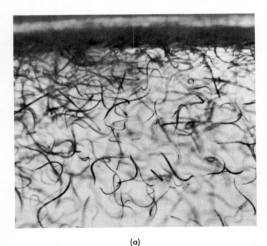

(a)

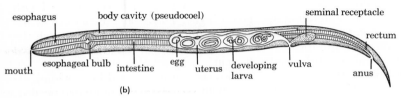

(b)

Fig. 1-13 *Turbatrix aceti*, the vinegar eel: (a) culture of these free-living, transparent nematode worms (about 2 mm long); (b) anatomy, showing digestive and reproductive systems. (a, courtesy of General Biological Supply House, Inc., Chicago; b, after R. W. Hegner and K. A. Stiles, *College Zoology*, 7th ed., Macmillan, New York, 1959.)

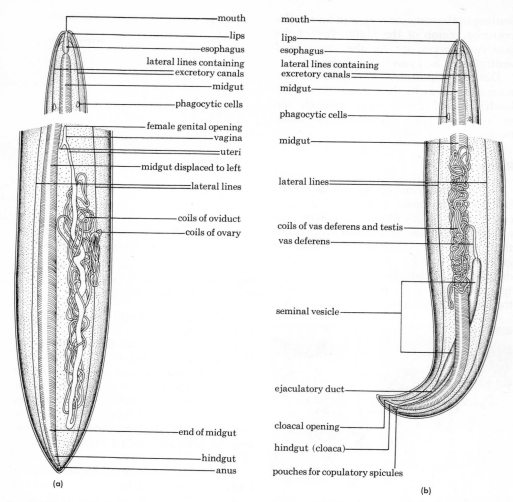

mouth
lips
esophagus
lateral lines containing
excretory canals
midgut
phagocytic cells

female genital opening
vagina
uteri
midgut displaced to left
lateral lines

coils of oviduct
coils of ovary

end of midgut

hindgut
anus

(a)

mouth
lips
esophagus
lateral lines containing
excretory canals
midgut

phagocytic cells

midgut

lateral lines

coils of vas deferens and testis
vas deferens

seminal vesicle

ejaculatory duct

cloacal opening
hindgut (cloaca)

pouches for copulatory spicules

(b)

Fig. 1-15 *Ascaris,* internal view: (a) female, with left reproductive organs omitted; (b) male, showing the curved posterior end. (After W. S. Bullough, *Practical Invertebrate Anatomy,* 2nd ed., Macmillan, London, 1958.)

debris. (Or add methyl cellulose, p. 112, to the slide to slow them down.) Notice the trochus, or ciliary disc, at the anterior end, and watch how the whirling motion of the trochus carries food material into the central mouth, which leads into a well-formed digestive tube. The mouth may be considered ventral since one side of the trunk of the rotifer is flattened.

If rotifers are fed green algae, such as *Chlorella,* the current created by the cilia is observable as the small green cells are

whirled about. The stomach and intestine become visible as they become filled with green algae. Focus on the mastax in the anterior third of the body, and watch the chewing action of the jaws contained therein.

Two cement glands aid in providing attachment for the foot organ when needed. At other times, this jointed tail is "telescoped," and rotifers may be found swimming rapidly across the field of the microscope.

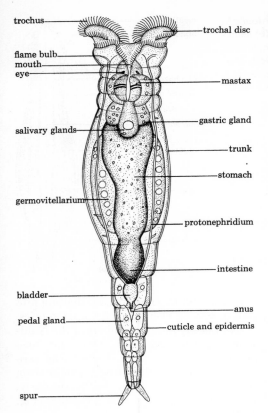

trochus — trochal disc
flame bulb
mouth
eye
— mastax
— gastric gland
salivary glands
— trunk
— stomach
germovitellarium
— protonephridium
— intestine
bladder
— anus
pedal gland
— cuticle and epidermis
spur

Fig. 1-16 *Hydatina*, a common rotifer: dorsal view of female, with circular and oblique bands of muscle omitted. (After R. D. Barnes, *Invertebrate Zoology*, Saunders, 1963.)

In the large body cavity, there is a fluid containing small granules, which probably acts as a circulating medium. There is also a large ganglion lying dorsal to the mouth and pharynx. Notice that the trunk region is enveloped in a clear lorica, which may be found as remains in very old cultures of rotifers.

The female is most often the form described; parthenogenetic females are found during the spring and summer. The large ovary of the female is connected to a vitellarium, or brood sac, which opens into the cloaca via an oviduct. The thin-shelled eggs are of two sizes. Except for a large testis, the male has reduced organ systems and is only a quarter the size of the female.

The males develop in the fall. Fertilized eggs are winter eggs; these hatch into females in the spring. Occasionally these fertilized eggs can be observed attached to the tail of the rotifer.

Gastrotrichs

Gastrotrichs are grouped with rotifers and roundworms in Phylum Aschelminthes. Gastrotrichs are found in pond water associated with *Ulothrix* and *Spirogyra*. Although the gastrotrichs resemble small ciliated protozoa, they are multicellular and have a distinguishable forked tail. The ventral surface is flattened and covered with two bands of cilia, which extend along the length of the body, while the dorsal region is more convex and covered with spine-like bristles (Figs. 1-17, 2-10b).

Chaetonotus is a commonly occurring gastrotrich with many species (well illustrated in *Fresh-Water Biology*[6]). Examine a specimen under a high-power microscope; look for four clusters of flagella extending from the anterior end. Try to locate the thick-walled esophagus, which leads from the mouth into a wide stomach. The intestine ends in a posterior anal opening; a pair of coiled nephridia terminate on the ventral surface. It will probably not be possible to distinguish the anterior ganglion, from which a pair of nerves extends ventrally along the body.

Fresh-water forms are parthenogenetic females.

Segmented worms

The Oligochaeta, a class in Phylum Annelida, are mainly fresh-water or terrestrial forms. They are segmented, with rows of setae, or bristles, which aid in classification. The smaller worms such as *Tubifex* (Fig. 1-22) and microscopic worms have anal gills. The Polychaeta, of which *Nereis* (Fig. 1-25) is a representative, are mainly marine forms.

[6] *Ibid.*, p. 415.

Fig. 1-17 *Chaetonotus,* a gastrotrich commonly found in old cultures of protozoa and algae: (a) side view; (b) details of anatomy. (a, from D. E. Beck and L. F. Braithwaite, *Invertebrate Zoology: Laboratory Workbook,* 2nd ed., 1962, courtesy of Burgess Publishing Co., Minneapolis; b, after T. J. Parker and W. A. Haswell, *A Textbook of Zoology,* Macmillan, London, 1930.)

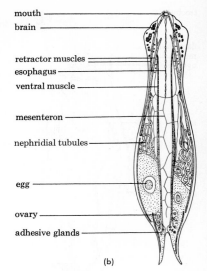

mouth
brain

retractor muscles
esophagus
ventral muscle

mesenteron

nephridial tubules

egg

ovary
adhesive glands

tail forks

(a)

(b)

Many oligochaetes are available for study and dissection. Most forms are hermaphrodites, and the location of ovaries and testes in specific segments aids in identifying different species. The smaller, almost transparent, worms that are found in pond cultures associated with protozoa and rotifers are often members of the families Aeolosomatidae and Naididae. Budding is a common asexual method of reproduction among many of them. *Aeolosoma* (Fig. 1-18) contains oil globules that

are yellowish in some species, shading into pale green or even blue in other species.

Commonly found members of the Naididae are *Nais* (Fig. 4-3), *Dero,* and *Aulophorus.* In *Dero,* a mud-dweller, the posterior end is modified into a gill-bearing respiratory organ (Fig. 1-19); similarly, posterior gills are found in *Aulophorus* (Fig. 1-20).

In *Nais,* the ventral setae of segments 2 to 5 are better differentiated than the posterior setae—an identifying characteristic. The dorsal, needle-like setae may be single or bifurcated, depending on the species.

Enchytraeus (Fig. 1-21), a small, white, aquatic or terrestrial form (less than 25 mm), is easily maintained in the laboratory (it may be purchased in aquarium stores). This member of the family Enchy-

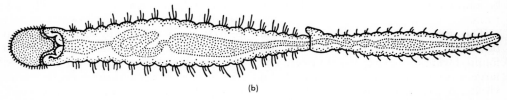

(a)

(b)

Fig. 1-18 *Aeolosoma,* a segmented aquatic worm (oligochaete) found in old cultures of protozoa: (a) the conspicuous globules, which are orange, are characteristic of this worm; (b) method of asexual reproduction by posterior budding. (b, reprinted with permission of the publisher from *College Zoology* by R. W. Hegner and K. A. Stiles, 7th ed., ⓒ The Macmillan Co., New York, 1959.)

traeidae may be cultured like earthworms (Chapter 12).

Tubifex

These fresh-water oligochaetes are commonly found in lakes and ponds containing organic matter. They form tubes of mud held firm by mucus secretions. *Tubifex* may also be purchased in aquarium shops, for they serve as live food for many aquarium fish; they are easily maintained in the laboratory (Chapter 12).

In the laboratory, *Tubifex* is often used in studies of regeneration (p. 482). However, they are useful to study as a representative of the segmented worms, for they are transparent enough to show active muscular contractions of the intestine, dorsal vessel, and aortic loop. Gills or respiratory branches are located in the terminal end; when cultured in an inch of mud, *Tubifex* may be seen to have its posterior end waving from mud tubes in which the anterior part of the worm is enclosed.

Each segment, or metamere, contains a portion of the body cavity, or coelom, filled with fluid. Bundles of setae are arranged segmentally; those along the more posterior end are in pairs on each side of a segment.

Mount a specimen in a drop of pond water, and examine it under low and high power. Students can observe first the activity of the worm—the means of locomotion, contractions, and ingestion. Examine the length of the worm, and note the arrangement of bristle setae on each segment. Either narcotize a specimen by adding a drop of magnesium sulfate solution (Epsom salts, p. 112) under the coverslip, or mount a worm in a very dilute agar medium (about 1 percent). In a quiet animal trace the internal organs (use Fig. 1-22 as a guide).

As in other annelids, ganglia concentrated in the anterior end are not easily seen in whole mounts, nor are the reproductive organs readily recognized.

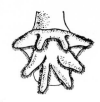

Fig. 1-19 *Dero*, a segmented aquatic worm (oligochaete): detail of posterior end, an identifying feature of the worm. (From H. B. Ward and G. C. Whipple, *Fresh-Water Biology*, 2nd ed., edited by W. T. Edmondson, 1959; courtesy of John Wiley & Sons.)

Fig. 1-20 *Aulophorus*, a segmented aquatic worm found in old cultures of protozoa: detail of posterior gills. (From H. B. Ward and G. C. Whipple, *Fresh-Water Biology*, 2nd ed., edited by W. T. Edmondson, 1959; courtesy of John Wiley & Sons.)

Leech

The leech is a brightly colored, rust or greenish, parasitic oligochaete, about 2 or 3 in. long, with a large posterior sucker. These hermaphrodites lack setae on the segments. As in other annelids, each somite consists of sections of the various systems (Fig. 1-22).

Somites 28 to 34 are almost fused and make up the caudal sucker, which is larger than the anterior sucker. Many ganglia are found in the posterior sucker.

With the aid of Fig. 1-23, dissect out the systems. Living leeches may be narcotized and killed, using methods described in Table 11-2.

Earthworm (Lumbricus)

When living earthworms are available, their behavior in response to many stimuli, such as light, touch, sound, and moisture, may be studied (see Chapter 7). Students will also want to study the external anatomy of the animals. When this has been done and the animal is ready for dissec-

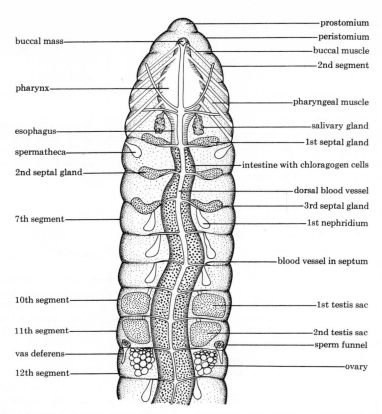

Fig. 1-21 *Enchytraeus,* a segmented white worm: dorsal view of organs in anterior end. (After W. S. Bullough, *Practical Invertebrate Anatomy,* 2nd ed., Macmillan, London, 1958.)

tion, gradually add chloroform to the water in a container holding several earthworms. Now stretch out one worm and pin it down firmly, dorsal surface up, through the first segment and the posterior end, in a waxed dissecting pan.

Cut through the skin of the back, beginning at the posterior end (Fig. 1-24).

Continue forward, cutting to one side of the midline. Cut the partitions, or septa, that hold down the body wall, and pin back the body wall. If pins are inserted through segments 5, 10, 15, and 20, the organs can be readily located. Start at the beginning of the alimentary canal, and locate the buccal region, the pharynx

Fig. 1-22 *Tubifex,* a segmented worm. (After R. W. Hegner, *College Zoology,* rev. ed., Macmillan, New York, 1926.)

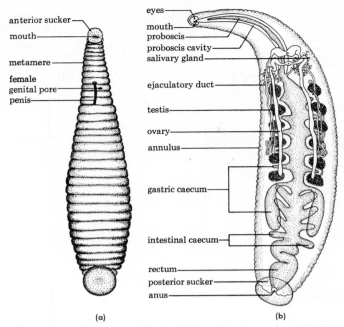

anterior sucker
mouth
metamere
female
genital pore
penis

eyes
mouth
proboscis
proboscis cavity
salivary gland

ejaculatory duct

testis

ovary

annulus

gastric caecum

intestinal caecum

rectum
posterior sucker
anus

(a)   (b)

Fig. 1-23 Leech: (a) ventral view; (b) internal anatomy, showing relative positions of digestive and reproductive systems. (After D. E. Beck and L. F. Braithwaite, *Invertebrate Zoology: Laboratory Workbook*, 2nd ed., Burgess, 1962.)

leading into the esophagus with its pouches in segments 10 and 13; and look for the glands in segments 11 and 12. In the vicinity of segment 15, locate the crop, and, further along, the muscular gizzard. Then trace the digestive tract (it continues as a yellow-colored intestine to the posterior end). Locate the large blood vessel dorsal to the stomach-intestine.

Carefully move the seminal vesicles aside, in order to trace the aortic arches, which branch from the dorsal blood vessel and circle around to connect with the ventral blood vessel (found under the digestive tube). In a freshly killed specimen pulsations of these swollen aortic "hearts" may be visible.

Blood is carried forward along the dorsal vessel to the five pairs of "hearts" in segments 7 to 11 and then into the ventral blood vessel. Here blood moves the length of the body (ventral to the digestive system), branching off to supply organs along the way. Along the ventral nerve cord you

may also locate the subneural blood vessel and the two lateral neural vessels. These also carry blood back along the length of the worm.

Now students may want to study the reproductive system. Remove the esophagus to get a clear view of the seminal vesicles, consisting of three lobes on each side of the esophagus. Also locate the two pairs of small testes on the septa of segments 10 and 11.

In segment 13 find a pair of ovaries, and look for the yellowish seminal receptacles in segments 9 and 10. Crush one in a drop of water on a slide and look for masses of filamentous sperms, received from another worm during copulation.

Also, each segment (except the first three and the last) contains a pair of nephridia opening into nephridiopores.

Finally, locate the two white cerebral ganglia above the buccal region in segment 3. Lift aside or carefully remove the pharynx to reveal the nerve ring, a pair of

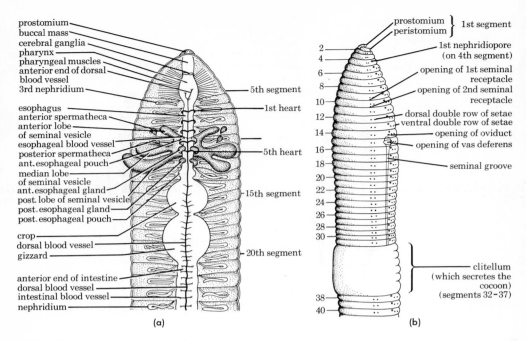

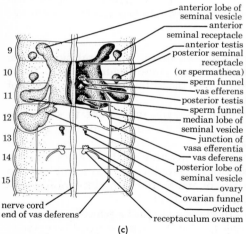

Fig. 1-24 *Lumbricus*, the earthworm: (a) general anatomy; (b) external view; (c) detail of reproductive systems. (After W. S. Bullough, *Practical Invertebrate Anatomy*, 2nd ed., Macmillan, London, 1958.)

subpharyngeal ganglia, and the long ventral double nerve cord.

Students may also want to study the parasites found in earthworms; see *Monocystis* and *Rhabditis* (p. 535).

Sandworm (Nereis)

The sandworm or clamworm, *Nereis,* is representative of annelids of the class Polychaeta. These marine forms, about 3 in. long, have a distinct head, bearing tentacles and four eyes; from each segment, or metamere, extends a pair of muscular parapodia which bear setae, or bristles, and some tactile receptors, cirri.

The bright red blood gives the more vascular regions a deep red color, which is visible through the greenish tint of the skin. Compare the external appearance with the terrestrial earthworm (Oligochaeta).

Look for the opening of the nephridia on the ventral side near each parapodium. Lay the worm, ventral side down, in a dissecting pan of water and cut through the skin, pinning it back. If the specimen is alive, it may be killed by adding chloroform to the water. Trace the buccal cavity from the mouth, the pharynx (segments 1 to 4), and the esophagus, which extends for the next five segments. Look for a pair of digestive (esophageal) glands—caeca—

that open into the esophagus; trace the intestine to the posterior region (Fig. 1-25).

Find the well-developed blood vessel which lies above the alimentary canal. Peristalsis may be seen in living specimens —blood is driven from the back to the anterior region of the worm. Try to find the ventral blood vessel that carries blood to the posterior end. Two pairs of looped, branching vessels in each segment connect the dorsal and ventral vessels; smaller branches of the looped, branching vessels extend into organs and tissues.

There are no branchiae, but respiratory exchange takes place through the rich blood supply in the parapodia. Nephridia, paired tubes, may be found in each metamere except in the most anterior and posterior segments.

Gonads may not be visible, for they are temporary organs that develop at breeding time as proliferations from the lining of the coelom. Although there are many varia-tions among the species of *Nereis,* the pair of testes (several pairs in some species) is usually found between segments 19 and 25. Ovaries are metamerically arranged, and eggs float in the coelom.

It may be possible to identify the ring of ganglia around the esophagus; also try to find the segmented, ventral nerve cord, which consists of masses of ganglia. Tracing forward from the nerve ring, look for a cerebral ganglion with branches leading to the eyes, tentacles, and palpi.

Where possible, compare *Nereis* with the leech (p. 35), or with another marine form, *Arenicola.*

Small fresh-water crustaceans

Daphnia

Daphnia is representative of the subclass Branchiopoda, forms that have flattened appendages used for both locomotion and

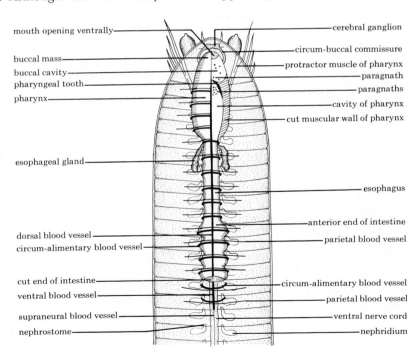

mouth opening ventrally
buccal mass
buccal cavity
pharyngeal tooth
pharynx
esophageal gland
dorsal blood vessel
circum-alimentary blood vessel
cut end of intestine
ventral blood vessel
supraneural blood vessel
nephrostome

cerebral ganglion
circum-buccal commissure
protractor muscle of pharynx
paragnath
paragnaths
cavity of pharynx
cut muscular wall of pharynx
esophagus
anterior end of intestine
parietal blood vessel
circum-alimentary blood vessel
parietal blood vessel
ventral nerve cord
nephridium

Fig. 1-25 *Nereis,* a marine segmented worm (polychaete): dorsal view of the anatomy; right halves of the buccal mass and the pharynx have been omitted. (After W. S. Bullough, *Practical Invertebrate Anatomy,* 2nd ed., Macmillan, London, 1958.)

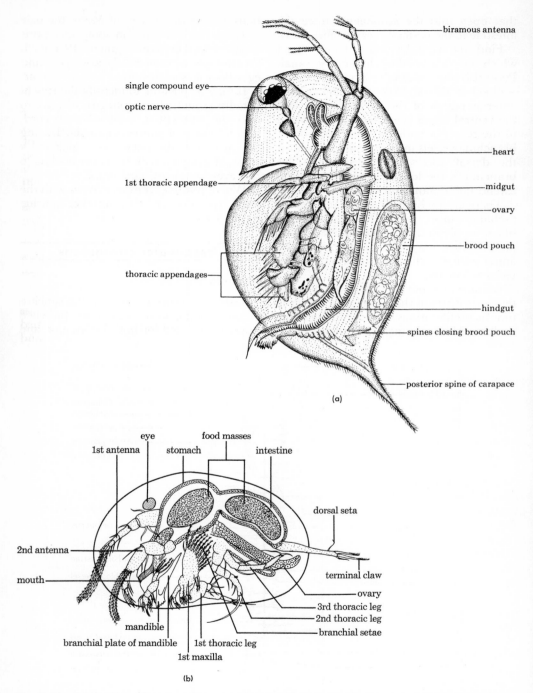

Fig. 1-26 (a) *Daphnia,* the water flea (a fresh-water cladoceran), lateral view of female; (b) *Cypris* (a fresh-water ostracod), lateral view of female. (b, from H. B. Ward and G. C. Whipple, *Fresh-Water Biology,* 2nd ed., edited by W. T. Edmondson, 1959; courtesy of John Wiley & Sons.)

respiration. Students examining pond water may find *Daphnia*, or water fleas, of the order Cladocera; these have a single compound eye, which is usually in active motion. Normally, there are five (or six) pairs of appendages attached to the thorax. Except for the head, the compressed body is enclosed in a flattened, transparent carapace. The sturdy second antennae, which are moved by large muscles, are the main organs of locomotion in Cladocera.

Use Fig. 1-26a to locate the organs in *Daphnia*. Although there are no blood vessels, circulating blood cells may be seen, for blood flows anteriorly from the pulsating heart and returns from the hemocoel and inner wall of the carapace, where gaseous exchanges occur—some exchange also takes place along the appendages.

Add single-celled algae such as *Chlorella* to the slide, and watch how the actively moving appendages create a current which carries in the algae. The food tube soon turns bright green.

Students often study changes in heartbeat in *Daphnia* caused by altered conditions: change in temperature, addition of hormones such as adrenin or thyroxin, addition of drugs such as caffein or thiouracil, or the addition of pesticides (see p. 245).

Most specimens are parthenogenetic females; eggs develop into miniature water fleas in the brood pouch. Under adverse conditions, sexual reproduction occurs. Some eggs hatch into males, and the females then produce some darkly colored, haploid eggs that require fertilization. These females are distinguished from the usual females by an altered, saddle-like shape of the dorsal valve that develops into an ephippium. At the next molt of the female, the eggs enclosed in the ephippium, or altered carapace, are left behind; they remain dormant until conditions become more favorable for development.

J. L. Brooks[7] reports that overcrowding seems to initiate production of males;

[7] *Ibid.*, "Cladocera."

sexual egg production seems to occur when there is a sudden decrease in food supply for females. Reported in the same source are methods of examination and preservation using a double coverslip. (Specimens are gradually transferred into full-strength glycerine jelly between two coverslips; this seal is mounted in resin on a slide.)

Cypris

Another small crustacean that bears a superficial resemblance to *Daphnia* is *Cypris* (Fig. 1-26b). However, this form is more opaque and is entirely enclosed in a bivalve carapace, which may be tinted

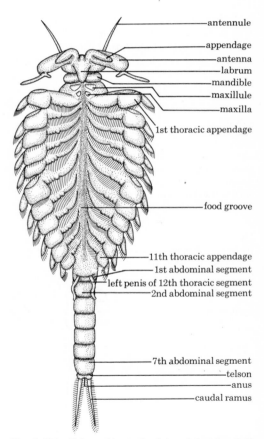

antennule
appendage
antenna
labrum
mandible
maxillule
maxilla
1st thoracic appendage

food groove

11th thoracic appendage
1st abdominal segment
left penis of 12th thoracic segment
2nd abdominal segment

7th abdominal segment
telson
anus
caudal ramus

Fig. 1-27 *Eubranchipus*, the fairy shrimp: ventral view of male. (After W. S. Bullough, *Practical Invertebrate Anatomy*, 2nd ed., Macmillan, London, 1958.)

greenish, orange, or yellowish brown. *Cypris* has seven pairs of thoracic appendages and is a member of the subclass Ostracoda.

Fairy shrimp (Eubranchipus)

Eubranchipus, classified in the order Anostraca, is also representative of the subclass Branchiopoda. Their occurrence is often sporadic, as a result of their being carried to temporary pools of rainwater or melting ice, where the nauplius larvae hatch. Another branchiopod, the brine shrimp *Artemia,* is found in saline (but not marine) basins such as the Great Salt Lake.

Using the microscope, find a quiet specimen of fairy shrimp, and examine the flattened appendages used for both locomotion and respiration. Notice that fairy shrimp lack a carapace around the cylindrical body (Fig. 1-27). Compound eyes are stalked on the prominent head region. The second antennae are enlarged, especially in the male. The number of appendages varies from 11 to 17 pairs, extending from as many segments of the thorax.

Hold up a container of fairy shrimp or brine shrimp, and watch how they swim on their backs, with their appendages oriented toward light.

A period of desiccation precedes hatching; in fact, it seems to initiate hatching. Students may purchase dry brine shrimp eggs and hatch out the nauplius larvae (pp. 354, 599).

Cyclops

Cyclops, a member of the subclass Copepoda, has a small body with five or six

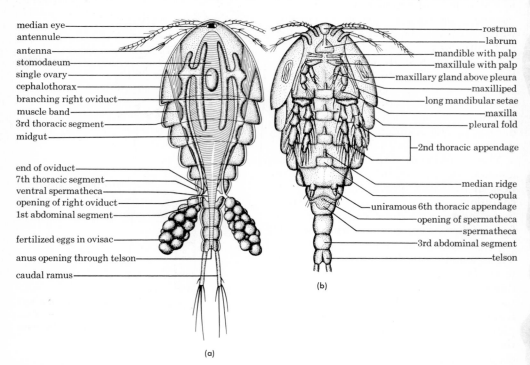

median eye
antennule
antenna
stomodaeum
single ovary
cephalothorax
branching right oviduct
muscle band
3rd thoracic segment
midgut

end of oviduct
7th thoracic segment
ventral spermatheca
opening of right oviduct
1st abdominal segment

fertilized eggs in ovisac

anus opening through telson

caudal ramus

rostrum
labrum
mandible with palp
maxillule with palp
maxillary gland above pleura
maxilliped
long mandibular setae
maxilla
pleural fold

2nd thoracic appendage

median ridge
copula
uniramous 6th thoracic appendage
opening of spermatheca
spermatheca
3rd abdominal segment
telson

(b)

(a)

Fig. 1-28 *Cyclops:* (a) dorsal view of female; (b) ventral view of female. (In both views, the 3rd, 4th, and 5th pairs of thoracic appendages have been omitted.) (After W. S. Bullough, *Practical Invertebrate Anatomy,* 2nd ed., Macmillan, London, 1958.)

pairs of thoracic appendages, the first four of which are biramous. The segments of the body are different sizes; this furnishes a distinguishing feature in identification and classification (Fig. 1-28).

Look for females carrying two posterior egg sacs, or possibly carrying capsules of sperm cells received from the male in mating. Fertilized eggs develop into nauplius larvae, which pass through several molts, a stage when they are difficult to identify unless the appendages are counted.

Gammarus

These small, laterally compressed, pond crustaceans belong in the subclass Malacostraca, with the fresh-water crayfish (and larger marine crabs and lobsters). They live in quiet lakes and ponds and can be maintained readily in the laboratory in large battery jars of green water; they often rest among the floating duckweeds.

Using Fig. 1-29, identify and compare the organs with those of the crayfish (Fig. 1-30).

Crayfish (Cambarus)

Examine the external anatomy (Fig. 1-30a) of this fresh-water decapod. Notice that the body is divided into two distinct parts, a rigid cephalothorax and a more flexible, segmented abdomen.

In a dissecting pan place a crayfish dorsal side up. Begin at the posterior dorsal end of the carapace, and cut forward to each side of the midline along each side of the outer edges of the thorax up to the rostrum. Now remove this top part of the carapace. In the dorsal part of the thorax find the pericardial sinus containing the angular, muscular heart. Look for the three pairs of ostia through which blood enters the heart.

Under the heart and a bit forward are the reproductive organs. In a mature female, find the bilobed ovary (containing eggs) in front of the heart and a single

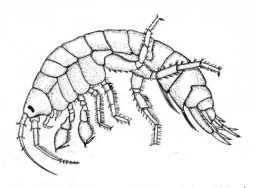

Fig. 1-29 *Gammarus,* the earwig. (After J. Needham and P. Needham, *A Guide to the Study of Fresh-Water Biology,* 4th ed., Comstock [Cornell Univ. Press], 1938.)

fused mass behind the heart. Locate the oviduct extending down on each side to the first segment of the third thoracic leg. In the male the white testis should be in a corresponding position, with two highly coiled vasa deferentia (sperm ducts). If these are filled with sperms, mount some in a drop of water, tease the ducts apart, and examine the unusual shape of the sperm cells. The vasa deferentia open on the first segment of the hindmost thoracic leg.

Under the roof of the head locate the stomach, a thin-walled sac. Digestive glands, which may be brown or greenish, are situated where the stomach joins the intestine. Trace the intestine along the length of the abdomen.

The kidneys, a pair of light green glands (sometimes called "green glands"), are located in front of and a bit below the stomach.

Next trace the nervous system. Find the large ganglion in back of the gullet, and trace two branches which unite into another large ganglion above the gullet, forming an esophageal collar. Moving backward along the animal, students may trace a double nerve cord which joins together pairs of segmental ganglia. Also look for branches of nerves extending into muscle on each side of the abdomen.

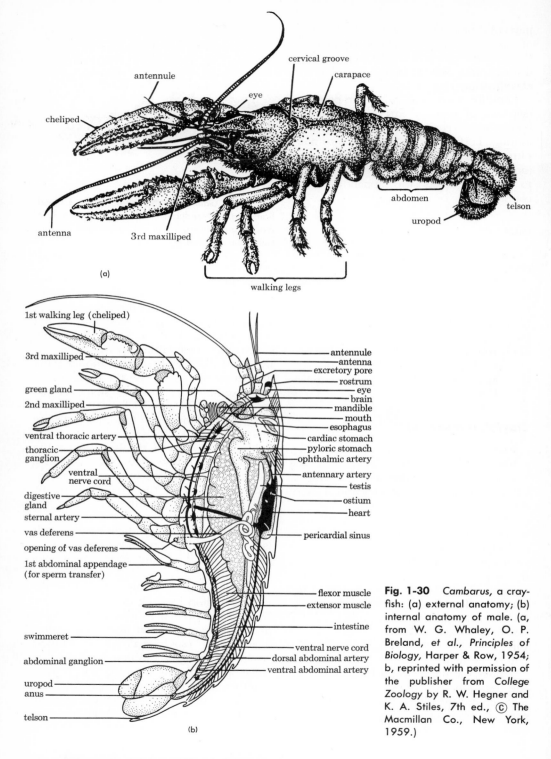

Fig. 1-30 *Cambarus*, a crayfish: (a) external anatomy; (b) internal anatomy of male. (a, from W. G. Whaley, O. P. Breland, *et al.*, *Principles of Biology*, Harper & Row, 1954; b, reprinted with permission of the publisher from *College Zoology* by R. W. Hegner and K. A. Stiles, 7th ed., Ⓒ The Macmillan Co., New York, 1959.)

Insects

Grasshopper (Melanoplus)

Remove the wings from a freshly killed or preserved grasshopper. Pin the speci- men with the dorsal side up in a dissecting pan (Fig. 1-31). Cut through each side of the top of the posterior abdomen; then remove this upper part of the abdomen— the heart, which is a delicate tube, may be

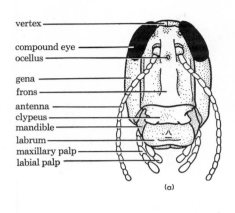

(a)

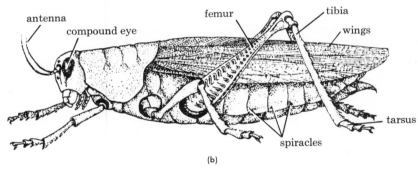

(b)

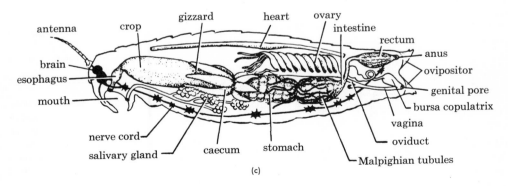

(c)

Fig. 1-31 *Melanoplus*, a grasshopper: (a) head; (b) external anatomy; (c) internal anatomy of female. (a, reprinted with permission of the publisher from *College Zoology* by R. W. Hegner and K. A. Stiles, 7th ed., ⓒ The Macmillan Co., New York, 1959; b, c, from W. G. Whaley, O. P. Breland, *et al.*, *Principles of Biology*, Harper & Row, 1954.)

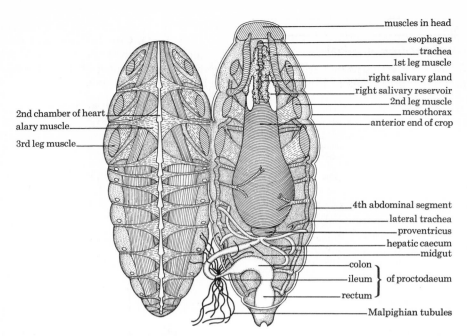

Fig. 1-32 *Blatta,* a cockroach: general dissection. The animal has been split lengthwise and spread open. (After W. S. Bullough, *Practical Invertebrate Anatomy,* 2nd ed., Macmillan, London, 1958.)

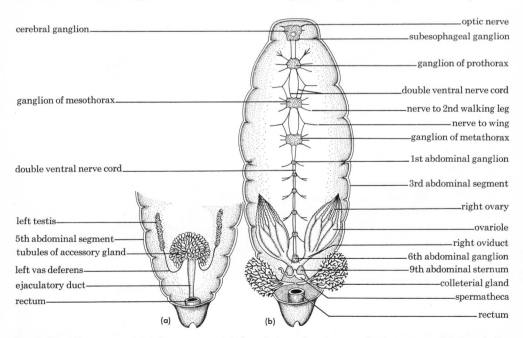

Fig. 1-33 *Blatta,* a cockroach, anatomy: (a) dorsal view of male reproductive system; (b) dorsal view of nervous system and female reproductive system. (After W. S. Bullough, *Practical Invertebrate Anatomy,* 2nd ed., Macmillan, London, 1958.)

attached to it. Notice the white air sacs and air tubes on each side of the abdomen.

If the specimen is a female, look for a mass of yellow sacs in the anterior end of the abdomen. Trace the oviducts from the two ovaries to an opening in the ovipositor. Push aside the eggs to find the dark-colored intestine.

When you remove the roof of the thorax, look for the muscles which move the wings. The crop, which contains hooked teeth, is in the prothorax. From this region locate the gastric caeca, which are cone-shaped pouches; then find the stomach, which leads into the intestine. Along the ventral region of the abdomen you may find enlargements of the nerve cord, the ganglia, from which many nerves branch to tissues.

Cockroach (Blatta)

For studying insect anatomy some teachers prefer the oriental cockroach *Blatta* or the larger *Periplaneta*. These as well as grasshoppers are available from biological supply houses.

Examine the external anatomy; if possible, compare it with that of the grasshopper. Identify the head and its appendages, the 3 segments of the thorax, and the 11 segments of the abdomen. The end segments are not easily distinguished, for they overlap and are reduced. Locate spiracles.

Pin the roach ventral side down in a waxed pan or on a corkboard (Fig. 1-32). Remove the leathery elytra and wings, and cut into the specimen along one side so that the dorsal section can be flapped over to expose the internal organs. Students can identify the dorsal "heart," a chain of node-like swellings which extends along the length of the roach. In the vicinity of the first leg locate the white salivary glands surrounding the esophagus. Trace the food tube; find the extended crop, thick-walled gizzard, hepatic caeca, colon, and rectum. The thread-like mass is composed of Malpighian tubules. Now remove the fatty tissue and the digestive tract, and trace the tracheal tubes, which branch throughout the body.

In the fourth and fifth segments of a male, locate the thin, coiled testes and the large accessory gland (Fig. 1-33). In a female look for the two ovaries with many-branched ovarioles, which probably contain clusters of eggs. Locate the spermatheca and the egg capsule in the seventh segment (which comes to a point). Finally, move the organs aside to locate the ventral double nerve cord and the chains of ganglia associated with it (Fig. 1-33).

Arachnids

This class of Phylum Arthropoda includes the "living fossil" (the king crab) and spiders, scorpions, mites, and ticks. There is a division of the body into a cephalothorax, to which six pairs of appendages are attached, and an abdomen. Antennae and mandibles are lacking; the first pair of appendages is modified as nippers, or chelicerae. Book lungs are characteristic of arachnids; there may also be tracheal tubes and spiracles.

Use Fig. 1-34 to explore the anatomy of a spider.

King crab (Limulus)

Limulus, the king crab (also called the horseshoe crab), may be found crawling through sand. The body consists of a large horseshoe-shaped carapace, a smaller shield, and a caudal spine, or telson. A pair of large eyes and a more anterior pair of smaller eyes are on the convex dorsal surface.

Turn the specimen over and locate the mouth at the base of the anterior walking legs; the first pair of appendages is the chelicerae (Fig. 1-35). The anus is posterior at the beginning of the long spine.

Examine the flattened appendages, the segmented body, and the specialized book gills. The bluish color of the blood is due to the copper-containing pigment hemocyanin.

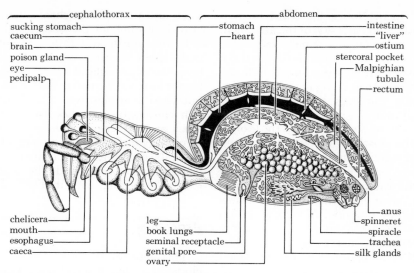

Fig. 1-34 Spider: internal anatomy. (After R. W. Hegner and K. A. Stiles, *College Zoology*, 7th ed., Macmillan, New York, 1959.)

A related ancient form, the scorpion, has a sting containing poison located at the end of the telson.

Some students may want to compare the two major lines in the evolution of the arthropods: the subphylum Chelicerata, which includes the horseshoe crabs, scorpions, and other arachnids, and the subphylum Mandibulata, comprising the crustaceans, centipedes, millipedes, and insects.

Mollusks

Phylum Mollusca comprises a large group of varied forms that are mainly marine—clams, oysters, snails, squids, nautiluses, and octopuses. The soft body may be enclosed in one shell, or in two, as in the bivalves. In the squid, a cartilaginous internal skeleton (called a pen) is present; the outer shell is lacking. Most mollusks have a foot, a mantle that is a fold of body wall that secretes a shell, and a coelom. Fresh-water forms such as pond snails (*Physa*, Fig. 1-38) and the fresh-water clam (*Anodonta*, Fig. 1-40) are available for dissection. A land snail, *Helix* (Fig. 1-39), and the garden slug may be ob-

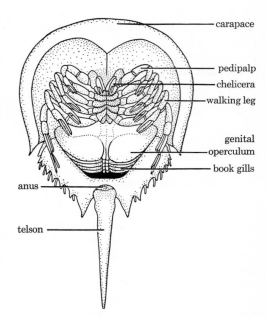

Fig. 1-35 *Limulus*, the horseshoe crab, or king crab: ventral view, showing walking legs and book gills. (Reprinted with permission of the publisher from *College Zoology* by R. W. Hegner and K. A. Stiles, 7th ed., © The Macmillan Co., New York, 1959.)

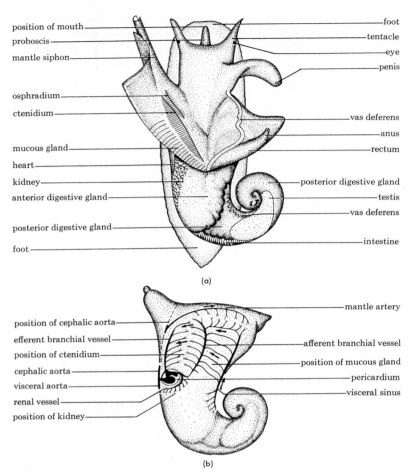

position of mouth — foot
proboscis — tentacle
mantle siphon — eye
— penis

osphradium —
ctenidium — vas deferens
— anus
mucous gland — rectum
heart —
kidney — posterior digestive gland
anterior digestive gland — testis
— vas deferens
posterior digestive gland —
foot — intestine

(a)

— mantle artery
position of cephalic aorta —
efferent branchial vessel —
position of ctenidium — afferent branchial vessel
cephalic aorta — position of mucous gland
visceral aorta — pericardium
renal vessel — visceral sinus
position of kidney —

(b)

Fig. 1-36 *Buccinum*, a whelk: (a) dorsal view of male, with shell removed and mantle cavity cut open to show the position of the organs; (b) part of the blood system as seen through the mantle. The anatomy of two other common whelks, *Busycon* and *Fulgar*, is similar. (After W. S. Bullough, *Practical Invertebrate Anatomy*, 2nd ed., Macmillan, London, 1958.)

tained. Pond snails are useful in studies of embryology (p. 351), and land snails are used in studies of behavior to explore responses to stimuli (p. 301).

The labeled diagrams are only an introduction to careful dissection. Students will want to use the guides that are listed in the bibliography at the end of this chapter.

In club work, or as special projects done at home, students may want to dissect one or more representative mollusks. The directions given for the fresh-water clam can be applied to mussels and oysters as well. Preserved *Pecten*, as well as other marine specimens, may be available. In large cities many of these marine forms can be obtained at fish markets that specialize in Italian fare. At one market, we have purchased squids, land snails, periwinkles, whelks, cuttlefish, and even an octopus.

Gastropods

Most of the gastropods are marine, for example, periwinkles, whelks (Fig. 1-36), and abalones (Fig. 1-37). The chiton and

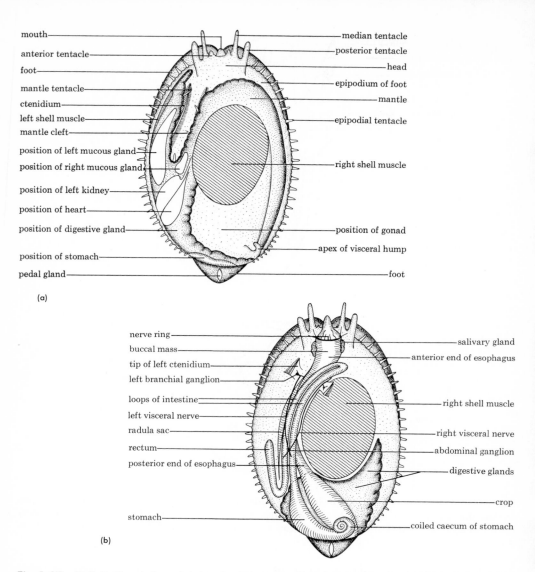

mouth
anterior tentacle
foot
mantle tentacle
ctenidium
left shell muscle
mantle cleft
position of left mucous gland
position of right mucous gland
position of left kidney
position of heart
position of digestive gland
position of stomach
pedal gland

median tentacle
posterior tentacle
head
epipodium of foot
mantle
epipodial tentacle

right shell muscle

position of gonad
apex of visceral hump
foot

(a)

nerve ring
buccal mass
tip of left ctenidium
left branchial ganglion
loops of intestine
left visceral nerve
radula sac
rectum
posterior end of esophagus
stomach

salivary gland
anterior end of esophagus

right shell muscle

right visceral nerve
abdominal ganglion
digestive glands

crop
coiled caecum of stomach

(b)

Fig. 1-37 *Haliotis,* the abalone: (a) dorsal view, with shell removed to show the position of the organs; (b) dorsal view, with respiratory, excretory, and reproductive organs removed to show the alimentary canal. (After W. S. Bullough, *Practical Invertebrate Anatomy,* 2nd ed., Macmillan, London, 1958.)

shipworm, *Toredo,* may also be available.

From an aquarium, students may obtain pond snails (Fig. 1-38); and from a garden or a fish market land snails, *Helix* (Fig. 1-39), are available for laboratory dissection. For a study of ciliated epithelial cells, see pp. 54, 107, 109.

Clam (Anodonta)

The directions given for this fresh-water clam can be applied to other bivalves—mussels and oysters.

Pry the two valves apart, and carefully cut through the large posterior adductor muscle below and behind the hinge. Then cut through the anterior adductor muscle. These two muscles keep the shell closed. Notice the mantle, a white membrane which lines each shell. Examine the posterior end, where the edges of the mantle meet to form the pigmented siphons. Notice too the papillae on the ventral siphon through which water and food enter, and the smooth edges of the dorsal, exhalant siphon (Fig. 1-40).

Turn back the mantle lobe. If you have a freshly killed specimen, scrape some cells from the edge of the mantle and from the gills. Look for ciliated cells (see p. 110). Notice that the gills on each side unite, forming a channel above the gills from which the dorsal siphon leads. The lower, or inhalant, siphon leads to the lower body cavity.

Now firmly grip the mantle lobe, and pull the body away from the dorsal margin. Under the hinge locate the heart. Carefully cut into this pericardial cavity, and find the yellowish ventricle, which may still be pulsating. A tube, the intestine, runs through the ventricle but has no connection with it. Look for an auricle on each side of the ventricle. Trace the anterior aorta (above the intestine) carrying blood forward, and the posterior aorta (below the intestine).

Posterior to the heart and in front of the posterior adductor muscle are the dark kidneys. The paired Keber organs are also part of the excretory system.

Next trace the entire digestive system. Pull down the anterior part of the foot, and locate the mouth (surrounded by ciliated labial palps) just back of the anterior adductor muscle, and in front of and above the foot. Behind the anterior adduc-

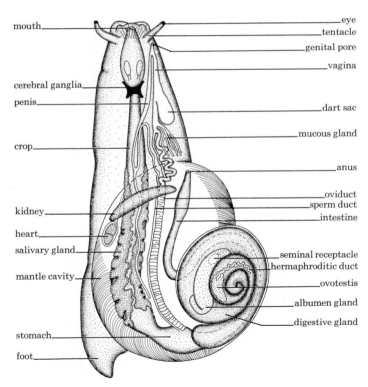

Fig. 1-38   Fresh-water snail: dorsal view, showing internal anatomy. (After R. W. Hegner and K. A. Stiles, *College Zoology*, 7th ed., Macmillan, New York, 1959.)

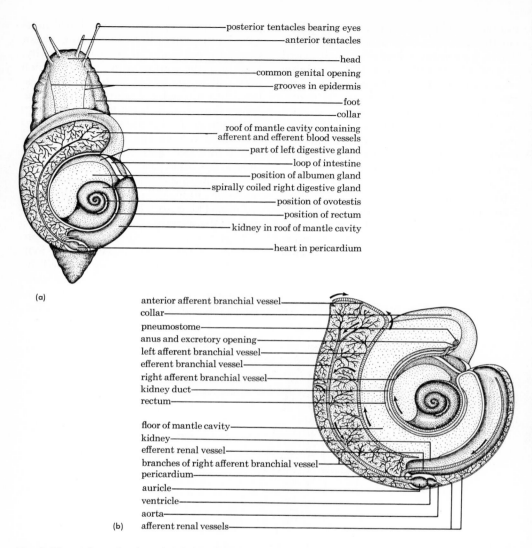

Fig. 1-39 *Helix,* a land snail: (a) dorsal view, with shell removed to show the visceral hump and the position of the organs within it; (b) direction of flow of blood through the body (mantle cavity exposed). (After W. S. Bullough, *Practical Invertebrate Anatomy,* 2nd ed., Macmillan, London, 1958.)

tor muscle find the darkly colored digestive gland, which surrounds the stomach. From here, trace the intestine through the heart and toward the excurrent siphon.

Carefully pare away the muscle of the foot to expose the ovary or testis in the posterior dorsal part. In a mature female, small, brown glochidia larvae (which soon attach themselves parasitically to the skin

of fish) may be found in the lamellae of the gills. Examine under the microscope.

It is difficult to locate the nervous system unless the specimen has been boiled or hardened in alcohol. Directly under the posterior adductor muscle try to find the two yellowish visceral ganglia encased by a thin membrane. From these ganglia, trace the nerves forward to the cerebral ganglia, near the surface at the mouth region (close to the base of the labial palps). Continue to follow the nerves into a pair of orange pedal ganglia deeply embedded in the foot (anterior to the gonad).

Microscopic examination of tissues.
When living clams are available, you can show ciliated epithelial tissue (Chapter 2). With forceps, peel thin sections from the edges of the mantle and gills, and mount them in the juice of the clam. Under high

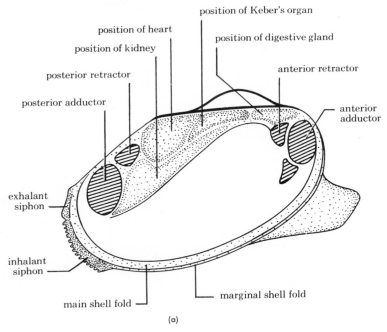

(a)

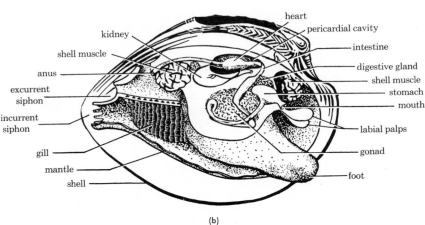

(b)

Fig. 1-40 *Anodonta,* a fresh-water clam: (a) details of mantle after removal or right shell valve; (b) internal anatomy, with shell, mantle, and gills on one side removed. (b, from W. G. Whaley, O. P. Breland, *et al., Principles of Biology,* Harper & Row, 1954; after R. W. Hegner, *Invertebrate Zoology,* Macmillan, New York, 1933.)

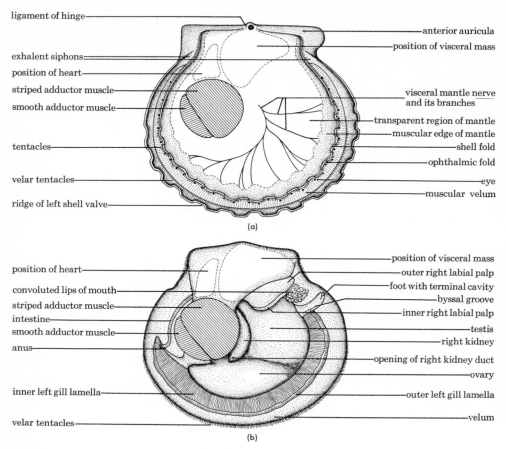

Fig. 1-41 *Pecten,* a scallop: (a) the right shell valve has been removed to show the right mantle; (b) internal anatomy, after removal of right mantle lobe. (After W. S. Bullough, *Practical Invertebrate Anatomy,* 2nd ed., Macmillan, London, 1958.)

power examine the rhythmic beating of the cilia; stain them with Janus Green B, methylene blue, or Lugol's solution (p. 121).

Students may also study the effects of different ions on ciliary motion (see p. 110).

In the aquarium snail *Physa,* a relative of the clam, ciliated epithelial cells line the intestine. Narcotize a snail in a warm solution of 1 percent magnesium sulfate. Mount its body in pond water on a glass slide, and dissect the intestine. With forceps peel off this lining, and mount it in a drop of pond water on a fresh slide; stain as above, if desired.

Scallop (Pecten)

Students may study a preserved specimen. Note that one side of the bivalve is more convex than the other. Study the ridges and the wide hinge. It may be possible to show by film the active, moving *Pecten* as it clicks through the water.

With the aid of Fig. 1-41, students may dissect and identify the main organs. Cut the adductor muscle with a knife to push back the convex shell on its hinge, and then remove it. Note especially the highly pigmented edge of the mantle with its blue "eyes." Remove the mantle in order to locate the organ systems.

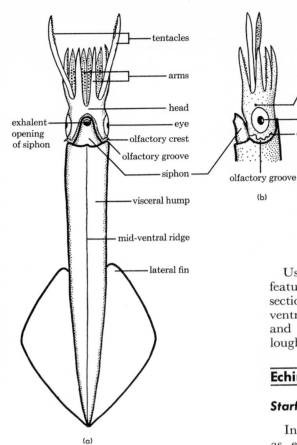

tentacles

arms

head

eye

exhalent opening of siphon

olfactory crest

olfactory groove

siphon

visceral hump

mid-ventral ridge

lateral fin

(a)

opening of aquiferous duct

eye

olfactory crest

olfactory groove

(b)

Fig. 1-42 *Loligo,* the squid: (a) ventral view, showing external anatomy; (b) lateral view of head region. (From W. S. Bullough, *Practical Invertebrate Anatomy,* 2nd ed., Macmillan, London, 1958, reprinted 1962.)

Squid and cuttlefish

Squids (*Loligo*) or the flatter cuttlefish (*Sepia*) are representative of the cephalopods. Both have ten arms and a reduced shell (the order Decapoda), while the octopus has eight arms and no shell. All these cephalopods have a highly developed head region with prominent eyes. (Students will want to make a careful study of the eye of the squid and compare it with a mammalian eye.)

Using Fig. 1-42, identify the external features of the squid. Then begin the dissection by cutting to one side of the mid-ventral ridge (Fig. 1-43). (For complete and excellent directions, refer to Bullough.[8])

Echinoderms

Starfish

In some areas many marine forms such as echinoderms are readily obtainable. When living starfish are available they may be narcotized by placing them in a small amount of water to which some crystals of magnesium sulfate (Epsom salts) have been added. If the starfish are put upside down when narcotized, the tube feet will probably remain extended. When the animals no longer respond to touch, they may be stored in 70 percent alcohol for future dissection.

Examine a starfish to locate the ventral mouth (the oral surface) and the ambulacral grooves containing tube feet (Fig. 1-44a). Each tube foot ends in suction cups —the means of locomotion and attachment. On the dorsal (aboral) side find the madreporite—part of the water-vascular system (Fig. 1-44c).

[8] *Op. cit.,* pp. 401–09.

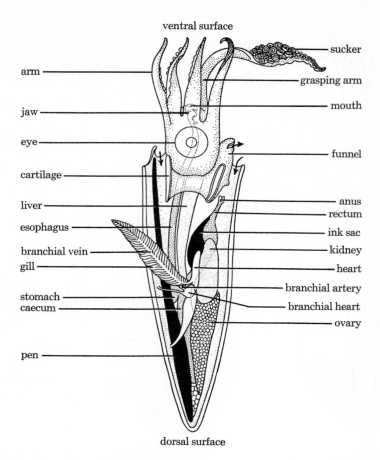

ventral surface

arm

jaw

eye

cartilage

liver

esophagus

branchial vein

gill

stomach
caecum

pen

sucker

grasping arm

mouth

funnel

anus
rectum

ink sac

kidney

heart

branchial artery

branchial heart

ovary

dorsal surface

Fig. 1-43 *Loligo,* the squid: internal anatomy, with body wall and arms removed from right side. (Reprinted with permission of the publisher from *College Zoology* by R. W. Hegner and K. A. Stiles, 7th ed., © The Macmillan Co., New York, 1959.)

Without breaking the madreporite, remove the skin from the dorsal region. Carefully separate the skin from the central disc, where parts of the digestive system are attached. Note the branched stomach divided into pyloric and cardiac parts (Fig. 1-44b). Ducts lead outward from the pyloric parts and branch into pyloric caeca containing digestive enzymes. Now remove the pyloric caeca from one arm to reveal the two retractor muscles connected to the cardiac regions of the stomach which pull in the stomach when it has been everted. Cut away these muscles, and find the tubes of the water-

vascular system. Below the madreporite, locate the calcareous stone canal. Water enters through the madreporite into the stone canal, which leads into the water-vascular ring around the mouth. This fluid moves along radial canals to ampullae, into tube feet. Tiedemann's bodies are thought to produce the amoeboid corpuscles found in the water-vascular system.

The paired reproductive organs, either testes or ovaries, are located at the junction of the arms.

The blood system and the nervous system are difficult to trace. Although you may find a nerve ring around the mouth,

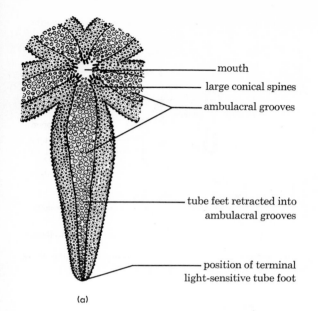

— mouth
— large conical spines
— ambulacral grooves

— tube feet retracted into
ambulacral grooves

— position of terminal
light-sensitive tube foot

(a)

Fig. 1-44 *Asterias,* a starfish: (a) external ventral (oral) view; (b) dorsal view of internal anatomy; (c) detail of part of the water-vascular system. (a, b, from W. S. Bullough, *Practical Invertebrate Anatomy,* Macmillan, London, 1950, reprinted 1954; c, reprinted with permission of the publisher from *College Zoology* by R. W. Hegner and K. A. Stiles, 7th ed., ⓒ The Macmillan Co., New York, 1959.)

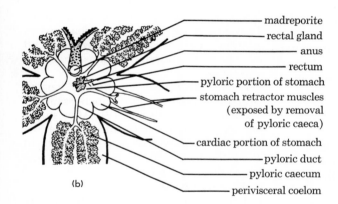

— madreporite
— rectal gland
— anus
— rectum
— pyloric portion of stomach
— stomach retractor muscles
(exposed by removal
of pyloric caeca)
— cardiac portion of stomach
— pyloric duct
— pyloric caecum
— periviseral coelom

(b)

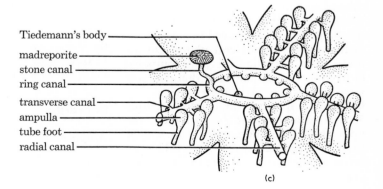

Tiedemann's body —
madreporite —
stone canal —
ring canal —
transverse canal —
ampulla —
tube foot —
radial canal —

(c)

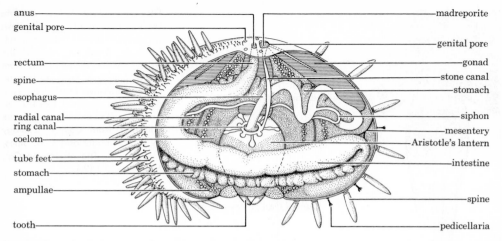

anus — — madreporite
genital pore — — genital pore
rectum — — gonad
spine — — stone canal
esophagus — — stomach
radial canal — — siphon
ring canal —
coelom — — mesentery
— Aristotle's lantern
tube feet — — intestine
stomach —
ampullae — — spine
tooth — — pedicellaria

Fig. 1-45 *Arbacia*, a sea urchin, showing both external structure and internal anatomy. (After R. W. Hegner and K. A. Stiles, *College Zoology*, 7th ed., Macmillan, New York, 1959.)

the radial nerves into the arms are not readily visible. Respiratory exchanges occur through thin-walled extensions of the body cavity that branch out through small openings in the skeleton on the dorsal surface of the arms. Both the outer and the inner surfaces of these dermal branchiae are ciliated; their movement produces a constant current.

Sea urchin

Arbacia (Fig. 1-45) may be available for study. The five-part radial symmetry is apparent in forms from which the spines have been removed. If the five rays of a starfish were turned backward and upward, the globular shape of this purple-colored sea urchin would result.

Look for tube feet among the long spines; these feet are moved by muscles attached to the ambulacral plates. On the oral surface, examine the jaws composed of five long, tooth-like radii that compose a structure called Aristotle's lantern.

Dissect a specimen by carefully cutting a circle around the globe to remove the upper half. The main systems are similar to those of the starfish. Use Fig. 1-45 as an introductory guide in locating the main organs.

Sea cucumber

These members of the class Holothuroidea have unusual powers of regeneration. They eviscerate through the cloaca when attacked, and the sac-like body crawls off to regenerate internal organs from the oral region.

Locate the mouth among the ten enlarged tube feet that form much-branched tentacles (Figs. 1-46, 12-19). After studying its external appearance, lay a sea cucumber on its side and carefully cut along the length of the animal to expose the organs; pin back the body wall and make a transverse cut through the body wall to flatten it. If you dissect in a pan of water, the branches of the two respiratory trees can float freely. Distinguish the coils of the digestive tube from the tubes of the one ovary or testis that is suspended from the dorsal wall mesentery.

Trace the water-vascular system using Fig. 1-46 as a guide.

Chordates

Amphioxus

Amphioxus, a member of the class Cephalochorda, is often studied as a representative of primitive chordates. Examine, under low power, prepared slides of a young lancet for a study of its total appearance: a fish-like shape with no specialized head or limbs. Find the notochord and locate the myotomes and gill slits. Reproductive organs are not apparent in these young forms that are mounted on slides.

If possible, examine preserved mature specimens (Fig. 1-47).

Dogfish shark

The basic chordate pattern is found in this cartilaginous fish, the dogfish shark,

Squalus. Related forms are skates and rays. These vertebrates can be collected, or obtained from supply houses and some local fish markets.

As a guide in dissection, use Fig. 1-48, plus a textbook such as R. W. Hegner and K. A. Stiles, *College Zoology.*[9]

Fish (generalized)

Hold the fish in one hand, begin the incision near the anal region, and cut forward for an inch or so. Then lay the fish in

[9] R. W. Hegner and K. A. Stiles, *College Zoology,* 7th ed., Macmillan, New York, 1959, pp. 368–73.

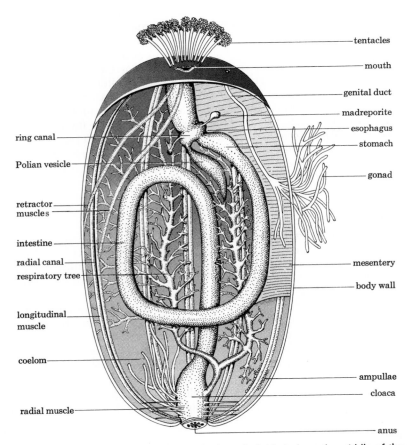

Fig. 1-46 Sea cucumber: internal anatomy, with body wall divided along the middle of the dorsal surface. (After T. J. Parker and W. A. Haswell, *A Textbook of Zoology,* Macmillan, London, 1930, and R. W. Hegner and K. A. Stiles, *College Zoology,* 7th ed., Macmillan, New York, 1959.)

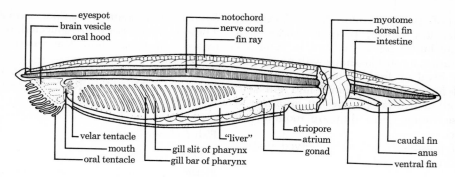

Fig. 1-47 *Amphioxus,* the lancet: internal anatomy. This organism, which measures about 2 in. long, illustrates the major characteristics of chordates. (Reprinted with permission of the publisher from *College Zoology* by R. W. Hegner and K. A. Stiles, 7th ed., © The Macmillan Co., New York, 1959.)

a waxed pan or on a corkboard, and cut forward along the midline. Be careful not to injure any organs. Find the heart in a pericardial sac beneath the pharynx (Fig. 1-49b). Blood moves from the auricle, into the ventricle, into the ventral aorta, along afferent branchial arteries, and into the gills. After the blood is aerated, it moves into efferent branchial arteries into the dorsal aorta. It returns through the veins into the sinus venosus, which empties into the auricle. Students may be able to trace some of these main blood vessels.

Food enters the mouth, passes along a short esophagus into a stomach, and then past the pyloric valve into the intestine. Look for the liver and the gall bladder. At the anterior end of the intestine you may find a large, red spleen.

Students should find the large air bladder in the dorsal part of the abdomen. High up in the dorsal region of the abdominal cavity look for the kidneys. Trace the ureters into a urinary bladder opening, and then into a urogenital orifice, posterior to the anus. Students should be able to distinguish ovaries from testes and trace the reproductive tubes into the urogenital opening.

Examine the respiratory system next. Find the paired gills, which are supported by gill arches. Notice the branchial filaments (double rows) on each gill and their rich supply of capillaries. Also note the gill

rakers, which act as sieves to hold back food particles.

Pare away the bony brain case with a scalpel, and locate the brain, which is composed of four distinct divisions: cerebrum, cerebellum, optic lobes, and medulla (Fig. 1-49a). Students may be able to find the olfactory lobes in front of the cerebrum. Try to trace the cranial nerves leading to the sense organs. In the neural arches of the vertebrae find the spinal cord with its branching spinal nerves.

A dissection of the eye of the fish will reveal a spherical lens.

Frog (Rana)

This description of the anatomy of the frog will hold for a study of frogs generally and for most toads and salamanders.

If you use preserved frogs, you cannot see a beating heart or study the contrast of colors of the organs. Try to have at least one freshly killed (pithed, p. 308; or anesthetized, p. 112) frog so that students can see the deeply colored auricles contract to pass blood to the ventricle, and the ventricle contract, changing from deep red to light pink as it sends blood into the dorsal aorta.

If you have live frogs, students can observe the breathing movements and eye reflex (p. 306); demonstrate circulation in the webbed foot (p. 246); or study fresh blood cells, ciliated epithelial cells from the roof of the mouth, intestinal protozoa, or sperms (see Chapter 2).

For dissection, lay a dead frog on its back in a waxed pan, and pin down the extended limbs. Use forceps to pick up the loose skin of the abdomen between the hind legs, and cut from there up to the lower jaw. Turn back the skin (notice its rich supply of blood vessels), and pin it down. Cut through the white abdominal muscles, being careful not to cut into the ventral abdominal vein or the organs beneath. At the pectoral region you will have to cut through bone. Now use both hands to spread out the pectoral region,

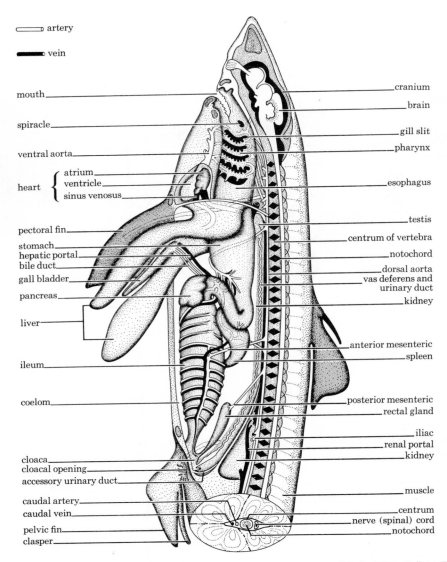

Fig. 1-48 *Squalus*, the dogfish: internal anatomy. (After R. W. Hegner and K. A. Stiles, *College Zoology*, 7th ed., Macmillan, New York, 1959.)

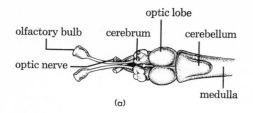

optic lobe
olfactory bulb   cerebrum   cerebellum
optic nerve
medulla
(a)

Fig. 1-49 Anatomy of the fish: (a) detail of brain; (b) lateral view. (From G. G. Simpson and W. S. Beck, *Life: An Introduction to Biology*, 2nd ed., Harcourt, Brace & World, Inc., 1965.)

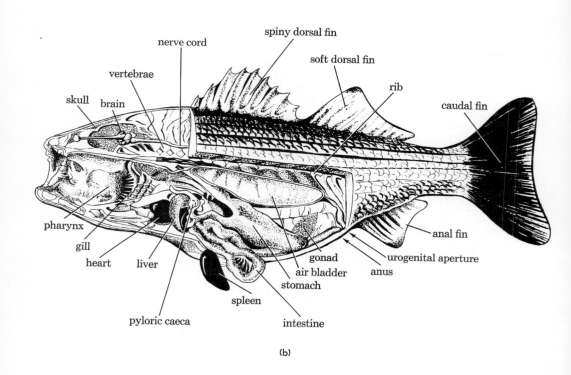

spiny dorsal fin
nerve cord
soft dorsal fin
vertebrae
rib
skull   brain
caudal fin
pharynx
gill
anal fin
heart   liver
gonad   urogenital aperture
air bladder   anus
stomach
spleen
pyloric caeca
intestine
(b)

carefully. Cut away any connecting mesentery, and pin the muscles back to the pan. Make incisions into each of the limbs so that the abdominal wall can be turned back and pinned. (See Fig. 1-50.)

Locate the heart in a pericardial sac (Fig. 1-51), and identify each organ system: respiratory, digestive, circulatory, excretory, and reproductive (Fig. 1-52). Blow into the glottis with a fire-polished glass tube to inflate the lungs. Use a blunt, flexible probe to push back into the mouth so that the gullet can be identified. The pancreas, which looks like a yellow cord, is in the loop formed by the stomach and intestine.

If your specimen is a female distended with eggs, one ovary and an oviduct should be removed. Trace the other oviduct to locate the funnel (ostium at the base of the lungs) where eggs which break away from the ovary migrate (by action of the ciliated cells lining the body cavity) to enter the oviducts. Locate, at the other end of the oviducts, the opening to the exterior, via the cloaca.

In the male locate two oval, orange or yellow testes. Trace the fine threads of the vasa efferentia connecting each testis to a kidney. Find the vas deferens, or ureter, leading from the outer edge of the posterior part of each kidney back to the cloaca.

Observe the yellow fat bodies, finger-like masses, connected to both the ovaries and testes. Near the rectum find the red, spherical spleen. Insert a flexible probe

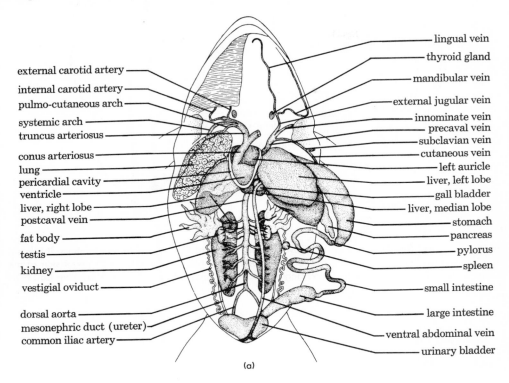

external carotid artery
internal carotid artery
pulmo-cutaneous arch
systemic arch
truncus arteriosus
conus arteriosus
lung
pericardial cavity
ventricle
liver, right lobe
postcaval vein
fat body
testis
kidney
vestigial oviduct
dorsal aorta
mesonephric duct (ureter)
common iliac artery

lingual vein
thyroid gland
mandibular vein
external jugular vein
innominate vein
precaval vein
subclavian vein
cutaneous vein
left auricle
liver, left lobe
gall bladder
liver, median lobe
stomach
pancreas
pylorus
spleen
small intestine
large intestine
ventral abdominal vein
urinary bladder

(a)

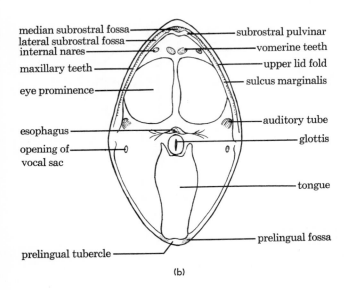

median subrostral fossa
lateral subrostral fossa
internal nares
maxillary teeth
eye prominence
esophagus
opening of
vocal sac
prelingual tubercle

subrostral pulvinar
vomerine teeth
upper lid fold
sulcus marginalis
auditory tube
glottis
tongue
prelingual fossa

(b)

Fig. 1-50 Anatomy of the frog: (a) internal anatomy; (b) structure of open mouth. (Courtesy of General Biological Supply House, Inc., Chicago.)

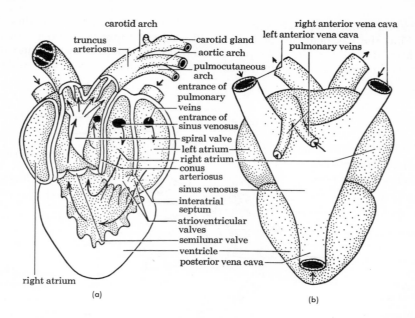

Fig. 1-51   Heart of the frog: (a) ventral view of heart, dissected to show the three chambers; (b) dorsal view of complete heart. Arrows indicate direction of flow of blood. (Reprinted with permission of the publisher from *College Zoology* by R. W. Hegner and K. A. Stiles, 7th ed., © The Macmillan Co., New York, 1959.)

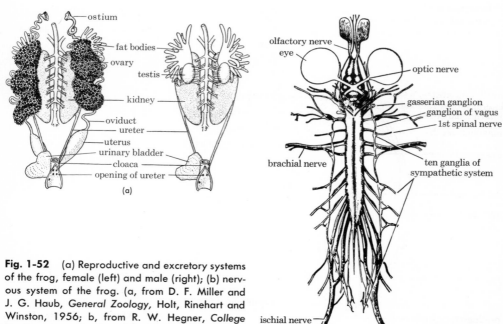

Fig. 1-52   (a) Reproductive and excretory systems of the frog, female (left) and male (right); (b) nervous system of the frog. (a, from D. F. Miller and J. G. Haub, *General Zoology*, Holt, Rinehart and Winston, 1956; b, from R. W. Hegner, *College Zoology*, rev. ed., Macmillan, New York, 1926.)

through the anus to locate the urinary bladder, or inflate the bladder by blowing through a fire-polished glass tube.

In a preserved specimen students may be able to study the nervous system in fine detail. Pare away the top of the skull with a scalpel until you find the two hemispheres of the cerebrum between the eyes. In front of the cerebrum you may be able to locate the two pear-shaped olfactory lobes, which lead out into the nasal region. Posterior to the cerebrum the optic nerves may be found extending from the optic lobes. Then trace the cerebellum and extension of the medulla into the spinal cord. When you lift up the organs within the abdomen locate the ganglia and branching spinal nerves extending from the spinal cord. Study the position of the spinal cord in relation to the spinal column. Follow the sciatic nerves into the muscles of the thighs.

Snake (generalized)

If the spinal cord of a freshly killed snake is not destroyed, the snake must be fastened securely to a board. (It is best, of course, to destroy the cord.) Lay it on its back and tack down the upper jaw and the tail end. Carefully pick up the skin at the throat, and make an incision (cutting slowly so as not to cut through the distended air sac which fills most of the body) extending to the anus; pin back the skin every few inches. Cut through the membranes across the rib section, avoiding blood vessels (Fig. 1-53).

Use glass tubing to inflate the stomach, then the right lung. Notice the air sac extending from the lung. Look for the undeveloped left lung found in some snakes at the end of the trachea.

Fig. 1-53 Internal anatomy of the snake. Note that there is a break in the drawing. Note also the enormous stomach. (After J. T. Saunders and S. M. Manton, A Manual of Practical Vertebrate Morphology, 2nd ed., Oxford Univ. Press, New York, 1949.)

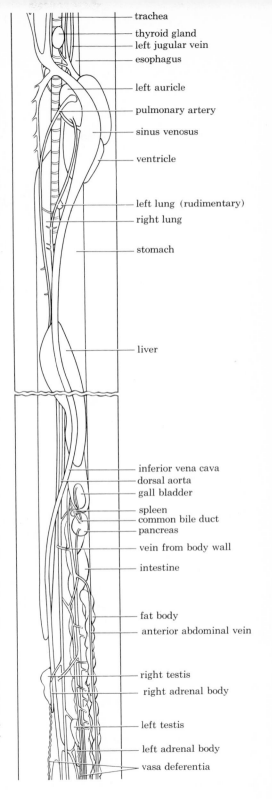

trachea
thyroid gland
left jugular vein
esophagus

left auricle

pulmonary artery

sinus venosus

ventricle

left lung (rudimentary)
right lung

stomach

liver

inferior vena cava
dorsal aorta
gall bladder

spleen
common bile duct
pancreas

vein from body wall

intestine

fat body
anterior abdominal vein

right testis
right adrenal body

left testis

left adrenal body

vasa deferentia

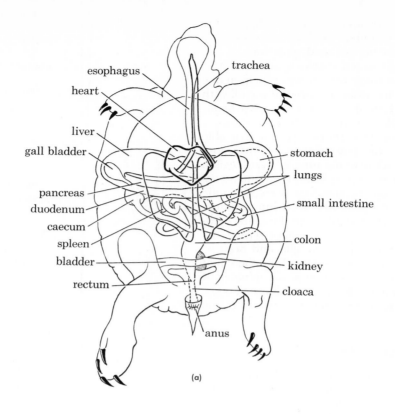

(a)

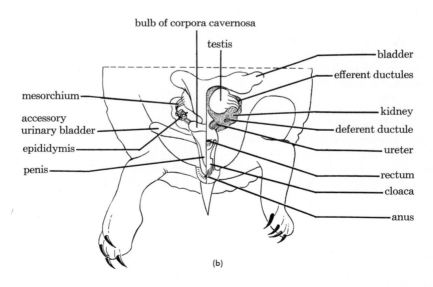

(b)

Fig. 1-54 Anatomy of the turtle: (a) digestive system, respiratory system, and heart; (b) male urogenital system.

Push aside the organs to find the two auricles and one ventricle of the heart lying within a pericardium. Perhaps students can trace the large aorta leading out of the ventricle between the auricles and bending around the gullet; follow its path posteriorly, or find some of the anterior branches.

Find the dark liver near the stomach, and the dark gall bladder and pale pancreas, both near the spleen. Trace the intestine along the length of the abdomen. Examine the ribbon-like fat bodies that extend posteriorly from the area of the intestine. In the posterior region of the abdomen find the whitish right testis lying anterior to the left testis; or, in the female, find pink masses of ovaries. Trace the pinkish oviducts or the smaller sperm ducts to the cloaca. Also look for the long adrenals located near the gonads. Posterior to the gonads, find the dark red, long, twisted kidneys. Notice that the right kidney lies a bit anterior to the left one. Trace the ureters leading out of the kidneys.

In a preserved specimen students might try to dissect the brain, as described for the frog (p. 65).

Turtle (generalized)

When preparing a fresh turtle for dissection, use a strong forceps to hold the head extended, and snip through the spinal cord at the back of the neck with strong clippers or shears. Then, with a hacksaw or coping saw, cut through the bridge which connects the carapace (upper shell) and the plastron (lower shell) on each side. Raise the plastron, and cut away the connecting membranes; then remove the plastron.

Cut through the membrane to reveal the internal organs, and examine the conspicuous heart (Fig. 1-54a). Near the heart find the liver and the gall bladder. The stomach is near or under the left lobe of the liver. Lift up the liver to locate the lungs. Try to inflate the lungs by blowing into them through glass tubing.

Follow the intestine from the stomach to the transverse vent under the tail. Locate the urinary bladder, and trace the ureters back to the kidneys. Masses of eggs may be found in the ovary of a female specimen. Trace the ducts leading from the gonads (ovaries or testes) to the outside.

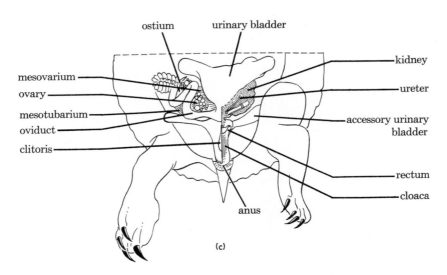

Fig. 1-54 (continued)   Anatomy of the turtle: (c) female urogenital system. (a, b, c, after L. M. Ashley, *Laboratory Anatomy of the Turtle*, Wm. C. Brown, 1955.)

Pigeon

Before any bird is dissected it should be dipped into hot water so that the feathers can be plucked off easily.

Stretch the wings and legs out on a corkboard or dissecting pan, and fasten them securely. Cut forward from the posterior end of the keel, or breastbone, revealing the crop (where food is macerated). Pull the crop forward to see how it is connected to the stomach, or proventriculus (Fig. 1-55). Then identify the trachea, with its cartilage rings. Students may also find the jugular vein on each side of the neck.

Try to insert glass tubing into the glottis, and blow into it to inflate the air sacs in front of the breastbone. Students may want to experiment by blowing air into a broken humerus and tracing the path of air through the hollow bones from the windpipe (Fig. 1-56). Air which passes from the glottis into the trachea finally enters nine large, thin-walled air sacs, situated mainly along the dorsal sides of the body cavity. When air is exhaled, the muscles of the thorax and the abdomen contract, sending air out of the air sacs into the lungs and finally out through the trachea.

A structure peculiar to birds is the syrinx, a vocal organ found at the base of the trachea where the trachea divides to form two bronchi.

Now cut through the skin of the abdomen, and continue posteriorly to the

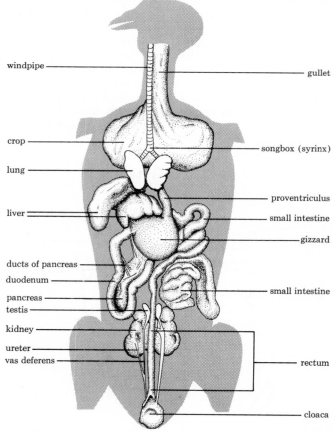

Fig. 1-55 Internal organs of the pigeon (heart has been removed). (After A. J. Thomson, *The Biology of Birds,* Sidgwick and Jackson Ltd., 1923.)

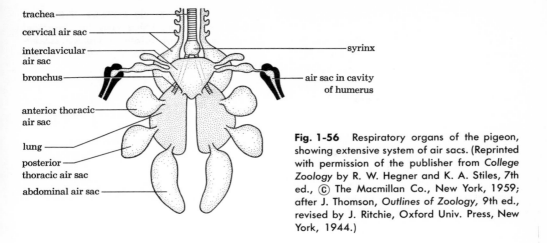

trachea
cervical air sac
interclavicular air sac
bronchus
anterior thoracic air sac
lung
posterior thoracic air sac
abdominal air sac
syrinx
air sac in cavity of humerus

Fig. 1-56 Respiratory organs of the pigeon, showing extensive system of air sacs. (Reprinted with permission of the publisher from *College Zoology* by R. W. Hegner and K. A. Stiles, 7th ed., Ⓒ The Macmillan Co., New York, 1959; after J. Thomson, *Outlines of Zoology*, 9th ed., revised by J. Ritchie, Oxford Univ. Press, New York, 1944.)

anus. Abdominal air sacs may be seen if they are inflated. Students should locate the heart in front of the large liver. Since the heart is comparatively large, students may dissect it to locate the four chambers.

Also find the pink lungs attached to the back of the animal. To the left of the body cavity, find the hard gizzard extending from the stomach. Trace the intestine from the gizzard. Cut into the gizzard and explain its structure. Find the pink-colored pancreas in a loop of the duodenum of the intestine. Toward the end of the intestine, near the rectum, find two side branches, caeca, and finally trace the intestine to the cloaca.

In the hollow of the pelvis find the lobed kidneys. Below these, students may look for the sciatic nerves which lead from the spinal cord into the hind legs.

Next, students should find the two testes or the single ovary (the left one) in front of the kidneys. Trace the one oviduct or the sperm ducts to the cloaca. There is no urinary bladder in birds.

Dissect the leg to show how a bird remains perched when asleep. Show that in a relaxed position the toes grasp the perch.

The dissection of the brain is a more difficult task. With a sharp knife pare away the bone of the skull to expose the soft brain. Then try to identify the parts of the brain.

Fetal pig

In some areas fetal pigs may be more readily available for dissection than rabbits, cats, or white rats. If you purchase a pig's uterus for a study of the relationship of the embryo, placenta, and umbilical cord (p. 366) there will be several fetuses available for dissection.

Place the pig fetus on its back in a waxed pan. Tie a cord around one forelimb, and bring it around underneath the pan to fasten back the other forelimb. Spread apart the hind limbs in the same way. Then make an incision in the midregion of the chest; cut through the skin laterally and along the medial line toward the umbilical cord. Use Fig. 1-57a as a guide in dissecting around the umbilical cord. Fold back the thick skin and note the peritoneum. Locate the liver; note the umbilical vein that runs from the umbilical cord to the liver. Fold over and pin down the body wall. (The umbilical vein must be cut before you can fold over the part of the body wall that contains the umbilical cord.) In the umbilical cord stump find the allantoic stalk, one umbilical vein, and two umbilical arteries.

Identify the liver, stomach, intestine, and diaphragm. Carefully remove the pericardium, and trace the blood vessels leading from the heart, especially the

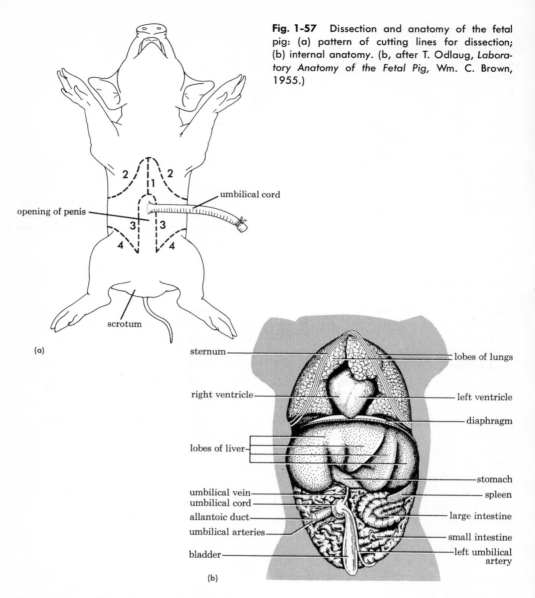

Fig. 1-57 Dissection and anatomy of the fetal pig: (a) pattern of cutting lines for dissection; (b) internal anatomy. (b, after T. Odlaug, *Laboratory Anatomy of the Fetal Pig,* Wm. C. Brown, 1955.)

umbilical cord

opening of penis

scrotum

(a)

sternum

right ventricle

lobes of liver

umbilical vein
umbilical cord
allantoic duct
umbilical arteries

bladder

lobes of lungs

left ventricle

diaphragm

stomach
spleen

large intestine

small intestine

left umbilical artery

(b)

dorsal aorta. The thymus gland is found ventral to the heart, extending forward.

Now trace the digestive system, including the gall bladder, bile duct, and pancreas. You will also find the spleen, a flattened, dark red organ near the stomach.

Examine the lungs, and trace their connection to the trachea.

Then begin a careful study of the urogenital system (Fig. 1-58). In a female, first

trace the urinary system. The kidneys lie in the dorsal part of the abdomen. From these trace the ureters toward the urinary, or allantoic, bladder. Can students find that one end of the bladder extends as the allantoic stalk into the umbilical cord? On top of the kidneys find the long, narrow adrenal glands. Then locate the small, oval ovaries a little posterior to the kidneys. Study the position of the Fallopian (uterine) tubes and the uterus. Now cut through the muscle in the pelvic girdle, and locate the white line of fusion, the pubic symphysis. At this point, split the girdle with a scalpel to reveal the rest of the reproductive system: the urethra (a duct leading from the urinary bladder to the urogenital sinus), the vagina, and the urogenital sinus.

In the male, separate the penis from the ventral body wall posterior to the umbilical cord. In older specimens the testes may have already descended into scrotal sacs. Look for the vasa deferentia, the thin sperm ducts from the testes, and find how they loop over the ureters. Locate the crescent-shaped epididymis enclosing the inner side of each testis, and see how each is connected with the vas deferens. Also trace the urethra from the bladder to the penis.

Finally, students may try to trace the nervous system and make a study of the structure of the eye. A hemisection of the brain may be more practical.

Rat

Lay the animal ventral side up and fasten with cords as described for the fetal pig. Make an incision in the skin along the midline from the base of the neck to the pelvis. Fold back the skin and fasten it with stout pins to a board or dissecting pan. Carefully cut through the muscular abdominal wall up to the breastbone. When the chest is opened, the lungs and heart collapse (Fig. 1-59). If you inflate the lungs to their normal size, students may see that the lungs fill the chest cavity and nearly encircle the heart. Study the lobes of the lungs and examine the muscles of the diaphragm.

For the most part, the rest of the anat-

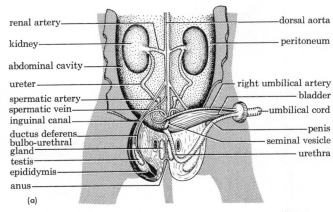

renal artery
kidney
abdominal cavity
ureter
spermatic artery
spermatic vein
inguinal canal
ductus deferens
bulbo-urethral gland
testis
epididymis
anus

dorsal aorta
peritoneum
right umbilical artery
bladder
umbilical cord
penis
seminal vesicle
urethra

(a)

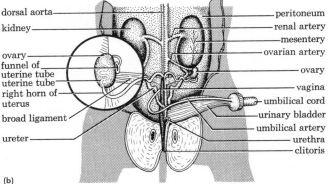

dorsal aorta
kidney
ovary
funnel of uterine tube
uterine tube
right horn of uterus
broad ligament
ureter

peritoneum
renal artery
mesentery
ovarian artery
ovary
vagina
umbilical cord
urinary bladder
umbilical artery
urethra
clitoris

(b)

Fig. 1-58 Urogenital systems of the fetal pig: (a) male; (b) female. (After T. Odlaug, *Laboratory Anatomy of the Fetal Pig*, Wm. C. Brown, 1955.)

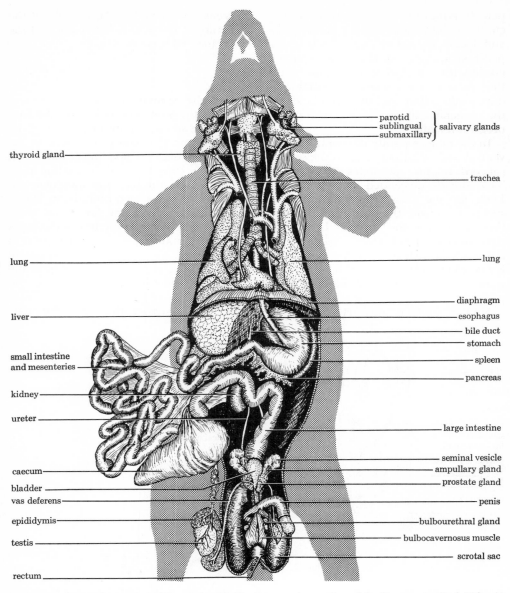

parotid
sublingual } salivary glands
submaxillary

thyroid gland

trachea

lung — lung

diaphragm
esophagus

liver
bile duct
stomach

small intestine
and mesenteries
spleen
pancreas

kidney

ureter

large intestine

seminal vesicle
ampullary gland
caecum
prostate gland

bladder
vas deferens
penis

epididymis
bulbourethral gland

testis
bulbocavernosus muscle

scrotal sac

rectum

Fig. 1-59 Internal anatomy of the rat (male). The heart and a portion of the liver are omitted. (After M. Ulmer, R. Haupt, and E. Hicks, *Comparative Chordate Anatomy*, Harper & Row, 1962.)

omy can be identified from previous dissections. Students should be able to trace the path of food through the alimentary canal and identify the pancreas and liver.

Locate the main blood vessels leaving the heart; trace the branches of the aorta carrying blood to the head, brachial vessels, and the main stem supplying the body organs. With care, students may trace the pulmonary circulation; notice the large veins, inferior and superior venae cavae, which enter the right auricle.

Now move the digestive organs aside and trace the urogenital system. Find the kidneys (with adrenal glands on top) and the ureters leading to the bladder; identify the ovaries or testes and their ducts. Prepare a slide of sperm suspension from a testis macerated in Ringer's solution (mammalian, p. 665).

Raise these organs and find the dorsal spinal column with spinal nerves leading out of the spinal cord. Dissect a limb and trace the sciatic nerve into it; examine the relationships of muscles, nerves, and bones in the leg.

Cat

In dissecting a cat, use Fig. 1-60 as a preliminary guide. Complete dissection guides are available; some are indicated in the Bibliography.

Seed plants

Leaves

Although leaves take many shapes, their function is the same—food-making for the plant. Six types of leaves are shown in

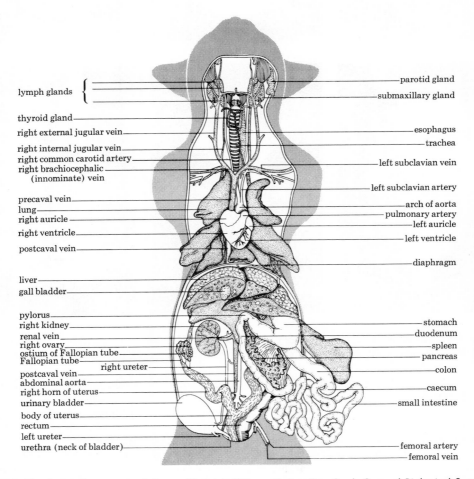

Fig. 1-60 Internal anatomy of the cat (female). (After a Turtox Key Card, General Biological Supply House, Inc., Chicago.)

(a) (b) (c)

(d) (e) (f)

Fig. 1-61 Types of leaves: (a) entire margin (mountain laurel); (b) serrate (birch); (c) dentate (chestnut); (d) lobed (chestnut oak); (e) cut (black oak); (f) compound (black locust). (Photos: A. M. Winchester.)

Fig. 1-61. Some leaves are only a few cells thick (elodea, p. 106), while others have elaborate adaptations (Fig. 1-65).

You may have students make some crude, freehand sections of a leaf, or you may prefer to purchase prepared slides. To make a freehand cross section, sandwich a piece of leaf between two halves of elder-berry pith or balsa wood (p. 113). With a

sharp razor, slice thin sections at an angle; mount in a drop of water. Some sections cut in this way will have thin edges; under a microscope the cellular structure of the leaves can be seen (Fig. 1-62).

Students may be able to identify the upper epidermis (which lacks chloro-plasts). Under this layer look for the palisade layer (or layers) filled with chloro-

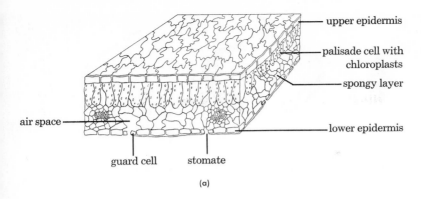

upper epidermis

palisade cell with chloroplasts

spongy layer

lower epidermis

air space

guard cell   stomate

(a)

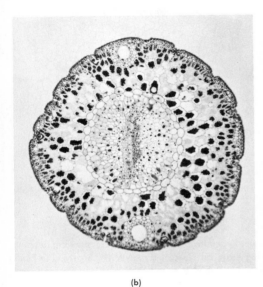

(b)

Fig. 1-62 Cross section of leaves: (a) typical green leaf (angiosperm); (b) pine leaf (*Pinus monophylla*), magnified 80×, reduced by 27 percent for reproduction. (b, photomicrograph by Larrance M. O'Flaherty and slide by LaRea J. Dennis, Herbarium, Dept. of Botany, Oregon State Univ., Corvallis, Oreg.)

plasts. This is the layer in which most of the food-making occurs. Below this is the loosely packed, spongy layer; the cells are interspersed with many air spaces. Below this layer students should find the lower epidermis, the layer throughout which the stomata and their guard cells (containing chloroplasts) are usually found. It is through these stomata that air containing carbon dioxide enters. The carbon dioxide dissolves in the water within the air spaces around the cells, diffuses through the membranes of cells, and finally reaches the actively photosynthesizing palisade cells.

Students may study guard cells and

Fig. 1-63 Dogwood. The colored bracts of dogwood are modified leaves; flower clusters are in the center. (Courtesy of U.S. Forest Service.)

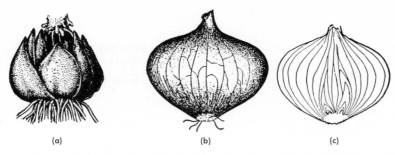

(a)   (b)   (c)

Fig. 1-64 Specialization of leaves for storage: (a) lily bulb; (b) and (c) onion bulb. (From W. H. Brown, *The Plant Kingdom*, 1935; reprinted through the courtesy of Blaisdell Publishing Co., a division of Ginn and Co.)

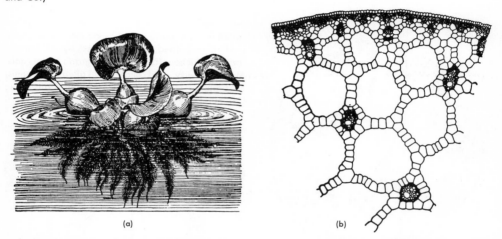

(a)   (b)

Fig. 1-65 Specialization of leaves for floating: (a) water hyacinth (*Eichhornia crassipes*), showing petioles modified as floats; (b) cross section of petiole, showing large air spaces. (From W. H. Brown, *The Plant Kingdom*, 1935; reprinted through the courtesy of Blaisdell Publishing Co., a division of Ginn and Co.)

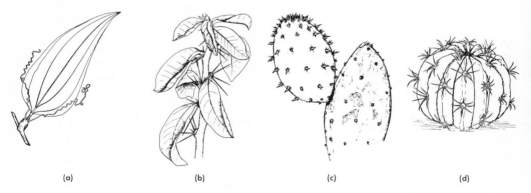

(a)   (b)   (c)   (d)

Fig. 1-66 Other modified leaves: (a) tendrils of *Smilax*; (b), (c), and (d) examples of spines on cactus. (a, from W. H. Brown, *The Plant Kingdom*, 1935, reprinted through the courtesy of Blaisdell Publishing Co., a division of Ginn and Co.; b, c, d, from C. L. Wilson and W. E. Loomis, *Botany*, rev. ed., Dryden [Holt, Rinehart and Winston], 1957.)

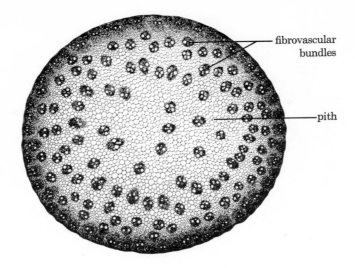

Fig. 1-67 Cross section of monocotyledonous stem (*Zea mays*). Fibrovascular bundles are scattered in the pith; the cortex is a layer only one or two cells thick between the epidermis and the outermost series of vascular bundles. (From W. H. Brown, *The Plant Kingdom*, 1935; reprinted through the courtesy of Blaisdell Publishing Co., a division of Ginn and Co.)

stomata in the lower epidermis of some leaves. Peel the lower epidermis near a midvein of a lettuce leaf, or from geranium, *Kalanchoë*, or *Tradescantia*, or an onion shoot (p. 175). Mount the lower epidermis in a drop of water, and examine it to find the numerous kidney-shaped pairs of guard cells surrounding the stomata.

Specialization of leaves. Some leaves are highly specialized organs. The familiar pink and white "petals" of dogwood are really modified leaves (Fig. 1-63). Students may remember that the leaves of onions and other bulbs are adapted to store food materials (Fig. 1-64). Still other leaves, such as those of the Venus flytrap, are modified to capture insects (Fig. 11-20b). Plants such as the water hyacinth have leaves containing large air spaces which keep the plants afloat (Fig. 1-65). The tendrils of *Smilax*, the garden pea, and many other plants are also modified leaves, as are the spines of cactus (Fig. 1-66). The leaves of *Bryophyllum* are specialized; the notches of the leaves are primordia of new plants (Fig. 9-33b).

Stems

Students can also make freehand sections of stems. Compare cross sections of monocotyledonous and dicotyledonous stems (Figs. 1-67, 1-68).

The rise of colored liquids through the fibrovascular bundles of celery, carrots, or growing seedlings may be demonstrated (p. 232). Still better, fibrovascular bundles can be studied in the orange-flowered jewelweed, or touch-me-not (*Impatiens*), found in swampy regions in late summer. Its stems are practically transparent and show the fibrovascular bundles leading up to the leaves. Students may collect these stems in August, cut them into short sections, and preserve them in 70 percent alcohol until they are needed.

They may also study the stem of a gymnosperm (Fig. 1-69). This is similar to the dicotyledonous stem, except that conifers have no companion cells in the phloem—the xylem of conifers contains only tracheids. Since vessels are lacking, the tracheids carry on conduction of water.

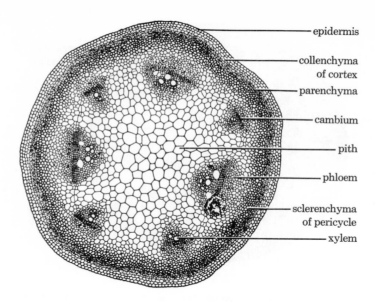

epidermis

collenchyma
of cortex

parenchyma

cambium

pith

phloem

sclerenchyma
of pericycle

xylem

Fig. 1-68 Cross section of dicotyledonous stem (*Aristolochia elegans*). (From W. H. Brown, *The Plant Kingdom*, 1935; reprinted through the courtesy of Blaisdell Publishing Co., a division of Ginn and Co.)

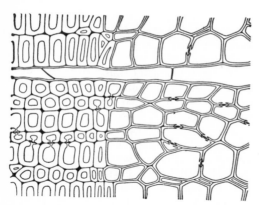

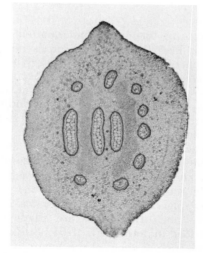

Fig. 1-69 Cross section of small portion of stem of a gymnosperm (redwood tree), showing xylem (tracheids) on both sides of a pith ray. (From W. H. Brown, *The Plant Kingdom*, 1935; reprinted through the courtesy of Blaisdell Publishing Co., a division of Ginn and Co.)

Fig. 1-70 *Pteridium,* a fern: cross section of rhizome. (Photo by Hugh Spencer.)

The pattern of stem growth, either alternate or opposite, is a means for identifying families of plants.

While the main function of stems is to support leaves and flowers and their fruits, many stems have specialized adaptations for other work. For instance, some stems may be storage regions, such as the fleshy stem of the white potato (tubers), the thickened stem of kohlrabi, or the rhizome of ginger. In fact, the wide spates of green which make up a cactus plant are not leaves, but stems which have taken over the work of carrying on photosynthesis.

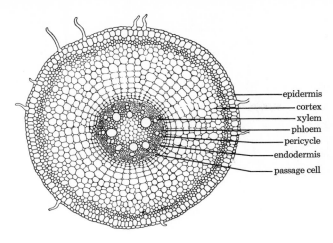

epidermis
cortex
xylem
phloem
pericycle
endodermis
passage cell

Fig. 1-71 Cross section of monocot root (*Commelina*). (From W. H. Brown, *The Plant Kingdom,* 1935; reprinted through the courtesy of Blaisdell Publishing Co., a division of Ginn and Co.)

Some stems are modified into twiners, for example, in the morning glory. Some stems, such as those of dodder and some epiphytes, are specialized to absorb water and minerals, thus taking on the usual function of roots. Runners and rhizomes (Fig. 1-70 and p. 475) may also serve a reproductive function in that they produce new plants without using seeds.

Roots

Perhaps students have already studied the root hairs on growing seedlings (p. 232) or measured the elongation of young growing roots (p. 453).

Cross sections of carrots or of the large castor-bean roots will give a fair picture of their structure. Prepared slides of roots may also be purchased (see Appendix C for supply houses). There is a close similarity between the structure of monocotyledonous and dicotyledonous roots, although there is a distinct difference in their stem structure (Figs. 1-71, 1-72).

While roots serve primarily for anchorage and absorption of soil water, some roots show wide modifications. Compare the fleshy storage root of the carrot, sweet potato, beet, radish, and dahlia. Brace roots and prop roots are found in the Indian rubber plant and the strangling fig tree. New plants can be grown from some of the fleshy roots; thus these roots serve a reproductive function too. Climbing plants, such as ivy, cling by means of adventitious roots.

Representatives of other members of the tracheophytes (the gymnosperms and ferns), the mosses and liverworts (Bryophyta), and bacteria, algae, and fungi are described in other chapters where their inclusion seemed more appropriate for a teacher (see Chapters 2, 8, 9, and 13).

Flowers, fruits, and seeds

The flower, as a specialized organ for reproduction, is described in Chapter 8. At this point, the anatomy of a typical flower is described (Fig. 1-73). Furthermore, special adaptations of flowers, seeds, and fruits are shown (Figs. 1-74, 1-75, 1-76).

Complete flowers have sepals, corolla, stamens, and pistils. Incomplete flowers lack one or more of these parts. For example, sepals and petals are missing, or reduced to hairs or scales, in the sycamore, willow, alder, grasses, and sedges. Imperfect flowers lack one of the essential structures; they may be either staminate or pistillate flowers. Willows, poplars, hop,

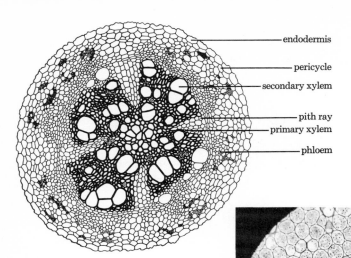

- endodermis
- pericycle
- secondary xylem
- pith ray
- primary xylem
- phloem

(a)

Fig. 1-72 Cross sections of dicot roots: (a) mungo bean; (b) *Ranunculus*, the buttercup. (a, from W. H. Brown, *The Plant Kingdom*, 1935, reprinted through the courtesy of Blaisdell Publishing Co., a division of Ginn and Co.; b, courtesy of Ward's Natural Science Establishment, Inc., Rochester, N.Y.)

(b)

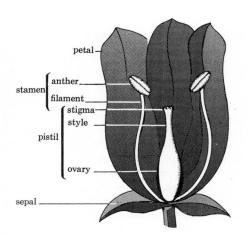

- petal
- stamen
  - anther
  - filament
- pistil
  - stigma
  - style
  - ovary
- sepal

Fig. 1-73 Diagram of a generalized flower, showing essential reproductive organs.

and the date palm are some examples of plants having imperfect flowers. Some species of maple and ash bear three kinds of flowers on one tree: staminate, pistillate, and perfect. Other plants, such as maize, may have staminate flowers as tassels, and ears composed of pistillate flowers.

Flowers are identified in a number of ways, for example, as hypogynous, perigynous, or epigynous; students may need to consult a botany text for explanations of such terms. Some representative types of flowers and fruits that identify common families of flowers are given in Figs. 1-74 and 1-75.

With a hand lens, examine the texture and patterns of pigmentation in petals, the sticky or feathery stigma, and the powdery pollen (black in a tulip, more often yellow in other flowers) that covers the stamens.

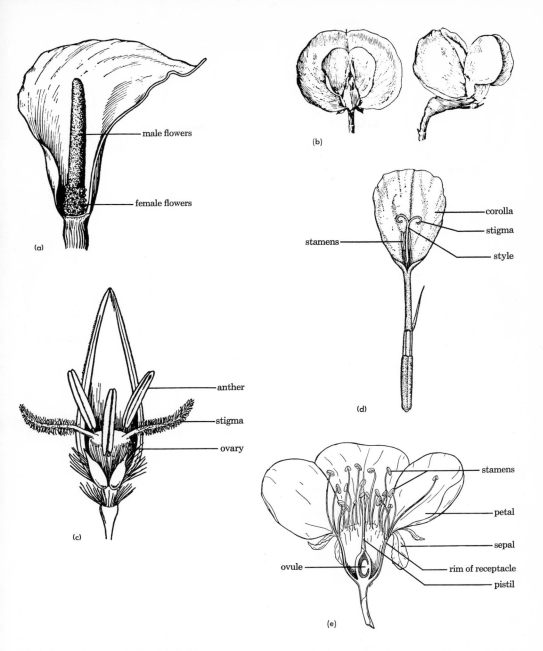

male flowers

female flowers

(a)

(b)

corolla

stigma

stamens

style

(d)

anther

stigma

ovary

(c)

stamens

petal

sepal

ovule

rim of receptacle

pistil

(e)

Fig. 1-74 Variety among flowers, showing inflorescences characteristic of families of flowers: (a) calla lily, showing spadix with female flowers below and male flowers above; (b) garden pea: left, front view; right, lateral view (stamens and pistil are largely hidden by petals); (c) *Avena sativa*, common oats, showing a single flower of a panicle; (d) marigold, showing a single flower; (e) perigynous flower of sour cherry (*Prunus cerasus*). (a, b, d, from W. H. Brown, *The Plant Kingdom*, 1935, reprinted through the courtesy of Blaisdell Publishing Co., a division of Ginn and Co.; c, e, from R. M. Holman and W. W. Robbins, *A Textbook of General Botany*, 3rd ed., 1934, courtesy of John Wiley & Sons.)

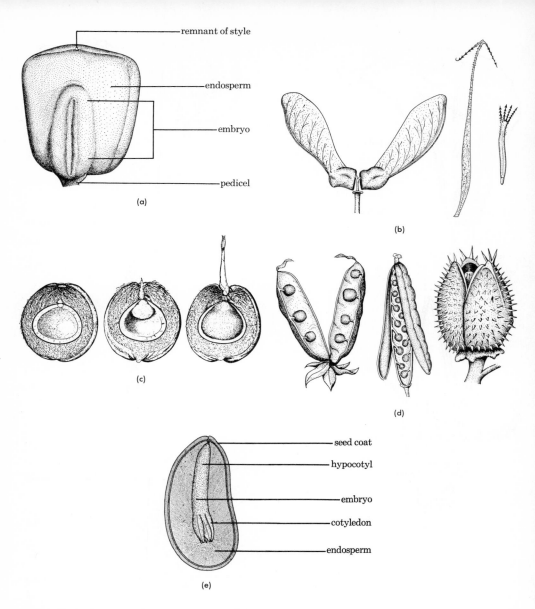

Fig. 1-75 Seeds and fruits, showing special adaptations: (a) Fruit (grain) of corn (*Zea mays*), external view. The small projection at the upper end of the grain is the base of the strand of corn silk. (b) Left, samara of maple (*Acer*); right, achenes of *Cosmos* and *Bidens* (beggar-ticks). (c) Germination of coconut. The large central meat is endosperm. In the left-hand drawing the embryo is still very small; the cotyledon, which is modified as an absorbing organ, is in the endosperm, while the remainder of the embryo projects up into the husk. In the middle drawing the modified cotyledon has enlarged, while the shoot appears through the husk. In the right-hand drawing the cotyledon fills the cavity in the kernel. (d) Left to right, pod of pea (*Pisum*); silique of mustard (*Brassica*); capsule of Jimson weed (*Datura*). (e) Longitudinal section of pine seed. (a, d, after R. M. Holman and W. W. Robbins, *A Textbook of General Botany*, 3rd ed., Wiley, 1934; c, e, from W. H. Brown, *The Plant Kingdom*, 1935, reprinted through the courtesy of Blaisdell Publishing Co., a division of Ginn and Co.; b, after Holman and Robbins and Brown.)

Distinguish the outer circles of sepals and petals; look for the common basic pattern of three, four, or five, or multiples thereof. Remove the sepals, which may be green or possibly the color of the petals (as in the tulip), and then carefully remove the petals. Notice whether or not the stamens are attached to the base of the petals; this fact is used in identification of flowers.

Identify the parts of the pistil. With a razor or scalpel, make a transverse cut through the ovary to examine the number of locules, or compartments, holding ovules. Are the locules in threes or fives? Notice how the ovules are attached to the wall of the ovary. For an examination of the macrospores to be found within the ovules, students will need prepared slides.

Dust some of the pollen grains on a slide, and examine in a drop of alcohol. Methods of germinating pollen grains are described in Chapter 8. However, if the flowers look as if they have been ripe for several days, look for germinating pollen grains on the stigma. Try to cut thin slices of the stigma, or carefully crush a portion of the stigma on a slide and mount in water. Examine under low power for evidence of extensions growing out of the pollen grains; these would be pollen tubes.

CAPSULE LESSONS: Ways to get started in class

1-1. On occasion you may bring to class an "unknown" plant or animal. Have students decide upon procedures to use to determine relationships. They should think of studying anatomy through dissection and examination of cells under the microscope.

1-2. In a biology club or a specialized curators club, students may want to dissect and compare the anatomy of many kinds of animals and plants. Perhaps they'll go further and examine tissues under the microscope (as developed in Chapter 2). The students may prepare and mimeograph guide sheets for dissection of several animals. Some teachers have worked with students to prepare a film of a dissection. Have students make an exhibit of adaptations among leaves for different kinds of functions in the plant. Exhibit these for all students to see.

1-3. Some teachers begin with dissections to introduce the relationship among living things. The class may be divided into smaller groups; each of these will dissect a fish, a frog, or a snake (or other vertebrate). The specimens can be tagged and stored in alcohol or formalin until the next day. Then students may switch places and study similarities and differences in the organisms they did not dissect the previous day.

1-4. Many ideas for student projects arise from a comparative study of plants and animals. For instance, students may make a phylogenetic collection of animals, preserved and mounted in jars (see Fig. 11-33). This collection may be used in the study of evolution. Or they may plant a plot of ground with examples of the various phyla of plants. Or they may try to culture some kinds of protists (see pp. 578–90, 622–38).

1-5. Some students may form the nucleus of a club whose purpose is to take field trips to learn the ecology of different locales in the community. They may want to organize a nature center which might supply schools nearby with living materials (from protists to mammals and flowering plants). This type of club might also supply student speakers for other schools, young people who have become proficient in some skill or technique.

1-6. Take a vicarious field trip, using 2×2 in. slides of a wide variety of flowers (Fig. 1-76). Have students identify the parts of flowers; or use slides of stems and roots, and examine their special adaptations. (See catalogs of color slides in most areas of biology, available from biological supply houses. Addresses are given in Appendix C.)

1-7. Use hall cases or possibly cabinets in the classroom to display examples of families of plants or animals. Specimens embedded in plastic may be used for display (Fig. 14-4), and models may be useful for review (Fig. 1-77). Herbarium specimens also make excellent displays; these are available from such supply houses as General Biological, Ward's, and Carolina (Appendix C).

1-8. Careful dissections of such organs as the heart of a cow, pig, or sheep (Fig. 1-78), or the brain of a sheep, cow, pig, cat, or dog (Fig. 1-79) may be part of club work or a special laboratory activity. Groups of four students can work together with one preserved specimen and share the chores of dissection and identification, using laboratory manuals. Fresh calf, chicken, or lamb hearts are available from local butchers.

(a)

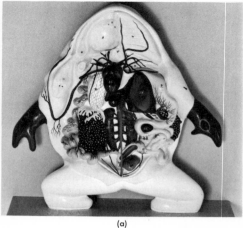

(a)

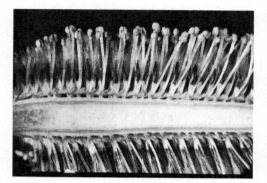

(b)

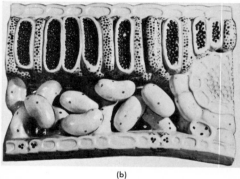

(b)

(c)

(c)

Fig. 1-76 Examples of lantern slides available from biological supply houses: (a) corn (*Zea mays*), detail of stamens on tassel; (b) willow (*Salix discolor*), detail of staminate catkin (flowers with paired stamens); (c) detail of flower of *Tradescantia*. (Courtesy of General Biological Supply House, Inc., Chicago.)

Fig. 1-77 Examples of models available for classroom study and review: (a) frog dissection; (b) typical leaf; (c) generalized flower (dicot). (a, photo courtesy A. J. Nystrom & Co.; b, c, courtesy of Clay-Adams, Inc., New York City.)

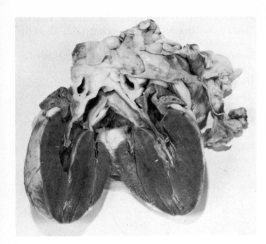

Fig. 1-78 Preserved sheep's heart, dissected to show chambers and valves. (Courtesy of Ward's Natural Science Establishment, Inc., Rochester, N.Y.)

PEOPLE, PLACES, AND THINGS

It may be possible for you to meet with teachers in different schools to plan joint club meetings, science fairs, seminars, and field trips. Students can combine their efforts and publish a science journal. This has two advantages: Students meet to exchange views and stimulate each other's learning, and they can afford to publish a more ambitious journal since a large number of students will share in the expenses.

Students in a central school might take responsibility for developing a culture center that would supply other schools with all sorts of living materials—plants and animals. Films and projectors may be shared between schools and with nearby colleges.

In addition, there may be in your community aquarium shops, nurseries, experimental laboratories and farms, or a zoological park. The helpful people in these places will offer both teachers and students their good services.

BOOKS

You will find books on ecology, and fieldbook guides for identification listed in Appendix A, the comprehensive Bibliography. The general texts in the major fields of biology will guide students in dissections and in physiology of animals and plants.

The following is a selected list of references

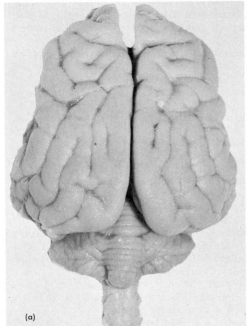

(a)

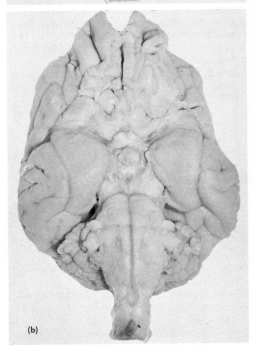

(b)

Fig. 1-79 Preserved dog's brain: (a) dorsal view; (b) ventral view. (Courtesy of Carolina Biological Supply Co.)

especially pertinent to the subject area of this chapter. We have drawn upon some of them heavily and have referred to them in context.

Beck, D. E., and L. F. Braithwaite, *Invertebrate Zoology: Laboratory Workbook,* 2nd ed., Burgess, Minneapolis, 1962.

Benson, L., *Plant Classification,* Heath, Boston, 1957.

Berman, W., *How to Dissect,* Sentinel, New York, 1961.

Borradaile, L., *Manual of Elementary Zoology,* 14th ed., edited by W. Yapp, Oxford Univ. Press, New York, 1963.

Brandwein, P. F., *et al., Teaching High School Biology: A Guide to Working with Potential Biologists,* BSCS Bull. 2, AIBS, Washington, D.C., 1962.

Buchsbaum, R., and L. Milne, in collaboration with M. Buchsbaum and M. Milne, *The Lower Animals: Living Invertebrates of the World,* Doubleday, Garden City, N.Y., 1960.

Bullough, W. S., *Practical Invertebrate Anatomy,* 2nd ed., Macmillan, London (St. Martin's, New York), 1958.

Cochran, D. M., *Living Amphibians of the World,* Doubleday, Garden City, N.Y., 1961.

Dugdale, C., *Manual for Dissection of the Cat,* Burgess, Minneapolis, 1949.

Du Porte, E. M., *Manual of Insect Morphology,* Reinhold, New York, 1959.

Eaton, T. H., Jr., *Comparative Anatomy of the Vertebrates,* 2nd ed., Harper & Row, New York, 1960.

Eddy, S., C. Oliver, and J. Turner, *Guide to the Study of the Shark, Necturus, and the Cat,* 3rd ed., Wiley, New York, 1960.

Edmondson, W. T., *see* Ward, H. B., and G. C. Whipple.

Esau, K., *Plant Anatomy,* 2nd ed., Wiley, New York, 1965.

Fisk, E., and W. Millington, *Atlas of Plant Morphology,* 2 portfolios: Portfolio I, *Photomicrographs of Root, Stem, and Leaf,* 1959, Portfolio II, *Photomicrographs of Flower, Fruit, and Seed,* 1962; Burgess, Minneapolis.

Florkin, M., *Unity and Diversity in Biochemistry,* Pergamon, New York, 1960.

————, and H. Mason, eds., *Comparative Biochemistry,* 7 vols., Academic, New York, 1960–64.

Foster, A., and E. Gifford, *Comparative Morphology of Vascular Plants,* Freeman, San Francisco, 1959.

Francis, E., *The Anatomy of the Salamander,* Oxford Univ. Press, New York, 1934.

Gans, C., and J. F. Storr, *Comparative Anatomy Atlas,* Academic, New York, 1962; dogfish, *Necturus,* cat.

Gilmour, D., *Biochemistry of Insects,* Academic, New York, 1961.

Greulach, V., and J. E. Adams, *Plants: An Introduction to Modern Botany,* Wiley, New York, 1962.

Harrison, B., *Dissection of the Shark,* Burgess, Minneapolis, 1949.

Harrow, B., and A. Mazur, *Textbook of Biochemistry,* 8th ed., Saunders, Philadelphia, 1962.

Hegner, R. W., and K. A. Stiles, *College Zoology,* 7th ed., Macmillan, New York, 1959.

Holmes, S., *The Biology of the Frog,* 4th ed., Macmillan, New York, 1927.

Hyman, L., *The Invertebrates,* 5 vols., McGraw-Hill, New York, 1940–59.

Jackson, F., and P. Moore, *Life in the Universe,* Norton, New York, 1962.

Roscoe B. Jackson Memorial Laboratory, staff, *The Biology of the Laboratory Mouse,* ed. by G. D. Snell, Blakiston, Philadelphia, 1941; Dover, New York, 1956.

Lanham, U., *The Fishes,* Columbia Univ. Press, New York, 1962.

Lockhart, R., G. Hamilton, and F. Fyfe, *Anatomy of the Human Body,* Lippincott, Philadelphia, 1959.

Meyer, B., D. Anderson, and R. Böhning, *Introduction to Plant Physiology,* Van Nostrand, Princeton, N.J., 1960.

Moore, J., *Principles of Zoology,* Oxford Univ. Press, New York, 1957.

Morrow, C. A., *Biochemical Laboratory Methods for Students of the Biological Sciences,* Wiley, New York, 1927.

National Science Foundation Course Content, *Science Course Improvement Projects,* NSF 62-38, National Science Foundation, Washington, D.C., 1962.

Pennak, R., *Fresh-Water Invertebrates of the United States,* Ronald, New York, 1953.

Popham, E. J., *Some Aspects of Life in Fresh Water,* 2nd ed., Harvard Univ. Press, Cambridge, Mass., 1961.

Prosser, C. L., and F. A. Brown, Jr., *Comparative Animal Physiology,* 2nd ed., Saunders, Philadelphia, 1961.

Richards, A. G., *The Complementarity of Structure and Function* (BSCS laboratory block), Heath, Boston, 1963.

Schwab, J., and P. F. Brandwein, *The Teaching of Science* (Inglis and Burton lectures), Harvard Univ. Press, Cambridge, Mass., 1962.

Simpson, G. G., *Principles of Animal Taxonomy,* Columbia Univ. Press, New York, 1961.

——, and W. S. Beck, *Life: An Introduction to Biology,* 2nd ed., Harcourt, Brace & World, New York, 1965.

Sinnott, E., and K. Wilson, *Botany: Principles and Problems,* 6th ed., McGraw-Hill, New York, 1963.

Smith, R. F., *Guide to the Literature of the Zoological Sciences,* 6th ed., Burgess, Minneapolis, 1962.

Torrey, T., *Morphogenesis of Vertebrates,* Wiley, New York, 1962.

Villee, C., *Biology,* 4th ed., Saunders, Philadelphia, 1962.

Wallace, G., *Introduction to Ornithology,* 2nd ed., Macmillan, New York, 1963.

Ward, H. B., and G. C. Whipple, *Fresh-Water Biology,* 2nd ed., edited by W. T. Edmondson, Wiley, New York, 1959.

Waterman, T. H., ed., *The Physiology of Crustacea,* Vol. I, *Metabolism and Growth,* Academic, New York, 1960.

Weisz, P., *The Science of Biology,* 2nd ed., McGraw-Hill, New York, 1963.

——, and M. Fuller, *The Science of Botany,* McGraw-Hill, New York, 1963.

Welty, C. J., *The Life of Birds,* Saunders, Philadelphia, 1962.

Witherspoon, J., and R. Witherspoon, *The Living Laboratory,* Doubleday, Garden City, N.Y., 1960.

Yapp, W., *see* Borradaile, L.

FILMS, FILMSTRIPS, AND SLIDES

This listing is intended to be only a guide for the selection of new films, filmstrips, and transparencies. Send for catalogs from distributors of films; catalogs of biological supply houses list extensive offerings of 2 × 2 in. slides. The addresses of distributors are given in Appendix B.

In general, the cost of color films averages $150; black and white films cost about $75. Newer films are made in varying lengths—time sequences of 1 min, 5 min, 11 min, or 30 min; the cost of films therefore varies. We have indicated the lengths of films so that lessons can be planned with the films in mind.

Sound films are available in color or black and white; rental rates are available from film libraries or by direct inquiries to producers. (Filmstrips are not available on a rental basis; the purchase price is usually about $6 to $7 for those in color.)

Amphibians (11 min), Coronet.

Angiosperms: The Flowering Plants (21 min), EBF.

Animals with Backbones (11 min), Coronet.

Arthropods: Insects and Their Relatives (11 min), Coronet.

Birds of the Countryside (11 min), Coronet.

Birds of the Dooryard (11 min), Coronet.

Cat Skeleton Preparation (filmstrip), Carolina.

Continuity of Life: Characteristics of Plants and Animals (10 min), Indiana Univ.

Crayfish Anatomy (11 min), Indiana Univ.

Demonstrations in Biology (set of 11 single-concept films, 8 mm), EBF; about the frog.

Diversity of Animals (set of 12 films, 28 min each), Part VIII of *AIBS Film Series in Modern Biology,* McGraw-Hill.

Diversity of Plants (set of 12 films, 28 min each), Part VII of *AIBS Film Series in Modern Biology,* McGraw-Hill.

Earthworm (11 min), Dowling.

Earthworm Anatomy (11 min), Indiana Univ.

Earthworm: Anatomy and Dissection (11 min), Coronet.

Echinoderms: Sea Stars and Their Relatives (17 min), EBF.

The First Many-Celled Animals: The Sponges (17 min), EBF.

Fish and Their Characteristics (11 min), Coronet.

Flatworms: Platyhelminthes (16 min), EBF.

Flower Structure (28 min), in Part III of *AIBS Film Series in Modern Biology,* McGraw-Hill.

Flowers at Work, 2nd ed. (11 min), EBF.

The Frog, 2nd ed. (11 min), EBF.

Frog Anatomy (17 min), Indiana Univ.; dissection of the bullfrog.

The Grasshopper: A Typical Insect (6 min), Coronet.

Gymnosperms (17 min), EBF.

The Honeybee: A Social Insect (6 min), Coronet.

Introduction to Animal Dissection (11 min), International Film Bur.

Introduction to Insects and Their Allies (filmstrip), Communicable Disease Center; free loan.

The Jointed-Legged Animals: Arthropods (19 min), EBF.

Life, Time, and Change (set of 12 films, 28 min each), Part X of *AIBS Film Series in Modern Biology,* McGraw-Hill.

Linnaeus (18 min), Swedish Film Center.

Living Birds (transparencies), Scientific Supplies.

Living Flowers (transparencies), Scientific Supplies.

Living or Non-Living (29 min), Indiana Univ.

Marine Invertebrates of the Maine Coast (34 slides), Ward's.

Monocot Plant Anatomy (filmstrip), Carolina.

Orders of Insects (20 min), Thorne.

Parasitic Plants (transparencies), Carolina.

Plant Associations (transparencies), Carolina.

Protozoa: One-Celled Animals (11 min), EBF.

Reptiles (14 min), EBF.

Reptiles (transparencies), Scientific Supplies.

Reptiles and Their Characteristics (11 min), Coronet.

Segmentation: The Annelid Worms (16 min), EBF.

Social Insects: The Honeybee (24 min), EBF.

Stinging-Celled Animals: Coelenterates (17 min), EBF.

Time Lapse Flower Magic from Hawaii (11 min), Hayes Spray Gun; free.

Virus (set of 8 films, 29 min each), Indiana Univ.

What Is a Bird? (17 min), EBF.

What Is a Reptile? (18 min), EBF.

Also refer to film listings at the end of Chapters 2, 4, 9, and 11.

LOW-COST MATERIALS

You may want to order from General Biological the set of *Turtox Service Leaflets* and subscribe to *Turtox News*. Also subscribe to *Ward's Bulletin* and receive Ward's leaflets on culturing plants and animals. Carolina, Welch, and Cenco also distribute bulletins to subscribers; Aloe Scientific offers a small catalog called *Animal Care*.

Aids and Sources for the Science Teacher lists apparatus, films, and new projects under way. It is obtainable from Am. Chemical Soc. (Washington 6, D.C.) and other sources, such as Am. Institute of Biological Sciences (AIBS).

Obtain offprints of the splendid papers in *Scientific American* from W. H. Freeman and Co., 660 Market St., San Francisco 4, Calif., at a cost of $0.20 per offprint.

A listing of supply houses, with addresses, will be found in Appendixes C and D. Many of the materials are free. Although we recommend the material for use in the classroom, we do not necessarily sponsor the products advertised.

Publications of the U.S. Dept. of Agriculture include its Yearbooks, which cost from $2.00 to $2.25 (free through the courtesy of your Congressman). Other publications, such as *Poisonous Snakes of the World* (D 201.2:Sn 1, $2.00), cost from $0.05 to $4.00 or so. Listings of publications will be mailed to you monthly if you request it; write to Supt. of Documents, U.S. Govt. Printing Office, Washington 25, D.C.

Chapter 2

Cells and Tissues

Some teachers begin their study of cells with tissue cells; others prefer to introduce the microscope with a study of motile microorganisms. Regardless of the mode of approach, the unifying concepts developed serve a fundamental conceptual scheme in biology—the complementarity of structure and function. Thus the cell is the unit of structure and function. Organization of life, in the biological sense, may be at the level of a unicellular organism,[1] or at that of one cell operating in a many-celled organism.

While many individual techniques are described in context in other chapters, the basic tools and techniques for preparing slides for study under the microscope are developed in this chapter. An introduction to the biochemistry of cells is also offered in the sense that several techniques relating to the action of enzymes on components of cells are described.

Possible ways to introduce lessons on the organization of organisms are offered in the "Capsule Lessons" at the end of the chapter. As in other chapters, suggestions for projects that students of varying

[1] Although many unicellular organisms have become highly specialized, the distinction between "plant-like" and "animal-like" is often difficult. Some biologists prefer to use the term "Protista" to include bacteria, flagellates, slime molds, and protozoa (see G. G. Simpson and W. S. Beck, *Life: An Introduction to Biology,* 2nd ed., Harcourt, Brace & World, New York, 1965). A further distinction is sometimes made in which those forms that lack a nuclear membrane are classified as Monera. Bacteria and blue-green algae would be grouped as Monera. Other algae, fungi, slime molds, and protozoa would then be grouped as Protista (see P. Weisz, *The Science of Biology,* 2nd ed., McGraw-Hill, New York, 1963).

abilities may undertake are offered wherever they seem appropriate.

Use of the microscope

Students enjoy as well as profit from using the microscope (Fig. 2-1), and they should be trained to use it correctly. Many teachers have found the following routines useful, and students should be encouraged to practice them.

1. Carry the microscope upright by holding the arm of the microscope with one hand and supporting its weight with the other hand. Set it down gently.

2. Align the low-power objective with the tube until it clicks into place.

3. Move the mirror to find the best light, using the concave side if the microscope has no condenser. It is not good practice to have direct sunlight fall on the microscope or reflect up through the mirror to the eye.

4. Watch the low-power objective (the shorter one, usually) as it is turned down near the slide, so that the working distance of the objective from the slide can be noted. Learn to lower the objective so that it is at a working distance—about ¼ in. from the slide. Many microscopes are adjusted to lock in this point.

5. While looking through the eyepiece, or ocular, slowly raise the barrel or tube by turning the coarse adjustment screw. Practice keeping both eyes open to avoid eye strain. Learn to move the slide with the

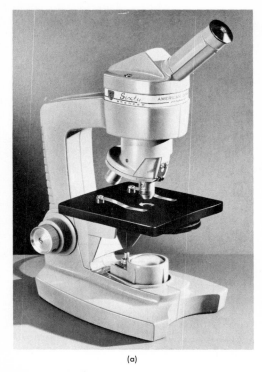

(a)   (b)

Fig. 2-1   Microscopes: (a) monocular; (b) binocular (stereoscopic). (a, courtesy of American Optical Co.; b, courtesy of Bausch & Lomb, Inc., Rochester, N.Y.)

free hand to locate the material to be examined (preparation of slides, p. 99).

6. Center the specimen found under low power; then slowly turn the fine adjustment screw a few degrees in each direction until the object is focused clearly. (It is important that the specimen be centered before switching to high power, for in high power a smaller part of the field is viewed and enlarged.)

7. Switch to high power (this can be done without changing the working distance if the microscope is parfocal). Adjust the light again by moving the mirror, and use the fine adjustment screw to focus the object clearly. The fine adjustment screw should be used with care. If the object was in focus under low power, students need make no more than a 10° turn in either direction to get a sharp image. Open and close the iris diaphragm until the light is comfortable to the eye.

Teachers may want to look into the advantages of a zoom optical system giving an "in focus" magnification continuing from $25\times$ to $100\times$ (Fig. 2-2). This magnification range and one varying from $50\times$ to $200\times$ are offered in inexpensive models by Bausch & Lomb (these zoomscopes cost $50.00).

Much of the beginning student's difficulty results from failure to clean the lenses of the microscope. Cloudiness of the surface of the glass lenses may be removed by breathing on them and then quickly wiping them with lens paper. If this does not remove the film, students should wipe the surface with lens paper moistened with 95 percent alcohol and then dry the surface. Dried glycerin, blood, or other albuminous materials may be removed by lens paper moistened with water to which a drop of ammonia has been added; the lenses should then be wiped dry. Balsam, paraffin, or oily substances may be wiped away rapidly with lens paper moistened with xylol. Since lens glass is softer than ordinary glass, care should be taken to avoid scratches from dust or other fine particles. If there are black specks which obscure the view, check to see where they are by turning the ocular. If the specks also turn, remove the ocular and brush the inner glass surface with a camel's-hair brush or lens paper. Do not use cloth, ordinary paper, or fingers (which are oily).

Oil immersion lens

In general, an oil immersion lens is needed for the examination of bacteria and other minute objects. (The substage diaphragm must be opened whenever the oil immersion lens is used.) Place a drop of cedar oil on the slide (or on the lens), and slowly and carefully lower the oil immersion objective so that it just touches the oil drop. Then focus with the fine adjustment screw. (Stained slides may be examined with or without a coverslip. Wet mounts must be covered with a coverslip.)

If the oil remains on the objective after a study is completed, it will harden; xylol will remove it, but xylol is detrimental to the oiled parts of the microscope. Therefore, students should immediately wipe the oil off the objective and the slide with lens paper.

Binocular (stereoscopic) microscope

The compound monocular (and the double ocular) microscope provides a two-dimensional image for observation; some appreciation of depth or thickness may be realized by using the fine adjustment screw. One thus sees sections or slices as if cut from the top, the middle, and down to the bottom of a cell.

The binocular stereoscopic microscope (Fig. 2-1b) has two oculars and gives a three-dimensional image. The true stereoscopic microscope is really two microscopes, for it has two oculars and two objectives. The oculars can be adjusted to the user's eyes so that a stereoscopic effect

is achieved. In this microscope, the image is not reversed or inverted as it is in the monocular compound microscope. Of course, these microscopes give less magnification, since they are used to locate objects and to provide a wide field of gross examination.

Degree of magnification

With a compound microscope, magnification is expressed in diameters. The total magnification may be determined by multiplying the magnification of the ocular lens by the magnification of the objective lens. For low power this is generally $10\times$ (ocular) by $10\times$ (objective) and for high power, $10\times$ (ocular) by $43\times$ (objective). Check the numbers on the objectives of your microscope.

Fig. 2-2 Zoomscope microscope utilizing continuously variable magnification. (Courtesy of Bausch & Lomb, Inc., Rochester, N.Y.)

On the low-power objective there is usually another number, which indicates the actual diameter of the field viewed under the low power, often 1.6 mm. The diameter under high power is usually 0.4 mm.

A rough estimate of the size of an organism or cell under low power may be made; for example, if a cell is one-fourth the diameter under low power, it would be ¼ of 1.6 mm, or 0.4 mm (which is 400 microns, since 1 mm = 1000 microns).

Students can prepare laboratory drawings with some indication of appropriate scale. If a student makes a drawing that measures 4 cm long, or 40,000 microns (μ) and the length of the actual object is 100 μ, his drawing is magnified 400×.

$$\text{magnification} = \frac{\text{size of drawing}}{\text{size of specimen observed}}$$

Students can also measure microorganisms and cells if they know the *area* of the field of vision. If they lay a transparent plastic ruler along the diameter of the field and observe the number of millimeters across the diameter, they can calculate the area by using πr^2, the formula for the area of a circle. This procedure may be used for low or high power or for oil immersion lenses. The volume of a cell can also be determined by using the formula for a sphere ($\frac{4}{3}\pi r^3$) or, if the cell is a cylinder, the formula $\pi r^2 h$.

For more careful measurements an ocular micrometer is used.

Ocular micrometer. The length and width of an organism can be measured with a small micrometer scale that is inserted into the ocular of a microscope. The scale, which may be subdivided into 50 units (or, more practically, into 100 equal divisions), is engraved on a glass disc that fits into the eyepiece. (When inserting the scaled disc, place the ruled side down into the diaphragm of the ocular.) Micrometers which are permanently mounted in the ocular are less adaptable for interchanging different kinds of counting discs.

The arbitrary divisions on an eyepiece micrometer must be calibrated against

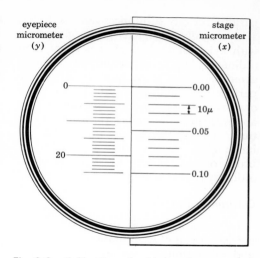

Fig. 2-3 Calibration of arbitrary divisions of an ocular micrometer against known divisions of a stage micrometer scale. Range of 20 divisions on an ocular micrometer would be the same as 80 μ; hence, each ocular micrometer division equals 4 μ.

some known unit, such as those found on a stage micrometer slide. Each eyepiece micrometer must be calibrated for low power, high power, or other combinations of objective and ocular, and for the specific tube length of each microscope.

The known units that are used on the stage micrometer may be either in the metric or the English system. Using the metric scale, the divisions may range from 1 to 2 mm, and be subdivided into tenths or hundredths of a millimeter. For example, a typical stage micrometer may have a distance scale of 2 mm measured off into 200 parts; each space is then ¹⁄₁₀₀ mm. The unit for measurement of microscopic objects is the micron (μ), which is 0.001 mm, or ¹⁄₂₅,₀₀₀ in. Students soon learn that the space of 1 mm on a ruler is equivalent to 1000 μ; 0.01 mm would then be 10 μ. Each unit on the stage micrometer described above would be 10 μ.

To calibrate the eyepiece micrometer, clip the stage micrometer slide under the microscope; rotate the ocular micrometer, and simultaneously move the stage micrometer slide so that the scales are superimposed in one plane (Fig. 2-3). Move the slide so that the beginning of 1 large-scale division on the ocular micrometer corresponds to the beginning of 1 large-scale division on the stage micrometer. The number of small divisions on the stage micrometer slide corresponding to 1 large-scale division on the eyepiece micrometer can then be counted and multiplied by 10 μ—the value of each small division on the stage micrometer in this example. This will give the value of 1 large ocular unit in microns.

Several readings should be made and then averaged. The fine calibration of the subdivisions of the scale and variations in the thickness of the slide affect the accuracy of measurements.

The calibration constant equals the true distance, as seen on the stage micrometer slide, divided by the number of divisions on the eyepiece micrometer.

$$C = \frac{\text{distance on stage micrometer slide}}{\text{divisions on eyepiece micrometer}}$$

$$= \frac{x}{y}$$

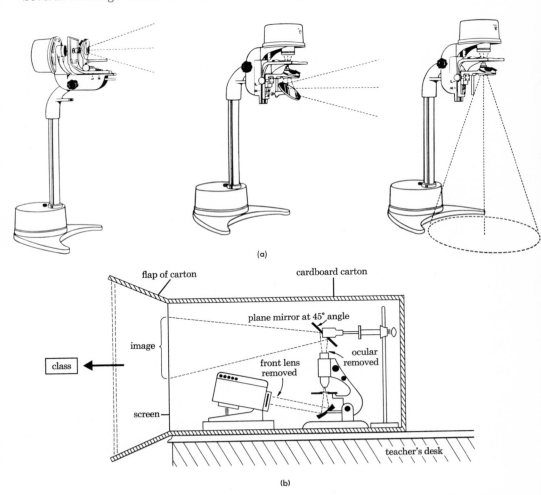

(a)

(b)

Fig. 2-4 Microprojectors: (a) Bausch & Lomb Tri-Simplex Micro-Projector; (b) homemade microprojector, shown for use in daylight. Without the cardboard container, the same homemade projector can be used to throw an enlarged image on a screen (or, without the plane mirror, on the ceiling) in a darkened room.

(C and x are in microns.) If a cell under the microscope covers 12 divisions on the scale of the eyepiece micrometer, its size can be found by multiplying the number of divisions it measures by the constant C.

Separate calibrations must be made for each lens.

Other methods for measuring size under the microscope. In lieu of a stage micrometer slide, the known dimensions of the squares of a counting chamber may be used to gauge the dimensions of objects under the microscope. However, counting chambers and micrometers are fairly expensive if each student needs the equipment. An inexpensive, thin circular disc with a measured series of saw-teeth on the rim is available from Bausch & Lomb. This micron disc is for use in a 31-15-09 $10\times$ Huygenian eyepiece only. With a $10\times$ 16 mm objective, the point-to-point value is 50 μ; with a $43\times$ 4 mm objective, the value is 12 μ; and with a $97\times$ 1.8 mm objective, 5 μ.

Microprojection

At times sufficient numbers of microscopes are not available (due to lack of funds, perhaps). And sometimes it is desirable to demonstrate to the class as a whole just which microorganism or portion of a slide should be studied. A microprojector is highly useful in such circumstances; the entire class, or groups of students, can view a fairly large image at the same time.

One commercial microprojector is shown in Fig. 2-4a. Detailed, careful directions for operation come with all commercial projectors. However, an essential requirement for efficient microprojection is a darkened room—the darker the better. If dark shades are not obtainable or do not darken the room sufficiently, you may want to devise a box with a screen, as shown in Fig. 2-4b. In this box is shown a "homemade" microprojector which can also be used in the standard way. Directions for both uses follow.

Homemade microprojector

If you have a microscope and a lantern-slide projector you can prepare a microprojector at no cost. Simply remove the eyepiece lens from the microscope and the front lens from the projector. Now focus the light beam from the projector at the mirror of the microscope. Place a slide on the stage and adjust the mirror until a spot of light shows on the ceiling overhead. Then adjust the position of the projector until the cone of light just fills the mirror. (You can check this by tapping chalk dust from an eraser to show the Tyndall cone.) Now focus on the specimen, using the low-power objective. A clear, enlarged image will form on the ceiling. If you wish, you can use a pocket mirror, clamped at a 45° angle, as shown, to throw the image forward to a screen.

If dark window shades are not available, set up the whole apparatus, with the mirror, inside a large carton. One side of the carton should be opened, leaving four flaps, and a translucent screen of tracing paper or tracing cloth attached across the opening. (The completely assembled apparatus is shown in Fig. 2-4b.) The flaps of the box can be braced open to act as a shadow shield, as shown. The carton can be placed on the teacher's desk, with the screen facing the class; a large, clear image will be seen on it.

Microscope screen

When students work with microscopes, it is not always possible to know whether they are all seeing what you hope they are seeing. An inexpensive substitute for a microprojector is a viewing device that consists of a microprojection hood that can be slipped over the eyepiece tube of a microscope. In one model, a real image is projected to a 3 × 3½ in. ground glass screen 7 in. above the microscope. A concentrated light source consisting of a ventilated lamp housing and a 100-watt projection lamp is needed.

Another model offers a viewing screen 4 × 5 in., which comes with an adapter

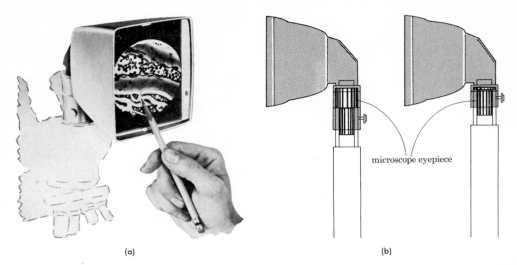

Fig. 2-5 Microprojection hood: (a) Welch microprojection screen with (b) adapter that fits regular and long-focus microscope oculars. (Courtesy of The Welch Scientific Co., Skokie, Ill.)

so that it fits practically all microscope draw tubes, both vertical and inclined (Fig. 2-5).

First microscope lesson

Some teachers like to train students in the use of the microscope by beginning with a wet mount or prepared slide of newspaper print. Cut out a single letter, say an *e*, from a newspaper and mount it in a drop of water. Apply a coverslip. Students should locate the *e* under low power and then center it. They should notice in what direction the letter is on the slide, then compare its position under the microscope (the image). Is it right side up, reversed, or upside down? Later, students may center the letter and switch to high power. If they move the slide a bit to the left, in what direction does the letter move under the microscope?

Some teachers make a number of permanent, prepared slides containing three crossed threads of different colors. The students find the bottom thread, and then find the center and upper ones by proper change of focus. Students enjoy this opportunity to practice the art of focusing.

Other teachers find that students learn quickly that the image of living material on a slide is reversed and inverted and start with a study of living materials immediately. For example, when students move the slide to follow a motile form, they notice in what direction the slide must be moved. In fact, some teachers like to use rich cultures of mixed protozoa—gray *Paramecium*, shapeless *Amoeba* (Fig. 2-6a), green *Stentor* (Fig. 2-6b), and *Spirostomum* (Fig. 2-6c). With slides of a mixed culture such as this, students can see a variety of distinctive organisms at the same time. Or you may want to begin with living algae such as *Chlorella*, or with epithelial cells from the cheek (Fig. 2-16), or with *Daphnia*, *Cyclops*, or a microscopic worm such as *Dero* or *Tubifex* (culturing, Chapter 12; also see Figs. 1-26, 1-28, 1-21, 1-22).

This first lesson often may be made exploratory by giving students some freedom in selecting the material to be used. Have several cultures available for study in a first lesson. Later, more careful studies of specific organisms can be made.

In summary, teachers may approach the study of cells from one or another level of organization. Two commonly used patterns often develop in this way:

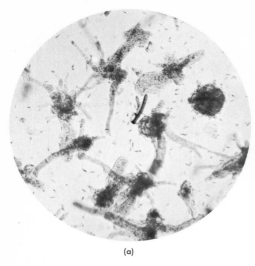

(a)

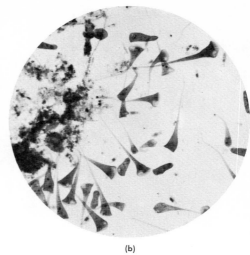

(b)

(c)

Fig. 2-6 Living protozoa: (a) *Amoeba proteus;* (b) *Stentor;* (c) *Spirostomum.* (Courtesy of General Biological Supply House, Inc., Chicago.)

Similarly, how do unicellular plant cells such as *Chlorella* and yeast cells compare?

From this approach, students may move into a comparison of many-celled organisms, or investigate the components of cells (p. 134).

Pattern 2: organization of many-celled organisms

Begin with an exploratory study of *Daphnia* (p. 39), *Hydra* (p. 22), segmented worms (p. 33), or rotifers (p. 31). After such an introductory examination, students are ready to study the specific organisms and their life functions. For example, they may examine epithelial cells from the lining of the cheek (p. 105), from an onion bulb, and from an elodea leaf (p. 106). In these three investigations, students learn the basic uniformity of cells in plant and animal tissues; they also discover the differences between plant and animal cells. Further, they learn that plant cells are diverse; for example, some have chloroplasts, but those that do not are unable to photosynthesize food materials.

Pattern 1: unicellular organization

Begin with a laboratory study of a mixed culture of protists—protozoa, algae, yeast cells, and possibly bacteria found in the protozoan culture. Spend subsequent laboratory periods on detailed study of *Paramecium* (p. 102), and/or *Amoeba* (p. 104), or other protozoa, comparing their similarities and diverse adaptations.[2] Explore methods of ingestion (Chapter 4), locomotion, reproduction, and so forth.

[2] F. Doflein, *Lehrbuch der Protozoenkunde,* G. Fischer, Jena, Ger., 1916.

When freshly killed frogs have been dissected, microscopic examination of tissues of the frog may follow (p. 107). Hopefully, there may also be time to look for parasites in frogs (see p. 534).

In either pattern of approach, students come to accept a basic concept: living things are made of cells. Cells have diversity or special adaptations, depending on their specific function.

Both patterns lead into a study of the composition of cells: the nature of protoplasm, possibly cyclosis (p. 106), diffusion gradients and density of cells (p. 241), or ion antagonism (p. 110). Finally, students need to realize the difference between their concept of the cell as viewed under the average compound microscope, magnifying some 430×, and the current general concept as based on observations using an electron microscope, which may magnify some 750,000× (Fig. 2-7).

Readings about cells

Students can begin to read *Scientific American* as their introduction to the functions of DNA, messenger RNA, ribosomes, and mitochondria, and the nature of grana in chloroplasts. As background to current reading, students might begin with the September 1961 issue, which is devoted to cells—their structure, growth, reproduction, functions in contraction and communication, and energy transformations. Further reading might include a perusal of the following from *Scientific American:* "The Mitochondrion" (January 1964); "Flower Pigments" (June 1964); "Cilia" (February 1961); "Amoeboid Movement" (February 1962); "The Membrane of the Living Cell" (April 1962); "Pumps in the Living Cell" (August 1962), dealing with excretion of sodium; "The Control of Biochemical Reactions"

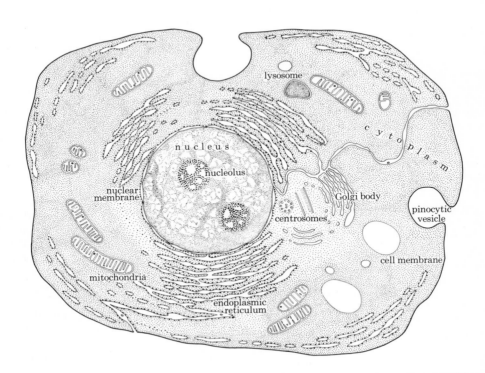

Fig. 2-7 Diagram of a cell based on image seen in electron micrograph. (After B. Tagawa, *Sci. Am.,* September 1961, p. 55.)

(April 1965); "Heart Cells in Vitro" (May 1962); "Genetic Code" (October 1962, March 1963); "Messenger RNA" (February 1962); "Hybrid Nucleic Acids" (May 1964); "Chromosome Puffs" (April 1964); and "Genes Outside the Chromosomes" (January 1965).

Some students may be interested in reading Leeuwenhoek's lucid and refreshing descriptions of his discoveries of microorganisms in the 1700's. Some teachers reproduce copies for all their students. By way of example, a short description from Dobell's collection follows.

I have a very little Cabinet, lacquered black and gilded, that comprehendeth within it five little drawers, wherein lie inclosed 13 long and square little tin cases, which I have covered over with black leather; and in each of these little cases lie two ground magnifying-glasses (making 26 in all), every one of them ground by myself, and mounted in silver, and furthermore set in silver, almost all of them in silver that I extracted from the ore, and separated from the gold wherewith it was charged; and therewithal is writ down what object standeth before each little glass.

This little Cabinet with the said magnifying-glasses, as I may yet have some use for it, I have committed to my only daughter, bidding her send it to You after my death, in acknowledgement of my gratitude for the honour I have enjoyed and received from Your Excellencies.[3]

On October 4, 1723, Leeuwenhoek's daughter, Maria, fulfilled her father's request and delivered the "little Cabinet" to the Royal Society.

Leeuwenhoek described his discovery of "animalcules" in a letter dated September 7, 1674, from Delft.

About two hours distant from this Town there lies an inland lake, called the Berkelse Mere, whose bottom in many places is very marshy, or boggy. Its water is in winter very clear, but at the beginning or in the middle of summer it becomes whitish, and there are then little green clouds floating through it; which, according to the

[3] From *Antony van Leeuwenhoek and His "Little Animals"* (p. 96), coll., trans., and ed. by Clifford Dobell. Published by Dover Publications, Inc., New York 14, N.Y., 1960, and reprinted through permission of the publisher.

saying of the country folk dwelling thereabout, is caused by the dew, which happens to fall at that time, and which they call honey-dew. This water is abounding in fish, which is very good and savoury. Passing just lately over this lake, at a time when the wind blew pretty hard, and seeing the water as above described, I took up a little of it in a glass phial; and examining this water next day, I found floating therein divers earthy particles, and some green streaks, spirally wound serpent-wise, and orderly arranged, after the manner of the copper or tin worms, which distillers use to cool their liquors as they distill over. The whole circumference of each of these streaks was about the thickness of a hair of one's head. Other particles had but the beginning of the foresaid streak; but all consisted of very small green globules joined together; and there were very many small green globules as well. Among these there were, besides, very many little animalcules, whereof some were roundish, while others, a bit bigger, consisted of an oval. On these last I saw two little legs near the head, and two little fins at the hindmost end of the body. Others were somewhat longer than an oval, and these were very slow a-moving, and few in number. These animalcules had divers colours, some being whitish and transparent; others with green and very glittering little scales; others again were green in the middle, and before and behind white; others yet were ashen grey. And the motion of most of these animalcules in the water was so swift, and so various, upwards, downwards, and round about, that 'twas wonderful to see: and I judge that some of these little creatures were above a thousand times smaller than the smallest ones I have ever yet seen, upon the rind of cheese, in wheaten flour, mould, and the like.[4]

In this letter, Leeuwenhoek is probably referring to the spirally arranged alga *Spirogyra;* if so, this is considered the earliest description of this alga. The animalcules are probably different kinds of protozoa; the ones with legs near the head and fins were possibly many-celled rotifers. *Euglena* is a green flagellate, and *Euglena viridis* is somewhat whitish at each end; this may be the species that Leeuwenhoek was describing in his letter to Mr. Oldenburg, who was secretary of the Royal Society in London.

[4] From *ibid.,* pp. 109–11.

Preparing material for the microscope

The techniques used to prepare slides depend upon the purpose and the materials available. Only a few of the commonly used methods can be described here. Detailed guides for specialized techniques can be found in texts listed at the end of the chapter.

Following is an outline of the procedures described in this chapter for handling plant and animal tissues and microorganisms.

I. Preparation of temporary slides
   A. Examining pond water
   B. Studying a variety of protists
     1. Protozoa
     2. Algae
     3. Bacteria
     4. Yeast cells
     5. Molds
   C. Tissue cells
     1. Onion bulb cells
     2. Epithelial tissue
     3. Elodea (*Anacharis*)
     4. Cyclosis
     5. Striated muscle in beef
     6. Human blood cells
     7. Tissues of a frog
     8. Tissues of *Hydra*
     9. Ciliated epithelial tissue and ion antagonism
    10. Chromatophores
    11. Enzyme activity of mitochondria

II. Accessory techniques
   A. Preventing swelling of cells
   B. Retarding evaporation
   C. Concentrating the number of organisms in a culture
   D. Slowing down the motion of protozoa
   E. Anesthetization
     1. Anesthetizing a frog
   F. Freehand sectioning
   G. Macerating tissue
     1. Using nitric acid
     2. Using potassium hydroxide
     3. Using formalin
     4. Using Jeffrey's fluid
   H. Mounting small forms
     1. In syrup
     2. In glycerin jelly
     3. In gum arabic
     4. In balsam
     5. In dioxane
   I. Repairing microscope slides
   J. Smear slide techniques
     1. Salivary chromosome smears
     2. Smear of onion root tip
     3. Blood smears
     4. Bacterial smears
   K. Staining temporary mounts; vital stains

III. Preparation of a hanging drop and a depression slide
   A. Hanging drop
   B. Cultures in depression slides

IV. Permanent stained slides
   A. Protozoa
     1. Concentrating protozoa
     2. Fixation
     3. Staining
     4. Other stains
   B. Tissue cells (histological techniques)
     1. Fixation
     2. Dehydrating, clearing, and embedding
     3. Sectioning and staining
     4. Other solutions needed in histological work
   C. Chick embryo

Preparation of temporary slides

Living protozoa and several tissue cells of plants and animals are best examined in the natural state—without the artifacts that may develop in fixing and staining cells. Occasionally, fixed, stained preparations are needed, especially in studies of components of cells (see the section on "Permanent Stained Slides"); protozoa are stained for the study of silver line systems, attachment of flagella and of cilia,

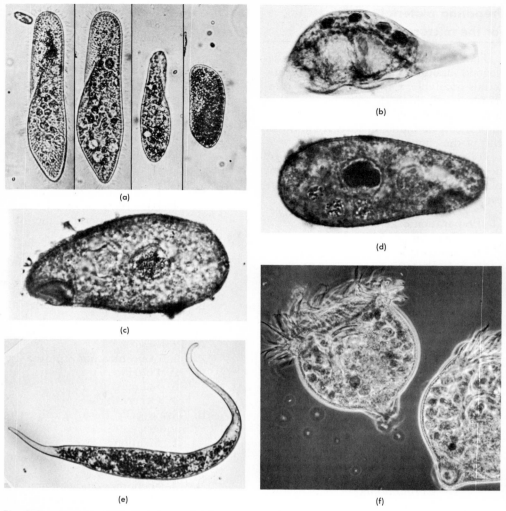

Fig. 2-8 Common ciliated protozoa: (a) four species of Paramecium (left to right: Paramecium multi-micronucleatum, P. caudatum, P. aurelia, and P. bursaria; all four photographs at the same magnification); (b) Blepharisma, stained to show chain of nuclei; (c) Tetrahymena pyriformis, stained; (d) Colpidium, stained; (e) Dileptus, a ciliate with a long proboscis; (f) Vorticella. (a, c, d, courtesy of Carolina Biological Supply Co.; b, courtesy of Ward's Natural Science Establishment, Inc., Rochester, N.Y.; e, f, photos by Walter Dawn.)

stages in nuclear divisions, and cell inclusions.

Additional descriptions of the preparation of temporary slides are given in context in other chapters (for example, *Monocystis* from earthworms, p. 535; flagellates in termites, p. 537).

Examining pond water. When collect-ing pond water, include some submerged leaves; possibly floating leaves of pond lilies, water hyacinths, and others; and some bottom mud. In the laboratory, divide the material into battery jars and several finger bowls (for more accessible study). Add a few rice grains or boiled wheat seeds, and store in a cool spot. Cover with several layers of cheesecloth, or loosely cap with aluminum foil. (For collecting material, see also Chapter 11.)

A gross examination of a large area of the finger bowl can be made with a binocular microscope (magnifying some 30×). Explore the contents: focus on the surface for several species of *Paramecium* (Fig. 2-8a), for free-swimming stages of *Vorticella* (Figs. 2-8f, 2-9d), and for *Stylonychia* (Fig. 2-9c). Focusing down, you may

also find *Euplotes, Blepharisma, Colpidium,* and *Tetrahymena* (Figs. 2-8, 2-9). Stalked *Vorticella* may be attached to submerged leaves or to the bottom of the dish. *Paramecium, Colpidium, Amoeba* (Figs. 2-6a, 2-9a), and some many-celled microorganisms (gastrotrichs and rotifers [Figs. 1-16, 1-17, 2-10]; and crustaceans such as *Cypris, Cyclops,* and *Daphnia,* pp. 39–43) may be found in the same region, feeding on bacteria around decaying leaves. Microscopic

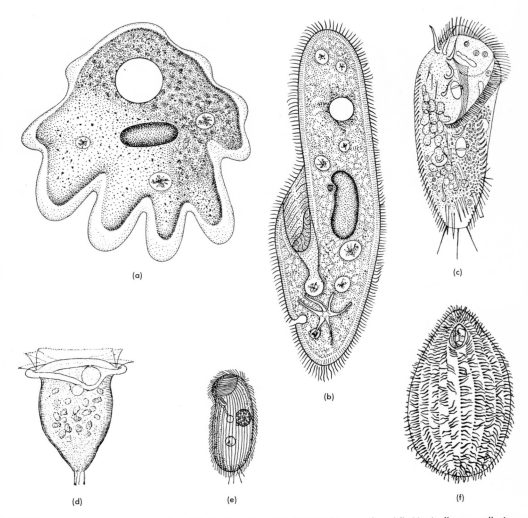

(a)

(b)

(c)

(d)

(e)

(f)

Fig. 2-9 Protozoa: (a) *Amoeba;* (b) *Paramecium;* (c) *Stylonychia mytilus;* (d) *Vorticella convallaria;* (e) *Colpidium colpoda;* (f) *Tetrahymena pyriformis.* (c, d, e, f, from R. Kudo, Protozoology, 4th ed., 1954; courtesy of Charles C. Thomas, Publisher, Springfield, Ill.)

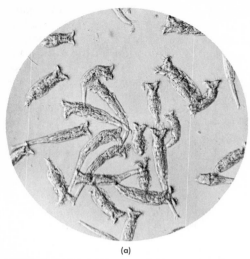

(a)

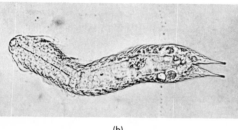

(b)

Fig. 2-10 (a) Rotifers; (b) *Lepidodermella*, a gastrotrich found among debris in collections from pond water rich in algae. While *Lepidodermella* is about the same size as rotifers, it lacks the whorls of cilia at the head end. (a, courtesy of General Biological Supply House, Inc., Chicago; b, courtesy of Carolina Biological Supply Co.)

worms such as *Aeolosoma* or *Dero* (Figs. 1-18, 1-19) may also be found.

Transfer samples from different levels to slides for more careful examination under low ($100\times$) and high power ($430\times$).

Pure cultures may be made of different genera of protozoa by isolating them in fresh-culture medium (culturing, Chapter 12). Some students will become interested enough to study samples over a week or a month to discover the cycles of these organisms; changes in succession, and alteration in population density of specific forms (Chapter 9).

Studying a variety of protists

PROTOZOA. On a clean slide center a small drop of culture medium. Apply a clean coverslip by pulling it along the slide at a 45° angle until it touches the edge of the drop, then carefully lower it over the drop so that the fluid is spread out evenly between the two surfaces of glass. If air bubbles occur, raise the coverslip and reapply to eliminate them. Methods for slowing down protozoa, and for preventing evaporation are described on pp. 111–12. Staining with vital dyes may be desired (pp. 121–22).

Examine the slide under the low-power objective. Center the slide and lower the objective to about ¼ in. above the slide or until it locks in place. Then use the coarse adjustment screw to raise the tube until the protozoa are in focus. With one hand move the slide to inspect the organisms in the drop; use the other hand to sharpen the image by slowly moving the fine adjustment screw a bit to the left or right.

Center the specimen under low power, then switch to high power for careful examination. We shall assume that *Paramecium* is ubiquitous enough to be in the culture under study (Fig. 2-8). There may be *Paramecium caudatum* or *P. bursaria* (green because of the symbiont *Chlorella*, or *Zoochlorella*, in the cytoplasm).

Wichterman[5] reports that the spiraling of *Paramecium* to the left is due to the oblique motion of cilia. In a summary of findings of others, he further reports that monovalent cation salts and hydrates tested induced reversal of ciliary motion, except $(NH_4)_2SO_4$ and $NH_4C_2H_3O_2$. None of the bivalent and trivalent cation salts tested induced reversal, except $CaHPO_4$ and $MgHPO_4$.

Watch the alternating contraction of the two contractile vacuoles of *Paramecium caudatum*, especially evident as the fluid evaporates and the coverslip presses down on the organisms. Active cilia in the oral groove create a current of water. Add

[5] R. Wichterman, *The Biology of Paramecium*, Blakiston (McGraw-Hill), New York, 1953, pp. 128–34.

some *Chlorella* to the slide, or add a dilute suspension of carmine powder, India ink, or yeast cells that have been dyed in carmine powder. Students can observe the ingestion of these materials into colored food vacuoles in the cytoplasm. Numbers of these *Chlorella* cells can be seen accumulating in a mass in the oral groove; the mass reaches a certain size and is then pushed into the cytoplasm. (For changes in pH in food vacuoles, see p. 200.) Further discussions of *Paramecium* and other protists are: reproduction, p. 345; responses, p. 299; vital stains, p. 121; preparation of permanent slides, p. 126; and silver line preparations, p. 125.

There may be other ciliates in the pond water. With the aid of Fig. 2-8, and the section on classification of protozoa (p. 588), try to identify the available specimens. *Blepharisma* is a pink or lavender form, the color being due to the presence of granules containing zoopurpurin. This pigment is light sensitive and bleaches out in strong light; it is soluble in alcohol and acetone. The pigment sten-

(a)

(b)

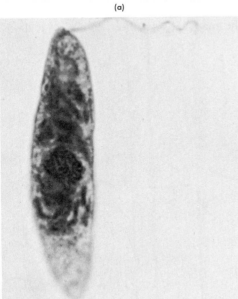

(c)

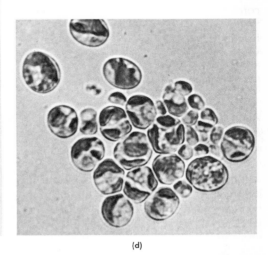

(d)

Fig. 2-11 Algae: (a) *Spirogyra*; (b) *Ulothrix*, a filamentous alga; (c) *Euglena*, a flagellate (stained); (d) *Chlorella*. (a, b, c, courtesy of General Biological Supply House, Inc., Chicago; d, photo by Walter Dawn.)

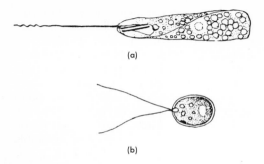

(a)

(b)

Fig. 2-12 Flagellates: (a) *Peranema trichophorum;* (b) *Chlamydomonas globosa.* (From R. Kudo, *Protozoology,* 4th ed., 1954; courtesy of Charles C. Thomas, Publisher, Springfield, Ill.)

torin, found in the ciliate *Stentor,* is not light sensitive and is not readily soluble. *Stentor* is large, bell-shaped when irritated but trumpet-like in its free-swimming form (Fig. 2-6b).

If amoebae are available, add a drop of yeast stained in carmine suspension, or a small drop of dilute India ink. Add some bristles or sand grains to raise the coverslip so that the amoebae can move freely. Try to observe pseudopodia in action (Figs. 2-6a, 2-9a).

When laboratory strains of protozoa and algal flagellates of known genetic composition are needed for a research project, pure cultures may be obtained from laboratory centers. Refer to L. Provasoli, "A Catalog of Laboratory Strains of Free-living and Parasitic Protozoa" (*J. Protozool.* 5:1–38, 1958).

If the desired forms do not arise in laboratory cultures from samples of pond or lake water, cultures for class study may be obtained from biological supply houses (Appendix C).

ALGAE. Soupy green water from an aquarium or culture jar standing in strong light contains a vast variety of algae. Prepare wet mounts of samples from different levels in the container. Under high power, find the bright green, cigar-shaped flagellate *Euglena* (Fig. 2-11c), with its prominent red eyespot containing the pigment astaxanthin. Several species of *Euglena*

may be found; some have a very flexible pellicle resulting in what is often called "euglenoid motion," a change in form. When the population density increases, *Euglena* encysts.

Peranema (Fig. 2-12a) is another flagellate, somewhat similar to *Euglena.* Also look for *Chlamydomonas* (Fig. 2-12b), a much studied flagellate in genetics research.

Small, spherical *Chlorella* cells may be found (Fig. 2-11d). They have a cup-shaped chloroplast that is visible only when the cells are at a certain angle; otherwise they appear uniformly green.

There may be desmids of rare beauty, and some filamentous algae such as *Spirogyra* or *Ulothrix* (Fig. 2-11a and b). Or students may find the blue-green alga *Oscillatoria,* which contains the pigment phycocyanin, along with the usual chlorophyll pigments of green algae.

BACTERIA. Temporary wet mounts may be made of dilute yoghurt (1:10), water containing decaying beans, or pickle or sauerkraut juices. Transfer a small drop of the material to a slide which has a dried

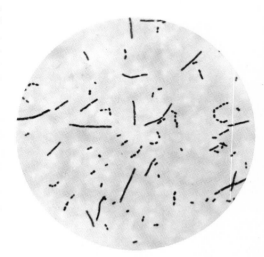

Fig. 2-13 Smear of yoghurt, showing both *Lactobacillus bulgaricus* (rods) and *Streptococcus lactis* (chain-like forms). (Courtesy of General Biological Supply House, Inc., Chicago.)

drop of methylene blue stain (p. 121) on it (culturing, p. 630).

Cocci and bacilli will be found in several preparations; spirilla are less frequently found. Especially common in yoghurt are *Lactobacillus bulgaricus* (rods) and *Streptococcus lacti* (strings of spheres), as shown in Fig. 2-13. Bacteria are visible in the clear areas amid the dark globules of coagulated milk products.

In sauerkraut or pickle juice, wild yeast cells are also found in abundance.

Cultures of pure bacteria may be purchased for class use; cultures of *Bacillus subtilis* (rod), *Sarcina lutea* (sphere), and *Rhodospirillum rubrum* (spiral) are especially useful. General Biological recommends this living collection of three harmless bacteria that are large enough to be seen under high power by students.

YEAST CELLS. Transfer a drop of actively fermenting sugar medium containing yeast cells (cultivation, p. 642) to a clean slide which has a dried drop of methylene blue stain on it. As the stain slowly diffuses into the cells, examine them under high power. Look for budding cells and whole colonies of undisturbed cells.

MOLDS. With a dissecting needle transfer a very small quantity of black mold, or green mold from a culture of mold (cultivation, p. 638), to a drop of glycerin on a clean slide (glycerin is used instead of water to reduce air bubbles).

As a first step in identifying the molds, examine under low and high power and compare the color and arrangement of sporangia with Fig. 2-14. Parasitic molds might be studied at this time (p. 641).

Tissue cells

ONION BULB CELLS. Slice a raw onion and cut the onion rings into ¼-in. sections. With forceps or with your nail remove the *inner* transparent membrane and flatten it in a drop of water. Examine under low power; note how difficult it is to distinguish the components of these rectangular cells. Stain the cells with Lugol's iodine solution: Apply a folded piece of lens paper at one side of the coverslip to draw

Fig. 2-14 *Penicillium notatum* (Fleming strain), mutant from Squibb culture (magnified 865×, reduced by 27 percent for reproduction; oil immersion technique). (Courtesy of U.S. Dept. of Agriculture, Peoria, Ill.)

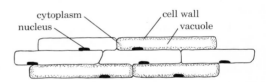

Fig. 2-15 Schematic diagram of onion cells (stained with Lugol's iodine solution so that nuclei are visible).

out the water under the slide; add a drop of the stain to the opposite side of the coverslip, and watch how the stain flows under the coverslip. Students will recognize the value of using stains in examining cells (Fig. 2-15).

EPITHELIAL TISSUE. Gently scrape the lining of the cheek with the flat end of a toothpick, and transfer these scrapings into a small drop of Lugol's iodine solution on a slide (or, if you prefer, use dilute methylene blue stain, p. 121). Stir the material in the stain with the toothpick so that the

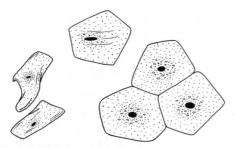

Fig. 2-16 Epithelial cells from lining of the cheek (stained with Lugol's iodine solution).

cells will not be clumped together; apply a coverslip and examine under low power (Fig. 2-16). Center some cells and switch to high power. Notice how the nucleus of each cell stains brown; the cytoplasm granules stain a very light brown with Lugol's stain. Other materials of sputum, such as free myelin globules and bronchial epithelial cells, may be found in addition to large, scattered masses of stained squamous epithelial cells.

Compare these epithelial cells with those found in skin of frog and in ciliated epithelial cells (Fig. 2-18a and b), also with ciliated cells of a snail (p. 51) and of the scallop, *Pecten* (p. 53).

ELODEA (ANACHARIS). Chloroplasts may be studied in algae, such as *Spirogyra* or *Euglena* (Fig. 2-11a and c), or they may be examined in cells of a leaf of the aquarium plant *Anacharis*. The cells of elodea leaves are only a few layers thick—students will need to learn to focus on different levels of cells. If students focus on the upper level of a cell, they will see chloroplasts scattered throughout. A shift in focus down to the middle of the cell will show chloroplasts forming a ring around the cells (Fig. 3-19).

After introducing chloroplasts, some teachers discuss the composition of chlorophyll, using chromatograms (pp. 137, 183), and then begin a study of photosynthesis (Chapter 3).

In young cells of *Anacharis* (especially in leaves near the growing tip of the plant), cytoplasm may be streaming along the periphery of cells; this gives the impression

of chloroplasts in motion, although the plastids are more probably floating in the moving cytoplasm. This streaming of cytoplasm is called "cyclosis" (a more extensive discussion follows).

CYCLOSIS. In many living cells mounted in salt solutions or aquarium water, students are likely to find cytoplasm streaming around the cell. In green plant cells they may often see chloroplasts moving around the border of the cell. (In some cases, it is believed that the chloroplasts have motility, aside from floating in streaming cytoplasm. Sauvageau[6] has shown that chloroplasts of the alga *Saccorhiza*, on exposure to strong light, show contractions and dilations when they move.)

Many aquatic plants illustrate this circulation of cytoplasm, or cyclosis. Mount a leaf of elodea (*Anacharis*), *Nitella*, *Chara*, or *Vallisneria* on a clean slide. In *Nitella* and *Chara*, focus especially on the internodal cells. In elodea, use the growing tips and focus on the midrib cells. Keep the uppermost layer of cells facing upward on the slide. Place the leaves in warmed water, or bring a warming light near the container to stimulate cyclosis. In some techniques, a bit of thiamine hydrochloride (vitamin B_1) is added to the water. The rate of streaming may be from 3 to 15 cm per hr and as high as 45 cm per hr at a temperature around 30° C (86° F). At times it is sufficient to heat the slide in the palm of the hand.

In *Nitella*, the many nuclei in the internodes also move. However, the chloroplasts are fixed within the inner surface of the cell walls and therefore do not move.

Students may find cytoplasmic streaming in other living material. Mount several threads of the mycelium of the bread mold *Mucor* in water or glycerin. Cytoplasm may be seen to be streaming up one side of the thread and down the other side.

Strip the epidermis from one of the inner

[6] As described in A. Guilliermond, *The Cytoplasm of the Plant Cell*, Chronica Botanica, Waltham, Mass., 1941.

scales of an onion bulb, and look for streaming in unstained cells mounted in water. Streaming may be stimulated by warming the water slightly.

Examine the unicellular hairs on the roots of *Tradescantia* seedlings, or the staminate hairs in the flower. Mount a filament of the stamen, which has several hairs attached, in warm water on a slide. Granules may be seen moving from the strands around the nucleus along the wall to another strand leading to the nucleus. At times the direction of streaming reverses.

Streaming is studied easily in one of the large amoebae—*Pelomyxa,* sometimes called *Chaos chaos* (Figs. 4-2, 12-2). Mount in a drop of culture solution (p. 579). Include a small bristle or broken coverslip in the drop to avoid crushing the specimen when the coverslip is applied. Students should see the many vacuoles and the actively streaming cytoplasm, which changes from sol to gel. The clearer ectoplasm and the denser endoplasm illustrate this change.

STRIATED MUSCLE IN BEEF. In a drop of water on a clean slide tease apart a minute piece of raw, lean beef. With dissecting needles separate the fibers, then transfer a few to a slide with a dried drop of dilute methylene blue stain. Add a small drop of water and apply a coverslip. Examine under low power, center the slide, and switch to high power. Careful focusing with the fine adjustment screw and proper illumination should reveal clear cross striations in this voluntary muscle tissue (Fig. 2-17b). Compare with a similar preparation from the thigh muscle of a freshly killed frog (p. 308); also refer to Figs. 2-17 and 2-18.

HUMAN BLOOD CELLS. Preparation of wet mounts of human blood in mammalian Ringer's solution, together with safety precautions involved when students prick their fingers, are described in the chapter on the circulation of blood (Chapter 5). For staining of blood smears with Wright's stain, see p. 118.

Students will want to compare nucle-ated red cells of the frog's blood with non-nucleated red cells of human blood.

TISSUES OF A FROG. *Epithelial tissue.* A live frog placed in a jar (with about 1 in. of water) will desquamate in about 24 hr. Mount small pieces of the sloughing skin in Lugol's solution or methylene blue (p. 121) on a clean slide with a coverslip. These mounts last for several hours if rimmed with petroleum jelly (Fig. 2-18a).

Students can also prepare temporary mounts of ciliated epithelial cells from the lining of the mouth of a freshly killed frog (Fig. 2-18b, also p. 323). After a careful study, show the effect of different ions on ciliary action (p. 110).

Dissect out small pieces of the stomach, intestine, and lining of the mouth of a freshly pithed frog; place in a macerating solution, 5 percent chloral hydrate (*caution:* poison). After 24 to 48 hr tease them apart with dissecting needles and mount in a drop of the solution. Students should be able to identify the cell membrane, cilia, cylindrical cells, and goblet cells.

Nerve cells. Dissect out small pieces of the spinal cord of a freshly killed frog and transfer to a clean slide. Gently press a second slide against this, mashing the tissue; hold the two slides parallel and pull apart, leaving a smear on each. Nerve cells, nuclei, cell contents, and bits of connective tissue should be seen. You can make permanent slides by drying these smears in air; then treat and stain them like fixed tissue (p. 124). They may show mitochondria, fat globules, nuclear constituents, and intracellular fibrils.

Blood cells. Study a wet mount of fresh frog blood in a drop of Ringer's solution (p. 665). Prepare blood smears and stain with Wright's blood stain (pp. 118, 251); compare red and white corpuscles (Fig. 2-18c).

Muscle cells. Tease apart a bit of macerated tissue from the stomach (see above) and from the thigh of a freshly killed frog. Stain with methylene blue or Lugol's solution (p. 121). What is the difference between these muscle cells?

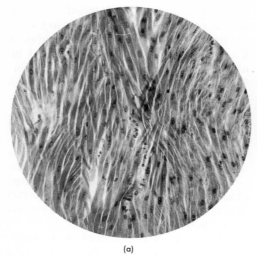

(a)

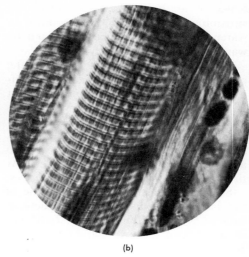

(b)

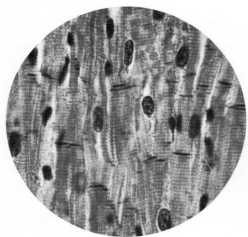

(c)

Fig. 2-17 Prepared slides of muscle and nerve cells: (a) smooth muscle as seen in section; (b) striated muscle under high magnification; (c) heart muscle, stained to show intercalated discs; (d) motor nerve cells; (e) motor nerve endings in muscle tissue. (Courtesy of General Biological Supply House, Inc., Chicago.)

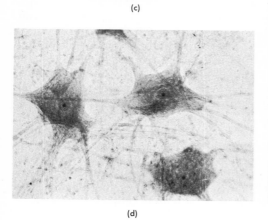

(d)

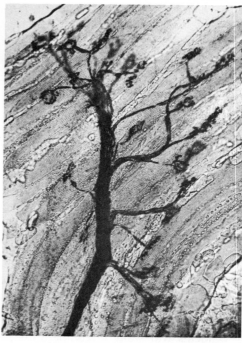

(e)

Sperm cells. Mount crushed testes of freshly killed frogs (p. 358).

TISSUES OF HYDRA. Place a hydra, preferably a green one, on a slide containing pond water, and tease it apart with dissecting needles. This maceration will release the symbiotic green zoochlorellae.

Under high power examine the large epithelio-muscle cells of the living animal. If you want students to stain the cells, first fix the cells, then add stain and macerate the stained cells as follows. Prepare a solution by mixing 1 part of glycerin and 1 part of glacial acetic acid with 2 parts of water. Then add 2 drops of this fixative to a clean slide; add the hydra. After 3 min add 1 drop of methyl violet stain (p. 131). Leave this for a few minutes. Then draw off the stain and wash the slide in a bit of water. Now macerate the stained hydra with dissecting needles in a drop of water. Add a coverslip and examine under high power (Fig. 1-1).

To study the nematocysts of a hydra, mount a hydra in pond water on a slide. Include small bristles to prevent the coverslip from crushing the specimen. If a small drop of safranin stain or a dilute acid is added to the wet mount, the nematocysts will be released (Fig. 2-19).

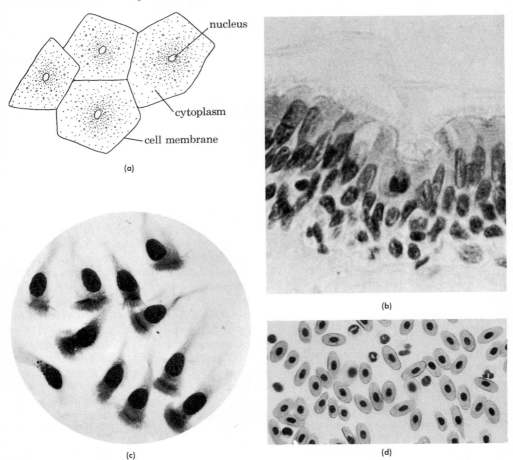

Fig. 2-18   Microscopic views of tissues of frog: (a) epithelial tissue (skin); (b) ciliated epithelium; (c) macerated (separated) ciliated epithelial cells; (d) blood cells stained with Wright's stain. (b, c, courtesy of General Biological Supply House, Inc., Chicago; d, courtesy of Carolina Biological Supply Co.)

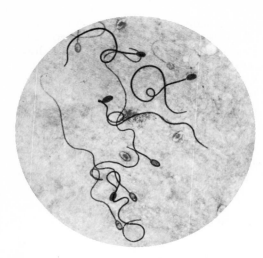

Fig. 2-19 Prepared slide showing glutinants, one kind of nematocyst of *Hydra*. (Courtesy of General Biological Supply House, Inc., Chicago.)

CILIATED EPITHELIAL TISSUE AND ION ANTAGONISM. Ion antagonism may be studied in the ciliary motion of epithelial tissue from the gills of a living clam, mouth of a living frog (p. 323), or intestine of an aquarium snail (p. 54). Dissect out several pieces of the gill, and put these in 10 ml of sea water in small watch glasses. Under low power watch the free ends of the cells, which show vigorously beating cilia. Then transfer the pieces of gill to the following solutions in watch glasses, and examine each under the microscope. First, put them into 0.9 percent potassium chloride solution; then transfer them into a 1 percent calcium chloride solution; finally, return the tissue to Ringer's solution. Compare the movement of cilia in each solution (refer to effect of ions on frog's heart, p. 244).

CHROMATOPHORES. If scales of live fish are mounted in Ringer's solution, chromatophores (pigment-containing cells) may be observed. Students may watch the contraction and expansion of these cells. First, draw off the Ringer's solution on the slide by putting filter paper on one side of the coverslip as a drop of chloretone (p. 112) is put on the opposite side. After students have examined these mounts, use the same procedure to draw off the chloretone, and add a drop of adrenalin or potassium chloride to the same mount. Watch the contraction of the chromatophores. In this way, by altering the size of the chromatophores, such animals as fish, amphibians, and several kinds of invertebrates are able to simulate the coloration or the varied intensity of the shadows in their background. Light is usually the original stimulus for the change in chromatophores.

ENZYME ACTIVITY OF MITOCHONDRIA. If you want to investigate the functions of cells further, have students observe mitochondria as centers of enzyme activity. In the following demonstration[7] Janus Green B is used to dye mitochondria of cells in celery, and then the dye is bleached by enzymes produced by these bodies.

Cut across two "strings" (collenchyma) of a fresh celery stalk; then make a second transverse cut 1 cm from the first. Transfer this 1-cm length, *inner surface up,* to a drop of 5 percent sucrose solution on a clean slide. With a razor blade carefully cut away the two strings of the strip, leaving the transparent center section containing two or three layers of cells. For ease in cutting, use two razor blades, as follows. Hold the two blades close to and parallel with the string. Keep the inner blade stationary and draw the outer blade along the edge of the string to make the cut. The stationary blade keeps the section from twisting. Repeat for the string on the other side. Now seal with a coverslip and look for cytoplasmic streaming in the epidermis and the uppermost, subepidermal layer. Also find the green plastids; clear nucleus; and small, moving spheres and rods, which are the mitochondria. Apply filter paper to one end of the coverslip, and at the same time add several drops of 0.001 percent Janus Green B

[7] Paper by H. Du Buy and J. Showacre, in National Institutes of Health, staff, "Laboratory Experiments in Biology, Physics, and Chemistry," mimeo pamphlet, U.S. Dept. of Health, Education, and Welfare, National Institutes of Health, Bethesda, Md., 1956.

solution to the opposite side. Watch the mitochondria dye blue; within minutes they will be decolorized by enzymes (dehydrogenases) on the mitochondria. The slides may be used for several days if they are placed in Petri dishes lined with moist toweling and stored in a refrigerator.

Directions for studying the mitochondria of a living fly or bee can be found in the BSCS laboratory block by A. G. Richards, *The Complementarity of Structure and Function* (Heath, 1963).

Accessory techniques

Preventing swelling of cells. Cells that have a fairly high salt concentration may swell when mounted in tap or aquarium water. This swelling can be prevented by putting the cells in a 10 percent aqueous solution of glycerin. Then add a coverslip. On the other hand, when the mounting fluid has a higher salt or sugar concentration than the cell, the cell loses water, resulting in a shrunken or plasmolyzed state.

Retarding evaporation. Students may ring a wet mount with petroleum jelly to slow down evaporation. Dip the mouth of a test tube into petroleum jelly. Apply this circle of petroleum jelly around the drop of material on the slide, and place a coverslip so that its edges are sealed to the slide.

Concentrating the number of organisms in a culture. Rich cultures of protozoa (*Paramecium, Blepharisma*) should be available for microscopic study. Many students will also want to examine flagellates and other motile algae (culturing, pp. 587, 627). At times a culture may be too dilute, that is, there may be too few organisms in the culture (only a small number of specimens would be picked up in a drop of fluid for a temporary slide). Students can increase the concentration of organisms in a volume of fluid by centrifuging 10 to 20 ml of the culture. Or pour a portion of the culture into vials or test tubes, and cover all but the top third of the tubes with carbon paper. Or prepare a short length of glass tubing; insert it into a one-hole stop-

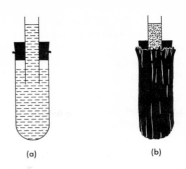

Fig. 2-20 A simple way to concentrate a culture of protozoa or of flagellates: (a) fit culture vial with stopper and short glass tube; (b) then cover vial with carbon paper.

per in a vial filled with culture, as in Fig. 2-20a; and cover the vial with carbon paper. The protozoa concentrate in the uncovered portions (the tops of the test tubes, or the tubing in the stopper, Fig. 2-20b), since most of them are positively phototactic or negatively geotactic, or gather where the concentration of oxygen is greatest (at the top surface). Then prepare temporary mounts containing large numbers of organisms for microscopic study.

Slowing down the motion of protozoa. Protozoa, especially the ciliated forms, move too rapidly for high school students, especially beginners, to follow under the microscope. There are several ways to slow down the motion of ciliates for close, careful study.

Most experienced workers simply prepare several slides ahead of time and let the fluid begin to evaporate. As evaporation continues, the weight of the coverslip is enough to impede the organisms. *Paramecium, Blepharisma, Spirostomum,* and *Stentor* have conspicuous contractile vacuoles which are best observed when the slides begin to dry—the pressure of the coverslip flattens the organisms. Under a darkened field (reduce light by closing the diaphragm a bit), the rhythmic pulsations of the vacuoles may be seen clearly.

You can also tease apart lens paper and apply the fibers on the slide containing a drop of rich protozoa culture. However, there is a danger that cilia and the larger cirri will be damaged with this crude method.

The following solutions may also be used to retard motion.

1. GUM TRAGACANTH. Get tragacanth at a drugstore or supply house, and grind some in a mortar to a fine powder. Add cold water to this to make a thick jelly. Dilute the material to the proper viscosity by placing a drop of protozoa culture and a drop of the jelly on a slide. Students may want to try out different dilutions.

2. METHYL CELLULOSE. Prepare a solution by dissolving 10 g of methyl cellulose in 90 ml of water. Place 1 drop of this syrupy solution with 1 drop of protozoa culture on a slide.

3. GELATIN. Prepare a 2 to 3 percent solution of clear gelatin by stirring in cold water and then heating gently until dissolved. Allow to cool to room temperature, and add 1 drop of this to a slide along with 1 drop of the culture containing the microorganisms.

You may also want to try a bit of isopropyl alcohol or chloretone, but these alter the structure and physiology of the organism.

Anesthetization. At times you may want to examine flatworms, a hydra, *Daphnia*, or some form having contractile tissue. Or students may want to anesthetize a large animal in order to dissect out a bit of tissue for examination, for example, ciliated epithelial tissue from the roof of the mouth of a frog (p. 323). The following materials and methods of narcotization or anesthetization are referred to throughout this book. Some experimentation may be necessary to meet specific needs.

1. MAGNESIUM SULFATE. A saturated solution of magnesium sulfate (Epsom salts) can be added, drop by drop, to the water containing the specimen. This procedure is useful for anesthetizing starfish and hydra; a 1 percent solution can be used to slow down some protozoa.

2. BUTACAINE SULFATE. A freshly prepared 0.1 percent solution of butacaine sulfate can be added, a drop at a time, to a container of protozoa and rotifers.

3. CHLORAL HYDRATE. Small crystals of chloral hydrate (*caution:* poison) can be added to the water containing planarians, small worms, and leeches. Or add small amounts of a 2 percent solution of chloral hydrate to the water containing the specimens.

4. MENTHOL AND CHLORAL HYDRATE. About 1 tsp of menthol crystals and a medium-sized crystal of chloral hydrate (*caution:* poison) when added to a small amount of aquarium water will anesthetize and extend small contractile animals before they are to be fixed in formalin or Bouin's solution.

5. MENTHOL. A few crystals of menthol on the surface of the habitat water will anesthetize planarians.

6. URETHANE. A 1 percent solution of urethane (*caution:* poison) may be used to anesthetize planarians and *Daphnia,* and a 5 percent solution can be applied to frogs to anesthetize them (see below). Mammals such as guinea pigs, rats, rabbits, and mice may be injected with 10 percent solution of urethane.

7. CHLORETONE. A solution of 0.1 percent chloretone can be added to water containing goldfish or frogs. One drop of chloretone of this dilution added to a large drop of culture of protozoa, hydras, or rotifers acts as an anesthetic. A frog may be anesthetized by immersion in a solution composed of 1 part of 0.5 percent chloretone and 4 parts of Ringer's solution for 15 to 30 min. (Use a 1 percent solution to kill planarians.)

8. ETHYL ALCOHOL (70 percent) may be used, drop by drop, to relax and anesthetize earthworms.

9. TRICAINE METHANESULFONATE prepared as a weak solution—0.1 g up to 1.0 g added to a gallon of water containing the specimens—is a useful anesthetizing agent for fish, frogs and frog larvae, and planarians.

10. ETHER OR CHLOROFORM. Spray ether or chloroform on water containing such specimens as *Daphnia* or worms, and then

cover the container. Mammals such as rabbits, rats, guinea pigs, and mice may be anesthetized by placing them in a container to which cotton saturated in ether or chloroform is added. (Quantity of anesthetic and the time factor can be altered so that these animals may be killed in this way. Refer to humane methods of handling animals, Chapter 1.)

ANESTHETIZING A FROG. There are times when you may want a frog anesthetized for a demonstration, such as circulation in the webbed foot. You may immobilize a frog by putting it in a covered jar with some cotton saturated with ether, but the immobilization will not last very long.

A more effective way[8] is to inject a 2.5 percent solution of urethane (ethyl carbamide; *caution:* poison) into the ventral lymph sac. As a guide, inject about 0.1 ml for every 10 g of the frog's weight.

There may be other occasions when you want to inject drugs of various sorts into a frog. Drugs are usually injected into the anterior lymph sac,[9] which is found in the floor of the mouth. Hold the animal in your hand so that its ventral surface is toward you. Draw the drug into the hypodermic syringe. Open the frog's mouth; avoid the tongue, and point the needle toward the floor of the mouth. When you press the needle into the skin, the needle enters the lymph sac; then inject the drug.

Freehand sectioning. Most tissues are not rigid enough to slice into thin sections for examination under the microscope. The usual procedure is to embed tissue in some paraffin mixture. A more rapid method is to freeze the tissue with carbon dioxide from a carbon dioxide cartridge, and then to cut sections.

Nevertheless, some plant leaves and some woody stems may be prepared as temporary mounts simply by inserting them between lengths of elderberry pith or fresh carrots and then cutting slices.

Slice a length of elderberry (or sunflower) pith or a raw carrot in half, and

[8] D. Pace and C. Riedesel, *Laboratory Manual of Vertebrate Physiology*, Burgess, Minneapolis, 1947.
[9] *Ibid.*

Fig. 2-21 Section razor and hand microtome for making freehand sections of plant tissue.

between the two halves sandwich a piece of a leaf, possibly of privet, or of a woody stem. Wrap these halves together tightly, and soak in water for a few hours. The pith and the enclosed tissue specimen will expand and become rigid enough to cut.

Then the material may be sliced with a sharp razor blade, at a slight angle, into thin sections. Keep the blade as well as the tissue wet, and float the sections on water so they will not curl. A better method, of course, is to insert the carrot-and-leaf preparation into a hand microtome and slice with a microtome razor blade (Fig. 2-21); the microtome, which can be adjusted for thickness, enables students to cut uniformly thin slices. Thin sections may be cut with a razor blade alone if a hand microtome is not available. The preparation is the same.

Staining is optional; methylene blue, eosin, or Lugol's iodine solution may be used.

Macerating tissue

USING NITRIC ACID. Mix 80 ml of water with 20 ml of concentrated nitric acid. (*Caution:* Pour the *acid* slowly into the *water.*) Place fresh muscle tissue from a frog or mammal in a glass dish containing this dilute nitric acid. In 1 to 3 days (at room temperature) this reagent should dissolve the connective tissue, leaving the muscle fibers isolated. Different sections of muscle may require varying periods of time for tissue breakdown. Shake the container to see the rate of maceration. Isolate and tease apart the fibers on a slide with dissecting needles; pour off the nitric acid; and wash the muscle tissue. You may want to have students stain the fibers with such stains as methylene blue and finally mount in glycerin or glycerin jelly.

When it is desired to keep the specimens for several days or indefinitely, pour the water off before staining and add a half-saturated solution of alum, prepared as follows. Add 100 g of alum to 500 ml of water and heat in an agate dish (all the alum should dissolve). Let the solution cool. Some alum will crystallize out; decant the resulting cold, saturated solution. From this prepare a 50 percent saturated solution by adding 100 ml of the saturated solution to 100 ml of water. (This alum preparation is also desirable if the specimens are to be stained and mounted in glycerin.)

USING POTASSIUM HYDROXIDE. A weak solution of potassium hydroxide will dissolve cells; a strong solution will separate the cells but will not destroy them. Prepare a solution by warming 35 g of potassium hydroxide in 100 ml of distilled water until it dissolves. Let the solution cool to room temperature.

Dissect out small pieces of tissue from the leg of a frog (striated muscle), from the stomach or intestinal wall (smooth muscle), and from the heart (cardiac muscle). Place these in separate containers of potassium hydroxide solution for 15 to 30 min. On separate slides mount a piece of each kind of tissue in a drop of solution. Tease apart the tissue with dissecting needles and apply a coverslip (see Fig. 2-17).

USING FORMALIN. Prepare formalin in a normal salt solution by adding 2 ml of formalin (40 percent solution of formaldehyde, p. 132) to 1000 ml of a normal salt solution (p. 659). This serves as a good dissociating agent for brain tissue and for all kinds of epithelial cells. While it acts quickly, it also preserves delicate cilia of epithelial tissue (Fig. 2-18) and the ependymal cells of the brain. Place small pieces of tissue in this solution. Two hours to 2 days later, depending on the size of the specimens, isolate nerve cells from the brain tissue. Examine a sample under the microscope to study how fast the process

is taking place. Often, epithelial tissue from the intestines and trachea may be isolated within 2 hr; stratified epithelial cells from the mouth and skin may require up to 3 days.

Finally, tease apart cells (in a drop of the solution) with dissecting needles on a clean slide. If you wish to have stained mounts, stain with a 1 percent aqueous solution of eosin. Draw off the formalin solution with filter paper at one end of the slide, and at the same time apply stain to the opposite side. For a more lasting preparation, mount in glycerin.

USING JEFFREY'S FLUID. This solution is used for plant tissues. In a Syracuse dish mix together equal parts of a 10 percent solution of nitric acid and a 10 percent solution of chromic acid. (*Caution:* Do not inhale fumes; work in a well-ventilated room.) Drop thin sections of plant tissue into this fluid. Mount small bits of tissue in water on a slide to examine the rate of maceration. When the cells separate from each other fairly readily, pour off the fluid and wash the tissue in water. Mount the cells in water and press down with the coverslip to separate them. Stain with eosin, methylene blue, or similar dyes. (Avoid putting metal forceps into the macerating fluids.)

Mounting small forms. Whole small forms may be mounted more or less permanently in several media. Forms which contain water need to be dehydrated before they can be embedded in balsam, or the balsam will become cloudy. On the other hand, some whole mounts may be embedded in glycerin jelly or in Karo syrup without dehydration.

Synthetic resins have been found to be superior in many ways to natural resins for mounting tissue on slides. Some of the natural resins often developed acidity and were variable in composition. Among the widely used synthetic resins, soluble in both xylol and hydrocarbon solvents, are Permount (from Fisher Scientific), Kleermount (from Ward's), and Piccolyte (from General Biological).

Some of the aqueous mounting media

may contain sugars to increase the refractive index. Many are recommended for small insects or other small invertebrates that are transferred from aqueous media or that have a high content of water.

We shall describe only a few of the simpler techniques that seem practical for the classroom.

IN SYRUP. Small forms such as *Drosophila,* ants, fleas, mosquito larvae, *Daphnia, Artemia,* and *Gammarus* may be mounted in a large drop of clear syrup, such as Karo.

Dehydration is not necessary. After the organism is oriented in the syrup, add a bristle so that the coverslip will not crush the specimen. Then gently lower the coverslip at a 45° angle from one side; let it sink slowly into the syrup so that no air bubbles form. Should air bubbles form, they may be broken with a dissecting needle. Then remove the excess embedding fluid with wet lens paper.

Some teachers use a medium containing fruit pectin in addition to the Karo syrup. Spread a thin layer of the following medium on a slide, and arrange small insects or worms that have been transferred from glycerol.

clear Karo syrup	50 ml
Certo fruit pectin	50 ml
water	30 ml
thymol (as a preservative)	small amount

IN GLYCERIN JELLY. Such organisms as roundworms, insects, small crustaceans, and plant specimens may be transferred into glycerin jelly from alcohol or formalin, as follows. Add glycerin to the alcohol or formalin in which the specimen is contained, until 10 percent of the storage fluid is glycerin. Cover the container with a bit of gauze, and allow evaporation to concentrate the glycerin. After the glycerin has become concentrated (this may take several days), transfer the specimens to a clean slide with a drop of glycerin jelly, prepared in the following way. Soak 10 g of gelatin in 60 ml of distilled water for about 2 hr. Then add 70 ml of glycerin and 1 g of phenol. Heat the solution in a water bath, and then let it cool. When ready to embed specimens on a slide, soften the mounting medium by heating to about 40° C (104° F) in a water bath. The temperature should not be allowed to rise above 40° C (104° F), or the colloid will no longer solidify.

Put a coverslip on the slide and ring it with balsam to retard evaporation. These slides need not be stained. They must be handled with care, for the glycerin jelly melts near room temperature. Examine under the microscope with reduced light intensity.

IN GUM ARABIC. Forms containing a high water content may also be transferred to a medium composed of gum arabic. Either of the following media offers successful slides.

1. FARRANT'S MEDIUM. Dissolve 200 g of gum arabic (lumps, not powdered form) in 200 ml of warm water. Add the following ingredients and carefully stir[10]:

glycerol	100 ml
carbolic acid	0.5 g

Keep in a water bath when used. (Carbolic acid is a preservative.)

2. BERLESE MOUNTING MEDIUM. Dissolve gum arabic in warmed distilled water. Add the chloral hydrate and glycerol; finally, filter.

gum arabic (lumps)	125 g
distilled water	200 ml
saturated aqueous solution of chloral hydrate	150 ml
glycerol	25 ml

IN BALSAM. Scales of fish and snakes, wings of insects, hair of mammals, feathers, and similar dry specimens may be embedded in balsam. They do contain some water, so in making permanent slides it is safer to dehydrate the specimens first in xylol for several hours (see p. 128); then mount them in a drop of balsam and seal with a coverslip.

Whole small insects must first be dehydrated before mounting in balsam. For a

[10] G. Humason, *Animal Tissue Techniques,* Freeman, San Francisco, 1962, p. 119.

rapid method, put them into glacial acetic acid for dehydration and then mount them directly in a drop of balsam. Or the dead insects may be put in a carbol-xylol solution made by adding 1 part of carbolic acid crystals to 2 parts of xylol. After 1 to 2 hr transfer to pure xylol for 6 to 24 hr; then mount in balsam and apply a coverslip.

IN DIOXANE. Permanent slides of small forms such as molds may be prepared for comparative studies during the school year. Transfer with dissecting needles a very small quantity of mold plants with fruiting bodies into a small drop of dioxane (available from biological supply houses) on a clean slide. Then carefully arrange the mold in the dioxane solution, and cover it with an additional drop of dioxane. Seal the slide to prevent evaporation by circling the drop of dioxane with balsam applied with a glass rod. Apply the balsam as a ring around the drop of dioxane, about ¼ in. away from the dioxane. Carefully apply a coverslip and press down gently with the needles so that a rim of balsam seals the edges of the slide.

In all these cases, leave the slides flat for several days until the balsam, Karo, or other medium hardens; later stack the slides in a slide box.

The preparation and staining of whole chick embryos in balsam is described on p. 132.

Repairing microscope slides. Eventually, a specimen on an expensive slide may slip, or the coverslip may be crushed. With some patience and practice, these slides can be repaired.

To repair a broken coverslip, soak the slide in xylol until the broken coverslip falls off by itself; avoid pushing, since the specimen may be torn. Apply fresh mounting medium and seal on another coverslip. Clear nailpolish may be used to seal the edges of the coverslip.

Protozoa will likely be washed off the slides in the first step in removing the broken coverslip. Draw off the xylol and collect the concentrated protozoa in the bottom of the dish.

A specimen that has slipped to an edge may be reoriented if the mounting fluid is heated slightly over an alcohol lamp. With a sliver of a tongue depressor soaked in xylol, the specimen may be moved in the softened mounting medium. Apply a bit more fresh medium if needed, and reapply a coverslip.

If specimens of tissue or of protozoa are faded, they may be restained. The process is laborious, since the material must be run down the series of alcohols, then destained. After restaining and dehydration, the material is moved along into fresh mounting medium.

Smear slide techniques. Basic techniques for making three types of smears on slides are described here: smears of dividing cells to show chromosomes, blood films, and smears of bacteria.

SALIVARY CHROMOSOME SMEARS. The salivary glands of fruit flies are excellent for the study of chromosomes; these insects have "giant" chromosomes, about 100 times as large as ordinary ones. (See larva, Fig. 8-20c.)

Demerec and Kaufmann[11] have published detailed instructions on preparing stained smears of salivary glands of the larvae of fruit flies. They advise that larvae be reared at a temperature between 16° and 18° C (61° and 64° F) with extra yeast added to the culture bottles (preparation of media, p. 415). Females laying eggs should be transferred to fresh bottles of medium every 2 days. In this way only a few eggs are laid in each bottle, and the larvae have a chance to grow without overcrowding. Full-grown larvae are collected as they emerge and begin to pupate. (Male larvae can be distinguished because they have larger, more conspicuous gonads, which can be seen through the transparent skin in the posterior third of the body.)

[11] M. Demerec and B. P. Kaufmann, *Drosophila Guide: Introduction to the Genetics and Cytology of Drosophila melanogaster,* 7th ed., Carnegie Institute of Washington, Washington, D.C., 1961 ($0.50). For a similar technique, using chromosomes of black flies (*Simuliidae*) having only six chromosomes as the diploid number, see D. Barley, "Salivary Gland Chromosomes of Black Flies," *Turtox News* 42:12, December 1964.

Dissect out the salivary glands of the larvae in 0.7 percent sodium chloride solution. On a clean glass slide add 2 drops of acetocarmine (see stains for protozoa, p. 122) to a dissected gland. Crush the glands on the slide by pressing a coverslip against them. Examine under low and high power.

Bensley and Bensley[12] described a similar technique using *Chironomus* flies (midges). The larvae may be found in tubes in mud or in water. They are soft, worm-like, and often red, and they may be found in the autumn and early spring. Collect them and culture them on decaying vegetation or in earth.

The recommended technique for handling *Chironomus* larvae is to mount an intact larva on a clean slide and gently press down with the coverslip to smear it. When examined under high power, the nucleolus and nuclear contents are visible. This material may be stained with neutral red (p. 121) or acetocarmine.

SMEAR OF ONION ROOT TIP. As a group activity (in class or in your science club) you may want students to stain the dividing cells in the growing tip of onion roots so that chromosomes are visible (Figs. 2-22, 2-23). The method described here is modified from one in *Stain Technology*.[13]

Germinate seedlings or place an onion bulb in water so that roots begin to grow. With a razor blade cut off the root tips and place them in a saturated aqueous solution of *p*-dichlorobenzene for 3 hr at 12° to 16° C (54° to 61° F). Next, transfer the root tips to a Pyrex test tube containing a mixture of 2 percent aceto-orcein solution and 1 *N* hydrochloric acid in the ratio 9:1. Heat this for a few seconds until it just reaches the boiling point; pour into a Syracuse dish or watch glass and let it cool for 5 min. Then transfer the root tips to a drop of 1 percent aceto-orcein solution on a clean slide.

[12] R. Bensley and S. Bensley, *Handbook of Histological and Cytological Technique*, Univ. of Chicago Press, Chicago, 1938.
[13] A. Sharma and A. Mookerjea, "Paradichlorobenzene and Other Chemicals in Chromosome Work," *Stain Technol.* 30:1–7, 1955.

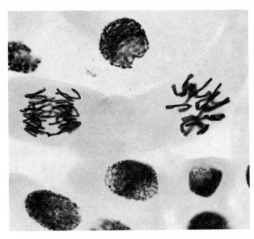

Fig. 2-22 Onion root tip, squash preparation. (Courtesy of General Biological Supply House, Inc., Chicago.)

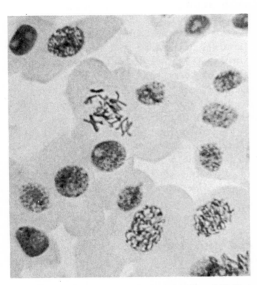

Fig. 2-23 Onion root tip, treated with colchicine. (Courtesy of General Biological Supply House, Inc., Chicago.)

Carefully cut off the deeply colored region of the root tip; discard the rest of the material. Now apply a coverslip and press uniformly with a dissecting needle along the coverslip so that the material is squashed. Place filter paper over the coverslip to squeeze out the excess stain. Ex-

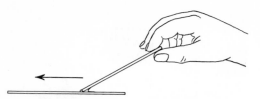

Fig. 2-24 Making a blood smear.

amine under high power for mitotic figures. (See also pp. 472–73.)

If these smear slides are sealed with a ring of paraffin they will last up to 15 days. The slides can be made permanent if they are placed in 10 percent acetic acid until the coverslips fall off. Then pass the slides through the following series of solutions: acetic acid and alcohol (half and half), absolute alcohol, then xylol (2 to 3 min in each solution). Mount in balsam.

BLOOD SMEARS. Clean the finger with cotton moistened in alcohol. Use a disposable, sterile lancet. These practices are similar to those described for making a slide of blood (p. 250) and for typing blood (p. 253). Students may be required to bring parental consent notes; regulations vary in different communities.

Puncture the finger and put a drop of blood on one end of a clean slide. Place a second slide, with one end at a 30° angle, on the first slide (Fig. 2-24). Bring the upper slide up to the drop of blood until the blood spreads along the narrow end of the slide, forming a uniform layer. *Push* the upper slide rapidly toward the opposite end of the bottom (level) slide to form a thin film. The greater the angle of the top slide, the thicker the film that is formed. Let the slide dry in the air. A good film should be smooth and not show wavy surfaces.

Chemically clean glassware is a requirement in preparing good blood smears. Use new slides and coverslips, or place them in a beaker of 95 percent alcohol, wipe dry, and flame over a Bunsen burner before use.

Wright's blood stain. Stand the dried blood smear slide across a Syracuse dish,

or support it across a wide cork in a Petri dish. Cover the dry blood film completely with Wright's stain for 1 to 3 min. This fixes the blood film. Next, add distilled water, or in another technique, a buffer solution, with a pipette, drop by drop, until a metallic, greenish scum forms on the surface of the slide. (The buffer solution is prepared by adding 1.63 g of monobasic potassium phosphate and 3.2 g of dibasic sodium phosphate to 1 liter of distilled water.) Continue to add water or the buffer until you have diluted the stain by half. Let this stand for 2 to 3 min.

The staining time varies with each batch of slides and should be predetermined by some trial slides. After the first application of distilled water or buffer to the original stain, place the slides in distilled water for 2 to 3 min. Now examine under the microscope; the granules in the basophils should stain deep blue, in the eosinophils bright red, and in the neutrophils lilac (Fig. 2-25). Wash the slide with

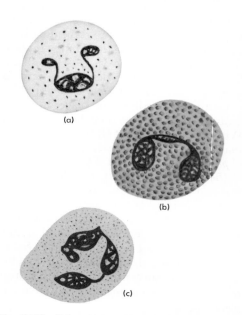

Fig. 2-25 White blood cells stained with Wright's stain: (a) basophil; (b) eosinophil; (c) neutrophil. (After W. Hamilton, J. Boyd, and H. Mossman, *Human Embryology*, 2nd ed., Williams & Wilkins, 1952.)

water continuously, until it is lavender-pink. Then stand the slides on edge to dry, or blot them dry with filter paper. Inspect under the microscope, without a coverslip or balsam, for the desired stain. If it is too dark, decolorize by returning the slides to distilled water.

Students may study these slides under an oil immersion lens (p. 91). The immersion oil should be added directly to the stained, dried blood film.

Giemsa stain. This is a fine stain for both blood smears and bacterial smears. A stock solution of Giemsa stain may be purchased, or prepared as follows. Dissolve 0.5 g of Giemsa powder in 33 ml of glycerin (this will take 1 to 2 hr). Then add 33 ml of acetone-free absolute methyl alcohol. To use the stain, dilute it by adding 10 ml of distilled water to each milliliter of stock solution of stain.

Let the blood smear dry in the air, and then fix the film by standing the slide in a Coplin jar (Fig. 2-26) with 70 percent methyl alcohol for 3 to 5 min. Dry in the air, or blot dry with filter paper. Next transfer the slide to a second Coplin jar containing dilute Giemsa stain for 15 to 30 min. Finally wash the slide in distilled water and dry it.

BACTERIAL SMEARS. Motility and fission in bacteria can be studied by using a hanging drop (Fig. 2-29). To a lesser extent, motility and fission may be seen in a wet mount, using high power or an oil immersion lens. A stain such as methylene blue may also be used in a wet mount (p. 99). However, "Brownian motion" will be seen and should not be confused with motility.

On the other hand, when bacteria are to be stained as permanent mounts, they must first be fixed to the slide as follows. Transfer a loopful (see preparation of loop, p. 634) of the bacterial suspension to a clean slide. Then, with the wire loop, spread this small drop into a thin film and let it dry in the air. (When the bacteria are transferred from a solid agar culture, use a sterile needle and transfer a small sample of a colony into a loopful of sterile water on a slide.)

Fig. 2-26 Coplin jar. Slides are placed in grooves for staining. (Courtesy of Ward's Natural Science Establishment, Inc., Rochester, N.Y.)

When the film is dry, pass the bottom of the slide through a flame three times to fix the bacteria to the slide, so that they will not be washed off in the subsequent staining procedure. There are many stains, generally basic aniline dyes, that can be used. These may be prepared and stored as *stock* solutions in saturated alcoholic solution. Prepare one or more of the following stains.[14]

1. CRYSTAL VIOLET. Add 13.87 g (or slightly more) of the dye to 100 ml of 95 percent ethyl alcohol. Let this stand for about 2 days, stirring frequently; filter and store.

2. BASIC FUCHSIN. Add 8.2 g of basic fuchsin to 100 ml of 95 percent ethyl alcohol. Let it stand for 2 days, stirring frequently; then filter and store.

3. SAFRANIN. Add 3.41 g to 100 ml of 95 percent ethyl alcohol and follow the procedure given above.

4. METHYLENE BLUE. Add 1.48 g of the dye to 100 ml of 95 percent ethyl alcohol and follow the procedure given above.

Before using, *dilute* the stock solution of

[14] Procedures are those suggested in U.S. War Dept., *Methods for Laboratory Technicians,* Tech. Man. 8-227, U.S. Govt. Printing Office, Washington, D.C., 1941.

the stains. For each 10 ml of the stock solution, add 90 ml of distilled water.

Apply the dilute stain to a bacterial smear on a slide (which has been fixed in a flame) for 2 to 4 min; then wash off with water and blot dry. Examine under the microscope. The time for staining varies with the thickness of the film, the concentration of the stain, and the kinds of bacteria under study. With practice, students will learn how many minutes it takes to stain the bacterial film. Coverslips need not be used. For permanent slides, however, add 1 drop of balsam when the slide is thoroughly dry, and then apply a coverslip.

Methods for culturing bacteria are described in Chapter 13, along with special techniques for plating and making transfers. Bacterial smears can be made from sauerkraut juice, yoghurt, timothy hay or beans left to decay in water, or the scrapings from one's teeth.

Gram's stain. Gram's stain is used as one step in distinguishing bacteria; some bacteria are "Gram-positive," others are "Gram-negative." Gram-positive organisms are stained violet or blue with this differential stain; Gram-negative organ-isms give a pink to reddish stain. Gram's stain involves the application of a stain, then a decolorizing agent, Gram's iodine, and finally a counterstain. In this technique, Gram-positive organisms do not become decolorized after treatment with a dye and iodine, while Gram-negative organisms lose the stain. Gram-positive bacteria remain stained because the cell membrane has different properties, or because the bacteria possess different chemicals, or for some other reason. In general, cocci are Gram-positive (except meningococci, gonococci, and catarrhalis group). Spirilla and spirochetes, as well as most bacilli (except the organisms that cause diphtheria), acid-fast bacteria, and many of the forms that produce spores are Gram-negative (Fig. 2-27).

Following are the stains that need to be purchased or prepared for Gram's stain technique.

1. CRYSTAL VIOLET, AMMONIUM OXALATE SOLUTION. Mix together 5 ml of crystal violet (stock solution) and 5 ml of 95 percent ethyl alcohol. To this add 40 ml of a 1 percent aqueous solution of ammonium oxalate.

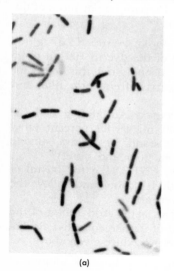 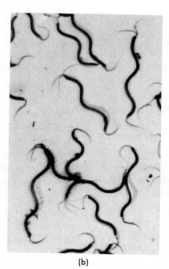 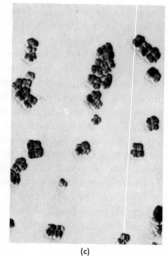

| (a) | (b) | (c) |

Fig. 2-27 Three kinds of bacteria: (a) bacillus, or rod-shaped (*Bacillus subtilis*); (b) spirillum, or spiral (*Spirillum rubrum,* or *Rhodospirillum*); (c) coccus, or spherical (*Sarcina lutea*). (Courtesy of General Biological Supply House, Inc., Chicago.)

2. GRAM'S IODINE. Mix together 2 g of potassium iodide and 1 g of iodine in 300 ml of distilled water. (For careful work, substitute 240 ml of distilled water and 60 ml of a 5 percent aqueous solution of sodium bicarbonate.)

The usual staining technique is as follows. Let a thin film of bacterial suspension dry on a slide, and then fix it in a Bunsen flame (wave the slide three times through the flame). Let the slide stand in crystal violet stain for 1 min. Pour off the excess stain. Then add Gram's iodine for 1 min. This fixes the Gram-positive organisms. Wash off in water, and decolorize with 95 percent ethyl alcohol. This is the step where the Gram-negative organisms lose the stain. Then wash in several changes of decolorizer until no more stain washes off. This may vary from less than 1 min to 3 min. Then wash in water again.

Now counterstain the slide with safranin (p. 119) for about ½ min. Wash off in water and blot dry.

Staining temporary mounts; vital stains. There may be times when you want students to stain living cells to illustrate specific details of cell structure; dyes other than Lugol's solution must be used. (Lugol's iodine solution kills the cells rapidly.)

Living cells may be stained to show cilia, flagella, or structures within the cells. Some stains destroy cells immediately, while others, called "vital stains," kill the organisms slowly. The organisms absorb these stains and continue to carry on their life functions for some time.

Place a drop of the stain (several are listed below) on a clean slide, and let it dry as a uniform film on the slide. Prepare several of these dried films of stain on slides, and keep them stored in a clean slide box. When they are to be used, just add a drop of protozoa culture, bacteria, yeast culture, or tissue cells to the slide. The stain will dissolve slowly into the drop of material on the slide.

Vital stains are usually prepared in absolute alcohol solutions. Basic dyes are less toxic than acidic ones. Prepare one or more of the following to have in readiness.

1. METHYLENE BLUE is used in dilutions of 1 part of stain to 10,000 or more parts of absolute alcohol. The stain is absorbed by the nucleus and by granules in the cytoplasm.

2. NEUTRAL RED is dissolved in absolute alcohol (1:3000–30,000). As an indicator, it is yellowish red in alkali, cherry red in weak acid, and blue in strong acid. The nucleus is stained lightly, especially the macronucleus of *Paramecium*.

3. CONGO RED may be used in dilutions of 1:1000 of absolute alcohol. Or, as an indicator, it may be diluted in water. In the presence of weak acids the indicator turns from red to blue (uses of indicators in food vacuoles, p. 200).

4. JANUS GREEN B, prepared as a saturated solution in absolute ethyl alcohol, is a specific vital stain for chondriosomes in protozoa in dilutions as weak as 1 part of the saturated solution to 500,000 parts of water. Furthermore, mitochondria in fresh frog's blood can be stained with dilute Janus Green B (1:10,000). Add a small drop of the vital stain to a small drop of blood on a clean slide. Or prepare slides in advance so that there is a dried drop of the vital stain on the slide. Then add a small drop of blood; the stain will slowly diffuse into the cells. Mitochondria stain bluish green. More permanent slides can be made by removing the coverslip before the components of the cells pick up all the stain and air-dry. Then apply balsam and seal with a coverslip. Lay the slides flat to dry. Janus Green B may also be used to stain bacterial inclusions and Golgi bodies in protozoa. For this reason, Janus Red is often used to stain chondriosomes—it is more specific than Janus Green B.

5. SUDAN III stains neutral fats red. Prepare a 2 percent stock solution, using absolute alcohol. When ready to use, dilute the stock solutions equally with 45 percent alcohol.

6. LUGOL'S SOLUTION reveals flagella and cilia, and stains glycogen reddish brown.

Kudo[15] suggests a preparation of the solution using 1.5 g of potassium iodine in 25 ml of water, then adding 1 g of iodine.

7. NILE BLUE SULFATE stains the macronucleus of protozoa green, and food vacuoles blue. Prepare dilute solutions (1 : 10,000–15,000).

8. NIGROSIN, although not a vital dye, can be applied to smears of protozoa. Nigrosin can be purchased as a 10 percent solution. Apply a drop to a smear of protozoa; dry the slide in air and examine under low and high power to see the pattern of rows of cilia in the pellicle of ciliates. (For silver line impregnation, see p. 125).

9. METHYL GREEN ACETIC ACID can be prepared by saturating a 1 percent solution of acetic acid with methyl green (p. 123) and filtering. Add 1 drop of a rich culture of protozoa to 1 drop of the methyl green acetic acid solution. The cytoplasm of the cells should remain clear while the nucleus is stained.

10. ACETOCARMINE is used to differentiate the nucleus. Add carmine powder to a boiling, 45 percent solution of acetic acid until it is saturated (*caution:* work in a hood). Filter. Add 1 drop of a protozoa culture to 1 drop of the stain on a clean slide.

Manwell reports the work of Ball (1927), who tried varying dosages of many vital stains on mortality of *Paramecium caudatum* (Table 2-1).

Preparation of a hanging drop and a depression slide

Hanging drop. If a drop of water on a slide were examined under the microscope, light would be reflected in several directions (Fig. 2-28). To avoid this we flatten out the drop with a coverslip. However, in so doing, we reduce the motility (if any) in organisms by the "crush" of the coverslip. A hanging-drop preparation enables students to study motility (especially in bac-

[15] R. Kudo, *Protozoology,* 4th ed., Thomas, Springfield, Ill., 1954, p. 892.

teria), fission in protozoa, germination of pollen grains, and similar subjects. Use a clean slide and coverslip, so that the surface tension of the water is not reduced (or a drop will not form). Place a small drop of the culture medium on a coverslip. Then apply petroleum jelly to the rim of the depression in a welled, or depression, slide. Place the inverted slide over the coverslip so that the rim of the coverslip and slide are sealed. Quickly invert the whole preparation; it should look like the one in Fig. 2-29.

Cultures in depression slides. Temporary hanging drops dry out quickly. A preparation which is easier for high school students to make, and more lasting, is one that uses the concavity, or depression, in the slide for the culture. Shallow concavity slides are for more temporary use, and straight-walled, deep-welled slides are used for more long-range work (but they must be used with care under the microscope). The straight-walled slides have a concavity that measures 3 mm deep and 16 mm in diameter.

Place 2 or 3 small drops of the culture to be examined in the concavity; ring with petroleum jelly; and seal with a coverslip. Avoid smashing the coverslip when focusing under high power.

Students may study the rate of reproduction of isolated protozoa or algae in depression slide cultures (p. 442). A thin agar preparation (1 percent) may be put in the bottom of the concavity and inoculated with mold spores. Algae may be grown on the surface of agar streaked with a transfer needle. In a way, these may be considered miniature Petri dishes for individual students, and they need little storage space. The succession of microorganisms in these sealed microaquaria may be examined over a period of 10 to 15 days; concepts of interdependence of organisms can be clearly developed (using controls). Liberation of oxygen bubbles in photosynthesis by algae, the effect of chemicals on growth, and the effect of other altered environmental conditions on

TABLE 2-1

Dyes staining cytoplasm of normal living Paramecium*

dye	minimal concentration for cytoplasm staining	percent mortality in 1 hr	hours needed for destaining cytoplasm
Bismark brown	1:150,000	0	7
Methylene blue	1:100,000	5	7
Methylene green	1:37,500	5	4
Neutral red	1:150,000	3	9
Toluidin blue	1:105,000	5	9
Basic fuchsin	1:25,000	30	9
Safranin	1:9000	30	1½
Aniline yellow	1:5500	0	1
Methyl violet	1:500,000	20	2
Janus Green B	1:180,000	40	7

* Reproduced, with adaptations, from *Introduction to Protozoology* by Reginald D. Manwell, St. Martin's Press, Inc., New York, and Edward Arnold Limited, London. Copyright © 1961 by St. Martin's Press, Inc. After G. Ball, "Studies on *Paramecium*, III: The Effects of Vital Dyes on *Paramecium caudatum*," *Biol. Bull.* 52:68–78, 1927.

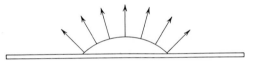

Fig. 2-28 Curved surface of a drop of water on a slide. Such a surface reflects light in many directions.

Fig. 2-29 Hanging drop.

microorganisms can be studied in depression slide cultures. This technique is often referred to in specific context in other chapters.

To retard evaporation, even in sealed slides, place the slides in Petri dishes lined with moist filter paper.

Streaming cytoplasm and vacuoles in root hairs of germinating radish or small grass seeds may be observed under the microscope. After soaking the seeds overnight, germinate them in depression slides. When germination has occurred, fill the depression slide with water. A portion of the roots in a flat mount may also be examined.

Permanent stained slides

Protozoa

CONCENTRATING PROTOZOA. A number of methods for concentrating protozoa may be used in preparing protozoa for fixation and staining. The following three methods are in common use.

1. Spread a thin film of Mayer's albumen (p. 132) on each of several clean coverslips. Then float some of these coverslips on the surface of a rich culture of protozoa; place others at the bottom of the culture jars. After 12 to 24 hr remove the coverslips with clean forceps and place them in a fixative.

2. Draw out a pipette in a Bunsen flame to form a capillary pipette (p. 677). Use the pipette to squirt a drop of a rich culture onto a slide or coverslip which has been spread with a film of Mayer's albumen. Quickly place the slide or coverslip into a jar of fixative.

3. This method permits the fixing of large quantities of protozoa in bulk. Centrifuge a culture slowly for some 30 sec. (Small hand centrifuges are available from biological supply houses.) Draw off the fluid quickly and pipette the concentrated culture of protozoa from the bottom into a

small container of fixative. When they are to be transferred out of the fixative into alcohol, concentrate them again by centrifuging.

FIXATION. While Schaudinn's fixative is recommended especially for fixing protozoa, others, such as Bouin's and Zenker's fixative solutions (pp. 127, 128), are also useful. Prepare Schaudinn's fixative, so that it is on hand, as follows. Prepare a saturated aqueous solution of mercuric chloride. To 2 parts of this solution add 1 part of absolute alcohol. Just before it is to be used, add 1 ml of glacial acetic acid.

Protozoa should stand in Schaudinn's fixative for 10 to 30 min (the average time is about 15 min). The time requirements are different for other fixatives. Then centrifuge the protozoa and transfer them into a small amount of 50 percent alcohol. After 10 min, transfer into 70 percent alcohol to which a few drops of a concentrated solution of iodine in alcohol has been added, so that the alcohol is slightly brown. After a few minutes, transfer to 90 percent alcohol, and finally into absolute alcohol, always keeping the quantity of alcohol as small as possible.

STAINING. Spread a film of albumen on slides or coverslips (see Mayer's albumen, p. 132); squirt a drop of previously fixed protozoa onto a slide or coverslip with some force so that the protozoa adhere to the albumen, and return the slides to absolute alcohol. Now the protozoa are on the slides, ready for staining. There are several suitable stains. Hematoxylin and the Feulgen nuclear reagent are chromosome stains. Borax carmine may be used for whole mounts (p. 130); or try Borrel's stain, which stains nuclei red and cytoplasm green. Many of these stains for protozoa may be found in textbooks of protozoology (some are included here under stains, p. 121). Delafield's or Ehrlich's hematoxylin may be used in place of Heidenhain's (p. 131).

The procedure for staining with hematoxylin is as follows. Remove the albumen-coated slides of protozoa from absolute alcohol and transfer through the series of alcohols to water. For instance, leave the slides (or coverslips) for about 2 min in each of these: from absolute alcohol to 90 percent, to 70 percent, to 50 percent, to 30 percent, and finally to water. Then place the slides in hematoxylin for 5 to 15 min so that the nuclei become stained (inspect under the microscope during the process); wash off the stain in water. Transfer the slides to ammonia water (about 1 drop of concentrated amonium hydroxide added to 250 ml of tap water).

Next, counterstain with a cytoplasmic dye such as eosin (1 percent alcoholic solution). After some 5 to 10 min transfer to water, then to 30 percent alcohol. Now follow with transfers at 2-min intervals into an upward series of alcohols: to 60 percent, 70 percent, 90 percent, and finally absolute alcohol. Next, transfer the stained slides into xylol.

The slides should be inspected under the microscope before adding balsam for final mounting. Should the nuclei be stained too deeply with hematoxylin, decolorize by placing the slides into 70 percent alcohol to which a small amount of 1 percent hydrochloric acid has been added. Then place the slides into alkaline 70 percent alcohol again to regain the blue color. (Prepare alkaline alcohol by adding 1 drop of 1 percent ammonium hydroxide to a Coplin jar full of the alcohol.) Continue into 90 percent alcohol, into absolute alcohol, and into xylol. Finally, mount with a coverslip by adding a drop of Canada balsam to each slide. Then apply coverslips. (Or add the balsam to the slide and cover with a stained coverslip, if that has been the procedure.)

OTHER STAINS. Kudo[16] recommends some special techniques, such as using Delafield's hematoxylin of a stock solution diluted 1:5 or 1:10 in order to achieve a slow, progressive staining of protozoa. Should slides become overstained, they can be decolorized in a 0.5 percent solution of hydrochloric acid in water or in

[16] *Ibid.*, p. 897.

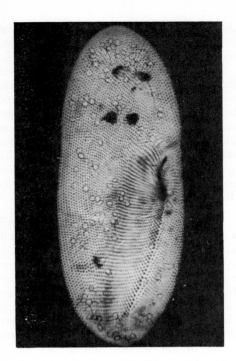

Fig. 2-30 *Paramecium* stained with silver nitrate solution to show silver line system. (Courtesy of General Biological Supply House, Inc., Chicago.)

alcohol. Then mount the protozoa in neutral mounting medium.

Mayer's paracarmine is recommended as a useful stain. It is prepared in slightly acidified 70 percent alcohol solution. As before, should protozoa become overstained, they may be decolorized in a 0.5 percent solution of hydrochloric acid in alcohol.

Silver impregnation method. Students can examine the silver line system of ciliates (Fig. 2-30); they may use a silver nitrate solution to impregnate the ciliary apparatus and its derivates on the pellicle of certain ciliates.

Manwell[17] reports on Lund's modification of the original Klein method. Centrifuge a culture of *Paramecium,* for example, to concentrate its numbers. Then follow

these steps, using centrifuge tubes or small vials.

1. Fix (3 min) in corrosive sublimate-formalin mixture (prepared: 95 parts of concentrated aqueous solution of mercuric chloride to 5 parts of formalin).
2. Wash twice in filtered water.
3. Impregnate (3 min) in 2 percent silver nitrate solution.
4. Reduce in distilled water by direct sunlight for 8 min.
5. Wash five times in distilled water.
6. As the first step in dehydration, add 9 ml of 95 percent alcohol slowly to each 1 ml of aqueous suspension of organisms.
7. Continue dehydration by passing from 95 percent alcohol to absolute alcohol; transfer into xylol to clean; mount in balsam.

Or students may wish to try the following modification of Klein's silver line system described by Kozloff.[18] Place a drop of concentrated ciliates on a slide and allow it to dry. Keep the forms from drying as a mass in the center of the slide by gently stirring the drop as it dries with a clean dissecting needle. Immerse the dried films in freshly prepared 2 percent solution of silver nitrate in distilled water. After 15 min, either expose the slides in the solution to sunlight or strong artificial light; or remove the slides from the silver nitrate solution, transfer them to a dish of distilled water, and expose this to sunlight to reduce the silver nitrate. After a few minutes in light, the slides turn brownish black. Remove a slide after 10 min and inspect under low power of the microscope. Some trial and error is necessary here because of variations in the thickness of the ciliated forms, their concentration, content of saline or protein, and so on.

After the silver impregnations have been adequately reduced, the slides should

[17] R. Manwell, *Introduction to Protozoology,* St. Martin's, New York, 1961, p. 584.

[18] E. Kozloff, "A Simple and Rapid Method for Demonstrating the Arrangement of Kinetosomes in Ciliates," *Carolina Tips* 27:3, March 1964. See also his reference text: A. Galigher and E. Kozloff, *Essentials of Practical Microtechnique,* Lea & Febiger, Philadelphia, 1964, pp. 366–83.

be washed in distilled water and blotted dry. Films may be mounted in neutral balsam or in synthetic resin media (p. 131) and coverslips applied.

Kozloff recommends the method for smaller ciliates such as *Colpoda* or *Tetrahymena,* and for symbiotic ciliates. (Avoid saline solutions, since the silver chloride precipitates out.)

Tissue cells (histological techniques). Unicellular organisms or bits of tissues are placed in a fixative for two reasons. First, the cells must be killed rapidly so that the cell contents are preserved and the cells closely resemble living cells. Second, a fixative hardens the tissue cells so that they can be cut into thin, transparent sections for examination under the microscope.

The fixative must be washed out of the cells before they can be processed for staining. The type of fixative used determines how the cells should be washed. When tissues have been killed in a fixative containing mercuric chloride (*caution:* corrosive sublimate, poisonous) or picric acid, the cells must be washed for at least 1 hr in 70 percent alcohol. On the other hand, when the fixative contains potassium dichromate, the tissues should be washed for at least 1 hr in water.

In general, the procedure for staining and mounting permanent slides is as follows.

1. Fix the tissue and harden.
2. Dehydrate through the series of alcohols.
3. Clear tissue.
4. Embed in paraffin.
5. Section with microtome.
6. Dissolve paraffin with xylol.
7. Pass through the series of alcohols into distilled water.
8. Stain and counterstain.
9. Dehydrate by moving up the series of alcohols into xylol.
10. Mount in balsam.

After discussing fixation, we shall follow through the next step, dehydration, and then proceed with subsequent procedures in preparing tissues for permanent, stained slides.

FIXATION. There are several widely used fixatives which teachers and students will wish to prepare. These fixatives are listed below in alphabetical order. Consult textbooks on histological techniques, such as those listed at the end of this chapter, for more detailed studies. Note that smaller quantities than those given can be prepared by cutting the "recipe" in half, or in thirds.

1. ALCOHOL (ABSOLUTE). Johansen[19] suggests a method for preparing absolute alcohol. (Absolute alcohol is, theoretically, 100 percent alcohol, with all the water removed.) Begin with 95 percent ethyl alcohol. Heat some crystals of cupric sulfate until only a white powder remains. Add this anhydrous form to the 95 percent ethyl alcohol. As long as there is still some water in the alcohol, the copper sulfate turns blue. Therefore, continue adding anhydrous sulfate until the solution no longer turns blue. Then filter the alcohol quickly into a dry stock bottle, and cork securely so that no moisture from the air enters the bottle. Apply petroleum jelly to the cork as an additional precaution. In fact, some workers keep a small bag of anhydrous sulfate suspended in the bottle to keep the solution free of water.

A sensitive test for water in absolute alcohol can be made by adding a few drops of the alcohol to a solution of liquid paraffin in anhydrous chloroform. If moisture is present in the alcohol, the solution becomes clouded.

2. ALCOHOL (ETHYL). A 70 percent solution of ethyl alcohol is a common preservative for small forms and tissue specimens. (For ways to dilute alcohols to lower percentages, see p. 658.)

3. ALLEN'S FLUID is recommended especially as an all-round general fixative. Combine the following:

[19] D. Johansen, *Plant Microtechnique,* McGraw-Hill, New York, 1940, p. 15.

water	75 ml
formalin (40% formaldehyde)	15 ml
glacial acetic acid	10 ml
picric acid	1 g
chromic acid	1 g
urea	1 g

Fix small pieces of tissue for 24 hr, and then wash in 70 percent alcohol until there is no loss of color.

4. BOUIN'S FIXATIVE is an excellent fixative for general use with both plant and animal tissue; however, it is difficult to wash out of tissues before staining. Its main advantage is that specimens may be stored in it for long periods of time. Combine the following:

glacial acetic acid	5 ml
picric acid (saturated aqueous)	75 ml
formalin (40% formaldehyde)	25 ml

Leave the tissue in the fixative for 24 to 48 hr; then wash in 70 percent alcohol until the color is removed.

5. CARL'S SOLUTION is an excellent preservative for insect forms. It is often used to kill and preserve small, soft-bodied forms such as mites, centipedes, and millipedes (see Table 11-2). A small amount of glycerin may be added to the solution to prevent hard-bodied insects from becoming brittle in the preservative. Combine the first three of these materials:

alcohol (95%)	170 ml
formalin (40% formaldehyde)	60 ml
water	280 ml
glacial acetic acid	20 ml

Add the glacial acetic acid to the solution just before using.

6. CARNOY'S FLUID is an especially useful fixative for tissue to be used for chromosome studies. Prepare small quantities of the solution just before it is to be used. (See p. 473.)

absolute alcohol	3 parts
chloroform	1 part
glacial acetic acid	1 part

7. CARNOY AND LEBRUN'S FLUID is recommended for hard-shelled specimens, such as some arthropods, for it has high penetrability. It also dissolves fat. (*Caution:*

highly poisonous and flammable.) Combine the following materials just before the fluid is to be used:

glacial acetic acid	33 ml
absolute alcohol	33 ml
chloroform	33 ml
mercuric chloride	to saturate about 25 g

8. FLEMMING'S FIXATIVE is recommended especially for small bits of tissue in preparation for careful histological study. Specimens are kept in the fixative for at least 24 hr. Combine these materials:

osmic acid (1%)	10 ml
chromic acid (10%)	3 ml
water	19 ml
glacial acetic acid	2 ml

Wash out the fixative with water for 24 hr; then transfer to 70 percent alcohol.

9. FAA (FORMALDEHYDE, ALCOHOL, AND ACETIC ACID) is a good hardening agent for plant tissues; in fact, plant materials may be stored in this preservative for several years. Pieces of leaf tissue should be killed and hardened in this fluid for 12 hr; thicker tissues, such as small stems and thick leaves, should be kept in it for 24 hr; and woody twigs should remain in it for a week. Tissues need not be washed after preservation in FAA. Many small animal forms can also be fixed in this fluid. The alcohol content counteracts the swelling effects of formalin and glacial acetic acid. Combine these materials:

ethyl alcohol (95%)	50 ml
glacial acetic acid	2 ml
formalin (40% formaldehyde)	10 ml
water	40 ml

10. GATES' FLUID, mainly used for plant tissues, is sometimes recommended to show chromosomes in root tips. Combine the following:

chromic acid	0.7 g
glacial acetic acid	0.5 ml
water	100 ml

Leave the specimens in this fixative for 24 hr. Then wash out the fixative with running water.

11. GILSON'S FLUID is an excellent fixative for careful histological work, and

especially recommended for the beginner. Avoid inhaling fumes of the fluid, for it is poisonous. Also, keep steel instruments out of this fixative, or out of any other fixative that contains mercuric chloride (corrosive sublimate).

Specimens are kept in this fluid for 24 hr; they can remain in it for a month or so. Be sure to wash out the fixative in 70 percent alcohol before dehydrating. Combine these materials:

alcohol (95%)	10	ml
water	88	ml
mercuric chloride	2	g
nitric acid	1.8	ml
glacial acetic acid	0.4	ml

12. KLEINENBERG'S FIXATIVE is recommended especially as a fixative for chick embryos (p. 132) and many small marine organisms. Prepare as follows. To a 2 percent aqueous solution of sulfuric acid, add picric acid until it is saturated.

13. LAVDOWSKY'S FIXATIVE is a general fixative especially useful in botanical work. Combine these materials:

potassium dichromate	5	g
water	100	ml
mercuric chloride	0.15	g
glacial acetic acid	2	ml

Again, avoid inhaling fumes of any of these fixatives that contain mercuric chloride (corrosive sublimate), and do not introduce steel instruments, such as tweezers, into the fluid.

Tissue may remain in the fixative for 24 hr. Then transfer tissue specimens into 70 percent alcohol for at least 1 hr.

14. ZENKER'S FIXATIVE is widely used in histological work. It is also a fine preservative for small marine forms. Since it is not very stable, only small amounts should be made at any one time, or add the glacial acetic acid just before using the fixative. Mix together:

potassium dichromate	2.5	g
water	100	ml
mercuric chloride	5	g
glacial acetic acid	5	ml
sodium sulfate	1	g

Fix the specimens for 24 hr. Wash out the fixative in 70 percent alcohol for the next 24 hr. (Avoid using steel instruments, and definitely avoid inhaling the fumes.)

There are several other preservatives which serve special uses in the laboratory or classroom. Some have found favor with technicians handling algae or fleshy fungi.

15. PRESERVATIVE FOR UNICELLULAR ALGAE. Sass[20] recommends the following preservative for storing unicellular algae, fleshy fungi, liverworts, and mosses. Combine:

water	72 ml
formalin (40% formaldehyde)	5 ml
glycerin	20 ml
glacial acetic acid	3 ml

16. PRESERVATIVE #1 FOR GREEN PLANTS is recommended often to prevent bleaching of chlorophyll in preserved plant specimens. Mix these ingredients:

alcohol (50%)	90	ml
formalin (40% formaldehyde)	5	ml
glacial acetic acid	2.5	ml
glycerin	2.5	ml
cupric chloride	10.0	g
uranium nitrate	1.5	g

The specimens may be stored in this preservative permanently, or they may be immersed in this solution temporarily before transferring to another preservative.

17. PRESERVATIVE #2 FOR GREEN PLANTS also prevents bleaching of chlorophyll in plant tissues. You can use the FAA fixative (p. 127) by adding enough copper sulfate to make a saturated solution, or you may prefer to make up this solution to have on hand, as follows. Dissolve 0.2 g of copper sulfate in 35 ml of water. To this solution, add a solution composed of:

ethyl alcohol (95%)	50 ml
formalin (40% formaldehyde)	10 ml
glacial acetic acid	5 ml

Specimens may be stored in this preservative indefinitely. Occasionally transfer specimens into fresh liquid.

DEHYDRATING, CLEARING, AND EMBEDDING. In the fixative the tissues were killed

[20] J. Sass, *Botanical Microtechnique*, 2nd ed., Iowa State Univ. Press, Ames, 1951, p. 100.

quickly and also hardened. The next step is dehydration of the tissues. After water is removed the tissues are embedded in paraffin to give the support needed to cut thin sections for examination under the microscope. Usually tissues are transferred from the fixative into 70 percent ethyl alcohol, but the process depends on the kind of fixative in which the tissue has been killed. Very delicate tissues are first washed in water and then transferred gradually, first into 30 percent alcohol, then into 50 percent, and finally into 70 percent alcohol (so that diffusion currents do not distort the tissues). Keep the tissues in each solution in the alcohol series for about 1 hr, with the exception of the 70 percent alcohol (2 to 6 hr). (If the fixative contained picric acid, remove it quickly by adding a few grains of lithium carbonate to the 70 percent alcohol used in washing.) Transfer the tissues to 95 percent, and into absolute alcohol. Transfer them into a clearing agent to prepare for embedding in wax. A clearing agent must be miscible with alcohol, for it replaces the alcohol in the tissues. Xylol is a rapid clearing agent. Keep the tissues in xylol for 1 to 3 hr. If the xylol becomes cloudy, return the tissues to absolute alcohol (the cloudiness indicates that the tissue has not been completely dehydrated). Incidentally, the best agent for removing water in the stock bottle of absolute alcohol is anhydrous copper sulfate (technique, p. 126).

When the tissues have been cleared, they are ready for embedding in melted paraffin. Keep the wax melted in a paraffin oven set at a temperature a few degrees higher than the melting point of the paraffin. Insert the specimens into square paper boats (Fig. 2-31) containing melted paraffin, and leave the boats in the oven for 2 to 3 hr. Change the specimens into fresh wax within this period of time. When the wax is ready to be cooled, remove the boats from the oven. Float the paper boats in a container of cold water (do not submerge them), and blow on the surface of the paraffin until a film begins to form. Or let the surface paraffin solidify first; then submerge the boat in cold water, and let the paraffin remain to harden throughout. Remove the paper and trim the block of paraffin so that it fits on the holder of the microtome.

SECTIONING AND STAINING. To attach the block of paraffin to the holder of the microtome, melt the wax at one end of the block and press it against the holder. Spread melted wax around the edges. The cutting knife and block should be set so that sections from 6 to 10 μ can be cut (1 μ equals $\frac{1}{25,000}$ in.). As you cut, a ribbon of wax sections forms; lift off the sections with a spatula and float them on slightly warmed water which has been spread on slides already prepared with a thin film of albumen (see Mayer's albumen, p. 132). The temperature of the water must be below the melting point of wax. Finally, dry the slides on which the sections are floating for about 24 hr in an incubator set at 37° C (99° F). If this time is shortened there is danger that the tissues may be washed off the slides in subsequent procedures.

At this point, the prepared slides consist of slices of tissue embedded in paraffin. In the next step, the wax must be removed before the tissue can be stained. First, briefly warm the slides with the paraffin sections over an alcohol flame, and then insert the slides into a jar of xylol to dissolve the wax. After 5 min, transfer the slides to absolute alcohol for 3 min, then for about 2 min in each of these alcohols: 70 percent, then 50 percent, then 30 percent. Finally, the slides are put into distilled water for 1 min. Since most stains are aqueous, the sections are brought gradually through the alcohols into water.

Now the slides containing the sections of tissue (with the paraffin dissolved away) are ready to be stained. There are many stains which may be used. We shall take Harris' hematoxylin, a nuclear stain, as an example of the general staining technique. First, add 2 drops of glacial acetic acid to the Harris' hematoxylin (preparation, p. 131). Immerse the slides in the stain for

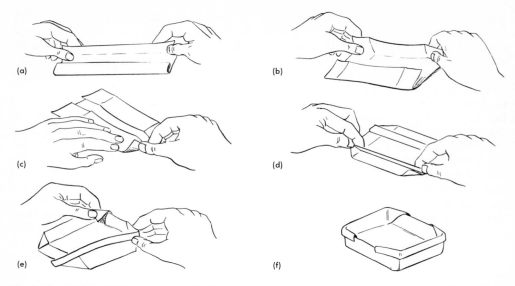

Fig. 2-31 Making a paper or aluminum foil boat in which to embed bits of tissue in paraffin. (After P. Gray, *Handbook of Basic Microtechnique*, McGraw-Hill, 1952.)

1 to 2 min. Transfer into tap water to wash out the acid (notice that the color fades). Wash for 5 min; then transfer slides to each of the following alcohols for 2 min: 30 percent, to 50 percent, to 70 percent, to 90 percent. To counterstain with eosin, a cytoplasmic stain, transfer the slides to eosin for 2 min and rinse in 95 percent alcohol. Move the slides into xylol, where they may be left for some time. Finally, mount in Canada balsam.

STAINS. Many stains may be used. Certain stains penetrate tissues better after use of a particular fixative. There are basic (alkaline) dyes, which stain acid structures within cells, and, conversely, acid dyes, which stain basic structures. For example, acid dyes are held by plasma, cilia, and cellulose structures. Chromosomes, centrosomes, nucleoli, cork, cutinized epidermis, and xylem tissues of plants retain basic dyes.

Some basic dyes in frequent use are safranin, hematoxylin, methyl green, gentian violet, and Janus Green B. Among the acid dyes are eosin, methyl orange,

orange G, fast green, and light green. The stains described below are listed in alphabetical order. Since this is an introduction to histological techniques, only a few are given. Refer to the specialized texts listed at the end of the chapter for more detailed procedures with special stains, and for general histological methods used with plant and animal tissue.

1. ACID ALCOHOL is included here since it is used to decolorize stains. (It can also be used to clean coverslips.) Acid alcohol is prepared by adding 1 ml of 1 percent hydrochloric acid to 90 ml of 70 percent ethyl alcohol.

2. BORAX CARMINE is prepared by mixing together 4 g of borax in 100 ml of distilled water. (Reduce the amounts if smaller quantities are needed.) Then add 3 g of carmine and boil for about 30 min. After this solution has cooled, dilute with 100 ml of 70 percent alcohol. Let this stand for a few days and then filter.

Specimens may be transferred into this stain directly from 70 percent alcohol.

Often the slides become overstained in this dye, but they may be decolorized by placing them in acid alcohol (see above).

3. CARBOL FUCHSIN is a stable stain at room temperature. Prepare in two parts; combine:

basic fuchsin	1 g
absolute alcohol	10 ml

Then mix with:

phenol	5 g
distilled water	100 ml

4. CONKLIN'S HEMATOXYLIN is recommended especially as a stain for whole mounts of chick embryos (p. 132). Prepare as follows:

Harris hematoxylin	1 part
water	4 parts

To this add one drop of Kleinenberg's picrosulfuric acid (p. 128) for each milliliter of fluid.

5. DELAFIELD'S HEMATOXYLIN is prepared by dissolving 4 g of hematoxylin in 25 ml of absolute alcohol. Add this to 400 ml of a saturated aqueous solution of ammonium alum. Expose the solution to light for a few days in a cotton-stoppered bottle; filter. Then add the following:

methyl alcohol	100 ml
glycerin	100 ml

The stain must be ripened for 2 months at room temperature before it is ready for use. Finally, store in well-stoppered bottles. Wash the specimens in water before transferring them into this stain. Delafield's, Ehrlich's, and Harris' (see below) hematoxylin stains may be purchased ready-made.

6. EHRLICH'S HEMATOXYLIN is also ripened at room temperature for about 2 months. Transfer specimens from water to the stain, which is prepared by mixing:

hematoxylin	2 g
absolute alcohol	100 ml

Then add these substances in the order presented:

glycerin	100 ml
distilled water	100 ml
glacial acetic acid	10 ml
potassium alum	in excess

7. EOSIN Y is a plasma stain often used for whole mounts or for contrast. Make up a 0.5 percent solution in distilled water.

8. ETHYL EOSIN is prepared as a 0.5 percent solution in 95 percent ethyl alcohol.

9. HARRIS' HEMATOXYLIN is prepared as follows. First combine:

hematoxylin	1 g
absolute alcohol	10 ml

Then prepare as a separate solution (heat is needed to dissolve the alum):

potassium alum	20 g
(or ammonium alum)	
distilled water	200 ml

Mix the two solutions and bring to a quick boil; add 0.5 g of mercuric oxide. Cool quickly.

10. METHYL GREEN SOLUTION is a good nuclear stain for general use. Prepare a 1 percent aqueous solution by dissolving 1 g of the dye in 1 ml of glacial acetic acid. Then dilute with distilled water to make 100 ml.

11. METHYL ORANGE is widely used as an indicator. Prepare a 0.1 percent solution by dissolving 0.1 g of methyl orange in 100 ml of distilled water.

12. METHYL RED may also be used as an indicator. Prepare a saturated solution in 50 percent ethyl alcohol.

13. METHYL VIOLET may be used to dye amphibian or human blood cells. Prepare as follows:

sodium chloride	100 ml
(0.7% solution)	
methyl violet	0.05 g
glacial acetic acid	0.02 ml

When human blood cells are to be stained, prepare the stain in 0.9 percent sodium chloride solution.

OTHER SOLUTIONS NEEDED IN HISTOLOGICAL WORK. At times there may be reason to use some of the following solutions (fixatives, stains, and so forth). Many of these have been referred to in other sections of this book.

1. BALSAM (NEUTRAL). The slight acidity of samples of balsam is well known. This acidity is an advantage when balsam is

used as the mounting fluid after acid stains, but detrimental when the basic hematoxylin stains are used. Balsam may be neutralized by adding a small quantity of sodium carbonate. Let the fluid stand for about a month. The supernatant balsam should prove to be slightly alkaline.

2. FORMALIN is a 40 percent solution of formaldehyde gas in water and is a stock solution.

A 10 percent solution of formalin, a common fixative for small forms, is prepared by adding 10 ml of stock formalin to 90 ml of water. This solution is an ingredient of many fixatives; it may also be substituted in certain fixatives for glacial acetic acid.

Sometimes a 4 percent solution of formalin is required. In such cases, add 4 ml of commercial formalin to 96 ml of water. (This is really a 1.6 percent solution of formaldehyde.)

Avoid inhaling the fumes of any formalin solutions, for they irritate the mucous membranes.

3. LOEFFLER'S METHYLENE BLUE SOLUTION is a frequently used stain for bacteria. Add the stain to a slide for ½ to 3 min. Then rinse off in water. The stain, which keeps indefinitely, may be prepared by combining:

methylene blue (saturated alcoholic)	30 parts
potassium hydroxide (1:10,000 aqueous solution)	1000 parts

4. MAYER'S ALBUMEN. A dilute solution of albumen is spread as a film on a clean slide so that tissue sections or protozoa adhere to the slide throughout the transfers from one liquid to another. Combine:

egg albumen	50 ml
glycerin	50 ml
thymol	1 crystal

(You can substitute for the thymol 1 g of sodium salicylate; they are both antiseptics.) Shake the mixture vigorously so that the air bubbles become trapped in the solution. When they rise to the surface, remove the foamy mass, and then store the clear fluid. It may be kept from 2 to 4 months without spoiling.

When the solution is to be used, add 3 drops of it to 60 ml of distilled water; with your finger, spread a very light film on a clean slide.

5. MOUNTING FLUID. At times this fluid may be used as a substitute for Canada balsam. However, it is poisonous and must be handled with caution. Mix together 2 parts of chloral hydrate to 1 part of phenol. Heat slightly until the mixture becomes fluid.

Chick embryo. Since a 72-hr chick embryo is the easiest to fix and mount whole on a slide, we will give a general description of the preparation of a slide of this stage. Then students may want to work back to the 24-hr stage. Complete directions may be found in Rugh's embryology manual.[21]

One way to incubate fertilized chicken eggs and to float early embryo stages in bowls of warmed saline solution is described on p. 363. Break open a 72-hr egg and float the embryo in warmed saline solution. Grasp the chalazae with forceps to float the embryo to the top of the mass of yolk. With sharp scissors cut outside the area vasculosa, and float this disc free of the underlying yolk (Fig. 2-32).

Lower a glass slide under the saline solution, and float the blastodisc, yolk side uppermost, onto the slide. Flatten out the blastodisc on the slide and, to hold the embryo in position, put a ring of filter paper large enough to encircle the area pellucida over that region. Then lift the slide out of the saline solution, and gently pipette a fixative onto the slide. Use Bouin's or Kleinenberg's picrosulfuric solution (pp. 127, 128). Place several such slides in Petri dishes of fixative for 8 to 10 hr. Then transfer the slides into a graded series of alcohols in this order: 30 percent, to 50 percent, to 70 percent. Leave the slides in each alcohol for 1 hr.

[21] R. Rugh, *Experimental Embryology: Techniques and Procedures,* 3rd ed., Burgess, Minneapolis, 1962, pp. 405–27.

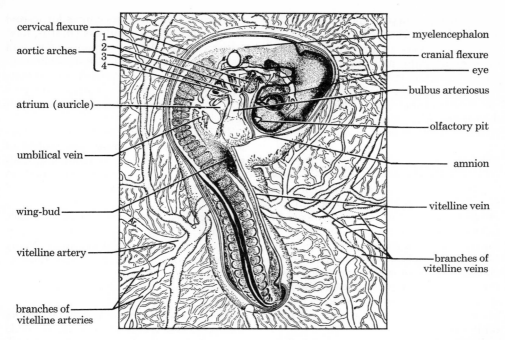

cervical flexure

aortic arches
1
2
3
4

atrium (auricle)

umbilical vein

wing-bud

vitelline artery

branches of
vitelline arteries

myelencephalon

cranial flexure

eye

bulbus arteriosus

olfactory pit

amnion

vitelline vein

branches of
vitelline veins

Fig. 2-32 Chick embryo with adjacent portion of area vasculosa, with 35 pairs of somites (about 72 hr). (From R. S. McEwen, *Vertebrate Embryology*, 3rd ed., Holt, Rinehart and Winston, 1949; after F. R. Lillie, *The Development of the Chick*, 2nd ed., Holt, Rinehart and Winston, 1919.)

To the 70 percent alcohol add a bit of lithium carbonate (or ammonium hydroxide, 3 percent by volume) to decolorize the bright yellow picric stain. Then transfer the specimens back to fresh 50 percent, then 30 percent alcohol, and finally into distilled water in preparation for staining.

Stain a 72-hr embryo with Conklin's hematoxylin (p. 131) for 5 min. For younger embryos less staining time is needed—possibly 2 min for a 24-hr embryo.

Then transfer the slides into tap water for a few minutes. Follow this with transfers into 30 percent alcohol, then to 50 percent; leave the slides ½ hr in each. Examine the slides under the microscope. Should the slides be overstained, destain them in acidified 70 percent alcohol (p. 130). Then follow with a washing in slightly ammoniated 70 percent alcohol

(p. 130). The slides may remain in pure 70 percent alcohol for several hours. Finally, transfer to 95 percent alcohol and follow with two changes in absolute alcohol.

When the embryos are ready for mounting, transfer the slides from absolute alcohol into pure cedar oil for clearing. After 24 hr in cedar oil, the embryos should appear translucent. Then make two transfers into xylol, ½ hr in each container. Mount the embryos in balsam with small chips of glass tubing or pulled-out tubing (so that the embryos will not be crushed). Add more balsam if needed. Flame a coverslip to dry any moisture on it, and lower it into the balsam. Let the slides remain flat for several days until the specimen hardens in place. Later remove the excess balsam with xylol.

This is one of many methods. There are several fixatives and stains used in preparing chick embryos. The blastodiscs also may be fixed in Zenker's fixative (p. 128). Other stains may be preferred, and some workers clear the mount in oil of wintergreen. Alternate techniques may be found in several of the specialized texts listed in the Bibliography. (See Chapter 14 for a description of a method for staining the skeleton of older chick embryos, p. 672.)

Cell contents: macromolecules

The chemical substances of which cells are composed may be examined in terms of basic subunits: amino acids, sugars, and nucleotides. From these subunits, the functioning macromolecules are assembled; these include polysaccharides, long chains of thousands of sugar molecules. By the elimination of molecules of water, proteins are formed from amino acids, and nucleic acids from combinations of nucleotide units.

Each nucleic acid is composed of four repeating nucleotides. Nucleotides consist of three simpler molecules: a nitrogen-containing organic base, a sugar residue, and phosphoric acid. The nitrogen-containing bases all fall into two groups—one containing the purine ring, and the other the pyrimidine ring. Ribose and deoxyribose are the two main nucleic acid sugars; thus, the two major groups of nucleic acids are the ribonucleic acids (RNA) and the deoxyribonucleic acids (DNA). The bases found in RNA are adenine and guanine (both purines), and cytosine and uracil (both pyrimidines). DNA contains adenine, guanine, cytosine, and thymine (a methyl derivative of uracil). In the organism, nucleic acids are attached to proteins, forming nucleoproteins.

Isolated nucleotides may be identified by chromatography. One method described by Wald and his colleagues[22] for a freshmen biology laboratory at Harvard is given on p. 140.

Students probably know the fairly

[22] G. Wald *et al., Twenty-six Afternoons of Biology,* Addison-Wesley, Reading, Mass., 1962, p. 23.

routine, elementary procedures for testing samples of cellular products (that is, food nutrients) for glucose, proteins, fats, and starches (Chapter 4). These tests have become part of the experience of students through the elementary science and general science courses in the junior high schools.

In the biology laboratory students may learn some of the basic techniques of separation, extraction, centrifuging, dialysis, filtration, and hydrolysis used to identify lipids, amino acids, proteins, and the like.

Hopefully, these procedures will give students some insight into the methods used in a biochemistry laboratory. Many laboratory procedures, and understanding of the nature of organic macromolecules may accrue from analyses of an egg or milk.

Teachers will want to refer to laboratory manuals; two excellent ones are G. Wald *et al., Twenty-six Afternoons of Biology* (Addison-Wesley, 1962), and B. Harrow *et al., Laboratory Manual of Biochemistry,* 5th ed. (Saunders, 1960). Or refer to such basic references as P. Hawk, B. Oser, and W. Summerson, *Practical Physiological Chemistry,* 13th ed. (Blakiston [McGraw-Hill], 1954), B. Walker, W. Boyd, and I. Asimov, *Biochemistry and Human Metabolism,* 3rd ed. (Williams & Wilkins, 1957), and J. Todd and A. Sanford, *Clinical Diagnosis by Laboratory Methods,* 13th ed., edited by I. Davidsohn and B. Wells (Saunders, 1962).

Separation of molecules by dialysis

You may want to have students combine equal quantities of several solutions and suspensions containing molecules of different sizes. These can all be placed in dialyzing tubing (available from supply houses).

Tie off each end of the tubing and immerse in a beaker half-filled with water. Stir the water occasionally to distribute the diffusing molecules uniformly. One possible series of compounds to test for diffusion through the membrane includes starch, glucose, a chloride (such as sodium chloride), and an amino acid.

After 5 min test for the presence of specific molecules in the water—evidence that these molecules have diffused through the membrane. Repeat after 15 min. In order for students to learn the standard identifying tests, a solution of each compound should also be tested separately. Use Lugol's iodine solution to identify starch; heat with Benedict's solution to show the presence of glucose; a white precipitate is formed if silver nitrate is added to a solution of sodium chloride; a solution containing an amino acid turns blue if heated with ninhydrin reagent (see p. 136).

Pentose sugars in plants

Leaf fragments or pure sugars may be used to demonstrate two tests for pentose sugars.[23]

1. BENZIDINE TEST. Compare the reaction of glucose (a six-carbon sugar) with xylose (a five-carbon sugar). Prepare 0.25 percent solutions of each sugar. To 0.5 ml of one solution of sugar in a Pyrex test tube, add 2 ml of 4 percent solution of benzidine in glacial acetic acid. (*Caution:* Work under a hood.) Heat to the boiling point and cool immediately in cold water. Repeat the procedure with the other sugar solution. The presence of a pentose is indicated by a cherry-red color. Hexoses give a yellow or brownish color.

2. BIAL'S TEST. Prepare Bial's reagent (using 0.2 percent orcinol solution in concentrated HCl). Add 2 ml of the reagent to an equal volume of a xylose sugar. (Also run a test on glucose, as described above in the benzidine test.) Heat in a boiling water bath. (Furfural is formed from the sugar and reacts with orcinol.) Add a few drops of ferric chloride (1 percent solution). A deep green color indicates the presence of pentose.

Students may use these tests on pectins of freshly pressed apple juice. Preparation of the pectins is described by Machlis and Torrey.[24]

Benedict's test for hexose sugars and aldehydes of sugars is described in Chapter 3.

Tests for proteins

Using bean meal extract. At some time you may want to have students learn something of the chemistry of proteins—the tests for sulfhydryl groups, benzene rings with hydroxyl (OH) groups, and so forth.

Prepare solutions of several pure compounds such as 10^{-3} M tyrosine, 10^{-3} M tryptophan, 10^{-3} M cysteine hydrochloride, 2 percent gelatin, and bean meal extract. Prepare the bean meal extract by shaking together 5 g of the bean meal and 50 ml of distilled water. Allow this to settle, and decant the supernatant which is to be used to run the tests to show the chemical nature of proteins.[25]

1. BIURET TEST. Most substances containing 2 carbamyl groups ($-CONH_2$) joined directly or through another C or N atom give a purple color reaction in the Biuret test. To 3 ml of the test solution, add 1 ml of 10 M NaOH and 1 drop of a 1 percent solution of copper sulfate.

2. MILLON'S TEST. Proteins containing compounds with a benzene ring with OH groups attached, such as tyrosine, give a positive reaction in this test. To 3 ml of the solution to be tested, add an equal volume of 10 percent mercuric sulfate in 10 percent sulfuric acid. Gently boil for 1 min. Then cool under the tap and add 1 drop of 1 percent solution of sodium nitrite. Warm this gently. A red precipitate is formed, due to the red mercury combinations of nitrophenol derivatives.

3. NITROPRUSSIDE TEST. The sulfhydryl groups among the proteins give a deep purple-violet color in the nitroprusside

---

[23] Adapted from L. Machlis and J. Torrey, *Plants in Action*, Freeman, San Francisco, 1956, p. 97.

[24] *Ibid.*, p. 98.
[25] Tests adapted from *ibid.*, p. 125.

test. To 2 ml of the solution to be tested, add 1 ml of 2 percent sodium nitroprusside —$Na_2Fe(CN)_5NO$—solution, and then add 2 drops of 10 N sodium hydroxide.

4. TRYPTOPHAN REACTION. The amino acid tryptophan can be identified by the deep purple color reaction in this test. To 2 ml of the test solution add 2 drops of commercial formalin (40 percent) that has been diluted 1:500. Then add 2 drops of 10 percent mercuric sulfate in 10 percent sulfuric acid. Mix this well. Add 2 ml of concentrated sulfuric acid. Mix again and look for the deep purple color which is a positive test for tryptophan.

5. NINHYDRIN TEST. This test may be considered quantitative: the intensity of the purplish color in a positive reaction is proportional to the concentration of the amino acid. Heat to boiling 0.5 ml of a 0.1 percent ninhydrin (triketohydrindene hydrate) with 3 ml of the solution of amino acid to be tested. (*Caution:* Avoid inhaling, poisonous fumes.) Or test hydrolyzed proteins for the presence of amino acids.

Ammonia and carbon dioxide are also evolved in the reaction (R stands for a hydrocarbon residue).

$$R \cdot CH(NH_2)COOH \longrightarrow$$
$$R \cdot CHO + NH_3 + CO_2$$

Refer also to the use of ninhydrin in the techniques of paper chromatography (p. 139).

Using seeds. Meyer, Anderson, and Swanson[26] suggest protein tests that may be made on slides. Thick slices of lima bean seeds that have been soaked long enough to be easily cut are used. Since the directions given by Meyer, Anderson, and Swanson vary from those described above, we include here the three tests that may be more successful when run directly on microscope slides holding soft lima beans. Additional microchemical tests that may be made on slides of tissues of plants are the osazone test for sugars, the phloro-

glucin test for lignin, the iodine–sulfuric acid test for cellulose, or the Sudan III test for fats and oils.[27]

Line up slides containing slices on a sheet of white toweling, and run the series of tests.

1. BIURET TEST. This is a specific test for the presence of peptide linkage. Proteins containing these compounds give a pink-to-violet color reaction in the Biuret test. Add a drop or two of 5 percent copper sulfate solution to a section of soft lima bean; add a coverslip and let it stand in a moist chamber for ½ hr. (A moist chamber may be made by lining a Petri dish with moist blotting paper or filter paper.) Remove the cover glass and wash the slice with distilled water. Blot dry with filter paper and add a drop of 50 percent potassium hydroxide solution. Does a pink or violet color appear?

2. MILLON'S REACTION. To a second section of lima bean on a slide, add a drop or two of dilute nitric acid (2 percent). Let it stand for a minute and blot off the excess nitric acid. Now add a drop of Millon's reagent. (Prepare the reagent by pouring 20 ml of mercury into a 500-ml Pyrex beaker. Stand the beaker under a hood; carefully pour 180 ml of concentrated nitric acid down the side of the beaker. [*Caution:* This is a violent reaction—some acid may be thrown out of the beaker.] After the mercury has dissolved, pour the solution into an equal volume of distilled water.) The reaction may be hastened by gently heating the slide. Proteins give a deep red color. Since the reaction depends on the presence of the monohydroxybenzene nucleus, some compounds other than proteins may give a similar color reaction. Therefore, this is not a conclusive test for proteins.

3. XANTHOPROTEIC REACTION. To the slice of lima bean, add a drop or two of concentrated nitric acid. A yellow color is a positive test for proteins. Since the reaction depends on the presence of the benzene nucleus, several other compounds

[26] B. Meyer, D. Anderson, and C. Swanson, *Laboratory Plant Physiology*, 3rd ed., Van Nostrand, Princeton, N.J., 1955, p. 142.

[27] Sass, *op. cit.*, pp. 96–98.

give the same reaction. Blot the excess nitric acid with filter paper and add a drop of strong ammonium hydroxide or sodium hydroxide to the slice. Proteins turn a deep yellow or orange-brown color in this check step.

Principles of chromatography

Slightly different solubilities of like compounds permit such compounds to be separated in water and an organic solvent. If a strip of Whatman #1 or #3 filter paper is introduced as a wick into a small sample of such a mixture or into a closed atmosphere saturated with the mixture, the compounds will migrate up or across the paper at different rates. The factors affecting the separation of pigments along the filter paper (or packed column of adsorbent, Fig. 2-33) are a combination of adsorption, ion exchange, and partition. Thus, unknown compounds can be identified by measuring the relative speed at which they migrate up the filter paper.

The distances traveled by any compound are dependent on the rate of migration of molecules along a solvent front that is produced by the rise of the solvent along the length of the filter paper. As these compounds move along the filter paper, they distribute themselves between the solvent and the cellulose fibers of the paper. (Compare the results of using Whatman #1 and #3 paper.) In general, the more soluble a compound is in water compared with the organic solvent, the slower it travels; it moves even more slowly if the pigment is also absorbed by the fibers of the paper.

Since the fastest molecules will travel the greatest distance, or to the highest point along the strip in ascending chromatography (or will be located at the outer circle of the concentric rings of the circular filter paper), the relative distances can be measured, and the rate of flow, or migration, of the molecules (R_f) can be calculated.

$$R_f = \frac{\text{distance substance (solute) traveled}}{\text{distance solvent traveled}}$$

In column and paper chromatography, substances are distributed in a one-dimensional row. In some cases, compounds overlap and a second separation step is required, producing a two-dimensional chromatogram. In the latter method, developed by two British chemists, the Nobel Prize winners A. J. Martin and R. L. Synge, the procedure followed in the first separation is repeated, using a different solvent traveling at right angles to the first spread, or separation; as a result, the closely associated compounds separate out in a second dimension.

In contrast to pigments that can be separated and perceived from chloroplasts or petals, most compounds such as sugars and amino acids are colorless. Of course, if radioactive tracers are included in their synthesis the compounds may be identified if the chromatographic paper is placed in contact with X-ray film for a few days. (Several techniques for partitioning pigments of chloroplasts are developed in Chapter 3.)

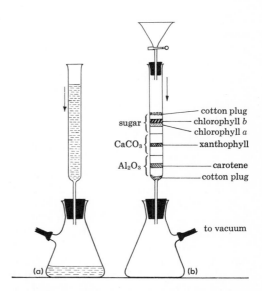

Fig. 2-33 Column chromatography: (a) column of one adsorbent; (b) use of different adsorbents and order of adsorption of pigments of chloroplasts. (After B. Harrow *et al.*, *Laboratory Manual of Biochemistry*, 5th ed., Saunders, 1960.)

TABLE 2-2

Rate of flow of molecules (R_f) of selected amino acids*

amino acid	R_f of 125 mm at 25° C (77° F) after 3 hr
Alanine	0.62
Arginine HCl	0.53
Aspartic acid	0.25
Glutamic acid	0.39
Glycine	0.49
Histidine HCl	0.81
Isoleucine	0.85
Leucine	0.86
Lysine HCl	0.41
Methionine	0.74
Phenylalanine	0.87
Proline	0.87
Serine	0.33
Threonine	0.57
Tryptophan	0.81
Tyrosine	0.53
Valine	0.82

* From B. Harrow, *et al.*, *Laboratory Manual of Biochemistry*, 5th ed., Saunders, Philadelphia, 1960.

One-dimensional chromatography of proteins. In the separation of proteins, such as the amino acids, the ninhydrin test for amino acids (p. 136) is used. In a laboratory demonstration, students may compare the migration rate of pure amino acids with a table of standards (Table 2-2), and then try to identify individual amino acids in an unknown containing a mixture of several amino acids. A suggested procedure follows.[28]

Obtain several pure amino acids, such as isoleucine, threonine, and aspartic acid, and prepare solutions of equal dilution. Fit each of several test tubes with strips of Whatman #1 paper (as shown in Fig. 3-18a), and make a pencil mark about 6 mm from the end of the papers to indicate the place to spot the papers. Use a capillary pipette to transfer a small drop of one

[28] Adapted from B. Harrow *et al.*, *Laboratory Manual of Biochemistry*, 5th ed., Saunders, Philadelphia, 1960, pp. 25–26.

of the amino acid solutions to the paper. Using separate pipettes spot the other papers. After drying, apply a second drop to the papers. After the spots are dry, insert each strip into a separate test tube; each test tube should contain 0.4 ml of butanol–acetic acid solution. (Prepare by adding 250 ml of water to 250 ml of *n*-butanol; shake well and add 60 ml of glacial acetic acid. Withdraw the bottom layer; store the butanol layer in the presence of the aqueous layer so that esterification will be reduced.)

In each test tube, only the tips of the papers are immersed in the solution. Keep a record of time and rate of the solution migration until the Whatman paper is wetted to within 5 mm of the top of the paper (about 2½ hr). Then remove the papers and, with a pencil, mark the upper front; dry for about 3 min at 110° C (230° F). Then spray with a 0.25 percent solution of ninhydrin (*caution:* poisonous) in water-saturated butanol. Again, dry at 110° C (230° F). As the color appears circle the spots. For each case, measure the distance that the solvent front migrated and the distance the amino acid migrated; then estimate the R_f value for each amino acid (refer to p. 137 for equation).

Use Table 2-2 to identify the amino acids and to check the skill of students in measuring the R_f of the molecules in solutions. Then prepare mixtures of several amino acids and repeat the procedure.

Students knowledgeable in this simple technique can design investigations of altered conditions of diet, drugs, or hormones on blood or urine, or compare the contents of cells and tissues, or the composition of the food residues in seeds.

Two-dimensional chromatography of proteins. Fold a sheet of Whatman #1 paper into a cylinder, and staple the two edges together. Apply a spot of a sample of amino acid (or of blood, urine, or organic acids) about 1.5 cm from the edge of one end of the cylinder; stand the cylinder in a 1-cm layer of a solvent for the test sample. After separation, unfold the cylinder and dry at 110° C (230° F). Turn the paper 90° and prepare a cylinder in this direction (hold it together with a staple). Place the cylinder, spot end down, into a 1-cm layer of a second solvent for this sample. After the second solvent has risen along the cylinder, remove and flatten the paper, and dry in hot air or at 110° C (230° F). Dip the sheet into a 0.25 percent acetone solution of ninhydrin; this may easily be done by sliding the paper in a wide tray in the manner used in photographic development.

After several hours, the purple spots should be well developed. Compare each spot of the experimental chemicals with standard chromatograms which have been spotted under similar conditions.

Separation of products of protein hydrolysis. Wald et al.[29] suggest a comparative chromatogram study of an unhydrolyzed protein, the protein hydrolysate, one unknown amino acid, and several known amino acids (such as alanine, histidine, lysine, methionine, and/or aspartic acid) prepared on one sheet of Whatman #1 paper.

On one edge of a square sheet of filter paper, mark off dots (with a lead pencil) 1.5 cm from the edge and at an equal distance from each other (Fig. 2-34). The number of dots is the same as the number of test spots to be applied. Apply different materials with fine capillary pipettes; allow the paper to dry; then reapply amino acids at least four times, and hydrolysate about six times. Label the spots with lead pencil.

Fold the paper into a cylinder, handling it only at its edges, and staple or sew it together so that the cylinder can stand in about 1 cm of solvent. Prepare 30 ml of solvent by combining 10 parts of formic acid, 70 parts of isopropanol, and 20 parts of water. Stand the paper cylinder in this solvent and allow the solvent front to rise to nearly 0.5 cm from the top. Remove the

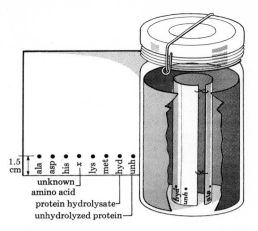

Fig. 2-34 Apparatus for comparative chromatogram of products of protein hydrolysis. (After G. Wald et al., *Twenty-six Afternoons of Biology*, Addison-Wesley, 1962.)

cylinder and allow it to dry. Open the cylinder and dip the sheet of paper into ninhydrin-acetone reagent (*caution:* avoid inhaling, poison). Remove and allow to dry in hot air. Circle the colored spots.

Measure the distances from the starting mark to the front for each of the test substances and for the solvent, and prepare a table of results. Compare with Table 2-2 and identify the amino acids in the hydrolyzed protein.

A protein that is easily obtained for hydrolysis and analysis is casein in milk. Casein may be purchased or extracted from skim milk.[30] Transfer about 0.2 g of casein into a test tube, and add 2 ml of 0.1 percent pancreatic extract dissolved in 0.1 *M* phosphate buffer at pH 7. Maintain in a water bath at 40° C (104° F). It is advisable to add a small crystal of thymol to inhibit growth of bacteria. The casein hydrolysate should be applied twice with a micropipette to Whatman #1 paper.

A chromatogram may also be prepared using the protein hydrolysate obtained from the extraction procedure described in the flow chart in Fig. 2-35.

[29] From Wald et al., *op. cit.*, pp. 21–22. For a similar demonstration, see Machlis and Torrey, *op. cit.*, pp. 121–23.

[30] See Wald et al., *ibid.*, pp. 144–45.

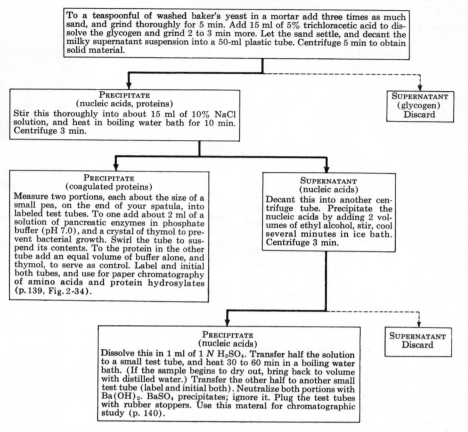

PRECIPITATE
(nucleic acids, proteins)
Stir this thoroughly into about 15 ml of 10% NaCl solution, and heat in boiling water bath for 10 min. Centrifuge 3 min.

SUPERNATANT
(glycogen)
Discard

PRECIPITATE
(coagulated proteins)
Measure two portions, each about the size of a small pea, on the end of your spatula, into labeled test tubes. To one add about 2 ml of a solution of pancreatic enzymes in phosphate buffer (pH 7.0), and a crystal of thymol to prevent bacterial growth. Swirl the tube to suspend its contents. To the protein in the other tube add an equal volume of buffer alone, and thymol, to serve as control. Label and initial both tubes, and use for paper chromatography of amino acids and protein hydrosylates (p. 139, Fig. 2-34).

SUPERNATANT
(nucleic acids)
Decant this into another centrifuge tube. Precipitate the nucleic acids by adding 2 volumes of ethyl alcohol, stir, cool several minutes in ice bath. Centrifuge 3 min.

PRECIPITATE
(nucleic acids)
Dissolve this in 1 ml of 1 N H_2SO_4. Transfer half the solution to a small test tube, and heat 30 to 60 min in a boiling water bath. (If the sample begins to dry out, bring back to volume with distilled water.) Transfer the other half to another small test tube (label and initial both). Neutralize both portions with $Ba(OH)_2$. $BaSO_4$ precipitates; ignore it. Plug the test tubes with rubber stoppers. Use this materal for chromatographic study (p. 140).

SUPERNATANT
Discard

Fig. 2-35 Flow chart for extraction and hydrolysis of nucleic acids. (After G. Wald et al., *Twenty-six Afternoons of Biology*, Addison-Wesley, 1962.)

Nucleic acids

Identifying nucleic acids (and nucleotides)

USING PAPER CHROMATOGRAPHY. Prepare an extract of nucleic acids of yeast cells according to the flow chart in Fig. 2-35 (or read complete directions in Wald's laboratory manual[31]). This extract will contain a large amount of RNA and a small amount of DNA. Another method of extracting RNA from yeast, as described by Hawk *et al.*, is given on p. 142.

The hydrolyzed extract contains nucle-

otides, even hydrolyzed nucleotides: adenine, cytosine, uracil, and guanine from the hydrolyzed RNA. Guanine is quite insoluble and its identity will not be attempted in this description. Since such small amounts of thymine are present, it will not be identified either.

In short, a chromatogram of three known nitrogenous bases—adenine, uracil, and cytosine (1.0 mg per ml solutions of each)—should be run on one sheet of Whatman #1 paper. Then the separation of macromolecules of the unknown extract of hydrolyzed yeast materials can be partitioned, and a comparison made between the standards and the unknown.

[31] *Ibid.*, Chapters 3, 4.

Apply the spots 1.5 cm from one end of the paper equidistant from each other. Reapply five times with a fine micropipette; allow each spot to dry before the succeeding application. Fold the paper into a cylinder, and staple or baste together with thread (Fig. 2-34).

Stand the paper cylinder (or cylinders, for, hopefully, several were prepared by groups of students) in a tray or jar of solvent composed of 15 parts of acetic acid, 60 parts of butanol, and 25 parts of water. Prepare to stop the process when the solvent front nears the point 1 cm from the top of the cylinder. Dry the cylinder; then flatten the sheet by cutting the threads or removing the staples.

These partitioned regions of nucleic acid components are not stained, but are exposed to ultraviolet light (as from a germicidal lamp with a wavelength of 260 mμ). The nitrogenous bases absorb ultraviolet light and appear as dark spots. (*Caution:* Use safety glasses; do not look into ultraviolet light, and avoid exposure of the skin for more than a few seconds.)

Measure the distance each component migrated along the solvent front, and compare with standard chromatograms prepared from solutions of pure nitrogenous bases (equation, pp. 137, 138).

USING PAPER ELECTROPHORESIS. In chromatography, separation of substances is based on the different rates of molecular movement through an adsorbing material. Electrophoresis is based on the fact that charged ions of colloidal substances migrate along paper (or other substances) toward one of the poles in an electric field. The distance through which a charged particle moves is dependent upon time and is proportional to the potential gradient at the spot where the particle is located. This last factor depends on the current, the conductivity of the solvent, and the cross-sectional area of the solution at this point. Several zones develop if the colloidal solution is impure, each zone composed of particles having similar rates of electrophoretic migration.

Paper electrophoresis is a simple adaptation of the original principle used by Tiselius, who obtained five distinct fractions composing plasma—globulins (alpha, beta, and gamma), albumin, and fibrinogen. Later workers have obtained subdivisions of the globulins. Only small samples of colloidal material are needed in paper electrophoresis. In essence, filter paper soaked in an electrolyte is connected to a suitable source of high-voltage direct current (about 4 to 8 volts per cm). Fractions of substances are made visible by staining.

The pH at which an amino acid has no tendency to migrate toward either the negative or positive electrode is its isoelectric unit.

The electrically charged nature of a protein molecule depends on its existing pH and the composition of the amino acids. At a pH that is acid to its isoelectric point, a protein molecule will contain an excess of NH_3^+ groups and will react as a positive ion and migrate toward the cathode. When the pH is alkaline to the isoelectric point, the protein molecule moves to the anode, since it has an excess of COO^- groups and behaves as a negative ion.

Proteins differ in their isoelectric points; as a result, they have net charges of different degrees at any specific pH. In brief, in a mixture of proteins, as in blood plasma, the components migrate at different velocities in an electric field.

If the pH of the protein mixture is adjusted with a buffer to a point that is alkaline to the isoelectric points of all the components, then all these constituents of proteins will carry negative charges. When an electric current is passed through an electrolyte, these proteins will all migrate to the anode, but at different rates. The constituents are thereby partitioned out.

In the paper electrophoresis of blood, for example, a spot of blood may be applied to filter paper and subjected to an electric current. Moisten the paper in an electrolytic solution (such as 1 tsp of sodium chloride and ¼ tsp of sodium bicarbonate in 300 ml of tap water). Immerse each end of the paper strip in two jars con-

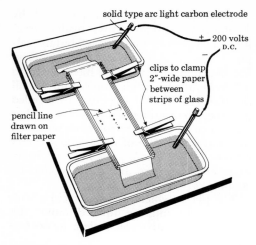

solid type arc light carbon electrode

+ 200 volts D.C.
−

clips to clamp 2"-wide paper between strips of glass

pencil line drawn on filter paper

Fig. 2-36 Simple apparatus for paper electrophoresis. (After C. L. Stong, *The Amateur Scientist,* Simon and Schuster, 1960.)

taining the electrolyte. Set the jars about 6 in. apart. Support the paper by sandwiching it between two glass rectangles so that only the ends of the papers are free to extend into the solutions (Fig. 2-36).

Attach electrodes in the containers. After the components have been separated in the electric field, ninhydrin (or another suitable stain such as bromphenol blue) is applied. (See test for amino acids, p. 135.) Graphite electrodes may be used with a current of some 10 milliamperes and a voltage of about 200 volts.[32] Use pure D.C. "B" batteries, or highly regulated D.C. power supply, $\pm\frac{1}{2}$ volt regulation.

The various components of blood that have negative charges migrate toward the positive electrode at varying rates.

Students will find some ideas for projects, such as studies of electrophorized serum of several species of tadpoles and the changes in quantity of serum albumins as metamorphosis progresses, clearly described in E. Frieden, "The Chemistry of Amphibian Metamorphosis," in *Scientific American* (November 1963). Also read S. Clevenger's paper on flower pigments in

the June 1964 issue and E. Hadorn's paper on fractionating fruit flies in the issue of April 1962. These papers suggest many avenues for possible inquiry. (See also the Bibliography for references by Prosser and Brown, by Florkin, and by Hawk, Oser, and Summerson.)

For detailed information on techniques of chromatography and electrophoresis, see F. Feldman, *Techniques and Investigations in the Life Sciences* (Holt, Rinehart and Winston, 1962), Chapter 5, and C. L. Stong, *The Amateur Scientist* (Simon and Schuster, 1960). The latter book is a compilation of many projects that have been described in *Scientific American.*

Nucleoproteins from yeast cells. Hawk, Oser, and Summerson[33] describe this extraction of RNA from yeast cells. Dilute 50 ml of 1 percent sodium hydroxide with 250 ml of water. To this solution add 30 g of dry yeast. Heat for ½ hr on a water bath, stirring on occasion. Then filter. Allow to cool and slowly add acetic acid to the filtrate until it is slightly acid to litmus. Filter again; then evaporate the solution to 100 ml. After cooling to 40° C (104° F) or below, vigorously stir the solution and quickly pour it into 200 ml of 95 percent alcohol to which 2 ml of concentrated HCl has been added. Allow this to settle in a tall container, and then decant. Wash twice with 95 percent alcohol; follow this with two washes with ether. Transfer to filter paper and allow to dry at room temperature. (*Caution:* Ether is explosive; keep away from flames and have good ventilation.)

The extracted nucleoproteins may be tested for solubility in cold and hot water, alcohol, acid, and alkali. What is the effect of heat on nucleoproteins? Do these proteins give a positive Biuret test?

A small portion may be boiled with 10 ml of 10 percent sulfuric acid for about 2 min. Then divide this into three portions.

[32] See C. L. Stong, *The Amateur Scientist,* Simon and Schuster, New York, 1960. Stong recommends the use of a rectifier from a junked radio receiver (p. 156).

[33] P. Hawk, B. Oser, and W. Summerson, *Practical Physiological Chemistry,* 13th ed., Blakiston (McGraw-Hill), New York, 1954, p. 213.
RNA is available also from biochemical supply houses (Appendix C).

To one portion, apply a carbohydrate test, such as Molisch's reaction (see p. 218). On the second portion, test for purine bases—add an excess of ammonia and then some silver nitrate solution. The third portion can be tested for the presence of phosphate by adding ammonia in slight excess; then add enough nitric acid to acidify the solution. Follow this with molybdate solution, and then warm the total solution. (See also partitioning of nucleic acids, p. 140.)

DNA from thymus glands. Cottony threads of DNA are available commercially. Or as a special club activity, students may wish to extract DNA from freshly minced and frozen thymus glands. Extractions are carried on at 0° C (32° F). Homogenize about 100 g of frozen tissue in 200 ml of cold, 0.15 M NaCl solution, which is also 0.01 M in sodium citrate, in a blender; then centrifuge the mixture at 2000 rpm. Repeat the centrifuging three times, discarding the supernatants each time.[34]

Disperse the residue in 200 ml of water in a blender at high speed. Bring the concentration up to 0.5 percent by adding enough 5 percent sodium dodecyl sulfate and enough NaCl to make the solution 1 M with respect to sodium chloride. Stir constantly. Store overnight at 5° C (41° F). Filter through a pad of Celite on a Buchner funnel. Pour the filtrate, when clear, into 2 volumes of ethanol, stirring constantly. Collect the fibrous precipitate of sodium deoxyribose nucleate, and wash several times with 70 to 95 percent alcohol.

Test the DNA for carbohydrate, purine bases, and phosphate (as described above for yeast nucleoproteins). Specific tests for purine and pyrimidine bases and derivatives can be found in Hawk, Oser, and Summerson, *Practical Physiological Chemistry,* or similar references. Methods for chromatographic separation of purine and pyrimidine derivatives are also given in Hawk *et al.*

DNA from testes. Rugh[35] suggests the extraction method used by Pollister and Mirsky. Grind up a large number of testes of frogs in a small volume of 1 M sodium chloride solution. To this mash, add 10 parts of 1 M NaCl; a viscous mixture results, due to the release of nucleoproteins. Centrifuge this mixture at 10,000 rpm and add 6 volumes of distilled water. Notice that the nucleoproteins precipitate out as fibrous matter. Pour off the supernatant fluid.

Wash the precipitate in 0.14 M NaCl and redissolve the precipitate in 1 M NaCl. Again centrifuge at a high speed. Precipitate the nucleoproteins again by adding 6 volumes of distilled water.

Now carefully stir with a glass rod; twist the rod and notice how the fibers of nucleoproteins wind around the rod.

Cell activity: enzymes

Cells contain thousands of enzymes. These catalysts are protein in composition, and are active within a limited pH range; they show a specificity for a given substrate.

Enzymes that catalyze the breakdown, or hydrolysis, of complex polysaccharides and of proteins through the addition of molecules of water are described in Chapter 4—amylases, lipase, and proteases, such as pepsin and trypsin of pancreatin extract. In cellular respiration many enzymes—especially dehydrogenases and phosphorylases—are involved in oxidation reactions.

Many examples could be given of demonstrations showing the action of enzymes involved in cellular respiration; however, only a selected few are described. Others may be found in laboratory manuals and in textbooks in biochemistry (see the bibliography at the end of this chapter).

Catalase

Students may explore the action of the enzyme catalase, which is found in plant

---

[34] *Ibid.,* p. 214.

[35] *Op. cit.,* p. 325.

and animal cells. The action of catalase in decomposing poisonous hydrogen peroxide (H_2O_2) in tissues into water and oxygen can be demonstrated using liver, blood, or freshly ground meat or plant cells. (Catalase found in liver has been isolated and can be obtained in crystalline form for quantitative work. See Appendix C for supply houses.)

$$2H_2O_2 \longrightarrow 2H_2O + O_2$$

Students, working in groups, may use different living tissues in different series of test tubes: minced raw potatoes, raw ground meat, yeast cells, fresh blood from a frog or human, or ground leaves from bean seedlings some 3 weeks old.

Fill test tubes one-third full of fresh hydrogen peroxide (3 percent), and add each tissue to be tested to a separate test tube; leave one tube of H_2O_2 as a control. Other controls will be suggested by students: boiling the tissue before adding to the test tube of peroxide, or testing the effect of temperature variations on the activity of the catalase.

Besides viewing the production of bubbles, which indicates tremendous activity of the catalase, some students may want to collect the oxygen gas by displacement of water, measure it, and test with a glowing splint to identify it (Fig. 2-37).[36] Or a volumeter of the type shown in Fig. 3-14 may be used to measure the quantity of oxygen produced.[37]

Students may also study some quantitative aspects of enzyme reactions, using the technique of titration with potassium permanganate solution. In an acid solution, hydrogen peroxide decolorizes potassium permanganate within a few minutes. In a beaker mix 1 ml of phosphate buffer and 9 ml of fresh hydrogen peroxide (3 percent). Record the time at which 1 ml of catalase solution is mixed with the perox-

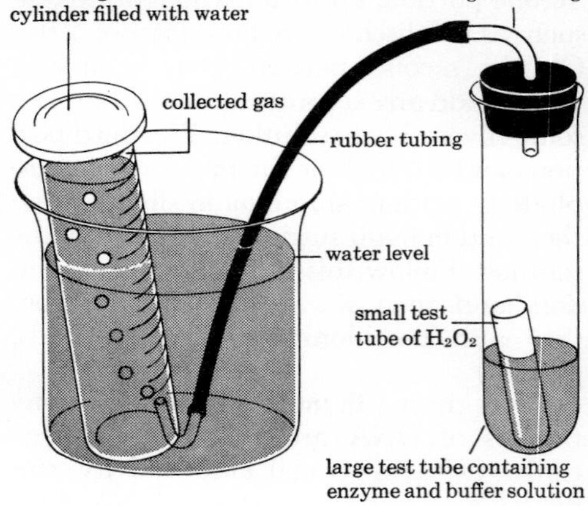

Fig. 2-37 Apparatus for collecting oxygen released from hydrogen peroxide as a result of the activity of catalase. (After L. Machlis and J. Torrey, Plants in Action, Freeman, 1956.)

ide-plus-buffer solution. Allow the action to go on for 10 min; then add 2 ml of sulfuric acid.

Now students can determine the quantity of hydrogen peroxide remaining in the solution after 10 min of reaction with the enzyme catalase. Count the number of drops of $KMnO_4$ needed to reach the end point where 1 drop of permanganate causes a pink color to persist. Allow a minute to elapse, and swirl between additions of drops of permanganate. Estimate the volume of permanganate needed (use a burette or graduated micropipette). What quantity of the H_2O_2 is destroyed in 10 min?

$$5H_2O_2 + 2KMnO_4 + 4H_2SO_4 \longrightarrow$$
$$2KHSO_4 + 2MnSO_4 + 8H_2O + 5O_2$$

Other students should titrate control solutions, that is, solutions containing the same chemicals, with water (1 ml) in lieu of the catalase solution. What is the effect of boiling on the activity of the enzyme?

Can students devise a demonstration to study the effect of metallic salts on the action of the enzyme? What is the effect of mercuric chloride?

---

[36] From Machlis and Torrey, op. cit., p. 113.

[37] Or use the technique described by Wald et al., op. cit., p. 35, using a volumeter to determine the quantity of oxygen liberated from Serratia cultures through activity of catalase.

Machlis and Torrey[38] suggest determining the effect of pH on catalase activity. Use bean leaves from ten seedlings (about 3 weeks old) ground in a mortar and added to 40 ml of distilled water; the apparatus used is the same as that shown in Fig. 2-37. At zero time the test tube is tipped so that the enzyme-buffer solution and the peroxide solution are mixed. Begin with a trial run using a buffer at 7.2 pH, and let the reaction continue for 5 min, that is, a time sufficient to fill the 10-ml cylinder that is used. Then carry on the other runs using different pH buffers (3.6, 4.8, 5.4, 6.0, and so forth).

Role of dehydrogenase

The removal of hydrogen in oxidation is catalyzed by a dehydrogenase which transfers hydrogen from an energy-yielding substrate to a hydrogen acceptor. Methylene blue can be used to show the action of a hydrogen acceptor. In this technique[39] the source of the dehydrogenase is fresh, finely minced muscle from a rat.

First prepare a phosphate buffer $(M/15)$ at 7.0 pH. Into each of a set of three test tubes pour 5 ml of the phosphate buffer, and then add 3 drops of 0.01 percent methylene blue solution. Into test tube 1 place finely minced, fresh muscle of rat. After mixing thoroughly, quickly add a layer of mineral oil over the surface to prevent air from reaching the dye.

Prepare tubes 2 and 3 in a similar manner, except for the treatment of the muscle tissue: To tube 2 add *boiled* muscle; to tube 3 add fresh muscle that has been *washed* several times in distilled water. Place the set or sets of tubes in a water bath maintained at 37° C (99° F). Have students explain the changes in the tubes.

Role of hydrogen acceptors. Students can examine the role of a vital dye such as Janus Green B in accepting hydrogen atoms in the oxidative process in cells of a developing chick embryo. Hydrogen is transferred from organic molecules to the Janus Green B dye, thereby reducing the dye. In the oxidized state the dye is green; in the reduced state it is red.

"Seat" a fertilized egg that has been incubated for 48 hr in a shallow depression of modeling clay so that it will not roll during the operation. Start to chip and carefully cut away a circle of shell from the top surface of the incubated egg. Use a forceps to lift off carefully the chipped circle of shell; with a pipette draw off some of the more fluid albumen to expose the surface of the embryo which is lying in the center of the egg.

Place a small ring of filter paper over the egg so that the embryo is in the center. The inner diameter of the filter paper ring should approximate the diameter of the blood sinus around the embryo; the outer diameter of the paper should fit into a depression slide. Carefully cut the membrane, using the outer rim of the paper ring as a guide. The embryonic tissue should adhere to the paper ring. Transfer this preparation into a Petri dish containing warmed (37° C, 99° F) physiological saline solution (0.9 percent sodium chloride). Several such preparations should be made. Then transfer each embryo to individual depression slides containing some warmed Janus Green B vital dye. (Prepare the solution by adding 0.1 g of Janus Green B to 500 ml of physiological saline solution.)

In about 15 min the embryo should have taken on a greenish stain. Draw off the vital stain with a pipette and replace the fluid with warm saline. Now the stained embryo is in warm saline solution in a depression slide. Seal the slide with a coverslip and make it airtight with petroleum jelly; avoid getting the petroleum jelly under the slide near the embryo. Use low-power microscope, and watch for the appearance of a reddish color. Keep a record of the areas that become red, and in what order in time.

Succinic dehydrogenase. This hydrogen-transferring enzyme is found in mitochon-

[38] Adapted from Machlis and Torrey, *op. cit.*, pp. 113–14.

[39] Adapted from "Laboratory Experiments in Elementary Human Physiology," mimeo sheets, prepared under the auspices of the Am. Physiological Soc., 1962.

dria of cells and is active in catalyzing one step in the Krebs cycle—the oxidation of succinic acid to fumaric acid.

In this procedure,[40] beef heart is used as a source of succinic dehydrogenase. In this oxidation reaction, methylene blue is reduced (and decolorized) upon acceptance of two hydrogen atoms from succinic acid. Cut up a small bit of beef heart (about the volume of a marble) into a test tube, and wash it several times with water. Prepare sets of the following test tubes: (1) no meat, only an equivalent volume of distilled water plus 3 drops of succinic acid (0.5 M), plus 7 drops of 0.01 percent methylene blue solution, plus 9 drops of water to bring the total volume to that of the other test tubes; (2) meat, plus 3 drops of succinic acid, plus 9 drops of distilled water; (3) meat, plus 7 drops of methylene blue solution, plus 12 drops of distilled water; (4) meat, plus 3 drops of succinic acid, plus 7 drops of methylene blue solution, plus 9 drops of distilled water; and (5) boiled meat (2 min), plus 3 drops of succinic acid, plus 7 drops of methylene blue solution, plus 9 drops of distilled water.

Down the side of each tube pour a layer of mineral oil to prevent oxygen of air from entering the tubes. Stand all the test tubes in a water bath maintained at 37° C (99° F). In which tubes are there changes in color? Vigorously shake the contents of test tube 4 to show oxidation of reduced methylene blue.

At some time, students may use the technique described in the laboratory manual of Machlis and Torrey[41] in which germinating white beans are the source of the succinic dehydrogenase.

Blocking effect of a competitive inhibitor

Succinic and malonic acids. A classic yet simple case of competitive inhibition is illustrated by the action of succinic acid

and malonic acid. Note the similarity of their molecular structures.

$$H_2C\text{—}CH_2COOH \qquad H_2C\text{—}COOH$$
$$\phantom{H_2C\text{—}}COOH \qquad \phantom{H_2C\text{—}}COOH$$

succinic acid malonic acid

Students, working in groups, may examine first-hand the blocking effect of malonic acid, which combines with the succinic dehydrogenase, thereby binding the enzyme so that succinic acid is not oxidized to fumaric acid. This example is actually a case of competitive inhibition, since addition of succinic acid can reverse the inhibition by malonic acid. The frequency of collisions of enzyme surfaces and substrate molecules depends on the concentration of the substrate.[42]

Beef heart may be used (as described earlier) in the demonstration. Prepare the sets of five test tubes described above, and add the following additional test tubes to the series: (6) meat, plus 3 drops of succinic acid, plus 3 drops of malonic acid (1 M), plus 7 drops of methylene blue, plus 6 drops of distilled water added to equalize the volume of all the tubes (as done for the first five tubes); and (7) meat, plus 9 drops of succinic acid, plus 3 drops of malonic acid, plus 7 drops of methylene blue.

Maintain these tubes in a water bath at 37° C (99° F), and watch for changes. What is the explanation of the results in test tubes 6 and 7?

PABA and sulfonamide. Teachers may want to have students report on competitive inhibition in the action of sulfonamides in destroying bacteria. Certain bacteria require p-amino benzoic acid (PABA) for the production of folic acid. Since the sulfonamides are structurally similar to PABA, bacteria are subject to the inhibition of activity of enzymes needed for normal metabolism. If sulfonamide is present, it becomes attached to the enzyme, thereby preventing the enzyme from

[40] From Wald *et al., op. cit.,* pp. 26–27.
[41] *Op. cit.,* pp. 108–09.

[42] For further reading, consult A. Giese, *Cell Physiology,* 2nd ed., Saunders, Philadelphia, 1962; Wald *et al., op. cit.;* and Hawk, Oser, and Summerson, *op. cit.*

carrying on the normal metabolism of PABA in the cell. Refer to any text in bacteriology for the mode of action of drugs used in chemotherapy, for example, P. Carpenter, *Microbiology* (Saunders, 1961).

Cytochrome oxidase

In the presence of molecular oxygen, cytochrome oxidase oxidizes reduced cytochrome c. Nadi reagent, a mixture of α-naphthol and dimethyl-p-phenylenediamine, reduces cytochrome c. These two components of the Nadi reagent are oxidized to the blue pigment indophenol blue by cytochrome oxidase. Several techniques are described in the literature.

Wald[43] uses chick embryos of different ages. Remove the chick embryo from the yolk and transfer to warm saline solution. Replace with warm Nadi reagent and record the time. Then watch for the first sign of blue color to appear and record its location at 3-min intervals. The preparation of Nadi reagent for this demonstration, as given by Wald, follows. Prepare a 0.01 M solution of α-naphthol by dissolving 1.44 g of α-naphthol in 1 liter of 0.9 percent NaCl solution; heat to dissolve. Prepare a 0.01 M solution of dimethyl-p-phenylenediamine by adding 1.36 g to 1 liter of physiological saline. To prepare the buffer, two salt solutions are needed. Make up 9.5 g of Na_2HPO_4 in 1 liter of distilled water. Also prepare 9.07 g of KH_2PO_4 in 1 liter of distilled water. The phosphate buffer is a mixture of these two salt solutions in the proportion of 7.8 ml of Na_2HPO_4 solution to 92.2 ml of KH_2PO_4 solution. Nadi reagent is now made by combining equal parts of the α-naphthol solution, the dimethyl-p-phenylenediamine solution, and the phosphate buffer.

Hawk[44] suggests adding Nadi reagent (5 drops of α-naphthol solution plus 5 drops of p-phenylenediamine hydrochloride solution) to a test tube containing

[43] *Ibid.*, p. 124.
[44] *Op. cit.*, p. 323.

5 ml of potato extract. The potato extract is prepared by grating a raw, peeled potato into cheesecloth that is suspended in a beaker of 200 ml of distilled water. Use the filtered water extract in the test. In this reaction involving potato oxidase, indophenol is produced from the α-naphthol and phenylenediamine.

To examine the action of an inhibitor—an azide—on the terminal oxidase, Machlis and Torrey[45] offer a method using fresh bean extract. Remove the seed coats from ten germinating white bean seeds, and grind the seeds in 20 ml of ice-cold, 0.4 M sucrose solution. When the preparation is a soft mash, centrifuge at 500 $\times$ g for 5 min to remove the cellular fragments. (This extract contains cytochrome c and active cytochrome oxidase.) Add 3 ml of this bean extract to each of two test tubes. Then to test tube 1 add 1 ml of 0.4 M sucrose solution; to test tube 2 add 1 ml of 10^{-3} M NaN_3 (sodium azide). To each test tube add also 5 drops of each of the following components of the Nadi reagent: 1 percent α-naphthol in 95 percent ethyl alcohol, and 1 percent aqueous solution of dimethyl-p-phenylenediamine HCl. After shaking the tubes, allow them to stand at room temperature. What changes occur in the next half-hour?

Machlis and Torrey also suggest a demonstration to show the action of oxidase of potato on tyrosine. This action results in the gradual formation of melanin, and takes a few hours to complete.

Phosphorylase

Phosphorylases split glycogen by introducing phosphoric acid between glycolic linkages. Machlis and Torrey[46] suggest a method for preparing phosphorylase. However, the extraction involves the use of poisonous potassium cyanide, so the extraction cannot be recommended for high

[45] *Op cit.*, p. 112.
[46] *Ibid.*, pp. 85–86. For a complete description of methods of extraction of phosphorylase, refer also to *High School Biology: Blue Version—Laboratory, Part I,* BSCS, Univ. of Colorado, Boulder, 1961 (experimental use, 1961–62), pp. 149–52.

school students. Refer to Wald[47] for a demonstration using phosphorylase from a potato tuber. A Waring blendor and sodium fluoride (*caution:* poison) are needed.

Cholinesterase and acetylcholine

The activity of the enzyme cholinesterase can be measured by titration of acetic acid liberated in the following reaction.

acetylcholine + H_2O $\longrightarrow$

choline + acetic acid

A demonstration that illustrates this action was described by G. Cantoni.[48] Acetylcholine is available from CIBA; the enzyme cholinesterase is obtainable from Winthrop Laboratories. The indicator used is cresol red solution, and titration is with 0.004 N sodium hydroxide.[49]

Amylase

In their laboratory manual, Machlis and Torrey[50] give directions for extracting amylase from 5-day-old barley seedlings. Grind about 15 seedlings in 50 ml of distilled water and clean sand. Pour off the amylase extract fluid and centrifuge for 5 min at 500 $\times$ g. Test a series of dilutions of the amylase on 0.4 percent soluble starch solution. Every half-minute, test a sample from each amylase-starch mixture for the disappearance of starch.

Invertase

In this demonstration of enzymatic action, invertase from yeast may be used. Mix half a package of yeast with 40 ml of distilled water. After this has been stand-

ing for 20 min, centrifuge for 5 min at 500 $\times$ g to obtain a crude solution of the hydrolytic enzyme invertase.

Add 10 ml of the crude invertase solution to 25 ml of a 2 percent solution of sucrose, and incubate for 1 hr at 38° C (100° F). Prepare a control containing similar quantities of the enzyme and sucrose; but first heat to boiling, then cool, the enzyme extract. Prepare another control containing 2 percent sucrose alone; another with enzyme alone.

Test 2-ml samples of the solution in the test tubes with 5 ml of Benedict's solution. Heat in a water bath for 5 min. In which test tube does a cuprous oxide precipitate form, indicating a positive test for reducing sugars?

Test each solution for fructose by adding a pinch of resorcinol and 6 drops of 25 percent HCl; then warm the solution. This is the Selivanoff test for ketoses; a red color indicates the presence of fructose. Run the test first on 1 ml of 0.5 M solution of fructose to show the color reaction. Avoid acid hydrolysis of the sucrose by running the test on a control tube containing a solution of sucrose alone. (See also tests for sugars, pp. 135, 218, 660.)

(You may want to refer to a comparison of anaerobic and aerobic metabolism in yeast cells, Chapter 6.)

Luciferin and luciferase

The emission of light by many invertebrates—for example, fireflies, marine dinoflagellates such as *Noctiluca,* and the ostracod crustacean *Cypridina*—and by luminous fish, bacteria, and fungi is known as bioluminescence and is the result of the oxidation by an enzyme of some substrate. Luciferin, a heat-stable substrate, is oxidized in the presence of oxygen, water, and the heat-sensitive enzyme luciferase. Cultures of luminous bacteria (such as *Achromobacter fischeri*) may be purchased from supply houses (Fig. 2-38). Some fish have a specific, complex organ—a photophore—that is regulated through nerve

[47] *Op. cit.,* pp. 29–31.

[48] Paper by G. Cantoni, in "Laboratory Experiments in Biology, Physics, and Chemistry," *op. cit.,* as demonstrated for members of the National Science Teachers Assoc. by the staff of the National Institutes of Health, Bethesda, Md., March 16, 1956.

[49] Refer also to L. Goodman and A. Gilman, *The Pharmacological Basis of Therapeutics,* 2nd ed., Macmillan, New York, 1955, pp. 389–418.

[50] *Op cit.,* pp. 88–89.

Fig. 2-38 *Photobacterium fischeri:* motile, Gram-positive, slightly curved rods; aerobic, facultative; luminescent. (Photographed in total darkness using only light emitted by the bacteria.) (Courtesy of Carolina Biological Supply Co.)

and hormone control; others have only symbiotic bacteria which inhabit an organ that is set below each eye of the fish. The luminous organ in the firefly is the lantern, which is regulated by the nervous system; it is a fat body supplied with oxygen from small tubules of the tracheae.

Luciferin is soluble in water and in higher alcohols. Quick drying of organisms with acetone to remove water is the usual method for preserving luciferin. Refer to Giese[51] for a comprehensive treatment of chemical composition and methods of extraction of luciferin from *Cypridina,* luminous bacteria, and fireflies. (Giese also discusses the extraction of luciferase and the mechanism of bioluminescence,

and gives a valuable bibliography.) Also consult a monograph in photobiology.[52]

This may be the time that you want to have individual students observe the brilliant blue light that results when a bit of powdered marine crustacean, *Cypridina,* is placed in the palm of the hand and a few drops of water added. The powder contains both luciferin and luciferase. The results are more effective in a darkened room. Small quantities of pulverized Japanese marine crustaceans in vials of ½-dram quantities can be obtained from supply houses (Appendix C).[53]

Investigations may be planned to study the step-by-step reactions and the rate of luminescence in a solution containing luciferin and luciferase. The concentration of enzyme, pH, and water, and the presence of heavy metals or drugs can be investigated. The effects of these variables follow the usual patterns characteristic of enzymatic reactions. At what time should oxygen be added to the system to affect maximum intensity of luminescence? If luciferin and luciferase are mixed before oxygen is added, maximum intensity occurs 0.02 sec after oxygen has been added to the system. However, if oxygen is added separately to the enzyme and to the substrate, and then they are mixed, maximum intensity is reached after 0.6 sec. This seems to indicate step-by-step reactions (see equations in Giese).

CAPSULE LESSONS: Ways to get started in class

2-1. Begin a study of cells by having students examine epithelial tissue from the lining of the cheek under low power. When they are ready to examine the slide that they have prepared (stained with dilute Lugol's solution, as described in this chapter), teach them to use the low power of the microscope. Then examine the slide under high power; students will begin to ask questions about the nature of the nucleus and the granular texture of the cytoplasm.

2-2. Introduce the parts and the manipulation of a microscope by having students examine a variety of protists in a drop of pond water. Or

use a culture of *Tubifex* worms; since they can be seen with the naked eye, students can locate the organisms easily on their slides. Examine under low power. Which way is the slide moved to follow a motile form, or to examine the length of a worm? Later, teach how an image is centered under low power before the microscope is switched to high power.

[52] H. H. Seliger and W. D. McElroy, *Light: Physical and Biological Action,* "A.E.C. Monograph Series on Radiation Biology and Industrial Hygiene," Academic, New York, 1965.

[53] General Biological suggests that a ½-dram quantity ($2.50) will supply 25 students with the small amount needed for individual demonstrations.

Ask about Carolina's ATP-firefly kit ($6.00).

[51] *Op. cit.,* Chapter 20, "Bioluminescence."

2-3. You may wish to start by asking how cells in the shoot of an onion compare with those in the bulb. Prepare slides and then diagram the two kinds of cells. Indicate those structures that both kinds of cells have in common. What additional structures are found in the cells of the green shoots?

2-4. At some time, you will want to use a block of modeling clay to represent a cell. Insert a marble in the clay. What does the marble represent? The clay? Enclose the clay in a sheet of saran wrap or other clear plastic. What does this layer represent? As an outer boundary of a cell what special functions might it have? Is this a typical plant or animal cell? Now cover the plastic with a sheet of paper. If this is a plant cell, what does the outer layer represent?

A student could prepare a gelatin block to which a bit of phenolphthalein solution had been added. Now enclose this in a membrane. Bring a solution of ammonium hydroxide close to this gelatin block, or suspend the block over the NH_4OH. How do students account for the reddish color in the gelatin? What is the nature of diffusion? Initiate a discussion of the nature of exchange of materials through a membrane. What materials in the environment of a cell diffuse into the cell? Do all materials diffuse readily out of cells?

2-5. Show a replica of Leeuwenhoek's microscope. Some students may be asked in advance of this lesson to prepare a talk to the class about Leeuwenhoek's findings. Read from his letters to the Royal Society in London. Compare the structure and differences in magnification of modern microscopes with Leeuwenhoek's. What effect did Leeuwenhoek's discovery of a new world of microorganisms have on the "science" of his day? What advances could be made with his new tool? What were the contributions of Hooke, Brown, Schleiden and Schwann, Dujardin, and others, to the accumulating knowledge of cells? To answer such questions, use a history of biology text such as one of the following: C. Singer, *A History of Biology*, 3rd ed. (Abelard-Schuman, 1959); E. Nordenskiöld, *The History of Biology* (Tudor, 1960); or J. Needham, *A History of Embryology*, 2nd ed. (Abelard-Schuman, 1959).

2-6. Plan several laboratory lessons in which students explore several cultures of living organisms. Students can fill the blackboards with drawings of what they see. Then develop the basis of classification of protozoa: whether the forms are ciliated, where cilia are, or whether forms are flagellated. Where is the oral, or

mouth, region located? Or are the organisms many-celled—such as rotifers, worms, and gastrotrichs?

If several kinds of *Paramecium* are present, develop the idea of genus and species.

2-7. You may want to have students use vital stains or begin to study ingestion in microorganisms. Refer now to activities given in Chapter 4.

2-8. Elicit from students how one might measure the length of a cell or of a protozoan. See pp. 91–94 and "Useful Formulas," Appendix F.

2-9. Use several models of cells and/or protozoa such as those in Figs. 2-8, 2-9, 2-11, and 2-18. Have students use the models to identify parts and then to describe the functions of the components of cells.

2-10. Use cells such as *Spirogyra* or *Anacharis* (elodea) in a wet mount. How could one determine that a cell has depth and is not two-dimensional?

2-11. Which would live longer on a slide, cells of the green alga *Chlorella* or *Scenedesmus*, or a drop of *Paramecium?* If the slides were sealed with petroleum jelly, could students watch the slides at intervals during the day? Introduce the many life functions of plant and animal cells; begin to build facts into concepts, concepts into conceptual schemes—for example, organisms are dependent on one another, and on their environment.

2-12. As a vicarious microscope lesson, show a film that goes beyond the available materials of a high school. You may want to preview some of the films in the AIBS series *Cell Biology* (McGraw-Hill). Some of the films in the AIBS series *Genetics* (McGraw-Hill) may be applicable at this time for your classes. (See other films listed at the end of this chapter.)

2-13. How has the electron microscope changed our concepts of the components of cells? Have students duplicate a diagram such as that on p. 55 of the September 1961 issue of *Scientific American* (Fig. 2-7 is a reduced version of that diagram). Or use offprints of the entire article (J. Brachet, "The Living Cell") for the class, plus later issues of *Scientific American*. What knowledge has been acquired since 1961? What is the structure of mitochondria? What is their role in the cell's activities? How are pigment molecules arranged in grana of chloroplasts? What is the site of enzyme production, and of protein synthesis? How are chromosomes or DNA molecules duplicated when cells reproduce? What do we know of the genetics of microorganisms?

2-14. Begin a lesson by showing dry yeast in a

package. Into one large test tube pour about 10 ml of fresh hydrogen peroxide (3 percent). Sprinkle dry yeast into the test tube. Students will begin to ask questions about the tremendous activity and the bubbles pouring out of the test tube. Elicit the observation that a gas seems to be liberated. How could it be collected and identified? Have students devise an apparatus for collecting a gas (see Fig. 2-37). Give the formula for hydrogen peroxide (H_2O_2). What kind of substance is present in yeast cells that could break down a poisonous substance so that it would not accumulate in cells? In subsequent lessons (laboratory periods) collect the oxygen liberated, or measure the amount of hydrogen peroxide left in the test tube after 10 min (by titration with $KMnO_4$, as described under "*Catalase*" in this chapter). The amount of peroxide not destroyed is in inverse relationship to the activity of the enzyme catalase.

Also use slides of potato or raw meat, as described in this chapter.

2-15. At some time, in preparation for laboratory work, show the short BSCS techniques film *Paper Chromatography* (Thorne). Plan to teach students to partition the components of hydrolyzed proteins as a laboratory technique that may be used for some long-range investigation.

2-16. In a series of laboratory lessons, students may study the effect of a dehydrogenase found in mitochondria—succinic dehydrogenase. Study the effect of temperature on enzymatic activity; plan to study also the blocking effect of a competitive inhibitor such as malonic acid.

2-17. Plan several laboratory lessons in which students can devise their own methods of testing the amount of enzyme in relation to the reaction produced. They can learn methods of dilution— what is the effect of an enzyme such as amylase diluted some 10,000 times? What is the effect of pH on enzyme activity? What kinds of substances inhibit the activity of enzymes? Why?

2-18. Show that molecules of different sizes travel through a membrane at different rates. Dialysis is a method of separating molecules of different sizes. Prepare a suspension of several compounds—amino acids, glucose, xylose, sodium chloride, and starch. Pour this into a length of dialyzing tubing (prepared as shown in Fig. 4-9).

Students can become familiar with the standard reagents used in tests for the identification of several cell components by testing the water in the beaker. Use Benedict's test; Lugol's iodine reaction; benzidine reaction or Bial's reaction; Biuret, Millon, and xanthoproteic tests. These and others are described in this chapter. (See also Molisch's reaction, p. 218.)

2-19. What is the effect of radiation on tissue cells? On microorganisms? Look into *Scientific American* (September 1959) and also several texts listed in the bibliography at the end of this chapter.

2-20. Many activities with the microscope are suggested throughout the book in discussions of such topics as growth, reproduction, modes of nutrition, behavior, photosynthesis, and respiration. What is the next problem for study for your classes leading out of the study of tissue cells and unicellular organisms?

No doubt you will want to look into suggestions offered in the BSCS laboratory blocks such as *Life in the Soil, Field Ecology, The Complementarity of Structure and Function, Plant Growth and Development, Animal Growth and Development, Animal Behavior*, or one in preparation, *Metabolism*. Also look into the fine laboratory manual *Twenty-six Afternoons of Biology*, by G. Wald *et al.* (Addison-Wesley, 1962).

2-21. Some teachers use free, exploratory laboratory periods to introduce some of the major topics to be studied in the year ahead. In these periods students might examine with the microscope the structure and behavior of such organisms as *Daphnia* (notice especially its heartbeat), *Tubifex* worms, and brine shrimps and their larvae (notice especially their mode of ingestion).

Questions continually arise about facts students learn from books and from their first-hand observations of those facts. For example, what changes the size of chromatophores in the skin or scales of fish? What kinds of microorganisms can be found in a handful of soil? What is the effect of different ions on the action of cilia?

2-22. Plan to have several microscope lessons to accompany the dissection of the frog. Prepare slides of tissues of the frog. Also hunt for protozoan parasites in the contents of the intestine. Are there worm parasites too (see p. 534)?

PEOPLE, PLACES, AND THINGS

Many of the techniques described here are specialized, and practice is needed to gain skill. Students who want help in gaining this skill may call upon a laboratory technician, a doctor, a member of a research staff of a hospital, or a teacher in another high school or in a college biology department. You will find that these resource people are happy to help youngsters who have interests similar to theirs.

We know of several students who have been given access to a histology laboratory in which they have been helped by specialists. These youngsters have gladly washed glassware and done other chores for this privilege.

BOOKS

The following are only a few of the books that are pertinent to the work discussed in this chapter. These and many other references are given in Appendix A.

Bloom, W., and D. Fawcett, *A Textbook of Histology,* 8th ed., Saunders, Philadelphia, 1962.

Brieger, E. M., *Structure and Ultrastructure of Microorganisms,* Academic, New York, 1963.

Brock, T., ed., *Milestones in Microbiology,* Prentice-Hall, Englewood Cliffs, N.J., 1961.

Bullough, W. S., *Practical Invertebrate Anatomy,* 2nd ed., Macmillan, London (St. Martin's, New York), 1958.

Butler, J. A., *Inside the Living Cell,* Basic Books, New York, 1959.

Carpenter, P., *Microbiology,* Saunders, Philadelphia, 1961.

Cheldelin, V., *Metabolic Pathways in Microorganisms,* Wiley, New York, 1961.

Clark, G. L., ed., *Encyclopedia of Microscopy,* Reinhold, New York, 1961.

Conn, H. J., *Biological Stains,* 7th ed., Williams & Wilkins, Baltimore, 1961.

———, M. Darrow, and V. Emmel, *Staining Procedures Used by the Biological Stain Commission,* 2nd ed., Williams & Wilkins, Baltimore, 1960.

Corliss, J., *The Ciliated Protozoa,* Pergamon, New York, 1961.

DeRobertis, E., W. Nowinski, and F. Saez, *Cell Biology,* 4th ed., Saunders, Philadelphia, 1965.

Dobell, C., *see* Leeuwenhoek, A. van.

Frobisher, M., *Fundamentals of Microbiology,* 7th ed., Saunders, Philadelphia, 1962.

Gabriel, M., and S. Fogel, eds., *Great Experiments in Biology,* Prentice-Hall, Englewood Cliffs, N.J., 1955.

Gerard, R. W., *Unresting Cells,* new ed., Harper & Row, New York, 1949; Harper Torchbook, 1961.

Giese, A., *Cell Physiology,* 2nd ed., Saunders, Philadelphia, 1962.

Gray, P., *The Microtomist's Formulary and Guide,* Blakiston (McGraw-Hill), New York, 1954.

Gurr, E., *Encyclopedia of Microscopic Stains,* Williams & Wilkins, Baltimore, 1960.

Hawker, L. E., *et al., An Introduction to the Biology of Microorganisms,* St. Martin's, New York, 1961.

Hollaender, A., ed., *Radiation Biology,* Vols. I, II, McGraw-Hill, New York, 1954.

Hooke, R., *Micrographia, or Some Physiological Descriptions of Minute Bodies—Observations by Robert Hooke,* orig. ed., 1665; Dover, New York, 1961.

Humason, G., *Animal Tissue Techniques,* Freeman, San Francisco, 1962.

Hutner, S., and A. Lwoff, eds., *Biochemistry and Physiology of the Protozoa,* 2 vols.: Vol. I, ed. by A. Lwoff, 1951, Vol. II, ed. by S. Hunter, 1955; Academic, New York.

Jensen, W., *Botanical Histochemistry,* Freeman, San Francisco, 1962.

Kavanagh, K., ed., *Analytical Microbiology,* Academic, New York, 1963.

Kudo, R., *Protozoology,* 4th ed., Thomas, Springfield, Ill., 1954.

Leeuwenhoek, A. van, *Antony van Leeuwenhoek and His "Little Animals,"* coll., trans., and ed. by C. Dobell, Harcourt, Brace & World, New York, 1932; Dover, New York, 1960.

McElroy, W., *Cellular Physiology and Biochemistry,* Prentice-Hall, Englewood Cliffs, N.J., 1961.

Manwell, R., *Introduction to Protozoology,* St. Martin's, New York, 1961.

Paul, J., *Cell and Tissue Culture,* Williams & Wilkins, Baltimore, 1960.

Richards, A. G., *The Complementarity of Structure and Function* (BSCS laboratory block), Heath, Boston, 1963.

Rugh, R., *Experimental Embryology: Techniques and Procedures,* 3rd ed., Burgess, Minneapolis, 1962.

Sager, R. A., and F. Ryan, *Cell Heredity,* Wiley, New York, 1961.

Schmitt, F. O., ed., *Macromolecular Specificity and Biological Memory,* M.I.T. Press, Cambridge, Mass., 1962.

Spector, W., ed., *Handbook of Biological Data,* Saunders, Philadelphia, 1956.

Strehler, B., *Time, Cells, and Aging,* Academic, New York, 1962.

Swanson, C., *The Cell,* 2nd ed., Prentice-Hall, Englewood Cliffs, N.J., 1964.

Tartar, V., *Biology of Stentor,* Pergamon, New York, 1961.

Waddington, C., ed., *Biological Organisation, Cellular and Sub-cellular,* Pergamon, New York, 1959.

Weesner, H., *General Zoological Microtechnique,* Williams & Wilkins, Baltimore, 1960.

This listing is intended to be only a guide for the selection of new films, filmstrips, and transparencies. Send for catalogs from distributors of films; catalogs of biological supply houses list extensive offerings of 2 × 2 in. slides. The addresses of distributors are given in Appendix B.

In general, the cost of color films averages $150; black and white films cost about $75. Newer films are made in varying lengths—time sequences of 1 min, 5 min, 11 min, or 30 min; the cost of films therefore varies. We have indicated the lengths of films so that lessons can be planned with the films in mind.

Sound films are available in color or black and white; rental rates are available from film libraries or by direct inquiries to producers. (Filmstrips are not available on a rental basis; the purchase price is usually about $6 to $7 for those in color.)

Ameboid Organisms (28 min), in Part II of *AIBS Film Series in Modern Biology,* McGraw-Hill.

The Analytical Balance (13 min), Fisher; free.

Animals in a Micro-Universe (11 min), Colburn.

Bacteria: Laboratory Study (15 min), Indiana Univ.

Biology in Today's World (11 min), Coronet.

Carbon 14 (12 min), EBF.

Cell Biology (set of 12 films, 28 min each), Part I of *AIBS Film Series in Modern Biology,* McGraw-Hill.

Cell Division (12 min), Phase.

Cell Division and Growth (13 min), Abbott; free.

Cell Division: Mitosis (single-concept film, 8mm, 2½ min), EBF.

The Cell: Structural Unit of Life (11 min), Coronet.

Cells and Their Functions (14 min), Contemporary.

Colorimetry (14 min), Fisher; free.

Compound Microscope (10 min), Bausch & Lomb; free.

Conquest of the Atom (22 min), International Film Bur.

Continuity of Life: Characteristics of Plants and Animals (10 min), Indiana Univ.

Diversity of Animals (set of 12 films, 28 min each), Part VIII of *AIBS Film Series in Modern Biology,* McGraw-Hill.

Diversity of Plants (set of 12 films, 28 min each), Part VII of *AIBS Film Series in Modern Biology,* McGraw-Hill.

Enzymes: The Key to Life (29 min), Indiana Univ.

The Eternal Cycle (13 min), in *The Magic of the Atom* series, Handel; free loan from U.S. Atomic Energy Comm. Film Library.

Fallout (15 min), Cenco.

The Fine Structure and Pattern of Living Things (2 reels, 50 min), U.S. Dept. of Agriculture, Graduate School; P. Weiss, narrator.

The Frog: Pithing (single-concept film, 8 mm, 2 min), EBF.

Fungi (16 min), EBF.

Genetics (set of 12 films, 28 min each), Part VI of *AIBS Film Series in Modern Biology,* McGraw-Hill.

Histological Techniques (9 min), BSCS techniques, Thorne.

How to Use the Microscope (filmstrip), Carolina.

The Life and Death of a Cell (21 min), Univ. of California.

Life in a Drop of Water (11 min), Coronet.

Life of the Molds (21 min), McGraw-Hill.

Life, Time, and Change (set of 12 films, 28 min each), Part X of *AIBS Film Series in Modern Biology,* McGraw-Hill.

Living Biology (set of films), R. Vishniac, Yeshiva Univ., 110 W. 57 St., New York 19, N.Y.

Microbiology (set of 12 films, 28 min each), Part II of *AIBS Film Series in Modern Biology,* McGraw-Hill.

The Microscope (11 min), McGraw-Hill.

Mitosis (24 min), EBF.

The Multicellular Animal (set of 12 films, 28 min each), Part IV of *AIBS Film Series in Modern Biology,* McGraw-Hill.

Multicellular Plants (set of 12 films, 28 min each), Part III of *AIBS Film Series in Modern Biology,* McGraw-Hill.

Paper Chromatography (14 min), BSCS techniques, Thorne.

Patterns of Life (set of films, 30 min each), Indiana Univ.

Preparing Slides of Plant Mitosis (filmstrip), Carolina.

Protozoa (set of 12 transparencies), Carolina.

Protozoa: One-Celled Animals (11 min), EBF.

Radiation (26½ min), National Film Board of Canada.

Simple Plants: Algae and Fungi (14 min), Coronet.

Simple Plants: Bacteria (14 min), Coronet.

Simple Plants: The Algae (18 min), EBF.

The Single-Celled Animals: Protozoa (17 min), EBF.

Slime Molds (set of 3 films: *Life Cycle* [30 min], *Collection and Use* [19 min], and *Identification* [24 min]), State Univ. of Iowa.

Smear and Squash Techniques (6 min), BSCS techniques, Thorne.

The Stuff of Life (29 min), No. 3 in *Virus* series, Indiana Univ.

The Unity of Life (28 min), Indiana Univ.

Weighing Procedure (single-concept film, 8 mm, approx. 3½ min), EBF.

What Is Life? (28 min), Indiana Univ.

The Worlds of Dr. Vishniac (20 min), in *Horizons* series, Educational Testing Service.

LOW-COST MATERIALS

A listing of suppliers of free and low-cost materials, with addresses, will be found in Appendix D. Although we recommend the material for use in the classroom, we do not necessarily sponsor the products advertised.

Am. Optical Co., Public Relations Dept., *AO Reports on Teaching with a Microscope; Evolution of the Microscope,* Am. Optical.

Guides to the Out-of-Doors (listing of *Audubon Nature Bulls.*), National Audubon Soc.

Johnston, I. M., *The Preparation of Botanical Specimens for the Herbarium,* Harvard Univ., Arnold Arboretum, distributed by Cambosco Scientific.

Microscope: Its Applications, Leitz Optical.

Microtechnique, Ward's *Curriculum Aids,* Nos. 4, 6, Ward's.

Native Grasses (set of six booklets in color), Phillips Petroleum Co., Bartlesville, Okla.; free.

Palmer, C. M., *Algae in Water Supplies* (illustrated manual in color), U.S. Dept. of Health, Education, and Welfare, U.S. Govt. Printing Office, Washington, D.C.; $1.00.

The Theory of the Microscope, Bausch & Lomb; $0.25.

Chapter 3

Producers:
Capture of Light Energy

The substance of this chapter is concerned with the specific conditions under which photosynthesis occurs. In addition, the major concepts related to modes of nutrition among organisms are developed briefly, especially in relation to the evolution of the photosynthesizing apparatus among autotrophic plants. Many of the investigations described reveal the methods of scientists: building models (some symbolic) to focus on underlying concepts; developing hypotheses; and testing possible predictions of these hypotheses in order to focus on experimental design.

There are the usual basic laboratory demonstrations which form the body of this book, and, of course, investigations which enable students to see for themselves. An additional feature, a laboratory block of investigations, develops problems which aim at a more advanced level of work and requires students to design somewhat sophisticated experiments. The series of problems in the block emphasizes inquiry. The problems may be pursued by individual students or by groups working in the laboratory or at home.

The inquiries offered in this chapter suggest studies in these areas[1]:

1. Evidence of photosynthesis: formation of sugar and starch
2. Role of light and wavelength in photosynthesis
3. Absorption of carbon dioxide in light
4. Oxygen evolved in light
5. Role of chloroplasts (and their analysis)
6. Growth as further evidence of photosynthetic activity

The suggestion is also made that algae such as *Chlorella* be offered as experimental plant forms useful in the high school classroom.

Photosynthesis as a process

It is estimated that 90 per cent of all photosynthesis is carried on by marine and fresh-water algae, and the remaining 10 percent by cultivated and wild land plants. Put another way, some 200 billion tons of carbon are fixed annually by photosynthesizing plants of the oceans and fresh waters and by land plants.

The overall process of photosynthesis can be represented by the following chemical equation:

[1] Naturally, in accordance with the basic notion that teaching is a personal invention, we expect that a teacher will select those demonstrations that are applicable to his class situation. If you care to see how teachers from all over the country have organized studies in biology, look at three versions of high school biology developed by the Biological Sciences Curriculum Study (BSCS) at the Univ. of Colorado (Boulder) from 1960 to 1963. These are available under the following titles: *High School Biology* (Green Version), Rand McNally, Chicago, 1963; *Biological Science: Molecules to Man* (Blue Version), Houghton Mifflin, Boston, 1963; and *Biological Science: An Inquiry into Life* (Yellow Version), Harcourt, Brace & World, New York, 1963. A laboratory manual is available for each textbook (in the Blue Version it is bound into the textbook and is called "Laboratory Investigations"). Also look into the BSCS laboratory blocks (published by Heath), and *Research Problems in Biology: Investigations for Students,* Series One–Four, Anchor Books, Doubleday, Garden City, N.Y., 1963, 1965.

$$6CO_2 + 12H_2O + 673{,}000 \text{ cal} \longrightarrow$$

low energy   from energy
level   of sun

$$C_6H_{12}O_6 + 6H_2O + 6O_2$$

high energy
level

Since the source of oxygen is water, at least 12 molecules of water must take part in the reaction for each molecule of hexose that is formed. The product, a six-carbon sugar, has built into it some 673,000 cal per mole of hexose. This energy is then available to those organisms that can reverse the photosynthetic process in biological oxidation (oxidation is discussed in Chapter 6).

The equation given for photosynthesis is deceptively simple. In 1905, Blackman demonstrated that there were two distinct phases in photosynthesis, only the first phase requiring light and chlorophyll. The first phase, finally explained by Hill in 1937 and named after him, is a photochemical phase in which the energy of light is captured by chlorophyll molecules and is used to split molecules of water in a photochemical lysis of water.

The second phase, or Blackman reaction, is a dark phase in which there is a thermochemical reduction of carbon dioxide. At low light intensity, the photochemical reaction is the limiting reaction in photosynthesis. The rate at which chlorophyll molecules absorb and transmit quanta of light is not increased by raising temperature. In order to increase the overall rate of photosynthesis, however, two factors must operate: (1) light intensity must be increased and (2) temperature must be increased to optimum.

It has been shown that intact chloroplasts isolated from cells carry on three distinct functions: (1) Hill reaction; (2) photosynthetic phosphorylation that occurs in the absence of oxygen, the process in which adenosine monophosphate is converted into adenosine triphosphate (ATP); and (3) the dark reaction, or the fixation of carbon dioxide.

Florkin,[2] in a review of current theories of photosynthesis, offers this summary of the total process:

$$3CO_2 + 9 \text{ ATP} + 5H_2O + 6 \text{ DPNH} + 6H^+ \longrightarrow$$

$$1 \text{ triose-P} + 9 \text{ ADP} + 6 \text{ DPN} + 8P$$

The introduction of a molecule of carbon dioxide into triosephosphate requires three molecules of ATP and four electrons (or four reduction equivalents). According to the proposals of Calvin, these substances are provided in the priming reaction of photosynthesis (the transformation of water in the presence of light energy) which furnishes much of the reduced triphosphopyridine nucleotide (TPNH) and adenosine triphosphate (ATP). The dynamics of photosynthesis is a means for providing ATP and DPNH. (The probable reactions involved in photosynthesis usually show phosphoglyceric acid as the first stable intermediate formed. It was this that led to the notion that photosynthesis involved a phosphorylating mechanism.)

In photosynthesis there is first a quantum absorption converting water into a reducing agent and an oxidizing agent:

$$H_2O \xrightarrow{h\nu} [H] + [OH]$$

The reduction of carbon dioxide is the second step:

$$CO_2 + [H] \longrightarrow (CH_2O)x$$

Hill's investigations confirmed these two steps (the liberation of oxygen in isolated chloroplasts when an H acceptor was present). Tagged oxygen (heavy oxygen-18) showed that oxygen came from water (Rubin and Kamen). Since the oxygen is derived from the decomposition of water, two molecules of water need to be decomposed for each molecule of oxygen that is evolved:

$$CO_2 + 2H_2O^{18} \xrightarrow[\text{chlorophyll}]{\text{light}}$$

$$(CH_2O) + O_2^{18} + H_2O$$

[2] M. Florkin, *Unity and Diversity in Biochemistry*, Pergamon, New York, 1960.

$$CO_2 + 4H \xrightarrow[\substack{112 \text{ kcal} \\ \text{per mole} \\ \text{of } CO_2}]{\text{reduction}} C(H_2O)_2$$

(one molecule) (four H atoms)

Since 1 quantum is equal to 40 kcal, the minimum of 4 quanta seems necessary (although some experiments show that closer to 8 quanta may be needed).

When radioactive tracer methods were developed in about 1945, they were used to trace the "first" organic compounds formed in the thermochemical stage of photosynthesis. Radio-carbon (first C^{11}, then C^{14}) appeared in organic materials after only *seconds* of exposure to light, mainly in the three-carbon phosphoglyceric acid (Fig. 3-1). After only 5 sec of illumination, killed cells contain C^{14} in traces of phosphoglyceric acid, alanine, aspartic acid, and malic acid. Further, after 15 sec of light, minute traces of sucrose, some glucose-phosphate, and fructose-phosphate are found.

One hypothesis advances this model: a pentose sugar is built up to which $C^{14}O_2$ is added, and this immediately splits into two molecules of phosphoglyceric acid. What is the possible scheme for the formation of the original pentose? It may be that hexose phosphate combines with a triose, forming a pentose and a tetrose sugar. Then the tetrose sugar might combine with a hexose and this new compound might split into a heptulose (a seven-carbon sugar) and a triose. If a triose combines with heptulose, two molecules of pentose could be formed so that a constant source of pentose would be available to fix carbon dioxide in the thermochemical step in photosynthesis.

Arnon and his coworkers[3] at the Univ. of California used isolated chloroplasts so that they could study photosynthesis separated from the complications of respiration. They found that isolated chloroplasts carry through not only the Hill reaction,

[3] D. Arnon, "Chloroplasts and Photosynthesis," in *The Photochemical Apparatus: Its Structure and Function,* Brookhaven Symposia in Biology, 11, Brookhaven National Laboratory, Upton, N.Y., 1959, p. 226. Also, D. Arnon, F. Whatley, and M. B. Allen, "Assimilatory Power in Photosynthesis," *Science* 127:1026–34, 1958.

but the complete process of photosynthesis, the conversion of water and carbon dioxide in light into carbohydrates and oxygen. Green plants cells could produce energy-rich ATP without oxidation of their stored carbohydrates in this process of photosynthetic phosphorylation.

The central notion seems to be that energy absorbed from a photon of light by molecules of chlorophyll raises the electrons from their usual energy level to a higher level in the molecule. These "excited" electrons from chlorophyll molecules pass along carriers and give off energy in small bits in changing ADP and ATP molecules of the photosynthesizing cell. (Calvin calls the structural units in grana of chloroplasts that convert light into chemical energy "quantasomes.")

The energy of ATP is then used to drive the reduction of carbon dioxide to sugar. The main function of light seems to be its use in the synthesis of ATP—a series of steps called photophosphorylation (Fig. 3-2). This process has nothing to do directly with the assimilation of carbon dioxide or the release of oxygen. The phosphorylation mechanism in chloroplasts (and in analogous particles in photosynthesizing bacteria) is independent of molecular oxygen, while mitochondrial phosphorylation requires an oxygen supply.

Arnon believes that the key problem in photosynthesis, that is, the conversion of light into useful chemical energy, need not be enmeshed in the dark enzymatic reactions responsible for synthesis of carbohydrates. In some of his experiments, substantial amounts of DPNH and ATP were obtained in light in the absence of carbon dioxide. At the end of the light period, radio-carbon C^{14} in the form of carbon dioxide was supplied in the dark. The newly formed radioactive compounds were identified by paper chromatography and by radioautography. Arnon considers the green portion of chloroplasts to be superfluous. The enzymes responsible in the chloroplast for carbon assimilation are present in the chlorophyll-free, water-solu-

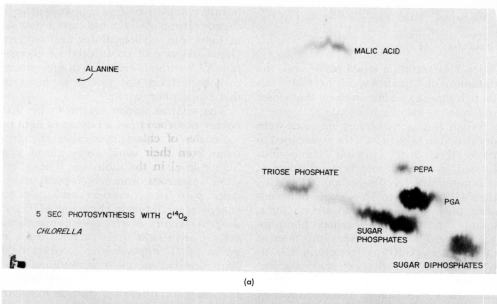

(a)

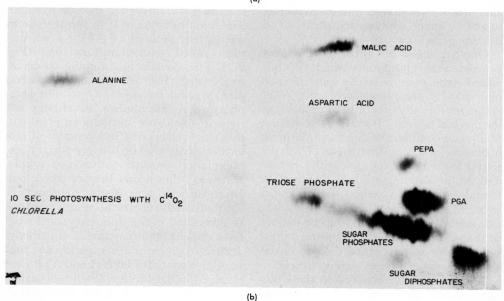

(b)

Fig. 3-1 Three radioautographs revealing compounds containing radioactive carbon that were produced by *Chlorella* in light: (a) during 5-sec exposure to light; (b) during 10-sec exposure to light.

ble part of the chloroplasts, the stroma of the chloroplasts. In short, it was to this chlorophyll-free suspension that radio-carbon dioxide was added, along with soluble enzymes, sugar, phosphate, and the two components of the "assimilatory power" obtained from the light reaction (reduced TPN and ATP). The chlorophyll-free extract was able to assimilate carbon dioxide in the dark. This is a relatively new concept —chemosynthesis in chloroplasts, rather than photosynthesis.

Additional evidence links protein synthesis more closely to assimilation of

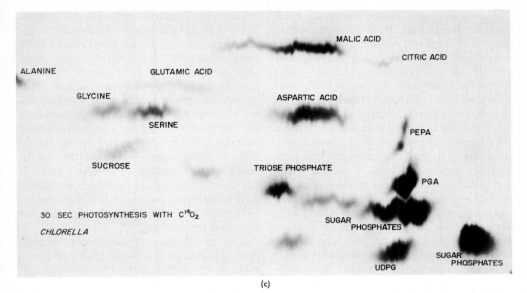

(c)

Fig. 3-1 (continued)   Radioautograph revealing compounds containing radioactive carbon that were produced by *Chlorella* in light: (c) during 30-sec exposure to light. (a, b, c, from J. Bassham, "The Path of Carbon in Photosynthesis," *Sci. Am.*, June 1962, p. 95.)

carbohydrates. Bassham[4] has found that alanine shows up labeled by C^{14} as rapidly as does a carbohydrate. Some 30 percent of carbon taken up by algae in Bassham's studies was incorporated directly into amino acids. When the full role of chlorophyll is revealed, the chloroplast may be found to be the site for production of all materials needed for the life of a green plant cell.

Evolution of photosynthesizing systems

Our present understanding of the chemical evolution of organic substances is the result of work of many scientists in different disciplines: Haldane, Oparin, Bernal, Fox, Urey, Miller, Calvin, and Arnon, and others. According to the hypothesis based on their ideas, methane and other hydrocarbons were probably formed by chemical abiogenesis during the thousands of millions of years before the origin of life on earth.

[4] J. Bassham, "The Path of Carbon in Photosynthesis," *Sci. Am.*, June 1962, pp. 88–100.

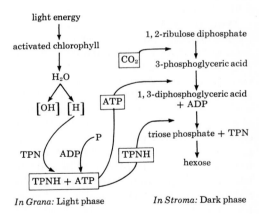

Fig. 3-2 Photosynthesis in isolated chloroplasts. In the light, grana catalyze the lysis of water, forming oxygen and liberating reduced triphosphopyridine nucleotide (TPN) and adenosine triphosphate (ATP). The assimilatory power (ATP) is then used in the stroma during the dark phase. (After D. Arnon, "Chloroplasts and Photosynthesis," in *The Photochemical Apparatus: Its Structure and Function*, Brookhaven Symposia in Biology, 11, Brookhaven National Laboratory, 1959.)

Those who are concerned with these theories hold that hydrocarbons could have reacted with reducing gases of the atmosphere—such as water vapor, ammonia, and hydrogen sulfide—when driven by ultraviolet radiation of the sun or by electric discharge, such as results from a discharge of lightning.

In the laboratory, successful attempts have been made to reproduce these conditions and, in fact, to produce complex organic molecules. Miller[5] passed spark and silent discharges for a week through mixtures of methane, ammonia, hydrogen, and water vapor. He obtained, along with some of the original gases, carbon monoxide, carbon dioxide, nitrogen, and several amino acids (glycine, alanine, and about 25 other amino acids in smaller amounts).

Calvin and his group[6] have experimentally confirmed methods suggested by Haldane and Oparin for duplicating the first production of organic molecules from inorganic materials in primeval seas. (Calvin's group has also irradiated two-carbon substances and has obtained succinic acid, a four-carbon metabolite of living cells.)

With no living organisms to consume them, these organic molecules might have accumulated, and, as Bernal has suggested, they might well have become adsorbed on clays in the seas' bottom, forming clusters at the bottom or along the edges of the seas. In such close contact, polymerization of amino acids and nucleotides could have occurred. Oparin calls the swarms of organic materials that precipitate out of solution "coacervate droplets." (These droplets can be duplicated in the laboratory by mixing solutions of gum arabic and gelatin; coacervates may also be formed by mixing together various proteins such as casein, egg or serum albumin, or RNA and other materials.[7])

[5] S. Miller, "Production of Some Organic Compounds Under Possible Primitive Earth Conditions," *J. Am. Chem. Soc.* 77:2351–60, 1955.
[6] M. Calvin, "Chemical Evolution and the Origin of Life," *Am. Scientist* 44:248–63, 1956.
[7] Work of B. de Jong, as described by A. Oparin in *Life: Its Nature, Origin, and Development,* trans. by A. Synge, Academic, New York, 1962, p. 69.

Since the quantity of molecular oxygen in the atmosphere was not sufficient to support life, the first living things must have been anaerobic heterotrophs, obtaining their food materials ready-made from the organic "soup" of the primeval seas. This hypothesis stating that early organisms were anaerobic heterotrophs is based on two criteria relating to known metabolic function: (1) contemporary living things have a metabolism based on the use of ready-made organic molecules; and (2) they also obtain energy through metabolic pathways based on anaerobic degradation of organic molecules (usually glucose), not by using free oxygen.

The metabolism of autotrophic organisms is based on the same chemical pathways as the metabolism of heterotrophs. It would seem that a special autotrophic mechanism—now apparently accepted as the common one, photosynthesis—has been added as a superstructure on a heterotroph foundation.

With the evolution of the photosynthetic mechanism, greater quantities of molecular oxygen became available in the atmosphere. In time, respiration became a more dominant process than the anaerobic fermentation that had first evolved. This new pathway in respiration also seems a superstructure that supplements a phase of anaerobic respiration in contemporary organisms. Oparin summarizes the virtually universal chain of reactions underlying lactic acid and alcoholic fermentation in microorganisms, glycolysis in animals, and similar respiration in higher plants. In subsequent biological selections in the evolution of early organisms, the three major coenzymes of living things—DPN, ATP, and coenzyme A—developed to supplement the catalytic activities of the protein enzymes. These enzyme systems may first have developed some billion years ago as metabolic pathways in primitive organisms having fermentation.

A central point in Arnon's thinking is that photosynthetic phosphorylation is compatible with the premise that emergence of photosynthetic organisms was achieved by their acquiring a pigment system capable of decomposing water by means of light. He surmises that the early capacity for utilizing light energy was probably more closely related to the synthesis of ATP than to the assimilation of carbon dioxide (photosynthesis). Arnon's reasoning is based on the close structural association in both chloroplasts and bacterial chromatophores of the phosphorylating activity with the chlorophyll pigment system. In contrast, those enzymes responsible for assimilation of carbon dioxide can easily be dissociated from the chlorophyll pigment system.

A further advance in the evolution of photosynthetic systems might well have been the conversion of only a part of the captured energy into ATP, with the remainder used to generate a reductant for carbon dioxide assimilation. Some mechanism had to be developed to prevent the regeneration of water after it had been split, so that free hydrogen would be available for the fixation of carbon dioxide. This would have to have been accomplished by diversion of the OH radical; such a mechanism was possibly dependent in the beginning on an external hydrogen donor, as it still is today among the photosynthesizing bacteria.

In a later stage of evolution, in higher green plants, this mechanism was converted into an enzyme system for the liberation of molecular oxygen. In this way, photosynthetic phosphorylation by chlorophyll (independent of molecular oxygen) could occur before oxidative phosphorylation by mitochondria, which need an oxygen supply. It is also interesting to note that after the evolution of a separate mechanism, independent of oxygen, for generating ATP in light, green plants also share with the non-green organisms (consumers) the evolution of an oxygen-dependent synthesis of ATP by oxidative phosphorylation in mitochondria. This is the basis of Oparin's assumption that photoautotrophic organisms evolved later than heterotrophic forms.

It is important for students to recognize that many aspects of photosynthesis are still not understood. In order to develop in students the ability to recognize that discoveries in science are the result of a collaboration throughout history—one scientist building on the work of others—some teachers use a case history approach such as Conant[8] developed at Harvard University. Some teachers duplicate materials drawn from original sources so that students read the contributions of van Helmont in the seventeenth century; of Priestley's findings about plants and "good" and "fixed" air in the eighteenth century; and the elegant work of Jan Ingenhausz elucidating the role of light in the photosynthetic process. Teachers may then make a rapid switch to Blackman in 1905 and his law of limiting factors in the process of photosynthesis. Many fine papers may be culled from M. Gabriel and S. Fogel, *Great Experiments in Biology*.[9] Melvin Calvin's speech in Stockholm accepting the Nobel Prize was printed in *Science*[10] and may be duplicated for reading and analysis by students in a class discussion. Freeman publishes offprints of many papers from *Scientific American*. Some of the relevant papers are listed in the bibliography at the end of this chapter. These readings provide the "reach" that is greater than most students' "grasp."

Modes of nutrition

Organisms that manufacture their energy-rich organic matter from inorganic sources are producers, or autotrophs. The driving energy for the involved syntheses may be from chemicals or from sunlight. On this basis, autotrophs are grouped as chemosynthetic autotrophs and photoautotrophs.

Chemosynthetic autotrophs manufacture their organic metabolites from carbon

[8] J. Conant and L. Nash, eds., *Harvard Case Histories in Experimental Science*, reprint ed., 2 vols., Harvard Univ. Press, Cambridge, Mass., 1957.

[9] M. Gabriel and S. Fogel, eds., *Great Experiments in Biology*, Prentice-Hall, Englewood Cliffs, N.J., 1955.

[10] M. Calvin, "The Path of Carbon in Photosynthesis," *Science* 135:879–89, 1962.

dioxide, and nitrate or ammonia, deriving energy to do so, not from light, but from the oxidation of an inorganic chemical. They lack photosynthesizing pigments and form a small group that includes iron bacteria; the colorless sulfur bacteria; *Nitrosomonas,* the aerobic bacteria that oxidize ammonia to nitrite; and *Nitrobacter,* which oxidizes nitrite to nitrate compounds.

Photoautotrophs convert light energy into chemical energy for their metabolism. This is the main group of autotrophs, and includes the higher plants, algae, colored sulfur bacteria, and the pigmented non-sulfur bacteria.

Photoautotrophs fall into three subgroups characterized by these generalized reactions[11]:

1. Green plants:

$$CO_2 + H_2O \xrightarrow{\text{light}} (CH_2O) + O_2$$

2. Pigmented sulfur bacteria:

$$CO_2 + H_2S \xrightarrow{\text{light}} (CH_2O) + S$$

3. Pigmented non-sulfur bacteria:

$$CO_2 + \text{succinate} \xrightarrow{\text{light}}$$

$$(CH_2O) + \text{fumarate}$$

Van Niel used the following generalized formula for all these types in a light reaction:

$$CO_2 + H_2A \xrightarrow{\text{light}} (CH_2O) + A$$
$$\text{(hydrogen acceptor)}$$

Evidence of photosynthesis

Formation of sugar and conversion to starch

Simple sugars are the first products of photosynthesis. Most green plants, namely most dicotyledons, support further reaction in the leaves; in the leaves the sugar is converted to starch. However, some green

[11] These equations are not balanced since not all reactants and products are indicated.

plants do not form starch in their leaves. The fleshy leaves of the onion, leaves of corn, sugar beets, and many other monocotyledonous plants (especially members of the Liliaceae) may be tested to show the presence of stored glucose rather than starch. In corn, the path of glucose from leaves may be traced to the seed, where glucose is stored. In sugar beets, glucose formed in the leaves is transported to the beet root to be stored as cane sugar. However, in the bean, and in most dicotyledonous plants, the conversion of glucose to starch continues in the leaves.

Sugar in green leaves. Sprouted onion bulbs grown in light show the presence of simple sugars in the shoots. Cut 1-in. lengths of the green shoots into a Pyrex test tube; add about 1 in. of either Benedict's or Fehling's solution (p. 660). Then bring the contents of the tube to a boil over a Bunsen burner or alcohol lamp. A change in color from blue, to shades of green, to reddish orange indicates the presence of simple sugars. This color reaction (as a test for simple sugars) may be checked by demonstrating the change in color when glucose solution or molasses is boiled with Benedict's solution.

Test for cane sugar in plants. Before you test for the presence of cane sugar in a plant, you may want to demonstrate the sugar test (using pure cane sugar). To a small quantity of cane sugar in a test tube add about 15 ml of distilled water. Next, add 1 to 2 ml of a 5 percent solution of cobalt nitrate, prepared by adding 5 g of cobalt nitrate to 100 ml of distilled water. Then add a small quantity of strong sodium hydroxide solution, prepared by adding 50 g of sodium hydroxide sticks (*caution:* caustic; use tongs) to 100 ml of distilled water. A violet color reaction is a positive test for cane sugar.

Note: A weak solution of cane sugar (sucrose) may be changed to glucose by adding a few drops of concentrated hydrochloric acid, and boiling gently for a minute or so. (Saliva will also change cane sugar, $C_{12}H_{22}O_{11}$, to glucose, $C_6H_{12}O_6$, on standing.) Neutralize the hydrochloric acid with sodium carbonate until no further effervescence occurs. Then test with Benedict's or Fehling's solution for the presence of glucose.

Role of light in photosynthesis

The need for light, the intensity of light, and the efficacy of specific regions of the spectrum can be studied by measuring: (1) rate of formation of starch; (2) changes in pH of the medium of aquatic organisms as CO_2 is absorbed in the presence of light; (3) number of bubbles of gas produced as oxygen is liberated, or distance a trapped bubble moves in a volumeter; and (4) rate of growth of cells, especially among algal forms.

Each of these methods for testing the rate of photosynthetic activity will be described. Where possible, examples will be given using land plants such as geraniums; one-celled water forms such as *Chlorella*, filamentous algae, and *Spirogyra;* or many-celled water plants such as *Anacharis* (elodea).

Starch-making by green plants. Select healthy leaves from a geranium or Golden Beta Yellow coleus plant which has been placed in the sunlight for several days. First, boil the leaves for a few minutes in water in a small Pyrex beaker on an *electric* hot plate. This preliminary step softens the leaves by breaking down cell walls. Next, transfer the softened leaves to a beaker half full of ethyl or methyl alcohol, and warm on the hot plate. Where a hot plate is not available, precautions must be taken to avoid igniting alcohol fumes. Alcohol may be heated in a water bath (Fig. 3-3) with the alcohol in a small beaker or long test tube standing in a larger Pyrex beaker of boiling water. (An electric immersion heater may also be used with a thermometer to make a fireproof alcohol-heating device.)

Chlorophyll is soluble in alcohol, and the pigment should be extracted completely from the leaves in about 5 min. For thick leaves, decant the alcohol and re-

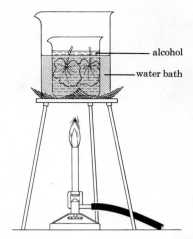

Fig. 3-3 Alternate apparatus for removing chlorophyll from green leaves. When an electric hot plate is not available, a water bath and Bunsen burner may be used as shown here.

place with fresh alcohol. Then heat again until all chlorophyll is removed. Wash the blanched leaves; spread them flat in open Petri dishes and cover with a dilute aqueous iodine solution such as Lugol's solution (p. 664). After 2 to 5 min, rinse off the excess iodine solution; hold the leaves up to the light to show the blackish areas which indicate the presence of starch. It may take as long as 15 min for all the iodine to react with the starch in the leaf.

You may want to show the reaction by first adding iodine solution to starch paste in a test tube. Students might show the absence of this reaction with other pure nutrients, such as sugar, fat, or proteins. However, the absence of starch is not a positive test that photosynthesis has not occurred; the end product may be glucose.

Testing the role of light in starch-making by green plants. Select geranium plants that have been in the dark for at least 24 hr. At the start of the experiment, some of the leaves should be tested for starch (since leaves originally free of starch will serve as one control). Have students cover several leaves of the plants with aluminum foil or carbon paper, or pin thin discs cut from cork to the top and bottom surfaces

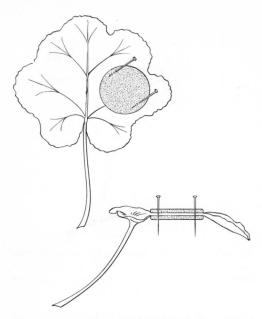

Fig. 3-4 Test of the role of light in starch-making in green plants. Part of a leaf is covered with two thin cork, carbon paper, or aluminum foil discs.

of several leaves (Fig. 3-4). Incidentally, to avoid denuding the plants of their leaves, only a portion of each leaf need be covered; the rest of each of the leaves (exposed) serves as a control. Then place the plants in sunlight, or artificial light from a 75-watt bulb about 2 ft from the plant.

After 24 hr or so, remove two or more covered leaves and an equal number of leaves exposed to the sunlight (be sure to identify them); place them in boiling water, extract the chlorophyll from the leaves in heated alcohol, and finally add diluted iodine as already described—all this to test for starch. Only the exposed leaves or areas should show the dark blue-black indicative of starch formation.

Or you can get results in 15 min by using a 500-watt lamp, 3 ft away from the plant. Make certain that the plants are not overheated. Use blue cellophane to arrest part of the heat.

Starch grains. You may want to have students examine the shape and size of

starch grains, since these are characteristic of specific plant species. In fact, starch grains are used to identify many plants. For example, in the potato the starch grains are irregularly oval in outline, and seem to consist of alternating dark and light lines in the grains. Compare these with starch grains of other plants (Fig. 3-5).

Prepare a mount of starch grains by lightly scraping the cut surface of a raw white potato with a scalpel; mount the scrapings in water, and apply a coverslip. Then allow a drop of dilute Lugol's solution (p. 664) to run under the coverslip as filter paper is placed near the opposite edge of the coverslip to draw up some of the water. Or press a clean slide over the cut surface of a raw potato. Add a drop of very dilute Lugol's iodine solution and apply a coverslip.

Starch-making in algae. It is especially difficult to identify the presence of starch in the small grains produced by green algae. When dilute Lugol's solution is applied, one is not completely certain as to whether the resulting color is dark brown or blackish.

Whitford[12] suggests digestion of starch as a means of distinguishing algae of the Chlorophyceae (which contain chlorophyll *b* and starch) from Xanthophyceae (which lack both starch and chlorophyll *b*).

Place a small quantity of algae on a nearly dry slide. Place a second slide over the algae and macerate the cells (so that enzymes may penetrate the cells faster) by pressing the slides between the fingers. After the algae are thoroughly ground, remove the top slide and apply a freshly prepared diastase solution. Then apply a coverslip. Digestion of the free starch is rapid.

Wavelength and photosynthesis. Evidence of a relationship between photosynthesis and chlorophyll is obtained by comparing the absorption spectrum of chlorophyll and the action spectrum of

[12] L. A. Whitford, "On the Identification of Starch in Fresh-Water Algae," *Turtox News* 37:62–63, 1959.

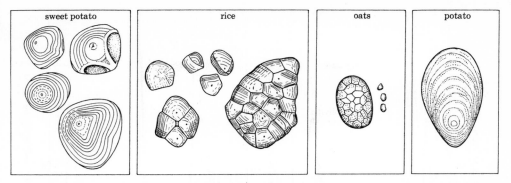

sweet potato	rice	oats	potato

Fig. 3-5 Starch grains from four plants. (After C. Wilson and W. Loomis, *Botany*, rev. ed., Holt, Rinehart and Winston, 1957, and T. Weier *et al.*, *Botany: A Laboratory Manual*, 2nd ed., Wiley, 1957.)

photosynthesis (Figs. 3-6, 3-7). Chlorophyll absorbs wavelengths of light, especially in the red and blue regions of the spectrum; these are also the regions in which the highest rate of photosynthetic activity occurs. (Refer to nature of chloroplasts, p. 182, and partitioning of pigments of chloroplasts, pp. 183–86.)

In describing light falling on a plant, three factors must be known: (1) intensity, (2) quality, and (3) duration of illumination. Intensity refers to the number of quanta falling on a given surface area. Quality describes the kind of wavelength making up that light. For example, plants growing under fluorescent light get more

blue than red wavelengths of light, while light from tungsten lamps is rich in red and poor in blue light. The third factor, duration, is a time factor.

In terms of photons, or quanta of light, the greatest radiant energy value is in the violet wavelength region of 3900 to 4300 Ångströms (Å). The red end of the visible spectrum at 6500 to 7600 Å is the region of photons of least radiant energy value.

Chlorophyll *b* extends the absorption range of chlorophyll *a* in the far red end of the spectrum, so that total absorption of chlorophyll occurs in the red and blue-violet ends of the spectrum. This is the "enhancement effect," or "Emerson effect."

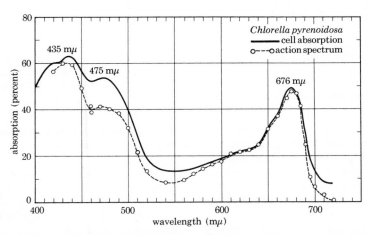

Fig. 3-6 Action spectrum of photosynthetic activity in *Chlorella pyrenoidosa*. (Adapted from M. B. Allen, ed., *Comparative Biochemistry of Photoreactive Systems*, Academic, 1960; from F. T. Haxo, "The Wavelength Dependence of Photosynthesis and the Role of Accessory Pigments," unpublished.)

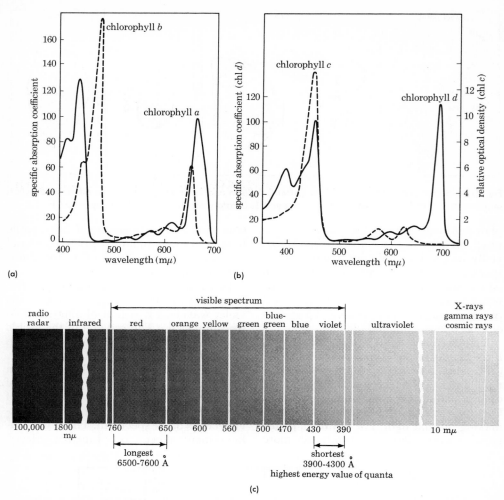

Fig. 3-7 Absorption spectra: (a) of chlorophylls *a* and *b*; (b) of chlorophylls *c* and *d*; (c) total spectrum of radiant energy (1 mμ = 10 Ångström units). (a, b, adapted from M. B. Allen, ed., *Comparative Biochemistry of Photoreactive Systems*, Academic, 1960; after J. H. C. Smith and A. Benitez, in K. Paech and M. V. Tracey, eds., *Modern Methods of Plant Analysis*, Springer, Berlin, 1955.)

Carotenoids absorb in the blue-violet and ultraviolet end of the spectrum; this is the region of greatest energy value.

In short, the range of the spectrum through which photosynthetic activity extends lies between 4000 to 7300 Å, with a peak at 4400 Å.

Chlorophyll also shows the characteristic of fluorescence, that is, it can emit energy at a different wavelength (see p. 164).

Absorption of carbon dioxide in photosynthesis

Among green plants, the rate of photosynthesis exceeds that of respiration by a factor anywhere from 10 to 100 times. In photosynthesis, only the dark phase is thermochemical and affected by temperature. Respiration, on the other hand, is a chemical reaction the rate of which increases some 2 to 4 times for each 10° C

(50° F) rise in temperature within the optimal range. Further, the rate of respiration is affected by several factors in the environment, including the supply of oxygen and a decrease in available nitrogen, as well as hereditary factors.

The study of exchange of gases is difficult under classroom conditions since photosynthesis masks respiration during periods of light. The carbon dioxide that green cells liberate in respiration is quickly assimilated in light for photosynthesis; conversely, the oxygen released in photosynthesis is used by green cells in respiration.

In short, studies of respiration in green plants must be made of plants maintained in darkness. The "apparent rate" of photosynthesis is that obtained before one makes corrections for the simultaneous, underlying process of respiration. The "true rate" of photosynthesis is established after corrections are made for respiration. This can be accomplished by keeping some samples of plants in the dark before, and after, a period of light. Then the mean value of the two dark periods can be used as the value applying during the light period.

Using indicators. Both Osterhaut and Haas (1919) used the resulting change in alkalinity as a means for measuring the rate of photosynthesis of water-living plants such as *Spirogyra* and *Ulva*. Changes in alkalinity of the suspensions in which the experimental plants were maintained could be compared against a series of buffer solutions of known alkalinity. And correction could be made for the observed rate of photosynthesis as compared with the amount of respiration that had occurred. Although they did not recommend this method for accurate research work, it is useful for high school students.

As an illustration, let us assume that the pH of a suspension containing some water-living plant or algae has been ascertained by adding an indicator and then comparing the color of the suspension with a series of known buffer solutions (p. 172). Suppose this suspension containing an indicator held at the end point is now placed in light for 20 min. If the suspension is now found to be more alkaline, we assume that the gas has been absorbed that previously maintained the culture at a lower pH. Our assumption, based on accumulated evidence in the domain of science, is that this gas is carbon dioxide. Thus we proceed, assuming that photosynthesis is an on-going process in light and that in this process a gas, carbon dioxide, is rapidly absorbed and removed from the medium. Similarly, the acidity of the culture medium should increase in darkness, for at this time carbon dioxide, constantly liberated in respiration, can be measured since photosynthesis has been stopped.

Phenolphthalein and brom thymol blue are suggested as useful indicators.

The quantity of carbon dioxide absorbed should equal the amount of oxygen liberated in photosynthesis, since the photosynthetic quotient (PQ) is 1, that is,

$$PQ = \frac{\text{volume of } O_2 \text{ liberated}}{\text{volume of } CO_2 \text{ absorbed}} = 1$$

The change in quantity of carbon dioxide is easier to measure, especially with indicators, since a change is quickly produced in 0.04 percent of a substance present in the medium. In photosynthesis, when carbon dioxide is rapidly absorbed, the acidity of the medium should decrease. (In respiration, the acidity should increase quickly.)

Indicators can be used only with media that have not been buffered, since buffered solutions can resist relatively large changes in pH. Indicators are usually considered weak organic acids or bases whose molecules give one color in the non-ionized state, while their anions or cations give a different color.

Phenolphthalein is a useful indicator in titration of weak acids (such as carbonic) and changes color in a slightly alkaline medium. Its pH range is from 8.3 to 10.

Absorption of carbon dioxide by algae. Cultures of *Chlorella* to which the indicator is added at the start give dramatic results in 5 to 20 min (well within the class period). Stock cultures of *Chlorella* vary in alkalinity, showing an increase in pH as cultures grow older and remain in light (permitting active photosynthesis). There are also variations in pH of the solutions depending on the stage in the growth cycle of these algal cells (Fig. 9-2). Some preliminary testing is needed to find cultures that are close to the end point for phenolphthalein. Test 10-ml samples of culture media by adding 1 or 2 drops of 0.1 percent phenolphthalein solution. If all cultures on hand are quite acidic, raise the pH by adding a small pinch of calcium carbonate.

Prepare several test tubes of Erlenmeyer flasks (125 ml) with measured amounts, such as 50 ml, of deep green suspensions of *Chlorella*. To each tube add 1 to 3 drops of the very dilute phenolphthalein solution (0.1 percent) so that the slightest pink tinge appears and then disappears on shaking the tube. Maintain several such tubes or flasks of equal cell density in light; keep the others in darkness or cover with aluminum foil. In bright light there is a deepening pink color within 5 to 20 min (depending on light intensity and cell density). Compare results with the controls kept in darkness. Other algae, or *Anacharis* (elodea), may be used, but *Chlorella* is especially recommended.

When stock cultures of *Chlorella* are limited, cultures may be prepared *without* the indicator. Prepare the tubes or flasks as already described, but not including the indicator solution; keep some tubes in light and some in darkness. At intervals up to 12 hr, add drops of indicator to *samples* drawn from the tubes or flasks. After 12 hr as little as 2 drops of indicator may turn the 10-ml samples of suspension a pink-rose (more alkaline). The addition of some 20 drops of indicator may not change the color of the samples from those suspensions maintained in darkness. The latter become more acidic over a period of hours, due to failure of carbon dioxide to be absorbed in darkness, and also to the liberation of carbon dioxide in respiration.

Similar materials may be used in the study of respiration (see Chapter 6). If those suspensions to which the indicator was added (so that they became pink-rose in light) are now shifted to darkness, the color fades within 10 min, and the cultures become colorless in 20 min. So sensitive is the test—if cell density and dilution of the indicator are appropriate—that students can observe changes in the color of the indicator in the tubes as the day advances. There is a fading as daylight begins to diminish; as light appears the next morning, the color deepens from pink to rose again. Students, in school or at home, can follow this over several days, and can keep a log of the time intervals.

Some students may want to measure the quantity of carbon dioxide that has been absorbed out of solution. The basis of colorimetry is given on p. 171, buffer solutions on p. 172, and a method of determining micromoles of carbon dioxide on p. 170.

In the laboratory or at home, students may investigate the relative amount of photosynthetic activity of *Chlorella* under red, blue, green, and orange wavelengths of light (p. 187). Also refer to the series of investigations beginning on p. 188.

Absorption of carbon dioxide by a water plant. The fact that carbon dioxide is absorbed in the presence of light is indirect evidence of its use in the process of food-making. One of the simplest methods of demonstrating this uses the absorption of carbon dioxide from water by such water plants as *Anacharis* (elodea).

USING BROM THYMOL BLUE WITH ELODEA. The procedure to be described is a modification[13] of a titration method adapted originally by H. Munro Fox. In this method the absorption of carbon dioxide from water by green plants in light is revealed through the change in the color of an indicator. (The use of phenolphthalein in studies with *Chlorella* has been described above.)

[13] P. F. Brandwein, "A Method for Demonstrating the Use of Carbon Dioxide by a Plant," *Teaching Biologist* 7:76, 1938.

Brom thymol blue[14] is the indicator used; it is blue in alkaline solution and yellow in acid medium. It has a fairly narrow range in pH, from 6.0 (yellow) to 7.6 (blue), so that slight changes in hydrogen-ion concentration show up quickly. Thus, a slight increase in acidity, as when carbon dioxide is added to the solution, will change the blue color (alkaline) to yellow. When carbon dioxide is absorbed, as in photosynthesizing plants, the yellow color (due to carbon dioxide in solution) is changed back to blue.

You may prepare a 0.1 percent *stock* solution of brom thymol blue by dissolving 0.5 g of brom thymol blue in 500 ml of water. To this add a trace of ammonium hydroxide (1 drop or so per liter) to turn the solution *deep blue*. In a beaker dilute the 0.1 percent *stock* solution with aquarium water in which elodea has been growing. Be certain that the solution is a deep blue. If it is not, turn it blue by adding just enough of the dilute ammonia. Then breathe through a straw into the indicator solution until the color *just* turns yellow. Now students might pour the brom thymol yellow into large test tubes. Prepare these demonstrations, and cork the tubes: (1) one set with no plants, exposed to light; (2) another set, each tube with a sprig of a vigorously growing elodea plant, each having an end bud, also exposed to light; and (3) a series like (2), except that each test tube is covered with aluminum foil or kept in the dark.

Those tubes containing plants, placed in sunlight, show a change from yellow to blue within 30 to 45 min (depending upon the intensity of sunlight). On cloudy days the series of tubes might be placed near a 75-watt bulb. Results similar to those in sunlight will take about 45 min. Since the (1) and (3) tubes show no color change, the color change in the tubes in the (2) series may be taken as indirect evidence of the absorption of carbon dioxide by green plants in the presence of light.

[14] Dictionaries usually list this indicator as "bromothymol blue"; we are following common scientific usage.

Phenol red, another indicator with a narrow range of pH (6.8 acid, to 8.4 alkaline), may be substituted for brom thymol blue or phenolphthalein (described earlier in the chapter). As the carbon dioxide is absorbed, the color change is from yellow (acid) to red (alkaline).

USING BROM THYMOL BLUE WITH CHLORELLA. Since brom thymol blue has such a narrow range above and below neutrality, it is obviously a good choice to indicate sensitive, slight changes from neutrality into either acidity or alkalinity. Students will find that they can readily relate absorption of carbon dioxide to intensity of light. Also have students compare suspensions of *Chlorella* of different cell densities with the rate of absorption of carbon dioxide in light (or use phenolphthalein, p. 168).

The green color of the suspensions of *Chlorella* often obscures the color changes from blue to green or yellow; therefore, the indicator is best used with samples of the medium after the cells have settled. On the other hand, if the cultures are not shaken, algae settle quickly so that a color change would be apparent in the clear fluid in test tubes or flasks.

Acidify some dilute brom thymol blue (preparation, above) so that it just reaches the turning point and remains yellow. If too much carbon dioxide is used, the time required for algae or any water plants to absorb it out of solution and show a change in color will exceed a class period. In bright light, photosynthesis is the on-going process, and tubes of algae containing the indicator show the change to blue as the medium becomes more alkaline. This is one demonstration showing that photosynthesizing cells absorb carbon dioxide in light.

Or count the small number of drops of brom thymol yellow needed to turn a sample of the culture blue. Compare with a control in the dark. Why is it that samples from the dark do not turn blue, but remain the yellow of acidified brom thymol?

Students will want to apply these techniques using indicators to test the effects of an altered environment on photosynthesis (and on respiration, too). Some investigations are suggested on p. 188 in the proposed block of activities.

Photolysis of water (simulating the Hill reaction). Wald[15] describes a demonstration that simulates the Hill reaction, the photolysis of water in the presence of chloroplasts. In the Hill reaction, the coenzyme TPN (triphosphopyridinenucleotide) is the electron or hydrogen acceptor A in the following reaction:

$$H_2O + A \xrightarrow[\text{chloroplasts}]{\text{light}} H_2A + \tfrac{1}{2}O_2$$

The dye 2,6-dichlorophenol-indophenol is used as an electron acceptor which is reduced with the liberation of oxygen. The blue color of the dye is bleached as the dye is reduced:

$$\text{dye (blue)} + H_2O \xrightarrow[\text{chloroplasts}]{\text{light}}$$

$$\text{dye} - H_2 \text{ (colorless, reduced)} + \tfrac{1}{2}O_2$$

In this technique, spinach chloroplasts are prepared by homogenizing spinach leaves with 0.5 M sucrose solution at 0° C (32° F) for 30 sec in a Waring blendor. After filtration through two layers of cheesecloth the filtrate is centrifuged at 50 × g for 10 min; the supernatant is decanted; and the filtrate is recentrifuged for 10 min at 600 × g. The supernatant from this second centrifugation is discarded. The pellet of chloroplasts at the bottom is resuspended in 0.5 M sucrose. The chloroplasts must be kept at 0° C (32° F) throughout this demonstration.

In carrying out this demonstration, follow this procedure. In each of two test tubes, mix the following: 2 ml of phosphate buffer (0.1 M, pH 6.5); 2 ml of dye solution (2,6-dichlorophenol-indophenol, 2.5 × 10^{-4} M); 0.1 ml of chloroplast suspension (about 2 drops); and 6 ml of distilled water. Mix by swirling the tubes.

Expose one tube to bright light; cover the other immediately with aluminum foil.

[15] G. Wald *et al.*, *Twenty-six Afternoons of Biology*, Addison-Wesley, Reading, Mass., 1962, p. 52.

After 10 min compare the two tubes. (It may be necessary to place the tube exposed to light in a tumbler of water to protect it from heat of the light source.)

Wald suggests that experiments be devised to show that the dye and chloroplasts must be illuminated together.

Measuring micromoles of carbon dioxide. Suppose several flasks of suspensions of *Chlorella* or other algae, or of water plants such as *Anacharis,* were on a ledge exposed to sunlight and we now plan to determine the concentration of carbon dioxide in the medium preliminary to some other investigation. Instead of a visual inspection (colorimetry, p. 171), students may devise a quantitative measure of the amount of carbon dioxide. Since carbon dioxide is absorbed by sodium or potassium hydroxide, a titration method can be used to calculate the amount of carbon dioxide in a given volume of suspension containing an indicator. In this procedure, alkali is added, drop by drop, to the medium. Equivalent amounts of equal concentrations of alkali and of acids react; the amount of carbonic acid (CO_2 in water) will be equal to the amount of hydroxide that is added to the fluid that is tested.

$$2NaOH + H_2CO_3 \longrightarrow$$
$$Na_2CO_3 + 2H_2O$$

Use freshly prepared 0.04 percent solution of sodium hydroxide (0.4 g to 1 liter of distilled water). The value in using this percentage solution is that each milliliter can combine with 10 micromoles of carbon dioxide.

Determine the content of carbon dioxide in the medium containing the algae, such as *Chlorella*, by pouring 100 ml of medium into a flask, and then adding 5 drops of 0.5 percent phenolphthalein indicator. To this add, drop by drop, the alkali—0.04 percent sodium hydroxide. The amount should be measured to 0.1 ml; use a graduated milliliter pipette, a hypodermic syringe, or a graduated burette. Add several drops, then shake the test flask. Continue this until a pink color appears; then carefully add 1 drop at a time to reach the end point where the pink color remains. Record the number of milliliters (in tenths) of alkali used to reach this end point. This is the amount of alkali equivalent to the carbonic acid in 100 ml of unknown medium.

Compute the number of micromoles of carbon dioxide in an unknown by multiplying the number of milliliters of sodium hydroxide by ten. (Recall that each milliliter of alkali will combine with 10 micromoles of CO_2.) Students will also need to compute the number of micromoles of CO_2 in a control medium such as Knop's solution.

In specific studies of photosynthesis, the number of micromoles of gas in the control equals the original amount of CO_2 in the containers at the start of the experiments. Subtract the amount obtained from the flask in light, in order to find the amount of carbon dioxide absorbed in the specific process of photosynthesis under study.

Furthermore, if the amount of carbon dioxide in the control is subtracted from the amount found for the flask kept in the *dark,* students will also have an estimate of the amount of carbon dioxide produced by green cells in the dark when respiration is the dominant on-going process.

For a clear, cogent description of a technique using radioactive barium carbonate in a study of carbon dioxide fixation, refer to Wald.[16] It goes without saying that teachers should not attempt these demonstrations unless they have received special training in handling radioactive materials.

COLORIMETRY. On occasion there may not be time to make titrations in the laboratory. In such cases color comparisons can be made, which provide a rough estimate of pH of the unknown. The pH of the unknown solution can be found by adding a measured volume of an indicator such as phenolphthalein to 10-ml samples of the unknown. The same amount of in-

dicator is added to a series of buffer solutions (10 ml each) of known hydrogen-ion concentration in steps of 0.2 pH units (or other units, as desired; see Tables 3-1, 3-2). Then the color of the unknown is matched against the series of labeled buffer solutions to determine the pH of the unknown. Test tubes of the same diameter and material must be used. Color is matched in a comparator block, or other device that may be made in the high school laboratory. The block should contain a lamp of specific wattage for uniform comparison tests.[17]

Simple colorimetry is based on a visual matching of colored solutions. This is different from photometry or spectrophotometry, where a photoelectric cell is used. While all colored solutions absorb some light, the relationship between the intensity of the incident and transmitted light depends on the thickness of the layer of solution that light passes through (Lambert's law), and also on the amount of monochromatic light the solution absorbs (Beer's law).

Colored solutions follow the Lambert–Beer law:

$$\log_{10} \frac{I_0}{I} = kbc$$

where I_0 = intensity of incident light
I = intensity of transmitted light
k = constant whose value depends on wavelength, temperature, solvent, and so on
b = thickness of layer of solution through which light passes
c = concentration of colored solution

Colorimetric apparatus for a research laboratory is expensive; yet, knowing the basic principles involved, students can prepare a set-up to allow light from a tungsten lamp to pass through some standard test tubes of known pH.

[17] Kits are available from supply houses such as LaMotte Chemical Products Co., Chestertown, Md. An especially useful reference for colorimetry is P. Hawk, B. Oser, and W. Summerson, *Practical Physiological Chemistry,* 13th ed., Blakiston (McGraw-Hill), New York, 1954.

[16] *Ibid.,* pp. 111–13.

TABLE 3-1

Proportions for preparing buffer solutions*

pH	add	to	dilute to
3.0	20.40 ml HCl	50 ml KHC_8H_4O	400 ml
4.0	0.40 ml NaOH	50 ml KHC_8H_4O	400 ml
5.0	23.85 ml NaOH	50 ml KHC_8H_4O	400 ml
6.0	45.40 ml NaOH	50 ml KHC_8H_4O	400 ml
7.0	29.55 ml NaOH	50 ml KH_2PO_4	400 ml
8.0	4.00 ml NaOH	50 ml H_3BO_3-KCl	400 ml
9.0	21.40 ml NaOH	50 ml H_3BO_3-KCl	400 ml
10.0	43.90 ml NaOH	50 ml H_3BO_3-KCl	400 ml

*Adapted from B. Meyer, D. Anderson, and C. Swanson, *Laboratory Plant Physiology*, 3rd ed., Van Nostrand, Princeton, N.J., 1955.

A wide range of pH test paper ("Hydrion") for each full unit or for each 0.5 unit can also be purchased. Then students can devise rapid methods (to be performed in one class period) to match colors of solutions when more accurate titration is not possible. The pH of the initial suspension used in demonstrations to show increase in carbon dioxide in respiration (Chapter 6) can be determined by using pH paper or by matching (after an indicator has been added) against the standard test tubes in the student-made colorimeter.

BUFFER SOLUTIONS. The preparation of a series of buffer solutions may be undertaken by groups of students. The choice of suitable indicators (and their preparation) may be guided by consulting Hawk, Oser, and Summerson.[18]

Or prepare the series of 0.2 M solutions suggested by Meyer, Anderson, and Swanson in *Laboratory Plant Physiology*. They use 0.2 M hydrochloric acid, sodium hydroxide, potassium acid phthalate, potassium dihydrogen phosphate, and boric acid-potassium chloride (use 0.2 M of each in the same solution). They further suggest that these reagents be mixed in the proportions shown in Table 3-1, and then diluted to 400 ml to get the pH range indicated for these buffer solutions.

[18] *Ibid.*

Machlis and Torrey, in *Plants in Action*, give a series of phosphate buffers at intervals of 0.2 pH which may be useful in the high school laboratory. Prepare these two *stock* solutions: (1) $M/15$ dibasic sodium phosphate (9.47 g of dry Na_2HPO_4 dissolved and diluted in 1 liter of distilled water) and (2) $M/15$ monobasic potassium phosphate (9.08 g of dry KH_2PO_4 dissolved in a liter of distilled water). Mix these two stock solutions as indicated in Table 3-2.

Preparations of a more acidic and a more alkaline range are also given in the appendix of *Plants in Action*.

Absorption of carbon dioxide by a land plant. Plants such as geranium or coleus may be used to show absorption of carbon dioxide through their leaves. The amount of carbon dioxide in the air reaching the plant may be modified, removed, or increased to demonstrate a change in the rate of starch-making.

First, reduce the amount of carbon dioxide around the plant in the following manner. Place a healthy geranium plant on a tray or large sheet of glass together with an open, wide-mouthed jar or beaker of solid KOH or pellets or sticks of NaOH (*caution:* caustic; use tongs).[19] Cover with a bell jar; seal the bottom of the jar to the

[19] Solid KOH or NaOH should be handled with forceps; for safety and first-aid procedures, see Chapter 14.

TABLE 3-2

Buffer solutions*

pH	15 M Na$_2$HPO$_4$ (ml)	15 M KH$_2$PO$_4$ (ml)
5.6	10.0	190.0
5.8	16.5	183.5
6.0	25.0	175.0
6.2	36.0	164.0
6.4	53.5	146.5
6.6	74.5	125.5
6.8	99.0	101.0
7.0	122.0	78.0
7.2	143.0	57.0
7.4	161.0	39.0
7.6	172.5	27.5
7.8	182.5	17.5
8.0	189.0	11.0

* From L. Machlis and J. Torrey, *Plants in Action*, Freeman, San Francisco, 1956.

Fig. 3-8 Apparatus for reducing the amount of carbon dioxide in the air surrounding a green plant. The beaker contains pellets of KOH or NaOH.

sheet of glass with petroleum jelly to make it airtight (Fig. 3-8). The hydroxide removes the carbon dioxide from the enclosed jar. Then set up a similar demonstration, but in this one omit the hydroxide. Place both preparations in moderate sunlight. After several days, the leaves may be tested for their starch content. The plant with the carbon dioxide absorbent will show little or no starch.

A more elaborate means for reducing the amount of carbon dioxide in the air is shown in Fig. 3-9. Begin with a geranium plant which has been in the dark for at least 24 hr, and which gives a negative starch test. An aspirator draws air through bottles containing barium hydroxide or strong soda lime in solution or pellet form. In this way the air that enters the bell jar lacks carbon dioxide. Later, test the leaves for the presence of starch.

The effect of an increased amount of

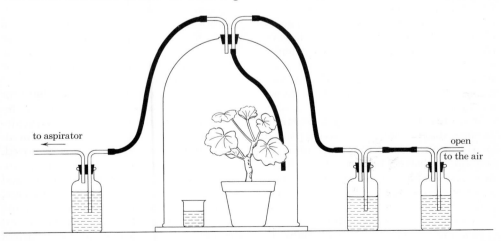

Fig. 3-9 More elaborate apparatus for reducing the amount of carbon dioxide in the air surrounding a green plant. The three bottles and the beaker contain Ba(OH)$_2$ or soda lime.

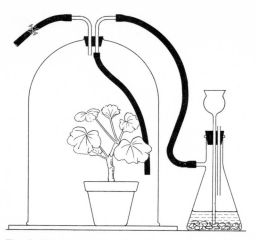

Fig. 3-10 Apparatus for increasing the amount of carbon dioxide in the air surrounding a green plant. Dilute acid and marble chips ($CaCO_3$) react in the flask to produce CO_2. Note the safety tube in the flask.

carbon dioxide in air on the rate of photosynthesis can also be shown (Fig. 3-10). Select a geranium plant which gives a negative starch test, and enclose it in the bell jar. Connect a delivery tube from a carbon dioxide generator (or a flask containing dry ice) into the bell jar. (Leave the clamp open while carbon dioxide is entering the bell jar; notice the safety tube on the generator.) Later, test these leaves for starch. There should be a significant increase in the amount of starch produced in the leaves of the plants grown in an atmosphere rich in carbon dioxide. However, this is a somewhat unsatisfactory demonstration because the relative difference in starch content is difficult to detect unless a comparison of the dry weight of the two sets of leaves is made.

Stomates, the air passages of leaves.
To show that stomates serve as air passages in the leaves, students may press the petioles of leaves between their fingers while the blades of the leaves are immersed in hot water. The air in the leaves will expand in hot water and escape as air bubbles through the stomates.

Floating water plants will show the stomates on the upper surface only, while most leaves will show them on the lower side. In the sunflower, clover, daffodil, and grasses, bubbles of air can be seen to escape from both surfaces of the leaves, indicating that stomates are found on both surfaces.

The epidermal layers may be peeled off the leaves and mounted for microscopic examination to confirm these observations (see p. 175). The fleshy leaves of *Peperomia*, types of *Crassula*, and lettuce allow easy removal of the lower epidermis (see also p. 241).

ENTRY OF CARBON DIOXIDE INTO LEAVES THROUGH STOMATES. At another time, you may want to show that stomates are the regions through which carbon dioxide enters the leaves of land plants. Select thin-leaved plants, such as geranium or coleus, which have stomates on the lower surface of the leaves. (Other useful plants are listed in Table 3-3; the average number of stomates on the upper and lower epidermal layers is indicated.)

Paint the lower surface of leaves of geranium with benzene. If the plant has been in sunlight, benzene enters the open stomates, resulting in the appearance of translucent areas. If some leaves had been covered, their stomates would likely be closed, producing no translucence.

In another method, petroleum jelly is used. Select two plants, geraniums for example, which have been kept in the dark overnight, or until the leaves give a negative starch test. Coat both sides of several leaves with melted petroleum jelly to close the stomates. On other leaves, coat the upper surface only; on still others, the lower surface only.

After the plants have been in light for several hours, wipe away the film (or remove it with ether or carbon tetrachloride). Boil the leaves in water, heat in alcohol until the chlorophyll is removed, and finally test for starch. Where petroleum jelly has clogged the stomates and prevented the entrance of carbon dioxide, students should find that starch-making has been arrested. The success of this demonstration depends on very careful techniques, for example, complete clogging of the stomates. Through the use of

TABLE 3-3

Number of stomates per sq mm*

Leaves with no stomates on upper surface

	lower surface
Balsam fir	228
Norway maple	400
Wood anemone	67
Begonia (red)	40
Barberry	229
Rubber plant	145
Black walnut	461
Lily	62
White mulberry	480
Golden currant	145
Lilac	330
Nasturtium	130

Leaves with few stomates on upper surface

	upper surface	lower surface
Swamp milkweed	67	191
Pumpkin	28	269
Tomato	12	130
Bean	40	281
Poplar	55	270
Bittersweet	60	263

Leaves with stomates more nearly equal on both surfaces

	upper surface	lower surface
Oats	25	23
Sunflower	175	325
Pine	50	71
Garden pea	101	216
Corn	94	158
Cabbage	219	301

*From B. Duggar, *Plant Physiology*, Macmillan, New York, 1930.

radioactive carbon, it has been found that carbon dioxide is absorbed through the upper epidermal region in some leaves that lack stomates in this upper layer.

EXAMINING STOMATES UNDER THE MICROSCOPE. Such leaves as *Bryophyllum, Sedum, Echeveria, Sempervivum, Kalanchoë, Peperomia,* "hen and chicks," *Tradescantia, Zebrina,* lettuce leaves, or Boston fern may be used for this demonstration. Tear the leaf to-ward the main vein. Then with a forceps pull off strips of the thin membrane that is the lower epidermis, and mount in water; examine under the microscope (Fig. 3-11).

Especially desirable for the purpose of examining stomates is the epidermis of certain species of *Tradescantia* whose cells (vacuoles) contain a pink pigment. The guard cells bounding each stomate contain chloroplasts and can be easily located in the pink background.

Where the epidermis is difficult to pull off, blanch the leaves in boiling water for a few minutes. Then the epidermis may be loosened easily. In this condition, as well as in the fresh state, the cells may be stained with methylene blue or dilute Lugol's iodine (pp. 121, 664).

Where a study of the other layers of cells in leaves is planned, it may be desirable to make freehand cross sections of leaves. This procedure is described in Chapter 2.

TURGOR IN THE GUARD CELLS SURROUNDING STOMATES. How do guard cells serve to regulate the size of stomates? Students will want to strip off the lower epidermis of a plant (see above) that has been standing in bright sunlight. Mount a bit of this in a drop of water. Examine the guard cells under high power. Now place a drop of a

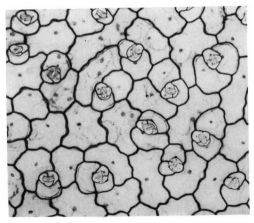

Fig. 3-11 Photomicrograph of epidermis from leaf of *Sedum*, showing stomates and guard cells. (Courtesy of General Biological Supply House, Inc., Chicago.)

0.4 M calcium chloride or sodium chloride solution (solutions, p. 658) near the edge of the coverslip, and draw off the water from the opposite end of the coverslip with filter paper. Examine the guard cells and the size of the stomates again. Notice the reduced size of the stomates as the guard cells lose turgor, because of removal of water by the sodium or calcium chloride. (For an explanation of plasmolysis and diffusion, with additional examples, see pp. 240–43.)

Students may return the tissue to water and note the increase in turgor of guard cells and in size of the opening of stomates (that is, if the cells have not been killed).

Oxygen evolved in photosynthesis

No demonstration satisfactory for use in the high school classroom has yet been elaborated to show the evolution of oxygen during photosynthesis. It would be a contribution if a suitable demonstration, simple enough to be used by students, were devised. However, until one *is* devised, the following demonstrations may be useful.

Rate of liberation of bubbles of gas in light as evidence of photosynthesis. Counting the bubbles of liberated gas as a method for estimating the amount of oxygen evolved during photosynthesis dates back in the literature to 1837. However, there are many inaccuracies in the method, as described by Miller.[20] He cites such variables as differences in size of air bubbles and different oxygen content of bubbles. Yet, for high school students, there is some value in the following two demonstrations. In flasks of rapidly photosynthesizing cells, bubbles of gas can be seen rising in the cultures; in fact, in strong sunlight, clumps of *Chlorella* rise to the surface rather quickly as they are buoyed up by a froth of bubbles. This is often apparent in a pond where masses of *Spirogyra* are brought to the surface by air bubbles.

[20] E. Miller, *Plant Physiology,* McGraw-Hill, New York, 1938.

In the other demonstration cut stems of *Anacharis* show fluctuations depending on the intensity of light.

While it is difficult to test for the presence of oxygen, we assume that the gas liberated in sunlight is oxygen. (This assumption is based on substantial experimental evidence with other plants, and on the research with manometric techniques, and semi-micro analysis.)

COUNTING BUBBLES IN A MICROAQUARIUM OF ALGAE. The rapid appearance of bubbles in media under controlled conditions can be used as a rough indication of rate of photosynthetic activity of algae. *Each student* may prepare his own series of slides to compare the numbers of bubbles produced under altered environmental conditions.

Depression slides or deep-walled concavity slides may be used by students. Place 3 to 5 drops of a dense culture of *Chlorella* (deep green) in the concavities so that they are completely filled. Then slide coverslips over the fluid in each concavity so that no air bubbles are trapped. For these short-range studies it is not necessary to seal the slides with petroleum jelly. With a hand lens, students can check that each slide contains no air bubbles at the start.

Place slides in sunlight or about 2 to 3 ft from a 100-watt bulb; inspect every 3 min and count the number of bubbles that appear over a 15-min period. Compare with slides maintained in the dark. Also prepare some microaquaria with medium of varying cell density. Is there a relationship between cell density and the number of bubbles produced? What is the effect of using cellophane coverslips of different colors—green, red, blue? (See also p. 187.)

Slides that are left for an hour or more are often useless, for they show one large air bubble; the small ones have coalesced.

COUNTING BUBBLES OF GAS FROM A CUT STEM. Bubbles of gas escape from the cut stems of elodea sprigs placed in bright sunlight. On the basis of evidence from research, we know this gas is oxygen. (See the list of possible inaccuracies, above.)

To show this, invert a 3-in. tip of a

vigorously growing elodea sprig into a test tube or beaker containing aquarium water to which about 2 ml of a 0.25 percent solution of sodium bicarbonate has been added for every 100 ml of aquarium water (which has been boiled to drive off dissolved gases, and then cooled). The bicarbonate will provide a source of carbon dioxide, since the small quantity of carbon dioxide ordinarily found in aquarium water acts as a limiting factor in photosynthesis. Tie the sprig to a glass rod immersed in the container so that it will be held down in place (Fig. 3-12).

The plant should be exposed to a bright light source. Soon, the number of bubbles of gas escaping from the cut stem per minute may be counted. You may want students to compare the rate of production of gas in relation to light intensity by moving the test tube or beaker varying distances from the light source. Use a light meter to check different intensities of light. Compare this with a control placed in relative shade or in darkness.

Fogg[21] reminds us that it is necessary to make comparisons for equal numbers of quanta since photochemical effects of light

[21] G. E. Fogg, *The Metabolism of Algae*, Wiley, New York, 1953.

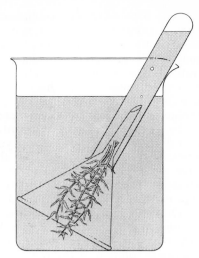

Fig. 3-13 Funnel method of collecting bubbles of gas liberated by sprigs of *Anacharis* (elodea).

quantities remain the same although energy per quantum varies with wavelength.

"Funnel" method for collecting oxygen. In this method the escaping gas is collected so that it can be tested and identified. Fill a large beaker or battery jar with aquarium water to which sodium bicarbonate solution (see technique above) has been added. Then cut the basal ends of five to ten sprigs of fresh elodea, and arrange them in a glass funnel so that their cut stems lie within or near the stem of the funnel. Force the wide mouth of the funnel to the bottom of the beaker (Fig. 3-13). Then cover the upper stem of the funnel with a test tube completely filled with water. Finally, set the preparation in bright light for a number of hours.

The water in the test tube should be displaced, more or less completely, by bubbles of gas rising from the cut stems of the elodea. Test the gas collected in the test tube for oxygen by inserting a glowing splint. Ganong[22] has criticized this method on the grounds that there is not enough oxygen produced to cause the splint to

[22] W. F. Ganong, "The Erroneous Physiology of the Elementary Botanical Textbooks," *School Sci. Math.* 6:297, 1906.

Fig. 3-12 Apparatus to show sprig of *Anacharis* (elodea) liberating bubbles of gas in light.

flare up. However, the probability of getting a positive test for oxygen is increased if the first ½-in. sample of gas, which usually has a good deal of air in it, is discarded. If the water has been boiled and then cooled, most of the dissolved gases will have been removed.

Indigo carmine test for oxygen production by leaves. Palladin[23] describes a demonstration which shows that oxygen is liberated by water plants such as elodea in the presence of light. The reagent, devised by Shutzenberger, is a solution of indigo carmine decolorized by $NaHSO_3$. The solution is yellow when prepared, but turns blue in the presence of oxygen. Place elodea plants in large test tubes containing a dilute solution of the reagent; then expose to light. Within a few minutes the reagent turns blue.

However, this test is extremely difficult to do because the reagent turns blue in air. This is the case with most reagents which turn color in the presence of oxygen; they turn color in air as well. We do not recommend this test for general use, but it might furnish the basis of a project—the use of indigo carmine as a test for oxygen.

Change in volume of gases. One can measure the exchange of gases in photosynthesis (and in respiration) by means of indicators, if buffered solutions are not used in the media. However, other methods are possible—students may observe changes in the volume of gases in a closed system such as a volumeter (Fig. 3-14; also Fig. 6-18, without cotton applicator). A change in volume, due perhaps to the liberation of oxygen during photosynthesis would affect pressure in the system so that the direction of a trapped air bubble or a trapped drop of fluid could be followed as it moved along a measured length of capillary tubing of known bore. Again, we assume the identity of the gases liberated in photosynthesis (or respiration).

In the research laboratory both photosynthesis and respiration in cells can be measured precisely with manometers and Cartesian divers.[24] In fact, Warburg in 1919 first used manometric applications in studies of photosynthesis in *Chlorella*. He was able to measure the increase in pressure in a closed system as oxygen was given off by algae; the pressure of carbon dioxide could be held constant with the use of buffers. Since oxygen is less soluble in water than is carbon dioxide, the increased volume of oxygen raised the pressure in the system and shifted the level of mercury in a capillary U-tube.

Methods using student-made manom-

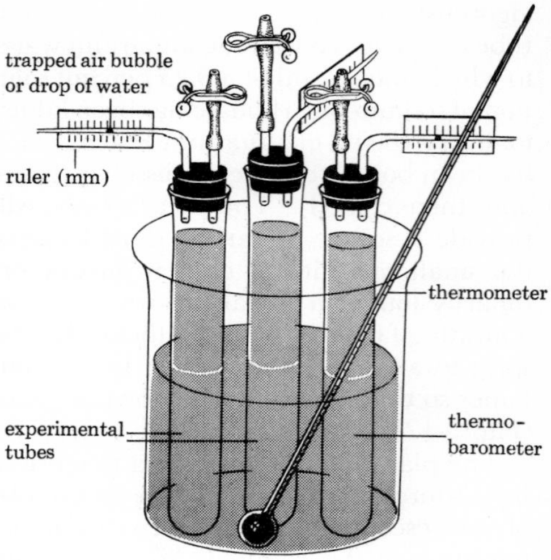

Fig. 3-14 Volumeter to demonstrate changes in volume resulting from exchanges of gases in photosynthesis. The apparatus may be used in studies of respiration as well as photosynthetic reactions. The thermobarometer acts as a control that records changes caused by temperature or barometric pressure. (After R. E. Barthelemy, J. R. Dawson, Jr., and A. E. Lee, *Innovations in Equipment and Techniques for the Biology Teaching Laboratory* [BSCS], Heath, 1964.)

[23] V. Palladin, *Plant Physiology*, Blakiston (McGraw-Hill), New York, 1926.

[24] Descriptions of chemical methods, and of volumetric and manometric techniques, are available in these, among many other texts: A. Giese, *Cell Physiology*, 2nd ed., Saunders, Philadelphia, 1962; H. Davson, *A Textbook of General Physiology*, 2nd ed., Little, Brown, Boston, 1959; W. Umbreit, *Manometric Techniques*, Burgess, Minneapolis, 1957; and D. Van Slyke and J. Plazin, *Micromanometric Analyses*, Williams & Wilkins, Baltimore, 1961.

eters with *Chlorella* or other water-living plants are difficult in the high school laboratory. Furthermore, the sensitive research instruments needed are expensive. (However, student-made manometers made of narrow-bore capillary tubing can be used effectively to show the change in volume of carbon dioxide produced by rapidly growing seedlings and growing yeast cells.)

USING A VOLUMETER. Measuring a change in the volume (in a volumeter, Fig. 3-14) is easier than measuring a change in pressure of minute amounts of gases. It is well to remember that the volume of oxygen given off in photosynthesis, per unit time, is some 10 to 30 times the volume of oxygen that is used in respiration.

In the volumeter, a change in volume causes a movement of a trapped air bubble, or drop of fluid, in a capillary tube. This provides some notion of a semi-quantitative test if a millimeter ruler is used to measure the distance the drop or bubble travels along the capillary arm. Such factors as barometric pressure and temperature must be kept constant by placing all the tubes in a water bath, and setting up a control using an equal volume of distilled water in place of the suspension of cells such as *Chlorella*. This is a thermobarometer (Fig. 3-14); students will need to maintain the same volume of gas above the medium in this control as in their experimental tubes.

When a gas is absorbed in the closed system of the volumeter, the drop or bubble should move toward the experimental tube; when a gas is liberated (such as oxygen in photosynthesis) at a faster rate than one is absorbed, the bubble should move away from the tube due to the increase in volume within the tube. Students will recall that carbon dioxide is more soluble in water than is oxygen. In the light, quantities of oxygen gas are liberated, changing the volume, so that students can readily gather data on oxygen production in photosynthesis.

Prepare several replicas of the apparatus shown in Fig. 3-14, using large test tubes or Erlenmeyer flasks of 125 ml capacity. All the joints must be air-tight. Draw out capillary tubing of 4.5 mm diameter over a Bunsen flame and allow it to cool. (Refer to the section on manipulating glass tubing, p. 677.) Then snap off the tubing at a region with a bore sufficient to permit entry of a hypodermic needle. The needle is used to inject a colored fluid into the tubing to obtain a trapped air bubble or drop. The thinner the tubing, the greater the distance the drop or air bubble will move, and the more sensitive the reading will be. To increase visibility, red ink may be used to color the fluid, but nontoxic brom thymol blue is more satisfactory in the event that some of the fluid should be drawn into the culture medium. Add a trace of detergent to the fluid to permit free flow in the tubing.

After the capillary tubing has been inserted into the rubber stopper, loosely apply the stopper to the test tube or flask. Then add the colored fluid as a drop, or trap an air bubble between drops of fluid. If a hypodermic needle is not advisable, use a medicine dropper and apply fluid to the wider bore of the tubing that is inserted into the stopper. Then deftly apply the stopper to the container so that the drop is positioned about midway along the horizontal arm of the capillary. If a two-hole stopper is used with tubing and a clamp (Fig. 3-14), pressure can be equalized more easily by opening and closing the clamp. Finally, fasten a plastic ruler under the bubble with plastic tape so that the distance in millimeters can be recorded.

Place all volumeters and the control (thermobarometer) in a water bath maintained at 25° C (77° F). Corrections may be made for changes found in the control due to pressure and temperature. Allow the apparatus to remain some 5 min in sunlight to reach equilibrium before taking readings. Then take readings every 5 min for a period of about 20 min.

USING A COMPENSATING FLASK. A varia-

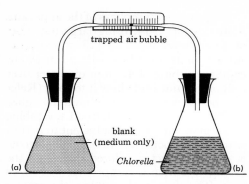

trapped air bubble

blank
(medium only)

Chlorella

(a)   (b)

Fig. 3-15 Modified volumeter: (a) compensating (control) flask; (b) experimental flask containing *Chlorella*.

tion of this volumeter is shown in Fig. 3-15. The extra space that this piece of apparatus requires in the laboratory is justified by its sensitivity. When the apparatus is placed in bright light, a trapped bubble moves rapidly from the experimental flask toward the compensating (control) flask. In sunlight, air in the closed system is warmed and expands; the use of a compensating flask equalizes this effect so that changes due to heat may be ignored in readings.

Teflon tubing (¼ in. bore) may be substituted to make a more flexible set-up. Apply petroleum jelly to make all joints air-tight.

DESIGNING INVESTIGATIONS. Using these techniques, students may devise various investigations. A teacher's role is to ask the right questions to stimulate possible approaches. Some variables that students can study using volumeters or indicators are: the effect of light on rate of photosynthesis; the effect of light of varying intensity; the effect of light of varying wavelength; the effect of flashing light as opposed to continuous light; the effect of varying temperature; and the effect of varying quantities of carbon dioxide available to the cells. Students may also see how the law of limiting factors applies in these experiments. (For further investigations, see pp. 188–93.)

Other methods for determining oxygen content. The volumeter described above

(Fig. 3-14) may also be utilized to absorb oxygen from the atmosphere in the closed system. Measurements may be made of the amount of oxygen liberated in photosynthesis by absorbing the oxygen with pyrogallol and taking readings of the movement of a trapped air bubble.

Add bicarbonate to the suspensions of *Chlorella* (p. 177). In this way a constant level of carbon dioxide in the solution and in the air above the solution is maintained. The bicarbonate solution liberates carbon dioxide as it is needed; as carbon dioxide is produced by cells, the solution absorbs the gas. As a result the exchange of carbon dioxide does not affect the pressure in the system; only changes in oxygen affect the pressure or create substantial differences in volume.

Apply freshly prepared 2 percent solution of pyrogallol to a cotton-tipped applicator and place it in the flask so that the cotton stands above the medium (Fig. 6-18). Insert applicators with oxygen absorbent into flasks containing a suspension of *Chlorella* (or other algae) and into the controls. Also prepare controls without the pyrogallol for comparison in interpreting readings. Maintain at a constant temperature, and take readings every 3 min for a 20-min period after the apparatus has come to equilibrium.

A substantial part of the 20 percent content of the atmosphere is absorbed at the start by the pyrogallol solution, and the resulting decrease in volume of the confined gases must be considered in subsequent readings. As photosynthesis is renewed in the light, quantities of gas are produced in the closed system, as shown by the movement of the trapped air bubble in the capillary.

Some preparations of pyrogallol are alkaline (containing potassium pyrogallate), and the oxygen absorbed forms an insoluble precipitate. Since there is an excess of hydroxide in this preparation, some carbon dioxide will also be absorbed, and this factor must be taken into consideration in the readings.

Prepare potassium pyrogallol solution

in two parts[25]: (1) dissolve 1 g of pyrogallate in 5 ml of water and (2) dissolve 5 g of potassium hydroxide in 25 ml of water. When you are ready to use the potassium pyrogallol solution, mix these two parts.

WINKLER METHOD. The Winkler method for determining the content of oxygen of a pond, lake, or aquarium is described in Chapter 6, and in many sources in the literature of biology.[26]

Evolution of oxygen by land plants. Many variations of the following method have been described. The best-known technique is probably the one in which a plant, such as a geranium, and a burning candle are placed under a bell jar. The bell jar is then sealed with petroleum jelly to a glass plate. The candle flame is eventually extinguished, presumably because the oxygen has been exhausted, and the demonstration is then allowed to stand overnight. The fact that the plant has produced oxygen may be tested for in several ways, for example, by inserting a glowing splint or by having a cigarette lighter ignite a candle inside the bell jar. A general criticism of this method is offered by several investigators.[27] It has been shown that a flame in an enclosed space will not consume all the oxygen. "An ordinary flame will not burn, as a rule, over about 3 percent of the oxygen from a confined space before it goes out."[28] That there is a residual amount of oxygen has been demonstrated. A piece of lens paper, cotton, or a match head can be ignited by a spark from a spark gap within the bell

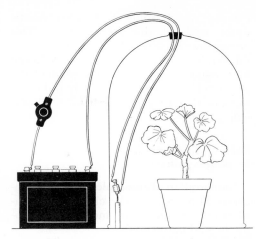

Fig. 3-16 Apparatus to demonstrate that green plants give off oxygen. An automobile cigarette lighter wired through a switch to a 6-volt battery is used to ignite a candle. The control apparatus for comparison is the same except that it does not contain the plant, or contains different types of plants.

jar after the candle flame has been extinguished. However, the candle enclosed with a geranium plant within a bell jar, used with a control, can be valuable to show the difference in quantity of oxygen evolved by green plants.

An electric hot wire in the form of an automobile cigarette lighter with a 6-volt storage battery may be used to ignite a candle in this method. Prepare the apparatus shown in Fig. 3-16, but do not connect the battery. Use a healthy geranium plant. Place the demonstration on a glass plate so that petroleum jelly can be used to seal the rim of the bell jar to the glass. Set up the control in the same way, but omit the green plant. Cork the opening in the top of each bell jar. When the demonstrations have been in the sunlight for several days, connect the storage battery in series to the wires hanging outside of the bell jars. This battery supplies current for the cigarette lighter.[29] Then

[25] Adapted from R. E. Barthelemy, J. R. Dawson, Jr., and A. E. Lee, *Innovations in Equipment and Techniques for the Biology Teaching Laboratory* (BSCS), Heath, Boston, 1964, p. 82.

[26] H. B. Benton and W. Werner, *Workbook for Field Biology and Ecology*, Burgess, Minneapolis, 1957; P. Welch, *Limnological Methods*, Blakiston (McGraw-Hill), New York, 1948; and Giese, *op. cit.*

[27] B. C. Gruenberg and N. E. Robinson, *Experiments and Projects in Biology*, Ginn, Boston, 1925: A. Raskin, "A New Method for Demonstrating the Production of Oxygen by a Photosynthesizing Plant," *Sci. Educ.* 21:231, 1937; Ganong, *op. cit.*; J. Glanz, "The Infamous Candle Experiment," mimeo (author, 87–81 145 St., Jamaica, N.Y.); and H. Alyea and F. Dutton, eds., *Tested Demonstrations in Chemistry*, 5th ed., *J. Chem. Educ.*, Easton, Pa., 1962, p. 18.

[28] Ganong, *ibid.*

[29] F. Vaurio in "Photosynthesis Apparatus," *Sci. Educ.* 22:309, 1938, comments on the standing joke about cigarette lighters and suggests a 6-volt transformer, copper wire, and a nichrome wire loop to ignite the candle, or the use of a storage battery or several dry cells.

compare the burning time of the candle inside each of the enclosed jars. This demonstration is usually not very successful; but you may be able to turn a failure into a success by assigning a student the project of either explaining why it did not work or designing modifications so that it will.

Role of chlorophyll in photosynthesis

Composition of chloroplasts. The photochemical apparatus, the chloroplast, contains several pigments, including chlorophylls and carotenoids.

Through the pioneer work of Willstatter and Fischer, the chemical nature of the chlorophyll molecule was established at the beginning of the twentieth century. Chlorophylls are magnesium porphyrins. A porphyrin molecule is composed of four basic nitrogen-containing pyrrole rings linked together by carbon and hydrogen atoms. In the center of the molecule is an atom of magnesium linked to an atom of nitrogen in each of the four pyrroles. Chlorophyll b differs from chlorophyll a due to the replacement of a methyl group in position 3 by a formyl group; chlorophyll b then has two less atoms of hydrogen and one more atom of oxygen:

chlorophyll a $C_{55}H_{72}O_5N_4Mg$
                 (blue-green color)

chlorophyll b $C_{55}H_{70}O_6N_4Mg$
                 (yellow-green color)

There are some ten different chlorophylls known, but in the higher green plants and the green algae the most common forms are chlorophyll a and b. Brown algae contain chlorophylls a and c; red algae have chlorophylls a and d. The structure of chlorophylls c and d is not clearly known.

Besides the chlorophylls found in those photosynthesizing plants that liberate oxygen, there are other groups of chlorophylls: (1) bacteriochlorophyll found in the purple bacteria; and (2) two kinds of chlorobiochlorophyll found in the green sulfur bacteria. These organisms do not liberate oxygen in photosynthesis.

Carotenoids are also pigments found among the green algae and higher green plants. The yellow and orange pigments of these plants contain four main carotenoids: β-carotene, lutein, violaxanthin and neoxanthin, and one or more of the xanthophylls. It has been postulated that carotenoids tend to protect light-sensitized chloroplasts from self-destructive oxidation. The Hill reaction may be inhibited if carotene is removed from the suspension of chloroplasts. Carotenoids of photosynthesizing organisms are of two types: (1) hydrocarbon carotenoids, or carotene; and (2) oxidized derivatives, the xanthophylls.

carotene $C_{40}H_{56}$ (orange-red)
xanthophylls $C_{40}H_{56}O_2$ (light yellow)

The central thought is that all photosynthesizing plants that produce molecular oxygen contain chlorophyll a, while chlorophyll b and associated carotenoids vary in their distribution among plant cells. Furthermore, β-carotene is almost as universal as chlorophyll a in those plants that evolve oxygen.

The number of molecules of chlorophyll per chloroplast seems to be the constant $\sim 1 \times 10^9$ for many photosynthesizing cells studied. The highly lamellar structure of chloroplasts consists of lipid lipoprotein layers; the pigments are thought to be on the surface of these lamellae, while the less dense structures, the stroma, may consist of aqueous protein.

Among the higher plants, each chloroplast, which may be from 2 to 20 μ in diameter, contains some 50 grana. These grana are arranged like a stack of coins, each consisting of chemicals arranged in a specific spatial pattern (Fig. 3-17). Several million molecules of chlorophyll are estimated to be found in individual grana.

Analysis of chloroplasts. Pigments of chloroplasts may be separated by several

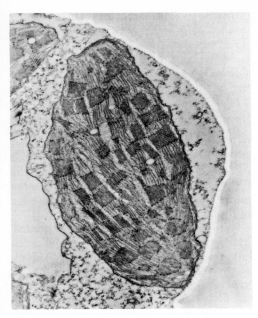

Fig. 3-17 Chloroplast of corn. An electron micrograph showing the complex lamellar system and the denser regions (grana), where chlorophyll pigment is found. (Courtesy of A. E. Vatter, Univ. of Colorado Medical School; and the Upjohn Co.)

extractions in a separatory funnel using different solvents, or by adsorption on a column of some adsorbent such as magnesium oxide powder. A third, more rapid and simple procedure for the classroom is the partitioning of pigments on filter paper to form a chromatogram.

SEPARATION OF PIGMENTS. Many solvents or combinations of solvents for extracting chlorophyll are described in the literature; several use acetone for the initial extraction of the pigments of chloroplasts. Further separation is based on differences in solubility of the pigments; for example, chlorophyll *a* is soluble in petroleum ether, while chlorophyll *b* is more soluble in methyl alcohol.

An extract of chloroplast pigments may be prepared by mixing 3 g of powdered leaves in 50 ml of 80 percent acetone (*caution:* poisonous fumes) for about 5 min (those pigments soluble in acetone will be

extracted).[30] Leaf powder may be prepared in advance and kept on hand for several weeks under refrigeration, but it tends to deteriorate after a time. Leaves of spinach, thin leaves of certain trees and shrubs, or bean or squash leaves can be dried thoroughly in an oven and stored in a closed container. Or students can begin with 15 g of fresh or frozen spinach leaves and grind them in a mortar with fine, clean sand in 50 ml of 80 percent acetone. Allow to stand some 10 min, grind again, and add more acetone. A deep green color results. To separate the sand from the extract, filter through a Buchner filter, using several layers of filter paper to obtain a filtrate that is an acetone extract of the chlorophylls, carotenoids, and other breakdown products of chloroplasts.

FLUORESCENCE. Students will want to observe this extract of chloroplasts in a beam of strong light (as from a projector) and distinguish the color of light as transmitted or reflected through the extract. Hold a beaker of the extract in the beam of light; by transmitted light observe the greenish color. Note the blood-red color of reflected light.

If the crude extract is further separated into chlorophyll *a* and chlorophyll *b* (as described on p. 185), students may observe the blue-green color of a solution of chlorophyll *a*, and the yellow-green color of chlorophyll *b* in transmitted light. Chlorophyll *b* looks brown-red in reflected light, while chlorophyll *a* gives a blood-red color.

Examine chromatograms of chlorophyll pigments under ultraviolet light (*caution:* avoid looking into the light). Note the fluorescence of the pigments.

FILTER PAPER CHROMATOGRAPHY. For a crude separation of carotenoids from chlorophylls, each student may place a 1 in. wide strip of Whatman #1, or prefer-

[30] These techniques have been adapted from several described in the literature, especially from B. Meyer, D. Anderson, and C. Swanson, *Laboratory Plant Physiology*, 3rd ed., Van Nostrand, Princeton, N.J., 1955, and also from L. Machlis and J. Torrey, *Plants in Action*, Freeman, San Francisco, 1956.

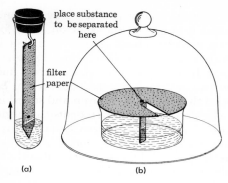

place substance to be separated here

filter paper

(a)   (b)

Fig. 3-18 Two techniques for paper chromatography: (a) strip filter paper method; (b) circular filter paper method.

ably the thicker #3, filter paper as a wick into a small container of a sample of the acetone extract of chloroplasts. Within 5 min the vivid yellow band will appear separated from the greens of the chloroplasts. Be sure to have the room well ventilated.

More careful partitioning of pigments can be done by hanging a strip of Whatman #1 or #3 paper in a closed container with a saturated atmosphere of solvent (Fig. 3-18a). Or a disc may be preferred with a segment cut ⅛ in. wide to be turned down into the solvent as a wick (Fig. 3-18b). Apply a drop of the extract of chloroplasts with a micropipette a few millimeters from the end of the strip and let this dry; apply a second time and, after drying, a third time, so that the drop becomes quite green as the pigment is concentrated on the strip. Notice in Fig. 3-18a that the strip is cut so that it tapers at one end. Cut the strip of filter paper so that it fits into the test tube without touching the sides of the tube; otherwise a clear separation of bands will not occur.

Now immerse the end of the strip in the saturated atmosphere of a test tube containing 1 in. or so of a mixture (9:1) of petroleum ether and acetone (*caution:* both poisonous). Use a curved pin or thumbtack to fasten the strip to the stopper, and close the tube or flask. If circular filter paper is used, place this over a Petri dish or

finger bowl so that the wick dips into the solvent, and cover with a similar dish or small bell jar. Apply the spot of pigment to the center of the filter paper so that, as the solvent rises by capillarity along the wick, the pigments spread out in concentric circles along with the advancing front of the solvent, all equidistant from the center.

Within 10 min students should see the bands of different colors—orange and shades of green, then shades of orange that range from yellow to orange. Remove the filter paper from the solvent when the pigments reach within an inch of the edge of the paper.

With a millimeter ruler and pencil, mark the position of the center of the dot of chloroplast pigments on the strip. Then measure the distance of the edge of the advancing front from the starting point on the filter paper; also measure the distances traveled by the different solutes or pigments and use the formula for Rf developed in Chapter 2 (p. 139) to establish the rate of migration of molecules. Should the Rf of a solute be 0.5, this would mean the solute moved 50 percent of the distance that the molecules of the solvent moved.

Using petroleum ether and acetone as suggested, students should find that carotene travels fastest, followed by xanthophyll, chlorophyll *a,* and then by slow-moving chlorophyll *b.*

After learning the technique, students may compare the rate of development of pigments in growing leaves of seedlings, or compare the development of pigments in leaves of buds of the maples on the lawn with the pigment content of full-grown leaves as they develop over a period of weeks. Compare the pigments in the varieties of Japanese maples, as well as the Japanese variety of the Norway maple. Of course, pigments of flowers can be tested with different solvents. Which solvents are best for carotenoids? For the anthocyanins of violets or lavendar? (Also refer to "Flower Pigments," *Sci. Am.,* June 1964.)

CHEMICAL EXTRACTION. The pigments of the acetone extract of chloroplasts may be separated, more tediously, by chemical extraction in a separatory funnel. As a first step, separate two solutions, one containing chlorophyll *a* and carotene and the other containing chlorophyll *b* and xanthophyll.

Into a separatory funnel pour 60 ml of petroleum ether (*caution:* petroleum ether is explosive; avoid flames); add 40 ml of green acetone solution (see p. 183). Carefully rotate the separatory funnel; hold the stopper with one hand and the stopcock with the other hand; then slowly rotate and invert the funnel. Open the stopcock to release the gas pressure. (Work in a well-ventilated room.) Close the stopcock and again slowly rotate the funnel. Add 80 ml of distilled water down the side of the funnel; rotate and invert several times. Open the stopcock again to release the gas pressure.

A separation of two layers should be visible—the upper layer, a deep green color, is soluble in petroleum ether and shows fluorescence. Draw off the bottom layer (an acetone-water layer) and discard. Then wash the petroleum ether layer again with 40 ml of distilled water; rotate the funnel and again note the separation of two layers. Draw off and discard the lower water layer. Repeat this procedure three or four times to obtain a clean, green petroleum ether solution.

Now this solution is ready to be separated into the two sets of pigments. To this petroleum ether solution, add 50 ml of 92 percent methyl alcohol solution (*caution: poisonous fumes*). Rotate in the funnel; pour the two layers that appear into separate bottles. The lower layer, the methyl alcohol layer, contains chlorophyll *b* and xanthophyll. The upper, petroleum ether, layer contains chlorophyll *a* and carotene. (How do chromatograms of these differ from each other?)

The acetone (containing some water) is a solvent for lipids. The chlorophylls are further separated (as is the carotene from the xanthophyll) by means of their different solubilities in a series of solvents. A subsequent step, using methyl alcohol and potassium hydroxide solution, saponifies the methyl and phytyl alcohol groups of the chlorophyll pigments so that they are water-soluble.

FURTHER SEPARATION. Into a separatory funnel pour 50 ml of the methyl alcohol solution (containing xanthophyll and chlorophyll *b*), and add 50 ml of ethyl ether.[31] Mix slowly by rotating the funnel. Add 5 ml of distilled water along the side of the funnel, and rotate; repeat four or five times until two layers separate out. Discard the lower, methyl alcohol layer.

Now two solutions are on hand: an ethyl ether solution and the earlier petroleum ether solution (the upper layer originally obtained in the separatory funnel). Obtain two large test tubes; pour 30 ml of the petroleum ether solution into one test tube, and 30 ml of the ethyl ether solution into the second test tube. Down the side of each test tube carefully pour 15 ml of freshly prepared 30 percent methyl alcohol-potassium hydroxide solution. Shake the tubes and observe for 10 min. Now add 30 ml of distilled water to each tube and shake until the two layers separate in the tubes. Prepare the methyl alcohol–KOH solution by diluting 92 ml of absolute methyl alcohol with water to 100 ml. To this 100 ml of methyl alcohol, add 30 g of KOH.

These pigment layers may be separated into clean test tubes and used in studies of the absorption spectrum of each of the four pigments. Also examine each solution of pigments in reflected and transmitted light to observe evidence of fluorescence (p. 183). Place the tubes against a black background and use a strong beam of light from a projector or some other source.

COLUMN ADSORPTION CHROMATOGRAPHY. Assemble the apparatus shown in Fig. 2-33a. The Buchner filter flask is to be connected to a mild vacuum; the chromatographic tube is of Pyrex glass, about 20 cm long with an outer diameter of 3 cm. Insert glass wool into the tube, which has first been inserted into a rubber stopper. Clamp the tube to a ring stand and use a large-

[31] Technique described by Machlis and Torrey, *ibid.,* pp. 136–41.

bore glass funnel in the packing process. Pack magnesium oxide powder into the tube, tap the side of the tube, and after each addition of a small quantity of powder, pack the column with a packing rod (the rod should have a diameter that just fits the bore of the tube so that the powder is evenly packed). Fill the tube to within 5 cm of the top. Then transfer the tube to the Buchner filter flask and apply gentle suction to aid in packing the powder.

Prepare a petroleum ether solution of pigments of chloroplasts (as described on p. 185). Pour this deep green solution down the side of the tube so that the magnesia becomes wet; this should nearly fill the tube. Then apply to the vacuum source so that the solution is drawn down through the length of the packed column. Different pigments are partitioned along the column due to differences in adsorptive abilities. As the extract moves along the column, add a second solution to help spread out the pigment bands so that they become easily distinguished. Prepare a 1:1 mixture of petroleum ether and benzene; pour this down the packed column. Students should find the following distribution of pigments, from *top to bottom:* chlorophyll *b*, a yellow-green layer; the dark green chlorophyll *a*; the xanthophylls; and the carotenes—a dark orange-yellow band of B-carotene, and the lowest layer, the bright yellow zone of A-carotene.

Several modifications of this technique may be tried. Some references suggest using a slurry of dry cornstarch in petroleum ether. In another technique,[32] several different adsorbents are used: a bottom layer of cotton, then a layer of aluminum oxide, a layer of calcium carbonate, a layer of powdered sugar, and another layer of cotton (Fig. 2-33b). Wet the column by running in petroleum ether from a dropping funnel, keeping a gentle vacuum to establish a filtrate flow rate of about 2 drops per sec. Pour in the petro-

leum ether extract of chloroplast pigments. Develop or spread out the pigment bands by adding a 4:1 mixture of petroleum ether and benzene.

When dried, the column can be cut between the colored zones and the pigments dissolved in ethyl ether to which a little methyl alcohol has been added. Where a spectroscope is available, students may examine the absorption spectrum of each of the chloroplast pigments. A chlorophyll solution is used as an absorption filter.

On occasion, students may also want to use a technique based on the electrostatic attraction of compounds or molecules—electrophoresis (p. 141). Here, too, the migration of complex biological compounds along a moist filter paper strip can be traced. The advantage of electrophoresis over paper chromatography is the greater speed of molecular migration and the wider separation of compounds.

Role of chloroplasts in photosynthesis. Variegated coleus leaves showing white regions or silver-leaf geranium leaves may be used to show that chlorophyll is needed for starch-making. First, make a diagram of the leaves showing their pattern of green and white. Then students may test several leaves for starch, using the standard method which has been described, that is, transferring leaves through boiling water, into alcohol, then into iodine (p. 163). The parts that previously contained chlorophyll show the presence of starch (these stain blue-black); the original white parts of the leaves, which lack starch, stain the brown of iodine.

If you compare the absorption spectrum of chlorophyll and the action spectrum of photosynthesis, you will notice a relationship between photosynthetic activity and chlorophyll. Chlorophyll absorbs wavelengths of light especially in the red and blue regions of the spectrum (Fig. 3-6). These are also the regions of highest photosynthetic activity (Fig. 3-7). (See section on light and wavelength, pp. 164–66, 176, 180.)

Although other pigments also absorb light for photosynthesis, the fluorescent

[32] Adapted from B. Harrow *et al., Laboratory Manual of Biochemistry,* 5th ed., Saunders, Philadelphia, 1960, p. 19.

spectrum obtained on illumination is that of chlorophyll *a*; this suggests that the light energy absorbed by other pigments may be transferred to chlorophyll, which in turn splits molecules of water. Chlorophyll *b* extends the absorption range of chlorophyll *a* in the far red end of the spectrum so that the total chlorophyll absorption occurs in the red and blue-violet ends of the spectrum.

USING COLORED CELLOPHANE. The action spectrum for the chlorophyll pigments of *Chlorella* (or other algae) can be approximated by maintaining suspensions of *Chlorella* under different color filters. Students will be able to measure the wavelength at which the rate of photosynthesis is high, as estimated by the amount of oxygen liberated. Estimate the amount of oxygen liberated by counting bubbles of gas in a microaquarium (p. 176), or by determining the change in alkalinity as carbon dioxide is absorbed out of the medium (p. 166).

When microaquaria covered with colored strips of cellophane or colored index tabs are not used, sets of tubes containing suspensions of *Chlorella* of similar cell density can be wrapped in several layers of different-colored cellophane, especially red, orange, blue, and green. Prepare several replicas, including some tubes covered with clear cellophane and some wrapped in aluminum foil. Maintain all the tubes in light for 3 to 4 days. They may be suspended in 0.25 percent bicarbonate solution if an indicator is not to be used to determine a change in alkalinity. Students can use a photometer (p. 445) to measure changes in turbidity; or cell counts may be made of suspensions in tubes, or of cells in microaquaria. If light green suspensions were originally used, there may be a clearly visible deepening of the green color. These are evidences of growth which result from synthesis of new protoplasm, an indirect measure that photosynthesis has occurred.

USING PETRI DISHES. Sterile dishes that have been seeded with *Chlorella* may be covered with colored cellophane. Either

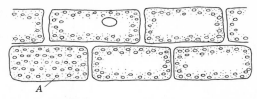

Fig. 3-19 Microscopic view of cells of *Anacharis*, showing chloroplasts lining the cell walls. Cell A is a view near the top of the cell, showing the layer of chloroplasts just under the top wall; the other cells are viewed through the center, showing distribution of chloroplasts near the cell walls.

seed the cooling, sterile agar before plating the dishes, or streak the dishes after the agar has solidified. Count the number and size of clones that arise when grown under different-colored (and clear) cellophane and aluminum foil.

In another method,[33] small, corked test tubes may be placed in water baths, each containing a different food color dye. A light meter may be used in all these techniques to establish uniform intensity of light.

Illuminated chloroplasts. At some club activity, or as a small project, students may want to show the reduction of 2,6-dichlorophenol-indophenol solution (0.1 percent) by illuminated chloroplasts. The demonstration is described in detail by Machlis and Torrey.[34] (See also p. 170.)

Chloroplasts in cells of leaves: Anacharis. One of the best plants in which to examine chloroplasts is a healthy elodea plant, *Anacharis* (Fig. 3-19). Mount a leaf in a drop of aquarium water (wet mounts, p. 106) and examine under low and high power. When the leaves from young, growing tips are examined, the chloroplasts may appear to be moving in the cytoplasm of the cells. In reality, it is the cytoplasm that is circulating (cyclosis), and the chloroplasts are being carried in the moving "stream" of cytoplasm (p. 106).

Chloroplasts in algae: Spirogyra. In *Spirogyra* the chloroplast is spiral (Fig.

[33] Described in *Biological Science: Molecules to Man* (BSCS Blue Version), *op. cit.*, "Laboratory Investigations," p. 54.
[34] *Op. cit.*, p. 141.

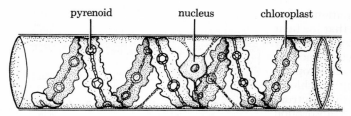

Fig. 3-20 Diagram of *Spirogyra* cell, showing pyrenoids. When stained with Lugol's iodine solution, pyrenoids appear black, which demonstrates that they contain starch. (From C. L. Wilson and W. E. Loomis, *Botany*, rev. ed., Dryden [Holt, Rinehart and Winston], 1957.)

pyrenoid   nucleus   chloroplast

3-20). Mount one or two threads in a drop of water. One species of *Spirogyra* has one spiral chloroplast; another fairly common species has a double spiral chloroplast. When Lugol's iodine solution (p. 664) is added to a wet mount, the pyrenoids along the chloroplasts stain darkly, showing that they contain starch.

Growth of organisms as further evidence of photosynthesis

When dilute Lugol's iodine solution is added to a slide of *Chlorella* that has been maintained in bright light, the cells appear much darker than cells that have been kept in darkness and stained with Lugol's. This difference is especially noticeable when dense cultures are used in preparing slides, which causes a darkening of the entire field.

Initial growth and subsequent reproduction of cells are both evidences of photosynthesis, for without the production of high-energy organic molecules, cells could not produce new protoplasm. The methods for measuring growth are applicable in studies of photosynthesis. (See growth, Chapter 9.)

If quantities of *Chlorella* are available, students might use uniform, known weights (grams per liter suspension) and compare the original weight with a possible increase in weight after photosynthesis has progressed for several weeks or a month. Or if graduated centrifuge tubes are used, students may centrifuge the media in flasks and measure increase in volume of cells per volume of liquid. (Also refer to pp. 442–48.)

Concept seeking and concept formation through a "block laboratory approach"

Students cannot learn all there is to know through inquiry, interpreted as laboratory research. Nor can they—through laboratory work—develop all knowledges necessary for future life. Inquiry does not mean only laboratory or experiment; it relates to general ways of seeking knowledge through personal effort, through expenditures of energy as a result of individual initiative.

The purpose of this "block" or "resource unit" is to help teachers to develop certain of the pervasive conceptual schemes in biology by means of many investigative approaches. Students need to know science as *process* (investigative approach or inquiry) as well as to evaluate the *product* of science. The products are the conceptual schemes that characterize biology and make it a special discipline. A discipline can become such only if there is an ordering of, or disciplined approach to, its understandings (the product), and its efforts to enlarge its understandings (the process). Biology, and science in general, lends itself to this discipline by learning by inquiry—this method of concept seeking and of concept formation.[35]

An organism, any organism, provides the focal point in the study of biology. For in its study are embodied all the understandings man has of life and of the nature of living in the biological sense. In the

[35] P. F. Brandwein, F. G. Watson, and P. E. Blackwood, *Teaching High School Science: A Book of Methods*, Harcourt, Brace & World, New York, 1958, Chapter 6.

study of a single organism in biology, the discipline emerges in coherence.

For example, the green alga *Chlorella* fulfills the requisites for an experimental organism for studying the biological processes. This autotrophic green plant is easily maintained, its environment can be controlled, it has uniform physiological properties and is nonpathogenic, and it offers opportunity for study as an individual organism or as a population. *Chlorella* has been the tool of the research laboratory; as a result, there is much well-established literature concerning it. For these reasons, *Chlorella* is used as a focus for the activities to be suggested for laboratory and classroom.

Perhaps we need to clarify a bit what we mean by the diverse methods used by scientists in seeking knowledge. Brandwein[36] suggests that we confuse the scientist's "concise report" of his work as given in his paper with the "manner of his work." Conant[37] distinguishes two orders of operations: (1) varied methods by which broad working hypotheses and grand conceptual schemes originate in the minds of men; and (2) fairly fixed patterns of testing and confirming these hypotheses.

In short, experimentation by itself does not produce science. Conant continues with his operational definition of science: "As a first approximation, we may say that science emerges from the other progressive activities of man to the extent that new concepts arise from experiments and operations, and the new concepts in turn lead to further experiments and observations."[38]

Brandwein also reminds us:

What we have tried to indicate is that problem solving begins even before the problem is stated; it begins perhaps unconsciously. The specific

[36] P. F. Brandwein, "Elements in a Strategy for Teaching Science in Elementary School," The Burton Lecture, in J. Schwab and P. F. Brandwein, *The Teaching of Science* (Inglis and Burton lectures), Harvard Univ. Press, Cambridge, Mass., 1962, p. 115.
[37] J. Conant, *On Understanding Science,* Yale Univ. Press, New Haven, Conn., 1947.
[38] *Ibid.,* Mentor ed., New Am. Library, New York, 1951, p. 37.

problem may be clarified when the scientist begins the empirical or the experimental investigative portion of his work; *but a great deal of his work goes on before he states his problem* in a formal way. The scientist solves problems; but this is not to say, we repeat, that he begins his investigation with a stated problem. It would seem, from our observations of the way scientists work, that the problem filters out of a mixture of observations, flashes of insight ("eurekas"), vague, muddled dissatisfactions, tentative essays into empiricism, reading, consultation with others, thinking (conscious or unconscious), mental scanning, and so forth. Eventually the problem does filter out, and an attempt to state it clearly is made. . . .

. . . We want, for our purposes, to distinguish between the way of the scientist in problem solving (a creative act) and the routine of the usual problem doing of the student taking an established course, in which the solution is foreordained, because there is a rigid schedule (perhaps 40 minutes) and standard equipment. Do we want our students to engage in problem solving or in problem doing? We think both.[39]

A list of some aspects and approaches to inquiry that reveal diversity of methods, yet a unity of purpose, in the efforts of scientists might well include the use of:

> the "educated guess," that is,
>   of hypothesis
> speculation
> the "leap in the dark"
> experimental design
> special laboratory techniques
>   (and invention)
> literature (the meaning of fact)
> criticism (confirmation)
> reports
> theory

Each of the learning situations that follow is created by asking a question that alters the conditions so as to cast doubt on the validity of the conceptual scheme. Whether the investigation is limited to a week or to a year depends clearly on what the individual student or the class as a group does.

[39] Brandwein, Watson, and Blackwood, *op. cit.,* pp. 26–27.

"Preludes to inquiry"
for individuals or groups[40]

1. *Chlorella* can be seeded into an agar medium so that colonies or clones form. Is there a limit to the size of a colony of *Chlorella?* How do these colonies compare with colonies of *Rhizopus,* or of *Saccharomyces?*

2. If light is necessary for photosynthesis, will twice as much light double the rate of photosynthesis?

This inquiry is meant to illustrate different levels of investigation. It is not necessary to go to the laboratory to do an "experiment" at all times. Library research is also a valid method of a scientist, leading to experimental design.

Students can inquire into what has been learned about the nature of light. They may devise an experiment to duplicate the kinds of thinking that they have read about in references. Here they will learn, no doubt, that there is all the difference in the world between reading the concise report of a scientist, and going through extensive experimentation with its endless replication to ensure statistically valid interpretation of experiments and controls before one can reach a tentative conclusion.

3. In using the "historical" approach to inquiry, a teacher may describe the elegant experiment that Engelmann[41] designed to show the production of oxygen by plant cells in certain regions of a microspectrum. Or students may read Engelmann's paper.[42]

What was known about photosynthesis at the time? What was the climate of thinking, the grand conceptual scheme used to explain how green plants made food?

Special problems for investigation often arise from focus on such a productive paper. For example, did Engelmann influence the thinking of van Niel, or of Hill? What would the next steps be after Engelmann's work? How did Hill's experiments with isolated chloroplasts affect van Niel's hypothesis about the origin of oxygen in photosynthesis? How did Arnon's hypotheses modify the whole conceptual scheme of photosynthesis?

4. A teacher may begin by asking what quantity of carbon dioxide is in the air. If 0.04 percent is present, what effect will 1.0 percent have on the rate of photosynthesis?

5. What part does *Chlorella* play in the economy of a pond? (If *Chlorella* did not exist, what other organism might take its place in the pond?)

6. What are some of the reasons for land becoming depleted so that the yield of food crops is diminished?

Using *Chlorella* as a "crop," determine whether the culture medium, the environment of *Chlorella,* remains constant from week to week.

7. What is the spectrum necessary for development of chlorophyll in *Chlorella?* Is the same spectrum necessary for photosynthesis?

Note how one aspect of inquiry—hypothesis—arises out of this problem posed to students. Many hypotheses are offered in class: (a) the wavelength has no effect on formation of chlorophyll; (b) certain wavelengths are more effective than others in stimulating the formation of chlorophyll; (c) photosynthesis is independent of wavelength of light; (d) chlorophyll formation and photosynthesis are related processes; (e) formation of chlorophyll and the process of photosynthesis are independent; and (f) certain wavelengths of light are more effective than others in photosynthesis.

8. "*Chlorella* cakes," especially rich in proteins as well as in lipids and carbohy-

---

[40] For examples of other kinds of inquiry, refer to J. Schwab, Supervisor, *Biology Teachers' Handbook* (BSCS), Wiley, New York, 1963.

[41] Described in W. P. Pfeffer, *The Physiology of Plants,* 2nd ed., Vol. I, Clarendon Press (Oxford Univ. Press), Oxford, 1900, pp. 342–56.

[42] Reprinted in Gabriel and Fogel, *op. cit.*

drates, may be a valuable supplement to diets of undernourished peoples.

9. In the lowest zone of the littoral region are found red algae; green algae live at the highest region; and in the middle zone are the brown algae.

Engelmann[43] offered a theoretical explanation for the chromatic adaptation of algae. Yellow light predominates in sunlight, but blue-green light is transmitted to the greatest depth in clear sea water. Pigmentation of algae is complementary to quality of the light of the position which the algae occupy.

Here students have an opportunity to test a theory. What predictions can be tested?

10. A mutation was found in *Chlorella* that resulted in its inability to synthesize sugar although it did have chlorophyll. How could such an organism be used in designing an experiment to test some factors in photosynthesis?

11. Other mutations—cells that lack chlorophyll—have been found in *Chlorella*. How might these organisms be used in the study of the light and dark reactions in photosynthesis? How could students demonstrate that the organism is the product of heredity and environment?

12. Glycollic acid is a product excreted by *Chlorella*. Under what conditions is there a maximum quantity of glycollic acid? What effect does this metabolic product have on the typical microecological system in a pond?

13. A student observed that when the growing tip of *Anacharis* was introduced into a microaquarium containing *Chlorella,* the algae increased in number to a greater extent than did controls lacking the tip of *Anacharis*. Methods for measuring growth of microorganisms are described in Chapter 9.

Clue: Growing tips are a rich source of auxins. This "prelude" offers a student an opportunity in hypothesis formation, and is fruitful of predictions that are susceptible to testing in the laboratory.

14. Does polarized light affect the rate of photosynthesis? Is it possible that certain phases in the growth cycle are more affected than others?

15. Can *Chlorella* synthesize any food materials in the dark if glucose is provided? Would *Chlorella* then be considered a heterotroph? Would this be an inherited adaptation? (How could a gain in organic matter be measured?)

16. What is the role of light in photosynthesizing plants? How do the factors of light intensity, temperature, or concentration of carbon dioxide affect the process? Is the stage in the growth phase of cells critical in this study?

17. Ionizing radiations include alpha, beta, gamma, X-ray, proton, and neutron radiations. The beta, gamma, and X-rays produce ionization of molecules in the protoplasm of plant cells. Neutrons and alpha particles cause electronic excitation. X-rays and other ionizing radiations produce their greatest effects on actively growing cells containing a high volume of water. Water is ionized and some OH forms H_2O_2, an oxidizing agent which may inhibit or interfere with certain metabolic processes.

Is there a difference in the effects produced by a given dose of X-rays on a culture of *Chlorella* in the exponential growth phase, as compared with the stationary phase? Does the quantity of oxygen affect irradiated cells? Students can devise ways to expose cultures to radiations.

18. Is the usual biological succession of organisms in a micro-lake (microaquarium) affected by altering the light intensity? Could such alterations (if any) affect the survival of *Chlorella*? Or of other organisms in the micro-lake?

19. If some substance altered the rate of growth of *Chlorella* (or any one auto-

[43] Pfeffer, *loc. cit.*

troph) in a pond, would there be effects on the other organisms—producers, consumers, or decomposers—in the pond?

Students might devise experiments using gibberellic acid, thiamine chloride, kinetin, or auxins.

20. What deficiency symptoms occur in *Chlorella* maintained in a culture lacking phosphates? Or lacking nitrates? Or lacking sulfates? Students may prepare Knop's solution (or others) and vary the chemical ingredients. Is the concentration of chlorophyll in cells affected? Is the rate of photosynthesis affected?

21. What is the effect of a change in the concentration of metallic ions on *Chlorella?* How do the ions of magnesium affect the concentration of chlorophyll in *Chlorella?*

Clue: It has been reported[44] that cell division stops in *Chlorella* in a magnesium-deficient medium, with the subsequent formation of large cells. In this case, the method of measuring optical density (turbidity, p. 445) is misleading since there is no growth. It has further been shown that manganese, zinc, copper, and cobalt inhibit growth and synthesis of chlorophyll in *Euglena,* while magnesium, molybdenum, and boron stimulate synthesis of chlorophyll. Do these metallic ions have similar effects on *Chlorella?*

Maintaining uniform genetic material. A technique used in bacteriological studies may be used for refining the design of some experiments with *Chlorella.* What kinds of problems could be posed, and solutions sought, if one knew a simple method of maintaining *Chlorella* of the same heredity, that is, uniform genetic material?

Lederberg developed a simplified technique, called replica plating, for continuing colonies of bacteria of the same heredity. A piece of velour was used to cover a cylinder of wood or other material, and held securely in place. The diameter of the cylinder and velour cover was that of a Petri dish.

When colonies are grown on a Petri dish, each colony theoretically has arisen from one organism. All the cells in that colony must have the same heredity (barring mutations); such a colony is a clone.

When Lederberg and his assistants inverted a Petri dish containing clones over the velour and pressed down lightly, a pattern of the clonal growth was formed on the velour. Now a series of fresh, sterile Petri dishes of agar could be inverted and touched to the "pattern" on the velour. When incubated, new clones formed at the points of contact with the original colonies. If several dishes were placed on top of each other, the same pattern of distribution of clones could be matched. What problems relating to *Chlorella* might now be *testable* with this method?

(Other "preludes" to investigations are offered in Chapter 11, relating to the "Web of Life," in which photosynthesis is the continuing thread.)

In several of the investigations presented here, emphasis is placed on the role of the altered environment in the *selection* of organisms, that is, the evolution of organisms. It becomes apparent that the organism is the product of its heredity, its interrelations with other organisms, and its environment. Adaptations of an organism may be explored, and the possibility of genetic continuity of such adaptations may be tested. Is it possible, for example, to grow *Chlorella* in highly alkaline medium? Or can *Chlorella* become adapted to an environmental change in salinity, or pH, and pass on this adaptation to successive generations?

Finally, succession of organisms leads to the problems inherent in the evolution of communities of organisms.

Further problems arise that lead into current areas of research in the laboratories. Students and teachers may select areas of interest from *Research Problems in Biology: Investigations for Students.*[45] These

---

[44] Finkle and D. Appleman, "The Effect of Magnesium Concentration on Growth of *Chlorella*," *Plant Physiol.* 28:664–73, 1953.

[45] Originally selected and edited by the Gifted Student Committee of the BSCS, Univ. of Colorado, Boulder, 1960–62; now Anchor Books, published by Doubleday, Garden City, N.Y., 1963, 1965.

investigations, presented in four series, were suggested by biologists over the country for long-range investigation by individual students who have ability and perseverance and are committed to biology.

Developing skills in understanding the reports of scientists

Along with the skills of observing, demonstrating, and planning experiments, students need practice in another skill, the ability to read and interpret the writings of scientists. Many of the experiments or demonstrations that we have described here are a part of the history of biology and the fascinating story of how green plants make their food supply. In some classes you may want students to read small sections of the original papers of some of the historic bench marks of the workers in this field: van Helmont, Ingenhousz, Priestley, Sachs, or Blackman, among the hundreds who have contributed in this area. Current reading should include work of Calvin, Arnon, Fox, Urey, Wald, and Oparin.

In fact, some of the experiments may be elicited "afresh" from students as they think through the need for large numbers of experimental plants and for controls. In this way, students will come to appreciate the design of experiments.

You may want to use a reading taken from a science journal or textbook, then prepare suitable questions based on the reading for students to answer. This is a valuable activity in giving practice in interpretation and reasoning.

Such a reading may be used as an introduction to a new topic; it might raise questions or give data related to the classwork; or it might introduce the need for experimental design in undertaking laboratory work. This is also one of the many ways a teacher can identify students with a high level of ability and interest in science. Such students read these passages with greater facility, appreciation, and understanding. In some classes, readings may be included on an examination as a test for under-

standings and application in a new view.

Here, for example, is an historical reading that one teacher assigned to his students: Jan van Helmont's statement of his classic, seventeenth-century experiment. Following the reading is a list of questions based on it that the students were to answer.

I took an earthen vessel, in which I put 200 pounds of earth that had been dried in a furnace, which I moistened with rainwater, and I implanted therein the trunk or stem of a willow tree, weighing 5 pounds, and at length, 5 years being finished, the tree sprung from thence did weigh 169 pounds and about 3 ounces. When there was need, I always moistened the earthen vessel with rainwater or distilled water, and the vessel was large and implanted in the earth. Lest the dust that flew about should be co-mingled with the earth, I covered the lip or mouth of the vessel with an iron plate covered with tin and easily passable with many holes. I computed not the weight of the leaves that fell off in the four autumns. At length, I again dried the earth of the vessel, and there was found the same 200 pounds, wanting about 2 ounces. Therefore 164 pounds of wood, bark and roots arose out of water only.[46]

The teacher asked the students to answer the following questions:

1. How long did Helmont continue his experiment?
2. What was he trying to find out?
3. What was the gain in weight of this 5-pound willow stem?
4. How did van Helmont account for the increase in weight?
5. What would he have found if he had included the weight of the leaves over four years?
6. What kind of control would you suggest to find out whether new plant tissue grows from water alone?
7. How do we explain the increase in weight of the plant nowadays?

You may find readings for similar purposes in scientific journals such as *Science,*

[46] From L. Nash, ed., *Plants and the Atmosphere,* "Harvard Case Histories in Experimental Science," 5, Harvard Univ. Press, Cambridge, Mass., 1952, p. 15.

a publication of the Am. Assoc. for the Advancement of Science (1515 Massachusetts Ave., N.W., Washington 5, D.C.); *American Scientist* (Society of Sigma Xi, 33 Witherspoon St., Princeton, N.J.); and *Scientific American* (415 Madison Ave., New York 17, N.Y.).

The two-volume set of the Harvard Case Histories in Experimental Science is a valuable source of readings in the history of science. Also, you may want to look into the readings edited by Gabriel and Fogel called *Great Experiments in Biology*.[47] This and others listed at the end of this chapter are rich sources of reading materials.

Tests of reasoning

You may also want to plan tests based on observation and reasoning rather than upon strict recall. By way of example, three sets of questions follow.

Experiment 1: A girl took five test tubes containing brom thymol blue and put elodea plants into three of them, as shown in Fig. 3-21. All the tubes were put in the dark. The next day she found that in the tubes containing the elodea (tubes 3, 4, and 5), the brom thymol blue had turned to yellow. (Brom thymol blue changes to yellow when enough CO_2 is added.)

[47] *Op. cit.*

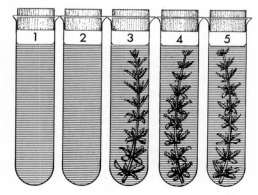

Fig. 3-21 Setup for Experiment 1. (Drawing by Marion A. Cox. From *Experiences in Biology*, by Evelyn Morholt and Ella Thea Smith, copyright, 1954, by Harcourt, Brace & World, Inc., and reproduced with permission of the publisher and the artist.)

1. What is the best explanation of the girl's results?
   a. Brom thymol blue turns yellow in the dark.
   b. The plants gave off oxygen in the dark.
   c. The plants gave off carbon dioxide in the dark.
   d. There is insufficient evidence for any explanation.

2. Why were tubes 1 and 2 necessary or not necessary?
   a. They were not necessary, because we know the action of brom thymol blue.
   b. They were needed to match colors.
   c. They were necessary to show whether any other factors change the color.
   d. None of these.

3. What should the girl do to change the brom thymol yellow in tubes 3, 4, and 5 back to blue again?
   a. Put the test tubes in the light.
   b. Put a goldfish into the tubes.
   c. Blow into the brom thymol yellow.
   d. None of these.

4. What process, going on in the green plants in the dark, accounts for the color change?
   a. assimilation
   b. oxidation
   c. photosynthesis
   d. transpiration[48]

Experiment 2: In the closed system of flasks and tubing shown in Fig. 3-15, there is a trapped air bubble in the tubing.

1. If the green algae (*Chlorella*) in flask (b) give off a gas, and there is no change in flask (a), the drop of water in the tubing will move along the tubing due to which of the following?
   a. increase in volume of gases in flask (b)
   b. decrease in volume of gases in flask (a)
   c. decrease in volume of gases in flask (b)
   d. increase in temperature of gases in flask (b)

2. In which direction will the trapped air bubble move along the tubing if the apparatus is in light?
   a. toward flask (b)
   b. away from flask (b)

[48] Experiment 1 adapted from *Experiences in Biology*, new ed., by Evelyn Morholt, copyright, 1954, © 1960, by Harcourt, Brace & World, Inc., and reprinted with their permission.

c. oscillate between the flasks and take a central position

d. unpredictable

3. What is the gas that is liberated if the apparatus is in light?

    a. carbon dioxide

    b. nitrogen

    c. oxygen

    d. water vapor

4. If the apparatus were placed in darkness, what would the trapped air bubble probably do?

    a. remain stationary

    b. move toward flask (b)

    c. move toward flask (a)

    d. contract due to a change in pressure

5. Phenolphthalein is an indicator that turns from rose to pale pink to colorless when the quantity of carbon dioxide increases in the medium.

If equal amounts of phenolphthalein were added to the flasks (a) and (b), after several hours of darkness we would likely find that

    a. flask (b) would be colorless

    b. flask (b) would be rose

    c. flask (a) would be rose

    d. flasks (a) and (b) would be colorless

6. The on-going process in algae maintained in the dark is

    a. transpiration

    b. oxidation

    c. photosynthesis

    d. growth

7. The color of the indicator would be markedly changed if the apparatus were

    a. maintained at a higher temperature

    b. kept agitated (stirred)

    c. placed in the light

    d. kept in darkness for several hours longer

8. Suppose the cells in flask (b) were yeast cells rather than *Chlorella*. What would you predict in the following situation?

a) In darkness, there would be a greater liberation of

    a. carbon dioxide

    b. nitrogen

    c. oxygen

    d. water vapor

b) This would be due to the fact that

    a. photosynthesis goes on faster than respiration

    b. respiration is the on-going process

    c. an appreciable quantity of alcohol is formed

    d. respiration goes on faster than photosynthesis

9. Were the culture of yeast cells placed in light there would be an increase in the quantity of

    a. carbon dioxide

    b. nitrogen

    c. oxygen

    d. water vapor

10. Yeast cells are considered

    a. symbionts

    b. autotrophs

    c. heterotrophs

    d. free living

11. Algae like *Chlorella* are considered

    a. symbionts

    b. autotrophs

    c. heterotrophs

    d. free living

Experiment 3: Fig. 2-29 shows a microscope slide with a concavity in the middle. Notice that a drop of pond water hangs from a coverslip inverted over the space of the slide.

Let us consider this drop of pond water as a microaquarium—a micro-lake. In this micro-lake are amoebae, motile *Euglena* that are green and have flagella, green algae called *Chlorella*, and two species of *Paramecium*, some many-celled animals called rotifers, and some bacteria.

1. The primary food source, the producers, are

    a. rotifers and *Chlorella*

    b. *Euglena* and *Chlorella*

    c. *Paramecium* and *Amoeba*

    d. *Euglena* and bacteria

2. The one consumer or heterotroph among the following is

    a. *Chlorella*

    b. *Euglena*

    c. green algae

    d. *Amoeba*

3. The usual role of the bacteria in this micro-lake is to act as

    a. producers

    b. decomposers

    c. symbionts

    d. parasites

4. Paramecia and rotifers show a preference for *Chlorella* as food. All things being equal, the limiting factor which determines the increase in population of paramecia in this micro-lake is the number of

    a. paramecia

    b. *Chlorella*

    c. *Euglena*

    d. amoebae

5. If two species of paramecia live in this micro-lake, a special problem may arise. This problem is competition
   a. for food between the two species of paramecia and the amoebae
   b. for food between the two species of paramecia and rotifers
   c. between the two species of paramecia
   d. between paramecia and bacteria

6. If *Euglena* should die off
   a. populations of both species of paramecia would also die off
   b. neither population of paramecia would die off
   c. population of one species of paramecia would die off
   d. population of *Chlorella* would die off

7. If the micro-lake is kept in sunlight, or near a light source, many bubbles of gas become visible. The gas is probably rich in
   a. ammonia
   b. carbon dioxide
   c. hydrogen
   d. oxygen

8. Which organisms are liberating this gas in light?
   a. rotifers
   b. bacteria
   c. *Chlorella*
   d. amoebae

9. The process which captures light energy and then converts the energy into organic matter is
   a. respiration
   b. oxidation
   c. assimilation
   d. photosynthesis

10. Of the following pairs of processes, the one which is a pair of *opposite* processes is
   a. photosynthesis-growth
   b. respiration-photosynthesis
   c. ingestion-digestion
   d. oxidation-locomotion

11. If red cellophane were placed over the micro-lake in the light, more bubbles of gas would occur than if green cellophane were used. This is because green algae
   a. absorb light of red wavelengths
   b. absorb light at the green end of the spectrum
   c. reflect red light
   d. carry on respiration faster in green light

12. The process by which these organisms reproduce (except rotifers) is called
   a. budding
   b. sporulation
   c. propagation
   d. fission

13. The flagellates, *Euglena,* become bleached when subjected to streptomycin. The structures which are disrupted are the
   a. mitochondria
   b. chloroplasts
   c. cell membrane
   d. vacuoles

14. These bleached *Euglena* can live in the darkness (or in light) provided the following is added to the micro-lake:
   a. starch
   b. glucose
   c. lipids
   d. minerals

15. Of the following pairs of terms, the pair that may best be applied to *Euglena* is
   a. heterotroph-autotroph
   b. heterotroph-symbiont
   c. autotroph-parasite
   d. autotroph-symbiont

16. The fact that populations of organisms in a micro-lake may change over a few weeks is due to the principle called
   a. adaptation
   b. food chain
   c. succession
   d. growth phases

CAPSULE LESSONS: Ways to get started in class

3-1. In a laboratory activity, have students prepare a micro-lake (see microaquarium, p. 176) containing a given density of *Chlorella*. Prepare several and cover with index tabs of different colors—red, green, blue, orange. Be sure there are no air bubbles at the start. Leave in the light and compare the number of bubbles in each slide after 5 min, then 10 min. What process is going on? Do bubbles form in the dark?

If yeast cells are used instead of *Chlorella,* are the observations the same? Are the explanations similar? (Other suggestions are in this chapter.)

3-2. Seal a snail in a soft glass test tube of pond water (Fig. 6-1). Ask the class: "How long might the snail live? Why?" By suitable questions lead the discussion toward the notion that green plants make food. And then the class is involved in a good discussion of how a green plant makes food and what materials it needs for food-making.

3-3. At the start of the period have, clearly in view, test tubes in a rack near a light source. Use dense suspensions of *Chlorella* to which phenolphthalein has been added so that they are at the turning point, just *slightly* pink. After 10 min students should observe that the medium becomes deeper rose color as it becomes more alkaline (due to the absorption of CO_2 from the medium). Of course, some preliminary experimentation is needed to discover the correct cell density, light intensity, and dilution of indicator (pp. 167–69).

Or use sprigs of *Anacharis* with phenolphthalein or brom thymol blue (p. 168).

3-4. Or begin by having students design an experiment to show that green plants use carbon dioxide in light (techniques, p. 176).

3-5. Begin by testing a silver-edged geranium leaf for starch. Elicit reasons why the white portion does not contain starch. From here, you may have students design experiments to test the importance of light, of carbon dioxide, and so forth, as factors in food-making.

3-6. Develop a case study around the topic of photosynthesis, using a reprint of E. Rabinowitch's paper "Photosynthesis" (*Sci. Am.,* March 1949, November 1953).

3-7. Introduce van Helmont's experiment with a willow twig, and proceed from there. (See P. F. Brandwein *et al., You and Science,* new ed. [Harcourt, Brace & World, 1960].)

3-8. Begin with laboratory work. Have the class examine cells of the onion bulb under the microscope (Fig. 2-22). Compare these with elodea cells. Lead into a discussion of the function of chloroplasts.

3-9. Do green plants give off oxygen? Collect a tubeful of the gas which bubbles out of the cut stem of elodea twigs in the light (described on p. 177).

3-10. Perhaps you may prefer to begin with a film such as *Gift of Green* (New York Botanic Gardens), *Leaves* (EBF), or *Photosynthesis* (EBF).

3-11. Using a crude acetone extract of chloroplasts of spinach, students may prepare chromatograms of the pigments, as described in this chapter. (*Caution:* Have good ventilation with a minimum concentration of acetone fumes.)

3-12. Use a roll of cellophane or a clear plastic box to model a cell. Elicit from students information concerning the structures typical of all cells. As they suggest structures, build the cell. For instance, as your students suggest the structure "nucleus" insert a nucleus (made of model-ing clay). Small balls of green clay might be chloroplasts, or a ribbon of green blotting paper the spiral chloroplasts of *Spirogyra.*

You may have models of cells made by students (in previous years) to show new students. Models can be made of plaster of Paris or papier-mâché (model-making, p. 667). This making of models may be done in a club or as a student project. An attractive visual aids collection may be developed in a short time.

3-13. Other suggestions for lessons may be found in these laboratory manuals and workbooks: *Student Laboratory Guide* to *Biological Science: An Inquiry into Life* (BSCS Yellow Version, Harcourt, Brace & World, 1963); the "Laboratory Investigations" at the end of *Biological Science: Molecules to Man* (BSCS Blue Version, Houghton Mifflin, 1963); E. Morholt, *Experiences in Biology,* new ed. (Harcourt, Brace & World, 1960); J. H. Otto, J. Towle, and E. Crider, *Biology Investigations* (Holt, Rinehart and Winston, 1963); and B. B. Vance, C. A. Barker, and D. F. Miller, *Biology Activities* (Lippincott, 1963).

3-14. Lessons may be devised by starting with any one of the demonstrations suggested in this chapter.

3-15. Keep a log of successful lessons. As these are revised each year (in the light of variations in classes and students) you will have a valuable piece of action research. (See the companion volume, P. F. Brandwein, F. G. Watson, and P. E. Blackwood, *Teaching High School Science: A Book of Methods* [Harcourt, Brace & World, 1958].)

3-16. Examine carefully and select feasible demonstrations from the rich source of materials prepared by the BSCS in the laboratory blocks (Heath), and C. Lawson and R. Paulson, eds., *Laboratory and Field Studies in Biology: A Sourcebook for Secondary Schools* (Holt, Rinehart and Winston, 1960).

3-17. Use molecular model sets to build basic concepts of linkage of carbon to carbon and carbon to hydrogen, to nitrogen, and so forth.

3-18. Develop the equation for photosynthesis. Students can report on the use of heavy oxygen and radio-carbon to learn the source of oxygen liberated in photosynthesis.

Develop the role of ATP in the cycle of photosynthesis.

PEOPLE, PLACES, AND THINGS

There are times when a teacher may need help—equipment on loan, a special chemical for an experiment, an idea for a student project for a youngster with a high interest in science. Likely as not, scientists in a nearby college, university, or hospital will want to be of help. Your colleagues in a nearby junior or senior high school may be of considerable help, too. Graduate students in nearby universities are often generous with suggestions for projects and new demonstrations. Many high school students have been sponsored in their project work in this way, and have thus been encouraged to continue in science.

You can get assistance from college librarians in hunting out an old or a new reference book on some specialty. A museum of natural history, botanic garden, or zoological park may have the information or equipment you need. And, of course, the parents of your students (doctors, engineers, scientists) may be resource people in some aspects of special work going on in your school. (Avoid, however, having students write to a busy scientist asking him to send samples, or to explain all he knows about photosynthesis.)

BOOKS

References on a subject as fundamental as photosynthesis soon become outdated. A sampling of papers from *Scientific American* is included here as supplementary reading to give students an introduction to the exciting range of current research. The papers are written by scientists about their own special interests and elegant experiments. (Offprints of some 500 articles are now available from W. H. Freeman and Co., 660 Market St., San Francisco 4, Calif., at $0.20 per copy.)

Symposia offer current research, but at a highly technical level; yet these are included for reference and for their splendid bibliographies, which are guides to the current literature of a specialized area of biology.

Allen, M. B., ed., *Comparative Biochemistry of Photoreactive Systems*, Academic, New York, 1960.

Blanchard, J. R., and H. Ostvold, *Literature of Agricultural Research*, Univ. of California Press, Berkeley, 1958.

Bonner, J., and A. Galston, *Principles of Plant Physiology*, Freeman, San Francisco, 1952.

DeRobertis, E., W. Nowinski, and F. Saez, *Cell Biology*, 4th ed., Saunders, Philadelphia, 1965.

Ehrensvärd, G., *Life: Origin and Development*, Univ. of Chicago Press, Chicago, 1962.

Fernald, M., *see* Gray, A.

Florkin, M., *Unity and Diversity in Biochemistry*, Pergamon, New York, 1960.

Fogg, G. E., *The Metabolism of Algae*, 2nd ed., Wiley, New York, 1964.

Frere, M., *et al., The Behavior of Radioactive Fallout in Soils and Plants*, National Academy of Sciences, Washington, D.C., 1963.

Galston, A., *The Life of the Green Plant*, 2nd ed., Prentice-Hall, Englewood Cliffs, N.J., 1964.

Gray, A., *Manual of Botany*, 8th Centennial ed., rewritten and expanded by M. Fernald, American Book, New York, 1950.

Heftmann, E., ed., *Chromatography*, Reinhold, New York, 1961.

Lewin, R., ed., *Physiology and Biochemistry of Algae*, Academic, New York, 1962.

Machlis, L., and J. Torrey, *Plants in Action*, Freeman, San Francisco, 1956.

Meyer, B., D. Anderson, and R. Böhning, *Introduction to Plant Physiology*, Van Nostrand, Princeton, N.J., 1960.

Moment, G., ed., *Frontiers of Modern Biology*, Houghton Mifflin, Boston, 1962; includes D. Arnon, "Photosynthesis as an Energy Conversion Process."

Perlman, P., *General Laboratory Techniques*, Franklin, Englewood, N.J., 1964.

The Photochemical Apparatus: Its Structure and Function, Brookhaven Symposia in Biology, 11, Brookhaven National Laboratory, Upton, N.Y., 1959.

Steward, F. C., ed., *Plant Physiology: A Treatise*, Vol. IB, *Photosynthesis and Chemosynthesis*, Academic, New York, 1960.

Webster, G., *Nitrogen Metabolism in Plants*, Harper & Row, New York, 1959.

Wolken, J., *Euglena: An Experimental Organism for Biochemical and Biophysical Studies*, Rutgers Univ. Press, New Brunswick, N.J., 1961.

A sampling from *Scientific American:* "How Cells Make Molecules" (September 1961), "The Role of Light in Photosynthesis" (November 1960), "The Path of Carbon in Photosynthesis" (June 1962), "The Living Cell" (September 1961), "Radiation and the Cell" (September 1959), "Control of Growth in Plant Cells" (October 1963), "Effect of Ionizing Radiation on Plants" (in "Amateur Scientist" section, December 1963), "Flower Pigments" (June 1964), "Life and Light" (October 1959), "Autobiographies of Cells" (August 1963), "How Cells Transform Energy" (September 1961).

FILMS, FILMSTRIPS, AND SLIDES

This listing is intended to be only a guide for the selection of new films, filmstrips, and transparencies. Send for catalogs from distributors of films; catalogs of biological supply houses list extensive offerings of 2 × 2 in. slides. The addresses of distributors are given in Appendix B.

In general, the cost of color films averages $150; black and white films cost about $75. Newer films are made in varying lengths—time sequences of 1 min, 5 min, 11 min, or 30 min; the cost of films therefore varies. We have indicated the lengths of films so that lessons can be planned with the films in mind.

Sound films are available in color or black and white; rental rates are available from film libraries or by direct inquiries to producers. (Filmstrips are not available on a rental basis; the purchase price is usually about $6 to $7 for those in color.)

All Flesh Is Grass (28 min), Indiana Univ.
The Atomic Greenhouse (13 min), in *The Magic of the Atom* series, Handel; free loan from U.S. Atomic Energy Comm. Film Library.
The Atomic Zoo (13 min), in *The Magic of the Atom* series, Handel; free loan from U.S. Atomic Energy Comm. Film Library.
Carbon 14 (12 min), EBF.
Chlorophyll (28 min), in Part III of *AIBS Film Series in Modern Biology,* McGraw-Hill.
Energy Relations (28 min), in Part IX of *AIBS Film Series in Modern Biology,* McGraw-Hill.

How Green Plants Make and Us Coronet.
Leaves (28 min), in Part III of *AIBS* *in Modern Biology,* McGraw-Hill.
Making the Most of a Miracle (27 min), Na Plant Food Institute.
Paper Chromatography (14 min), BSCS techniques Thorne.
Patterns of Energy Transfer (28 min), in Part I of *AIBS Film Series in Modern Biology,* McGraw-Hill.
Photosynthesis (21 min), EBF.
Photosynthesis: Chemistry of Food-Making (14 min), Coronet.
The Riddle of Photosynthesis (13 min), in *The Magic of the Atom* series, Handel; free loan from U.S. Atomic Energy Comm. Film Library.
Titrating with Phenolphthalein (single-concept film, 8 mm, 3 min), EBF.
Vitamins in Fact and Fancy (29 min), Indiana Univ.
Water Movement in Soils (27 min), Washington State Univ.

LOW-COST MATERIALS

Leaflets describing methods of culturing algae (and higher plants) are available from biological supply houses: General Biological, Carolina, Ward's, and Welch (addresses are in Appendix C).

Among Organisms

Ingestion

The devices used by animals, as food-takers (or heterotrophs), to capture or to ensnare their food are an essential part of the study of life functions of plants and animals.

The means of ingestion by protozoa, by small fresh-water invertebrates (including insects), and by some amphibia may be studied directly. Organisms to be studied in class might be cultured in the laboratory and thus be available throughout the year. The cultivation and maintenance of these animals are described in Chapter 12. Where facilities favor support of a marine aquarium (p. 600), interesting studies can be pursued on how ingestion takes place among small echinoderms, squids and other mollusks, crayfish, and others. Many suggestions for student projects are inherent in the subject area of this chapter.

Protozoa

Paramecium. To a drop of a concentrated culture of *Paramecium* add a small drop of a thick suspension of *Chlorella*, a unicellular green alga. Watch the formation of green food vacuoles and especially the food mass in the oral groove. Or add a small pinch of carmine powder to a thick culture of *Paramecium* (or other ciliate). Students may prepare wet mounts of the mixture (pp. 95, 102) and watch how cilia in the oral groove create a current of water. They will note also the formation of

a food ball in the oral groove and the passage of the dark red food mass into the cytoplasm, where a food vacuole is formed. (You may substitute a drop of India ink for the carmine powder. In this case, the food vacuoles appear black.)

As a special project a student might study digestion within the food vacuoles using indicators. Stain a few drops of milk with a few grains of the indicator Congo red (indicators, p. 121), and then add this to an equal volume of a culture of *Paramecium*. Prepare wet mounts and examine the ingestion of red butterfat globules as food vacuoles. In a number of cases the food vacuoles become bluish as acid is secreted into the vacuoles, and finally change to red (alkaline).

A boiled suspension of Congo red and yeast may be preferred. Prepare this indicator-yeast suspension by adding a pinch of Congo red powder to a thick suspension of dry yeast that has been mixed with water. Bring to a gentle boil for 5 min. After the suspension is cooled, dip a toothpick into it and transfer this small quantity to a drop of pond water containing *Paramecium*. Congo red is blue at a pH of 3, and orange-red at about pH 5.

Or add a drop of *Paramecium* to a drop of dilute neutral red (0.02 percent) on a slide and seal the coverslip with petroleum jelly for a prolonged examination. On the acid slide (pH 6 and below), neutral red is deep red, at pH 7 the color is rose, at pH 8 the indicator turns orange, and then yellow at pH 9.

Kudo[1] reports on the use of indicators to render food vacuoles visible and also to test the pH of the food vacuole.

Dissolve any one of the following indicators in 100 ml of distilled water: 50 mg of neutral red, 100 mg of phenol red, 150 mg of litmus, or 50 mg of Congo red. To 1 drop of the culture on a slide add 1 drop of the indicator solution.

An interesting but unsolved problem may be examined by some students as a project. Do protozoa continue to take in carmine particles which cannot be digested, or do they "learn" to select their food types? One method of attack might be to count the number of food vacuoles formed during measured time intervals. Students may think of many more approaches. We have found this a good project.

Pelomyxa. Prepare several slides of thick cultures (see Chapter 12) of *Stentor*, *Paramecium*, *Blepharisma*, or *Chilomonas* (Fig. 4-1). To each of these slides add some *Pelomyxa* (sometimes called *Chaos chaos*), a large amoeba (Fig. 4-2). These may be purchased from a biological supply house (Appendix C). Watch to see how a food cup is formed as the amoeba's pseudopodia close around the prey. Food vacuoles form, of different colors depending upon the food source (red with *Blepharisma*, blue-green with *Stentor*). At times you may see the captured prey moving within the vacuole. Circle the coverslips with petroleum jelly to prevent evaporation of the water. Students may study the change in the size of the food vacuoles of the carnivorous amoebae over a period of several hours.

Blepharisma. Some *Blepharisma* are cannibalistic and contain deep red food vacuoles of ingested, smaller *Blepharisma*. Mount several drops of *Blepharisma* culture for microscopic examination.

Trumpet-shaped *Stentor* (Fig. 4-1c), mounted in a rich culture of *Blepharisma*, will soon show pink food vacuoles within the blue-green body.

[1] R. Kudo, *Protozoology,* 4th ed., Thomas, Springfield, Ill., 1954, pp. 103–12.

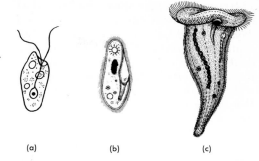

Fig. 4-1 Three common protozoa: (a) *Chilomonas;* (b) *Paramecium;* (c) *Stentor.* (a, from G. G. Simpson and W. S. Beck, *Life: An Introduction to Biology,* 2nd ed., Harcourt, Brace & World, Inc., 1965; b, c, courtesy of General Biological Supply House, Inc., Chicago.)

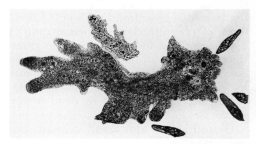

Fig. 4-2 A large amoeba: *Pelomyxa,* sometimes called *Chaos chaos.* (Contrast size with size of slipper-shaped paramecia in photo.) (Courtesy of Carolina Biological Supply Co.)

Other microscopic invertebrates

For maintaining these invertebrates in the laboratory, see Chapter 12.

The microscopic annelid worm *Dero* or *Nais* (Fig. 4-3) may be mounted in a thick suspension of *Paramecium* for examination under the microscope. Watch how it ingests particles of food.

You may also want students to place some planarians (unfed for several days) in Syracuse dishes along with such annelid worms as *Dero* or *Aulophorus* or brine shrimp (*Artemia*). The action of the proboscis of planarians during ingestion may be observed with a binocular microscope or hand lens (see also p. 91).

The action of tentacles and stinging organs such as those in *Hydra* (Figs. 1-1,

12-9) makes interesting study. Select hydras which have not been fed for several days, and add them to a Syracuse dish containing *Artemia, Daphnia,* or *Dero.* Students may be able to see the hydras grasp the food organisms and paralyze them with the nematocysts on the tentacles. Watch the movement of the food toward the hypostome. The gastrovascular cavity becomes distended, and the captured organisms may be seen through the thin walls of this cavity. Under a binocular microscope or small dissecting lens you may see several organisms ingested in rapid succession. (For hatching brine shrimp eggs, see p. 354.)

Students can show the role of glutathione in the feeding response in hydras that was demonstrated by Loomis.[2] Body fluids of prey attacked by nematocysts of hydras are said to contain reduced glutathione, which stimulates a feeding response in the hydras. According to Loomis, hydras reject dead food, for reduced glutathione becomes oxidized in dead organisms.

Glutathione is also found in yeast; students may add dry yeast to some hydras in a watch glass and observe the feeding response. Or they may extract reduced glutathione from yeast cells by grinding dry yeast with a little water, and then centrifuging and decanting the liquid portion which contains the glutathione. Then dip a bit of dead food or a minute bit of filter paper into the chemical, and place this next to some hydras in a watch glass or depression slide; examine the ingestion reflex under the microscope.

Insects

Mealworms (*Tenebrio molitor*) may be raised in the laboratory (see p. 603). You may want to show the insect's mandibular mouth parts; use a hand lens.

In the fall, students may collect egg cases of praying mantis (Fig. 4-4), which

[2] W. F. Loomis, "Glutathione Control of the Specific Feeding Reactions of Hydra," *Ann. N.Y. Acad. Sci.* 62:209–28, 1955.

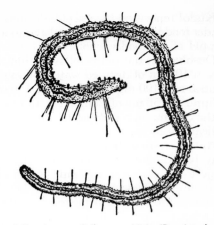

Fig. 4-3 An annelid worm: *Nais.* (Reprinted with permission of The Macmillan Co., New York, from *Invertebrate Zoology* by R. W. Hegner. Copyright 1933 by The Macmillan Co., renewed 1961 by Jane Z. Hegner; after Leunis.)

Fig. 4-4 Egg mass of praying mantis. (Photo by Hugh Spencer.)

appear as a frothy mass of tan material attached to twigs of trees and shrubs. Or egg masses may be purchased from supply houses or botanical gardens. They should be placed in a vivarium or large jar when they are received. When the mass is kept at room temperature, hundreds of nymphs hatch and are available for study. Place several nymphs in small vials and watch the action of their forelegs and mandibles.

In contrast, the sucking mouth parts and the extensible and flexible proboscis of adult fruit flies may be observed by placing adults in slender vials of media (preparation, p. 415). Study the action of labrum, labium, and hypopharynx which terminate in a pair of soft, pad-like, labellar lobes. On the other hand, the mouth parts of the larvae show only strong mouth hooks; the usual mouth parts are missing.

Some vertebrates

The rapid, lashing tongue movements of frogs, toads, and salamanders may be demonstrated by placing one of these animals in a small container. Then empty a bottle of fruit flies into the container. (Earthworms may be substituted as food.) Similar demonstrations may be prepared using chameleons, which feed upon fruit flies (Fig. 4-5).

In some classes, this may be the time to introduce broad notions of modes of nutrition among consumers, or heterotrophs. Refer to pp. 533–39 for demonstrations of holozoic nutrition, saprophytism, parasitism, and mutualism. Students will probably also ask about insectivorous plants (Figs. 11-20, 11-21).

Digestion

The food tube

In microscopic many-celled organisms the path of food through the digestive tract may be observed, as well as ingestion. By using indicators (p. 662) it is possible to

Fig. 4-5 Chameleon, with tongue extended for feeding. (Courtesy of the American Museum of Natural History, New York.)

see changes in color as food moves along the food tube. Prepare wet mounts of rotifers, *Tubifex, Dero, Nais,* or the crustacean *Daphnia.* (Culture methods are described in Chapter 12.)

When cells of *Chlorella* are added to a culture of *Daphnia* or rotifers, the digestive tract becomes conspicuously green. Or you may want students to use carmine powder suspended in pond water, or to use indicator dyes (such as those described earlier in this chapter for *Paramecium,* for example, Congo red boiled with yeast cells). Add a drop of the Congo red-yeast suspension to a drop of thick culture of rotifers, *Cyclops, Daphnia,* or some other microorganism on a slide. Seal with petroleum jelly and examine for the next 15 to 30 min. Watch the change in the food tube as the color changes from red to blue (blue at pH 3.0, and orange-red at pH 5.0).

A more graded series of changes in acidity may be traced with a very dilute solution of neutral red (0.02 percent). Add a drop of a thick suspension of microorganisms to a drop of the indicator and seal with petroleum jelly. At pH 9 the indicator is yellow; at pH 8 orange; rose at pH 7; deeper red below pH 6.

What change in color is visible as the food mass moves through the pharynx, stomach, and intestine of the microscopic multicellular organisms?

Also demonstrate the path of food along the digestive tract of frogs. First, pith to destroy the brain and the spinal cord (Fig. 7-9). Then pin down the frog on a dissection board and remove the skin and muscular layer to expose the digestive tube (dissection, Chapter 1). Keep the organs bathed in Ringer's solution (p. 665). Note the slow, rhythmic contractions along the length of the intestine and rectum. These contractions will be more apparent if the frog has been fed recently.

You may also want to demonstrate the ciliary action along the sides of the jaw and roof of the mouth of the frog. Use the same pithed frog; dissect away the lower jaw and the floor of the mouth. Then remove the viscera, leaving as much of the esophagus as possible. Sprinkle fine particles of cork or filings of lead or iron on the anterior part of the roof of the mouth. Watch how the particles are carried along the roof and through the esophagus. This is accomplished by ciliary motion. The particles reappear in the cut part of the esophagus and move into the coelom, since the stomach has been cut away. (Keep the tissue moist throughout.)

You may also prepare exceedingly interesting demonstrations by mounting bits of ciliated epithelial tissue from the roof of the mouth or the sides of the frog's jaw in Ringer's solution (Fig. 2-18). Examine under high power and look for the rhythmic ciliary motion. You may, if you wish, slow down the motion with ice. Seal the coverslip with petroleum jelly if you plan to study the motion over some time. And after a study of cilia you may try a vital dye to bring out details (pp. 323, 121).

A comparison of the adaptive devices among several organisms—small invertebrates as well as vertebrates—may be made through dissection followed by a careful study of anatomy (Chapter 1).

The need for digestion: diffusion through a membrane

We have studied ingestion among several heterotrophs, and we have compared their digestive systems. What happens to the food they consume? What is digestion and why is it necessary? Food materials must be made soluble; that is, they must be in a form which can pass through membranes. The process of diffusion may be shown in many ways. Probably all of us have used one or another of these methods to show diffusion of molecules from place to place.

You may want to drop a small piece of copper sulfate or several potassium permanganate crystals into water and leave them undisturbed. To do so, insert a few dry crystals into a section of glass tubing. Hold a finger over the top of the tubing as you insert the other end into water. When the tubing is at the bottom of the container of water, release the finger; the crystals will fall to the bottom. Each day students will find that molecules of copper sulfate (or other substances used) have moved from their place of greatest concentration and are becoming distributed throughout the water solvent.

Another striking demonstration of diffusion is the movement of a gas such as ammonia in another gas (air). Wet a circle of filter paper with phenolphthalein and insert it into the bottom of a large test tube. Now invert the test tube over a bottle of ammonium hydroxide. Students should be ready to explain the rapid change of the filter paper to red. The color change of the alkaline ammonia on the indicator phenolphthalein may be shown in a test tube. (Use controls with filter paper soaked in water.) Diffusion of gases may be shown in the spreading of illuminating gas, or of the volatile materials in perfume, around a room. Molecules move generally from their point of greatest concentration to a region of lowest concentration. This may also be demonstrated by carefully pouring red ink or a dye along one side of a container into water. Here it is demonstrated that a liquid diffuses through a liquid.

Now we may also show that some molecules diffuse through a membrane. This makes an impressive class demonstration. To a large test tube or cylinder of water add a few drops of a 1 percent solution of phenolphthalein in alcohol (preparation, p. 664). Cover the mouth of the tube with a wet goldbeater's membrane or wet cellophane (do not use the kind that covers cigarette packs, since an additional protective agent is added) or use dialyzing tubing. Fasten with a rubber band and invert over a bottle of ammonium hydroxide (Fig. 4-6). A pink stream will move rapidly up through the test tube. Since phenolphthalein turns deep red in an alkaline medium (ammonia in this case), a stream of molecules in motion is visible.

Diffusion of iodine through a membrane. Pour dilute Lugol's iodine solution

phenolphthalein in water

NH₄OH

Fig. 4-6 Apparatus to demonstrate diffusion through a membrane. The water in the test tube contains phenolphthalein and turns red as molecules of ammonia evaporate and diffuse up through the membrane.

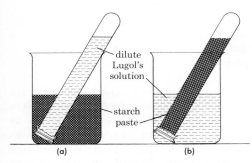

dilute Lugol's solution

starch paste

(a)   (b)

Fig. 4-7 Apparatus to demonstrate diffusion through a semipermeable membrane. Molecules of Lugol's iodine solution pass through the membrane, while starch does not. Thus, the characteristic blue-black color appears in the starch solution and not in the Lugol's solution: in the beaker in (a) and in the tube in (b).

(p. 664) into a large test tube, cover with a wet goldbeater's membrane, and secure with rubber bands. Invert the test tube into a beaker containing a 1 percent starch paste (Fig. 4-7).

Then prepare another test tube, but this time pour starch paste into the test tube and fasten the wet membrane over the mouth of the tube. Invert this test tube into a beaker half full of dilute Lugol's iodine solution. Note that molecules of the iodine solution will pass through a membrane in either direction, whereas molecules of starch will not pass through at all. Starch is insoluble; it needs to be digested to diffuse through membranes.

Diffusion of glucose through a membrane. Fill large test tubes with glucose solution, molasses (dilute), corn syrup, or honey. Into other test tubes pour starch paste (1 percent). Cover each test tube with a wet goldbeater's membrane. Fasten with a rubber band and invert each tube into an individual beaker containing a small volume of water. Within 15 min test for starch by adding dilute Lugol's iodine solution to the water in which the inverted tubes of starch paste stand. Then test the water in the other beakers for glucose with Benedict's solution (pp. 218, 662).

It may be desirable at this time to set up a test tube containing starch paste to which saliva has been added. Demonstrate with dilute Lugol's iodine solution that starch has not diffused, but that there is now some simple sugar which has passed through the membrane (test with Benedict's solution). Where did the simple sugar come from? Of course, both the starch paste and the saliva should be tested for simple sugar before saliva is added and the test tube prepared. Also refer to p. 218; the end product is maltose.

Osmosis and osmotic pressure. Fill the bulb of a thistle tube with heavy molasses while holding the finger over the tube opening. (Instead of covering the tube opening with a finger, some teachers connect a short piece of rubber tubing with a clamp.) Next cover the bulb with a wet semipermeable membrane (goldbeater's) or cellophane tubing and immerse in a beaker of water (Fig. 4-8). Soon the level of the liquid inside the thistle tube will rise. This special diffusion of molecules of water through the membrane is called

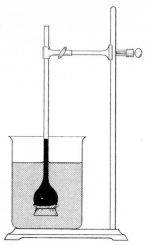

Fig. 4-8 Apparatus to demonstrate diffusion through a membrane. Water in the beaker diffuses rapidly through the membrane, causing the level of molasses in the thistle tube to rise.

osmosis.[3] As the water level rises in the tube, its pressure (hydrostatic pressure) will eventually halt the further upward diffusion of water. Some teachers connect extensions of glass tubing with rubber tubing to the stem of the thistle tube. You may get a considerable rise of fluid in the tube due to the diffusion of water molecules into the thistle tube. Also refer to demonstrations with a raw potato and with a raw egg (Figs. 5-15, 5-16).

Diffusion through the intestine of a frog. Dissect out the stomach and intestine of a freshly killed frog (dissection, p. 60). Clean out its contents by rinsing in Ringer's solution. Be certain the intestine is not perforated. Tie one end of the intestinal tube with thread; into the other end pour a solution of molasses diluted with Ringer's solution. Then tie this end off with thread and wash off the intestinal tract to remove excess molasses. Now suspend the intestinal tract in a beaker of water, as shown in Fig. 4-9, by tying the ends to a wood splint or tongue depressor. Within a half-hour, test the water in the beaker with

[3] *Osmosis* should not be confused with *diffusion*. Osmosis is a special case of diffusion, where pure water passes through a membrane permeable only to the water.

Benedict's solution for glucose. For rapid, positive results use a small quantity of water in the beaker.

Intestinal casing purchased from a butcher or slaughterhouse may be used instead of the frog's intestine. Tubular lengths of semipermeable cellophane (dialyzing tubing) are available from some supply houses for the same purpose. Absorption of other soluble food products through the intestine is demonstrated using intestinal casing or tubular cellophane on p. 207.

Diffusion through a collodion membrane. Use large, clean, thoroughly dry test tubes. Into each test tube pour about 5 ml of collodion from a stock bottle. Rotate the test tube so that the collodion completely lines the test tube; then pour the collodion back into the stock bottle. Allow the film of collodion to harden slightly. Then loosen one edge of the film along the mouth of the tube with your fingers. Along this edge let water run down between the glass and the film of collodion so that more of the film is loosened (Fig. 4-10). Rotate the tube. By careful manipulation you can remove the collodion film intact. Test the bag of collodion for leakage by filling it with water. Then let it soak in water for a short time. Repeat the procedure until several collodion membranes are made.

You may fill these cylindrical bags with starch paste (as a control), glucose, or a 50 percent solution of corn syrup or molasses. Close the tops securely by twisting and tying with thread. Wash off the overflow materials and immerse the bags in small beakers of water as shown in Fig. 4-11. After 10 min or so test the fluid in the beakers at intervals for molecules of glucose diffused into the water.

Chemical digestion in plants and animals

The independent green (holophytic, or autotrophic) plant is the ultimate source of food, because, generally speaking and especially for our purposes, the green

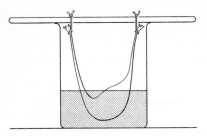

Fig. 4-9 Apparatus to demonstrate diffusion through the intestine of a frog. If the intestine of a frog, other intestinal casing, or semipermeable cellophane tubing is filled with molasses and suspended in water, molasses diffuses into the water and water diffuses through the membrane into the molasses.

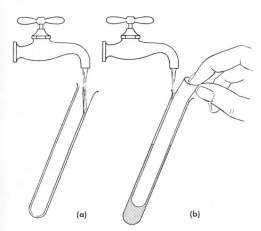

(a)   (b)

Fig. 4-10 Making a collodion bag: (a) pouring water between the top of the collodion film and the test tube; (b) removing the film from the tube.

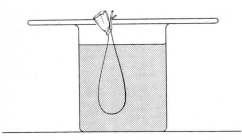

Fig. 4-11 Apparatus to demonstrate diffusion through a collodion membrane. A collodion bag is filled with starch paste, glucose, or solutions of corn syrup or molasses and is suspended in water.

plant is the major converter of light energy. Green plants make carbohydrates, fats, and proteins and use minerals in building foodstuffs. While a few animals (and some fungi and bacteria) can synthesize vitamins, most vitamins are derived from green plants.

Food nutrients must, in turn, be broken down chemically (digested) within the bodies of the plants and animals that consume them. (The need for digestion has been established earlier; see p. 204.) The nutrients are then built into protoplasm in the pattern of the individual organism.

Digestion must take place within the green plant as well as in other organisms, for stored foods must be digested before they can be distributed and utilized within a plant. Digestion may be studied in seeds, which contain stored food materials.

Controlled experiments are described for investigating the role of enzymes and their specificity in the digestion of starch, proteins, and fats.

Digestion in plants

Hydrolysis of starch in plants. Demonstrate the hydrolysis of storage starch by an amylase active in germinating seedlings. Soak some corn grains and plan three groups for testing: (1) dry corn grains; (2) grains soaked for 2 days; and (3) corn grains that have been germinating for a week or 10 days. Grind the grains (include the shoots in those for the third group) in separate mortars with fine sand in 10 ml of water. Filter each set and test the filtrates for simple sugar with Benedict's solution (p. 660). Under what conditions is the amylase active? What is the advantage to the growing plants of having a supply of soluble glucose rather than starch?

EXTRACTING DIASTASE FROM SEEDLINGS. Students may want to extract this starch-splitting enzyme as a project. Thoroughly grind up 25 germinating barley seeds with about three times as much water. Filter the preparation after it has been standing for about half an hour.

In the meantime, prepare a thin starch paste (0.1 percent) by dissolving 0.1 g of arrowroot starch in a little cold water, and adding this paste to boiling water; use 100 ml of water in all. (Stir to avoid lumps.) If it is easier to weigh out 1 g of starch, prepare a 1 percent solution with 1 g of starch and 100 ml of water, as above; then dilute with water an appropriate quantity of this solution to 0.1 percent.

Add 5 ml of barley extract to an equal amount of cooled starch paste. Let this stand in a warm place. Heat some barley extract to boiling, cool, and add to another test tube containing starch paste. This serves as a control (boiling destroys the activity of the enzyme). At intervals, test samples from each tube for starch with Lugol's solution. Starch disappears gradually as the diastase, extracted from the germinating barley seeds, completes the digestion of arrowroot starch. The test tube containing starch paste and boiled barley extract does not show this reaction; starch paste remains in the test tube.

As a project, a student or small group might explore this problem: Is there a difference in the quantity of the enzyme present during the first few days of germination and during the second week of germination, or is the same quantity produced at all times? Students might design an experiment such as this: Test dried seeds, then test seeds after 1 day of germination, 2 days, and so forth. They may use the procedure already described for extracting diastase, as long as they use uniform quantities of water and seeds.

You may prefer to vary the demonstration of the enzymatic hydrolysis of starch by germinating seedlings in the following way, suggested by the BSCS.[4] Into several sterile Petri dishes pour a thin layer of starch agar (5 g of agar plus 7.5 g of arrowroot starch brought to a quick boil, with constant stirring, in 500 ml of water).

After cooling, scatter on the surface of the agar some radish or mustard seeds that

[4] *Teachers' Guide* to *High School Biology: Blue Version— Laboratory, Part I*, BSCS, Univ. of Colorado, Boulder, 1961 (experimental use, 1961–62), p. 65.

have been soaked overnight. Let the seeds germinate for 2 to 3 days. Then remove the seeds and pour a thin film of Lugol's iodine solution over the surface; tilt the dishes to spread the iodine evenly. What is the explanation of the clearer areas that do not stain blue-black with iodine? How could you test whether simple sugar is present in these clearer areas?

USING COMMERCIAL DIASTASE. The concept of digestion as the breaking down of a substance by an enzyme may be demonstrated with commercial powdered diastase. Students may perform the experiment individually, or you may demonstrate it to the class as a whole.

Prepare a 0.1 percent solution of arrowroot starch as described in the preceding activity. While the starch paste is cooling, prepare a 0.1 to 0.2 percent diastase solution in water. (The enzyme is available from supply houses, and it is always ready to use; simply dissolve a small quantity in a large amount of water.)

Each student needs a clean glass slide or a small, square piece of glass. A student laboratory assistant in each row or group adds drops of different fluids to each student's slide in this order: (1) 2 separate drops of dilute starch paste; (2) 1 drop of very dilute Lugol's solution (light brown in color) to each drop of starch paste so that both drops become very light blue in color; and (3) 1 drop of the diastase solution to only 1 drop of the combined starch and Lugol's solution. The other drop on the slide acts as a control throughout this demonstration.

Within a few minutes students should notice how the blue color of one drop disappears. Why? Why does the other drop remain blue? In this experience students gain a notion of how rapidly an enzyme acts in splitting starch paste.

As a demonstration for the class as a whole, using two *large* test tubes, prepare the same dilutions of the starch paste, diastase, and Lugol's solution as for the individual student work. Turn starch paste in each of the two test tubes light blue with dilute Lugol's solution. Then add to one tube a volume of diastase solution equal to one-third the volume of solution already in the tube. The blue color should disappear within 1 to 15 min, depending on the quantities used and the temperature. Why doesn't the control test tube lose its blue color?

Some teachers prefer to add the enzyme to the starch paste, wait a short time, and then add iodine solution. This fails to produce a blue color, since starch is no longer present in the solution. This method is described on p. 211 in a demonstration in which students test their own saliva.

There are many extensions of this activity which can be used by students as practice in designing an experiment. For example: (1) How could you show that this same change in starch goes on within the body (use saliva instead of diastase)? (2) How could you find out what the starch has been changed into? (3) How dilute a solution of the enzyme can digest the starch paste? (4) Is digestion independent of temperature, or would digestion go on faster at body temperature? (5) Does it make any difference whether large particles or finely chopped or chewed particles of food material are acted upon by an enzyme?

Students may use similar experimental designs using pancreatin (pp. 214–16).

INVERTASE IN YEAST. Add 1 ml of a thick yeast suspension to 5 ml of sucrose (1 percent). As a control, test a sample of the sucrose solution with Benedict's solution or with a Clinitest tablet to show a negative test for the presence of glucose. Allow the yeast-sucrose suspension to remain in a warm place, or incubate at 37° C (99° F), for 1 to 2 days. Test the suspension for glucose; also test a sample of sucrose solution, and of yeast suspension.

TESTING THE EFFECT OF DILUTION ON ENZYME ACTIVITY. You may want students to demonstrate how powerful the action of an enzyme is. This might be a small project. Use the enzyme diastase which was prepared earlier (p. 208). Have the students dilute a given volume of diastase in solution with an equal volume of water so

that it is diluted in half; to a given quantity of this dilution have them add an equal volume of water so that the diastase is diluted to one-fourth its original strength. They may continue to dilute it many times. Is a negative starch test obtained (starch disappeared) when their most dilute diastase solution is added?

For another project, students may design a plan for an experiment to discover whether diastase acts as readily on starch paste at 4° C (40° F) or 38° C (100° F) as it does at 18° C (65° F, approximately room temperature). (See also p. 211.) They may contrive many questions which can be answered by carrying through an activity in the laboratory or at home.

Proteins. Fresh pineapple juice contains the protein-splitting enzyme bromelin. (Another proteolytic enzyme, papain, may be purchased as a product derived from papaya.[5]) Students may make a mash of some slices of a fresh pineapple and collect the juice. Add a given quantity to two test tubes containing some chopped hard-boiled egg white. As a control, boil a similar volume of the pineapple juice and add this boiled juice to chopped egg white in two other test tubes. Add 2 to 4 drops of toluene to each test tube to inhibit spoilage. Notice the ragged, partially digested bits of egg white in the test tubes containing fresh pineapple juice.

Digestion in animals

Enzymes. According to the abilities of their students, teachers may want to introduce the biochemistry of digestion—the enzyme reactions in the hydrolysis of starch into simple sugars. You probably used chemical equations in developing concepts of photosynthesis (Chapter 3). The role of the respiratory enzymes in glycolysis, the Krebs cycle, and the cycles of ATP formation are taught in oxidation (Chapter 6).

We may demonstrate in several ways that starch is changed into reducing sugar through the action of saliva.

[5] This product is sold as a meat-tenderizing agent.

DIGESTION OF STARCH BY SALIVA IN A TEST TUBE. Collect saliva from a student into a large test tube (filter if necessary). Salivation can be increased by chewing soft paraffin or clean rubber bands.

Prepare two or three test tubes each containing 5 ml of a 0.1 to 1 percent starch paste suspension and 5 drops of saliva. Roll the test tubes between the hands to ensure thorough mixing. You may want to use a water bath with a thermometer for precise work (Fig. 4-12). At timed intervals, remove samples of the saliva and starch paste mixture and test with Lugol's solution for the presence of starch. Changes will occur gradually, until the blue-black reaction does not take place. At one point a reddish reaction may occur as the intermediate dextrins form. Finally, the color remains the light brown tint of the dilute iodine. Now the starch has disappeared. What has happened to it?

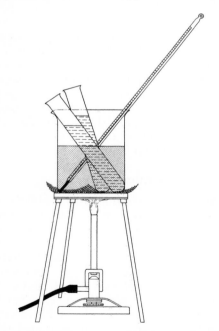

Fig. 4-12 Apparatus to demonstrate digestion of starch by saliva. Test tubes of starch paste and saliva are heated in a water bath at a constant temperature of 37° C (98° F). A hot plate may be substituted for the Bunsen burner—especially when younger children are using the apparatus.

TABLE 4-1
Blackboard diagram for salivary digestion demonstration

	result	interpretation
Starch paste		
plus Lugol's ———→		
plus Benedict's ———→		
Saliva		
plus Benedict's ———→		
Starch paste and saliva		
plus Benedict's ———→		

At this point, test the mixture for a reducing sugar with Benedict's solution (or Fehling's reagent). A change from blue to green to reddish-orange gives evidence of the presence of varying amounts of reducing sugar (maltose or glucose).

Of course, students will want to demonstrate that neither the original starch paste nor the saliva contained glucose (or maltose). Test both the starch paste and a sample of the saliva with Benedict's solution. These two tubes serve as controls. Other useful controls are two or three similar test tubes containing starch plus saliva.

Students often use the form shown in Table 4-1 to put on the blackboard as a guide in recording their observations.

The activity of the enzyme ptyalin can be plotted against time by measuring the amount of light transmitted by a test tube containing starch and dilute Lugol's (very dilute, as an indicator). Stong[6] gives a fine description and illustrations for making a photometer using a projector for a source of light and a silicon solar cell. Or students may make the photometer described on p. 445.

Also test the effect of pH on the activity of ptyalin, and test the effect of temperature. Both of these factors are critical for the activity of specific enzymes.

What is the effect of an inhibitor such as 0.1 percent mercuric chloride on the rate of hydrolysis of starch by ptyalin?

[6] C. L. Stong, "The Amateur Scientist," *Sci. Am.,* January 1963, pp. 147–54.

DIGESTION OF A CRACKER IN THE MOUTH. This procedure may be used as a demonstration or an individual laboratory exercise. Collect saliva from a student into a clean test tube. Then ask the student to chew slowly on a piece of soda cracker (which was tested previously to ascertain that reducing sugar was absent, or present in only small amounts). Use a cracker that is all white; tan or brown patches may be indications of dextrins.

Have the student place this chewed mixture into a clean test tube. While this part of the experiment is under way, you may want a student to test the saliva for reducing sugar with Benedict's solution, for the student may have eaten sweets previous to this activity. Also test a whole cracker for the presence of starch (as another control).

Clinitest tablets may be used in testing the end product of salivary digestion. Clinistix strips may be used to distinguish glucose from maltose.

Within 5 min the chewed cracker may be tested with Benedict's solution; a color change from green to orange will be observed. Digestion goes on more rapidly here than in the preceding demonstration, since the process goes on at body temperature.

You may want to divide the chewed cracker mixture originally obtained from the student so that half could be tested for reducing sugar and the other half for starch (should a student ask if all the starch had been changed to simple sugar). In most cases some starch remains. The quantity varies inversely with reaction time.

STUDENTS TEST THEIR OWN SALIVARY DIGESTION. In this *laboratory* exercise students work alone or in pairs, depending upon available materials. Here we shall assume that students work in pairs. A small vial of dilute starch paste and a drop bottle of dilute Lugol's solution (with a pipette stopper) are given to each pair of students. Then *each* student gets a pipette, a microscope glass slide, and an empty vial. Each student now takes 10 drops of

starch paste from the stock vial and places it into his own empty vial. Then each student puts 1 drop of the starch paste on the extreme left end of his microscope slide. To this he adds 1 drop of iodine (Lugol's) to check the starch content. Now each student adds his own saliva to the 10 drops of starch paste in his vial. Students warm the vial by rolling it between the palms of the hands. At 2-min intervals the student puts successive drops along the length of the glass slide. Each is tested with iodine solution.

Gradually, the blue-black color fails to appear as successive drops of the starch-saliva suspension are tested. The reddish color of dextrins may appear, and finally there is no color change. (Starch and dextrin intermediates do not generally appear.) A few tests may show no change in the starch reaction. (A very small percentage of the population seems to lack ptyalin, the salivary amylase. This is believed to be an inherited trait.)

One or two students should test the remaining saliva-starch mixture of the vials for reducing sugar. As controls, saliva as well as a sample from the original starch paste stock should be tested for reducing sugar. The main advantage in this procedure is that each student has the opportunity to test his own saliva, and each set of "experiments" acts as a control for the others.

EFFECT OF TEMPERATURE ON ENZYME ACTIVITY. You may want to demonstrate the action of salivary amylase under different conditions. A committee of students may undertake this activity as a small project. Prepare six test tubes, each with about 5 ml of a 0.5 percent arrowroot starch suspension prepared in 0.25 percent NaCl solution. Leave two tubes at room temperature. Chill two tubes in a beaker of water to about 5° C (41° F); place two more in a water bath at 40° C (104° F).

Add an equal quantity of saliva to one of the test tubes at each of the three different temperatures. To each of the other matched tubes add an equal quantity of saliva which has first been boiled for 10 min. Test all the tubes for the presence of reducing sugar.

If you prefer to use sucrose, it is possible to show the effect of invertase on the hydrolysis of sucrose to simple reducing sugars. Add Benedict's solution (p. 660).

EFFECT OF DILUTION OF ENZYME ON DIGESTION. Students may want to demonstrate the minute quantity of enzyme required to produce the hydrolysis of starch.

Dilute filtered saliva in the following manner. Into each of four test tubes pour 9 ml of water. To test tube 1, add 1 ml of saliva; shake the tube thoroughly. With a clean pipette transfer 1 ml of this diluted solution from tube 1 into tube 2. (Tube 1 is now a 1:10 dilution; tube 2 is a 1:100 dilution.) With a clean pipette transfer 1 ml of fluid from the tube 2 into tube 3; continue until tube 4 will have a dilution of 1:10,000. Of course, further dilutions may be made as time and materials permit. To each of the tubes of saliva of different dilutions, add 1 ml of a dilute starch suspension composed of 0.5 percent starch prepared in 0.25 percent sodium chloride solution. Incubate all the tubes in a water bath maintained at 38° C (100° F) or in an incubator.

Test samples from each test tube after 1, 5, 10, and 15 min for the amount of starch present, using Lugol's iodine solution (p. 664), and/or test for the presence of simple sugars using Benedict's solution (p. 660). Be sure to test a control (dilute saliva), as well as the starch suspension, for the presence of simple sugars. Repeat in 30 min. Note that time is a factor in the hydrolysis of starch. Given more time, more digestion of starch will occur. Therefore limit the period of testing to 15 or 30 min. Record the time it takes for digestion to occur. Note the quantity of sugar produced in a given period of time. With the same technique, students may test the effect of pH on reaction time of this hydrolysis.

SPECIFICITY OF ENZYMES. At this point, a small committee or a student interested in an individual project may want to extend the work on enzymes (see procedure above). In so doing they will have the opportunity to learn how to design simple experiments with controls. Students may plan to investigate the effect of the amylase in saliva on starch, sugar, fat, and protein, as well as on foods containing many nutrients (for example, milk). Similarly, experiments may be devised to discover whether pepsin reacts with starch, sugar, fat, and protein.

Students can develop the following simple plan into a small project. Does saliva act only on starch? Dilute about 24 ml of saliva with an equal amount of water. Prepare eight test tubes so that each contains about 6 ml of the diluted saliva. Suppose we set up two test tubes for each kind of nutrient so that one may act as a check for the other. To the first set of two test tubes add 10 ml of 1 percent starch paste. To the second set of two tubes add 10 ml of a freshly prepared 3 percent cane sugar solution. To the third set of tubes add 2 drops of olive oil with 10 ml of 1 percent sodium carbonate. To the last pair add a pinch of casein (or chopped hard-boiled egg white) with 10 ml of 0.2 percent hydrochloric acid. Now put all the test tubes in an incubator or in a 40° C (104° F) water bath for half an hour. Students may now test the starch paste and cane sugar solutions for reducing sugar with Benedict's solution. Then they may want to test the casein-containing tubes for peptones (with the Biuret test, p. 213).

Tests for proteins

A GENERAL TEST FOR A PROTEIN. Hawk, Oser, and Summerson[7] describe a technique for testing the component elements of proteins. These are carbon, nitrogen, hydrogen, and sulfur. In a dry test tube, heat some powdered egg albumin (see preparation below). Suspend a strip of wet red litmus paper in the tube. Across the mouth of the test tube place a strip of filter paper which has been moistened in lead acetate solution. If hydrogen and nitrogen

[7] P. Hawk, B. Oser, and W. Summerson, *Practical Physiological Chemistry*, 13th ed., Blakiston (McGraw-Hill), New York, 1954, p. 168.

are present, ammonia fumes are produced, and the litmus will be turned blue. The presence of sulfur is indicated by the blackening of the lead acetate paper. If the powder chars you know that carbon is present. If the powder and the test tube are dry at the beginning of the test, you can see drops of moisture condense on the sides of the test tube. This indicates the presence of hydrogen and oxygen (as water).

PREPARATION OF ALBUMIN. Hawk, Oser, and Summerson also suggest this preparation of albumin. If you need albumin solution for immediate use, beat the white of an egg with about 8 volumes of water. Strain through cheesecloth, and filter if necessary.

To prepare powdered egg albumin, beat the egg white with about 4 volumes of water, and filter. Then evaporate the water from the filtrate on a water bath at about 50° C (122° F). Powder the residue in a mortar.

SPECIFIC TESTS FOR PROTEINS

1. XANTHOPROTEIC TEST. In demonstrating this test, carefully add 1 ml of concentrated nitric acid to a few bits of hard-boiled egg white or to about 3 ml of an egg albumin solution. Gently heat the test tube over a Bunsen flame and note the yellow color (an indication of protein) resulting. When the solution has cooled, add an excess of sodium hydroxide or ammonium hydroxide to the test tube. Watch the yellow color deepen into orange.

2. BIURET TEST FOR PROTEOSES AND PEPTONES. Peptones and proteoses give a pink color reaction in this test. In a test tube place fresh egg white. To this add an equal volume of 10 percent sodium hydroxide solution and shake to mix. Then carefully add a 0.5 percent copper sulfate solution (about 2 to 5 drops of copper sulfate solution in a test tube of water), drop by drop, into the test tube. (Or use prepared Biuret reagent, p. 136.) Shake the tubes as each drop is added and watch for a violet or pink-violet color.

3. MILLON'S REACTION. This is especially useful in testing solid proteins. Dilute Millon's reagent (p. 136) with 2 or 3 volumes of distilled water. To chopped egg white or egg albumin solution add a few drops of Millon's reagent; then heat slowly. Notice the protein precipitates, which turn red. In this test the color reaction is due to the hydroxyphenyl group in the protein molecule. Any phenolic compound such as tyrosine, thymol, or phenol will also give this reaction.[8]

Gastric digestion

DIGESTION OF PROTEINS BY PEPSIN. Take a hard-boiled egg and separate the albumin, or "white," which is almost pure protein, from the yolk. Chop the egg white and place equal amounts into each of eight test tubes. Label the test tubes and add the following substances: (1) in two tubes, about 10 ml of water; (2) in two more tubes, about 10 ml of 0.5 percent commercial pepsin suspension; (3) in two more tubes, about 10 ml of 0.2 percent hydrochloric acid; and (4) in the last two tubes, about 10 ml of the pepsin suspension and 2 drops of the hydrochloric acid to make artificial gastric juice (lacking rennin, however). Or you may want to use another formula for the artificial gastric juice.[9]

Place all the test tubes in an incubator or into a water bath maintained at 37° to 40° C (99° to 104° F) for a period of 24 hr. At the end of this time you will find evidence of digestion in (2) and (4). However, the increased activity of the enzyme in an acid medium (note the rapid digestion) should indicate the effect of an acid pH on the action of pepsin.

A more effective test of conditions affecting the action of pepsin can be shown if you add two more test tubes to the demonstration described above. To the two test

[8] *Ibid.*, pp. 169–70.
[9] Bennett suggests this formula for preparing artificial gastric juice: Mix together 0.5 g of pepsin with 0.7 g of hydrochloric acid and 0.2 g of lactic acid. Add these ingredients to 100 ml of distilled water. H. Bennett, ed., *Chemical Formulary*, Chemical Publishing Co., New York, 1951.
Artificial gastric juice may also be prepared by dissolving 750 mg of pepsin (U.S.P.) in 100 ml of 0.1 N hydrochloric acid. Hawk, Oser, and Summerson, *op. cit.*, p. 369.

tubes, containing chopped egg white, add about 10 ml of 0.5 percent commercial pepsin suspension and 5 ml of 0.5 percent sodium carbonate solution. Compare the rate of digestion in the test tubes containing pepsin and acid with those containing pepsin and alkali.

EFFECT OF TEMPERATURE ON PEPSIN DIGESTION. In this activity students should place 10 ml of a solution of pepsin and hydrochloric acid into each of eight or more test tubes. These are prepared by adding a few drops of 0.2 percent HCl to each test tube of 0.5 percent pepsin solution. Students should immerse at least two test tubes in ice water; another set should be kept at room temperature; and a third set of two tubes should be placed in an incubator or water bath at 40° C (104° F). Have students boil the contents of another set of two test tubes for a few minutes, cool, and then place in the water bath. To each test tube, they should add equal amounts of hard-boiled egg white or albumin solution. Students can observe the rate of digestion in all the test tubes.

Fibrin may be substituted for egg albumin or egg white (boiled) whenever a practically pure source of protein is needed for testing digestion. Fibrin may be obtained from freshly drawn blood, as a clot forms, or commercial preparations may be purchased. Out-of-date whole blood stored in blood banks can also be used.

For a more quantitative test of digestion of egg white by pepsin suitable for a project, Hawk, Oser, and Summerson[10] describe the preparation of Mett tubes. In this technique egg white is drawn up into narrow-bore glass tubing. Subsequently, measured lengths of this tubing are cut and added to flasks of digestive juice. The amount of digestion in the tubes can be measured in millimeters. Normally, human gastric juice digests about 2 to 4 mm of albumin in tests with Mett tubes.

ANTIPEPSIN ACTION OF SOME WORMS.[11] Students may want to examine antipepsin

action. Try to get intestinal worms from a slaughterhouse; grind a few worms, such as *Ascaris,* in a mortar with sand. To this add some 0.9 percent sodium chloride solution; this should be mixed well and filtered.

Into each of two test tubes pour 10 ml of a solution of pepsin and hydrochloric acid as described earlier, along with some fibrin or egg white. To one test tube add 4 ml of the worm extract, and to the other tube add 4 ml of water. Keep both tubes at 37° to 40° C (99° to 104° F). As the rate of digestion is studied the next day, it will be noted that the worm extract inhibits digestion of proteins.

ACTION OF RENNIN ON MILK. You may want to demonstrate the coagulating action of this enzyme. Crush a Junket tablet and add it to a test tube of milk. Warm, but do not boil, the milk. Notice how rapidly the milk solidifies in the test tube. Compare this with a duplicate test tube in which the milk has first been boiled for 5 min. Notice the destructive action of heat on enzyme activity. Some teachers prefer to prepare a rennin solution first by crushing a Junket tablet in water, and add this to milk.

Digestion in the small intestine. You may want to have students perform demonstrations of digestion of proteins, carbohydrates, and fats.

DIGESTION BY TRYPSIN. You may want to show that proteins are digested by trypsin in the small intestine. Commercial pancreatin powder is useful for this purpose. Pancreatin contains the enzyme trypsin. Digestion and the optimum pH for the activity of trypsin can be demonstrated. Add chopped hard-boiled egg white or albumin solution (p. 213) to six test tubes, then add pancreatin to the tubes in the proportion of a pinch to a half test tube of water. Set two tubes aside, and to two others add 2 drops of phenolphthalein solution. Add a 0.5 percent solution of sodium carbonate, drop by drop, until the first pink color appears. This will result in a pH (p. 662) close to 8.3 (alkaline). To two other test tubes add 5 ml of a 2 percent

[10] Hawk, Oser, and Summerson, *ibid.,* p. 347.
[11] *Ibid.,* p. 237.

boric acid solution, which will result in a pH close to 5. For more rapid results use a water bath kept at 40° C (104° F), or keep the test tubes at the same temperature in an incubator.

Or you may want to have some students use casein of milk and a more carefully prepared solution of "pancreatic juice."

Prepare 0.25 g of pancreatin in 50 ml of water; neutralize (use litmus) using 0.05 M sodium bicarbonate. To this add 5 g of casein and 50 ml of 0.1 M sodium carbonate. Preserve with toluol and incubate at 38° C (100° F). (*Caution:* Avoid flames.)

Compare the action of a dilute solution of pancreatin with the action of such plant proteases as papain or other meat tenderizer, or fresh pineapple juice, which contains the enzyme bromelin. Incubate casein, albumin, or hard-boiled egg white with the plant proteolytic enzyme in a water bath or incubator maintained at 38° C (100° F).

AMYLASE ACTIVITY OF PANCREATIC JUICE. Also test the amylase activity of pancreatic juice on 0.5 percent arrowroot starch suspension.

Prepare a pancreatin solution as described above. Then prepare two test tubes containing about 10 ml of 0.5 percent starch paste, and add 5 ml of "pancreatic juice" to each. Keep the tubes at 40° C (104° F) in a water bath for about half an hour. Test the contents of one test tube for simple sugar with Benedict's solution or Clinitest tablets; test the other tube for starch with Lugol's solution. Is there an amylase in pancreatin? The digestion of starch begins in the mouth and is completed in the small intestine.

By following the pattern of experiments already described for salivary digestion, you may want also to show the effect of temperature on pancreatic digestion of carbohydrates.

INVERTASE IN THE SMALL INTESTINE. Also demonstrate that sucrose is hydrolyzed to glucose and fructose. Test the end product with Benedict's solution (other tests, pp. 136, 218, 162).

EMULSIFICATION OF FATS. In a test tube shake together a small quantity of water with a few drops of olive oil; an emulsion forms. On standing (for a short time), the oil separates from the water. To another test tube containing about 10 ml of water add 5 drops of a 0.5 percent sodium carbonate solution. To this add 3 drops of olive oil and shake well. On standing, this emulsion does not separate as quickly (since an alkaline medium exists). To a third test tube of water, add a 5 percent solution of bile salts such as sodium taurocholate (or add a soap solution). Then add a few drops of olive oil and shake well. Compare this more permanent emulsion with the preceding transitory ones. The size of the oil drops may be compared by placing a drop of fluid from each test tube on a slide for examination under low power of the microscope. Students may readily see the small size of the oil drops in an emulsion; in discussion these observations may be elicited, and the significance of the small size of the drops as a preliminary step in fat digestion by lipase in the small intestine should be apparent.

BILE SALTS AND SURFACE TENSION. Students may want to demonstrate to the class that bile salts lower surface tension. Add 5 ml of water to one test tube; to another tube, add 5 ml of bile suspension. Then sprinkle powdered sulfur on the surface of each test tube. What is the value of reduced surface tension?

HYDROLYSIS OF FAT BY LIPASE IN THE SMALL INTESTINE. A small committee of students might use olive oil in this hydrolysis. Prepare three test tubes containing 10 ml of water and a few drops of olive oil. To two of these tubes add 2 ml of a soap solution or 5 percent bile salt solution (sodium taurocholate). Then prepare a pancreatin solution. Dissolve the following ingredients in 100 ml of distilled water:

pancreatin	1	g
CaCl$_2$ (1% solution)	1	ml
K$_2$HPO$_4$ (0.2 M solution)	25	ml
NaOH (0.2 M solution)	11.8 ml	

Add 5 ml of this pancreatin solution to

one of the test tubes with soap solution or bile salts and to the one without this solution. Thus we have three test tubes: (1) a tube containing water, olive oil, soap solution, and pancreatin solution; (2) a control containing water, olive oil, and soap solution; and (3) another control containing water, olive oil, and pancreatin solution.

At the start of this demonstration, check the pH with litmus or phenolphthalein or hydrion paper. The original solutions must be neutral, not acid, for the objective is to show that fat digestion results in the production of fatty acids. Demonstrate the formation of acid with litmus paper, or other indicator, after the test tubes have remained in a water bath at 40° C (104° F) for at least half an hour.

HYDROLYSIS OF CREAM BY PANCREATIC LIPASE. Compare the effect of lipase and bile salts in pancreatic digestion. Prepare pancreatin solution as already described. Pour equal amounts of cream into three test tubes. To tube 1 add pancreatin solution; to tube 2 add an equivalent amount of pancreatin solution, plus a pinch of bile salts; to tube 3 add an amount of water equal to the solution used, plus a pinch of bile salts. Maintain all the tubes in a water bath at 37° C (99° F). Committees of students can prepare replicates. After a few seconds, begin to test the rate of hydrolysis by adding 5 drops of blue litmus solution to each test tube. Prepare litmus solution by adding 1 g of blue litmus powder to 100 ml of water; show the action of blue litmus by adding drops to a small quantity of dilute acid.

Is pancreatin effective in hydrolysis of fats in cream? Also test pure fats or oils such as olive oil (p. 215). What is the apparent role of bile salts? The effect of temperature on the activity of enzymes? Are there inhibitors of any sort? How dilute may the enzyme preparation be and still hydrolyze fats? As assignments, students can devise demonstrations to test the validity of their predictions; they can later perform their demonstrations in the laboratory.

LITMUS MILK TEST USING PANCREATIN. You may want to show that pancreatin contains digestive enzymes which split fats into fatty acids and proteins into amino acids. Litmus can be used as an indicator to show the resulting acid reaction after digestion. First, prepare litmus milk powder: to 50 parts of dried milk powder add 1 part of blue litmus powder. Then dissolve 1 part of prepared litmus milk powder in 9 parts of water.

To each of two test tubes add 10 ml of litmus milk powder solution, and to one test tube add a pinch of pancreatin powder. To the other test tube add only water or boiled pancreatin solution. If the tubes are kept in a water bath, students may observe the change in litmus to pink, indicating the presence of acid.

Absorption in the small intestine. We have described demonstrations illustrating diffusion of substances through a membrane (pp. 204, 242). Although the need for digestion was established in diffusion demonstrations, students may advantageously engage in this project. Students may fill a length of sausage casing or cellophane dialyzing tubing[12] (without holes—even pinpricks—or the demonstration will fail) with a variety of foods and pancreatin. Be sure to include sources of the common nutrients. For example, include molasses, egg white and yolk, salt, bread or a cracker, and some milk. (Mix these together in a mortar first, if necessary.) Tie each end of the tubing with thread and attach each end to a support as illustrated in Fig. 4-9. Wash off the outside of the tubing to remove traces of nutrients, and insert into a beaker of water. Place in a water bath at 40° C (104° F).

The next day students may test samples of the water in the beaker for reducing

[12] Dialyzing tubing made of unseamed cellophane may be purchased in rolls. When cellophane is substituted for semipermeable animal membranes or for dialyzing tubing, the nonwetting coating on the cellophane covering of packaged products needs to be removed. Immerse cellophane for 3 to 5 min in 95 percent ethyl alcohol (or in isopropyl for some 10 min). Under a faucet of warm water, wash and wipe off the film.

sugar, simple proteins, salt, and fatty acids. Use Benedict's solution or other reagents to test for simple sugar (p. 660); the Biuret test for simple proteins (p. 213); litmus paper for acids. Add 1 or 2 drops of silver nitrate to a water sample; the formation of a white precipitate will indicate generally the presence of a chloride.

ABSORPTION OF POTASSIUM IODIDE. You may demonstrate the speed of absorption of certain substances through the small intestine (or possibly the stomach) on *yourself,* to the amusement of your students. Have a pharmacist prepare 0.2-g gelatin capsules of potassium iodide.[13] Swallow one of these capsules before class begins. The iodide is absorbed quickly and appears in the salivary glands.

The saliva should be tested before the capsule is swallowed, or immediately afterward, by adding it to a starch paste suspension. Then test the saliva every few minutes. Shortly the blue-black color reaction occurs.

Student projects

There are many opportunities for individual student research work of varying levels of complexity in this area.

1. Individual students may want to learn to extract digestive juices from an organ. A club group may find this an introduction to the chemistry involved in the extraction of such proteins as enzymes. A reference such as Hawk, Oser, and Summerson's *Practical Physiological Chemistry*[14] gives concise steps for more elaborate undertakings by a group working under a teacher's supervision. One technique is given here as an example.

Cut up a sheep's or pig's pancreas from which the fat was trimmed. Grind it in a meat chopper or a Waring blendor. Put the ground pancreas in a 500-ml flask, add 150 ml of 30 percent ethyl alcohol, and let it stand for 24 hr, shaking it occasionally.

Then strain this alcoholic extract through cheesecloth and filter it. Begin to neutralize the filtrate with potassium hydroxide. Use litmus as an indicator; near the end point slow down, by using 0.5 percent sodium carbonate solution.

The pancreatic juice extract may be added to hard-boiled egg white or albumin solution to test whether it has retained its potency.

2. One student might investigate the kind of digestion and enzymes involved in insectivorous plants which trap flies or digest bits of meat. (See culture methods, Chapter 13.)

3. A student might devise a way to study extracellular digestion by molds, such as bread mold. In this case, enzymes are secreted into the bread (or other substrate), where starches are changed to simple sugars. These sugars then diffuse through the rhizoids of the mold. Is it possible to extract the enzyme? Is it possible to demonstrate that rhizoids do secrete enzymes?

4. Here is an advanced research problem. A student may want to learn more about the way some molds and bacteria develop certain enzymes, adaptive enzymes, depending upon the kind of medium in which they are growing. For example, starch-splitting enzymes are developed if the plants are grown in a medium containing starch; sugar-splitting enzymes are developed by the molds or bacteria if they grow in a medium that is rich in a certain sugar. Look up the subject of adaptive enzymes in a book on physiology of bacteria. (See the references at the end of this chapter.)

Nutrients, nutrient tests, and nutrition

Most teachers think it desirable for students to understand how food is tested. While few consumers stop to test the nutrient content of the family diet, students should understand the general idea

---

[13] G. Du Shane and D. Regnery, *Experiments in General Biology,* Freeman, San Francisco, 1950.
[14] *Op. cit.*

of *testing* that scientists use, and nutrient testing furnishes a very useful device. For example, iodine may be added to all nutrients in the pure form (in solution, of course). This establishes the specificity of the iodine test. Similarly, all tests should be tried on all nutrients.

Tests as tools for experimental work are therefore emphasized in this section. "Food nutrients" is the area selected for teaching the *idea* of testing.

Summary of nutrient tests

Starch. Add a few drops of a dilute iodine solution such as Lugol's solution (p. 664) to the substance to be tested for starch. A change to a blue-black color is a positive test. Dextrins give a red color with iodine. (*Note:* Test pure nutrients to establish specificity.)

Carbohydrates. In the presence of non-oxidizing acids, carbohydrates undergo dehydration, producing furfural or compounds of furfural. If a strong reagent is used, all carbohydrates can be made to give a positive test. However, a specific test can be obtained with more dilute solutions. You may want to refer to the many tests for carbohydrates that are given in P. Hawk, B. Oser, and W. Summerson, *Practical Physiological Chemistry,* 13th ed. (Blakiston [McGraw-Hill], 1954) and in A. Cantarow and B. Schepartz, *Textbook of Biochemistry,* 3rd ed. (Saunders, 1962).

SUGARS. Only three of the many tests for sugars are given here.

1. BENEDICT'S TEST. A quantitative test for reducing sugars such as glucose can be made by heating the sample with an alkaline reagent containing cupric copper. In the reaction, copper is reduced to red cuprous oxide.

In a Pyrex test tube boil 5 ml of Benedict's or Fehling's solution (p. 660) with a sample of the unknown. In a positive test for simple sugars (such as glucose or fructose), Benedict's or Fehling's solution is reduced so that a series of color changes results: green to yellow to orange.

(*Note:* Pure sucrose—ordinary table sugar —is not a reducing sugar and will not give a positive test with Benedict's solution; see p. 660.)

Clinitest tablets,[15] used for testing sugar in urine of diabetics, do not require heating. The sample to be used should be in solution.

2. MOLISCH'S REACTION. In this test, α-naphthol is the reagent. Molisch's test is not specific for glucose; positive reactions can be obtained with sucrose, aldehydes, and such acids as lactic, citric, and oxalic acids. Prepare the reagent by dissolving 10 g of α-naphthol in 100 ml of 95 percent ethyl alcohol. In the test, add 2 drops of Molisch's reagent to 5 ml of sugar solution in a test tube; mix well. Slant the test tube, and carefully pour 3 ml of concentrated H_2SO_4 down the side of the tube (*caution*). The acid forms a layer under the sugar, and a reddish-violet zone appears between the two layers.

You may prefer to substitute thymol for the α-naphthol since it does not deteriorate as quickly. In the test, use 3 or 4 drops of a 5 percent alcoholic solution of thymol to the test tube of sugar solution.

3. SELIVANOFF'S TEST. Ketose sugars give a bright red color when heated with HCl and resorcinol. This is a test for fructose; hydrolyzed sucrose yielding fructose also gives a positive test. Prepare Selivanoff's reagent by dissolving 0.05 g of resorcinol in 100 ml of dilute (1:2) HCl.

In this test, add 5 drops of fructose solution to 5 ml of Selivanoff's reagent in a test tube, and heat to boiling. Note the red color that is found in a positive test. There may or may not be a brownish-red precipitate. If there is a precipitate, it should dissolve in alcohol, giving a bright red color.

Proteins. The xanthoproteic test consists of adding concentrated nitric acid to the unknown in solution. When proteins are present the unknown turns yellow. As a

---

[15] These tablets, as well as Clinistix, are available from Ames Co., Inc., Elkhart, Ind.

further check, you may pour off the nitric acid and add a small amount of ammonium hydroxide. The yellow color should change to a deep orange. Or use Millon's reaction (p. 213). For partially digested proteins, use the Biuret test (pp. 136, 213).

Fats and oils. A simple test for fats and oils is to place some of the unknown substance on unglazed paper, such as wrapping paper. A permanent translucent spot through which light can pass indicates the presence of fats or oils. (Water spots will dry, but fat spots remain translucent.)

Water. Insert the sample of food into a dry test tube. Heat gently over a Bunsen flame; look for drops of moisture which condense on the sides of the test tube in a positive test.

Minerals. Heat a sample of food such as bread or milk in a test tube or evaporating dish until it forms an ash or until the liquid evaporates. A whitish ash indicates that minerals are present. Students may be guided (in their club activities) into procedures in qualitative analysis. A pattern for such a procedure is suggested here for the qualitative analysis of bone ash.

Qualitative analysis of bone ash

Hawk, Oser, and Summerson[16] give a number of qualitative analyses of organic materials which can serve as a guide. A procedure such as this one may be worked through by a small committee or club. Add a small quantity of dilute nitric acid to a beaker containing about 1 g of bone ash. Stir until most of the ash is dissolved; then add an equal amount of water, and filter. Now add ammonium hydroxide until the solution is alkaline and the phosphates precipitate out. Filter off the phosphates, and conduct tests for chlorides, sulfates, phosphates, iron, calcium, and magnesium. For example, the addition of 1 percent silver nitrate will cause a precipitation of silver chloride if chlorides are present. For further qualitative tests, see the above-mentioned reference.

[16] *Op. cit.,* p. 254.

Examining phosphates and magnesium in muscle of frog

The same authors[17] describe Hürthle's method. Use a dissecting needle to tease apart a small bit of muscle of frog on a glass slide. Then expose the slide to ammonia vapor for a few minutes. Cover with a coverslip and examine under the microscope. You should find large amounts of ammonium magnesium phosphate crystals distributed throughout the muscle fiber.

Vitamin content of foods

How have scientists learned to test for the vitamin content of foods? Students may be able to design experiments with animals to discover whether foods contain vitamins B_1 or D. Are there chemical tests for vitamins in foods? We shall first deal with experiments using animals to show the results of deficiencies, and then go on to demonstrate chemical tests for vitamin content of foods.

Effects of vitamin D on rats. Two groups of rats of the same age (about 25-day-old albino rats, 4 to 6 days past weaning) might be used. At least two rats in each group should be used, even for a comparatively short study of the effect of differences in vitamin D content. One group is fed a vitamin-D-free diet, while the second group receives the same diet with viosterol added. The diet which has been recommended is taken from *Turtox Service Leaflets.*[18] Both groups are fed this diet:

yellow corn	76 parts
gluten flour	20 parts
sodium chloride	1 part
calcium carbonate	3 parts

For one group, supplement this diet with 20 drops of viosterol per 1000 g of diet. Both groups should get adequate supplies of water daily. They should be housed in separate cages, and the cages should be kept away from direct sunlight.

[17] *Ibid.,* p. 288.
[18] *Turtox Service Leaflet,* No. 49, General Biological Supply House, Chicago.

Within 3 weeks, differences in weight should be apparent. As the animals are weighed every day and records kept, it should be found that the group lacking vitamin D in the form of viosterol stops growing. They show a generally unhealthy appearance, they are somewhat wobbly in gait, and their fur appears ruffled.

The differences listed should be seen within 30 days, the maximum period the vitamin-D-free diet should be used. When the animals on the deficient diet develop symptoms of the deficiency disease they should be placed on the more complete diet; that is, viosterol should be added to the original diet. Almost immediately the affected animals will gain weight.

From the time the affected animals are given viosterol, weigh them daily and record the rapid gain in weight and the progressive change to a healthy appearance.

Vitamin B₁ requirements of pigeons. When healthy young pigeons are fed a diet deficient in thiamine (B₁) they develop polyneuritis (beriberi).[19] Students who maintain pigeons at home may try this experiment, or it may be carried on in school.

Separate cages are needed for the experimental and the control pigeons. The cages should be kept free of droppings, or pigeons will consume this along with food. Feed all the birds equal amounts of white polished rice, finely cracked egg shells, and water.

One group gets, in addition to this diet, a mixture of barley, hemp seeds, yellow corn, and some fresh vegetables such as shredded cabbage or lettuce.

In about 10 days the pigeons without the "extras" will develop a paralytic condition, since nerve-muscle connections are impaired. Then add the "extras" to the diet and watch how quickly the pigeons recover their health and activity.

Chemical test for vitamin C. Ascorbic acid, vitamin C, is a strong reducing agent with the ability to reduce indophenol, iodine, silver nitrate, and methylene blue (in the light). A simple, quantitative test for the presence of vitamin C in foods may be done as a class demonstration. Even better, it may serve as an excellent laboratory experience—a controlled experiment involving simple measurement.

Indophenol (2,6-dichlorophenol indophenol) is an "indicator" which is blue in color, and is reduced in the presence of ascorbic acid or juices containing ascorbic acid.[20] Prepare a 0.1 percent solution of indophenol in water by dissolving 1 g of indophenol in 1000 ml of water.[21]

The test should first be demonstrated with pure ascorbic acid and later repeated with several fruit juices. You can purchase a 10 percent ascorbic acid solution in a drugstore under the trade name Cecon, and dilute it further with water as necessary.

Since this is a sensitive test, the ascorbic acid should be diluted enough so that the number of drops of ascorbic acid added to the indophenol may be counted accurately. If the ascorbic acid is too strong, 1 drop will immediately bleach the indophenol, so that a comparative study is difficult to make. About 10 ml of the indicator is added to a test tube. Diluted (1 percent) ascorbic acid is added to this, drop by drop; count the number of drops needed to change the color from blue to colorless. (The intermediate pink stage should be disregarded.)

Now, diluted fruit and vegetable juices —canned, fresh, and frozen—may be tested. Test the juices beforehand so that you may dilute them sufficiently for students to add about 10 to 20 drops of juice

[19] Diets deficient in specific nutrients may be obtained from Nutritional Biochemicals Corp., Cleveland 2, O., and General Biochemicals Inc., Chagrin Falls, O.

[20] Indophenol can be purchased from the Chemical Division of Eastman Kodak Co., Rochester, N.Y., or from biological supply houses.
[21] A BSCS laboratory activity recommends the use of a 10 to 20 percent solution of Lugol's iodine as a substitute for indophenol. The color of iodine bleaches until the end point is reached. *High School Biology: Green Version—Laboratory, Part II*, BSCS, Univ. of Colorado, Boulder, 1961 (experimental use, 1961–62), p. 13-3-1.

to a given quantity of indophenol. Dilute them all equally, so that the ascorbic acid content of canned juices may be compared with that of fresh juices. In comparative tests the quantity of indophenol used should be standardized.

Some juices may be boiled, or left exposed to air for several hours and then tested again. The data collected may then be compared with the original readings.

Bicarbonate of soda, which is often added to vegetables to preserve their green color while cooking, may be added to samples of the same juices. Note the loss of ascorbic acid under these conditions. (The larger number of drops of juice needed to bleach a given quantity of indophenol is a measure of the small amount of ascorbic acid remaining in the juices.)

There may be errors in the titration of vitamin C, due to the presence of other reducing compounds formed during the heat-processing of food or while food substances are stored. Hawk, Oser, and Summerson[22] describe procedures for correcting such errors, as well as means of determining the ascorbic acid content of whole blood, urine, or plasma, or plant tissues; colorimetric methods are included.

They also suggest biological methods for assaying for vitamin C—experiments with guinea pigs fed on standard scurvy-producing diets.

Microbiological assay of riboflavin. Hawk, Oser, and Summerson[23] describe the method of Snell and Strong for determining riboflavin content by measuring the stimulation of growth of *Lactobacillus casei*. The acid produced by the bacteria is determined by titration with sodium hydroxide.

Vitamin requirements of plants

Problems for small groups of students or for a club might revolve around a study of the vitamin needs of plants—the vitamin K requirements of *Euglena*; the use of

22 *Op. cit.*, pp. 1233–41.
23 *Ibid.*, pp. 1160–64.

Euglena, Neurospora, and other microorganisms—mainly bacteria—in microbiological assay of vitamin and amino acid content of foods. Two possible demonstrations, using thiamine, which is readily available, are given. (See also p. 462.)

Thiamine and growth of Phycomyces. Show that *Phycomyces blakesleeanus* (having a much-branched mycelium with many nuclei) requires thiamine for growth. Prepare a thiamine-deficient medium for the mold: To 1 liter of distilled water, add 100 g of glucose, 10 g of asparagine, 15 g of KH_2PO_4, and 5 g of $MgSO_4 \cdot 7H_2O$. Adjust the pH to between 4.0 and 4.5.

Now prepare a solution of thiamine chloride and dilute to varying concentrations: 5 mg per liter, 2.5 ml per liter, 1.0 mg per liter, 0.5 mg per liter, 0.1 mg per liter. Pour an equal amount of medium (in milliliters) into 12 test tubes. Leave 2 as controls, and to pairs of test tubes add 1 ml of the different dilutions of thiamine chloride solution. In short, there should be 2 tubes, or replicates, of each dilution and 2 controls. Loosely stopper with cotton plugs, or cover with aluminum foil, and sterilize for 15 min at 15 lb per sq in. pressure. Inoculate each tube with an equivalent amount of mycelia of *Phycomyces blakesleeanus;* incubate or let stand for 2 weeks, and compare the rate of growth in each test tube. Can one estimate the vitamin requirement for growth of these molds? Is there growth in the controls? In assay work, the mat of mold would be weighed; possibly a rough index might be made. Does one dilution show twice the growth of another dilution? Does the mat weigh about twice as much?

Deficient strains of Neurospora. Students may repeat, on a small scale, the procedures developed in 1941 by Beadle and Tatum for identifying biochemical mutations after irradiation of the red mold *Neurospora crassa* (Fig. 8-71).

Normal strains of this red mold synthesize the requirements for growth, provided a minimal diet is supplied: a source of carbon in the form of a carbohydrate (for this mold is a heterotroph); some inorganic salts, including nitrate; and one vitamin, biotin. Some mutants have lost the ability to synthesize a specific vitamin or amino acid; apparently the enzyme is missing due to the loss or modification of a gene. As a result, these vitamins or amino acids must be added to the mineral medium to support growth.

Purchase a normal strain of *Neurospora* and a deficient strain—possibly a thiamine-deficient strain that cannot make its own thiamine, or a strain unable to make pyridoxine and choline, or arginine.

Transfer a bit of the normal and the deficient strains to separate sterile test tubes or Petri dishes containing a minimal medium; also transfer each strain to individual test tubes that have the minimal medium plus some of the required thiamine (or arginine or choline, as the case may be).

The preparation of a minimal medium for *Neurospora* is suggested in the *Teachers' Guide* to the BSCS Blue Version.[24] To 1 liter of water add 15 g of sucrose, 5 g of ammonium tartrate, 1 g of ammonium nitrate, 1 g of monopotassium phosphate, 0.5 g of magnesium sulfate, 0.1 g of calcium chloride, 0.1 g of sodium chloride, 10 drops of 1 percent ferric chloride, 10 drops of 1 percent zinc sulfate, and 5 mg of biotin. Sterilize in an autoclave for 10 min at 15 lb per sq in. Dehydrated *Neurospora* culture agar is available from Baltimore Biological Laboratory and from Difco Laboratories (see Appendix C). Media for assays of pyridoxine and choline are also available. (Agar may be added to the liquid medium given above.) See p. 400.

Students may devise ways to test for the quantity of choline or thiamine or other essential substances in foods by using specific deficient strains of *Neurospora*. Or they may be given an unknown mutant strain and be asked to identify it. See also genetics of *Neurospora* (pp. 398–402).

A well-developed series of laboratory activities is described in the BSCS labora-

[24] *Teachers' Guide* to High School Biology: Blue Version—Laboratory, Part II, BSCS, Univ. of Colorado, Boulder, 1961 (experimental use, 1961–62), pp. 153–60.

tory block *Microbes: Their Growth, Nutrition, and Interaction* by A. S. Sussman (Heath, 1964).

Mineral requirements

Effects of nitrates, phosphates, and sulfates may be demonstrated using growing seedlings or algae such as *Chlorella* (pp. 458–62).

Calories

At this point, or later in a study of respiration (Chapter 6), you may want to develop a discussion of the need for calories in the diet. There are several approaches to this work, involving the use of the blackboard, charts, or paper-and-pencil techniques.

For example, you may put a chart such as Table 4-2 on the blackboard. After a study of the chart, students often ask why boys and girls have a higher caloric requirement than men and women. A minimum average of 1500 Cal seems to be needed to keep a person alive.

For what kinds of activities does a person need more calories? (See Table 4-3.)

Why are some people overweight or underweight?

An activity that creates much interest is this one. Ask students to keep a list of all the foods they have eaten during 3 school days. Then, in class, distribute charts which give calories, mineral content, vitamin content, and the amount of proteins of many foods. One example of such a chart is the fine pamphlet *Facts About Foods*, which is distributed by H. J. Heinz Co., Pittsburgh, Pa.

In this booklet, or a similar one, students may look up the number of calories, proteins, vitamins, and minerals in the foods they have consumed. Then total the list and divide by three; this is an average for 1 day. What number of calories is needed on the basis of body weight?

Students might compare their own diet with the recommendations given in Table 4-2. Are they getting an adequate diet?

TABLE 4-2
Recommended daily dietary allowances*

	weight (lb)	height (in.)	calories	proteins (g)	calcium (g)	iron (mg)	vitamin A (IU)	thiamine (mg)	riboflavin (mg)	niacin (mg)	ascorbic acid (mg)	vitamin D (IU)
Boys												
12–15	98	61	3000	75	1.4	15	5000	1.2	1.8	20	80	400
15–18	134	68	3400	85	1.4	15	5000	1.4	2.0	22	80	400
Girls												
12–15	103	62	2500	62	1.3	15	5000	1.0	1.5	17	80	400
15–18	117	64	2300	58	1.3	15	5000	0.9	1.3	15	70	400
Men												
18–35	154	69	2900	70	0.8	10	5000	1.2	1.7	19	70	
35–55	154	69	2600	70	0.8	10	5000	1.0	1.6	17	70	
Women												
18–35	128	64	2100	58	0.8	15	5000	0.8	1.3	14	70	
35–55	128	64	1900	58	0.8	15	5000	0.8	1.2	13	70	

* Adapted from *Recommended Dietary Allowances*, 6th ed., *Report of Food & Nutrition Board* (Public. 1146), National Academy of Sciences–National Research Council, Washington, D.C., 1964.

TABLE 4-3
Approximate energy expenditure in a variety of activities*

activity	time (hr)	man rate (Cal/min)	man total	woman rate (Cal/min)	woman total
Sleeping	8	1.1	540	1.0	480
Sitting, driving a car, bench work	6	1.5	540	1.1	420
Standing, or limited walking	6	2.5	900	1.5	540
Walking, purposeful, or outdoors	2	3.0	360	2.5	300
Occupational activities involving light physical work; spasmodic week-end swimming, golf, picnics†	2	4.5	540	3.0	360
			2880		2100

* Adapted from *Recommended Dietary Allowances*, 6th ed., *Report of Food & Nutrition Board* (Public. 1146), National Academy of Sciences–National Research Council, Washington, D.C., 1964.

† Week-end swimming, tennis, and so forth may use 5 to 20 Cal per min for a limited time.

A balanced one? Plan a diet for a week for a person who needs to gain weight. Similarly, suggest a diet for a person who wishes to lose weight. Why should dieting of any kind be undertaken with the assistance of a doctor? Nibblers will be interested in the caloric contents of some common snack foods (Table 4-4).

Using the library in the study of nutrition and health

Gathering facts from authoritative sources is one of the methods used by scientists. Students in science classes need practice in developing this skill. Intelligent use of the facilities of the library requires the same kind of planning that is needed in preparing for a field trip. Learning to find and organize information is time well spent.

Suppose the unit topic is "The Foods We Need for Good Health." The problems for study in this area may be elicited from students in class. Students' questions might be listed on the board. For example: What is food used for? Where do we get our food? Is it all right to go without breakfeast? Why do some people gain weight? What are calories? What good are vitamins? Which vitamins do we need? Can your diet cause pimples? What's wrong with taking vitamin pills? What is a balanced diet? Must you have one every day? What should you eat to lose weight?

Then these questions may be selected by groups of three to six students working as committees for library "research." Allow time, possibly 10 min, for students to break up into their small groups to interpret among themselves the specific problems about which they want information. At the end of this time, the students reassemble and a spokesman in each group defines the areas his group will investigate. The rest of the class can add suggestions.

TABLE 4-4

Caloric values of some "snack" foods*

	amount or average serving	calories
Fruits		
Apple	1 3-in.	90
Banana	1 6-in.	100
Grapes	30 medium	75
Orange	1 2¾-in.	80
Pear	1	100
Candies, etc.		
Almonds or pecans	10	140
Cashews or peanuts	10	60
Cheese crackers	10 2-in. diameter	220
Chocolate bars, small size		
Plain	1 1¼-oz	190
With nuts	1	275
Caramels		
Plain	1 ¾-in. cube	35
Chocolate-nut	1 ¾-in. cube	60
Potato chips	10 2-in. diameter	110
Sweets		
Ice cream		
Plain vanilla	1 ⅙-qt serving	200
Chocolate and other flavors	1 ⅙-qt serving	230
Ice cream soda, chocolate	1 10-oz	270
Sundaes, small chocolate-		
nut with whipped cream	1 average	400
Midnight snacks for icebox raiders		
Chicken leg	1 average	88
Ham sandwich	1 ½ oz ham	350
Hamburger on bun	1 3-in. patty	500
Peanut butter sandwich	1 2 tbsp peanut butter	370
Mouthful of roast	1 ½ x 2 x 3 in.	130
Brownie	1 ¾ x 1¾ x 2¼ in.	300
Beverages		
Chocolate malted milk	1 10-oz	450
Milk	1 7-oz	140
Soft drinks (soda, root beer, etc.)	1 6-oz	80
Tea or coffee, straight	1 cup	0
Tea or coffee, with 2 tbsp cream and		
2 tsp sugar	1 cup	90
Desserts		
Pie		
Fruit (apple, etc.)	1 average serving (⅙ pie)	560
Custard	1 average serving (⅙ pie)	360
Lemon meringue	1 average serving (⅙ pie)	470
Cake		
Iced layer	1 average serving	345
Fruit	1 thin slice (¼-in.)	125

* Adapted from "Caloric Values for Common 'Snack' Foods," published by Smith, Kline & French Laboratories, Philadelphia.

You may want to bring textbooks, magazines, and vertical files of current clippings to class. Many teachers prefer to bring the students to the library for a period (or two) of reading.

While students are engaged in the library, seeking references (on the stories of vitamin discoveries, of the "basic seven," of calories and food intake), there is opportunity to observe how well the students use a card catalog, "readers' guide," or the index of a book, how they select a pamphlet they can read, how fast they read, and how well they take notes.

In sharing what they have learned with others, students may use many ways to present the information to their classmates. A panel group may elicit (by questioning) the facts from the class, another committee may dramatize a story of a vitamin discovery, another group may devise an experiment in nutrition, another committee may plan a review for the class on the unit of nutrition—particularly in preparation for an examination.

Several other areas lend themselves to this fruitful research by the group of students working in the library: "How Living Things Behave," or "Ancient Life on Earth," or "Diseases: Their Cause and Methods of Control," or some similar unit on maintaining health. In the last-mentioned unit, for example, students learn to classify diseases according to their causes or kinds, or glean the stories of how diseases have been brought under control, discover the men associated with major advances in medicine, vaccination, antibiotics, and so on. In this way students get practice in organizing and assimilating new reading material. The process is fruitful in that this kind of reading gives rise to a new awareness, and thus new questions arise for study in class.

CAPSULE LESSONS: Ways to get started in class

4-1. You may want to use this demonstration. In one test tube seal a snail, in another a green water plant such as *Cabomba* or *Anacharis* (elodea) along with a snail (p. 265). After developing the idea of their interdependence, you might compare the snail's dependence on plants with man's dependence on green plant foods. What is an autotroph? A heterotrophic organism?

4-2. Have two students stretch out some 20 ft of clothesline. This simulates roughly the length of the intestine. How does food move along the digestive tract? What changes take place along the way? How does food material get out of this intestine into the blood stream, finally to the cells?

4-3. You may want to begin with a laboratory lesson. If you have many microscopic organisms available, students may examine food-getting in several protozoa (using *Chlorella* or carmine powder); also examine microscopic worms and crustaceans. Then lead to a discussion of food-getting.

If you have living frogs or chameleons, examine their way of ingesting food. Elicit from students discussion about the adaptations for food-getting found in several kinds of animals. Develop a discussion concerning the uses of food to the body.

4-4. Food-making by plants and food-taking by animals, along with digestion of insoluble materials, may be part of a broad study in interdependence among living things, as developed in some activities in Chapter 11 (especially those concerning food chains). You may want to begin this study of digestion with activities in that chapter.

4-5. In a laboratory lesson, have students dissect frogs (see Chapter 1) so that they may trace the path of food through the body. Compare this with the digestive system of man (as shown in charts, or use a manikin). Further, how do frogs' parasites get their food?

4-6. Perhaps you may want to begin the work with a comparison of the kinds of heterotrophic modes of nutrition: holozoic, predator-prey, parasitism, and saprophytism. From here, develop a discussion of the need for digestion and the uses of food in the body.

4-7. Use one of the demonstrations of diffusion of materials through a membrane to develop the notion that foods taken into the digestive system are not "part of us" until they diffuse out of the digestive system into the blood.

4-8. You may want to begin with a film which reveals many of the ways animals get their food. Have you seen some of the films in the AIBS series *The Multicellular Animal* (McGraw-Hill)? Or perhaps you may want to show *Patterns of Nutrition* (Indiana Univ.) or *Partnerships Among Plants and Animals* (Coronet).

4-9. Begin, sometimes, with a discussion of molds and how they get their food supply. How are they able to live on bread? Then you are into a discussion of interrelations among plants and animals and the dependence on green plants as the food source. Refer to soil microfauna and microflora. How do they get their food?

4-10. Or set up a demonstration to show that a mixture of starch paste and Lugol's solution loses its blue color when a diastase solution or saliva is added (see techniques in this chapter). What has happened to the starch? Elicit discussion concerning the need for digestion.

4-11. As a laboratory lesson have students test the action of their own saliva on starch paste, as described in this chapter. Or have one or two students demonstrate how saliva changes the starch in starch paste or a chewed cracker into soluble form.

4-12. Why do plants need a starch-splitting enzyme? Demonstrate the change from starch to sugars in germinating seedlings. Use non-germinating seeds as controls.

4-13. Have a committee of students report on the classic experiments in digestion done by Spallanzani, by Réaumur, and by Dr. Beaumont (on his patient Alexis St. Martin). Lead into a discussion of the role of enzymes in making food materials soluble so they can diffuse through membranes out of the digestive system.

4-14. Have a student report on the way termites get their cellulose food supply digested by flagellates (p. 537). If possible, examine living flagellates from the intestine of termites.

4-15. You may find it useful to have students summarize the work of the digestive system on the blackboard in this way.

name of organ	role in digestion
a. -----------------	-----------------
b. -----------------	-----------------
c. -----------------	-----------------
d. -----------------	-----------------
e. -----------------	-----------------
f. -----------------	-----------------
g. -----------------	-----------------

4-16. Perhaps you may want to plan a laboratory lesson on the examination of epithelial cells such as those in the lining of the mouth. Directions are given for scraping the lining of the cheek to get epithelial cells for mounting in Lugol's solution (p. 664).

4-17. If you have not already demonstrated diffusion of molecules through membranes, you may want to begin this work on digestion with either starch paste or glucose in dialyzing tubing. Fasten each end to a tongue depressor or splint, and suspend over a beaker of water. Or use a starch paste and saliva mixture (or diastase and starch paste). If you do not have this cellophane tubing, use test tubes across the mouth of which wet goldbeater's membranes are stretched as described previously (Fig. 4-7). Test the water in the beakers to find out which molecules pass through a membrane. Why didn't starch go through the membrane? Then you are in the topic of digestion.

4-18. Ask students to list all the food they consumed the previous day. Develop the notion of nutrients (and balanced diet) and lead into a discussion of the "basic seven," then into tests for nutrients. Discuss animal experimentation and the use of chemical tests for the nutrients in foods.

4-19. At some other time, begin with a laboratory lesson testing the vitamin C content of some fruit juices (fresh and canned) with indophenol (description in this chapter).

4-20. Occasionally, you may want to have a supervised study lesson. Have you tested the reading and comprehension level of your students by having them read a chapter in their text? Or use supplementary readings to discover their interests. (See the companion volume, P. F. Brandwein, F. G. Watson, and P. E. Blackwood, *Teaching High School Science: A Book of Methods* [Harcourt, Brace & World, 1958].) With slow readers you may want to try a chapter in a general science textbook (see the list of general science texts in this same volume).

4-21. Discuss the significance of the Federal Food, Drugs, and Cosmetics Act. As an aid in this study have students examine and paste into their notebooks labels from food packages and from empty medicine bottles. What may be the danger in using additives?

4-22. Describe the diet of an average individual in the United States, Mexico, India, the Arctic, Spain, South America, China, Greece, and Germany. (See R. W. Gerard, ed., *Food for Life* [Univ. of Chicago Press, 1952].) Compare the amount of proteins, minerals, and vitamins as well as the other nutrients contained in these diets. Develop a discussion of what makes an adequate diet.

4-23. You might want to start a lesson this way: Let's examine what makes a good or a poor meal. A boy had this for breakfeast: black coffee with a sugared doughnut. How would you improve this meal?

For lunch a girl had spaghetti, rolls and butter, custard pie, and black coffee with sugar. What's wrong with this lunch? What nutrients are in excess? Which nutrients should be added—and in what foods?

Ask the class to plan a good diet, and develop a discussion of the kinds of food needed for a good diet.

4-24. Or prepare a demonstration to show which components of gastric juice digest proteins (described in this chapter). Better still, students may design the demonstration themselves; further, they may investigate factors that affect the activity of enzymes.

4-25. With a photometer, introduce techniques for measuring growth of microorganisms. Plan several projects relating to growth (see Chapter 9). Which minerals are critical for stimulating and maintaining growth?

4-26. A student might introduce a study of vitamins by reporting some historic discovery, for example, the story of how Dr. Goldberger discovered the cause of pellagra. Here there are many opportunities for group work in the library.

4-27. You may want to introduce laboratory work on microbiological assay using *Neurospora* or *Euglena*. Students can compare growth in minimal and supplemental media. Have you shown the BSCS techniques film on *Neurospora* (Thorne)?

4-28. Perhaps you plan to discuss food-making in crop plants as part of the need for soil conservation. What good farming practices are needed for a more nutritious food supply? Have you used the film made available by the Beet Sugar Development Foundation called *Feed the Soil and It Will Feed You*? Or the films *Hunger Signs* and *Life of the Soil* distributed by the National Fertilizer Assoc. (A directory of film distributors is given in Appendix B; a directory of distributors of low-cost or free materials is given in Appendix D.)

4-29. Or you may show a film describing how radioactive tracers are used in plant nutrition. Show *The Eternal Cycle, The Atomic Greenhouse,* or others in *The Magic of the Atom* series available from the U.S. Atomic Energy Comm. What has been learned about nutrition in plants as a result of using tracers?

PEOPLE, PLACES, AND THINGS

Students may often get help in planning a project in nutrition from the nutrition or biology department of a nearby college or hospital.

Living materials may be collected in the field or borrowed from other schools; organisms for nutrition experiments, from protists to rats, may be purchased from biological supply houses (Appendix C).

There may be a canning plant or a food-processing plant in your community to which you may take your classes. Or possibly someone may come to speak to your classes on some aspect of nutritional research.

Your school dietician is another resource person. Perhaps she will describe to the class how balanced meals are planned in the school kitchen.

When a student needs assistance in a research problem—either technical help or equipment—there may be an opportunity for him to work (washing glassware) in a laboratory in plant or animal physiology at a nearby college or university. There may be a part-time position in the college library for a student who wants to read in advanced areas of science. A government agricultural station or a biological testing laboratory may be in your community. In these places you will find resource people who are willing to provide chances for mutual assistance between the junior or senior high school and the college. Perhaps you will plan visits to these places.

BOOKS

Areas described in this chapter are as broad as life: nutrition, modes of nutrition, ingestion, and digestion of food materials. These are only a few of the references pertinent to the work; also refer to books listed in Chapters 1, 2, and 11, as well as to Appendix A, the comprehensive Bibliography.

Asimov, I., *The Wellsprings of Life,* Abelard-Schuman, New York, 1961; Signet, New Am. Library, New York, 1962.

Bates, M., *The Forest and the Sea,* Random, New York, 1960; Signet, New Am. Library, New York, 1961.

Bonner, J., *The Cellular Slime Molds,* Princeton Univ. Press, Princeton, N.J., 1959.

Borek, E., *Atoms Within Us,* Columbia Univ. Press, New York, 1961.

Cantarow, A., and B. Schepartz, *Textbook of*

Biochemistry, 3rd ed., Saunders, Philadelphia, 1962.

Chandler, A., and C. Read, *Introduction to Parasitology,* 10th ed., Wiley, New York, 1961.

Christensen, C., *Common Fleshy Fungi,* Burgess, Minneapolis, 1955.

Duddington, C. L., *Micro-Organisms as Allies,* Macmillan, New York, 1961.

Dukes, H. H., *et al., The Physiology of Domestic Animals,* 7th ed., Comstock (Cornell Univ. Press), Ithaca, N.Y., 1955.

Eaton, T. H., Jr., *Comparative Anatomy of the Vertebrates,* 2nd ed., Harper & Row, New York, 1960.

Esau, K., *Plant Anatomy,* 2nd ed., Wiley, New York, 1965.

Foster, J., *Chemical Activities of Fungi,* Academic, New York, 1949.

Frisch, K. von, *Ten Little Housemates,* Pergamon, New York, 1960; insect pests.

Gerard, R. W., ed., *Food for Life,* Univ. of Chicago Press, Chicago, 1952.

Greulach, V., and J. E. Adams, *Plants: An Introduction to Modern Botany,* Wiley, New York, 1962.

Harrow, B., and A. Mazur, *Textbook of Biochemistry,* 8th ed., Saunders, Philadelphia, 1962.

Hegner, R. W., and K. A. Stiles, *College Zoology,* 7th ed., Macmillan, New York, 1959.

Hyman, L., *The Invertebrates,* 5 vols., McGraw-Hill, New York, 1940–59.

Manwell, R., *Introduction to Protozoology,* St. Martin's, New York, 1961.

Meyer, B., D. Anderson, and R. Böhning, *Introduction to Plant Physiology,* Van Nostrand, Princeton, N.J., 1960.

Nickerson, W., *Biology of Pathogenic Fungi,* Ronald, New York, 1947.

Oginsky, E., and W. Umbreit, *An Introduction to Bacterial Physiology,* 2nd ed., Freeman, San Francisco, 1959.

Prosser, C. L., and F. A. Brown, Jr., *Comparative Animal Physiology,* 2nd ed., Saunders, Philadelphia, 1961.

Rogers, W. P., *The Nature of Parasitism,* Vol. II of J. F. Danielli, ed., "Theoretical and Experimental Biology," Academic, New York, 1961.

Schmidt-Nielsen, K., *Animal Physiology,* 2nd ed., Prentice-Hall, Englewood Cliffs, N.J., 1964.

Simpson, G. G., and W. S. Beck, *Life: An Introduction to Biology,* 2nd ed., Harcourt, Brace & World, New York, 1965.

Singer, C., *A History of Biology,* 3rd rev. ed., Abelard-Schuman, New York, 1959.

Sistrom, W. R., *Microbial Life,* Holt, Rinehart and Winston, New York, 1962.

Villee, C., *Biology,* 4th ed., Saunders, Philadelphia, 1962.

Walsh, E. O., *An Introduction to Biochemistry,* Macmillan, New York, 1961.

Weisz, P., *Elements of Biology,* McGraw-Hill, New York, 1961.

———, *The Science of Biology,* 2nd ed., McGraw-Hill, New York, 1963.

Welsh, J., and R. Smith, *Laboratory Exercises in Invertebrate Physiology,* Burgess, Minneapolis, 1960.

Westcott, C., *Plant Disease Handbook,* 2nd ed., Van Nostrand, Princeton, N.J., 1960.

Witherspoon, J., and R. Witherspoon, *The Living Laboratory,* Doubleday, Garden City, N.Y., 1960.

Also refer to Yearbooks of the U.S. Dept. of Agriculture: *Food* (1959), *Animal Diseases* (1956), *Soil* (1957), *Insects* (1952), and *Plant Diseases* (1953).

Offprints are available of many of the fine papers in *Scientific American;* write to W. H. Freeman and Co., 660 Market St., San Francisco 4, Calif., for a listing ($0.20 each).

FILMS, FILMSTRIPS, AND SLIDES

This listing is intended to be only a guide for the selection of new films, filmstrips, and transparencies. Send for catalogs from distributors of films; catalogs of biological supply houses list extensive offerings of 2 × 2 in. slides. The addresses of distributors are given in Appendix B.

In general, the cost of color films averages $150; black and white films cost about $75. Newer films are made in varying lengths—time sequences of 1 min, 5 min, 11 min, or 30 min; the cost of films therefore varies. We have indicated the lengths of films so that lessons can be planned with the films in mind.

Sound films are available in color or black and white; rental rates are available from film libraries or by direct inquiries to producers. (Filmstrips are not available on a rental basis; the purchase price is usually about $6 to $7 for those in color.)

Antibiotics (14 min), EBF.
Bacteria: Friend and Foe (11 min), EBF.
Bacteria: Laboratory Study (15 min), Indiana Univ.
Balance in Nature (15 min), Filmscope.
Between the Tides (22 min), Contemporary.
Biology: Human Anatomy Series 30020 (overhead teaching transparencies, two units), EBF.

Birds of the Seashore (11 min), Canadian Travel Film Library; free.

Carnivorous Plants (10 min), Moody.

Cell Biology (set of 12 films, 28 min each), Part I of *AIBS Film Series in Modern Biology,* McGraw-Hill.

Classification of Invertebrate Animals (set of 11 filmstrips), EBF.

Classification of Plants (set of 9 filmstrips), EBF.

Economic Botany (set of 57 transparencies in color), Carolina.

The Eternal Cycle (13 min), in *The Magic of the Atom* series, Handel; free loan from U.S. Atomic Energy Comm. Film Library.

Flight of the Sea Birds (10 min), Utica Duxbak; free.

Fungi (16 min), EBF.

The Human Body: Digestive System (14 min), Coronet.

The Human Body: Nutrition and Metabolism (14 min), Coronet.

Human Digestion (10 min), Contemporary.

The Insects: Harmful and Useful (filmstrip), EBF.

Insects That Live in Societies (filmstrip), EBF.

Life Cycles of Insects (filmstrip), EBF.

Life of the Molds (21 min), Sterling; free.

Louis Pasteur (30 min), Sterling.

Microbiology (set of 12 films, 28 min each), Part II of *AIBS Film Series in Modern Biology,* McGraw-Hill.

Microorganisms: Beneficial Activities (15 min), Indiana Univ.; nitrogen cycle.

Microorganisms: Harmful Activities (15 min), Indiana Univ.; includes Koch's postulates.

Minerals of Life (29 min), Indiana Univ.

The Multicellular Animal (set of 12 films, 28 min each), Part IV of *AIBS Film Series in Modern Biology,* McGraw-Hill.

Orders of Insects (set of 8 filmstrips), EBF.

Parasitic Plants (set of 21 transparencies), Carolina.

Partnerships Among Plants and Animals (11 min), Coronet.

Patterns of Nutrition (29 min), Indiana Univ.

Simple Plants: Bacteria (14 min), Coronet.

Slime Molds (set of 10 transparencies), Carolina.

Virus (set of 8 films, 29 min each), Indiana Univ.

Vitamins and Your Health (17 min), National Vitamin Foundation.

Wings on the Wind (14 min), Wilcox.

Wood Decay by Fungi (20 min), Syracuse Univ., College of Forestry.

World Within (28 min), Univ. of California; parasites.

LOW-COST MATERIALS

A listing of suppliers of free and low-cost materials, with addresses, is given in Appendix D. Although we recommend the material for use in the classroom, we do not necessarily sponsor the products advertised.

The Basic 7 Food Groups; food charts, General Mills.

Care and Feeding of Laboratory Animals, Ralston Purina.

The Challenge of Health Research, in "Health Hero Series," Metropolitan Life Insurance.

Food Facts; How to Conduct a Rat Feeding Experiment, Wheat Flour Institute.

Food Values Charts, National Livestock and Meat Board.

U.S. Atomic Energy Comm., *Atomic Energy and the Life Sciences,* U.S. Govt. Printing Office, Washington, D.C.

U.S. Govt. Printing Office, *Nutrition Charts* (set of ten charts in color, 19 × 24 in.), Washington, D.C.; $0.75 per set.

Inquire of the large livestock, milk, citrus fruit, and milling companies (listed in Appendix D) for the low-cost materials that may be available. Vitamin charts and booklets are often available from Heinz, Merck, Upjohn, and Eli Lilly.

Add your name to the mailing list of the U.S. Govt. Printing Office (Washington 25, D.C.) to receive listings of the latest publications as they are prepared.

Chapter 5

Distribution of Materials

Each cell has requirements like those of the single cell in the flowing stream. The cells of our bodies, however, are shut away from any chances to obtain directly food, water, and oxygen from the distant larger environment, or to discharge into it the waste materials which result from activity. These conveniences for getting supplies and eliminating debris have been provided by the development of moving streams within the body itself—the blood and lymph streams.[1]

For heterotrophs, or consumers, the "supplies," or organic molecules, must be ingested, or absorbed, and then digested. On the other hand, in organisms whose nutrition is autotrophic, the raw materials —water, minerals, and gases—must reach cells. These materials also move in a stream—the transpiration stream in the xylem of trees. Further, the products of photosynthetic activity and assimilation must also be distributed along another channel moving in an opposite direction— the assimilate stream—carrying manufactured materials from leaves downward through plant tissues. Only in rapidly growing regions of a plant are the two streams moving in the same direction.

The transport of organic substances in the phloem is still under study. In recent studies of the physiology of transportation in trees, analysis of sieve-tube exudate has been made by paper chromatography. Zimmerman[2] describes Hartig's classic method of collecting sieve-tube exudate; he then describes the newer, stylet method, which takes its name from the procedure which uses the stylets of aphids. It has been found that aphids do not suck juices but are fed by the internal pressure of the sieve tubes. As a result, plant exudate flows through the stylets of the aphids, provided the aphids have inserted their stylets into the sieve tubes of the phloem. Aphids can be anesthetized and cut away from their mouth parts; exudation continues from the sieve elements so that the exudate can be collected with a pipette for hours or even days. Such material can be analyzed for its organic composition—that is, kinds of sugars and nitrogenous and phosphorus compounds—at different seasons, at times of defoliation, and after specific experimental treatment.

The upward transport of water and minerals, or the transpiration stream, is described next. The remainder of the chapter deals with techniques to demonstrate the circulatory stream of animals, both invertebrates and vertebrates.

Movement of water in plants

How does water rise in plants? The generally accepted explanation of water transport offered is the transpiration-cohesion-tension theory. Simply stated, this theory implies that a water column rises in the xylem tubes because of the cohesive tendency of water. Thus, a column of water, or transpiration stream, is continuous from roots to leaves. Since

[1] W. Cannon, *The Wisdom of the Body*, Norton, New York, 1932, p. 28.
[2] M. Zimmerman, "Movement of Organic Substances in Trees," *Science* 133:73–79, 1961.

evaporation of water goes on through stomates of leaves (during transpiration), there is a loss of water from the top of the water column of the plant. A tension is set up through which the column of water is lifted up through the plant to the leaves.

Students can demonstrate the rise of water in special tubes or vessels of the plant, and also the occurrence of transpiration in leaves. As a project, a student or a committee of students might study the lifting power of transpiration in plants. A related but seemingly independent action, that of root or exudation pressure, is described here. Students might study this and also examine water absorption by seeds.

The techniques and materials in this section deal with those processes in plants commonly classified as conduction, transpiration, translocation, and diffusion. The exchange of gases (carbon dioxide and oxygen) through stomates of leaves or through the membranes of algal cells is described in the chapters relating to photosynthesis and respiration (Chapters 3, 6).

Transpiration stream

Fibrovascular bundles. Examine the fibrovascular bundles of jewelweed (*Impatiens*). Since the stem is clear, the bundles may be seen by holding the plant up to light. Plants removed from the field may be maintained in the laboratory. In fact, they can be grown from seed in the laboratory. However, the seeds must first be subjected to freezing temperatures for about 2 weeks, then dried for another 2 weeks. Germination and rapid growth take place in moist or wet soil. Seeds or cultivated specimens of *Impatiens* may be purchased from florists.

When the plants are to be used for demonstrating conduction of water, remove the plants—roots and all—from the soil. Then wash the roots and immerse them in a colored solution such as dilute fuchsin, phenol red, eosin, methylene blue, or red ink in water. The path of the colored solution may be traced along the fibrovascular bundles.

It may be convenient to collect jewelweed in season; cut the stems into 3-in. sections and preserve them in a mixture of equal parts of 95 percent alcohol and glycerin. Then, throughout the school year, material will be available to show xylem ducts. Cut the stems into longitudinal sections to reveal these ducts. Thin cross sections of stems may also be made by bundling a few stems tightly; then section them by hand with a razor blade (p. 113) for examination under a microscope. After mounting in water, parenchyma cells and epidermal cells, as well as the spiral markings of the xylem ducts, may be seen. Directions for the preparation of temporary and permanent stained slides are given in Chapter 2.

If jewelweed is not available, other stems may be used effectively. For example, the cut stalks of fresh celery, cornstalks, or bean seedlings, to suggest a few, may be immersed in water colored by adding a few crystals of eosin, fuchsin, or red ink. After a few hours the dye will have moved up to the stems and colored the fibrovascular bundles and the veins of the leaves. Shoots of beans or the leaves of celery or lettuce show red venation. Cross sections of the stems will show the red-colored fibrovascular bundles.

Root hairs. You may also want to have students grow seedlings to show root hairs. Grow soaked seeds of radish in covered Petri dishes for a few days. Or pin a soaked seed to the cork of each of several small vials (Fig. 5-1). (Add a few drops of water before closing the vials.) Each root will grow down into the vial if the vials are placed upright, and the root hairs can be examined with a magnifying glass without removing them from the vials. In this way the root hairs will not dry out as many students examine them during the day.

In a summary exercise, students should be ready to trace the entry of water and soluble salts from the soil into root hairs, up the stems of the plants to the leaves through the fibrovascular bundles.

Fig. 5-1 Apparatus for germinating a seed in a vial to show extensive growth of root hairs.

Use of radioactive compounds and radioisotopes

Many teachers use radioactive substances to show how materials are rapidly absorbed and transported in living things. Current techniques utilize Geiger counters, scintillation counters, incineration techniques, and radioautographs to locate minute quantities of radioactive tracers.

Many of the procedures are not practical at the high school level for lack of equipment, lack of skill, or inability to provide safety precautions. Specially trained teachers are putting into practice activities of the type described here.

Tracing circulation in a stem with uranium nitrate. In this demonstration students can trace the rate of upward movement of a water solution of uranium nitrate by means of a Geiger counter. Prepare a 10 percent solution of uranium nitrate (which may be purchased from a chemical supply house). Stand the stems of geranium plants, stalks of fresh celery, or germinating bean seedlings in the solution.

Before using the Geiger counter, record the background count, which is the rate of clicking that is produced when the Geiger tube or probe is *not* brought near anything in particular. (This background count is due to cosmic rays and radiation from rocks and soil.) Then bring the counter near the plant after 1 hr, 2 hr, and 4 hr. Students should find a high count in the leaves after a few hours. The correct count at any time is the registered count less the previously noted background count.

Radioautograph of a leaf. If a Geiger counter is not available, the demonstration can still be carried out; a radioautograph of the leaf can be made to show the accumulation of uranium nitrate. Dental X-ray film is excellent for this purpose, or use photographic film. Place a leaf from a plant which has been standing in a radioactive solution against unopened film (wrapped in lightproof paper—not aluminum foil). Similarly, place an untreated leaf against the same type of film. Hold the leaves flat against the paper with a weight, such as a block of wood, and keep them in a dark place for 2 weeks. Then remove the leaves and develop the film by placing it in developer, washing it in 3 percent acetic acid (or pure white vinegar), and standing it in photographic hypo for 20 min. Students should see the outline of the treated leaf, with the veins the more prominent regions, in the radioautograph produced (Fig. 5-2).

Using radioactive isotopes as tracers. The Atomic Energy Comm. makes available to high schools certain radioactive isotopes: potassium and sodium phosphate (radiophosphorus-32) and iodine chloride

Fig. 5-2 Radioautograph of leaf of coleus plant that has taken up radioactive phosphorus-32. (Brookhaven National Laboratory.)

(radioiodine-131). When the ordinary caution suggested by the AEC is used, there is no hazard in using these radio-isotopes, for their strength is approximately 10 microcuries, weaker than doses used in medicine.

Young tomato plants may be grown in a hydroponic solution (see p. 458), and radioactive phosphate may be substituted for the ordinary phosphate in preparing the solution. Use the method described above with uranium nitrate to trace the uptake and upward path of the radioisotope with a Geiger counter or the radio-autograph technique.

You may want to write to the U.S. Atomic Energy Comm. in Washington, D.C., to learn the location of distribution centers for your community. The Atomic Energy Comm. provides suggested aids and rules for handling radioactive materials; in fact, a number of publications giving practical demonstrations are available.[3] You may want to try some of the techniques given, such as using a Geiger counter; making radioautographs of a fish skeleton, a leaf, or colonies of bacteria and molds; the uptake of phosphorus-32 by goldfish. Safety precautions and health hazards are discussed in the appendix of the booklet edited by Schenberg.

Evaporation of water through leaves (transpiration)

There are many techniques for showing that water is given off by leaves, specifically, through the stomates in the leaves.

Using collodion to locate the stomates. Spread a thin film of collodion over the upper surface of some leaves, and on the lower surface of other leaves, of a healthy plant. Do not remove the leaves from the plant. After several hours students should see that collodion remains transparent over the regions of the leaves which remain dry, whereas moisture due to transpiration turns the film of collodion an opaque, whitish color. (Also refer to the use of benzene, p. 174.)

The different patterns of distribution of stomates may be discovered by examining upper and lower epidermal layers under the low power of a microscope (see Table 3-3).

Using watch crystals to observe release of water.[4] Fasten small watch crystals to both sides of several different types of green leaves so that part of each leaf is sandwiched between the crystals (Fig. 5-3). These can be fastened with cellophane tape. Seal the glass edges on both sides of the leaves with a film of petroleum jelly. Since the leaves are not detached from the plant, photosynthesis continues and the leaves are not injured. Water vapor, released by the stomates in transpiration, will condense on the cool surfaces of the watch crystals and form drops of moisture. Students should be able to see which surface of the leaf releases more water. Then they may compare, by microscopic examination, the relative number of stomates in the upper and lower epidermis of the different leaves studied (see Table 3-3).

Using cobalt chloride paper to demonstrate loss of water. Transpiration may be shown in leaves on a plant in yet another way. Place a square of dry cobalt chloride paper (blue) on the upper and lower surfaces of the leaves. (Cobalt chloride paper is blue when dry but turns pink when moist.) Then fold a strip of cellophane over the paper (across the leaves) to hold it in place and to protect it from moisture in the air. Fix both with a clip as shown in Fig. 5-4. Elicit from students which surface of the leaves shows a greater degree of change in color of the paper.

[3] S. Schenberg, ed., *Laboratory Experiments with Radioisotopes for High School Science Demonstrations*, rev. ed., U.S. Atomic Energy Comm., U.S. Govt. Printing Office, Washington, D.C., 1958 ($0.35).

An additional reference is H. Miner, R. Shackleton, and F. Watson, *Teaching with Radioisotopes*, U.S. Atomic Energy Comm., U.S. Govt. Printing Office, Washington, D.C., 1959 ($0.40). Also refer to *Radioisotopes*, a booklet of demonstrations prepared by General Biological Supply House, Chicago.

[4] W. J. V. Osterhout, *Experiments with Plants*, Macmillan, New York, 1905.

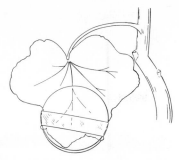

Fig. 5-3 Apparatus for demonstrating transpiration. Water vapor released by the stomates found on the lower surface of the leaf condenses on the lower watch crystal.

Fig. 5-4 Alternate apparatus for demonstrating transpiration. Water vapor released by the stomates on the lower surface of the leaf changes the blue cobalt chloride paper to pink.

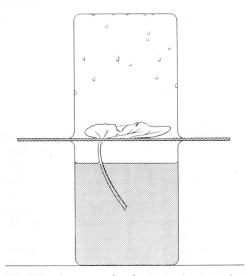

Fig. 5-5 Apparatus for demonstrating transpiration of cut leaf. Compare this with a control in which the leaf is covered with petroleum jelly.

Cobalt chloride paper is easy to prepare. Soak strips of filter paper in a 3 percent cobalt chloride solution. The water solution is red, and the filter paper is red when wet. Dry the strips of filter paper (they will turn blue); then store the prepared paper in a closed container. Before using the paper, it may be necessary to dry it (to turn it blue) in an oven or over a flame, since the paper turns pink in the presence of minute amounts of moisture in the air.

Loss of water from excised leaves. Under a small bell jar or other closed container, place a handful of leaves from geraniums, *Sempervivum,* maple, *Tradescantia,* or other plants. Let the container stand in bright sunlight. Use another small bell jar without leaves to serve as a control. Within 15 to 30 min, students may see moisture resulting from transpiration on the inner surface of the glass.

Or you may use single leaves. Punch a small hole in the center of a piece of cardboard. Then remove a leaf with its petiole from a plant, and insert the petiole through the hole in the cardboard. Place this over a beaker of water so that the petiole is immersed in water (Fig. 5-5). Plug the hole in the cardboard in which the petiole is inserted with petroleum jelly or paraffin to prevent circulation of water vapor. Then cover the leaf with another beaker as shown. As a control, have a student prepare a similar demonstration, but cover both surfaces of the leaf with petroleum jelly. After several hours water vapor condenses on the inside of the upper glass tumbler of the experimental setup (the untreated leaf).

Loss of water from a growing plant. Cover the soil of a small, healthy plant with rubber sheeting or plastic material so that the moisture from the soil and from the surface of the pot cannot escape by evaporation. Then place the plant under a bell jar and keep it in bright sunlight. Or it may be possible to enclose a small plant in a plastic bag. As a control, use a covered pot of soil, with no plant, under another bell jar. Within 15 min, a film of moisture should appear on the inner surface of the bell jar covering the green plant.

Water loss in a whole plant. In this method, a whole plant is weighed periodically for a week. Grow sunflower or other rapidly growing plants (which have an active rate of transpiration) in glazed earthenware crocks or glass jars. Cover the surface of the soil with rubber sheeting or plastic to prevent evaporation from the surface of the soil.

You may want to have students weigh the whole plant in its crock or jar at the beginning of the preparation. Weigh the plant or plants at 24-hr intervals for a week. *Note:* Students might arrange to test many plants in this way so that they have a basis for comparison. (If the plants are watered during the experiment, the weight of the water added should be considered.)

Loss of water from twigs of trees. Get two freshly cut woody twigs and remove the leaves from one of them. With a cork-borer, punch a hole in two corks just large enough to permit the passage of the stems. Cork two flasks with one of the twigs inside each of them, and the cut ends protruding from both. Spread paraffin or petroleum jelly around the joints of the cork and twig so the joints are airtight. Immerse the cut ends of the stems in water. A comparison may be made of the amount of condensation within each flask;[5] Fig. 5-6 shows the experimental flask, with the leafy twig.

Loss of weight in a leaf due to transpiration. Remove at least two leaves, approximately equal in size, from a rubber plant (*Ficus elastica*).[6] When the flow of latex has stopped, slip the petiole of each leaf halfway through a 1-in. length of tightly fitting rubber tubing. Fold over the ½-in. excess of rubber tubing (Fig. 5-7), and wire each securely to avoid evaporation from the petiole.

Coat the lower epidermis of one or more leaves with petroleum jelly; coat the upper

Fig. 5-6 Apparatus for demonstrating transpiration of cut twigs. Considerable water vapor condenses on the inside surface of the flask. Compare this with a control containing a twig with the leaves removed. (Photo: A. M. Winchester.)

epidermis of the other leaves. Now weigh each leaf with its attached tubing and mark each leaf for future identification. Hang these leaves by means of the wire in a dry room or outside the window. After a few hours, weigh both leaves again and compare the rate of transpiration. Students might return to class to weigh these leaves several times during the day. Remember that the stomates in *Ficus elastica* are found in the lower epidermis. Other succulent leaves may be substituted in this demonstration.

Potometer method for measuring water loss. Connect the bottom ends of two burettes with a short rubber tube, as shown in Fig. 5-8. Then fill the whole apparatus with water and plug one bu-

---

[5] B. C. Gruenberg and N. E. Robinson, *Experiments and Projects in Biology,* Ginn, Boston, 1925.
[6] F. Darwin and E. H. Acton, *Practical Physiology of Plants,* Cambridge Univ. Press, New York, 1925.

Fig. 5-7 Leaf of rubber plant, with petiole sealed with rubber tubing, to demonstrate loss of weight due to transpiration through stomates.

rette with a one-hole stopper through which a woody stem has been inserted. Now seal the plant in place with paraffin. Invert a beaker over the other burette to prevent evaporation of water. The water level in this burette should be marked at the start of the demonstration so that as transpiration continues the amount of

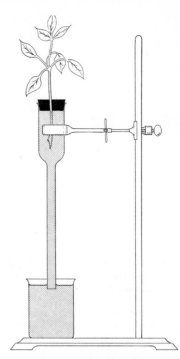

Fig. 5-9 True potometer. An air bubble is introduced at the bottom of the tube and its rate of movement up the tube is measured.

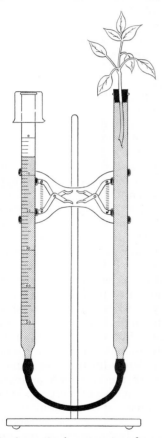

Fig. 5-8 Improvised potometer for measuring water loss in transpiration. As the plant absorbs water, the level in the burette at the left will change; the change can be read from the scale.

water absorbed may be measured by the change in the water level.

The method above is a modification of a true potometer method. Where a potometer is available, a quantitative measurement of transpiration may be made. Place a potometer, filled with water, into a small beaker of water (Fig. 5-9). Keep the bottom of the tube about 1 cm below the surface. Hold a leafy stemmed plant under water (to avoid air bubbles) and cut the stem near the bottom. Insert this stem into a one-hole rubber stopper and fit this tightly into the potometer bowl. Support the potometer with a stand and clamp. No air should remain in the potometer.

After 15 to 20 min introduce the indicator bubble into the potometer tube by raising the end of the potometer out of the beaker and waiting (for a few minutes) until a large bubble appears at the opening. Then lower the end into the water again and measure the rate of movement

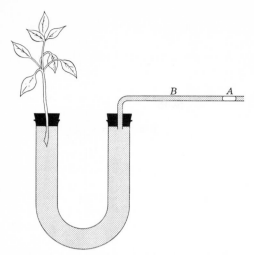

Fig. 5-10 U-tube potometer. As transpiration occurs, the air bubble in the tube will recede a measurable distance from A to B.

of the bubble in the transpiration stream. Calibrate the tube, or attach a millimeter ruler with plastic tape so that the distance the air bubble moves may be measured.

A simple U-tube to measure water loss. This method[7] for measuring the amount of transpiration is simpler than those previously described. Place the cut end of a leafy shoot into a one-hole stopper, and fasten this securely in one side of a water-filled U-tube. Plug the other side of the tube with a stopper through which a right-angle glass bend extends as shown (Fig. 5-10). Attach a millimeter ruler to the side arm to measure the distance the bubble moves. Clamp the preparation to a stand to support it. Wait 15 min before taking readings. As transpiration of water from the leaves of the shoot occurs, measure the movement of water in the extended delivery tube, by measuring how far the air bubble moves.

Students, working in groups, can compare the rate of transpiration under a variety of conditions: in light in still air; in light in a stream of moving air (as from an electric fan); in the dark, with both moving air and still air. Is the rate of transpira-

[7] G. Atkinson, *A College Textbook of Botany*, Holt, Rinehart and Winston, New York, 1905.

tion different if the temperature is lowered some 10°? Or raised 10°?

Lifting power of transpiration

Mechanics of the lifting power of leaves. A porcelain cup (called an atmometer cup) is used to simulate evaporation of water from leaves. The evaporation from the cup creates the conditions under which water is lifted up through thin tubes by capillarity. First, boil the porcelain cup in water to remove all air bubbles. Then insert a 30-in. length of 5-mm glass tubing into a one-hole stopper to fit the atmometer securely. Fill the atmometer and glass tubing with cooled, boiled water so that no air bubbles appear. Cover the end of the tubing with one finger and invert into a container of mercury (Fig. 5-11a). Or use the method shown in Fig. 5-11b, which uses water only. In either method, clamp the preparation to a stand to support it. Then fan the porcelain cup or use an electric fan to stir the air to speed evaporation (water loss) from the porous cup. In (a) the mercury will soon rise in the tube; it will continue to rise higher than normal atmospheric pressure would lift it. The additional rise of the column of mercury is the result of a tension set up by rapid evaporation of water through the porous cup. In (b) the amount of water loss can be read from the calibrated burette. A similar demonstration using living plants may be developed (as below).

The lifting power of leaves in a living plant. Hold a plant under water to cut off a healthy leafy shoot; insert the shoot into rubber tubing. Then connect a section of glass tubing into the other end of the rubber tubing to make an airtight connection. Now completely fill the glass tubing with water, cover the end with a finger, and lower it into a dish of mercury. When the plant is supported by a clamp, the setup is similar to the previous demonstration (Fig. 5-11a), but in place of the atmometer, a living plant is used. Have students watch the gradual rise of mercury in the tube. In this demonstration a sizable rise

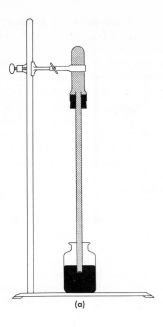

(a)

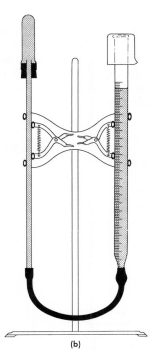

(b)

Fig. 5-11 Atmometers: (a) porous cup (porcelain) atmometer with water and mercury; (b) porous cup (porcelain) atmometer with water-filled burette from which amount of evaporation can be read.

in the mercury column is usually not observable because air bubbles often interfere.

A simple U-tube to show lifting power. You may want to have students prepare the apparatus shown in Fig. 5-12. Here a leafy shoot is fitted into a stopper, airtight. Add some water to this arm of the U-tube; to the other arm add mercury so that the level is equal in both arms of the "U." In a short time (within the hour) students may see the mercury level displaced in the direction of the tube containing the shoot.

Exudation pressure (root pressure). You may have noticed that when a plant is cut off at the ground level, water exudes from the cut stem because of root pressure.

Plants grown in the open may be used, but potted plants such as the geranium are more convenient for use in the classroom. Cut off the stem close to the soil level, and use a short rubber tube to attach a long glass tube to the rooted part of the stem.

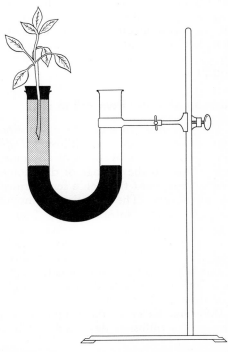

Fig. 5-12 U-tube apparatus demonstrating the lifting power of the transpiration stream.

Clamp the setup to a stand to keep it upright (Fig. 5-13). Now pour a small amount of water into the tube so that the stem remains moist. Within a short time (10 min or so) water begins to rise in the tube.

Palladin's method to show conduction of water up a stem.[8] Both the ascent of liquids and the possible effect of transpiration on the rise of fluids in plants may be studied in this method (Fig. 5-14).

Select a plant with a forked stem or side branches; a branch of a bush, or a tree with a side shoot will do nicely. Remove the leaves from the side shoot and attach the shoot to rubber tubing which leads to a glass tube. Be sure all joints are watertight. Then fill the glass tube with water and immerse it in mercury. For the best results, the shoot should be thin and the glass tube of narrow bore.

Guttation. When transpiration is retarded, and there is a free intake of water through the roots, exudation of water from the leaves occurs. This can be shown best by growing cereal seeds under a small bell jar. In this moisture-laden air, transpiration decreases, and *drops* of water appear on the leaves (the process is known as guttation).

Passage of water from cell to cell

First, we need to make a clear distinction between osmosis and diffusion. *Diffusion* refers to the movement of any molecule. *Osmosis* is the passage or diffusion of molecules of *water* through a semipermeable membrane. The direction of motion (in non-ionized substances) is from a region of higher to one of lower concentration of the particular molecule under study. Thus, diffusion refers to the distribution of molecules of ammonia through a classroom as well as to the passage of glucose through a membrane.

Diffusion. Refer to the simple demonstrations of diffusion described in Chapter 4.

[8] V. Palladin, *Plant Physiology,* Blakiston (McGraw-Hill), New York, 1926.

You may want to set up a thistle tube as described in Chapter 4 and shown in Fig. 4-8, but with a 20 to 30 percent sucrose solution in the thistle tube. This may be considered a model of the content of the cell sap of a cell. (It is not a good model of a cell, however, since it cannot show changes in turgidity.) Just as the water enters the thistle tube, so water molecules are thought to enter root hairs and diffuse from cell to cell along a diffusion gradient.

Membranes of cells. Cells are separated from each other by membranes through which certain molecules are able to pass; the membranes, however, prevent the diffusion of other molecules. Hence these membranes are semipermeable. Further, membranes also have the characteristic of selective permeability; uncharged particles pass through more readily than charged ones, and fat-soluble molecules seem to diffuse regardless of size.

The movement of water in and out of cells is more complex than indicated in the following activity. For example, permeability to water may vary with the physiological state of cells, as well as problems offered by the overall positive charge of cell membranes.

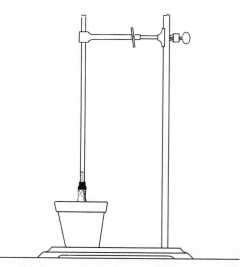

Fig. 5-13 Apparatus for demonstrating exudation pressure of roots.

For more advanced study you may want to have students investigate the factors involved in the diffusion of molecules of glucose or essential minerals, as well as of water, through membranes of cells. They may want to study the description of Donnan equilibrium in a general physiology text such as *Comparative Animal Physiology,* and also in *Plants in Action.*[9]

Some students may want to learn more about active transport through membranes and possibly may demonstrate for the class the active transport of glucose, using the gut of a freshly killed hamster.[10]

Active transport requires a supply of energy from ATP for a pumping process which transports molecules against the diffusion gradient from a region of lower concentration to one of higher concentration. (Refer to A. Giese[11] for a full chapter on active transport through the cell membrane. Examples described include the accumulation of potassium in cells of some water plants such as *Nitella* and *Chara,* root pressure in plants, vacuolar function in protozoans, nephric organs in marine fish, and a possible mechanism of active transport.)

Also investigate the role of living kidney fragments from a goldfish. Treated with phenol red and examined under a microscope, the lumen of a tubule may be found to have concentrated the phenol red from the surrounding solution.[12]

Osmotic pressure of cells. Select leaves of *Anacharis* (elodea) that are in the same or associated whorls near the growing tip to show the effect of placing cells in hypertonic solutions, that is, more concentrated solutions. As water moves out of the cells,

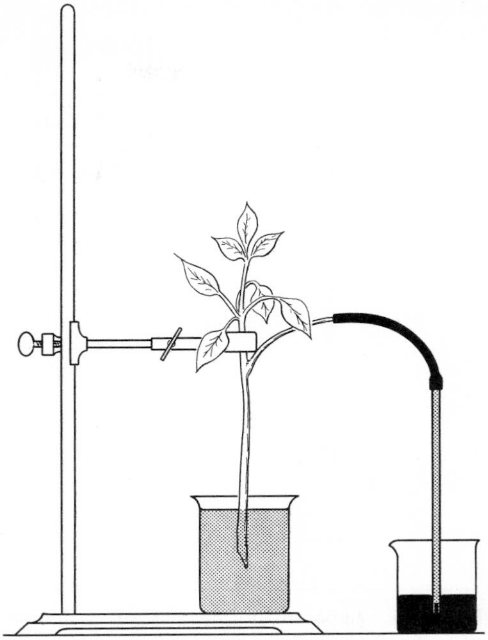

Fig. 5-14 Apparatus for demonstrating movement of water in a woody twig.

the cell contents shrink and the cell is plasmolyzed.

It is possible to study the average osmotic pressure of the cell sap of these cells. Place the leaves in separate Syracuse dishes, each containing one of the following concentrations of cane sugar: $M/2$, $M/3$, $M/4$, $M/5$, $M/6$, $M/7$, and $M/8$. Place similar leaves in Syracuse dishes each containing one of the following concentrations of potassium nitrate: $M/2$, $M/3$, $M/4$, $M/5$, $M/6$, $M/7$, $M/8$, $M/9$, and $M/10$.

Under the microscope look for evidence of plasmolysis. Students can determine which concentration of cane sugar and of KNO_3 is isotonic, that is, the concentration that just fails to cause plasmolysis in most cells.

More readily visible to students are the results obtained using cells with cytoplasm containing anthocyanin. Use epithelial tissue from rex begonia, cells of the red onion bulb, or tissue from the lower epidermis of some varieties of *Tradescantia*

[9] C. L. Prosser and F. A. Brown, Jr., *Comparative Animal Physiology,* 2nd ed., Saunders, Philadelphia, 1961. B. Machlis and J. Torrey, *Plants in Action,* Freeman, San Francisco, 1956, Chapter 3.

[10] See the excellent description in G. Wald *et al., Twenty-six Afternoons of Biology,* Addison-Wesley, Reading, Mass., 1962, Chapter 17.

[11] A. Giese, *Cell Physiology,* 2nd ed., Saunders, Philadelphia, 1962, Chapter 13.

[12] This demonstration is described in *Biological Science: Molecules to Man* (BSCS Blue Version), Houghton Mifflin, Boston, 1963, "Laboratory Investigations," pp. 102–03.

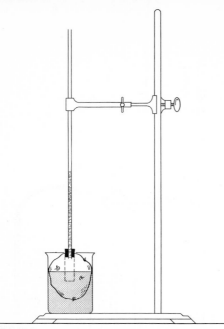

Fig. 5-15 Apparatus for demonstrating diffusion of water through the cells of a raw potato. The cavity in the potato is filled with sucrose solution; water diffuses in through the membranes of the cells in the potato to the cavity and rises in the tube, carrying with it some of the sucrose.

(purplish pigment on the undersurface of the leaves).

Some teachers use descriptions of the experiments of Stephen Hales on water transport in *Vegetable Staticks* (1727). For additional activities on calculating osmotic pressure of cells, or, in fact, the whole area of absorption of water, look into *Plants in Action*[13]; or refer to the other references given in the bibliography at the end of this chapter.

Osmosis through a raw potato. The living cells of a potato (or a carrot) may be used to show the passage of water through the semipermeable membranes which surround the cells (Fig. 5-15). You may want to use an apple corer to remove a center cylinder of a raw white potato. Do not plunge the corer through the entire length of the potato, but leave about ½-in. thickness at the bottom. Slowly pour a concen-

[13] *Op. cit.*, Chapter 3.

trated sucrose solution into the core, and close with a one-hole rubber stopper through which a piece of glass tubing has been inserted. Place this in a beaker of water; use a clamp to hold the tube upright. Then seal the stopper with melted paraffin to prevent leakage.

Osmosis through a raw egg. This technique may be of interest to some teachers, although the results may not be worth the time consumed in its preparation. Carefully crack and remove the shell at the blunt end of an egg to expose the underlying membrane. This is a semipermeable membrane (it permits the passage of water only). You may dissolve the shell with dilute hydrochloric acid instead of cracking the egg. At any rate, the cut region of the shell must be smooth to avoid puncturing the membrane. Invert the egg and fit it into a bottle of water so that the egg is supported by the neck. Next, carefully break through the other end of the egg to make a narrow hole to fit a piece of glass tubing, which is to be inserted into the egg contents. Clamp the tubing in place, and seal it to the egg with melted paraffin (Fig. 5-16). You may need to replace the water in the bottle if it falls below the level of the egg membrane.

Amy Applegate,[14] a teacher who has corresponded with us, has offered a much simpler technique. Stand an egg in the mouth of a bottle. Fasten a short length of glass tubing to the pointed end of the egg with sealing wax. Reach through the glass tubing with the end of the handle of a deflagrating spoon to make a small hole in the tip of the egg. By means of a short length of rubber tubing connect a longer length of glass tubing and support this on a ring stand. It is only necessary to crack the blunt end of the egg so that water from the container can diffuse through the membrane; this reduces breakage of the membranes because the shell does not have to be chipped away or dissolved away with

[14] A. Applegate, Bloom Township High School, Chicago Heights, Ill., personal communication, June 1958.

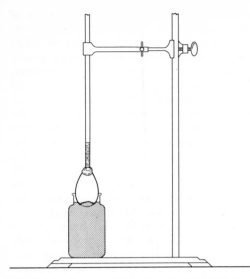

Fig. 5-16 Apparatus for demonstrating diffusion of water through the membrane of a raw egg. Water diffuses through the exposed shell membrane at the blunt end, up through the egg, out the hole in the top, and up the tube, carrying with it some of the egg contents.

acid. Students soon can observe the contents of the egg rising in the glass tube.

(For a demonstration of selectivity in the absorption of ions in nutrient solutions, see hydroponics, pp. 458–62.)

Absorption of water by seeds: imbibition and swelling

The rapid absorption of water by the colloids found in the seed may be readily illustrated. Fill a tumbler one-third full with dry beans or peas, and the rest with water. Within a few hours students may see swollen seeds; they should notice that the seeds now almost fill the tumbler.

Imbibition pressure of seeds. Make a cardboard box mold about 1 in. in depth. Fill it with a thick paste of plaster of Paris; drop in a handful of dry oats, wheat, or peas, and allow the paste to harden. After an hour or so, cut away the cardboard box and place the plaster block in water.

The small capillary spaces in the plaster block absorb water, so that water is carried

to seeds. The block is burst as a result of the swelling of the seeds due to imbibition. *Note:* You may want to refer to the structure of leaves, stems, and roots at this time. Freehand sections for study under the microscope may be attempted (p. 113); or prepared, stained slides may be purchased. A description of the parts of plants is given in Chapter 1. The preparation of tissue for slide-making is described in Chapter 2.

Circulation in animals

You may want to demonstrate the heartbeat and the circulation of blood in several animals, invertebrate and vertebrate. There are also many opportunities for individual laboratory experiences and for group demonstrations. In this section some activities related to circulation and the heart, blood cells, and blood typing are described.[15]

Recording arterial pulse

Students learn easily to take their own pulse. They should place the index and middle finger of one hand on the inner side of the other wrist. About 1½ in. above the wrist joint, near the thumb, the pulse may be felt in the radial artery. Then students may want to take each other's pulse. They should count the pulse for several 15-sec periods, find the average, and multiply this by four to get the average pulse rate per minute.

Next, compare this normal pulse rate with the rate after exercise. Students might hop in place, or do knee-bending exercises, or run up a flight of stairs. Then count the pulse again. (*Note:* When teachers select students for any exercise they should be apprised first of their students' medical records. These may be obtained from the

[15] A teacher must, of course, select those techniques that are best suited to his classroom situation. As a guide you may want to examine the laboratory activities in the *Student Laboratory Guide* to *Biological Science: An Inquiry into Life* (BSCS Yellow Version), Harcourt, Brace & World, New York, 1963; or in the "Laboratory Investigations" section of *Biological Science: Molecules to Man* (BSCS Blue Version), *op. cit.*

school's health education department.)
Also refer to the effect of increased activity
on rate of production of CO_2 (p. 268).

The heart

Heartbeat in the frog. Anesthetize a
frog with chloretone or urethane (p. 112),
or pith a frog (Fig. 7-9). Then open the
frog to expose the heart. Remove the pericardial sac, and bathe the heart (throughout the demonstration) in Ringer's solution (p. 665). In this way the heart will
continue to beat for several hours.

For class demonstration you can am-
plify the contractions of the heart by
making a lever arrangement (Fig. 5-17).

ALTERING THE HEARTBEAT. The effect of
temperature may be shown by bathing the
frog's heart with water warmed to 40° C
(104° F), then with iced water.

The action of adrenalin may be shown
by adding a minute quantity (0.01 per-

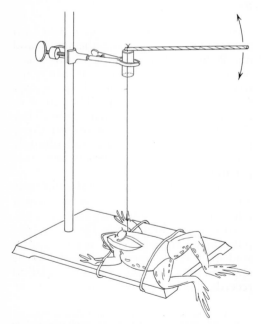

Fig. 5-17 Apparatus for demonstrating heartbeat of an anesthetized frog. The tip of the ventricle
is pierced by a bent pin tied to a thin thread that is
attached to a straw supported by a hollow glass
tube.

cent) to the Ringer's solution. A bit of
experimentation may be necessary to
determine the quantity needed to affect
the particular frog's heart under study.

EFFECT OF MINERALS ON CONTRACTION.
Ringer's solution is composed of ions of
sodium, potassium, and calcium in definite
proportions. You may want to show the
effect of these ions when applied separately
to the beating frog's heart. Suspend the
heart of a freshly killed or pithed frog by
means of a bent pin (like a fishhook) and
thread attached to a tongue depressor in
a beaker of Ringer's solution. Count the
number of times the heart beats in a
minute. Now transfer the heart into a
beaker of 0.7 percent sodium chloride
solution. Count the number of beats now.
Keep the solutions at room temperature.
You will notice that the heartbeats are
weaker and slower. Now replace the heart
in Ringer's solution and note the recovery
in rate of heartbeat.

Next, transfer the heart into a beaker of
0.9 percent potassium chloride solution.
Again count the heartbeats until the heart
muscle stops contracting. If the heart is
now transferred into a 1 percent calcium
chloride solution you will find the heart
contracting again. Replace the heart in
Ringer's solution and watch the return to
normal rate of heartbeat. The ions in the
Ringer's solution are balanced to effect
the normal contraction of the frog's heart.

Heartbeat in the clam. In many areas
students may get hard-shelled clams from
a fish market. Open a live clam by gently
cracking one shell; then cut the anterior
and posterior adductor or shell muscles
with a sharp knife. Remove the mantle
around the viscera and observe the beating heart in the pericardial cavity, which
lies below the dorsal hinge ligament (Fig.
1-40). Bathe the heart in clam juice or
Ringer's solution. Watch the contraction
of the ventricle.

Heartbeat in embryo snails. Students
may have the snail *Physa* (or other egg-
laying snails) in their aquaria at home.
Tease apart a developing egg mass in a
drop of aquarium water on a microscope

slide and examine under low power (without a coverslip). Students will find a beating heart in an embryo in the early stages when the shell is still transparent. (Also note the periodic muscular contractions of the body.)

Heartbeat in Daphnia. *Daphnia* (water fleas) may be obtained from aquarium stores or from lakes and ponds (culturing, Chapter 12). Mount several for examination under low power of the microscope. Add a bristle to the slide before applying a coverslip, to prevent crushing the specimen. Look for the rapidly beating heart, which is situated in the dorsal region behind the eye (Fig. 1-26a). You may change the rate of the heartbeat by warming the slide or by adding a trace of adrenalin (about 0.01 percent). Since the heartbeat is very rapid (almost 300 beats per min), this effect may not be simple to detect. Students will need a stopwatch for these counts.

A strobe disc made of cardboard with a single slit and turned by a 300-rpm motor can be used in front of the light source to slow the heartbeat's apparent motion.

Heartbeat in aquatic worms. Mount any one of these annelid forms, which may be obtained from an aquarium store or a lake or pond: *Dero, Nais, Aulophorus,* or *Tubifex* (culturing, Chapter 12). Add a bristle to the mount before the coverslip is added. In *Tubifex* look for the single pair of aortic arches or loops, and note the movement of the dorsal blood vessel, which pumps blood.

The worms may be anesthetized first by placing them in a drop of water to which a drop of very dilute formaldehyde has been added. (Other methods of anesthetizing small organisms are given in Table 11-2 and p. 112.)

Dissecting the heart. Students may get an untrimmed heart of a beef, dog, sheep, or hog (see Fig. 1-78) from a butcher or a biological supply house. They can locate the auricles, ventricles, aorta, venae cavae, and pulmonary vessels (Fig. 5-18). Insert a soft metal probe, or use a fire-polished section of glass tubing as a probe, to ex-

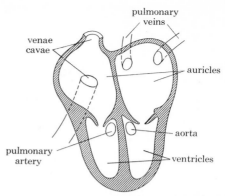

Fig. 5-18 Schematic diagram of the structure of the mammalian heart.

plore the point of origin of blood vessels which lead into and leave the chambers of the heart. Note the semilunar valves in the blood vessels, and determine the direction of blood flow.

Dissect the chambers of the heart. Notice a diagonal deposit of fat along the lower two-thirds of the heart to be dissected. Use this as a guideline that marks the wall between the two ventricles. Make an incision into each of the two ventricles. Spread the heart open and use a straw or a glass rod to probe the position of the bicuspid and tricuspid valves, which separate the ventricles from the auricles.

If possible, direct a stream of water (through tubing from a faucet) through the ventricles to examine the operation of the valves.

Compare the thickness of the walls of the auricles and ventricles. Note especially the thickness of the left ventricle, through which blood is pumped to the entire body. Look for the coronary arteries from the aorta and trace their paths as they spread over the heart. Students might cut cross sections of the aorta and a vena cava to compare the elasticity and thickness of the walls.

Circulation in capillaries

Circulation in the tail of a goldfish. You may want to show circulation in the capillaries, arterioles, and venules in the almost

Fig. 5-19 Demonstrating circulation of blood in the tail of a goldfish.

transparent tail of a small goldfish. Wrap the body of the goldfish in wet absorbent cotton so that only the tail is exposed (Fig. 5-19). Place it in a Petri dish with a small amount of water. Cover the tail with a glass slide to hold it flat and expanded. Examine under low and high power of the microscope.

Watch the pulse and the swiftly moving blood in the arterioles. Contrast this with the more slowly moving blood which circulates in the opposite direction in the venules. Many criss-crossing capillaries are visible. Look for blood cells floating in the blood stream. Incidentally, irregular black patches of pigment in the skin (chromatophores) may be visible in the specimens (see p. 110).

Have students set up several similar demonstrations so that everyone may observe circulation within a 15-min period. Then return the fish to the aquarium. At times the fish flip their tails vigorously, and the demonstrations must be adjusted. When desired, the fish may be anesthetized with chloretone or urethane (p. 112). However, circulation is sluggish in anesthetized fish.

Circulation in the gills and tail of a tadpole. When very young tadpoles of am-
phibia are available, students can observe circulation of blood in the external gills. Later the gills are internal as epithelial cells of the skin grow over them. Nevertheless, in these older forms circulation can be observed in the almost transparent tail.

Mount a tadpole on a depression (welled) slide in a drop of aquarium water. Omit the coverslip or add a bristle to prevent the coverslip from crushing the specimen. Watch the circulation in capillaries under the low and high power.

Circulation in the frog

IN THE FOOT. The webbed skin between the toes of the frog is rich in capillaries. Wrap a live frog in wet cheesecloth, paper toweling, or absorbent cotton, and hold fast in a plastic bag. Lay the animal on a frog board made of cork, so that only one foot is extruded. Or you may want to anesthetize the frog by injecting it with about 5 ml of a 2.5 percent solution of urethane (p. 112). Inject into the lymph sacs under the loose skin along the sides and back of the frog. Frogs can be anesthetized without injection by putting them into a solution of chloretone. Add 1 part of a 0.5 percent solution of chloretone in water, to 4 parts of Ringer's solution. Keep the frog in this solution for about half an hour. Avoid longer immersion.

Whether the frog is untreated or anesthetized, pin it to the board by fastening it with cellophane tape. Punch a hole in one corner of the corkboard and spread the web of one foot over this hole; pin in place or wind thread around the toes and secure it with pins to keep the toes immobile (Fig. 5-20). Keep the frog's skin wet during the demonstration.

Place the board on the stage of the microscope and focus on the webbed foot. Observe capillaries, arterioles, and venules and the rate of blood flow. Notice the direction of the flow of blood and the pulsations of the blood vessels. Estimate the diameter of the capillaries (see below). Under high power compare the movement of red and white corpuscles. Notice the appearance and disappearance of capillaries.

It may be possible to find, using high power, several white corpuscles moving through capillary walls.

A committee of students may want to study the effect of varying conditions on circulation of blood. For example, place a small piece of ice on the web and observe the effect on circulation. Then alternate with warm water, 35° to 40° C (95° to 104° F), and observe the results. Watch the action of chemicals by adding a drop of dilute adrenalin (0.01 percent) to the web. The following solutions are recommended[16] to study the effects of dilators and constrictors on capillaries: lactic acid solution (0.1 g per liter), histamine acid phosphate solution (0.1 g per liter), acetylcholine (0.05 g per liter), adrenalin chloride solution (0.1 g per liter), nicotine solution, ethyl alcohol (95 percent), and sodium nitrite solution (0.1 g per liter). (Each solution should be washed away before the next solution is applied.)

[16] *Biological Science: Molecules to Man* (BSCS Blue Version), *op. cit.,* "Laboratory Investigations," p. 88.

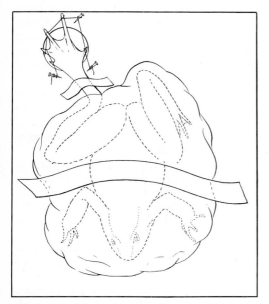

Fig. 5-20 Demonstrating circulation of blood in the webbed foot of a frog. The frog is wrapped in wet gauze in a small plastic bag and secured on a corkboard with cellophane tape.

IN THE TONGUE. Anesthetize a frog by placing it in chloretone or injecting it with urethane (p. 113). Wrap the body in wet toweling, cheesecloth, or absorbent cotton; and place the animal, dorsal side up, on a corkboard. Spread the unattached tip of the tongue over the hole in the corkboard and secure with pins. Or cover the tongue with a strip of cellophane and fasten to the corkboard with pins.

Place the board on the stage of the microscope, keep the tongue well lighted, and observe under low power a section revealing blood vessels of different sizes. Clamp the board to the stage of the microscope with a rubber band.

Keep the tongue wet with Ringer's solution. Watch the pulsations in the arteries, the different direction of blood flow in arteries and veins, and the capillaries connecting the larger blood vessels.

Then under high power try to distinguish red and the larger white corpuscles. Estimate the diameter of the capillaries by using red corpuscles as reference. In the frog, these cells average 22 μ in length, 15 μ in width, and 4 μ in thickness.

Compare these effects on circulation: Apply slight pressure to the tongue by stroking it with a dissecting needle. Next apply, drop by drop, a small quantity of urethane to a small area of the tongue. Watch the changes in the capillaries. Then tie a string around a portion of the tongue to close off circulation for a short time. Observe the effect and watch the changes as the thread is released. Also apply a small amount of adrenalin (0.01 to 0.1 percent) to a small area of the tongue, and watch the changes that the adrenalin produces.

IN THE LUNG. Anesthetize (p. 113) or pith (Fig. 7-9) a frog. Then pin the frog to a corkboard or a waxed dissection pan and dissect to expose the lungs (dissection, Chapter 1). Inflate the lungs by blowing through fire-polished glass tubing inserted into the glottis of the frog. Keep the lungs bathed in Ringer's solution. Examine the inflated air sacs and the rich network of capillaries in a bright light using a hand lens or binocular microscope. Observe circulation in the capillaries.

A small section of inflated lung can be examined under a compound microscope too. Spread a thin section over the hole in a sheet of cork similar to the one described above.

Note: You may also spread the capillary-rich mesentery supporting the intestine across the opening in the sheet of cork for observation of circulation. This is a dramatic demonstration.

IN THE URINARY BLADDER. The study of peripheral circulation in the urinary bladder of the leopard frog has also been recommended.[17] In this preparation, blood vessels are not obscured by pigment cells, and since the vessels are confined to a single plane, complete circuits can be observed through arterioles, capillaries, and venules without refocusing the microscope.

A frog is prepared for the demonstration by a preliminary injection of 5 ml of Ringer's solution into the dorsal lymph sinus. This pretreatment results in a distension of the bladder. A half-hour later the frog is anesthetized by an injection of 3 to 6 ml of 5 percent urethane (p. 113) into the dorsal lymph sinus. When the frog is anesthetized (withdrawal reflex has ceased), a 2-cm incision is made parallel to the midline and a bit dorsal to the ventral edge of pigmentation. Start the incision about 1 cm anterior to the hind limb. The distended urinary bladder can be extruded through the incision and prepared for study under a binocular microscope with transmitted light. (If the bladder is not distended, Ringer's solution may be inserted through the cloaca.)

McNeil *et al.* also suggest observation of vasodilator effects of histamine (1 or 2 drops of 0.001 histamine) or of 1 drop of 1 percent acetic acid. Wash off the chemicals between applications. Also apply 1 drop of 0.001 M solution (or more dilute) epinephrine to show the vasocontractor effect on arterioles.

Circulation in the earthworm. Secure several live earthworms. Suspend them, anterior end down, for a few minutes or twirl them around so that the blood is concentrated into the anterior end. Anesthetize the worms by placing them in a shallow dish with enough water to cover them. To this add a small amount of alcohol, sufficient to immobilize the worms but *not* to kill them. Next wash off in water, and pin each worm to a waxed dissection pan or a sheet of cork. Make a lengthwise cut from the clitellum to the prostomium (tip of the head), cutting to one side of the mid-dorsal line (dissection, Chapter 1). Cut through the body wall but avoid cutting into the viscera. Watch the beating of the five pairs of aortic loops, or "hearts." With a hand lens or a binocular microscope students may observe the circulation of blood more clearly.

Circulation may also be observed in *Tubifex* (p. 245).

Blood vessels. When possible have students get lengths of arteries and veins from a butcher and bring them to class. Compare their thickness in cross section and the difference in elasticity. Cut along the length of a vein, and open it to show the cup-like valves. Hold it upright and try to fill it with water to show the action of the valves. Valves in lymph vessels may also be studied as a prepared slide (Fig. 5-21).

ACTION OF THE VALVES.[18] Students may prepare the model shown in Fig. 5-22.

Attach a thin strip of rubber sheeting to a one-hole stopper with thumb tacks. This represents a valve in the one cylinder. Fill the glass cylinders with red ink. As the bulb is squeezed the increased air pressure forces the water through A into B. Why doesn't the water flow back into A? Then compare this with the flow of blood from the extremities back to the heart.

DEVICES TO ILLUSTRATE BLOOD VESSELS. At times we may establish a concept with remarkably simple visual devices. Because

[17] C. W. McNeil *et al.,* "The Use of the Urinary Bladder of the Leopard Frog in the Demonstration of Peripheral Circulation," *Turtox News* 36:8, August, 1958.

[18] P. E. Blackwood, *Experiences in Science,* 2nd ed., Harcourt, Brace & World, New York, 1955, p. 24.

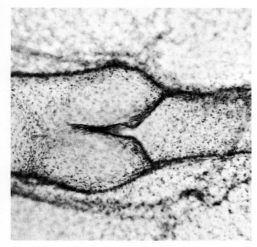

Fig. 5-21 Photo of valve in lymph vessel as shown on a prepared microscope slide. (Courtesy of General Biological Supply House, Inc., Chicago.)

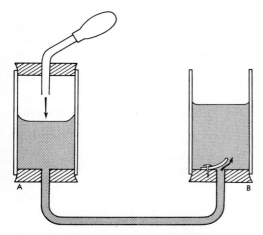

Fig. 5-22 Model of valves in blood vessels. (After E. Morholt, *Experiences in Biology*, new ed., Harcourt, Brace & World, 1960.)

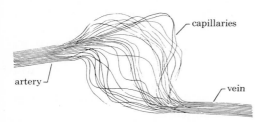

Fig. 5-23 Demonstrating a capillary bed, using a piece of frayed cord.

they are immediately contrived before the eyes of youngsters they have additional impact.

Have you ever used these? Pull apart cheesecloth or gauze. This will represent the appearance of capillary beds. Or have you separated out the central strands of a coarse hemp cord (Fig. 5-23)? Dip one unfrayed end of the cord into red ink, the other into blue. Isn't this a standard representation of arteriole, capillaries, and venule?

Remove the insulation from the center section of a 4-in. length of electric wire. Spread apart the individual fine copper wires in the center. Imagine that these separated wires represent capillaries. Then the ends might represent an artery and vein.

When you want to show capillary circulation around an air sac in the lungs, strands of cord wrapped around a Florence flask may give students the general idea. Or with a glass-marking pencil draw "capillaries" on the flask. And, in similar fashion, show circulation in a villus of the small intestine. Wrap cord around a small test tube. Or draw capillaries with a glass-marking pencil. Now insert this into a larger test tube. This will serve as a villus with an outer epithelial layer. The smaller test tube within the larger one represents a lacteal.

The blood

Differences between blood in arteries and in veins. The effect of the amount of oxygen supply in the blood in arteries as compared with veins may be demonstrated. This technique is described in the section on respiration of animals (Chapter 6).

Blood cells. Fresh mounts of blood for examination under the microscope will be described here. Staining techniques using Wright's blood stain or Giemsa stain are described in Chapter 2.

HUMAN BLOOD CELLS. You may want to have students examine their own blood cells under the microscope.

Note: Whenever activities call for taking samples of students' blood, the following procedure is advisable:

1. Get consent notes from parents if your community or school requires them. A note might be obtained for the term or year to cover such activities as making blood smears and typing blood.
2. Apply alcohol to the area of the finger to be pricked. Allow the alcohol to evaporate.
3. Use individually wrapped sterile, disposable lancets.
4. After puncturing the skin and obtaining blood, clean the finger with alcohol again. If bleeding continues, have the student apply pressure by clenching a piece of sterile cotton or a square of gauze.

After following this procedure, apply the blood to a clean slide. Add a drop of Ringer's or Locke's solution (pp. 664, 665) and a coverslip. Compare the red corpuscles, which are present in vast number, with the white blood cells. When red blood cells show crenation (wavy, irregular outline), the mounting fluid is hypertonic and should be diluted with distilled water. Introduce a drop of 10 percent sodium chloride solution under the slide. What is the effect on red blood cells? What is an isotonic solution?

White blood cells. White corpuscles may be studied more readily after red corpuscles have been destroyed. Prepare the slide as described above. Then put a drop of acetic acid at the edge of the coverslip so that it diffuses into the mounting fluid. Note the clear view of white blood cells that remains.

Blood counts. A small group of students, for example, a club, may want to learn to count red blood cells. For this technique micropipettes and special counting chambers are needed.

First, a measured volume of normal blood is taken from the finger and is drawn into a micropipette. This is diluted to a definite volume with a solution such as Hayem's (p. 662). Then the red blood cells become diluted and visible for count-

ing. The blood in the pipette is mixed well and a drop is placed in the counting chamber. Specific details of these techniques may be found in such references as those by Hawk, Oser, and Summerson and Todd and Sanford (listed at the end of this chapter). Or a doctor may demonstrate the technique. (See p. 443 for a modification—counting algal cells.)

Facts about human blood. For students who are interested in doing blood counts and similar work, the information in Table 5-1 and Fig. 5-24 may be useful.

You may also want students to observe the action of catalase of blood in decomposing hydrogen peroxide (Chapter 2).

BLOOD CELLS OF A FROG. Mount a drop of fresh frog's blood on a clean slide. Touch one end of a second slide to the drop; hold the second slide at a 30° angle to the first slide. Push the second slide along the first so that a thin film of blood is spread the length of the slide (Fig. 2-24). Let the slide dry. Examine without a coverslip, or later stain with Wright's blood stain (p. 118) and mount the slide in balsam with a coverslip.

A drop of blood may be examined in a drop of Ringer's solution as a wet mount instead of a dry smear. Look for red cells that are oval, nucleated discs.

PHAGOCYTOSIS BY LEUCOCYTES OF A FROG. Students may observe the manner in which leucocytes of frog's blood ingest materials. Add a small amount of India ink or carmine powder to some physiological saline or Ringer's solution for frogs (p. 665).

Anesthetize a frog or use a pithed frog; insert a hypodermic needle into a dorsal lymph sac and draw off a small quantity of lymph. On a clean slide add a drop of lymph to a drop of saline plus ink (or carmine particles). Add a bit of broken coverslip or a piece of straw before placing a coverslip over the preparation; this will prevent the coverslip from flattening the cells. Seal the edges of the coverslip with petroleum jelly.

Inspect the slide under low power and then switch to high power and refocus the

TABLE 5-1

Normal conditions of human blood

Red blood cells	4,750,000–5,400,000 per cu mm
White blood cells	7000–10,000 per cu mm
Platelets	300,000 per cu mm (approx.)
Hemoglobin	90–105%
Color index	0.9–1.0
Specific gravity	1.055–1.066
Coagulation time	3½–5 min
Differential white cell count	
Polymorphonuclears	60–70%
Small lymphocytes	20–25%
Large lymphocytes	5–8%
Eosinophils	1–2%
Basophils	1–2%
Pulse rate	
Birth	130–150 per min
6–10 years	90–100 per min
10–14 years	80–90 per min
Adulthood	72 (average)
Old age	67 (average)

light to look for leucocytes engulfing particles suspended in the fluid.

MODEL OF PHAGOCYTOSIS. The movement of phagocytic cells, such as some kinds of white corpuscles and amoebae, is partially explained on the basis of surface tension; you may want to simulate such an action by preparing a "model." Place a drop of clean mercury in a bit of dilute acid in a Syracuse dish. Next to the mercury put a small crystal of potassium dichromate. Note the movement of the mercury, and explain in terms of changes in surface tension.

BLOOD CELLS OF A MAMMAL. Occasionally you may want students to compare the red blood cells of a frog with those of a mammal. Blood may be obtained from a slaughterhouse; usually ox blood is available. To prevent clotting, add 0.1 g of potassium or sodium oxalate for every 100 ml of ox blood. Or you may prefer to add 1 part of 2 percent sodium citrate solution to 4 parts of blood. Both the oxalate and citrate are anticoagulants.

Then mount a drop of the ox blood as a wet mount (or prepare a blood smear as for frog's blood, above). Both white and red cells are visible. You may use petroleum

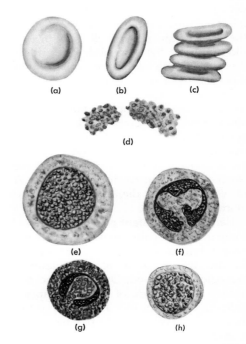

Fig. 5-24 Human blood cells: (a)–(c) red blood cells; (d) platelets; (e)–(h) leucocytes—(e) eosinophil, (f) neutrophil, (g) basophil, (h) lymphocyte. (After W. G. Whaley, O. P. Breland, et al., *Principles of Biology*, Harper & Row, 1954.)

jelly on the mount to prevent evaporation (p. 111).

AMOEBOID MOVEMENT OF WHITE CORPUS-CLES. In this technique, described by Bensley and Bensley,[19] vital dyes are added to blood cells on a slide. As the stains are absorbed, the cells gradually die. However, the structures in the cells may be studied.

Prepare a number of slides with a coating of the stains described as follows: First prepare a *stock* solution of neutral red in absolute alcohol by adding 100 mg of neutral red to 10 ml of absolute alcohol. Then make a *dilute* solution from this stock solution by adding 10 ml of absolute alcohol to 4 ml of the stock solution. To this dilute solution add 1 or 2 drops of a saturated solution of Janus Green B in distilled water.

Now support a chemically clean, dry slide on a Syracuse dish, and flood the slide with the diluted neutral red solution containing Janus Green B (Fig. 5-25). Allow the excess stain to flow off by tilting the slide into the Syracuse dish. Set aside to dry. Prepare a supply of such stained slides. (Protect them from dust.)

Mount a drop of fresh frog's blood on this stained slide. Apply a coverslip. Examine under low power and under high power. As the dyes slowly dissolve, notice that the mitochondria of the cells stain green and the cell inclusions become red. Amoeboid movements may be studied. As the cells gradually die, their nuclei absorb more of the stain.

Hemoglobin content of human blood. A comparison of color may be used as a simple, but somewhat inaccurate, method for the determination of the amount of hemoglobin in blood. One scale, the Tallquist scale, is made of strips of paper of different shades of red. Each strip is perforated in the center. A drop of blood from a finger which has just been pricked (see p. 250) is placed on special white absorbent paper which is provided with the Tall-

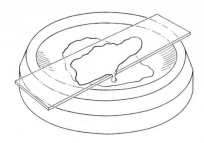

Fig. 5-25 Flooding a slide with stain. The Syracuse dish acts as a catch basin.

quist scale, and the resulting color is matched against the color scale. The approximate percentage of hemoglobin may be read from the scale. Students may be able to bring to class demonstration material of this type, for doctors use several of these scales for rough estimates of hemoglobin content.

A more accurate method, involving the use of a hemocytometer, is described in physiology texts, such as those by Todd and Sanford and Hawk, Oser, and Summerson listed at the end of this chapter. These books give complete, detailed accounts of the commonly used techniques for determining the percentage of hemoglobin.

Hemin crystals in blood. This might be an activity for a small group of students interested in a more extensive study of blood.[20] Take a drop of blood from the fingertip (technique, p. 250), mount it, and allow it to dry in the air. To the dried blood add a crystal of salt, a drop of water, and finally a drop of glacial acetic acid. Now apply a coverslip. Gently heat one end of the slide until bubbles escape. Under high power locate the reddish brown rhombic crystals of hemin.

Coagulation time. The coagulation time is the time it takes blood to begin to form fibrin, that is, to clot. The average coagulation time varies from 2 to 8 min, depending on the method used and the amount of

[19] R. Bensley and S. Bensley, *Handbook of Histological and Cytological Technique,* Univ. of Chicago Press, Chicago, 1938.

[20] D. Pace and C. Riedesel, *Laboratory Manual for Vertebrate Physiology,* Burgess, Minneapolis, 1947. For a variety of activities also look into P. Hawk, B. Oser, and W. Summerson, *Practical Physiological Chemistry,* 13th ed., Blakiston (McGraw-Hill), New York, 1954.

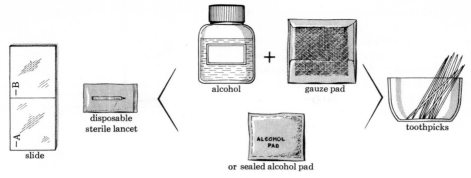

Fig. 5-26 Materials needed for A-B-O typing of blood.

blood. Here are two ways in which you may show how rapidly blood clots.[21]

USING A GLASS SLIDE. This is a convenient method. Obtain a few drops of blood from the finger (technique, p. 250), and place them on a clean slide. (Note the time at which the blood was taken.) Slowly pull a clean needle through the drop of blood at ½-min intervals. When a fine thread of fibrin can be pulled up by the point of the needle, coagulation has begun. Record the time interval between the flow of blood and the formation of fibrin—this is the coagulation time.

USING CAPILLARY TUBING. For this method, pull out narrow-bore capillary tubing into capillary pipettes of even bore over a wingtop Bunsen burner. Clean the finger with alcohol, puncture the finger (see p. 250), and fill a 4-in. length of pipette by placing it near the drop of blood exuding from the finger. Blood rises in the fine tubing by capillarity. At ½-min intervals break off small sections of the capillary pipette. When fibrin forms at the broken edge, coagulation has begun. Record the time intervals as in the slide method above. (Also look into BSCS laboratory activities.)

Fibrin clot and plasma. Try to get fresh blood from a slaughterhouse or meat-packing house, or "out-of-date" whole human blood from a hospital blood bank.

If possible, keep it in a vacuum bottle. Some of the blood may clot in transit, but it may be possible to get fresh blood to class. Pour some into a beaker and whip it with straws or small sticks. Fibrin will form shortly. When some of the blood is allowed to stand undisturbed, a clot also forms. As the clot contracts from the sides of the container, the clear, straw-colored plasma is visible.

Typing blood

A-B-O blood typing. Students may type their own blood in the classroom. Laboratory space is not essential for this activity.

Divide the class into five or six groups and supply each group with a tray or box containing the materials illustrated in Fig. 5-26. Thus, each student has access to a glass slide, which he divides into two halves with a glass-marking pencil. There are also cotton supplies; individual, disposable needles or lancets; toothpicks; and a bottle of alcohol or—better still—individual, sterile, disposable alcohol pads.[22] These alcohol pads wrapped in aluminum foil eliminate the need for special precautions to keep gauze or cotton sterile, and there is no hazard in having alcohol spilled from tilted bottles.

[21] U.S. War Dept., *Methods for Laboratory Technicians,* Tech. Man. 8-227, U.S. Govt. Printing Office, Washington, D.C., 1941.

[22] Absorbent pads saturated with 70 percent iso-propyl alcohol are available from U.S. Hospital Supply Corp., 838 Broadway, New York 3, N.Y., or from Becton, Dickinson & Co., Rutherford, N.J. (Cost of about 100 pads is $1.50.)

First, demonstrate the technique for the students. Divide a slide in half with a glass-marking pencil. At the left, place 1 drop of anti-A serum; on the right, 1 drop of anti-B serum. Make this a uniform practice to avoid confusion. Since the serum is expensive and youngsters may tip over the vials, it is advisable for the teacher or one student to put the 2 drops of serum on each student's slide.

Blood serum should be refrigerated when not in use. However, it cannot be stored indefinitely. It may be purchased already prepared from supply houses and certain biological laboratories (Appendix C). Dyes are added to distinguish the two serums. Powdered serum is also available for purchase and is accompanied by explicit directions for preparing solutions. But it is easier to purchase the anti-A and anti-B serums already prepared, in solution.

The pattern of inheritance of blood groups is described in Chapter 8.

Demonstrate for the students how to obtain blood (technique, p. 250). Use an alcohol-saturated pad to clean the skin; remove the sterile lancet from its wrapping, and make a quick, painless jab of the finger. With one end of the toothpick, transfer a drop of blood to the drop of anti-A serum and stir it a bit. Now with the *other end* of the toothpick transfer a second drop of blood into the drop of anti-B serum. Tilt the slide back and forth a bit and note the immediate reaction (unless the blood is type O). Students may want to check their observation by examining the slide under a microscope or hand lens. Check the results against those illustrated (Fig. 5-27).

Students readily understand the principle upon which blood typing is based. The four blood groups are classified on the basis of the presence or absence of the two agglutinogens, A and B, found in the red blood cells. Both of these antigens are present in type AB blood and both are absent in type O blood. However, when type A blood is added to type B blood it clumps. Therefore the plasma of type B

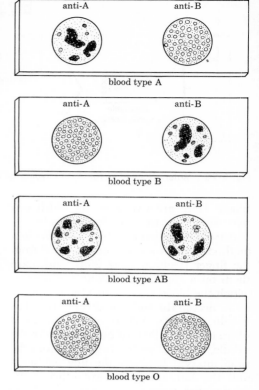

Fig. 5-27 The four possible results from adding a drop of "unknown" blood sample to each drop of serum on a slide.

blood contains a chemical which clumps "A" cells, and this is known as anti-A serum (see Table 5-2). In summary, type B blood has a "B" chemical (agglutinogen) in the red blood cells and an anti-A chemical (agglutinin) in the plasma. It is obvious that a person of B blood type couldn't live with anti-B in his own plasma, for it would clump his own blood.

A person of A blood type has "A" chemical in his red blood cells and anti-B in his plasma. (The anti-B serum which you use in blood typing in class has, of course, been extracted from blood of a person of type A.)

When students find that a drop of their "unknown" blood is placed into a drop of anti-A serum and clumping occurs, the unknown blood must be type A, as shown in Fig. 5-27. Similarly, if the unknown

TABLE 5-2
Summary for blood typing

blood group	agglutinogen in red blood cells	agglutinin in plasma
A	"A" chemical	anti-B
B	"B" chemical	anti-A
AB	"A" and "B" chemical	no anti-serum present
O	no chemical present	both anti-A and anti-B present

blood clumps in anti-B serum, the unknown blood belongs to group B. Should the unknown blood clump in both drops of serum, anti-A and anti-B, the unknown blood belongs to type AB. O type blood, which lacks the two antigens in the red blood cells, will not clump in either serum.

After students have typed their blood, compare the frequency of each blood group in the class (see Table 8-5). Type O is the most frequent, very roughly 44 percent (the percentage varies among racial groups within a population). Type A follows with about 39 percent (note there are several subtypes); type B is found in some 13 percent of occidentals; type AB is the rarest in the western hemisphere among the white population, about 4.5 percent. (We repeat: these figures are approximations.) Table 8-5 also gives some indication of the several subtypes of these four groups. Fig. 5-28 includes distribution of the Rh factor among a sampling of 100 donors. (See Tables 8-5, 8-6, 8-8.)

A description of the genetics of blood types and other factors in blood is given in Chapter 8.[23]

Rh blood typing. Slide or tube techniques are used for determining Rh blood factor as in A-B-O blood typing, but false positive reactions often occur unless directions are carefully observed for using a specific anti-Rh typing serum.

Slides are prepared as for A-B-O typing, but the slides must be maintained at about

37° C (99° F) and not above 47° C (117° F). Place 1 drop of anti-Rh serum at the center of the slide. Adjacent to this, but not touching, place an equal amount of a suspension of red blood cells. With a toothpick mix the 2 drops and agitate or tilt the slide. Clumping is macroscopic and occurs almost immediately, completed within 2 min, if there is to be clumping. If the test blood clumps in anti-Rh the blood is Rh positive; if there is no clumping, the test blood is Rh negative.

Be certain to follow the directions. When some serums are used, the blood is not to be washed with saline; in other preparations, blood is to have saline added. Oxalated blood is further specified in some techniques.

Typing serum, with specific directions or manuals of directions, are available from blood-donor services.

(See inheritance of blood factor Rh in Chapter 8. See Table 8-8.)

M-N blood types. Through a pair of allelic genes, M and N, three possible genotypes are inherited: MM, NN, and MN. In testing for these antigens, a slide is divided in half with a glass-marking pencil as described previously. A drop of anti-M serum is placed on half of the slide, a drop of anti-N serum on the other end. To each drop of serum on a slide add a

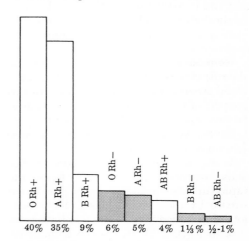

Fig. 5-28 Distribution of four main blood groups and Rh factor among a sampling of 100 donors.

[23] Students may want to read E. Zuckerkandl, "The Evolution of Hemoglobin," *Sci. Am.,* May 1965, pp. 110–18.

TABLE 5-3

*Blood chemistry**

Glucose (0.09–0.12%)	90.0–110 mg per 100 ml
Nonprotein nitrogen	25.0–30.0 mg per 100 ml
Urea nitrogen	12.0–15.0 mg per 100 ml
Uric acid	1.5–3.0 mg per 100 ml
Creatinine	1.0–2.0 mg per 100 ml
Calcium	10.0–12.0 mg per 100 ml
Chlorides (NaCl)	0.45–0.5%
Carbonate content (CO_2)	53–77 vol. %
Alkali reserve (RpH)	8.5
Hydrogen-ion concentration (pH)	7.4

*D. Pace, B. McCashland, and C. Riedesel, *Laboratory Manual for Vertebrate Physiology*, 3rd ed., Burgess, Minneapolis, 1964.

drop of 5 percent suspension of test cells in normal saline. Cells must be washed twice in saline before use in M-N typing.

If test cells clump in anti-M, the blood type is M; clumping in anti-N indicates blood type N, and clumping in both serums indicates blood type MN. Clumping should occur within 5 min at room temperature.

The blood stream

Homeostasis. Blood maintains, within a small range of limits, a constant concentration of glucose and salts, and a constant pH and temperature (Table 5-3). The role of the kidneys and of hormones in maintaining a stable internal environment are mentioned in Chapters 6 and 7.

Hormones. You may prefer to include the role of hormones which circulate in the blood stream throughout the body in this context of the role of the blood stream in homeostasis rather than in a study of behavior. We describe hormones and their functions in Chapter 7.

Using a new laboratory tool, radioactive tracers, it is possible to follow the rapid uptake of iodine from the blood stream by the thyroid gland. Techniques for using radioactive substances have been described (p. 233). In essence, radioactive iodine is combined into an iodide compound. This is added to a sugary solution and fed by pipette to a white rat or rabbit three times daily for 2 days. Over the next day or days a high concentration can be located only in the region of the thyroid gland when a Geiger counter is moved over the body. (See also "Some Radioactive Isotopes," Appendix F.)

In most studies, however, the experimental animals must be sacrificed and assays made of microincinerated tissues.

Body defenses. In some classes, there may be the opportunity to study the immunological reactions of blood more thoroughly. For example, students may want to report to the class on the role of plasma cells, a specialized group of white blood cells (plasmocytes) that produces antibodies. (Refer to Chapter 24 in W. Umbreit's *Modern Microbiology* [Freeman, 1962] for an illustrated description of the complement-fixation test.)

Using the blackboard as a device for summary

As a summary, students might outline the components of blood and their functions on the blackboard. Then a diagram such as Fig. 5-29 might be drawn on the board in colored chalk. A short review of the functions of blood might be established by this presentation "in a new view" at this time. What kind of blood vessel brings blood to the cells (that is, what is *A*)? What

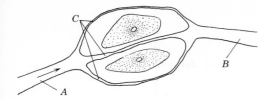

Fig. 5-29 Schematic diagram showing two tissue cells surrounded by a network of capillaries.

TABLE 5-4

What does the blood gain and lose in circulating through these organs?

organ	blood gains	blood losses
small intestines	. . . .	. . . .
liver	. . . .	. . . .
kidneys	. . . .	. . . .
lungs	. . . .	. . . .
bone marrow	. . . .	. . . .
muscle cell	. . . .	. . . .
thyroid gland	. . . .	. . . .
all cells	. . . .	. . . .

kind of blood vessel is represented by *B*? By *C*? Then complete a chart such as Table 5-4 in class.

CAPSULE LESSONS: Ways to get started in class

5-1. Demonstrate how to prepare a goldfish for microscopic study of the circulation in its tail (see technique in this chapter). Then have several students prepare additional demonstrations so that all students may examine blood circulating through the arteries, capillaries, and veins of the tail. In which blood vessels does the blood move more quickly, and in spurts? Which are the fine connecting vessels? What are the large vessels in which blood flows the other way?

5-2. Set fresh celery or bean seedlings or carrots into red-colored water. Ask students to trace the entrance of water up into the leaves. What is the water to be used for? How did the water get into the roots?

5-3. Bring in a fresh specimen of a beef's heart for dissection in class. Elicit discussion about each chamber of the heart and of the valves. How many times does the heart beat in an hour if the average rate is 70 heartbeats per min? Develop a discussion, with diagrams, of a closed circulatory system from the heart and back again. Follow, the next day, with a development of the story of what materials the blood gains and loses in its trip around the body.

5-4. Have students examine fresh celery, jewelweed (*Impatiens*), or carrots so they may see fibrovascular bundles. Very thin sections of a carrot can be cut with a razor blade; if the edges of the slices are very thin, they can be examined under the microscope. Or study prepared, stained slides.

5-5. At some point show a film on the composition of blood. You may want to show *Work of the Blood* (EBF) or, in review, the complete circulation of blood as in *Hemo, the Magnificent* (Bell System) or *The Red Blood Cell* (Moody).

For a more leisurely development, show an 8-mm, silent, single-concept film—*The Heart in Action* (3 min) or *Circulation: The Flow of Blood* (3 min), both EBF—or a filmloop on the frog's heartbeat (see the listings at the end of this chapter).

5-6. Have students take each other's pulse (see Table 5-1). Develop the idea that the beating heart pumps blood through the body by way of blood vessels. Then compare the pulse after some kind of exercise; now the class is into a good discussion of the role of the heart in pumping blood and other aspects of circulation.

5-7. Perhaps you want to begin with the story of Harvey and his study of the circulation of blood. With advanced students you may want to read Harvey's original paper on circulation (obtainable from Regnery Publishing Co., Chicago, Ill.). Have students draw their conception of how blood circulates around the body.

5-8. Explore some of the approaches given in the BSCS laboratory block on metabolism (in preparation).

5-9. Distribute vials containing seedlings which show root hairs. Have students examine them with a magnifying lens. Ask them what advantage there may be in the increased surface provided by the extensive root hairs. What process goes on through these cell extensions?

5-10. Provide these materials for students: several plants, cobalt chloride paper, petroleum jelly, and pins. Ask students to design a demonstration to reveal which surface of the leaves, the upper or lower surface, contains more stomates.

5-11. You may want to set up several demonstrations to show transpiration. Put them in the classroom, where they may be watched, and wait for students to ask questions. You are then into the topic "How Is Water Transported in a Plant?"

5-12. Have a laboratory lesson in which students examine the underside of such a leaf as *Sedum, Tradescantia,* or lettuce. Teach students how to strip off the lower epidermis for examination under the microscope as a wet mount (see p. 175 for wet mounts).

5-13. Have a student report to the class on the theories which have been advanced to explain the rise of water in a tall tree. Almost any botany text will have a description of the different theories (see the list of books at the end of this chapter). Further, what happens to the products of photosynthesis in leaves?

5-14. Perhaps you may want to show the film *The Growth of Plants* (EBF) or one or more in the AIBS series of 12 films *Multicellular Plants* (McGraw-Hill). Use the film as a device for raising questions about the way plants get water and how it is transported through the plant. Other times, use a film as a review of the topic.

5-15. Begin with a demonstration of diffusion or any other activity described in this chapter.

5-16. Open a bottle of ammonia or perfume in class for a few seconds. Have students raise their hands when they can smell it. Elicit the facts that the liquid changed into a gas and the molecules diffuse among the molecules of gases in the air. Then lead into a discussion of diffusion of materials through a membrane. Refer to the passage of water through root hairs, and from cell to cell.

5-17. Sometimes you may want to include in a test one of the demonstrations of work done in class, with questions based on that principle or concept. For example, you might repeat a demonstration or, better still, show a modification of it (several are described in this chapter) which illustrates the same principle. Then make up five questions which test both observation and reasoning. This might be used as one test question. A sample of this type of question may be found on p. 194.

5-18. Some teachers begin with a study of the composition of the blood. (In most school systems, parental consent notes are necessary.) Study either a drop of blood in Ringer's solution or a dried blood smear. A biology club might want to stain smears for the entire class (with Wright's blood stain, p. 118). What is the effect of an isotonic solution on blood cells? Of a hypertonic solution?

5-19. Have students report to the class on some blood disturbances (such as types of anemia or leukemia) and some heart conditions. Elicit a discussion concerning the factors which affect the circulatory system. Remember that the highest death rate is caused by circulatory diseases. (There are many inexpensive booklets available from the Red Cross or the Am. Heart Assoc.)

5-20. As a laboratory lesson have students prepare wet mounts of the water flea, *Daphnia*. These are available from an aquarium shop. Examine the beating heart found in the anterior top end of the animal. Develop a discussion of the role of the heart in circulation. You may also see blood circulating in such microscopic worms as *Dero, Nais,* or the *Tubifex* worms students may be able to get from an aquarium shop (culturing, Chapter 12). What are some factors that affect the heartbeat?

5-21. A very successful laboratory lesson is blood typing. Have students type their own blood. Demonstrate the technique and have students around the desk as you do it. Get anti-A and anti-B serums and prepare the materials listed in this chapter. Have a student report on the distribution of types A, B, AB, and O in the total population. Compare with the findings in your class. Warn students not to consider this exercise in blood typing as official, for this class work is subject to all the usual pitfalls of novices.

5-22. Begin the lesson with a question which arouses good discussion. These may be possibilities: How many know that food materials are sometimes injected into a person who is very ill? What kind of fluid is inserted into a vein? Why? Into which kind of a blood vessel? (Students might trace the path of this glucose solution through the body.)

5-23. You may want to use this little story as a means for getting attention focused on the day's lesson. A bad-tasting but harmless drug was injected into the leg of a man. In less than a minute he sensed a bitter taste on his tongue. Explain. From this elicit a discussion of circulation time around the body, the path of circulation, and the need for blood's circulation.

5-24. Ask students if they know of the disease diabetes. Some students know a great deal about insulin injections and the precautions necessary. Here the topic of circulation is approached through the study of hormones. Of course, many other hormones might be used in this manner: role of thyroxin in metabolism; what causes a giant or a midget; a story of the adrenals and their role in stress. The role of the blood in

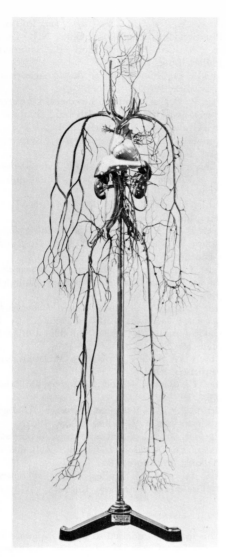

Fig. 5-30 Wire model of circulatory system. (Courtesy of Clay-Adams, Inc., New York City.)

homeostasis is developed in this manner. Or perhaps students may want to read from Cannon's classic *The Wisdom of the Body*.[24]

5-25. This may be the place you want to introduce the use of radioactive isotopes as tools in a study of functions of the body. Students might report to the class on how radioactive iodine, radioactive sodium chloride, and other

[24] *Op. cit.,* pp. 28, 302–03.

isotopes are used (see the comprehensive Bibliography, Appendix A).

5-26. At some time, students may read a reprint from *Scientific American* such as S. Biddulph and O. Biddulph, "The Circulatory System of Plants" (February 1959, pp. 44–49). The subject of the paper may be used as an introduction to a topic, as a review, or as a basis for a test of understanding.

5-27. Fill six quart-bottles with water. This is about the amount of blood in the human body. Develop the notion of a closed system of tubes for circulation of blood. Use a plaque or a wire model such as that in Fig. 5-30 to have students voice their own concepts of how the blood circulates around the body.

5-28. Use a fresh haslet, a heart-lung model, or a heart model such as that in Fig. 5-31. Ask students to identify as many parts as they can. What is the advantage of the close proximity of lungs to heart? What is the advantage of thick walls in the ventricles? What is the function of the thin-walled small auricles?

PEOPLE, PLACES, AND THINGS

Do students in a club need the assistance of a technician in learning how to make a blood count? Do you need advice as to where to take students on a field trip for living materials? Where can you get live frogs any time of the year? A beef's heart? A film on circulation?

There are many resource people in your community: a lab technician at a hospital, or a parent. Neighboring high schools and colleges

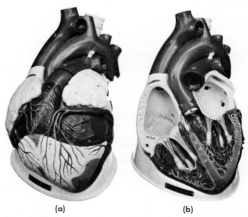

(a) (b)

Fig. 5-31 Take-apart model of heart: (a) closed view; (b) open view. (Photo courtesy A. J. Nystrom & Co.)

may provide you with a film or living material and possibly some suggestions for fruitful locations for a field trip. Are you near an experimental station, an agricultural college, a nursery, or a farm? An aquarium shop and the local butcher can supply some materials suggested in this chapter.

If possible, try to be involved in the landscaping of the school grounds. Refer to some uses of the school grounds for field trips and work throughout the year in soil erosion, reproduction, interdependence, and other aspects of conservation (see Chapter 11).

A visit to an experimental station and several short field trips over the year around the school grounds to interpret some aspect of class work give meaning to the work at hand.

At the end of the book you'll find a listing of supply houses and a directory of film distributors. Are you looking for some information? Perhaps the books listed here or in the Bibliography include what you want.

BOOKS

These are only a few of the books that are pertinent to the work discussed in this chapter. These and many other references are given in Appendix A, the comprehensive Bibliography.

Best, C., and N. Taylor, *The Physiological Basis of Medical Practice,* 7th ed., Williams & Wilkins, Baltimore, 1961.

Blanchard, J. R., and H. Ostvold, *Literature of Agricultural Research,* Univ. of California Press, Berkeley, 1958.

Bonner, J., and A. Galston, *Principles of Plant Physiology,* Freeman, San Francisco, 1952.

Bourne, G., ed., *Medical and Biological Problems of Space Flight,* Academic, New York, 1963.

Carlson, A., V. Johnson, and H. M. Cavert, *The Machinery of the Body,* 5th ed., Univ. of Chicago Press, Chicago, 1961.

Cassidy, H., *Fundamentals of Chromatography,* Interscience (Wiley), New York, 1957.

Davidsohn, I., and B. Wells, *see* Todd, J., and A. Sanford.

Faires, R., and B. Parks, *Radioisotope Laboratory Techniques,* 2nd ed., Pitman, New York, 1960.

Fisk, E., and W. Millington, *Atlas of Plant Morphology,* 2 portfolios: Portfolio I, *Photomicrographs of Root, Stem, and Leaf,* 1959, Portfolio II, *Photomicrographs of Flowers, Fruit, and Seed,* 1962; Burgess, Minneapolis.

Foster, A., and E. Gifford, *Comparative Morphology of Vascular Plants,* Freeman, San Francisco, 1959.

Galston, A., *The Life of the Green Plant,* 2nd ed., Prentice-Hall, Englewood Cliffs, N.J., 1964.

Giese, A., *Cell Physiology,* 2nd ed., Saunders, Philadelphia, 1962.

Gilmour, D., *Biochemistry of Insects,* Academic, New York, 1961.

Hawk, P., B. Oser, and W. Summerson, *Practical Physiological Chemistry,* 13th ed., Blakiston (McGraw-Hill), New York, 1954.

Heftmann, E., ed., *Chromatography,* Reinhold, New York, 1961.

Lockhart, R., G. Hamilton, and F. Fyfe, *Anatomy of the Human Body,* Lippincott, Philadelphia, 1959.

Machlis, L., and J. Torrey, *Plants in Action,* Freeman, San Francisco, 1956.

Meyer, B., D. Anderson, and R. Böhning, *Introduction to Plant Physiology,* Van Nostrand, Princeton, N.J., 1960.

Prosser, C. L., and F. A. Brown, Jr., *Comparative Animal Physiology,* 2nd ed., Saunders, Philadelphia, 1961.

Todd, J., and A. Sanford, *Clinical Diagnosis by Laboratory Methods,* 13th ed., edited by I. Davidsohn and B. Wells, Saunders, Philadelphia, 1962.

Umbreit, W., *Modern Microbiology,* Freeman, San Francisco, 1962.

Villee, C., *Biology,* 4th ed., Saunders, Philadelphia, 1962.

Wald, G., *et al., Twenty-six Afternoons of Biology,* Addison-Wesley, Reading, Mass., 1962.

Walker, B., W. Boyd, and I. Asimov, *Biochemistry and Human Metabolism,* 3rd ed., Williams & Wilkins, Baltimore, 1957.

Weisz, P., and M. Fuller, *Science of Botany,* McGraw-Hill, New York, 1963.

FILMS, FILMSTRIPS, AND SLIDES

This listing is intended to be only a guide for the selection of new films, filmstrips, and transparencies. Send for catalogs from distributors of films; catalogs of biological supply houses list extensive offerings of 2×2 in. slides. The addresses of distributors are given in Appendix B.

In general, the cost of color films averages $150; black and white films cost about $75. Newer films are made in varying lengths—time sequences of 1 min, 5 min, 11 min, or 30 min; the cost of films therefore varies. We have indicated the lengths of films so that lessons can be planned with films in mind.

Sound films are available in color or black and white; rental rates are available from film libraries or by direct inquiries to producers.

(Filmstrips are not available on a rental basis; the purchase price is usually about $6 to $7 for those in color.)

The Blood (16 min), EBF; includes typing and Rh factor.
Circulation (17 min), United World.
Circulation of Blood, Am. Heart Assoc.; free.
Circulation: The Flow of Blood (single-concept film, 8 mm, 3 min), EBF.
Dynamics of Phagocytosis (28 min), Campus.
Endocrine Glands, 2nd ed. (11 min), EBF.
Flow of Life (20 min), Educational Testing Service.
Frog Heartbeat (6 min), BSCS techniques, Thorne.
Frog Heartbeat (filmloop, 3 min), Ealing; adapted from BSCS techniques film.
The Growth of Plants (21 min), EBF.
The Heart and Circulatory System (29 min), Moody.
Heart Disease: Its Major Cause, Am. Heart Assoc.; free.
Heart: How It Works, Am. Heart Assoc.; free.
The Heart in Action (single-concept film, 8 mm, 3 min), EBF.
Hemo, the Magnificent (2 reels, 25 min each), Bell System; free.
The Human Body: Circulatory System (14 min), Coronet.
Human Physiology Series (filmstrips), Soc. for Visual Education.
Infectious Diseases and Man-made Defenses (11 min), Coronet.
Infectious Diseases and Natural Body Defenses (11 min), Coronet.
Microorganisms That Cause Disease (11 min), Coronet.

The Multicellular Animal (set of 12 films, 28 min each), Part IV of *AIBS Film Series in Modern Biology,* McGraw-Hill.
Multicellular Plants (set of 12 films, 28 min each), Part III of *AIBS Film Series in Modern Biology,* McGraw-Hill.
Principles of Endocrine Activity (16 min), Indiana Univ.
Radiation in Biology: An Introduction (14 min), Coronet.
The Red Blood Cell (33 min), Moody.
Work of the Blood (14 min), EBF.

LOW-COST MATERIALS

This is only a partial listing of inexpensive material available to the teacher. Many of these materials are distributed to teachers without charge. While we recommend the material for use in the classroom, we do not necessarily sponsor the products advertised.

About Your Blood, John Hancock Mutual Life Insurance.
Rheumatic Fever; Your Heart, Metropolitan Life Insurance.
The Story of Blood, Am. National Red Cross; ask for additional materials for class use.
U.S. Atomic Energy Comm., *Some Applications of Atomic Energy in Plant Science.* U.S. Govt. Printing Office, Washington, D.C.
U.S. Dept. of Agriculture, Forest Service, *How a Tree Grows* (poster, 16 × 21 in.), U.S. Govt. Printing Office, Washington, D.C.
Your Heart and How It Works (chart), Am. Heart Assoc.; ask for additional pamphlet information.

Chapter 6

Use of Materials

An organism, whether a plant or an animal, draws materials from its environment and returns other materials to the environment in the processes of maintenance, growth, repair, and reproduction.

Oxidation of glucose

The laws of thermodynamics hold for living cells as well as for inanimate objects. Living cells constantly need a supply of free energy; this energy is obtained from chemical bonds in foods. Photosynthesizing cells store energy in energy-rich bonds of carbohydrates. Reversing the arrow in an equation for photosynthesis gives this generalized equation for respiration in cells:

$$C_6H_{12}O_6 + 6O_2 + 6H_2O \longrightarrow$$
$$6CO_2 + 12H_2O + energy$$

When glucose is oxidized in cells, one molecule of carbon dioxide is evolved for each molecule of oxygen used in the process (respiration quotient R.Q. = 1). When fatty acids are oxidized, the R.Q. is less than 1; substances richer in oxygen than is glucose have an R.Q. greater than 1.

A small amount of energy is liberated in each step-wise oxidation of glucose, and it is stored in energy-rich phosphate bonds. The overall efficiency is about 60 percent, in that 38 high-energy bonds are formed per molecule of glucose that is completely oxidized. Only a small amount of energy is liberated in the steps leading to pyruvic acid. It is the oxidation of pyruvic acid that yields the major energy supply of cells.

The oxidation-reduction reactions involved in the breakdown of glucose are coupled with the formation of energy-rich bonds in ATP molecules. The process of oxidative phosphorylation occurs in the mitochondria of plant and animal cells (for demonstrations with mitochondria, see Chapter 2). Energy from the oxidation of glucose that might be lost as heat is transferred to form a third phosphate group for an ADP molecule, forming a molecule of ATP—the energy currency for all the life functions of cells. The energy for the phosphorylation of ATP itself comes from the loss of hydrogen atoms by pyridine nucleotides. Thus powerful reductants are formed: reduced pyridine nucleotide DPNH (diphosphopyridine nucleotide) or TPNH (triphosphopyridine nucleotide). As a result, hydrogen atoms may be "forced" on other molecules in a reduction process. In the synthesis of ATP, some TPNH is oxidized (losing hydrogen atoms).

In brief, pyridine nucleotides and ATP provide the energy for the conversion of carbon dioxide to carbohydrates in cells; respiration is the reverse of this photosynthetic process.

In the oxidation of glucose into two molecules of pyruvic acid, hydrogen is removed and transferred to a hydrogen acceptor; these intermediate steps are common to both glycolysis and fermentation. The subsequent fate of pyruvic acid depends on whether anaerobic or aerobic respiration occurs. In anaerobic systems, pyruvic acid is converted into ethyl alcohol and carbon dioxide through this series of reactions:

1. Pyruvic acid $\longrightarrow$
   acetaldehyde $+ CO_2$
   (The catalyst is carboxylase, which removes CO_2.)
2. Acetaldehyde $+ 2H \longrightarrow$
   ethyl alcohol $+ CO_2$
   (The catalyst, an alcohol dehydrogenase, transfers H ions.)

In aerobic respiration, pyruvic acid is converted to CO_2 and 2H with a large yield of energy. Dehydrogenases remove hydrogen from the compounds, and CO_2 is removed by carboxylases.

In the overall series of oxidation-reduction reactions in cells, pairs of hydrogen atoms are removed from the intermediate nutrient fragments, and transferred by means of the dehydrogenases (with coenzyme I and II) to hydrogen acceptors, or flavoprotein enzymes. Each hydrogen becomes a hydrogen ion in the fluids of cells. The electrons of the hydrogen ions are transferred to the cell pigments found in all aerobic cells—the cytochromes. There are usually three cytochromes in cells (a, b, and c) involved in hydrogen transfer. (Cytochromes are iron porphyrins which have a general structure similar to the heme part of hemoglobin, and also to a chlorophyll molecule except that the chlorophyll molecule contains magnesium rather than iron.) Finally, cytochrome oxidase passes along the electrons to oxygen and, as a result, water is formed. (For a demonstration of the activity of cytochrome oxidase, see Chapter 2.)

Not all the pyruvic acid molecules that result from glycolysis are shunted into a Krebs cycle. Some of the pyruvates are intermediates in the synthesis of amino acids, or are precursors for the synthesis of fatty acids (via acetyl coenzymes A), or become part of the high-energy phosphate bond enzymes. The metabolism of sugars, fats, and nitrogenous compounds are all linked in the Krebs cycle.

(For a brief discussion of the evolution of heterotrophs, see Chapter 10.)

Enzymes: space and time relationships

Further study: a reading. There is a definite order in which processes occur in the living cell; that is, there is a space and time relation. This relationship is described by Oparin in the following passage from his book *Life: Its Nature, Origin, and Development.* The passage might be the basis for a reading and discussion by a class at some time during the study of enzymatic reactions or in a study of differentiation of embryonic tissue (Chapter 9).

Owing to the specificity of their structure, present-day protein-enzymes are extremely efficient chemical agents in protoplasm. With their help, not only are the biochemical reactions necessary for the rapid flow of the vital processes accelerated, but they are directed along particular tracks. This happens because enzymes are unlike the other catalysts of the inorganic world. In the first place they have an extremely powerful catalytic effect; in the second place they have a very great catalytic specificity; in the third place they are markedly labile so that the strength of their catalytic activity can vary within very wide limits under the influence of a great variety of external influences and internal factors.

This may be clarified by reference to the following simple scheme. Let us suppose that the organic substance A can be converted into substances B, C, D, etc. In the accompanying diagram these transformations are represented by radial vectors the length of which correspond with the rates of the respective reactions.

Thus the rate of the reaction A $\longrightarrow$ B is seven times that of the reaction A $\longrightarrow$ D while this reaction proceeds only half as fast as A $\longrightarrow$ C. Naturally, when a given time has passed and all the substance A which was originally present has disappeared, the resulting mixture will consist of 70 per cent B, 20 per cent C and 10 per cent D. If we make some non-specific alteration which accelerates all the reactions equally, the changes will proceed more quickly but the proportions of the products obtained at the end will remain the same. As before, substance B will predominate. If, however, we introduce into the original mixture an enzyme which will accelerate the reaction A $\longrightarrow$ D by many millions of times without affecting the rates of the other reactions, almost all of substance A will be converted into substance D and the other possible reactions will have no practical significance.

The substance D which has thus arisen, like any other organic compound, has many chemical potentialities, but, in the presence of a new specific enzyme, it will only follow the one path laid out for it by that enzyme. Thus there arises a chain of consecutive reactions coordinated with one another in time. But such a chain is only

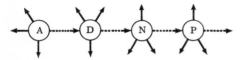

possible where there is a whole series of enzymes each of which is specific for the corresponding link in the chain. And, in fact, in Buchner's juice, which is extracted from yeast, we may find a large collection of enzymes which, acting concurrently, can reproduce the chain of reactions by which sugar is converted into alcohol and carbon dioxide, though essentially in a one-track way, without any branching of the chain.

When we are dealing with a branched chain of reactions rather than with a simple one we must take into account, not only the presence of all the necessary enzymes, but also the relationship between their catalytic activities. In such a case the intermediate products formed at the points where the chain branches follow, not one, but at least two paths. The proportion following each path depends on the relative catalytic activity of the specific enzymes involved in these paths.

In the living body, with its extremely elaborate network of metabolism, the matter is even more involved. Here there must be a precise, yet very versatile, coordination of the catalytic activities of all the enzymes in the well-stocked arsenal of the protoplasm. This coordination is largely based on the high lability of enzymes, the

great variability of their activity under the influence of many external and internal agents.

There is no physical or chemical factor, no organic compound or inorganic salt which cannot, in one way or another, affect the course of enzymic reactions. Any rise or fall of temperature, any change in the acidity of the medium, its oxido-reduction potential, the salts dissolved in it or its osmotic pressure, will interfere with the relationship between the rates of the enzymic reactions and thus alter their mutual connections in the network of metabolic reactions. In this respect the development within protoplasm of a new sort of specific activators or inhibitors, which selectively strengthen or weaken the activity of one or another of the protoplasmic enzymes, is very important.

We must also bear in mind the exceptional part played by the spatial organisation of protoplasm in biological metabolism. Enzymes are mainly concentrated on interfaces and on the structural formations of protoplasm. They are, as one might say, rationally "assembled" here into "production lines" each of which fulfils its own biological function. Any alteration in these lines may, not merely raise or lower the rate of the reactions catalysed by the enzymes, but may displace the stationary equilibrium of chemical reactions in the direction of synthesis or breakdown. This will naturally be of decisive importance for the stability of the whole living system, even for its survival under the given environmental conditions.[1]

This may be the time to introduce the "one gene-one enzyme" notion of Beadle and Tatum and to consider the effect of a mutant on the network of enzymatic reactions. Further, in tissue located spatially in one part of the anterior of the gastrula, what conditions determine that the tissue shall become an eye at a certain time? After its determination in time, a change of location does not alter its destiny. Further, in the evolution of early "protoplasm," or the coacervate system (or whatever name one wishes to use), how were enzymes and substrate evolved? By chance? Hardly. Oparin describes his hypothesis of the evolution of abiogenic

[1] From A. Oparin, *Life: Its Nature, Origin, and Development*, trans. by A. Synge, Academic, New York, 1962, pp. 76–79. Reprinted by permission of Oliver & Boyd Ltd. and the author.

organic substances; students will want to read Oparin's entire book.

Respiration in animals and plants

Metabolic pathways are similar in most respects for microorganisms and higher plants and animals.

Teachers begin in many ways, and they vary their approaches with individual classes during the day. Indeed, as we have reason to state at different times, teaching is a personal invention and the procedures vary with the individual. One teacher may begin with the concept of the interrelationship of living things, as demonstrated in a sealed test tube or in a "microaquarium" in a depression slide. Another teacher may begin with a comparison of exhaled and inhaled air of the students themselves, or of changes produced in an enclosed atmosphere by plants, seeds, insects, or a mouse. Several of these approaches will be described. The class and teacher can plan together the direction in which to move.

(*Note:* Demonstrations concerning the role of respiratory enzymes in living tissue are described in Chapter 2.)

Making a microaquarium

You may want to call students' attention to a large aquarium in the classroom (preparation, p. 605). What is the need for green plants? From where do the animals get their oxygen supply? Or you may want to have students make microaquaria for examination over the next month.

In a test tube. Begin by asking students to predict what might happen if you were to seal a snail in a test tube. Seal a snail in one tube (Fig. 6-1a). Into another, place a snail and a green aquarium plant (Fig. 6-1b). You may prefer to "seal" each tube with a rubber stopper; cover with tape or wax to prevent an exchange of gases.

However, the demonstration seems to be more effective when the test tubes are sealed in a flame. Hold the ends of a large, soft-glass test tube (not Pyrex) and rotate

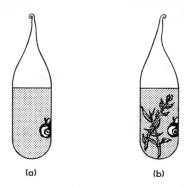

(a)   (b)

Fig. 6-1 Sealed microaquaria: (a) a snail (*Physa*) in a sealed tube of aquarium water; (b) a snail plus a green aquarium plant in a sealed tube of aquarium water.

it in a Bunsen flame. As the central area of the test tube softens, remove it from the flame and pull it out quickly so that a 1½-in.-long section is made (Fig. 6-2a). Make several of these and set aside until cool (3 min or so). Then add aquarium water to each test tube (not more than half full). To some of the tubes add green aquarium plants and a snail. Into others place a snail without the plants.

To seal each test tube, hold the tube at a 45° angle and heat the narrow part of the tube until the glass softens (Fig. 6-2b). Then remove it from the flame and pull it out quickly to form a fine strand of glass (still open for exit of gases). Place the center of this strand in the flame again in order to seal the tube. Be careful not to heat the liquid as you heat the glass tube, since the liquid will boil and expanded gases will cause the tube to crack. Stand the sealed microaquaria in a jar or test-tube rack in good light but not in direct sunlight.

You may want to use plastic bags instead of test tubes. Such bags may be obtained in pet shops or by the roll in neighborhood supermarkets.

On a glass slide. The existence of the oxygen-carbon dioxide cycle among microorganisms in a microaquarium on a sealed depression slide may be studied.[2]

[2] M. Rabinowitz, "A Balanced Microaquarium," *Teaching Biologist* 7:5, 1938.

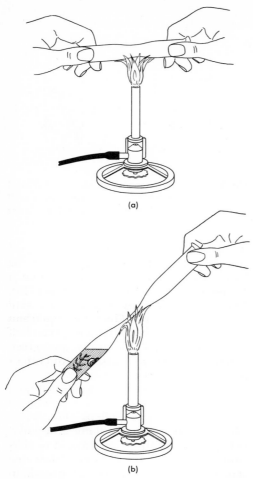

Fig. 6-2 Making a sealed microaquarium: (a) constrict an empty soft glass tube by heating and pulling; (b) after placing materials in tube, make constriction very narrow. After expanded gases have escaped, completely seal off tube.

Into the depression place the growing tip and a small segment of an aquarium plant such as *Nitella* in several drops of a rich culture of mixed protozoa and possibly some microscopic worms (*Nais, Dero*) and a crustacean, *Daphnia*. Allow some space for several air bubbles to be trapped in the microaquarium as you apply the coverslip. Seal the coverslip to the slide with petroleum jelly or paraffin (to prevent evaporation). Also prepare controls without the *Nitella*.

Keep the microaquaria in moderate light, not direct sunlight. These slides may be studied under a binocular or standard monocular compound microscope or with a microprojector. In which slides do the worms or protozoa live longer?

Teachers who culture their own protozoa during the year know that cultures thrive better when algae such as *Chlorella* are introduced into the same culture dishes. Students may examine this relationship by preparing several microaquaria either in depression slides or as hanging drops (preparation, p. 122). Introduce six to eight paramecia onto the slides in a drop of Brandwein's solution A (p. 581). To half of the slides add a drop of a thriving culture of *Chlorella*. (Culture methods for all these microorganisms are in Chapter 12.) Maintain the slides in moderate light. During the next 2 to 3 weeks, which slides show vigorously active, and reproducing paramecia? Why is this so?

Burning a butter candle

It is sometimes useful to draw a crude analogy between the burning of food in the body and the burning of a candle, a peanut, or a butter candle as described here. Yet this is only a starting point for probing the differences in the two processes. Several introductory demonstrations of the action of respiratory enzymes are given in Chapter 2, with references to further study in the literature.

We can compare the products of the burning candle and the products of foods that are burned in the body in some of the following activities. Several show the liberation of carbon dioxide by plants and animals (including humans); others show the consumption of oxygen by living organisms in the process of respiration (or oxidation).

Shape a mound of butter into a candle, in an evaporating dish or a Pyrex beaker, and insert a cotton string as a wick halfway into it (Fig. 6-3). Then ignite the wick. Heat is produced.

Fig. 6-3 A butter candle to illustrate burning of food in the body.

Hold a cold glass plate or a mirror over the butter candle and look for drops of condensed water on the surface. Wipe dry. Then a student might breathe on the cold glass plate.

Test the air around the burning candle for carbon dioxide. To do this pour some limewater into the evaporating dish or beaker and cover with a glass plate. Shake slightly and note the cloudy appearance of the limewater caused by the precipitation of calcium carbonate. Or use brom thymol blue instead of limewater; the color of the solution changes from blue to yellow (indicators, pp. 662–64).

Students recognize that water, heat, and carbon dioxide were released in oxidation. Do we exhale carbon dioxide? What instrument is used to measure the temperature of the body?

Also refer to the demonstration of luminescence, using dried *Cypridina,* described in Chapter 2.

Oxidation in living cells

In the presence of oxygen the indicator methylene blue is blue in color. When oxygen is removed methylene blue is bleached and becomes colorless. This change in color may be observed and utilized in the study of oxidation in living cells. Here are two methods described in the literature.

In the first method[3] saturate an isotonic solution (0.7 percent sodium chloride)

with methylene blue. Then inject about 2 to 3 ml into the dorsal lymph sac of a living frog. Within 1 hr after injection place the frog under anesthesia (p. 113), dissect, and examine the organs for the presence of methylene blue.

Since the cells use oxygen rapidly, the methylene blue will be bleached. When cells are exposed to air again, or when the cells die, the blue color returns. In this way it is possible to estimate the speed at which oxidation stops in the different organs of the frog (or the rate at which tissues die). (Other demonstrations are in Chapter 2.)

A more advanced project is the study of oxidation in muscle cells.[4] The fact that methylene blue is reduced so that it becomes colorless in living tissues is used in this project.

Add a few drops of sodium hydrosulfite to a methylene blue solution. Note the disappearance of the blue color. In the presence of hydrogen peroxide or in shaking with air (so that oxygen is added) methylene blue is oxidized back to the blue color. This method may be used to show the disappearance of oxygen in muscle cells.

The student will need to dissect out two sartorius muscles from the leg of an anesthetized or freshly killed frog (p. 308). Stain these muscles with methylene blue–Ringer's solution (as described above) and then mount each muscle strip on a slide. Apply a coverslip; use petroleum jelly around the rim to prevent evaporation. Within a short time, the center region of the muscle tissue may be examined; the methylene blue is colorless.

As the coverslip is raised, exposing the muscle tissue to air, the blue color will return. When the coverslip is replaced, the methylene blue will again be reduced to the colorless form in the muscle. One of the slides may be steamed (over a beaker of boiling water) until the tissue is killed. After this slide has cooled, lift off the cover-

[3] D. Pace and C. Riedesel, *Laboratory Manual for Vertebrate Physiology,* Burgess, Minneapolis, 1947, p. 214.

[4] R. Root and P. Bailey, "A Laboratory Manual for General Physiology," 2nd ed., mimeo, City College, New York, 1946.

slip on both slides. Note any difference in the rate of reappearance of the blue color.

Oxidation-reduction illustrated

The fact that oxidation-reduction is a reversible process may be demonstrated by mixing a solution of thionine dye with a solution of ferrous sulfate.

In the presence of intense light, the color of the dye disappears within a second. In the dark, the color reappears immediately. In the presence of light the ferrous iron reduces the dye to the colorless form, and is itself oxidized at the same time to ferric iron; in the absence of light, the process is reversed. What is the applicability of this to living tissue?

Carbon dioxide released in exhalation

Students may breathe through a straw or glass tube into a beaker of limewater or water containing an indicator (such as brom thymol blue). When limewater is used, it becomes cloudy (as calcium carbonate is formed).

Since carbon dioxide added to water results in a weak acid (carbonic acid), we may use several indicators to reveal the increase in acidity of the liquid. For example, pour dilute brom thymol blue (slightly alkaline and light blue in color) into a beaker (preparation, p. 169). When a student breathes into this it turns a light yellow or straw color (see Fig. 6-4).

At other times you may use phenolphthalein made slightly alkaline with sodium carbonate (or sodium hydroxide) so that it is pinkish (or red). When carbon dioxide is breathed into this, the medium becomes colorless as the acidity increases. Phenol red may also be made slightly alkaline with sodium carbonate. When exhaled air is bubbled through this indicator it turns from a red to a yellowish color (as the acidity is increased). For other indicators, see Fig. 14-1.

Students may use short-range hydrion test paper to make determinations of pH at home. Liquid indicators are not necessary.

Effect of exercise on production of carbon dioxide. In this demonstration, six volunteers (boys and girls) first exhale into water containing an indicator (alkaline), and records are kept of the time (in seconds) needed to decolorize the indicator. Then the same volunteers exercise vigorously for 1 min—by doing knee bends or by running around the hall—and at a signal breathe into a second beaker or container of the same indicator; timekeepers record the time (in seconds) that it now takes to decolorize the indicator. Students should be ready to explain why the time factor has been reduced by half (or thereabouts). (Rate of breathing, p. 288.)

Some teachers have students titrate these solutions to determine the carbon dioxide content (or increased acidity).

The solutions are prepared as follows. Add 3 to 5 drops of phenolphthalein to 100 ml of tap water. (Prepare the solution of indicator by adding 1 g of phenolphthalein powder to 40 ml of ethyl alcohol; add this to 120 ml of distilled water.) Now make the solution slightly alkaline by adding a few drops of 0.04 percent NaOH solution (4 g to 1000 ml of water) until a rose color appears and persists. Test the solution by breathing into a sample; have enough NaOH so that it takes at least 30 to 45 sec for a change in color to occur.

Prepare two similar flasks with straws for each of the volunteers taking part in the demonstration. Cork or add an extra drop of NaOH to the second flask of each pair, since it will otherwise be exposed to air before the student uses it.

Good planning requires that you examine the students' health records to be sure that volunteers for any special class activity, such as this one involving vigorous exercise, are not over-strained by it.

At a signal have each volunteer exhale into the first flask of the pair that each has at hand. How many seconds are needed to bleach out the pink color? (The solution becomes colorless when the pH shifts to the acid side.) After the 1-min exercise session students breathe into the second flask of the pair. Record the time (in seconds) needed to bleach out the pink color.

Now each flask can be titrated to determine the number of micromoles of carbon dioxide in the flasks of each volunteer. Each milliliter of 0.04 percent solution of NaOH will combine with 10 micromoles of CO_2. (A mole of CO_2 weighs 44 g [C = 12, O_2 = 32]; a micromole is a millionth of a mole.)

With a graduated pipette or burette tube, add, drop by drop, a 0.04 percent solution of NaOH to the flasks containing exhaled air. Swirl the flasks as each drop is added. How many drops are needed to reach the end point and maintain the pink color? How many milliliters were used? To get the number of micromoles of carbon dioxide exhaled in a given time, multiply by 10 the number of milliliters of 0.04 percent NaOH used to reach the end point.

A comparison of exhaled and inhaled air

Students may know that there is a small percentage of carbon dioxide in air at all times. They may question the conclusion that limewater (or some other indicator) turned milky because of the presence of exhaled air and claim the change indicated only that some carbon dioxide was in the air at all times.

A more valid way to show that exhaled air contains more carbon dioxide than inhaled air is diagramed in Fig. 6-4. To make this device connect a glass Y-tube to two sections of rubber tubing. Then fit one short and one longer piece of glass tubing into each of two two-hole rubber stoppers which fit two Erlenmeyer flasks.

Attach the rubber tubing to the *short* section of glass tubing in one flask and to the *longer* piece in the other flask.

Into each flask pour 100 ml of water to which alkaline brom thymol blue (p. 663) has been added so that the water is colored light blue. Compare your setup with the

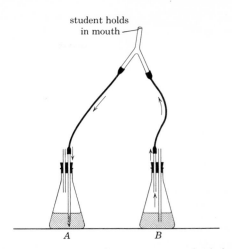

Fig. 6-4 Apparatus for comparison of inhaled and exhaled air. Student inhales (through mouth) air from flask B, exhales into flask A. An indicator in the water will show one difference in the two types of air.

diagram again. Note that only one tube to which the rubber tubing is attached extends below the level of the fluid. Insert a short section of a straw into the mouth piece of the Y-tube. Have students inhale and exhale continuously without removing the lips from the straw. Watch the change in color of the indicator in flask A, the "expirator." Of course, phenol red, phenolphthalein, or limewater may be substituted for the brom thymol blue.

The principle upon which this "inspirator-expirator" works is relatively simple for students to understand. During inspiration, air from flask B is inhaled. As air is removed from this flask the air pressure inside it is reduced, and more air is pushed into it and bubbles through the indicator. In exhalation, air is breathed into the indicator in flask A, and the indicator changes color rapidly.

Respiration in small land animals

The increase in carbon dioxide content of exhaled air of land animals may be measured. Make the apparatus shown in Fig. 6-5; be sure that the long and short

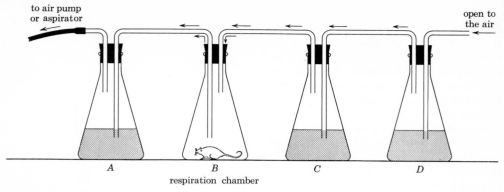

to air pump
or aspirator

open to
the air

A   B   C   D

respiration chamber

Fig. 6-5 Apparatus for measuring respiration in small animals. Flasks A, C, and D contain water with a CO_2 indicator.

tubes are in the right place (note that they are reversed in the respiration chamber). Pour dilute brom thymol blue, slightly alkaline phenolphthalein, or limewater into three flasks, skipping the respiration chamber. Place some fruit flies, grasshoppers, or beetles, or a small frog, toad, or white rat into the respiration chamber. (Germinating seeds can also be used; see p. 274.) Be sure to fire-polish the ends of glass tubing before inserting them into the rubber stoppers.

After connecting the apparatus, make it airtight. Connect an exhaust pump or aspirator to flask A. If an aspirator is used, begin the flow of air through the flasks by starting the siphon of the aspirator and letting the water drain into a sink or pail. As the pressure is reduced in flask A, air will be forced from the respiration chamber into this flask. If the organisms have been in the respiration chamber for a short time before the apparatus is used, carbon dioxide should have accumulated in the chamber. As this air is forced into flask A the indicator changes color. Thus, brom thymol blue becomes yellow, phenolphthalein becomes colorless, or limewater becomes milky. (The time in which this takes place will vary, of course, with the animal used—particularly in relation to its size—so that a change in the indicator may occur in an hour, or it may take several hours.)

In this same way, reduced air pressure in the respiration chamber will cause air

from flasks D and C to enter flask B, the respiration chamber. Watch for a slight change in the color of the indicator in flasks C and D as the small amount of carbon dioxide in air gradually shows a cumulative effect.

After 20 to 30 min terminate the demonstration and titrate the solution in flask A with 0.04 percent NaOH (see p. 271 for details). Multiply the number of milliliters of NaOH by 10 to estimate the number of micromoles of carbon dioxide exhaled by the mouse, or by insects or other experimental animals, in a given period of time.

In a BSCS Blue Version laboratory manual[5] it is suggested that the effect of chemicals on rate of respiration of mice might be studied after the "normal" rate of respiration has been obtained. Teachers might want to inject 0.5 ml of a 0.025 mg per ml solution of chlorpromazine into the thigh of a mouse and record the effect on the rate of respiration. Also, inject into another mouse 0.5 ml of adrenalin chloride solution.

In the same laboratory manual it is suggested that breathing rate may be studied in quiet and excited rats using a cellulose strip as a lever attached to the abdominal wall of a rat; or a kymograph drum might be used (see Chapter 7 for preparation of drum with smoked paper).

[5] *High School Biology: Blue Version—Laboratory, Part II*, BSCS, Univ. of Colorado, Boulder, 1961 (experimental use, 1961–62), pp. 359–60.

Respiration in small aquatic animals

Add enough brom thymol blue to a bottle or large test tube of water so that a light blue color results. To this add a goldfish, guppies, or several tadpoles. Then stopper the container. Within 20 min (as carbon dioxide is excreted) a color change becomes apparent; the blue color changes to yellow. Similar preparations, without the experimental animals, should be used as controls.

As a short review in a new context, ask students how to reclaim the yellow brom thymol blue so that the blue color is regained. Some students may recall that green water plants carry on photosynthesis. What would happen if the plants were added to the yellow brom thymol solution and placed in strong light? In this process, carbon dioxide would be absorbed out of solution; thus, the liquid would become less acidic and turn back to a blue color as it becomes slightly basic.

Rate of respiration

Of aquatic animals. Divide a sample of pond water into two containers. To one container, add a given weight of many of the same kind of organism (for example, test organisms might be strained *Daphnia,* a few guppies, or tadpoles). If students work in groups, samples of water may be tested for micromoles of carbon dioxide (p. 272) at different time sequences— 20 min, 30 min, 1 hr, 2 hr, and so forth. At the end of the designated time, add an indicator to each container (about 3 drops of phenolphthalein solution, as used previously, to a 100-ml sample).

Titrate 0.04 percent sodium hydroxide to the end point until a pink color remains in each container. Estimate the number of micromoles of carbon dioxide by multiplying by 10 the number of milliliters of sodium hydroxide needed to reach the end point. (Each milliliter of 0.04 percent sodium hydroxide solution contains 10 micromoles of carbon dioxide.)

If we let N_c equal the number of micro-moles of CO_2 evolved by the control, N_e the number of micromoles of CO_2 evolved by the experimental animals, W the weight of the animals (in grams), T the time (in hours), and R the rate of respiration (number of micromoles of CO_2 per gram of organism per hour), then we can calculate R as follows:

$$R = \frac{N_c - N_e}{W \times T}$$

In a carefully prepared series of activities for a 2-hr laboratory session, Verduin[6] describes procedures for measuring respiration in small aquatic organisms such as tadpoles, snails, crayfish, or small fish. He also describes measurement of rate of photosynthesis using filamentous algae.

In this technique, 0.02 N sodium hydroxide and phenolphthalein are added to the pond water at the start rather than after a given time lapse as in the method already described. Then 50 ml of the stock solution, distinctly pink in color, is transferred into a 100-ml graduated cylinder. Students can determine the volume of the fish or tadpoles by displacement of water in the graduated cylinder. This 50-ml sample plus the organisms are transferred to one container; in another container pour 50 ml of the fluid without the animals. At the end of a half-hour titrate each solution with 0.02 N sulfuric acid. Apply the same formula as given in the previous activity. In this case, the volume of animals (V, in milliliters) is used instead of weight.

$$R = \frac{N_c - N_e}{V \times T}$$

= micromoles of H_2CO_3 evolved per milliliter of organism per hour

Here N_c refers to the number of drops needed to titrate the control, and N_e refers to the number of drops needed to titrate the experimental.

[6] J. Verduin, "Simple Measurement of Respiration and Photosynthesis in Aquatic Organisms," *Turtox News* 41: 234–37, 1963.

The procedure recommended for measuring the rate of *photosynthesis* in this same laboratory period is similar to those already described in Chapter 3. Verduin adds the phenolphthalein indicator to the pond water and bleaches it if necessary by exhaling into the solution with a straw. Students measure out 50 ml of this solution in a 100-ml graduated cylinder. They then introduce a small quantity of filamentous algae such as *Spirogyra, Cladophora,* or *Zygnema* into this solution and measure the volume of the algae by displacement. This solution with plants is transferred into one container; a control solution is also prepared with algae and placed in the dark. Begin to titrate the experimental solution after 30 min in light. Titrate this (and the control) with 0.02 N NaOH until a permanent pink color appears. Since each drop of NaOH will react with 1 micromole of H_2CO_3 or bicarbonate in the water in the same way that algae remove the bicarbonate in photosynthesis, the following computation of rate of photosynthesis can be made:

$$\frac{N_d - N_l}{V \times T} = \begin{array}{l} \text{micromoles of } CO_2 \text{ absorbed} \\ \text{per milliliter of algae} \\ \text{per hour} \end{array}$$

In this case, N_l refers to the number of drops needed to titrate the sample kept in the light, N_d to the sample kept in the dark. V refers to the volume of algae, and T to time.

A small pond might be sampled many times in the course of a day, and the rates of respiration and photosynthesis can be compared.

Of unicellular algae (*Chlorella*). Many of the procedures described in Chapter 3 can be used in tests of rate of respiration in *Chlorella*.

In one technique for measuring the rate of photosynthesis in *Chlorella* a suspension of cells plus 0.1 percent phenolphthalein and enough NaOH (if needed) to make the suspension pink was prepared (p. 168). In the light the medium rapidly turned deep rose as CO_2 was absorbed and the medium became more alkaline. A similar preparation now placed in darkness or covered with aluminum foil shows a rapid bleaching of the rose color. Individual students may prepare several test tubes with a rich (deep green) suspension of *Chlorella* plus an indicator at the sensitive end point so that slight changes in pH in light and dark can be quickly obtained. In fact, so sensitive is the test that if students observe the tubes kept in light they should find a fading of color as daylight begins to diminish.

If the stock of *Chlorella* is limited, it may be more practical to draw off samples of medium as cultures are exposed to different light conditions. Then count the number of drops of indicator needed to turn a 10-ml sample. (The pH range of phenolphthalein is from 8.3 to 10.)

Brom thymol blue, with a pH range from 6.0 to 7.6, may also be used. Because of its narrow range, it is obviously a good choice to indicate slight shifts from neutral into either acidity or alkalinity. The indicator is best used with samples of medium since the green color of *Chlorella* interferes with perception of color changes.

Also refer to measurement of micromoles of gas produced in light and dark (see pp. 273, 167).

Production of carbon dioxide

By an aquatic plant

You will recall that the indicator brom thymol blue is blue when alkaline; when carbon dioxide is added, the solution becomes more acid and turns yellow. This indicator was used previously to show the absorption of carbon dioxide by a green plant in photosynthesis (p. 168). Now we shall use the blue solution again. If aquatic plants produce carbon dioxide during oxidation, the solution should become yellow. (This must be done in the dark where photosynthesis will not obscure the effect.)

Prepare a 0.1 percent solution of brom thymol blue in tap water (p. 169). Add about 20 ml of this to some 50 ml of aquarium water. Set up several test tubes containing this blue solution; to each add a sprig of elodea (*Anacharis*) with a growing tip. As controls, prepare other test tubes without elodea plants. Then place some of the tubes in the dark (or cover them with aluminum foil). Let one control (containing the indicator, without a plant) remain in light. Which tubes show a color reaction? How long does it take for the solution to turn from blue to yellow as a result of oxidation?

Or use phenolphthalein and titrate using the procedure above.

By a leafy plant

In the presence of light, green plants absorb carbon dioxide in food-making. They also continuously give off small amounts of carbon dioxide as a result of oxidation, but this effect is masked by the large amounts absorbed for photosynthesis. Therefore a green plant must be put in darkness in order to demonstrate or ascertain the speed of production of carbon dioxide in oxidation.

Place a healthy, leafy green plant, such as a geranium or coleus plant, and a small beaker of limewater on a sheet of glass or cardboard. Cover this with a bell jar, and seal the glass rim with petroleum jelly. As a control, prepare a similar demonstration omitting the green plant. Then cover both bell jars with black cloth for about 24 hr. At the end of this time the limewater with the green plant should be whiter than that with the control, due to the formation of carbon dioxide by the plant.

When a bell jar is not available, immerse cut stalks of plants in water in a large bottle or jar containing water. As shown in Fig. 6-6, suspend a vial of limewater from the tightly fastened stopper. Prepare a control with branches or stalks from which the leaves have been removed, and place both containers in the dark or cover with black cloth.

By roots

The excretion by roots of carbon dioxide as carbonic acid can be demonstrated by

Fig. 6-6 Apparatus for demonstrating that leaves give off CO_2 in the dark. The suspended vial contains an indicator (brom thymol blue or limewater).

using an indicator in this way. Support several vigorously growing seedlings (such as lima beans) on wet absorbent cotton with their roots submerged in test tubes containing water and a dilute indicator.

One indicator may be phenolphthalein (add a bit to the water, then introduce the small amount of sodium hydroxide needed to turn the water slightly pink). At other times you may vary the demonstration by using another indicator, such as litmus powder or brom thymol blue (turn the litmus powder blue by adding a small amount of limewater).

Within a few days, watch the change as the seedlings grow with their roots in the water (with indicator). The phenolphthalein solution will turn from pink to colorless, the litmus solution from blue to red.

You may improve upon this demonstration if you wish by using a solid medium. Prepare some unflavored gelatin. Before it solidifies, add a small quantity of litmus and limewater to tint the gelatin blue. (Or add phenolphthalein and enough sodium hydroxide to color the gelatin red.)

Fill test tubes about one-third full of the liquid gelatin. Then insert the roots of vigorously growing seedlings, such as sunflower or lima beans, into the cool, solidifying gelatin. Loosely cork the test tubes. Within 2 days you should detect increased acidity. The litmus should be red; the phenolphthalein should now be colorless.

(Of course, brom thymol blue might be used; in this case, the change in color would be from blue to yellow.)

By germinating seeds

Rapidly oxidizing tissues such as those of growing seeds produce appreciable amounts of carbon dioxide.

Place a dozen soaked peas, beans, or corn kernels into a test tube, or fill a bottle not more than one-third full of soaked seeds. As shown in Fig. 6-7a, connect this to another test tube or bottle containing limewater, brom thymol blue, or other indicator. Where space is limited, use the apparatus in Fig. 6-7b as a substitute. In this case, a small test tube containing seeds is placed on a support inside a larger test tube containing the indicator solution.

It is advisable to first disinfect the seeds by putting them in dilute formalin (1:500) or in dilute potassium permanganate for about 20 min. Then wash the seeds and soak them (see also p. 456). Controls can be set up by using seeds killed with formalin (p. 661) to prevent decay or mold.

While the previous demonstration is effective, a simpler one may be used. In fact, this demonstration as well as the previous one may be used to show respiration in flower buds, tips of growing stems, small insects, or insect larvae. Put a grain

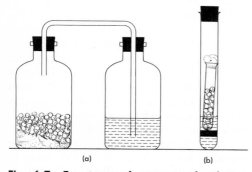

Fig. 6-7 Two types of apparatus for demonstrating that germinating seeds give off CO_2: (a) bottle on left contains germinating seeds, bottle on right contains water with an indicator; (b) a small tube of seeds is supported inside a large tube containing an indicator.

of barium hydroxide (or limewater, brom thymol blue, or other indicator) into a test tube containing approximately 10 drops of water. Then put a small wad of cotton in the tube about 1 in. above the water level. On this cotton surface lay six germinating seeds of peas or beans, or about two dozen oat seeds; seal the test tube. Then set up several such tubes using different seedlings, as well as controls using either killed seeds or none. The rapidity with which a precipitate forms (or the color changes) is a rough indication of the rate of carbon dioxide formation.

Throughout this section a variety of indicators as well as techniques for demonstrations are given. The teacher can select from these and use many different means for bringing variety into classroom demonstrations. You can ascertain whether students have memorized the results instead of understanding basic principles by using demonstrations which have not been described in the textbook or workbook used.

Titrating the solution. Some students may want to titrate the solution containing liberated carbon dioxide. Pour 50 ml of 0.2 N solution of sodium hydroxide into a flask and close with a rubber stopper. Measure out a quantity of germinating pea seedlings and tie them in a cheesecloth bag; suspend the bag into the flask over the sodium hydroxide solution, holding the bag in position by means of a string caught in the rubber stopper.

Maintain this flask at 35° C (95° F) in an incubator if necessary; also, prepare controls without seeds, or with killed seedlings. After 2 days, remove the bag of seedlings from the flask and quickly stopper again. Many variations are possible; some students may investigate how temperature affects the rate of aerobic respiration in germinating seeds. Maintain some flasks at temperatures varying from 5° to 25° C (41° to 77° F), as well as the one at 35° C (95° F).

By titration, students can now estimate the amount of carbon dioxide given off. First, precipitate the carbon dioxide absorbed by the hydroxide by adding 5 ml of barium chloride solution (20 g of barium chloride to 100 ml of water) to a sample, say a 10-ml sample, of the hydroxide from the flask. Add 3 drops of indicator such as phenolphthalein (prepared by adding 1 g of phenolphthalein powder to 50 ml of 95 percent ethyl alcohol and then adding 50 ml of water) to the solution to be titrated.

Now titrate with 0.1 N HCl until the rose color is bleached. Also titrate the control flasks (no seeds, dead seeds, or seeds that have been maintained at different temperatures). Then subtract the value (in milliliters) obtained from the flask with seeds from the value in the control flask and multiply by 5 to get the total amount of acid equivalent to the amount of carbon dioxide liberated by the seedlings.

Compare the value (in milliliters) of HCl equivalent to the liberated CO_2 in the flasks maintained at different temperatures.

The respiratory quotient of germinating seedlings may be measured by using a technique[7] that demonstrates the principles of manometry.

Measuring the volume. Now that the gas has been identified as carbon dioxide, you may want to measure the volume produced in respiration. Several techniques are described below. Frequently students want to find answers to questions which involve a need for effective ways to measure volumes of a gas absorbed or a gas released in a specific life activity. The suggestions described may serve as guides in planning the tools to use in a project.

USING WATER. Place a quantity of germinating seeds such as beans or peas, or kernels of corn, on a pad of moist absorbent cotton in a bottle or test tube. Connect the bottle to a delivery tube that runs to another bottle (Fig. 6-8). Set up as controls similar demonstrations using seeds which have been killed. Within a few hours students will see a substantial rise in the level of water in the delivery tube. The rise is a measure of the volume of the carbon diox-

[7] L. Machlis and J. Torrey, *Plants in Action*, Freeman, San Francisco, 1956, p. 99.

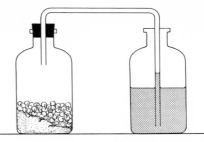

Fig. 6-8 Apparatus for measuring CO_2 given off by germinating seeds. CO_2 produced in seed-containing bottle is absorbed by water in second bottle; hence, decreased pressure in the bottle of seeds causes atmospheric pressure to push water up the tube.

ide produced by the seeds, since carbon dioxide is soluble in water.

USING MERCURY. A committee of students might want to use this device to demonstrate the evolution of carbon dioxide. Soak a quantity of seeds and let them begin to germinate. Then place a wad of wet cotton on the bottom of a bottle and add the germinating seeds until the bottle is about one-third full. Seal off one end of a short piece of glass tubing. Into the open end of the tubing insert a pellet of potassium hydroxide (*caution:* use tongs); hold this in place with a bit of cotton. As in Fig. 6-9, insert this into a two-hole stopper. Into the other hole insert one end of a delivery tube. Place the stopper in the bottle and seal it with paraffin. Immerse the other end of the delivery tube into a dish of mercury.

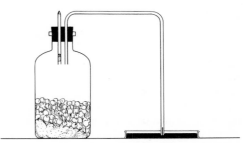

Fig. 6-9 Apparatus for measuring CO_2 given off by germinating seeds. Pellet of KOH absorbs CO_2; as air pressure in bottle is consequently greatly reduced, mercury rises in capillary tube.

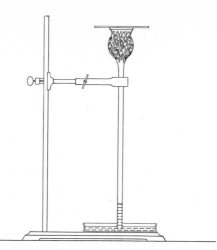

Fig. 6-10 Thistle tube apparatus for measuring CO_2 given off by germinating seeds. KOH solution in dish absorbs CO_2; as the air pressure is reduced, liquid rises in the tube.

As carbon dioxide is produced in respiration, it is absorbed by the potassium hydroxide pellet. As a result, air pressure within the bottle is reduced and normal air pressure pushes the mercury up the delivery tube arm. This gives an approximate measure of the volume of carbon dioxide produced by germinating seeds.

The negligible amount of carbon dioxide within the bottle at the start of the demonstration can be measured in a control; the same materials are used except that the seeds have been killed.

USING A THISTLE TUBE. Soak some wheat seeds or oat seeds for 24 hr in water and allow them to germinate in folds of damp paper. Then fill the bulb of a thistle tube with these germinating seeds. Support the tube upright so that the stem of the tube stands in a strong solution of potassium hydroxide (*caution:* one stick of caustic potash in two-thirds of a glass of water). This hydroxide will absorb CO_2. Red ink may be added to the liquid to increase visibility in a large classroom. Cover the bulb with a glass plate and seal with petroleum jelly (Fig. 6-10). Prepare several of these devices, using different types of seeds, as well as killed seeds as controls.

Within a few hours students may see the rise of potassium hydroxide solution in the tube. In about 6 hr the liquid will rise halfway up the tube. This serves to demonstrate that some gas in the thistle tube has been absorbed and that more carbon dioxide has been formed. In turn, the carbon dioxide was absorbed by the potassium hydroxide.

USING A BURETTE. This technique is similar to the previous one. Place some germinating seeds into the constricted end of a 100-ml burette. Hold them in place with wet cotton. Now invert the burette into a large test tube of saturated potassium hydroxide solution (Fig. 6-11). Adjust the level of the solution to the zero mark. When the burette clamp is closed, there

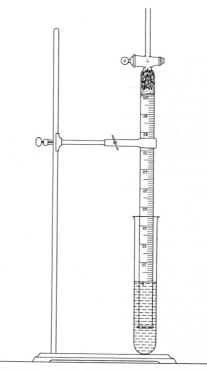

Fig. 6-11 Burette apparatus for measuring CO_2 given off by seeds. Saturated KOH solution in test tube absorbs CO_2, and the liquid consequently rises in the burette; the volume of CO_2 produced is determined by reading the level of the solution in the burette.

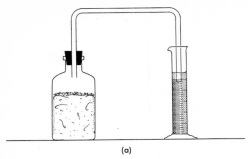
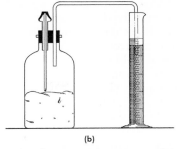

Fig. 6-12 Apparatus for demonstrating role of yeast in bread-making: (a) CO_2 is given off by yeast sprinkled on top of flour-and-water "dough"; (b) CO_2 given off in a well-mixed bread dough of flour, water, and yeast may be trapped but can be released by punching the dough several times with the glass rod.

will be 100 ml of air in contact with the germinating seeds. Students might prepare a control with dead seeds or without seeds. (*Caution:* Saturated KOH is corrosive.)

There will soon be a difference in the rise of potassium hydroxide in the two burettes. This will be a rough measure of the carbon dioxide evolved in respiration, since carbon dioxide was absorbed by the hydroxide.

USING A RESPIRATION CHAMBER. The demonstration using a series of flasks with a respiration chamber and an aspirator described for animal respiration (Fig. 6-5) applies here as well. Fill the respiration chamber two-thirds full of germinating seeds. Then prepare a control with either killed seedlings or no seedlings.

USING A VOLUMETER. Use the volumeter shown in Fig. 6-17 and the procedure described on p. 281. Measure the distance the trapped drop of water or air bubble moves. Compare the volume of carbon dioxide evolved by germinating seeds with the volume produced by dead seeds.

By fruits and tubers

You may want to use the apparatus shown in Figs. 6-5 and 6-6 to show production of carbon dioxide in respiration of tissues of such tubers as potatoes, and in such fruits as apples. Place a portion of the apple or potato in a jar or tube, and line the jar with moist filter paper. Then use an indicator or limewater to show that carbon dioxide is evolved in respiration.

By fungi

You may want to use a rich culture of yeast or bread mold to demonstrate the evolution of carbon dioxide.

Such forms as the edible mushroom (*Agaricus campestris*) give excellent results, because they do not readily decay. Prepare the demonstration as in Fig. 6-6. Or use a container with a delivery tube extending from a one-hole stopper into a smaller container of brom thymol blue or limewater. Within an hour enough carbon dioxide is produced to change brom thymol blue to yellow (or to make limewater milky). Include a control in which the fungi are omitted.

Role of yeast in bread-making. Students may want to show that dough rises because bubbles of carbon dioxide are trapped within it. Mix flour with water, place this in a jar, and sprinkle the paste with yeast (Fig. 6-12a). The indicator (brom thymol blue or limewater in the cylinder) will show the evolution of carbon dioxide. (The addition of sugar to the dough will speed up fermentation.)

Prepare a similar jar; this time thoroughly *mix* the yeast with the dough. Note how quickly the dough rises (temperature is a factor); but the limewater or brom thymol blue does not react readily since the carbon dioxide is trapped within the dough. Now arrange a glass rod as in Fig. 6-12b so that the dough can be punctured as it rises. Place the rod in a glass tube and make the connection airtight with clay or with wired rubber tubing. Then fit the glass tube containing the rod into the rubber stopper. (Prepare a control similar in all details except that yeast is not added.) Why does the indicator change when the dough is punctured?

Effect of temperature on fermentation. Students can devise a piece of apparatus to demonstrate that carbon dioxide is liberated by yeast cells in fermentation. A delivery tube may be carried into a beaker containing an indicator such as brom thymol blue or phenol red. Or one arm of a thin-tube manometer may be attached to a delivery tube from one test tube or flask containing a glucose-yeast solution, and the second arm of the manometer attached to a delivery tube from a flask containing only a glucose solution. Maintain both tubes in a water bath. Prepare duplicates and maintain at different temperatures. Does the liberation of carbon dioxide vary with changes in temperature?

Some teachers prefer to set up a container of molasses and yeast cells and extend a delivery tube into a jar of a culture of *Chlorella*. A control is also prepared, using boiled molasses and yeast cells. After 2 to 3 days in the light, the experimental is many shades greener than the control. What gas do yeast cells liberate? In what process do *Chlorella* cells use this gas? (Cell counts may be made; see pp. 443–44.)

Measuring the volume. Prepare the apparatus shown in Fig. 6-17, using either test tubes or flasks. Into one tube or flask pour molasses (or glucose) and a suspension of yeast cells to within 2 in. of the top of the flask. Into the other, the thermobarometer, pour water to occupy the same volume. A control may be used (a volume of *boiled* molasses and yeast cells equal to that of the experimental). Maintain at 30° C (86° F) in a water bath.

There is a very rapid consumption of

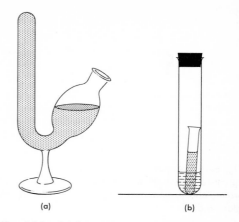

Fig. 6-13 (a) Fermentation tube containing yeast cells in a molasses (or 10 percent glucose) solution; (b) yeast culture growing under anaerobic conditions—oxygen in outer tube is absorbed by potassium pyrogallic acid.

oxygen; the liberated carbon dioxide may be measured. If a layer of cotton is applied over the suspension of yeast cells and sodium hydroxide is spread over this, the carbon dioxide will be absorbed and the volumeter will measure the rate of consumption of oxygen. Have students take readings at 1-min intervals of the distance that the trapped drop or bubble moves. Of course, this is also a measure of the rate of release of carbon dioxide.

Fermentation tubes. Fill the closed tubular portion of a fermentation tube with a dilute solution of molasses (Fig. 6-13a). Leave the bulb side partly empty. Prepare another fermentation tube in which yeast is added to the molasses. Then plug with cotton and keep the tubes in a warm place. Within ½ hr students may see bubbles of gas rise in the fermentation tube that contains yeast cells. There should be a considerable accumulation of gas in the upper arm of the fermentation tube. Now a small piece of potassium hydroxide may be put in the liquid, where it will soon dissolve. Watch the liquid rise again into the tubular arm as the carbon dioxide gas is absorbed by the caustic potash.

Alcohol production by yeast cells. Fermentation results in two main products, carbon dioxide and alcohol. This method indicates the production of alcohol. Since fermentation proceeds rapidly under anaerobic conditions the apparatus shown in Fig. 6-13b may be preferable to fermentation tubes.

Prepare a rich culture of yeast plants by adding yeast cells to a solution made of equal parts of molasses and water, or to a 10 percent glucose solution. Let the yeast culture grow for at least 24 hr, and then pour it into the inner tube (see diagram). Then fill the outer tube to the level indicated with alkaline pyrogallic acid (prepared by mixing equal volumes of 5 percent solutions of pyrogallic acid and potassium hydroxide) or any other oxygen absorbent; stopper tightly. From time to time, test the contents of the inner tube for the presence of alcohol, as follows.

TESTING FOR ALCOHOL PRODUCTION BY YEASTS. In anaerobic respiration or fermentation, yeast plants ultimately break down sugars into carbon dioxide and alcohol. The odor of alcohol is evident in both of the previous demonstrations. If the odor of alcohol is not characteristic in itself, the presence of ethyl alcohol may be detected in this way: Pour about 5 ml of the fermenting molasses or glucose solution into a clean test tube. To this add 4 drops of a 10 percent solution of sodium hydroxide. Then add iodine potassium iodide solution (Lugol's solution, p. 664), drop by drop, until the liquid remains faintly yellow. Let it stand for 2 min and then shake the tube. Note the separation of a layer of iodoform settling to the bottom of the tube. Ethyl alcohol, acetone, and acetaldehyde and ketones containing the acetyl group give the iodoform test.

Comparison of anaerobic and aerobic metabolism in yeasts. Prepare a quantity of glucose-yeast suspension and divide it between two flasks. Into two other flasks place only glucose solution. Prepare two flasks with a one-hole stopper to maintain anaerobic conditions; use two-hole stoppers in the other two flasks to provide an incoming source of air (oxygen supply).

Maintain a flask of glucose-yeast solution and one containing only glucose under anaerobic conditions, and keep similar (the other two) flasks under aerobic conditions, that is, with the source of air through the second inlet tube in the flasks.

Incubate at 37° C (99° F) for several hours or a day. Analyze the contents of the several solutions by fractional distillation.[8] Alcohol is the fraction given off between 78° and 79° C (172° and 174° F). (Use the iodoform test, which is described above.)

The equation for fermentation in yeast is:

$$C_6H_{12}O_6 \longrightarrow$$

$$2C_2H_5OH + 2CO_2 + 2 \sim P$$

(ethyl alcohol)   (two high-energy phosphate groups)

The equation for respiration in yeast is:

$$C_6H_{12}O_6 + 6O_2 \longrightarrow$$

$$6CO_2 + 6H_2O + 38 \sim P$$

(approximate)

Absorption of oxygen

By germinating seeds

Students might prepare this demonstration (Fig. 6-14). Place a quantity of germinating seeds on a cushion of moist absorbent cotton in a 250-ml bottle so that it is about one-third full of seeds. Make the long arm of the delivery tube about 50 cm long. Insert the short arm into a one-hole stopper, with a small vial of 10 percent potassium hydroxide suspended from the stopper by means of a string and a bent

[8] Described in *High School Biology: Blue Version— Laboratory, Part I,* BSCS, Univ. of Colorado, Boulder, 1961 (experimental use, 1961–62), pp. 145–47.

For information concerning the transformations of glucose to alcohol in yeast extracts, look up the Meyerhoff–Embden system of reactions in texts in cell physiology.

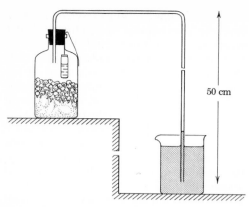

Fig. 6-14 Apparatus for indirectly measuring O_2 absorbed by germinating seeds. CO_2 evolved by the seeds is absorbed by KOH in the vial; since the seeds absorb a volume of O_2 equal to the volume of CO_2 evolved, the decrease in pressure, indicated by a rise of water in the tube, is an indirect measure of O_2 absorbed.

pin, and stopper the bottle. Immerse the other end of the tube in a beaker of water colored with a few drops of red ink.

Then set up a control using killed seeds. Keep both preparations at the same temperature. Potassium (or sodium) hydroxide absorbs carbon dioxide immediately, and since there is an approximate

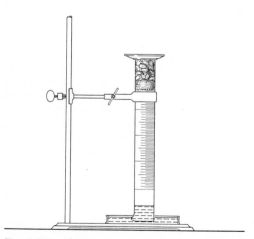

Fig. 6-15 Alternate apparatus for indirectly measuring O_2 absorbed by germinating seeds. CO_2 given off (equal in volume to O_2 absorbed) is absorbed by KOH solution, and atmospheric pressure forces liquid up the graduated cylinder.

1:1 ratio between the amount of oxygen taken in and the carbon dioxide given off in respiration, the rate of oxygen consumption will be indicated by the water level in the delivery tube.

Or students might set up the demonstration in Fig. 6-15 using sprouting seeds held in place with moist cotton. The graduated cylinder should then be inverted into potassium hydroxide or sodium hydroxide. Prepare a duplicate demonstration using killed seeds. As in the previous preparation, seedlings absorb oxygen, and an equal quantity of carbon dioxide is produced. Here, the carbon dioxide is absorbed by the hydroxide, which rises in the cylinder.

Burning splint. In this method you may want to show that oxygen is used up in a container in which germinating seeds are growing. The decrease in oxygen content may be shown by testing with a burning splint. Soak bean, wheat, or oat seeds or corn kernels. Fill several jars about one-third full of germinating seeds; cork the jars securely. Also prepare replicas as controls using seeds which have been killed with formalin.

After 48 hr insert a burning splint into the bottles. The splint will burn more brightly in the control bottles than in the jars containing living, germinating seeds.

This demonstration is generally unsatisfactory because there is little difference in the glow of the burning splint in the two jars.

Using fermentation tubes. Place a quantity of soaked seeds (wheat, oats, or corn kernels) into the tube-part of each of four fermentation tubes. Then fill this tube-part with water.[9] Generate carbon dioxide and let it pass into the first tube; add oxygen to the second tube, and nitrogen (if available) to the third tube. Let the gases displace nearly all the water in each tube. Then insert a rubber stopper into the open end under water. Take out the tube, placing the corked end into a beaker con-

[9] L. J. Clarke, *Botany as an Experimental Science*, Oxford Univ. Press, New York, 1931.

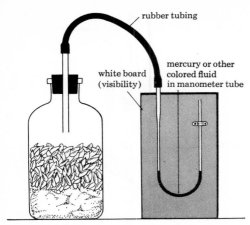

rubber tubing

white board (visibility)

mercury or other colored fluid in manometer tube

Fig. 6-16 Apparatus for illustrating that O_2 is consumed by germinating seeds.

taining water so that air is prevented from entering. In the control tube, the fourth tube, place seeds in a small quantity of water and stopper the fermentation tube. Within 24 hr germination will be found in the control tube and in the tube containing oxygen.

By different seeds

A small group of students might undertake a project to compare the quantity of oxygen absorbed by seeds rich in fats with that absorbed by seeds rich in starch. In other words, is there a 1:1 ratio of oxygen to carbon dioxide when fats are oxidized? For instance, students might germinate soybeans and wheat seeds which have been soaked for 24 hr in advance. Prepare the device shown in Fig. 6-16, using soybeans (rich in fats) in one setup, and wheat seeds (rich in starch) in the second. Put mercury (or other colored fluid) into the U-tube in each preparation and seal the bottles so they are airtight. Students should observe some differences in 24 hr. Does the level of fluid in the two arms of the U-tube manometer remain about the same in the case of the wheat seeds? If so, this indicates that the volume of oxygen consumed equals the volume of carbon dioxide given off. What happens in the device using soybeans? The additional

amount of oxygen consumed is indicated by the higher column of fluid in the tube nearer the bottle.

Using a volumeter. Germinating pea seedlings (3 to 4 days old) consume oxygen at a rapid rate; this can be measured in a volumeter (Fig. 6-17). One molecule of carbon dioxide is formed for each molecule of oxygen absorbed so that no change in the gas volume occurs (Avogadro's law). Sodium hydroxide is added to absorb the carbon dioxide formed in respiration.

Large test tubes or flasks may be used. Fill one tube almost to the top with germinating seedlings of peas; then insert a layer of cotton. On the top of the cotton place a 1-in. layer of soda lime (sodium hydroxide) to absorb the carbon dioxide liberated. Fill the second tube with water so that the same volume is occupied; this tube will be the thermobarometer. Insert an escape tube with clamp into each two-hole stopper, and insert a 1-ml calibrated pipette in the second hole of each stopper. Or pull out capillary tubing and apply a plastic ruler as a scale (as in Fig. 6-17). With a fine capillary or a hypodermic syringe insert the indicator drop or trapped air bubble; adjust the escape

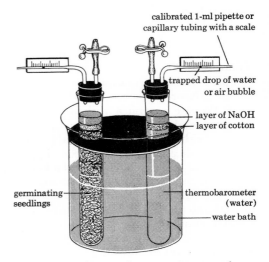

calibrated 1-ml pipette or capillary tubing with a scale

trapped drop of water or air bubble

layer of NaOH
layer of cotton

germinating seedlings

thermobarometer (water)

water bath

Fig. 6-17 Volumeter for measuring rate of consumption of O_2 (or rate of liberation of CO_2) by germinating seedlings.

clamp and set the drop or bubble at the far end of the pipette. Let the apparatus and the control reach equilibrium (place in a water bath, if needed); then take readings of the temperature and the distance that the trapped bubble or drop moves toward the flask or test tube. Subtract the readings in the thermobarometer from those of the experimental until the rate of change in the experimental becomes constant.

This is the rate of consumption of oxygen by germinating pea seedlings in the flask. What is the rate of liberation of carbon dioxide?

By algae

Using a volumeter. A volumeter was used to test the amount of oxygen given off in photosynthesis of algae such as *Chlorella* (Fig. 3-14). Determine the respiration rate in the dark; let light fall on the suspension of cells and estimate the net photosynthesis.

To measure the amount of oxygen absorbed, the carbon dioxide must be removed from the system. A volume of oxygen is simultaneously being absorbed equal to the volume of carbon dioxide released from cells (R.Q. = 1).

Use identical samples of suspensions of *Chlorella* in two flasks or tubes (Fig. 6-18). In one flask add 10 percent KOH to absorb the CO_2 in the flask and the CO_2 liberated during respiration, thereby keeping the pressure of carbon dioxide at a low level. Prepare a replica without *Chlorella* as a thermobarometer. Also prepare a control with the KOH. Use a cotton-tipped applicator that has been dipped into a freshly prepared solution of alkali—either 10 percent potassium or sodium hydroxide. If the cotton applicator is now dipped into phenolphthalein it turns rose-colored; stand this applicator in the flask so that the cotton tip is exposed to the air above the level of the fluid. Watch how quickly the color is bleached as the alkali absorbs the carbon dioxide in the air over the fluid. Since the volume of oxygen ab-

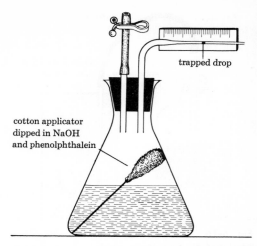

Fig. 6-18 Flask for measuring rate of consumption of O_2 by *Chlorella*. Flask contains a dense culture of *Chlorella* and a cotton applicator dipped in CO_2-absorbent NaOH and phenolphthalein indicator. (Large test tubes may be used instead of flasks.)

sorbed in respiration is to be measured, the flasks must be covered so that photosynthesis ceases. Cover the flasks with aluminum foil and maintain at 25° C (77° F) in a water bath (as shown in Fig. 4-12). Students should take readings every 2 to 3 min for a period of 15 min. (Avoid handling the flasks, and let the system come to equilibrium before beginning to take readings.)

Record in millimeters the distance traveled by the drop (or trapped air bubble) toward the experimental flasks; in the case of the controls (including the thermobarometer) record any change in position of the drop.

Concentrated populations of cells of *Chlorella* can be studied, and by using dilutions (see p. 441), students can show the diminished rate of exchange of gases with a decrease in population density.

How might these changes in acidity affect other organisms growing in a pond? (See investigations suggested in Chapter 11.) Does a change in acidity of pond water affect succession of other microorganisms in the pond?

If a supply of oxygen is needed for

respiration, would three times the supply of oxygen increase the rate of respiration threefold? Does gibberellic acid, or one of the auxins such as indoleacetic acid, affect the rate of respiration in plants? Is the rate of respiration the same at every phase of the growth cycle of plant cells (see Chapter 9)?

By aquatic plants

A simple modification of Warburg's manometric method (used previously to show that oxygen is given off in photosynthesis, Fig. 6-16) may be used here to show that oxygen is absorbed during respiration.

By small land animals

After students understand the techniques for measuring the consumption of oxygen (or the liberation of carbon dioxide, as previously described), they may plan investigations that focus on the effect which changes in temperature or pressure have on the consumption of oxygen; the

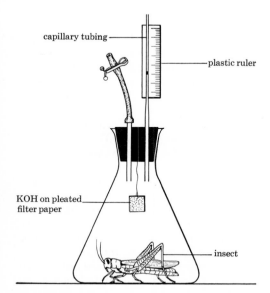

Fig. 6-19 Volumeter for measuring consumption of O_2 by an insect or other air-breathing organism. KOH absorbs CO_2; as O_2 is consumed, the trapped bubble in the tubing moves toward the bottle.

capillary tubing

plastic ruler

KOH on pleated filter paper

insect

effects of treatment with narcotics, drugs, or hormones such as thyroxin and adrenalin; and the effect on breathing of animals that are quiet or in an agitated state (also see Fig. 6-5).

Using a volumeter. Small insects, frogs, land snails, or mice are convenient experimental animals for this demonstration. Prepare a volumeter like that shown in Fig. 6-19; insert thin capillary tubing (0.2 ml in diameter) into one hole of a rubber stopper and secure a plastic ruler along its length with plastic tape; or use a 1-ml calibrated pipette. Insert a short piece of glass tubing and rubber tubing with a clamp into the other hole of the rubber stopper; this is an escape tube. Fasten a square of filter paper that has been pleated to increase its surface area to the rubber stopper; or place it or cotton in a small vial that can be suspended from the rubber stopper. Soak the filter paper or cotton with 10 percent potassium hydroxide. Since KOH is caustic, avoid having the experimental animals come in contact with it. Prepare a control without the test animal; also prepare a thermobarometer. In all the setups insert a drop of colored water containing a trace of detergent in the length of capillary tubing or pipette. Maintain in a water bath at 30° C (86° F), and wait until the flasks (or bottles) have reached the temperature of the water bath. Introduce the test animals. Take readings at 1-min intervals (unless the drop moves very slowly); also record the temperature.

Carbon dioxide liberated by the respiring organisms will be absorbed by the potassium hydroxide. Compare the experimental with the controls. Have groups of students compare their results. As oxygen is absorbed by the organisms, the change in volume within the containers will result in the trapped drop moving down into the container.

Using a manometer. Hammen[10] suggests a modification that uses the common

[10] C. Hammen, "Oxygen Consumption as a Laboratory Exercise," *Turtox News* 38:296–97, 1960.

land snail, *Helix,* for measurements in which a U-tube manometer is used instead of a volumeter. (*Helix* can be obtained in fish markets.) Assemble the apparatus shown in Fig. 6-20; one container accommodates the large snail (or earthworms or crayfish) along with a vial containing 20 percent KOH to absorb the carbon dioxide liberated by the animal. Insert a pleated piece of filter paper to increase the surface of the KOH solution; suspend the vial by a thread as shown. Also prepare the empty container to act as a thermobarometer. Bend glass tubing and assemble the thin U-tube capillary containing colored ink. Immerse both containers in a water bath and allow the apparatus to reach equilibrium. The level of the liquid in the manometer can be adjusted by manipulating the stopper on the empty container. The syringe filled with air is inserted through the rubber stopper. Take readings of the change in the level of the manometer (using an attached plastic millimeter ruler) as oxygen is consumed by the test animal. Restore the original level by releasing the syringe to introduce air into the container. Students should take several readings, restoring the air at the end of each reading. Average the readings. The volume of oxygen that is used is the difference between these readings and the original volume. Students can weigh the animal (or animals) to calculate the rate of respiration in millimeters of oxygen consumed per gram per hour. Some students may want to calculate the difference in rate of consumption at 21° C (70° F) and at 31° C (88° F). Plot the different rates at the two temperatures.

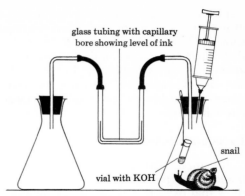

Fig. 6-20 Constant-pressure respirometer for measuring consumption of O_2 by a land snail. Both flasks are immersed in a constant-temperature water bath, which may be an aquarium or a large finger bowl. (After C. Hammen, "Oxygen Consumption as a Laboratory Exercise," *Turtox News* 38:296, 1960.)

Measuring the amount of dissolved oxygen in pond water

Winkler method. This method is especially useful in testing the concentration of oxygen in pond water that has only a small amount of organic matter, nitrates, and iron in it.

Collect a 100-ml sample of water to be tested; avoid agitating the water while collecting it. Several chemical reactions are involved in this rather long procedure, and separate stock solutions need to be prepared beforehand.

1. To the 100-ml sample of water add 0.5 ml of solution of manganous sulfate. (Prepare by adding 400 g of $MnSO_4 \cdot 2H_2O$ to 1 liter of water.) The oxygen which is present oxidizes some manganous ions to manganic ions.

2. Add 1.5 ml of a mixture of potassium hydroxide–potassium iodide solution. (Prepare by adding 500 g of sodium hydroxide to 1 liter of water; then add 150 g of potassium iodide.)

3. Shake the sample with the test solutions.

4. Add strong acid—0.5 ml of undiluted sulfuric acid. A brown precipitate forms in the acid solution; the amount of iodide that is changed to iodine is in proportion to the number of manganic ions present. The color of the solution is a rough indication of the quantity of oxygen present, since the amount of iodine is equal to the amount of oxygen in the water.

5. Titrate iodine with sodium thiosulfate (hypo). (Prepare the thiosulfate by adding 3.1 g to 1 liter of water.)

Use starch as an indicator in this titration. (Boil a cornstarch suspension—1 tbsp in a cup of water.)

To a 100-ml sample of pond water containing the iodine, add a few drops of starch suspension so that a blue-black color results. Titrate with thiosulfate until the sample turns colorless. The number of milliliters of thiosulfate needed to turn the sample colorless is equal to the parts per million of oxygen dissolved in the sample of pond water.[11]

Effect of shortage of oxygen on seedlings

Using potassium pyrogallate to remove oxygen. Soaked oat, wheat, or other small seeds or corn kernels should be placed on moist absorbent cotton in a jar or wrapped in a small cheesecloth bag, tied with string and suspended as shown in Fig. 6-21. Prepare two demonstrations. Pour fresh potassium pyrogallate solution (prepared by mixing equal volumes of 5 percent solutions of pyrogallic acid and potassium hydroxide) into the two bottles. Then prepare two replicas as controls, using water instead of potassium pyrogallate solution. Students may observe the small amount of germination in the bottles which contain potassium pyrogallate. There may be a small residue of air containing oxygen trapped within the seed coats so that a little germination occurs; yet there should be an appreciable difference in germination between the jars with water and those with potassium pyrogallate solution.

Drowning the seedlings. The fact that drowned seedlings stop growing is indirect evidence for the lack of oxygen. (Although this approach is described in the literature, it is not satisfactory to our way of thinking.) In this method, bean seeds or corn kernels are germinated until the root tips are about 10 mm long. Then the root tips are marked at uniform lengths with India

[11] Further information is available in P. Welch, *Limnological Methods*, Blakiston (McGraw-Hill), New York, 1948, pp. 207–11.

Fig. 6-21 Apparatus for demonstrating that germinating seeds require O_2. Seeds suspended over potassium pyrogallate solution should not grow; those over water should grow.

ink. Test tubes or jars are lined with blotting paper, and the seedlings are inserted between the blotter and the glass walls. Let several jars stand with seeds drowned by water. In other jars, only moisten the blotting paper. Then cover all the jars to prevent evaporation. Measure the length of the root tips after 1 day and after 2 days.

Immersing the seedlings in mercury. In this method mercury is used to displace air which would otherwise be available to seedlings. Completely fill a large test tube with mercury and invert this into a dish of mercury; fill another test tube nearly full with mercury and invert it into the same dish. By this procedure, the second test tube contains a small amount of air. Now support both test tubes with clamps on a stand (Fig. 6-22).

Students will need peas or beans that have been germinated until the roots are about 25 mm long. After first removing the outer seed coats (to eliminate air pockets) measure the lengths of the roots. Then insert six of the seedlings into each of the two inverted test tubes so that they float to the top. Place this preparation in a warm place and measure each root after 24 hr. Students should learn from the results that seeds which receive some air show greater growth.

Overcrowding. You may want students to soak a large quantity of beans, peas,

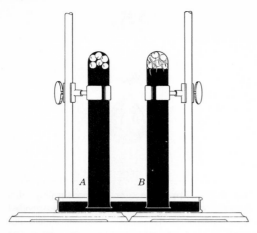

Fig. 6-22 Alternate apparatus for demonstrating that germinating seeds require O_2. The seeds in tube B, with an air pocket, should grow; those in tube A should not.

corn kernels, or oat seeds in water. Place some of the seeds in a small bottle, then pack the remainder of the bottle with moist sand so that every space is filled; stopper this bottle tightly. Then prepare another bottle with about the same quantity of soaked seeds. Fill this bottle only halfway with moist sand; do not seal this bottle.

Students may compare the degree of germination of seeds in both bottles after the bottles have been exposed to the same temperature conditions for a few days.

Using the blackboard as a device for summary

Students might diagram a cross section of a leaf. (You may want to examine leaves, stems, and roots at this time; see Chapter 1.) Then students might compare photosynthesis and respiration, perhaps in the form of Table 6-1.

Production of heat by living plants

The evolution of considerable heat by seedlings during respiration may be demonstrated by preventing its escape, and noting the substantial difference in tem-

perature (Fig. 6-23). Fill a pint vacuum bottle nearly full with germinating beans, corn grains, peas, wheat, or oats on a bed of moist absorbent cotton. Insert a thermometer through a one-hole stopper and seal it in place with modeling clay to make it airtight. Prepare a control containing seedlings killed in formalin. Unless the seeds in the control are properly killed and disinfected (p. 456), there will be evidence of temperature change due to heat generated through bacterial decay.

When vacuum bottles are not available an effective substitute can be made by using two bottles with thermometers as above (one experimental, one control) and packing them into glass-wool insulation or excelsior. Also refer to biological luminescence, Chapter 2.

Respiratory enzymes and mitochondria

The role of respiratory enzymes may be demonstrated using some of the techniques presented in Chapter 2.

TABLE 6-1
Photosynthesis vs. respiration

	photosynthesis	respiration
Check one column:		
Process in which food is used		
Process in which food is made		
Process which uses energy		
Process which releases energy		
Fill in one or both columns, as appropriate:		
Gas taken in		
Gas given off		
Gas entrance		
Gas exit		
Water entrance		
Water exit		
Time of day process goes on		

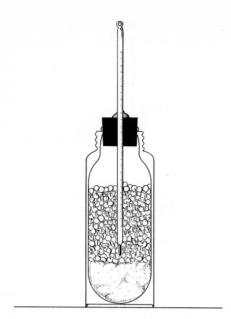

Fig. 6-23 Apparatus for demonstrating that germinating seeds generate heat. The thermometer in this vacuum bottle containing germinating seeds should show a temperature rise within 24 hr; the same setup, without seeds or with killed seeds, should not.

Mitochondria, obtained from preparations of celery stalks (see p. 110), may be examined under the microscope. Mitochondria may also be obtained from a living fly or bee, as described in a BSCS laboratory block.[12] The procedure recommended is this.

Open the thorax of a live fly or bee, after first cutting off the head. Remove a bit of flight-muscle tissue to a slide and tease it apart in physiological saline solution (p. 121) to which a saturated solution of Janus Green B has been added. Apply a coverslip and examine under high power; look between the muscle fibers to locate hundreds of floating mitochondria, which are round or slightly oval. With Janus Green B the mitochondria should appear green or bluish green.

[12] A. G. Richards, *The Complementarity of Structure and Function* (BSCS laboratory block), Heath, Boston, 1963, pp. 66–69.

Mechanics of breathing

Students may want to show that certain aspects of respiration are "mechanical." Prepare the bell-jar model shown in Fig. 6-24. Insert a glass Y-tube into the opening of a rubber stopper which fits into the top of the bell jar. Fasten two balloons of the same size by means of string or rubber bands to the two arms of the Y-tube. Next cut a circle of thin rubber sheeting large enough to fit over the bottom of the bell jar. In the center of the sheeting pinch out a piece, insert a small cork, and tie it off as a "holdfast." Tie the sheeting in place. Check to see that there are no breaks in the sheeting.

Now the balloons represent lungs and the rubber sheeting demonstrates the action of the diaphragm. When students pull out or push in the rubber "diaphragm" as shown, a change takes place in the size of the balloon "lungs."

What happens when the rubber sheeting is pulled down? Since we have in-

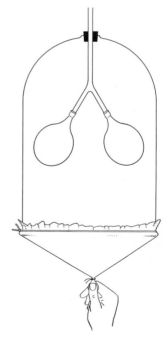

Fig. 6-24 Bell-jar model to show mechanics of breathing.

creased the volume within the bell jar by pulling down the "diaphragm," there is less pressure on the "lungs." Thus atmospheric pressure pushes outside air into the "lungs." When the sheeting is pushed upward the air pressure in the interior of the bell jar is increased, the pressure on the "lungs" is increased, and the air in the "lungs" is forced out. Then the balloon "lungs" collapse.

Of course, in the body the size of the chest cavity is also changed through the action of intercostal muscles between the ribs. Students may want to make a movable model of wood with hinges to show these changes in chest cavity.

Capacity of the lungs

A rough estimate may be made of the volume of air exhaled. Make right-angle bends in glass tubing (½-in. bore) as illustrated (Fig. 6-25). Then fill a gallon bottle about four-fifths full with water to which some red ink or dye has been added. Insert the tubing into the two-hole stopper and seal the bottle with paraffin.

Cover the mouthpiece with paper toweling and have students exhale into it. The amount of water displaced into the cylinder will be equal to the volume of air exhaled. Use a graduated cylinder to facilitate measuring the volume of displaced water.

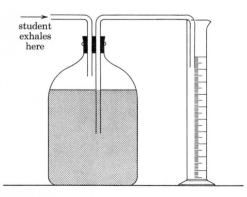

student
exhales
here

Fig. 6-25 Apparatus for estimating capacity of human lungs.

Effect of excess carbon dioxide on the rate of breathing

Students may determine their normal breathing rate while at rest. Count the number of times breathing occurs in half a minute and then multiply by two. In general, respiration rate varies with age:

at birth	30–50 per min
1st year	25–35 per min
4–15 years	20–25 per min
adulthood	16–18 per min

Next, have students breathe into a paper bag held closely for several minutes over the nose and mouth. Determine the rate of breathing while breathing into the bag, then again after breathing ordinary air. The increase in carbon dioxide in the air stimulates the medulla, which in turn regulates the rate of breathing. However, this "experiment" yields very "rough" data. (Also refer to effect of exercise on production of carbon dioxide, p. 268.)

Oxygenated and deoxygenated blood

Get several pints of fresh blood from a slaughterhouse, and add 1 part of sodium oxalate to 9 parts of blood to prevent it from clotting.

Prepare two flasks, each with a short piece of glass tubing and a right-angle bend of glass tubing inserted in a rubber stopper. A piece of rubber tubing leads from each bent tubing—to an oxygen generator for one flask and to a carbon dioxide generator for the other (Fig. 6-26).

Note the froth and the brilliant red color of oxygenated blood and the darker red color of blood rich in carbon dioxide.

Clamp the delivery tubes and reverse the attachments so that oxygen now flows into the flask of deoxygenated blood. Watch the change to bright red as oxyhemoglobin is formed in the blood.

In lieu of a generator of carbon dioxide, the gas may be supplied by a small piece of dry ice or by chalk dust added to a carbonated beverage. Oxygen may be sup-

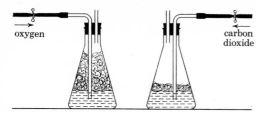

Fig. 6-26 Apparatus for demonstrating effects of O_2 and CO_2 on oxalated blood.

plied (if no generator is available) by adding 1 ml of hydrogen peroxide to 15 ml of citrated blood. (These alternatives are suggested in several of the BSCS laboratory techniques in the 1961 preliminary editions.)

Comparing respiratory systems in animals

How are different animals adapted for permitting a rapid interchange of oxygen and carbon dioxide? Students may study how surface membranes are increased by means of different kinds of gills (as in worms, fish, and amphibian tadpoles), tracheae in insects, and book lungs in some spiders and in king crabs (dissections, Chapter 1).

Among vertebrate groups, examine gills of fish, which can be obtained from a fish market. Also study the external gills of early amphibian tadpoles, and the lungs of mature frogs and toads. Observe the breathing movements of amphibia by watching the external nares and throat movements.

In the freshly killed frog, the lungs may be inflated by inserting a straw or glass tubing into the glottis and blowing gently as the small lungs begin to expand. Perhaps students have already studied the capillary circulation in the lungs as described in Chapter 5.

Study of mammalian lungs

Have students bring to class the lungs and windpipe (called a haslet) of a cow or sheep, which they may get from a slaugh-terhouse, a local butcher, or a biological supply house.

Drop the untrimmed lungs into a bucket of water and have students explain why they float. Dissect carefully to reveal the two bronchi and smaller bronchial tubes. Also examine the arrangement of cartilage rings in the trachea, or windpipe.

Many biological supply houses also have preserved plucks of dogs or sheep. These plucks, which are large and easy to handle, include heart, lungs, liver, gall bladder, pancreas, part of the diaphragm, and the duodenum. Such a specimen might serve as a fine review after dissecting a frog or rat.

Lamore[13] recommends the use of a dog pluck and includes a labeled diagram of the pluck (Fig. 6-27).

Excretion

We have already described several indicators which may be used to show that carbon dioxide is excreted as a waste product of respiration. In brief, the method involves blowing into a water solution of an indicator in a beaker or other container. You may use the following indicators, which will undergo these changes: Phenolphthalein, which has been made slightly alkaline (to a pink color) by adding sodium carbonate (or sodium hydroxide), will be decolorized; brom thymol blue will turn yellow; phenol red (alkaline solution) will turn yellow. We have described devices for enclosing small animals or plants in a chamber so that gaseous exchanges in the air may be tested.

Excretion of water

To show that water is also excreted by lungs during exhalation, have students breathe against the blackboard or on a mirror. (Compare this with water vapor given off in the burning of a butter candle, described early in this chapter.)

[13] D. Lamore, *Carolina Tips* 21:35, November 1958. (*Carolina Tips* is a publication of Carolina Biological Supply Co.)

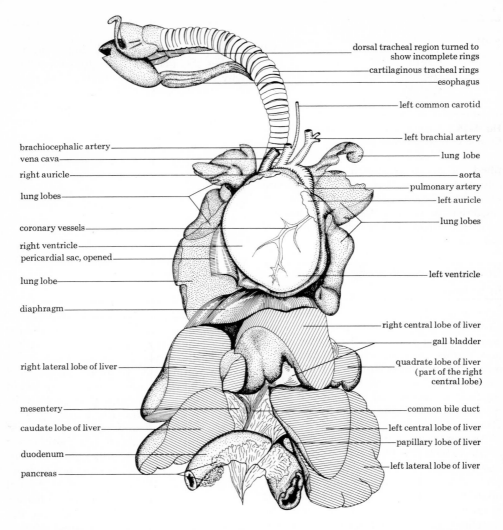

dorsal tracheal region turned to
show incomplete rings

cartilaginous tracheal rings

esophagus

left common carotid

left brachial artery

lung lobe

aorta

pulmonary artery

left auricle

lung lobes

left ventricle

right central lobe of liver

gall bladder

quadrate lobe of liver
(part of the right
central lobe)

common bile duct

left central lobe of liver

papillary lobe of liver

left lateral lobe of liver

brachiocephalic artery

vena cava

right auricle

lung lobes

coronary vessels

right ventricle

pericardial sac, opened

lung lobe

diaphragm

right lateral lobe of liver

mesentery

caudate lobe of liver

duodenum

pancreas

Fig. 6-27 A dog pluck, for study of mammalian lungs. (After D. Lamore, *Carolina Tips* 21:35, November 1958.)

Study of kidneys

From a butcher, students may get an untrimmed kidney of a hog, sheep, or cow. Study the fat capsule which envelops the kidney. Remove this sheath and slice the kidney lengthwise. Distinguish the region rich in tubules and capillaries (cortex) and the collecting funnel which leads into the ureter.

Speed of action of an enzyme, urease

Urease is an enzyme which transforms urea into ammonia:

$$CO(NH_2)_2 \xrightarrow{\text{urease}} CO_2 + NH_3$$

Both urea and urease may be purchased from a supply house.

Dissolve some urea in water and divide the solution into two beakers. To each beaker add a bit of phenolphthalein so that the solutions are slightly milky. Then add a few crystals of urease to one beaker. Watch how quickly the solution changes to red (alkaline due to the presence of ammonia).

Or this procedure may be preferred: Add 4 drops of 0.04 percent phenol red solution to 3 ml of 0.1 M urea. If the solution is not pink at the start, add sodium hydroxide solution (0.1 M) drop by drop. Then neutralize by adding 0.1 M acetic acid until the end point is reached and the color disappears (at pH 7).

To this solution, add pure urease and shake. Explain the rapid change. (When urease is not available, add 0.5 g of soybean meal and shake. Let this stand for a while. The results are not immediate, as they are when pure urease is used.)

Study of a nephron

When studying the role of the kidney in maintaining the stability of blood, students may be able to examine the basic unit, the nephron, comprising a glomerulus and tubule. In an activity described in the BSCS Blue Version[14] it can be shown that a dye, phenol red, is carried across membranes of the tubules by active transport. The energy for this transport is provided by ATP. The kidney of a goldfish is fragmented by expelling it through a hypodermic needle. Fragments are put into several Petri dishes containing physiological saline solution. Each dish is aerated by tubing connected to a vacuum pump. In essence, the investigation involves a microscopic examination of kidney tubules that have been kept for a half-hour in saline solution to which 1 ml of phenol red has been added. Small fragments are then teased apart on a slide for study under low power. The lumen of some of the tubules should show a concentration of phenol red.

[14] *Biological Science: Molecules to Man* (BSCS Blue Version), Houghton Mifflin, Boston, 1963, "Laboratory Investigations," pp. 101–04.

This accumulation of dye against the diffusion gradient is an example of active transport of molecules across a membrane. Certain drugs, such as dinitrophenol, interfere with the formation of ATP required for active transport of molecules. This disruption of ATP by a drug can be demonstrated by maintaining tubules in saline to which 1 ml of dinitrophenol has been added. Then 1 ml of phenol red is introduced into the medium. After a half-hour, the treated tissue is examined under a microscope; it shows no concentration of phenol red in the tubules. (See pp. 240–41 for further work on active transport.)

Using the blackboard as a visual device

Standard charts or models of a cross section of skin are useful but elaborate. A diagram drawn with the students participating may be as effective.

This is one suggestion. Draw a straight line on a blackboard to indicate the surface of the skin. From an opening on the surface (a pore), diagram a long shaft ending in a coiled tube. Elicit from the students the kind of blood vessel which would be found encircling small groups of cells. Then, in the diagram, enmesh this sweat gland with capillaries. By means of arrows (use chalk of a different color) indicate the waste substances that the kidney or the skin receives from the blood (see Table 6-2).

Cooling effect of evaporation

Students can easily demonstrate the cooling effect of evaporation of sweat. They may wet some absorbent cotton with alcohol. Quickly apply it to one hand of each of several students in class. Do students notice that the hand wet with alcohol feels colder? When a liquid evaporates, it takes heat from the surface of the skin, thus resulting in a cooling effect. Using this demonstration as a guide, students can explain how evaporation of sweat cools the body. Why is it that evaporating alco-

TABLE 6-2
Composition of urine*

Average amount (24 hr)	1200–1500 ml
Reaction to litmus	faintly acid
Color	clear amber
Specific gravity	1.015–1.022
Number of grams of the following in a 24-hr specimen:	
Urea	20–30 g
Uric acid	0.6–0.75 g
Ammonia	0.5–0.7 g
Chlorides	10.0–15.0 g
Creatinine	0.3–0.45 g
Phosphate	2.0–4.0 g
Sulfur	1.0–3.5 g
Total solids	50.0–70.0 g

* D. Pace and C. Riedesel, *Laboratory Manual for Vertebrate Physiology*, new ed., edited by D. Pace *et al.*, Burgess, Minneapolis, 1958.

hol cools the skin more than evaporating water?

Test for glucose in urine

Place 5 ml of Benedict's qualitative solution (p. 660) and exactly 8 drops of urine in a test tube. Boil for 1 to 2 min and allow to cool. A precipitate will form when there is about 0.3 percent glucose. You may also want to demonstrate the use of commercial tablets used for testing sugar in the urine that do not require boiling over a flame. Show how Clinitest (or Galatest or other such tablets) and Clinistix[15] are used by diabetics to test their own urine.

Test for albumin in urine

There are many tests for albumin in urine. The method described here is Heller's nitric acid ring test. Pour 5 ml of concentrated nitric acid into a test tube. Tilt the tube, and use a medicine dropper to add urine so that it flows slowly down the side of the test tube. Watch for the

[15] Clinitest tablets and Clinistix strips are available from Ames Co., Inc., Elkhart, Ind. Combistix is a dip-and-read strip that tests for protein, glucose, and pH. Results for all three of these tests are given in seconds.

stratification of liquids with a white region of precipitated protein at the point of contact.

Some workers suggest that the urine be diluted in this test since concentrated urine will form a white ring due to the presence of uric acid. At times, bile pigments or other substances may give a colored ring, but the albumin ring is white. (See also other tests cited in Hawk, Oser, and Summerson.[16])

Phenylketonuria

There is also a dip-and-read test to detect preventable mental deficiency in infants. Phenistix (available from Ames Co.) is a strip of paper impregnated with a reagent, ferric chloride (see also Chapter 8).

Introducing a discussion about a reading in science

At some time you may want to use a reading—a passage from a book or magazine—as a test or as a device for starting a discussion in some classes. At times this may be part of a symposium in a club. For example, you may want to select Chapter 6, "Molecular Traffic—Metabolism" from *Unresting Cells.*[17] Offprints of many of the fine papers in *Scientific American* are available. Also take a reading from *American Scientist* (Society of Sigma Xi, 33 Witherspoon St., Princeton, N.J.) or from *Science* (Am. Assoc. for the Advancement of Science, 1515 Massachusetts Ave., N.W., Washington 5, D.C.).

The following are some of the articles from *Scientific American* relevant to this chapter. The September 1961 issue was devoted to cells—"How Cells Transform Energy," "How Cells Move," "How Things Get into Cells," "How Cells Make Molecules." Look into "Membranes of

[16] P. Hawk, B. Oser, and W. Summerson, *Practical Physiological Chemistry*, 13th ed., Blakiston (McGraw-Hill), New York, 1954, pp. 830–31.
[17] R. W. Gerard, *Unresting Cells*, new ed., Harper & Row, New York, 1949; Harper Torchbook, 1961, pp. 154–85.

Living Cells" (April 1962), "The First Breath" (October 1963), "Heart Cells in Vitro" (May 1962), "Autobiographies of Cells" (August 1963), "Blood Vessel Surgery" (April 1961), "The Master Switch of Life" (December 1963), "Bioluminescence" (December 1962), and "Enzymes in Medical Diagnosis" (August 1961).

Students might be asked to interpret these papers, or a committee of students might prepare five or ten questions based on one of the papers.

CAPSULE LESSONS: Ways to get started in class

6-1. Prepare a demonstration ahead of time, using elodea plants and brom thymol blue in test tubes placed in the dark (p. 272). Ask students to explain why the brom thymol blue turned yellow in the dark. This should raise many questions. Then have students set it up again as an experiment with controls. This may help clarify one difference between photosynthesis (in which carbon dioxide is absorbed) and oxidation, or respiration (in which CO_2 is given off).

6-2. Use a Florence flask to represent one air sac in the lungs. With a glass-marking pencil draw interweaving capillaries on the glass. What materials would pass into the air sac from the blood? What would the blood gain from the air sacs? Develop a discussion of the difference between inhaled and exhaled air and the path of air into the air sacs.

6-3. You may want to begin a lesson with a demonstration showing that growing seedlings give off enough heat to raise the temperature above that of the room. Elicit from students a discussion of the process going on: respiration. What materials are used? What happens when food is oxidized?

6-4. Fasten a paraffin candle to the bottom of a battery jar by heating the end of the candle. Pour a bit of brom thymol blue into the jar. Light the candle and cover the jar. Have students watch the candle become extinguished. (There is value in observing and thinking in silence.) Then shake the jar to show that the brom thymol indicator has changed to yellow. Have a student blow into a small beaker of brom thymol blue. Now ask the class what the two cases seem to have in common. Develop a discussion of the details involved in burning, or oxidation.

6-5. Recall this activity, which you may have used before, and introduce a new relationship. Seal an aquarium plant and a snail in a large test tube (p. 265). Have students describe the oxygen–carbon dioxide cycle. Extend this to life on land as well as plant-animal relationships in a lake or ocean.

6-6. Begin with the inspirator-expirator apparatus described in this chapter (Fig. 6-5) and elicit an explanation for the color change in one flask from brom thymol blue to yellow. From this point on develop the need for, and the materials used in, respiration.

6-7. What effect does strenuous exercise have on breathing? (Refer to the activity on p. 268.) Why is this so? From this point on, develop a discussion of the materials exchanged in respiration. Trace the path of oxygen throughout the body.

6-8. Ask students to hold their breaths for some time. Why will they be forced to exhale? Develop the need for respiration.

6-9. Have a student make the bell-jar model as a project. Demonstrate its use, and have the class develop from the animated model the mechanics of breathing.

6-10. Start a lesson with a discussion of untrimmed lungs or—better still—a haslet, which a student may bring to class. Also refer to a model of lungs (supply houses, Appendix C). Have students trace the path of oxygen into the body and the path of carbon dioxide out of the blood and into air sacs.

6-11. You may want to introduce the role of respiratory enzymes using a speck of dried *Cypridina* containing both luciferin and luciferase (Chapter 2). Add a few drops of water. Compare this brilliant blue color with the flame of a candle, or of a burning butter candle. What is the difference?

6-12. For a laboratory activity plan to demonstrate the active transport of phenol red into the lumen of tubules of a kidney from a goldfish, as described in this chapter. Develop the role of the kidneys and the active part played by membranes.

6-13. Provide the class with jars and germinating seeds. You may want to ask students to design an experiment, after they have established the fact that respiration may be measured by testing either the amount of oxygen used or the amount of carbon dioxide given off.

6-14. Some other time begin with a film. There are several to select from in the field. Are you familiar with *Enzymes: The Key to Life* (Indiana Univ.), or *Mechanisms of Breathing,* 2nd ed. (EBF), or the films in the AIBS series (McGraw-Hill)?

6-15. You may want to begin with a reading from a book or from *Scientific American,* and either ask questions or develop the meaning in class.

6-16. You may want to begin with a film such as *Seeds and How They Grow* (Cox). Or show *Life of a Plant* or *Plant Growth* (both EBF). Use the film to focus attention on the life functions of plants. Send for catalogs from companies; there are new films and filmstrips produced each year. Preview them on a loan basis.

6-17. Show the cooling effect of evaporation of perspiration from the skin in the following way. Stroke the blackboard with cotton soaked in water and also with cotton soaked in alcohol (or ether). Notice which streak evaporates faster. Next wet one of a student's arms with water and the other with alcohol. Repeat on several students. Which arm is cooler? Why?

Ask for an explanation which should include the fact that heat must be removed from a surface in evaporation. Why do we feel cool on a windy day? Why are we uncomfortable on a humid day?

6-18. Use a film to review the uses of the skin, kidneys, and lungs in excretion of liquid and gaseous wastes. You may want to use *The Human Body: Excretory System* (Coronet), a film which discusses these three organs.

6-19. Have a student bring in a fresh kidney, and dissect this in class. Or use a model and trace the path of nitrogenous wastes from the kidney tubules into the ureters and urinary bladder.

6-20. Demonstrate the action of the enzyme urease on urea. Where does this occur in the body? What would happen if urea accumulated?

PEOPLE, PLACES, AND THINGS

There are many people who may be of help. A doctor may be invited to speak on how the body is affected in disease. A hospital technician may help a group of students in a demonstration. Students may solicit the help of a butcher for fresh materials such as organs or blood. Nurserymen can supply young plants.

Other science teachers in nearby schools and colleges may have models or living materials which you may borrow. There may be an opportunity to place a capable student in a position in a college or hospital laboratory on a part-time basis.

Models, slides, charts, and other illustrative aids are available from supply houses listed at the end of the book. Enzymes are available from biochemical houses.

Talented students can make models or other devices which can be used in subsequent years for classroom demonstrations (Chapter 14). For instance, a model of a circulatory system that traced the path of a drop of blood along a "lighted" path was made by a student knowledgeable in electrical wiring.

BOOKS

You will want to refer to Appendix A, the comprehensive Bibliography, for references in biochemistry. Such texts are listed under "Physiology of Plants and Animals" and "Microbiology and Health."

A few titles are listed here to give the flavor of the field.

Bonner, J., *The Ideas of Biology,* Harper & Row, New York, 1962; Harper Torchbook, 1962.
———, and A. Galston, *Principles of Plant Physiology,* Freeman, San Francisco, 1952.
Borradaile, L., *Manual of Elementary Zoology,* 14th ed., edited by W. Yapp, Oxford Univ. Press, New York, 1963.
Brown, J. H., ed., *Physiology of Man in Space,* Academic, New York, 1963.
Burnet, F. M., *The Integrity of the Body: A Discussion of Modern Immunological Ideas,* Harvard Univ. Press, Cambridge, Mass., 1962.
Butler, J. A., *Inside the Living Cell,* Basic Books, New York, 1959.
Cannon, W., *Bodily Changes in Pain, Hunger, Fear, and Rage,* 2nd ed., Branford, Newton Centre, Mass., 1953; Harper Torchbook, Harper & Row, New York, 1963.
———, *The Wisdom of the Body,* rev. and enl. ed., Norton, New York, 1939; Norton Library, 1963.
Carlson, A., V. Johnson, and H. M. Cavert, *The Machinery of the Body,* 5th ed., Univ. of Chicago Press, Chicago, 1961.
Cellular Regulatory Mechanisms, Cold Spring Harbor Symposia on Quantitative Biology, 26, Cold Spring Harbor Laboratory of Quantitative Biology, Cold Spring Harbor, N.Y., 1961.
Cheldelin, V., *Metabolic Pathways in Microorganisms,* Wiley, New York, 1961.
Crocker, W., and L. V. Barton, *Physiology of Seeds,* Chronica Botanica, Waltham, Mass. (Ronald, New York), 1953.

Davson, H., *A Textbook of General Physiology,* 3rd ed., Little, Brown, Boston, 1964.

Florkin, M., *Unity and Diversity in Biochemistry,* Pergamon, New York, 1960.

Gerard, R. W., *Food for Life,* Univ. of Chicago Press, Chicago, 1952.

Giese, A., *Cell Physiology,* 2nd ed., Saunders, Philadelphia, 1962.

Hawk, P., B. Oser, and W. Summerson, *Practical Physiological Chemistry,* 13th ed., Blakiston (McGraw-Hill), New York, 1954.

Machlis, L., and J. Torrey, *Plants in Action,* Freeman, San Francisco, 1956.

Meyer, B., D. Anderson, and R. Böhning, *Introduction to Plant Physiology,* Van Nostrand, Princeton, N.J., 1960.

National Academy of Sciences, National Research Council, *Recommended Dietary Allowances,* 6th ed., Report of Food & Nutrition Board, Public. 1146, National Academy of Sciences, Washington, D.C., 1964; $1.00.

Oparin, A., *Life: Its Nature, Origin, and Development,* trans. by A. Synge, Academic, New York, 1962.

Prosser, C. L., and F. A. Brown, Jr., *Comparative Animal Physiology,* 2nd ed., Saunders, Philadelphia, 1961.

Umbreit, W., R. Burris, and J. Stauffer, *Manometric Techniques,* Burgess, Minneapolis, 1957.

Van Slyke, D., and J. Plazin, *Micromanometric Analyses,* Williams & Wilkins, Baltimore, 1961.

Wald, G., *et al., Twenty-six Afternoons of Biology,* Addison-Wesley, Reading, Mass., 1962.

Yapp, W., *see* Borradaile, L.

FILMS, FILMSTRIPS, AND SLIDES

This listing is intended to be only a guide for the selection of new films, filmstrips, and transparencies. Send for catalogs from distributors of films; catalogs of biological supply houses list extensive offerings of 2×2 in. slides. The addresses of distributors are given in Appendix B.

In general, the cost of color films averages $150; black and white films cost about $75. Newer films are made in varying lengths—time sequences of 1 min, 5 min, 11 min, or 30 min; the cost of films therefore varies. We have indicated the lengths of films so that lessons can be planned with the films in mind.

Sound films are available in color or black and white; rental rates are available from film libraries or by direct inquiries to producers. (Filmstrips are not available on a rental basis; the purchase price is usually about $6 to $7 for those in color.)

Biological Transformation of Energy (2 reels, 54 min), U.S. Dept. of Agriculture, Graduate School; A. Szent-Györgyi, narrator.

Cell Biology (set of 12 films, 28 min each), Part I of *AIBS Film Series in Modern Biology,* McGraw-Hill.

Elimination (13 min), United World.

Enzymes: The Key to Life (29 min), Indiana Univ.

The Fine Structure and Pattern of Living Things (2 reels, 50 min), U.S. Dept. of Agriculture, Graduate School; P. Weiss, narrator.

The Flow of Life (20 min), Educational Testing Service.

The Fuel of Life (29 min), Indiana Univ.

Healthy Lungs (11 min), Coronet.

The Human Body: Excretory System (14 min), Coronet.

The Human Body: Respiratory System (14 min), Coronet.

Kidneys, Ureters, and Bladder (10 min), Bray.

Mechanisms of Breathing, 2nd ed. (11 min), EBF.

The Multicellular Animal (set of 12 films, 28 min each), Part IV of *AIBS Film Series in Modern Biology,* McGraw-Hill.

Nutrition: The Chemistry of Life (2 reels, 58 min), U.S. Dept. of Agriculture, Graduate School; J. Mayer, narrator.

Understanding Vitamins (14 min), EBF.

The Unity of Life (30 min), in 23-film *Of Science and Scientists* series, Indiana Univ.

LOW-COST MATERIALS

A listing of suppliers of free and low-cost materials, with addresses, will be found in Appendix D. Although we recommend the material for use in the classroom, we do not necessarily sponsor the products advertised.

We have mentioned the availability of monthly or periodical bulletins from the biological supply houses. At times pharmaceutical firms such as Upjohn print excellent booklets in current biochemistry which you may draw upon for classroom work.

Carolina Tips (periodical), Carolina.

Turtox News (monthly); *Turtox Service Leaflets* (set of "How to Do" pamphlets), General Biological.

U.S. Atomic Energy Comm., *Some Applications of Atomic Energy in Plant Science,* U.S. Govt. Printing Office, Washington, D.C.; use of radioisotopes in study of life activities.

Ward's Bulletin (periodical), Ward's.

The Welch Biology and General Science Digest (periodical), Welch.

Chapter **7**

Behavior and Coordination

Responses of organisms—plants, animals, and humans—are considered first, for these are what many students can observe together. Many of the stereotyped plant reactions or tropisms are described, as are taxes in lower animals. These cases of orientation to a specific environment are followed by demonstrations of conditioning, habit formation, and other kinds of learning such as trial and error, and cognitive, or insight, behavior.[1]

In the second half of the chapter the structural bases for behavior are developed in terms of coordination of the animal organism through a nervous system, or coordination of behavior through hormones in plants and in animals.

Tropisms and taxes

The bending of plants toward light was first called a "tropism" by De Candolle in 1832. Now most investigators restrict the term to the growth responses of plants and the orientation movements of some sessile animals. The term "taxis" is preferred for protozoa. One speaks of phototropism in plants, but of phototaxis in protozoa.

Responses to light

Phototropisms

LEAVES. Observe plants growing in the field. Note the mosaic of leaves formed as the leaves are oriented in relation to the angle at which light strikes. Few leaves are

---

[1] You may want to examine H. Follansbee, *Animal Behavior* (BSCS laboratory block), Heath, Boston, 1965.

shaded by others. Notice the leaves of vines in particular. Indoors, healthy, full-grown geranium and coleus plants show the same tilting of leaf petioles as a positive response to light.

STEMS. Shoots of growing seedlings show a positive phototropism more readily than fully grown plants. Soak seeds of radish, oats, wheat, or bean, or kernels of corn overnight and plant them just under the surface of moist, clean sand in paper cups or flowerpots. Students may place several of these cups of seeds under a box to exclude light. Arrange a second box covering similar cups of seeds, but with a slit on the side at about the level of the cups, and set the boxes where moderate light can enter the slit.

Within a few days, depending on the kind of seeds used, a marked growth or bending of the stems of the shoots should be apparent in the box in which light enters from one side. In the other box, seedlings grow upright in random fashion.

A more elaborate version of this demonstration can be constructed which provides standard conditions of humidity and ventilation. Students might build inside a suitable wooden box a partition reaching to within ¼ in. of the top of the box. They can also make a hinged cover for convenient handling if the box is to be a permanent piece of apparatus. Apply black paint to the inside surfaces.

Two variations are shown: In Fig. 7-1a a slit has been cut in the side of the box; in Fig. 7-1b small flashlight bulbs have been wired to the side of one compartment and to the top of the other. In this way an equal

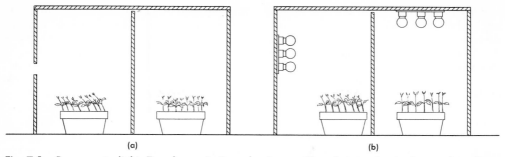

(a)   (b)

Fig. 7-1 Response to light. Two demonstrations showing positive phototropism in shoots of seedlings.

amount of light (and heat) is received by both sets of seedlings. Light the bulbs for about 5 min every half-hour or so.

LIGHT AND THE GROWTH OF SHOOTS. Individual students or small groups may want to repeat some of the classic experiments of Boysen-Jensen and Went on the production of auxin (growth hormone) by the tips of oat coleoptiles (also see growth-promoting substances, p. 326). The tips of the coleoptiles, which are the early sheaths that enclose the shoots, are especially sensitive to small amounts of light (when they are only a few centimeters long). Small cups of aluminum foil may be placed over the tips (work in the dark or in red light), or students may cut off the tips. In such cases there is no bending toward light (Fig. 7-2b), since sufficient auxin is no longer produced. Thus, the tip of the coleoptile is the region which produces auxin, and it is the section which is light sensitive. Light (*one* factor among many) seems to deflect the flow of auxin to one side, the shaded side of the shoot (see Fig. 7-26b). Therefore the cells on the shaded side, those with the greater concentration of growth hormone, elongate faster and the stems bend toward the light. A physiological regeneration usually takes place within 24 hr, so tips may have to be cut off again. Some students may want to try some of the techniques using auxin. Knowing the effect of auxins will help students explain the bending of stems toward light, and the negative response of roots (see below). Also look into photoperiodism and flowering (p. 332).

ROOTS. Roots show a negative phototropism, that is, they grow away from light. Soak radish or mustard seeds overnight. Spread cheesecloth across several tumblers of water and fasten with a rubber band, allowing some slack so that the cheesecloth remains wet. Then sprinkle seeds on the surface of the cheesecloth.

Enclose some of the tumblers in a darkened box. Cut a small slit in a similar box and cover other preparations with this box. In this way light enters the box from one side (as described above for shoots of seedlings).

You may find it more convenient to make small black paper boxes which fit snugly over individual preparations (aluminum foil may be used instead of black paper). Just cut a small slit along one side of some of the paper boxes. After the response of roots (and stems) is apparent, rotate the position of the slit 180°. Notice the change in the growth of stems and roots over the next few days.

Responses of animals

The fact that animals respond to light may be shown in several ways.

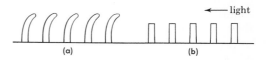

(a)   (b)

Fig. 7-2 Light and production of auxin: (a) the tips of oat coleoptiles, which are light-sensitive, bend toward the light source; (b) oat coleoptiles with tips removed do not respond to light.

PROTOZOA. Most protozoa are indifferent to light, although moderate intensity is preferred. However, some forms, such as *Blepharisma* (Fig. 2-8b), are especially light sensitive (see p. 336).

EUGLENA.[2] Pour a concentrated culture of the green flagellate *Euglena* into a finger bowl or Petri dish (culturing, p. 587). Cover half the dish with a library card, so that light shining down on the dish falls on one half of the dish. Within 10 min remove the cover. Observe the concentration of these organisms in large numbers in the lighted half of the dish. This is a splendid example of positive phototaxis.

Students may show this phototactic response under the microscope. Cut a library card to the size of a glass slide. In the center cut out a slit about $1/16$ in. wide. Prepare a wet mount of *Euglena*, focus under low power, and place the library card under the slide. Examine the organisms that are visible through the narrow slit and quickly remove the library card. Observe that the majority of the flagellates congregate in the region of the narrow slit in the card through which light passes.

In fact, flagellates show a positive or directed taxis. The rate of swimming increases to ~0.18 mm per sec until it reaches the light saturation point.[3] Light affects the eyespot of *Euglena* and the organism is oriented with its axis in line with light rays. Some chemicals may alter this directed response, or phototaxis.

Some students may also want to examine the phototactic action spectrum of *Euglena*.[4]

FLATWORMS (PLANARIA). Planarian worms are found in streams on the underside of rocks or leaves of water plants. In the laboratory they thrive best when kept in shallow, darkened jars of pond water containing some small stones under which they can hide (culturing, p. 592). Remove the cover of the containers holding planarians and watch their movement away from a source of light. They exhibit a negative response to light.

EARTHWORMS. Students may keep earthworms in wet sphagnum moss in a dark box or covered dish. Or they may use a layer of humus to line the container.

Suddenly lift the cover and direct a beam of light from a flashlight on the anterior region of the worms. Observe the way they avoid the light by contracting or burrowing into the moss.

FRUIT FLIES. Transfer a fairly large number of adult fruit flies (*Drosophila*) from a culture bottle into a long test tube; stopper the test tube with cotton. Then cover one half of the test tube with black paper or aluminum foil. Direct a beam of light at one end of the tube and observe the attraction of the flies to the light. This is a positive response to light. If the position of the foil or paper is reversed and the beam of light is shifted to a new position, you will find that the flies again orient themselves in the lighted part of the tube.

Also refer to the effect of light on the development of fruit fly larvae (photoperiodism, p. 332). Many additional interesting activities relating to phototaxis in fruit flies and other insects may be found in H. Kalmus, *101 Simple Experiments with Insects*.[5] Also, there is a BSCS Green Version laboratory activity,[6] for ascertaining whether fruit flies are attracted to light of one intensity in preference to another intensity.

DAPHNIA. What intensity of light affects migration of *Daphnia?* Use a quantity of water fleas, wide, shallow containers, and a light source, the intensity of which can be varied. What happens when light of low intensity is directed at one side of the container? What reaction occurs when the light intensity is increased?

Furthermore, are *Daphnia, Artemia,* or

---

[2] You may prefer to classify *Euglena* as a member of the plant kingdom.

[3] J. Wolken, *Euglena: An Experimental Organism for Biochemical and Biophysical Studies,* Institute of Microbiology, Rutgers, The State Univ., New Brunswick, N.J., 1961.

[4] *Ibid.*

[5] H. Kalmus, *101 Simple Experiments with Insects,* Doubleday, Garden City, N.Y., 1960.

[6] *High School Biology: Green Version—Laboratory, Part II,* BSCS, Univ. of Colorado, Boulder, 1961 (experimental use, 1961–62), Exercise 17-2.

other small organisms affected by the wavelength of light? Suppose that red light or green, blue, or yellow filters are used. Is the behavior of the organism the same under each wavelength of light?

OTHER ORGANISMS. Perhaps pond snails, land snails, earthworms, or *Tubifex* can be obtained easily. Can students use these organisms to design their own investigations on different taxes, on the effects of alternating light and dark periods, on the effects of polarized light, as well as light of different wavelengths? How do these responses compare with growing oat seedlings? Do seedlings bend in response to all wavelengths of light?

Response to electricity

In an electric current, most protozoa move toward the cathode. This may be due to a direct effect on cilia (or other organelles), for the cilia on the side of the cathode beat forward, while those on the side of the anode beat backward. As a result, the organisms are turned to face the cathode. Even rhizopods put out pseudopodia on the side of the cathode.

The electrical charge on the surface of microorganisms may be demonstrated in this way. Partially fill a U-tube or a flat-sided glass container with a thick culture of *Paramecium*. Connect two dry cells in series and insert the two wires from the cells into opposite ends of the container or U-tube containing the protozoa (Fig. 7-3).

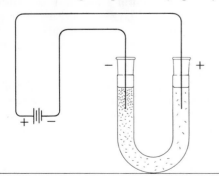

Fig. 7-3 Response of protozoa to electricity. Paramecia move toward the negative arm of the U-tube.

Notice the clustering of the microorganisms around the negative pole. This seems to indicate that the surface of the microorganisms is positively charged.

You may throw a shadow picture of this reaction on a screen or wall. Use a flat-sided container which fits into the slide compartment of a lantern-slide projector or a projection cell. Show the clustering effect. Then reverse the current and show the migration of the organisms to the other electrode, which is now the negative pole.

Response to chemicals

Chemotaxis seems universal in protozoa and other small organisms; they avoid most chemicals, and exhibit a positive taxis in some weak acids. In fact, paramecia collect around a bubble of carbon dioxide in preference to a bubble of air.

Chemotaxis

POSITIVE CHEMOTAXIS IN PARAMECIUM. As food materials diffuse through the culture medium, *Paramecium* responds by swimming toward the food. Mount a drop of culture fluid containing some debris. Examine under the microscope. Protozoa will be clustered around the food material. This response to "food chemicals" is considered a positive chemotaxis.

NEGATIVE CHEMOTAXIS IN PARAMECIUM. Prepare a wet mount of a concentrated culture of *Paramecium*. Add a drop of acetic or hydrochloric acid (1 drop of concentrated hydrochloric acid in 20 drops of water) or 1 drop of Waterman's ink[7] to one side of the coverslip. If the acid is strong enough, *Paramecium* will move away from the incoming acid. Often the trichocysts of *Paramecium* are extruded.

Another way to show negative chemotaxis in protozoa uses a bit of cotton thread soaked in dilute hydrochloric acid. This thread is placed across a drop of thick culture of *Paramecium* on a slide. Hold the slide against a dark background or view it

[7] Not every ink contains enough tannic acid to give this response. Some preliminary experimentation may be necessary.

under the microscope. Students will see a cleared area around the thread. The animals have moved away from the acid.

RESPONSE OF HYDRA TO ACID. Mount several specimens of *Hydra* in a drop of the aquarium or pond water in which they were found. They will elongate when not disturbed. Place a bristle under the coverslip so that the organisms will not be crushed by the weight of the coverslip. Then introduce a drop of weak acetic acid to one side of the coverslip. Observe the discharge of nematocysts on the tentacles of the animals as the acid diffuses through the water. This makes an especially fine demonstration of nematocysts (p. 109). (This reaction is not considered a taxis.)

RESPONSE OF PLANARIA TO CHEMICALS. Place small strips of raw liver in the culture dishes of planarians and watch the rapid aggregation of these forms around the liver. In fact, you may collect planarians from a lake or brook by suspending a bit of raw liver in the water for a few hours. If planarians are present large numbers will be found collected around the liver. This may be considered a positive chemotaxis.

Place a drop of ammonium hydroxide or acetic acid near the planarians and watch the negative response.

EARTHWORMS. What is the effect of bringing a cotton applicator dipped in ammonium hydroxide near an earthworm?

Chemotropism

POLLEN GRAINS. For an investigation on how fragments of the pistils of flowers stimulate germination of pollen grains, see W. Rosen, "Studies on Pollen Tube Chemotropism."[8]

Response to contact

You may want to show how large protozoan forms such as *Spirostomum* or *Stentor* respond to contact or disturbance. When a drop of a rich culture of these organisms

[8] W. Rosen, "Studies on Pollen Tube Chemotropism," *Am. J. Bot.* 48:889–95, 1961.

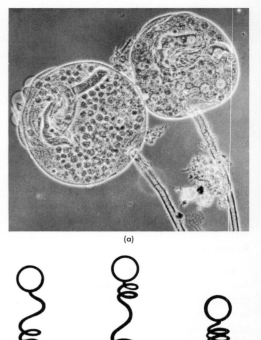

(a)

(b)

Fig. 7-4 *Vorticella:* (a) living stalked *Vorticella* (magnified 1200×, reduced by 45 percent for reproduction); (b) diagrams of contraction of *Vorticella* stalk: left and center, various types of relaxation after contraction; right, a spontaneous contraction. (a, photo by Walter Dawn; b, from L. V. Heilbrunn, *An Outline of General Physiology*, 3rd ed., Saunders, 1952.)

is mounted on a slide and the slide is tapped, these large organisms will contract. This is visible to the naked eye. You need a microscope, however, to see similar contractions in *Vorticella*, a stalked form (Fig. 7-4).

Students may recall that whenever *Paramecium* or other microorganisms run into each other they respond by altering the direction in which they are moving. This, too, may be considered a taxis, a thigmotaxis.

Similarly, when hydras, earthworms, insect larvae, planarian worms, and many microscopic worms are touched with a dissecting needle they contract.

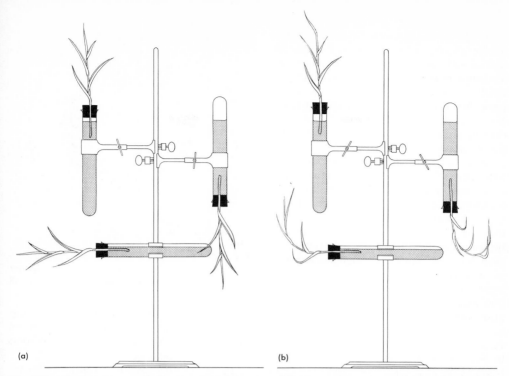

(a)   (b)

Fig. 7-5   Response to gravity (geotropism): (a) sprigs of *Tradescantia* are inserted into stoppered tubes of water, clamped in various positions; (b) negative geotropic response about 6 hr later.

Response to gravity

Geotaxis

PARAMECIUM. These forms show a negative geotaxis; they swim up to the top of a container. This is especially apparent when carbon dioxide accumulates in the culture medium. Using a hand lens, watch how they swim, anterior end upward, to the top of the container. You may show this by pouring a thick culture of the animals into a stoppered test tube or vial. Students may want to check whether light or other factors also influence this response. Set up several types of controls. For example, several tubes might be prepared; some might be placed in a horizontal position, some two-thirds covered with carbon paper, and so on. Do the protozoa respond to light? To gravity?

You can put this response to use in concentrating a culture when you plan a laboratory lesson on *Paramecium* (see Fig. 2-20; also, culturing, p. 585).

HELIX. Live land snails are available from a fish market. A pound of snails will be sufficient for many investigations in behavior. Place some snails in a tall battery jar (with a cover, or they may soon be walking up the walls). Why do they quickly move upward? What happens when the jar is placed on its side? (Also test the effect of other stimuli: a bristle near the antennae; a bit of lettuce; a lighted match; ammonium hydroxide near, but not touching.)

Pond snails may also be used, but their responses are much slower.

Geotropism in stems. Healthy shoots of *Tradescantia* are excellent materials for demonstrating the negative response of shoots to gravity. Tropic responses can be observed within 1 hr. Clamp to a ringstand three test tubes with stoppers containing shoots of *Tradescantia* in the positions shown in Fig. 7-5a. Be sure to seal the stopper of each test tube with melted paraffin to avoid leakage of water. (Germinating seedlings may be used in place of the shoots of *Tradescantia,* but the response will not be as rapid.) Of course, seedlings may be grown in paper cups of sand or vermiculite (some vertical and some horizontal) instead of in test tubes. You may want to place some controls in the dark so that light as a possible factor in this response is eliminated.

Here is another way to use seedlings when fast-growing shoots such as *Tradescantia* are not available. Line large test tubes (about 10 in. long) with blotting paper and insert a plug of absorbent cotton into the bottom of each tube. Wet the blotting paper thoroughly. Between the glass and the blotter arrange soaked seeds of oats, radishes, or wheat. Fill the tubes again with water; pour off the excess. If you attach string or wire to both the top and bottom of each of the test tubes you have a loop through which they can be suspended in different positions from hooks around the classroom. When the test tubes hang at a slant, the growth of stems shows a negative geotropic response. Change the position of the tubes and watch for the change in direction of the growth of stems. In addition, set up replicas covered with aluminum foil or hang some in the dark.

Students soon begin to ask questions about the nature of these responses.

Geotropism in roots. Some of the preparations made to show negative response of stems to gravity are useful here. For example, when using seedlings in place of *Tradescantia,* students will notice that the roots respond positively to the stimulus of gravity. Regardless of their original position roots bend and respond to gravity.

Also have students observe the growth of roots of seedlings planted between the glass tube and blotter and those suspended from hooks in the classroom. The stems show a negative, the roots a positive geotropic response.

If students previously germinated seeds in paper cups of sand to show negative geotropism in stems, they can now demonstrate the direction of root growth. After

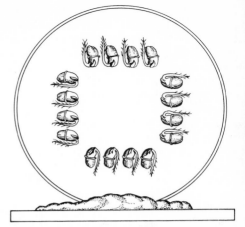

Fig. 7-6 Response to gravity (geotropism). Germinating seedlings are attached to a moist blotter lining a Petri dish, which is held upright by a ball of clay.

3 or 4 days of growth, loosen the seedlings from the moist sand by shaking them gently, and look for the bending of roots toward a vertical position.

Here are some other techniques to show responses of plants; these give variety to demonstrations.

USING A PETRI DISH. Select soaked seeds of beans, oats, wheat, or peas which have begun to germinate so that a few centimeters of roots are visible. Then fasten seeds to a blotter which has been cut to fit the bottom of a Petri dish (Fig. 7-6). Thread fine wire through the cotyledons of the seeds to fasten them to the blotter; arrange the seeds so that the roots face the compass points; wet the blotter; and cover the Petri dish. Prepare several of these Petri dishes. Then stand the dishes in various positions by inserting one edge into a bit of modeling clay.

When roots show a positive response to gravity, shift the position of the dish and watch the change in the direction of growth of each root. Similar dishes may be kept in the dark to show that this is not phototropism.

With a glass-marking pencil, trace the path of each growing root on the surface of the Petri dish if the growth pattern of the roots themselves is not apparent.

USING A POCKET GARDEN. Students may find a pocket garden easier to handle than Petri dishes. Prepare a "garden" by soaking a sheet of blotting paper and laying it over a glass square. Then scatter soaked radish seeds evenly over the surface of the blotter. Cover this with a second glass square and secure with adhesive tape or rubber bands. Now stand one edge of the pocket garden in a finger bowl containing some water. When the roots of the seedlings have grown sufficiently and demonstrate positive geotropism, invert the pocket garden. Within a few days, note the changed direction of growth of the roots. Students can make several preparations and place some of them inside jars lined with black paper or cover them with a box. If students ask whether light had an effect they may study the controls which were placed in the dark.

USING A BATTERY JAR. In a variation of this demonstration, neither pocket gardens nor Petri dishes are required. Pin seedlings to a corkboard and place the board in a darkened, covered battery jar, lined with moist blotting paper. You may turn the corkboard from time to time and show the persistent tendency of roots to grow downward as a positive geotropic response to gravity.

USING A CLINOSTAT. Some students can show that when roots are constantly revolving there is no response to gravity. (This may be a project for students especially interested in the "esoteric" responses of plants and their possible causes.) Roots continue to grow in the directions in which they were originally placed. Make a homemade clinostat by attaching the preparation described above, in which seedlings were wired to a blotter in a Petri dish, to the minute-hand shaft of an old alarm clock (in an upright position) so that a continuous revolving motion is produced. A small, low-geared electric motor can also be used.

At another time a small committee of students may try to fasten seedlings to a water wheel. A much faster rotation may be effected with a water wheel and students may show that roots fail to respond to gravity in this condition.

RESPONSES AGAINST OBSTACLES. At some time, have students demonstrate how persistent is the response of roots to gravity, even when obstacles exist. For example, roots grow into solidified agar or into mercury.

Line the sides of several finger bowls with sheets of cork. Into the bowls pour a layer of mercury ½-in. deep (or use a layer of agar prepared by boiling together 1 part of agar with 4 parts of water). Pin several germinating seedlings, which have a primary root about ½ in. long, around the cork lining so that the roots are in a horizontal position and almost touch the mercury (or agar). Add a film of water to the surface of the mercury to cover the roots. After the finger bowls have been covered and sealed, if necessary, with petroleum jelly, put some in the dark and others in the light. After a day or so, depending upon the kind of seedlings used, notice how the roots curve and grow into the mercury. Here roots respond to gravity and illustrate an expenditure of energy, for the mercury or agar offers resistance.

Although the use of mercury to demonstrate the expenditure of energy by seedlings is effective, there are two drawbacks: (1) a large quantity of mercury is needed, and (2) mercury frequently poisons the roots of the seedlings.

REGION OF SENSITIVITY IN ROOTS. Again place seedlings with roots about ½ in. long in the directions of the compass points as described earlier. Then cut off the tips of the roots with a razor. Prepare replicas in which the root tips are left intact. Keep all the preparations in the dark. Students will notice that the roots of the tipless seedlings do not show the normal geotropic response (until a physiological regeneration of the root tips occurs). Thus we demonstrate that it is the tips of roots that receive the stimulus, and an uneven growth of one side of the root results in a bending. Compare this with the effect of auxin on bending in stems (Fig. 7-2). Which tissue is inhibited by auxin, that from the stems or

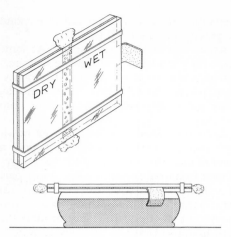

Fig. 7-7 Response of roots to water (hydrotropism). Soaked fast-growing seeds are placed in the channel of the glass-blotter-and-rubber-band pocket garden. Garden is kept moist by means of blotter wick extending into a finger bowl of water.

from the roots? Also refer to the region of greatest growth in young roots (Fig. 9-13).

Response to water

Hydrotropism. There are several ways to show responses to the stimulus of water (hydrotropism). You may want groups of students to try several methods.

POCKET GARDEN. Use two squares of glass to fashion a pocket garden. Place two thicknesses of blotting paper between the glass squares so that a clear channel remains in the center (Fig. 7-7). Along the center arrange small seeds, such as radish, mustard, or lettuce, which have been previously soaked. Then plug the ends of the row of seeds with cotton so that they will not fall out. Fasten the glass plates together with rubber bands. Now stand one end of the preparation in water until one blotter becomes soaked. Then attach a strip of filter paper to the edge of this wet blotter to form a wick. Rest the pocket garden on a finger bowl of water, and immerse the wick as shown. Be sure the other blotter remains dry. Within a few days observe the roots of the germinating seeds growing in the direction of the wet blotter rather than toward the dry blotter. Since the preparation is in a horizontal position, the stimulus of gravity will not interfere with the responses.

HANGING GARDEN. Shape a ball of sphagnum moss (or use a sponge), soak it in water, and then hold it together with cord. Insert soaked seeds of oats, corn grains, bean, radish, or wheat seeds at different locations: top, sides, and bottom. Hang this ball from a hook somewhere in class. Within a few days (depending on the kind of seed) the seeds will sprout. Students will see that roots grow *into* the sphagnum rather than downward in response to gravity. That is, the roots show a positive hydrotropism. The stimulus of water is greater than the stimulus of gravity. Should students suggest that the roots grow inward as a negative response to light, have them devise a plan for a control experiment (the same preparation growing in the dark) so they may refute their own suggestions.

GLASS PANEL. You may find this method practical for showing roots growing toward water. Fill a glass tank (or box with a glass panel) with soil. Plant seedlings near the glass. Insert a small flowerpot into the box. Cork the bottom opening of the flowerpot and half fill it with water. Water will diffuse out into the soil through the porous pot. As time passes, watch the roots bend and grow in the direction of the supply of water around the flowerpot rather than toward drier soil. The same preparation may be made by inserting a small flowerpot into a larger one, but the roots and their direction of growth cannot be seen unless the seedlings are lifted out of the soil.

Response to touch and heat

Climbing roses and many vines show a positive thigmotropism, a response to touch that enables them to encircle a trellis. However, there are other cases of sensitivity in plants which probably are not tropisms. Turgor movements, which are reversible, are more rapid than growth movements or tropisms.

We may show sudden responses in such plants as the sensitive plant *Mimosa* (Fig. 7-8). The responses are the result of loss of turgor of certain cells found in a swelling, called a pulvinus, situated at the base of each cluster of leaflets. Bring a lighted match close to the tip of a cluster of leaflets. Watch the sudden folding and drooping of the entire cluster of leaflets.

Students may show how these plants respond to mechanical stimulus, touch. Gently tap the terminal leaflets with a pencil or with your finger, or pinch the leaflets with a forceps. See how quickly the end leaflets fold over each other.

This excitation of cells is followed by an action current that spreads from the tactile hairs on the leaves of *Mimosa* along the sieve tubes down through the petiole into the pulvinus. Here the permeability of the cells of the lower region of the pulvinus is affected so that the cells lose water. As a result of the loss of turgor, the stem drops.

If possible, study the response of the Venus flytrap (Fig. 11-20b). This too is a case of a response due to excitation created by the fly touching tactile hairs on the leaf and resulting in a wave of excitation that causes some cells to decrease their permeability due to a change in turgidity. The protoplasm of plant cells transmits excitation accompanied by changes in electrical potential that affect the permeability of cells.

Other movements of leaves or flowers (as in *Oxalis,* and in beans, peas, and other legumes) are sometimes called sleep movements. These also rely on changes in turgidity rather than changes due to growth (tropism).

A summary demonstration of several tropisms

You may want students to prepare the demonstrations which were first described by Coulter[9] over 80 years ago. (These are especially good for a "new" view of the

[9] J. M. Coulter, "The Influence of Gravitation, Moisture, and Light upon the Direction of Growth in the Roots and Stems of Plants," *Science* 2:5–6, 1883.

(a)

(b)

Fig. 7-8   Response of plants to touch: (a) *Mimosa pudica,* undisturbed; (b) response of *Mimosa pudica* to touch. (Courtesy of General Biological Supply House, Inc., Chicago.)

topic.) Students can prepare balls of sphagnum moss with soaked seeds and devise experiments to test whether light, gravity, water, and so forth, affect the growth of stems and roots. Use fast-growing seedlings such as oats, barley, radishes, or corn grains.

Moisten masses of sphagnum moss and scatter soaked seeds into the moss. Then shape the moss into balls about 4 to 5 in. in diameter. Tie each mass together with string so each retains its shape. Prepare some seven to ten balls of moss, so that the following demonstrations can be prepared. Where possible, prepare duplicates.

Suspend one sphagnum ball in the classroom so that it receives light from all sides. Place a second ball on a glass tumbler which has some water in it, about 1 in. above the level of water. Prepare a third in the same way over a tumbler, but cover the glass with black paper or aluminum foil. Then cut a thin slit in the paper or foil so that light enters from one side.

Insert a fourth ball part way into another tumbler as though the ball were a stopper. Thus one half of the ball is outside the tumbler and the other half is inside. Then prepare a fifth ball in the same way, but place the tumbler in a horizontal position. With two more balls (the sixth and seventh) prepare setups similar to the fourth and the fifth, but cover the tumbler with black paper so that light is excluded as a factor in the response.

Mount the eighth ball on a spindle, such as a knitting needle, which runs through its center. Then attach the spindle to the minute hand of an upright alarm clock so that the spindle forms a continuation of the minute hand of the clock.

Watch the direction of the growth of shoots and roots. On the basis of these experiments students may be able to generalize about the direction of tropic responses in plants.

Responses in higher animals

Reflexes in the frog

A living frog is an excellent specimen for the study of reflexes. Here are a few reflex responses which can be shown in class.

Blinking and other reflexes. Hold a live frog securely in one hand by grasping the hind legs. Bring a blunt probe or a glass rod near one eye. Students should observe that the frog blinks. In fact, the response at times is so strong that the eye may be pulled into the throat region. Observe, also, the lids covering the eyes.

A student may elicit the blinking response in another way. Dip a cotton-tipped toothpick into ammonium hydroxide and bring it near the eye of the frog.

Touch the "nostrils" and watch the response. Stroke the throat and belly regions. When these regions are first stroked in males, watch the distinct clasping reflex in the forelimbs. When the frog is placed on its back and stroked longer, students will not fail to notice the quieting, almost hypnotic effect on the frog. Stroke the back and sides of a male frog to make it croak. Watch the extension of the vocal sacs and note the closed "nostrils" which prevent the escape of air.

"Scratch" reflex. Demonstrate the scratch reflex in this way. Grasp the head and hold the forelimbs securely in one hand. Wash off any mucus present on the back of the frog. Now touch its dorsal skin with a cotton-tipped toothpick or glass rod dipped in dilute acetic acid. Notice how the hind leg attempts to brush off the irritant. With a stronger acid, a more violent response occurs. In fact, the animal may try to use both legs to brush off the irritating substance. Wash off the acid with water. Which part of the nervous system is the center of this reflex, the brain or the spinal cord? (You may also want to show the same response in a spinal frog at this time. See Fig. 7-9 and p. 307.)

Other responses. Place a normal frog in water. Notice the resting position. Watch the normal breathing movements. Suddenly tap the jar and watch its reactions. Now watch the motion of the limbs in swimming.

Perhaps a student will demonstrate changes in the size of chromatophores in the frog's skin in response to light. Put one frog in the dark (several would be preferable) and set others in the light for several hours. Which frogs are darker in color? These responses are due to reactions of the nervous system as well as the effect of a hormone, called intermedin, from the intermediate lobe of the pituitary. (See also chromatophores of fish, p. 110.)

Show the feeding reactions of a live frog by dangling earthworms or mealworms before it. Note the position of the eyes in swallowing.

Locating responses. Which region of the nervous system is responsible for these reflexes? If we destroyed only the brain of a frog, would the blinking reflex or the scratch reflex remain in this "spinal frog"? What if we also destroyed the spinal cord? Do any reflexes remain in this "pithed frog"? For destroying (pithing) the brain and spinal cord of a frog, see the photographs and description in Fig. 7-9. A longitudinal section of the frog's head is shown in Fig. 7-10, as a guide to demonstrating the procedure.

THE "SCRATCH" REFLEX IN A SPINAL FROG. Is the brain of the frog necessary for the scratch reflex to occur? Is the spinal cord necessary? If the brain is the center for the scratch reflex, then the frog should lose this response when the brain is destroyed. Use a spinal frog (brain removed or destroyed) for this demonstration.

Destroy the brain of a frog as shown in Fig. 7-9; avoid injuring the spinal cord. In fact, some teachers prefer to cut off the entire head to be sure they have removed the brain; insert sharp dissecting scissors into the mouth and cut off the head by cutting behind the tympanic membranes; place a bit of cotton over the cut end. (*Note:* This should not be done in front of the class.) Whether the brain is destroyed or severed from the body, suspend the frog from a ring stand by clamping the lower jaw (Fig. 7-11).

Wash off the mucus covering, which may be quite thick, but remember to keep the skin moist. If spinal shock has occurred the animal gives no response; wait 5 or 10 min for recovery before proceeding. Now dip a glass rod into dilute acetic or nitric acid and apply it to the back of the frog. The scratch reflex takes place even though the brain has been destroyed. Wash off the irritating acid by raising a battery jar of water up to the suspended frog and immersing the frog's body in the water. Repeat.

Release the frog from its suspended position; now destroy the spinal cord (see Fig. 7-9). Suspend the frog once more and apply acid to the body again. Why is there no response this time? Apparently the spinal cord is the center for this body reflex.

You may now want to trace the path of the stimulus in the body. For example, when acid is placed on the skin of the back, why is the response given by the hind leg? Develop a simple pattern of a reflex arc on the blackboard as a summary activity (Fig. 7-12). Students might represent sensory neurons with one color of chalk and motor neurons with another color. Now trace the path of a stimulus through a reflex arc.

Reflexes in man

Have students observe their own reactions in the following situations.

Blinking. Have one student stand and hold a sheet of glass in front of his face. Then ask a second student to throw cotton balls or crumpled paper at the glass. The student cannot avoid blinking even though he is protected by the glass. Some teachers prefer to use a pane of clear plastic or wire screening rather than glass.

Contraction of the iris. Have students cover one eye for a minute. When they remove the hand have them look at once into a mirror. Note the dilation of the pupil of the eye that was covered. As light strikes the eye observe the change in size of the pupil (as the iris contracts in bright light). Students will need to try this several times.

Or darken the room for a minute. Have students observe their eyes in a mirror as the light is suddenly turned on. The contraction of the iris is marked. Also note the size of the pupil when the eye accommodates for close and distant vision.

Flow of saliva. Slice a lemon in front of the class, or describe its sour taste. Recall

Spinal frog. Hold the frog securely with the left hand (if you are right-handed) with its legs extended. Bend its head down by pressing its snout with the index finger.

Rest the dissecting needle on a line bisecting the head. Slide the needle down until the point is at the cranial opening (foramen magnum) just below the cranium and above the spinal column (see Fig. 7-10); this is about in the middle of the posterior line of the tympanic membranes.

 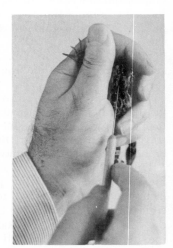

Insert the needle without changing its direction, and move it quickly from side to side to sever the brain from the spinal cord (hold the frog firmly, for it will wriggle); tilt the needle forward, insert it into the brain case, and move it around to destroy the brain. You now have a "spinal frog": Its spinal cord is intact, but its brain has been destroyed. (*Note:* We call this a "spinal frog" for ease of reference; different authors apparently accept different terms. For example, some people refer to a "double pithed frog" when both the brain and spinal cord are destroyed.)

Fig. 7-9 Pithing a frog. *Note:* The frog should be kept moist throughout. (Stanley Rice.)

Pithed frog. A pithed frog has not only its brain but also its spinal cord destroyed. Take a spinal frog and insert the needle into the same opening, but this time tilt it downward through the length of the spinal column, and scrape a bit. Observe the extension and relaxation of the hind legs. Notice, in the tank above, the difference between a spinal frog (swimming peculiarly) and a pithed frog (seemingly dead).

One reflex of a spinal frog. When the toes of a spinal frog are pinched, its legs jerk up close to its body. See the text for other reflexes to test in spinal and pithed frogs.

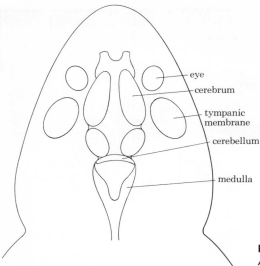

Fig. 7-10 Head of a frog. Schematic diagram of dorsal view, showing parts of brain in relation to tympanic membranes.

eye
cerebrum
tympanic membrane
cerebellum
medulla

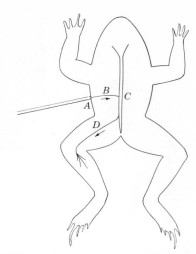

Fig. 7-12 Reflex arc in a frog. Acid applied at A is the stimulus, received by sense receptors in the skin; the impulse is carried by sensory neuron B to the spinal cord, C; motor neuron D transmits an impulse to the muscles, and the leg contracts to brush off the irritant.

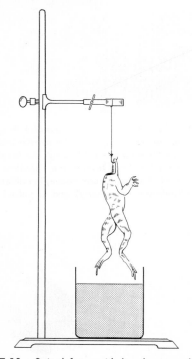

Fig. 7-11 Spinal frog, with head removed, suspended from a clamp by lower jaw for testing "scratch" reflex. Alternatively, the frog may be held by hand, using a secure grip on the forelimbs.

a favorite food by means of a full description (or pictures). Ask students to describe the results of a reflex that occurs within them.

Cilio-spinal reflex. Watch the pupils of the eyes of several students when the skin of the back of the neck is pinched.

Patellar reflex (knee jerk). Have a student sit on a chair or a table with legs crossed or freely suspended. When the subject is completely relaxed, strike a blow just below the patella bone with a rubber hammer or with the side of the hand.

Nerve-muscle preparations: contraction of muscle

You can readily show how a nerve transmits a stimulus to a muscle and how this muscle tissue contracts to produce movement. Use a double-pithed frog. Dissect the hind limb to expose the gastrocnemius muscle, or calf muscle, the largest muscle between the knee and ankle.

Either pull the skin from the gastrocnemius muscle as shown in Fig. 7-13 or cut the skin around the waist of the frog and pull off the skin (as you would pull off rubber gloves). Be sure to bathe all tissues in *cold* Ringer's solution (p. 665). Continue to dissect the sciatic nerve from the dorsal side of the thigh, back into its origin in the spinal cord. Cut the nerve from the spinal cord and snip off, very carefully, all the branch nerves along the thigh. Transfer the nerve-muscle preparation into Ringer's solution. This preparation may now be used to simulate the discovery of Galvani, or a kymograph may be employed to record muscle twitches.

Galvani's discovery. If copper wires are inserted into the muscle, as shown in Fig. 7-13e, and if the wires are touched to a dry cell, the current will stimulate the con-

traction of the muscle. Or strips of different metals can be used with Ringer's solution to form a battery—a wet cell.

Prepare a "galvanic" forceps by soldering a strip of copper and a similar length of zinc (about $3 \times \frac{1}{2}$ in.) at one end to form a forceps. Stimulate the nerve or muscle by placing the nerve-muscle preparation on blotting paper (soaked in Ringer's solution). With this salt solution acting as an electrolyte, touch the galvanic forceps to the muscle. Students should see the muscle twitch whenever it is touched. Now touch the nerve. What would happen if the nerve were crushed?

The surfaces of muscles and of nerves normally carry a positive charge; when

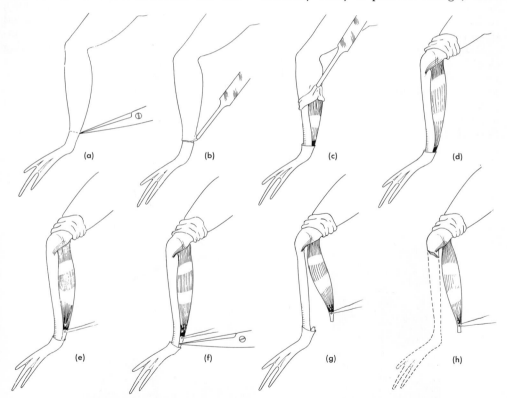

Fig. 7-13   Dissection of the gastrocnemius muscle of the frog in preparation for study: (a) the skin is carefully cut with sharp scissors near the insertion of the Achilles' tendon; (b), (c), and (d) the loose skin is rolled back with tweezers; (e) a hole is made with a dissecting needle through the tendon and a copper wire is inserted; (f) and (g) the tendon is cut *below* the wire; (h) the bone and other muscles are cut away. The loose skin can be pulled down over the muscle to keep it from drying. (After D. Pace and C. Riedesel, *Laboratory Manual for Vertebrate Physiology*, Burgess, 1947.)

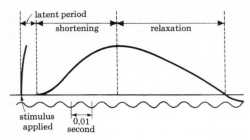

Fig. 7-14 Periods of a single twitch of a muscle of frog: latent, shortening (contraction), and relaxation. (After A. Carlson and V. Johnson, *The Machinery of the Body*, 4th ed., Univ. of Chicago Press, 1953.)

dry cells or galvanic forceps are brought in contact with the tissue, negative charges are produced and conducted along the surface of the membranes. Then the positive charge is restored.

When a frog's muscle is given a single electric shock, the response is a sudden twitch which lasts about 0.1 sec. Students should read further about the separate periods of a single twitch: latent, shortening (or contraction), and relaxation periods (Fig. 7-14).

There is a brief recovery period during which oxygen is consumed by muscle tissue. There must be a recovery period between stimulations or the muscle will become fatigued and the twitches cease.

However the immediate source of energy for contraction is from ATP. After the contraction of a muscle, glycogen is broken down into lactic acid, which is oxidized in the Krebs citric acid cycle. This reaction provides energy for the resynthesis of ATP and phosphocreatine.

Using a kymograph. Muscle tissue, such as a ventricle or the gastrocnemius muscle of the hind limb of a frog, may be attached to the lever of a kymograph. It is then possible to trace the normal pattern of contraction of the muscle (in saline solution), especially when the muscle is stimulated electrically, or when accelerators or inhibitors are applied to the contracting muscle.

There are many models of kymographs. Essentially, the apparatus consists of a rotating drum around which a sheet of smoked paper has been wrapped (Fig. 7-15). The speed of rotation can be ad-

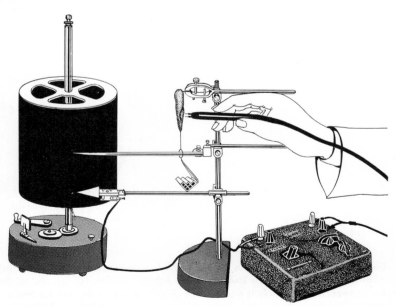

Fig. 7-15 Kymograph. Typical setup for recording the effect of electrical stimulus on skeletal muscle preparation. (After catalog of Harvard Apparatus Co., Dover, Mass.)

justed. When the fine point of a stylus attached to a movable lever is brought near the smoked drum, the point traces a pattern on the smoked paper.

Accessories for the kymograph are available from supply houses. Especially useful is the catalog distributed by Harvard Apparatus Co., Dover, Mass., that offers descriptions of apparatus for teaching physiology; also included are several teaching kits.

If the large calf muscle of a frog is kept moist it will continue to contract for hours. A record may be made of the contractions. The muscle may be clamped at its point of origin (the end that was near the knee of the frog) with a fixed hook or clamp. The other end of the muscle (near the Achilles' tendon), the insertion end, can be hooked to a lever that has a pointed stylus at its tip.

The stylus is later placed so that it just touches the frictionless surface of the smoked paper on the kymograph drum. As the muscle contracts the duration and vigor of the contraction are recorded.

Some teachers attach additional styli to mark the time each stimulus is given. This may also be marked by hand.

Highly glazed paper is wrapped around the drum of the kymograph and this is rotated in a dense smoke. The source of the smoke may be a Bunsen burner with a fishtail or the oily smoke of a kerosene smoker lamp. (Work under a hood to avoid "smoking" the walls of the room.) Apply a thin layer of smoke. Where the stylus moves against the drum a clear line is traced. This record may be preserved by spraying the surface with lacquer. Smoked paper is less expensive than ink. However, some teachers prefer writing with ink on a drum to avoid smudged results and the problem of storing smoked paper. It is also possible to get fairly good smoked paper using a candle flame.

Using different kinds of muscle. Some students may compare the contraction of striated muscle with smooth muscle of the frog. For example, cardiac muscle has a very long refractory period; each heart-beat represents a single twitch. In general, smooth muscle may remain relaxed or tightly contracted for varying periods. While the calf muscle may relax and contract in 0.1 sec, smooth muscle takes from 3 to 160 sec, and cardiac muscle may take from 1 to 5 sec.

Perhaps students can compare the smooth muscle from some invertebrates with the striated muscle from an invertebrate such as an arthropod.

Myosin, actin, and ATP. Hawk, Oser, and Summerson[10] describe several preparations to demonstrate the formation of muscle fibrils and the contraction that occurs when myosin, actin, and ATP are combined.

Role of acetylcholine. Sympathetic ganglia liberate acetylcholine, which functions at some synapses. The "hormone" is synthesized from acetate and choline in the presence of ATP and calcium ions. The activity of acetylcholine depends on a chemical electrogenic action, and acetylcholine is rapidly hydrolyzed by acetylcholinesterase. An idea of the effect of acetylcholine on a frog's heart may be obtained from the activity described on pp. 244 and 325. Some students will want to try to reproduce the simple, elegant demonstration of Loewi.

You will want to have students read further about excitation of nerve cells and fibers and the contraction of individual fibers and muscles. Many students can comprehend a text in physiology such as A. Carlson, V. Johnson, and H. M. Cavert, *The Machinery of the Body*, 5th ed. (Univ. of Chicago Press, 1961), or C. Villee, *Biology*, 4th ed. (Saunders, 1962).

More advanced reading, profitable for some students, might be recommended for

[10] P. Hawk, B. Oser, and W. Summerson, *Practical Physiological Chemistry*, 13th ed., Blakiston (McGraw-Hill), New York, 1954, pp. 284–88.

For further reading, see H. Weber and H. Porzehl, "The Transference of the Muscle Energy in the Contraction Cycle," *Progr. Biophys.* 4:60–111, 1954; A. Szent-Györgyi, *Chemistry of Muscular Contraction*, 2nd ed., Academic, New York, 1950; and also T. Hayashi, "Contractile Properties of Compressed Monolayers of Actomysin," *J. Gen. Physiol.* 36:139–52, 1953.

a study of the chemistry of excitation and contraction. A teacher might suggest A. Giese, *Cell Physiology,* 2nd ed. (Saunders, 1962), and C. L. Prosser and F. A. Brown, Jr., *Comparative Animal Physiology,* 2nd ed. (Saunders, 1961).

Reprints of papers from *Scientific American* are available for individuals or for a reading and discussion by the whole class. Teachers also have available the BSCS laboratory block *The Complementarity of Structure and Function* by A. G. Richards (Heath, 1963). The block offers a rich source of material for individual investigations and for class laboratory work.

Learning ("educated" responses)

Students are interested in human behavior and that of animals. Many have trained a puppy, or have guppies, or have played with the young of farm animals.

At one time, behavior was considered to be a rather simple function of reflexes, instincts, conditioning, habit formation, and conscious learning. Current research (for example, on "imprinting" in young ducks, RNA and memory, social deprivation in the young organism, and the role of imitation in developing the so-called instincts) has raised fundamental questions about behavior.

After introductory studies on conditioning and habit formation, teachers may want to use offprints of articles from *Scientific American* concerning behavior. Only a sampling of papers is included here. Freeman (San Francisco) has a catalog of offprints available at $0.20 each.

Beginning with the experimental aspect of behavior of organisms, students will want to read "Protopsychology" (February 1963), which is an article about planarians and learning; also read "Messenger RNA" (February 1962). "Electrically Controlled Behavior" (March 1962) describes studies on chickens. The rationale for using teaching machines is presented by B. Skinner in "Teaching Machines" (November 1961). And work with pigeons described by D. Blough in "Experiments in Animal Psychophysics" (July 1961) should be read along with the paper by Skinner.

Factors that influence learning in humans are described in many issues; these may be topics for discussion in club work as well as supplementary work for the classroom. Students will be interested in "A Study of Aspirations" (February 1963), "The Measurement of Motivation" (May 1963), "The Nature and Measurement of Anxiety" (March 1963), and "Early Experience and Emotional Development" (June 1963).

Behavior of animals is described in "The Social Life of Baboons" (June 1961), "The Evolution of Behavior in Gulls" (December 1960), "Population Control in Animals" (August 1964), "Experimental Narcotic Addiction" (March 1964), "Vision in Frogs" (March 1964), "Sound Communication in Honeybees" (April 1964), "Learning in the Octopus" (March 1965), "The Synapse" (April 1965), "The Navigation of the Green Turtle" (May 1965).

Conditioning in planarians

Students who wish to learn more about planarians might begin by reading "Research on Learning in the Planarian" by A. Jacobson and J. McConnell (*Carolina Tips* 25:7, September 1962); a bibliography is included. Then read J. Best, "Protopsychology" (*Sci. Am.,* February 1963) and A. Jacobson, "Learning in Flatworms and Annelids" (*Psychol. Bull.* 60:74–94, 1963).

Methods for culturing planarians are described on p. 592, and the techniques for bisecting and regenerating head and tail ends of these worms are described on p. 482. (If any investigations involve the use of ribonuclease, this enzyme may be purchased from Worthington Biochemical Corp., Freehold, N.J.)

Conditioning in earthworms

With some patience, students may show that "learning" of a sort takes place among

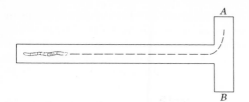

Fig. 7-16 T-maze for training earthworms. Place rich garden soil at *A*; place coarse sandpaper, weak acid, or a weak electric current at *B*.

earthworms. They can build a T-shaped maze of wood or other materials (Fig. 7-16). At one end of the T-bar, place a rich sample of organic soil. At the other end place a piece of filter paper or cotton soaked in a dilute acid, or cover the area with coarse sandpaper. Place a hungry earthworm at the beginning of the maze and watch its trail along the length of the path. As a project students might keep a record of the number of times the same worm (or worms) turned toward the humus and thus avoided the acid or sandpaper. Identify the worms with India ink (or some similar means) so that the records of trials and errors may be identified. Repeat the experiments daily. It may take some 50 trials before the earthworms "learn" to avoid the irritating substances and turn to the humus, where the food is located.

Conditioning in goldfish

Goldfish, unlike earthworms, have a definite brain in the acknowledged sense of the term. If you feed the fish from one end of the aquarium regularly, and at the same time introduce another stimulus, the fish will associate the two stimuli. Flash a light each time the fish are fed, or tap on the tank. Students might take over this activity as a learning experience. It may take 3 weeks or more to condition the fish. After conditioning, the fish will rise to one end of the aquarium when the tank is tapped (or when a light is on), although food may not be given. The original stimulus, food, has been removed. The fish respond to a substitute stimulus (tapping or light) associated with the food stimulus.

We may consider the original response to food a reflex, an inborn behavior pattern. As a result of conditioning, a conditioned reflex has been established and the inborn response is now given to a substitute stimulus.

Learning in the white rat

Construct a rat maze using wood and window screening (Fig. 7-17). Trace out a maze and nail down strips of wood topped with screening (6 in. high) so that the animal moves along channeled paths.

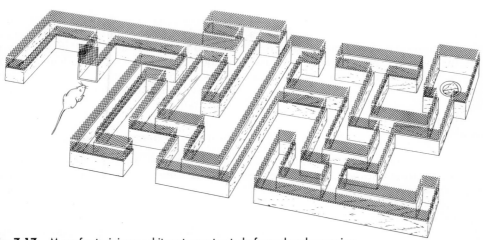

Fig. 7-17 Maze for training a white rat, constructed of wood and screening.

Select a hungry white rat for training. Lead it through the maze several times. When it is successful in running the maze, reward it by giving it food and fondling it. In this way the rat comes to associate food and petting with a completed task.

Keep a record day to day or several times a day, as the animal (when hungry) is put through the maze. Students might want to compare the number of trials needed for learning the maze among several rats, some young, some older.

A committee of students may want to study the rate of learning of rats which have a thiamine (or other) deficiency. Perhaps they may want to study the effect of drugs (caffeine) on speed of learning. (Normal controls should be used for comparison.)

Learning seems to go on faster when a reward rather than punishment accompanies the learning task at hand. What would happen if we slapped the rat each time it failed to complete the maze? (Do not try this in class, for the animal may become vicious.)

Conditioning in human beings

You may want to demonstrate how conditioning occurs. Stand at the rear of the classroom so that your students are not facing you. Ask them to try an experiment with you and to follow what you say. Direct them to mark a tally line on a sheet of paper each time you say "write." Use a ruler to tap on the blackboard or on a desk each time you say "write," so that the two stimuli are associated with the students' response (the drawing of a line on the paper).

After some 20 times of repeating these signals (about two every second) continue to rap with the ruler but stop using the stimulus word "write." Many students will continue to draw lines at the sound of the rapping. They have been conditioned temporarily so that they draw a line in response to the stimulus (the sound of the tap). Students will vary in this experiment. Some stop immediately when you omit the oral command. Others may continue to write as many as ten extra lines when results are compared with the rest of the students. Have students explain the factors needed for conditioning. How is this type of behavior "unconditioned"?

Refer to the classic work of Pavlov in conditioning dogs.

Habits in human beings

Show the value of making certain kinds of behavior habitual. Here are two methods to show that habits are time-saving and require little conscious thought.

Have students write their full names in the usual manner as often as they can in half a minute. Then have them write their names with the other hand at least five times. Record the time it takes. Why does it take so long to write with the other hand? What is one value of a habit?

Direct students to copy as quickly as possible an oral paragraph of material. Read at a fair pace and record the time it takes for students to copy the material you dictate. Then read another paragraph with one change in directions. Students are not to dot any *i* or cross any *t* in the words they copy. Select material with many words containing *i*'s and *t*'s; read at the same fair pace. Have students score the number of dotted *i*'s and crossed *t*'s.

Have a boy describe in detail how he ties a shoelace. Have someone describe in detail the kinds of houses, trees, store signs, and so forth that he passes on the way to school. Why must the students stop to think, even though they pass the same way to school daily? Even though they tie shoelaces daily?

Learning in human beings

Considerable time should be spent on gaining some experimental evidence with young people. Many students lack proper study habits; they fail to get the "big idea" in their assignments, in reading, in work in class. Some of these "experiments" may help to change students' work habits.

Comparing the rate of learning under differing conditions

LEARNING SENSE AND NONSENSE RHYMES.
On the blackboard list in two vertical columns the words in column A and in column B. Cover the lists so that students cannot see them.

column A	column B
thing	Its
whatever	a
a	very
as	odd
odd	thing
eats	as
its	odd
very	as
Miss T	can
can	be
Miss T	that
odd	whatever
as	Miss T
be	eats
turns	turns
that	into
into	Miss T[11]

Now uncover the list of words in column A. Direct students to memorize the list in vertical order. Record the time. Ask students to raise their hands when they have memorized the list. Allow this activity to go on for 5 min.

Now uncover column B, and have students memorize it. Keep a record of the time and the number of students who quickly raise their hands when they have memorized the words. Students readily explain that column B made sense and that column A was meaningless. From here it is not difficult to guide the class toward the need for understanding main ideas so that most work and reading in school does not fit into the category of "nonsense."

LEARNING UNDER DISTRACTION. Have students copy a stanza of poetry which is dictated. Better still, distribute mimeographed sheets containing two stanzas of a poem.

Ask students to memorize the first stanza as quickly as possible. Students might raise their hands when they have memorized the stanza. Keep a record of the time it takes for the first hand to be raised until some dozen or so students raise their hands.

After complimenting them on their success, ask them to memorize the second stanza. However, occupy yourself by making distracting noise. Turn on a portable radio and find a nonmusical program. Also slam drawers or metal lockers from time to time. Keep a record of the time it takes for the first student to indicate success in memorizing the second stanza. Count others. Ask for an explanation of the relative slowness or even failure to memorize the second stanza. There is an opportunity here to stress the need for a quiet place for work and study.

Learning by trial and error. One of the simplest but most time-consuming ways to learn something is by trial and error. Give students a simple puzzle to solve, without pictures as a guide. Use the exact shape shown in Fig. 7-18 as a guide. Or use a simpler one, a large letter T or R cut into three or four pieces. Draw it on onion-skin paper, and have a group of students volunteer to cut out sets of four cardboard or oaktag pieces from this pattern. Place four of these pieces in each one of enough envelopes for all the students in class.

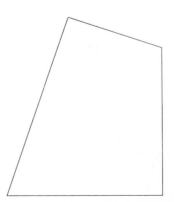

Fig. 7-18 Learning by trial and error. Cut four pieces of this shape and have students learn how to put them together to form a perfect square.

---

[11] Walter de la Mare, "Miss T"; reprinted by permission of the Literary Trustees of Walter de la Mare and the Society of Authors as their representative.

Distribute an envelope to each student (note the time). Tell the students to put together the four pieces to form a perfect square, with no spaces. Allow 10 min. When a student is successful have him raise his hand; keep a record of the time. Then have him take apart his puzzle and put the pieces back in the envelope. After 5 min, tell each of the successful students to try solving the problem again.

Why is it that most students have difficulty in solving the puzzle? Why is it that some students who successfully completed the puzzle the first time cannot repeat their performance? Elicit from students that this is hit-or-miss learning, as they say, or trial-and-error learning.

Show that when something has meaning, learning goes on faster. Have a student who was successful show the rest of the class, in detail, how to put the puzzle together. Then time the students in their performance in learning to solve the puzzle.

These experiences help the students gain insight into some methods of improving study habits. Many who worry about their poor study habits appreciate this help. Naturally, these are suggestions only and need to be personalized through individual guidance.

What is meant by a "gestalt"? What is field theory as applied to learning? (Refer to a text in psychology, such as E. R. Hilgard, *Introduction to Psychology,* 3rd ed. [Harcourt, Brace & World, 1962].)

The nervous system and the sense organs

The brain

Structure of the brain. Students may get the brain of a sheep or calf from a butcher. In class, identify the cerebrum, cerebellum, and medulla. A student may report on a comparison of brains in different groups of animals. Develop the notion that the cerebrum has become larger in the higher animals, with more convolutions in the cortex. A series of models might be used by students in a comparative study of brains of vertebrates.

A student might diagram the human brain on the blackboard (Fig. 7-19). Then associate these activities with the region of the brain or spinal cord which controls them. For example:

1. Region where thinking occurs
2. Controls sneezing and swallowing
3. Coordinates body movements and equilibrium
4. Center for sight, hearing, and speech
5. Controls heartbeat
6. Center for reflexes of the body, such as pulling away from a hot object
7. Controls breathing
8. Center for decision-making and other voluntary behavior

If you study sense organs at this time, you may want to study the structure of the eye of a sheep (p. 319), or that of a frog. Students may try to trace the optic nerve to the brain. Some teachers like to use a sagittal half of a sheep's head. (Refer to Chapter 1 for instructions on dissecting a frog and other vertebrates if you want to study the nervous system in more detail.)

Dissection of a brain. The brain of a sheep or calf may be obtained locally from a butcher. This material, as well as fetal pigs, may also be obtained from biological supply houses (addresses in Appendix C).

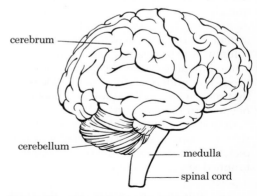

cerebrum

cerebellum

medulla

spinal cord

Fig. 7-19 Schematic drawing of human brain. (From J. W. Clemenson, T. G. Lawrence, et al., *Your Health and Safety,* 4th ed., Harcourt, Brace & World, Inc., 1957.)

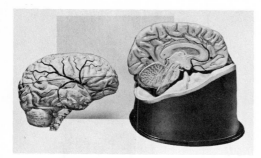

Fig. 7-20 Take-apart model of brain. (Courtesy of Clay-Adams, Inc., New York City.)

If fetal pigs are available, their brains may be dissected.

If a fetal pig is used, carefully cut each occipital bone by inserting scissors into the foramen magnum and then cutting down on each side; then cut forward. Lift off the central, dorsal portion and locate the outermost tough membrane of the brain, the dura mater. Remove all the bone fragments, then cut into the dura mater and observe how the membrane separates the cerebral hemispheres from the cerebellum.

Use Fig. 7-20 as a guide in completing this dissection. Locate the olfactory lobes, the sulci and gyri of the brain, the olfactory nerve (first cranial nerve), the cerebellum, medulla oblongata, and choroid plexus.

Lift the brain gently out of the skull, and identify the structures that can be seen from the ventral view: the optic nerves, optic chiasma, and the infundibulum, to which the pituitary may still be attached. Try, also, to find the cranial nerves and identify them.

The brain may also be cut in half sagittally, so that a median view may be examined (Fig. 7-20).

The eye

Sheep's eye. Obtain the eye of a sheep or calf from a butcher or biological supply house. Examine the eyelids and the third, or nictitating, membrane. Look for the optic nerve at the back of the eye, and examine the arrangement of muscles that move the eyeball. (Have ammonia inhalants on hand, for the fixed "stare" of the eye has an eerie effect on many students.)

Locate the iris and pupil. Using Fig. 7-21 as a guide, dissect away the lids, fatty tissue, and muscles, but leave the optic nerve. Students may readily identify the thick, white, tough sclera; the black choroid; and the whitish retina, which is probably shriveled and may have fallen into the eyeball cavity. The sclera continues along the front of the eye as the cornea. Of course, in a living animal the cornea, lens, and vitreous chamber are transparent.

Locate the large vitreous humor, or body, that holds the retina[12] in place in the living animal. Find the lens and study its shape and attachment to the ciliary body. Students can also focus images with this lens.

Eye of *Limulus*. The arachnid eye of the horseshoe crab, or king crab, is often studied for comparisons with the eye of a mammal.

Some teachers may find the occasion to study the biochemistry of the visual pigment (red in the *Limulus* eye, formed by attachment of retinene to opsin). It may

[12] Students may want to read about color vision and the detection of polarized light in animals as compared with vision in man. See C. L. Prosser and F. A. Brown, Jr., *Comparative Animal Physiology*, 2nd ed., Saunders, Philadelphia, 1961.

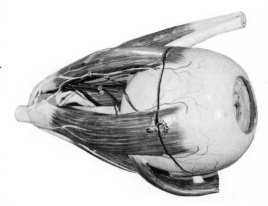

Fig. 7-21 Take-apart model of eye. (Courtesy of Clay-Adams, Inc., New York City.)

also be interesting to stimulate the eye of a living king crab with brief flashes of light. One cotton wick "electrode" is placed on the cornea and the second electrode is placed through a small hole made behind the eye in a study of the retinal generator potential. The duration of the dark adaptation, the pattern of nerve impulses of the optic nerve, and electronic recording equipment needed are well described by G. Wald *et al.*[13]

The ear

First use a model of an ear such as that in Fig. 7-22. (Models may be purchased from a biological supply house.) Then examine a prepared slide of the cochlea of the ear under low power. How many times is the coiled tube of the cochlea sectioned? Use high power and try to locate the organ of Corti (Fig. 7-23), the basilar membrane, and hairs; then look for the auditory nerve.

Especially try to observe fibers of different lengths in the basilar membrane. Explain how sound waves are received, and transmitted to the auditory nerve.

Refer to two papers in *Scientific American:* H. Kalmus, "Inherited Sense Defects" (May 1952) and G. von Békésy, "The Ear" (August 1957).

[13] G. Wald *et al., Twenty-six Afternoons of Biology,* Addison-Wesley, Reading, Mass., 1962, Chapter 20, "Electrical Activity of a Sense Organ: The *Limulus* Eye."

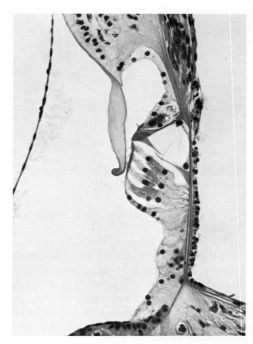

Fig. 7-23 Organ of Corti in internal ear. Photomicrograph of a prepared slide. (Courtesy of General Biological Supply House, Inc., Chicago.)

Autonomic nervous system

A model of the nervous system such as that in Fig. 7-24 may be available for study. Pretend to give an unannounced test; distribute your "traditional test paper." Wait a moment; ask, with a smile, that students describe their physical state. Why does the heart beat faster? Explain the causes of hyperglycemia, which accompanies examination anxiety. Also refer to coordination by adrenals and other hormones, pp. 323–26.

Further reading

These references are only to *Scientific American,* which is readily available for high school students. One or several of the following papers might be read and discussed over several sessions in class or in club work: "The Analysis of Brain Waves" (June 1962), "The Great Cerebral Com-

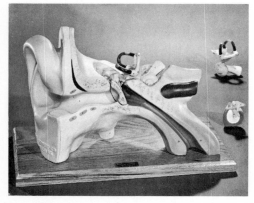

Fig. 7-22 Take-apart model of ear. (Courtesy of Denoyer-Geppert Co., Chicago.)

missure" (January 1964), "The Human Thermostat" (January 1961), "The Perception of Pain" (February 1962), and "Truth Drugs" (March 1960).

Articles dealing with the biochemistry of how we see make interesting reading: "The Visual Cortex of the Brain" (November 1963), "Visual Pigments in Man" (November 1962), and "Stabilized Images on the Retina" (June 1961). "Inhibition in Visual Systems" (July 1963) describes "eyes" of scallops and photoreceptive systems of other invertebrates. Also look into "Afterimages" (October 1963), "The Origin of Form Perception" (May 1961), and "Aftereffects in Perception" (January 1962).

Theories about amoeboid movement, contraction of muscle, and the beating of cilia are described in "How Cells Move," in the September 1961 issue. (This issue is devoted entirely to work on cells.) In the same issue, see "How Cells Receive Stimuli" and "How Cells Communicate."

An introduction to the long years of work of J. T. Bonner on slime molds is given in "How Slime Molds Communicate" (August 1963). See also Fig. 9-39.

Teachers may also want to select activities from the BSCS laboratory block developed by H. Follansbee, *Animal Behavior* (Heath, 1965). Studies in the behavior of carpenter ants, *Daphnia*, mosquito fish, and chameleons are included.

Sense receptors of the skin

The senses for taste, touch, sight, hearing, and smell are clusters of nerve endings that form sensory receptors. These keep the organism in touch with the outer environment through mechanisms of "feedback."

You may want to have students map the location of several of the sense receptors in the skin (and errors in sensing the location of these receptors) in the following ways.

Localization of touch. Have a student keep his eyes closed throughout the experiment. Show another student how to touch

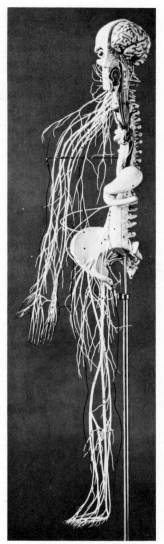

Fig. 7-24 Wire model of nervous system. (Courtesy of Clay-Adams, Inc., New York City.)

the skin of the hand of the subject with the pointed end of a soft pencil (leaving a mark). Remove the pencil, give the subject a blunt probe, and ask him to locate the place on the skin where he received the stimulus. Use a millimeter ruler to measure his error in locating where the stimulus was applied. Try this on several subjects.

Here is another method for localizing these receptors. Insert two pins closely spaced in a cork, or use pointed calipers. Then gently bring the pinpoints to the surface of the hand, forearm, and fingertips. When the pins are placed closely the sensation received by the subject is that of one pinpoint. When the distance between the pinpoints is increased a bit the subject receives two sensations. This approximates the distance between the two sense receptors in the skin. Are the receptors grouped more closely on the forearm or at the fingertips?

Temperature contrast. Have students immerse one finger in water at 40° C (104° F) and at the same time put a finger from the other hand into water at 20° C (68° F); after 30 sec they should transfer both fingers into water at 30° C (86° F). What is the sensation? Use the same procedure but this time have students first immerse one finger in water at 45° C (113° F) and the other finger in water kept at 30° C (86° F), and finally shift both fingers into water at 10° C (50° F). Have them describe the sensation.

Temperature response. Draw out a glass rod to form a blunt point. Prepare several, fire-polish the ends, and allow them to cool. Chill the rods in ice water and then have students apply one to the skin of the back of the hand, and also to the forearm. Locate the receptors. Use washable ink to mark the receptors on the skin.

Repeat the procedure, but this time with a rod warmed in hot water (*caution:* not too hot). Locate the receptors which sense the stimulus of heat. Mark these receptors on the skin with different colored ink.

Confusing the senses. Ask students to cross the middle finger and the index finger of one hand. Then have them roll a small pill or bean in the palm of the other hand with the crossed fingers. The students should be able to describe the sensation.

Estimation of weight. A blindfolded student might be asked to compare the weight of two graduated cylinders each holding a given volume of water. Then add small volumes of water to one cylinder; estimate at what increase in weight (1 ml of water weighs 1 g) a student can sense a difference in weight. Shift the cylinder with increased weight from hand to hand at random during the time of the test.

Blind spot. The blind spot is the area in the eye in which there are no visual receptors because it is the point at which the optic nerve leaves the retina. Wald *et al.*[14] describe a demonstration that each student may try. Draw a small cross on a large sheet of white paper, a bit left of center. Have one student close the left eye and stare at the cross with the right eye holding the paper 30 cm from the eye. A partner of each subject should now bring a pencil point into the subject's field of vision, starting some 2 to 4 in. to the right of the cross on the paper. At what point does the pencil disappear? Mark this point on the paper; repeat, bringing the pencil from another angle. By bringing the pencil from different directions toward the cross, one can plot the boundary of the blind spot.

Chemical response. Map the areas of the tongue which are sensitive to salty, sweet, sour, and bitter substances. Apply the solutions with a sterilized camel's-hair brush, a toothpick, or a glass rod drawn out to a blunt point and fire-polished. Or apply small squares of filter paper that have been soaked in a solution to different regions of the tongue. (Use a forceps to handle the filter paper.)

Wash out the mouth between tastings. First use a solution of 2 parts of water to 1 part of vinegar, or use 1 percent solution of acetic acid. Apply it to the tip, the sides, the center, and the back of the tongue. Locate the area of the tongue that is sensitive to sour substances.

In testing for the salt-sensitive area of the tongue, use a 10 percent solution of sodium chloride. Apply it to the same regions of the tongue as before. Remember to wash out the mouth between tests.

For a bitter substance, use a weak solution of aspirin in water, or quinine sulfate.

[14] *Ibid.,* pp. 132–33.

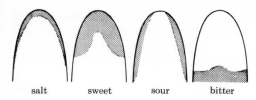

salt   sweet   sour   bitter

Fig. 7-25 Areas of the four kinds of taste receptors in the tongue.

Use a 5 percent cane sugar solution as a sweet substance. Use coded bottles so that students do not know the true contents. Also test students with tap water to eliminate the factor of suggestion. Compare your findings with the areas shaded in Fig. 7-25.

Also refer to inherited taste sensitivity, p. 383. Some students may want to report back to the class after reading Chapter 11, "Chemoreception," in Prosser and Brown.[15]

Many ideas for investigations, or for short-range projects, may be found in the books by Prosser and Brown, and Hawk, Oser, and Summerson.[16]

An engaging book, *101 Simple Experiments with Insects,* by H. Kalmus[17] describes how to make an olfactometer, a T-shaped glass tubing device by which one can measure responses of fruit flies, bees, and butterflies.

Sense of smell. Blindfold a student and have him hold his nose. Now offer cubes of raw potato, apple, and onion. Have him chew and identify the materials. (A solution might stimulate nerve endings associated with taste or smell or even heat.)

Coordination of responses in animals

Ciliated epithelial tissue in coordination

While ciliated epithelial tissue may have been observed earlier in studies of tissues of a frog, this is a "new view" in the context of coordination of responses.

Ciliated epithelial tissue covers the surface of the roof of the mouth and pharynx of the frog. Examine the action of cilia. Cut away the lower jaw of a pithed frog, and lay it on its back. Place a drop of India ink or carmine powder at the anterior part of the mouth; watch the rate at which the particles move toward the rear of the mouth.

Cut off a small bit of tissue, and fold it over in a drop of saline solution. Under high power the active current set up by the cilia can be observed at the edge of the small bit of tissue. Cut down the light under high power until this current is apparent.

J. Longley[18] describes a procedure that may readily be duplicated in class. Demonstrate the movement of a gross object, such as a glass tube, by cilia. Seal one end of a glass tubing that is several centimeters long and has a bore of about 0.5 cm. Place this over a vertically held hat pin or similar device on which it can freely rotate. Attach a light-weight pointer—a thread of glass—to the upper end of the glass tube. This may be done by bringing a heated piece of thin glass tubing close to the upper end of the tube, which is also hot; then draw out a fine thread about 4 to 5 cm long. Around the glass spindle wrap a strip of ciliated epithelial tissue (keep it moist), keeping the ends of the strip secure with threads. Cilia should rotate the glass spindle, and the pointer should be seen to move at a readily detected rate.

Coordination through hormones

How do the hormones of ductless glands affect behavior? Most students have heard of diabetes, and can describe the conditions under which insulin is given. They can describe their feelings in "stage fright." Someone can report on the work of Hans Selye and describe stress and the

15 *Op. cit.*
16 *Op. cit.*
17 *Op. cit.*

18 J. Longley, "The Activity of Ciliated Epithelia," in National Institutes of Health, staff, "Laboratory Experiments in Biology, Physics, and Chemistry," mimeo pamphlet, U.S. Dept. of Health, Education, and Welfare, National Institutes of Health, Bethesda, Md., 1956.

alarm reaction (see *Sci. Am.,* March 1949).

Students may feel tissue of the thyroid gland by lightly putting their fingers against their windpipes. Then they should swallow and try to locate the thyroid tissue, which seems to move up and down in front of the windpipe. Many can describe a basal metabolism test from their own experience.

Which gland is overactive in producing a giant? What are some kinds of midgets? Which is the master gland that coordinates all the ductless glands? What controls the pituitary?

Use a manikin or a chart to show the position of the ductless glands in the body. Many organs may be brought to class by students; for example, the pancreas and possibly the adrenals are available from a butcher. You may plan to dissect a frog, rat, pigeon, or cat to locate some of the ductless glands (dissections, Chapter 1).

In season, a group of students may treat tadpoles with thyroxin or an iodine salt, or use pituitary hormones to induce ovulation in the frog (as described on p. 360).

Perhaps there is occasion to do a rat thyroidectomy; complete directions are available in F. D'Amour and F. Blood, *A Manual for Laboratory Work in Mammalian Physiology* (Univ. of Chicago Press, 1954).

Students may report on the classic work of Bayliss and Starling, "The Mechanism of Pancreatic Secretion," as reproduced in M. Gabriel and S. Fogel, eds., *Great Experiments in Biology* (Prentice-Hall, 1955). Perhaps there is time to read and discuss Otto Loewi's paper, "On the Humoral Transmission of the Action of Heart Nerves," reproduced in the same publication. (Also refer to p. 325 for an investigation using a frog's heart.)

Which hormones are involved in pregnancy tests? What are the experimental animals? What is the evidence of a positive test for pregnancy?

What is the effect of adrenalin on melanophores of fish scales (p. 110)? On the heartbeat of *Daphnia* (p. 245)?

Students should surely know something of the way hormones affect skin color in frogs (*Sci. Am.,* July 1961). Further reading in texts by Prosser and Brown, by Patton, and by Turner (see end-of-chapter bibliography) may be useful. These have excellent coverage of endocrine mechanisms in invertebrates as well as in higher forms.

How do hormones affect metamorphosis of insects? Duplicate the work with *Cecropia* pupae as described in the texts mentioned, as well as in C. Williams, "Metamorphosis of Insects" (*Sci. Am.,* April 1950), and in V. Wigglesworth, "Metamorphosis and Differentiation" (*Sci. Am.,* February 1959).

In the BSCS laboratory block by F. Moog, *Animal Growth and Development* (Heath, 1963), there are investigations for the class on the effect of gonadotrophic hormones on sexual development in cockerels. In these studies chicks are injected, so students must learn to use hypodermic syringes. The height and length of the comb are measured and chicks are weighed. Daily injections are made over a week. Other studies involve using female chicks; these are injected with testosterone to see if male characteristics can be elicited in female chicks.

Other studies are described: the normal development of the chick embryo, regeneration of the frog's tail, the effect of a thyroid inhibitor (thiourea) on a chick's development, ovulation in frogs (pituitary), and the effect of thyroxin-feeding on tadpoles. Ovulation in frogs is described here in Chapter 8.

Effect of feeding thyroxin to frog tadpoles. The hormone thyroxin stimulates the metamorphosis of the tadpole into an adult frog. Bullfrog tadpoles with the hind limbs appearing give the best results in this work. Set up finger bowls containing different concentrations of thyroxin. Use a concentration of 1 part of thyroxin to 5 million parts of water in one series. In another set, use 1 part of thyroxin to 10 million parts of water. (The hormone is absorbed directly through the skin of the tadpoles.)

Put individual large bullfrog tadpoles or five to ten small frog tadpoles in separate finger bowls. Set aside one series as controls (use spring or pond water without thyroxin); establish two series using different concentrations of thyroxin.

Rugh[19] suggests preparing stock solutions of thyroxin by dissolving 10 mg of thyroxin (crystalline) in 5 ml of 1 percent sodium hydroxide. Dilute to a 1-liter volume with distilled water. This 1:100,000 concentration should be refrigerated until it is used. Dilute this stock solution to prepare the solutions needed for the series of finger bowls.

Feed the tadpoles on alternate days with small amounts of hard-boiled egg yolk, and change the water each week to prevent fouling of the medium.

Rugh also recommends another method which makes use of thyroxin tablets. When these are available, crush and dissolve five 2-grain tablets in 5 ml of distilled water in a mortar. Weigh out an equal amount of whole-wheat flour and grind it up with the thyroxin tablets. Then spread this paste in a thin layer on glass squares and leave to dry. Finally, powder the dry mixture and store it in closed bottles in a refrigerator. Use 50 mg of this wheat-thyroxin powder per tadpole daily for 1 week. Feed the tadpoles, both controls and the experimental animals, with parboiled spinach or lettuce. Change the medium daily to prevent fouling.

Effect of feeding iodine to frog tadpoles. Since the main fraction in thyroxin is iodine, students may devise a project investigating the effects of iodine on the metamorphosis of amphibian tadpoles.

Here again work with bullfrog tadpoles is most successful, but any species of frog, or the salamander *Necturus* may be used. No injections are necessary, since iodine is absorbed directly through the skin. Select tadpoles in which the hind limbs are just becoming visible, 1 mm or so. Set up several finger bowls with a maximum of 20 tadpoles of small frogs such as leopard frogs, or one tadpole of a bullfrog, in about

30 ml of prepared medium. Prepare several finger bowls with animals in spring water or pond water as controls.

In the experimental dishes use the following medium. First, prepare a stock solution: Dissolve 0.1 g of iodine (crystalline) in 5 ml of a 95 percent solution of alcohol; then dilute to 1 liter with distilled water. This is a stock solution of a concentration of 1:10,000.

Dilutions as weak as 1:500,000 and 1:1,000,000 have been successful in stimulating metamorphosis. Each week change the medium to prevent fouling due to decay of food material. Treat the tadpoles for 7 to 14 days. However, avoid more frequent changes, for too much iodine will result in such accelerated growth that tadpoles may die. On alternate days feed the controls as well as the experimental animals with small amounts of hard-boiled egg yolk or partially cooked lettuce. Rugh recommends 1 sq in. of parboiled spinach leaf per tadpole. Remove uneaten food.

Keep a record of the changes in the length of tail and hind limbs and the time of appearance of forelimbs in both the experimental and the control animals. Also note the changes in the shape of the head and body. Normally, larvae of *Rana pipiens* (leopard frog) reach stages of metamorphosis about 75 days after fertilization of eggs (maintained at 23° to 25° C [73° to 77° F]).

If students plan a project in the field of experimental embryology, or for club work in this area, they can learn many techniques from a study of Rugh's fine text-manual *Experimental Embryology,* cited above. Rugh also gives excellent directions for preparing glass operating needles and for making transplants of tissue from donor to host tadpole. The way a transplant of a future eye or gill or limb can be made is also described. Students should be forewarned that the techniques are difficult and require much practice before any degree of success can be expected (see also Chapter 9).

Hormonal control of the frog's heart. As a laboratory activity you may want to

[19] R. Rugh, *Experimental Embryology: Techniques and Procedures,* 3rd ed., Burgess, Minneapolis, 1962, p. 315.

have students observe the effects of adrenalin and of acetylcholine on the contraction of the heart. A beating heart, suspended from a support in Ringer's solution, was described on p. 244.

If a kymograph is available (description, p. 312), the following preparation may be used to compare the action of several hormones and drugs. Lay a freshly killed frog on a frog board and dissect out the beating heart (cut through the pericardial membrane to expose the heart). Keep the heart bathed in Ringer's solution.

Fasten a fine thread to the tip of the ventricle or pass a needle and thread through the tip of the ventricle. Attach the extension of one thread to the lever of a kymograph. Adjust the level of the frog and the position of the lever; the needle should just touch the smoked paper on the drum.

Get a reading of the heartbeat when the heart is bathed in Ringer's solution. Apply 1 or 2 drops of a solution of acetylcholine ($2 \times 10^{-4} M$). Record the change in the heartbeat. Then rinse the heart in Ringer's if it stops beating. Also, what is the effect on the heartbeat of a solution of adrenalin chloride or epinephrine ($2 \times 10^{-4} M$)? What is the role of the sinoauricular node?

Further reading. Some teachers use offprints from *Scientific American* in teaching the nature of experimental design. Studies of hormones, both plant and animal, provide good examples for high school students. Tests of understanding can be designed asking students to distinguish the underlying working hypothesis, the problems that arose, and the predictions that could be tested.

Look into such papers from *Scientific American* as "The Thyroid Gland" (March 1960), "The Parathyroid Gland" (April 1961), "The Thymus Gland" (November 1962), "Insects and the Length of the Day" (February 1960), "Cortisone and ACTH" (March 1950), "Plant Growth Substances" (April 1957), and "The Control of Flowering" (May 1952).

Coordination of responses in plants

Phytohormones

Auxins, or phytohormones, have some regulatory control in many activities of plants. Auxins inhibit root elongation and the growth of lateral buds, and they stimulate cell elongation and thus elongation of stems at internodes (possibly by changing the properties of cell walls so that more water can enter and the walls can become elongated). Auxins also promote initiation of roots, flower buds, and fruit-set. These growth-regulating substances also delay the falling of leaves (abscission of petioles).

Indoleacetic acid (IAA) is the key compound in the family of phytohormones. Auxins consist of a number of indolyl derivatives (for example, indole-3-acetaldehyde, indole-3-glycollic acid, indole-3-pyruvic acid, and alpha indole-3-butyric acid); these affect plants at the organ level. Apparently algae have indoleacetic acid, but it doesn't appear to play a key role.

Individual students, or groups, may want to repeat some of the classic experiments of Darwin, Boysen-Jensen, and F. W. Went. Many students are fascinated to read an excerpt from the paper "Sensitiveness of Plants to Light: Its Transmitted Effects" by Charles Darwin and his son. In this paper they describe using tin foil caps over tips of shoots to investigate which part of the shoot is sensitive to light, with the result that only the illuminated shoot bends toward the light source. This excerpt is reprinted in M. Gabriel and S. Fogel, eds., *Great Experiments in Biology* (Prentice-Hall, 1955). In the same book are included excerpts from subsequent papers that are classics in the early investigations of plant hormones: P. Boysen-Jensen, "Transmission of the Phototropic Stimulus in the Coleoptile of the Oat Seedling"[20] and F. W. Went, "On Growth-accelerating Substances in the Coleoptile of *Avena sativa*."

[20] Or select sections from Boysen-Jensen's book, *Growth Hormones in Plants*, McGraw-Hill, New York, 1936.

An excellent source of descriptions of test methods (with many photographs) used by research scientists is available: *Test Methods with Plant-regulating Chemicals,* by leaders in the field.[21] There are at least 45 different tests, among which are tests for root induction and fruit size; the split pea stem test; the cambium test; tests for abscission, angle, and elongation of branches; and radioautograph tests of translocation of tagged regulators.

Two BSCS laboratory blocks rich in investigations for individual and group laboratory work are W. P. Jacobs and C. E. LaMotte, *Regulation in Plants by Hormones* (Heath, 1964), a study of experimental design; and A. E. Lee, *Plant Growth and Development* (Heath, 1963). Each block is suggested for a 6-week sequence.

Many teachers have already discovered certain fine laboratory manuals available, especially L. Machlis and J. Torrey, *Plants in Action* (Freeman, 1956) and B. Meyer, D. Anderson, and C. Swanson, *Laboratory Plant Physiology,* 3rd ed. (Van Nostrand, 1955). Clear, concise instructions are provided for reproducing effects of auxins on seedlings, stems, leaves, and roots.

Investigations using auxins

GROWTH OF SHOOTS. Students may germinate soaked oat seeds in paper cups of vermiculite or moist sand kept in the dark. When the coleoptiles emerge and are about 1 to 1.5 cm high, cut off about 3 mm of the tips of several with a sharp, sterile razor or the edge of a coverslip. Then apply a bit of lanolin paste to the cut surfaces. Decapitate an equal number of coleoptiles; this time apply lanolin to which 0.1 percent indoleacetic acid has been added. (Prepare this by dissolving 100 mg of indoleacetic acid in 2 ml of absolute ethyl alcohol. Add this solution to 100 g of lanolin paste and mix thoroughly so that the auxin is evenly distributed in the paste. Or melt the lanolin

[21] J. Mitchell, G. Livingston, and P. Marth, *Test Methods with Plant-regulating Chemicals,* Agriculture Handbook 126, U.S. Dept. of Agriculture, U.S. Govt. Printing Office, Washington, D.C., 1958 ($0.40).

and stir in IAA.) Set aside a third batch of coleoptiles from which the tips are not removed. Place all the shoots (that is, cut tips with lanolin, cut tips with lanolin plus indoleacetic acid, and normal tips) near a source of unilateral light. After 24 hr students should measure the angle of curvature in all three sets of plants. Which part of the coleoptile seems to produce a growth-regulating substance? Which part of the coleoptile seems sensitive to light?

GROWTH OF ROOTS. In a similar approach, students may repeat the previously described technique for growing oat seedlings. This time the aim is to show the inhibitory effect of growth-regulating substances on root growth. In this case students should cut off the 2-mm tip of several roots. Prepare untreated roots and treated roots with the lanolin paste and lanolin-plus-auxin preparation, respectively. Then set up the seedlings in Petri dishes for a geotropism demonstration as shown in Fig. 7-6. Students may observe in subsequent demonstrations that auxins initiate root formation, but, in this case, after the root primordia are formed and the roots elongate, auxin inhibits further growth of roots.

Some students may also plan experiments to find out whether terminal buds produce a substance that inhibits the growth of lateral buds, or they may plan activities to show the effect of leaves on stimulating root formation. Does root initiation proceed faster when stem cuttings have leaves?

Effect of leaves and buds on the growth of cuttings. Select healthy coleus, geranium, or other herbaceous cuttings which have several internodes. In some, remove all the leaves; in others, remove the terminal buds and the growing tips; in a third group, remove only the lower leaves. Set aside several others as controls. Then compare the rate of root formation after planting the cuttings in moist sand. Better results should be forthcoming if cuttings are grown in nutrient solutions (p. 459) to which thiamine, pyridoxine, and other vitamins of the B group are added.

Students may use soaked seeds—garden beans or lima beans—rather than cuttings. After the seeds have germinated in the dark for about 2 weeks, or until a pair of leaves is formed, cut off the roots, leaving only a stem portion about 8 to 10 cm long. Stand several of these young plants in containers of the following solutions: a nutrient solution such as Shive's or Hoagland's (p. 459) to which micronutrients have been added; nutrient solution plus 0.1 mg of indoleacetic acid; nutrient solution plus 1.0 mg of indoleacetic acid; and distilled water. Under which conditions does root initiation occur?

There is indication that root-forming substances may be made in buds and leaves and transported to the region where new growth is stimulated. Demonstrations may be devised in which there is a unilateral application of growth hormones. One side of the stem (the side to which the hormone has been applied) elongates to a greater extent than the other side (Fig. 7-26). This activity may be tried with the growth-promoting substances to be described. Of course, when the hormones are applied to cuttings they stimulate good root formation.

Growth-promoting substances

The growth-promoting substance which has been shown to be elaborated by the plant is indoleacetic acid, but there are substitutes for this substance that are less expensive and that are not quickly destroyed by the enzyme systems of plants. Two that can be used successfully are β-indolebutyric acid and α-naphthaleneacetic acid. In fact, small amounts of the latter are active in duplicating the entire range of activities initiated by the growth substances native to the plants. A general name for these growth-promoting substances is "auxin," which we shall use for a whole group of chemicals that stimulate growth or elongation of cells, not just one particular chemical compound.

Other related compounds have practical

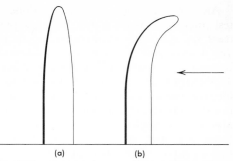

(a)   (b)

Fig. 7-26 Unilateral growth in plant shoot. Growth hormone applied to left side of shoot (a) causes elongation of cells on left side and consequent bending, as in (b). (Note: This effect is the same as if light were shining from the right side; light deflects auxins to the shady side.)

applications. In minute amounts, the herbicide 2,4-D (2,4-dichlorophenoxyacetic acid) is effective in reducing fruit drop in citrus fruits. Yet at higher concentrations it is toxic to broadleaved plants and thus acts as a selective herbicide for such weeds as plantain, dock, thistle, dandelion. In fact, Mitchell and Marth[22] offer explicit directions for preparing large quantities of 2,4-D to be used on lawns or fields. In their text they give tables to indicate the effect of 2,4-D on several hundreds of weeds, crop plants, and ornamental shrubs.[23]

In some communities actual field experiments may be undertaken to compare treated and untreated fields or lawns. Other projects may use refined techniques to find the stage in growth at which the chemicals are most effective. Perhaps some students want to demonstrate the specific physiological effects of auxins; for instance, it is thought that 2,4-D interferes with the sugar transport from leaves to stems and roots. How is 2,4-D transported through the plant?

Rooting preparations may be applied as liquids, powders, or pastes.

[22] J. Mitchell and P. Marth, *Growth Regulators for Garden, Field, and Orchard,* Univ. of Chicago Press, Chicago, 1947, pp. 7–27.
[23] Bio-assay of 2,4-D is a laboratory activity described in L. Machlis and J. Torrey, *Plants in Action,* Freeman, San Francisco, 1956, pp. 31–32.

Liquid form. Mitchell and Marth also suggest these concentrations for soaking stem cuttings. Dissolve ⅓ level tsp of indolebutyric acid in ½ tsp of grain alcohol. Then stir this into 5 qt of water, making a 0.01 percent solution. This solution is still effective after refrigeration for several months. The time needed for soaking stem cuttings varies with the kind of plant used (from 2 to 24 hr). For example, young leafy cuttings may take root quickly after soaking for some 1 to 2 hr in a 0.005 percent solution of indolebutyric acid. More woody cuttings of herbaceous plants may need to soak for 6 to 24 hr in the same dilution. Woody cuttings of trees such as dogwood or elm will root best if the leaves are still attached. They should be soaked for 1 to 2 hr in 0.01 percent indolebutyric acid. Consult the tables in Mitchell and Marth's text mentioned above.[24]

At times you may need a more rapid method. Dip the cuttings (about 1 in. of the stem) into a stronger solution (0.05 to 0.1 percent) for about 1 sec. Prepare the 0.1 percent solution by dissolving 1 oz by weight of indolebutyric acid in 1 oz by volume of 50 percent (100 proof) grain alcohol. Students may make more dilute solutions by adding more of the alcohol.

Powder form. Probably the easiest method is to dip the cuttings in a powder mixture. One method suggests rolling succulent stem cuttings in a mixture of 1 part of growth hormone powder to 1000 to 5000 parts of clay or talc. Stronger mixtures are needed for stems of woody plants, that is, 1 part to 500.

There are several commercial products on the market called by such names as "Auxilin," "Hormodin," "Quick-root," "Rootgro," and "Rootone" (among many others). "Hormodin" was developed at the Boyce Thompson Institute for Plant Research and is easy to use. Powder seems simpler to use than the liquid form, although both are available.

This experiment may be useful to

[24] Also refer to Mitchell, Livingston, and Marth, *op. cit.*

demonstrate how growth hormones stimulate rooting. Select several cuttings of privet or willow which are about 4 to 6 in. long. Wet the cut ends of several of the stems in water and dip them into Hormodin powder (or a similar preparation). Shake the twigs to remove excess powder; the amount of powder sufficient to produce results is that amount which just adheres to the wet tip. Plant the cuttings, the treated ones and several untreated controls, in separate pots of sand. Examine the growth of roots each week.

Paste form. A paste is made by mixing indolebutyric acid or other growth-promoting substance in melted lanolin. The advantage of a paste is that the substance can be placed on a localized region of the plant. First, prepare a solution (0.01 to 0.05 percent) of a growth-promoting substance such as indolebutyric acid or naphthaleneacetamide. Use the directions given above under "Liquid Form" for the preparation of a 0.1 percent solution, and dilute this. Then mix 1 tsp of the solution in 1 tbsp of melted lanolin. The small amount of alcohol used as a solvent for the growth substance evaporates in the heated lanolin. After a thorough mixing, apply this paste with a toothpick to different parts of several plants. For example, apply a circle of paste around the stem, to the upper leaves, to the lower leaves, and to cuttings. During the next few weeks, compare the treated parts with untreated controls. Keep both sets of plants under similar conditions of temperature and moisture. Indoleacetic acid may be used in some tests, naphthaleneacetic acid or naphthaleneacetamide in others.

Another easily prepared auxin formula suggested in the literature uses indoleacetic acid in this way (Fig. 7-27). Mix thoroughly 0.1 g of β-indoleacetic acid with 1 ml of 70 percent ethyl alcohol and then add this to 50 g of hydrous lanolin. A smooth paste should be made. Heat the lanolin a bit. If you wish to have a color for identification, add a pinch of carmine powder to tint the paste red.

(a)   (b)

Fig. 7-27 Effect of indoleacetic acid (IAA) on petioles of nasturtiums: (a) petiole before treatment; (b) petiole 1¼ hr after IAA paste was applied halfway up the petiole. (Courtesy of Carolina Biological Supply Co.)

In other project work, students may want to use β-naphthoxyacetic acid, which is effective in stimulating the formation of parthenocarpic fruits (seedless fruits, formed without pollination).

Other plant hormones

There are several biologically active substances besides the auxins.[25] Kinetin, a derivative of adenine, although not yet found in living cells, promotes cell division. It also interacts with auxins in effecting differentiation, for example, a high kinetin-low auxin ratio favors growth of buds, while a low kinetin-high auxin ratio enhances development of roots on cuttings.

The florigens are substances which affect flowering (in association with photoperiodism, p. 332). Several other substances affect sexual development of flowers, for example, it is possible to alter the sex of flower buds with naphthaleneacetic acid. A high level of auxin is associated with female flower bud tissue.

Capable and interested students may find further suggestions for individual

---

[25] Students interested in planning investigations may want to read J. Raper's paper "Hormones and Sexuality in Lower Plants," in H. K. Porter, ed., *The Biological Action of Growth Substances,* Symposia of the Soc. for Experimental Biology, 11, Academic, New York, 1957, pp. 143–65.

long-range investigations by reading Chapter 18 in E. Sinnott, *Plant Morphogenesis* (McGraw-Hill, 1960), and also Chapter 18 in B. Meyer, D. Anderson, and R. Böhning, *Introduction to Plant Physiology* (Van Nostrand, 1960). Also have students examine copies of *Science* and of *Biological Abstracts.* Sections of the latter may be purchased separately. In the section on plant physiology, students will encounter papers on current research with growth substances in plants.

Perhaps reviews of research are available. Three papers which give background material are F. Skoog and C. Miller, "Chemical Regulation of Growth and Organ Formation in Plant Tissues Cultured *in vitro*"[26] and two papers by K. V. Thimann, "Promotion and Inhibition: Twin Themes of Physiology" (*Am. Naturalist* 90:145–62, 1956), and "Growth and Growth Hormones in Plants" (*Am. J. Bot.* 44:49–55, 1957).

Some students may also want to investigate the newer ideas concerning antiauxins.

Gibberellins

Of the group of gibberellins (at least A_1, A_2, and A_3) the most active is gibberellic acid (A_3). The effects of gibberellic acid in rice attacked by a fungus (*Gibberella fujikuroi*) were originally observed and described in 1898. Young rice plants grew unusually tall and died of the "foolish seedling" disease. This effect was studied more thoroughly in 1926, and now the growth substance has been extracted from the fungi and its molecular formula is known. Possibly it is also present in higher plants.

While showing some of the same effects as auxins, gibberellic acid is not classified as an auxin since it does not give the usual oat coleoptile test, or show epinasty of leaves. It effects the elongation of stems and leaves in some plants (genetic composition is also a factor). (See Fig. 7-28.)

[26] *Ibid.,* pp. 118–31.

Seedlings of dwarf pea plants treated with gibberellic acid grow tall. As with auxins, this is due to the elongation of cells rather than an increase in number of cells. Does gibberellic acid supply some substance that has been lost through mutation in dwarf plants?

Unlike auxins, gibberellic acid does not show polarity of transport; it does not inhibit the growth of lateral buds or check abscission of leaves. Gibberellic acid also effects flowering (in association with photoperiodism, p. 332) in some plants (inhibits it in others), retards root growth, and breaks dormancy. There seems to be an increase in surface area of leaves; this results in increased assimilation of carbon.

Does gibberellic acid stimulate callus formation? With a cork-borer cut round plugs (2 mm × 8 mm) of tissue from carrots, citron, or other plant material. Or

(a)   (b)

Fig. 7-28 Effect of gibberellic acid on sunflower seedlings: (a) untreated control; (b) stem that was smeared with gibberellin paste just above the cotyledons 5 weeks before photograph was taken. (Courtesy of Carolina Biological Supply Co.)

use measured lengths of stems of plants to test growth of the stems in gibberellic acid. Treat material aseptically and grow in the medium suggested by P. White (p. 467). The medium requires thiamine, pyridoxine, nicotinic acid, and glycine.

Record the weight of the plugs and plant them in agar. In different trials, add varying amounts of gibberellic acid to Petri dishes of medium: 0.0, 0.5, 5.0, 25, 50, and 100 ppm. Also prepare medium to which both gibberellic acid and some auxin are added. Is there a synergistic effect, or a counteraction?

Application. Apply the gibberellic acid directly to seeds or to shoots of pea plant seedlings, and also to stems of ornamental plants. Does the substance affect seeds? Are there other factors such as wavelength or intensity of light, or is temperature a factor that affects the sensitivity of plants to gibberellic acid? Does gibberellic acid affect regeneration of such plants as *Bryophyllum* (see Fig. 9-33)? Is the growth rate of duckweed affected? Duckweed is a useful plant since new leaves are easily measured, and the plant requires little care or space (Fig. 13-26b).

Prepare a stock solution of gibberellic acid and then try applications of varying dilutions. Weigh out 100 mg of the powder, dissolve in a few milliliters of ethyl alcohol, and mix into 1 liter of distilled water. Or follow the instructions given on the preparations obtainable from supply houses. Use the solution when it is prepared; it may be refrigerated for a week or so, but then it begins to deteriorate.

Add a wetting agent so that the substance does not run off the plant surfaces. Dilute solutions of polyoxyethylene sorbitan monolaurate, such as the commercial preparation Tween 20, may be added (1 ml to 1 liter of gibberellic acid solution).

Apply the solution to leaves by using a fine capillary pipette and then using a glass rod to spread the solution evenly. Or use a cotton-tipped applicator on buds or stems. Gibberellic acid may also be added to the soil or applied as a spray. Seeds can

TABLE 7-1

Action spectrum
of some responses in plants*

	range (Å)	peak (Å)
Phototropism	3000–5000	4400–4700
(Flavoprotein)		blue light
Photoperiodism	5800–7200	6500
(Phytochrome)		red light
Photosynthesis	4000–7300	4400
(Chlorophylls)		blue light

* Adapted from A. G. Norman, "The Uniqueness of Plants," *Am. Scientist* 50:445, 1962.

be strewn on blotting paper or filter paper soaked in the chemical.

On initial treatment, rapidly growing leaves are a pale green, but they later develop a deeper color, especially if fertilizer has been added to the soil.

Many supply houses offer descriptive literature suggesting demonstrations and plants for successful experimentation.

Photoperiodism

Light affects several on-going processes in cells besides photosynthesis (see Table 7-1). One sensitive response is a rhythm of cell division which is timed by periods of darkness and light. This type of photoperiodism has been found in organisms at the cellular level as well as among higher plants and animals.[27]

In general, organisms "clock" their dark hours in a 24-hr cycle, not the light hours. For example, in plants that have been exposed to a brief light interruption midway in the period, the "dark-timer" starts timing again from this last light interruption, and disregards the previous period of darkness. Thus, a new growth

[27] A review of recent work may be found in R. Withrow, ed., *Photoperiodism and Related Phenomena in Plants and Animals,* Am. Assoc. for the Advancement of Science, Washington, D.C., 1959.

Also read the review of the effect of light on the reproductive cycles of animals in Prosser and Brown, *op. cit.*

Chapter 13 in E. Sinnott, *Plant Morphogenesis,* McGraw-Hill, New York, 1960, is a review rich in ideas for investigations.

stage begins after a certain number of hours of darkness. The region of the spectrum controlling photoperiodism in plants is situated between 5800 and 7200 Å (in the red region).

The original work of Garner and Allard of the U.S. Dept. of Agriculture in 1920 was based on the amount of daylight. They called these sequences photoperiods. This notion of long-day and short-day plants persists, although now we know that it is the hours of darkness that the plant clocks. (The term "long-night plants" would be more accurate than "short-day plants.")

Students may rear Cladocera (such as *Daphnia*) in different regimens. Raise some under a light period of 12 to 16 hr; illuminate others for only 8 hr per day. Several replicas should be maintained. Then count the number of juveniles and adults at the end of a week. Replace in fresh media; separate the adults from juveniles and keep counts at weekly observations.[28]

Among many male birds, the gradual increase in hours of light will stimulate activity of the testes; among ferrets, the increase in illumination brings them into estrus ahead of their normal reproductive cycle. Prosser and Brown[29] give several examples from fish to field mice and squirrels. Also, among immature ducks, red and orange light especially stimulates the pituitary gland, which triggers gonad activity. Light need not reach the eye in the duck; direct stimulation of the pituitary has been achieved with a quartz rod. When direct stimulation is used even blue light is effective. Furthermore, pituitaries of illuminated ducks have been transferred to immature mice; this increases gonadotropic activity in the mice. In the case of ferrets, light must reach the eye; in the sparrow this is not needed. However, there is no common pattern among vertebrates. Nerve connec-

[28] Read A. Giese, "Reproductive Cycles of Some West Coast Invertebrates," in Withrow, *ibid.,* pp. 625–37.

[29] Prosser and Brown, *op. cit.,* pp. 555–57.

tions between the pituitary and hypothalamus seem necessary for most gonadotropic activity; yet there are cases where gonadal activity seems to be the result of hormonal transmission only.

Where space is available (for example, a greenhouse or an outdoor plot of ground) students may find out first-hand the effects on flowering in plants when the number of hours of light and dark is altered. For example, they may grow short-day plants, such as chrysanthemum, ragweed, soybeans, cocklebur, or poinsettia, which need only about 10 hr of light in every 24 hr in order to flower. These plants need long periods of dark to flower normally; if this dark period is interrupted by even a short exposure to light, the plants remain in a vegetative stage. It is believed that the hormone which controls flowering is made in the leaves and transported to the sites of flower formation. An advanced project for a competent student might involve designing experiments to learn more about the effect of covering leaves, rather than the total plant; the possible effect of covered leaves and exposed leaves on flower development of grafted branches of shrubs or trees; or the effect of temperature on photoperiodism.

Similar projects may be undertaken using sturdy long-day plants (light for 14 hr) such as aster, rose mallow, spinach, or barley plants. Plan to use plants that are about 4 to 5 weeks old, and expose them to changed light conditions for a week. Students should provide for controls. Also use corn, snapdragon, cucumber, tobacco (most varieties except Maryland mammoth), and tomato plants as examples of plants not affected by fixed periods of light and dark.

Phytochrome

The existence of a light-sensitive blue or bluish green pigment in plants was announced by the U.S. Dept. of Agriculture in 1959, and isolated in 1961. Phytochrome must exist in minute quantities; while its bluish color is not apparent in albino plants, these plants respond to red and far red, revealing the presence of a blue pigment.

The existence of phytochrome was postulated when the action spectrum for sensitivity to photoperiodism was found to lie between 5800 and 7200 Å, and to be most effective in the red end of the spectrum. This did not coincide with the absorption spectrum of any pigment that was known at the time.

At Beltsville, Md., the scientists of the Agricultural Research Service built a spectrograph with a carbon-arc lamp that could throw a light beam, and with two prisms that dispersed light into a spectrum that formed a band (5 ft $\times$ 3 in.) long enough so that 14 small pots of plants could be illuminated simultaneously (Fig. 7-29). The focal length of the instrument was 33 ft.[30]

In the experiment described, short-day plants were used; cocklebur and soybean have been used extensively. All leaves were removed except one leaf on each of the potted plants; this leaf of each plant was fastened to a screen (Fig. 7-29). In each 24-hr cycle, plants received the number of "dark" hours known to be needed to initiate flowering; then they were given a brief interruption of light. As brief an exposure as 30 sec of *red* light (not other wavelengths) caused soybean and cocklebur to start clocking over again; they never registered enough hours of darkness to trigger "flowering."

Reversible shift. Seeds exposed to light in the region of 7000 to 8000 Å do not germinate. It has been established that some lettuce seeds can germinate when exposed to brief periods of red light, but if *far red* light is used the seeds fail to germinate. The light-sensitive substance reacts to the last kind of light signal. For example: (1) with seeds given red light, germination results; (2) with seeds given red, then far red light, there is no germination; (3) with seeds given red, far red, then

[30] U.S. Dept. of Agriculture, Agricultural Research Service, *Plant Light-Growth Discoveries*, ARS 22-64, January, 1961.

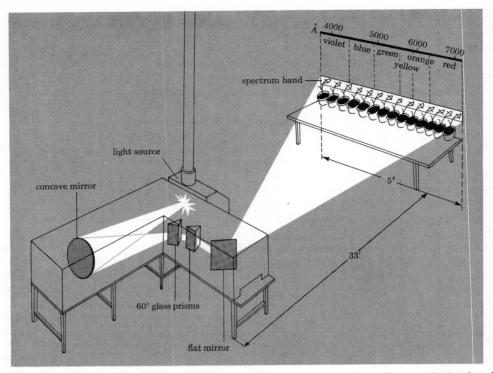

Fig. 7-29 Diagrammatic drawing to show details of ARS spectrograph and how plants can be irradiated. In the initial experiments that showed the importance of red light, each soybean plant, when ready to flower, had all foliage removed except one leaf. This leaf was fastened to a screen for brief irradiation during each night photoperiod to record the flowering response to separate colors of the spectrum. (After U.S. Dept. of Agriculture, Agricultural Research Service, *Plant Light-Growth Discoveries*, ARS 22-64, January 1961.)

red light, germination results; and (4) with seeds given red, far red, red, then far red light, there is no germination.

Students might duplicate this work. Seeds first are given dark storage at 20° C (68° F); then they are scattered on wet filter paper or a white blotter. Some lettuce seeds germinate in 2 days on wet blotter paper when given 30 sec of red light, then darkness. Most seeds sprout when exposed up to 16 min in red light.

The same reversible action of far red light (7200 to 8000 Å) has been found for flowering and elongation of stems.

In experiments with 6-day corn seedlings, the concentration of phytochrome was greatest in the upper region of the first internode and in the first sheath-like leaf folded around the stem in the corn shoot.

It may be that the floor of a forest receives more far red than red light (since red light is absorbed), so that far red would be the effective factor and would inhibit the germination of some seeds. In open fields, rapidly growing weeds need only a short flash of light to activate the phytochrome of their seeds; therefore the seeds of these weeds germinate faster than other seeds.

For use in greenhouses. Filters of colored cellophane can approximate pure color. Use a fluorescent lamp with a red filter to get almost pure red light, and an incandescent filament lamp with blue and red filters to get far red light.

Incandescent filament lamps resemble sunlight in providing red and far red light; they are effective in promoting flowering and the vegetative growth of plants, even though there is some far red. If a greenhouse is equipped with fluorescent lights only, add some incandescent lamps to supply far red if needed.

Source for red light. Wrap two layers of DuPont 300 MSC red cellophane around a 15-watt fluorescent tube.

Source for far red light. Wrap two layers of DuPont 300 MSC red cellophane and two layers of 300 MSC dark blue cellophane paper around an incandescent bulb.

Going further

The BSCS laboratory manuals provide activities on photoperiodism; also refer to the BSCS laboratory blocks.

Activities on the effect of periods of light on emergence of *Drosophila* are available in H. Kalmus, *101 Simple Experiments with Insects* (Doubleday, 1960). Excellent coverage is provided in textbooks by Giese and by Prosser and Brown (see end-of-chapter bibliography).

Students will also profit from certain articles in *Scientific American:* "Light and Plant Development" (December 1960), "Plant Growth Substances" (April 1957), "The Flowering Process" (April 1958), and "The Control of Flowering" (May 1952). These offer background information; also refer to the current issues.

Ionizing radiation

There is much literature on the effects of ionizing radiations on cells. There are texts, publications of the Atomic Energy Commission, reviews in textbooks of physiology or heredity, and current papers in journals (these are easily found in *Biological Abstracts*). Only a few guides are suggested here: R. F. Kimball, "The Effect of Radiation on the Genetic Mechanisms of *Paramecium aurelia*" (*J. Cellular Comp. Physiol.* 35: supplement, 157–69, June 1950) and "Non-genetic Effects of Radiation on Microorganisms" (*Ann. Rev. Microbiol.* 11:199–220, 1957); H. W. Schoenborn, "Protection Against Lethal Damage Induced by Ultraviolet Radia-

tion" (*J. Protozool.* 3:97–99, 1956); and R. Wichterman, *The Biology of Paramecium* (Blakiston [McGraw-Hill], 1953) and "Biological Effects of Radiations on the Protozoa" (*Bios* 28:3–20, 1957).

A textbook by A. Giese (see end-of-chapter bibliography) has an excellent summary of effects of radiation. See also the following papers by Giese: "Protozoa in Photobiological Research" (*Physiol. Zool.* 26:1–22, 1953); "An Intracellular Photodynamic Sensitizer in *Blepharisma*" (*J. Cellular Comp. Physiol.* 28:119–27, 1946); and, with E. Zeuthen, "Photo-oxidations in Pigmented *Blepharisma*" (*J. Gen. Physiol.* 32:525–29, 1949).

Send for publications of the Atomic Energy Comm. and of the National Research Council of the National Academy of Sciences, Washington, D.C. (the latter are not free). Look into "Long-Term Effects of Ionizing Radiations from External Sources," Public. 849, 1961; "The Biological Effects of Atomic Radiation," summary reports, 1960; and "The Effects of Inhaled Radioactive Particles," 848, 1961 (all from the National Research Council).

Among specialized textbooks on radiations, look into D. E. Lea, *Actions of Radiations on Living Cells,* 2nd ed., revised by L. H. Gray (Cambridge Univ. Press, New York, 1955); R. Zirkle, *Biological Effects of External X-Rays and Gamma Radiation* (McGraw-Hill, 1954); A. Hollaender, ed., *Radiation Biology,* Vols. I, II (McGraw-Hill, 1954); B. Duggar, *Biological Effects of Radiation,* Vols. I, II (McGraw-Hill, 1936); and M. Kamen, *Radioactive Tracers in Biology* (Academic, 1947).

Protozoa

Radiation slows down cell division among protozoa, and flagellates seem more sensitive to ionizing radiations than ciliates. For example, *Paramecium aurelia* can survive an exposure to 150,000 roentgens; yet, even in apparently normal paramecia, the effects may show up later.

In general, the physiological state of protozoa and other cells affects the severity of the effects of radiations.

Photodynamic sensitization

When the pink protozoan *Blepharisma* (Fig. 2-8b) is maintained in moderate light or in darkness, it becomes deeply pigmented, almost purple (p. 585). These forms are especially sensitive to light; when exposed to intense light they quickly die. This photosensitization occurs only when oxygen is present. Students may want to compare the effect of intense light on light-grown forms with those maintained in darkness.

Further, it is possible to use color filters to determine whether light of a specific wavelength is more critical in producing death among these protozoa. Is it possible to reduce the effects of photosensitization? Is cannibalism among *Blepharisma* affected by light? (Refer to Giese's textbook listed in the end-of-chapter bibliography, and also to B. Christensen and B. Buchmann, eds., *Progress in Photobiology* [Elsevier, Amsterdam, 1961].)

CAPSULE LESSONS: Ways to get started in class

7-1. Ask students to name activities their bodies perform with no thought involved (that is, reflexes). List examples of reflexes on the blackboard. Develop a definition of a reflex. What is the path of the reflex arc within the body?

7-2. Or begin by asking the class what behavior a child has from birth. List the reflexes on the board. As students mention them, ask how these activities differ from the behavior involved in dressing, riding a bike, dancing, or writing a composition.

7-3. Or ask a student to hold up a square of screening or clear plastic in front of his face. Have another student throw paper balls at him. Ask the class to note the student's behavior. Why does he blink even though he is protected? Develop the idea that some behavior is inborn and takes place without any thought.

7-4. In a panel discussion, students might describe some of the problems facing adolescents in school, in getting a job, and in getting along with others.

7-5. You may want to show a film on hormones such as *Endocrine Glands* (EBF). This might serve as a summary of the work in that area. How do ductless glands affect behavior?

7-6. Begin by asking students how they would train a new puppy to answer to its name. Develop the steps in habit formation: desire, repetition, and satisfaction.

7-7. You may want students to list the traits they admire in other people. Then have them check themselves against this yardstick. Develop a discussion of ways to improve oneself. What are the characteristics of a pleasant personality?

7-8. Ask students to list their fears. Discuss the questions: "Are fears inherited?" "How are fears developed?" Perhaps some students can recall what situations were associated with the development of their fears.

7-9. Start a discussion of study habits by using the activity described in this chapter wherein students memorize stanzas of poetry in quiet and with distractions. Or use a list of nonsense words first, and let students develop the conclusion that problems can be solved faster when one knows what the problem is, and how to tackle it.

7-10. Use the puzzle referred to in this chapter. Students take from 5 to 15 min to solve it. After 10 min, have one student explain how to put the pieces together. Then time the class again. Why is it easier to solve the puzzle now? Keep a tally of the number of students who solve the puzzle in each case. How does this apply to doing a school assignment? How do past experiences affect our insight into new problems?

7-11. You may have time in class to read or discuss a student's report on some of the booklets published by Science Research Associates, 259 E. Erie St., Chicago, Ill. These, for instance, may be useful: *You and Your Mental Abilities, Getting Along with Brothers and Sisters, Why Stay in School?, What Is Honesty?,* and *Choosing Your Career.*

Students in high school give much time to the problems discussed in these booklets; they often wonder if they are normal or if others feel the way they do. The stress should be on the fact that all young people growing up have similar problems to face and to solve. What are the best ways to solve problems of living?

7-12. A student might want to give a report of the kind of learning that W. Kohler described in his book *Mentality of Apes* (Harcourt, Brace & World, 1927). Was the putting together of two sticks to make a longer one for reaching food an example of an instinct or insight? Also compare mentality of apes with behavior in sharks, ducks, baboons, and other forms, as described in issues of *Scientific American*.

7-13. You may want students to describe instincts in animals, such as nest-building, spinning of a spider's web, migration of some birds and fish, behavior in a beehive, and many others. How do these behavior patterns differ from problem-solving behavior in man? (A description of an ant-observation case is given on p. 604.)

7-14. What are some successful ways to study? Students might contribute answers from their own experience. Then you may want to try some of the "experiments" described in this chapter. Include a discussion of the role of insight in the students' problem-solving activities.

7-15. Begin with a film showing instinctive behavior, such as *Dance of the Bees* (a description of von Frisch's work), available from Wilner Films and Slides. What is an instinct? Or show *Behavior in Animals and Plants* (Coronet). Also preview films in the AIBS series on behavior (McGraw-Hill).

Have you seen *Man Is a Universe* and *Stress* (both produced by the National Film Board of Canada)?

7-16. Have the class plan a field trip around the school grounds to look for examples of tropisms among plants. Students may search for examples of plants responding to environmental conditions. Give one or two examples; then set them free for 10 min to find more. Bring the class together again; all of the students will be working together to check the examples they have found.

7-17. Ask students to bring in some *Tradescantia* sprigs and supply equipment to set up the demonstration in Fig. 7-5. Get the students to plan the demonstration themselves. Ask what might happen if the plant were reversed (turned upside down). Why do stems grow up and roots grow into the soil?

7-18. Ask students why roots of plants grow downward. Is it a response away from light or a response to gravity (or something else)? Establish how an experiment might be designed so that only one factor is tested at a time. This is a good

device for teaching the elements of a controlled experiment.

7-19. In planning a test, you may want to include some questions with this pattern:

1. In which part of the nervous system is each of these behavior patterns centered:

Learning to drive a car, pulling away from a hot object, swallowing, breathing, touch-typing, solving a puzzle, dilation of the iris of the eye, doing your math homework, dancing, suspending judgment until you have the facts.

2. Identify a possible cause of each kind of behavior:

Increased flow of adrenin, sluggish metabolism, sudden increase in glucose in the blood, rapid growth of long bones in the body, flow of saliva when you see a picture of certain foods, fear of the dark, finding the right key among those on a key ring, ducking when a ball is thrown your way, ducking a clothesline which has recently been removed.

You may also want to include a reading from the classic works of William James, W. Cannon (especially *The Wisdom of the Body,* rev. ed. [Norton, 1939], which describes homeostasis, or chemical equilibrium), or Bayliss and Starling. Then prepare several questions based on the reading.

7-20. Have a student describe how to knot a tie without using gestures. Why is the description slow and stumbling? Develop the notion that our habits are activities done without thinking.

7-21. Have students write their names five times. Then have them shift their pens into the other hand and again write their names five times. Why is it so much more difficult and time-consuming? Elicit from students that each letter must be thought through in spelling a word. What is the value of having a habit?

7-22. You may want to bring to class an example of some tropism described in the chapter, and then to elicit explanations from the students as to the cause of this response.

7-23. As independent research, or a group project, students may want to learn more about auxins (see p. 328) and plant growth. They may want to go further into the whole field of polarity. Is it possible for roots to grow at the growing tip of a shoot? Try using cuttings of fast-growing plants—willows, coleus, and similar plants.

7-24. You may want to show a film as an introduction or as a review of behavior in plants. Be sure to check through the films in the AIBS film series (McGraw-Hill). Or show *Rhythmic Motions of Growing Plants* (Syracuse Univ. College of Forestry).

7-25. There are many films on learning in rats, cats, chimpanzees, and children. Some of these may be suitable for class. For instance, *Behavior Patterns at One Year* (EBF), *An Analysis of the Forms of Animal Learning, I and II* (Pennsylvania State College Psychological Cinema Register), *Fears of Children* (International Film Foundation), *Problem Solving in Monkeys* and *Problem Solving in Infants* (International Film Bur.), *Baboon Behavior* (Univ. of California), *Self-Conscious Guy* (Coronet), *Shyness* (McGraw-Hill), and others.

7-26. You may want to show a film on the structure of the nervous system and the function of such sense organs as the eye and ear. Examples include *Nerves: Demonstrations and Experiments* (EBF), *Eyes: Their Structure and Care* (Coronet), and *Gateways to the Mind* (Bell System).

7-27. You may want to begin by asking students to describe the conditions under which a diabetic must live. Or ask students how they feel when they are frightened. From these questions, elicit the uses of insulin and of adrenin in the hormonal coordination of the body. Develop the notion of some regulator, such as the pituitary.

7-28. Elicit an experimental design to investigate how the effects of a ductless gland are discovered. It may be necessary to describe an experiment such as that of Gudernatsch, who removed thyroid glands from tadpoles, and also added thyroid extract to the diet of tadpoles. Or read together the work of Banting; or show the film that describes an experiment, *Endocrine Glands* (EBF).

7-29. Do you plan to use films to clarify or summarize the work on hormones? You may want to use *The Hormones: Small but Mighty* or *Principles of Endocrine Activity* (both Indiana Univ.).

7-30. There may be time to demonstrate a technique. A 5-min film is available, *Frog Skeletal Muscle Response* (Thorne). Then develop the theories of nerve excitation and muscle contraction.

PEOPLE, PLACES, AND THINGS

Where in your community can you get living material to demonstrate behavior patterns? Ask a teacher in another school or a member of the biology department of a nearby college for fruitful places to take a class on a field trip.

Living materials which are out of season may be purchased from a biological supply house. Then you may culture many kinds of living things the year around in the laboratory (see methods in Chapters 12 and 13).

Students may bring in organs of animals, such as a brain, endocrine glands, or an eye, which they can get from a butcher. You may need to borrow a model of an eye or a brain from a college department, a doctor, or another school. Perhaps you may be able to borrow films from a nearby school or college (psychology department).

When possible, invite a guest speaker from a bureau of child guidance to talk about getting started toward a vocation, or about setting goals. It might be advisable to inform students early about the requirements for college entrance and the requirements for specific vocational opportunities.

In addition, it may be possible to visit a nursery or an experimental station where work is in progress on the effects of hormones on plant growth.

Visit a nearby woods if possible, or plan a field trip near the school. Solicit the aid of nurserymen and of parents in landscaping a small area of the school grounds so that you can take classes to this spot to teach the many aspects of interrelationships of plants and animals, and aspects of conservation (see pp. 545–52).

Students might prepare several terraria in the classroom. These may be made from leaking aquarium tanks, large mayonnaise jars, or other similar jars. Plan to have terraria which illustrate different environmental relations: plants native to a swamp, a desert, and a wooded area (see p. 611 for making terraria).

BOOKS

The references that follow give an indication of the scope of work in behavior. More complete coverage is given under "Physiology of Plants and Animals" in Appendix A.

Audus, L. J., *Plant Growth Substances,* 2nd ed., Interscience (Wiley), New York, 1960.

Buddenbrock, W. von, *The Senses,* Univ. of Michigan Press, Ann Arbor, 1961.

Bugelski, B., *The Psychology of Learning,* Holt, Rinehart and Winston, New York, 1956.

Bullough, W. S., *Vertebrate Reproductive Cycles,* 2nd ed., Wiley, New York, 1961.

Carlson, A., V. Johnson, and H. M. Cavert, *The Machinery of the Body,* 5th ed., Univ. of Chicago Press, Chicago, 1961.

Carthy, J. D., *An Introduction to the Behavior of Invertebrates,* Macmillan, New York, 1958.

Dethier, V. G., and E. Stellar, *Animal Behavior,* 2nd ed., Prentice-Hall, Englewood Cliffs, N.J., 1964.

Eiduson, S., *et al., Biochemistry and Behavior,* Van Nostrand, Princeton, N.J., 1964.

Follansbee, H., *Animal Behavior* (BSCS laboratory block), Heath, Boston, 1965.

Giese, A., ed., *Photophysiology,* Vol. II, *Action of Light on Animals and Microorganisms; Photobiochemical Mechanisms; Bioluminescence,* Academic, New York, 1964.

Gorbman, A., and H. Bern, *A Textbook of Comparative Endocrinology,* Wiley, New York, 1962.

Gray, J., *How Animals Move,* Cambridge Univ. Press, New York, 1959; Pelican, Penguin, Baltimore, 1964.

Guilford, J. P., *Personality,* McGraw-Hill, New York, 1959.

Harlow, H., and C. Woolsey, eds., *Biological and Biochemical Bases of Behavior,* Univ. of Wisconsin Press, Madison, 1958.

Harvey, O. J., D. E. Hunt, and H. M. Schroder, *Conceptual Systems and Personality Organization,* Wiley, New York, 1961.

Hillman, W., *The Physiology of Flowering,* Holt, Rinehart and Winston, New York, 1962.

Jenkins, J., *Studies in Individual Differences,* Appleton-Century-Crofts, New York, 1961.

Jenkins, P., *Animal Hormones: A Comparative Survey,* Part I, *Kinetic and Metabolic Hormones,* Pergamon, New York, 1962.

Patton, R., *Introduction to Insect Physiology,* Saunders, Philadelphia, 1963.

Penfield, W., and L. Roberts, *Speech and Brain Mechanisms,* Princeton Univ. Press, Princeton, N.J., 1959.

Pierce, J., and E. David, Jr., *Man's World of Sound,* Doubleday, Garden City, N.Y., 1958.

Prosser, C. L., and F. A. Brown, Jr., *Comparative Animal Physiology,* 2nd ed., Saunders, Philadelphia, 1961.

Rugh, R., *Experimental Embryology: Techniques and Procedures,* 3rd ed., Burgess, Minneapolis, 1962.

Savory, T., *Instinctive Living: A Study of Invertebrate Behavior,* Pergamon, New York, 1959.

Smythe, R. H., *Animal Vision: What Animals See,* Thomas, Springfield, Ill., 1961.

Thorpe, W., and O. Zangwill, *Current Problems in Animal Behaviour,* Cambridge Univ. Press, New York, 1961.

Turner, C. D., *General Endocrinology,* 3rd ed., Saunders, Philadelphia, 1960.

Vogel, H., V. Bryson, and J. Lampen, eds., *Informational Macromolecules,* Academic, New York, 1963.

Waring, H., *Color Change Mechanisms of Coldblooded Vertebrates,* Academic, New York, 1963.

Wigglesworth, V. B., *The Physiology of Insect Metamorphosis,* Cambridge Univ. Press, New York, 1954.

Withrow, R., ed., *Photoperiodism and Related Phenomena in Plants and Animals,* Am. Assoc. for the Advancement of Science, Washington, D.C., 1959.

Also refer to Freeman's listing of offprints of articles from *Scientific American;* these offprints are available from W. H. Freeman and Co., 660 Market St., San Francisco 4, Calif. ($0.20 each).

FILMS, FILMSTRIPS, AND SLIDES

This listing is intended to be only a guide for the selection of new films, filmstrips, and transparencies. Send for catalogs from distributors of films; catalogs of biological supply houses list extensive offerings of 2×2 in. slides. The addresses of distributors are given in Appendix B.

In general, the cost of color films averages $150; black and white films cost about $75. Newer films are made in varying lengths—time sequences of 1 min, 5 min, 11 min, or 30 min; the cost of films therefore varies. We have indicated the lengths of films so that lessons can be planned with the films in mind.

Sound films are available in color or black and white; rental rates are available from film libraries or by direct inquiries to producers. (Filmstrips are not available for rental; purchase price is usually about $6 to $7 for those in color.)

Baboon Behavior (31 min), Univ. of California.

Behavior in Animals and Plants (11 min), Coronet.

The Brain and Nervous System (30 min), National Educational Television.

The Discovery of Insulin (19 min), International Film Bur.

Endocrine Glands (11 min), EBF.

Eyes: Their Structure and Care (11 min), Coronet.

Feeling of Hostility (27 min), in National Film Board of Canada *Mental Mechanisms* series, McGraw-Hill.

Feeling of Rejection (23 min), in National Film Board of Canada *Mental Mechanisms* series, McGraw-Hill.

Frog Heartbeat (filmloop, 3 min), Ealing; adapted from BSCS techniques film.

Frog Skeletal Muscle Response (5 min), BSCS techniques, Thorne.

Fundamentals of the Nervous System (16 min), EBF.

Gateways to the Mind (2 reels, 25 min each), Bell System; free.

The Hormones: Small but Mighty (29 min), Indiana Univ.

The Human Body: Nervous System (14 min), Coronet.

The Human Brain (11 min), EBF.

Learning and Thinking (30 min), National Educational Television.

Life Science: Response in a Simple Animal (11 min), Film Assoc.; amoeba.

Life, Time, and Change (set of 12 films, 28 min each), Part X of *AIBS Film Series in Modern Biology,* McGraw-Hill.

Man Is a Universe (11 min), National Film Board of Canada; brain and nervous system.

The Man Who Beat Death (27 min), Post; tennis star W. F. Talbert and diabetes.

Mother Love (27 min), Carousel; baby Rhesus monkeys.

The Multicellular Animal (set of 12 films, 28 min each), Part IV of *AIBS Film Series in Modern Biology,* McGraw-Hill.

Multicellular Plants (set of 12 films, 28 min each), Part III of *AIBS Film Series in Modern Biology,* McGraw-Hill.

Nerves: Demonstrations and Experiments (11 min), EBF.

Plant Motions: Roots, Stems, Leaves (11 min), EBF.

Removing Frog Pituitary (filmloop, 3 min), Ealing; adapted from BSCS techniques film.

Rhythmic Motions of Growing Plants (11 min), Syracuse Univ., College of Forestry; free.

Sense Perception (set of 2 films, 30 min each), Moody.

Stress (11 min), National Film Board of Canada film, McGraw-Hill.

Symptoms of Our Times (set of 6 films), Am. Osteopathic Assoc.; alcoholism, addiction, arthritis, and so forth, free.

"Thinking" Machines (20 min), Educational Testing Service.

A Time of Migration (14 min), Potomac; alewife's struggle.

Visual Perception (20 min), Educational Testing Service.

Voice of the Insect (27 min), Carousel.

We, the Mentally Ill (28 min), Association.

Wonder of the Senses (27 min), Moody.

Many of these materials are distributed to teachers without charge. Where there is a small fee, the cost is usually indicated, although prices are subject to change. Although we recommend the material for use in the classroom, we do not necessarily sponsor the products advertised.

"Career Series" (many titles), New York Life Insurance.

Carolina Tips (periodical), Carolina.

Emotions and Physical Health; other titles, Metropolitan Life Insurance.

Free and Low-Cost Materials for Science Clubs of America; Thousands of Science Projects, Science Service.

New York State Dept. of Mental Hygiene, *School Days; Teen Time; Your Job;* many other titles, "Guideposts to Mental Health Series," Albany.

Redl, F., *Understanding Children's Behavior,* Bur. of Publications, Teachers College, Columbia Univ., New York; $0.75.

Study Your Way Through School; Understanding Yourself; many other titles, Science Research Associates; $0.60 each.

Turtox Leaflets (some 60); *Turtox News* (periodical), General Biological.

U.S. Dept. of Health, Education, and Welfare, Children's Bureau, *Your Child from Six to Twelve,* U.S. Govt. Printing Office, Washington, D.C.; $0.20.

U.S. Govt. Printing Office, *Catalog of Mental Health Pamphlets,* Washington, D.C.; $0.25.

———— , *Publications on Wildlife, Conservation, and Agriculture* (catalog), Washington, D.C.

Chapter **8**

Continuity of the Organism

One of the significant and basic conceptual schemes developed in biology concerns itself with the continuity of the organism. It may be stated this way: An organism is the product of its heredity and environment.

In this chapter we concern ourselves with reproduction as the mechanism transmitting genetic material from organism to organism in a given life span. In a later chapter (Chapter 10) we deal with continuity of organisms through evolution.

We begin with a study of Redi's classic investigations.

Classic experiments disproving spontaneous generation

Redi's experiment

Redi was among the early investigators who suspected that the familiar stories of spontaneous generation of living things were false. In 1668, he set up three jars to discover whether or not maggots and flies came from decaying meat. In season, this experiment may be duplicated (Fig. 8-1). Leave one jar open, another covered with cheesecloth (close mesh), and a third sealed with a sheet of plastic, as a substitute for Redi's parchment. Place a piece of fresh meat in each jar and set the jars out of doors overnight. You will find eggs of blow flies on the meat in the open jar and on the cheesecloth of the second jar, and no trace of flies on the jar from which the odor could not escape. In fact, any time

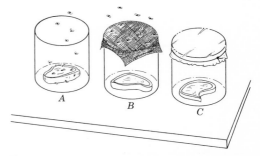

Fig. 8-1 Procedure simulating Redi's experiment. Flies have direct access to meat in jar A; jar B is covered with cheesecloth; jar C is covered with clear plastic or glass.

throughout the year you may use fruit flies and a medium of ripe banana (p. 415) in place of decaying meat. If fruit flies are used to repeat this classic experiment, cover sets of half-pint bottles containing the banana medium (some covered with cheesecloth, others with cellophane or plastic, still others open) with an open-top bell jar (Fig. 8-2). Then select a thriving stock bottle of fruit flies and invert the mouth of the bottle into the opening in the bell jar. Quickly place a stopper to close the bell jar. The flies within have the opportunity to select among the bottles. Within a week students should see evidence that eggs have been laid and larvae are burrowing in the bottles that have been kept open.

Spallanzani's experiment

For a time Redi's work seemed at least to have dispelled the idea that spontane-

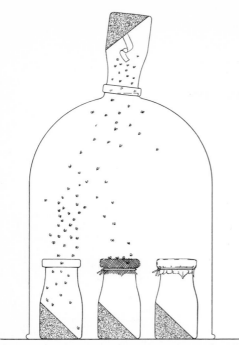

Fig. 8-2 Apparatus using fruit flies to simulate Redi's experiment. Flies are introduced to a bell jar containing bottles of banana medium; some are open, some are covered with cheesecloth, some are sealed with plastic.

ous generation occurred in large forms. Then with the invention of the microscope and the discovery of the world of microorganisms doubts again arose. Could such minute organisms have arisen from previously existing, equally tiny organisms? You may want to develop some of the case histories involving Needham and Spallanzani, Schwann, Pouchet, and Pasteur (to mention only a few who took sides in this controversy which lasted for some three centuries). Pasteur finally gave evidence of the propagation of microscopic forms from previously existing forms.[1]

Spallanzani's experiments disproving

[1] J. Conant, ed., *Pasteur's and Tyndall's Study of Spontaneous Generation,* "Harvard Case Histories in Experimental Science," 7, Harvard Univ. Press, Cambridge, Mass., 1953. Also see J. Conant, ed., *Pasteur's Study of Fermentation,* 1952, case study 6 in the series. The studies have been compiled into the two-volume set *Harvard Case Histories in Experimental Science,* reprint ed., Harvard Univ. Press, Cambridge, Mass., 1957.

spontaneous generation of microorganisms can be repeated in class. You may recall that he boiled broth and quickly sealed the flasks. To prepare similar containers use nine test tubes pulled out in the center (see Fig. 6-2a). Make the constriction about ¾ in. in length. As the tubes cool, prepare beef broth (p. 631) and fill the lower part of each tube with broth. Then arrange the tubes this way (Fig. 8-3):

Boil the broth in three tubes (*A*) in a water bath or double boiler and leave them unsealed.

Boil the broth in another three tubes (*B*). Plug the tubes with sterile cotton. When the test tubes are cool enough to handle, hold them at a 45° angle over the Bunsen burner and heat the constricted area. Pull out the top and seal each tube (see Fig. 6-2b).

Seal three tubes of broth (*C*) without any preliminary heating.

In a few days students should find that tubes *A* and *C* contain turbid broth while the broth in tubes *B* remains clear. The tops of the sealed tubes may be broken and sample drops of the broth examined under the microscope under high power. Use a sterile wire loop to transfer a drop of the broth onto a clean glass slide (see Fig. 13-11a) and mount with a coverslip.

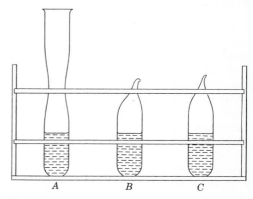

Fig. 8-3 Procedure simulating Spallanzani's experiment. The broth in tube *A* is boiled and left open to the air; broth in tube *B* is boiled and tube is sealed; broth in tube *C* is not boiled but tube is sealed.

Flame the loop after each dip into the broth. You may want to stain dry smears with methylene blue (p. 119).

Modification of Pasteur's experiment

Pasteur devised many experiments on spontaneous generation. He designed a series of flasks like the one shown in Fig. 8-4a. To simulate his experiments you may want to prepare tubes of broth with an S-shaped delivery tube (Fig. 8-4b). As the test tubes of broth are heated, water vapor will collect in the trap in the delivery tube. In this way Pasteur permitted dust-free (bacteria-free) air to enter the tubes. The heated broth should remain clear, but the controls set up as in Fig. 8-4c should become turbid. Students may examine the bacteria in the broth to confirm their observations.

Asexual reproduction in animals and plants

Fission

Many one-celled plants and animals divide into two halves. In well-fed, healthy cultures of protozoa, bacteria, and many algae, organisms undergoing fission can be found (Fig. 8-5). Techniques for preparing such cultures are described in Chapters 12 and 13.

On occasion, you may want students to study the rate of fission in some protozoans or algae. They may isolate individual protozoans under a binocular microscope with a micropipette, that is, glass tubing that has been pulled out to a fine bore (in a Bunsen flame). The students should count the number of protozoans they put into a small quantity of medium in a Syracuse dish or welled slide (see p. 442 for technique). In 24 hr they may find twice as many or possibly four times the original number. Then isolate each of these in fresh medium in other dishes or slides.

If you lack a stereoscopic microscope, place a row of small drops of the protozoa

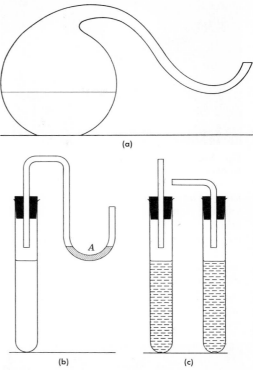

(a)

(b)   (c)

Fig. 8-4 Flasks for Pasteur's experiment: (a) one of Pasteur's flasks; (b) simple apparatus which works like Pasteur's flasks, forming at A a water trap to catch dust particles; (c) control tubes to be used with tube in b.

Fig. 8-5 Fission in *Paramecium*. (Courtesy of Carolina Biological Supply Co.)

culture on a slide, and examine each drop. In a random sampling you should find a drop with only a few animals. This drop may be transferred to a drop of culture medium. After a week or so the number of organisms will have increased greatly.

You may want to have students study the rate of division among several organ-

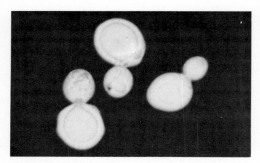

Fig. 8-6 Budding in yeast. (Photo: Dr. Carl Lindegren, Southern Illinois Univ.)

isms. They may compare the effects of varying temperatures or pH values, or adding vitamins, hormones, or antibiotics to the medium (see Chapters 3 and 11).

Use a solution of methyl cellulose or gum tragacanth (p. 112) for slowing down the protozoa. Vital stains (for temporary mounts) and staining techniques (for permanent mounts) are described in Chapter 2.

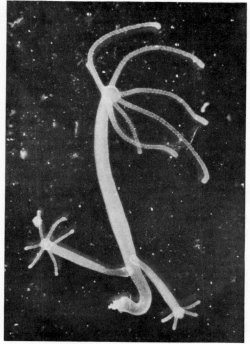

Fig. 8-7 Budding in *Hydra*. (Photo: A. M. Winchester.)

Budding

Prepare yeast cultures as described in Chapter 13. You should find many examples of small buds protruding from the parent plants (Fig. 8-6).

Use a vital stain (such as neutral red) to make the cells easily visible. Add a drop of a 1 percent solution of neutral red to one side of the coverslip of a mounted specimen. Remove the fluid in which the yeast cells were cultured by applying a piece of filter paper to the opposite side of the coverslip. Note how the volutin vacuoles and the granules of the living cells take up the neutral red stain. Dead cells, on the other hand, stain a uniform red color.

Examine also a culture of *Hydra* and look for budding: small replicas of *Hydra* may be found growing out of the stalk (Fig. 8-7). Culture methods are described in Chapter 12; see also Chapter 1.

Spore formation

Spore formation may be demonstrated readily, using molds such as bread mold (*Rhizopus,* Fig. 13-14b), *Penicillium* (Fig. 2-14), *Aspergillus* (Fig. 13-13), or *Neurospora* (Fig. 8-44). Under unfavorable conditions yeasts also may form spores, but these are difficult to show. While sporulation is found to some extent among protozoa (especially the malarial *Plasmodium*), examples are difficult to demonstrate unless prepared slides are used.

Prepare cultures of molds as described in Chapter 13. Mount the fruiting bodies (sporangia) of the molds in a drop of glycerin or alcohol to prevent the formation of air bubbles in the mounted specimens.

You may also want to study spore formation in forms such as the mildews or even in yeasts and mushrooms (p. 105), or alternation of generations in mosses or ferns (Figs. 8-45, 8-46). Students can observe dispersal of fern spores. Macerate sori (clusters of sporangia) on a dry slide without a coverslip. By the time students have focused under low power, the sporangia will be ejecting spores.

You may want to simulate the mechanical dissemination of spores from a mold spore case. Almost fill a round black balloon with balls of cotton no larger than ¼ in. in diameter.[2] Then put the neck of the balloon over one end of a 3-ft length of ½-in. glass tubing with fire-polished ends. Inflate the balloon by blowing through the open end of the glass tubing. Plug the open end with modeling clay so that the air will not escape. Sink this end into a mass of modeling clay so that the upright "sporangium and sporangiophore" of a bread mold are simulated. The mass of clay may be considered the substrate. A network of glass tubes may be partially embedded in the clay to represent mycelia.

Burst the "sporangium" by dropping one or two drops of xylol on the balloon, or casually stroke the balloon after wetting your fingers in xylol. Students can watch the bursting and the release of the cotton "spores," and notice that the cotton balls are expelled to fairly great distances around the room.

(For examples of vegetative propagation, see Chapter 9.)

Sexual reproduction in animals and plants

Historic papers that were bench marks in the twentieth-century development of genetics are available as reprints in J. Peters, ed., *Classic Papers in Genetics* (1959), and M. Gabriel and S. Fogel, eds., *Great Experiments in Biology* (1955), both from Prentice-Hall.

After reading Gregor Mendel's paper (in Peters),[3] students will want to read Van Beneden, "Researches on the Maturation of the Egg and Fertilization"; Wilson, "The Chromosomes in Relation to the Determination of Sex in Insects"; and Sutton, "The Chromosomes in Heredity"

(all in the paperback edited by Gabriel and Fogel). Sutton's paper and papers by Morgan ("Sex-limited Inheritance in *Drosophila*") and Bateson and Punnett ("Experimental Studies in the Physiology of Heredity") are in Peters' compilation.[4]

These are the basic studies that introduced problems concerning our understanding of reduction division (meiotic division) and our understanding of the concept of individuality of chromosomes.

Evolution of sexuality

Basically, algae and higher plants are considered to have evolved from the flagellates. *Chlamydomonas,* a primitive alga that is intermediate between flagellates and many of the higher green algae, holds a pivotal position. In many studies of variations in sexual patterns, *Chlamydomonas* (Figs. 2-12b, 8-73) and *Ulva* (Fig. 8-8) are experimental forms. (One may also study sexuality in bacteria and yeasts.[5])

Among *Chlamydomonas* there are two physiologically different types, plus and minus mating types (isogametes). Meiosis occurs in the zygote, forming a tetrad, then four haploid cells—two plus and two minus mating types (Fig. 8-73).

Much work with alternation of generations has been done with the marine green alga *Ulva,*[6] the sexual plants are haploid (Fig. 8-8). *Oedogonium* (Fig. 8-42) is a widely studied green alga that shows highly developed sexual reproduction; it has small, motile sperm cells and large, nonmotile eggs (heterogametes).

Differentiation of sexes seems to have arisen independently in several unrelated groups of algae.

Among filamentous algae, *Spirogyra* (Fig. 8-39) has been a classic for study of conjugation between active and passive cells; these are isogametes, morphologically similar.

---

[2] P. F. Brandwein *et al.,* "Laboratory Techniques in Biology," mimeo, 1940.

[3] Also in the appendix of E. Sinnott, L. C. Dunn, and T. Dobzhansky, *Principles of Genetics,* 5th ed., McGraw-Hill, New York, 1958.

[4] Also refer to *Biological Science: An Inquiry into Life* (BSCS Yellow Version), Harcourt, Brace & World, New York, 1963, pp. 532–35.

[5] Also, the life cycle of yeast, *ibid.,* p. 240.

[6] *Ibid.,* pp. 254–55.

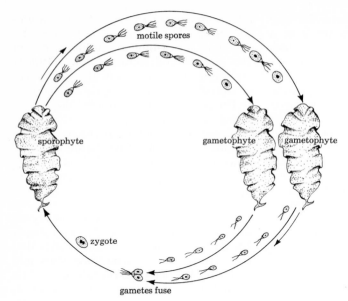

motile spores

sporophyte

gametophyte

gametophyte

zygote

gametes fuse

Fig. 8-8 Alternation of generations in a marine alga, *Ulva*. (After *Biological Science: An Inquiry into Life* [BSCS Yellow Version], Harcourt, Brace & World, 1963.)

Acetabularia (Fig. 8-76), a beautiful, branching green alga that has, when fully differentiated, an umbrella-like top composed of a whorl of sporangia (elongated cells), has been the subject of some elegant investigations into polarity and differentiation. Aplanospores are formed within sporangia; after a rest period, these produce within them a number of biflagellated isogametes, which in turn fuse to form zygospores that grow into new branched plants with umbrella tops. Some students may want to read of J. S. Hämmerling's grafting experiments showing the role of the nucleus in differentiation.[7] (See also p. 404.)

In studies of the mating of protozoa, mating types of *Paramecium aurelia* and of the green form, *P. bursaria,* are available from biological supply houses (Appendix C). *Paramecium bursaria* (Fig. 8-14) is green due to the presence of symbiotic algae—*Chlorella,* or *Zoochlorella.*

These paramecia conjugate more often when they are raised on dried lettuce leaf

powder (added to distilled water and then autoclaved).

Mix together cultures of *Paramecium bursaria* of two mating types. Initially, several very large clumps form that slowly separate into paired conjugating forms (in 5 to 6 hr; see Figs. 8-13, 8-14).

Mating types among other protozoa have also been studied: for instance, several other species of *Paramecium; Euplotes;* and *Tetrahymena pyriformis,* which seems to have some 40 mating types.[8]

For a study of evolution of sexuality among fungi, many teachers use *Neurospora* (Fig. 8-70) or bread mold, and then continue with a study of the complex cycles of the rusts (involving several hosts). Alternation of generations and the reduction of the gametophyte accompanied by the appearance of a larger, more conspicuous sporophyte generation are found among

[7] J. S. Hämmerling, "Nucleo-cytoplasmic Relationships in the Development of *Acetabularia,*" *Internat. Rev. Cytol.* 2:475–98, 1953.

[8] A. M. Elliott, "A Quarter Century Exploring *Tetrahymena,*" *J. Protozool.* 6:1–7, 1959. See also J. R. Preer, "Genetics of the Protozoa," *Ann. Rev. Microbiol.* 11:419–38, 1957, and R. A. Sager, "Genetic Systems in *Chlamydomonas,*" *Science* 132:1459–65, 1960.

Also look into a general text in protozoology, such as R. Manwell, *Introduction to Protozoology,* St. Martin's, New York, 1961.

mosses (Bryophyta) and the ferns and seed plants (Tracheophyta).

Meiosis

Students will first need to understand the concept that among the higher organisms an egg and a sperm each contains half of the DNA content of the species. It follows that, at fertilization, the normal number of chromosomes for the species, or the quantity and kind of DNA, is restored. This genetic code is now available to the fertilized egg cell for regulating differentiation and the potentialities of an organism in a specific environment. This may be the right time in class to raise queries and to describe experiments concerning the role of hormones in regulating production of different kinds of RNA, and thereby effecting differentiation in the cytoplasm. Students may want to read E. Davidson, "Hormones and Genes" (*Sci. Am.* June 1965).

In meiosis there are two successive divisions accompanied by only one division of chromosomes. The stages in meiosis share the same names as those in mitosis, although the long prophase in meiosis is divided into five specific substages: leptotene, zygotene, pachytene, diplotene, and diakinesis. During the prophase there is a pairing of homologous chromosomes, with an interchange of hereditary material. The chromosomes have already duplicated themselves so that the doubled chromosomes lying in close pairs (synapsis) look double, or bivalent. In the first metaphase (of the first meiotic division), the nuclear membrane disappears and a spindle can be seen. At this step different pairs of homologous chromosomes align themselves independently at the equator. In a subsequent stage in meiosis, the first anaphase, homologous chromosomes separate. Then follow the first telophase and the interphase, or interkinesis, and the second metaphase. At the second anaphase the centromeres divide so that each chromatid may be considered as a separate chromosome. As a result of the two

meiotic divisions four cells, each containing the haploid number of chromosomes, are formed. Diagrams such as those in Fig. 8-9 may clarify the movement of chromosomes, so that students may be able to follow the stages in prepared slides such as those of a section through the testis of an insect (Fig. 8-10).

Time of meiosis. In the life cycle of different organisms the time of meiosis may differ, although the pattern is constant for a species. Among the many-celled animals, meiosis occurs during gametogenesis (as in the generalized schema in Figs. 8-9, 8-11, 8-12). This pattern is also found in some of the lower plants, such as the brown alga *Fucus,* and in certain protozoa. Some of the lower plants have retained a more primitive pattern: meiosis occurs in the zygote, immediately after fertilization, at the first cleavage stage, resulting in an adult stage that is haploid. Most green algae, for example, are haploid in the body cells and in gametes since meiosis occurs in the first two divisions of the fertilized egg (see *Chlamydomonas,* Fig. 8-73). Among the red algae, the nonsexual plants are diploid, arising from carpospores (fertilized eggs). These diploid plants have a meiotic division in the formation of sexual spores, called tetraspores, giving rise to the subsequent haploid sexual stage.

The higher plants that have an alternation of generations have a complete stage or generation occurring between meiosis and the production of gametes. In the diploid sporophyte meiosis occurs forming spores that give rise to a haploid gametophyte generation (Figs. 8-45, 8-46, 8-53). This gametophyte generation is greatly reduced among the seed plants so that it is not independent but is confined within the sporophyte. You will remember that the anther and ovule are sporangia containing microspores and a megaspore, respectively. The megaspore develops into an embryo sac, the equivalent of the gametophyte generation, and the sac contains the egg cell. The microspores, or pollen grains, give rise to two nonmotile gametes.

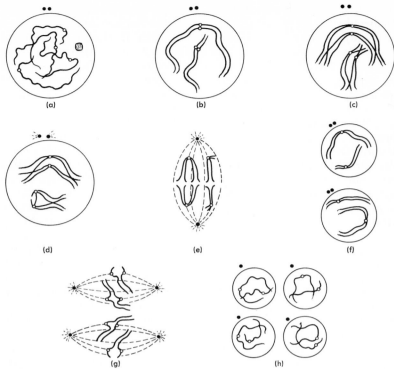

Fig. 8-9 Diagrammatic representation of stages in meiosis illustrated by two pairs of chromosomes: (a) interphase; (b) pachytene, or zygotene (homologous threads come together in synapsis, forming bivalents); (c) diplotene (each chromosome has duplicated itself so that four homologous chromatids form a tetrad); (d) diakinesis; (e) anaphase I; (f) interphase; (g) anaphase II (centromeres divide so that each chromatid is now a separate chromosome); (h) the four haploid products. (After E. Sinnott, L. C. Dunn, and T. Dobzhansky, *Principles of Genetics*, 5th ed., McGraw-Hill, 1958.)

Also see inheritance in the fungus *Neurospora* (Figs. 8-44, 8-70, 8-71, 8-72), inheritance in *Chlamydomonas* (Figs. 8-74, 8-75), and sexual reproduction of this alga (Fig. 8-73). Stages in mitotic divisions in cells are given in Chapter 9.

Chiasmata. The homologous chromatids break in one or several places and exchange parts so that the new chromatids that are formed consist of sections of the two chromosomes that were in synapsis. This results in chromosomes combining both maternal and paternal parts of the original chromosomes and provides the opportunity for diversity in heredity among offspring. This exchange occurs at some time between the zygotene and diplotene stages. These cross-shaped ar-

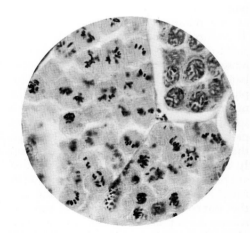

Fig. 8-10 Prepared section through testis of a grasshopper, showing spermatogenesis. (Courtesy of General Biological Supply House, Inc., Chicago.)

rangements can be studied on prepared slides of gametogenesis. There may be one or several crossover regions. As a result, in the first anaphase when homologous maternal and paternal chromosomes separate they are different from the original chromosomes that went into synapsis. Two of the chromatids of the homologous pair have remained unchanged while two of the chromatids exchanged sections. Finally, the two centromeres that unite the chromosomes (there is only one in mitosis) separate and move to opposite poles of the cell—maternal centromere to one pole, and paternal centromere to the other. This is the end of the first anaphase. To recapitulate, in the second metaphase the chromosomes are arranged on the equatorial plate; the centromeres divide, with the result that each chromatid can now be considered a chromosome (see last stages in Figs. 8-9, 8-11, 8-12).

Modeling chromosomes. While bright students do not need to manipulate models of chromosomes to understand the difference between mitosis and meiosis, students who lack a facility with abstract concepts will find models helpful. Give each student some clay of various colors, with enough of each color to make pairs of "chromosomes." Then on a sheet of paper have students draw circles to represent stages, or cells. Ask them to place the right number and kind of chromosomes in each kind of cell as it undergoes mitosis, and as it undergoes two divisions in meiosis. This will help clarify the difference between the two processes.

Some teachers use Poppits (beads that are easily detached from and reattached to one another), or they cut plastic sponges into strips. Pipe cleaners, wire, wooden beads, and imagination produce excellent results. (Also refer to the double helix concept of DNA molecules, p. 428.)

Be sure to have students show how chromatids are held attached by means of a centromere. Demonstrate synapses and crossover of chromatids (use Figs. 8-11 and 8-12 as models). Compare this process with mitosis (Chapter 9).

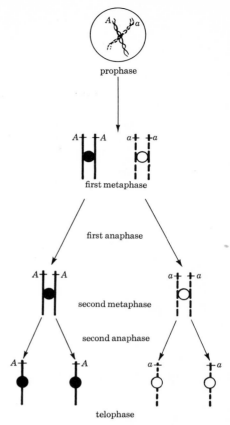

Fig. 8-11 Meiosis: the segregation of genes of a pair of homologous chromosomes resulting in a 1:1 ratio of meiotic products. (After R. Levine, Genetics, Holt, Rinehart and Winston, 1962.)

A listing of the number of chromosomes in some plants and animals is given in Table 8-1.

Sexual reproduction in animals

We shall first investigate reproduction among Protozoa, and then go on to other invertebrate forms and the vertebrates.

Conjugation in protozoa. The sexual fusion of two cells or two organisms which superficially are structurally alike is called conjugation. Fertilization, on the other hand, is thought of as the union of two cells which are structurally different, that is, an egg and a sperm cell. (However, in strict

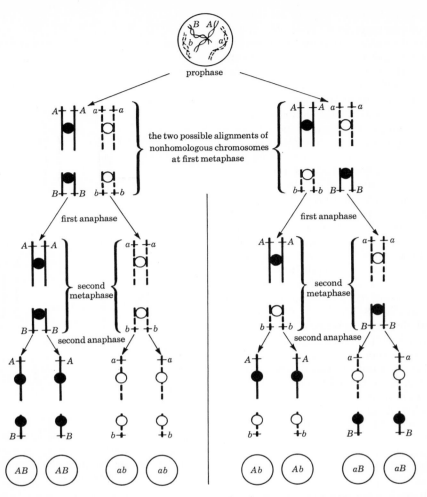

Fig. 8-12 Independent assortment of two pairs of nonhomologous chromosomes at meiosis. During metaphase I of meiosis any two pairs of nonhomologous chromosomes are independent of each other when they align at the equator of the cell; further, each pair of chromosomes moves independently through meiosis forming gene combinations AB, Ab, aB, and ab in a 1:1:1:1 ratio. (After R. Levine, *Genetics*, Holt, Rinehart and Winston, 1962.)

accuracy, fertilization is the union of sperm and egg nuclei.)

The fact that there are complicated chemical differences among *Paramecium*, for example, has been described by Sonneborn and many other workers (Fig. 8-91). You may want to purchase the different mating types to show conjugation. Or in a well-fed, old, overcrowded culture you may find cases of this union of two protozoans (Figs. 8-13, 8-14, 8-91).

Selected references for reading more about the heredity of mating types are listed at the end of this chapter (see recent books on heredity). Spector[9] gives a summary of mating types. Also see p. 424 here.

Sexual differentiation in Hydra. *Hydra* is a choice organism for study of budding (Fig. 8-7), regeneration (p. 480), and differentiation of gonads at certain times in

[9] W. Spector, ed., *Handbook of Biological Data*, Saunders, Philadelphia, 1956, pp. 98–99.

TABLE 8-1

Number of chromosomes in some plants and animals*

Apple	34, 51	Housefly	12
Artemia salina	42	Hydra	32
Ascaris	4	Liverwort (*Riccia*)	8
Ash (American)	46	Man	46
Barley	14	Monkey (*Rhesus*)	48
Begonia carminata	42	Mosquito	6
Bullfrog	26	Moss (*Bryum*)	20
Cabbage	18	Mouse	40
Cat	38	*Neurospora*	14
Chameleon	34	Oak (scarlet)	24
Cherry	32	Onion	16
Chicken	78	Pear	34, 51, 68
Corn	20	Pig	40
Crab grass	36	Pine (yellow)	24
Daphnia pulex	20	Platyfish	48
Dog	78	Plum	48
Fern (*Pteris*)	64	Potato	48
Fruit fly		Rabbit	44
Drosophila		Rat	42
melanogaster	8	Rice	24
virilis	12	Sheep	54
pseudoobscura	10	Shrimp (*Eupagurus*)	254
Garden pea	14	*Spirogyra*	24
Goldfish	94	Strawberry	56
Guinea pig	64	Tobacco	48
Horse	66	Tomato	24

* Data from W. Spector, ed., *Handbook of Biological Data,* Saunders, Philadelphia, 1956; C. D. Darlington and A. P. Wylie, *Chromosome Atlas of Flowering Plants,* 2nd ed., Macmillan, New York, 1956; and E. Sinnott, L. C. Dunn, and T. Dobzhansky, *Principles of Genetics,* 5th ed., McGraw-Hill, New York, 1958.

a laboratory medium (Fig. 8-15). Students may devise experiments to show the effect of starvation, overfeeding, and accumulation of carbon dioxide on differentiation of ovaries and testes in these organisms. The brown hydra *Hydra littoralis* is dioecious.

Perhaps the paper on sex gas in *Hydra* in *Scientific American* (April 1959) will be useful; also refer to Chapters 1, 2, and 12 for further work on hydras. H. M. Lenhoff[10] suggests staining male hydras with Feulgen reagent since this is a specific stain for DNA. (Also see Chapter 2 for descriptions of the use of Feulgen stain.)

Eggs and sperms of snails. If the aquatic snail *Physa* is kept in an aquarium tank, material will be available for the study of

[10] H. M. Lenhoff, "Laboratory Experiments Using Hydra," *Carolina Tips* 23:8, October 1960.

eggs and sperms. Dissect out the multiple-lobed ovotestis located in the uppermost portion of the spiral of a fair-sized snail and macerate it a bit in a drop of aquarium water on a clean glass slide. (Also see dissections, Chapter 1.)

Sperms as well as the spheroid eggs are visible under the high-power objective if the material is at the right stage. Students can also segregate fertilized eggs found on leaves of aquarium plants or the aquarium glass into finger bowls and watch the early stages of cleavage under a microscope or with a hand lens.

Prepared slides of cleavage stages in mollusks or echinoderms (Fig. 8-16) may be studied at this time.

Fertilization in echinoderms. When sea urchins or starfish are available, students

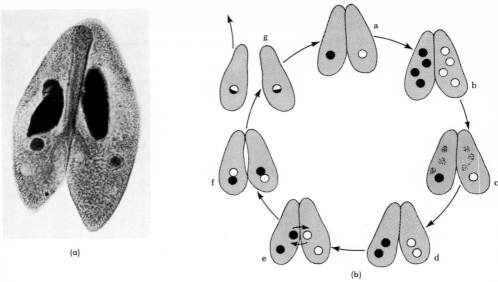

(a)

(b)

Fig. 8-13 Reproduction in *Paramecium*: (a) stained preparation of conjugation in *Paramecium cauda-tum*, showing a small micronucleus and a large macronucleus; (b) diagrammatic flow chart of conjuga-tion in paramecia: In a and b, the single micronucleus in each conjugant divides twice, producing four haplonuclei, of which three disintegrate, c; in d, each remaining micronucleus divides in two; in e, the cells exchange one of their micronuclei; in f, the micronuclei in each individual fuse; and g shows the two exconjugants. (a, courtesy of Carolina Biological Supply Co.; b, after E. Sinnott, L. C. Dunn, and T. Dobzhansky, *Principles of Genetics*, 5th ed.; copyright, 1958, McGraw-Hill Book Co. Used by permission.)

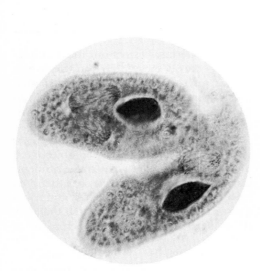

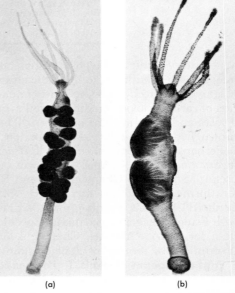

(a)   (b)

Fig. 8-14 Conjugation in *Paramecium bursaria*, stained to show details of spindles and micro-nuclear activity. (Courtesy of General Biological Supply House, Inc., Chicago.)

Fig. 8-15 *Hydra*: (a) with spermaries; (b) show-ing amoeboid ova. (Courtesy of Ward's Natural Science Establishment, Inc., Rochester, N.Y.)

may have the opportunity to study fertilization under the microscope.[11] Clean glassware and dissecting needles are needed. Collect unfertilized eggs by stripping the females and placing the eggs in a finger bowl of sea water. Dissect out testes from males and store them in another finger bowl with a small amount of sea water (see anatomy, Chapter 1). Put a few eggs on a slide with sea water. Mount pieces of straw or broken coverslips with them so that they are not crushed by the coverslip. Locate some eggs under low

[11] G. Wald *et al.*, in *Twenty-six Afternoons of Biology* (Addison-Wesley, Reading, Mass., 1962, Chapter 23), suggest passing a weak electric current through sea urchins or injecting a small quantity of potassium chloride solution. Refer to this activity for a possible laboratory lesson on cleavage stages in sea urchins.

power. Then with a clean dissecting needle or a micropipette, pick up some sperm suspension (from the finger bowl containing testes) and place this at the edge of the coverslip. As the sperms swim under the coverslip, move the slide around to locate a sperm drawn into the fertilization cone formed by the egg. Watch how the other sperms are pushed aside by the fertilization membrane which arises from the egg's surface after one sperm has entered (Fig. 8-17). Place a coverslip (with petroleum jelly applied) on the slide to prevent drying, and watch early cleavage stages (Fig. 8-18).

Films that show an egg fertilized by a sperm cell are also available (see film listing at the end of this chapter).

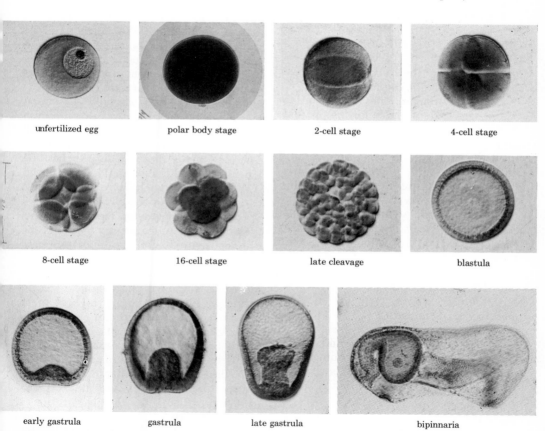

| unfertilized egg | polar body stage | 2-cell stage | 4-cell stage |

| 8-cell stage | 16-cell stage | late cleavage | blastula |

| early gastrula | gastrula | late gastrula | bipinnaria |

Fig. 8-16 Prepared slides of early cleavage stages of starfish egg (magnified 165×). (Courtesy of General Biological Supply House, Inc., Chicago.)

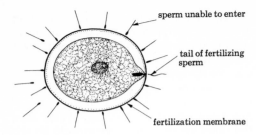

sperm unable to enter

tail of fertilizing sperm

fertilization membrane

Fig. 8-17 Diagram of sperm entry. A fertilization membrane raised around the egg immediately after one sperm comes in contact with the surface of the ovum prevents entry of other sperms.

Some students will want to read about photoperiodism in invertebrates, as well as vertebrates and plants. Refer to R. Withrow.[12]

Eggs of *Daphnia*. The summer parthenogenetic eggs or the winter fertilized eggs may be found in the brood pouch of *Daphnia* (Fig. 1-26a) if you examine the organism under the microscope as a wet mount (culturing, Chapter 12).

Eggs of *Artemia*, the brine shrimp. These eggs remain viable for long periods in the dry state. Students may get them at aquarium shops. The eggs may be developed as follows. In a wide, flat pan add 4 tbsp of rock salt to a gallon of tap water. Introduce ½ tsp of dried brine shrimp eggs. If the eggs are put in a fish feeding ring they will not scatter. Keep the pan covered, but leave an opening large enough for light to enter at one end. Larvae are attracted to light, and as they hatch they accumulate at this end of the pan. Then you can transfer them to a fresh salt solution. (Also refer to methods described in Chapter 12.)

The developing forms may be observed in depression slides under the microscope (Fig. 8-19). If the mounts are rimmed with petroleum jelly the slides may be used for hours; placing them in a moist chamber

will preserve them even longer. You may want to arrange a simple chamber by placing slides across Syracuse dishes containing water; then cover all these with a small bell jar, or place the dishes in a covered plastic box.

Metamorphosis of *Drosophila*. Culture fruit flies as described on pp. 415–16. When flies are cultured in many small vials of medium they will lay eggs which students may examine closely with a hand lens or microscope. In fact, students may streak banana mash (or other medium) in depression slides and insert them into bottles of fruit flies. Then some flies will lay eggs on the surface of these slides. Observe that the eggs are white, with two filaments attached to one end (Fig. 8-20). After examination, return the slides to the breeding bottles. In time, pupa formation may be observed on these slides and on the walls of the culture bottles. If medium is spread on plastic spoons instead of depression slides, transfer and handling of early stages of fruit flies are facilitated.

Egg masses of tent caterpillar. In winter, you may find egg masses in shiny, almost metallic, black clusters circling small branches of trees, especially wild cherry. Enclose some of these small branches in a glass jar with a fine screen or mesh cover. The eggs will hatch into larvae; these caterpillars build a tent (Fig. 8-21).

Egg masses of praying mantis. Students may find the tan, foamy egg masses of the praying mantis. Or you may purchase them most readily from botanical gardens. When these masses are put into a terrarium they will hatch out in early spring. Study the nymph stage of these insects which undergo an incomplete metamorphosis, that is, egg, nymph, and adult. Students may compare this with complete metamorphosis in such insects as fruit flies, tent caterpillars (see above), beetles, and butterflies and moths. (Also see Chapter 12.)

Cocoons and chrysalises. Examine pupal stages of moths and butterflies by cutting open the protective coverings. Commonly found forms include the bag-

[12] R. Withrow, ed., *Photoperiodism and Related Phenomena in Plants and Animals,* Am. Assoc. for the Advancement of Science, Washington, D.C., 1959.

Also read A. Giese, "Annual Reproductive Cycles of Marine Invertebrates," *Am. Rev. Physiol.* 21:547–76, 1959.

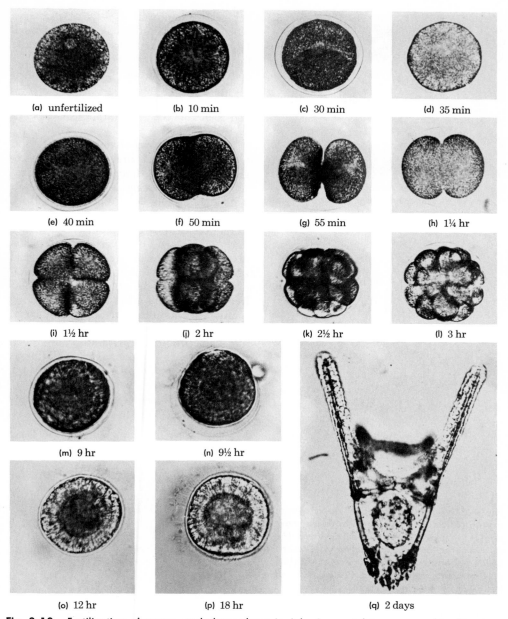

(a) unfertilized   (b) 10 min   (c) 30 min   (d) 35 min

(e) 40 min   (f) 50 min   (g) 55 min   (h) 1¼ hr

(i) 1½ hr   (j) 2 hr   (k) 2½ hr   (l) 3 hr

(m) 9 hr   (n) 9½ hr

(o) 12 hr   (p) 18 hr   (q) 2 days

Fig. 8-18   Fertilization, cleavage, and pluteus larva in *Arbacia punctulata,* a sea urchin. (Courtesy of Dr. Ethel Browne Harvey.)

worm (the small, spindle-shaped form that is covered with twig fragments and dangles on a stalk, Fig. 8-22). Other common forms are pupae of *Cecropia, Polyphemus, Vanessa,* and *Promethea.* Let other pupae remain at room temperature in a terrarium with a bit of moisture. Watch them hatch into adults. Students can study or make models of the life cycles of the silkworm. Or if you lack pupae of butterflies or moths, students may want to study the honey bee (Fig. 8-23).

Fig. 8-19 Larva of *Artemia*, the brine shrimp. (Courtesy of General Biological Supply House, Inc., Chicago.)

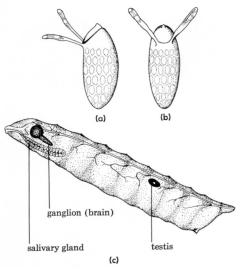

(a)   (b)

ganglion (brain)

salivary gland   testis

(c)

Fig. 8-20 Eggs and larva of *Drosophila melanogaster*: (a) side view of egg; (b) dorsal view of egg; (c) lateral view of third-instar larva. (After M. Demerec and B. P. Kaufmann, *Drosophila Guide: Introduction to the Genetics and Cytology of Drosophila melanogaster*, 7th ed., Carnegie Institution of Washington, 1961.)

Earthworms. Directions for dissection are given in Chapter 1. Holmes and Smith[13] suggest using mature earthworms as a ready source for live sperm cells and ova. Inspect the dorsal region of the worm to locate the swollen, prominent seminal

[13] De F. Holmes and J. Smith, "Earthworms: A Source of Living Gametes," *Turtox News* 40:8, August 1962.

vesicles (segments 9 to 12); also observe whether the clitellum is fully formed. Draw fluid with a hypodermic needle from the seminal vesicle and place in cold Ringer's solution; view this fluid in dimmed light under high power. To obtain ova, make an incision into the segments of the body containing ovaries (13 to 14).

Eggs and patterns of cleavage. Students will find excellent diagrams and photographs, and lucid descriptions of fertilization and types of eggs in insects, mollusks, echinoderms, and vertebrates in B. I. Balinsky, *An Introduction to Embryology* (Saunders, 1960). The patterns of cleavage and "fate maps" of invertebrates, *Amphioxus*, fish, and birds are given; also read about the role of organizers in the developing embryos—the classic work of Holtfreter, Mangold, and Spemann, among many. Techniques for handling eggs, artificial fertilization, and parthenogenesis, and for inducing ovulation are well described in R. Rugh, *Experimental Embryology: Techniques and Procedures*, 3rd ed. (Burgess, 1962).

Introductory descriptions of these processes are given (Figs. 8-16, 8-18, 8-26, 8-31, 9-22).

Fig. 8-21 Nest of larvae of the eastern tent caterpillar. (Courtesy of U.S. Dept. of Agriculture.

Fig. 8-22 Bags, adult, and pupa of *Thyrodopteryx efemoriformis*. The pupal stage is known as the bagworm. (Courtesy of U.S. Dept. of Agriculture.)

Amphioxus. Prepared slides of successive series of cleavage stages in the small protochordate *Amphioxus* are available for students to study under the microscope.

Ovaries and testes of fish. During the spring, smelts or perch provide good specimens for dissection in the classroom.[14] Dissect the fish from the anal region to-

[14] Refer to the effect of photoperiods on reproduction in fish in Withrow, *op. cit.*

Also refer to W. S. Bullough, *Vertebrate Reproductive Cycles*, 2nd ed., Wiley, New York, 1961, and P. Korringa, "Relations Between Moon and Periodicity in the Breeding of Marine Animals," *Ecol. Monographs* 17:347–48, 1947.

ward the head (dissection, Fig. 1-49). Remove the body flap to reveal the ovaries or the testes. Trace the tubes (oviducts or sperm ducts) which carry the eggs or sperms to the cloacal region. Use a hand lens to examine the eggs in the roe or ovaries. Caviar illustrates variations in the size of fish eggs; cod roe and salmon eggs are available as canned food. Also examine the eggs of some tropical fish.

Development of fresh-water medaka (Oryzias latipes). When these interesting fish are given the right amount of food and light, they lay eggs daily, shortly before

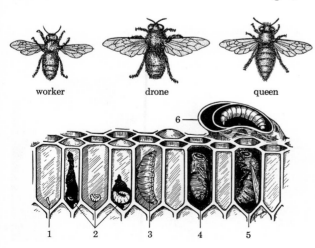

worker   drone   queen

Fig. 8-23 Life cycle of honey bee: (1) egg; (2) three stages of larva; (3) fully developed larva; (4) pupa; (5) adult queen ready to emerge; (6) queen larva in queen cell. (Courtesy of General Biological Supply House, Inc., Chicago.)

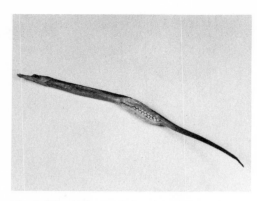

Fig. 8-24 Male pipefish with eggs (or young) in brood pouch. (Courtesy of Carolina Biological Supply Co.)

freshly killed or preserved frogs (dissection, Fig. 1-50). In season, the two many-lobed ovaries (covered by a membrane) contain large black and cream-colored eggs and occupy the greater part of the female's abdominal cavity. Remove the ovaries, locate the many-coiled oviducts, one on each side of the abdomen, and trace them forward to their origin, under the heart region, high in the chest cavity. Also follow the oviducts along to the cloacal region. Each egg breaks out of the ovaries, is pushed along the abdominal cavity by cilia, and is brought to the opening of the oviducts. In passage along the oviducts, each egg receives three layers of jelly; the jelly swells when it comes in contact with water.

dawn. Then the eggs are fertilized externally and remain attached to the female for several hours. Students may isolate the eggs in small Syracuse dishes with aquarium water and trace their development through early cleavage stages. They hatch in 6 days. The first cleavage stage occurs 1 hr after fertilization, the second 1½ hr later, and the third 2 hr later.

Keep several fish in a large tank. Feed them, on alternate days, such worms as *Tubifex* or *Enchytraeus,* or mixed dry food.

Varieties of fish may be studied: the pipefish (Fig. 8-24) and the sea horse, to show the breeding pouch of the male; the skate, to show the peculiar egg case (Fig. 8-25). Students may also trace the embryology of the goldfish (Fig. 8-26).

Ovaries and testes of frogs. Dissect

In the male frog, locate two yellow testes lying ventrally on the kidneys (Fig. 1-52a). The thread-like sperm ducts may be traced to the cloaca or followed back to their connection with the kidneys. Notice the finger-like, yellow fat bodies found in the vicinity of the ovaries and testes. Their function is thought to be food storage during hibernation and during the active reproductive period.

When fresh material is dissected, the testes may be removed and crushed in water on a slide, to show living sperm cells.

Sperm cells of frogs. Place the testes of a freshly killed frog into a clean Syracuse dish containing some 10 ml of Ringer's solution (p. 665) or spring water. Tease the organs apart with clean forceps.

Fig. 8-25 Embryo and yolk sac of a skate in the egg case. (Mr. Franklin C. Daiber, University of Delaware.)

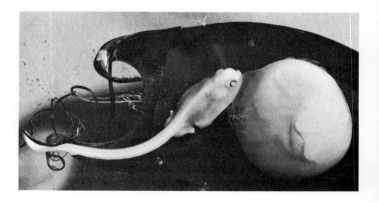

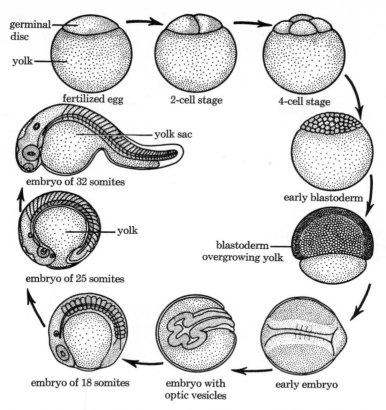

Fig. 8-26 Stages in the early embryology of a bony fish, the goldfish. (Reprinted with permission of the publisher from *College Zoology* by R. W. Hegner and K. A. Stiles, 7th ed., © The Macmillan Co., New York, 1959.)

Mount a drop of this suspension on a clean glass slide and apply a coverslip.

At first the sperm cells do not move, for the concentration and accumulation of carbon dioxide is high. In a few minutes, however, the oscillating heads of sperms may be observed under high power. A drop of methylene blue (p. 121) added to the slide will stain the flagellum of each sperm cell, but the cells will be killed by the stain.

Egg cells of frogs. Dissect freshly killed frogs to show the large masses of eggs. Examine the black and cream-colored eggs with a hand lens. Or collect egg masses in the spring from lakes and ponds.

When live eggs have been fertilized, they rotate within the layers of jelly so that all show the black area uppermost. When the cream-colored part is still uppermost, the eggs are dead or have not been fertilized.

If possible, follow the cleavage stages of living cells with a hand lens or binocular microscope. Compare with the normal stages in the development of a frog's egg (Figs. 8-27, 8-28).

If you have collected frogs you can stimulate ovulation. This and artificial fertilization are described later in this chapter.

EFFECT OF TEMPERATURE ON THE DEVELOPMENT OF FROGS' EGGS. Before students study the rate of cell division or cleavage, and the rate of differentiation in frogs' eggs at different temperatures, the rate of

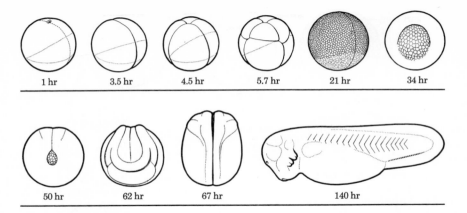

Fig. 8-27 Diagram of development of a frog's egg (*Rana pipiens*); time in hours when temperature is 18° C (64° F). (Drawings by Dr. J. F. Mueller, from Ward's Natural Science Establishment, Inc., Rochester, N.Y.)

cleavage at some constant temperature should be established. (Cell division, or mitotic division, is described in Chapter 9.) Keep one finger bowl of fertilized eggs in Ringer's solution at 15° C (59° F) as a control. Set up similar finger bowls and keep at 10° C, 20° C, and 30° C (50° F, 68° F, and 86° F). Then examine the eggs at intervals. Students are likely to find that gastrulation begins in approximately 27 hr at 15° C, in about 72 hr at 10° C (compare with Figs. 8-27 and 8-28).

Procedures for laboratory work are well developed by Rugh.[15]

[15] R. Rugh, *Experimental Embryology: Techniques and Procedures,* 3rd ed., Burgess, Minneapolis, 1962, Chapter 11, "Temperature and Embryonic Development."
Also refer to F. Moog, *Animal Growth and Development* (BSCS laboratory block), Heath, Boston, 1963. Many activities are also provided in the *Student Laboratory Guide* to *Biological Science: An Inquiry into Life* (BSCS Yellow Version), Harcourt, Brace & World, New York, 1963.

Induced ovulation in *Rana pipiens*. With most animals the eggs and cleavage stages can be studied only at a specific time of the year. However, certain amphibians can be stimulated to ovulate out of season when they are injected with the material from the anterior lobe of the pituitary glands.

In the late summer and fall, large doses of pituitary glands are needed, but from October into spring the dose decreases from eight glands down to just one. There is no qualitative difference between the glands in the two sexes, but the female's is larger and probably contains more of the follicle-stimulating hormone. Many teachers prefer to buy pituitary extract, which is readily available in drugstores (it need not come from a frog).

First, dissect out the required number of pituitary glands; then crush them in Ringer's solution (p. 665) and inject this

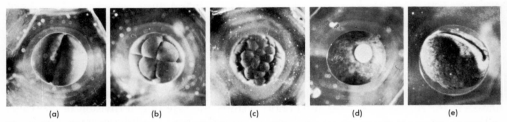

Fig. 8-28 Photographs of stages in development of frogs' eggs: (a) 2-cell; (b) 8-cell; (c) 32-cell; (d) late yolk plug; (e) early neural groove. (Courtesy of Carolina Biological Supply Co.)

material under the skin of a female frog with a hypodermic needle. A description of the technique follows.

The pituitary gland is located just posterior to the level of the eyes of the frog (Fig. 8-29). It is necessary to cut off the upper head and jaw region of a living frog to reach the pituitary gland. Insert sharp scissors into the mouth at right angles to the jaw, and quickly sever the cranium at its junction with the vertebral column. Then cut the base of the skull (the upper palate) on either side by inserting the point of the scissors through the foramen magnum (the opening into the cranium); turn the point toward the underside of the orbit. Turn back the base of the skull and hold it in position with the left thumb. A small, ovoid, pink pituitary gland adhering to the turned base of the back of the skull should be apparent. Frequently the gland is surrounded by flocculent tissue. Remove the gland, and store it in 70 percent alcohol if it is not to be used immediately.

Mash the necessary number of whole glands in 2 ml of 0.1 percent Ringer's solution. Use a needle of rather wide bore to pick up this material and inject it into the abdominal cavity lymph spaces lying directly under the loose skin. Inject along the side of the animal—be careful to avoid injury to other organs. Rub the finger over the injected area after you remove the needle to keep the injected material from running out. Keep the injected female in a battery jar at 15° C (59° F) in about 1 in. of water. If the frogs are kept at a higher temperature, say 23° C (73° F), you will find that the eggs are often abnormal (in addition, the reserve food materials are too rapidly consumed).

If enough frogs are available, you may want to have students dissect one of the ovulating females to study eggs in all stages of development. Cilia which line the coelom propel the eggs to the ostium. There are also cilia on the edges of the liver, and on the pericardium and mesovarium. No cilia are found in males or immature females. Possibly, the presence

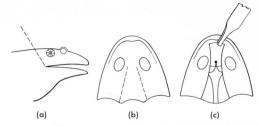

Fig. 8-29 Locating the pituitary gland in the frog: (a) cut off the upper jaw and the head; (b) view of upper palate: insert scissors into back of cranial cavity as far to the right as possible, and cut through bone; repeat on left side; (c) pull up flap with forceps, exposing pituitary gland (black ball). (After *Carolina Tips*, Carolina Biological Supply Co.)

or absence of cilia may be considered a secondary sex characteristic.

Artificial fertilization of frogs' eggs. After 36 hr at 15° C (59° F) the eggs of the pituitary-injected female may be stripped from her body. It is certain, however, that after 72 hr most of the eggs will be in the uterus. (At 15° C the eggs can be left in the body for 5 days after injection.) Gently hold a female and apply pressure to both sides of the abdomen and press in the direction of the cloaca. Strip the eggs into large, dry dishes so that the eggs are well separated. Be sure that all glassware is biologically clean. Then with a *clean* pipette, wash the eggs with 0.1 percent Ringer's solution for a few minutes. Finally, draw off the Ringer's solution, and pipette previously prepared sperm suspension over the eggs. Now the eggs have been artificially fertilized.

Make the sperm suspension earlier so that it can stand for about 15 min, to give the sperms time to become activated. To obtain the suspension, kill two males (save their pituitary glands in 70 percent alcohol) and then dissect out the testes. Roll the testes on toweling to remove blood and body fluids; wash them in 0.1 percent Ringer's solution.

Macerate the four testes in 50 ml of 0.1 percent Ringer's solution and let this stand. This amount of sperm suspension will be sufficient to fertilize about 1000 eggs. If it is necessary to save the testes you can keep them for a few days in wet cotton at a temperature of from 4° to 8° C (39° to 46° F).

Pipette the sperm suspension over the eggs and shake the dish occasionally. Let this stand for about 20 min. Finally, wash the eggs with more 0.1 percent Ringer's and flood the dish so that the eggs float freely. After half an hour students will be able to see which eggs have been fertilized, because the fertilized eggs rotate within the jelly layers so that the black surfaces are uppermost. When the jelly is completely swollen (after about an hour), cut the clusters of eggs apart with scissors and separate them into smaller finger bowls. Put about 50 eggs in each finger bowl in the same Ringer's solution and keep at 15° C (59° F). Discard the unfertilized eggs before they begin to decay. Transfer the eggs into fresh Ringer's solution twice a week.

Artificial parthenogenesis in frogs' eggs. Certain eggs can be made to begin cleavage without the entrance of a sperm cell. Rugh[16] describes a technique for causing unfertilized frogs' eggs to develop. Segregate the female to be used in this experiment from males for several days prior to the work. Boil the instruments and glassware or wash them in alcohol to be sure that they are clean and free of sperm cells. If the work is to be done at a time other than the breeding season, you will first need to induce ovulation.

When the eggs are available, parthenogenesis can be induced in this way: Strip the eggs in single file along the length of clean glass slides. Prepare several slides with rows of eggs. The slides may be kept in a moist chamber for about an hour, if necessary, until ready for use. A suitable moist chamber may be prepared by setting the slides over Syracuse dishes of water and covering with a bell jar. Now pith a nonovulating female frog (Fig. 7-9). Dissect it to expose the heart. Cut off the tip of the ventricle and let the blood flow into

the coelomic fluid. Next, dissect out a strip of abdominal muscle and dip it into the mixture of blood and coelomic fluid. Now streak each egg so that a film of blood–coelomic fluid remains on each egg. Then prick each egg with a glass or platinum needle within the animal hemisphere (that is, the black area), just off center. Finally, immerse the slides with the eggs on them into spring water or distilled water. Keep the eggs in Petri dishes with just enough fluid to cover the eggs. Keep them in a cool place and examine the eggs hourly under a binocular microscope, dissecting microscope, or hand lens for signs of cleavage stages.

In addition to Rugh's text–laboratory manual,[17] teachers will want to refer to some activities in the BSCS laboratory block developed by F. Moog, *Animal Growth and Development*.[18] Activities relate to the effect of injecting hormones, and the measurement of comb development in chicks.

Organizers in frogs' eggs. For more advanced work (especially with a small group) certain operative techniques may be developed. Especially useful are those for making grafts of small squares of tissue from the early gastrula stage of a frog's egg to early tadpole stages (before muscles develop so that tadpoles can flex the body).[19] Glass microneedles are needed with tips of glass thread bent at a 45° angle. These are fashioned in a microburner. Hair loops are also needed (Fig. 9-44).

Techniques for preparing needles and the methods for making the grafts are given in Rugh's fine *Experimental Embryology*, cited above. See also pp. 484–88.

Also try duplicating Vogt's classic experiments in which the regions of the blastula are stained with vital stains in an attempt to trace the path of the presumptive germ layers during invagination (also in Rugh's book). See p. 483.

---

[17] *Ibid.*
[18] *Op. cit.*
[19] Also refer to Chapters 27–29 in *Biological Science: An Inquiry into Life* (BSCS Yellow Version), *op. cit.*

[16] *Ibid.*, Chapter 20, "Artificial Parthenogenesis."

Ovaries and testes in the pigeon or chicken. Students may be able to bring to class the ovary of a chicken in order to show the different-sized egg cells, or "yolks," which look like a bunch of grapes. Or a freshly killed pigeon may be dissected according to the instructions given in Chapter 1.

Look for the ovary, oviduct, and enlarged shell gland near the cloacal region. In the male find the testes and trace the sperm ducts to the cloaca. Study the relationship between the testes and the kidneys.

Living chick embryos. It is a fascinating study to watch a chick develop from the apparently amorphous egg. When fertilized eggs are obtainable, students may incubate them in the laboratory and examine them while they develop over a 3-week period. (Students can build an incubator if you cannot afford to buy one for school.) Some companies guarantee eggs up to 90 percent fertility, but seasonal variations lower this figure. You cannot store fertilized chicken eggs in the laboratory before incubation unless you can get temperatures as low as 10° to 15° C (50° to 59° F).

Keep the incubator at 37° C (99° F) and insert the bulb of the thermometer at the level of the eggs, not high up in the incubator, for there may be a difference as great as 10° C between the two places. Include a pan of water in the incubator to keep a uniform humidity. A relative humidity of 60 percent is optimum. Do not wash the egg surface; washing will remove a protective film which reduces bacterial infection. Turn the eggs daily to prevent adhesion of membranes. Students might mark the eggs with a pencil so that they know which surfaces to turn.

There seem to be two periods when mortality is high, the third day and just before hatching time. Some mortality must be expected, for some hens produce fewer viable eggs than do others. And you will want to allow for variations in handling.

After eggs have been incubated for about 36 hr they are relatively easy to handle for examination. The number of hours of incubation is not a true index of the age of the embryo, for cleavage started in the oviduct before the egg was laid. As you remove an egg from the incubator hold it in the same position so that the blastoderm will be floating on top and the heavier yolk underneath.

Crack the egg on the edge of a finger bowl which contains slightly warmed saline solution (or Locke's solution, p. 664). If you place finger bowls of solution in the incubator beforehand, they will be at the proper temperature (the embryo will not run the danger of being chilled). Let the egg contents flow into the saline solution so that the embryo is submerged in solution. Notice how the chick blastoderm rotates to the top position. When the embryos are larger they do not rotate so easily but must be moved with a forceps by pulling on the chalazae, the thickened albumen "ropes" on each side of the egg yolk mass (Fig. 8-30).

Use a hand lens or a binocular microscope to examine this early embryo. You should find the head beginning to curve to the left side (Fig. 8-31b). Three vesicles of the brain should be visible and the neural tube should be closed at this time. In a 40-hr embryo you should begin to see the heart beat, and a complete blood circulation should be discernible in a 45-hr embryo. If you stagger incubation of the eggs

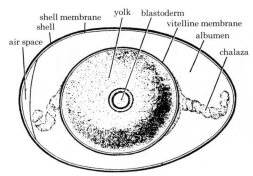

Fig. 8-30 Cross section of fertilized chicken's egg. (From J. W. Mavor, *General Biology*, 3rd ed., Macmillan, 1947.)

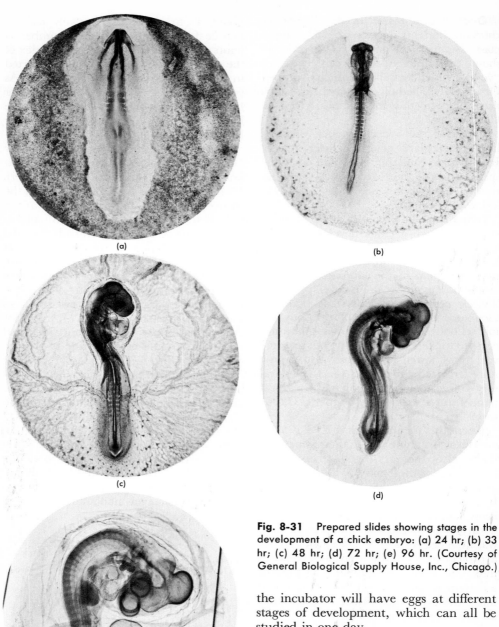

Fig. 8-31 Prepared slides showing stages in the development of a chick embryo: (a) 24 hr; (b) 33 hr; (c) 48 hr; (d) 72 hr; (e) 96 hr. (Courtesy of General Biological Supply House, Inc., Chicago.)

the incubator will have eggs at different stages of development, which can all be studied in one day.

Students in a microscopy club or a biology research club may want to make permanent stained slides (technique, Chapter 2) of 24-hr, 36-hr, 48-hr, and 72-hr chick embryos. Older embryos become too opaque for examination under the microscope.

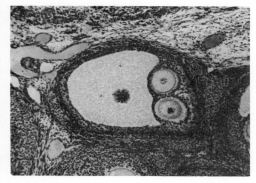

Fig. 8-32 Prepared slide of a section through an ovary, showing Graafian follicles of a cat. (Courtesy of General Biological Supply House, Inc., Chicago.)

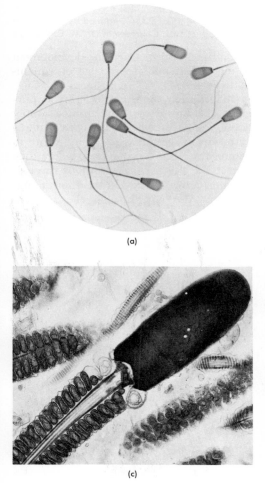

(a)

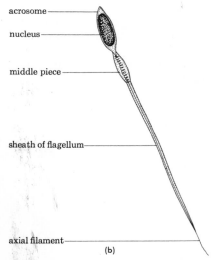

(c)

Ovaries and testes of a mammal. Dissect a freshly killed rat or other small mammal which has been etherized by being placed in a closed container with some ether-soaked absorbent cotton. Check the reflexes to make certain that the animal is dead before you begin to dissect it. Make an incision in the ventral body wall at the posterior end and cut forward along one side of the median line. Cut through the sternum to the shoulder or pectoral girdle. Then cut back the abdominal wall flaps and pin them back to the dissection pan (dissection, Chapter 1).

In the female, identify the ovaries, oviducts, and uterus. Cut into the uterus to look for developing embryos or fetuses if the female is pregnant. Then with a blunt probe examine the uterine lining, and trace the oviducts forward, and posteriorly to the vagina. You may want to show a microscope slide of a section of the ovary (Fig. 8-32). In the male, dissect the scrotal

acrosome

nucleus

middle piece

sheath of flagellum

axial filament

(b)

Fig. 8-33 Sperm cells of mammals: (a) prepared slide of sperm cells of a bull (magnified $1380\times$, reduced by 49 percent for reproduction); (b) diagram of typical sperm cell of mammal; (c) electron micrograph of sperm cell of a bat (enlarged 21,500 diameters, reduced by 56 percent for reproduction). (a, courtesy of General Biological Supply House, Inc., Chicago; c, courtesy of Dr. D. W. Fawcett and Dr. S. Ito, Harvard Medical School.)

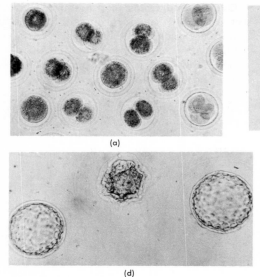

(a)

(b)

(c)

(d)

Fig. 8-34 Prepared slides showing stages of development in eggs of a rabbit: (a) 1 day; (b) 2 days; (c) 3 days; (d) 4 days. (Each photograph is magnified 93×, reduced by 29 percent for reproduction.) (Courtesy of Dr. M. C. Chang.)

sac to expose the testis. Examine a cross section of the testis; trace the epididymis, vas deferens, vesicles, and the urethra in the penis.

If there is time, students may identify other organs (dissections, Chapter 1). Also refer to studies on photoperiodism and the role of the pituitary gland in the reproductive cycle (Chapter 7).

Students will want to examine prepared slides of sperm cells of mammals (Fig.

8-33a) and also to trace the development of the almost microscopic egg of a mammal (Fig. 8-34).

Dissecting a pig's uterus. If you purchase the uterus of a pig from a supply house you will receive a bifurcated uterus with several fetuses. You may want to demonstrate this to your class before they dissect the uterus. Carefully cut through the uterine wall to reveal the chorion and amnion. Also cut through the amnion,

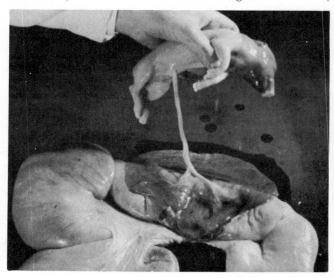

Fig. 8-35 Bifurcated uterus of pig, showing one fetus dissected out. (Photo from *Reproduction Among Mammals*, Encyclopaedia Britannica Films Inc.)

since it contains amniotic fluid. Students will easily identify the placenta and umbilical cord as you gently lift out the fetus (Fig. 8-35). If a sufficient supply is purchased, every student may dissect a portion of the uterus. The fetuses can be cut out and preserved in alcohol or formalin in a museum jar, or saved for future dissection.

Pregnancy and delivery. Many specimens of uteri and mammalian embryos are available from biological supply companies (Fig. 8-36). Placental types available include diffuse (pig), cotyledonary

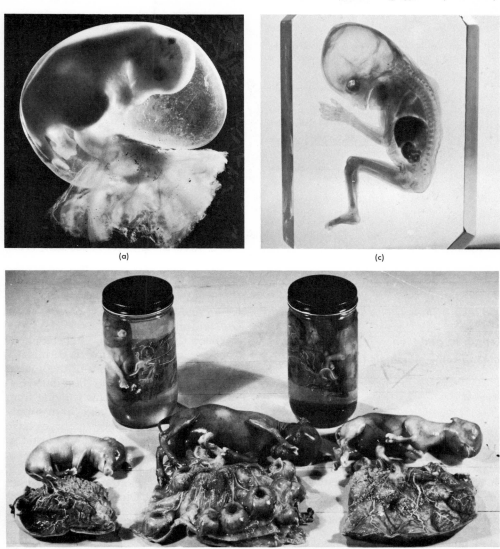

(a)

(c)

(b)

Fig. 8-36 Some specimens of uteri and embryos available for study: (a) pig embryo in amnion; (b) placental types (left to right): diffuse (pig), cotyledonary (sheep and cow); (c) human embryo, bisected along median line, cleared and mounted in plastic. (a, b, courtesy of Carolina Biological Supply Co.; c, courtesy of General Biological Supply House, Inc., Chicago.)

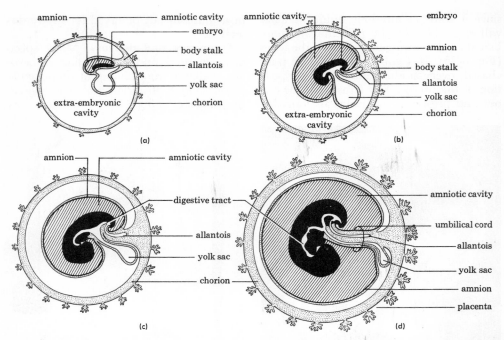

Fig. 8-37 Stages in the early weeks of development of human umbilical cord and body form: (a) 15 days; (b) 20 days; (c) 30 days; (d) 33 days. (After C. Villee, *Biology*, 4th ed., Saunders, 1962.)

* Adapted from W. Spector, ed., *Handbook of Biological Data*, Saunders, Philadelphia, 1956.

TABLE 8-2

Length of gestation among some mammals (in days)*

mammal	gestation period	mammal	gestation period
Ape	210	Fox	52
Baboon	210	Guinea pig	68
Bat	35	Hamster	16
Bear	208	Horse	336
Bobcat	50	Kangaroo	38
Buffalo	275	Man	274–280
Camel	315–410	Mouse	19
Cat	63	Rabbit	31
Cattle	281	Rat	21
Chipmunk	31	Sheep	151
Dog	63	Squirrel	44
Elephant	624	Whale	360

(cow and sheep), zonary (cat), and discoid (rabbit). Also available is a set of comparative uterine types: duplex, bipartite, bicornate, and simplex.

Where the social and intellectual atmosphere in the community is permissive, a study of the development of the human embryo in the uterus may be made (Fig. 8-37). Embryos of humans ranging from 3 to 4 months are available from supply houses (Fig. 8-36c). These embryos have been cleared and mounted in plastic so that they can be examined easily in class and can be readily stored.

There are excellent models of human reproduction, the Dickinson-Belskie models, available from General Biological. There is a set of models which is useful in teaching different stages of fertilization and pregnancy; also a set of models useful in teaching parturition, as shown in Fig. 8-38.

Refer to Table 8-2 for a comparison of length of gestation among mammals.

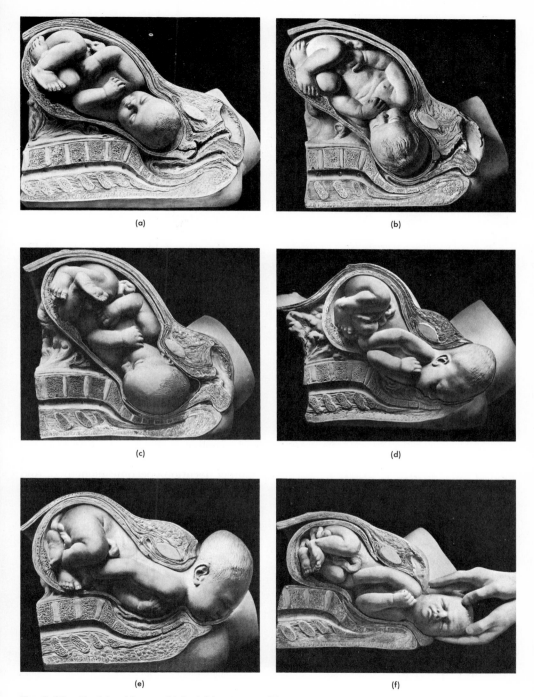

Fig. 8-38 Models of human birth: (a) beginning; (b) uterus contracting, cervix dilating; (c) head deep in birth canal; (d) head emerging; (e) head turning upward; (f) baby born. (Reproduced with permission from the *Birth Atlas*, published by Maternity Center Association, New York City.)

Fig. 8-39 Conjugation in *Spirogyra* (magnified 435×, reduced by 43 percent for reproduction). Notice the bridge along which one entire cell, as a gamete, moves into the passive cell, forming a zygospore. (Courtesy of General Biological Supply House, Inc., Chicago.)

Sexual reproduction in plants

Spirogyra. In a form such as *Spirogyra* the gametes which unite look alike (as in *Rhizopus*, fusion is called conjugation). Trace the process and note that the active cell moves across the "bridge" to fuse with the contents of the passive (female) cell, or gamete (Fig. 8-39). A zygospore is formed as the result of each union. Examine prepared slides of conjugation; techniques for culturing *Spirogyra* are described in Chapter 13.

Fucus, a brown marine alga. With the aid of preserved specimens (if fresh material is not available) and slides, students may trace the life cycle of this alga (Fig. 8-40c).

Vaucheria. This tubular green alga, often found on flowerpots and called "green felt," is available for study of sexuality in plants (Fig. 8-41; also see culturing methods, Chapter 13).

Chlamydomonas. Refer to Fig. 8-73 for a study of sexual reproduction; also see heredity in *Chlamydomonas* (p. 402) and meiosis (Fig. 8-75).

Oedogonium. This is a common, unbranched, green filamentous alga which reproduces by unlike gametes (Fig. 8-42). These gametes are produced in special cells called gametangia, which are developed from vegetative cells of the filaments. The female gametangia are oval cells within which an egg develops. Motile ciliated sperms are liberated from the male gametangia (antheridia); they swim to the egg through an opening, or pore, formed by the breakdown of a small area of the wall of the oogonium.

Oedogonium also reproduces asexually by forming ciliated zoospores, which are transformed vegetative cells. These cells with a collar of cilia break away from the filament and swim off to take hold at another location, become elongated, and carry on cell division. Thus they form a new filament. Stain a wet mount with Lugol's solution (p. 664) to show cilia.

Molds. Among *Rhizopus* hyphae there are no visible differences between the two mating types, but a physiological difference exists. This type of sexual union is referred to as conjugation. Hyphae of opposite mating types (plus and minus) grow toward each other, and the tips of the hyphae fuse. The plus and minus strains of *Rhizopus* and of certain other

Fig. 8-40 Reproductive organs and cycle of *Fucus*, a marine brown alga: (a) male conceptacle, showing antheridia; (b) female conceptacle, showing oogonia; (c) reproductive cycle. (Courtesy of General Biological Supply House, Inc., Chicago.)

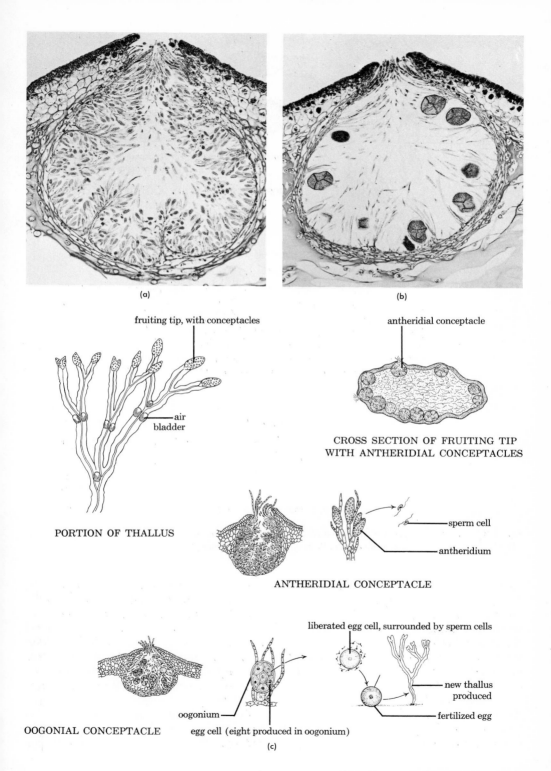

(a)

(b)

fruiting tip, with conceptacles

air bladder

PORTION OF THALLUS

antheridial conceptacle

CROSS SECTION OF FRUITING TIP
WITH ANTHERIDIAL CONCEPTACLES

sperm cell

antheridium

ANTHERIDIAL CONCEPTACLE

liberated egg cell, surrounded by sperm cells

new thallus produced

fertilized egg

OOGONIAL CONCEPTACLE

oogonium

egg cell (eight produced in oogonium)

(c)

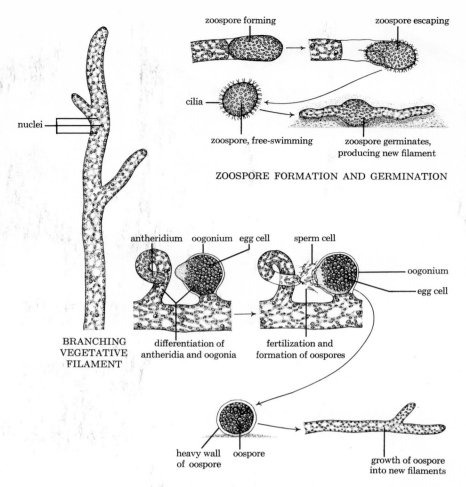

ZOOSPORE FORMATION AND GERMINATION

SEXUAL STAGES AND GERMINATION OF OOSPORE

Fig. 8-41 Cycles of asexual and sexual reproduction in *Vaucheria,* a tubular green alga (green "felt"). (Courtesy of General Biological Supply House, Inc., Chicago.)

molds such as *Phycomyces blakesleeanus* can be obtained from supply houses or college laboratories. "Seed" opposite sides of moistened bread (culturing, p. 639) with the plus and minus strains. Examine under the microscope the dark zygospores which eventually form along the middle of the bread (Fig. 8-43).

NEUROSPORA. Obtain sexual strains of *Neurospora* from a biological supply house. This pink mold grows well; however, it readily contaminates other cultures in the laboratory unless care is taken. If possible, make transfers in another room; handle the tube cultures as though they were bacterial cultures. Flame the cotton stoppers and mouths of the tubes; use a moist loop to transfer spores from an old culture to new agar slants. Some workers use an aqueous suspension of spores to make transfers.

Prepare slanted agar medium (Difco has a prepared medium, or see p. 400) in wide-mouthed test tubes or small jars

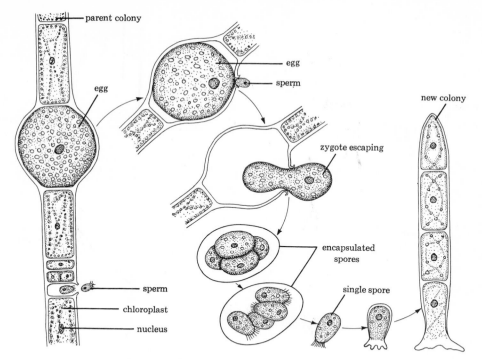

Fig. 8-42 Sexual reproduction in *Oedogonium*, a filamentous green alga. Notice the four swarm spores resulting from meiotic divisions of the zygote. (After D. Marsland, *Principles of Modern Biology*, 3rd ed., Holt, Rinehart and Winston, 1957.)

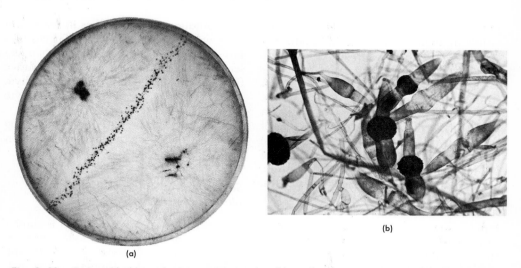

Fig. 8-43 Fusion of plus and minus gametes of molds to form zygospores: (a) *Phycomyces blakesleeanus* mycelia fuse, forming zygospores at the juncture of plus and minus strains; (b) zygospores of *Rhizopus nigricans,* formed as a result of conjugation (magnified 172×, reduced by 61 percent for reproduction). (a, courtesy of Carolina Biological Supply Co.; b, courtesy of General Biological Supply House, Inc., Chicago.)

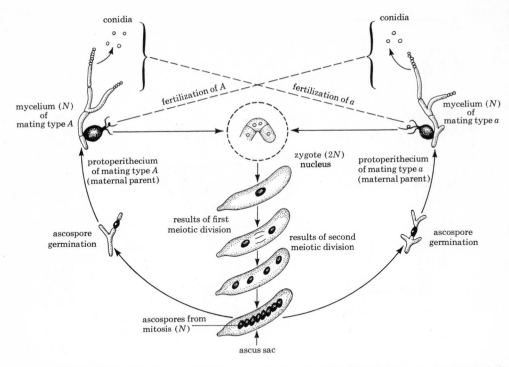

Fig. 8-44 Sexual reproduction in *Neurospora*, a bread mold. (After R. P. Wagner and H. K. Mitchell, *Genetics and Metabolism*, Wiley, 1955.)

plugged with cotton. Perithecia become visible in about 1 week and are mature in 2 weeks (Fig. 8-44; see also Figs. 8-70, 8-71, 8-72).

Crosses may be made between two strains, for example, wild type and albino. Inoculate spores at different regions of a prepared Petri dish, or at opposite ends of the slants in test tubes, and watch the growth of the mycelia and the meeting of the mycelial fronts forming sexual spores.

Alternation of generations in mosses. Compare the life history of the moss (Fig. 8-45) with that of the fern (Fig. 8-46). The same pattern exists, although the gametophyte stage in the moss is conspicuous and the sporophyte grows from the gametophyte (some botanists consider this "parasitic").

Try to grow the spores of mosses, following steps similar to those for growing fern spores (pp. 645–51).

Alternation of generations in ferns. You may want to collect spores of ferns and grow them according to the method described on p. 648. Use Fig. 8-46 to trace the life history of a fern (the gametophyte generation alternates with the sporophyte generation).

Also examine prepared slides showing sperm cells entering the archegonia of fern prothallia.

Sexual reproduction in higher plants—seed plants

Fertilization in higher plants is similar to that in animals, but the mode of transfer of the gametes, sperm and egg, is different. A reduction division, similar to the division in animals, occurs in the maturation divisions in the spore mother cells in the anthers, and within the ovules. Each gamete, then, is haploid, so that when a

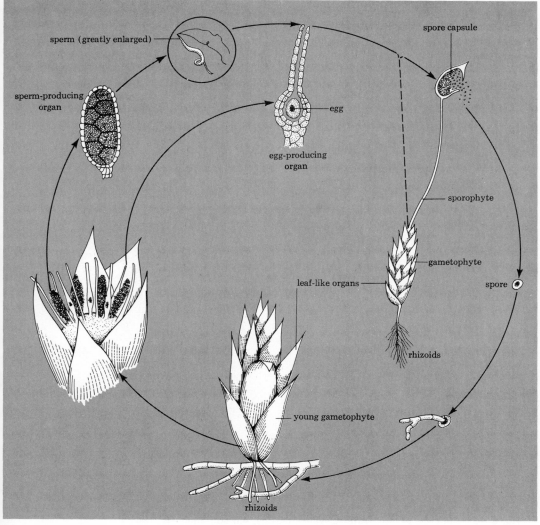

Fig. 8-45 Life cycle of moss. (After C. Villee, *Biology*, 4th ed., Saunders, 1962.)

pollen-grain nucleus unites with the nucleus of an ovule the original diploid number is restored.

A desirable procedure, we believe, is to take students out of doors to study flowers of shrubs and trees in season as they come to blossom (see Chapter 11). (Refer also to the effect of hormones, and of photoperiods, Chapter 7.) *Forsythia,* maple, cherry, magnolia, dogwood, oak, elm, and others blossoming at about the same time

may be studied during walks around the school grounds. Students may bring to class tulips, daffodils, crocuses, and pussy willows. Around the lawns you should find dandelions and other flowers. Later in the school year students may collect and identify fruits and seeds. Where the school is in a suburban or rural community, it is fairly easy to find living material.

Students will learn that not all flowers have petals and that some flowers contain

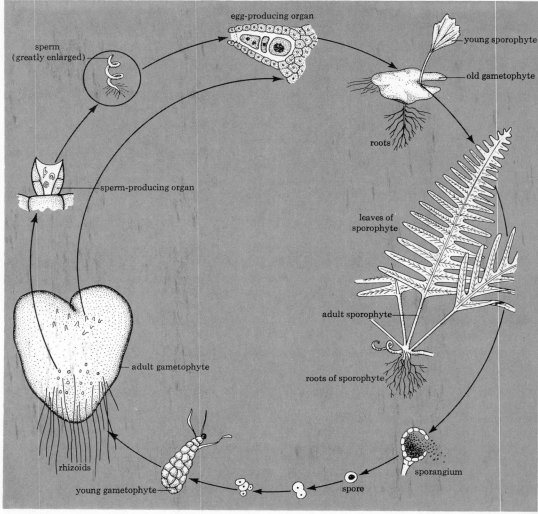

sperm
(greatly enlarged)

egg-producing organ

young sporophyte

old gametophyte

roots

sperm-producing organ

leaves of
sporophyte

adult sporophyte

roots of sporophyte

adult gametophyte

sporangium

rhizoids

spore

young gametophyte

Fig. 8-46   Life cycle of fern. (After C. Villee, *Biology*, 4th ed., Saunders, 1962.)

only stamens, while others have only the female organs, the pistils.

Parts of a flower. Tulips, daffodils, and gladioli are excellent forms to show an almost diagrammatic flower (Fig. 1-73); the pistil and stamens are clear. Students should also see a variety of flower shapes and realize that these shapes help to identify families of plants. They might learn to recognize a few, such as composite flowers, members of the Compositae (dan-delions, asters, daisies, and so forth, Fig. 8-47); members of the pea family, the Leguminosae (sweet pea, wistaria, locust trees, clover, Fig. 8-48); and the rose family, the Rosaceae (apple, pear, and cherry trees, as well as roses, Fig. 8-49). These families each have a different flower shape or arrangement. Some students who show interest in classifying plants may want to learn how to use a flower key (study the pattern, pp. 79, 549). Fig. 8-50

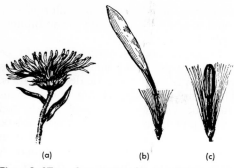

(a)   (b)   (c)

Fig. 8-47 Composite flower *Aster novae-angliae*: (a) composite head; (b) ray flower; (c) tubular flower. (From G. F. Atkinson, *Botany*, 2nd ed., Holt, Rinehart and Winston, 1905.)

Fig. 8-48 Typical flower shape of pea family: *Erythrina fusca*. (From W. H. Brown, *The Plant Kingdom*, 1935; reprinted through the courtesy of Blaisdell Publishing Co., a division of Ginn and Co.)

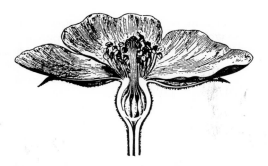

Fig. 8-49 Typical flower shape of rose family: rose flower. (From W. H. Brown, *The Plant Kingdom*, 1935; reprinted through the courtesy of Blaisdell Publishing Co., a division of Ginn and Co.)

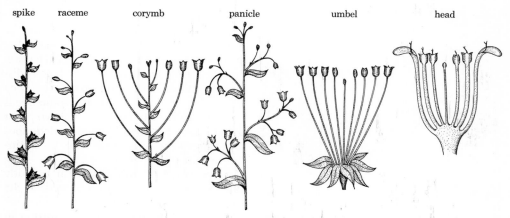

spike   raceme   corymb   panicle   umbel   head

Fig. 8-50 Types of inflorescence. (After J. W. Mavor, *General Biology*, 3rd ed., Macmillan, 1947.)

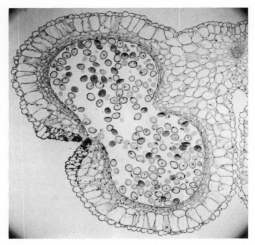

Fig. 8-51 Prepared slide of cross section of anther of lily, showing pollen grains in the pollen chamber. (Courtesy of General Biological Supply House, Inc., Chicago.)

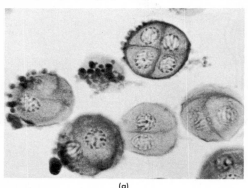

(a)

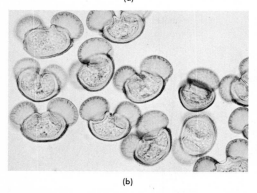

(b)

Fig. 8-52 Examples of pollen: (a) lily; (b) pine. (a, courtesy of Carolina Biological Supply Co.; b, courtesy of Ward's Natural Science Establishment, Inc., Rochester, N.Y.)

shows the main kinds of inflorescence. Some flower keys are listed in the Bibliography (Appendix A).

To study flowers, students should be provided with hand lenses or dissection microscopes, and razor blades or scalpels. They can then dissect out the stamens after the exterior whorls of sepals and petals have been studied. Anthers may be examined under a hand lens, and then dissected to show the pollen chambers. A prepared slide of an anther may be examined, or pollen grains may be mounted on a slide and examined under high power of the microscope (Fig. 8-51). The sculpturing and the different shapes of pollen grains from different plants will be noted. In fact, flowers may be identified by means of their pollen grains (Fig. 8-52). On occasion, students may want to germinate pollen grains (p. 382).

Similarly, cross sections and lengthwise sections of the ovary may be studied (Fig. 8-53). Students will locate the ovules and note that the number of chambers in the ovary equals (or is a multiple of) the number of petals and stamens. Floral parts are mainly in "fives," or "threes" (especially among the monocots).

During the year short field trips may be taken to show fruits and seeds on trees and shrubs around the school. Students may bring to class the fruits they collect. They may want to prepare a display of some of the special adaptations which facilitate seed dispersal (Fig. 8-54).

Furthermore, seeds themselves may be studied. Students can soak some large seeds such as beans, peas, or the corn grain (really a fruit containing a seed) and dissect them to learn their structure. They may also try to germinate some seeds (p. 450) and study how the embryo plant grows. In this way, monocotyledonous and dicotyledonous seeds can be distinguished in their pattern of growth.

Some students may also wish to study pine cones and the relation of seeds to the scale-like bracts in the cones. The life history of pollination and fertilization in gymnosperms (Fig. 8-55) may also be

Fig. 8-53 Prepared slide of cross section of ovary of *Fritillaria,* showing megaspore mother cell. (Courtesy of General Biological Supply House, Inc., Chicago.)

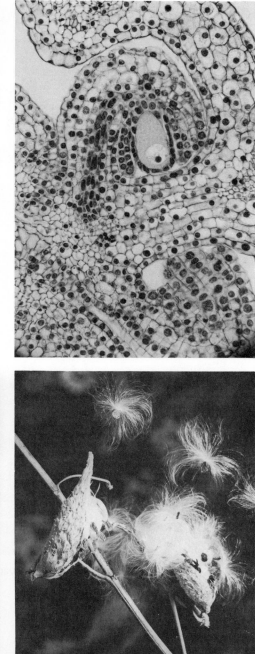

(a)

(b)

Fig. 8-54 Adaptations for seed dispersal: (a) cocklebur; (b) milkweed. (Photos by Hugh Spencer.)

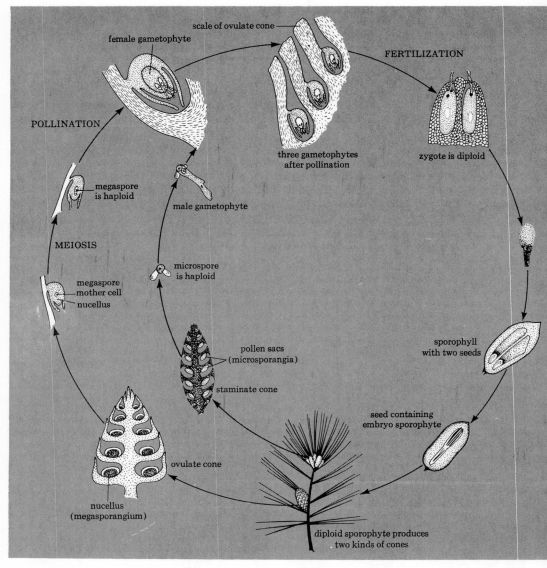

Fig. 8-55 Life cycle of gymnosperm. (After C. Villee, *Biology*, 4th ed., Saunders, 1962.)

studied and compared with the life cycle of angiosperms (Fig. 8-56).

Artificial pollination. You may want to demonstrate how a breeder would pollinate flowers which he selects for mating. Hybridization techniques of this kind are commonly used to get new varieties of flowers through recombinations in the off-

spring of different genes. The flower to be pollinated should be selected in the bud stage, that is, before its own pollen becomes ripe. Open the flower bud carefully and remove the unripened stamens by cutting them out with scissors. Wash with alcohol all the equipment you use, to prevent any foreign pollen from entering the

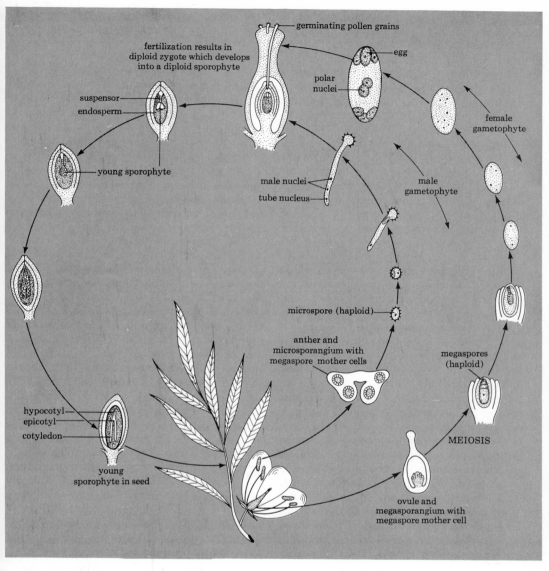

Fig. 8-56 Life cycle of angiosperm. (After C. Villee, *Biology*, 4th ed., Saunders, 1962.)

plant blossom you have selected. Then cover the flower bud with a small bag (or square of plastic) until you are ready to pollinate that flower. When other plants show ripened stamens, select the special pollen (from the same kind of plant or one closely related) by removing the stamens from those flowers; then touch them to the stigma of the selected flowers. Again cover the flowers you have now pollinated to prevent other pollen grains from fertilizing the ovules. Plan to pollinate several flowers in one operation. Students can plant the seeds that result, and, if possible, study the inheritance of flower color of the next generation.

Examining pollen grains. Prepare slides of many kinds of pollen grains. Students can discover the variety of shapes and patterns of pollen grains by mounting some in a drop of water or in xylol. (The motion of the pollen grains in water is Brownian movement.) Or you may want to mount the pollen grains in melted glycerin jelly (see Fig. 8-52).

Students may want to study the kinds of pollen grains in the air at a particular season. Lightly coat slides with petroleum jelly and suspend the slides out of doors. They will catch pollen grains (and dust as well).

Staining pollen grains. To a small amount of aniline oil add a bit of crystal violet so that a light purple tint results. Mount pollen grains in a few drops of the tinted aniline oil on a slide. Then hold the slide over a Bunsen flame until the pollen grains stain deeply, but do not let the slide become warm to the touch.

Let the slide cool and remove the excess oil with filter paper. Wash several times by adding drops of xylol to the slide, then drawing it off with filter paper until the excess stain has been removed. Then add a drop of balsam and a coverslip (staining techniques, Chapter 2).

Germinating pollen grains. There are several ways to grow pollen grains. These grains germinate in the sticky, sugary exudate of the stigma. Students can brush ripe anthers against a ripe stigma (of the same kind of flower), then crush the stigma in water on a slide. Include a few bristles or pieces of broken coverslip before adding the coverslip, to prevent the grains from being crushed; or make a hanging-drop preparation (Fig. 2-29). The slides can be kept in a moist chamber between each examination. Maintain at a temperature of 21° C or 25° C (70° F or 77° F). Students can study pollen grains of narcissus, lily, daffodil, tulip, or sweet pea in this way. Pollen grains can also be sprinkled into a sucrose (cane sugar) solution. Different pollen grains require different kinds of sugar content. Where the stigma of the flower is quite sticky, pollen grains from this species of flower require more concentrated sugar solution for germination. In general, a majority of pollen grains grow in solutions varying from 2 to 20 percent; most at 10 percent. However, many composites require 30 to 45 percent sugar solution. The pollen grains of many cycads grow tubes in 2 to 3 days when grown in a 10 percent cane sugar solution.

Students may use more precise methods (described by Johansen[20]). To do this they should boil 1 g of sugar and 0.5 g of agar in 25 ml of water. This medium is cooled to about 35° C (95° F); then 0.5 g of powdered gelatin is added and stirred until it melts.

Keep the solution at 25° C (77° F) in a water bath or on a hot plate. Put a thin film of this solution on clean slides. Dust the slides with pollen grains and keep them in a moist chamber without applying coverslips. Examine the slides under the microscope for growth of pollen tubes. (Try crushing anthers of *Tradescantia* in a film of this sugar solution.)

Also refer to chemotropism of pollen grains in relation to ovarian tissue (p. 300 and footnote 8 in Chapter 7). For a study of genetic differences between two types of pollen grains, see p. 395 and Fig. 8-68.

Heredity: the genetic code

Heredity in man

Since most youngsters are interested in their own heredity, it is a good idea to begin with a study of heredity in human beings. Youngsters enjoy looking into the inheritance of maleness and femaleness, color blindness, freckles, musical ability, hair color, shape of ears, attached or free ear lobes, curly hair, ability to taste specific chemicals, and blood types. Students can learn the laws of heredity by using examples from human traits as well as those from traditional plant material (garden

[20] D. Johansen, *Plant Microtechnique,* McGraw-Hill, New York, 1940, pp. 476–77.

peas, corn, or *Neurospora*), and other animal material (*Drosophila* or Protozoa). They will want to study Mendel's work, and demonstration crosses using rats, peas, maize, sorghum, fruit flies, and inheritance of RNA in planarians. Students may also wish to model in three dimensions the replication of a chain of DNA molecules.

Tasters vs. nontasters. About seven out of every ten human beings taste PTC (phenylthiocarbamide) as a salt, sweet, sour, or bitter substance. To others, PTC is tasteless. Paper soaked in this harmless chemical may be purchased from the Am. Genetic Assoc.[21] Or you may want to prepare it yourself as described later in this chapter.

Cut the paper into ½-in. squares. (Do not explain what kinds of tastes will be obtained; otherwise some students will imagine a taste.) Distribute the paper and have all the students taste it at the same time. Then notice their reaction. Tally the results—tasters and nontasters. Ability to taste PTC paper seems to be a dominant trait.

Students might use symbols to illustrate some possible crosses, as in Table 8-3. Is it possible for one child to be a taster and his sibling a nontaster? Have students calculate these possibilities beginning with the known facts which are given:

nontasters must have genetic make-up   *tt*
tasters must have at least one gene   *T*

Work back to find the parents' genetic make-up. For example, both parents might be hybrid for the trait (Table 8-3*b*); then 75 percent of the offspring would be tasters, 25 percent nontasters. Or one parent might be hybrid and the other a nontaster (Table 8-3*c*); then 50 percent of the offspring would be tasters and 50 percent nontasters.

Elicit from students that there are six possible crosses whenever *one* pair of genes is under study:

[21] Located at 1507 M St., N.W., Washington 5, D.C. (Purchase of 50 or more, $0.02 each.)

TABLE 8-3

Inheritance of the trait of tasting PTC paper: some possible crosses

Key: *T* = taster, *t* = nontaster

(a)	*Parents:*	*TT*	×	*tt*
	Gametes:	*T*		*t*
	Offspring:		*Tt*	
			(hybrid taster)	
(b)	*Parents:*	*Tt*	×	*Tt*
	Gametes:	*T t*		*T t*
	Offspring:		*TT Tt Tt tt*	
			(75% tasters,	
			25% nontasters)	
(c)	*Parents:*	*Tt*	×	*tt*
	Gametes:	*T t*		*t*
	Offspring:	*Tt*		*tt*
		(50% tasters,		
		50% nontasters)		

1. Pure dominant × pure dominant (*TT* × *TT* = 100% *TT*)
2. Pure recessive × pure recessive (*tt* × *tt* = 100% *tt*)
3. Pure dominant × pure recessive (*TT* × *tt* = 100% *Tt*)
4. Hybrid × pure dominant (*Tt* × *TT* = 50% *Tt*; 50% *TT*)
5. Hybrid × pure recessive (*Tt* × *tt* = 50% *Tt*; 50% *tt*)
6. Hybrid × hybrid (*Tt* × *Tt* = 50% *Tt*; 25% *TT*; 25% *tt*)

You may be able to provide enough paper for students to take home to test their parents' reactions as well. Then students might compile all the data.

The Am. Genetic Assoc. also supplies tablets of mannose, a sugar which is tasted differently by different people. But the genetic basis for these differences is not as well known as in the case of PTC tasting.

PREPARING PTC PAPER. Gradually dissolve 500 mg of phenylthiocarbamide in 1 liter of water. At room temperature this will take about 24 hr. Then soak sheets of filter paper in this solution and hang them

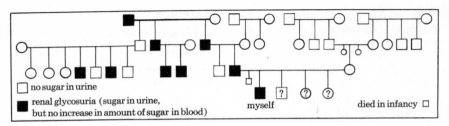

Fig. 8-57 Family pedigree of a student who suffered from renal glycosuria, excess of sugar in the urine. (Drawing by Marion A. Cox. From *Experiences in Biology*, by Evelyn Morholt and Ella Thea Smith, copyright, 1954, by Harcourt, Brace & World, Inc., and reproduced with permission of the publisher and the artist.)

up to dry. Cut small pieces, about ½-in. squares, and store them in envelopes for distribution to students. (At higher concentrations, nontasters will also react.)

A family pedigree. Students may be interested in tracing the inheritance of a trait in their own family. They may study the inheritance of eye color, hair color, freckles, PTC tasting or nontasting, and many other traits. They may want to model their "pedigree" after the one diagramed in Fig. 8-57. See also Table 8-9.

Blood groups. There are four basic blood types in man, a classification based on the distribution of the two antigens found in human red corpuscles. (See Tables 5-2 and 8-4 and the accompanying discussion for the facts underlying blood types and the typing of blood.)

The inheritance of blood types is ex-plained on the basis of multiple alleles. However, it is not necessary to involve students with these complex details. The following explanation has proved success-ful in the classroom.

1. Type A person can have genetic make-up *AA* or *OA*.
   (*A* is dominant over *O*.)
2. Type B person can have genetic make-up *BB* or *OB*.
   (*B* is dominant over *O*.)
3. Type AB person can have genetic make-up *AB* only.
   (Neither *A* nor *B* is dominant over the other.)
4. Type O person can have genetic make-up *OO* only.
   (Neither *A* nor *B* can be present.)

In crossing blood types, note that group A (or B) never appears in the offspring

TABLE 8-4

Inheritance of blood groups

blood type of parents	possible genetic combinations of offspring	blood type of offspring	blood types that do not occur in offspring
O × O	*OO*	O	A, B, AB
O × A	*OA, OO*	A, O	B, AB
O × B	*OB, OO*	B, O	A, AB
O × AB	*OA, OB*	A, B	O, AB
A × A	*AA, OA, OO*	A, O	B, AB
A × B	*OA, OB, AB, OO*	A, B, AB, O	———
A × AB	*AA, OA, OB, AB*	A, B, AB	O
B × B	*BB, OB, OO*	B, O	A, AB
B × AB	*OA, BB, OB, AB*	A, B, AB	O
AB × AB	*AA, BB, AB*	A, B, AB	O

TABLE 8-5

Blood groups (A_1-A_2-B-O)

group	frequency (percent)	phenotype and frequency		genotype and frequency	
O	43.5	O	43.5	OO	43.5
A	39.2	A_1	31.0	A_1O	25.1
				A_1A_2	2.3
				A_1A_1	3.6
		A_2	8.2	A_2O	7.9
				A_2A_2	0.3
B	12.7	B	12.7	OB	11.9
				BB	0.8
AB	4.5	A_1B	3.4	A_1B	3.4
		A_2B	1.1	A_2B	1.1

unless it was present in at least one parent. What are the possible mating combinations (Table 8-4)?

Students readily contribute this information and can predict what blood group a parent could not have if the child had a known type. You may want to develop several hypothetical cases to have students apply this information to solve the cases. If you have not already typed blood in class, you may want to have students do this now (pp. 253–56).

Since the discovery of A-B-O, it has been found that antigen A has two forms: A_1 and A_2. If we consider that multiple alleles are the basis for explaining heredity of blood types, it is clear that genes A_1, A_2, B, and O belong to the same locus. As a result, ten possible genotypes may result from recombinations of the four alleles (Table 8-5).

HUMAN MIGRATION. There may be occasion for students to read further about the use of blood groups to trace the patterns of migration of peoples over the face of the earth. Study Table 8-6. The alleles I^A, I^B, and i represent the gene for the production of the antigen A, for the antigen B, and for neither antigen, respectively. (Sometimes the alleles are indicated as L^A, L^B, and l, in honor of Landsteiner, the discoverer of these blood groups.)

Read further in such references as W.

Boyd, *Genetics and the Races of Man* (Little, Brown, 1950); also refer to the discussion of migration of genes in population "pools" in Chapter 10.

Secretors and nonsecretors. Some students will want to investigate A, B, and H antigens in water-soluble form in body fluids as well as saliva, tears, sweat, urine, semen, and gastric juice. Substance A is secreted by a member of blood group A; substance B by a member of blood group B. The antigenic substance H is secreted by members of group O. (Anti-H serum is used to detect H; all secretors have H in their saliva regardless of A-B-O blood groups.)

Let us assume that S is the gene for secretor, and s is the recessive allele. Secretors are either SS or Ss. This gene is inherited independently of the A-B-O antigens. Some association has been found between this S-gene and the Lewis blood group system. Table 8-7 is a summary of recent knowledge of human blood antigens (or agglutinogens).

Rh factor. Although there are many complexities involved in the inheritance of the Rh factor, we shall assume that the trait is inherited as a simple dominant. This assumption is based on the finding of a substance in the red corpuscles of 85 percent of the sample white population tested. This type of blood is called Rh positive. Fifteen percent of the population lack this substance and are designated as Rh negative. There are no natural antibodies against this substance in human serums, but the introduction of Rh antigen into blood that is Rh negative stimulates antibody production. However, there are several Rh antigens. This blood factor now seems to be inherited as a set of multiple alleles (14 or more alleles). See Table 8-8; also refer to typing blood, pp. 253–56.

You may want to have students work out several possible crosses to show the pattern of inheritance of the Rh factor, assuming—for purposes of simplicity—that it is a single gene trait. The following questions may serve as samples. A woman with Rh-negative blood marries a man

TABLE 8-6

Frequencies in different populations of the alleles i, I^A, and I^B, giving rise to the blood groups O, A, B, and AB*

population	number of persons tested	i	I^A	I^B
Americans (white)	20,000	0.67	0.26	0.07
Icelanders	800	0.75	0.19	0.06
Irish	399	0.74	0.19	0.07
Scots	2,610	0.72	0.21	0.07
English	4,032	0.71	0.24	0.06
Swedes	600	0.64	0.28	0.07
French	10,433	0.64	0.30	0.06
Basques	400	0.76	0.24	0.00
Swiss	275,644	0.65	0.29	0.06
Croats	2,060	0.59	0.28	0.13
Serbians	6,863	0.57	0.29	0.14
Hungarians	1,500	0.54	0.29	0.17
Russians (Moscow)	489	0.57	0.25	0.19
Hindus	2,357	0.55	0.18	0.26
Buriats (N. Irkutsk)	1,320	0.57	0.15	0.28
Chinese (Huang Ho)	2,127	0.59	0.22	0.20
Japanese	29,799	0.55	0.28	0.17
Eskimos	484	0.64	0.33	0.03
American Indians (Navajo)	359	0.87	0.13	0.00
American Indians (Blackfeet)	115	0.49	0.51	0.00
W. Australians (aborigines)	243	0.69	0.31	0.00
African Pygmies	1,032	0.55	0.23	0.22
Hottentots	506	0.59	0.20	0.19

* From E. Sinnott, L. C. Dunn, and T. Dobzhansky, *Principles of Genetics,* 5th ed., McGraw-Hill, New York, 1958; data from W. Boyd, *Genetics and the Races of Man,* Little, Brown, Boston, 1950, and A. Mourant, *The Distribution of the Human Blood Groups,* Thomas, Springfield, Ill., 1954.

who has Rh-positive blood. Will her children all have the Rh-positive factor since this trait is dominant? Explain. What are the chances of her having an Rh-positive child? An Rh-negative child? Explain. If both parents were Rh negative, what would be the chances of having an Rh-positive child?

At some time in this work students might report to the class on the role of the Rh factor in pregnancy and in cases of blood transfusions. Have them explain how Landsteiner and Wiener came to identify an "Rh" factor.

Adding to our understanding of the complexity of inheritance of the Rh factor is the use of two systems of symbols. Wiener proposed the use of Rh_0 and rh'; Fisher and Race use alleles C and c and genes D and E (Table 8-8). The two systems are based on differences in interpretation of serological mechanisms.

Students can read an excellent account of blood systems in L. H. Snyder and P. R. David, *The Principles of Heredity,* 5th ed. (Heath, 1957), in A. Srb, R. Owen, and R. Edgar, *General Genetics,* 2nd ed. (Freeman, 1965), and in E. Sinnott, L. C. Dunn, and T. Dobzhansky, *Principles of Genetics,* 5th ed. (McGraw-Hill, 1958), Chapter 9.

Inheritance of M and N blood types. Antigens M and N are inherited as a pair of alleles resulting in three genotypes: *MM, NN,* and *MN.* Natural antibodies

TABLE 8-7

Nine independent systems of human blood agglutinogens*

system	blood factors
A-B-O	A, B, O
Rh	D (Rh$_0$), d (Hr$_0$)
	C (rh'), c (hr')
	E (rh''), e (hr'')
MNSs	M, N
	S, s
P	P
Lewis†	Le^a, Le^b
Lutheran	Lu^a
Kell	K, k
Duffy	Fy^a, Fy^b
Kidd	Jk^a, Jk^b

* From *Hyland Reference Manual of Immunohematology,* 3rd ed., Hyland Laboratories, Los Angeles, 1965.

At the present time there are nine clearly defined blood group systems. The Diego and Sutter systems may become the tenth and eleventh, but their exact place has not yet been established. Also, a number of "private" and "public" antigens, described in the literature, may eventually become a part of existing or new blood group systems.

† There is evidence that Lewis and Lutheran may be linked.

TABLE 8-8

Common genotypes of eight Rh types*

	Rh type of the cells		approxi-mate % (Caucasian)	most common genotype		approxi-mate %	some other possible genotypes†		
Rh negative (15%)	rh	(cde)	14.4	cde/cde	(rr)	14.40			
	rh'	(Cde)	0.46	Cde/cde	(R'r)	0.46			
	rh''	(cdE)	0.38	cdE/cde	(R''r)	0.38			
	rh'rh''	(CdE)	rare	Cde/cdE	(R'R'')	rare			
Rh positive (85%)	Rh$_0$	(cDe)	2.1	cDe/cde	(R$_0$r)	2.0	cDe/cDe	(R$_0$R$_0$)	0.06%
	Rh$_0$'	(CDe)	50.7	CDe/cde	(R$_1$r)	31.68	CDe/cDe	(R$_1$R$_0$)	2.09%
				CDe/CDe	(R$_1$R$_1$)	16.61	CDe/Cde	(R$_1$R')	0.8%
	Rh$_0$''	(cDE)	14.6	cDE/cde	(R$_2$r)	10.96	cDe/cdE	(R$_2$R$_0$)	0.72%
				cDE/cDE	(R$_2$R$_2$)	2.5	cDE/cdE	(R$_2$R'')	0.34%
	Rh$_0$'Rh$_0$''	(CDE)	13.4	CDe/cDE	(R$_1$R$_2$)	11.5	CDe/cdE	(R$_1$R'')	0.97%
				CDE/CDe	(R$_z$R$_1$)	0.2			

* From *Hyland Reference Manual of Immunohematology,* 3rd ed., Hyland Laboratories, Los Angeles, 1965. Both Wiener (Rh) and Fisher (CDE) systems are included in the table.

† Rare genotypes, frequently less than 0.1 percent, are omitted.

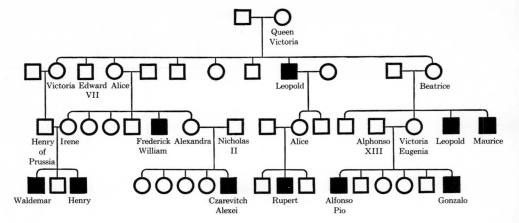

Fig. 8-58 Partial pedigree of royal families of Europe, showing transmission of the gene for hemophilia. (From L. H. Snyder and P. R. David, *The Principles of Heredity*, 5th ed., Heath, 1957.)

against them are not found in the serum of humans. (Blood typing is described in Chapter 5.)

Parents who have the genotype *MM* produce 100 percent *MM* offspring; *NN* × *NN* results in 100 percent *NN*; *MM* × *NN* results in 100 percent *MN*; *MN* × *MN* gives a probability of 25 percent *MM*, 25 percent *NN*, and 50 percent *MN* offspring.

Sex-linked genes in man. Students can study the transmission of hemophilia,

traced to a gene on the X-chromosome. Hemophiliacs lack a specific anti-hemophilic globulin in their plasma, a substance that is present in normal individuals. While hemophilia has been associated with royalty (Fig. 8-58), the disease occurs in about 1 in 25,000 males (and in 1 in 625 million females). A female would need the combination X^hX^h to have hemophilia (Fig. 8-59).

Also try the Ishihara test for one kind of color blindness, the familiar "red-green"

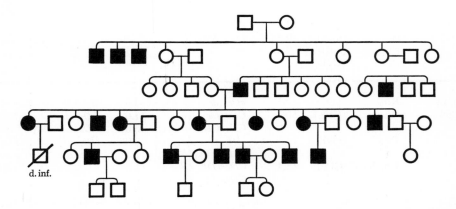

Fig. 8-59 Pedigree showing the occurrence of hemophilia in women. (From L. H. Snyder and P. R. David, *The Principles of Heredity*, 5th ed., Heath, 1957.)

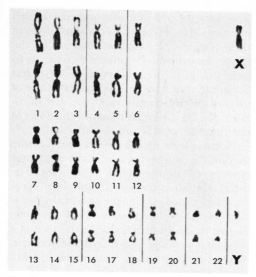

Fig. 8-60 Chart of human chromosomes: 22 pairs of autosomal chromosomes, plus X and Y. (Courtesy of J. H. Tjio.)

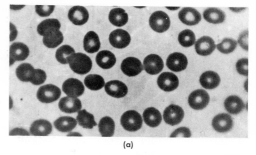

(a)

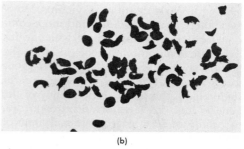

(b)

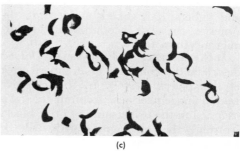

(c)

Fig. 8-61 Photomicrographs of human red blood cells: (a) normal cells from person of genotype *SS*; (b) cells from a heterozygous individual of genotype *Ss*, showing sickling of some red blood cells; (c) cells from a homozygous individual of genotype *ss*, showing complete sickling. (Courtesy of the Univ. of Michigan Heredity Clinic, Dr. James V. Neel, Director.)

color blindness.[22] About 8 percent of males and 0.5 percent of females show this trait, which has been traced to the X-chromosome. Why doesn't a father who is color-blind pass the gene for color blindness to his sons?

Autosomal traits. These are traits caused by genes found on pairs of chromosomes other than the sex chromosomes (see human chromosomes, Figs. 8-60 and 8-63, Table 8-9).

SICKLE-CELL ANEMIA. Red corpuscles of clinically healthy people (of heterozygous genotype) shrivel to a sickle shape when exposed to reduced oxygen tension (Fig. 8-61). Those who are homozygous "sicklers" (*SS*) suffer from severe anemia. The disease generally results in early death.

The change in shape of the corpuscles is due to the solubility of the reduced sickle-hemoglobin. In those who are "sicklers," either homozygous (*SS*) or heterozygous

(*Ss*) hemoglobin crystallizes out; water is also lost under conditions of reduced oxygen absorption.[23]

Hemoglobin of "sicklers" differs in only one amino acid of the nine that make up the specific peptide of normal hemoglobin.

---

[22] You may purchase a set of 13 color slides for testing color blindness (2 × 2 in. slides) from General Biological; or refer to the color chart on p. 97 in L. H. Snyder and P. R. David, *The Principles of Heredity*, 5th ed., Heath, Boston, 1957, or on p. 289 in the *Student Laboratory Guide* to *Biological Science: An Inquiry into Life* (BSCS Yellow Version), *op. cit.*

[23] L. Pauling *et al.*, "Sickle-Cell Anemia: A Molecular Disease," *Science* 110:543–48, 1949.

In this case, valine replaces the normal glutamic acid. In an electrophoresis apparatus, normal hemoglobin migrates to one pole, and sickle hemoglobin moves to the opposite pole; heterozygotes for the trait (*Ss*) have hemoglobin that separates into two fractions, each migrating to an opposite pole.

THALASSEMIA FACTOR. The gene for thalassemia major, when homozygous, causes Cooley's disease, a severe anemia leading to early death of its victims. However, there is a less deadly, and more common form, thalassemia minor, which seems to be due to a dominant gene (lethal in the homozygous condition). The heterozygous forms show varying degrees of anemia in their blood picture. The frequency of the gene is high in Mediterranean countries, especially in certain geographical areas of Italy; here some regions may have as high a frequency as 10 percent or 20 percent, and others only about 1 percent. In America, the gene is found mainly in immigrants from Italy.

ALCAPTONURIA. The physician Garrod described this condition as an inborn error of metabolism (1909). In this recessive trait, homogentisic acid accumulates and is excreted in the urine, turning urine black. Normally the homogentisic acid would be broken down into acetoacetic acid by a gene that produces an enzyme in blood serum. There is also a hardening and darkening of cartilage in the body of the affected individuals. An increase in the consumption of tyrosine and phenylalanine by alcaptonurics leads to an increase in homogentisic acid since there are blocks in the normal sequence of enzyme-regulated breakdowns of the amino acids into acetoacetic acid and final conversion to carbon dioxide and water. Use Fig. 8-62 to trace the complexity of this metabolic series that also includes tyrosine and albinism, and phenylalanine and phenylketonuria.

PHENYLKETONURIA. Another recessive causes a block in the metabolism of phenylpyruvic acid or of phenylalanine, resulting in extreme mental retardation.

The lack of the specific enzyme which handles the increased consumption of these amino acids can be detected by the large amounts of phenylpyruvic acid excreted in urine. There is a strip test available that can be used to test urine of infants for this condition;[24] the diet of infants can be corrected to avoid the onset of the mental retardation, or idiocy.

Normal individuals possess the enzymes to oxidize phenylpyruvic acid into precursors of homogentisic acid (see Fig. 8-62). These errors in metabolism have revealed much of the picture of degradation of phenylalanine and of tyrosine. Notice that albinism is due to the lack of a gene that produces an enzyme effective in converting a precursor into melanin (skin color pigments).

DEAF-MUTISM. Two different dominant genes are needed to produce normal hearing. For example, *D* and *E* interact to produce normal hearing. Any other combination of genes—such as *Ddee*, *DDee*, *ddEE*, or *ddEe*—results in children deaf at birth. They are mutes since they are deaf (unless they have special education). In this type of inheritance two "normals" may have a child who is deaf; two deaf parents may have a normal child. Most people have at least one dominant of each pair of genes, and are normal. Assume normal parents, *DdEe* × *DdEe*; could these parents give birth to a deaf child (*DDee*)?

SKIN COLOR. As many as three or four sets of genes may act to produce a multiple, cumulative effect of pigmentation. However, for the sake of simplicity, let us assume that only two pairs of genes are involved. In this multiple factor inheritance let *AABB* represent Negro skin color and *aabb* represent white, with mulatto having the genotype *AaBb*. Any three of the color-producing genes would produce dark skin; two would produce medium coloring; and one gene would produce a light skin. What variations may

[24] Phenistix is the trademark of the "dip-and-read" test for phenylketonuria in infants (Ames Co., Inc., Elkhart, Ind.).

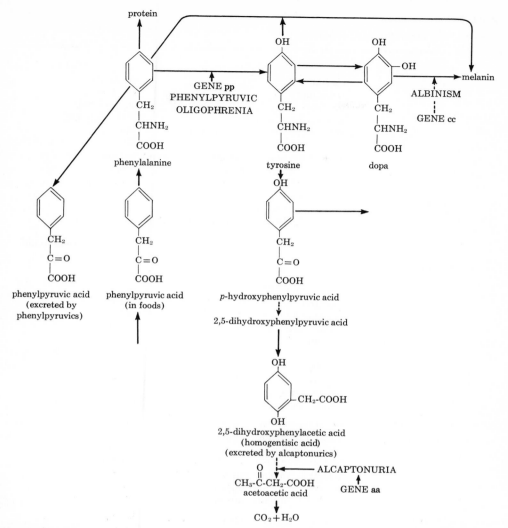

Fig. 8-62 Phenylalanine-tyrosine metabolism in man. The metabolic steps assumed to be blocked in homozygotes for the three genes indicated are shown by arrows. (After E. Sinnott, L. C. Dunn, and T. Dobzhansky, *Principles of Genetics*, 5th ed.; copyright, 1958, McGraw-Hill Book Co. Used by permission.)

occur in skin color of offspring of two mulattos, $AaBb \times AaBb$?

OTHER INHERITED TRAITS IN HUMANS. Examine Table 8-9 for other traits that students may want to read about and report back to the class. Further references in genetics are found at the end of this chapter. Students will find good reviews in some of the older issues of *Scientific American*

as well as in the current ones. For example, "Color Blindness" (March 1951), "Mongolism" (February 1952), "Inherited Sense Defects" (May 1952), "Radiation and Human Mutation" (November 1955), "Chemistry of Hereditary Diseases" (December 1956), "The Explanation of Twins" (May 1951), "Sex Differences in Cells" (July 1963), "Radiation and the

TABLE 8-9

Some human traits that are inherited*

	dominant	*recessive*
Hair, skin, etc.	dark hair	blond hair
	nonred hair	red hair
	curly hair	straight hair
	abundant body hair	little body hair
	early baldness (dominant in males)	normal
	white forelock	self-color
	piebald (skin and hair spotted with white)	self-color
	pigmented skin, hair, eyes	albinism
	black skin (two pairs of genes, dominance incomplete)	white skin
	ichthyosis (scaly skin)	normal
	epidermis bulbosa (sensitiveness to slight abrasions)	normal
	normal	absence of sweat glands
Eyes and features	brown	blue or gray
	hazel or green	blue or gray
	"Mongolian fold"	no fold
	congenital cataract	normal
	nearsightedness	normal
	farsightedness	normal
	astigmatism	normal
	free ear lobes	attached ear lobes
	broad lips	thin lips
	large eyes	small eyes
	long eyelashes	short eyelashes
	high, narrow bridge of nose	low, broad bridge
Skeleton and muscles	short stature (many genes)	tall stature
	dwarfism (achondroplasia)	normal
	midget (ateliosis)	normal
	polydactyly (more than five digits)	normal
	syndactyly (webbed fingers or toes)	normal
	brachydactyly (short digits)	normal
	progressive muscular atrophy	normal
Some conditions affecting systems	hereditary edema (Milroy's disease)	normal
	hypertension	normal
	normal	hemophilia (sex-linked)
	normal	sickle-cell anemia
	normal	diabetes mellitus
	tasters (PTC)	nontasters (PTC)
	normal	congenital deafness
	normal	amaurotic idiocy
	migraine	normal
	normal	dementia praecox (several pairs of genes)
	normal	phenylketonuria

* Adapted from C. Villee, *Biology,* 4th ed., Saunders, Philadelphia, 1962.

Human Cell" (April 1960), "Distribution of Man" (September 1960), and the entire issue of September 1959 on ionizing radiations.

Examining human chromosomes. In 1956 Tjio and Levan[25] reported that the number of chromosomes of humans is 46, not 48; this was soon confirmed by others.

Slides are available for the study of chromosomes in human leucocytes (magnification of at least 900× is needed for study of details). White blood cells are cultured and stimulated to divide by using an extract of red kidney beans, called phytohemagglutinin. As cells undergo mitosis, colchicine is added to arrest mitosis so that the chromatids do not separate; metaphase cells accumulate as a result. Cells are then placed in a hypotomic sodium citrate solution so that the cells become swollen as water enters; the chromosomes are then spread apart in the cells. Centrifuging then results in scattering of the chromosomes in the cells. Finally, the cells are fixed, dropped on slides, and rapidly dried in warm air. The procedure causes the swollen cells to collapse, and (after staining) chromosomes can be counted in one plane. Each chromosome is actually two chromatids that are joined by a centromere (Figs. 8-11, 8-63).[26]

Photomicrographs are made and the chromosomes are numbered according to a standardized system, developed in 1959, which describes shape, length, and arrangement of arms and the distance of the chromosomes from the centromere. As a result, female somatic cells would show 23 pairs (22 autosomal pairs and one pair of XX); there would be 22 pairs of autosomes in a male plus one asymmetrical pair of XY sex chromosomes (Fig. 8-60).

Several hereditary disorders have been

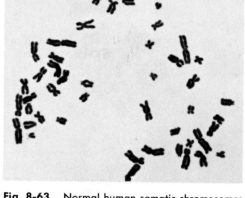

Fig. 8-63 Normal human somatic chromosomes from an *in vitro* culture of cells biopsied from an adult female; colchicine metaphase stage. (From J. H. Tjio and T. T. Puck, "The Somatic Chromosomes of Man," *Proceedings of the National Academy of Sciences* 44:1229–37, 1958.)

found to be associated with an abnormal number of chromosomes; Mongolism is an example in which the chromosome number is 47, due to the presence of an extra autosomal chromosome.

For additional information concerning the technique of preparing such slides refer to H. Spence and C. Grefer, "A Simple Method for Demonstration and Enumeration of Human Chromosomes" (*Am. J. Med. Technol.* 29:281–90, 1963).

SEXUAL DIMORPHISM IN BODY CELLS. A structural difference found in *undividing* cells is useful in distinguishing male and female genotypes. Davidson and Smith (1954) described a dimorphism in polymorphonuclear leucocytes of blood films from females and males. A bulb-like projection attached by a thread-like fiber to the lobed nucleus is found in cells of females; these bulb-like structures, called drumsticks, are absent in blood cells of males (Fig. 8-64).

Some workers place more confidence in the observation of Barr bodies (named after Barr, who described them in 1949) in cells than in the appearance of drumsticks in leucocytes, since the drumsticks have been found to be absent in some cases. The Barr body is a dark-staining body that lies

[25] J. H. Tjio and A. Levan, "The Chromosome Number of Man," *Hereditas* 42:1–6, 1956.

[26] Summarized from H. Edgerton and R. Flagg, "Human Chromosomes, Nuclei, and Sex," *Carolina Tips* 27:1–2, January 1964.

Also refer to *Mental Retardation,* published by Am. Medical Assoc., 535 N. Dearborn St., Chicago, Ill. ($0.40).

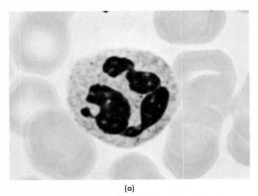
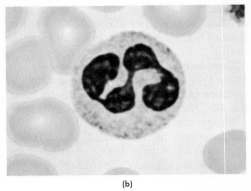

(a)　　　　　　　　　　　　　　　　(b)

Fig. 8-64 White blood cells of normal humans: (a) cells from a female, showing drumstick on nucleus; (b) cells from a male. (Both are magnified 2200×.) (Courtesy of Eastman Kodak Co.)

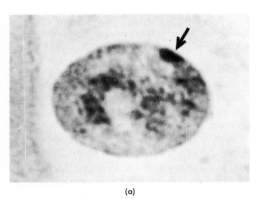
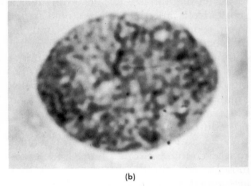

(a)　　　　　　　　　　　　　　　　(b)

Fig. 8-65 Nucleus of normal squamous epithelium in humans: (a) cell of female, showing Barr body (arrow); (b) cell of male. (Both are magnified 1750×.) (Courtesy of Carolina Biological Supply Co.)

against the membrane in the resting nucleus forming a distinctive pattern of heterochromatin material of the X chromosomes in females; it is used as a sex indicator (Fig. 8-65). Skin biopsies are made to secure deeper layers of epidermis, but more frequently oral smears are used. In deep skin cells of females, Barr bodies are found. Prepared slides of epithelial cells of males and females may be examined (Fig. 8-65).

Heredity in some plants

Corn (maize). Purchase hybrid seeds of corn containing recessive genes for al-binism from the Texas Agricultural Experiment Station, College Station, Tex.; Meyers Hybrid Seed Corn Co., Hillsboro, O.; biological supply houses; and Genetics Laboratory Supplies, Clinton, Conn. (among others). When germinated, these seeds should show 75 percent green shoots and 25 percent albino shoots, or a 3:1 ratio (Table 8-10). These seedlings show variations in the rate of formation of chlorophyll; further, they should be grown in sunlight so that the 75 percent show development of chlorophyll.

GENETIC EARS OF CORN. Ears of corn offer excellent opportunities for counting segregating traits. Large numbers of kernels

TABLE 8-10
Law of segregation in maize*

Parents:	Gg	$\times$	Gg	
Gametes:	G	g	G	g
Offspring:	GG	Gg	Gg	gg

	G	g
G	GG	Gg
g	Gg	gg

* Hybrid green plants, when inbred, produce off-spring in the ratio of three green to one albino (25 percent pure green, 50 percent hybrid green, and 25 percent albino).

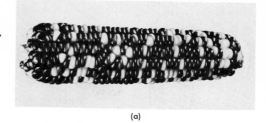

(a)

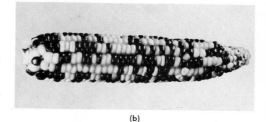

(b)

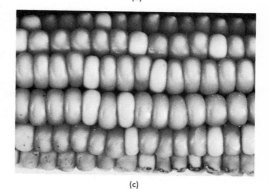

(c)

per ear give numbers that are close to theoretical Mendelian ratios. In many school situations, study of corn may be more practical than culturing and counting fruit flies.

Many examples of genetic crosses are available from biological supply houses. Study the 3:1 ratio of purple aleurone color to yellow aleurone (Fig. 8-66). These are the F_2 of hybrids resulting from crossing colored $\times$ colorless (yellow) aleurone: $RR \times rr \longrightarrow (F_1)\ Rr \longrightarrow (F_2)$ 1 RR : 2 Rr : 1 rr, or three purple to one yellow. Or show the results of a backcross of F_1 hybrid Rr to recessive parent— $Rr \times rr \longrightarrow$ 1 purple (Rr) : 1 yellow (rr).

Or count F_2 progeny for starchy-waxy endosperm (three starchy to one waxy) resulting from inbreeding hybrids in the F_1 progeny of a cross: $SS \times ss \longrightarrow Ss$.

Independent assortment is shown in the 9:3:3:1 ratio which results from crossing dihybrids with traits for purple and yellow aleurone and the starchy-sweet endosperm. You may also want students to work out crosses as in Fig. 8-67 to illustrate linkage and the percentage of crossover in genes for colored and colorless aleurone and for the endosperm trait full or shrunken.

WAXY-STARCHY POLLEN GRAINS. Preserved tassels of corn plants that are

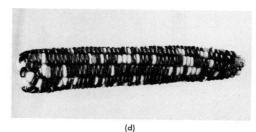

(d)

Fig. 8-66 Genetic ears of corn, showing segregation of factors: (a) 3:1 ratio of purple aleurone to yellow aleurone, a second generation cross from $RR \times rr$; (b) 1:1 ratio of purple to yellow resulting from $Rr \times rr$; (c) starchy-waxy endosperm (3:1), demonstrating segregation of recessive endosperm character, waxy; (d) 9:3:3:1 ratio of segregation of two independent factors, purple: yellow–starchy : sweet, from dihybrid parents with genotypes $RrSs \times RrSs$. (Courtesy of Carolina Biological Supply Co.)

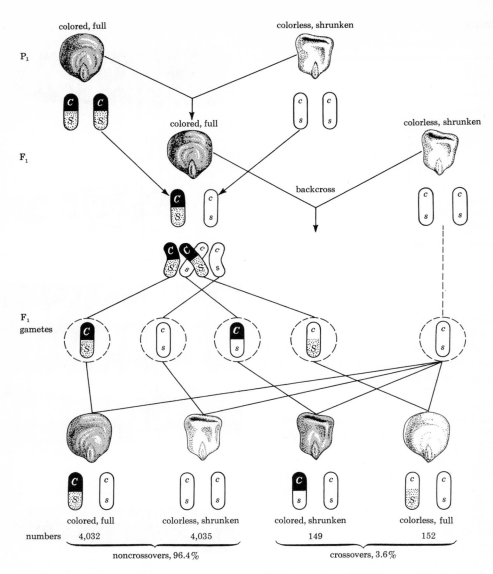

Fig. 8-67 Chromosome linkage and crossover, showing the behavior of genes for colored and colorless aleurone and for shrunken and full endosperm in corn. (After E. Sinnott, L. C. Dunn, and T. Dobzhansky, *Principles of Genetics,* 5th ed.; copyright, 1958, McGraw-Hill Book Co. Used by permission.)

heterozygous for non-waxy (*Wxwx*) are available; slides of pollen can be made in class. When Gram's iodine solution is added to the pollen on a slide, some pollen grains (starchy) show a distinct dark blue color, and others do not stain with iodine

(waxy). Students can count the proportion of both types on the slide. Since pollen grains of the heterozygous plants are gametophytic, a gamete ratio of 1:1 results.

STAINING KERNELS. If kernels of similar

heterozygous corn plants are cut and Gram's iodine applied to each, a ratio of three dark (starchy) to one waxy results— 1 *WXWX* : 2 *WXwx* : 1 *wxwx*.

(Also refer to enzymatic differences in wrinkled and smooth peas, that is, ability to change sugar to starch. See below.)

Soybeans. Obtain seeds that will produce seedlings in which the heterozygous plants (whether pure or hybrid) show a phenotypic difference: Homozygous shoots are green; heterozygous shoots are light green; and homozygous recessives are yellow. This gives a ratio of 1:2:1, not 3:1.[27]

Seeds germinate within 10 days. The green and yellow variations are distinct soon after germination, but after about 7 more days, the green color becomes distinct so that the dominant heterozygous and dominant homozygous are differentiated.

Peas. You may also obtain F_1 seeds from a cross between pure tall and dwarf peas. They will develop into hybrid tall pea plants. At the same time also purchase the seeds that are a result of a cross between two hybrid tall pea plants. Germinate the seeds in paper cups containing moist sand. As the seedlings grow, students will see the difference in height; a ratio of three tall to one short or dwarf will result.

What is the result when gibberellic acid is applied to dwarf pea plants? (Refer to the discussion of gibberellic acids in Chapter 7.)

GENETIC VARIATIONS IN PEAS. Wrinkled and smooth (or round) peas show variations in the type of starch grains they contain and in the rate of starch formation from a source of glucose-1-phosphate (dependent on the action of the enzyme starch phosphorylase).

Students can add scrapings from a cut surface of a wrinkled pea to a drop of Lugol's iodine solution on a microscope slide; prepare a slide of scrapings from a round pea in a similar manner. Notice,

under the microscope, the oval starch grains of round peas and the grooved, round grains of wrinkled peas.

Also prepare fluid extracts of chopped or ground dried peas of the two types. Be sure to wash the food grinder, blender, or mortar and pestle when preparing crushed peas of both round and wrinkled types. Add about 50 ml of water for every 10 g of peas used. Centrifuge if necessary to separate the fluid containing the enzymes.

Test the extracts for potency (and/or quantity) of the starch-forming enzyme by examining how rapidly a glucose solution is converted to starch by each type of extract. Use a small quantity of glucose-1-phosphate (about 0.5 g) added to 100 ml of distilled water. Use Lugol's iodine to test sample drops of the glucose solution on a slide at 10-min intervals. The enzyme extract from wrinkled peas shows a darker bluish black color. (Use a white background for the slides.)

You may prefer to add 2 g of plain agar to the glucose-1-phosphate and distilled water. Bring this to a boil. Then plate Petri dishes by pouring the extracts on separate dishes containing a film of glucose and agar. Test with Lugol's solution for the appearance of starch.

Sorghum. Red-stemmed sorghum is dominant over green-stemmed (when grown in the presence of sunlight). Seeds produced from inbred hybrid plants may be obtained from supply houses as well as the Brooklyn (N.Y.) Botanic Garden. These seedlings will germinate in about 5 days, but the differences in stem color are most obvious in about 7 to 10 days. When large numbers of seeds have been germinated, the 3:1 ratio of red-stemmed to green-stemmed plants results.

Also notice the difference in pollen heterozygous for waxy (Fig. 8-68).

Tobacco. Germinate seeds of tobacco— normal and albino—on agar plates. Carolina Biological recommends using a 3 percent agar to which carbon black is added in the proportion of 6 tsp of carbon black to 1 liter of water. You may want to add a mold inhibitor such as Tegosept M or

---

[27] Such seeds are available from Carolina Biological Supply Co., Burlington, N.C.

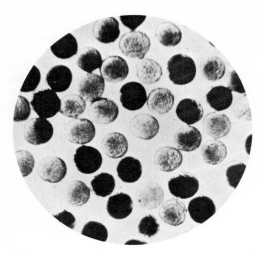

Fig. 8-68 Normal and "waxy" pollen grains in sorghum (magnified 270×, reduced by 39 percent for reproduction). When pollen grains from sorghum heterozygous for the waxy gene are treated with iodine, half the grains (the normal grains) stain blue (dark in the photo) and half (the waxy grains) stain red. (Courtesy of R. E. Karper, Texas Agricultural Experiment Station, from *J. Heredity* 24:6, June 1933.)

Methyl Parasepts (see p. 415); there are many such inhibitors under different trade names. Plug the flasks and sterilize them.

Devise a demonstration to show that the appearance of chlorophyll is due both to heredity and to the proper environment, in this case, light.

Irradiated seeds. Supply houses, such as Carolina Biological and General Biological, have available seeds that have been irradiated with doses ranging from 15,000

to 30,000 roentgens.[28] Many of the young seedlings will show variations resulting from radiation (Fig. 8-69). To determine whether mutations have occurred, at least two subsequent generations of the plants would have to be studied.

Neurospora. A series of laboratory activities may be planned to study heredity in this pink mold. Inoculate one side of an agar film (containing a minimal medium, described on p. 222) in a Petri dish with plus type *Neurospora crassa*. On the other half of the film add the opposite mating type, minus (these two types can be called *A* and *a*). A growth of asexual haploid hyphae result with haploid asexual spores, or conidia. However, conjugation, sexual fusion, or fertilization occurs when either haploid filaments or nuclei of conidia of opposite mating types fuse. As a result, a diploid fruiting body (a sac-like perithecium) forms (see Fig. 8-44).

Mount a perithecium on a slide in a drop of alcohol. Use fine jeweler's forceps or very thin platinum wire transfer needles. In alcohol, the perithecium softens and the long, tubular asci separate in the perithecium. The diploid fusion nucleus in each ascus undergoes two divisions in meiosis resulting in four haploid nuclei; these undergo a mitotic division so that eight haploid ascospores are lined up in each ascus in the perithecium.

[28] Clear descriptions of laboratory activities for measuring effects of radiations on chromatin and on cytoplasmic changes are given in D. Halleck, "Nuclear and Cytoplasmic Abnormalities Due to Gamma Radiation: Measurement Techniques," *Turtox News* 42:8, August 1964.

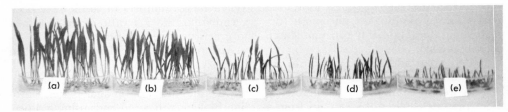

Fig. 8-69 Seedlings of 7-day-old Himalayan barley (*Hordeum vulgare*) germinated in water after X-irradiation with the following doses: (a) no irradiation; (b) 10,000 R; (c) 20,000 R; (d) 30,000 R; (e) 50,000 R. All these dosages (b through e) are sufficient to produce mutations. (Photo by H. Luippold, courtesy of Oak Ridge National Laboratory.)

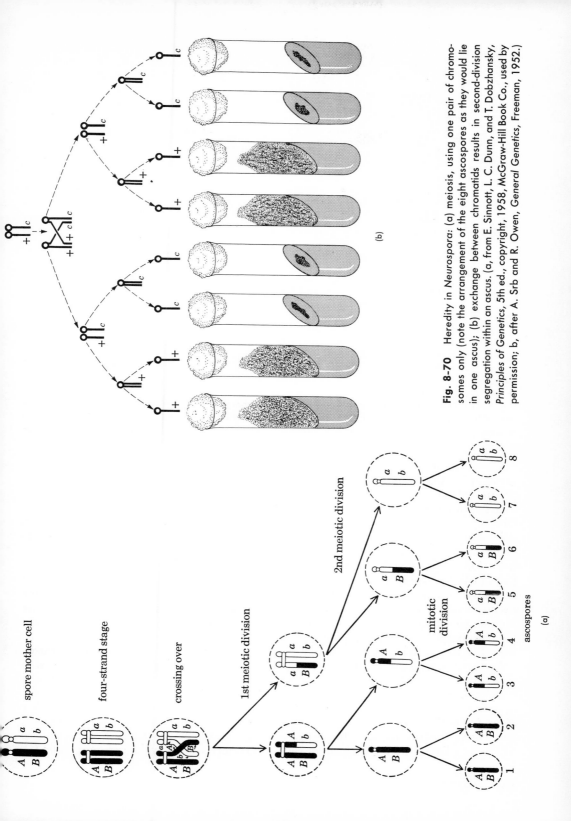

Fig. 8-70 Heredity in *Neurospora*: (a) meiosis, using one pair of chromosomes only (note the arrangement of the eight ascospores as they would lie in one ascus); (b) exchange between chromatids results in second-division segregation within an ascus. (a, from E. Sinnott, L. C. Dunn, and T. Dobzhansky, *Principles of Genetics*, 5th ed., copyright, 1958, McGraw-Hill Book Co., used by permission; b, after A. Srb and R. Owen, *General Genetics*, Freeman, 1952.)

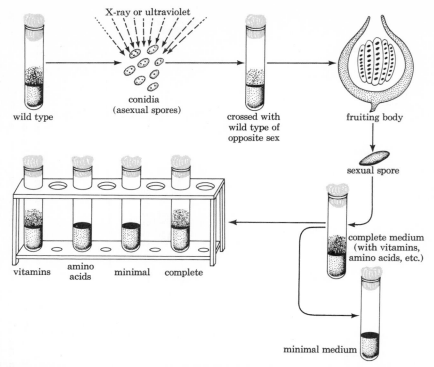

Fig. 8-71 Detection of biochemical mutants in *Neurospora*. The mutant lacks the ability to synthesize one of the vitamins. Notice that the mutant does not grow on minimal medium, or on minimal medium to which amino acids are added. It does grow when vitamins are added. (After G. W. Beadle, *Science in Progress,* Yale Univ. Press.)

Dodge showed that each ascospore is lined up so that four at one end are the product of one mating type, and the other four the product of the other mating type. The type of genes they contain will vary, depending on whether the alleles segregated at the first or second meiotic division. Lindegren showed that when two pairs of linked genes are crossed over, an exchange occurs in any one region between two of the four chromatids (between the one chromatid of one homologue and the chromatid of the other homologue) as shown in Fig. 8-70.

Students interested in special crosses, such as normal wild *Neurospora* with albino, may practice opening a fruiting body or perithecium, and separating out a single ascus. Carefully separate each of the eight ascospores, keeping track of the order in which they lay in the ascus. Transfer each spore with a transfer loop of sterile water (to avoid loss of dry ascospores or contamination) to an agar slant or a Petri dish containing the special nutrient medium needed for *Neurospora* to grow. Heat in a water bath at 55° C (131° F) for half an hour; cool and then incubate. Complete details are provided in the fine laboratory manual of N. Horowitz, *Chemical Genetics* (Division of Biology, California Institute of Technology, Pasadena, 1960). Also refer to the BSCS film *Neurospora Techniques* (Thorne).

DETECTION OF BIOCHEMICAL MUTANTS. The steps used by Beadle and Tatum to detect nutritional mutants are illustrated in Figs. 8-71 and 8-72. *Neurospora* normally synthesizes its amino acids and other materials from a minimal diet. This diet, which may be purchased from Difco (see directory of supply houses, Appendix C), may be prepared by combining the following:

ammonium nitrate	1	g
ammonium tartrate	5	g
monopotassium phosphate	1	g
magnesium sulfate	0.5	g
calcium chloride	0.1	g
sodium chloride	0.1	g
ferric chloride	10	drops of 1% solution
zinc sulfate	10	drops of 1% solution
sucrose	15	g
biotin	5	mg
distilled water	1	liter

The medium should be sterilized at 15 lb pressure for 10 min; since sucrose is present, avoid overheating. If solid medium is needed, pure agar can be added; the agar should be a select grade to avoid contamination of the medium with other minerals. It is preferable to grow strains on liquid media (see minimal diet above) to avoid unknown contaminants in the agar.

When a nutritional deficiency due to a mutation had occurred, Beadle and Tatum could locate it quickly because the spores did not grow on the minimal medium. If the spores could grow on a minimal medium to which one amino acid or one vitamin had been added as a supplement, this was the factor that the mutant could no longer synthesize; apparently it had lost the enzyme to synthesize this product. Ascospores from treated (by radiation or chemicals) and from untreated fungi are placed in individual test

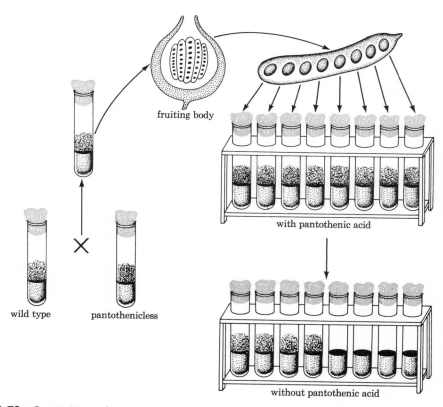

fruiting body

wild type   pantothenicless

with pantothenic acid

without pantothenic acid

Fig. 8-72 Determining inheritance of a biochemical mutant type in *Neurospora*. Conidia are transferred from medium containing pantothenic acid to medium lacking the test substance. Notice the segregation of the mutant in the test tubes lacking pantothenic acid. (After G. W. Beadle, *Science in Progress*, Yale Univ. Press.)

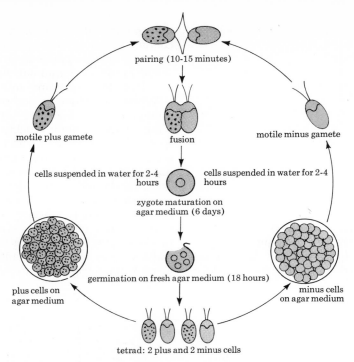

pairing (10-15 minutes)

motile plus gamete

fusion

motile minus gamete

cells suspended in water for 2-4 hours

cells suspended in water for 2-4 hours

zygote maturation on agar medium (6 days)

plus cells on agar medium

germination on fresh agar medium (18 hours)

minus cells on agar medium

tetrad: 2 plus and 2 minus cells

Fig. 8-73 Sexual reproduction in *Chlamydomonas*. (Adapted from R. Levine, *Genetics*, Holt, Rinehart and Winston, 1962, and R. Manwell, *Introduction to Protozoology*, St. Martin's, 1961.)

tubes containing a complete medium; then they are allowed to produce mycelia so that the plant is established stock. Transfers are then made to tubes with a variety of minimal media.

This tedious, yet elegantly simple, technique was the basis for the "one gene, one enzyme" theory that led to the discoveries of biochemical processes that gained for Beadle and Tatum, along with Lederberg, the 1958 Nobel Prize.

Seven linkage groups have been found in *Neurospora crassa*. Laboratory work may lead to fairly complex studies for small groups of highly motivated students. Extreme caution must be practiced in the laboratory, for *Neurospora* can easily contaminate the whole laboratory (the spores are highly resistant).

Also read a description of a cross in *Neurospora* that shows independent assortment (genes on different chromosomes) of two sets of traits, a pattern of spreading growth, and its allele a pattern of colonial

growth; these alleles are on chromosome 1. Another pair of alleles, found on chromosome 2, is for color—pink or orange, and albino conidia. Students will want to read further in D. Bonner and S. Mills' fine little book, *Heredity*, 2nd ed. ("The Foundations of Modern Biology Series," Prentice-Hall, 1964). For further information, see the text by Sinnott, Dunn, and Dobzhansky, and that of Srb, Owen, and Edgar (see the end-of-chapter bibliography).

Chlamydomonas. The biochemical genetics of sexuality has been widely studied in this flagellated green alga (Fig. 8-73). Haploid cells reproduce by fission. Diploid zygotes are formed in some species under specific conditions and then undergo meiotic divisions resulting in four haploid cells.

Moewus and his group (1950), working with *Chlamydomonas reinhardi*, showed that, under gene control, chemicals were produced that influenced potencies or valences of sexuality. They described five valences of maleness and five of femaleness. Furthermore, they postulated that certain hormone-like substances initiated growth of flagellae, control of conjugation, or determination of valence of sexuality, and so on.

There are two physiologically different mating types among *Chlamydomonas*. Members of a clone arising from one cell are all "plus" (mt^+) type; members of another clone are "minus" (mt^-) mating type. Pairing occurs when algae of opposite mating types are brought together in a test tube of medium, and the fusions result in diploid zygotes. After meiotic divisions haploid cells (or gametes, if you will) are formed, the typical *Chlamydomonas* cells. Two cells of the tetrad are mt^+ and two are mt^- (Fig. 8-73). These cells show no structural differences, as all cells of *Chlamydomonas* look alike, but there is a physiological difference, so they may also be considered gametes.

One trait that has been studied is the wild, or normal, flagellated form as compared with paralyzed (nonmotile) flagella. When a mating type of the strain producing normal flagella is crossed with the opposite mating type of a strain having paralyzed flagella, zygotes that are heterozygous are formed. As a result of meiosis, four cells are formed—two that have flagella and two that lack motile flagella. In a liquid medium in a test tube, the paralyzed cells sink to the bottom of the test tube so that the two types can be separated, thereby obtaining individual cells from which pure clones can be developed (barring mutations). Like most unicellular forms and fungi, the predominating stage in the life cycle of *Chlamydomonas* is haploid.

Levine and Ebersold[29] describe segregation of a pair of alleles. Wild-type green *Chlamydomonas* can carry on photosynthesis

in a minimal medium containing carbon dioxide, water, and inorganic salts. A yellow biochemical mutant can grow only on a minimal medium supplemented with those specific substances that it cannot synthesize. For example, one mutant cannot synthesize the amino acid arginine. When opposite mating types of the two strains, "wild" green and arginine-deficient, are mixed, pairing occurs and fusion results in a diploid zygote. In meiosis, a tetrad is formed, 50 percent carrying the normal gene ($+$) and 50 percent deficient in arginine synthesis (*arg*). If large numbers of these are mixed together, a cross of $+mt^+ \times arg\ mt^-$ occurs; and in about 15 min, pairing and zygotes result. Zygotes are transferred to minimal agar medium to which arginine has been added. Replica plating (a technique developed by Lederberg) is used to reproduce pure cultures, thereby avoiding transfers by needles or loops (Fig. 8-74). Further data is available in Levine's *Genetics*.

In Fig. 8-75, note the independent assortment of two pairs of alleles in tetrads of haploid *Chlamydomonas*. Levine describes a cross between an arginine-requiring strain (*arg*) and an acetate-requiring strain (*ac*): $arg+ \times +ac$. Products of meiosis are separated and grown on medium supplemented with both arginine and acetate; the colonies which grow can be transferred to fresh media by means of replica plating (Fig. 8-74). Transfers are made to four different kinds of media: minimal, minimal plus arginine, minimal plus sodium acetate, and minimal plus both supplements. Study the possible crosses which result, depending on whether crossover occurs in the first division or not.

Taking a specific case, as in the test analysis of cross $arg+ \times +ac$ in haploid *Chlamydomonas*, there are three classes of tetrads formed (Fig. 8-75). Two of the products have the genotype of one parent, and two show the genotype of the other parent, with the result that the parental types, *PD*, are segregated in a ratio of 1:1. Also note that the second possibility gives

[29] R. Levine and W. Ebersold, in *Exchange of Genetic Material*, Cold Spring Harbor Symposia on Quantitative Biology, 23, Cold Spring Harbor Laboratory of Quantitative Biology, Cold Spring Harbor, N.Y., 1958; also in R. Levine, *Genetics*, Holt, Rinehart and Winston, New York, 1962.

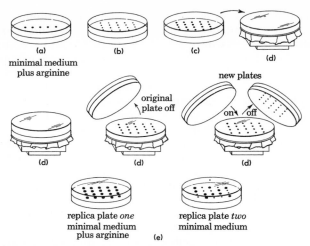

Fig. 8-74 Tetrad analysis of a cross of arginine (*arg*) and normal gene (+) in *Chlamydomonas*: (a) transfer by replica plating of mature zygotes to minimal medium supplemented with arginine; (b) meiosis and germination; (c) mitosis; (d) replica plating; (e) resulting plates. (From R. Levine, *Genetics*, Holt, Rinehart and Winston, 1962.)

minimal medium plus arginine

new plates

original plate off

on / off

(d)

replica plate *one*
minimal medium
plus arginine

replica plate *two*
minimal medium

(e)

a 1:1 ratio of new combinations of genotypes: *arg ac* and + +; these are *NPD*, or nonparental types. In the third possible set of tetrads (as shown in Fig. 8-75b) there are four different genotypes in a ratio of 1:1:1:1 of tetratypes, *T*; two show the parental types *arg*+ and +*ac*, and two are recombinations *arg ac* and + +. Locate where the segregation occurred for *arg* and its + allele, and for *ac* and its + allele.

Acetabularia. There may be the opportunity to show that the nucleus of a cell is the source of genetic control and that the cytoplasm is affected by the genetic composition of the nucleus. *Acetabularia* (Fig. 8-76) is a one-celled marine alga differentiated into an umbrella-like cap on a stalk, which has a holdfast, or rhizome. (Remember this is a one-celled organism.) The nucleus of this alga is in the rhizome. Regeneration experiments by Hämmerling (see p. 346) showed that when the stalk and cap were removed, the rhizome containing the nucleus regenerated a new cap and stalk. Further, notice in Fig. 8-77 that grafting experiments exchanging reciprocal parts of two species of *Acetabularia* resulted in producing the shape of cap controlled by the genetic contents of the nucleus of the holdfast. Whatever substance or substances emanate from the nucleus pass through the intervening cytoplasm of a different species and affect the shape of the cap.

Yeasts and bacteria. Plus and minus mating strains are found in some yeasts. You may want to obtain yeast cultures that ferment galactose (producing carbon dioxide that can be measured) and cross these with cells that are unable to ferment this sugar. Students will want to refer to the classic work of Lindegren on yeasts (as well as on *Neurospora*).[30]

The literature on sexuality in bacteria is extensive, as is the study of mutations among bacteria. Some activities are well described in *Biological Science: Molecules to Man* (BSCS Blue Version, Houghton Mifflin, 1963). In the "Laboratory Investigations" section of this text, see Investigations 20 and 21 for work with bacteria.

Also refer (in genetics texts) to work that has been done with *Pneumococcus* and transformation, studies of transduction,

[30] C. Lindegren, *The Yeast Cell: Its Genetics and Cytology*, Educational Publishers, St. Louis, 1949.

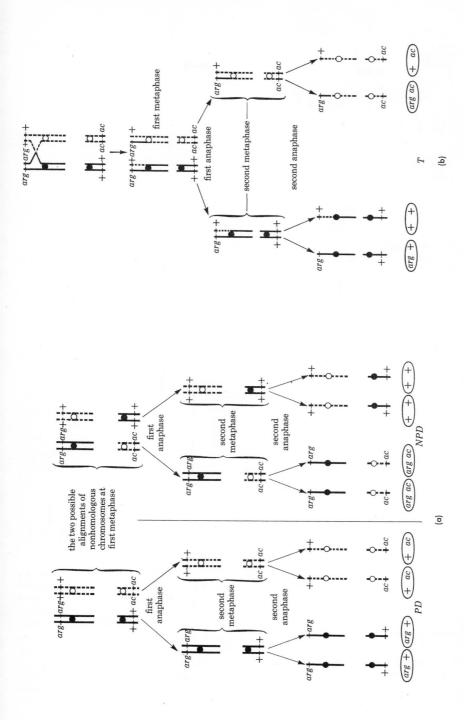

Fig. 8-75 Independent assortment of two pairs of alleles in tetrads of *Chlamydomonas*: (a) tetrad analysis of possible alignments of nonhomologous chromosomes at the first metaphase, producing either parental ditypes (*PD*) or new combinations, nonparental ditypes (*NPD*); (b) formation of tetratypes (*T*), a third possible recombination of alleles in a ratio of 1:1:1:1. Segregation of *arg* and its + allele and its + allele. (After R. Levine, *Genetics*, Holt, Rinehart and Winston, 1962.)

Fig. 8-76 *Acetabularia*, a unicellular marine alga. (Courtesy of General Biological Supply House, Inc., Chicago.)

and work in phage and *Escherichia coli.* Students should be aware of the precautions used in handling all bacteria. Pathogenic bacteria should not be used in high school biology (see Chapter 9).

Mutations. We have mentioned the possibility of studying the effects of radiation and of chemical treatment on the growth of seedlings, of bacteria, of *Neurospora,* and others. Some students may wish to demonstrate the results of treating plants with colchicine (*caution:* poisonous and especially dangerous to the eyes). Squash smears may be prepared of root tips of onions to show the doubling of chromosomes due to irregularities in spindle formation and the production of new cell walls. See Chapter 9 for the preparation of squash slides to show mitosis and effects of colchicine (also see Fig. 2-23).

Slides of mounted specimens of mutant forms of fruit flies are available for study, or flies may be prepared by mounting them in Karo syrup (see Chapter 2).

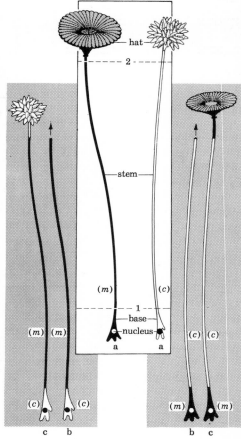

Fig. 8-77 Grafting experiments in *Acetabularia.* Grafting *Acetabularia mediterranea* stem to the nucleated base of *A. crenulata* (dark, left-hand panel) results in the generation of a deeply indented hat characteristic of *A. crenulata* (center panel, left). The reciprocal graft (dark, right-hand panel) produces an umbrella-shaped hat consistent with the character of *Acetabularia mediterranea* (center panel, right). (After A. Srb, R. Owen, and R. Edgar, *General Genetics,* 2nd ed., Freeman, 1965.)

Law of probability

It is important to illustrate the fact that a 3:1 ratio among the offspring of hybrids is the result of random combinations of eggs and sperms. There are several ways to show this symbolically. Among these are the following.

1. Have students toss two similar coins simultaneously on the desk and keep tallies of the number of combinations of two heads, two tails, and a head-and-tail toss. Have students toss coins 10 times, then 50 times, and finally 100 times. Compile all the results and show that only when large numbers are used will the 1:2:1 ratio come out with some accuracy. The two sides of a coin represent two possibilities similar to the segregation of a pair of genes in a hybrid, such as $TD \times TD$. Each pair TD separates into T and D. Thus, in recombination at fertilization this ratio appears: 25 percent TT, 50 percent TD, and 25 percent DD.

2. Mix together in a large bowl 100 green and 100 yellow dried peas. Or use black and white marbles, or beads. Have a student pick out two at a time, without looking. His partner may keep a tally of the number of times each of the three possible combinations appears. Each pea represents an egg or sperm (carrying one gene of a pair). Two together are symbolic of a union, a fertilization. For instance, approximately 25 percent of the times a student will pick up two green peas; 25 percent of the times, two yellow ones; and 50 percent of the times he will select a combination of green and yellow peas. Of course, a large number of selections is needed to approximate this 1:2:1 ratio.

You may want to have some students use the chi-square method for determining goodness of fit (below). Also refer to the Hardy-Weinberg law, and other work on population genetics as it is referred to in tracing the migration of alleles for blood groups among humans (Chapter 12).

Chi-square (χ^2) method. Chi-square is a criterion for testing the discrepancy between a set of observed values and the corresponding theoretical values expected on the basis of some hypothesis concerning a population.

In a random "picking" of yellow and green peas, or in the tossing of coins (as described above), results do not exactly follow the theoretical expected numbers. How great is the difference between the theoretical results (based on a proposed hypothesis) and the actual data obtained? Can the difference in results still be explained on the basis of chance, or is there a larger difference, significant enough, so that one may suspect that some factor other than chance is involved in the deviations observed? (*Note:* These are qualitative differences; quantitative differences in a population are not tested with this chi-square method.)

To test how often chance alone can be expected to account for the deviation, the formula

$$\chi^2 = \sum \frac{d^2}{c}$$

is used (d = deviation, c = the theoretical expectation, and Σ = the summation, or sum of). This simplified formula is derived from

$$\chi^2 = \sum \left[\frac{(o - c)^2}{c} \right]$$

where o = observed data. The deviation squared is

$$d^2 = (o - c)^2$$

Suppose that a student picks up at random several handfuls of yellow and green peas mixed thoroughly in a container. Upon counting his sample of 104 peas he does not find 52 yellow and 52 green as would be expected on the hypothesis of a 1:1 ratio. Instead he finds 56 yellow and 48 green peas. The deviation of the observed results from the antici-

TABLE 8-11
Table of Chi-square*

degrees of freedom	P = 0.99	0.95	0.80	0.50	0.20	0.05 (1 in 20)	0.01 (1 in 100)
1	0.000157	0.00393	0.0642	0.455	1.642	3.841	6.635
2	0.0201	0.103	0.446	1.386	3.219	5.991	9.210
3	0.115	0.352	1.005	2.366	4.642	7.815	11.341
4	0.297	0.711	1.649	3.357	5.989	9.488	13.277
5	0.554	1.145	2.343	4.351	7.289	11.070	15.086
6	0.872	1.635	3.070	5.348	8.558	12.592	16.812
7	1.239	2.167	3.822	6.346	9.803	14.067	18.475
8	1.646	2.733	4.594	7.344	11.030	15.507	20.090
9	2.088	3.325	5.380	8.343	12.242	16.919	21.666
10	2.558	3.940	6.179	9.342	13.442	18.307	23.209
15	5.229	7.261	10.307	14.339	19.311	24.996	30.578
20	8.260	10.851	14.578	19.337	25.038	31.410	37.566
25	11.524	14.611	18.940	24.337	30.675	37.652	44.314
30	14.953	18.493	23.364	29.336	36.250	43.773	50.892

* From A. Srb and R. Owen, *General Genetics,* Freeman, San Francisco, 1952 (abridged from Table III of *Statistical Methods for Research Workers* by R. Fisher, Oliver & Boyd, Edinburgh).

pated is $+4$ and -4. Substitute in the formula:

$$\chi^2 = \sum \frac{d^2}{c} = \frac{(4)^2}{52} + \frac{(-4)^2}{52}$$

$$= \frac{16}{52} + \frac{16}{52} = 0.31 + 0.31 = 0.62$$

Now that the value of χ^2 in a given experiment is known, the next step is to calculate how often, on the basis of chance alone, this value χ^2 is likely to occur; that is, what is the probability that the deviation obtained will occur by chance alone. The number of degrees of freedom referred to in Table 8-11 relating to χ^2 values is one less than the number of classes in the ratio. For example, in the example given above either green or yellow seeds will occur; if green is counted, there is only one freedom, the chance of yellow. Hence, in this case the degree of freedom is 1. Notice that χ^2 value of 0.62 falls between probability, or P, values of 0.20 and 0.05. Therefore it is not unusual for a sampling to have 56:48 instead of the theoretical 52:52 of the total of 104 in the sample.

When two coins are tossed the expected results are 1 (tail-tail) : 2 (head-tail) :

1 (head-head). In an actual tally of 200 flips of the two coins, the expected results should be 50:100:50. Can the deviation of the actual data be accounted for on the basis of chance? A test of a 1:2:1 ratio would have two degrees of freedom. (A 3:1 ratio would have one degree of freedom.) Calculate the χ^2 value and use Table 8-11, knowing the number of degrees of freedom at which to enter the table. Notice that the probability decreases as the value of χ^2 increases. (A probability of 0.05 is 1 in 20; a probability of 0.01 is 1 in 100.)[31]

Students may want to read the paper "Probability," as well as others in the issue of *Scientific American* for September 1964, which is given over to mathematics in the modern world.

Galton apparatus. Some students may be skillful enough to make a pinball

[31] There are excellent chapters on statistics useful in genetics in both A. Srb, R. Owen, and R. Edgar, *General Genetics,* 2nd ed., Freeman, San Francisco, 1965, and Sinnott, Dunn, and Dobzhansky, *op. cit.*
Also refer to the *Student Laboratory Guide* to *Biological Science: An Inquiry into Life* (BSCS Yellow Version), *op. cit.,* and to the "Laboratory Investigations" section of *Biological Science: Molecules to Man* (BSCS Blue Version), Houghton Mifflin, Boston, 1963.

machine of the type that Galton used so that when lead shot, small marbles, or beads are released from a chamber they pass through a grid of nails and fall in random assortment into a curve of distribution showing the laws of probability.

E. Dodson described the construction of such a "Galton apparatus" (Fig. 8-78) in this way:

> The machine is based upon a sheet of 1″ plywood 20″ × 36″, and the whole sheet surrounded by strips of 1″ × 2½″, flush with the lower surface of the plywood. Thirteen columns for the reception of shot were made by cutting 12 slots in the baseboard equidistant (1⁹⁄₁₆″ apart), 12″ long, and ¾″ deep. Into these, strips of sheet metal 1¼″ wide and 12″ long were driven, thus making 13 columns ½″ deep and 12″ long on the surface of the board. (See figure to visualize the arrangement.) 2″ above these columns, a row of 1″ finishing nails was driven into the baseboard so as to project ½″ above the board. The nails were spaced 1″ apart. 9 such rows in all were used, each separated by 1″ from the rest, and the nails of successive rows alternating (see figure). The storage chamber for shot was separated from this grid by triangular dividers made of 1″ board 5″ on the base and 9½″ on each of the longer sides, thus leaving an opening of about 1¼″ at the center. A quantity of ⅜″ lead shot was then put into the shot chamber, and a sheet of glass put over the machine and held in place by metal clips. By raising one end of the machine, the shot can be run through the grid. While every run is a little different from the last one, in general the pattern of shot gives a beautiful approximation of a normal probability curve, as shown in the accompanying figure.[32]

A reading from Mendel's work

There may be times when a reading from an original source is appropriate. Such a translation as the one given below (of Mendel's paper) might be mimeographed and developed in class as part of a case history in heredity. This example shows several methods used by scientists.

[32] From E. Dodson, "The Galton Apparatus for Demonstration of the Laws of Probability," *Turtox News* 36:2, February 1958. Reprinted by permission of General Biological Supply House, Inc., Chicago, and the author.

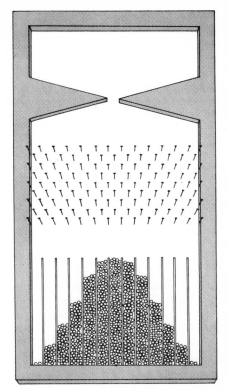

Fig. 8-78 Galton apparatus for demonstrating the laws of probability. (After E. Dodson, "The Galton Apparatus for Demonstration of the Laws of Probability," *Turtox News* 36:2, February 1958.)

Or you may want to add questions based on the reading and test the ability of students to interpret what they read.

THE GENERATION BRED FROM THE HYBRIDS

In this generation there reappear, together with the dominant characters, also the recessive ones with their peculiarities fully developed, and this occurs in the definitely expressed average proportion of three to one, so that among each four plants of this generation three display the dominant character and one the recessive. This relates without exception to all the characters which were investigated in the experiments. The angular wrinkled form of the seed, the green colour of the albumen, the white colour of the seed-coats and the flowers, the constrictions of the pods, the yellow colour of the unripe pod, of the stalk, of the calyx, and of the leaf venation, the umbel-like form of the inflorescence, and the dwarfed stem, all reappear in the numerical proportion given, without any essential alteration. *Transitional forms were not observed in any experiment.*

Expt. 1. Form of seed—From 253 hybrids 7,342 seeds were obtained in the second trial year. Among them were 5,474 round or roundish ones and 1,850 angular wrinkled ones. Therefrom the ratio 2.96 to 1 is deduced.

Expt. 2. Colour of albumen—258 plants yielded 8,023 seeds, 6,022 yellow, and 2,001 green; their ratio, therefore, is as 3.01 to 1.

Expt. 3. Colour of the seed-coats—Among 929 plants 705 bore violet-red flowers and grey-brown seed-coats; 224 had white flowers and white seed-coats, giving the proportion 3.15 to 1.[33]

The entire translation is a splendid example of reporting the results of experiments and the presenting of a working hypothesis to explain inheritance in garden peas.

Demonstration Punnett square

Students may want to construct this do-it-yourself device by screwing cup hooks into a large square of plywood as in Fig. 8-79. Two small hooks close together represent the diploid condition of the parent organisms; single hooks represent the gametes, or the haploid possibilities. Below this, show a Punnett square with four boxes. In each box show two hooks, representing a diploid organism. Along the top and left side of the square, insert single hooks, representing the gametes.

Then students might make models of flowers out of plywood or cardboard and color them with crayon. Also model small animals such as guinea pigs or rats, black and white ones. Use small circles of color to indicate the genes in the gametes. Make

[33] From Gregor Mendel, *Experiments in Plant-Hybridisation,* trans. by Royal Horticultural Soc. of London, Harvard Univ. Press, Cambridge, Mass., 1933, p. 321. Reprinted by permission of the publisher.
Students may find Mendel's paper reproduced in the appendix of Sinnott, Dunn, and Dobzhansky, *op. cit.*

holes in these models and hang them on the hooks. Youngsters may try to make a variety of crosses in front of the class. Errors are easily detected as they occur.

Heredity in rats, mice, and hamsters

You may be able to purchase pedigreed hooded rats from an experimental station or college. Cross hooded with albino rats. Hoodedness seems to be dominant (with a few white, or albino, offspring appearing in the F_2, or second, generation after inbreeding the hybrid hooded F_1 generation). This assumes the original cross was pure hooded × albino.

Pedigreed mice may be purchased from several supply houses (Appendix C). Directions for possible crossings will be supplied by the biological supply house. (For ways to maintain mice or rats in the laboratory, see Chapter 12.) Make crosses between pure black mice and albino mice (Table 8-12); the F_1 offspring will be all black. These are hybrid black, each animal carrying a recessive gene for albinism.

When these F_1 offspring are inbred (Table 8-13), the second filial generation, F_2, should show the ratio of three black to one white.

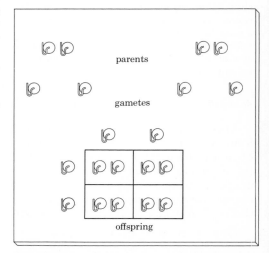

Fig. 8-79 Demonstration board, made of plywood and cup screws, for simple genetic crosses.

TABLE 8-12

Crossing pure black and albino mice*

Key: B = black, b = albino

Parents:	BB	×	bb
Gametes:	B		b
Offspring:		Bb	

* Offspring are hybrid black.

TABLE 8-13

Crossing two hybrid black mice*

Parents:	Bb	×	Bb
Gametes:	B b		B b
Offspring:	BB Bb Bb bb		

	B	b
B	BB	Bb
b	Bb	bb

* Offspring show a ratio of three black to one white (25 percent pure black, 50 percent hybrid black, and 25 percent white).

TABLE 8-14

Test cross: mating a black mouse of unknown genetic make-up to an albino*

Parents:	$B?$	×	bb
Gametes:	B $?$		b
Offspring:	Bb		$?b$

	b			b
B	Bb	or	B	Bb
B	Bb		b	bb

* If the mouse was pure black, there can be no white mice in the litter; if the mouse was hybrid, the offspring will show a 1 : 1 ratio.

Students may make test crosses (back-crosses) by mating the black offspring in the second generation with an albino parent in order to determine whether the black animal is pure black or hybrid black (Table 8-14).

Any of these ratios, of course, can be expected to hold only when large numbers of offspring are produced. In practice, a male would be mated to a large number of genetically similar females in order to get many offspring.

R. Whitney[34] recommends hamsters as subjects for genetic study. Some mutations that have been studied are partial albino (white coat, dark ears, red or black eyes), cream (yellow coat, black ears and eyes), microphthalmia (minute eyes), tawny (tan coat with black markings, black eyes and ears), brown or cinnamon (cinnamon coat, peach-gray ears, and eyes a rich red color), and white band (coat with a broad white band circling the body just behind the shoulders).

Genetics of fish

There is an extensive literature on heredity in the guppy, *Lebistes reticulatus*. (For ways to maintain tropical fish, see Chapter 12.) Blond and golden are alleles of the wild type and illustrate Mendel's laws.

For detailed studies see H. B. Goodrich *et al.,* "The Cellular Expression and Genetics of Two New Genes in *Lebistes reticulatus*" (*Genetics* 29:584–92, 1944). Also look into O. Winge and E. Ditlevsen, "Color Inheritance and Sex Determination in *Lebistes*" (*Heredity* 1:65–83, 1947).

Some students may have pure strains of platys—unspotted and the dominant, spotted. Swordtails show readily observable traits such as body color and color pattern of the wagtail.

Useful pamphlets, prepared by M. Gordon,[35] are available.

[34] R. Whitney, "Hamsters as Genetic Subjects," *Turtox News* 40:4, 1962.

[35] M. Gordon, *Platies, Siamese Fighting Fish,* and *Swordtails,* Am. Museum of Natural History, New York ($0.25 each).

Also refer to M. Gordon, "Distribution in Time and Space of Seven Dominant Multiple Alleles in *Platypoecilus maculatus*," *Advances in Genetics* 1:95–132, 1947.

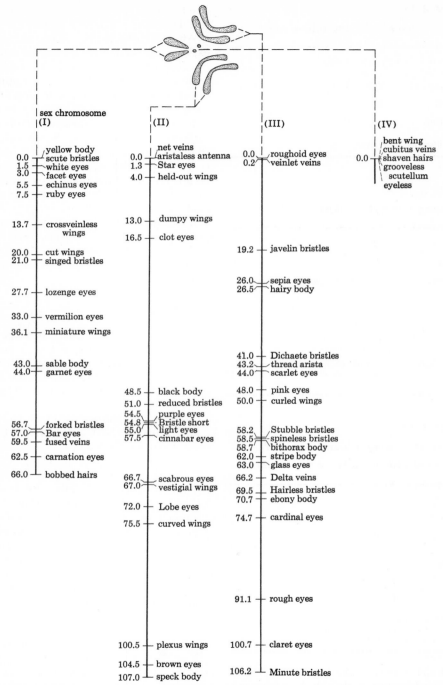

Fig. 8-80 Map of the four linkage groups, or chromosomes, of the fruit fly, *Drosophila*, showing loci of genes. (After E. Sinnott, L. C. Dunn, and T. Dobzhansky, *Principles of Genetics*, 5th ed., McGraw-Hill, 1958.)

Heredity in fruit flies

Pure stocks of *Drosophila* may be obtained from a supply house or college laboratory. Some suggested stocks for work in heredity are these (see the map of chromosomes, Fig. 8-80).

1. *Wild:* These are the typical forms found in nature, with red eyes, gray body, and normal wings.

2. *Vestigial wings:* This trait is recessive to normal wings. Mate winged males to virgin vestigial-winged females (since the vestigial forms are unable to fly). See Fig. 8-85.

3. *Curly wings:* This is dominant over wild-type normal wings. The wings curl up at the tips, making it difficult for the insects to fly.

4. *White eyes:* This is a sex-linked trait found in the X chromosome. Make reciprocal crosses; that is, cross red-eyed females with white-eyed males, and white-eyed females with red-eyed males (Figs. 8-86, 8-87). Flies must be pure for these traits.

5. *Yellow body:* This is a sex-linked trait, found in the X chromosome.

6. *Black body:* Flies with this characteristic can be used to show complete linkage and crossing over in one region since the black (or the wild-type) body gene is located in the same chromosome as vestigial (and/or long, normal wild-type) wings. See map, Fig. 8-80. Cross black-bodied, long-winged flies with gray-bodied, vestigial-winged flies. Also cross black-bodied, vestigial-winged flies with normal gray-bodied, wild-type flies having normal long wings (Table 8-16). (See linkage, pp. 418–21.)

Mating fruit flies. Students should refer to the map of the chromosomes of *Drosophila* for an indication of the position of the genes on the chromosomes of the animal (Fig. 8-80). In this way students will discover which genes are linked on the same chromosome.

Virgin female flies are needed when mating flies of two different types, for example, red eyes × white eyes. Since females retain sperms for a considerable time, females must be isolated as they emerge from pupae.

Many workers isolate virgin female flies in separate vials for 3 days. Flies that have been so "aged" are ready to lay eggs after mating. Larvae will hatch within a day and consume any mold that may have been carried into the new culture bottles.

When pupae appear, remove the parents by etherizing them and then placing them in a "morgue," that is, a stoppered bottle containing some mineral oil.

Usually females do not mate for some 12 hr after hatching. A more casual method is to remove the original parents as pupae begin to form in the culture bottles. Then the emerging flies will be virgin for the first 12 hr.

Parent flies should be removed in any case, as they may create errors in determining the results (when counting offspring). When the flies are kept at 20° C (68° F), counts of offspring may be continued up to the twentieth day. Thereafter, the second generation will emerge and errors in count will occur.

Examining and crossing fruit flies. Students will quickly learn to identify male and female flies (Fig. 8-81). The abdomen of the male is bluntly rounded with a wide band of dark pigment; the ventral posterior end of the abdomen shows a dark spot which is easily seen under low magnification, or with the naked eye. The female's abdomen is elongated, with thinner pigment bands. Females distended with eggs are easily recognized. With a hand lens it is possible to distinguish seven abdominal segments in the female and five in the male. The females have an ovipositor; the males possess sex combs (ten short black bristles) on the end of the tarsal joint of the first pair of legs.

It is difficult to look for banding in newly hatched flies, for pigment is not well developed. Also, dark-colored flies need to be identified principally by possession of sex combs and by shape of the body.

Etherizing flies. The flies must be etherized so that students may examine them

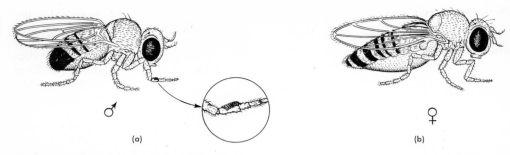

Fig. 8-81 *Drosophila*, fruit flies: (a) male (note position of sex combs on first pair of legs, inset); (b) female. (After W. G. Whaley, O. P. Breland, *et al.*, *Principles of Biology*, Harper & Row, 1954.)

with ease. To make an etherizing container, obtain a bottle having the same circumference as the mouth of the stock bottles. Insert a nail 1 in. long into the cork stopper of the vial. Then cover the nail with layers of cotton and tie the cotton to the nail with cord (Fig. 8-82b). Put a few drops of ether on the cotton when you plan to etherize the flies.

To transfer the flies, first tap the culture bottle on the table. When the flies fall to the bottom (this will be but for a moment), quickly remove the cover and hold the mouth of the etherizer closely against the mouth of the stock bottle (Fig. 8-82a). Invert the stock bottle and tap it slightly, so that the flies move into the etherizing bottle. Since the flies show a positive phototaxis, and a negative geotaxis, an electric light may be placed near the etherizer to encourage the flies to move into the empty bottle. Then quickly separate the bottles, and stopper both (Fig. 8-82b). The flies will be anesthetized within a few seconds. Guard against an excess of ether or too rapid etherization to avoid killing the flies. (When the wings stand out at an angle, the flies are dead.) A better device to use is the plastic anesthetizer shown in Fig. 8-83.

Spill the anesthetized flies out onto white paper for examination. If the flies recover before the examination is completed, they must be re-etherized. Have on hand a Petri dish cover in which a strip of filter paper is held fast with adhesive tape. Add a drop of ether to the filter paper and place the dish over the flies for a few seconds.

Since the flies are fragile, use a camel's-hair brush or a copper strip to handle them during examinations and to transfer them from place to place. When you plan to cross two different genetic types, select

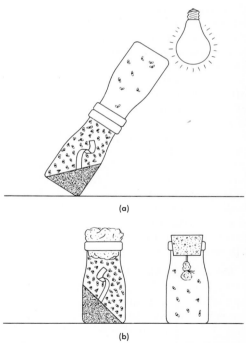

Fig. 8-82 Technique for transferring and etherizing fruit flies: (a) flies show phototaxis and leave culture bottle for new one; (b) old stoppered bottle with cotton plug, new one with etherizing stopper.

several pairs of flies of the stock with which you are dealing and place them in a small cone of paper. Then introduce them into new culture medium for stock. Otherwise, anesthetized flies dropped into the medium will adhere to the medium and drown.

The stock cultures should be kept at a temperature range of 20° to 25° C (68° to 77° F). At 25° C the life cycle takes 10 days; at 20° C the cycle is lengthened to 15 days. Larvae crawl up to the dry sides of the bottle or onto the paper strip inserted into the bottle. Here the larvae pupate. At 25° C the pupa stage lasts about 4 days. Lower temperatures prolong the life cycle. High temperatures increase sterility and reduce viability. The temperature within the bottles will be a bit higher than the room temperature. This is because of fermentation of the medium.

Subculture or transfer flies to new stock bottles with fresh medium each month, or more frequently, depending upon the temperature and the amount of evaporation of fluid from the medium. Prepare several duplicate stock cultures. Label the bottles and record the date of transfer.

Establishing stock cultures. Fruit flies are attracted to soft grapes, plums, bananas—in fact, to any fermenting fruits. Larvae feed freely on yeast and other microorganisms in the fermenting fruit juices; clearly, a fermenting medium must be prepared. For rapid, temporary cultures, where little handling will occur, the simplest medium is prepared by dipping a piece of ripe banana into a suspension of yeast (made from a quarter of a package of yeast dissolved in 100 ml of water). Insert this piece of banana along with a strip of paper toweling into a clean glass vial or bottle. In season, this may be left open to attract fruit flies, or you may introduce flies into the bottles. Then plug with cotton wrapped in cheesecloth, or with milk-bottle caps. However, this medium is not recommended for careful work. In the 2-week life span of fruit flies the medium described above will become a mash with

Fig. 8-83 Plastic anesthetizer for fruit flies. The funnel end fits different-sized containers; a wick in the chamber can be saturated with ether. (Courtesy of Carolina Biological Supply Co.)

the flies embedded in the soft material. Also, molds will grow in profusion.

Demerec and Kaufmann[36] describe several media which use agar to solidify the medium so that bottles may be inverted; they may then be used with ease in making transfers. You may want to prepare one or more of these media. Use about 50 to 60 ml of medium for a half-pint culture bottle.

After any one of the four media has been prepared, you may want to add a trace of a mold inhibitor. (A minute quantity of Methyl Parasepts[37] in 0.15 percent solution can be added; in excess any inhibitor will reduce the growth of yeast and slow down the development of the flies.)

Heat the medium again so that it comes close to boiling. Then quickly pour the medium into half-pint milk bottles or glass vials to a depth of ½ in. It is safer to sterilize the bottles before introducing the medium. Then insert a strip of paper toweling into the medium while it is soft; this will provide additional surface for egg-laying and pupation. Cover the bottles with cotton wrapped in cheesecloth or

[36] M. Demerec and B. P. Kaufmann, *Drosophila Guide: Introduction to the Genetics and Cytology of Drosophila melanogaster,* 7th ed., Carnegie Institution of Washington, Washington, D.C., 1961 ($0.50).
[37] One preservative, Methyl Parasepts (*p*-hydroxybenzoic acid esters), may be purchased from Heyden Newport Chemical Corp., 342 Madison Ave., New York 17, N.Y. Another preservative, Tegosept M, may be obtained from Goldschmidt Chemical Corp., 153 Waverly Place, New York 14, N.Y.

with milk-bottle caps. Tilt the bottles against a ledge to increase the surface, and allow the medium to cool; a completed bottle looks like the one in Fig. 8-84b. You may want to pour medium from the cooking pot through a funnel into the bottles so that the medium does not spill along the sides of the bottles. Store the bottles in a refrigerator until the flies are to be introduced. Just before using the bottles to accept the flies, add 2 or 3 drops of a rich yeast suspension to the surface of the solid medium. Or add a pinch of dried yeast; this will dissolve in the fluid on the surface.

1. BANANA MEDIUM. Dissolve 15 g of agar in 480 ml of water by bringing it to a boil; stir well. To this add 500 g of banana pulp made by mashing a banana with a fork or by putting it through a strainer.

2. CREAM OF WHEAT MEDIUM. This preparation eliminates the agar. Measure out 775 ml of water, 115 ml of molasses or Karo, and 100 g of Cream of Wheat. Add the molasses to two-thirds of the water; bring to a boil. Mix the Cream of Wheat with the remaining third of cold water and add this to the boiling mixture. Continue to stir, and cook for 5 min after boiling begins. Add mold inhibitor and pour the medium into sterilized bottles, add strips of toweling, stopper the bottles, and tilt them, as before.

3. CORNMEAL MEDIUM. This medium uses agar. Dissolve 15 g of agar in 500 ml of water and heat. After this comes to a boil, add 135 ml of corn syrup (Karo) or molasses. Combine 100 g of cornmeal in 250 ml of cold water and add to the heated mixture. Boil this *slowly* for about 5 min. Add mold inhibitor. Then pour this medium into sterilized bottles or vials, insert toweling as before, and plug the bottles with cotton or cover with caps. This quantity will fill 25 culture bottles.

4. CORNMEAL – MOLASSES – ROLLED-OATS MEDIUM. This recipe requires no agar. Measure out 730 ml of water, 110 ml of molasses or Karo, 150 g of cornmeal, and 16 g of rolled oats (not quick-cooking). Add rolled oats and molasses to two-thirds

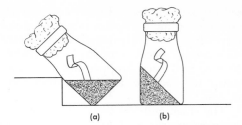

Fig. 8-84 Culture bottles for fruit flies: (a) bottle containing soft medium and strip of paper toweling, left slanted to cool; (b) bottle with solidified medium, ready for flies.

of the water and bring this to a boil. Mix the cornmeal with the remaining cold water and introduce this into the boiling mixture. Cook for a few minutes until it thickens but still can be poured. Then add the mold inhibitor. You may want to add agar to the medium to prevent it from softening with a rise in room temperature.

In fact, for work during the summer months, agar should be included in any medium you use. Cover the bottles of medium after you insert toweling strips and tilt the bottles to increase the surface area.

Should molds appear in the cultures of *Drosophila,* paint the surface of the culture medium with 95 percent alcohol or with a solution made from 1 part of carbolic acid to 8 parts of water. (*Caution:* Remove the flies from the bottle before this treatment.)

Life cycle. Some teachers have students keep their own vials of flies to study the entire life cycle of *Drosophila.* To accommodate large numbers of students, prepare the medium and use a small paper or plastic spoon to hold a small quantity of medium. The small spoon is then inserted into a vial. If several different cultures are given to the students, they will have materials to make crosses of newly emerging flies. They may transfer several pairs of flies into vials with fresh medium (on a new spoon).

Since the spoons have handles, the handling of transfers is greatly facilitated; further, less cooking is needed.

Some demonstration crosses. Types of

fruit flies available from supply houses are described on p. 413.

LAW OF DOMINANCE. Cross a wild-type fruit fly (that is, gray-bodied, long-winged) with a black-bodied, long-winged fly. The first generation should be 100 percent hybrid gray-bodied flies if both parents were pure for these traits.

Cross pure long-winged with vestigial-winged flies to get all long-winged offspring (hybrid).

Or cross pure long-winged flies with curly-winged ones to get 100 percent curly-winged flies. In all these crosses students may observe the appearance in the offspring of one of the pair of contrasting traits. The trait which appears is the dominant trait; the hidden trait is the recessive one.

LAW OF SEGREGATION. Cross the offspring; that is, inbreed the offspring of each of the crosses made above to illustrate dominance. When these hybrids are crossed, students should find that the second generation reveals a 3:1 ratio. Seventy-five percent will show the dominant trait and 25 percent will show the recessive trait. Students should see that the recessive trait which was not apparent in the F₁ hybrids has reappeared in the second generation due to the recombination of genes at fertilization.

TEST CROSS, OR BACKCROSS. In order to determine whether an organism showing the dominant gene is pure or hybrid, the "unknown" should be crossed with the recessive type. The answer lies among the offspring of this cross, provided there are enough of them. For example, if we wanted to find out whether some long-winged flies are hybrid and carry a recessive gene for vestigial, we would cross them with vestigial-winged flies. If they were pure long-winged, with two genes for long wings, then there could be no vestigial-winged flies among the offspring. However, if the long-winged flies were hybrid for the trait and they were crossed with vestigial, 50 percent of the offspring should show the recessive trait, vestigial wings (Fig. 8-85). In a two-factor backcross, the

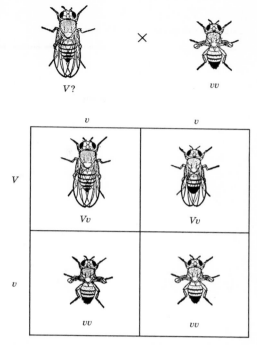

Fig. 8-85 Test cross, or backcross, in fruit flies. Long-winged flies of unknown genetic composition are mated with vestigial-winged flies; if some vestigial-winged offspring appear (as shown), the unknown was hybrid for long wings.

unknown type would be crossed with the double recessive. In an actual cross, apply the chi-square method, pp. 407–08.

LAW OF INDEPENDENT ASSORTMENT OF FACTORS. When two pairs of genes are located in different pairs of chromosomes, they are inherited independently of each other (see Fig. 8-80). Each gene behaves as a unit since the chromosomes segregate in maturation of eggs and sperms (meiosis), and are recombined in fertilization (see meiosis, p. 347).

Students may examine this type of cross first-hand (in club or laboratory work). At least two factors must be studied. For example, cross fruit flies pure for wild or gray body and pink eyes with flies pure for black body and red eyes (Table 8-15). (Note in Fig. 8-80 that these genes are on different chromosomes, not within the same chromosome.) Let *B* stand for a gene

TABLE 8-15
Independent assortment in fruit flies*

Key: B = gray body; b = black body; P = red eyes; p = pink eyes

(a)   Parents: $(BBpp)$ × $(bbPP)$

Gametes: Bp bP

Offspring: $(BbPp)$

(b)   Parents: $(BbPp)$ × $(BbPp)$

Gametes: BP Bp bP bp BP Bp bP bp

	BP	Bp	bP	bp
BP	BBPP gray body red eyes	BBPp gray body red eyes	BbPP gray body red eyes	BbPp gray body red eyes
Bp	BBPp gray body red eyes	BBpp gray body pink eyes	BbPp gray body red eyes	Bbpp gray body pink eyes
bP	BbPP gray body red eyes	BbPp gray body red eyes	bbPP black body red eyes	bbPp black body red eyes
bp	BbPp gray body red eyes	Bbpp gray body pink eyes	bbPp black body red eyes	bbpp black body pink eyes

* (a) Pure gray-bodied, pink-eyed flies are mated with pure black-bodied, red-eyed flies; the offspring are all dihybrid, with gray bodies and red eyes; (b) if these dihybrids are inbred, independent assortment occurs; the genetic and physical make-up of the offspring are shown in the Punnett square.

for gray body; b, a gene for black body; P, a gene for red eyes; and p, a gene for pink eyes.

Upon inspection of the Punnett square in Table 8-15, students should be able to locate the $9/16$ of the flies that have gray body with red eyes; $3/16$, gray body with pink eyes; $3/16$, black body with red eyes; and $1/16$, black body with pink eyes. Students should find similar results if they actually count large numbers of fruit flies. A dihybrid cross such as this one yields a $9:3:3:1$ ratio. (In a trihybrid cross, multiply the $9:3:3:1$ ratio by a $3:1$ ratio, which represents the added hybrid, so that the offspring would yield a ratio of $27:9:9:$

$3:3:3:1$. In this case, $27/64$ of the offspring should show the combination of three dominant traits and $1/64$ should show the three recessive traits.)

Apply the chi-square test; remember that with four classes of offspring you enter the table (Table 8-11) at three degrees of freedom.

SEX LINKAGE. After students have learned how maleness or femaleness is inherited on the basis of XY- or XX-chromosome pairs, some students may want to cross red-eyed female fruit flies with white-eyed males (Fig. 8-86). The gene for red or white eye color is located on the sex chromosome (see Fig. 8-80). In this cross the offspring will all

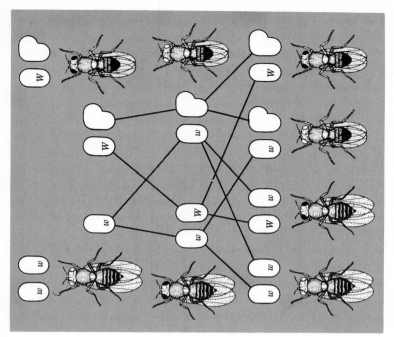

Fig. 8-87 Sex linkage in fruit flies: reciprocal cross showing white-eyed female mated with red-eyed male. (After E. Sinnott and L. C. Dunn, *Principles of Genetics*, 2nd ed., McGraw-Hill, 1932.)

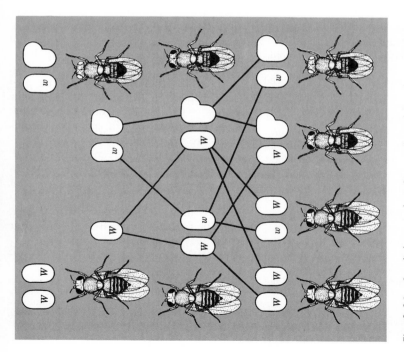

Fig. 8-86 Sex linkage in fruit flies: red-eyed female crossed with white-eyed male. (After E. Sinnott and L. C. Dunn, *Principles of Genetics*, 2nd ed., McGraw-Hill, 1932.)

be red-eyed. But in the reciprocal cross (Fig. 8-87), white-eyed females crossed with red-eyed males, the females in the first generation are all red-eyed and the males are white-eyed. Trace the results as shown in both cases in the second generation. This is similar to the inheritance of color blindness and hemophilia among human beings (p. 388).

Some teachers have their students use the experiments outlined in the pamphlet by Demerec and Kaufmann, *Drosophila Guide*. Some instructions from the guide for demonstrating sex linkage in fruit flies are reproduced below. The letter w represents recessive white eyes; m, recessive miniature wings; and f, recessive forked bristles. The plus sign (+) is read "wild type."

DEMONSTRATION OF LINKAGE BETWEEN CERTAIN GENES AND SEX

A. With wild-type females

1. Mate wild-type virgin ♀ ♀ with wmf ♂ ♂. (In the two sex-linkage crosses only one character need be followed, and white has been chosen for these instructions. Results of the same type can be obtained by following either of the other two genes.)
2. Remove parents after 7 to 8 days, and transfer to a fresh culture bottle.
3. Determine the types of F_1 flies. They should be wild type.
4. Mate several pairs of F_1 flies. ♀ ♀ need not be virgin.
5. Remove F_1 after 7 to 8 days, and transfer to a fresh culture bottle.
6. Examine the F_2 flies, and count the numbers of white-eyed and wild-type red-eyed flies by sexes. What is the proportion of w^+ to w flies? What is the sex of all w flies?

Record and explain results.

B. The reciprocal cross

1. Mate virgin wmf ♀ ♀ with + ♂ ♂.
2. Remove parents after 7 to 8 days, and transfer.
3. Determine the types of F_1 flies by sexes.
4. Mate several pairs of F_1 flies. ♀ ♀ need not be virgin.
5. Remove F_1 after 7 or 8 days, and transfer, providing four bottles in all.
6. Examine the F_2 flies, and count the num-

bers of white-eyed flies and red-eyed flies by sexes. (Save the flies for the next experiment.) The proportion of w^+ to w flies should be quite different from that resulting from the reciprocal cross in A above.

Record and explain the results of these two crosses, making diagrams to show the transmission of the sex chromosomes.[38]

AUTOSOMAL LINKAGE. Since there are thousands of genes in *Drosophila* and only four linkage groups, or chromosomes, many genes are found together or linked on the same chromosome. We focus now on autosomal linkage, not linkage on the X chromosome.

To show the linkage of two genes found on the same chromosome students might try this cross. Cross a pure black-bodied, normal long-winged fly with a gray-bodied, vestigial-winged fly. (See chromosome map, Fig. 8-80, to find on which chromosomes the genes are located.) The offspring of pure parents should all be gray-bodied and long-winged (Table 8-16a).

Students who are interested may want to go further. They may show that there usually is not a complete linkage in the female since there is a crossing over of chromosomes. In the *male* fruit fly, however, there is *no crossover* of chromosomes. As a result, different ratios will be produced, depending upon whether a male or a female of the above F_1 generation is used. If the male F_1 is used (Table 8-16b) the offspring are 50 percent gray-bodied, vestigial-winged, and 50 percent black-bodied, long-winged. The expected types of crossover, gray-long and black-vestigial, do not appear. However, when the reciprocal cross is made (F_1 female with black, vestigial males, Table 8-16c), the four expected types appear.

In 1919 Morgan showed that the per-

[38] From M. Demerec and B. P. Kaufmann, *Drosophila Guide: Introduction to the Genetics and Cytology of Drosophila melanogaster,* 7th ed., Carnegie Institution of Washington, Washington, D.C., 1961, 3rd printing, 1964, pp. 37–38. Reprinted by permission of the publisher.

TABLE 8-16

Linkage in fruit flies*

Key: B = gray body; b = black body
         V = long wings; v = vestigial wings
         ♂ = male; ♀ = female

(a)

Parents: $(bV)(bV)$ × $(Bv)(Bv)$

Gametes: bV Bv

Offspring: $(bV)(Bv)$

(b)

Parents: F₁ ♂ $(bV)(Bv)$ × $(bv)(bv)$ ♀

Gametes: bV Bv bv

Offspring: $(bV)(bv)$ $(Bv)(bv)$

(c)

Parents: F₁ ♀ $(bV)(Bv)$ × $(bv)(bv)$ ♂

Gametes: bV Bv BV bv bv

Off-
spring: $(bv)(bV)$ $(bv)(Bv)$ $(bv)(BV)$ $(bv)(bv)$

---

* (a) Pure black-bodied, long-winged × pure gray-bodied, vestigial-winged; (b) F₁ male (gray-bodied, long-winged) × double recessive female (black-bodied, vestigial-winged); (c) reciprocal cross to b: F₁ female × double recessive male. Note that, while the usual way of symbolizing the parents would be $(bbVV)$, since the genes are linked we use the symbols $(bV)(bV)$ instead.

---

centages from such a cross (as in c) were as follows:

	noncrossovers
gray, vestigial	41.5%
black, long	41.5%
total	83.0%

	crossovers
black, vestigial	8.5%
gray, long	8.5%
total	17.0%

Heredity in wasps

Habrobracon. P. W. Whiting has used the small parasitic wasp *Habrobracon* (Fig.

8-88a and b) for studies in genetics. The wasp is about the size of the fruit fly. It is also recommended by Whiting[39] for studies of complete metamorphosis, and to show parthenogenesis of eggs of unmated females. The effect of environment on the development of body color can also be observed readily by students; at higher temperatures (30° C [86° F]) the body color is light honey yellow, and black at lower temperatures (20° C [68° F]).

Since *Habrobracon* is parasitic on the caterpillar stage of its host, the flour moth —*Ephestia*—must also be cultured (Fig. 8-88c).

The life cycle of *Habrobracon* is 10 days at 30° C (86° F), about 14 days at room temperature. The larvae of the host, *Ephestia,* are easily reared. A thin layer of yellow cornmeal is placed on the bottom of a pasteboard box and kept at room temperature. Moths are released into the box, which is then covered. A high humidity stimulates egg-laying; this may be arranged by keeping a container of water in the box.

In about 3 weeks the larvae are almost half grown; they may then be transferred to new containers to avoid overcrowding. Larvae are ready for use in about 6 weeks.

Males of *Habrobracon* feed on honey, as do the females, but females do not lay eggs until they have fed on the host larvae. Students may observe the wasps kept in shell vials. When the host larvae are inserted into the vials students may observe the host-parasite relationship.

Both host and parasite can be refrigerated and maintained for months without the preparation of media customary in the maintenance of such animals as fruit flies.

PARTHENOGENESIS. For a study of parthenogenesis, the host larvae, after being stung, may be separated into individual vials so that the eggs of the wasp may be observed as they develop. As the pupae gradually mature within the cocoons, their

[39] P. W. Whiting, "The Parasitic Wasp, *Habrobracon,* as Class Material," *Ward's Natural Sci. Bull.* February 1943.

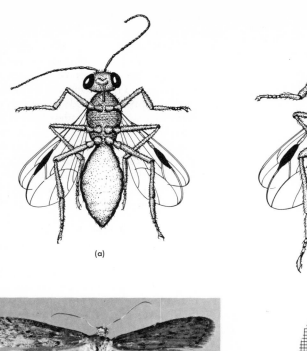

(a)

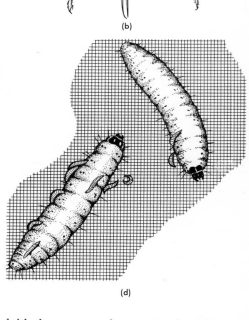

(b)

(c)

Fig. 8-88 Parasitic wasp *Habrobracon* and its host moth *Ephestia:* (a) male (note long antennae and small size of body); (b) female; (c) adult *Ephestia;* (d) *Habrobracon* maggots feeding on larvae of its host, *Ephestia*. (c, courtesy of Canada Dept. of Agriculture.)

(d)

sexes may be observed before they emerge. If only females are desired, destroy the males before they emerge. Insert a needle into the cocoons to destroy all insects (males) that have long antennae and lack the ovipositor sting.

Unmated females that may be heterozygous for a trait produce sons according to their *gametic* ratios.

MAKING CROSSES. Females are diploid, males are haploid. Since males are hap-

loid, there cannot be a cross of two heterozygous wasps.

Whiting suggests a distinctive cross for students to make to show dominance and segregation. Mate virgin females of a stock having orange eyes with wild-type black-eyed males. In F$_1$ pupae, observe that all male pupae have orange eyes (like their mothers) and that female pupae have the dominant trait, black eyes like their father.

Independent assortment is illustrated by using stocks of small-winged, white-eyed females (*swsw whwh*) that are diploid (as shown) with wild-type males (*Sw Wh*). F_1 males all have small wings and white eyes, like the female parent; F_1 females show the wild type (*Swsw Whwh*). What kinds of offspring, all males, are produced from unmated F_1 females? There will be equal numbers of wild-type (*Sw Wh*); small-winged (*sw Wh*); white-eyed (*Sw wh*); and small-winged, white-eyed (*sw wh*).

When you are planning materials for large numbers of students, place orders about 8 weeks in advance so that host larvae (*Ephestia*) can develop and be ready for shipment when they are half grown. Teachers need only place these larvae in boxes on yellow cornmeal. The wasps are shipped in shell vials in time so that a teacher can breed enough materials for use 3 weeks later.

Wild-type wasps have black eye color, dark brown ocelli, and honey yellow body color with a pattern of black that becomes more extensive if the temperature is lowered. Among mutants are specimens having white eyes; eyeless; red eyes that are dark at higher temperatures, brighter at lower temperatures; orange that varies to pink and red; carrot that pales to white at room temperature; very black body color; very minute wings and white eyes (double mutants) (Fig. 8-89); and many other mutations.

Mormoniella. This wasp, smaller than fruit flies, is parasitic on pupae of a fly. Males are haploid, so gametic ratios are shown directly in the offspring of unmated females.[40]

Carolina Biological maintains stocks of *Mormoniella* and of the host, the pupa stage of which is parasitized; they use one genus of flies, called *Sarcophaga* (Fig. 8-90a). Directions are available from the supply company for maintaining both wasps and *Sarcophaga* pupae.

[40] P. W. Whiting, "A Parasitic Wasp and Its Host for Genetics Instruction and for General Biology Courses," *Carolina Tips* April 1955.

Fig. 8-89 A double mutant of *Habrobracon* available for genetic studies, a small-winged, white-eyed male. The ocelli as well as the compound eyes are white.

The life cycle covers 10 days at 27° C (81° F). There is a minimum of preparation of culture media, for, as with *Habrobracon* and its host, the larvae of the host fly can also be refrigerated and maintained for months. Cultures are kept in small shell vials stoppered with cotton.

There are many stocks available, a good number with easily observed mutants of eye color.

Heredity in protozoa

Those students who become deeply interested in heredity of microorganisms may investigate the elegant experiments of Jennings, Sonneborn, Kimball, Chen, Kidder, and others on the identification of the mating types in some protozoa (Fig. 8-91). There are some 30 mating types in *Paramecium aurelia;* in *P. bursaria* (the green one), Jennings, Chen, and others have described 6 varieties containing 23 mating types. In *Tetrahymena* some 40 mating types have been found; there may be more. In fact, one variety has at least 9 mating types. Kimball has described at least 6 mating types in *Euplotes patella*. The alga-

(a)

Fig. 8-90 *Mormoniella*, a small parasitic wasp, and its larger host, *Sarcophaga*. (a) Heads of host, *Sarcophaga*: left, head of male; right, head of female. (Note the wider space between the eyes in the female.) (b) *Mormoniella*, female (white eyes). (c) *Mormoniella*, male (wild type). (d) *Mormoniella vitripennis* stinging *Sarcophaga* pupa. (a, b, d, courtesy of Carolina Biological Supply Co.; c, courtesy of George B. Saul 2nd.)

(c)

(b)

(d)

like flagellate, *Chlamydomonas*, has already been described (pp. 345, 402–04, Fig. 8-75). Also refer to extranuclear inheritance (Fig. 8-92).

Refer to Chapter 12 ("Genetics"), then Chapter 11 ("Reproduction") in R. Manwell, *Introduction to Protozoology* (St. Martin's, 1961). Also read Chapter 6 ("Heredity") and then the splendid Chapter 5 ("Reproduction") in R. Kudo, *Protozoology*, 4th ed. (Thomas, 1954). Both books offer excellent end-of-chapter bibliographies.

Conjugation vs. autogamy. These two processes that occur in paramecia emphasize the many advantages in using *Paramecium* as an experimental organism in studies of heredity in microorganisms. In conjugation of different pairs of mating types, the nuclei of each conjugant undergo meiosis, and ultimately three of the four resulting nuclei disintegrate. The one remaining haploid nucleus in each conjugant now undergoes a mitotic division, thereby forming two genetically identical haploid nuclei. An exchange of one nucleus of this pair of nuclei between the two conjugants restores the diploid number and gives each conjugant a nucleus of the other conjugant. These exconjugants now have a nucleus of genetically diverse material and in subsequent rapid fissions duplicate this new diversity.

In autogamy, on the other hand, each paramecium becomes more homozygous and thereby provides the experimenter with a rapid method of gaining genetically similar material. In this nuclear reorganization, the nuclei of *unpaired* individuals undergo meiosis. One of the nuclear products of meiosis survives and the other three disintegrate. Then this one haploid nucleus undergoes mitosis, producing two identical nuclei that now fuse. This "self-fertilization" results in the diploid number being restored and produces homozygosity. Recessive genes cannot long remain hidden in organisms that undergo autogamy.

Killer trait in *Paramecium*. Students may want to report on Sonneborn's work in which he found that some paramecia secrete a substance that is lethal to other paramecia. Killer paramecia have very fine particles of a substance called "kappa" in their cytoplasm; there also is a pair of genes necessary for the development of the killer trait (Fig. 8-93). Kappa particles produce a poison, paramecin, which destroys the "sensitive" varieties of paramecia. Kappa particles are composed of DNA and duplicate themselves in the cytoplasm.

Some interesting comparisons have been made between kappa particles, or "plasmagenes" (Sonneborn's term), and viruses.

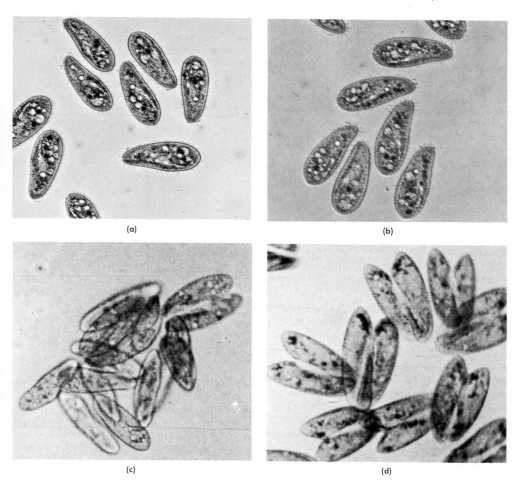

(a)

(b)

(c)

(d)

Fig. 8-91 *Mating behavior of* Paramecium aurelia: *(a) and (b) the two complementary mating types of* P. aurelia, *stock 51; (c) clumping, an early stage in the mating reaction; (d) formation of pairs, following the clumping reaction. (Courtesy of T. M. Sonneborn.)*

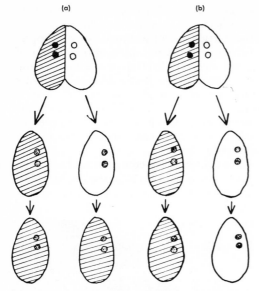

(a)   (b)

Fig. 8-92 Cytoplasm transfer at conjugation may contribute to differences in inheritance between two groups of varieties of *Paramecium aurelia:* (a) phenotypes of the two members become identical; (b) differences prior to conjugation persist. In both, the micronuclei are identical. (Reproduced from *Introduction to Protozoology* by Reginald D. Manwell. Copyright © 1961 by St. Martin's Press, Inc.)

Some 20 inherited traits have been described for *Paramecium aurelia* (due to single genes). There may be as many as 30 to 40 chromosomes making up the diploid number (Table 8-17).

Antigens in Paramecium. When paramecia are repeatedly injected into a rabbit over the period of a week, the rabbit develops antibodies. Then this immune serum is collected, diluted, and used to test the presence of similar or different antigens in other paramecia. Antigens appear to be associated with the cilia or with the surface membrane of the protozoa. A variety of antigens has been found in all populations of *Paramecium aurelia.*

Kidder[41] and his group have described

[41] G. Kidder and V. Dewey, "Studies on the Biochemistry of *Tetrahymena*," *Arch. Biochem.* 20:433–43, 1949.

TABLE 8-17
Number of chromosomes in some protozoa*

protozoa	number of chromosomes
Chlamydomonas	10 (haploid)
Euglena viridis	30 or more
Phacus pyrum	30–40
Trichonympha campanula	52, or 26 doubles
Amoeba proteus	500–600
Endamoeba histolytica	6
Actinophrys sol	44 (diploid)
Didinium nasutum	16 (diploid)
Paramecium aurelia	30–40
Paramecium caudatum	about 36
Stentor coeruleus	28 (diploid)
Uroleptus halseyi	24 (diploid)
Stylonychia pustulata	6
Euplotes patella	6 (diploid)
Vorticella microstoma	4
Carchesium polypinum	16 (diploid)

* Adapted from R. Kudo, *Protozoology*, 4th ed., 1954; courtesy of Charles C. Thomas, Publisher, Springfield, Ill.

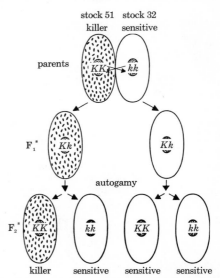

Fig. 8-93 Heredity of the killer trait in *Paramecium aurelia*. Without the dominant Mendelian gene (*K*) in the nucleus, kappa particles in the cytoplasm disappear and will not reappear, even with *K*, unless kappa is reintroduced from the cytoplasm of another killer organism. (Reproduced from *Introduction to Protozoology* by Reginald D. Manwell. Copyright © 1961 by St. Martin's Press, Inc.)

differences in biochemical reactions, especially antigens, in six strains of *Tetrahymena pyriformis*.

Additional facets of heredity

Genes and the effect of environment. Genes have their effect within a given environment. In our desire to make the study of heredity simple we sometimes put stress on genes for a specific trait. We speak, for example, of the gene for vestigial wings in fruit flies. Yet if the eggs and the hatching larvae develop at a higher temperature, longer wings result, not the vestigial condition. Recall the Himalayan rabbit: When white fur is plucked from the body and that area is subjected to very low temperatures, the hair which grows back is not white but black in color. We minimize the effect of the environment in which the genes are acting.

In another example, we know there are genes for the development of chlorophyll. But sunlight must be present, or the genes do not "reveal" themselves. As a result, chlorophyll does not appear. Students may show this condition in the following way. Grow bean seeds in the dark. Note that the chlorophyll does not develop in the dark. Then remove some of the containers of seedlings to a well-lighted location and watch the development of chlorophyll within a few days. The plastids are always present, but light is needed for the expression of the green color of chlorophyll.

Also refer to the raising of pupae of *Habrobracon* at 20° C (68° F) and 30° C (86° F), and the effect of these temperatures on the development of body color (p. 421).

(*Note:* In the high school classroom it is difficult to demonstrate that a trait is the result of the expression of many genes all acting together. Usually a trait is not the expression or the effect of a single gene.)

In more advanced seminars, discussions of complementary genes, epistasis, cumulative effect, and sex-limited effects can be studied.

Variations. Two major concepts in biology need development and illustration: (1) The individual is the product of its heredity and environment; and (2) the constant variability among offspring provides the raw materials for emerging new species. (This concept is further developed in Chapter 10.)

Teachers use several methods for demonstrating variability (aside from direct work in the classroom). Bulletin boards and hall cases are used to display collections made by students. We have seen students' collections of the following:

1. Variations in shells of a given species, such as might be found on a beach—scallop shells, periwinkles, jingle shells, or others—or variations among species within a given genus of shells (Fig. 8-94).

2. Variations in leaves of one tree—sassafras, or a species of maple, oak, or linden.

3. Variations in potato beetles or among ladybird beetles.

4. Variations among seeds—of sunflowers or others.

5. Graphs of variations in height among students of a given age; of their grades; of athletic skills; and so forth.

Fig. 8-94 Some variations among species of moonsnail shells. (Courtesy of the American Museum of Natural History, New York.)

6. Collections showing effects of light—for example, the reddish color of barberry leaves compared to the green of the hidden leaves.

7. Collections showing patterns of submerged leaves of plants that grow near the edge of a pond, or with roots in water.

Other ideas for demonstrating these concepts are offered in Chapter 11. Mounted specimens showing variations are available from supply houses.

Artificial pollination and breeding. When a new type of plant has been developed as a result of careful breeding, or of a mutation, more plants of the same genetic make-up may be produced by using some method of vegetative propagation (Chapter 11). You may want students to explore the method commonly practiced by the breeder (Chapter 9).

Interested students will find provocative reading in the chapters on breeding plants and domestic animals in L. H. Snyder and P. R. David, *The Principles of Heredity*, 5th ed. (Heath, 1957).

Also look into the 1936 and 1937 Yearbooks of the U.S. Dept. of Agriculture, which describe experiments on breeding plants and animals.

Models of DNA. It is interesting to note that the ideas that R. Goldschmidt developed in the 1930's concerning the importance of the entire length of the chromosome as a physiological unit come close to the heart of current thinking. Unfortunately, Goldschmidt was not able to provide a *model* to fit his hypothesis.

The recent work of Wilkins, Watson, and Crick is useful in presenting a model of a "chromosome"—a model of DNA (Fig. 8-95).

Teachers may use many devices to offer a visual model of DNA. (See Chapter 9 for further descriptions and activities on mitosis.)

Some students have used tongue depressors and strips of stiff plastic to form a double helix and to show the arrangement of adenine-thymine and guanine-cytosine. It may also be possible to use a magnetic board; small blocks are used to represent the bases and to show attachments at various positions. The magnetic model has the advantage of remaining an animated model that can be juggled around in front of the class; students can show the unzippering of the helix and replication.

Directions for take-apart models are given in C. M. Hutchins, *Life's Key: DNA* (Coward-McCann, 1961). An especially clear explanation for making models of DNA is provided in a small pamphlet by M. Kory, *Biology at the Molecular Level* (distributed at one time by Eli Lilly and Co., Indianapolis, Ind.). In their model, microscope slides are used as nucleotides, so that a giant DNA molecule is likened to a stack of microscope slides with each slide (or nucleotide) separated from its neigh-

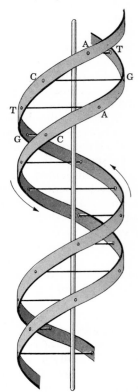

Fig. 8-95 Watson-Crick model of DNA. Two phosphate-sugar chains (represented by strips), held together by bases (bars, here), represent the double helix form of DNA.

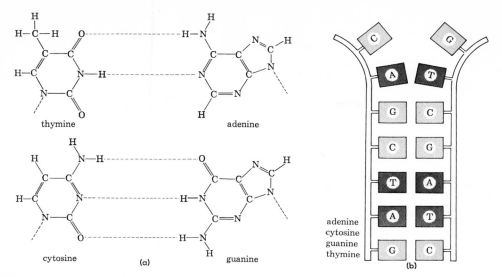

Fig. 8-96 (a) Pairing of bases of DNA: Hydrogen bonds (dotted lines) link adenine to thymine, and guanine to cytosine. (b) Sequence of bases: The sequence of bases in one strand of DNA determines the order in the second strand, since adenine pairs with thymine, and guanine pairs only with cytosine. (After J. P. Burnett, *The Synthesis of Proteins,* Eli Lilly and Co.)

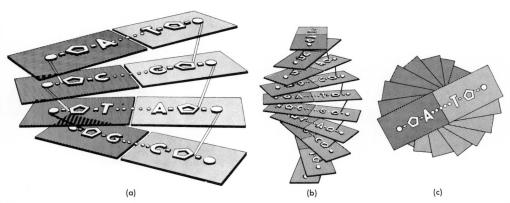

Fig. 8-97 Model of DNA: (a) demonstration of the twist generated by rotation of each nucleotide in its own plane; (b) side view of one complete revolution made per 10 nucleotide pairs; (c) top view of twist. (After M. Kory, *Biology at the Molecular Level,* Eli Lilly and Co.)

bors by a fixed distance. Some idea of this DNA model may be gleaned from a study of Figs. 8-96, 8-97, and 8-98.

A set of paper cut-outs that can be constructed to form a model of DNA is available.[42]

Chemical supply houses offer nucleic

[42] V. Potter, *DNA Model Kit,* Burgess, Minneapolis, 1959 ($1.00).

acids and their derivatives, as well as DNA-ase. The samples of white, cottony threads of DNA elicit great interest from students.

Fruit fly chromatograms. As a long-range investigation some students may want to try to identify amino acids in fruit flies using the methods of paper chromatography (see pp. 137–39). E. Hadorn de-

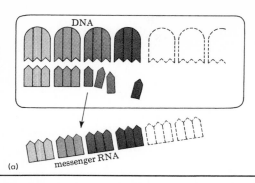

(a) messenger RNA

$$\text{amino acid} + \text{ATP} + \text{enzyme} \rightleftharpoons \text{enzyme} \cdot \text{AMP} \cdot \text{amino acid complex} + \text{PPi}$$

(b)

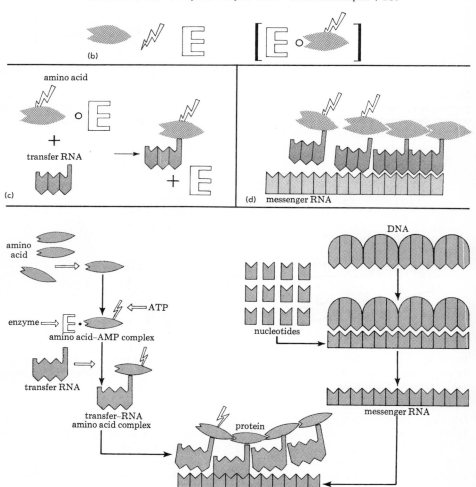

(c)

(d) messenger RNA

(e)

scribes his own work in "Fractionating the Fruitfly" (*Sci. Am.*, April 1962).

Also refer to C. L. Stong's description and wonderful diagrams of techniques for studying the genetics of fruit flies with chromatograms viewed by an ultraviolet lamp ("The Amateur Scientist," *Sci. Am.*, June 1965).

Background reading. To read papers in which scientists describe their own work is a rare privilege for the high school student. Refer to J. Hurwitz and J. Furth, "Messenger RNA" (*Sci. Am.*, February 1962), S. Benzer, "The Fine Structure of the Gene" (*Sci. Am.*, January 1962), and

F. H. Crick, "The Genetic Code" (*Sci Am.*, October 1962).

Salivary chromosomes. Techniques for making slides of smears from *Drosophila* larvae are described in Demerec and Kaufmann's *Drosophila Guide*;[43] the techniques are very briefly summarized here in Chapter 2. Other techniques for demonstrating chromosomes in mitosis and the effect of colchicine are given in Chapter 2.

Population genetics. Some references to the Hardy-Weinberg law, the migration of alleles in a population, and techniques of "negative eugenics" are given in Chapter 10.

CAPSULE LESSONS: Ways to get started in class

8-1. Hold up a hen's egg (*precaution:* hard-boiled) and elicit the fact that it is a hen's egg. What will it give rise to? A chicken? Under what conditions? And why a chick? Focus this introductory discussion on the notion that an organism arises from a "fertilized" egg. How many eggs do animals produce? Compare codfish with chicken. Elicit differences between external and internal fertilization. What are the odds against each egg of a codfish becoming fertilized? Begin to touch upon the evolution of sexuality from protozoa, algae, yeasts, and bacteria to the flowering plants and the mammals.

8-2. The questions that students raise probably lead to the next lesson: What is in a fertilized egg? What governs the genetic continuity? How does the fertilized egg develop? Use models, films, and slides, and combine them with other approaches that are listed among these suggested lessons. Show the film *The Rabbit's Development* (International Film Bur.).

8-3. At some time, plan a reading of offprints from *Scientific American* (catalog available from Freeman, San Francisco). There may be

seminars planned where students each give a 10-min talk and answer questions of the class concerning such topics as hormones and reproduction; photoperiodism and its effect on flowering and on reproductive cycles of animals; role of messenger RNA in protein synthesis; replication of DNA; patterns of cleavage among eggs of mollusks, worms, echinoderms, and vertebrates; differentiation in time and space of tissues from early gastrula; polarity and regeneration; heritable metabolic disorders; radiation and mutations; genetic drift; and what a species is.

8-4. Begin with the approach used in the *Student Laboratory Guide* to *Biological Science: An Inquiry into Life* (BSCS Yellow Version, Harcourt, Brace & World, 1963), in which fertile chick eggs that have been incubated for 48 hr, 72 hr, and further are used. Locate the beating heart and try to trace the circulatory system; discuss the origin of blood cells, somites, anlage of eyes, limbs, and other organs and organ systems.

8-5. Collect frogs' eggs and watch their devel-

[43] *Op. cit.*

Fig. 8-98 (a) Transfer of information: Messenger RNA is synthesized under the direction of DNA. (b) An activating enzyme and ATP "activate" amino acids. In these diagrams ATP stands for adenosine triphosphate, AMP stands for adenosine monophosphate, and PPi stands for inorganic pyrophosphate. (c) Symbolic representation showing transfer RNA, specific for each amino acid, bringing activated amino acids to a messenger RNA template, where bonding of amino acids in a specific sequence occurs. (d) Triplets of nucleotides in transfer RNA's may be attracted to complementary triplets in messenger RNA, thereby enabling each amino acid to be properly aligned on the messenger RNA template to form a polypeptide chain. (e) Summary of protein synthesis: DNA of chromosomes directs the synthesis of messenger RNA, which in turn directs the synthesis of protein on the ribosomes by means of transfer RNA's (adapter molecules). (After J. P. Burnett, *The Synthesis of Protein*, Eli Lilly and Co.)

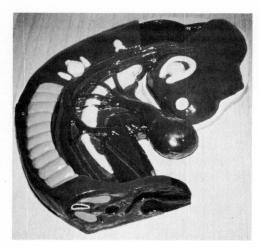

Fig. 8-99 Model of 72-hr chick embryo. (Courtesy of The Welch Scientific Co., Skokie, Ill.)

opment. A copy of Schumway's chart giving a timetable of development of frogs' eggs is reproduced in the *Student Laboratory Guide* cited above. Some students may want to test the regenerative power of the tail of the early tadpole stages (and also to examine prepared slides of chromosomes, pp. 473, 483; see also Fig. 9-22).

8-6. At some other time, try the approach in the "Laboratory Investigations" section of *Biological Science: Molecules to Man* (BSCS Blue Version, Houghton Mifflin, 1963), using Investigation 29, which is on the regeneration of plants. Also look into the activity on growing pollen grains.

8-7. Have students use a series of models and describe the stages in the development of the frog or the chick (as in Fig. 8-99). Also use models of stages in cleavage of a fertilized egg and have students arrange them in chronological order. Many students have made models of papier-mâché or of special materials, as shown in Fig. 14-2. Or use commercial models to develop stages in cleavage; then, to show what goes on within the nucleus, show stages of mitosis. Compare with models showing stages in meiosis of cells in the ovaries and testes or in the ovary and anthers of flowers. Transparencies are available, and sets of slides for the microscope are useful for first-hand, close examination of details of chromosomes and of sperms and pollen grains.

8-8. It may be desirable to construct a magnetic board, consisting of a sheet of iron and cut-outs of nucleotides. These nucleotides may be made of colored cardboard backed with small magnets that can be moved and positioned on the upright sheet of iron to show models of DNA, the process of mitosis, and comparisons with meiosis in reproductive organs. The models must be light-weight so that they do not slide off the tilted board. Have students use the materials in developing the concepts.

8-9. Without a preliminary explanation, draw the pedigree of sickle-cell anemia on the blackboard (diagramed in the *Student Laboratory Guide* cited above). Have students describe the type of inheritance.

8-10. After students know how to use symbols for pedigree charts, draw several pedigrees such as the one in Fig. 8-100; or copy others from the problems at the ends of chapters in genetics textbooks such as those by Srb, Owen, and Edgar and by Sinnott, Dunn, and Dobzhansky (see the end-of-chapter bibliography). Draw these on the board, or reproduce some as part of homework assignments. Have students explain the modes of heredity.

8-11. Select one laboratory activity that will require time for follow-up at home or during free time, if there is not time in class for a 6-week laboratory block. Look into the BSCS laboratory block *Animal Growth and Development*, by F. Moog (Heath, 1963)—excellent, detailed descriptions with illustrations are provided so that students may supplement classwork with independent researches, several of which may be done at home.

8-12. Some students may plan to try to cross different strains of guppies or platys. Or it may be possible to get mating types of *Paramecium bursaria* or of *Chlamydomonas* and study, first-hand,

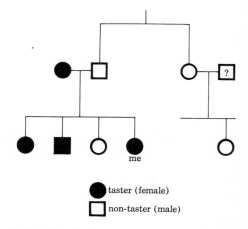

Fig. 8-100 Pedigree of a student who is a taster of PTC.

the behavior and heredity of living organisms.

8-13. Begin by showing the short silent film *Fertilization* (EBF). Watch how a living sperm cell seems to be drawn into an egg. Notice that only one sperm enters, while the others are rejected after a fertilization membrane lifts off the surface of the egg. Elicit from the students that a sperm carries the "chemicals of heredity," the genes; so does an egg. From this fusion the next generation arises. The class is now into a study of where egg and sperms "originate." What is the structure of sperms and eggs? How does a fertilized egg develop?

8-14. On a field trip around the school grounds, examine the parts of a flower on a tree or shrub. Notice that some flowers lack petals; some are staminate only. Students may notice that flower parts come in multiples of three or five. Key out a specimen; students seem to be very interested in keying out an unknown plant (see Chapter 11).

8-15. Plan several lessons with a microscope to show fission in protozoa, or budding in yeast cells, or sporulation in some fungi. Develop the notion that, in general, asexual reproduction results in offspring the heredity of which is identical to that of its parents. How does this compare with reproduction by seeds, or reproduction in higher animals?

8-16. Have students prepare a bulletin board showing methods of asexual and sexual reproduction in plants. Where possible, exhibit actual specimens that students have grown. After a breeder has developed a new type of tulip, how may he propagate more of the same kind?

8-17. You may want to refer to the interrelationships of living things and indicate the conditions needed for reproduction. What conditions make for success in survival?

8-18. Encourage students to bring in specimens to keep in the aquaria and terraria in the classroom, and also to maintain such specimens at home. Some may want to undertake the fascinating study of germinating seeds. They may also study the devices for dispersing seeds. In the study of reproduction, there are rich opportunities for laboratory work on individual plants or animals. At the end of the period have students bring together the work they did and share their experiences with the rest of the group. Reference books can be made available in the classroom.

8-19. Where possible, dissect a pig uterus and study the way a fetus is nourished through the placenta. Discuss the role of the amnion in birds or reptiles with that in mammals. Develop the

reasons mammals can perpetuate themselves even though they produce so few eggs; compare with fish or amphibia.

8-20. In season, crush stigmas of flowers on a slide, and under the microscope look for the growth of pollen tubes. (Perhaps a student may try to work out concentrations of sugar solutions which could serve as an artificial medium for germination of pollen grains of a given flower.) From here, have students describe how a sperm nucleus in the pollen grain reaches the nucleus of an egg cell in the ovule so that the egg cell is fertilized. Compare fertilization in plants with that of animals.

8-21. Have students examine prepared slides showing reduction division in testes of insects or other forms. Sets of slides are available. Also see the film *Meiosis in Spermatogenesis of Grasshopper* (Phase).

Examine living sperm cells if frogs are available. Or a club may want to study chromosomes; they may make smears of salivary glands of fruit flies or onion root tips (pp. 116–18, 472–73).

8-22. As a research project, a student might want to study polarity in the regeneration of some worms or *Hydra*. Will the anterior parts of cut pieces always regenerate a "head" part, or can they grow a "tail," or posterior portion?

If a shoot of a plant is turned upside down, will a new shoot regenerate at the anterior part, or may roots grow on the portion which was originally in the most anterior (terminal) part?

8-23. Show the BSCS films on techniques: *Removing Frog Pituitary* (if you plan to stimulate ovulation in female frogs), or *Neurospora Techniques,* or *Genetics: Techniques of Handling Drosophila* (all from Thorne). Many laboratory activities for class or club develop from these films.

8-24. As a summary, show the color transparencies *Flowers and Fruit* and *Plant Life Cycles* (Scientific Supplies). Or you will want to select some of the 12 films in the AIBS series *Reproduction, Growth, and Development* (McGraw-Hill). At some time, show *Amphibian Embryo: From Fertilization to Hatching* (EBF). Students might summarize the facts developed in the films.

8-25. Since reproduction is a means for passing on inherited traits to the next generation, students (in committee) may want to report to the class on the development of eggs of different animals and plants. One committee could describe the reproductive habits of salmon or of tropical fish. Have students bring in living fish for examination. Fish may be purchased for dissection (if class time is available). In season collect frogs' eggs and watch them develop.

Other students may raise pigeons or chickens; they can bring in eggs. If possible, incubate fertilized eggs, and examine different stages in development. Compare the similarities and differences in reproduction among many groups of animals: How many eggs are laid? Are the eggs fertilized internally or externally? How long does it take for the embryo to develop? How large is the egg? What factors determine the size of the egg? How can a mammal's egg be microscopic?

8-26. Use clay of different colors to illustrate chromosomes, and show them going through maturation. With clay, have students demonstrate reduction division and compare this division with mitotic division. Also thread a few wooden beads of different colors on wires to represent genes on pairs of chromosomes. Use the wire "chromosome" device (Fig. 9-26) to show mitosis. At some time, show the film *Mitosis and Meiosis* (Indiana Univ.). Then extend the lesson into a discussion of DNA and replication of these molecules.

8-27. Plan a field trip, in season, to collect eggs of animals. Also study flowers as trees and shrubs begin to blossom. Establish the similarity between sexual reproduction in plants and animals; that is, fertilization consists of the union of an egg nucleus and a sperm nucleus.

8-28. Examine prepared slides of cleavage stages in whitefish eggs or mitotic figures in dividing cells in the tip of the onion root. Trace the stages in the reduplication of chromosomes in mitosis, where each new cell formed receives an equal amount of chromosome material. Have students explain how all the body cells are replicas of the fertilized egg. How does differentiation occur?

8-29. Ask students how to grow plants without using seeds. Develop the methods of asexual reproduction. Have students in committee bring in specimens and watch them grow in the classroom. Start sweet potatoes, white potatoes, bulbs, cuttings, and so forth. Develop the notion that cells divide and that an equal distribution of chromosome material is effected in mitosis. Show the short film *Plants That Grow from Leaves, Stems, and Roots* (Coronet).

8-30. Ask the class how an apple tree could produce five different kinds of apples. When possible have students demonstrate how a graft is made. Have students establish the value of vegetative propagation.

8-31. Distribute squares of PTC paper and tally the number of students who taste the chemical and those who do not. Have some students test their parents too. How is the trait inherited? Develop the concept of the pairing of factors and their segregation. You may want to follow this lesson with a committee report on Mendel's work. Since the need for reduction division of chromosomes has been developed, students may develop the proof for the law of dominance and the law of segregation. (A student may read from Mendel's paper.)

8-32. There may be a student in class who has been breeding tropical fish. Perhaps he can be encouraged to become an expert in one phase of the genetics of fish. There is a fine paper by O. Winze and E. Detlevsen, "Color Inheritance and Sex Determination in *Lebistes*" (*Heredity* 1:65–83, 1947). The magazine *Your Aquarium* carries many articles on genetics of fish.

8-33. Show culture bottles of fruit flies. Ask the students to predict the approximate number of males and females in the culture. Develop the concept of XX and XY chromosomes and how males and females are produced.

8-34. When students know the pattern of inheritance of X and Y chromosomes, develop how color blindness is inherited in humans. Do students understand that the gene is on the X chromosome? That a colorblind female is thought to have two genes for color blindness? What is the proof that the gene is not on the Y chromosome? Have students also diagram the way hemophilia is inherited.

8-35. Are some students searching for ideas for long-range "research"? Look into *Research Problems in Biology: Investigations for Students,* Series One–Four (BSCS, Anchor Books, Doubleday, 1963, 1965).

8-36. Have students type their own blood (as a laboratory lesson, p. 253). Before you begin, be certain to get notes of permission from parents. Develop the basic laws of heredity from the data secured. For example, if a student has type A blood what possible blood types could his parents have? What blood types could the parents not have? How might a child of B type parents have O type blood?

8-37. What is the difference between identical and fraternal twins? When students know that identical twins develop from one fertilized egg and thereby have the same genetic composition, they should be able to describe in what ways these twins resemble each other. What kinds of traits may be different in a set of identical twins? Students should recognize that learned behavior patterns result in differences between identical twins.

8-38. At some time in the work a committee of students might report on the inheritance of the

Rh factor in human beings. While there are many genes involved in the inheritance of this blood factor, simple examples of dominance may be used. If 85 percent of the population tested in the United States have a factor in their red corpuscles called an Rh factor, and about 15 percent of the people lack this factor and are called Rh negative, the Rh-positive factor is considered dominant. What percentage of the offspring might be Rh negative, if both the parents are hybrid for Rh positive?

8-39. Students might make a "flip book" to show how the two processes mitosis and reduction division differ. What effect might irregularities in movement of chromosomes have on the heredity of the offspring? Also have students report on radiation damage to chromosomes.

8-40. Committees of students might breed hooded and white rats, fruit flies, or *Mormoniella*. Or they might grow seeds resulting from a cross of two hybrid sorghum plants or corn plants. Students can gain skill in reasoning on the basis of the evidence. Sets or kits of materials for making crosses are available from biological supply houses.

8-41. Organize a committee of students with some creativity and some degree of dexterity to make models of synthesis of proteins from the DNA code in the nucleus of a fertilized egg. Hall cabinets might well be used to exhibit "flow charts" and models of "egg to organism."

Fine assembly programs have been organized around such materials; furthermore, students may report to the student audience about their work. This gives students some sense of audience reaction, and a chance to stimulate the thinking of the younger students in the school.

8-42. You may want to show a film such as *DNA: Molecule of Heredity* (EBF), narrated by G. W. Beadle. Also show *Genetics: Mendel's Laws* (Coronet), which illustrates several cases of dominance. Be sure to select films from the set of nine films on *Heredity* (Indiana Univ.), and films from the AIBS series on *Genetics* (McGraw-Hill).

8-43. You may want to arrange to have a panel of student "experts" discuss the effects of radiation of genes and chromosomes. Students might get their information from books listed in the Bibliography. Also, what are some chemical mutagens?

8-44. When possible, have students visit (or report on a previous visit to) a county or state fair. How have new kinds of plants and animals been developed?

8-45. If possible, show color slides on variations in cattle (beef and dairy) and in dogs; these may be purchased from supply houses. How do these forms differ from the ancestral wild type? How have these new types been developed? Perhaps some student in class breeds dogs, or a parent breeds horses, cattle, cats. Have the student describe some pedigrees. (See L. H. Snyder and P. R. David, *The Principles of Heredity,* 5th ed. [Heath, 1957].)

8-46. Show a new plant or animal. You may easily obtain pink grapefruits, double-petaled flowers, and so forth. Elicit the idea of a mutation as a change in a gene. What might be some causes of a change in the chemistry of a gene? What kinds of irregularities might occur among chromosomes which could result in a different, a "new" organism? Students might report on their reading of a text in genetics (see Bibliography). Use several strings of wooden balls or beads that can be plugged together (Poppits) to show translocation, crossover, and nondisjunction.

8-47. Report on the work of Beadle and Tatum. What was their "one gene, one enzyme" hypothesis? What predictions could be tested? Why was *Neurospora* a good organism for experimentation?

Plan to show the BSCS film on techniques for handling *Neurospora* (Thorne), or the filmloop on the subject (Ealing).

8-48. Students may want to report to the class about improvements which have been made in beef and dairy cattle, horses, sheep, food crops, fruits, dogs, bees, and so forth. How does a breeder select points for breeding? What are desirable traits for the consumer? They may consult the Yearbooks of the U.S. Dept. of Agriculture (Washington, D.C.) for 1936 and 1937 on improvements in plants and animals through breeding. Also use the Yearbook for 1943–47, *Science in Farming.* These books give actual case histories of new types of plants and animals. The texts by Snyder and David and Srb, Owen, and Edgar (see the end-of-chapter bibliography) provide chapters on the genetics of domestic plants and animals. (Human family histories are also analyzed.) Develop the values of inbreeding, hybridization, and selective breeding.

8-49. Have students keep seedlings in the dark. Do they notice that chlorophyll is lacking? What is the proof that these genes need specific environmental conditions before an effect is produced in the body of the plant? How long does it take, when the seedlings are put in light, for the shoots to turn green? Develop the interaction of genes in a given environment. Plan to show the film *Heredity and Environment* (Indiana Univ.).

8-50. Or you may want to develop the notion

this way. Use several coleus plants of the same genetic make-up (obtained from cuttings from the same plant). Place some in moderate light, others in bright light. How does the environment affect the development of color in coleus plants? Students may find many examples of similar conditions around the school grounds.

8-51. You may want to pose the following problem for solution. A group of science students has purchased a black male guinea pig for a study of heredity. How may the students (limited to one generation of guinea pigs) discover whether this animal is pure black or hybrid black? Elicit from students that the male might be mated to several white females (recessive). Should a white-furred offspring appear, the unknown must have been a hybrid; for if the unknown were pure black, then all the offspring should be black-furred (hybrid black).

8-52. A student might report to the class on exceptions to the "law" of dominance—blended inheritance. For example, when red and white four-o'clock flowers are crossed, the hybrid offspring have pink flowers. Many examples may be found in reference books in heredity. What is the heredity of roan cattle?

PEOPLE, PLACES, AND THINGS

Perhaps you have a student in class who breeds dogs, tropical fish, or pigeons. Or there may be an aquarium shop or experimental station or farm in your vicinity. These students, as well as the scientists, may be resource people in developing the work in heredity and reproduction. Zoology laboratories in colleges (or supply houses) may have stocks of fruit flies, white rats and hooded ones, protozoa, or invertebrates, which you can purchase.

On a field trip around the school you may find many examples of variations among plants of the same species. Perhaps you will want to explore the differences in environmental factors which might affect the differences in growth, time of flowering, color variations, and so forth.

Along a beach it may be possible to collect shells showing variations; also collect *Fucus* and egg cases of whelk, dogfish, or whatever there may be.

Perhaps you may be able to share equipment such as glassware, tanks, films, or living materials with teachers in other schools or colleges. It may be possible to set up a nature center in one school in which student curators (Chapter 14) learn to cultivate many living things all year round for schools in the community.

Are there experts you could invite as guest speakers? Could you plan a visit to study some interesting work going on with plants and animals: an experimental farm, an agricultural station, a government conservation project, a fish hatchery?

BOOKS

The following list indicates the diversity in scope and depth of the references for students in the area of transmission of the genetic code through reproduction. Also refer to the comprehensive Bibliography, Appendix A.

Allard, R., *Principles of Plant Breeding,* Wiley, New York, 1960.

Anfinsen, C., *The Molecular Basis of Evolution,* Wiley, New York, 1959.

Asimov, I., *The Genetic Code,* Orion, New York, 1963; Signet, New Am. Library, New York, 1963.

Auerbach, C., *The Science of Genetics,* Harper & Row, New York, 1961.

Balinsky, I. B., *An Introduction to Embryology,* Saunders, Philadelphia, 1960.

Bonner, D., and S. Mills, *Heredity,* 2nd ed., "The Foundations of Modern Biology Series," Prentice-Hall, Englewood Cliffs, N.J., 1964.

Boyd, W., *Genetics and the Races of Man,* Little, Brown, Boston, 1950.

Brachet, J., *The Biochemistry of Development,* Pergamon, New York, 1960.

Carter, C., *Human Heredity,* Pelican, Penguin, Baltimore, 1962.

Darlington, C. D., and A. P. Wylie, *Chromosome Atlas of Flowering Plants,* 2nd ed., Macmillan, New York, 1956.

Dunn, L. C., *Heredity and Evolution in Human Populations,* Harvard Univ. Press, Cambridge, Mass., 1959.

Elliott, F., *Plant Breeding and Cytogenetics,* McGraw-Hill, New York, 1958.

Fuller, J., and W. Thompson, *Behavior Genetics,* Wiley, New York, 1960.

Goldschmidt, R., *Theoretical Genetics,* Univ. of California Press, Berkeley, 1955.

Hadorn, E., *Developmental Genetics and Lethal Factors,* Wiley, New York, 1961.

Hamburger, V., *A Manual of Experimental Embryology,* rev. ed., Univ. of Chicago Press, Chicago, 1960.

Hanson, E., *Animal Diversity,* Prentice-Hall, Englewood Cliffs, N.J., 1961.

Harris, H., *Human Biochemical Genetics,* Cambridge Univ. Press, New York, 1959.

Lenz, W., *Medical Genetics,* trans. by E. Lanzl, Univ. of Chicago Press, Chicago, 1963.

Levine, R., *Genetics,* Holt, Rinehart and Winston, New York, 1962.

Li, C. C., *Human Genetics: Principles and Methods,* McGraw-Hill, New York, 1961.

Lwoff, A., *Biological Order,* M.I.T. Press, Cambridge, Mass., 1962.

Meeuse, B. J., *The Story of Pollination,* Ronald, New York, 1961.

Neel, J. V., and W. J. Schull, *Human Heredity,* Univ. of Chicago Press, Chicago, 1954.

Peters, J., ed., *Classic Papers in Genetics,* Prentice-Hall, Englewood Cliffs, N.J., 1959.

Raven, C. P., *Morphogenesis: The Analysis of Molluscan Development,* Pergamon, New York, 1958.

Rothschild, N., *Fertilization,* Wiley, New York, 1956.

Rugh, R., *Experimental Embryology: Techniques and Procedures,* 3rd ed., Burgess, Minneapolis, 1962.

———, *Vertebrate Embryology: The Dynamics of Development,* Harcourt, Brace & World, New York, 1964.

Schmitt, F. O., ed., *Macromolecular Specificity and Biological Memory,* M.I.T. Press, Cambridge, Mass., 1962.

Sinnott, E., L. C. Dunn, and T. Dobzhansky, *Principles of Genetics,* 5th ed., McGraw-Hill, New York, 1958.

Snyder, L. H., and P. R. David, *The Principles of Heredity,* 5th ed., Heath, Boston, 1957.

Srb, A., R. Owen, and R. Edgar, *General Genetics,* 2nd ed., Freeman, San Francisco, 1965.

Stern, C., *Principles of Human Genetics,* 2nd ed., Freeman, San Francisco, 1960.

Strauss, B., *Chemical Genetics,* Saunders, Philadelphia, 1960.

Timakov, V., *Microbial Variation,* Pergamon, New York, 1959.

Waddington, C., *New Patterns in Genetics and Development,* Columbia Univ. Press, New York, 1962.

Withrow, R., ed., *Photoperiodism and Related Phenomena in Plants and Animals,* Am. Assoc. for the Advancement of Science, Washington, D.C., 1959.

Wollman, E., and F. Jacob, *Sexuality and the Genetics of Bacteria,* Academic, New York, 1961.

FILMS, FILMSTRIPS, AND SLIDES

This listing is intended to be only a guide for the selection of new films, filmstrips, and transparencies. Send for catalogs from distributors of films; catalogs of biological supply houses list extensive offerings of 2 × 2 in. slides. The addresses of distributors are given in Appendix B.

In general, the cost of color films averages $150; black and white films cost about $75. Newer films are made in varying lengths—time sequences of 1 min, 5 min, 11 min, or 30 min; the cost of films therefore varies. We have indicated the lengths of films so that lessons can be planned with the films in mind.

Sound films are available in color or black and white; rental rates are available from film libraries or by direct inquiries to producers. (Filmstrips are not available on a rental basis; the purchase price is usually about $6 to $7 for those in color.)

Amphibian Embryo (16 min), EBF; frog and salamander.

Asexual Reproduction (10 min), Indiana Univ.

Basic Nature of Sexual Reproduction (15 min), Indiana Univ.

Biography of the Unborn (17 min), EBF.

Bountiful Heritage (21 min), Ferry-Morse; free.

The Chick Embryo: From Primitive Streak to Hatching (13 min), EBF.

Chicks (set of 12 transparencies), Carolina.

DNA: Molecule of Heredity (16 min), EBF; G. W. Beadle, narrator.

Development of the Chick Embryo (6 min), Coronet.

The Double Life of the Dragonfly (11 min), Film Pharos.

The Embryonic Development of Fish (28 min), National Film Board of Canada.

The Fish Embryo: From Fertilization to Hatching (12 min), EBF.

Flowers and Fruits (color transparencies), Scientific Supplies.

Flowers at Work, 2nd ed. (11 min), EBF.

Flowers: Structure and Function (11 min), Coronet.

Frog Eggs (set of 16 transparencies), Carolina.

Gene Action (16 min), EBF.

Genetic Corn (set of 12 transparencies), Carolina.

Genetic Investigations (12 min), Indiana Univ.

Genetics (set of 12 films, 28 min each), Part VI of *AIBS Film Series in Modern Biology,* McGraw-Hill.

Genetics (2 reels, 53 min), U.S. Dept. of Agriculture, Graduate School; G. W. Beadle, narrator.

Genetics: Improving Plants and Animals (14 min), Coronet.

Genetics: Mendel's Laws (14 min), Coronet.

Genetics of Mendelian Populations (30 min), in *Genetics* series, McGraw-Hill.

Genetics: Techniques of Handling Drosophila (3 min), BSCS techniques, Thorne.

Growth of Flowers, 2nd ed. (11 min), Coronet.

Handling Drosophila (filmloop), Ealing; adapted from BSCS techniques film.

Heredity (set of 9 films, 30 min each: *Fact or Fallacy, Many Pairs of Genes, Sexuality and Variation, The Sex Chromosomes, It Runs in the Family, Heredity and Chromosomes, Reproduction and Heredity, Heredity and Environment,* and *Mendel's Experiments*), Indiana Univ.

Hereford Story (27 min), Farm Film Foundation; free.

The Human Body: Reproductive System (14 min), Coronet.

Human Reproduction (21 min), McGraw-Hill.

Hydra Development (set of 20 transparencies), Carolina.

Laws of Heredity (16 min), EBF.

Life Before Birth (27 min), Carousel.

Life Cycle of a Malaria Parasite (12 min), Univ. of California.

Life of the Molds (21 min), Sterling; free.

Magnolia (set of 5 transparencies), Carolina.

Meiosis in Spermatogenesis of Grasshopper (19 min), Phase.

Meiosis: Sex Cell Formation (16 min), EBF.

Mr. Stickleback (10 min), Almanac.

Mitosis (24 min), EBF.

Mitosis and Meiosis (16 min), Indiana Univ.

Neurospora Techniques (8 min), BSCS techniques, Thorne.

The Newt (10 min), International Film Bur.

The Onion (10 min), International Film Bur.

Perpetuation of Life (29 min), Indiana Univ.

Pin Mold (10 min), International Film Bur.

Plant Life Cycles (transparencies), Scientific Supplies.

Plants That Grow from Leaves, Stems, and Roots (11 min), Coronet.

The Rabbit's Development (29 min), International Film Bur.

Radiation in Biology (14 min), U.S. Atomic Energy Comm.; free.

Radiation in Biology: An Introduction (14 min), Coronet.

Removing Frog Pituitary (1 min), BSCS techniques, Thorne.

Reproduction, Growth, and Development (set of 12 films, 28 min each), Part V of *AIBS Film Series in Modern Biology,* McGraw-Hill.

Reproduction in Animals (11 min), Coronet.

Reproduction in Plants (14 min), Coronet.

The Reproductive Processes of the Frog (45 min), R. Rugh, Columbia Univ., 630 W. 168 St., New York 32, N.Y.

Salmon: Catch to Can (14 min), U.S. Fish and Wildlife Service, Bur. of Commercial Fisheries; free.

The Salmon's Struggle for Survival (26 min), National Film Board of Canada.

Sea Urchin (8 min), United World.

Seed Germination (15 min), EBF.

Span of Life, Upjohn; free.

The Story of the Breeds (25 min), Texas Co.

Syngamy and Alteration of Generation in Allomyces, A Water Mold (20 min), Phase.

The Thread of Life (2 reels, 25 min each), Bell System; free.

Many of the biological supply houses publish periodical bulletins of suggestions for using plants and animals in the classroom. These offer excellent sources of ideas for the teacher and resourceful students. Also refer to Appendix D, the directory of distributors of free and low-cost materials.

The Biological Effects of Atomic Radiation, National Academy of Sciences.

Demerec, M., and B. P. Kaufmann, *Drosophila Guide: Introduction to the Genetics and Cytology of Drosophila melanogaster,* 7th ed., Carnegie Institution of Washington; $0.50.

Holstein-Friesian Cow and Bull (charts), Holstein-Friesian Assoc. of America.

Potter, V., *DNA Model Kit,* Burgess, Minneapolis; a kit for preparing a three-dimensional model, $1.00.

Story of Cereal Grains, General Mills, Education Section.

The Story of Meat Animals; The Story of Plants, Swift, Education Division.

U.S. Govt. Printing Office, *Human Conservation: The Story of Our Wasted Resources,* Washington, D.C.; $0.20.

Your Heredity, Science Research Associates; $0.60.

The Am. Genetic Assoc. provides upon request pedigree charts and lists of articles on heredity of domesticated animals.

Look into H. Deason, ed., *A Guide to Science Reading* (Signet, New Am. Library), an annotated guide to more than 900 paperbound science books ($0.60).

"Modern Biology Series" (Holt, Rinehart and Winston) includes a number of fine paperbacks that cost less than $2.00 each. "The Foundations of Modern Biology Series" (Prentice-Hall) also has a wide selection of splendid paperback books costing less than $2.00 (available in hard covers at a higher price). "Fundamentals of Botany Series" (Wadsworth) is another fine paperback series ($1.00 each for instructors).

Chapter 9

Growth and Differentiation

Among the topics discussed in this chapter are studies of growth curves in populations of microorganisms, in germinating bean or pea seedlings, in young mice, in chicks, and in other animals; rate of growth of organs, such as roots, leaves, and stems of seedlings; mitotic divisions of nuclear material; and growth of eggs of animals, especially anuran and uredele eggs, and cleavage stages.

A very brief description is offered of the tools and techniques for investigating the potencies of the dorsal lip of the blastopore of the amphibian egg, and the method of making transplants of tissue from the dorsal lip region (organizer) to the flank or belly region of an older host.

Many of the solutions used in studies of mineral needs of growing plants can be made available to students interested in hydroponics. Some suggestions are offered for culturing bacteria and for testing the effectiveness of antiseptics and antibiotics on bacteria.

The final portion of the chapter deals with studies of regeneration of missing parts in plants and animals. We there discuss tissue that has retained the qualities of the embryonic state, that is, tissue capable of differentiating into leaves, or roots and shoots in plants, or of regenerating missing parts in such animals as planarian worms, hydra, and protozoans.

The measurement of growth

Growth, in a general sense, is often identified by an increase in height. In a biological sense, growth is associated with an increase in the mass of the living substance. The distinction is not a simple one; an increase in height or length may accompany an increase in mass. In either sense, growth is a dynamic indicator of metabolism of cells. Furthermore, each organism seems to have a specific cycle of growth, and even providing an abundance of nutrients does not permit a continued accelerated rate of growth. It is often imperative that an enzyme be present too.

Students can measure the increase in weight of an organism as an index of its growth. Plotting the increase in weight against time gives a sigmoid curve (Fig. 9-1) which is a measure of the absolute increase from starting to final weight regardless of other factors.[1] In short, a sigmoid growth curve gives a graphic illustration of absolute increase in weight and shows the greatest absolute increase in weight per unit of time in the middle part of the curve. Now, if increases in mass are measured for equal time intervals, then the amount of increase at different periods in the life cycle of one organism may be known. An absolute increase is not a true picture of the rate of growth at *different periods* of life (Fig. 9-1), nor does it offer a valid basis for comparing the rates of growth of *different organisms* (Fig. 9-2).

[1] B. I. Balinsky, *An Introduction to Embryology*, Saunders, Philadelphia, 1960, offers a splendid chapter on growth on the cellular and intercellular levels and on the organismic level. Balinsky also interprets growth curves and aging, as well as disproportional growth in organs. Some of the material presented here has been gleaned from Balinsky's Chapter 15.

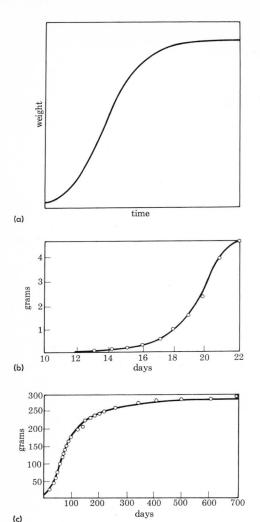

Fig. 9-1 Sigmoid growth curves: (a) ideal sigmoid curve; (b) intra-uterine growth of white rat; (c) postnatal growth of white rat. (In b and c, the open dots indicate actual measured values.) (After B. I. Balinsky, *An Introduction to Embryology*, Saunders, 1960.)

Clearly, if the same absolute increase in a given time is shown by a large and a small animal, their rate of growth is not the same; the smaller animal had to grow at a faster rate.

Growth, then, is an increase in geometric progression, an exponential process.

The increase is proportional to the initial quantity of growing material (if the initial quantity is exceedingly small and if the initial size is negligible). The general formula for exponential growth is $W = e^{vt}$, where W is the weight of an animal at any time t, v is the observed rate of growth, and e is the base of natural logarithms (equal to 2.71828 . . .).

Students may plot exponential growth curves of several organisms, and of populations of organisms. Analyzing growth at the intracellular level, however, is more complex. The overall rate of synthesis of molecules depends on the quantity of molecules of enzymes at hand, and the quantity that is available for each step in the chain of syntheses. Possibly the enzymes that control growth of protoplasm exist as only a few molecules in a cell and must be multiplied before organic syntheses in growth can get under way.[2]

Investigate the factors affecting the change in the height of a tree. Is this growth pattern similar to that of a definitive structure such as a leaf or a fruit? Is the growth pattern of an individual cell or a multicellular plant or animal similar to the growth of a population of cells, such as a culture of yeast cells, or of *Chlorella*, or of *Paramecium?* Does the logistic theory of population growth hold for all kinds of organisms? (For some limitations, see H. Andrewartha and L. Birch, *The Distribution and Abundance of Animals* [Univ. of Chicago Press, 1954] and more recent references.)

Logarithmic growth curves

Populations of such cells as *Chlorella, Paramecium,* and yeasts have been carefully studied (Fig. 9-2); populations of these organisms exemplify distinct growth phases (Figs. 9-2a, 9-3). The first phase is a lag phase, in which the organisms become adapted to their medium or environment. The second phase is an exponential growth phase, in which rapid growth of

[2] *Ibid.* Also read Balinsky's chapters on differentiation and regeneration in the 2nd ed., 1965.

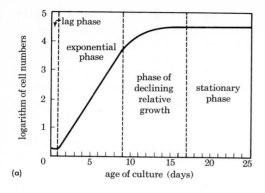

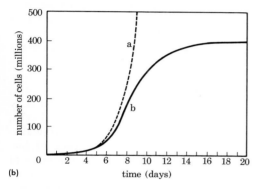

(a)

age of culture (days)

(b)

time (days)

Fig. 9-2 Sigmoid growth curves of different organisms: (a) growth of *Chlorella;* (b) growth of yeast cells, showing theoretical rate of multiplication with unlimited space and food (curve a) and actual rate of multiplication in a definite limited amount of nutrient solution (curve b). (a, after G. E. Fogg, "Famous Plants: 4, *Chlorella*," in M. L. Johnson, M. Abercrombie, and G. E. Fogg, eds., *New Biology*, Vol. XV, Penguin, Harmondsworth, Eng., 1953; b, after D. Marsland, *Principles of Modern Biology*, rev. ed., Holt, Rinehart and Winston, 1951.)

viable cells occurs. When young cultures are used in studies of growth, counts of cells may be made regularly and plotted against time. The points will be found to lie closely along a straight line. How long can the exponential growth rate continue in a given culture medium? Are there physical conditions of density of cells per volume of medium that have detrimental effects on growth? What deficiencies in the medium may arise? What materials may accumulate in the medium?

The third phase of growth is a declining growth phase, or maximum stationary period. Cells are viable, but there is little growth. The duration of this phase may depend upon alterations in the environment. Final phases 4 and 5 are stationary or decline phases, which may lead to the death of the population (or to spore formation among cells that have this ability).

Measuring growth of microorganisms

Students may either estimate the rate of growth of individual cells (that is, the rate of fission) or measure the growth of a population of similar organisms.

Making serial dilutions

Usually one must dilute suspensions of microorganisms in order to count or isolate cells.

Prepare a series of small test tubes or vials of the culture medium in which the cells are growing, for example, Knop's solution if the cells are *Chlorella*. When an entire class is engaged in a laboratory activity, small quantities may be used.

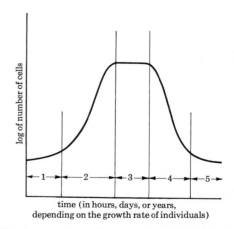

time (in hours, days, or years, depending on the growth rate of individuals)

Fig. 9-3 Logarithmic growth curve. Phase 1 is the initial lag phase; phase 2 includes the exponential phase of rapid growth; phase 3 is the stationary phase; phase 4 is an exponential death phase; phase 5 is a survival phase.

Add 9 ml of Knop's solution to each of a series of small test tubes. With a micropipette of 1-ml volume, or a calibrated micropipette, remove 1 ml of the algal suspension from the main culture and add it to the first of the series of test tubes. Agitate this tube, draw up 1 ml of the mixed dilution, and add it to the second test tube. Agitate this tube, draw up 1 ml of this, and add it to the third test tube; and so forth.

TUBE 1: 1 ml of algal suspension
    + 9 ml of Knop's = 1:10 dilution
TUBE 2: 1 ml of 1:10 dilution
    +9 ml of Knop's = 1:100 dilution
TUBE 3: 1 ml of 1:100 dilution
    + 9 ml of Knop's = 1:1000 dilution
TUBE 4: 1 ml of 1:1000 dilution
    + 9 ml of Knop's = 1:10,000 dilution

Students may make counts of cells in each dilution. Or they may streak solidified agar in Petri dishes or welled slides.

Growth of individual cells

Individual cells isolated in welled slides, or in a hanging drop, may be used for studies of the rate of fission. The same technique can also be used for developing a clone, or a group of organisms of similar heredity (barring mutations). Pick up individual cells with capillary tubing pulled out to a fine thread. Or use the following method, which was originally used by Spallanzani. Place 1 small drop of culture on a slide; near it, place 1 small drop of fresh culture medium. With the tip of a capillary pipette make a connecting channel of fluid between the 2 drops. Observe under a microscope. When a motile cell moves into the drop of fresh medium, destroy the channel, pick up the isolated cell, and transfer to a welled slide or hanging drop. Seal the coverslip with petroleum jelly.

Nonmotile cells may be isolated on a solidified medium, such as agar. Dilute a portion of the culture; then, with a transfer loop or needle, streak the surface of the cooled medium with a diluted culture (Fig. 9-4). Or a row of small drops of the dilution can be placed along a slide, and each drop can be inspected for isolated cells.

These techniques are useful for students working with yeast cells, *Chlorella, Chlamydomonas,* or motile protozoa. The nonmotile forms that can be grown on agar, such as bacteria, yeasts, and *Chlorella,* may be thought of as isolated, individual cells, so each colony that forms is, theoretically, a pure colony—a clone of cells of like heredity. The change in the diameter of the colonies may be taken as an estimate of the rate of fission of the cells. (For other techniques for handling bacteria, see Chapter 2 and p. 119.)

Growth of populations of microorganisms

Viable populations of cells increase exponentially: 1, 2, 4, 8, 16, 32, 64, . . . (that is, 2^0, 2^1, 2^2, 2^3, 2^4, 2^5, 2^6 ,. . .; see Fig. 9-1). Notice in the growth curve that there are limits to the theoretical curve of "com-

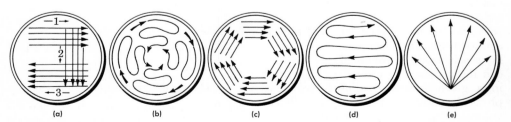

Fig. 9-4 Patterns for streaking the surface of sterile agar medium. (Adapted from *Student Laboratory Guide to Biological Science: An Inquiry into Life* [BSCS Yellow Version], Harcourt, Brace & World, 1963.)

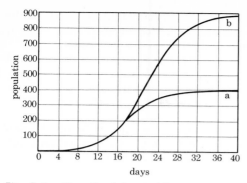

Fig. 9-5 Growth of *Drosophila*: curve a, in a ½-pt bottle; curve b, in a 1-pt bottle. (After D. Marsland, *Principles of Modern Biology*, rev. ed., Holt, Rinehart and Winston, 1951.)

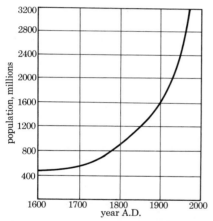

Fig. 9-6 Actual growth of human population of the earth.

pound interest"—limits set by such environmental factors as the quantity of food. Then the rate of growth falls off so that the number of cells reaches a constant value. Populations of other organisms show similar growth curves (Fig. 9-5). Present data of the human population of the earth indicates that the same general growth curve will be approximated (Fig. 9-6).

There are several methods of measuring growth suitable for the high school laboratory.[3] One good method uses *Chlorella* as an experimental organism because it is

[3] Also examine the many activities in A. S. Sussman, *Microbes: Their Growth, Nutrition, and Interaction* (BSCS laboratory block), Heath, Boston, 1964.

nonpathogenic, nonmotile, easy to maintain in the laboratory, and useful in studies of photosynthesis (Chapter 3) and respiration (Chapter 6). Yeast cells can also be used, but *Chlorella* is preferred because it can be inspected without the use of a microscope (the culture medium very soon begins to turn green).

Under optimum conditions, *Chlorella* has a very short lag period—about 1 day. It then moves rapidly into active growth for about 10 days, until the rate begins to level off. In the later, stationary phase, *Chlorella* cells are large and vacuolated and have thick walls. The generation time for *Chlorella* is about 15 hr, more than twice that for yeast cells or bacteria. (Culture media for *Chlorella* are described in Chapter 13.[4])

Counting cell populations. The number of cells in a given volume of suspension must be known at the start of experiments in which growth rate is estimated (for example, under varying conditions of light, minerals, or wavelengths). The number of cells of *Chlorella* (or any other nonmotile organism) in 1 cu mm of culture medium may be determined fairly rapidly in a counting chamber, using the same method as for counting red blood cells.

The counting chamber is a ruled glass slide with a 1-mm-square center section marked off into 25 squares. Each of these 25 squares is further subdivided into 16 squares, giving a total of 400 small squares (Fig. 9-7). On each side of the specially ruled section of the slide are raised supports that are exactly 0.1 mm high. When a coverslip is placed over the ruled slide there is a distance of 0.1 mm between the slip and coverslip, which establishes a given volume (0.1 cu mm) in which the cells can be counted.

Use a diluted suspension of cells. (The dilution factor must be recorded, because it enters into the calculation of the total number of cells [see p. 444].) If the cells

[4] Urea is a good source of nitrogen for *Chlorella,* since urea does not support the growth of contaminants in the culture medium; further, it does not change the pH.

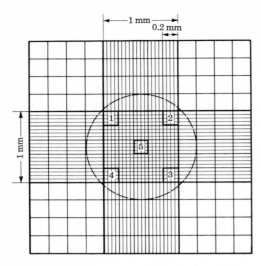

Fig. 9-7 Ruled section of chamber for counting cell populations. Portion within circle contains 25 major squares, each of which is subdivided into 16 minor squares.

tend to form into clumps, agitate the fluid.

First place the coverslip over the supports on the ruled slide. Then, with a Thoma pipette or a hypodermic needle and syringe, allow a small drop of the diluted suspension to flow by capillarity between the ruled surface and the coverslip. Let the slide stand for 2 or 3 min so that the cells will settle.

Count the cells in each of five of the large blocks of squares marked 1, 2, 3, 4, and 5 (Fig. 9-7). If there is a large discrepancy in the number of cells in each of the five blocks, the cells in the suspension probably are not evenly distributed. (Using the standards used in counting red blood cells, the difference should not exceed 15.) To avoid counting some of the cells twice, include those that touch the top of any square or the inner line on the right side, and exclude cells that touch the bottom or left side of any square.

Now calculate the number of cells in each cubic millimeter of suspension:

1. Total the count of cells in the five blocks of squares. For example, there may be $96 + 99 + 93 + 104 + 101 = 493$ cells.

2. The 80 small squares in which the cells were counted occupy $^{80}/_{400}$, or $\frac{1}{5}$ sq mm and contain the 493 cells.

3. Therefore, *1 sq mm* would contain $5 \times 493 = 2465$ cells.

4. Since the thickness of the cell layer is 0.1 mm, 1 cu mm would contain $10 \times 2465 = 24,650$ cells.

5. Multiply this number by the dilution factor (for example, 100). Hence, there are $100 \times 24,650 = 2,465,000$ cells in 1 cu mm of suspension.

In short,

$$\underset{\text{in 80 squares}}{\text{number of cells}} \times \underset{\text{(area)}}{5} \times \underset{\text{(volume)}}{10} \times \underset{\substack{\text{(dilution} \\ \text{factor)}}}{100}$$

$$= \text{number of cells in 1 cu mm} \\ \text{of suspension}$$

There are three main sources of error in this method: The cells may be clumped; there may be an overflooding of the slide; or the dilution of the suspension may be inaccurate.

Agitators for cultures of microorganisms. Some students may want to build agitators, which are used for keeping cultures of microorganisms stirred to facilitate distribution of nutrient materials and to promote growth. Some students have contrived an agitator from discarded parts of an old phonograph. The turntable and motor can be fastened to a wooden base, and to this may be attached wire test-tube baskets to hold a supply of test tubes or flasks.

Counting organisms. Umbreit[5] describes Breed's method for counting organisms. This method is especially useful for counting small cells such as bacteria in a sample, and can be done without a counting chamber or any special equipment. It has wide use for high school laboratory activities.

With a glass-marking pencil, mark off a 1-cm square on each of several microscope

[5] W. Umbreit, *Modern Microbiology*, Freeman, San Francisco, 1962, pp. 45–46.

slides. With a sterile standardized loop, transfer a small drop of suspension of bacteria onto a slide, and spread the drop evenly over this *known,* marked-off area. The volume of fluid transferred by a given loop is about the same for each loopful and can be determined by weighing and averaging 50 or 100 loopfuls. Probably a loopful will carry 0.01 ml of fluid. This amount of fluid, then, has been spread over an area of 1 sq cm.

After the smear is dried, fixed, and stained (see Chapter 2), the cells are counted under an oil immersion lens. This lens usually has a field of 0.16 mm, and the area of the smear seen under the field is about 1/5000 of the 1-sq-cm area over which the sample was spread. The area of the oil immersion field is $\pi r^2 = 3.14 \times 0.08 \times 0.08 = 0.02$ sq mm.

Consider that the smear was spread over 1 sq cm, or 100 sq mm; therefore there are 100/0.02, or 5000 microscopic fields in the 1 sq cm. Umbreit continues this calculation, estimating that each organism seen in one field represents 5000 in the area of 1 sq cm. If the loop carried over 0.01 ml of the sample of the fluid, this would represent 5000 per 0.01 ml, or 500,000 per ml. It can be calculated, in this case, that each

organism seen per field represents 500,000 per ml of the original sample.

Thus, in general, for any number of organisms N seen in a field, there are

$$N\left[\frac{100Y}{\pi(d/2)^2}\right]$$

organisms in the area of the slide, where d is the diameter of the microscopic field (in millimeters) and Y is the known area over which the total sample is spread (in square centimeters). To convert this to a general formula giving the number of organisms in 1 ml, multiply by $1/Z$, where Z is the total sample on the slide (in milliliters). Thus,

$$\text{organisms per milliliter} = \frac{400NY}{Z\pi d^2}$$

If, as in the preceding discussion, the sample (Z) is 0.01 ml and the given area is 1 sq cm, the equation can be reduced to

$$\text{organisms per milliliter} = N\left[\frac{4 \times 10^4}{\pi d^2}\right]$$

Turbidity as a measure of growth. A photometer of the type shown in Fig. 9-8 may be made by high school students. Fairly good approximations can be obtained of changes in turbidity (increase in number) of test tube cultures of *Chlorella* or

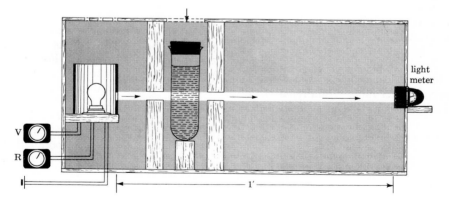

Fig. 9-8   Student-made photometer for measuring growth of microorganisms in a culture. The interior of the box must be lightproof, painted dull black, with black felt fitted around apertures. An aluminum shield on the light source concentrates light through a slit on one side, thus permitting light to pass in a direct line through the wood partitions and the culture to the light meter, placed 1 ft away from the light source. The voltmeter is inserted into the circuit to maintain a constant voltage, and the rheostat permits variation of the current through the light bulb. Ventilation holes above the light bulb prevent the culture medium from becoming overheated.

yeast cells, or of protozoa, bacteria, or *Euglena.* Measurements may have to be made over days or even weeks to note definite variations in relative rate of growth.

In the measurement of turbidity, a beam of light is passed through the culture medium, and the transmitted light is registered on the photosensitive surface of a light meter (or a photoelectric cell). This, in turn, causes a current and a corresponding deflection of the needle of the galvanometer; the greater the illumination, the greater the current. A dial can be calibrated to measure foot-candles of intensity of illumination on a surface.

Since the intensity of illumination is inversely proportional to the square of the distance, when the meter is placed *exactly 1 ft away* from the lamp, the reading in foot-candles will be equal numerically to the candlepower of the lamp:

$$\frac{\text{intensity of illumination}}{\text{(foot-candles)}} = \frac{\text{candlepower}}{d^2}$$

With increased turbidity, due to the multiplication of cells in the culture medium, less light is transmitted through the suspension, which reduces the reading on the meter (foot-candles in this case).

Students will need to compare the meter reading to a standard sample of culture medium. Also, blanks, or empty test tubes, should be measured (similar test tubes must be used in all the experimentals).

If three cultures of *Chlorella* of varying density, as determined visually, are placed in the light of a 100-watt lamp, the exposure meter readings might approximate the following order:

solution	reading (foot-candles)
blank tube	28
Knop's solution (control)	24
pale green suspension	20
light green suspension	12
deep green suspension	8

Counts of cells made of these three cultures in a counting chamber would give results of the following order:

solution	cell density (cells per cu mm)
pale green	12,400
light green	14,500
deep green	22,700

In this case, turbidity may be standardized in terms of the number of cells estimated in a counting chamber.

Increase in clones as a measure of growth. Students may count colonies of cells of algae in much the same way as clones of bacteria are examined in Petri dishes. A diluted suspension (see p. 441) of cells is added to cooling agar medium before the agar is plated into sterile Petri dishes. Or a measured sample of cells is poured into the Petri dish and cooled agar is then added.

The agar solution is a thin medium composed of about 1 to 1.5 percent agar prepared with Knop's solution or other culture medium. Use mineral agar rather than nutrient agar in order to hold down the bacterial contaminants that are in the cultures of algae. (Agar melts at 100° C [212° F] and solidifies below 42° to 45° C [108° to 113° F]. Algae may be added to the agar cooled to 42° to 45° C. Some practice is needed because, on the addition of the cool algae culture, the agar will solidify rapidly so that it cannot be poured.)

Petri plates may be stacked and kept in moderate light if green algae are to be cultured. After about 2 weeks, hold the plates up to the light or against a white background to locate small pinpoint colonies as they begin to appear.

A student who has some manual dexterity may construct a colony counter using a diagonal ledge of wood with a hole in it. A light bulb is placed on the underside of the board and a Petri dish is placed over the hole (Fig. 9-9). Complete directions for making a colony counter are available from the Am. Optical Co., Buffalo 15, N.Y. In the slanted side of a wooden light box, cut a hole the size of a Petri dish. Place a glass plate over the hole and rule vertical and horizontal guidelines about ½ in. apart on the glass. Position an

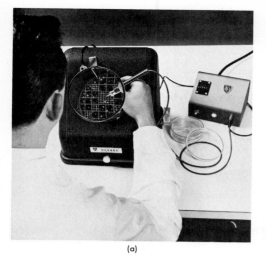

(a)

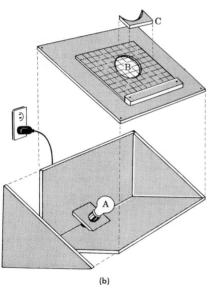

(b)

Fig. 9-9 Colony counters: (a) commercial counter with an electronic register, used in bacteriology laboratories; (b) plan for constructing colony counter (A, 25-watt light bulb; B, opening, the diameter of a Petri dish, covered by ruled glass; C, wooden block to hold Petri dish). (a, courtesy of American Optical Co.; b, after an American Optical Co. drawing.)

exposed Petri dish containing colonies on this glass plate, and, with a hand lens, count the colonies that are within each square drawn on the glass plate. When light shines through the Petri dish, even

small pinpoint colonies are observable.

In experimental work, students would inoculate the medium with a given volume of dilute algal suspension. When colonies arise, transfers may be made into small test tubes or welled slides to provide cells of the same heredity, that is, clones.

In a mixed culture of algae, *Chlorella* usually appears first in the colonies developing on agar. Prepare a slide for checking under the microscope.

The diameter of the colonies may be measured with a millimeter ruler, or the plates may be laid on graph paper and students can devise their own method for estimating numbers of colonies. Plastic Petri dishes that are already ruled for counting colonies are available.

If the colonies are evenly distributed over the surface of the Petri dish, students may count only those in a small section and, from this, calculate the clones in the total area of the sphere. Use a cork-borer with a known diameter to measure off a plug or cylinder of seeded agar in the Petri dish. Then calculate the volume of this small cylinder ($\pi r^2 h$) and count the number of clones in this volume of cylinder. From this, calculate the number of colonies in the known volume of agar plus medium used (possibly 15 ml plus 1 ml of culture). Welled slides with a straight-walled concavity usually are 16 mm in diameter and 3 mm deep.

Counting clones may be a useful measure, provided one remembers its limitations; the number of live cells is not determined. The count represents an estimate of the number of live cells that are capable of reproduction. A constant error in counting cells, particularly in methods which measure turbidity, is that dead cells cannot be distinguished from live ones.

Dispose of Petri dishes as though they were contaminated with bacterial cultures; cover with Lysol or other strong disinfectant.

Small prescription bottles with one flat side may be substituted for Petri dishes. Add 15 ml of agar medium to which a specific dilution has been added. Seed the

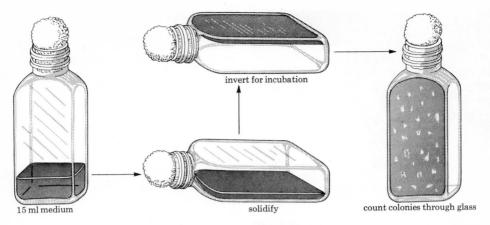

Fig. 9-10 Procedure for using prescription bottle cultures (in lieu of Petri dishes) for measuring growth of colonies of microorganisms. (After W. Umbreit, *Modern Microbiology*, Freeman, 1962.)

agar medium while it is cooling at 45° C (113° F). Shake the bottle, and allow the agar to solidify on one flat side (Fig. 9-10). When the bottles are to be incubated, invert them, as is also customary for Petri dishes. Inverting the bottles prevents the condensed drops of moisture that usually accumulate on the "roof" of the container from dripping back onto the growing colonies on the surface of the agar medium.

Students can count the colonies through the glass. Suspensions of *Chlorella,* yeast, or bacteria may be seeded in the agar medium in these bottles.

Measuring growth by a gain in volume (or weight). The growth of populations of organisms over several weeks (or even over a month) can be studied by estimating the increase in volume or weight of cultures that have been maintained under a variety of experimental conditions.

Either allow tubes of a given volume of culture to stand undisturbed so that cells become packed at the bottom of a graduated tube (bulk volume), or centrifuge 10-ml samples of agitated experimental cultures, as well as the controls, in calibrated centrifuge tubes at a given speed for a given time. Then calculate the volume of sedimented, packed cells in relation to the volume of fluid (milliliters of algae per 100 ml of fluid). The packed volume of cells in a sedimentation tube per milliliter

of culture may be determined daily. Or the cells from diluted samples of cultures can be counted, and the number of cells per milliliter of fluid can be estimated.

Wet weight may be obtained by running a sample collected from a centrifuge through a Buchner flask attached to a vacuum pump. Collect the cells on filter paper in the funnel and weigh. Or allow the filter discs to dry in a dust-free container or oven and weigh to obtain the dry weight of cells in 1 liter of fluid.

Replica plating

Lederberg developed a simplified technique, called replica plating, for continuing colonies of bacteria of the same heredity without making transfers with needles or loops. A piece of velour was used to cover a cylinder of wood or other material. The diameter of the cylinder and velour covering was the same as that of a Petri dish, and the velour was held securely in place.

When colonies are grown in a Petri dish, each colony may be considered as having arisen from one organism. Hence all the cells are a clone, having the same heredity. Lederberg and his group inverted a growing Petri dish over the velour and wood cylinder, pressing lightly, so that a pattern of the colonial growths was formed on the velour (see Fig. 8-74). Then a fresh, sterile Petri dish of agar was inverted and touched to the "pattern" on the velour. When incubated, new colonies formed at the points of contact with the original colonies. In fact, if two dishes were placed on top of each other, the same pattern of distribution of clones could be matched. This, then, is also a way of marking off, or of identifying similar clones without the tedious task of making transfers.

What kinds of investigations might be carried out when students can rely on having microorganisms of the same heredity?

Growth of yeast cells

Many laboratory activities can be planned with a package of dried yeast cells and dilute molasses solution. Use the methods of making cell counts or analyzing turbidity described for *Chlorella* or other nonmotile microorganisms (pp. 443–48).[6]

Growth of motile forms

Growth of flagellates such as *Euglena* or *Chlamydomonas* and of ciliates such as *Paramecium* can be measured by means of increased turbidity, and by counts of sample loopfuls in diluted culture media. They may be isolated by using the channeled-drop method (p. 442), or by examining a series of drops for individual organisms and then transferring these into hanging drops on microscope slides.

For investigations using microaquaria, see Chapters 3 and 11. Use of *Euglena* for assay of vitamin B$_{12}$ is described on p. 462. See BSCS laboratory activities on competition between *Paramecium aurelia* and *P. caudatum;*[7] methods for counting protozoa are also presented.

[6] Also consult the activities in Sussman, *op. cit.,* which include investigations of population growth of yeast cells.

[7] *High School Biology: Blue Version—Laboratory, Part I,* BSCS, Univ. of Colorado, Boulder, 1961 (experimental use, 1961–62), Exercise 10.

Conditions required for growth of bacteria

The optimum factors for growth of bacteria may be rediscovered by committees of students. We are thoroughly acquainted with the conditions bacteria need for growth, and we sometimes forget that for children this work presents a fresh experience in learning. Our emphasis is not so much on the end results, but on providing practice in devising experiments.

Where are bacteria found?

To determine whether bacteria are present in a given environment, optimum conditions for their growth must first be provided.

Prepare nutrient agar medium and plate several sterile Petri dishes (Chapter 13). Set aside a few unopened plates as controls. Then carefully lift the cover on each of several Petri dishes and inoculate in the following ways. Into one dish place several coins. Have a student touch the surface of the agar of another dish lightly with his fingertips. To another dish add a film of yoghurt, milk, sauerkraut, pickle juice, or water in which beans or peas are left to decay. A student might cough into still another dish. Expose the final dish to the air of the classroom. Incubate all the Petri dishes. Students should recognize that the absence of a full growth of bacterial colonies on a window sill is due not to the lack of bacteria, but to the absence of optimum conditions for growth.

Students may begin to ask how antibiotics interfere with the growth of bacteria, and inquire about the role of antiseptics in affecting bacterial growth. Many teachers relate these laboratory activities concerning antibiotics and antiseptics as man-made defenses to discussions of the control of diseases of plants, animals, and man.[8]

[8] For an especially readable discussion of diversity among microorganisms see *Biological Science: An Inquiry into Life* (BSCS Yellow Version), Harcourt, Brace & World, New York, 1963, Chapters 9–12.

Effect of temperature on growth of bacteria

Prepare a nutrient agar medium (Chapter 13). Plan to plate several Petri dishes (p. 634). Inoculate the cooling nutrient agar with a pure culture of harmless bacteria such as *Bacillus subtilis* or the bacteria in sour milk. Then plate the Petri dishes so that each dish contains about the same quantity of bacteria.

Place two or more dishes in a refrigerator at 4° to 8° C (39° to 46° F). Keep two at room temperature (20° to 25° C [68° to 77° F]), and put two in an incubator at 37° C (99° F). If you have a second incubator, keep two more Petri dishes at approximately 55° C (131° F). Check the temperature each day with a thermometer and observe the rate of development of visible colonies.

Effect of penicillin on growth of bacteria

Prepare several test tubes of nutrient agar culture medium (p. 630). While the medium is cooling (45° C [113° F]) but still liquid, inoculate it with a pure culture of harmless bacteria and agitate by rolling between the palms. Then plate the suspension into several sterile Petri dishes. Prepared this way, the agar medium will have a uniform distribution of bacteria. Inoculate some culture media with a pure culture of nonpathogenic, Gram-positive bacteria (*Bacillus lactis,* found in milk; or *B. subtilis,* found in hay and soil), others with nonpathogenic, Gram-negative bacteria (*Pseudomonas hydrophilus,* found on frogs suffering from red leg; *Achromobacter nitrificans,* found in soil; or *Escherichia coli* var. *acidilactici*).

Now set two inoculated Petri dishes aside as controls; to the other inoculated dishes add a few discs of filter paper (about ¼ in. in diameter), which have been soaked in a solution of penicillin prepared by dissolving a penicillin lozenge in cooled, sterile water; or soak the filter paper in a culture of *Penicillium* mold. After incubation of all dishes students should find abundant growth of colonies in the cultures. On the other hand, the Petri dishes containing discs of the antibiotic show a clear zone, a "zone of inhibition," around the discs in some cases. Penicillin shows greater activity against Gram-positive bacteria, mainly cocci.

You may want to compare the action of several antibiotics at one time, or to find the quantitative differences in the spectrum of a single antibiotic. Bacto-Unidisks[9] are sterile preparations containing specific amounts of commonly used antibiotics and sulfa drugs. Many other commercial preparations are also available.

Antagonism between two kinds of bacteria

Two kinds of bacteria may be grown in one Petri dish to show antibiosis, the suppression of growth of one bacteria by substances produced by the other bacteria (Fig. 9-11). Carolina Biological offers a kit of cultures of *Sarcina subflava* and of a strain of *Bacillus subtilis.* The following procedure is suggested in Carolina's catalog.

[9] Available from Difco Laboratories, Detroit 1, Mich., and other supply houses.

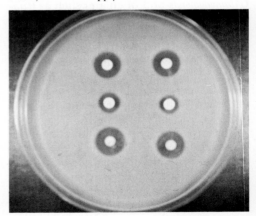

Fig. 9-11 Demonstration of zones of inhibition. Assay of varying concentrations of penicillin by application of saturated paper discs on the surface of sterile agar medium that has been seeded with *Staphylococcus aureus.* (Food and Drug Administration photo.)

First, melt a sterile tube of prepared nutrient agar medium and let it cool to 45° C (113° F). Inoculate this tube with a suspension of *Sarcina subflava* and rotate it between your hands to distribute the bacteria evenly in the culture medium. Pour this uniformly seeded suspension into a sterile Petri dish and set it aside to solidify.

Next, inoculate this seeded plate, either by streaking or by touching a transfer needle at several places on the agar surface, with a strain of *Bacillus subtilis*. Then incubate the plate or maintain it at room temperature. After 24 hr, inspect the plate; *Sarcina subflava* cultures should be prominent over the agar except at the places where *Bacillus subtilis* was inoculated. Notice the zones of inhibition (clear areas) around the areas where *Bacillus subtilis* grows; apparently *subtilis* produces an antibiotic (a bacteriostatic action) that diffuses through the agar and inhibits the growth of *Sarcina subflava*.

Effect of antiseptics on bacterial growth

You may also wish to demonstrate the bacteriocidal effect of antiseptics. Prepare nutrient agar culture medium (Chapter 2) and inoculate the cooling nutrient medium with a pure culture of harmless bacteria. Then plate the seeded agar in Petri dishes; in this way bacteria are distributed throughout the medium.

Now set aside several dishes as controls. To the other dishes add several selected antiseptics. To some dishes you may add a film of 2 percent tincture of iodine, or use filter paper discs that have been soaked in the antiseptic. To others add a film of 70 percent alcohol, to others a film (or discs) of full-strength Lysol, and to others diluted Lysol. You may want to include household mercurochrome (1 percent), as well as witch hazel and 3 percent hydrogen peroxide.

Incubate all the Petri dishes, including the control dishes. Look for the appearance of colonies. Students may compare the effectiveness of the various antiseptics used by counting the number of colonies and the size of each colony. Which chemicals were effective in a 24- to 36-hr period? Which chemicals were still effective after 72 hr?

Soil bacteria and utilization of starch

Many soil bacteria break down starches. Students may show this by preparing a nutrient agar medium to which cornstarch, potato starch, or arrowroot starch has been added before it is sterilized. Plate sterile Petri dishes and, after cooling, streak the surfaces of several plates with transfer needles that have been dipped into different samples of soil. Incubate the plates and, after 48 hr, spread a film of Lugol's iodine solution over the surface of each dish. What may be surmised when some areas show a clear zone?

Quantitative plating of bacteria

One way of counting the number of bacteria (or other microorganism) in a *diluted* suspension is to spread a thin film of the suspension on the surface of sterile agar in a Petri dish or to add the suspension first, covering it with a film of medium. We may assume that each bacterium trapped in the agar develops into a colony. Finally, count the number of colonies.

Begin with five or six dilution blanks, that is, with test tubes containing 9 ml of sterile diluting fluid and an accompanying sterile pipette for each blank. Dilute the suspension by adding 1 ml of sample to 9 ml of diluting fluid to get 10 ml of a 1 : 10 dilution. Then use a sterile pipette to transfer 1 ml of this 1 : 10 dilution (which has been well mixed) to the second tube containing 9 ml of fluid; this results in 10 ml of a 1 : 100 dilution. Again, transfer 1 ml of this to a 9-ml blank to obtain 10 ml of a 1 : 1000 diluted suspension; continue as required.

Plate known volumes of these samples. Introduce 1 ml of the diluted sample into the bottom of a sterile Petri dish and add 15 ml of the appropriate agar medium. Swirl the dish to mix thoroughly; cover and allow to cool until solidified. Then incubate.

Count the colonies using a hand lens; or, when feasible, use a colony counter as in Fig. 9-9. Compare the number of colonies present in suspensions of different dilutions.

Irradiation of bacteria

Wald et al.[10] suggest an activity in which a dilute suspension of *Serratia marcescens* is spread as a film over sterile agar. The dish is divided into four quadrants by marking the bottom of the dish with a waxed pencil. Each quadrant is successively exposed to ultraviolet light of wavelengths near 260 mμ for 90, 120, and 150 sec (and one quadrant is left unexposed as a control). *Serratia marcescens* is especially effective as a tool for this study of possible mutations since several mutants occur that lack the normal brilliant red pigmentation.

Atomic radiation. In general, microorganisms have been found to be unusually resistant to atomic radiation. A lethal dose for man is less than 1000 roentgens (r), whereas representative microorganisms have the following lethal doses:

Escherichia coli	10,000 r
yeast cells	30,000 r
Amoeba	100,000 r
Paramecium	300,000 r

Other investigations using bacteria

Wald et al.[11] describe a fine laboratory activity to show the transformation of *Pneumococcus*. A strain of nonpathogenic streptomycin-sensitive *Pneumococcus* is transformed into a streptomycin-resistant strain by introducing to its suspension DNA extracted from a streptomycin-resistant strain of *Pneumococcus*. Careful bacteriological techniques must be used and whole blood must be added to prepare a special medium. Although the strain of *Pneumococcus* that Wald suggests is nonpathogenic,

the procedure for demonstrating genetic transformation of bacteria is recommended for more advanced college laboratory work, not for high school students.

Students will want to read Chapters 7 and 8 in *Biological Science: Molecules to Man*[12] to learn more about these investigations changing the genetic composition of some bacteria.

We have described a procedure for extracting DNA from yeast cells in Chapter 2. A laboratory technique described by Rugh[13] for extracting DNA from testes of frogs has also been included in the studies of the biochemical nature of cells in Chapter 2.

Wald et al.[14] give lucid directions for demonstrating the attack on colonies of *Escherichia coli* by a bacterial virus, the T$_4$ bacteriophage. This work will be more meaningful if students have previously read about viruses; look into *Biological Science: An Inquiry into Life*[15] for the life cycle of a phage in a bacterial cell.

It is well to repeat the caution that work with microorganisms, however low in pathogenicity, requires exceeding care, and is more suitable for college students.

Also see Wald et al.[16] for a laboratory activity in which a search is made for antibiotic-resistant forms, that is, forms resistant to penicillin and to streptomycin.

An especially fine group of ideas and techniques for further investigations is a special issue, *Microbiology in Introductory Biology,* of *The Am. Biol. Teacher* (Vol. 22, No. 6, June 1960).

Also look into the laboratory activities accompanying the three BSCS high school biology texts. For students who are looking for long-range investigations using bacteria as experimental tools, examine *Research Problems in Biology: Investigations for Students,* Series One–Four (BSCS, Anchor Books, Doubleday, 1963, 1965).

[10] G. Wald et al., *Twenty-six Afternoons of Biology,* Addison-Wesley, Reading, Mass., 1962, pp. 39–40, 45.
[11] *Ibid.,* Chapter 8.

[12] *Biological Science: Molecules to Man* (BSCS Blue Version), Houghton Mifflin, Boston, 1963.
[13] R. Rugh, *Experimental Embryology: Techniques and Procedures,* 3rd ed., Burgess, Minneapolis, 1962, p. 325.
[14] *Op. cit.,* Chapter 9.
[15] *Op. cit.,* Chapter 9.
[16] *Op. cit.,* Chapter 7.

The BSCS laboratory block by A. S. Sussman, *Microbes: Their Growth, Nutrition, and Interaction* (Heath, 1964) provides a rich variety of investigations and laboratory activities. Each of the activities is designed to be carried out over a period of about 6 weeks.

Measuring growth of the total plant

Growth in height of a total plant

Use a ruler to measure the total vertical growth of a plant. For more accurate work you may want to use an auxanometer (Fig. 9-12). Measure growth in width of stems or roots with a caliper. Also weigh the total plant; an increase in weight is an index of growth. (If a truer index of growth is desired, the dry weight of the plant should be taken.)

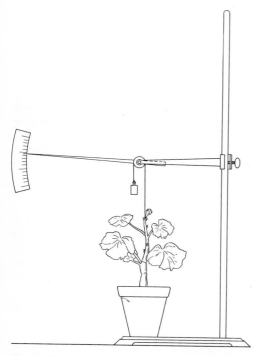

Fig. 9-12 Auxanometer for measuring vertical growth of a plant.

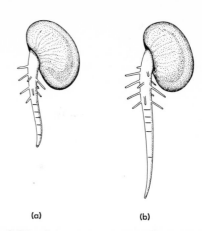

(a)　　　　　　　　(b)

Fig. 9-13 Demonstration comparing regions of growth in a young root: (a) marked root; (b) same root after growth. (After *Biological Science: An Inquiry into Life* [BSCS Yellow Version], Harcourt, Brace & World, 1963.)

The growing regions of a plant

Finding the region of greatest growth in roots. Germinate bean, pea, lupine seeds, or kernels of corn in a germinating dish. When the rudimentary roots are about 1 in. long, mark equal intervals along the length of the root with India ink. To do this, students should first dry the roots with blotting paper, then place a ruler alongside the root and mark off intervals ⅛ in. or 1 mm apart. Replace the seeds in the germinating dish and examine after 24 hr, 48 hr, and so on (Fig. 9-13).

Growth of stems. Shoots of seedlings of beans, oats, wheat, or peas are suitable for marking with India ink. Blot the stems dry and mark intervals ⅛ in. apart.

Growth of leaves. Germinate bean, lima bean, or pea seeds in moist sand or sawdust. When small leaves are visible, flatten them against a glass plate or similar support which has been marked off into boxes like graph paper. Trace these equidistant markings on the leaves with India ink (Fig. 9-14). Examine the leaves from day to day and watch the changes in the pattern of the boxes.

As an interesting variation of this demonstration, a student might grow these

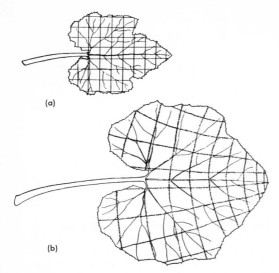

(a)

(b)

Fig. 9-14 Demonstration of growth of a squash leaf: (a) leaf marked into squares with India ink; (b) same leaf after growth. Note that some squares have enlarged more than others. (From R. M. Holman and W. W. Robbins, *A Textbook of General Botany*, 4th ed., 1939; courtesy of John Wiley & Sons.)

seedlings—with marked roots, stems, and leaves—under different colored lights or different conditions of moisture and atmospheric pressure, or in media lacking specific minerals (see pp. 458–61).

Further studies. For further studies and activities describing patterns of growth of leaves, stems, and roots, as well as internal changes in structure, refer to the BSCS laboratory block by A. E. Lee, *Plant Growth and Development* (Heath, 1963). The block also describes activities on metabolism and regulation by auxins and gibberellic acid.

Refer to Chapter 7 of this text for activities involving auxins and other growth substances and photoperiodism. Some students may want, at this time, to look into Sir D'A. Thompson's classic *On Growth and Form* (2 vols., Cambridge Univ. Press, New York, 1952). Is there a hereditary basis for the plane of growth of cells of leaves or of fruits that gives them their shape? Look into the chapters on "Cellular Basis of Growth" and "Symmetry" in E.

Sinnott, *Plant Morphogenesis* (McGraw-Hill, 1960). Also refer to the "Genic Control of Development" in E. Sinnott, L. C. Dunn, and T. Dobzhansky, *Principles of Genetics,* 5th ed. (McGraw-Hill, 1958).

Peculiarities in growth

The characteristic growth of twining plants is due to unequal distribution of cells, causing curvature in cylindrical organs. This pattern of growth is called nutation. Unequal distribution in growth of cells lying in one plane results in another kind of curvature—nastic curvature. If more cells are located at the lower layer of one plane, leaves or stems may curve upward giving a hyponastic curvature. In epinasty, the curvature is downward.

These growth characteristics are not regarded as tropic responses; tropisms are regarded as growth movements caused by asymmetrical conditions and resulting in a bending due to elongation of cells on one side (see tropisms, Chapter 7).

Height retardation

The growth of some plants is deliberately retarded to produce bushy, compact plants for ornamental purposes (Fig. 9-15). One commercially available growth retardant for certain plants is Phosfon (tributyl - 2,4 - dichlorobenzlphosphonium chloride).[17] Phosfon causes a shortening of the internodes of the plant, producing a thicker stalk with deeper green leaves. Phosfon is effective on such plants as chrysanthemums, Easter lilies, seedling red and silver maple, black locust, mimosa, coleus, azalea, alfalfa, some species of holly, and summer cypress. The comparative effects of the growth retardant Phosfon and the growth promoter gibberellin are shown in Fig. 9-16.

Phosfon is completely soluble in water, ethanol, isopropanol, and acetone. It per-

[17] Phosfon is available from Carolina Biological Supply Co., Burlington, N.C. Refer to the experiments described by S. Felton and C. Downing in *Carolina Tips* 26:3, March 1963; a good bibliography is included.

(a)

(b)

Fig. 9-15 Comparison of normal plants and plants treated with growth-retarding chemicals: (a) chrysanthemums (left, untreated; right, growth retarded by addition of Phosfon); (b) blossoms of Easter lilies (left, from treated plant; right, from untreated plant). (Courtesy of U.S. Dept. of Agriculture.)

sists in soil, so it retards height of plants for several years.

The substance is applied to the young growing stages of plants in a lanolin base or as an additive to soil. A spray is not used because the chemical destroys chloroplasts. Some plants, such as Black Valentine beans, show the effect of treatment in a few days; in other plants, 2 to 3 weeks

are needed to show results. Plants are reduced to one-third to one-half of their normal height.

Felton and Downing[18] suggest specific methods for retarding growth of several plants. For example, plant Black Valentine beans in small pots. Just before the terminal bud unfolds, use a wooden applicator to apply a circle of Phosfon in lanolin paste around the stems midway between the cotyledons and the primary leaf node.

To use Phosfon as a soil additive, dissolve 4 tsp of 10 percent Phosfon liquid in 6 qt of water. Mix 2 oz of this solution with soil for a 4-in. pot or 4 oz for a 6-in. pot. Plant seeds of garden annuals such as *Salvia,* cleome, and petunias in flats of untreated soil. Transfer young plants with good leaf growth into larger pots containing Phosfon-treated soil.

For rooted chrysanthemum cuttings, dissolve 1 tsp of 10 percent Phosfon liquid in 6 qt of water.

Competition among plants

Ecological factors affect the rate of growth of plants (see also Chapter 11).

[18] *Ibid.*

(a)   (b)   (c)

Fig. 9-16 Effect of plant-growth regulators on sunflower seedlings: (a) Phosfon-treated seedling; (b) untreated seedling; (c) gibberellin-treated seedling. The lanolin paste method was used with these seedlings. (Courtesy of Carolina Biological Supply Co.)

GROWTH AND DIFFERENTIATION   455

While many kinds of animals may exist in the same region and make different demands on the environmental supplies, green plants draw from a common, limited pool of materials: minerals, water, light, carbon dioxide, and oxygen. As a result, plants compete for materials and interfere with one another's growth.

J. Harper[19] describes the exponential rate of growth of fronds of duckweed (growing in 400-ml beakers of nutrient solution) as a population growth of 0.3 g per g per day. Were this to continue, the fronds would cover an acre in some 50 days. But this exponential growth rate stops, and a linear growth rate results, increasing by 15 mg *per beaker* per day. In the exponential phase, the limiting factor is the number of fronds; in the linear phase of growth, the limiting factor is some environmental circumstance. A third or stationary phase of growth occurs in about 7 to 9 weeks; as many fronds die off as new ones grow. The limiting factor in this case seems to be light; the fronds under the thick mass growing on the surface do not get light and thus die off. (Comparisons can be made with land plants, especially young plants and the relationship between their growth rate and the amount of shading they receive.)

What happens when two species of duckweed, *Lemna minor* and *L. gibba,* are grown in the same culture? *Lemna gibba* gains dominance and *L. minor* is eliminated; this seems to be due to the presence in *L. gibba* of air sacs in the fronds, which probably raise the plants to the top of the mat of growth of duckweeds.

(Ecologists are also concerned with the possibility that one species of land plant may produce toxins that inhibit growth of other plants in their proximity.)

Ways to germinate seeds

Seeds may be sown on paraffined cheesecloth drawn tautly over a container

[19] J. Harper, "Interference Between Plants," *The Times Sci. Rev.* 13:7–8, Autumn 1963.

of water, or in moist sand, sawdust, excelsior, or sphagnum moss. Methods vary with the purpose and the type of seeds. Several successful methods are described.

Growing seeds to show root hairs

This is a rapid method for use by students in the classroom. Students might line several test tubes with wet blotting paper and place radish seeds between the wet blotter and the glass walls of the test tubes. Enough test tubes for the entire class may be prepared. In a few days, students may examine the root hairs with a hand lens without disturbing their growth. Occasionally, add a few drops of water to the bottom of the test tubes; these will be absorbed in the blotter by capillarity.

When seeds larger than radish seeds are used, soak them first and pin the germinating seeds to the cork of a vial, one seedling per vial. When the vials stand upright the root with the root hairs grows down into the vial (Fig. 5-1).

Large-scale germination

Prepare a germinating dish with two 12-in. porous flowerpot saucers and a smaller one (Fig. 9-17a). Wash all the porous plates in hot water at the start to reduce fungus growth. Place the smaller plate containing germinating seeds inside one of the 12-in. plates; fill the peripheral moat with water. Now cover the large plate with the other 12-in. plate.

In some cases, you may want a glass germinating dish so that growth stages of seedlings are visible without the cover being removed. As shown in Fig. 9-17b, fit a small, flat dish within a larger glass dish and fill the outer moat with water. Then line the inner dish with moistened filter paper; two pieces of blotting paper are bent from the inner dish into the water of the outer dish to form a wick. Cover the dishes with a plate of glass or a bell jar. The whole device may be covered with dark paper, aluminum foil, or heavy cloth to aid germination.

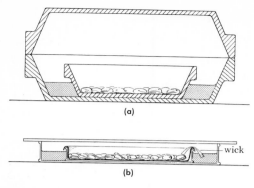

Fig. 9-17 Germination dishes: (a) large-scale dish made from three porous flowerpot saucers; (b) smaller dish made from two glass dishes, utilizing a blotting paper liner and filter paper wick in the smaller dish.

Test of viability in seeds

Viable seeds produce dehydrogenase, which reduces tetrazolium salts; dead seeds do not. As a result, tetrazolium salts —which are colorless in the oxidized form and colored in the reduced state—can be utilized to detect and separate viable seeds from dead seeds. For example, when 2,3,5-triphenyl tetrazolium chloride (TTC) is added to living tissues, it will be reduced by the dehydrogenases produced by the tissues and will change from colorless to red. (Other derivatives of tetrazolium may be black or blue in the reduced form.) The usual germination tests take up to 10 days to perform, but the use of TTC reduces the test time to 24 hr or less.

Soak 100 barley seeds in water for about 18 hr, and then germinate 50 seeds on moist filter paper in Petri dishes kept in the dark. Cut the remaining 50 seeds in half lengthwise; cut through the embryo plants as well (Fig. 9-18). Place these seeds in Petri dishes. Add a 0.1 to 1.0 percent solution of the TTC to the dishes so that the seeds are just covered. Let them soak for 2 to 4 hr in the dark at 20° C (68° F).

The tetrazolium salt is colorless, water-soluble, and nontoxic. By chemical or phytochemical action this indicator is re- duced to triphenylformazan—an insoluble, bright red dye. As germination starts, chemical action changes the colorless solution into the insoluble red dye. The viable seeds may be identified by the staining of at least half of the scutellum, the whole of the shoot, and the regions where the adventitious roots will later develop (Fig. 9-18).

Some days later, have students compare the number of viable seedlings which germinated in the untreated dish with the number obtained in the rapid test using the indicator.

Sterilization of surfaces of seeds

The conditions favorable for germination of seeds also enhance bacterial or mold germination. To prevent this, the surfaces of the seeds may be disinfected; three methods are suggested here.

1. Place seeds in a dilute formalin solution (1:500) for about 20 min. Then rinse off in water.

2. In a glass container prepare a 1 percent solution of bichloride of mercury (1 g to 100 ml of water). (*Caution:* This mercury salt is very poisonous to mucous membranes. Avoid having students handle it.)

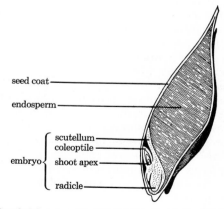

seed coat
endosperm
scutellum
coleoptile
embryo { shoot apex
radicle

Fig. 9-18 Lengthwise section through a barley seed. (From *Plants in Action* by Leonard Machlis and John F. Torrey. San Francisco: W. H. Freeman and Co., 1956.)

Soak the seeds in this solution for about 10 min and rinse in water.

3. Soak the seeds for about 15 min in a 1 percent solution of chlorine, prepared by adding 21 g of pure sodium hypochlorite to 100 ml of water. Or use Clorox (5 percent sodium hypochlorite); add 1 part of Clorox to 4 parts of water.

Growth of seedlings in relation to stored food

Generally, the growth of seedlings is dependent upon the amount of stored food available to the embryos. You may want students to germinate several bean seeds which have been soaked in water overnight. When the rudimentary roots, or radicles, become visible, students may select several seeds at the same stage of germination and treat them in the following manner:

1. Remove one cotyledon from several seeds.
2. Remove both cotyledons from other seeds.
3. As controls, leave the cotyledons intact on some seeds.

Suspend the seeds on a wire mesh with the roots in nutrient solution (p. 459), or plant them all in clean, moist sand. Students may watch their growth at regular intervals over the next few weeks.

This preliminary classroom or laboratory activity might serve to stimulate further activities which are developed in this chapter. Some of these questions will arise: Can we substitute nutrient solutions for the stored food in the seeds? How do roots grow? What substances are responsible for the growth of roots or the bending of stems? How do leaves grow into different patterns? What environmental factors affect the growth rate of parts of plants? Can we speed up the growth of roots on cuttings by using growth substances? Can we retard the height of plants? What conditions do bacteria need to grow? How does penicillin affect the growth of bacteria?

Effect of darkness upon the growth of seedlings

Light seems necessary for several of the formative processes as well as for photosynthesis. Seedlings in darkness generally grow tall and spindly, because they lack adequate supporting tissue and have long internodes. Most plants do not develop chlorophyll in the dark. Students may test this by growing some bean or pea seedlings in the dark and others in the light. In about a week the seedlings grown in the light should be sturdier, shorter (since there are short internodes), and greener than the etiolated plants grown in the dark (see Fig. 10-4). A few days of exposure to light should cause the development of chlorophyll in the pale leaves.

Mineral nutrition in plants

Although green plants can manufacture food (simple sugars) during photosynthesis, for the total growth of the plant certain minerals are necessary. As a series of laboratory investigations, students may grow plants in soilless cultures (hydroponics). They may also want to show the effects upon plants of a deficiency of certain minerals.

Although the term "hydroponics" was originally applied to water-culture methods, the term has been extended to water, sand, and gravel cultures (Fig. 9-19). While detailed care is not possible in the classroom, emphasis may be given to the need for balanced solutions for the growth of plants. In our experience, these activities are successful if the following conditions are observed.

Provision should be made for roots to receive adequate aeration (by using an aerator pump). Glazed crocks are excellent containers; enameled pails may also be used. Metal cans or Mason jars (soft glass) may be used if a paraffin or asphaltum lining is applied to the metal or glass to prevent the solution from making contact with the fairly soluble surface. Glass jars may be painted black or covered with

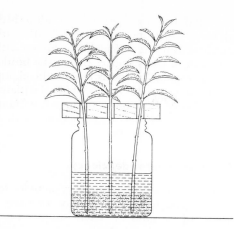

Fig. 9-19 Soilless culture for growing plants. Seedlings or cuttings supported by a perforated wooden or cork block are grown in a glass jar containing a layer of sterile sand and nutrient solution.

black oilcloth or aluminum foil to inhibit the growth of algae in the solution.

Some nutrient solutions

Plants seem to have a wide tolerance for different salts, so no one solution of nutrient salts serves all plants.

Some of these solutions date back to the original ones of Knop, Sachs, and Pfeffer, which were developed nearly 100 years ago. Other solutions show modifications, including varying quantities of trace elements. Some trial-and-error activity may be needed to learn which of these solutions is the most fruitful in a specific school situation. (Other culture solutions for algae are described in Chapter 13 and in Chapter 3, which includes buffers.)

1. KNOP'S SOLUTION. Weigh out the following materials and dissolve them in 1 liter of water.

$Ca(NO_3)_2 \cdot 4H_2O$	0.8 g
KNO_3	0.2 g
KH_2PO_4	0.2 g
$MgSO_4 \cdot 7H_2O$	0.2 g
$FePO_4$	trace

2. SACHS' SOLUTION. Dissolve the following salts in 1 liter of water. (The calcium phosphate is only partially soluble.)

KNO_3	1.0 g
NaCl	0.5 g
$CaSO_4$	0.5 g
$MgSO_4$	0.5 g
$Ca_3(PO_4)_2$	0.5 g
$FeCl_3$ (1% solution)	trace (1 drop)

3. PFEFFER'S SOLUTION. Weigh out and dissolve the following salts in 1 liter of water. (This is similar to Knop's solution.)

$Ca(NO_3)_2$	0.8 g
KNO_3	0.2 g
KH_2PO_4	0.2 g
$MgSO_4 \cdot 7H_2O$	0.2 g
KCl	0.2 g
$FeCl_3$	trace

4. SHIVE'S SOLUTION. Dissolve the salts listed below in 1 liter of water. The iron and phosphate salts should be added first to the water, since they precipitate readily. (Enough of the minerals will remain in the solution, however.)

$Ca(NO_3)_2 \cdot 4H_2O$	1.23 g
KH_2PO_4	2.45 g
$MgSO_4 \cdot 7H_2O$	3.70 g
$FePO_4$ (0.5% solution)	1.0 ml

5. HOAGLAND'S SOLUTION. Weigh out the following salts and dissolve them in 1 liter of water.

$Ca(NO_3)_2 \cdot 4H_2O$	0.95 g
KNO_3	0.61 g
$MgSO_4 \cdot 7H_2O$	0.49 g
$NH_4H_2PO_4$	0.12 g
ferric tartrate	0.005 g

Agricultural experimentation stations may vary the ingredients of some of these well-known solutions. Yet most of the nutrient solutions contain six of the essential elements for plant growth: phosphorus, nitrogen, sulfur, potassium, calcium, and magnesium.

One practical activity is growing seedlings or cuttings in Hoagland's solution minus the nitrogen compounds. Later test the petioles of plants for the presence of nitrates using diphenylamine sulfate reagent.[20] To test for nitrates, place thin sections on a slide in a few drops of the

---

[20] Dissolve 0.5 g of diphenylamine in 100 ml of concentrated sulfuric acid.

Refer to L. Machlis and J. Torrey, *Plants in Action*, Freeman, San Francisco, 1956, p. 120.

reagent; a blue color indicates the presence of nitrates.

Most workers in this field advise the addition of trace elements to all the nutrient solutions. These trace elements needed in minute quantity include boron, manganese, copper, zinc, and probably molybdenum. Curtis and Clark[21] recommend the addition of 1 ml of Haas and Reed's "A to Z" solution per liter of nutrient solution. This should fulfill the requirement of trace elements for plant growth.

6. HAAS AND REED'S "A TO Z" SOLUTION. Weigh out these ingredients and dissolve in 1 liter of water.

H_3BO_3	0.6 g
$MnCl_2 \cdot 4H_2O$	0.4 g
$ZnSO_4$	0.05 g
$CuSO_4 \cdot 5H_2O$	0.05 g
$Al_2(SO_4)_3$	0.05 g
KI	0.03 g
KBr	0.03 g
$Co(NO_3)_2 \cdot 6H_2O$	0.05 g
LiCl	0.03 g
TiO_2	0.03 g
$SnCl_2 \cdot 2H_2O$	0.03 g
$NiSO_4 \cdot 6H_2O$	0.05 g

Acidity of solutions

Students may want to learn to check the pH of the solution with an indicator (as described on p. 662), pH meter, or Hydrion pH paper. Most plants grow best within a range between 4.5 and 6.5 (acid); a pH over 7 (alkaline) may retard plant growth. Further, when the solution is slightly acid, iron salts are more soluble. Check the pH every few days, since the solution may become alkaline as nitrogen-bearing ions are absorbed. If potassium ions are absorbed rapidly, sulfate ions accumulate and produce an acid solution (see buffer solutions, Chapter 3).

To avoid these changes in pH, some workers prefer a drip method to immersion of the plants directly in the solution. If you cannot get an intravenous drop bottle like the kind used in hospitals to administer glucose, make one by filling a bottle with nutrient solution and sealing a siphon with a clamp to the stopper. Invert the bottle, and at regular intervals release the clamp to moisten the bedding in which the plants grow. When the plants are grown without bedding, it is easier to change the solution than to use a drip bottle.

Effect of mineral deficiency

This is a useful laboratory activity for many groups. Students may prepare a series of containers, each supporting a growing seedling held fast in a one-hole stopper. In each of the containers in the series, students might omit a different salt from the nutrient solution (see p. 459). Later, they may compare each seedling's growth with the control, that is, with seedlings grown in the total nutrient solution. They may also want to test the effect of different degrees of acidity on plant growth. For example, while potassium dihydrogen phosphate, KH_2PO_4, gives the solution a slightly acid reaction, potassium monohydrogen phosphate, K_2HPO_4, produces an alkaline solution. Do these salts have an effect on growth of plants?

There are several valuable keys for identifying nutrient deficiencies in plants. In these keys, the main traits described refer to changes in chlorophyll and to types of plant growth. All these observations may be made by students. In addition, the work has a practical application in some school areas. Useful keys may be found in Curtis and Clark's reference[22] and in S. Dunn's physiology text.[23] Or send for the catalog of color slides showing nutrient-deficiency symptoms in plants ($0.25 per slide) from the National Plant Food Institute, 1700 K St., N.W., Washington 6, D.C.; slides of crops, flowers, fruits and nuts, and weeds are available.

[21] O. F. Curtis and D. G. Clark, *An Introduction to Plant Physiology,* McGraw-Hill, New York, 1950, p. 384.

[22] *Ibid.*

[23] S. Dunn, *Elementary Plant Physiology,* Addison-Wesley, Reading, Mass., 1949, p. 85.

Excellent guides for students in preparing apparatus for hydroponics are given by Ellis and Swaney.[24]

Growing cuttings in solution

When single plants are used, plant them in wide, porous flowerpots, as shown in Fig. 9-20. Wash the pot thoroughly in hot water. Choose a jar into which 2 in. of the pot will fit. Fill the jar with the nutrient solution and fit the pot into this. Plant the cutting in the pot in excelsior, glass wool, or sphagnum moss. As the cutting grows, supports may be made of sections of plywood through which holes have been bored. Instead of wood supports you may find it useful to make supports for cuttings by molding paraffin sheets to fit as covers on the pot; then make holes in the paraffin by means of a hot iron rod.

Seedlings or cuttings need not be grown in solution alone; they can be grown in coarse sand cultures. For this method, use a mixture of peat moss (about 30 percent by volume) and clean sand. Peat moss (sphagnum) retains moisture and also reduces the alkalinity of the medium. Fill a small flowerpot with the sand–peat moss mixture and fit it into a container of solution. Run glass wool, placed in contact with the roots of the plants, through the drainage hole in the flowerpot, into the nutrient solution. This wick transports nutrient solution to the roots with less evaporation than would result if you watered the sand from the top of the mixture.

Although most plants can be grown by these methods in the classroom, certain plants are especially resistant to poor environmental conditions. Good results should be forthcoming with wheat, rye, corn, radish, sunflower, castor beans, peas, beans, pepos, tomatoes, and potatoes. (*Precaution:* Avoid growing these plants in a laboratory where illuminating gas or

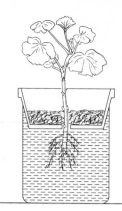

Fig. 9-20 Soilless culture for growing single plant or cutting. A porous flowerpot is immersed in a container of nutrient solution.

bottled gas is used, since this causes wilting and loss of leaves.)

Growing seedlings in solution

Begin with a simple method that is successful with small, fast-growing seeds—especially cereal plants, which require little support.

To prepare support for the seedlings, you may want to immerse a layer of cheesecloth or mosquito netting in hot paraffin and spread it across the top of a container of fluid. Allow a little slack in the center to dip into the solution. (The seeds will be planted in this depression.) You may prefer to add sterile sand or vermiculite to the bottom of the container. This gives some support to plants rooted in the sand. Or bore holes into a wooden block (or cork), which can be fitted over a container to give support to stems (Fig. 9-19).

Use zinc (or galvanized iron) mesh troughs to make a more substantial support for larger plants. These supports must be given several coats of paraffin or asphaltum to prevent toxicity. You may try to use nontoxic stainless steel mesh. As shown (Fig. 9-21), shape ¼-in. mesh into a trough with flanges to support it against the sides of the jar.

Most plants need a surface area of about

[24] C. Ellis and M. W. Swaney, *Soilless Growth of Plants,* 2nd ed., revised by T. Eastwood, Reinhold, New York, 1947.

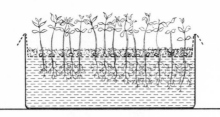

Fig. 9-21 Soilless culture for growing seedlings bedded in glass wool or excelsior in a paraffined zinc mesh trough. The roots grow down into nutrient solution.

4 sq in. of nutrient solution and an air space of about 2 sq in. When large numbers of seedlings are planted in crocks, use a fish-tank aeration pump to provide a stream of air through the solutions. Other specialized devices for ensuring aeration and circulation of nutrient solutions are described in texts in this area, for example, *Soilless Growth of Plants*.[25]

First, cover the mesh trough with a 2-in. layer of straw, glass wool, coarse sawdust, excelsior, or sphagnum moss. (Beforehand, you might want to test the possible toxicity of the bedding by germinating some seeds in the specific material you plan to use.) Next, plant the seedlings in the bedding.

Add enough solution to the container to cover half of the roots of the seedlings. Then cover the other half of the roots with another 2-in. layer of bedding. This second layer of bedding excludes light and provides a moist air space around the roots. Keep the jars in medium light. As the plants grow, support them with stands. Change the nutrient solution weekly.

Assay of vitamin B_{12}

The flagellate *Euglena* is one of many microorganisms that is unable to manufacture its own vitamin B_{12}, even though it is essential for growth. (Vitamin B_{12} is the anti-pernicious anemia factor.) Because of this dependence on an outside source of vitamin B_{12}, the rate of growth of *Euglena* may be used as a measure of the

[25] *Ibid.*

presence and concentration of vitamin B_{12} in a substance.

Begin with suspensions of *Euglena* of known cell density. Using a photometer, such as the student-made instrument shown in Fig. 9-8, students may measure the growth of *Euglena* in each suspension. They may then prepare a graph, plotting the concentration of cells on the Y-axis and the growth variable (time) on the X-axis. Then, the concentration of vitamin B_{12} in the unknown suspension may be estimated by determining the ratio of the Y-intercepts of the unknown suspension and the known suspension.

For culturing *Euglena*, see Chapter 12. Add glutamic acid, malic acid, and asparagine to the basic culture. For complete details for making a basal medium, and suggestions for research projects, refer to S. Hutner, M. Bach, and G. Ross, "A Sugar-containing Basal Medium for Vitamin B_{12}: Assay with *Euglena*—Applications to Body Fluids" (*J. Protozool.* 3:101-12, 1956). (Commercially, B_{12} is a byproduct of the *Streptomyces griseus* from which streptomycin is obtained.)

Nutritional deficiencies in Neurospora

The use of deficiency diets in studies of heredity of the pink mold *Neurospora* is described in Chapter 8.

Identifying amino acids

Amino acids can be partitioned by the methods of paper chromatography described in Chapter 3. The color of the amino acids can be brought out by spraying the chromatogram with a solution of 0.25 percent ninhydrin in acetone.

Growth in animals

Nutritional deficiencies in animals

Begin with two groups of young male and female rats or mice from the same litter, about 25 days old. *Turtox Service Leaflet* No. 49, *Nutrition Experiments,*[26] offers the following mineral-deficient diet for one group of animals, and the accompanying complete diet for the control group.

1. MINERAL-DEFICIENT DIET

casein	15 parts
lard	4 parts
cornstarch	74 parts
agar	6 parts
yeast	1 part
Viosterol	15 drops per 1000 g of diet

2. COMPLETE DIET FOR THE CONTROL GROUP

casein	14 parts
lard	3 parts
cornstarch	70 parts
agar	5 parts
yeast	1 part
salts (No. 51)	7 parts
Viosterol	15 drops per 1000 g of diet

Prepare the salts (No. 51) required in the control diet from the following:

calcium carbonate	1.5 parts
potassium chloride	1.0 part
sodium chloride	0.5 part
sodium bicarbonate	0.7 part
magnesium oxide	0.2 part
ferric citrate	0.5 part
monobasic potassium phosphate	1.7 parts

For an experiment that includes four animals, prepare about 3 lb of each diet. Animals that begin to show deficiency symptoms should be placed on a recovery diet that includes the salts listed above.

Growth of animals

You may wish to have students investigate the rate of growth of tadpoles. Compare growth rate on a normal diet with growth rate on a diet that includes iodine salts (Chapter 7). Or students may raise chicks, young pigeons, young mice, or hamsters on normal diets and diets that are deficient in minerals, vitamins, or

[26] A set of some 60 *Turtox Service Leaflets* is available to teachers from General Biological Supply House, Inc., 8200 South Hoyne St., Chicago, Ill.

essential amino acids. (Housing and maintenance of animals are described in Chapter 12.)

Other investigations for the nutrients in foods, including the test for vitamin C, are described in Chapter 6. Also plan to use *Euglena* in an assay of vitamin B_{12} (p. 462).

Growth of eggs

Students may subject a batch of fertilized eggs of fish, snails, or amphibians to different temperatures. Prepare several finger bowls containing an equal number of eggs and an equal volume of culture medium. Place the finger bowls in incubators and refrigerators and maintain them at temperatures of 4°, 10°, 20°, and 25° C (39°, 50°, 68°, and 77° F); also maintain one finger bowl at the fluctuating temperature of the laboratory.

Maintain fresh culture medium at the same temperature as the finger bowls so that the eggs can be transferred into fresh culture during the experiment without changing the temperature of the medium.

Using Fig. 9-22 as a guide for the stages of development in the amphibian egg (and the descriptions in Chapter 8 of fish and mollusk eggs), students should record the development of each group of eggs, noting the time, in hours, at which each stage occurs.

There may be occasion to refer to Rugh's *Experimental Embryology*[27] for studies of heteroploidy induced by transferring eggs immediately after insemination through extremes of temperature. Through diagrams and discussion, both Rugh and Balinsky[28] describe procedures for experimental embryology of fish and of birds. (See p. 483 for further reference to Rugh's text as a guide for the study of experimental embryology of the amphibian egg.)

Students may study the effect of several variables on development of the egg and embryo: temperature, crowding, additional oxygen supply (aeration), drugs,

[27] *Op. cit.*
[28] *Op. cit.*

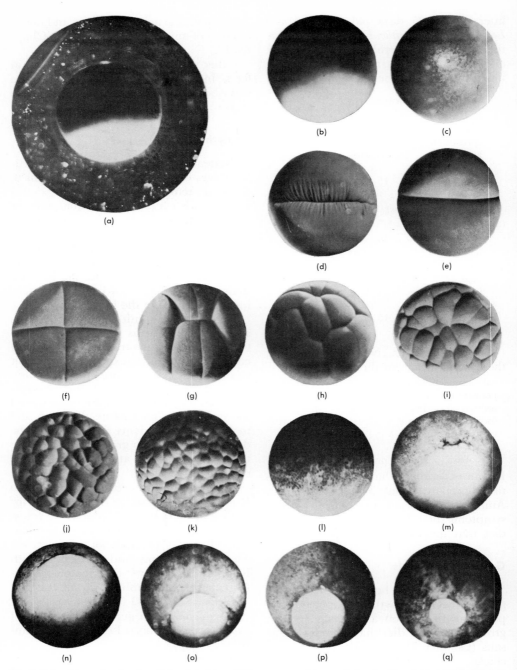

Fig. 9-22 Stages in the development of amphibian egg and larva: (a) unfertilized egg; (b) fertilized, gray crescent; (c) polar body formation; (d) and (e) first cleavage; (f) second cleavage, 4 cells; (g) third cleavage, 8 cells; (h) fourth cleavage, 16 cells; (i) fifth cleavage, 32 cells; (j) temporary morula, 64+ cells; (k) early blastula; (l) late blastula, epiboly; (m) early dorsal lip; (n) and (o) active gastrulation; (p) and (q) yolk plug formation.

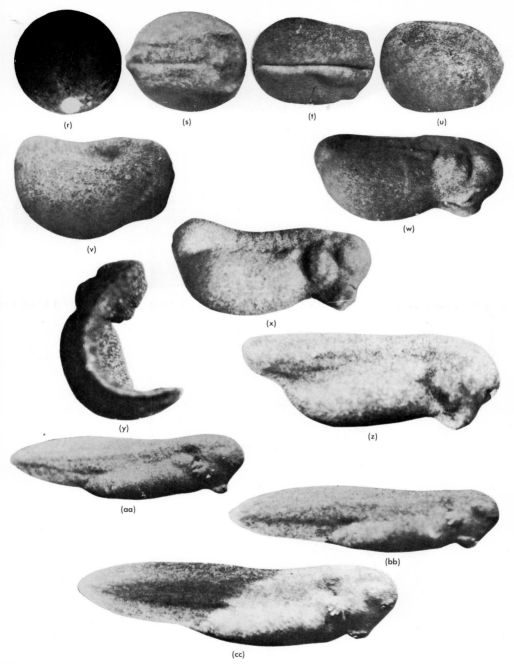

Fig. 9-22 (continued): (r) disappearing yolk plug; (s) neural folds; (t) neural groove, rotation; (u) neurula—neural tube closed; (v) early tail bud; (w) muscular response; (x) initial heartbeat, myotomes; (y) and (z) hatching, gill development; (aa) heartbeat; (bb) gill circulation; (cc) mouth open.

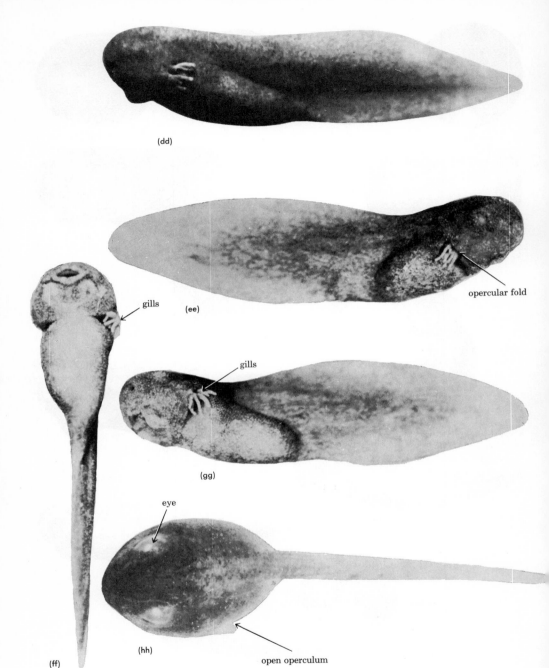

Fig. 9-22 (continued): (dd) tail fin circulation; (ee) opercular fold and teeth, right side; (ff) opercular fold and teeth, ventral view; (gg) opercular fold and teeth, left side; (hh) operculum complete. (a–cc, ee–hh, from R. Rugh, *Experimental Embryology: Techniques and Procedures,* 3rd ed., Burgess Publishing Co., 1962; dd, from *Vertebrate Embryology: The Dynamics of Development* by Roberts Rugh, © 1964, by Harcourt, Brace & World, Inc., and reproduced with their permission.)

and chemicals such as lithium chloride. Use Fig. 9-22 as a basis for comparison. Refer to the series of photos of abnormalities in the development of amphibian eggs and early embryos in Rugh's text. Also refer to Shumway's figures in Rugh's text.

Students should certainly read the relevant chapters in *Biological Science: An Inquiry into Life*.[29] Especially readable are Chapter 28, "The Development of Animals," and Chapter 29, "The Analysis of Development." Students will then be familiar with the work of Wilhelm Roux and of Hans Spemann.

The technique of inducing ovulation in frogs throughout the year is described in Chapter 4 of Rugh's *Experimental Embryology* and in Chapter 8 of this sourcebook.

Artificial parthenogenesis. There may be an opportunity for some students to demonstrate that eggs of frogs can be stimulated to develop without the entry of sperm cells. Chapter 20 in Rugh's *Experimental Embryology* gives directions for such a laboratory activity.

The materials required for this experiment include uterine eggs from an ovulating frog, plus a nonovulating frog and glass or platinum needles. The procedure, in brief, is as follows. Streak the surface of the eggs with a strip of abdominal muscle that has been dipped in blood and coelomic fluid of the nonovulating frog. Then prick the uterine eggs with the glass or platinum needle. For a more complete description, see pp. 361–62.

Experimental embryology of fish

Students interested in investigations in embryology will find that fish are splendid experimental animals. Rugh gives a complete description of techniques and a timetable for the embryo development of several fish: the Zebra fish (*Brachydanio*), the sea horse, the Japanese medaka (*Oryzias*), the multicolored platy (*Xiphophorus* or *Platypoecilus*), and *Fundulus*. He also describes techniques for vital staining of fish embryos and for culturing fish explants *in vitro*. Chapter 48 in Rugh's *Experimental Embryology* provides a good discussion on the genetics of fish.

Experimental embryology of chicks

Many students have incubated fertile chick eggs and watched the development of the embryo over 21 days (see Chapter 8). Some students may want to culture early chick embryos *in vitro*. Several media are available, and these and the techniques useful in explanting early chick embryos are described by both Rugh and Hamburger.[30] Chorio-allantoic grafting is also described.

Plant tissue-culture techniques

Some students may want to study aseptic tissue techniques using plant tissues. In this method the apical tips, about 2 to 5 mm of root tips of seedlings (tomato seedlings are desirable), are removed from Petri dishes in which the seedlings are grown and transferred, under sterile conditions, to sterile flasks of culture medium. Transfer methods used in bacteriological techniques are recommended (see p. 634). P. White's *The Cultivation of Plant and Animal Cells*[31] is a fine guide for this work and offers suggestions for student investigations.

Preparation of White's medium is time-consuming, but the medium is necessary; it contains macronutrients and micronutrients needed for growth. Nonphotosynthesizing parts also require a source of vitamins. (Many vitamins are made in shoots; therefore excised root tips require the vitamins in the medium.) White's medium, as described here or supplemented with coconut milk, may be purchased from Baltimore Biological Laboratory (see Appendix C for address).

[30] Rugh, *op. cit.,* Chapters 49–55. V. Hamburger, *A Manual of Experimental Embryology,* rev. ed., Univ. of Chicago Press, Chicago, 1960, Part III, pp. 143–73.

[31] P. White, *The Cultivation of Plant and Animal Cells,* Ronald, New York, 1954.

[29] *Op. cit.*

Begin by preparing the following five *stock* solutions, each in 1 liter of distilled water. Sucrose is added as a final step.

solution A

Ca(NO$_3$)$_2$	2.0	g
KNO$_3$	0.8	g
KCl	0.65	g
NaH$_2$PO$_4$	0.165	g

solution B

MgSO$_4$	36.0	g

solution C [32]

MnSO$_4$	0.45	g
ZnSO$_4$	0.15	g
H$_3$BO$_3$	0.15	g
KI	0.075	g
$\begin{bmatrix} \text{CuSO}_4 \\ \text{Na}_2\text{MoO}_4 \end{bmatrix}$	$\begin{bmatrix} 0.002 \text{ g} \\ 0.021 \text{ g} \end{bmatrix}$	

solution D (must be freshly prepared)

Fe$_2$(SO$_4$)$_3$	0.25	g

solution E

glycine	0.3	g
thiamine	0.01	g
pyridoxine	0.01	g
nicotinic acid	0.05	g

In preparing White's solution, use only small amounts of the stock solutions. For example, begin with 500 ml of distilled water; to this add 100 ml of stock solution A, and 10 ml each of solutions B, C, D, and E. After these have been thoroughly mixed by shaking, add 20 g of sucrose and make up the quantity to 1 liter by adding distilled water.

Now divide the solution by pouring 50-ml amounts into smaller flasks, plug with cotton; autoclave for 15 min at a pressure of 15 lb per sq in. These flasks may be stored for a few days in a refrigerator until transfers can be made. Since the solution contains organic compounds the solution cannot be stored indefinitely. Once they have mastered this tissue-culture technique, students may try to grow isolated sections of roots, stems, and leaves in nutrient solutions to examine how differentiation and growth take place.

[32] Machlis and Torrey suggest the salts of molybdenum and copper (the bracketed salts) as additions to the original formula of White. See Machlis and Torrey, *op. cit.*, pp. 46–47, 56.

Growth of excised roots

For studies of growth of excised roots, Machlis and Torrey[33] suggest beginning with seeds of a ripe tomato. Use sterile procedures to dissect the tomato and transfer the seeds to sterile Petri dishes. You may substitute dried tomato seeds if they are sterilized by being placed for 30 min in a 0.5 percent aqueous solution of sodium hypochlorite containing a small amount of a detergent.

Germinate the seeds in the dark, at 25° C (77° F), in Petri dishes lined with filter paper that has been soaked in sterile water. Machlis and Torrey describe how to excise 1-cm root tips from the germinating tomato seeds and how to transfer them to Erlenmeyer flasks containing White's medium. (Media may be solidified by adding 0.8 percent agar to the media.) Also transfer excised portions of stem to sterile medium.

The amount of growth of the roots may be observed after a week; measure the root lengths by holding a plastic ruler behind the flask.

Is there any differentiation of the excised stems into roots or leaves?

Growth of excised embryos

Use seeds that germinate rapidly, for example, radish, oats, peas, sunflower, or morning glory. Sterilize the surfaces of the seeds by placing them for 30 min in a 0.5 percent solution of sodium hypochlorite to which a little detergent has been added. Then soak seeds in sterile water to hasten germination. Be sure to observe aseptic techniques.

Excise the embryo of each seed by cutting near the hilum and then around the seed. Separate the embryo from the halves of the seed, or from the endosperm of the seed.

Use sterile transfer needles to place the embryo in a flask or test tube containing White's medium. If necessary, make a

[33] *Ibid.*, pp. 56–59.

small hook in the transfer needle to catch or secure the embryo tissue. Maintain the cultures in the dark, at 28° to 30° C (82° to 86° F), for 4 to 5 days.

Growth of excised leaves

Excised leaf primordia of ferns and of flowering plants such as the sunflower can be studied. Transfer the inner leaves of the apical buds to sterile medium, such as White's solution, which has been solidified by adding 0.8 percent agar.

Students may be successful in obtaining developing well-formed, although smaller, mature leaves. Some investigators recommend that 15 percent autoclaved coconut milk or acid casein hydrolyzate be added, in concentrations of 1:1000, to the sterile medium.

Further, to correct or avoid the usual chlorosis of leaves grown in culture, ammonium nitrate can be added to the medium.[34]

Planning investigations

Students may be interested in the classic papers by Skoog and Miller[35] and by Erikson and Goddard.[36] An abridged edition is available of Sir D'A. Thompson's classic text *On Growth and Form* (J. Bonner, ed., Cambridge Univ. Press, New York, 1961). See also E. Willmer, *Tissue Culture* (Wiley, 1958).

Some students may want to try an investigation proposed by White.[37] Others may wish to grow small, isolated discs of leaves in sterile culture medium to study metabolism, as described by Nieman.[38]

Animal tissue-culture techniques

Growing animal tissue *in vitro* requires skills and experience not commonly found among high school students, unless they are working directly under the supervision of a research worker in the field. Students may read the descriptions by Rugh[39] of methods for culturing isolated sections of early embryos (such as from the tail bud stage) to observe the differentiation of these primordia, or anlage, when grown in another embryo host or in a hanging drop. These techniques are difficult, and asepsis must be observed. (See our introductory description, pp. 483–88.)

For an introduction to the techniques of culturing mammalian tissues, refer to the description of the chemically defined medium developed by H. Eagle, "Nutrition Needs of Mammalian Cells in Tissue Culture" (*Science* 122:501, 1955). Also refer to E. Willmer, *Tissue Culture* (Wiley, 1958) and J. Paul, *Cell and Tissue Cultures* (Livingston, London, 1959).

In general, culture media for animal tissues are difficult to prepare and maintain. The temperature must be carefully regulated for vertebrates, and vertebrate tissue is very susceptible to infection.

Organ rudiments and tissue cultures are available for *in vitro* studies. Kidney and embryo tissue of mammals are among the tissues that can be purchased from Baltimore Biological Laboratory or Hyland (see Appendix C). For suggestions as to which of a wide variety of prepared sterile tissue-culture media to use and how to maintain tissues in the laboratory, refer to BBL's bulletin *Tissue Culture: A Manual of Materials and Methods.* Several modifications of Eagle's media, as well as those of Hanks, Earle, Puck, and Spinner, may be purchased already prepared. Also consult

[34] T. Steeves, H. Gabriel, and M. Steeves, "Growth in Sterile Culture of Excised Leaves of Flowering Plants," *Science* 126:350–51, 1957.

[35] F. Skoog and C. Miller, "Chemical Regulation of Growth and Organ Formation in Plant Tissues Cultured *in vitro*," in H. K. Porter, ed., *The Biological Action of Growth Substances,* Symposia of the Soc. for Experimental Biology, 11, Academic, New York, 1957, pp. 118–31.

[36] R. Erikson and D. Goddard, "An Analysis of Root Growth in Cellular and Biochemical Terms," *Growth* 15:symposium supplement 10, 89–116, 1951.

[37] P. White, "Nutrition of Excised Plant Tissues and Organs," in *Research Problems in Biology: Investigations for Students* (BSCS), Series Two, Anchor Books, Doubleday, Garden City, N.Y., 1963.

[38] R. Nieman, "Growth and Synthesis of Metabolites by Isolated Leaf Tissue," in *ibid.,* Series One, 1963.

[39] *Op. cit.,* Sections 27, 29.

Hyland Laboratories' publication *Tissue Culture Reagents*.

Mitosis

In regeneration of organs of plants and animals, new cells are formed as a result of fission, budding, or sporulation. In each case the nuclear content of the cells is distributed equally to each new cell; chromosomes and genes, or molecules of DNA, are replicated so that each new cell receives its proper genetic contents (see also Chapter 8).

You may want to have students study mitosis in actively growing tissues; root tips or fertilized eggs are the classic materials for study. These tissues need special staining to make chromosomes visible. Of course, chromosomes are not visible unless they are going through mitosis, or reduction division (Fig. 8-11).

Prepared slides of onion root tips (Fig. 9-23) or of cleavage stages in the eggs of starfish (Fig. 8-16), fish (Fig. 9-24), sea urchins (Fig. 8-18), or worms such as the roundworm *Ascaris* (Fig. 9-25) may be studied under the high power of the microscope. Some techniques are described here for staining chromosomes of fruit fly larvae

and onion root tips. You may want to try these in class or as a club activity, as a group or individual investigation (see also Chapter 2).

Wire "chromosomes"

The actual mechanics of mitosis (and of reduction division, or meiosis, Fig. 8-9) as a continuous series of chromosome changes is a difficult concept for students to understand. When they observe mitotic division under the microscope or when they study a chart, students see only a series of "still" shots. Try using insulated electric wire to show the dynamics of the splitting of chromosomes. (Actually, the chromosomes duplicate themselves rather than split, but the picture looks the same as the two parts of the chromosome move away from one another.) The technique is described in the legend of Fig. 9-26. Also refer to the suggestions in Chapter 8 for using glass slides and other materials to model molecules of DNA.

Students may prepare themselves for laboratory observations of mitosis in cells by reading Chapter 7 in *Biological Science: An Inquiry into Life*.[40]

[40] *Op. cit.*

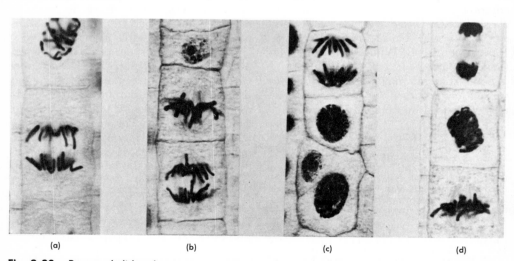

(a) (b) (c) (d)

Fig. 9-23 Prepared slides showing mitosis in onion root tips: (a) typical prophase, also anaphase; (b) typical metaphase, also anaphase; (c) late anaphase; (d) telophase, also metaphase. (Courtesy of Carolina Biological Supply Co.)

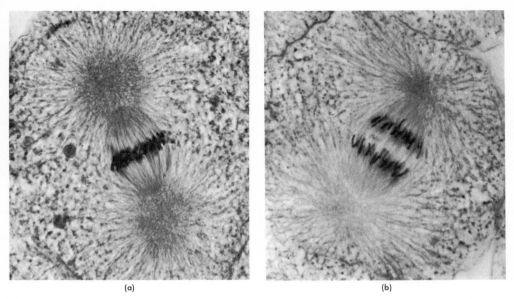

(a)   (b)

Fig. 9-24 Prepared slides, showing mitosis in the cleavage of whitefish eggs: (a) metaphase stage (chromosomes are lined up at the center of the spindle; note the spindle fibers and centrosomes); (b) anaphase stage (chromosomes move along the spindle fibers to opposite poles of the cell). (Courtesy of General Biological Supply House, Inc., Chicago.)

Examining chromosomes

Specific directions are offered in Chapter 2 for preparing smear or squash slides of root tips of onions and of salivary glands of larvae of *Drosophila, Chironomus,* and black flies, *Simulium.* At this point, we consider the preparation of slides of the tail

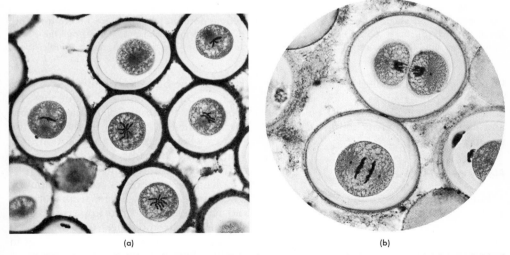

(a)   (b)

Fig. 9-25 Prepared slides, showing mitosis in the parasitic roundworm *Ascaris:* (a) large field of view; (b) two stages at higher magnification. (Courtesy of General Biological Supply House, Inc., Chicago.)

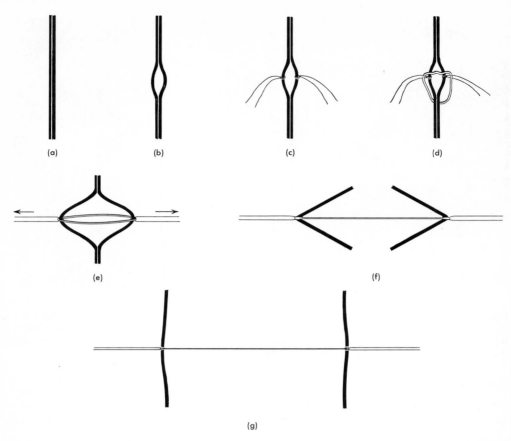

Fig. 9-26 Construction of wire model of chromosomes and spindle fibers during mitosis: (a) take a piece of insulated electrical cord; (b) separate the middle portion; (c) tie a piece of string to each piece of cord; (d) slip a rubber band around the cord, through the strings; (e) and (f) pull the strings apart slowly until (g) the wire splits in two, much as chromosomes are drawn out along spindle fibers.

fin of amphibian tadpoles. We also describe ways of preparing slides to show the effect of several chemicals (such as colchicine and caffeine) on altering mitotic divisions in onion root tips.

Ploidy in onion root tips

Fill a number of tumblers with water and, using toothpicks as props, balance an onion bulb over the mouth of each tumbler so that only the bottom regions of the bulbs are immersed in the water. Let stand in darkness for a few days until a good growth of roots, some 1 cm long, appears. Then

transfer some bulbs to containers of 0.01 to 0.02 percent colchicine (aqueous) for about 3 hr. (*Caution:* Colchicine is poisonous. To prevent damage to the eyes, use goggles when preparing the solution; also wear plastic or rubber gloves to avoid getting colchicine on the fingers.)

Colchicine suppresses cell division by preventing spindle fibers from forming and new cell walls from appearing. Consequently, chromosomes are duplicated and remain in the same cell, resulting in ploidy—double chromosome number.

Prepare cells for staining as follows. Cut off 2-cm tips of both untreated and colchicine-treated onion root tips, and place them in separate containers of fixative (such as Carnoy's or Navashin's, described below). After 12 to 24 hr snip off about 1 mm of the tips of these fixed lengths and place the 2-cm tips in a solution of equal parts of concentrated hydrochloric acid and 95 percent ethyl alcohol. After about 10 min, transfer the tips through 70 percent ethyl alcohol and into a drop of aceto-orcein stain on a slide. Cover with a coverslip and squash the tissue by pressing with the ball of the thumb in a rotary motion. Or apply pressure by rolling a glass rod over the slide; avoid crushing the tissue.

To prepare the aceto-orcein stain, begin by heating 45 ml of acetic acid. When the acid is hot, add about 2 g of orcein.[41] Allow the solution to cool and dilute it with 55 ml of distilled water. Apply a drop of the stain to a slide, and add the tissue. In some techniques, acetocarmine is substituted for aceto-orcein. To prepare acetocarmine, first boil 0.5 g of carmine in 100 ml of 45 percent acetic acid for about 3 min. Cool and filter the solution. In staining chromosomes, dilute 1 part of the solution with 2 parts of 45 percent acetic acid. The addition of 1 or 2 drops of ferric hydrate gives the stain a darker bluish tint.

Chromosomes that have been treated with colchicine are shorter and thicker than untreated chromosomes and are almost straight in one plane.

Preparation of special fixatives

Humason[42] gives the recipes for many fixatives, including Carnoy's and Navashin's; both are given as onion root tip preparations.

1. CARNOY'S FIXATIVE. Combine the ingredients listed below. Keep tissue in this fixative solution for 3 to 6 hr. Remove tissue from fixative and wash for 2 to 3 hr in absolute alcohol to remove chloroform.

glacial acetic acid	10 ml
absolute ethyl alcohol	60 ml
chloroform	30 ml

2. NAVASHIN'S FIXATIVE. Mix equal parts of solutions A and B given below. Keep tissue in the mixture for 6 hr, then transfer tissue into fresh solution for 18 hr. This fixative is especially useful for plant tissues.

solution A

chromic acid	1 g
glacial acetic acid	10 ml
distilled water	90 ml

solution B

| formalin | 40 ml |
| distilled water | 60 ml |

Caffeine and onion tips

Hennessey, Martin, and Carr[43] found that caffeine delays many stages in mitosis, and that in stronger concentration, metaphase chromosomes are especially distinct. It is less dangerous to treat onion root tips with caffeine than with colchicine. To do so, immerse young roots from onion bulbs for ½ hr in 0.1 to 0.5 percent caffeine.

Chromosomes in the tail fin of tadpoles

Two techniques for examining chromosomes in the tail fin are fully described by Rugh.[44] In one method, more appropriate for anuran tadpoles, whole tadpoles are fixed, and epidermal tissue is trimmed from the tail fin. The second method, more successful with urodele larvae (*Amblystoma*, and especially *Triturus*), does not destroy the tadpoles. Only the distal third of the tail fin is cut off. In practice, larvae with *regenerated* tails are preferred since these new cells do not contain as much pigment.

This second method is, briefly, the following. Narcotize tadpoles in chloretone, and cut off the distal third of the tails.

[41] G. Humason, *Animal Tissue Techniques,* Freeman, San Francisco, 1962.
[42] *Ibid.*

[43] T. Hennessey, M. Martin, and V. Carr, "A Cytological Study of the Effects of Uranium Nitrate and Caffeine on the Roots of *Allium cepa*," *Turtox News* 28:146, 1951.
[44] *Op. cit.,* pp. 128–31.

Transfer tail tips to Bouin's fixative, run them through alcohols, and stain them in Harris' acid hematoxylin. Finally, process the tips for mounting in clarite on slides.

Chromosome puffs

Students may want to read about the chromosome puffs found around the enlarged giant chromosomes in the cells of some insects, such as *Drosophila* and *Chironomus*. These puffs may be the loci for genes that proliferate messenger RNA, which passes into the cytoplasm to act as templates for the production of specific proteins.[45]

Differentiation and regeneration

How does a cell in a developing gastrula, having a complement of the same chromosomes as other cells, become individualized, developing specificity of function and structure? How does the seemingly undifferentiated tissue at the notches of leaves of *Bryophyllum* or in the "eye" of the potato become translated into "whole" new plants?

In this section we shall consider some of the classic work of Spemann and the "organizer"—the "time" and "place" relationships in the determination of the amphibian blastula and gastrula. Let us begin with studies of regeneration, with which students are better acquainted.

Regeneration among plants

Regeneration of missing parts is more easily studied in plants than in animals. Regeneration regularly occurs in wound healing, growth of adventitious roots, tumors, "witches' brooms," fasciations, and the many forms of propagation or vegetative multiplication.

Among higher plants most vegetative multiplication results from mitotic growth of primordia or dormant buds; at times, regeneration may be possible only in meri-

[45] Read W. Beermann and U. Clever, "Chromosome Puffs," *Sci. Am.,* April 1964, pp. 50–58.

stematic tissues. On the other hand, in regeneration among animals, there is more often a movement of cells, a reorganization and growth of cells.

The plants and animals resulting from regeneration bear the same heredity as the parent (barring mutations or chromosome aberrations) since growth results from mitotic divisions of cells (p. 470).

Regeneration may occur naturally, in which case it may be considered a form of asexual reproduction. Or plants may be multiplied from isolated parts of plants—leaves, stems, roots, or even buds of some desired variety of plant (usually obtained through breeding). The latter would be a purposeful, or artificial, method for propagating new plants.

Many of these studies of regeneration may be conducted by students in their homes as extensions of classroom discussions.

Natural methods in plants

ROOT STOCKS AND RHIZOMES. Many species of plants spread themselves by means of vigorous underground stems which grow out in all directions from the parent plant. Rootlets develop at intervals and stems shoot up at these nodes (Fig. 9-27). Two hardy forms that may be grown in the classroom and used to demonstrate this are varieties of snake plant (*Sansevieria*) and *Aspidistra*. Other common forms that reproduce similarly are Solomon's-seal, calla lilies, many ferns (especially New York, Boston, holly, hay-scented, and polypody), Johnson grass, and Bermuda grass (Fig. 9-28).

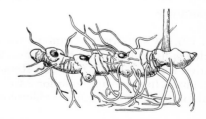

Fig. 9-27 Rhizomes of Solomon's-seal. (From W. G. Whaley, O. P. Breland, *et al., Principles of Biology,* Harper & Row, 1954.)

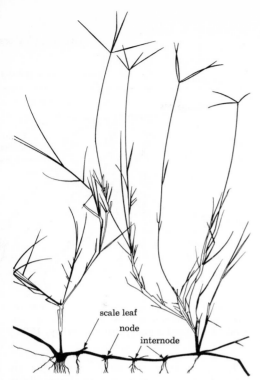

Fig. 9-28 Rhizomes of Bermuda grass. (After R. M. Holman and W. W. Robbins, *A Textbook of General Botany*, 4th ed., Wiley, 1939.)

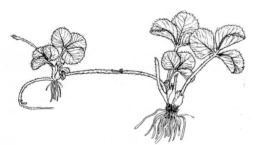

Fig. 9-29 Stolon of strawberry. (From W. G. Whaley, O. P. Breland, *et al.*, *Principles of Biology*, Harper & Row, 1954; after W. W. Robbins and T. E. Weier, *Botany: An Introduction to Plant Science*, Wiley, 1950.)

RUNNERS (STOLONS). Some plants produce trailing or reclining stems which take root at their ends or at nodes; erect stems grow from these nodes. Specimens of the hardy strawberry geranium (*Saxifraga*)

and the spider plant (*Anthericum liliago*) are available for classroom use throughout the year. Other useful plants are the strawberry (*Fragaria,* Fig. 9-29), cinquefoil (*Potentilla*), and *Cotyledon secunda*. Several aquarium plants, such as eel grass (*Vallisneria*), also produce long runners.

BULBS AND CORMS. A bulb is a compact group of fleshy scales or leaves surrounding a small, sometimes vestigial stem. In bulb forms like the lily the scales are narrow and loose, while in the onion, narcissus, amaryllis, daffodil, and hyacinth the scales are continuous and fit closely. Small bulbs, or bulblets, may develop around the parent bulb. These small bulbs are borne above the ground, for example, in the axils of the leaves of the tiger lily, or at the top of the flower stalk in the leek. Many other bulbs divide into two or more parts.

A corm, on the other hand, is solid throughout. Cormels are developed in the same way bulblets are formed. Good examples of corms are those of Indian turnip, or jack-in-the-pulpit (*Arisaema*), *Caladium,* crocus (Fig. 9-30), and gladiolus.

Students may grow many of these bulbs and corms in class. For example, when onions are partially submerged in water (with the root end just below the water surface), a good root system develops in a week. The onion may be supported by the rim of the container. When bulbs are not

Fig. 9-30 Corm of crocus. (From W. G. Whaley, O. P. Breland, *et al.*, *Principles of Biology*, Harper & Row, 1954; after R. M. Holman and W. W. Robbins, *A Textbook of General Botany*, 3rd ed., Wiley, 1934.)

placed in water they first send forth shoots, not roots. Also use a garlic and have many of the bulblets sprout.

Similarly, narcissus and hyacinth bulbs may be grown in water or gravel. Narcissus bulbs grow quickly and flower in a warm room in a month's time. Hyacinth bulbs, however, require a preliminary cooling treatment in a dark place for about 4 weeks, and flowers appear 2 to 3 months later. Tulip and spring crocus bulbs require a cool temperature (about 4° C [40° F]) for a preliminary "forcing." Cover the bulbs lightly with sandy loam; water them well; and keep them in a cool, dark place for about 9 to 10 weeks. The bulbs of Easter lilies, which are rather slow to flower, should be covered with some 2 in. of sandy loam, watered well, and kept in the dark until the shoots are a few inches long. After this, transfer them to a warm location, where it takes from 3 to 4 months for the lilies to flower.

TUBERS. These are thickened underground stems, in which starch is stored. The new plants arise from rather inconspicuous buds in the tubers, such as the "eyes" in the white potato. Cover tubers of the white potato, dahlias, *Anthericum,* or *Caladium* with ½ in. of wet sand, or set them in a tumbler of water so that about half of the tuber is covered (Fig. 9-31).

FLESHY ROOTS. If sweet potatoes, carrots, or radishes (supported by toothpicks) are placed in containers of water, new shoots will appear.

Methods used by man

CUTTINGS. A cutting is a part of the plant, either a stem or a leaf, which has been separated from the parent plant and which, under suitable conditions, will produce a complete plant.

Stem cuttings. Select the tip of a healthy, vigorous young stem of geranium, begonia, or *Tradescantia* to use for a cutting. Include several nodes in the cutting, usually a piece 1 to 4 in. long, depending on the plant and the length of the internodes. When vines are used, such as bittersweet (*Celastrus*), the older parts will root better.

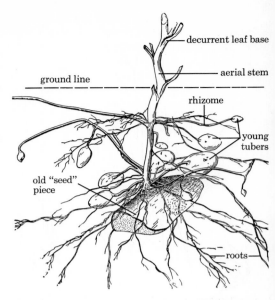

Fig. 9-31 Sprouting tubers of white potato. (From R. M. Holman and W. W. Robbins, *A Textbook of General Botany,* 3rd ed., Wiley, 1934.)

In preparing a stem cutting, cut squarely below a node with a sharp, clean knife. Pinch off the flower bud if there is one. Then cut off most of the leaves to reduce the rate of transpiration so that the cutting does not wilt. Plant the cutting 1 in. deep in medium-grained wet sand or sphagnum moss.

For good results with fleshy-stemmed plants such as geranium and *Dieffenbachia* (which should see use more often where stem cuttings are desired), the tip of a plant is not used. Instead, cut the stem into 2- or 3-in. pieces and plant them horizontally in moist sand. Allow a cover of about ½ in. of sand.

Other plants suitable for stem cuttings are coleus, ivy (either *Hedera* or *Parthenocissus*), privet (*Ligustrum*), willow (*Salix*), golden-bell (*Forsythia*), pineapple (*Ananas*), and cactus (*Opuntia* or other similar cacti). Succulent forms usually take root within 2 to 3 weeks, while woody types often require months. (See also polarity, Fig. 9-37.)

There are a few precautions to observe

in making cuttings. Use healthy stocks and work in a place that is not excessively hot or dry so that plants do not wilt rapidly. Keep the plant cuttings in moist newspaper until they are ready for planting.

The planting medium must be moist to keep the plants from wilting, and well drained so that they do not rot. Pack the medium around the cuttings to keep them upright. Use coarse-grained sand or vermiculite for good drainage. (You will find that round-grained sand such as sea sand does not pack well.)

After the cuttings have been planted in sand, water them well and cover with moist newspaper (or invert a glass tumbler over each plant to retain moisture). Remove the newspaper (or glasses) after a few days. Keep the young cuttings sheltered from strong sunlight.

A callus forms around the cut end and adventitious roots develop in about 2 to 3 weeks. Then set the small plants into 2-in. pots. The pots should first be washed and soaked in water for several hours so that they will not immediately absorb water from the potting soil. Prepare this soil by mixing 3 parts of loam (garden soil), 1 part of sand, and 1 part of humus. Crush the lumps so that the soil is smooth and fine textured. Then water the soil thoroughly about an hour before potting, and, finally, fill the pots with the soil.

Slip a broad knife or spatula under the cutting, lifting the sand along with the cutting to avoid injuring the roots. Make a hole in the soil with your finger and set the cutting in place. Avoid packing the soil too tightly. If the roots are long, hold the plant in place in the empty pot and gradually add the soil. When the plants show good growth, transfer them, with the same technique, to 4-in. pots.

In general, most cuttings, fleshy roots, and stems of herbaceous plants are less subject to bacterial or fungus attack when started in moist sand rather than plain water. Storage parts of plants containing sugar or starch are most susceptible to decay.

(A special note about geranium plants.

In the fall they should be cut back, repotted, and allowed a period of rest. Cut the plants back about two-thirds of their length and divide the plants so that each cutting has about three main stems. These cuttings should be repotted in a mixture of equal parts of loam, sand, and leaf mold. You may plant smaller cuttings in a large flowerpot around a partly submerged smaller pot, as in Fig. 9-32. Water the center pot so that equal drainage takes place with less chance for leaching of the minerals from the soil.)

Leaf cuttings. A new plant may be grown from part of a leaf. Use the leaves of *Bryophyllum* (Fig. 9-33), *Kalanchoë, Sedum,* begonias (especially *Begonia rex*), *Sansevieria,* or African violets. Lay the leaves of the first three plants on wet sand and hold them flat with a small stone or flat piece of glass placed on the leaf blade. For *Begonia rex* and *Sansevieria* a section some 2 in. long may be planted upright, about 1 in. deep in sand. Look for rooting and the appearance of shoots. When a leaf of *Begonia rex* is laid on sand, many shoots appear at the regions where the veins are severed (Fig. 9-34).

LAYERING. While layering occurs naturally, especially among wild roses, raspberries, and blackberries, it may be done with purpose. A branch is placed in contact with the moist soil and held fast so that roots and shoots are produced at the point

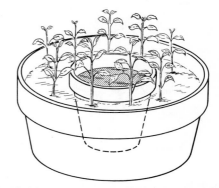

Fig. 9-32 Growing plants from stem cuttings. The smaller, inset flowerpot is kept full of water to provide equal moisture to the cuttings.

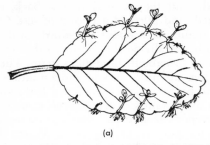

(a)

Fig. 9-33 Growing plants from leaf cuttings: (a) diagram showing small plants arising from notches along the edges of a leaf of *Bryophyllum pinnatum;* (b) photograph of new plants differentiated from tissue in notches of a leaf of *Bryophyllum.* (a, from W. H. Brown, *The Plant Kingdom,* 1935, reprinted through the courtesy of Blaisdell Publishing Co., a division of Ginn and Co.; b, Richard F. Trump, from R. F. Trump and D. L. Fagle, *Design for Life,* Holt, Rinehart and Winston, 1963.)

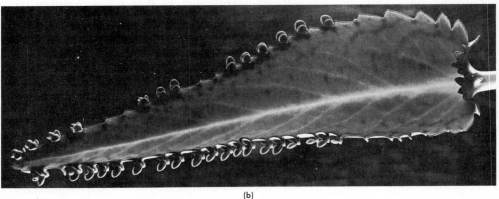

(b)

of contact with the soil (Fig. 9-35). These can be separated from the parent plant. This method is used when plants cannot be successfully grown from cuttings because they do not root readily. Layering is the commercial method for propagating such plants as magnolia, grapevines, and raspberries.

In class you may want to use such plants as *Vinca,* English ivy, and *Philodendron.* The small climbing rose and the small blackberry *Rubus villosus* are especially suitable for classwork.

Sometime you may want to use the Chinese, or pot method of layering. This is generally successful with rubber plants (*Ficus*) and castor-oil plant (*Croton*). Cut

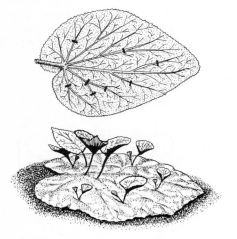

Fig. 9-34 Growing plants from leaf cuttings of *Begonia rex.* Shoots appear where veins have been severed. (After E. Sinnott, *Plant Morphogenesis,* McGraw-Hill, 1960.)

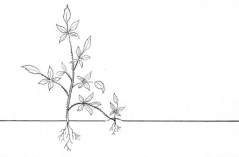

Fig. 9-35 Growing plants by layering. Where the tip of a branch is covered with soil, a new shoot grows.

into the bark of a branch about 1 ft away from the tip, where the wood is slightly hardened, and pack it around with sphagnum moss tied with raffia or cord. Keep the packing moist. Then cut a flowerpot in half lengthwise and shape each half around the wounded section of the stem. Hold it together with cord. Paper may be used instead of the flowerpot, but it is less effective. In 2 to 3 weeks, when the stem becomes rooted, cut off the branch from the tree.

GRAFTING. In this method of vegetative propagation the cut surfaces of two woody plants are placed together. If their cambium layers make contact, they will grow together. In our experience, useful plants for classroom work in grafting are the common privet and reciprocal grafts between the potato (*Solanum*) and the tomato (*Lycopersicum*). Thriving tomato plants furnish suitable scions to graft onto the potato stock.

Some precautions are necessary in grafting. Try only the simplest kinds of grafts, that is, stem grafts. Bud, saddle, and other types of grafts are not generally successful in the classroom. The work is fast enough to hold interest and students may practice these grafts with success, too. The scion and the stock should be similar in width. Remove all the leaves of both the stock and the scion when you graft. Make clean cuts with a sharp, clean knife (Fig. 9-36).

Fit the cut ends snugly and tie them together with raffia or similar material. Impregnate the binding material with beeswax or grafting wax (pp. 661–62). When possible, cut the stock about 1 or 2 in. from the soil line in order to eliminate many buds. On the scion there should be as many buds as possible, including the terminal bud.

Students may try bud grafts and tongue grafts with apple and with privet where these are available. Select a stem ½ in. in diameter. Also try grafts among a variety of cactus plants.

Effect of growth hormones. Students may also want to examine the effect of growth hormones on the rate of regenera-

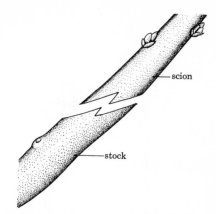

Fig. 9-36 A stem graft. The cut scion (which includes its terminal bud, not shown) is fitted into the rooted stock.

tion of new parts of the plant (see auxins, pp. 326–30).

They may repeat the classic study of Vöchting, showing polarity in willow shoots. What happens when a twig is inverted and kept moist? Do roots continue to be regenerated from the original, or morphologically basal, region? What growth occurs at the original apical region? If the twigs are subdivided into many smaller sections, is polarity retained? Refer to Fig. 9-37; note, however, that polarity is more complex than Vöchting anticipated (Fig. 9-38). Refer also to regeneration in *Acetabularia* (Fig. 8-76).

Students who want to investigate how "embryonic" tissue becomes differentiated, and what is meant by polarity will find lucid reading and illustrations in E. Sinnott, *Plant Morphogenesis* (McGraw-Hill, 1960). Read especially Chapter 9 on "Regeneration," then Chapter 20 on "Organization"; follow up with Chapter 8 on "Differentiation" and Chapter 6 on "Polarity."

For advanced reading, some students may profit from a perusal of the paper by G. L. Stebbins, "From Gene to Character in Higher Plants" (*Am. Scientist* 53:104–25, 1965).

Experimental work on differentiation among slime molds may be an area for long-range study. Learn how acrasin, a

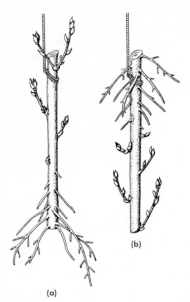

(b)

(a)

Fig. 9-37 Polarity in the willow: (a) stem suspended in moist air, producing roots and shoots; (b) stem in inverted position. (After E. Sinnott, *Plant Morphogenesis*, McGraw-Hill, 1960.)

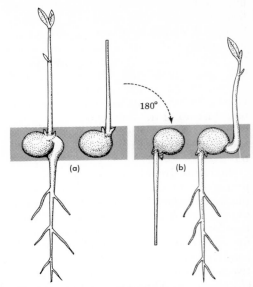

180°

(a)

(b)

Fig. 9-38 Inversion of polarity: (a) the epicotyl is decapitated from etiolated pea seedlings; (b) when the seedling is inverted and placed in water, roots grow out from the epicotyl and a shoot grows out from a cotyledonary bud. (After E. Sinnott, *Plant Morphogenesis*, McGraw-Hill, 1960.)

diffusible substance produced by some individual vegetative amoebae of slime molds, affects patterns of aggregation, development, and differentiation of amoebae into stalk-forming or spore-forming types. Read J. Bonner, "How Slime Molds Communicate" (*Sci. Am.*, August 1963). An introduction to Bonner's work on differentiation and polarity in slime molds is given in Fig. 9-39.

Regeneration among animals

Some animals are able to regenerate parts of their bodies. In this process new cells are formed as a result of differentiation and growth. When regeneration leads to the formation of new individuals, as it can with planarians and hydras, it may be considered a means of reproduction.

Regeneration in hydras. Place a well-fed hydra in a Syracuse dish containing aquarium or pond water. Cut the animal transversely with a sharp razor or with a coverslip. Try several kinds of cuts: one below the ring of tentacles, to remove the

hypostome and tentacles; one above the bud region; one below the bud region; or one at the base of the hydra. Isolate the pieces in welled slides containing aquarium or pond water, and seal the slides with petroleum jelly. Prepare several of these specimens and store them at 15° to 30° C (59° to 86° F). Diagram the types of cuts used, and record the observations made during the next week or so. Are there factors that enhance regeneration—thiamine chloride, growth substances, or others?

Of course, in asexual reproduction in hydras (Fig. 8-7), a small bud regenerates a complete organism.

Grafting hydras. H. Lenhoff has reviewed several techniques for making grafts on hydras, including the classic work of E. B. Harvey showing induction.[46]

Try to transplant a small portion of the

[46] H. Lenhoff, "Laboratory Experiments with *Hydra*," *Carolina Tips* 23:8, October 1960. E. B. Harvey's original paper (*J. Exptl. Zool.* 7:1, 1909) may be available for students to read.

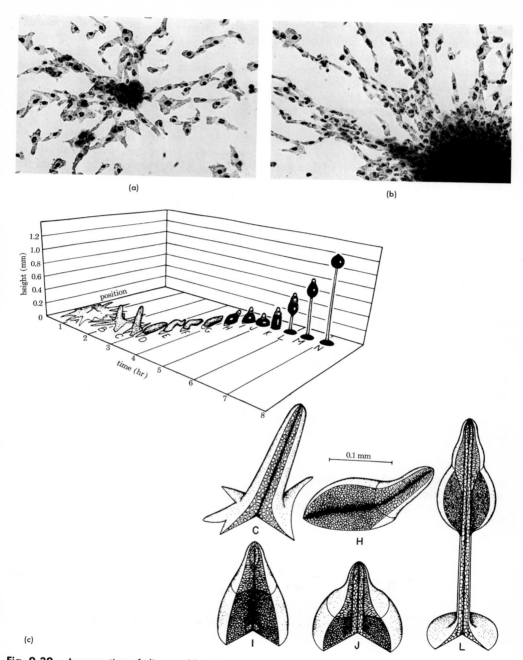

Fig. 9-39 Aggregation of slime molds: (a) photomicrograph showing individual organisms beginning to congregate as a fruiting body; (b) later stage in aggregation; (c) various stages in the process of aggregation: A–C, the fruiting body is formed by aggregation of hundreds of individual slime molds; D–H it crawls along the substrate for some time, until it grows a stalk (I–N), which lifts the spore-producing part off the substrate; finally, the spores are released. (a, b, photographs by J. T. Bonner; c, courtesy of J. T. Bonner, from *American Journal of Botany*.)

mouth or hypostome of a hydra, with tentacles attached, to the midregion of a recipient hydra. The hypostome acts as an "organizer" inducing, at this region, a new hydra. Lenhoff suggests that shallow Petri dishes lined with paraffin be used for the operations. Further, he suggests that if the hydras are introduced a day in advance into the waxed dishes, they tend to adhere to the wax and are easier to cut. E. B. Harvey used both green hydras and some that had been made albino; the latter were formed when they were placed in 0.5 percent glycerin, which apparently releases the symbiotic algae, the zoochlorellae, from the cells of the hydras.

Regeneration in Planaria. Most planarians may be used for regeneration studies, but *Dugesia dorotocephala* and *D. tigrina* are especially suitable experimental animals. (For culturing techniques, see Chapter 13.) Stretch out the animals in a drop of water on a glass slide or on a moist cork. Cut the worms, with a sharp razor or a coverslip, in the various ways shown in Fig. 9-40. Make sketches of the cuts made, and isolate the cut pieces into Syracuse dishes containing pond water or Brandwein's solution A (p. 581). Store the slides in a covered container in the dark (or in an enameled container) at 21° to 27° C (70° to 80° F). Change the water frequently to encourage rapid healing, but do not feed. Notice the colorless mass of tissue —a regenerated blastema—that protrudes from the cut edge.

The rate of regeneration in relation to the metabolic gradient (or degree of cephalization) may be studied as an extended laboratory activity. Some students may also be interested in reading C. M. Child's classic work in observing the rate of oxidation (using methylene blue and planarians sealed in anaerobic conditions).[47]

Refer to the papers previously cited (p. 314) concerning naive heads, trained heads, and regenerated heads in studies of memory, RNA, and learning, and to Best's[48] paper on planarian learning. Students may also want to read Balinsky's[49] discussion of regeneration and Bronsted's[50] fine review paper. Also see Turtox suggestions[51] about patterns and precautions for cutting worms, and investigations described by Hamburger.[52]

Regeneration in protozoa. You may want to have the class discuss experiments that have been done using *Stentor, Paramecium, Euplotes,* and *Blepharisma,* among other animals, to show regeneration of parts of the cell (provided the nucleus is intact).[53]

Regeneration in Tubifex worms. *Tubifex* is an annelid worm that you may purchase in aquarium or pet shops. In these forms posterior regeneration is most successful. Put the specimens in clean Syracuse dishes and make a transverse cut with a clean scalpel or razor blade. Then separate the sections into individual dishes and keep them at room temperature (10° to 16° C [50° to 61° F]). It takes 2 to 3 weeks for successful regeneration. A new anal opening is formed within 48 hr when the cut end heals.

In other demonstrations of regeneration use *Dero,* a microscopic annelid, or try using the earthworm. In the latter, ante-

[47] C. M. Child, *Patterns and Problems of Development,* Univ. of Chicago Press, Chicago, 1941.
[48] J. B. Best, "Protopsychology," *Sci. Am.,* February 1963, pp. 54–62.
[49] *Op. cit.,* pp. 476–502.
[50] H. V. Bronsted, "Planarian Regeneration," *Biol. Rev.* 30:65–125, 1955.
[51] *Turtox Service Leaflets,* No. 16, General Biological Supply House, Chicago.
[52] *Op. cit.,* pp. 181–91.
[53] See R. Manwell, *Introduction to Protozoology,* St. Martin's, New York, 1961.

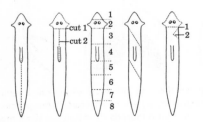

Fig. 9-40 Cutting patterns for demonstrating regeneration in *Planaria.* (Courtesy of General Biological Supply House, Inc., Chicago.)

rior-end regeneration seems to be more successful.

Regeneration of tails of tadpoles. When young tadpoles of frogs or salamanders are available, cut off the tips of their tails with a sharp razor. Refer to specific techniques in Hamburger's book.[54] Make drawings of the cuts and compare the regeneration of tails in the specimens under observation. To reduce mortality, maintain the tadpoles at a cool temperature. Keep a record of observations for a week or 12 days. (See also mitosis, using cells of tadpoles, p. 473, and refer to Balinsky.[55])

The embryology of the frog

Early amphibian development

Students may become familiar with the classic, elegant experiments of Roux, Vogt, Mangold, and Spemann by reading in a high school text, such as *Biological Science: An Inquiry into Life*.[56]

With a hot needle, prick one cell of a two-celled stage of frog or salamander. What are the subsequent cleavage patterns of the eggs? Are the results the same if the "killed" cell is removed from the healthy cell?

Rugh describes techniques for centrifuging eggs, for constricting early cleavage stages with ligature, and for providing a restrictive environment by partially embedding early embryos in agar. Also see Hamburger's[57] and Rugh's[58] descriptions of formation of double embryos and of Vogt's experiments in applying vital stains to trace the "fate maps" of developing eggs.

Embryonic induction

Mesodermal tissue underlying ectoderm in the late blastula stage of amphibians

[54] *Op. cit.*, pp. 192–96.
[55] *Op. cit.*, Chapters 18, 19.
[56] *Op. cit.*, Chapters 28, 29. Also refer to classic papers in M. Gabriel and S. Fogel, eds., *Great Experiments in Biology*, Prentice-Hall, Englewood Cliffs, N.J., 1955.
[57] *Op. cit.*, pp. 15–140.
[58] *Op. cit.*, Investigations 15, 16, 21, 23, 25.

influences or causes the differentiation of the overlying ectoderm into nervous system, eye, or other ectodermal derivatives. This is, in very elementary terms, the essence of embryonic induction. When Spemann transplanted tissue of the dorsal lip of the blastopore of a very early gastrula into a host in the gastrula stage, the transplanted section of dorsal lip caused organization and differentiation of a brain and spinal cord. Thus, the host had its own original organization plus an imposed "smaller embryo" growing on it. Spemann referred to the dorsal lip region as the "organizer" in the induction of tissues, leading to differentiation of organ systems.

Time is a factor in induction and differentiation. At a very early stage, there seems to be a faint surface etching of outlines of organs—tissue has a presumptive fate (Fig. 9-41). Yet, if such tissue from a presumptive, or future, eye region (anlage for eye) is transplanted to the belly region of a host in the tail bud stage (Fig. 9-22), the tissue differentiates into skin similar to the surrounding tissue. In the early gastrula, however, the presumptive fate is fixed, or determined, and it is boldly drawn so that, when transplanted, an anlage for an eye differentiates into an eye regardless of its environs.

Rugh[59] and Hamburger[60] describe specific techniques for investigations to test the organizing potencies of the dorsal lip of the blastopore. Urodele gastrulae withstand these operations more successfully than do the anuran embryos.

These techniques require considerable practice, maturity, and skill, and, although high school students enjoy reading these fascinating studies, few students seem to achieve success in these techniques.

Students may, however, subject eggs to different temperatures and pressures and to centrifuging. Or they may try to constrict eggs with a hair loop, possibly try to use vital stains, and to experience the "awe" of watching cleavage stages, gastru-

[59] *Ibid.*, pp. 197–203.
[60] *Op. cit.*, pp. 118–21.

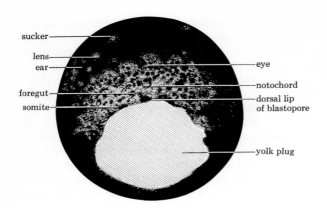

Fig. 9-41 Gastrula stage of frog, showing anlage for several organs. (After R. Rugh, *Experimental Embryology: Techniques and Procedures,* 3rd ed., Burgess, 1962.)

Labels: sucker, lens, ear, foregut, somite, eye, notochord, dorsal lip of blastopore, yolk plug

lation, and formation of a neural tube in a living mass of cells. Hamburger and Rugh (cited above) give splendid directions and illustrations for these many techniques, and for transplanting pieces of the dorsal lip and for eye and limb field operations. Other possible investigations utilize fish eggs, chick embryos, and even early mouse embryos.

To give a view of the delicate manipulations required in this work, we will briefly describe a technique for transplanting anlage for gills from an early embryo (Fig. 9-22) to an older host, such as part z in Fig. 9-22. All three germ layers contribute to the formation of the external gills.

Making a transplant

Select appropriate donor and host embryos. Remove the jelly and membranes as follows. Cut an egg apart from an egg mass with a pair of scissors. With jeweler's forceps, gently grasp an egg and roll it along a piece of toweling to remove most of the jelly. *Be gentle:* Do not injure the egg. Next, with a pair of finely pointed jeweler's forceps, hold onto the vitelline membrane and, with a second forceps, carefully pull or prick the membrane. When done correctly, the egg or embryo will pop out of the encasements of membrane and jelly; however, many embryos are jabbed or squashed in this procedure.

Now the delicate embryos must be handled and transferred with a hair loop (see below) or with a glass pipette (of wide enough bore; see Fig. 14-8). Operating dishes are either salt cellars or Syracuse dishes embedded with plain agar, wax to which lampblack has been added to reduce the glare, or Permoplast clay. With a glass ball, shape two depressions in the embedding material used in the operating dishes. With the hair loop or the glass pipette, transfer the donor and the host to the depression in the operating dish (Fig. 9-43). Add enough operating solution to the dish to cover the embryos.

First prepare the host embryo to receive the transplant by cutting out a rectangular piece of tissue with a glass needle. Remove ectoderm, with the underlying mesoderm, from either the tail bud region or the belly region. In fact, plan several operations, transplanting gill anlage into different sites in different hosts. With the aid of a hair loop, cut deeply enough to accommodate the rather thick transplant. The transplant must fit snugly into this area, and the operation must be done quickly since the cells are alive and mitosis proceeds quickly.

Using the glass operating needle and the hair loop, cut a similarly sized rectangle of tissue from the gill swelling in the donor. This gill swelling is located in line with the eye, directly below the first few somites (see Figs. 9-22, 9-42). The tissue transplanted from the donor should include ectoderm with underlying mesoderm; transfer it, without changing its orientation, to the host. Use both the needle and the loop to insert the transplant carefully so as not to injure the tissue. Then use a coverslip bridge to hold the graft in place for about 30 min (for preparation, see directions below). If any extra "fringes" of tissue remain around the wound after ½ hr, carefully trim them with a hair loop. Finally, transfer the host to fresh medium —which Rugh calls "growing medium" in contrast to "operating medium" (recipe, below)—and maintain in a cool room, or in a refrigerator at 10° to 20° C (50° to 68° F).

Also maintain the donors to examine the region from which the anlage for gills was removed; use the other side of the donor embryos as controls. Compare the rate of growth of the gills in the original donor with the rate of growth of the transplanted tissue in the host. Keep a record of observations, including drawings or photographs.

Solutions and operating tools

Although contamination by bacteria is not as great a danger in fast growing amphibian tissue maintained at cool temperatures, sterile procedures should be followed in handling the embryos. The agar in which the embryos are embedded

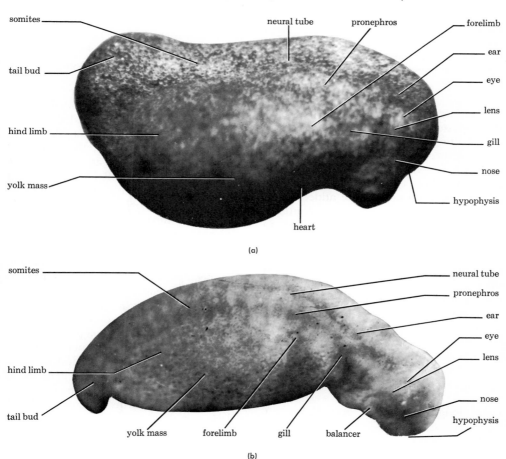

(a)

(b)

Fig. 9-42 Organ fields in amphibian embryos: (a) *Rana pipiens;* (b) *Ambystoma punctatum.* (From R. Rugh, *Experimental Embryology: Techniques and Procedures,* 3rd ed., Burgess Publishing Co., 1962.)

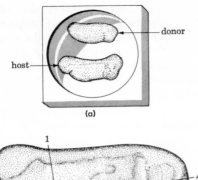

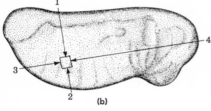

Fig. 9-43 Transplanting anlage to host: (a) donor and host in operating dish; (b) parallel cuts made to remove tissue from host so that transplanted anlage from donor can be inserted into belly or flank of host.

is plain agar, not nutrient agar; wax or Permoplast clay may be preferred.[61]

Operating solution.[62] Operating solutions are usually more concentrated than solutions in which embryos are left to grow. A dilute solution (0.5 percent) of sodium sulfadiazine may be added to the solutions to inhibit growth of contaminants; it has been found harmless to embryos.

FOR URODELES. Prepare quantities of the following stock solution; autoclave and store.

Great Bear Spring water[63]	10 liters
NaCl	70 g
KCl	1 g
$CaCl_2$	2 g

For an operating medium, dilute 2 parts of stock solution in 1 part of Great Bear Spring water. This is a hypertonic solution.

For longer periods for growth, prepare

[61] Permoplast is available from American Art Clay Co., Indianapolis, Ind.

[62] Rugh, *op. cit.*, p. 15.

[63] Great Bear Spring water is available from Great Bear Spring Co., 51 Madison Ave., New York, N.Y. *Conditioned* tap water may be substituted.

a more isotonic solution. Mix 2 parts of stock solution to 4 parts of Great Bear Spring water.

FOR ANURANS. For an operating and growing medium, use Holtfreter's solution, which is prepared as follows. To 1 liter of distilled water add

NaCl	3.5 g
KCl	0.05 g
$CaCl_2$	0.1 g
$NaHCO_3$	0.2 g

FOR FROG EMBRYOS. For an operating medium, prepare Holtfreter's solution *double strength;* autoclave and store.

Light source. A strong light source is needed for operating on early embryo stages. To reduce the heat of the lamp, place a round Florence flask, filled with water to which a few crystals of copper sulfate have been added, between the light and the operating dish. Support the flask on a ring stand.

Microburner. A microburner must be prepared in order to fashion the glass needles and hair loops needed for the operations. Select a 6-in. length of soft glass tubing (7 mm diameter), and pull out one end, in a flame, to a diameter of about 1 mm. As in Fig. 9-44, make a right angle bend in the center of the length of tubing. Allow to cool, and insert the more pointed end through a cork supported in a burette clamp on a ring stand. Arrange the cork and tubing so that it may be tilted in several directions for ease in making glass needles. Use rubber tubing to connect the large end of the glass tubing to a source of gas. Apply a screw clamp to the rubber tubing so that you can control the amount of gas passing through the thinner tubing. Light the microburner. It should be placed fairly close to the desk or work table since students will find it easier to twirl glass tubing and make 45° glass needles with both arms supported against a desk.

Hair loops. Prepare hair loops from 4-in. lengths of 6-mm-bore glass tubing. Pull out one end of the tubing in a microflame so that the opening is reduced to about 1 mm; close the opening in the other end.

Obtain small lengths of fine hair, preferably baby's hair, and thread each end of a 1-in. length into the tubing, leaving a very small loop (about ⅛ in.) protruding from the glass tubing. Secure the loop in place by dipping this end into warm wax (beeswax or paraffin); allow some of the wax to run up the capillary so that a good seal is made. Should an excess of wax adhere to the hair loop, touch the loop lightly to a warmed slide to remove the excess wax. Prepare several of these hair loops and store in a test tube with a cotton stopper to avoid contamination by dust in the laboratory.

Glass needles. Use 4-in. lengths of 6-mm glass rods to prepare several needle holders. In a flame, draw out the glass rod to a tapered end. Insert this tapered end into the flame again; the glass melts and balls up, forming a sphere. This is the region into which the needle (to be prepared next) will be inserted.

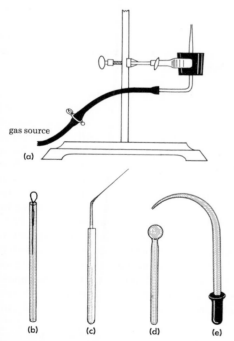

Fig. 9-44 Microburner and operating tools: (a) microburner; (b) hair loop; (c) glass needle; (d) glass ball roller; (e) transfer pipette.

In the microflame, pull out lengths of 4-mm soft glass rods to a thickness of 1 mm. Twirl 7- to 8-cm lengths of this fine thread of glass in the microflame to melt the center; support your arms against the work table and pull out a bit and then quickly downward so that the apex of a triangle is formed. Quickly break this thread with a quick brush near the flame. Allow to cool. If a small ball of glass has formed at the tip, break it off with your fingernail. For pointed needles, the pulled thread of glass should be pulled apart completely. Check the point of the needles under a microscope.

Rugh[64] suggests another method for preparing needles. Pull out lengths of glass rod to 1 mm thickness (as above), and put a hook or bend at the end of each length. Then hang this hook over the metal ring of a ring stand in such a way that the glass is directly over a 20 × 60 mm glass vial containing a small bit of cotton at the bottom. At a point 2 cm from the bottom of the hanging glass rod apply the microburner flame from one side. As the glass melts, the weight of the hanging rod will draw the end of a fine point and will cause the rod to drop into the glass vial.

Now the needles are ready to be inserted into the glass rod holders that were prepared earlier. Heat the ball at the tip of the holder and, when it is quite soft, insert the needle (held in a forceps) in a straight line with the holder. Allow to cool. Stand the completed needles temporarily in small balls of clay.

In use, the needles are broken very easily, so many needles should be prepared at one time. They can be stored in wide-mouthed vials containing a cushion of cotton. Stand the vials upright in a test-tube holder or in an appropriate block of wood. Some workers prepare and store the proper lengths of 1-mm diameter glass rods; then they prepare the needles in a microflame as they are needed.

Glass ball rollers. A length of glass rod with a round ball at one end is needed to

64 *Op. cit.,* p. 8

shape depressions in beeswax, paraffin, or agar in the operating dishes (Fig. 9-44d).

Glass transfer pipettes. Pull out several lengths of glass tubing to obtain pipettes with different sized bores. A slight twist in the bend of the pipette enables the operator to insert the pipette at an angle into a dish, thereby increasing the ease of manipulation. Therefore, as the glass becomes soft, twist it by spreading your hands or by turning your hands so that the palms are upward a bit more; this will bend the soft glass. Or let the weight of the glass produce a bend; then pull out a short length. Some pipettes should have bores wide enough to pick up eggs; others should have smaller bores for transferring tissue to be transplanted.

Glass coverslip bridges. Hold one edge of a clean coverslip in a microflame until it is softened along its center line. Then lay the soft glass over a thin length of glass rod, causing a slight bend in the coverslip. Allow to cool. The thickness of the glass rod, of course, determines the degree of bend. Only a slight bend is needed for applying pressure to a transplant to hold it in place as the graft grows together. Practice will illustrate how much of a bend is necessary. Remember that the embryo is in a depression in the operating dish, so that often a bend is not essential.

All glass materials may be stored in Petri dishes or vials containing cotton. They may be cleaned in alcohol and then dipped in operating solutions containing 0.5 percent sodium sulfadiazine.

Suggestions for long-range investigations

Students of high ability may be guided into long-range projects which can develop into "independent research." Techniques of plant and animal tissue culturing require patience and practice to avoid contamination, but these skills open the door to the creative ability of young people. Devising experiments to explore the role of special hormones, photoperiodism, differentiation, and polarity introduces students to areas of biology rich in problems.[65] Techniques described throughout this text are useful in finding answers and raising more questions for continued investigations.

A study of chlorophyll extracts or of the properties of chlorophyll in transmitted lights and in reflected light can be used for individual study or as part of a larger project. The separation of carotinoids from the leaves of plants, as well as the types of chlorophyll, may be done with paper chromatography.

Committees of students may devise experiments to test which wavelengths of light are most effective for photosynthesis and for growth, or for flowering. Small land plants or test tubes of *Anacharis* or microscopic algae such as *Chlorella* may be covered with different-colored sheets of cellophane. The plants and the control plants kept in sunlight and in the dark may later be tested for starch; or the rate of growth may be recorded on a graph. Have students develop a comparison scale for use in judging the amount of starch present in leaves. (It will be found that the yellow to red region of the spectrum is the most active one.)

Some students may study the causes of dormancy in seeds. They may devise ways to show how dormancy may be broken. Some hard surfaces of seeds may be filed or cut, others placed in acid, and so forth. Controls should be included in these experiments. Be sure to look into avenues for investigation on regeneration in animals and on the cleavage patterns of invertebrate and vertebrate eggs.

We have discussed many aspects of growth in animals and plants. Much of this work may be considered an extension of classwork. Topics in other chapters—such as reproduction, nutrition diseases, photosynthesis, and genetics—include

[65] For an interesting study of differentiation, see J. Bonner and M. Hoffman, "Evidence for a Substance Responsible for the Spacing Pattern of Aggregation and Fruiting in the Cellular Slime Mold," *J. Embryol. Exptl. Morphol.* 11:571–89, 1963.

some aspect of the work described in this chapter. The techniques and the subject areas may be coordinated effectively, as the teacher sees fit, into specific, related learning experiences in the laboratory and classroom.

CAPSULE LESSONS: Ways to get started in class

9-1. In planning laboratory activities using bacteria, be sure to consult the section on diversity in microorganisms in the *Student Laboratory Guide* to *Biological Science: An Inquiry into Life* (BSCS Yellow Version, Harcourt, Brace & World, 1963). Using some of these techniques, can students test several predictions within the framework of any of the hypotheses that may be raised in class?

9-2. In a laboratory activity dealing with population studies of yeast cells use the methods suggested in this chapter, or look into the *Student's Manual: Laboratory and Field Investigations* to *High School Biology* (BSCS Green Version, Rand McNally, 1963). When 1 g of dry yeast is added to 10 ml of water, the average number of cells per field under a high-power objective is about six. Refer also to Breed's method for estimating numbers of microorganisms (in this chapter).

9-3. Prepare several grids by marking off squares in India ink on rubber sheeting. On the sheet, outline the shape of a specific leaf, fruit, or fish. Stretch the sheet in several directions to show how altered patterns of the plane of growth might affect the shape or form. Elicit a discussion of ways that changes in the genetic code might alter the pattern or plane of growth or polarity; lead into a discussion of interrelated steps in genetic control. Also refer to studies of mitotic divisions and to the effects of hormones such as auxins and of retardants such as Phosfon (in this chapter).

9-4. When a batch of amphibian eggs is available (they may also be purchased), students may devise a whole series of investigations, from simple, careful observations of cleavage patterns, invagination, and formation of a neural tube to plans for a study of environmental factors that affect rate of growth (see this chapter).

9-5. At some time, raise the following questions. Are both cells in the two-celled stage of the developing egg equally totipotent? Can each develop an entire embryo? What would happen if one cell were killed by pricking it with a hot needle? What would happen if the two cells were separated from each other by constricting them with a hair loop? Each cell might then be isolated in a separate dish and left to develop.

9-6. As a special investigation, try to stain parts of eggs of a salamander or a frog with vital dyes. The stains may be applied to small bits of agar, which, in turn, are placed against regions of the eggs. R. Rugh and V. Hamburger (see the end-of-chapter bibliography) give complete details.

9-7. Many students may do experimental work at home; there they may try to regenerate planarian worms, limbs of salamanders, or tail fins of salamander tadpoles. Plan to have such students report to the class or at special seminars. Perhaps there is space in a hall cabinet to display the project.

9-8. Where perennials, shrubs, and trees are available in school plots or home gardens, vegetative propagation of woody plants may be investigated. Grafting procedures may also be attempted.

9-9. Many of the activities of this chapter are techniques which are useful in illustrating all the life activities of plants. As such, these skills may be developed through a variety of projects undertaken as club work, laboratory experiences, or home activities.

For example, begin with a handful of dried beans and another of germinating beans. You may develop a lesson on respiration and the use of the stored food supply. Or you may ask the following questions. What are the characteristics of living things? Are the dried beans living? This is one way to develop the notion of organization of materials in living systems.

You may want to show tropisms suggested by students. Or you may plan a study of plant functions beginning with seedlings, which are so adaptable for class demonstrations.

9-10. If you introduce the concept of a hormone as a substance made in one part of a body which produces an effect in other parts, you will find studies with auxins of plants a fruitful approach to the topic. Have the students offer explanations. Perhaps students will report to the class on some of the classic experiments in the field. You might want to demonstrate the effect of feeding iodine or thyroxin to tadpoles of salamanders or frogs.

9-11. Introduce the notion that bacteria are

present in the air and on the body (also in the body) by having students inoculate prepared Petri dishes. Develop the conditions needed for bacterial growth. Establish from this discussion ways to preserve food, that is, ways to prevent bacteria from growing in our food. For if we know the conditions for bacterial growth, we can protect our food supply by removing at least one of the conditions required for their optimal growth. Students may devise an experiment at home testing the effectiveness of vinegar, sugar, and salts as preservatives (see pp. 449–52).

9-12. Have students grow seeds of beans, sunflowers, squash, or other plants at home. Show them how to soak and place the seedlings between wet blotters and the glass walls of baby-food jars or other containers. This activity may lead into a study of growth hormones, the way plant embryos grow, or root hairs and the conduction system of seedlings.

These are the times when first-hand experience is easier to provide, and more meaningful and exciting, than a film showing a growing seed. However, there may be appropriate times for showing time-lapse photography of plant growth.

9-13. When possible, try to grow excised embryos of seedlings or pieces of roots or leaves as tissue cultures, as described in this chapter.

9-14. In lessons on control of diseases, demonstrate the effect of penicillin on the growth of bacteria; use Gram-positive and Gram-negative organisms. Also devise ways to test the effectiveness of antiseptics. Compare antiseptics and antibiotics and the role they play in control of diseases.

Be sure to show the zone of inhibition around discs soaked in antibiotics and applied to Petri dishes seeded with bacteria.

PEOPLE, PLACES, AND THINGS

Is there an agricultural experimentation station, a college laboratory, a plant nursery, or a botanical garden in your community? You may want to plan a field trip to one of these places to learn the kind of research being done in the areas of hormones, photoperiodism, soil-saving and conservation, soilless growth of plants, tissue-culture techniques, or growth rate measurements of a variety of microorganisms.

Students may glean ideas for individual research projects which may be carried on at home or on the school grounds (see Chapter 11). Possibly you may borrow from a nearby hospital or college bacteriology department sterile Petri dishes or a dye for staining slides. Many laboratories will supply grateful schools with frogs' eggs, *Drosophila, Chlorella, Hydra,* or other forms —if the requests are within reason.

You may want to invite a guest speaker to help your students discover what is being done in the field of antibiotics (or some other field of biology). Possibly a parent doing research of some kind will be willing to visit the class.

The librarian of a college may help you find some books (such as those in the Bibliography) which furnish information on specific subjects. Science teachers in other schools may have books, films, and other equipment to share with you. You will want to look into P. Goldstein's *How to Do an Experiment* (Harcourt, Brace & World, 1957) for ways to guide students.

Some universities have experimental stations from which you may get help in your specific problem. You will want to look into the opportunities for placing interested, able students in voluntary or part-time jobs in experimental stations, plant nurseries, experimental farms, or research departments of a hospital.

BOOKS

The references in the previous chapter on areas in heredity and reproduction should be consulted along with these in growth and differentiation. A comprehensive Bibliography is given in Appendix A.

Audus, L. J., *Plant Growth Substances,* 2nd ed., Interscience (Wiley), New York, 1960.

————, ed., *The Physiology and Biochemistry of Herbicides,* Academic, New York, 1964.

Balinsky, B. I., *An Introduction to Embryology,* 2nd ed., Saunders, Philadelphia, 1965.

Berrill, N., *Growth, Development, and Pattern,* Freeman, San Francisco, 1961.

Bonner, J., *The Evolution of Development,* Cambridge Univ. Press, New York, 1958.

————, ed., *see* Thompson, Sir D'A.

Butler, E., ed., *Biological Specificity and Growth,* Princeton Univ. Press, Princeton, N.J., 1955.

Eigsti, O. J., and P. Dustin, *Colchicine: In Agriculture, Medicine, Biology, and Chemistry,* Iowa State Univ. Press, Ames, 1955.

Evans, L. T., ed., *Environmental Control of Plant Growth,* Academic, New York, 1963.

Goldstein, P., *How to Do an Experiment,* Harcourt, Brace & World, New York, 1957.

Hadorn, E., *Developmental Genetics and Lethal Factors,* Wiley, New York, 1961.

Hamburger, V., *A Manual of Experimental Embry-*

ology, rev. ed., Univ. of Chicago Press, Chicago, 1960.

Jacobs, W. P., and C. E. LaMotte, *Regulation in Plants by Hormones* (BSCS laboratory block), Heath, Boston, 1964.

Lee, A. E., *Plant Growth and Development* (BSCS laboratory block), Heath, Boston, 1963.

Leopold, A. C., *Auxins and Plant Growth,* Univ. of California Press, Berkeley, 1955.

Machlis, L., and J. Torrey, *Plants in Action,* Freeman, San Francisco, 1956.

Miller, M. W., *The Pfizer Handbook of Microbial Metabolites,* McGraw-Hill, New York, 1961.

Mitchell, J., G. Livingston, and P. Marth, *Test Methods with Plant-regulating Chemicals,* U.S. Dept. of Agriculture Handbook 126, U.S. Govt. Printing Office, Washington, D.C., 1958.

Moog, F., *Animal Growth and Development* (BSCS laboratory block), Heath, Boston, 1963.

Needham, J., *A History of Embryology,* 2nd ed., rev. with A. Hughes, Abelard-Schuman, New York, 1959.

Rugh, R., *Experimental Embryology: Techniques and Procedures,* 3rd ed., Burgess, Minneapolis, 1962.

Seeley, H. W., Jr., and P. J. VanDemark, *Microbes in Action,* Freeman, San Francisco, 1962.

Sinnott, E., *Plant Morphogenesis,* McGraw-Hill, New York, 1960.

Soc. for the Study of Development and Growth, *Regeneration,* Symposium 20, ed. by D. Rudnick, Ronald, New York, 1962.

Soc. of Am. Bacteriologists, Committee on Bacteriological Technic, *Manual of Microbiological Methods,* ed. by H. J. Conn, McGraw-Hill, New York, 1957.

Sussman, A. S., *Microbes: Their Growth, Nutrition, and Interaction* (BSCS laboratory block), Heath, Boston, 1964.

Thompson, Sir D'A., *On Growth and Form,* 2 vols., Cambridge Univ. Press, New York, 1952; abridged ed. by J. Bonner, 1961.

Torrey, T., *Laboratory Studies in Developmental Anatomy,* Burgess, Minneapolis, 1962.

Umbreit, W., *Modern Microbiology,* Freeman, San Francisco, 1962.

Wain, R., and F. Wightman, eds., *The Chemistry and Mode of Action of Plant Growth Substances,* Academic, New York, 1956.

White, P., *The Cultivation of Plant and Animal Cells,* Ronald, New York, 1954.

Withrow, R., ed., *Photoperiodism and Related Phenomena in Plants and Animals,* Am. Assoc. for the Advancement of Science, Washington, D.C., 1959.

FILMS, FILMSTRIPS, AND SLIDES

This listing is intended to be only a guide for the selection of new films, filmstrips, and transparencies. Send for catalogs from distributors of films; catalogs of biological supply houses list extensive offerings of 2×2 in. slides. The addresses of distributors are given in Appendix B.

In general, the cost of color films averages $150; black and white films cost about $75. Newer films are made in varying lengths—time sequences of 1 min, 5 min, 11 min, or 30 min; the cost of films therefore varies. We have indicated the lengths of films so that lessons can be planned with the films in mind.

Sound films are available in color or black and white; rental rates are available from film libraries or by direct inquiries to producers. (Filmstrips are not available on a rental basis; the purchase price is usually about $6 to $7 for those in color.)

Age of Promise (10 min), Association.
And the Earth Shall Give Back Life (25 min), Squibb.
Antibiotics (14 min), EBF.
Are You Positive? (14 min), National Tuberculosis Assoc.
Bacteria: Friend and Foe (11 min), EBF.
Bacteria: Laboratory Study (15 min), Indiana Univ.
Bacteriological Techniques (5 min), BSCS techniques, Thorne.
Basic Biology of Bacteria (filmstrip), Communicable Disease Center; free.
The Challenge of Malaria (14 min), Contemporary.
The Complement Fixation Test (5 min), Univ. of California.
Culturing Slime Mold Plasmodium (7 min), BSCS techniques, Thorne.
Drugs: The Biochemist's Scalpel (29 min), National Educational Television.
Explorations in Laboratory Animal Care (22 min), National Soc. for Medical Research; free.
For More Tomorrows (28 min), Am. Cyanamid–Lederle; free.
Germfree Animals in Medical Research (19 min), Communicable Disease Center; free.
Growth of Seeds (14 min), EBF.
Immunization, 2nd ed. (11 min), EBF.
Infectious Diseases and Manmade Defenses (11 min), Coronet.
The Life and Death of a Cell (21 min), Univ. of California.
Measuring Techniques (14 min), BSCS techniques, Thorne; micrometer, hemocytometer.
Microbiology (set of 12 films; 30 min. each) AIBS, McGraw-Hill.

Phagocytosis (4 min), Univ. of California.

Rabies Can Be Controlled (14 min), Lederle.

Reproduction, Growth, and Development (set of 12 films, 28 min each), Part V of *AIBS Film Series in Modern Biology,* McGraw-Hill.

Science Detectives (28 min), Modern Talking Pictures; finding antibiotics, free.

Scientific Method in Action (19 min), International Film Bur.; Salk's methods.

Seed Germination (15 min), EBF.

Simple Plants: Bacteria (14 min), Coronet.

The Smallest Foe (20 min), Lederle; virus and Rickettsia.

A Tree Is Born (29 min), Syracuse Univ., College of Forestry; free.

Unseen Enemies (32 min), Shell; free.

Unseen Harvester (29 min), Du Pont; free.

Victory over Polio (27 min), McGraw-Hill.

Virus (set of 8 films, 29 min each), Indiana Univ.

Vitamins in Fact and Fancy (29 min), Indiana Univ.

The World Within (28 min), Univ. of California.

Your Health: Disease and Its Control (11 min), Coronet.

LOW-COST MATERIALS

A listing of suppliers of free and low-cost materials, with addresses, will be found in Appendix D. Among these are large pharmaceutical firms which manufacture antibiotics; write to them and ask for literature.

Write to W. H. Freeman and Co., 660 Market St., San Francisco 4, Calif., for their catalog of available offprints of papers from *Scientific American* ($0.20 each), many of which deal with aspects of the work discussed in this chapter. You may also want to look at *Microbiology in Introductory Biology,* a special issue of *The American Biology Teacher* (Vol. 22, No. 6, June 1960).

Publications we have mentioned earlier in other contexts contain material relevant to growth and differentiation as well:

Carolina Tips (periodical), Carolina.

Turtox Service Leaflets (set of "How to Do" pamphlets), General Biological.

U.S. Atomic Energy Comm., *Some Applications of Atomic Energy in Plant Science,* U.S. Govt. Printing Office, Washington, D.C.; use of radioisotopes in study of life activities.

Chapter 10

Evolution

There are two major conceptual schemes relating organisms and their environment:

1. An organism is the product of its heredity and environment.
2. An organism is interdependent with other organisms and with its environment.

This chapter is concerned with still another conceptual scheme:

3. Organisms and environments have changed over the ages.

A variety of techniques and procedures are available in this study, including laboratory and field work and demonstrations. Another useful technique is the "reading."

A reading: concerning the first living things

A passage from A. Oparin's *Life: Its Nature, Origin, and Development,* such as the one that follows, or from a different source may be useful. Through reading such passages, students become aware of current thinking in biological sciences.

THE FIRST LIVING THINGS
WERE ANAEROBIC HETEROTROPHS

By trying to detect in the tremendous variety of systems of metabolism in different organisms those similarities, those features of organisation which are most widespread among all living things, and therefore most ancient, we can, in the first place, establish two cardinal principles.

First, the metabolism of all contemporary living things is based on systems which are designed for the utilisation of ready-made organic substances as their primary building material for biosynthesis and as the source of that energy which is necessary for life although one might, theoretically, postulate many other satisfactory metabolic pathways.

In the second place, all contemporary living organisms have a biochemical system for obtaining energy from organic substances which is based on the anaerobic degradation of these substances although, with free oxygen in the atmosphere, as it is now, it would be perfectly rational for them to be oxidised directly.

It is self-evident and generally accepted that the overwhelming majority of the biological species which now inhabit our planet can only exist if they are constantly supplied with ready-made organic substances. This applies to all animals, both higher and lower, including most of the protozoa, the great majority of bacteria and all fungi. This fact alone is extremely suggestive. One can hardly imagine that all these evolved simply as Batesonian simplifications, as a complete loss of the autotrophic abilities which they once had. This is also contradicted by the intensive biochemical studies of the metabolic systems of these organisms. We do not find in them the least trace or vestige of those specific enzymic complexes or groups of reactions which are required by autotrophic forms of life while, on the other hand, the metabolism of autotrophs is always based on the same internal chemical mechanisms as that of all those other organisms which can only exist by consuming organic substances. The specific autotrophic mechanisms are merely superstructures on this foundation. It is just this sort of organisation of their metabolisms which allows autotrophs, under certain conditions, to revert entirely to the consumption of ready-made organic substances.

This can be demonstrated particularly clearly in the case of the least highly organised photo-autotrophs—the algae—both in natural conditions and in the laboratory. By means of such

493

'ong ago been shown that if
᠆ introduced into sterile
will assimilate these sub-
᠊s may go on at the same time
᠊ but, in some cases, photosyn-
᠊ altogether and the algae go over
᠊ely saprophytic way of life.

these conditions one gets a very
᠊nt growth of blue-green algae such as
᠊c and the diatoms as well as such green
᠊gae as, for example, *Spirogyra*. Many forms of
blue-green and other algae must obviously
assimilate organic compounds directly even
under natural conditions, when they live in dirty
ponds. This is suggested by the fact that they
grow especially luxuriantly in stagnant waters
and such places where organic substances
abound.

A heterotrophic basis for nutrition may be
found, not only in the algae, but also in higher
plants, although their photosynthetic apparatus
has here reached the acme of its development.
It is, however, only present in the chlorophyll-
bearing cells of higher plants; the metabolism of
all the rest of the tissues, which are colorless, is
based, like that of all other living things, on the
use of the organic substances supplied to them, in
this case, by the photosynthesizing organs.
Furthermore, even leaves revert to this form of
metabolism when they are without light.

Thus the metabolism of all higher plants is
entirely based on the heterotrophic mechanism
of assimilation of organic substances, but in the
green tissues this mechanism is accompanied by
a supplementary, specific superstructure which
has the task of supplying the whole organism
with ready-made organic substances. If these
substances reach the plant from outside in one
way or another, then it can exist even without its
photosynthetic superstructure, as may be observed under normal, natural conditions, especially in the germination of seeds. This may
also be demonstrated artificially in cultures of
vegetable tissues or in the growing of a complete
adult plant of sugar beet in the dark from roots
grown the previous year. In these cases either a
complete higher plant or its tissues live in the
total absence of any activity by their photosynthetic apparatus by the assimilation of exogenous
organic substances. If, however, one breaks even
one link in the chain of heterotrophic metabolism (by introducing a specific inhibitor, for
example) then all the vital activities of the plant
are brought to a standstill and it is destroyed.

Hence it is perfectly clear that the vital processes of the photoautotrophs are founded on the

original and ancient form of metabolism based
on the use of ready-made organic substances and
that they developed the power of photosynthesis
considerably later, as an accessory to their
earlier heterotrophic metabolic mechanism.[1]

You may also wish to refer to the work
on chemical evolution in Chapters 3
and 6.

A reading: concerning chemical evolution

You may want to introduce students to
the kind of speculation and experimentation under way in learning about the
primordial chemicals which became organized into "living" systems.

The following passage from a paper by
M. Calvin in *American Scientist* might be
used as a focus for class discussion; or a
committee of students may report on the
complete paper, much of which they will
understand. Or you may want to reproduce the passage and compose a series of
questions relating to it, as shown for the
selection by Darwin in the next section
and the one by van Helmont in Chapter 3.

The time element that is involved is a very
long one. By extrapolating the idea of evolution
to include nonliving systems as well as living
ones, we can go clear back to a time when the
universe and the stars were evolved and eventually an earth was formed. This time period
starts roughly about 5 to 10 billion years ago,
as far as the astrophysicists can tell us. Roughly
ten billion years ago the universe was formed
by an explosion of matter and the elements
were formed in an evolutionary pattern, a discussion of which would be beyond our present
scope, and which may more properly be called
"nuclear evolution." The next period that we
can characterize after the earth's formation is
the time for the formation of chemicals of various degrees of complexity upon the surface of the
earth, but before the appearance of systems that
we could call living—"chemical evolution." . . .

In order to begin the discussion of chemical
evolution, we have to decide what sort of an

---

[1] From A. Oparin, *Life: Its Nature, Origin, and Development*, trans. by A. Synge, Academic, New York, 1962,
pp. 100–02. Reprinted by permission of Oliver & Boyd
Ltd. and the author.

earth we had to work on. What sort of a chemical system did we have about 2½ billion years ago, when the earth first began to take its present form? I might point out that these various periods that I have tried to delineate are, of course, simply regions in time, and there is no sharp dividing line between them—they grade one into the other. One can say only that the earth gradually took shape, by some process—perhaps by aggregation, which is one of the modern cosmological theories. Regardless of what path its formation followed, we can say that at some period of time the earth had acquired very nearly its present form, and by this time chemical evolution was well under way—it had already begun. We should try to decide what sort of an earth, i.e., what sort of chemicals, we had to deal with, and what the earth was like at that time. Unfortunately, the geochemists can't agree on whether the atmosphere of the earth was an oxidized one or a reduced one, or some intermediate stage between.

For the present purpose it isn't necessary to know exactly what form the atmosphere of the earth had during that period. The reason is that there exist at least four different ways in which more or less complicated chemical compounds could have been formed in either condition — oxidized or reduced — although the reduced starting point seems the easier one to develop. These ways have been described as follows, and in this order. The first method by which larger molecules containing more than one carbon atom could have been formed from simple ones was suggested by J. B. S. Haldane about 1926, and has been experimentally checked. (We, among others, have checked it.) Under the influence of ultraviolet light from the sun, it is possible to make more or less complex substances, like the amino acids and heterocyclic compounds that are now found in biological materials, by simply illuminating aqueous solutions containing simple carbon compounds such as formic acid or formaldehyde (one-carbon compounds) and a nitrogen-containing material such as ammonia, nitric acid, or nitrates; and one can get fairly complex organic materials. A second possible method is the one that was suggested by the Russian biochemist Oparin. He had the idea that the earth cooled down from a hot miasma and that carbon was mostly in the form of metallic carbide which, upon being put in contact with water, formed acetylene; the acetylene under suitable catalytic influences such as rocks and minerals could polymerize and form large chains which could give rise to molecules of the type we now see in biological materials. The third way in which simple organic substances could have been formed in a world without life is by means of the action of very high-energy radiation, such as is produced by radioactive materials or comes to us from the stars in the form of cosmic rays. This we have also checked in an experimental way. We have taken solutions of carbon dioxide and water and irradiated them in the cyclotron and have gotten formic acid, and irradiation of formic acid produced oxalic acid (a two-carbon compound). My colleagues have irradiated a variety of other substances since then. They have irradiated two-carbon substances and obtained four-carbon compounds such as succinic acid, which are even now important metabolites in modern living organisms. . . . These three methods have permitted the building up of complex chains of atoms from simple ones, and this is essentially what we are trying to do: to devise ways and means of getting more complex substances from simple ones, without the intervention of living organisms, which today is the only way it occurs in nature, outside the laboratory.

The last method that has been suggested, and tested experimentally, is the one involving an electric discharge in the upper atmosphere, like a lightning discharge, when there are present methane, hydrogen, ammonia, and water, i.e., a reduced atmosphere. If an electric discharge is passed through such a mixture, a variety of compounds can be produced in which there are carbon atoms tied to each other, particularly compounds of the type of amino acids, which are the essential building blocks of proteins.[2]

Certain students, especially those with a high level of ability and an interest in science, may find a passage such as this one the springboard for an interest in learning more about the chemistry of living things. They may try to read with profit some of the books listed at the end of this chapter, for example, those by Simpson and Beck, Anfinsen, Huxley, Blum, Waddington (*The Nature of Life*), Lwoff, and Ehrensvärd.

Many of the methods used by scientists are included in Calvin's paper. There is a

[2] From M. Calvin, "Chemical Evolution and the Origin of Life," *Am. Scientist* 44:3, July 1956, pp. 248–51. Used with permission.

value, we believe, in bringing the reports of scientists who are actively engaged in research into the science classroom.

A reading: from *Origin of Species*

The following passage from Darwin's writings might be used to help explain his theory of natural selection. Students often enjoy this type of activity; they find that they gain skill in interpreting what they read.

The giraffe, by its lofty stature, much elongated neck, forelegs, head and tongue, has its whole frame beautifully adapted for browsing on the higher branches of trees. It can thus obtain food beyond the reach of the other *Ungulata* or hoofed animals inhabiting the same country; and this must be a great advantage to it during dearths. The Niata cattle in South America show us how a small difference in structure may make, during such periods, a great difference in preserving an animal's life. These cattle can browse as well as others on grass, but from the projection of the lower jaw they cannot, during the often recurrent droughts, browse on the twigs of trees, reeds, etc., to which food the common cattle and horses are then driven; so that at these times the Niatas perish if not fed by their owners . . . . Man has modified some of his animals, without necessarily having attended to special points of structure, by simply preserving and breeding from the fleetest individuals, as with the race horse and greyhound, or as with the gamecock, by breeding from the victorious birds. So under nature with the nascent [beginning] giraffe, the individuals which were the highest browsers and were able during dearths to reach even an inch or two above the others, will often have been preserved; for they will have roamed over the whole country in search of food. That the individuals of the same species often differ slightly in the relative lengths of all their parts may be seen in many works of natural history, in which careful measurements are given. These slight proportional differences, due to the laws of growth and variation, are not of the slightest use or importance to most species. But it will have been otherwise with the nascent giraffe, considering its probable habits of life; for those individuals which had some one part or several parts of their bodies rather more elongated than usual, would generally have survived.

These will have intercrossed and left offspring, either inheriting the same bodily peculiarities, or with a tendency to vary again in the same manner; while the individuals less favored in the same respects will have been the most liable to perish.[3]

Elicit from students a discussion of the main points developed by Darwin in his theory. You may want students to answer questions of the kinds that follow, or you may wish to compose your own. (If the material is mimeographed, students have time to study the passage and the related questions.)

A. Select the best answer.

1. According to Darwin, an advantage the giraffe has over other grazing animals is that
   a. it can eat more food.
   b. it can get along with less water.
   c. it can reach food on the higher branches of trees.
   d. it migrates from place to place.

2. According to Darwin, when food is plentiful the possession of a long neck is
   a. an advantage.
   b. a hazard.
   c. a nuisance.
   d. of no special advantage.

3. Darwin used the example of Niata cattle to show that
   a. changes in the environment have little effect on the welfare of animals.
   b. Niata cattle are better fitted than giraffes for survival.
   c. some animals have adaptations which fit them for survival.
   d. the environment stimulates animals to develop variations.

B. Indicate whether the following statements are true or false.

1. According to the passage, Darwin
   a. recognized that living things differ among themselves.

[3] From Charles Darwin, *On the Origin of Species by Means of Natural Selection, or the Preservation of Favoured Races in the Struggle for Life,* 1859.

b. thought that variations arose by mutation.

c. thought that breeders usually selected the fastest or the strongest animals for breeding purposes.

d. said that a small variation might spell the difference between survival or death for an animal.

e. stated that long-necked giraffes survived because they left more descendants than the short-necked giraffes did.

How fossils were formed

There are a number of materials and procedures to help students visualize fossils and their formation.

Plan field trips to fruitful areas—limestone, coal, shale, or sandstone deposits. Have students start a collection of fossil specimens (some may already have the beginnings of such a collection); small fossils may be purchased. You may want to obtain metal casts and ceramic models of prehistoric animals for display cabinets (Fig. 10-1).

Plan to show films and filmstrips on formation of fossils (see the listing at the end of this chapter), and 2 × 2 in. slides (available from supply houses) that trace the evolution of some kinds of horses, furnish reconstructions of prehistoric life in the seas, or illustrate the emergence of plants, insects, and amphibians on land.

Many students enjoy simulating formation of fossils and of sedimentary rock. Several procedures are given below. Organisms embedded in clear plastic may also serve as models of how fossils have been formed. Fine specimens are available from supply houses (Fig. 14-4). As an individual or group club project, students may learn the technique by which such specimens are made.

Simulating formation of fossils

"Fossils" in plaster. Make some imprints of leaves or shells in class to show how some fossils are formed. Students may embed leaves or shells in plaster of Paris. The plaster can be compared to the sediments that are deposited by a slow-moving stream or to a muddy embankment or swamp.

(a)

(b)

(c)

(d)

Fig. 10-1 Metal casts of some prehistoric animals: (a) *Dimetrodon;* (b) *Tyrannosaurus rex;* (c) *Stegosaurus;* (d) a plesiosaur. (Courtesy of the American Museum Supply Co.)

Prepare a mixture of plaster of Paris and water, the consistency of pancake dough; stir until smooth, and pour a layer 1 in. deep into a cake pan. Then coat a variety of leaves or shells with petroleum jelly and lay them on the plaster. Cover these with another layer of plaster of Paris. After it has hardened, students may remove the plaster from the pan and crack it open. They will find the casts or imprints made by the shells or leaves. The actual molds, especially of the shells, can be compared to what might be found when cracking open a fragment of sedimentary rock.

Students might make an imprint of the human hand, too. Coat the hand lightly with petroleum jelly and make an impression in soft plaster. Let the impression harden without adding another layer of plaster.

"Fossils" in ice. Students often have small figures of animals as toys or charms. Place these in water in the compartments of an ice tray in a refrigerator and let the water freeze. Each ice cube shows the "animal fossil" embedded in ice. Or use fresh pieces of fruit in the same way to show that no decay takes place when bacteria of decay do not have access to the material or when conditions are unfavorable for their growth.

"Fossils" in "amber." Melt a few pieces of resin over a low flame; avoid boiling. Then make small paper boxes of the type used to embed tissue in paraffin (see Fig. 2-31) about 1 in. square, stapled or clipped together. Now pour the molten resin into several paper boxes. Completely submerge an insect, preferably a hard-bodied one, such as a beetle, in the resin. If air bubbles appear in the resin, heat a dissecting needle and insert it into each of the air bubbles.

Let the resin harden at room temperature (rapid cooling causes cracking). After the resin has cooled, slough off the paper covering by soaking the box in water. The product is an insect "fossil" which simulates the manner in which ancient insects, mainly ants and small flies, became em-bedded in drops of resin which trickled down the bark of evergreens.

In fact, students may embed small insects in a drop of clear Karo syrup or in balsam on a glass slide to simulate the way insects were embedded.

"Fossils" in gelatin. Gelatin may be used as an embedding material instead of resin. Add ¼ cup of cold water to 1 tsp of gelatin. When the water has been absorbed, dissolve the mass in ¾ cup of hot water. Then pour this into paper boxes or Syracuse dishes. Embed small objects, such as fruits, in the gelatin before it hardens.

Simulating formation of sedimentary rock

You may also show how layers of sedimentary rock are formed. Have students partially fill large jars (pickle or mayonnaise jars) with fine sand, coarse gravel, small rocks, humus, and clay. If possible, try to include several different textures and colors. Include some shells or toy dinosaurs in this muddy mixture to simulate remains of organisms. Fill the container with water so that all this material represents a muddy stream. Shake the jar and then let it stand; note the size and kinds of materials which settle to the bottom first (Fig. 10-2). (Have students explain how shells might be found in sedimentary rock.)

Overproduction among living things

Collect the cones of pine trees, the pods of milkweeds, or the heads of sunflowers. On a trip around the school grounds have students count the number of seeds produced by several other plants. Count the number of seeds in each pine cone and estimate the number of cones on one tree. What would be the total number of seeds produced by one tree? What are the chances of survival for each seed? If students have estimated the number of seeds on a maple tree, have them return later to count the number of seedlings nearby.

Fig. 10-2 Sedimentation in a jar, showing the layers of different-sized particles that settle out of still water.

Why are there so few? Begin to relate these observations to the steps in Darwin's theory.

Students might count the number of seeds in one pod of a milkweed and calculate the total number of seeds one plant could produce. Similarly, how many seeds might one sunflower produce?

Examine the roe of one fish and the eggs of one frog; look up the number of eggs laid by one shrimp, one oyster, one insect, and so forth. What are the chances for each egg to be fertilized? To grow into an embryo? To reach adulthood?

What conditions operate against the development of a seedling or spore? The fertilization of an egg and its development into an embryo?

Overcrowding in seedlings

Students may study how space can be a limiting factor in the growth of seedlings. Soak overnight a large number of seeds, such as mustard, radish, oats, beans, or corn grains. Plant half of the seeds (of one kind) in a small flowerpot and the other half in a pot twice as large.

Keep the seeds in a dark place and water them regularly. Over the next 3 weeks, count the number of seedlings which grow and survive in each flowerpot.

Or you may want to try this variation. Plant soaked seeds in several flowerpots of the same size, but vary the number of seeds in each pot. For example, in one pot plant 10 seeds, in another plant 20 seeds, in a third plant 30, in another plant 50, and finally in another pot plant 100 seeds. Keep other factors (kind of soil, temperature, and water conditions) constant and compare the percentage of seeds that grow in each pot. Compare this with conditions of growth under a parent tree outdoors.

Overcrowding among fruit flies

Does overcrowding of organisms in one area affect their survival? Examine how the food supply can be a limiting factor in the survival of organisms.

Prepare fruit fly medium and pour into culture bottles (stock cultures, Chapter 8). This time pour the medium into several quart milk bottles and also into several small vials. Then introduce four pairs of fruit flies into each container. (Students should be reminded to sprinkle the surface with yeast before introducing the flies.) During a 3- to 4-week period, students may count the number of offspring surviving in each quart bottle and in each small vial.

A student may want to try this variation as an individual project. Prepare several ½-pt bottles, each containing food and four pairs of flies. In this way there is an increase in the total number of flies within a uniform environment. In about 2 months, when the population is at an ebb, count the total number of flies in all these bottles, distinguishing the dead ones from the living. The student should determine the total number of dead and living flies in all the bottles. Now he should transfer all of the living flies in each bottle into bottles of fresh culture medium. The total population expands each time a new food supply is added at the end of every reproductive cycle. Then counts might be made of the initial increase in population in each bottle with an increase in food. Again, after 2 months, when the food supply has reached a low point, count the number of living flies. Is there a point at which the food supply is not large enough to sustain a further increase in population? (Also see a population study, Fig. 9-5 and p. 513.)

Overcrowding among praying mantids

If egg masses of the praying mantis are available, they may be kept in a small terrarium. Provide a source of water (or sugar water), such as a small sponge inserted halfway into a vial of water. Students may examine and estimate the tremendous number of nymphs which hatch out of each egg mass. Each day watch the decrease in population as cannibalism increases. You may later provide fruit flies as food for the nymphs or release them in a garden or park, since they are beneficial insects.

Similarity and variation

Similarity in related plants

Students may observe first-hand the early embryology of some plants. In what ways are all monocots similar?

Monocots. Soak the seeds of monocotyledonous plants such as oats, rye, barley, bluegrass, and corn. Plant them in flats of moist sand or light soil and label the types of seeds. As the seedlings begin to grow observe the first leaves. Since there is a similar genetic pattern based on common ancestry these first leaves closely resemble each other.

Dicots. At the same time, show the similarity among the first leaves of dicotyledonous seedlings. For example, soak seeds of mustard, lettuce, pumpkin, marigold, melon, radish, pepper, tomato, pansy, and sweet peas. Also plant these seeds in flats of moistened sand and compare their first leaves. (Be sure to label the seeds, or students will not be able to recognize the plants.)

Gymnosperms. Scar or file seeds of pine, hemlock, spruce, arbor vitae, and fir; soak overnight and plant in moist, light soil.

They germinate slowly, but the results are rewarding. Compare the first leaves of the seedlings. Note their similarity. Students may watch the specific differences among the seedlings which develop later, showing that the seeds also contain different genes as well as the early similar patterns which first are expressed.

Mustard family. Students may also want to compare the seedlings of members of the same *family* of plants, for example, the mustard family. Plant soaked seeds of cabbage, brussels sprouts, kale, cauliflower, kohl-rabi, and collards. Label the seeds as they are planted. All of these forms are members of one family, descendants of the wild cabbage.

Variations in plants

Measuring seeds and leaves. In class, students might examine variations among seeds of one kind. For example, get several pounds of dried lima beans or other seeds; have each student measure the length of five seeds in millimeters. Arrange eight to ten large jars or test tubes of uniform size, and label the jars in millimeter intervals to correspond with the range of distribution you might find among the seeds. If students measure the lengths of the seeds and then place them in the approximately labeled jar or test tube (Fig. 10-3), a continuous normal curve of distribution re-

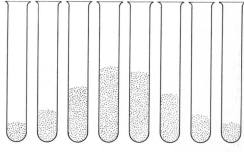

Fig. 10-3 Test tubes containing seeds that have been sorted according to length. Note that the seed levels approximate the normal distribution curve.

sults; a visualized histogram is produced. If these seeds are kept dry, they can be saved from year to year for this activity.

At other times you may want to distribute a few pounds of fresh green pea pods. Students might measure the length of each pod. (In addition they could compare the weight of each pod.) Find the average length of the pea pods and compare this average with the length of individual pods. Students might plot their data.

You may find sunflower seeds useful for studying other variations. Students might study the variations in stripings on the seeds. Measure seed clusters of linden (*Tilia*), the Chinese Tree of Heaven (*Ailanthus*), the tulip tree (*Liriodendron*), maples, the cones of evergreens, or the size of acorns for one species of oak.

Students might examine variations in the length of needles from one evergreen, or the diameter of leaves of broad-leaved plants. Collect a large number of needles from one kind of pine, or a large number of leaves from a single maple, oak, elm, or sassafras tree. With a centimeter ruler measure the width of the leaves from the one tree. Plot the frequency distribution of the sizes.

Percentage of germination of seeds. You may want to show some qualitative differences among seeds of the same species in the following way. Soak seeds overnight, and place them between pieces of wet blotting paper which have first been ruled off in squares. Place an equal number of seeds in each square, the number depending upon the size of the seeds. Or use flats for germinating seeds. Prepare several dishes or flats containing different kinds of seeds. Then keep a daily record of the rate of germination. Students may estimate the percentage of seeds which germinated in 24 hr, 36 hr, 48 hr, and so on. Also count the number of nonviable seeds (those which did not germinate).

Variations due to differences in environment. Members of the same species of plant may show variations due to the environmental factors under which the

organisms grow. Students may recall that development of chlorophyll in seedlings depends on light (pp. 427, 502). Seedlings grown in the dark are yellow and spindly (etiolated). Variegated coleus plants develop patterns of brilliant color in sunlight. What is the effect of sunlight on red corn?

On field trips students may find many of the following examples: (1) variations in color of leaves on barberry shrubs, dependent on differences in the amount of light which reaches the leaves; and (2) differences in the shape of leaves of many plants growing along a stream or lake. Many plants show variations depending on whether the leaves are growing in water or in air. Students may collect specimens of crowfoot (*Ranunculus*), several of which have filamentous leaves when growing in water. Also find some specimens which grow farther back from the water's edge, and examine their broad leaves. In beggar's-ticks (*Bidens Beckii*) the lower submerged leaves are deeply cleft, while farther up on the stem, well above the water level, the leaves are almost entire; whereas in the water parsnip the submerged leaves are feathery while the aerial leaves are almost entire. The submerged leaves of arrowhead (*Sagittaria*) are thin and linear, while the leaves above water are arrow-shaped. In addition many of the transitional stages can be found.

From these demonstrations students may come to recognize that changes in the environment affect the expression of the genetic pattern, thereby altering the appearance of plants. In other words, the appearance of an organism depends upon the combined effect of a specific heredity acting in a specific environment. (See also Chapters 8, 9, 13.)

In effect, an organism does not "become adapted" to an environment. Instead, it already *has* the special adaptation, pre-adaptation, or genetic variation. When the organism is in an environment where this variation is not a hindrance, the organism survives and reproduces more of its own kind. Thus the organism is said to be adapted to its environment. In fact, the variation may be beneficial and thereby have special survival value. Students might "explain" such examples of adaptations as (among many) cactus, sun-loving plants, shade-loving plants, and dandelions (see specialization of leaves, Chapter 1).

EFFECT OF A MOISTURE-LADEN ATMOSPHERE ON THE FORM OF GROWTH OF HERBACEOUS PLANTS. Dandelion plants, *Sempervivum,* and the like may be grown in a moist terrarium or under a bell jar in which the air has been saturated with moisture by placement of a wet sponge inside. In controls, develop dry conditions by placing the plants under a bell jar with an open glass container or calcium chloride (or concentrated sulfuric acid). Other plants may be maintained under conditions of usual humidity. The long internodes and broad leaf blades which develop under conditions of high moisture content should be apparent. In dry conditions students should find that the plants have short internodes and small leaf blades.

Among dandelion plants grown under normal conditions, leaves usually grow some 15 cm in length; in a moist atmosphere they may grow to 60 cm. Under moist conditions, *Sempervivum* loses its low growth form and becomes spindly with smaller leaves and a thinner cuticle.

EFFECT OF DARKNESS ON POTATO PLANTS. A group of students may try to reproduce the classic experiment performed by Vöchting in 1887, which Palladin[4] describes. Terminal buds of potato plants were enclosed in a darkened box supported by a ring stand; aerial tubers developed. Other plants were placed in darkened boxes with the upper branches exposed to light; here, potato tubers formed above the soil on the darkened stem portions.

You may also want to demonstrate the

Fig. 10-4 Effect of darkness on growth of potato plants. Etiolated plant (left) grown in dark, compared to normal plant of same age (right). (Photo by Dr. Carl L. Wilson, Dartmouth College.)

effect of darkness on the growth of potato plants (Fig. 10-4).

EFFECT OF MINERAL DEFICIENCIES ON PLANT GROWTH. To study the effect of mineral deficiencies on the growth and form of plants, students may grow the young shoots of one kind of plant in different solutions of chemicals (nutrient solutions, Chapter 9).

Similarity and diversity in animals

Comparative anatomy. In the laboratory, several vertebrates may be dissected in order to compare organ systems of the body. Among vertebrates, trace the evolution of the heart (Fig. 10-5), the brain (Fig. 10-6), or the respiratory or excretory systems.

[4] V. Palladin, *Plant Physiology,* Blakiston (McGraw-Hill), New York, 1926, p. 327.

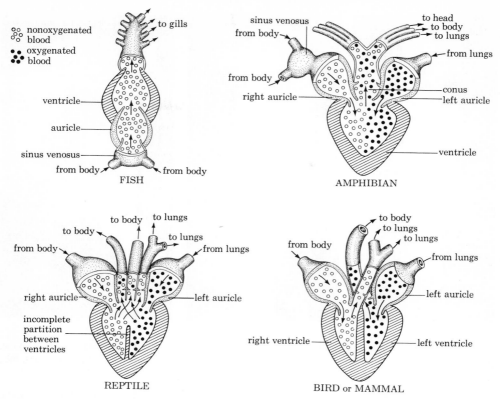

Fig. 10-5 Evolution of the vertebrate heart. (After R. W. Hegner and K. A. Stiles, *College Zoology*, 7th ed., Macmillan, New York, 1959.)

Similarly, make comparative studies of the evolution of conductive and supporting tissue in plants.

In both plants and animals, trace the evolution of sexuality from protista through the phyla to the flowering plants and, among the animals, to the chordates (also see Chapters 1, 2, 3, 7, 8).

Vestigial structures. Additional evidence of relationships through common ancestry is the possession of vestiges. Have students list the structures in their body for which there is no use. How many vestigial structures can they name? What is the genetic basis for vestigial organs? Name some in the horse, in the male frog, in birds, and in the whale (Fig. 10-7).

Comparative embryology. Refer to studies of developing eggs and embryos (Chapter 9). How can we explain the fact that eggs of animals go through similar cleavage stages, gastrulation, and differentiation? Further, students may examine preserved specimens of vertebrate embryos in jars that have the labels concealed. How many vertebrates can students recognize in the early stages (as in Fig. 10-8)? Why do closely related forms remain similar for the longest time? (Also refer to Chapters 8, 9, 11.)

Variations within a species

Variations among mollusks. It may be possible for students to collect, on a field trip, large numbers of mollusk shells of some one species, for example, scallop shells (*Pecten*), oyster shells (*Ostrea vulgaris*), or fresh-water mussels (*Anodonta mutabilis*). In class, students might measure the width

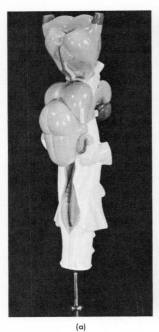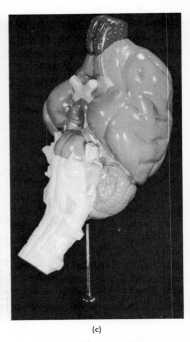

(a)   (b)   (c)

Fig. 10-6 Models showing evolution of the vertebrate brain: (a) shark (dogfish); (b) frog; (c) cat. (Models designed by Dr. Justus F. Mueller. Courtesy of Ward's Natural Science Establishment, Inc., Rochester, N.Y.)

in centimeters of the shells of one species. Then plot a graph of the frequency distribution, using the measurement as the abscissa and the frequency as the ordinate. Notice that while the basic pattern of genes is similar within a species, there are still some variations due to different recombinations of genes.

Variations among human beings. In class, students may want to study some differences among themselves. For instance, all the boys who are 15 years old might stand. The rest of the class could record the height and the weight of these boys. Then plot the data on a graph. Perhaps a small committee of students might gather the same facts—height and weight—for large numbers of 15-year-old boys in school. Then add these to the graph. Does a curve of normal distribution hold for human beings as well as for beans and shells? If we could plot a graph of the I.Q., or the grades in school, would they fall into a similar pattern when the assort-

ment was as random as in the study of the height of 15-year-old boys?

Variations among frogs. Moore[5] describes variations within a species; for example, he compares the characteristics of *Rana pipiens* from northern states with those of *Rana pipiens* from southern states. Students may also study variations within the population of one locality.

If possible, order *Rana pipiens* from different supply houses. Compare patterns of spots and coloring and anatomical details. Moore also compares adaptations of the globular shape of the egg masses of early spring egg-layers (in cold water), such as *Rana pipiens, R. palustris,* and *R. sylvatica,* with species such as *R. clamitans* and *R. catesbiana* that breed later (in warmer water) and lay eggs that float on the surface as a thin film. Attempt to duplicate and study the effects on eggs of a variation in oxygen supply by allowing some large

[5] J. Moore, "*Rana pipiens* and Problems of Evolution," *Ward's Natural Sci. Bull.* 31:3–7, Fall 1957.

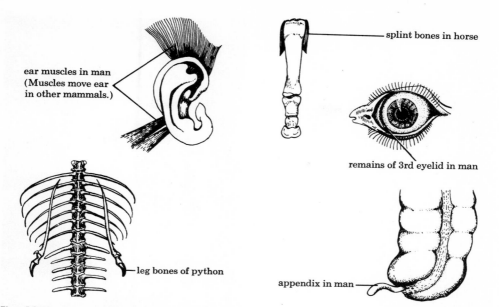

Labels in image:
- ear muscles in man (Muscles move ear in other mammals.)
- splint bones in horse
- remains of 3rd eyelid in man
- leg bones of python
- appendix in man

Fig. 10-7 Five examples of vestigial structures in animals. (Drawings by Marion A. Cox. From *Experiences in Biology*, new ed., by Evelyn Morholt, copyright, 1954, © 1960, by Harcourt, Brace & World, Inc., and reproduced by permission of the artist.)

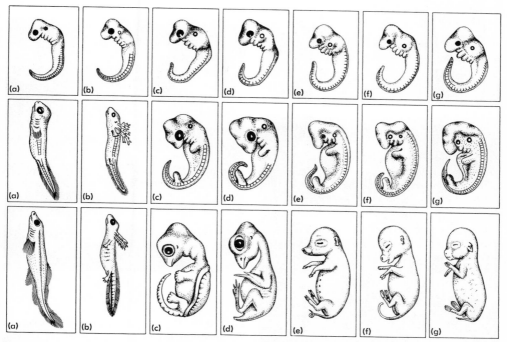

Fig. 10-8 Comparison of development of vertebrate embryos: (a) fish; (b) frog; (c) turtle; (d) bird; (e) pig; (f) sheep; (g) rabbit. Why are the embryos in the top row so similar? (After G. J. Romanes, *Darwin and After Darwin*, Open Court Publishing Co.)

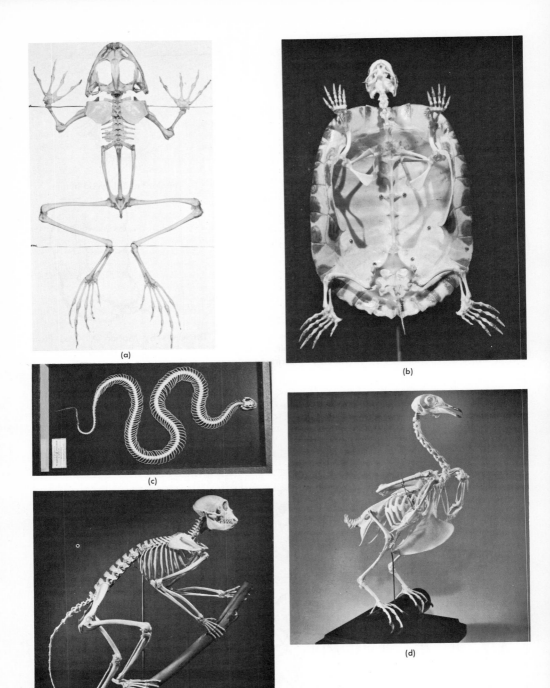

Fig. 10-9 Comparison of vertebrate skeletons: (a) bullfrog; (b) turtle; (c) snake; (d) pigeon; (e) monkey. (a, courtesy of Carolina Biological Supply Co., b–e, courtesy of Ward's Natural Science Establishment, Inc., Rochester, N.Y.)

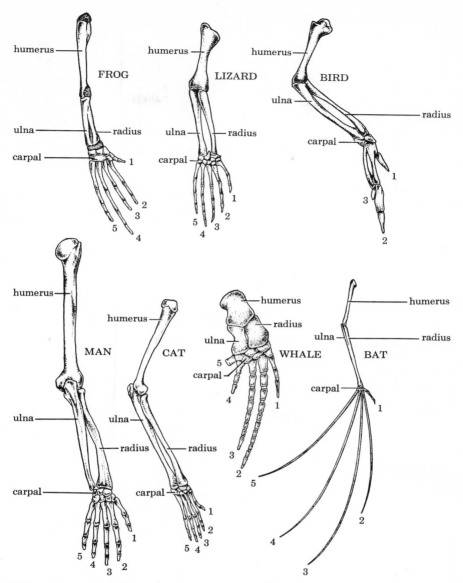

Fig. 10-10 Homologous structures: comparison of skeletal structure of forelimbs of vertebrates. (From C. A. Villee, *Biology,* 4th ed., W. B. Saunders Co., Philadelphia, 1962.)

egg masses to develop in battery jars; compare these with smaller masses of eggs left to develop in other containers.

An interesting sidelight to this study would be to compare the parasites of frogs from different regions. (See Figs. 11-9, 11-10, 11-11.)

Further evidences of relationships

Skeletons. If it is desirable, show students a series of vertebrate skeletons, such as those in Fig. 10-9, and elicit from the class possible explanations for similarities in bone structure. Also, what are the possible causes of variations?

What are possible explanations for the similarities in the forelimbs of vertebrates (Fig. 10-10)? Ask for explanations of the origin and survival value of the genes for these special adaptations.

Adaptation. How might such highly diversified fish as sturgeon or eel have developed (Fig. 10-11)? Have students prepare an exhibit case of other kinds of adaptations, such as mimicry as shown in the *Kallima* butterfly (Fig. 10-12). Or show monarch and viceroy butterflies or walking sticks, a few of the many cases of mimicry and camouflage. (*Note:* There is some doubt of the value of the mimicry of monarch and viceroy butterflies. Is there the possibility of an investigation here?)

Also use the example of gene frequencies in a population of moths in a wood containing pine and birch trees, as developed in BSCS materials. Refer to either *Biological Science: An Inquiry into Life* (BSCS Yellow Version, Harcourt, Brace & World, 1963) or *Biological Science: Molecules to Man* (BSCS Blue Version, Houghton Mifflin,

(a)   (b)

Fig. 10-12 *Kallima*, the dead-leaf butterfly, a classic example of adaptation of form and coloration: (a) upper surface of the wings, which is brightly colored; (b) underside of the wings, which has the shape and coloration of a dead leaf. (Courtesy of the American Museum of Natural History, New York.)

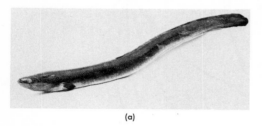

(a)

(b)

Fig. 10-11 Fish that show special adaptations: (a) common eel; (b) sturgeon. (Courtesy of Carolina Biological Supply Co.)

1963). Also read H. B. Kettlewell, "Insect Survival and Selection for Pattern" (*Science* 148:1290–96, 1965). Make use of the Hardy–Weinberg law (discussed later in this chapter), which gives the ratio of genotypes for a single gene difference as $p^2 : 2pq : q^2$, where p and q represent the frequency of the dominant and recessive alleles, respectively. Explain, in relation to the moths described, how both heredity and environment are responsible for the coloring of the moths. What is meant by a preadaptation?

A species' niche. Are there two species of birds living in the same area? Are there several species of frogs in a given pond? Or several species of insects in the same 5 sq ft of ground somewhere near the school? Or two species of *Paramecium* in the same culture? If both species can survive in the same setting, they must have different ecological niches. (Also refer to Darwin's finches, Fig. 10-16.)

How might out-of-doors investigations be planned for the study of frogs at a pond, of insects in a vacant lot, or of two species of *Paramecium?* Under what hypothesis are the students operating? What testable predictions can be made? What is the design of the experiment? (Be sure to refer to the studies of succession in Chapter 11, especially the use of microaquaria on a slide for studies of microorganisms; also see "Preludes to Inquiry" in the chapter, and in Chapter 3.)

Many studies of ecological relationships may be found in the laboratory manuals of the BSCS texts, and in E. Morholt, *Experiences in Biology,* new ed. (Harcourt, Brace & World, 1960). Be sure to examine the long-range investigations in *Research Problems in Biology: Investigations for Students,* Series One–Four (BSCS, Anchor Books, Doubleday, 1963, 1965).

Variations in structure and adaptation

What are possible explanations for diversified adaptations in beaks and in limbs of birds (Fig. 10-13)? For adaptive radiation (Fig. 10-14)? Students should also offer explanations for convergent evolution, as shown in Fig. 10-15.

Darwin's finches and adaptive radiation. Perhaps reprints of D. Lack's paper "Darwin's Finches" (*Sci. Am.,* April 1953) can be distributed to the class so that students may read and discuss this example of adaptive radiation. Or a student might report on the work as described in G. G. Simpson and W. S. Beck, *Life: An Introduction to Biology,* 2nd ed. (Harcourt, Brace & World, 1965) (Fig. 10-16), or in D. Lack, *Darwin's Finches* (Macmillan, 1947). Explain how finches varied. What are some possible explanations? What is meant by adaptive radiation?

Migration and isolating mechanisms. What factors affect migration of plants and animals? How may physical barriers, such as oceans, mountains, or a clearing, serve as isolating mechanisms that help to originate new species of plants or animals? (Examine the texts by Villee, Dobzhansky, and Simpson and Beck included in the end-of-chapter bibliography.)

What is the effect of introducing a species into a new area where it has no natural enemies?

Use data on the distribution of blood groups among humans of different racial

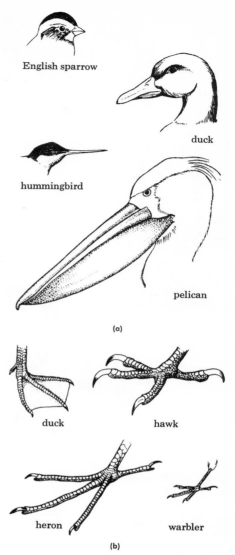

Fig. 10-13 Variation in beaks and claws of birds: (a) beaks; (b) claws. (Reproduced from *Experiences in Biology,* new ed., by Evelyn Morholt, © 1960, by Harcourt, Brace & World, Inc., with their permission.)

or national origin, as in Table 10-1, as the basis for a discussion of human migration and isolation. Refer also to relevant material in Chapter 8.

What is meant by a gene pool? What is genetic drift (see population genetics, discussed later in this chapter)?

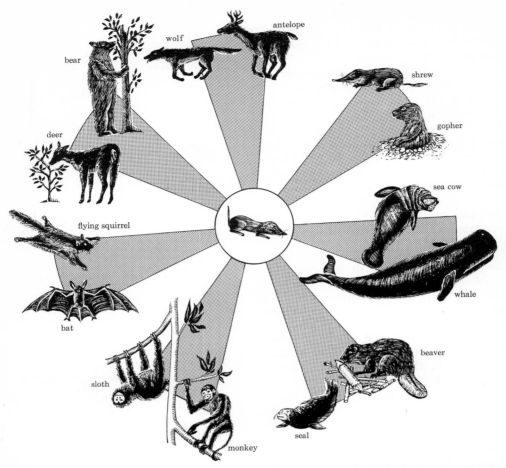

Fig. 10-14 An example of adaptive radiation among mammals. (After C. Villee, *Biology*, 3rd ed., Saunders, 1957.)

Classification of plants and animals

You may decide to bring to class a sampling of animals or plants. Better still, take your class outdoors to examine a biome, such as a pond, beach, desert, or wooded area. How do you help develop in your students the idea of order among the many kinds of plants and animals so that the students can identify them or classify them for reference? Some students may suggest grouping according to habitats, or habits of getting food. Certainly organisms differing as sharply as the animal inhabitants of a pond (fish, frogs, water striders,

Daphnia, protozoans, snails, planarians, water snakes, perhaps a beaver) cannot be placed or classified together. Students can readily name one outstanding trait that makes a fish different from a frog, a snake, a planarian, or other inhabitants of the pond.

Your students may compare the categories (kingdom, phylum, order, family, genus, species) with the categories used by a librarian to classify books. This is familiar ground. Here is a book entitled *The Fight to Live* by Raymond Ditmars. In what area of the library should it be placed? This book might be classified:

TABLE 10-1

Distribution of blood groups (percent)*

peoples	type O	type A	type B	type AB
A > 2B				
Swedes	34	51	10	5
English	46	43	7	2
Americans, white	46	41	8	4
Portuguese	38	52	6	3
Italians	36	51	9	4
Austrians	42	40	10	8
Greeks	38	42	16	4
Armenians	22	52	13	13
Philippine Negritos	48	33	14	4
A > B, <2B				
Russians	41	31	22	6
Poles	32	38	21	9
Uzbek Turks	29	34	27	10
Japanese	31	38	22	8
South Chinese	32	39	19	10
A < B, O < B				
North Chinese	31	25	34	10
Manchus	27	27	38	8
Hindus	30	24	37	8
Gypsies	34	21	39	6
Philippine Moros	25	18	45	12
A± = B±, O > B				
Sumatrans	44	23	29	4
Annamese	42	22	28	7
Congo Negroes	46	22	24	8
Senegal Negroes	43	23	29	5
Madagascans	45	26	24	4
O > A + B + AB				
American Indians, full blood	91	8	1	0
Filipinos	65	15	20	1
Melanesians	54	27	16	3
Australians, natives	57	38	3	1
Bechuana Negroes	53	19	24	4

* From A. L. Kroeber, *Anthropology: Biology and Race.* Copyright 1923, 1948 by Harcourt, Brace & World, Inc. Reprinted by permission of the publisher.

Area of library—Nonfiction
Section of Nonfiction Area—Science
Case in Science Section—Biology
Shelves in Biology Case—Zoology
Group on Zoology Shelves—Books
  About Animals
Although this is a simplification of the librarian's method, it illustrates the principle of sorting and classifying—putting together the things that are closely related.

Students may practice putting similar animals or plants into broad categories and then into smaller ones. For example, are all frogs in a pond the same species?

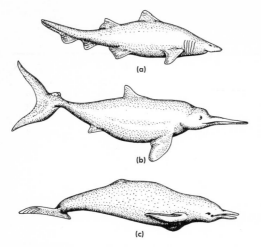

Fig. 10-15 An example of convergent evolution: (a) shark; (b) a fossil aquatic ichthyosaur (dinosaur); (c) dolphin (mammal). (After C. Villee, *Biology*, 3rd ed., Saunders, 1957.)

Using a key for identification and classification

When students' observation of fine details has been sharpened, you may want to take them outdoors to learn the nature of a key as a means for identifying or classifying a plant or an animal. Suppose we begin with plants, since they are available the year round.

An elementary approach might be used for a study of trees in winter. Have students examine trees and shrubs and select one *single*, obvious difference among them. Obviously, some trees have leaves and some do not. Begin to build a key by pairing off these evident single characteristics. Look more closely at the trees that do have leaves in the winter. Find one single characteristic that sets one tree or shrub apart from another. Perhaps one tree has clusters of needles, another has needles growing out in all directions from the stem, some have needles growing in one plant so

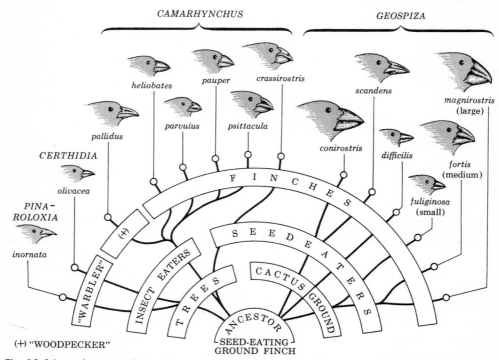

Fig. 10-16 Adaptive radiation among Darwin's finches on the Galápagos Islands. (After G. G. Simpson and W. S. Beck, *Life: An Introduction to Biology*, 2nd ed., Harcourt, Brace & World, 1965.)

that they look 2-ranked, another kind has especially small, fine needles (Fig. 10-17). Then have students examine one group more carefully, perhaps those with needles in clusters (the pines); what differences can be observed? Do some clusters have two needles? Others three or five needles in a bundle, or fascicle? Students can readily distinguish these differences within the genus: species differences (Fig. 10-18). Both Figs. 10-17 and 10-18 are highly simplified keys that only indicate one way that students may approach the problem of variations among living things and glean a basis for classification of organisms. Should you want to have students practice using a standard key, they may use such references as Graves' guide to trees and shrubs or an insect key such as Swain's. These and other guides are listed in the Bibliography.

A simplified key to Protozoa is given in Fig. 12-7 and to insects in Fig. 11-32; from these, students may recognize basic differences and begin to build their own simplified keys, which may then be compared with standardized keys in textbooks.

Population genetics

If dark hair is dominant over blonde, why doesn't blonde hair eventually disappear in a population? In random mating in a large population, the frequency of a gene remains constant; that is, the proportion of homozygous to heterozygous remains constant. However, in a small population that interbreeds, chance, rather than selection, may shift heterozygous gene pairs so that they become homozygous. If the homozygous gene pair is harmful, it may disappear from the small interbreeding population. This example of genetic drift is an exception to the principle developed by Hardy and Weinberg that holds for random mating.

Hardy–Weinberg law

If D and d make up a pair of alleles, the distribution of the genotypes in a large population is DD, Dd, dd. Hardy and Weinberg recognized this frequency as a quadratic equation:

$p = $ frequency of D gene

$q = $ frequency of d gene

$p + q = 1$ (since the allele of the pair will be either D or d)

Random mating in a large population (Table 10-2) results in the following algebraic sum:

$$p^2 (DD) + 2pq (Dd) + q^2 (dd)$$

This algebraic sum represents the frequency of the gene considered in any generation showing the stability of a pair of alleles in a population. (The assumption is that selection is not at work; see genetic drift, mentioned earlier.)

Villee[6] offers the following example. Consider albinism as a trait that occurs in about 1 in 20,000. Let a represent albinism and A represent normal pigmentation. Then the frequency of a is 1/20,000. Use a quadratic equation $p^2 + 2pq + q^2$. Since the square root of 20,000 is about 141, $q = 1/141$. Since $p + q = 1$, it follows that $p = 1 - q$, or $p = 1 - 1/141$. Then $p = 140/141$.

How many may be heterozygous for albinism (Aa)?

$$2pq = 2\frac{(140)}{(141)} \times \frac{(1)}{(141)}$$

Hence, Aa is about 1 in 70; that is, about 1 in 70 in the population is a carrier of albinism.

[6] C. Villee, *Biology*, 4th ed., Saunders, Philadelphia, 1962.

TABLE 10-2

Random mating in a large population according to the Hardy–Weinberg law

		sperms	
		p (D)	q (d)
eggs	p (D)	p^2 (DD)	pq (Dd)
	q (d)	pq (Dd)	q^2 (dd)

A. Leaves needle-like, mainly evergreen; seeds
   borne on scales of a cone: Gymnosperms

   B. Leaves shedding in fall
      C. Leaves needle-like or scale-like on
         slender twigs: *Taxodium* (bald cypress) ────────►

      CC. Leaves needle-like, many in
          cluster on short spur branches:
          Larix (larch) ────────►

   BB. Leaves needle-like, remaining on tree
       D. Leaves needle-like, in clusters
          of two to five with
          sheath at base: *Pinus* (pine) ────────►

                        See Fig. 10-18.

       DD. Needles borne singly, not in clusters

           E. Twigs roughened, cones hanging down

              F. Needles with short
                 leafstalks, in two rows

                 G. Needles stiff, sharp, extending
                    down twig: *Taxus* (yew) ────►

                 GG. Needles soft, blunt, not
                     extending down twig:
                     Tsuga (hemlock) ────────►

              FF. Needles without leafstalks,
                  4-angled, stiff, sharp, on all
                  sides of twig: *Picea* (spruce) ────►

           EE. Twigs smooth, cones upright:
               Abies (fir) ────────►

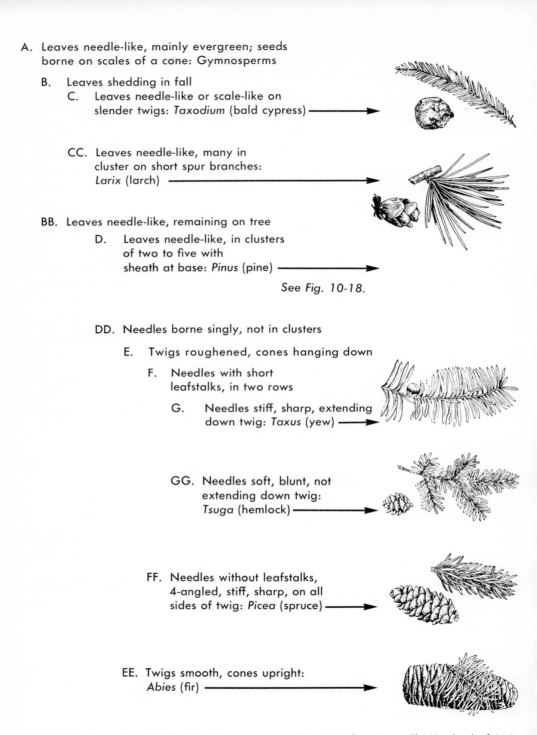

Fig. 10-17 Beginning classification of gymnosperms. (Drawings from *Trees, The Yearbook of Agri-culture,* U.S. Dept. of Agriculture, 1949.)

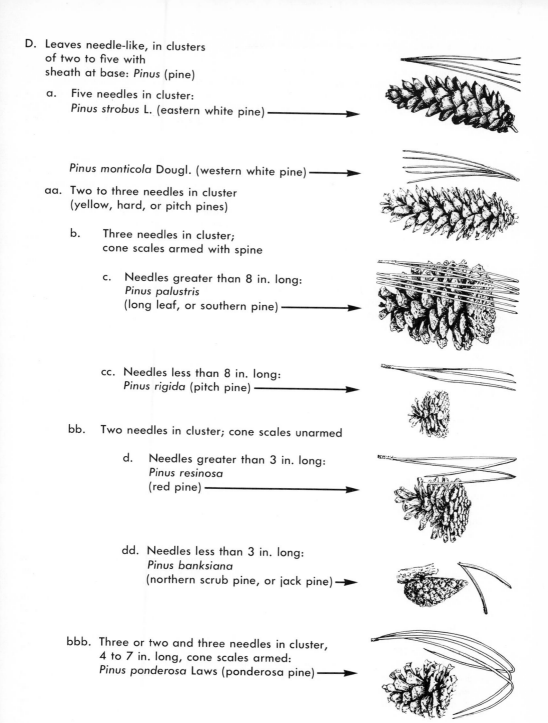

D. Leaves needle-like, in clusters
 of two to five with
 sheath at base: *Pinus* (pine)

    a. Five needles in cluster:
        Pinus strobus L. (eastern white pine) ⟶

        Pinus monticola Dougl. (western white pine) ⟶

aa. Two to three needles in cluster
    (yellow, hard, or pitch pines)

    b.   Three needles in cluster;
        cone scales armed with spine

        c.   Needles greater than 8 in. long:
            Pinus palustris
            (long leaf, or southern pine) ⟶

        cc. Needles less than 8 in. long:
            Pinus rigida (pitch pine) ⟶

    bb.   Two needles in cluster; cone scales unarmed

        d.   Needles greater than 3 in. long:
            Pinus resinosa
            (red pine) ⟶

        dd. Needles less than 3 in. long:
            Pinus banksiana
            (northern scrub pine, or jack pine) ⟶

    bbb. Three or two and three needles in cluster,
         4 to 7 in. long, cone scales armed:
         Pinus ponderosa Laws (ponderosa pine) ⟶

Fig. 10-18 Beginning classification of pines. (Drawings from *Trees, The Yearbook of Agriculture,*
U.S. Dept. of Agriculture, 1949.)

CAPSULE LESSONS: Ways to get started in class

10-1. You may want to have students demonstrate the way some fossils were formed. Use the techniques described in this chapter. Also make a sedimentation jar. Students may use plaster of Paris or ice as embedding materials. Establish the need for a quick burial, that is, removal from bacteria of decay, to inhibit the decay of the plant or animal. Discuss the other materials in which fossils have been found: ice, amber, tar, coal, and so forth. Plan a trip to a museum if possible.

10-2. Some students may be able to make clay models of dinosaurs (model-making, Chapter 14) or to develop a diorama. Use such devices as these to explore this question: How much of the reconstruction of prehistoric animals or plants is fact and how much is the imagination of the artist? Plan an exhibit of "Life One Million Years Ago." Or have students prepare a phylogenetic collection of plants and animals.

10-3. Plan a field trip to a museum to study fossil remains of ancient plants and animals. Or take your class through time by means of a series of Kodachrome slides, such as *Digging Up Dinosaurs, Changes in the Horse,* or *Ancient Man* (available from Am. Museum of Natural History, New York).

How has the horse changed through time? What were dinosaurs like? How did Neanderthal man differ from Cro-Magnon man?

10-4. In a laboratory lesson you may want to have students study variations found in the length of seeds in a few pounds of lima beans. Measure the length of seeds and graph the findings. Or give each student an envelope containing a pine cone full of seeds. If possible have the cones from a single tree. Again, have students measure the length of the seeds. Elicit the possible causes of the variations. Estimate the total number of seeds produced by a single tree. What are the chances that all the seeds from one tree will develop into new pine trees? Explain.

10-5. You may want to use a film as an introduction to the theories of evolution, such as the series on *Camouflage in Nature* (2 reels, Coronet). Or show one of the AIBS films (12 in the set) on *Life, Time, and Change* (McGraw-Hill). Develop a discussion around these questions: How did the different kinds of mimicry originate? How is it that an animal matches its environment in pattern, form, and color? What are some adaptations in plants?

10-6. How might Darwin have modified his theory if he had known of Mendel's work? This might be an approach to a review of the steps in Darwin's theory. You may want to discuss a reading from Darwin's *Origin of Species* to show that he was not certain as to how variations occurred. Compare this with the concept that inheritable changes in plants and animals are based on changes in genes.

10-7. To one committee of students give a variety of seeds of one family of plants, such as the mustard family; to another committee give seeds of cereal plants; to a third committee give seeds of different members of the legume family. Do this at least a week in advance of the classroom discussion. Then students may compare the shape, size, and time of appearance of the first leaves of the plants within each family. Have students notice the similarity. Develop the notion that these plants share a pool of genes from a common ancestor in that family group. Students may notice that variations exist in the rate of germination of seeds of the same kind. What is an advantage to the plant species of rapidly germinating seeds?

10-8. You might want to begin a discussion of theories relating to evolution. Although every female frog lays several hundred eggs each spring, the population of frogs in a certain pond remains relatively constant. Establish the facts about overproduction, the role of enemies, unfavorable environment, and so forth. What is meant by the "fittest" organisms?

10-9. At some time you may want to show the film *Heredity and Environment* (Coronet) as a review of the role of heredity in a new context, that is, in relation to changes in living things.

10-10. A committee of students might report to the class on evidence indicating relationships among plants or animals, for example, comparative embryology and vestigial remains. How have different species of oaks, maples, azaleas, and others developed?

10-11. You may want to have students plan short debates or a panel discussion on some of these topics: (1) the role of mutations and recombinations in evolution; (2) the role of barriers in the formation of new species; (3) genetic drift and its effect on populations of organisms; (4) the differences between natural selection and artificial selection (that is, man's use of selective breeding); and (5) the role of lethal mutations.

10-12. After a study of the classic theories of Lamarck, Darwin, and De Vries, you may want to have students apply these theories to an explanation of this adaptation: Assume that all bears were originally brown; how might the origin of white polar bears have been explained by Lamarck? By Darwin? By De Vries?

How might each of these men have accounted for the long neck of giraffes? How might they have explained camouflage in animals (walking sticks, moths the color of tree bark, green grasshoppers, and brown ones, too, among many other examples)?

10-13. You may want to plan a laboratory lesson with skeletons of several vertebrates (or study these in a visit to a museum). Students may compare the similarities and differences among the skeletons. How do we explain the similar pattern among the vertebrates? The differences? (See Fig. 10-9.)

10-14. It may be possible to plan a trip to a zoo to study primitive mammals such as the duck-billed platypus and the spiny anteater. And at a botanical garden students can examine an ancient gymnosperm, the last species of *Gingko*.

10-15. Show skulls of several mammals. Have students try to identify the animals. How are they adapted to their specific food habits? Elicit a discussion of the relationship between structure and function; between homologous and analogous structures.

Or you may wish to have students trace the evolution of the temporal and mandibular bones of the skull from fish to man (Fig. 10-19).

10-16. Students may want to trace human migration (Table 10-1). They may report on the evidences of relationship among groups of people based on their possession of certain blood types. Students will find material in W. Boyd, *Genetics and the Races of Man* (Little, Brown, 1950).

10-17. You may want to use the filmstrips based on the series of articles that appeared in *Life* magazine under the title "The World We Live In." Or a student may bring to class the book that is a compilation of those articles: editorial staff of *Life* and L. Barnett, *The World We Live In* (Time, 1955). Possibly pictures may be projected for the entire class to see.

10-18. Use the story of the moths as described in H. B. Kettlewell, "Insect Survival and Selection for Pattern" (*Science* 148:1290–96, 1965) or in the BSCS biology texts to develop a discussion of population genetics and the Hardy–Weinberg law. (See the *Student Laboratory Guide* to *Biological Science: An Inquiry into Life* [BSCS Yellow Version, Harcourt, Brace & World, 1963], Exercises 33–1, 33–2, 34–2, 36–1.)

10-19. Be sure to explore the suggestions for laboratory work developed in other chapters

EVOLUTION OF THE HUMAN SKULL
TEMPORO-MANDIBULAR BONES

This exhibit is designed to show ten structural stages in the evolution of the temporo-mandibular bones of the skull from fish to man, represented by the known fossil and recent specimens.

Fig. 10-19 Evolution of the temporal and mandibular bones of the skull from fish to man. Clockwise, beginning in the lower left-hand corner, the ten stages are as follows: Stage I, Fossil crossopterygian fish (*Rhizodopsis*), Devonian; Stage II, Embolomerous amphibian (*Eogyrinus*), Lower Carboniferous; Stage III, Cotylosaurian reptile (*Seymouria*), Permo-Carboniferous; Stage IV, Theromorph reptile (*Mycterosaurus*), Permo-Carboniferous; Stage V, Gorgonopsian reptile (*Scymnognathus*), Permian; Stage VI, Cynodont reptile (*Ictidopsis*), Triassic; Stage VII, Marsupial (Opossum), Recent (representing the primitive marsupials of the Upper Cretaceous); Stage VIII, Primitive primate (*Notharctus*), Eocene; Stage IX, Anthropoid (Chimpanzee), Recent; Stage X, Man, Recent. (Courtesy of the American Museum of Natural History, New York.)

(especially Chapter 11). Also refer to the splendid activities described in the laboratory manuals accompanying BSCS texts. Many suggestions are also provided in the BSCS laboratory blocks.

PEOPLE, PLACES, AND THINGS

Is there a museum of natural history near you? (Or is there one near enough to supply your school with traveling exhibits?) You may want to take your classes on a field trip through a section of the museum. In fact, you may be able to make arrangements to have your classes go behind the scenes; there scientists will explain how fossil bone preparations and other displays are made.

A member of a science department in another school or the teachers in a college biology department may help you by lending fresh or preserved materials for a study of interrelationships among groups of organisms. Perhaps you may be able to share films, color slides, and projection equipment.

Students may take frequent short field trips around the school grounds in search of the number of seedlings growing under a parent tree. An extension of a lesson might guide students to look, on their own, for variations due to environmental factors. (See Chapter 11 for ways to use a school lawn.) Students may collect some seeds and fruits; others may be purchased as you need them from a grocery store or seed distributor.

BOOKS

The study of the evolution of plants and animals in the interacting environments of the earth's crust is as broad as all of life. This following list of references is only a sampling; a more comprehensive bibliography is given in Appendix A.

Andrews, H. N., *Studies in Paleobotany,* Wiley, New York, 1961.

Anfinsen, C., *The Molecular Basis of Evolution,* Wiley, New York, 1959.

Blum, H., *Time's Arrow and Evolution,* 2nd ed., Harper Torchbook, Harper & Row, New York, 1962.

Bondi, H., *et al., Rival Theories of Cosmology: A Symposium and Discussion of Modern Theories of the Structure of the Universe,* Oxford Univ. Press, New York, 1960.

Colbert, E., *Evolution of the Vertebrates,* Wiley, New York, 1955.

Darlington, C. D., *Evolution of Genetic Systems,* rev. and enl. ed., Basic Books, New York, 1958.

Dobzhansky, T., *Mankind Evolving: The Evolution of the Human Species,* Yale Univ. Press, New Haven, Conn., 1962.

Dodson, E., *Evolution: Process and Product,* Reinhold, New York, 1960.

Dowdeswell, W. H., *The Mechanism of Evolution,* 2nd ed., Harper Torchbook, Harper & Row, New York, 1960.

Ehrensvärd, G., *Life: Origin and Development,* Univ. of Chicago Press, Chicago, 1962; Phoenix, Univ. of Chicago Press, Chicago, 1962.

Fisher, R. A., *The Genetical Theory of Natural Selection,* 2nd ed., Dover, New York, 1959.

Florkin, M., *Biochemical Evolution,* Academic, New York, 1949.

Goldring, W., *Handbook of Paleontology for Beginners and Amateurs,* Part I, *Fossils,* 3rd ptg., Paleontological Research Institution, Ithaca, N.Y., 1960.

Haber, F., *The Age of the World: Moses to Darwin,* Johns Hopkins Press, Baltimore, 1959.

Heusser, C., *Late-Pleistocene Environments of North Pacific North America,* Special Public. 35, Am. Geographical Soc. of New York, 1961.

Huxley, J., *Evolution in Action,* Harper & Row, New York, 1953.

Kroeber, A. L., *Anthropology: Biology and Race,* Harbinger, Harcourt, Brace & World, New York, 1963.

Life, editors of, and L. Barnett, eds., *The Wonders of Life on Earth,* Time, New York, 1960.

Lwoff, A., *Biological Order,* M.I.T. Press, Cambridge, Mass., 1962.

Mayr, E., *Animal Species and Evolution,* Harvard Univ. Press, Cambridge, Mass., 1963.

Moody, P., *Introduction to Evolution,* 2nd ed., Harper & Row, New York, 1962.

Nairn, A., ed., *Descriptive Palaeoclimatology,* Interscience (Wiley), New York, 1961.

Oparin, A., *Life: Its Nature, Origin, and Development,* trans. by A. Synge, Academic, New York, 1962.

Portmann, A., *Animal Camouflage,* Univ. of Michigan Press, Ann Arbor, 1959.

Romer, A., *The Vertebrate Story,* rev. ed. of *Man and the Vertebrates,* Univ. of Chicago Press, Chicago, 1959.

Simpson, G. G., and W. S. Beck, *Life: An Introduction to Biology,* 2nd ed., Harcourt, Brace & World, New York, 1965.

Thiel, R., *And There Was Light,* Knopf (Random), New York, 1957; Mentor, New Am. Library, New York, 1960.

Villee, C., *Biology,* 4th ed., Saunders, Philadelphia, 1962.

Waddington, C., *The Nature of Life,* Atheneum, New York, 1962.

————, *The Strategy of the Genes,* Macmillan, New York, 1957.

Wallace, B., and A. Srb, *Adaptation,* 2nd ed., Prentice-Hall, Englewood Cliffs, N.J., 1964.

FILMS, FILMSTRIPS, AND SLIDES

This listing is intended to be only a guide for the selection of new films, filmstrips, and transparencies. Send for catalogs from distributors of films; catalogs of biological supply houses list extensive offerings of 2 × 2 in. slides. The addresses of distributors are given in Appendix B.

In general, the cost of color films averages $150; black and white films cost about $75. Newer films are made in varying lengths—time sequences of 1 min, 5 min, 11 min, or 30 min; the cost of films therefore varies. We have indicated the lengths of films so that lessons can be planned with the films in mind.

Sound films are available in color or black and white; rental rates are available from film libraries or by direct inquiries to producers. (Filmstrips are not available on a rental basis; the purchase price is usually about $6 to $7 for those in color.)

Adaptations of Plants and Animals (14 min), Coronet.

Adaptive Radiation: The Mollusks (18 min), EBF.

Animals of the Ice Age (16 min), Northern.

Camouflage (10 min), Utica Duxbak; free.

Darwin's Finches (11 min), Film Assoc.

Darwin's World of Nature (2 filmstrips), *Life* Filmstrips.

The Earth in Evolution (10 min), AV-Ed.

Evolution (transparencies), Scientific Supplies.

Evolution of Vascular Plants: The Ferns (17 min), EBF.

The Fossil Story (19 min), Shell; free.

Fossils: Clues to Prehistoric Times (11 min), Coronet.

Genetics (set of 12 films, 28 min each), Part VI of *AIBS Film Series in Modern Biology,* McGraw-Hill.

How Living Things Change (11 min), Coronet.

Hunting Animals of the Past (22 min), Nebraska.

Journey into Time (14 min), Sterling.

Life, Time, and Change (set of 12 films, 28 min each), Part X of *AIBS Film Series in Modern Biology,* McGraw-Hill.

Mammals of the Rocky Mountains (11 min), Coronet.

Mammals of the Western Plains (11 min), Coronet.

Natural Selection (16 min), EBF.

Origin of Land Plants: Liverworts and Mosses (14 min), EBF.

Our Changing Earth (13 min), Film Assoc.

Prehistoric Animals (transparencies), Scientific Supplies.

Prehistoric Animals of the Tar Pits (21 min), Film Assoc. of California.

Prehistoric Times: The World Before Man (11 min), Coronet.

The Rival World (27 min), Shell; free.

Story in the Rocks (18 min), Shell; free.

The Story of Prehistoric Man (11 min), Coronet.

Underwater Reflections (15 min), Do All; fish-eye view of the struggle for existence.

The World Before Man (11 min), Coronet.

The World Is Born (20 min), Disney.

LOW-COST MATERIALS

This is only a partial listing of free and inexpensive materials available to the teacher. Many of these materials are distributed to teachers without charge. Suppliers are listed in Appendix D. While we recommend the material for use in the classroom, we do not necessarily sponsor the products advertised.

Adler, I., *How Life Began,* Signet, New Am. Library, New York.

Asimov, I., *The Wellsprings of Life,* Signet, New Am. Library, New York.

Classification of Animals (chart), Denoyer-Geppert.

Colbert, E., *Evolution of the Vertebrates,* Wiley, New York, 1955; Science Editions Paperbacks, Wiley, 1961.

Prehistoric Men, Chicago Natural History Museum.

Races of Mankind, Chicago Natural History Museum.

Romer, A., *The First Land Animals,* Am. Museum of Natural History.

Simpson, G. G., *The Meaning of Evolution,* Yale Univ. Press, New Haven, Conn., 1949; Yale Paperbounds, 1960.

Chapter 11

Web of Life:
Ecological Patterns

Throughout a study of biology we usually look at the individual organism—its mode of nutrition, pattern of regulation, continuity of special adaptations, and individuality through heredity.

Our purpose in this chapter is to examine *populations* of organisms in various ecological patterns. Two conceptual schemes guide our study:

1. The organism is the product of its heredity and environment.
2. Each organism is interdependent with other organisms and with its environment.

This chapter comprises two parts, each subdivided into the following major topics:

Part I
    A. Relation of organisms to their environment
    B. Interrelations among living things
    C. Preludes to inquiry: a study of a microecosystem (This focuses on the two major conceptual schemes given above.)

Part II
    A. Field work: exploring living things in their environment
    B. Collecting and identifying organisms
    C. Preserving animal specimens

PART I

Relation of organisms to their environment

Living things exist as populations: populations of deer, of *Daphnia,* of pond lilies, and so forth. All of the populations of plants and animals living in a given habitat (geographical area) make up a biotic community. Each population occupies its own niche; that is, it has a given role in a community.

But the interacting living populations (the biotic community) also interact with the nonliving environment, exchanging energy with it.

Furthermore, physical and chemical factors limit the range of populations, or bring adaptive mechanisms (which are genetic) into action.

Range of a species is influenced by physical factors in the environment, by limits of tolerance or ability to adapt to light, temperature, pressure, wind, gravity, texture of soil, water, water currents, ionizing radiation, and other physical conditions.

Some chemical factors that may limit the range of organisms are pH of water (or soil), salinity of water (or soil), amount of oxygen and carbon dioxide, and presence of nitrates, phosphates, and other chemicals.

A number of these physical and chemical factors are described in the following activities. Other interactions are discussed, by way of example, as they relate to one organism—*Chlorella* (see the "Preludes to Inquiry," or laboratory block, p. 542).

A third set of factors—biotic factors—is described in the next section, "Interrelations Among Living Things."

Soil

Living organisms in soil. What kind of plant and animal communities exist in soil? Have students collect 1-ft-block samples of soil from a garden, wood, meadow, vacant lot, or bank of a pond or lake. Pack each sample separately in plastic bags, labeling the source, moisture, and texture of the soil. Record the temperature and other factors that may be helpful later.[1]

Compare the texture and color of the top and the undersurface of the soil. Examine the soil for earthworms, sowbugs, slugs, insect larvae and pupae, millipedes, and other forms of life. Are there small snakes or buried eggs of small reptiles in it? With white enamel pans serving as backdrops, use a sieve to separate specimens from each sample. To separate out smaller specimens use a Berlese funnel (Fig. 11-1), which may be purchased or made by a student. Collect each type of organism in separate small containers (for example, small baby-food jars) so that a census may be taken after the organisms have been identified. Learn to use an identification key such as those in this chapter. Also refer to the survey of the plant and animal kingdom (pp. 549–52).

Transfer small portions of the soil samples to dishes containing a little water, and set them aside. After a week or 10 days, examine the soil, and record the plant forms that grow: seedlings, mosses, and fungi (including molds).

Also locate small worms, insects, and

[1] Also refer to D. Pramer, *Life in the Soil* (BSCS laboratory block), Heath, Boston, 1965. Included are studies for high school students of the microfauna and microflora of the soil and of activities and interrelations of soil organisms.

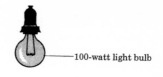

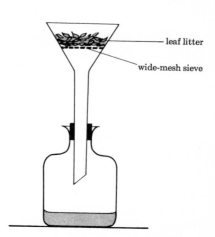

100-watt light bulb

leaf litter

wide-mesh sieve

Fig. 11-1 Berlese funnel for collecting small organisms in leaf litter.

larvae of some insects, and place them in finger bowls covered with aluminum foil. Examine the finger bowls daily with a hand lens or binocular microscope.

Also mount a bit of soil in a drop of water on a glass slide. Under the high-power objective of the microscope look for small worms, fungi, large bacteria, and protozoa. You may want to have students stain slides with methylene blue dye (p. 121) in an attempt to locate some of the larger bacteria.

Also streak a sterile transfer needle dipped into a sample of soil across the surface of nutrient agar in a Petri dish. Incubate, and examine after a few days.

List all the kinds of organisms found in soil. What is their role in increasing the fertility of soil? What is the pH of the soil (p. 522)?

Effect of nitrates on growing plants. Students probably know that plants need certain substances for proper growth (mineral cycle). It may be desirable to show the effect of mineral deficiencies on growing plants. Have students place seedlings in jars of nutrient solutions (pp. 458–62), each of which lacks one mineral substance —nitrate, phosphate, or sulfate. Compare the growth of seedlings in these deficient solutions with seedlings grown in complete nutrient solutions.

If a patch of sandy soil is available, an outdoor field experiment may be devised. Locate an area where the grass seems pale (due perhaps to a deficiency of nitrates in the soil). Mark off a section several square feet in area, and evenly scatter over it about 10 to 30 g of sodium nitrate. (Emerson and Shields[2] recommend about 30 g for a 10-sq-ft area.) Then water this area immediately so that the nitrates will be absorbed into the soil. During the next few weeks look for differences in color of the grass in the treated area and the surrounding, untreated area. Students should find that plants grow faster and that grass is greener than in the untreated control area.

You may want to relate this activity to crop rotation and the role of nitrogen-fixing bacteria in the soil and in the nodules of leguminous plants (Fig. 11-17; p. 538).

Binding force of roots. You can show in class that roots hold soil and thereby reduce erosion. Germinate some fast-growing seeds such as radish, oats, or mustard seeds in paper cups of moistened sand. Soak the seeds first to speed their germination. Let the seedlings grow for 2 weeks but water them sparingly. (Root systems will be more extensive when the plants have not been well watered.) With a firm tug on the shoots try to remove from the paper cups the entire sand mass bound by the roots. Students will find the sand particles held together by the root system so that the sand takes the shape of the container. Each student can germinate seeds to show this binding force. There may be time to study root hairs with a hand lens (p. 232).

[2] F. Emerson and L. Shields, *Laboratory and Field Exercises in Botany*, Blakiston (McGraw-Hill), New York, 1949, pp. 89–92.

Why are cover plants grown on a slope? Why is soil eroded quickly on a denuded hillside (Fig. 11-2)?

Water-holding ability of soils. Students may prepare three funnels or glass chimneys, with gauze taped or tied to the bottom of the mouth. Pack one with sand, another with clay, and the third with soil rich in humus. Then pour equal amounts of water into each funnel and watch the water run through the soil. If graduated cylinders are placed beneath the funnels, students can readily measure the amount of "runoff" water. Which soil permits water to run through most quickly? Which soil retains the water?

Capillarity of soils. The texture of the soil is a factor in determining how much water clings to particles of soil and is thereby readily accessible to the roots of plants.

Use the same materials as above but instead of pouring water into the funnels, put an equal amount of water into each of the graduated cylinders. Be sure that the stem of the funnel reaches the water level in each case.

Watch the rate of water absorption in each preparation. Through which kind of soil does water rise readily?

Soil for plants. Students may prepare several flowerpots in the following way. In some place sand, in others a clay soil, and in a third group a rich loam (clay, sand, and humus). Sow soaked seeds of fast-growing plants, such as mustard, oats, or radish, in each flowerpot. Compare the growth of seedlings in each kind of soil. In what ways do the texture and composition of soils affect plant growth? Which type of soil holds water best? In which soil does water move readily by capillarity?

Testing the acidity of soils. You may want to have students test the acidity of small samples of soil from different places, such as a weedy region, swamp area, beach, woods, burned-over area, well-tended garden, along a fence row, and also along a whitewashed wall. Cut a section of soil with a spade so that a slice about 5 in. deep is transported to class. These sections of soil might be wrapped carefully in newspaper and a plastic bag or transported in a pan.

Use litmus paper as an indicator. Acid turns litmus from blue to red; alkaline turns red litmus blue. Each student can test several samples of soil. Lay moistened strips of litmus paper, red and blue, side by side in several Syracuse dishes or on glass slides. Then put ½ tsp or less of a sample of soil on the strips of litmus paper. A drop of water may be added to moisten the soil. Then turn the glass dish or slide over and inspect the litmus paper. Use samples of soil from different levels of the section of soil and repeat the procedure. Are the top and bottom layers of the same pH? (See p. 662 for a discussion of pH.)

Desert soils are usually alkaline; grasslands of the central United States are approximately neutral; and the forest regions of the eastern section of the country tend to be acidic. Soils that hold much rainwater tend to have sodium, calcium, and magnesium leached away and become acidic.[3]

One student might demonstrate the effect of adding a bit of lime to a sample of acid soil. Repeat the litmus test. (In fact, soil may be prepared for demonstration. Add a bit of vinegar or other acid to a sample of soil when you want to show the test for an acid soil. Add lime—CaO—when you want alkaline soil samples.)

Committees of students might use different indicators (p. 662) as a check on their observations with litmus paper. When a liquid indicator such as brom thymol blue is used, a bit of soil to be tested is placed in a small evaporating dish or against a white background of some sort. Then the indicator is added and the dish is tilted to examine the color of the indicator as it flows from the moistened soil.

Several other indicators may be used for testing soil. For example, for testing the different pH ranges you can use brom

cresol green with a pH range of 3.6 to 5.2, brom cresol purple with a range of 5.2 to 6.8, brom thymol blue with a range of 6.0 to 7.6, and phenol red with a range of 6.8 to 8.4 (see Fig. 14-1).

For testing, select five or six thin layers of soil 6 to 8 in. from the top surface. Sprinkle the sample with rainwater (1:4) and let stand. Then add 1 drop of an indicator solution to a white dish and add 5 drops of the clear soil extract. Lapp and Wherry[4] suggest using only one indicator, brom cresol purple, because it has a wide range with sharp color shifts: dark purple at pH 7.5, reddish brown at 6, yellow at 5. Prepare the indicator by dissolving 0.2 g of brom cresol purple indicator powder in 50 ml of methyl or ethyl alcohol. Dilute with water to a total volume of 500 ml. Add a few drops of limewater until a red color appears. To test a soil sample, add 1 drop of indicator to 5 drops of the clear soil extract.

Students may recognize that the pH of the soil, one factor among many, affects the type of plant community existing in different soil. In fact, an experienced observer can estimate approximately the pH of the soil from the kinds of plants growing in a given region. Table 11-1 shows some of the plants that thrive in soil of three different pH ranges.

[4] W. Lapp and E. Wherry, "pH Preferences of Common Plants," *Sci. Teacher* 18:121–26, 1951.

TABLE 11-1
Acidity of soil in which certain plants thrive

pH 6–8 (neutral)	pH 5–7 (slightly acid)	pH 4–5 (very acid)
ailanthus	bayberry	azalea
ash	begonia	sweet pepper
barberry	bittersweet	bush
cherry	candytuft	maple-leaved
elm	chokeberry	viburnum
honeysuckle	chrysanthemum	stagger bush
locust	coleus	hydrangea
pear	goldenrod	lady slipper
sugar maple	mountain laurel	sheep laurel
	privet	many ferns
	wintergreen	

[3] Refer to U.S. Dept. of Agriculture, *Yearbook of Agriculture, 1957: Soil*, U.S. Dept. of Agriculture, Washington, D.C., 1957, pp. 67–71, "pH, Soil Acidity, and Plant Growth."

In what way do the kinds of plant communities change when the minerals are constantly leached out of the soil? A student looking for an interesting project may want to read further in a text on soilless growth of plants to learn how the ability of plants to absorb minerals is affected by the pH of the soil.

Profile of soil. How does the top layer of soil differ from subsoil and bedrock? If possible, have students examine an excavation under way, or a hillside cut through so that more than 5 ft of soil in profile is exposed. If you live in a desert area, compare the difference between the top layer of sand and the more granular, rocky layers several feet below the surface. What factors contribute to the formation of fine-textured sand?

In a study of so-called average soil, students may recognize the comparatively thin layer of dark topsoil. Under this layer, about 1½ to 2 ft down, is a layer of subsoil. This rests on a deep (below 5 ft) layer of bedrock. An examination of the content of the soil will reveal that living organisms are found only in topsoil (p. 521).

How does the texture of the usual topsoil differ from subsoil? Which kind of soil holds water best? You might want a committee of students to report on the way that bedrock, like granite and basalt, was formed. What factors change rock into soil? For example, what effect does each of the following conditions have on the breaking down of rock into soil: weathering (such as alternating heat and cold), winds, water in motion, plant roots, and ground water?

When they study fragmentation and exfoliation of rock, students can demonstrate some factors responsible for the breaking up of rock masses. Pour a dilute acid, such as vinegar, on limestone. This reaction might simulate the action of carbonic acid (carbon dioxide) dissolving in ground water. Have students bring to class bits of rock on which mosses or lichens are growing. Notice how readily the rock crumbles? Can students find examples to show how the outer layers of rock "peel off" after they have been subjected to alternating contraction and expansion due to cold and heat? If a rock were cracked, what effect would freezing water that is trapped in the cracks have on the rock?

Conversely, how does soil form sedimentary rock (see Fig. 10-2)?

How leaves add to the soil. Have students ever tested the sponginess of soil, particularly in a wooded region? Perhaps such a region is near enough for a short field trip. Examine the cushion-like layers of leaves on the ground. Notice that the top layers of these leaves can be identified. Lift up layers of leaves. Notice how only veins of the underlying leaves are apparent. In layers below this, only fragments of leaves can be found. Now squeeze a mass of this material in your hands. Do you feel the moisture? Students may have had similar experiences in observing the changes in a compost heap (made of grass cuttings and leaves) around their own homes.

Which organisms in the soil aid in the rapid change of dead leaves into humus, giving this absorbent quality to the topsoil? How would the nitrogen cycle (as well as cycles of other minerals) be affected if bacteria of decay and soil fungi disappeared?

Erosion and the slope of the land. Does soil erode faster when land is sloped or flat? Do cover crops protect the soil on a hillside and prevent the transportation of this soil distances away? Where it has been possible to plan a section of the school grounds for outdoor demonstrations and field trips (see pp. 545–48), a committee of students may undertake this demonstration in a natural setting using flat and sloped land, land covered with vegetation, and bare soil.

Or it may be necessary to construct boxes (at least three) about 1 ft deep and 3 ft square and filled to the top with a loam soil. It would be best to set this demonstration outdoors. The committee will need to plan this work well in advance, for one box needs a thick growth of

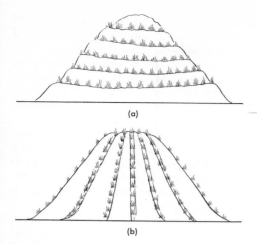

(a)

(b)

Fig. 11-2 Models of two ways to plow a hillside: (a) contour plowing; (b) plowing that increases erosion of the soil.

a cover crop such as grass, oats, barley, rye, or some other fast-growing crop.

In the demonstration students should let one box of loam lie flat; the other box of loam and the third box, containing a cover crop, should be tilted. In short, one box represents flat land, and two boxes show sloped land (one with vegetation). What will happen to the soil in each box in a rainstorm?

Have students direct a gentle spray of water from a garden hose on the three boxes of loam. Are there differences in the amounts of soil carried away from the flat box as the water overflows it and from the

tilted box? What is the effect of cover crops? Where these conditions occur naturally around school, students may visit these areas after a heavy rainstorm. In fact, they may want to take such measures as planting cover crops or terracing, and to keep a record of results. Do their corrective measures reduce soil erosion?

CONTOUR PLOWING. Another committee of students may show the effect of poor farming methods on promoting erosion of soil. For instance, they may build several mounds of topsoil and pack these firmly. In some mounds make furrows that run vertically downward from the top of the mounds like spokes of a wheel. In other mounds, make furrows that circle around the mounds (Fig. 11-2). Students might plant "crops" in the furrows and then pour equal quantities of water over each mound. How does the amount of runoff water (with soil) compare in the two "plowing" techniques?

USING THE BLACKBOARD TO SHOW TERRACING. Try this simple but effective way to show how soil absorbs water and helps to prevent floods. Dash some water on the blackboard and, as the small rivulets of water flow "downstream" along the board, dam these streams by pasting strips of wet toweling across them. Notice how the flow of water is stopped or slowed down (Fig. 11-3). This demonstration may represent terracing or strip cropping or the effect of a dam—whatever you wish.

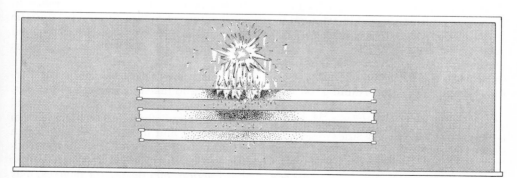

Fig. 11-3 Using the blackboard to demonstrate terracing. Water splashed on the blackboard runs down quickly until it meets the strips of paper toweling; each strip absorbs some water, breaks its stream, and slows it down.

STAKING OUT A CONTOUR LINE. If your school is in a farming community, a group of students may want to learn simple ways to lay out a contour line for contour plowing or strip cropping. An easy method is described in *Conservation: A Handbook for Teachers.*[5] In this method a carpenter's level is used by a "sighter"; a companion, the "lineman," marks the line. The sighter locates the level line along a slope or hillside where both he and the lineman are in position.

RAINDROP EROSION. What kind of soil is eroded easily by the force of raindrops? Students may set out discs of metal or wood on bare soil. After a heavy rainstorm they will probably find that the soil under the discs is at a slightly higher level than in the region of exposed soil (where some soil was carried away).

You may want to use a method described in *Conservation: A Handbook for Teachers* (mentioned above) in which the amount of soil splashed up on stakes covered with white paper is measured. In soil covered with vegetation there should be less splashed or transported soil than on uncovered soil. During what seasons of the year is it likely that more soil is transported from cultivated regions?

HOW MUDDY IS THE RUNOFF WATER? Another way to gain an understanding of how much soil is transported or eroded from any one place is to measure the amount of sediment in the runoff water. Water carries many minerals which have been leached from the soil in solution, but much of the soil is transported as a suspension.

After a rainstorm it may be possible for students to collect in individual large bottles samples of water running off from

different regions. Allow the suspended particles to settle, and note the amount of sediment and the size of the particles carried by runoff water from soil where corn is cultivated. Where wheat is growing. Where there is no cover crop. Where there is a gully. Where land is plowed following its contour. Along a roadbank.

What is meant by sheet erosion? Explain how silt is deposited at the foot of a slope. What is the value of cover crops in conserving soil? Explain how the continued existence of wells (and springs) is dependent upon cover crops.

STUDYING THE TREES IN YOUR COMMUNITY. Students have learned that trees and cover crops help prevent floods and soil erosion. Does your community plant trees or conserve them in other ways?

On individual field trips, as well as a group trip, students can make a count of trees, a tree census. What kinds of trees are there and how tall are they? Have youngsters identify the kinds of trees and shrubs in the neighborhood. Which kinds of trees are used for special purposes, for example, for attracting birds to the gardens or fields, as windbreaks for cultivated land, as pioneers in moving into a new area?

Students might list many ways in which trees aid in conserving the resources of the region in which they live.

Is the temperature of soil uniform? Certain plants and animals exist in sunny areas, while others seem to grow only in shaded regions. Is there a difference in temperature of soils in the shaded and the sunny regions which might affect the kind of community of living things that exists there?

Students may devise ways to use a thermometer and take readings of soil samples. For instance, check the readings of several thermometers beforehand by placing them in ice water and in boiling water. Keep a record of any degree of error in the readings among the different thermometers so that fractions of a degree (or degrees) may be added or subtracted from the readings students take. Now lay one thermometer so that the bulb touches soil in a shaded

---

[5] *Conservation: A Handbook for Teachers,* Vol. 45, No. 1 (September) of *Cornell Rural School Leaflets.* This booklet is available from the New York State College of Agriculture, Cornell Univ., Ithaca, N.Y. In it are references to many other useful *Cornell Rural School Leaflets.* The devices used in the demonstrations described in these leaflets are simple and effective: showing water penetration with the use of tin cans, making a wind card and a rainfall gauge, measuring moisture in the air with old photographic film which curls when dry, and so forth.

area, and a second on soil in a sunny region. Students may want to compare the temperatures of light-colored and dark-colored soil. Which surface warms up faster? Similarly, is there a difference between the temperature of soil on the surface layer and the soil several feet down? Tie a string on a thermometer and lower it into a hole which students have made so that a reading is taken of the soil several feet down. How does this compare with the temperature at the surface?

This demonstration may be done at any time in the year. For example, students might take readings of soil temperatures under a blanket of snow and on the surface of the soil.

Water

Well water. In a community where wells are common, students can trace the origin of well water back to precipitation, then to ground water. What is the relationship of different types of rock formation to well-drilling? What cautions should be observed in deciding on the location of a well? How can water spoilage be avoided?

What causes water to rise in a well? Insert a straw into a glass of water. Let the water level represent the water table in the soil. When air is sucked out of the straw, what causes water to rise in the straw? How deep can wells be drilled? What is the difference between a simple well and an artesian well? What is the community doing to prevent water waste and spoilage or pollution of water in streams, wells, and reservoirs?

Study of watersheds. From where does your water supply come? Maps of the watersheds that supply your community may be available from your state departments of agriculture (or geology). Students may bring maps of this area to class. Then trace the path of water from the watershed to the individual home. For instance, how is water made fit to drink? What are the requirements of a good watershed region? What precautions are in practice to con-

trol forest fires in this region? To control insect pests which damage forests? To prevent stream pollution?

Besides reading the excellent booklets published by state departments, students should learn to find information in the Yearbooks of the U.S. Dept. of Agriculture. The 1955 Yearbook, *Water,* is a mine of valuable information for students in all sections of the country.

Tracing the water cycle. You may want to have students demonstrate principles involved in evaporation and condensation; then their observations may be extended to an interpretation of the water cycle over the earth.

A student might streak the blackboard with cotton soaked in water. Where does the water go when the board becomes dry? Set up a simple demonstration as in Fig. 11-4. Place muddy water (or water colored with a dye) in the flask; boil this and collect water vapor in a cold container so that students can see drops of clear water condensing in the container.

Students can relate this work to changes in water in a water cycle: evaporation, condensation of water on dust particles, and cloud formation followed by precipitation. (A student might report on seeding clouds with iodide salts or dry ice). An extension of this work may develop into a discussion of ways to make muddy water (from streams in a watershed) fit for drinking. You may want to have students demonstrate settling (Fig. 10-2) and the addition of alum to muddy water. Discussion

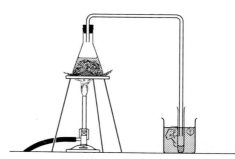

Fig. 11-4   Setup to demonstrate the water cycle: evaporation and condensation.

of filtration, aeration, the need for bacterial counts, and the addition of chemicals to purify water may develop from this demonstration.

Action of ground water. After precipitation, some water may run off into streams, but much of it soaks into the ground; the quantity depends upon the kind of soil (p. 522). What is the action of this ground water? Students may demonstrate how ground water in which carbon dioxide (carbonic acid) is dissolved helps to break down rock into soils. Have them add dilute hydrochloric or even acetic acid to limestone or marble chips (calcium carbonate). Notice the destructive action of the acid on the bits of "rock." This destructive action goes on in limestone beds where ground water percolates between layers of rock, forming caverns and sinkholes. Students will also recognize that the formation of stalactites and stalagmites in limestone caves illustrates the constructive action of ground water.

Interested students may be referred to texts in metallurgy and geology for explanations of the ways ores and precious stones are formed in rocks. A student may demonstrate tests for different ores. How are copper ores or iron ores identified? How does ground water supply artesian wells? Under what conditions does ground water form springs? What relation do these studies have to conservation?

Interrelations among living things

Substances are taken by organisms from the nonliving environment and, in turn, substances are returned from living things to the earth or air. As the interaction between organisms and their environment shifts, ecological successions occur. Thus, as the forest floor becomes progressively more shaded due to the increase in foliage above, a succession of invading seedlings may take hold, or changes in pH of soil may affect ability of some seedlings to grow, and so forth. Some shifts may be periodical, such as the temporary rain pools or pools due to melting snow. In addition, there may be a stratification from water to swamp to land, dominance of one species, climax forms, and so on.

A simple activity can be used to focus attention on these interchanges. Prepare a sealed test tube containing only pond water and a snail (for preparation, see Chapter 6). How long will the snail live? Then hold up a green aquarium plant sealed in a test tube of pond water. How long will it live? Students should have developed concepts which they can apply to an analysis of the situation. They will suggest that plants and the snail placed together will last longer. They can outline many of these cyclic transfers—oxygen-carbon dioxide cycle. Refer to photosynthesis-oxidation (Chapters 3, 4, 6), the laboratory block (pp. 190–92), the nitrogen cycle (nitrogen-fixing bacteria, p. 538), the mineral cycle (pp. 458–62), the energy cycle (or food chains), and the notion of a pyramid of producers and consumers (Chapters 3, 4).

Students can prepare diagrams of the relations in these cycles, such as the one in Fig. 11-5. Also see the laboratory block (p. 542) and the study of food chains (Chapters 3, 4, and this chapter).

These cyclic transfers between populations of plants and animals in a community and their nonliving environment are implied in the term "ecosystem." The role of the specific population of plants or animals in an ecosystem may be considered in terms of modes of nutrition (part of the energy cycle). Is the population a predator, a primary consumer, a tertiary consumer, a producer, a decomposer (Fig. 11-6)?

It is also possible to look at relations among populations: for example, competition for a given exploited area or mutualism, as exemplified by termites and flagellates (p. 537), green hydra (p. 591), *Paramecium bursaria* (p. 585), and lichens (p. 537). (Symbiosis is considered a general term for examples of mutualism and commensalism.)

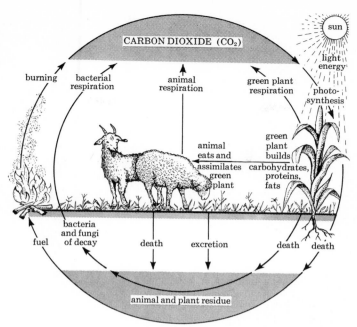

Fig. 11-5 The carbon cycle in nature. (After R. W. Hegner and K. A. Stiles, *College Zoology*, 7th ed., Macmillan, 1959.)

Web of life in a microaquarium

If a small snail were sealed in a test tube, how long would it live? Students already are aware that living plants and animals are dependent upon one another. They know that a snail cannot exist long without green plants. Sealing a snail and a water plant in a test tube is an effective demonstration or laboratory exercise. (For the technique, see Fig. 6-1.) You may save time, but the effect is less dramatic, if instead of sealing them in a flame you merely close several test tubes with rubber stoppers.

If students are familiar with the indicator brom thymol blue, you may want to add to the closed tubes some indicator solution that has been turned to a yellow color (by exhaling into it through a straw). Then have students explain why the brom thymol yellow in the tubes containing the green plants turns blue in the presence of light. Why does the blue color turn yellow during the night? They will recall that green plants absorb carbon dioxide from water during sunlight. Therefore the yellow color (yellow in the presence of excess carbon dioxide) of the indicator is changed back to blue since the reduction in carbon dioxide content of the water decreases the acidity of the water (oxygen–carbon dioxide cycle). Thus, the indicator is in a less acid condition; the more alkaline medium turns blue. During the night, the green plants *give off* carbon dioxide (in respiration) and since there is no light, photosynthesis is stopped. Therefore the yellow color results, since the plants release carbon dioxide into the aquarium water. But as soon as light appears, the carbon dioxide is absorbed by the plants, and the indicator becomes blue again (see the laboratory blocks of activities, pp. 190–92, 543–44).

Elicit from students a discussion of the kinds of habitats that will support a biotic community.

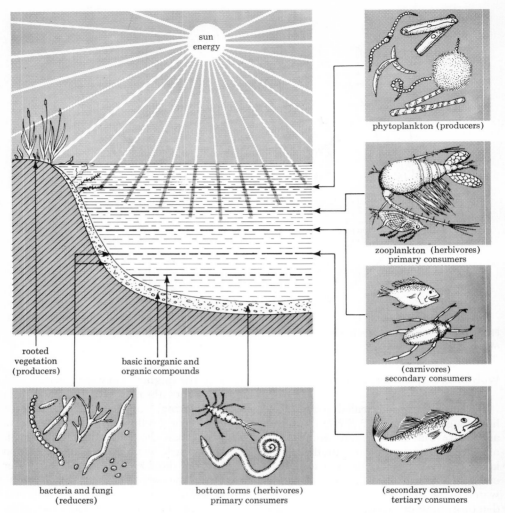

Fig. 11-6 An ecosystem, showing the roles of many plants and animals in the system. (After C. Villee, *Biology*, 4th ed., Saunders, 1962.)

The image contains the following labels: sun energy; phytoplankton (producers); zooplankton (herbivores) primary consumers; (carnivores) secondary consumers; (secondary carnivores) tertiary consumers; bottom forms (herbivores) primary consumers; bacteria and fungi (reducers); basic inorganic and organic compounds; rooted vegetation (producers).

Study of a biotic community

Take field trips around school lawns, along a stretch of beach at ebb tide, in an open meadow, in the woods, or in a burned-over region; students are certain to find some kinds of plants and animals living together, dependent upon one another (Fig. 11-23). (The animals are clearly dependent upon the green plants.) Any such community of living things is known as a biome. Students may be able to follow the succession of living things over a period of several years, as one kind of plant gives way to another in a forest or as plants begin to invade a lake and turn it into swamp and finally into dry land. As this happens, the kinds of animals and plants in the region change too (see field trips and collecting, pp. 545–59).

Several biotic relations are described here which may be useful for group investigations, either in the classroom or outdoors.

Study of a quiet pond. Many students find a study of the plant and animal relationships in a biome, such as a pond, a rewarding experience. Begin by listing the kinds of animals that are present in the pond. Are there forms which skim the surface, such as water striders? Are there free-swimming forms? Any attached to vegetation? Bottom dwellers and tube dwellers? Are they vertebrate or invertebrate?

With a dip net students might gather living specimens. If they spread the contents of the net out on a white cloth or white pan they will find it easier to identify the specimens. Some pictorial guides such as those by Needham and Needham, Ward and Whipple, or Pennak may be helpful (see the end-of-chapter bibliography).

Similarly, examine the vegetation. Are there floating forms—algae, duckweed, or water lilies? Submerged forms? What kinds? Seed plants or algae? Try to identify them (p. 552; also classification, p. 549).

A schematic drawing such as Fig. 11-7 might be made of a cross-sectional view of the pond, with the plants and animals at the proper level—floating, submerged, free-swimming, attached, and so forth.

What forces may alter the dynamic interactions among the organisms? How may changes in physical conditions affect life in the pond? Of course, the green plants make their food supply. What is the food source for the consumers and decomposers, for the insects, larvae, small fish, snails, or other inhabitants? Try to trace food chains. What might happen if one kind of organism were killed off or in some other way removed from the pond? Would there be an effect on the other organisms in the web of life in that pond? Which forms hold others in check so that they do not overreproduce?

You may want students to draw contour maps of the pond showing its depth, its elevation in relation to land, and its area and shape. Students might record their observations of other features as well. For example, what is the source of the water? The possibility of contamination? The temperature of the water? Would cold or warm water hold more oxygen and support active life? Is it clear water or turbid? What kind of bed is there—gravel, flat rock, mud, hard clay? What is the pH of the water?

Photographs might be taken of the pond each year. Over a period of years a comparison of photographs might show a suc-

Fig. 11-7 Cross section of a pond, an ecosystem showing typical life forms. (After R. Buchsbaum and M. Buchsbaum, *Basic Ecology*, Boxwood Press, 1957.)

cession of different plant and animal migrants. It may be possible to study an area where plants invade the pond; as their roots bind soil, a swamp may gradually result. What kinds of plants and animals survive in this new biome? Which are the pioneer plants? What is the pattern of succession?

Study of fast-moving streams. It may also be possible to have students examine the inhabitants of a fast-moving stream. Fast-moving streams are characterized by well-aerated water.

Study the life in the stream. Which organisms live under submerged rocks? Which are free-swimming? Which attached? Are there living things found on the banks and shallow part of the stream? Which are predators? Which are competitors? Which are found in the deeper water? Do organisms move with the current or face into it? How does the oxygen supply of the stream compare with that of pond water?

Finally, develop the food chains between plants and animals, and animals and animals. What factors may upset the balance in this web of life?

Find larvae of beetles, mayflies, stoneflies, flatworms, and snails, and then make a comparative study of the two biomes: the stream and the pond. Students may identify the plant and animal inhabitants of each and enter them in their proper location in a cross-sectional view, as shown in Fig. 11-7 for a pond. (See also Figs. 11-25, 11-31, 12-20, 12-21, 12-22, 12-23.)

On a contour map of the stream students might show the nature of the bed and banks. Also keep a record of the temperature, clarity, and relative velocity of the water in the stream, as well as the nature of the slope. Are the organisms active or slow-moving? A fast-moving stream often has a high oxygen content, which supports active organisms.

If students plan to transport living organisms from a stream to the classroom, they will have greater success in keeping the forms alive if they select organisms from a quiet stream (see collecting, discussed later in this chapter).

In your vicinity, it may be more appropriate to study a flood plain marsh, a bog, a desert, woods, a beach at ebb tide, a burned-over area, or an open meadow. Students may record the kinds of organisms and the physical features of each area. In these places they may want to test the acidity of the soil (p. 522) and examine the organisms living in a section of soil (p. 521). What natural enemies hold the existing organisms in check?

Study of a soil community. What kinds of plants and animals interact with one another—cooperatively or competitively —in the soil? How do organisms change the soil? (See also p. 528.)

Food chains

One form of life feeds upon another, and ultimately the food source upon which all other forms are dependent is the producer, the green plant. Students may have grasped this broad concept in a study of the biomes previously described or in a study of photosynthesis (Chapter 3). The notion of a pyramid with a large number of organisms supporting a smaller number of organisms is basic to an understanding of interrelationships upon which food chains and the web of life are based.

Study of plankton. With a dip net collect (directions, p. 552) free-floating surface organisms from a lake, a slow-moving stream, or an ocean. Many small crustaceans, algae, small mollusks, and microorganisms (protozoa, desmids, rotifers, larvae, and so forth) may be present. Turn the samples from a dip net into a white-bottomed tray for close examination. With a hand lens students may identify some forms; examine other specimens in class with a microscope.

When it is not feasible to go outdoors for this kind of study, a classroom aquarium can be used to develop the idea of the balance in nature. Students may prepare an aquarium (fresh-water, p. 605; marine, p. 600) or a microaquarium using test tubes (p. 265) or slides (p. 265). Trace the food chains and the interdependence of the living things in an aquarium. Then compare this with a lake or an ocean which contains many food chains, all of which relate back to an autotrophic plant.

Modes of nutrition: summary. Organisms that make their own food are called *autotrophic.* To produce food they need water, carbon dioxide, inorganic salts, and a source of energy. Among the autotrophs are the green plants and purple bacteria which are *photosynthetic autotrophs.* Some bacteria are *chemosynthetic autotrophs,* getting energy from the oxidation of inorganic molecules. Nitrite bacteria such as *Nitrosomonas* oxidize ammonia to nitrites; nitrate bacteria such as *Nitrobacter* oxidize nitrites to nitrates. Some sulfur bacteria oxidize hydrogen sulfide to sulfates; iron bacteria oxidize ferrous iron to the ferric form.

Organisms that get their food ready-made are called *heterotrophs.* All animals, fungi, and most bacteria are in this category. Animals that take in food as solid material that must be digested have a *holozoic* form of nutrition. Another form of heterotrophic nutrition is the *saprophytic* form—soluble materials are absorbed through the cell membrane—as in molds, most bacteria, and yeasts. *Parasitism* is a third type of heterotrophic nutrition.

Also among heterotrophs, one may classify those organisms that compete as *competitors;* interaction between some other species is *mutualism* (including commensalism), more generally known as *symbiosis.*

Students may examine the specialized ways in which organisms such as parasites (dodder, Fig. 11-8; frog parasites, Fig. 11-9) get their food. Try to collect specimens of fungus parasites such as corn smut or wheat rust (Fig. 13-16). Look for nematode worms, which infest plant crops.

Saprophytes, such as molds (p. 105), yeasts (p. 105), mushrooms, and shelf fungi, may be examined in class. Examples of symbionts (p. 537) are not difficult for students to find.

(a)

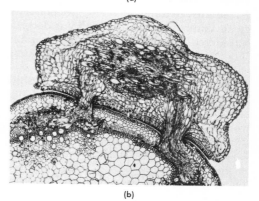

(b)

Fig. 11-8 Relationship of parasite and host: (a) dodder (*Cuscuta Gronovui*) on host plant; (b) microscope slide showing haustoria of dodder growing into stem of host. (a, from C. L. Wilson and W. E. Loomis, *Botany*, rev. ed., Dryden [Holt, Rinehart and Winston], 1957; b, courtesy of Carolina Biological Supply Co.)

As part of their study of interdependence of living things students have learned some facts about the use of foods in the body, that is, digestion, circulation, respiration, and assimilation of new tissues. They have learned the ways living things reproduce themselves, which traits have survival value, and how variations adapt certain organisms for their way of living. Students need to know that intelligent development of resources presupposes an understanding of basic life functions of both plants and animals.

Specialized ways of getting food

Parasitism

PARASITE AND HOST. *Dodder.* On a field trip look for dodder, an orange-colored seed plant which grows like a vine around a host plant such as goldenrod (Fig. 11-8; see also p. 653).

Dodder produces small clusters of white flowers and a large number of seeds. When students try to separate the parasite from its host plant, they'll notice that dodder has projections (haustoria) growing into the phloem tubes of the host (Fig. 11-8b). What would happen if dodder seeds germinated in a region where independent green plants were absent? Why is dodder considered a parasite, while Indian pipe (Fig. 11-14), another seed plant lacking chlorophyll, is a saprophyte?

Frogs and their parasites. When freshly killed frogs are dissected in class, students are likely to find examples of parasitic worms whirling around in the region of the lungs and liver or inside these organs. Look especially for a dark flatworm about ⅜ in. long, *Pneumonaeces* (Fig. 11-9b). These worms may be mounted in a drop of Ringer's solution (p. 665) and viewed under the microscope. Also inspect the contents of the urinary bladder for round-worms and flatworms. Examine the inside of the urinary bladder with a hand lens. Small trematodes with large ventral suckers may be present. *Gorgodera* (Fig. 11-9a) is a common inhabitant of frogs. Are there yellowish cysts in the mouth of the frog? Tease one open on a slide and look for encysted cercaria of the flatworm *Clinostomum.*

Students may undertake an interesting project: a comparative study of the parasites in frogs from different localities (parasites travel with their hosts). Order live frogs from such diverse areas as Vermont and Alabama.

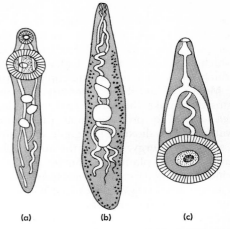

Fig. 11-9 Some parasitic flatworms found in frogs: (a) *Gorgodera* (in urinary bladder); (b) *Pneumonaeces* (in lungs and glottis); (c) *Diplodiscus* (in large intestine). Note the large posterior sucker of *Diplodiscus.*

Also look for protozoans living as parasites within the digestive system of frogs. Begin below the stomach, and remove a bit of the lower intestine and/or of the rectum. Mount the contents of the intestine in a drop of Ringer's solution. From this transfer a bit onto a clean slide as a dilute wet mount. Students may find the large, slow-moving protozoan *Opalina* (Fig. 11-10a), which lacks an oral groove and a contractile vacuole, along with *Balantidium*, a smaller form (Fig. 11-10b).

In the rectum of the frog another protozoan, *Nyctotherus* (Fig. 11-11), may be found. It is not definitely known whether this form is parasitic in the frog or whether it is free-living. *Nyctotherus* can be distinguished from *Opalina* by its laterally placed oral groove. It is often present in the intestine of cockroaches, centipedes, and millipedes as well.

PARASITES AND SEVERAL HOSTS. Unraveling the complete food chain of some parasites and their hosts has taken years of intensive study. Often one stage of a parasitic worm might be found in a bird, another in a fish, and a third in a snail. In each host the parasitic worm was in a different stage of its life cycle, and these were not recog-

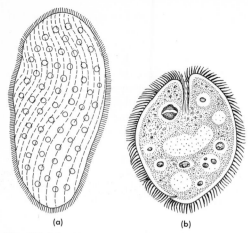

(a)   (b)

Fig. 11-10 Protozoans sometimes found in intestines of frogs: (a) *Opalina;* (b) *Balantidium.* (a, from D. F. Miller and J. G. Haub, *General Zoology,* Holt, Rinehart and Winston, 1956; b, after R. Kudo, *Protozoology,* 4th ed., Thomas, 1954.)

nized as developmental stages of one organism. Students may find many examples of parasites that have two or three hosts. You may want them to report on several examples, such as the malarial plasmodium, the broad tapeworm of man (*Diphyllobothrium*), the pork tapeworm (*Taenia solium*), and the Oriental liver fluke (*Clonorchis*).

What would be the effect on the food chain if a parasite destroyed its host? Explain why parasitic forms produce such large numbers of eggs.

Parasites of the earthworm. Monocystis, a gregarine parasite in the seminal vesicles

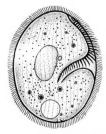

Fig. 11-11 *Nyctotherus,* a protozoan found in rectum of frogs or intestine of centipedes and cockroaches. (After W. S. Bullough, *Practical Invertebrate Anatomy,* Macmillan, London, 1954.)

of the sexually mature earthworm, is a member of the class Sporozoa among the protozoa.

After narcotizing an earthworm (by placing it in a 5 percent alcohol solution, increased to 10 percent to kill the worms), split open the body wall, between the 10th and 15th segments, and pin back the skin to expose the white seminal vesicles. Mount a seminal vesicle in 0.7 percent sodium chloride solution and crush the contents on a slide. Add a drop of dilute methylene blue stain, and examine under low power, then under high power. Look for elongated trophozoites (Fig. 11-12). Spores were probably ingested by earthworms; these were liberated as sporozoites, which penetrated the wall of the gut of each worm and traveled through the blood to reach the seminal vesicles. Here the cigar-shaped trophozoites destroy sperm mother cells. Dormant sporozoites in cysts may also be found.

Trophozoites later pair off and form gametes. After fertilization, new sporozoites are formed. Different species of large gregarines also inhabit the gut of grasshoppers and cockroaches. Slides may be prepared of the contents of the gut and mounted in saline at 22° C (72° F).

A nematode, *Rhabditis,* is a common parasite that may be found in the nephridia of the earthworm (Fig. 11-13).

PART-TIME PARASITES. Several living forms are partially parasitic in the sense that they are not permanently attached to or do not live continuously within a host. Mistletoe carries on photosynthesis and absorbs water and minerals from the host plant. Forms such as mosquitoes or leeches are also temporary parasites.

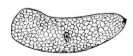

Fig. 11-12 Trophozoite state of *Monocystis,* a parasitic sporozoan often found in seminal vesicles of earthworms. (From R. Kudo, *Protozoology,* 4th ed., 1954; courtesy of Charles C. Thomas, Publisher, Springfield, Ill.)

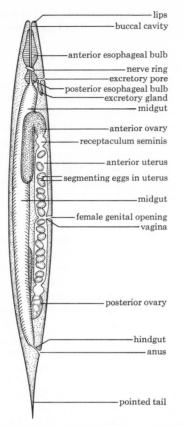

lips
buccal cavity
anterior esophageal bulb
nerve ring
excretory pore
posterior esophageal bulb
excretory gland
midgut
anterior ovary
receptaculum seminis
anterior uterus
segmenting eggs in uterus
midgut
female genital opening
vagina
posterior ovary
hindgut
anus
pointed tail

Fig. 11-13 *Rhabditis,* a parasitic nematode commonly found in nephridia of earthworms. (After W. S. Bullough, *Practical Invertebrate Anatomy,* 2nd ed., Macmillan, London, 1958.)

Saprophytes. Molds growing on bread or lemons probably have been studied previously in class (pp. 105, 344, 370). Or students may grow mold cultures on bread (p. 638).

You may want students to study the role of saprophytes in a food chain. Students may collect puffballs, mushrooms, shelf fungi, mildews, fungi, and bacteria of decay in soil, among many saprophytes which affect the storage of food crops.

You may want students to examine several specimens under a microscope (see field trips, p. 545; culturing, pp. 630–40).

What is the role of saprophytes? List some saprophytes beneficial to man in that they return minerals to soil. What would happen to the fertility of soil and the balance of life if bacteria of decay disappeared from the earth? What use has man made of bacteria of decay? Of yeast plants (p. 277)? Name several saprophytes that are harmful to man since they interfere with man's use of resources such as lumber or food crops. How does man attempt to hold these saprophytes in check?

INDIAN PIPE. Students may find examples of the white seed plant Indian pipe (Fig. 11-14) growing in shaded regions, sometimes pushing up under masses of fallen leaves. Indian pipe is a higher plant which lacks chlorophyll and is therefore unable to manufacture carbohydrates. It exists as a saprophyte, breaking down decaying vegetation and absorbing nutrient materials as mushrooms do.

A whole cluster, with the soil in which it is growing, may be transferred carefully to a terrarium. When bruised, the plants turn blackish.

Fig. 11-14 Indian pipe, a seed plant lacking chlorophyll, which lives as a saprophyte. (Photo by Harold V. Green.)

Symbiosis (mutualism)

LICHENS. Students may collect lichens from rocks or tree barks (Fig. 11-15). When a bit of the lichen is teased apart in a drop of water on a glass slide (wet mounts, p. 99) and examined under the microscope, the green algal cells enmeshed in filaments of fungi can be distinguished. In this mutual partnership of two plants the algae carry on food-making and the fungi absorb and hold water and minerals. Together these two plants can thrive on rocks, tree bark, and fence posts. Alone, each plant would have a limited range of existence (culturing, p. 642).

TERMITES AND THEIR PROTOZOA. Termites consume wood, yet they lack a cellulose-splitting enzyme to break down this "food." The large, slow-moving flagellates found in the intestinal tract of termites are symbiotic organisms that digest the cellulose consumed by the termites (Fig. 11-16).

On a field trip students may locate a termite colony in a tree stump. (This can be kept in the laboratory, p. 605.)

To prepare a wet mount of the flagellates, place a termite in a drop of physiological salt solution (0.5 to 0.7 percent) or Ringer's solution (p. 665) on a clean slide. Use forceps to hold down the abdomen on the slide. With another forceps pull off the termite's head. The intestine (with other viscera) will be attached to the head part. Remove the remains of the termite from the slide, keeping only the intestine. Macerate the intestinal contents, and then apply a coverslip. Under the high-power objective of the microscope the long flagella will be visible on these specimens. Preparations may be circled with petroleum jelly, covered, and used for several hours. Students may also want to add a vital dye (p. 121).

Many of the flagellates belong to the Kofoidea (Order Hypermastigina) or to the Pyrsonympha (Order Polymastigina). They reproduce asexually by longitudinal fission and encystment. These organisms possess a cellulose-splitting enzyme lacking

Fig. 11-15 *Physcia*, a lichen, shown growing on tree trunk. (Courtesy of General Biological Supply House, Inc., Chicago.)

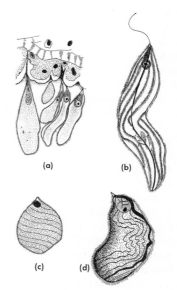

(a) (b)

(c) (d)

Fig. 11-16 *Pyrsonympha vertens*, flagellates found in intestines of termite: (a) cross section of the intestine of a termite, showing young *Pyrsonymphae* attached; (b) extended form; (c) contracted form; (d) form showing wood fibers inside. (After J. F. Porter, "*Trichonympha* and Other Parasites," *Bull. Museum Comp. Zool.* 1897–98.)

in termites. Cleveland[6] in 1925 showed that when termites were defaunated they died within a few days. However, when dying termites were reinfected with flagellates through oral feeding, digestion again took place.

When the termites molt, the chitinous lining of the gut is lost along with the flagellates. After molting, reinfection occurs. Some wood-eating beetles, sow bugs, woodroaches, and cockroaches have similar flagellates, amoebae, or bacteria serving the same function. *Trichonympha* is common in the woodroach (*Cryptocercus*). Encystment has been observed once a year when the woodroach molts.

What would happen to termites if they didn't become infected with flagellates immediately upon hatching?

NITROGEN-FIXING BACTERIA AND LEGUMES. In season, on a field trip, students may uproot and bring to class leguminous plants such as clover, peanuts, alfalfa, or beans. Look for swellings, or nodules, on the roots. Each nodule contains thousands of nitrogen-fixing bacteria which have invaded the root tissues but do little harm. This is a case of mutualism, or symbiosis.

In class, wash off the roots. Crush a nodule between two glass slides. Add a drop of water to each slide and examine each as a wet mount. For more careful study these slides may be stained in the following way. Place a drop of methylene blue stain (p. 121) on each of several slides. Let the stain dry; store the slides in a box so that they are ready for use. Add a drop of water containing nitrogen-fixing bacteria to the slide, apply a coverslip, and examine under high power. Notice how the stain gradually diffuses into the water and stains the bacteria.

Or you may prefer to make a smear of the bacteria and stain the smear to prepare a permanent slide, as follows. Make a thin film of the bacteria by spreading the

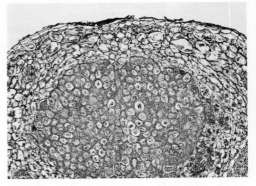

Fig. 11-17 *Rhizobium leguminosarum,* nitrogen-fixing bacteria. Slide shows cross section of nodule attached to root of leguminous plant. (Courtesy of Carolina Biological Supply Co.)

fluid of a wet mount along the slide with a dissecting needle or with another slide. Fix the bacteria to the slide by waving the slide through a Bunsen flame two or three times. Lay the slide across a Syracuse dish, and flood it with methylene blue for 1 min. Then rinse off the stain with water, let it dry, and examine under high power magnification (Fig. 11-17).

Why do farmers rotate their crops with leguminous plants? How do these nitrogen-fixing bacteria aid in replenishing the nitrate content of soil?

EFFECT OF NITROGEN-FIXING BACTERIA ON PLANT GROWTH. Students may demonstrate the difference in growth of plants grown in soil deficient in nitrates and soil rich in nitrates (see hydroponics, pp. 458, 521).

Sterilize four small flowerpots of soil in an autoclave or an oven. Then add nitrogen-fixing bacteria (purchased from a seed company) to the soil in two of the flowerpots. Add an equal quantity of soaked clover seeds to each of the four flowerpots. In which pots do students notice a more luxuriant growth of clover plants? What is the advantage of rotating nonleguminous crops with leguminous plants?

OTHER EXAMPLES. Students may come upon many cases of symbiosis and mutualism on field trips near school.

If there is a beach or bay in the vicinity,

---

[6] L. R. Cleveland, "The Effects of Oxygenation and Starvation on Symbiosis Between the Termite Termopsis and Its Intestinal Flagellates," *Biol. Bull.* 48:455, 1925.

students may find hermit crabs (Fig. 11-18), which inhabit deserted gastropod shells—possibly whelk, periwinkle, or moonsnail. To the shells are attached sea anemones (as shown in the figure), sea lettuce, or other forms. This is a type of camouflage. When the crabs are extricated from the shells, students can see the degenerated abdomen of each crab. Notice how the abdomen is twisted to fit into the spirals of the mollusk shell. The usually sessile anemones are carried about by the hermit crab to sources of food; and possibly the hermit crab has better chances of survival with these devices for camouflage. Students may also know of the harvestfish (Fig. 11-19), which swims among the tentacles of the Portuguese man-of-war.

In collecting specimens from ponds you may come upon green hydras or green paramecia (or they may be purchased from supply houses). In these animals live green algae in a symbiotic relationship. (See also *Paramecium bursaria* [p. 585] and green hydra [p. 591].)

Students may find ants and their "cows," plant lice. These aphids, or lice, are tended by some species of ants which "milk" the aphids of a sugar solution they excrete. This is a symbiotic relationship.

Insectivorous plants. Several insectivorous plants may be found in bogs (or purchased). Students may find sundew, pitcher plant, or Venus flytrap (Fig. 11-20).

Fig. 11-18 A hermit crab and its "home": an empty snail shell. (Note the sea anemone attached to the shell.) (Photo: © Douglas P. Wilson.)

Fig. 11-19 *Peprilus paru*, the harvestfish, which swims among the dangerous tentacles of the Portuguese man-of-war. (Courtesy of Carolina Biological Supply Co.)

In fresh water they may find *Utricularia*, a floating plant which has small bladder traps into which microorganisms swim and become captive (Fig. 11-21).

In all these cases the insectivorous plants are green and carry on photosynthesis. Yet there may be a deficiency in nitrates; the plants secrete enzymes which act upon the insects, and possibly this protein material supplies the sources of nitrogen for plant growth.

Students may try to establish these bog plants in a terrarium which duplicates the environment from which they were transferred. Add some bits of charcoal to the tank to absorb odors. Introduce fruit flies into the tank (p. 415).

When insectivorous plants are not available, students may report to the class on these unusual plants and their kind of nutrition; or you may choose to show a film (see listing at the end of the chapter).

Upsetting the balance in nature

How have certain plants and animals become "pests" in one area while they are part of a natural food niche in another location?

Students may report on the history of the invasion and spread of the European corn borer, Tussock moth, Japanese beetle, English sparrow, thistle, Mediterranean

(a)

(b)

(c)

Fig. 11-20 Insectivorous plants: (a) sundew; (b) Venus flytrap; (c) pitcher plant. (a, Roche photo; b, courtesy of Carolina Biological Supply Co.; c, courtesy of Chicago Natural History Museum.)

fruit fly and starling, among many others. What conditions enabled these immigrant insects, plants, or birds to overproduce? What is a natural enemy? What is the role of natural enemies in the web or balance of life?

This may be the time to elaborate upon the facts of overproduction in animals and plants (pp. 498–500, 508) and to clarify the concept of natural enemies. What other factors prevent the survival of each egg, seed, or spore produced by an animal or plant? Students may recall the unfavorable conditions, such as shade or crowding, under which seedlings germinate, or the lack of food and water and overcrowding among the young of a given species of animals which are all looking for the same conditions for survival.

Students can list the ways man has up-set the web of interrelationships in nature. In addition to introducing organisms into a new territory, man has upset the balance in nature by doing the following things: (1) starting forest fires through carelessness, (2) recklessly cutting down trees, (3) allowing sheep and cattle to overgraze the land, (4) overcultivating land, (5) contaminating waterways, (6) destroying certain animals, and (7) breeding new plants and animals.

Students may, in committee, report on several of these practices. They should come to realize the need for knowing the life cycles of insects (pp. 354, 602) so that they may be destroyed or reared with care, depending upon their role in relation to a balance in nature and the needs of man. Why is it more advantageous to control insect pests by introducing their natural enemies than by using chemical controls, such as DDT and other insecticides?

This may be the time for students to read an offprint of C. F. Cooper's paper "The Ecology of Fire" (*Sci. Am.,* April 1961). In addition to loss of vegetation and cover for animals, organisms in the streams are affected by the change in pH due to wood ashes washed into their environment.

What is the value of knowing the facts about reproduction and development of frogs, birds, plants of all kinds, fish, snails, and so forth?

How are the seeds of plant pests disseminated? How does man through his actions encourage the spread of seeds of plant pests?

Have living things become extinct as a result of man's careless use of resources? Students may describe many examples, such as the story of the passing of the passenger pigeon. What is the value of sanctuaries? Of closed seasons?

Going further: projects

Making a film. Members of a photography club or individual students interested in photography may develop a fine documentary. They might select nearby areas that are in need of repair—barren slopes, places that have been affected by raindrop erosion, and the like—and then show the effects on these areas of planting cover crops, terracing, or contour plowing. Or they might show in film all the intricacies of a few selected examples of plant and animal relations in a biome—for example, a woods, an open field, a fence post, or a pond.

Another committee of students might write the script for this film, which might be shown as an assembly program in school. Might it be useful to show the film at a PTA meeting?

Some camera enthusiasts may combine their interest in ecology and talents in photography by collecting their specimens

(a)

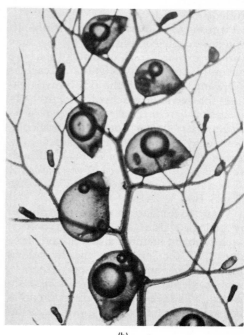

(b)

Fig. 11-21 *Utricularia,* the bladderwort: (a) view of aquatic, insectivorous plant; (b) enlarged to show bladder traps. (Photo by Hugh Spencer.)

Fig. 11-22 *Dryopteris marginalis,* marginal shieldfern: under surface, showing sori. (Courtesy of the American Museum of Natural History, New York.)

as photographs. They may get excellent results (such as Fig. 11-22) using close-up photography.[7] Such a collection of portraits of plants and animals might well become part of an identification key that students could devise.

Long-range investigations. Students who are deeply committed to biology may want to explore areas of biology that are currently under investigation. Teachers will want to look into the BSCS investigations for secondary school students—*Research Problems in Biology: Investigations for Students,* Series One–Four (Anchor Books, Doubleday, 1963, 1965). A guide accompanies the investigations—P. F. Brandwein *et. al., Teaching High School Biology: A Guide to Working with Potential Biologists* (BSCS Bull. 2, AIBS, 1962).

These investigations were contributed by biologists from all over the country. Each problem includes a description of the nature of the problem, the hypothesis under study, and some approaches. A brief guide to the literature is also included. The investigations are as broad as life, and the methods of approach or in-

[7] See C. Neidorf, "A Technique for Close-up Photography of Freshly Collected Fertile Fern Specimens," 2 parts: Part I, *Turtox News* 38:178–80, 1960, Part II, *Turtox News* 38:236–38, 1960.

quiry are as varied as the scientists who prepared the investigations.

A broad view. Students may find it interesting to trace the changes in geographical distribution of plants and animals over a period of time as the land masses changed, the climate changed, or rivers were diverted. These changes are continuing today, and the "meddling" of man is a part of this picture.

Review Darwin's writings on the effects of barriers upon the migration of plants and animals and the rise of new species. Students may go further into reading in population genetics and the importance of genetic drift in accounting for the kinds of plants and animals that exist anywhere. The work of Sewall Wright and G. G. Simpson, as well as the writings of Richard Goldschmidt, offer a broader view of the origin of new species.

Preludes to inquiry: a study of a microecosystem

In addition to field work, or when outdoor field work is not feasible, small groups of students may work on varying aspects of problems raised in the following series of investigations (a "block" activity). The organism selected was the green alga *Chlorella*. Similar blocks of activities may well be correlated around some other organism in activities designed to support the concepts that organisms are interrelated and also related to their environment.

Chlorella was selected because there is a wide research literature using this organism that can be consulted. Also, results of investigative work with *Chlorella* may be observed within a class period, or within a few days; changes in other organisms, such as seeds, frogs, and geraniums, take much longer to occur.

Somehow students are seldom really taught *how to inquire;* they are usually taught *how to confirm* what is known. They engage in "problem-doing," not "problem-solving."[8] For instance, consider the study of photosynthesis and the use of the geranium plant in the high school classroom. Students "find out" that starch is produced (problem-doing); the solution to the problem is known and the student is, indeed, guided irrevocably to the known solution. Hence the activity may be considered "problem-doing" not "problem-solving." On the other hand, problem-solving should involve a confrontation of a "new" situation without a readily available solution in the text.

The study of *Chlorella* in a block of class time follows. (See culturing of *Chlorella*, Chapter 13.) A series of questions for investigation is given, and a number of possible procedures are suggested, including some described in greater detail elsewhere in this book. (See also pp. 190–92.)

1. What is the average rate of reproduction of *Chlorella?*

Isolate a small number of cells in a microaquarium (pp. 122, 265, 442); make daily counts of the cells (p. 444).

2. What are the effects of adding addi-

tional quantities of minerals such as nitrates, magnesium, or phosphates on growth and physiology?

Prepare culture medium with *additional* amounts of these minerals in each stock preparation (see Knop's or others, p. 459). Conversely, what is the effect of nitrate *deficiency* or a lack of other substances on formation of chlorophyll? On rate of growth of *Chlorella?* Students may devise their own methods of investigation.

3. What is the effect of additional amounts of carbon dioxide on *Chlorella?*

Introduce a small bubble of carbon dioxide into the microaquarium by means of a micropipette attached to a source of CO_2, such as a bottle of soda pop or dry ice. Be sure to establish controls. (See techniques for measuring growth, p. 441; indicators, p. 662.)

4. Do yeast cells have any effect on *Chlorella?*

Prepare a flask or bottle of yeast cells in dilute molasses to which two dried peas (as a protein source) have been added. Attach a delivery tube from this flask to a flask of *Chlorella*. Or on a sealed slide—a microaquarium—add a small pipetteful of yeast cells to a drop of very dilute *Chlorella* suspension. Also prepare controls without the yeast cells. Is the difference between the controls and the *Chlorella* with yeast cells added measurable? For example, is there a greater intensity in the green color?)

5. For what organisms does *Chlorella* serve as a food niche?

When students introduce *Paramecium* into a microaquarium of *Chlorella* and *Euglena*, they find that *Paramecium* feeds on *Chlorella* in preference to *Euglena*. What factors affect this preference? What is the preference of other heterotrophs such as *Vorticella*, rotifers, *Daphnia*, and *Aeolosoma?* What happens when the supply of *Chlorella* is consumed and *Euglena* is present in great numbers? What happens to the heterotrophs when the primary producers disappear?

[8] P. F. Brandwein, "Elements in a Strategy for Teaching Science in Elementary School," The Burton Lecture, in J. Schwab and P. F. Brandwein, *The Teaching of Science* (Inglis and Burton lectures), Harvard Univ. Press, Cambridge, Mass., 1962.

6. How many *Chlorella* cells are needed to support a given number of heterotrophs? For example, to support 6 paramecia? Or 12 paramecia? Which multiplies faster, the food crop or the heterotroph? What factors affect balance in a microecosystem?

Students may plan several designs for investigation: slide 1—1 drop of rich *Chlorella* suspension added to 1 drop of a known number of *Paramecium;* slide 2—1 drop of rich *Euglena,* plus 1 drop of a known number of *Paramecium;* slide 3—same as slide 1, but in a dark room; slide 4—same as slide 2, but in a dark room; slide 5—1 drop of *Euglena,* plus 1 drop of *Chlorella,* plus three to six paramecia; slide 6—1 drop of *Chlorella,* plus known quantities of *Paramecium,* plus *Didinium* (protozoan that preys upon *Paramecium*); slide 7—1 drop of *Chlorella,* plus *Aeolosoma* (segmented worm, p. 594) and *Daphnia,* or other mixed culture (p. 100); slide 8—same as slide 7, but without *Chlorella* (or controls) and in darkness.

7. Does *Chlorella* thrive better in a pure culture? Or are there biological or physical advantages in the combination of *Chlorella* with certain heterotrophs? Or with other autotrophs?

Here the design of the experimental techniques is left for students. What possible hypotheses may be formed? What predictions can be tested?

8. Does *Chlorella* affect the pH of its environment? If so, does this affect the succession of other microorganisms in the habitat?

9. What gas is given off by *Chlorella* in sunlight? What factors affect this process?

Count the number of bubbles accumulating in the microaquarium in a period of 15 min when it is exposed to light; also when it is kept in the dark. Other investigations concerning photosynthesis and wavelengths of light may be proposed; see Chapter 3.

10. Is the usual biological succession of organisms affected (in a microaquarium or in a flask culture) by increasing the intensity of light? Or by decreasing the intensity of light? Could such alterations in the environment affect the survival of *Chlorella?* For example, growth of cover plants such as shrubs or trees might shade a small pond, or lightning might fell the shade trees.[9]

Students may design their own experiments. (See suggestions for measuring growth, Chapter 9; photosynthesis, Chapter 3; culturing solutions, p. 627.)

11. Does exposure to X-rays affect the exponential growth phase of a culture of *Chlorella?* Does the quantity of available oxygen affect irradiated cells?

Ionizing radiations include alpha, beta, gamma, X-rays, proton, and neutron radiations. Beta, gamma, and X-rays ionize the molecules in protoplasm of plant cells.

12. Can *Chlorella* synthesize any food materials in the dark?

13. Are there photoperiods, or rhythms of cell divisions, in *Chlorella?* Does light intensity affect reproduction of *Chlorella?*

14. What is the optimum temperature for growth of *Chlorella?* What is the effect of culturing the cells at 40° C (104° F) for 6 hr? Can they thrive at 10° C (50° F)? Is it possible to adapt *Chlorella* to different temperature ranges? Is photosynthesis affected by the different temperature ranges? (See counting the bubbles of gas, p. 176.)

Students might carry on these investigations at home. Use a double boiler to maintain a raised temperature (use a thermometer), or add ice cubes to lower the temperature to 10° C (50° F).

Students and teachers will surely think of many other problems that may be investigated by the methods of inquiry, among the many methods used by scientists.

[9] See J. Bonner, "The Upper Limit of a Crop Yield," *Science* 137:3523, 1962.

PART II

Field work: exploring living things in their environment

Field work is an essential part of biology.[10] Field trips may be brief—one period or less—or long—a day's planned excursion or a weekend at a nature camp. From such activities young people develop a deeper understanding and appreciation of the way living things function, their interdependence, the effect of the environment on their survival, and the selective forces which affect heredity and evolution. Field work provides students with the necessary opportunity to observe in life the organisms they read about in texts.

This section describes methods of collecting living things for a more or less immediate study in class, and ways to preserve some living things for a school museum. Chapters 12 and 13 describe many ways to keep alive, in class and laboratory, the plants and animals that have been collected in field work.

Many varieties of field trips (Fig. 11-23) are possible, depending on the purposes of the teacher and the location of the school. Whatever the nature of the trip, it should always be carefully planned. Are students going to gather water forms or land forms? Are they going to the seashore or to a wooded area? Proper clothes and equipment are necessary. The size of specimen bottles and the kind of nets required will differ with each trip. Bring along hand lenses for preliminary examination of specimens in the field. (Several types may be purchased from a supply house or an optical firm.) Students should remember to bring along fieldbook guides to help identify plants and animals (see the Bibliography at the end of the book).

Nature trails around school

It may be possible to develop nature paths as a class or club project. In such projects, some schools plant trees and shrubs which are especially useful in teaching. For example, reproduction in different flowers, variations, and evolutionary sequence may be shown on a field trip around the school grounds if you have magnolia, forsythia, varieties of cherry trees, flowering crabapple, maples of several varieties, dogwood, elm, oak, or barberry. If possible, include witch hazel, which blossoms in the fall. On the lawn, dandelions probably flourish. These complex flowers, which are really clusters of flowers, might be compared with the flowers of the primitive magnolia.

The trees and shrubs may attract insects and birds to the school grounds. Students might build birdhouses in shop classes. They may make attractive markers for a nature trail and erect a bulletin board showing a map of the trail or announcing interesting things to look for each month. Committees of students often prepare a mimeographed guide sheet.

Later, seedlings may be observed and aspects of the struggle for existence and the survival value of some seedlings may be investigated on the school grounds.

You may also find it possible to have students prepare specific associations of plants and animals to exemplify interdependence in a web of life. For example, in a shady area on the school grounds, with moist, slightly acid soil, some ferns,

[10] An ecological approach to the study of biology is given in *High School Biology* (BSCS Green Version), Rand McNally, Chicago, 1963. In the accompanying *Student's Manual: Laboratory and Field Investigations,* see Section 1, "The World of Life: The Biosphere" (chapters on "The Web of Life," "Individuals and Populations," and "Communities and Ecosystems"); Section 2, "Diversity Among Living Things"; and Section 3, "Patterns in the Biosphere" (patterns of life among microorganisms, on land, and in the water, and chapters on "Patterns of Life in the Past" and "The Geography of Life").

Another rich source of activities is E. A. Phillips, *Field Ecology* (BSCS laboratory block), Heath, Boston, 1964. Many teachers probably already know of C. Lawson and R. Paulson, eds., *Laboratory and Field Studies in Biology: A Sourcebook for Secondary Schools,* Holt, Rinehart and Winston, New York, 1960, a cooperative contribution of many teachers.

(a)

(b)

Fig. 11-23 Field trips can be made to many places: (a) a forest pool; (b) the school grounds.

(c)

Fig. 11-23 (continued) Field trips can be made to many places: (c) a slow-moving stream. (a, b, Hays from Monkmeyer; c, courtesy of Howard H. Michand, Purdue Univ.)

mosses, and spring flowers from other regions might be transplanted. Perhaps a decaying log with its "inhabitants" might also be transported.

A small pond might even be constructed near school and stocked with plants and animals taken from a nearby lake.

You may plan short field trips, possibly 15 min of one period, to study flowers, to measure variations in plants of the same kind, to get a sample of soil to test its pH, to look for examples of tropisms, to examine a compost pile of decaying leaves and cut grass in order to study the steps in the breakdown of vegetation through the action of bacteria of decay.

Students might search for examples of erosion and go about using corrective measures to reduce the loss of soil.

Using the school grounds

If it is not possible to develop nature trails, other ways may be found to use your school grounds. Here are some practices that have been used with success by many teachers.

1. Plant trees and shrubs that show a variety of flowers for students to study.

2. Take a census of insect pests in a measured area of land.

3. Examine and list all the kinds of organisms living in the top 2 in. of 1 sq ft of soil (p. 521).

4. Test a weed killer on one part of a lawn and leave the other part as a control (p. 328).

5. Add fertilizer (or inoculate the soil with nitrogen-fixing bacteria) to one portion of lawn and leave the other portion untreated (p. 521).

6. Practice soil-saving methods on a slope, such as terracing, contour plowing, or planting cover crops (pp. 521–27).

7. Study raindrop erosion. Set out discs of wood over a barren region. After a heavy rain, are there higher levels of soil under the discs than in the exposed regions? What happened to the soil in the exposed regions?

8. Plant seeds of flowering plants or use seeds of crop plants, so that the complete life cycle may be studied.

9. Illustrate methods of vegetative propagation; demonstrate the use of cuttings, bulbs, layering, or runners. It may be possible to make several grafts.

10. Test the effect of growth hormones in speeding regeneration of roots of cuttings.

11. Demonstrate breeders' methods of artificially pollinating flowers.

12. Investigate the effect of shortening or lengthening the period of light that a plant receives (photoperiodism).

13. Are there insect pests around the school grounds? Perhaps you may try the effect of specific insecticides. What stage do they attack in the life cycle of the insects?

Indoor field trips

It is useful to have a museum of living organisms in school. This may be a room or a portion of a laboratory, the size dependent upon the space you have available.

Students willingly bring living things to class; also, some living organisms may be purchased. Often some students know how to keep fish in an aquarium. Student curators may be trained to maintain both fresh-water and marine aquaria. They may be trained to prepare terraria of various kinds—one duplicating an arid environment with sand, cactus plants, horned "toads," and another containing a miniature woodland with humus in which ferns, mosses, lichens, and small wild flowers are kept. If students place a pan of water at one end of a terrarium they may include frogs' eggs, tadpoles, or adult frogs or salamanders. (Toads mess a tank, since they burrow.) A bog terrarium can be made from sphagnum moss and insectivorous plants. Add charcoal to absorb odors and reduce acidity (preparation of aquaria and terraria, pp. 600, 605, 611).

Turtles, snakes, small lizards such as chameleons, and small mammals such as rats and hamsters may be cared for by students trained to handle them (student curators and squads, Chapter 14; caring for animals, Chapter 12).

Students can be taught to culture protozoa, fruit flies, many worms, and other organisms useful in the year's work in biology. A living-material center may grow out of these activities. Then one school might be able to supply nearby schools with living materials (also some exhibits of life cycles of plants or insects which may be preserved).

You may have facilities for arranging a small greenhouse or, more simply, a sandbox in which seeds or cuttings of plants and other kinds of vegetative propagation may grow. Battery jars or bell jars can be used to cover young plants to increase moisture around them.

Many lessons may be taught in a museum of living organisms. In fact, you may find it useful to prepare a guide—a mimeographed sheet of questions which guide the observations and thinking of students. In this learning situation, students may study food chains, reproduction, behavior, interdependence of living things, camouflage, heredity, evolutionary relationships, and many aspects of nutrition.

A museum of plants and animals

If it is at all possible, students might take on the task of setting up a school museum containing examples of the major plant and animal phyla. A listing of the examples which might be collected is given below; of course, microscopic forms may be represented by models. Where possible, maintain living organisms or well-prepared preserved specimens in museum jars, as shown in Fig. 11-33. Also, guides for identification can be constructed, such as those for insects (p. 599), protozoa (p. 589), and conifers (p. 514). See also suggestions for the preservation of representative organisms (p. 360) and work of museum curators (p. 683).

A survey of the plant kingdom[11]

SUBKINGDOM THALLOPHYTA

Simplest plants, without true roots, stems, or leaves, composed of single cells or aggregations of cells; plants not forming embryos.

Phylum Cyanophyta	Blue-green algae; no distinct nuclei or chloroplasts
Phylum Euglenophyta	Euglenoids
Phylum Chlorophyta	Green algae; chloroplasts and nuclei
Phylum Chrysophyta	Golden-brown and yellow-green algae and the diatoms
Phylum Pyrrophyta	Dinoflagellates and cryptomonads
Phylum Phaeophyta	Brown algae (large seaweeds); multicellular
Phylum Rhodophyta	Red algae; multicellular
Phylum Schizomycophyta	Bacteria
Phylum Myxomycophyta	Slime molds
Phylum Eumycophyta	True fungi
Class Phycomycetes	Bread molds and leaf molds
Class Ascomycetes	Yeasts, mildews, and sac fungi
Class Basidiomycetes	Mushrooms, rusts, and smuts
Class Fungi Imperfecti	A miscellaneous assortment of fungi

SUBKINGDOM EMBRYOPHYTA

Plants forming embryos.

Phylum Bryophyta	Multicellular plants, usually terrestrial, with alternating asexual and sexual generations; prominent plants are gametophytes (sexual generation), on which the asexual sporophytes are dependent
Class Musci	Mosses
Class Hepaticae	Liverworts
Class Anthocerotae	Hornworts
Phylum Tracheophyta	Vascular plants
Subphylum Psilopsida	Leafless and rootless vascular plants
Class Psilophytinea	
Order Psilophytales	Fossil remains only
Order Psilotales	
Subphylum Lycopsida	Club mosses
Class Lycopodineae	Club mosses
Subphylum Sphenopsida	Horsetails
Class Equisetineae	
Subphylum Pteropsida	Plants with large leaves and complex conducting systems
Class Filicinae	Ferns
Class Gymnospermae	Conifers, cycads; no true flowers or ovules, seeds naked on cone scales
Subclass Cycadophytae	
Order Cycadofilicales	Fossil seed ferns
Order Bennettitales	Fossils

[11] Based on C. Villee, *Biology,* 3rd ed., Saunders, Philadelphia, 1957.

Order Cycadales	Cycads, the most primitive living seed plants
Subclass Coniferophytae	
Order Cordaitales	Fossil remains
Order Ginkgoales	Maidenhair tree (*Ginkgo*)
Order Coniferales	Common evergreen trees and shrubs
Order Gnetales	Small trees or climbing shrubs resembling angiosperms; tropical
Class Angiospermae	Flowering plants with seeds enclosed in an ovary
Subclass Dicotyledoneae	Most flowering plants; vascular bundles in rings, flowers in twos, fours, or fives, two seed leaves, or cotyledons
Subclass Monocotyledoneae	Grasses, lilies, and orchids; vascular bundles scattered, flower parts in threes or sixes, leaves with parallel veins, one seed leaf

A survey of the animal kingdom[12]

Phylum Protozoa	Unicellular or sometimes aggregate of cells, microscopic
Class Flagellata	Primitive forms with flagella
Class Sarcodina	Protozoa that move by pseudopods
Class Sporozoa	Parasitic protozoa, nonmotile, reproduce by spores
Class Ciliata	Protozoa that move by cilia
Class Suctoria	Protozoa with cilia in early stages, stalked adult stage, some parasitic
Phylum Porifera	Sponges (fresh-water and marine)
Phylum Coelenterata	Hydra, jellyfish, and corals; radially symmetrical forms with two layers of cells making up body wall, which surrounds a central gastrovascular cavity, stinging cells (nematocysts)
Phylum Ctenophora	Sea walnuts, or comb jellies; locomotion by means of eight bands of cilia, lack stinging cells
Phylum Platyhelminthes	Flatworms; bilateral symmetry, true nervous system, flame cells as excretory structures
Phylum Nemertea	Proboscis worms; marine, nonparasitic, protrusible proboscis equipped with a hook, blood vascular system
Phylum Nematoda	Roundworms; bilateral symmetry, free-living or parasitic in plants and animals, very large phylum
Phylum Acanthocephala	Hook-headed worms; parasitic
Phylum Chaetognatha	Free-living marine worms
Phylum Nematomorpha	Horsehair worms; about 6 in. long, brown or black, larvae parasitic in insects
Phylum Rotifera[13]	"Wheel animalcules"; microscopic, complete digestive system, flame cells (sometimes classified as Phylum Aschelminthes with gastrotrichs)
Phylum Gastrotricha	Microscopic, somewhat like rotifers, sometimes mistaken for protozoa

[12] Based on *ibid.*
[13] Or Phylum Trochelminthes , or sometimes classified with gastrotrichs as Phylum Aschelminthes.

Phylum Bryozoa	"Moss" animals; marine, microscopic, forming colonies
Phylum Brachiopoda	Lamp shells; marine
Phylum Phoronida	Marine, worm-like, live in a leathery tube
Phylum Annelida	Segmented worms, including earthworms and leeches; some marine forms, some small fresh-water and terrestrial worms
Phylum Onychophora	A few species of tropical forms, with respiratory system like insects and excretory system like segmented worms
Phylum Arthropoda	Segmented, with jointed legs, hard, chitinous skeleton, body divided into head, thorax, and abdomen
Class Trilobita	Fossil trilobites
Class Crustacea	Crabs, lobsters, barnacles, water fleas, and sow bugs; two pairs of antennae and gills
Class Chilopoda	Centipedes
Class Diplopoda	Millipedes
Class Arachnoidea	Scorpions, spiders, ticks, mites, and king crabs
Class Insecta	Insects; head with four pairs of appendages, thorax with three pairs of legs, usually two pairs of wings, respiration by tracheae
Phylum Mollusca	Soft-bodied, unsegmented, usually covered by a shell, ventral foot for locomotion, respiration by gills protected by a mantle
Class Amphineura	Chitons
Class Scaphopoda	Tooth shells
Class Gastropoda	Snails, whelks, abalones, and slugs; asymmetrical, with single spiral shell (or no shell)
Class Pelecypoda	Oysters, clams, mussels, and scallops; shell of two valves (bivalves)
Class Cephalopoda	Squids, octopuses, and cuttlefish
Phylum Echinodermata	Marine forms, radially symmetrical as adults and bilaterally symmetrical as larvae, water-vascular system, tube feet
Class Asteroidea	Starfishes
Class Ophiuroidea	Brittle stars
Class Echinoidea	Sea urchins and sand dollars
Class Holothuroidea	Sea cucumbers
Class Crinoidea	Feather stars and sea lilies; few living forms, mainly fossils
Phylum Chordata	Bilaterally symmetrical forms with a notochord and dorsal nerve tube
Subphylum Hemichorda	Acorn worms; marine
Subphylum Urochorda	Tunicates
Subphylum Cephalochorda	Amphioxus
Subphylum Vertebrata	Forms with backbone of vertebrae, well-developed brain, ventrally located heart, usually two pairs of limbs
Class Agnatha	Lampreys, hagfish, and fossil ostracoderms
Class Placodermi	Fossil spiny sharks

Class Chondrichthyes	Skates, sharks, rays, and chimaeras
Class Osteichthyes	Bony fishes
Class Amphibia	Frogs, toads, and salamanders
Class Reptilia	Lizards, turtles, snakes, crocodiles, fossil dinosaurs, and other forms
Class Aves	Birds; warm-blooded, feathers
Class Mammalia	Skin covered with hair, mammary glands, warm-blooded
Subclass Prototheria	Monotremes (duckbill platypus and spiny anteater)
Subclass Metatheria	Pouched mammals such as opossums, kangaroos, and wombats
Subclass Eutheria	Placental mammals

Collecting and identifying organisms

Collecting and examining water forms

Several kinds of dip nets are used for collecting small water forms such as insect larvae, crustaceans, and plankton. One type of net, used to brush through pond weeds and mud or sand at low tide (Fig. 11-24), should have a strong wire frame, a diameter of about 6 to 8 in. and strong netting, either bobbinet or nylon. It is well to have two kinds of nets which may be fastened to the same handle: a fine-weave net to catch the smaller specimens and a coarser net to catch larger forms (for example, crayfish, fiddler crabs, and mussels) and to brush through weeds. For ease in handling, the handle should not be more than 3 ft long.

Transfer the collected specimens into shallow white pans or place them on large sheets of paper and spread them out in the sun for inspection.

When pond water is collected, large, clean jars should be filled with the water and some of the mud from the bank added. When plants are collected, avoid packing too many in one container. Some submerged branches and aquatic plants should also be collected. These may be transported to the laboratory in wet newspaper or plastic bags. Within 5 hr transfer the pond water, twigs, and plants to large battery jars or small aquaria.

After the mud settles, identify the swimming specimens with a hand lens or binocular microscope. In the spring you are likely to find *Daphnia* and larvae of mosquito, dragonfly, damsel fly, and mayfly (Fig. 11-25). In samples of mud you may find caddice fly cases and *Tubifex* worms (Fig. 1-22).

Microorganisms congregate at different levels in the battery jars—some are surface forms, others bottom dwellers. For a rapid inventory of the microscopic forms that have been collected, place some clean coverslips on the bottom of the containers and float others on the surface of the water.

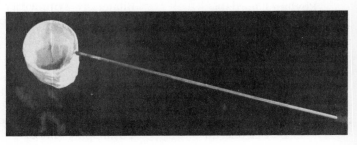

Fig. 11-24 One type of dip net for collecting aquatic forms. (Courtesy of General Biological Supply House, Inc., Chicago.)

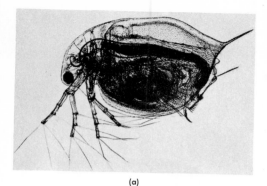

(a)

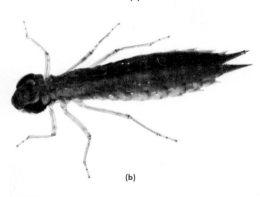

(b)

(c)

(d)

Fig. 11-25 Aquatic specimens that may be collected on a field trip: (a) *Daphnia*; (b) dragonfly nymph; (c) damsel fly nymph; (d) mayfly nymph. (a, b, courtesy of Carolina Biological Supply Co.; c, d, photos by Hugh Spencer.)

If these coverslips are left in these positions for several hours (or overnight), many microorganisms become attached to them. Carefully remove the coverslips with forceps, place them on a drop of water on clean slides, and examine under the microscope (see key to protozoa, p. 589). If you scrape the surfaces of submerged leaves and examine the scrapings under the microscope, you may find protozoa such as *Amoeba* (Fig. 12-1) and *Vorticella* (Figs. 2-8f, 7-4), and some flatworms such as planarians (Fig. 12-10), as well as insect eggs, rotifers, and some snails. Break apart swollen, rotting twigs which have been submerged. Look for snails, planarians, and hydras (Figs. 12-9, 12-10, 12-12) and, in clean water, specimens of the fresh-water sponge *Spongilla*.

Keep all these containers in moderate light. Maintain at a temperature below 24° C (75° F), and keep the containers covered to prevent evaporation. Plan to subculture certain organisms important to your work by isolating them in containers of aquarium water in which plants and animals have been growing (see culture methods, Chapters 12, 13).

Plant specimens may include diatoms, blue-green and green algae, elodea, duckweed, *Cabomba*, and other types of water plants. Spores, eggs, and encysted forms develop quickly in the laboratory. Other forms become dormant for a time and reappear within a period of a month or so. This is excellent material for studies of food cycles and ecological patterns in different biomes.

In some cases the collection of living material is not feasible, and the materials, such as soft-bodied specimens, might better be preserved in the field. Methods for preserving specific animal forms are given in brief, easy reference form in Table 11-2.

Collecting algae

While some algae may be collected on land, especially in damp or boggy places, most fresh-water algae are found in ponds and lakes. Marine algae may be studied immediately or kept for a short time in marine aquaria (pp. 600, 627),[14] but they cannot be maintained in class as readily as fresh-water forms.

When collecting fresh-water algae (Fig. 11-26), it is best to get specimens which thrive in slow-moving bodies of water; these forms have a better chance of surviving in the school aquarium. Avoid overcrowding the specimens; transfer them as soon as possible into large containers in the laboratory. If you can refrigerate the specimens, you will increase their viability.

In shallow water you may find the slow-moving, blue-green filaments of *Oscillatoria* (Fig. 11-26b). This form is also found in damp places (such as the outer black layer on damp flowerpots). The blue-black mats on the surface of damp soil in the pots are also likely to be *Oscillatoria*. *Euglena* (Fig. 11-26a) may be found in shallow pools, and you may well find desmids (Fig. 13-2) of several kinds in the greenish mud of very shallow ponds and at the edge of a lake. The silky threads which float in sunny spots of ponds, lakes, and ditches, and have a slippery feel between the fingers, are likely to be those of *Spirogyra* (Fig. 2-11a). And in the same locale you may find the water net *Hydrodictyon* (Fig. 11-26c). The colonial green algae *Volvox* (Fig. 11-27), composed of hundreds to thousands of cells, are also inhabitants of lakes and ponds. *Volvox* are green spheres, about the size of the head of a pin, which can be more easily recognized with a hand lens.

You may also recognize *Vaucheria*, which is found as a green felt covering on rocks in ponds and ditches or on flowerpots. On the other hand, a green, fuzzy or hairy covering on rocks in ponds and slow-moving rivers may be the simple filamentous alga *Ulothrix*. Note the difference be-

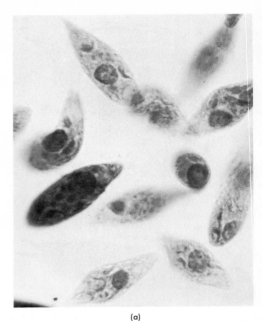
(a)

(b)

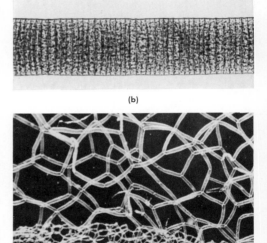
(c)

Fig. 11-26 Some fresh-water algae: (a) *Euglena*; (b) *Oscillatoria*, showing portion of single filament of this blue-green alga; (c) *Hydrodictyon*, showing two sizes of net. (a, photo by Hugh Spencer; b, c, courtesy of Carolina Biological Supply Co.)

[14] If students search for *Fucus* in the fall, they will find that fertilization is going on; antheridia are orange-red in color and the branches containing eggs are a dull green. Dry and preserve *Fucus* and other marine forms, such as *Ulva* and the delicate red algae, on cardboard or bristol board.

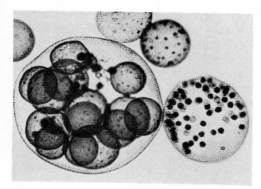

Fig. 11-27 *Volvox globator,* a colonial green alga, with sexual stages and several daughter colonies. (Courtesy of Carolina Biological Supply Co.)

tween *Vaucheria* (Fig. 8-41) and *Ulothrix* (Fig. 2-11b). *Vaucheria* has branching filaments which lack cross-cell walls; it is multinucleated, a syncytium.

In damp places on land (especially in greenhouses) or on damp bark, fences, and rocks, students will find a green covering made of *Protococcus* (or *Pleurococcus*). This simple form (Fig. 11-28) seems to be more likely a reduced form, rather than a primitive alga. Yet it is a useful plant form for class study.

In samples of pond water, along with desmids (which can be recognized by their symmetrical halves), you should find the primitive green alga *Chlamydomonas* (Fig. 2-12b). Under high-power magnification look for the chloroplast, two contractile

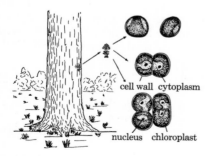

cell wall cytoplasm

nucleus chloroplast

Fig. 11-28 *Protococcus (Pleurococcus),* a simple green alga found on damp bark. (From J. W. Mavor, *General Biology,* 3rd ed., Macmillan, 1947.)

vacuoles, and two flagella in the anterior region. Some species also have a red eye-spot.

Collecting fungi, mosses, ferns, and seed plants

These plants may be transported in small boxes along with some of the soil in which they were growing. Refer to Chapter 13 for places to locate specific plants and descriptions of ways to maintain the plants in school.

Collecting insects

Collecting nets for capturing insects differ in size from the water nets previously described. One type, a light-weight collecting net, is used primarily for the collection of fragile insects such as butterflies, moths, and dragonflies. The net is fine mesh, possibly nylon, sewed to a rim about 1 ft in diameter. The net should be at least two times deeper than the diameter of the rim, since a twist of the net is made when a specimen is caught in the bottom of the net. This confines the specimen and makes it possible to transfer it to a collecting bottle. The handle should be light-weight and about 3 ft long.

Another type of net, a sweeping net, is used to brush or sweep through tall vegetation; different kinds of insects may be captured in this way, including many beetles. This net should have a stout wire rim and a stout handle. The length of the handle depends somewhat upon the individual, since increased length will need additional strength. In fact, some nets on the market have handles only about 6 in. long. The sweep net may be made of white muslin, nylon, or duck; the diameter of the wire rim should be about 1 ft. (Again, this depends upon the strength of the individual.) The depth of the net should be at least one and one-half to two times the diameter of the net.

When hunting insect specimens with a net, twist the net to enclose the insects in the bottom. Then transfer them into a killing jar (described below). When flying specimens or delicate-winged lepidoptera have been captured, their wings may be damaged as they struggle within the net; this can be avoided by placing a drop of ether or chloroform on the net. Then they may be transferred to a killing jar and later to an insect case.

The culturing of certain insects useful in classroom demonstrations and investigations is described in Chapter 12.

Experience in field work will result in the development of some ingenious devices for carrying the variety of bottles and jars which should be on hand. A sturdy knapsack is desirable; it is usually expandable so that it may hold jars of many sizes. Both hands then are free for work. Plastic vials and bottles are preferable to glass ones; these make the "pack" lighter and are practically unbreakable. If a section of the plastic containers is rubbed with sandpaper or emery cloth to roughen it a bit, it is possible to write on the bottles with pencil. These markings may be washed off later in the laboratory or classroom. (Some plastic bottles, available from supply houses, are coated with a special lining so that strong chemicals may be stored in them.)

Small vials are useful in collecting arachnids, larvae, and other soft-bodied forms that must be preserved immediately in alcohol. Two or three uniformly sized vials should be prepared as killing jars, others should contain alcohol (70 percent), and the remainder might be left as transfers, to carry specimens from the net to a large jar.

Out in the field you will also need medicine droppers (carry them in one envelope), as well as forceps, hand lenses, and prepared envelopes in which to pack lepidopterans, as described below.

To repeat, in the field, larval stages and soft-bodied insects should be dropped into vials of 70 percent alcohol. Methods for preserving these forms permanently are described later in this chapter. However, lepidoptera, green insects, most beetles, flies, and bees (and other Hymenoptera) should not generally be put into alcohol.

Killing jars. Prepare both large and small killing jars for collecting insects in the field. Since the crystals and fumes of the potassium cyanide contents of the standard killing jar are deadly and poisonous to handle, it is advisable to purchase the jars ready-made.

As the bottles are used, moisture accumulates in the bottom from the secretions of captured insects. To avoid this, cover the plaster with a thin sheet of blotting paper or cork sheeting.

A safer type of killing jar is made by packing rubber bands into the bottoms of jars of different sizes, then soaking the rubber with carbon tetrachloride (or Carbona). Cover this with absorbent cotton and pack tightly. Finally, cover this with a circle of cardboard or a thin disc of cork to hold the materials in place. Since the fumes are not as lingering as cyanide, lift the cork disc and add carbon tetrachloride from time to time.

When large, fragile-winged specimens are placed in killing jars, they may struggle furiously, since the fumes act slowly. Wet a bit of cotton with ether or chloroform (also carried in the knapsack) and put this into the killing jar, or dab it on the insect net before you transfer the captured animal to the killing jar, so that the wings will not be damaged.

Paper envelopes for large-winged insects. Dragonflies, damsel flies, butterflies, and Dobson flies should be put into special envelopes (Fig. 11-29) to prevent wing damage. In fact, these insects may be stored in the laboratory in these envelopes until you are ready to mount them. At that time they may be put into a relaxing jar, then placed on a wing spreader, and finally mounted (all three techniques are described later in this chapter). You can purchase these envelopes, made of cellophane, from supply houses; or take slips of paper of various sizes, the largest about the size of a postcard, and fold them as shown. Students should have a good supply of them before starting out on a field trip. As they are filled, label all col-

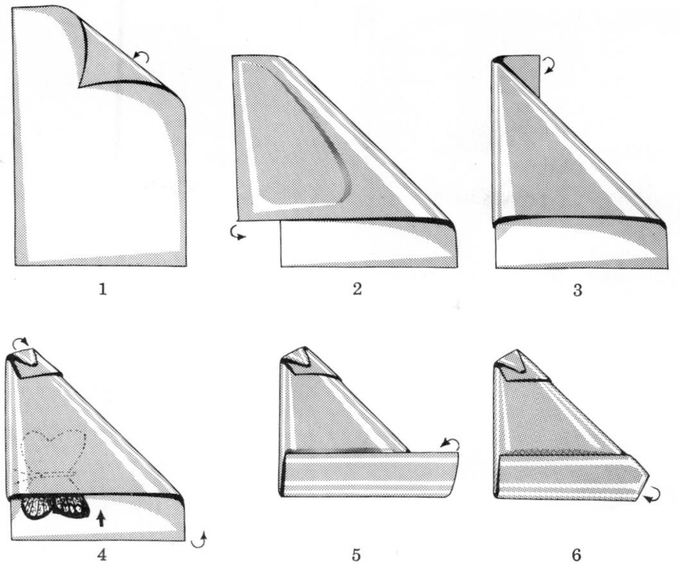

Fig. 11-29 Steps for preparing paper envelopes for transporting large-winged insects. (After F. E. Lutz, *Fieldbook of Insects,* 2nd ed., Putnam's, 1921.)

lected materials, giving location, date, time of day, and ecology of the collecting area.

Collecting other land animals

While we have thus far confined the discussion of collection of land forms to the insects, spiders and other arachnids may also be collected by net. Earthworms, even earthworm cocoons (Fig. 11-30), slugs, land snails, and insect eggs (as well as vertebrates such as salamanders), should be collected in cigar boxes or jars containing soil. Decaying tree stumps may be transported to the laboratory (in moist newspaper and plastic bags), and placed in covered containers (since termites may be in the decaying wood). The flagellates in the intestine of termites, sow bugs, and related forms are excellent for microscopic study. (See p. 537 for methods of preparing slides of symbiotic flagellates.)

Identifying insects

Prior to a field trip, the main orders of insects may be distinguished by studying illustrations in textbooks. On field trips many students like to be able to distinguish a beetle from a wasp or a cicada. An introduction to the notion of classifying animals and plants is given in Chapter 10, where the categorization of living things was compared to a library system of classifying books.

Students may devise their own simplified classification system to "key" out unknown insects when they know the basic patterns of differences among the orders. Broadly speaking, what kinds of variations distinguish the orders of insects? How would students classify the specimens in Fig. 11-31? Students might collect

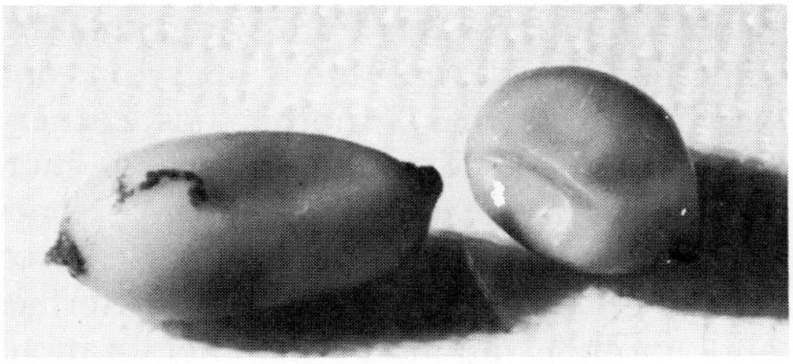

Fig. 11-30 Cocoons of an earthworm (cocoons are slightly larger than an ordinary wheat grain). (Courtesy of Carolina Biological Supply Co.)

Fig. 11-31 Representatives of different orders of insects: (a) praying mantis; (b) June beetle; (c) cabbage butterfly; (d) damsel fly (adult). (a, b, c, courtesy of Carolina Biological Supply Co.; d, photo by Harold V. Green.)

many samples of insects, then sort them into probable similar groups. How does a beetle differ from a butterfly, a dragonfly, a bee, a fly, or a grasshopper? What are the distinguishing features that set apart one order from others? While students will recognize that insects, in general, have three body parts, two pairs of wings, and three pairs of jointed legs, there are some large differences that are evident to students. One order has front wings that are scale-like (beetles), another order has one pair of wings (flies), another has leathery front wings and color-patterned hind wings which pleat below the upper ones (grasshopper group), another has colored wings of overlapping scales (butterflies and moths), another has membranous wings and a constricted "waist" (wasp and bee group). Representatives of these orders and a few others are shown in Fig. 11-32.

There are many, many orders of insects that have not been included in this much abbreviated introduction.

Now, *within* an order of insects, such as the Coleoptera, students should be able to choose the traits which differentiate ground beetles from cucumber beetles, weevils, longhorned beetles, or ladybird beetles. Similarly, they should be able to distinguish families within the order Hymenoptera. Why are wasps in a different family from ants or bees?

When students have learned the traits of each order and have a nodding acquaintance with the notion of families, they may look again with sharpened observation for the distinguishing differences *within a family*. Students may turn to a professional key such as those found in books on insects (see the Bibliography at the end of the book) and learn the genera which make up specific families of insects. Furthermore, *within one genus*, for example, among ladybird beetles or potato bugs, what variations may be counted?

At the same time you may want to use the studies of variations described among protozoans (Chapter 12) or evergreens (Chapter 10); or refer to other examples of variations and their causes as developed in Chapter 10.

Preserving animal specimens

You may want to refer to Table 11-2 and to the descriptions that follow for some notes on the preservation of animals

INSECTA

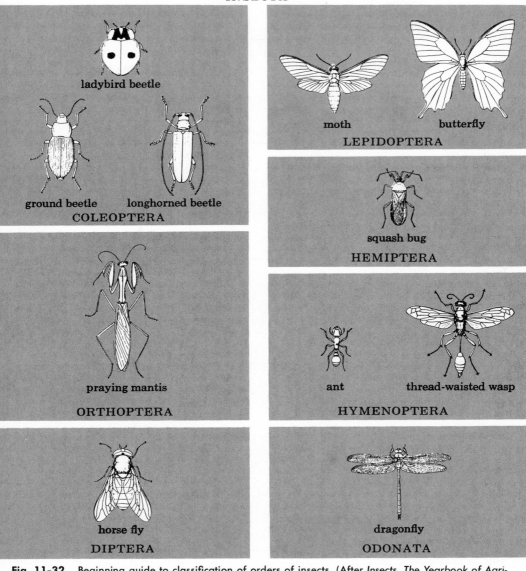

Fig. 11-32 Beginning guide to classification of orders of insects. (After *Insects, The Yearbook of Agriculture,* U.S. Dept. of Agriculture, 1952.)

TABLE 11-2
Condensed outline of methods of collecting and preserving animals*

animal	where found	special collecting devices	how to kill	fixative	preservative
Fresh-water sponges	Mid-summer in fresh water attached to branches and submerged wood	Flat-bladed knife or scalpel	70% alcohol, changed when it becomes discolored	70% alcohol	70% alcohol
Hydra	Lagoons, ponds, rivers, lakes, attached to vegetation, stones, fallen leaves	Flat-bladed knife or scalpel and pipette	Hot Bouin's solution flooded over specimens from base to peristome; or use menthol	Bouin's solution	70% alcohol
Fresh-water planarians	Fresh spring-fed streams, lakes, rivers	Fresh liver placed in water where *Planaria* are found	Use menthol crystal method; or extend on glass slide and submerge in hot Gilson's or corrosive sublimate†	Gilson's or corrosive sublimate†	Formalin or 70% alcohol
Tapeworms	Intestine of dogs, cats, rabbits, sheep	Scalpel and forceps	Relax in cold water; wrap animals around support to stretch them and immerse in 10% formalin	Bouin's or formalin	Alcohol or formalin
Ascaris	Intestines of pig, horse, cat or dog	Scalpel and forceps	Dip momentarily into 98° C (208° F) water	5% formalin or saturated corrosive sublimate†	5% formalin or alcohol
Rotifers	Plant material taken from ponds or lagoons	Pipette	Anesthetize with solution magnesium sulfate or menthol crystals	When cilia cease to move add few drops of osmic acid	Wash in water and store in 10% formalin

*From *Turtox Service Leaflet,* No. 2, General Biological Supply House, Inc., Chicago.
†*Caution:* poison.

animal	where found	special collecting devices	how to kill	fixative	preservative
Pectinatella and Plumatella (Bryozoa)	Attached to stems, rocks, leaves in streams, especially in late fall	Scalpel	When fully expanded, flood with boiling Bouin's	Bouin's	70% alcohol
Earthworms	In spring on rainy nights on golf courses or bluegrass lawns	Flashlight and pail	Anesthetize by slowly adding alcohol to water in which worms are (or see p. 112)	5% formalin	5% formalin
Leeches	Hand pick from hosts or with dip net among weeds in ponds and streams	Dip net	Anesthetize in warm chloretone or magnesium sulfate; or asphyxiate in closed jar	Inject with 10% formalin and submerge in same extended position	8% formalin
Crayfish	Streams, ponds, lagoons, in water or burrowing in mud	Dip net, seine, or spade	Drop alive into alcohol or 8% formalin	70% alcohol or 8% formalin	70% alcohol or 8% formalin
Ticks and mites	Cattle, dogs, horses, old cheese, decaying organic matter	White paper and brush for taking specimens from parasitized animals	Drop directly into 70% alcohol	70% alcohol	70% alcohol
Centipedes and millipedes	Under logs or stones	Forceps	Carl's solution	Carl's solution, injected into body cavity	Carl's solution

TABLE 11-2 (continued)*

animal	where found	special collecting devices	how to kill	fixative	preservative
Insects	Woods, fields, water, air—everywhere	Nets, forceps, and other equipment, depending on the kind collected	For drying, in killing jars; for liquid preservation, in alcohol	Alcohol, Carl's solution, chloral hydrate, and special solutions	70% alcohol, Carl's solution, or drying
Slugs	In damp places under leaves, logs, stones		Anesthetize in boiled water which has been cooled, and immerse in formalin or alcohol	Alcohol or formalin	70% alcohol or 8% formalin
Aquatic snails	Streams, ponds, lakes; most abundant among vegetation	Dip net, scraper net	Anesthetize in warm water by adding magnesium sulfate causing them to expand; then drop into 10% formalin	10% formalin	8% formalin
Clams	Streams, lakes, partly buried in the bottom	For large numbers, dredge or crowfoot hooks are used	Place wooden pegs between the two halves of shell and drop into 10% formalin	10% formalin	8% formalin
Lampreys	Occasionally may be taken from fish but for large numbers must be taken in breeding season in streams	Seine	Remove from water for few minutes and inject 10% formalin into body cavity	10% formalin	8% formalin
Fish	Streams, lakes	Nets, seines, or hook and line, depending on kind	Drop into full-strength formalin	10% formalin	8% formalin

*From *Turtox Service Leaflet*, No. 2, General Biological Supply House, Inc., Chicago.

animal	where found	special collecting devices	how to kill	fixative	preservative
Grassfrogs	In meadows or borders of marshy lakes	Net	Inject ether into body cavity or drop into 80% alcohol	Inject 5% formalin into body and place in 5% formalin	5% formalin
Grassfrog eggs	Shallow water of ponds in early spring when singing is started	Jars	Place in fixative	8% formalin	8% formalin
Salamanders	Damp places in woods, ponds, rivers, streams	Hook and line or nets	Inject ether into body cavity and drop into 80% alcohol	5% formalin	5% formalin injected into body cavity
Reptiles	Woods, fields, dunes, depending on kind	Snares for handling poisonous snakes; nets for capturing turtles and aquatic forms	Inject ether and drop into 70% alcohol	10% formalin	8% formalin injected into body cavity
Birds and small mammals	Most of world	For taxidermy, a 12-gauge shotgun and shells with fine shot (No. 8 or 12)*	Bird skins are most generally used for study or reference; body is removed; skin is then stuffed with cotton and dried		
Large mammals			Gas or drown if taken to laboratory alive	Embalm or inject 8% formalin into body and large muscles	8% formalin

*Some taxidermists prefer to use .22-caliber bird shot, which is less expensive; state and federal collecting permits are needed. Small mammals may be easily collected with mousetraps baited with rolled oats.

in the laboratory or in the field. Preserved specimens are useful in lessons on relationships among living things (classification and dissection) and identification of animals, when the weather does not permit outdoor field trips. Also, students may want to learn these techniques for project work.

Preserving soft-bodied specimens

In general, soft-bodied animals should be preserved first in 50 percent alcohol for a few days, then transferred to 70 percent, and finally to 90 percent alcohol. Formaldehyde may also be used as a preserving fluid, although it tends to make the specimens more brittle. Puncture the body of each animal at several widely separated places with a dissecting needle before you put it in the preserving fluid so that complete penetration of the preservative takes place. (See pp. 126, 360 for preparation of some fixatives, or preserving fluids.)

Later, the preserved specimens can be mounted in jars (Fig. 11-33). Each specimen can be supported against a glass slide and fastened to the glass with thread. Use uniformly sized, wide-mouthed jars and seal the jars with paraffin to prevent evaporation of the preserving fluid. Label the jars and cover the labels with plastic tape so that they will not get smudged in use.

In time, the specimens will bleach and fragment; that is a signal to get new specimens and fresh fluid.

In summary, *soft-bodied forms* that may be preserved in alcohol or formaldehyde (formalin, p. 132) include the following.

Jellyfish and *hydras* should be anesthetized before preserving so that they will remain in an extended position. Ways to anesthetize small forms are described on pp. 112 and 126. The specimens will become opaque in preserving fluid. See further instructions in Table 11-2.

Flatworms, roundworms, and *annelid worms* should be anesthetized before preserving. Flatworms can be flattened between two glass slides, which should then be inserted into alcohol. Further notes are given in Table 11-2.

Soft-bodied mollusks, such as squids, become opaque and pinkish when preserved.

Crustaceans, such as hermit crabs, fiddler crabs, and larger crabs, as well as barnacles and the shipworm (*Teredo*), may be preserved or mounted (see below).

Soft-bodied insects, such as grasshoppers, praying mantids, locusts, some beetles (and larvae and pupae stages), should be preserved. Barker's dehydration method[15] may be used to preserve insects (as well as soft-bodied specimens). In this method a fat solvent such as carbon bisulfide or benzol (*caution:* flammable) is added to 80 percent alcohol in the ratio of 1 part of solvent to 10 parts of alcohol. The specimens are punctured and placed in this fluid for 24 to 72 hr, depending upon their size. Then the specimens are transferred

[15] In *How to Prepare Insect Displays,* Ward's Natural Science Establishment, Rochester, N.Y., 1950.

Fig. 11-33 Museum jars showing preserved specimens of invertebrates. (Courtesy of General Biological Supply House, Inc., Chicago.)

to 95 percent alcohol for a similar period of time. The next transfer is to absolute alcohol to which a bit of calcium chloride has been added to absorb any moisture. This dehydrates the specimen. Finally, transfer to xylol and dry on a plaster block. Students may want to mount the insect forms on pins and display them in an insect box. Larvae, as well as insects, may be mounted on pins for display. For methods of pinning insects for display, refer to pp. 565–68.

Spiders, centipedes, millipedes, scorpions, mites, and *ticks* may be preserved in 70 percent alcohol.

Small fish, as well as their *eggs* and *fry,* may be preserved individually in 70 percent alcohol in small vials to make a life-history series.

Among *amphibia,* all the stages in the life history may be preserved (preferably in 70 percent alcohol).

Young bird embryos and *embryos of mammals* should be preserved in a fixative or transferred through a graded series of alcohol (70, to 80, to 90 percent) to prevent undue shrinkage.

Preserving hard-bodied specimens

Preserve the following, where possible, in the dry state.

Sponges may be mounted on bristol board after drying in the sun.

Starfish may be dried in the sun and weighted down in a flat position before complete drying, then mounted.

Sea urchins should be dried in the sun after removing the viscera; with a knife cut away the mouth parts (Aristotle's lantern), and pull away the attached internal organs.

Sand dollars are quite flat and may be completely dried in the sun.

Crabs may be preserved whole in alcohol. Or the carapace may be dissected away carefully, and then the crab dried completely and mounted.

Mollusk shells must be dried out completely (Fig. 11-34). Mount on bristol board or in boxes containing cotton (Riker mounts, perhaps).

Small horseshoe crabs, in various stages of development, may be collected and allowed to dry thoroughly. Then the surfaces can be shellacked (if the forms are not completely dry, the shellac will become opaque). Larger forms of horseshoe crabs must be eviscerated. Then the cleaned shells can be dried.

Young crayfish may be dried in the sun.

Treatment of specimens that have become brittle. When preserved specimens become brittle and fragmented, they should be replaced with new specimens. On occasion, however, you may have a form that is not readily obtainable. A method has been suggested for reclaiming dried specimens, mainly soft-bodied forms such as worms. In this method,[16] the specimens are first soaked in 0.5 percent trisodium phosphate solution, then warmed in this solution for about 1 hr at 35° C (95° F). After the forms are removed from the heat they should be kept in a closed container to avoid evaporation. Within 1 hr the specimens should appear normal; within 2 days they should be soft.

Special techniques for mounting and displaying insects

Mounting insects. While it is best to mount insects soon after collecting, this step may be postponed. Keep the larger insects in the paper envelopes which you brought from the field. However, as time passes insects become brittle, and antennae, legs, and wings break. Better specimens of lepidoptera are hatched from pupae rather than captured in nets.

Such insects as lepidoptera, dragonflies, and damsel flies are mounted with their wings fully extended. Grasshoppers are mounted with one wing outspread (Fig. 11-35) to show the color and pattern of the underlying wing, a point often required for identification. Frequently the wings of

---
16 H. Bennett, ed., *Chemical Formulary,* Chemical Publishing Co., New York, 1951, p. 106.

(a)

(b)

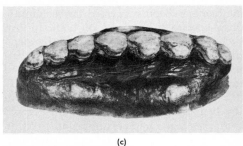

(c)

(d)

(e)

Fig. 11-34 Dried and mounted mollusk shells: (a) giant sea scallop (5 to 6 in. in diameter); (b) *Venus mercenaria*—the quahog, or marine clam; (c) *Katharina tunicata*, a large Pacific chiton (3 to 4 in. long); (d) *Urosalpinx cinerea*, the common Atlantic coast oyster drill that bores small holes in oyster shells, causing considerable damage to oyster beds; (e) channeled whelk, or conch. (a–d, courtesy of Carolina Biological Supply Co.; e, photo by Hugh Spencer.)

cicadas, lacewings, and Dobson flies are spread before they are mounted.

If insects have become brittle they may be softened in a relaxing jar, then placed on a spreading board if their wings are to be extended.

Relaxing jars. Prepare several jars of different sizes to accommodate insects of varying sizes. Into the bottom of each glass jar insert a wad of wet cotton to which a few drops of carbolic acid have been added (to inhibit the growth of molds). Cover this with a layer of blotting paper. (You may prefer to substitute moist sand at the bottom of the jar, in which case add a few drops of carbolic acid to the blotter.)

Students should now place the dried insects on the blotter; cover the jar. After 24 hr the insects should be soft enough to mount or spread on a spreading board. Nevertheless, they must be handled with some care, as they are not as pliable as they were when freshly killed.

Spreading the wings of insects. The

kind of spreader shown in Fig. 11-36 may be purchased from biological supply houses, or constructed by students. The spreader consists of two side sections made of soft wood, with a channel along the center just wide enough to accommodate the body of an insect such as a moth or grasshopper.

Students should learn to pin insects such as dragonflies, damsel flies, caddice flies, butterflies, and moths through the thorax and insert the mounting pin into the cork in the groove of the wing spreader. Use forceps to insert the pin into the cork so that the pin does not bend under the pressure of the fingers.

Arrange the legs of the insects as they are in life, and use forceps to spread out the wings. Pin down the outspread wings by inserting pins along the front marginal part of the wings. Use strips of paper as shown to hold the wings in position and fasten them with pins.

The wings will dry in a slightly tilted position, but their weight will cause them to droop a bit when the insects are finally mounted. You will find that the length of time needed for the insect to dry out depends upon the size of its body. It may vary from 1 day to 2 weeks.

Mounting insects for display. The usual method for mounting insects for observational study or exhibit is to set them on pins which can be inserted into a cork-lined box. Cigar boxes make fine display cases; even more attractive are clear plastic boxes with a bottom lining of pressed cork, corrugated paper, or balsa wood.

Use special black enamel insect-mounting pins about 1½ in. long. The commonly used grades are Nos. 1 and 3. There is a uniform method for pinning insects.

Most insects are pinned through the thorax. Moths and butterflies are pinned squarely in the center of the thorax. On the other hand, grasshoppers, flies, bees, wasps, squash bugs, and the like are pinned to the right of center in the thorax. Beetles are not pinned through the thorax at all, but through the right wing cover about a quarter of the way back.

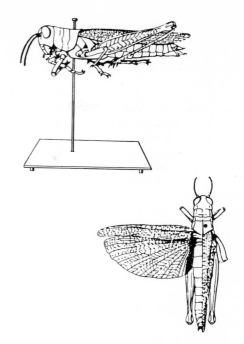

Fig. 11-35 Technique for mounting grasshoppers. Note that the wings on one side of the body are extended to show coloration of the underwings. (From *Insects, The Yearbook of Agriculture*, U.S. Dept. of Agriculture, 1952.)

Very small beetles, such as weevils, are too small to be pinned. First, fasten these small insects to small triangles of lightweight stiff paper, such as library card stock, using thinned white shellac. Then

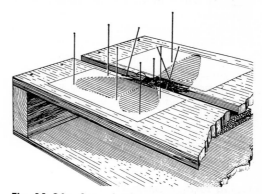

Fig. 11-36 Spreader for spreading and drying wings of such insects as butterflies, moths, and dragonflies. (From *Insects, The Yearbook of Agriculture*, U.S. Dept. of Agriculture, 1952.)

insert the pin through the broad base of the paper triangle. Small fragile insects such as mosquitoes may be first pinned to a bit of cork (Fig. 11-37).

To label insects, prepare a small strip of paper (about ¼ × ½ in.) carrying the name, location, and date of collection. Place this under the insect by transfixing it with the same pin.

Insects that have a high fat content leave brown stains at the base of the pin. You may avoid this by immersing the insect (except lepidoptera) in carbon tetrachloride for 2 to 3 hr before pinning it in the box.

Care of insect collections. Some small insects attack preserved insects. To keep these museum pests under control, heat a number of pins and insert them through small chunks of paradichlorobenzene. Pin the paradichlorobenzene in two corners of each box which houses your collection.

Mounting specimens preserved in alcohol. At times it is desirable to show the complete life history of a specific insect. This may be important in the life history of insect pests or of special beneficial insects. Place each stage (eggs, larvae, and pupae) into separate small vials of 70 percent alcohol or 5 percent formalin. Seal the vials with paraffin. Keep the adult stage or nymphs dried.

Mount the small vials and the adult stage in Riker mounts (Fig. 11-38) or in homemade display cases containing cotton. Students may make display cases by cutting away the paper cover of several boxes of the same size, leaving a ½-in. margin as a frame. Tape a sheet of glass or of plastic to the inside of the cut-away box cover, and line the box with cotton. Lay the specimens in the cotton and close the box with the glass cover. Insert pins on each of the four sides of the box to fasten the cover to the box.

Going further: projects

Some students may prepare valuable collections of a single species of insect. For example, a study may be made of varia-

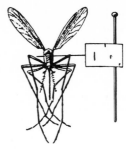

Fig. 11-37 Technique for pinning and labeling small insects. A mosquito is shown pinned with a "minuten nadeln" to a block of cork on a regular insect pin. (From *Insects, The Yearbook of Agriculture,* U.S. Dept. of Agriculture, 1952.)

tions of striping among potato beetles, or the number of spots in one species of ladybird beetles. Some students have successfully embedded specimens in polyester resins.[17]

Students and teachers who want more advanced information on handling of insects are fortunate, for there is a tremendous literature on collecting, preserving, exhibiting, and identifying insects. A book such as D. J. Borror and D. M. DeLong, *An Introduction to the Study of Insects,* rev. ed. (Holt, Rinehart and Winston, 1964) is a valuable guide, providing interesting details of each order of insects. Further, it gives detailed suggestions for projects concerning insects, for collecting and preserving them. Students will also enjoy looking into H. Kalmus, *101 Simple Experiments with Insects* (Doubleday, 1960).

[17] Most biological supply houses offer information for this technique.

Fig. 11-38 Riker mount, which is useful for displaying vials containing early stages of an insect's life cycle or whole mounts of the adult form. (Courtesy of General Biological Supply House, Inc., Chicago.)

For further information on insects that live in water or lead a partially aquatic existence, look into the chapters dealing with insects in H. B. Ward and G. C. Whipple, *Fresh-Water Biology*, 2nd ed., edited by W. T. Edmondson (Wiley, 1959) and R. Pennak, *Fresh-Water Invertebrates of the United States* (Ronald, 1953). These and other valuable books are listed in the Bibliography.

CAPSULE LESSONS: Ways to get started in class

11-1. Seal a snail in a test tube (Fig. 6-1) and ask how long it will live. Elicit from students the idea of the dependence of living things on green plants. Then develop different kinds of nutrition among plants and animals (saprophytes, parasites, symbiotic relations, and other special types as described in this chapter).

11-2. If possible, begin with a field trip to examine a specific plant-animal community or habitat: pond, desert area, beach, woods, meadow, or bog. Have students describe the food chains in the habitat. You will find R. Buchsbaum and M. Buchsbaum, *Basic Ecology* (Boxwood Press, 1957) a useful guide.

11-3. Have students bring in samples of soil from different regions and examine under a microscope the living organisms found in the soil. Develop the notion of the interdependence of these plants and animals. What would happen if one kind of organism disappeared?

11-4. There are soil-testing kits available for purchase. Demonstrate as described in this chapter the acidity or alkalinity of some soils around the school area. How would living things be affected if the acid soil of swamps or bogs became more alkaline? Use compost formed from decaying leaves for tests of acid soil.

Develop the idea that plants require certain conditions for growth, such as the proper pH. In fact, the degree of acidity of the soil determines, in some measure, the availability of minerals to the plants.

11-5. Draw a diagram of the primary producers, primary consumers, secondary consumers, and reducers (decomposers) in a pond (as in Fig. 11-6). Ask students to complete the picture and to explain the relationships. Trace food chains, cycles of oxygen and carbon dioxide supplies, and the nitrogen cycle.

11-6. Make *Chlorella* central to a lesson. Use the lesson as a basis for introducing students to a block of preludes to investigations concerning *Chlorella* (pp. 542–44, also pp. 190–92).

11-7. Ask a student to give an account of his experiences as a fisherman. What did he catch? Was it a predator or prey? Upon what did it feed? Trace the food chain back to a primary producer. Be certain to include the decomposers in the cycles. What effect did the removal of the catch from the stream have on the plant and animal community in the stream?

11-8. That minute bacteria have an essential role in the balance of nature may be developed by asking students to explain what would happen if all bacteria of decay disappeared from the earth. This is the time to develop the idea that not all bacteria are disease-producing; a good number are helpful.

11-9. Begin a study of the relationship between water and soil by demonstrating the ability of different soils to hold water. Use funnels of different soils as described in this chapter. Have students explain how cover plants and forests prevent floods and the erosion of soil.

11-10. You may plan to have a committee of students make a survey of good land practices in use in your community. They might also recommend conservation measures that could effectively be introduced. Perhaps they can write an article for a newspaper—either school or community—to report their findings and their suggestions.

Students should learn to explain in simple language the usefulness of these farming techniques: crop rotation, strip cropping, ways to spread ground water, terracing, use of green manure, irrigation, contour plowing, and the use of fertilizers and the proportion of sulfates, phosphates, and nitrates these should contain for specific crops.

11-11. Take a field trip around the school grounds and list places that show erosion of soil by wind, water, or ice. Decide the kind of conservation practices that could be used effectively in each area. Have students explain the value of each of their suggestions.

In some cases a class or a club might put some of these ideas into practice. They might terrace the lawn or a piece of land nearby to reduce a sharp angle of an existing slope. Or plant saplings or cover crops on a hillside or a burned-over area. If a photographic record is kept year by year, students in subsequent classes might continue the practices when they see how effective they have been.

11-12. Perhaps a debate may be planned around the proposition "Conservation Is Today's Most Pressing Problem."

11-13. Have students report on the destruction of soil by overgrazing. Also plan to have reports on the use of radioactive isotopes as tracers added to fertilizers to learn the amounts and the kinds of minerals that different crop plants take from the soil. Many reference materials are available from the U.S. Atomic Energy Comm. and from current magazine articles.

11-14. As a student or club project, the value of attracting birds to a garden (school grounds) may be investigated. For example, a survey of the kinds of birds attracted to a given area could be made. Then birdhouses can be built to attract birds to the school area. Although many may consume seeds, how effective in general are birds in holding down the insect population? Perhaps students can explain the practice of planting berry shrubs and trees along rows of cultivated crops.

While chicken hawks may consume some chickens, why is it a poor practice for a community to offer a bounty for dead hawks?

(Many colleges of forestry and departments of agriculture in colleges offer splendid publications. For example, you may send for a list of publications of the New York State College of Agriculture, Cornell Univ., Ithaca, N.Y., to learn the kinds of materials available for student work.)

11-15. Perhaps your school is located in a vicinity where you may solicit the help of a forester (or a county agent). He may, on a field trip, demonstrate how forests are handled as crops, the succession of trees in woods and climax forests, what kinds of seedlings are pioneers in a new region, and whether the prevailing winds carry light seeds or heavy seeds greater distances into new regions. Students might prepare questions in advance to ask the expert. What are the new practices in forest management?

You may ask a county agricultural agent to explain how to test crops and soil for mineral deficiencies.

11-16. In a broad view of human conservation, you may want to relate conservation of resources to constructive measures at work in the community such as improvement of man's neighborhood environment, human health, plans for hospitalization insurance, opportunities for scholarships and education, and play-grounds. What effect has the increased use of electrical power had on man's standard of living and his life span? What will the life span probably be in the year 2000? The population of the United States? The population of the world? Will there be land area to feed this population?

11-17. Can students recognize poisonous plants such as poison ivy or poison oak which may be found in your community? Do they know the harmful plant and animal pests? Can they recognize poisonous spiders, scorpions, and snakes in the community?

11-18. There are many fine films and filmstrips which enable you to take your classes on a "trip" when no other type of trip is possible. Many are listed at the ends of chapters.

11-19. There are excellent bulletins available from experimental stations and state and federal departments of agriculture. You may want to plan a class period of free reading at a time when you can make available large numbers of publications. Students will profit from a perusal of these publications which show scientists at work. Be certain to include the fine Yearbooks of the U.S. Dept. of Agriculture: *Insects* (1952), *Plant Diseases* (1953), *Marketing* (1954) (students may practice reading the many different kinds of graphs in the back of this yearbook), *Water* (1955), *Trees* (1949), *Science in Farming* (1943–47), *Food and Life* (1939), *Soils and Men* (1938), *Soil* (1957), *Seeds* (1961), *Land* (1958), *Food* (1959), and *After a Hundred Years* (1962).

11-20. Plan to study some kind of biome—the plants and animals living together in a given area. What are the needs of living plants? Of animals? Here you may develop the work for the entire year.

11-21. Collect living things for examination under the microscope. What role do microscopic forms of life play in the lives of the other plants and animals in the community?

11-22. Show organisms found in the soil, or the community life in a fallen log. Develop the concept of interrelations, and the checks and balances among living things. What are natural enemies? Look for examples of parasites, saprophytes, and symbionts.

11-23. Possibly as a club project, a section of the school grounds may be used to build a nature trail; or prepare the section for outdoor demonstrations and field work (see suggestions in this chapter).

11-24. Outdoors you may show variations in leaves, in fruits, in shells, in beetles, in the shaded undersurface of leaves compared with the top (barberry), or in the shape of leaves underwater

(a)

(b)

(c)

(d)

(e)

Fig. 11-39 Specimens of fruits and seeds:
(a) pistillate flower clusters and leaves of maple;
(b) leaves and fruit of sugar maple; (c) seeds of
goat's beard, or wild salsify tragopogon; (d) large
staminate cones of Southern long leaf pine; (e) female pine cones (*Pinus nigra*). (a, Jerome Wexler
from National Audubon Society; b, courtesy of the
American Museum of Natural History, New York;
c, Tet Borsig from Monkmeyer; d, courtesy of
U.S. Dept. of Agriculture; e, Roche photo.)

Fig. 11-40 Overhead delineascope in use. (Courtesy of American Optical Co.)

and their aerial shoots. Why do some trees of the same kind flower earlier than others?

11-25. You may want to introduce the topic "Reproduction in Plants" by taking students on a short walk around the school grounds. Students can study the floral structure and compare the flowers of many different plants in bloom without picking the flowers from trees or shrubs. Elicit through questions the essential organs needed for reproduction in flowering plants. How do pollen grains get to the stigma of flowers? How is a seed formed?

11-26. Study seed-dispersal mechanisms outdoors. What are the chances for survival of seeds around the parent plants? What is the rate of migration for plants? What effect does the survival of seeds have on other plants and animals? Examine lawns for seedlings of maple, linden, and others.

11-27. Or use the same motivation to elicit the major ideas of overproduction, variations, need for survival and the "fittest," which are concepts requisite to an understanding of heredity and evolution.

11-28. It may be possible to collect stages in the life cycle of an insect or in the development of frogs or the stages in development from pollination to seeds and fruits, or the distribution of seeds or young seedlings (Fig. 11-39). This might lead into a study of reproduction.

11-29. Field work is a requirement in any study of soil, water, and conservation. (See suggestions throughout Chapter 14.)

11-30. Sometimes it is necessary to bring a "field trip" into the classroom. By means of slides students can trace the destruction of a forest by sand dunes or examine plants of a tundra and communities in a forest, a desert, or a grasslands

biome. The biological supply houses carry series of slides on parasitic plants, behavior, adaptations, parasite-host interrelations, the cycles of oxygen, carbon dioxide, and nitrogen, and food chains.

In developing lessons in ecological relationships, some teachers also use labeled preserved specimens of plants of the Alpine Tundra, of marine algae, or of important crop and pasture plants. Delicate specimens may be displayed; for example, overhead projectors can be used to show fine structures of red algae. Using transparencies students may also trace changes in a pond, or succession in a forest (Fig. 11-40).

PEOPLE, PLACES, AND THINGS

Many teachers are expert in field work; some are not. Over several years, one may become competent in recognizing about a dozen of the trees and shrubs around the community, the orders of insects, and some representative forms of invertebrates and vertebrates in the area. These are the specimens that students bring into class. Someone at a botanical garden or zoological park, or another science teacher nearby may give help, or suggest a source of help. It is possible to send specimens of organisms, carefully packed, to an "expert" at a museum or an agricultural station.

There may be a student at school who has developed competence in some area of field work. You may want to have such a student lead a field trip. Is there a parent in the community, or a student, who has a hobby of bird-watching or insect-collecting? Here is a resource person students may turn to for help in suggesting projects for long-range study.

The work of this chapter is as broad as life. Many people can offer help to the classroom teacher. You may be in a vicinity where a forester can help you, or where a county agent can demonstrate useful practices.

A visit may be planned to a sanctuary, a fish hatchery, an experimental farm using radioactive isotope techniques, a commercial or a college laboratory which contributes to some aspect of the wise development of resources.

Your local conservation, wildlife, and agricultural agencies will be eager to help. Perhaps you may borrow a film or some equipment for a demonstration, or get pamphlets suggesting many demonstrations that students might undertake in class or as individual or club projects.

If your school grounds are limited, perhaps someone in the community has fields, or a lot, or a bit of woodland you may use for field trips. Many schools have made successful arrangements of this sort.

BOOKS

These are only a few of the references that are pertinent to the work discussed in this chapter. These and many other references are compiled in a comprehensive, subject-indexed listing (Appendix A) that includes field guides for identification of plants and animals. Also refer to books listed in Chapters 1, 2, 4, and 9.

Barclay, G., *Techniques of Population Analysis,* Wiley, New York, 1958.

Barnes, H., *Oceanography and Marine Biology,* Macmillan, New York, 1959.

Bates, M., *The Forest and the Sea,* Random, New York, 1960; Signet, New Am. Library, New York, 1961.

————, *Man in Nature,* 2nd ed., Prentice-Hall, Englewood Cliffs, N.J., 1964.

Borror, D. J., and D. M. DeLong, *An Introduction to the Study of Insects,* rev. ed., Holt, Rinehart and Winston, New York, 1964.

Buchsbaum, R., and M. Buchsbaum, *Basic Ecology,* Boxwood Press, Pittsburgh, 1957.

Bullough, W. S., *Practical Invertebrate Anatomy,* 2nd. ed., Macmillan, London (St. Martin's, New York), 1958.

Cain, S., and G. de Oliveira Castro, *Manual of Vegetation Analysis,* Harper & Row, New York, 1959.

Cameron, T., *Parasites and Parasitism,* Wiley, New York, 1956.

Darlington, P., *Zoogeography: The Geographical Distribution of Animals,* Wiley, New York, 1957.

Daubenmire, R. F., *Plants and Environment,* 2nd ed., Wiley, New York, 1959.

Dice, L., *Natural Communities,* Univ. of Michigan Press, Ann Arbor, 1952.

Edmondson, W. T., *see* Ward, H. B., and G. C. Whipple.

Elton, C. S., *The Ecology of Invasions by Animals and Plants,* Wiley, New York, 1958.

Faust, E., C. Beaver, and C. Jung, *Animal Agents and Vectors of Human Disease,* 2nd ed., Lea & Febiger, Philadelphia, 1962.

Frey, D., ed., *Limnology in North America,* Univ. of Wisconsin Press, Madison, 1963.

Hardy, A., *The Open Sea: Its Natural History—The World of Plankton,* Houghton Mifflin, Boston, 1956.

Hubbs, C. L., ed., *Zoogeography,* Am. Assoc. for the Advancement of Science, Washington, D.C., 1958.

Hutchinson, G. E., *A Treatise on Limnology,* Vol. I, Wiley, New York, 1957.

Hyman, L., *The Invertebrates,* 5 vols., McGraw-Hill, New York, 1940–59.

Kalmus, H., *101 Simple Experiments with Insects,* Doubleday, Garden City, N.Y., 1960.

Lack, D., *The Natural Regulation of Animal Numbers,* Oxford Univ. Press, New York, 1954.

Macfadyen, A., *Animal Ecology: Aims and Methods,* 2nd ed., Pitman, New York, 1963.

Moore, H., *Marine Ecology,* Wiley, New York, 1958.

Needham, J., and P. Needham, *A Guide to the Study of Fresh-Water Biology,* 4th ed., Comstock (Cornell Univ. Press), Ithaca, N.Y., 1938.

Nicol, J., *The Biology of Marine Animals,* Interscience (Wiley), New York, 1960.

Nikolsky, G., *The Ecology of Fishes,* trans. by L. Birkett, Academic, New York, 1963.

Odum, E., with H. Odum, *Fundamentals of Ecology,* 2nd ed., Saunders, Philadelphia, 1959.

Oosting, H., *The Study of Plant Communities: An Introduction to Plant Ecology,* 2nd ed., Freeman, San Francisco, 1956.

Pennak, R., *Fresh-Water Invertebrates of the United States,* Ronald, New York, 1953.

Phillips, E. A., *Field Ecology* (BSCS laboratory block), Heath, Boston, 1964.

Popham, E. J., *Some Aspects of Life in Fresh Water,* 2nd ed., Harvard Univ. Press, Cambridge, Mass., 1961.

Pramer, D., *Life in the Soil* (BSCS laboratory block), Heath, Boston, 1965.

Reid, G., *Ecology of Inland Waters and Estuaries,* Reinhold, New York, 1961.

Reitz, L. P., ed., *Biological and Chemical Control of Plant and Animal Pests,* Am. Assoc. for the Advancement of Science, Washington, D.C., 1960.

Sears, M., ed., *Oceanography,* Am. Assoc. for the Advancement of Science, Washington, D.C., 1961.

Storer, J. H., *The Web of Life,* Devin-Adair, New York, 1953; Signet, New Am. Library, New York, 1956.

Swain, T., ed., *Chemical Plant Taxonomy,* Academic, New York, 1963.

U.S. Govt. Printing Office, *Common Trees of Puerto Rico and the Virgin Islands,* A 1.76:249, Washington, D.C., 1964.

Ward, H. B., and G. C. Whipple, *Fresh-Water Biology,* 2nd ed., edited by W. T. Edmondson, Wiley, New York, 1959.

Welch, P., *Limnological Methods,* Blakiston (McGraw-Hill), New York, 1948.

This listing is intended to be only a guide for the selection of new films, filmstrips, and transparencies. Send for catalogs from distributors of films; catalogs of biological supply houses list extensive offerings of 2×2 in. slides. The addresses of distributors are given in Appendix B.

In general, the cost of color films averages $150; black and white films cost about $75. Newer films are made in varying lengths—time sequences of 1 min, 5 min, 11 min, or 30 min; the cost of films therefore varies. We have indicated the lengths of films so that lessons can be planned with the films in mind.

Sound films are available in color or black and white; rental rates are available from film libraries or by direct inquiries to producers. (Filmstrips are not available on a rental basis; the purchase price is usually about $6 to $7 for those in color.)

Arctic Wilderness (set of 6 filmstrips), EBF.
Arthropods of Public Health Importance (filmstrip), Communicable Disease Center; free loan.
Balance in Nature (19 min), Filmscope.
Battle of the Beetles (15 min), U.S. Forest Service; Engelmann spruce bark beetle, free.
Between the Tides (22 min), Contemporary.
Biology and Control of Domestic Mosquitoes (21 min), Communicable Disease Center; free.
Birds of the Prairie Marsh (9 min), National Film Board of Canada film, Contemporary.
Carnivorous Plants (filmstrip), Carolina.
The Cave Community (13 min), EBF.
Classification of Invertebrate Animals (except insects) (set of 11 filmstrips), EBF.
Classification of Living Amphibians and Reptiles (set of 4 filmstrips), EBF.
Classification of Living Birds (set of 5 filmstrips), EBF.
Classification of Living Fish (set of 4 filmstrips), EBF.
Classification of Living Mammals (set of 6 filmstrips), EBF.
Classification of Plants (set of 9 filmstrips), EBF.
The Community (11 min), EBF.
Conserving Our Forests Today (11 min), Coronet.
Conserving Our Soil Today (11 min), Coronet.
Cotton Fertilization (13 min), California Spray Chemical; free.
The Desert (22 min), EBF.
Ecology (set of 12 films, 28 min each), Part IX of *AIBS Film Series in Modern Biology*, McGraw-Hill.
Ecology Series (set of 5 films: *Above the Timberline:*

The Alpine Tundra Zone [15 min], *The Changing Forest* [19 min], *Life in the Woodlot* [17 min], *The Spruce Bog* [23 min], and *World in a Marsh* [22 min]), National Film Board of Canada films, McGraw-Hill.
Fallout (15 min), Cenco.
Fishing: Five Great Lakes (28 min), U.S. Fish and Wildlife Service; free.
500,000 to One (25 min), Sinclair Refining; destructive and helpful insects, free.
The Grasslands (17 min), EBF.
Gypsy in the Trees (22 min), Rothschild.
The High Arctic Biome (23 min), National Film Board of Canada film, EBF.
High Arctic: Life on the Land (22 min), National Film Board of Canada film, EBF.
The Housefly and Its Control (11 min), Coronet.
How to Collect and Preserve Plants (14 min), Illinois Natural History Survey; free (preference to Illinois schools).
How to Collect Insects (13 min), Illinois Natural History Survey; free (preference to Illinois schools).
In the Beginning (28 min), Modern Talking Picture; free.
Insect Mounting and Preserving (14 min), Dowling.
Introduced Animals (18 min), New Zealand Embassy; free.
Life of the Molds (21 min), McGraw-Hill.
Look to the Land (21 min), EBF.
Marine Animals and Their Food (8 min), Coronet.
Modern Mosquito Control (26 min), Modern Talking Picture; free.
Mysteries of the Deep (23 min), Disney.
Natural Enemies of Insect Pests (27 min), Univ. of California.
Nature's Half-Acre (30 min), Disney.
Nematodes: The Thief of the Soil (15 min), Modern Talking Picture; free.
Orders of Insects (set of 8 filmstrips), EBF.
Plankton and the Open Sea (19 min), EBF.
Plant Associations (sets of 50 transparencies), Carolina; sets on tundra, forest, grasslands, and so forth.
Plant-Animal Communities: Physical Environment (11 min), Coronet.
Plant-Animal Communities: The Changing Balance of Nature (11 min), Coronet.
Plants Interacting with Their Environment (set of 17 transparencies), Carolina.
The Pond (20 min), International Film Bur.
Population Ecology (21 min), EBF.
Radiation (26½ min), National Film Board of Canada.
Realm of the Wild (25 min), U.S. Forest Service; free.

The Rival World (25 min), Shell; free.
The Sea (26 min), EBF.
Secrets of the Ant and Insect World (15 min), Disney.
Secrets of the Plant World (15 min), Disney.
Secrets of the Underwater World (15 min), Disney.
Social Insects: The Honeybee (24 min), EBF.
Soil Conservation Series (4 parts, 10 min each),
U.S. Dept. of Agriculture; free.
Succession: From Sand Dune to Forest (16 min), EBF.
Survival in the Sea (set of 4 films, 30 min each),
Indiana Univ.
The Temperate Deciduous Forest (17 min), EBF.
This Vital Earth (10 min), U.S. Soil Conservation
Service; free.
The Tropical Rain Forest (17 min), EBF.
Valley and the Stream (16 min), Do All; free.
The Vanishing Prairie (set of 6 filmstrips), EBF.
The Voice of the Insect (27 min), Carousel.
The Web of Life Series (set of 2 films: *A Strand
Breaks* [17 min] and *The Strands Grow* [15
min]), EBF.
What Is Ecology? (11 min), EBF.
Wild Flowers of the West, Richfield Oil; free.
Winning the Weed Battle (25 min), Stauffer Chemi-
cal; free.
The World at Your Feet (22 min), International
Film Bur.
World of Insects (22 min), California Chemical;
free.

World of Microbes (30 min), McGraw-Hill.
Yours Is the Land (21 min), EBF.

LOW-COST MATERIALS

A directory of suppliers of free and low-cost
materials constitutes Appendix D. You will find
many useful hints in the free publications of the
biological supply houses. Many low-cost pamphlets are available from state and federal agricultural departments, and from agricultural and
forestry stations affiliated with colleges.

Do you have *Turtox Service Leaflets* and *Turtox
News* (General Biological), and *Carolina Tips* and
Carolina's bulletins on culturing plants and
animals? Also refer to *Ward's Bulletin* and to
Ward's series of curriculum aids on microtech-
niques. Welch distributes *Biology and General
Science Digest.*

Inquire of the U.S. Dept. of Agriculture about
their Yearbooks and other publications on conservation practices. The department's Forest
Service and Soil Conservation Service have films
available to schools on free loan; write to them
for lists of their films and pamphlets. Also write
to the Fish and Wildlife Service of the U.S. Dept.
of the Interior for a list of their publications.

Among other organizations that lend films to
schools without charge are National Wildlife
Federation and Southern Pine Assoc.

Section **THREE**

Special techniques

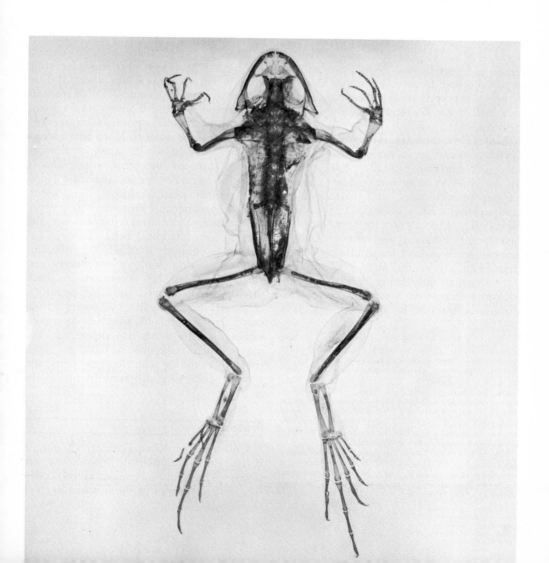

Chapter **12**

Maintaining Animals Useful
in the Classroom

The best methods of maintaining living things are those that reproduce the most favorable field conditions and eliminate natural enemies. This is, in essence, the principle upon which the construction of vivaria is based. For all forms—invertebrate and vertebrate—certain precautions must be taken. In all cases proper diet and temperature control are the factors most responsible for keeping the animals in a healthy state.

Many invertebrates and some classes of vertebrates can be cultured in the laboratory or classroom, providing living forms for activities in behavior, ecology, classification, reproduction, and variation, for comparative studies of organ systems and circulation of blood, and for observation of heartbeat.[1] Furthermore, many of these animals serve as food for small vertebrates that often are reared in the laboratory.

Students usually bring in materials for identification when a resourceful teacher stimulates interest in living things. If a teacher brings in eggs of frogs and sets them out in aquaria or bowls in the classroom, or if he brings in patches of different mosses, a fern or two, and a lichen for a terrarium, youngsters will ask questions.

In many schools there are a few experts among the students who can give short talks to classes or to clubs about the behavior of certain animals. There also are students who would like to care for living materials (student squads, Chapter 14). Many students want to learn more about how to identify shells, trees, or flowers; how to start an insect collection; how to set up a terrarium or an aquarium; how to get more plants without using seeds; how to photograph animals and flowers; how to make prints and enlargements; and how to breed tropical fish.

In this chapter, methods for collecting and cultivating both invertebrates and vertebrates will be described. As each phylum is mentioned, possible usefulness of these animals in the classroom is indicated. In Chapter 13, methods for maintaining living plants are described. You may want to refer now to Chapter 11 for specific details for planning field trips, and tips on collecting different living organisms.

---

[1] You will want to read the approaches to biology developed in the three versions of high school biology written by college and high school teachers in the Biological Sciences Curriculum Study (BSCS): *Biological Science: An Inquiry into Life* (Yellow Version), Harcourt, Brace & World, New York, 1963; *Biological Science: Molecules to Man* (Blue Version), Houghton Mifflin, Boston, 1963; and *High School Biology* (Green Version), Rand McNally, Chicago, 1963. A laboratory manual accompanies each of the textbooks (in the Blue Version it is bound into the textbook).

Also look into the BSCS laboratory blocks of activities (published by Heath). The blocks explore such topics as ecology, microbes, and growth of organisms. Each block of investigations is designed to be completed in about 6 weeks.

Protozoa

Culturing protozoa is relatively simple if certain fundamental precautions are taken. The techniques described below have proved useful to the authors in their efforts to maintain living things for use in the biology classroom and laboratory. First, we shall describe the necessary general conditions for a culture center. Then we shall discuss specific culture methods for the more common, easily obtained forms of protozoa.

Suggestions for maintaining a culture center

In general, the room or portion of the laboratory where cultures are to be kept should fulfill the following environmental conditions:

1. Keep cultures at a constant temperature within an optimum range of 18° to 21° C (64° to 70° F). You may maintain these temperatures during warm weather by stacking finger bowls in a metal container, which you can keep cool by putting it in a sink and circulating tap water to the level of the top dish. At temperatures above 25° C (77° F), cultures do not maintain themselves at maximum and may die off; below 15° C (59° F), development is very slow and the cultures may also die off.

2. Keep cultures away from fumes of concentrated acids such as nitric acid, hydrochloric acid, and sulfuric acid and such alkalies as ammonium hydroxide.

3. Try to keep the cultures at a hydrogen-ion concentration approximately neutral (pH 7).

4. Keep the cultures in medium light. Darkness is not detrimental. Direct sunlight is harmful, since the temperature of the culture may be raised above the optimum. (Of course, this does not apply to the culturing of green forms.)

5. Avoid sudden drafts near the cultures, since they may carry contaminants, such as some common chemicals in powdered form.

6. Keep glassware clean. Use low-suds detergent; rinse repeatedly. For a final rinsing, wash in distilled water if the tap water is not free of copper and chlorine.

The number of organisms in a beginning culture greatly affects the growth of the entire population. For this reason, when a limited number of organisms is available, begin cultures in Syracuse dishes. As the population density increases, transfer larger quantities into small finger bowls.

For pure strains of protozoa, consult L. Provasoli, chairman, "A Catalog of Laboratory Strains of Free-living and Parasitic Protozoa" (*J. Protozool.* 5:1–38, 1958).

Rhizopods

Amoeba. The amoeba most commonly used in the classroom, *Amoeba proteus,* is not as common in nature as other amoebae. It may be found in ponds or pools which do not contain much organic matter, where the water is clear and not too alkaline, and where exceedingly swift currents are absent. Both *Amoeba proteus* and *A. dubia* (Fig. 12-1) are found among aquatic plants such as *Cabomba,* elodea, and *Myriophyllum.* Often scrapings from the base of a stalk of the cattail (*Typha*) or from the underside of a leaf of the water lily, *Nymphaea,* will yield many amoebae.

Place small amounts of these plants on which amoebae may be present into finger bowls, large Petri dishes, or flat jars. Cover the material with pond water in which the plants had been growing, or add spring water. Be certain to maintain them at room temperature. To each container add

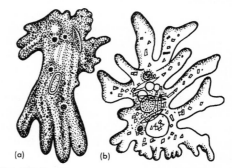

Fig. 12-1 Two species of Amoeba useful in the classroom: (a) *Amoeba proteus;* (b) *A. dubia* (not drawn to scale). (From G. G. Simpson and W. S. Beck, *Life: An Introduction to Biology,* 2nd ed., Harcourt, Brace & World, Inc., 1965.)

two to four uncooked rice grains. Amoebae generally appear in 1 week to 10 days in successful cultures. Examine with a binocular (stereoscopic) microscope to locate amoebae in the cultures.

Amoebae congregate on the bottom or sides of a container. They may be removed in either of two ways. Carefully pour off the excess fluid into another dish. Then with a pipette pick up the amoebae attached to the bottom and place them in fresh spring water to which a few uncooked rice grains have been added. When the rice grains begin to decay (by bacterial action), small animals such as *Chilomonas* (Fig. 4-1a) begin to increase. *Chilomonas* furnishes a good food supply for *Amoeba*.

If the animals are not congregated on the bottom, swirl the finger bowl with a rotating motion, thereby causing the heavier particles to fall into the center of the dish. You may then pick up the amoebae with a pipette.

The method we have just described is useful for *temporary* cultures in the laboratory. When amoebae are to be cultured continuously, the following methods have been successful.

METHOD A. This method involves the use of a hay infusion Halsey[2] has described a typical method. Place eight 1-in. lengths of timothy hay stalks in 100 ml of spring water. Boil this mixture for 10 min, and let it stand for 24 hr. Then add large quantities of *Colpidium* or *Chilomonas*. Let this medium stand for 2 to 3 days, then inoculate it with amoebae. As the culture develops, the number of food organisms (that is, the ciliated forms) decreases. When this happens, remove half of the culture medium, and add an equal amount of fresh hay infusion to which *Colpidium* or *Chilomonas* have been added. Add two grains of uncooked rice or boiled wheat, or four 1-in. lengths of boiled timothy hay for every 50 ml of culture medium which has been added. Cultures

may last as long as 6 months. However, if large amounts of organic matter with accompanying large quantities of bacteria are present, they tend to cause the death of the amoebae.

Many methods described in the literature are similar to that of Halsey (see papers by Jennings,[3] Kofoid,[4] Hyman,[5] Dawson,[6] and LeRoy and Ford[7]). In all these methods a medium that has no specific chemical composition is used. While these methods are successful in the hands of some workers, they offer pitfalls for the beginner. In fact, the very simplicity of the method is its undoing. The beginner often needs a method that takes care of all the variables that may cause failure, namely, medium, food, temperature, pH, and so forth. Such a preparation is described in method B.

METHOD B. Chalkley,[8] Pace,[9] Brandwein,[10] and Hopkins and Pace[11] have described methods which make use of synthetic pond water of a specific chemical composition, instead of natural pond water or hay infusion. (In some cases a buffer may be needed.) In our experience, these methods are superior to those described in the preceding paragraphs. Although more time is spent in preparing this culture medium, it is fully repaid by the quantity of animals found in each culture.

[3] H. S. Jennings, "Methods of Cultivating Amoebae and Other Protozoa for Class Use," *J. Appl. Microbiol. and Lab. Methods* 6:2406, 1903.

[4] C. A. Kofoid, "A Reliable Method for Obtaining *Amoeba* for Class Use," *Trans. Am. Microscop. Soc.* 34:271, 1915.

[5] L. Hyman, "Methods of Securing and Cultivating Protozoa: General Statement and Methods," *ibid.*, 44:216, 1925.

[6] J. A. Dawson, "The Culture of Large Free-living Amoebae," *Am. Naturalist* 62:453, 1928.

[7] W. LeRoy and N. Ford, "Amoeba," in Galtsoff *et al., op. cit.*

[8] H. Chalkley, "Stock Cultures of *Amoeba proteus*," *Science* 71:442, 1930.

[9] D. Pace, "The Relation of Inorganic Salts to Growth and Reproduction in *A. proteus*," *Arch. Protistol.* 79:133, 1933.

[10] P. F. Brandwein, "Culture Methods for Protozoa," *Am. Naturalist* 69:628, 1935.

[11] D. L. Hopkins and D. Pace, "The Culture of *Amoeba proteus* Leidy Partim Schaefer," in Galtsoff *et al., op. cit.*

[2] H. R. Halsey, "Culturing *Amoeba proteus* and *A. dubia*," in P. Galtsoff *et al.*, eds. (J. Needham, chairman), *Culture Methods for Invertebrate Animals*, Comstock (Cornell Univ. Press), Ithaca, N.Y., 1937, p. 80.

We have found the methods of Chalkley and the one of Brandwein especially successful. The method described here is our modification of existing methods. This is selected merely because of our prolonged experience with it; it has also been successfully used by other teachers and students. The method has been used with many other protozoa and small invertebrates. Both methods depend upon a synthetic pond water prepared as follows.

1. BRANDWEIN'S SOLUTION A. Weigh out the following salts, and dissolve them in distilled water to make 1 liter of solution:

NaCl	1.20 g
KCl	0.03 g
$CaCl_2$	0.04 g
$NaHCO_3$	0.02 g
phosphate buffer (pH 6.9–7.0)	50 ml

This is *stock* solution A. For use, it should be diluted 1:10 with distilled water. (For each milliliter of stock solution A, add 10 ml of distilled water.)

Rinse a number of finger bowls in hot water, then in cold. Next prepare a 1 percent aqueous solution of powdered nonnutrient agar in distilled water, or in solution A. Heat slowly until smooth, then pour, while fluid, a 1- to 2-mm layer into the bottom of the finger bowls. While the agar is still soft, embed five grains of rice in it.[12]

Introduce about 50 amoebae, together with 15 ml of the medium in which they have grown, into each bowl, and add about 30 ml of dilute solution A. During each of the next 3 days, add 15 ml of dilute solution A, until the total volume is about 90 ml. A few days after the cultures have been started, the layer of agar will separate from the bottom of the dish. Then amoebae may be found growing in layers on the upper and lower surfaces of the agar as well as on the glass surfaces.

After about 2 months of growth, the culture wanes and should be subcultured. This may be accomplished by dividing the

contents of each finger bowl into four parts. Prepare fresh finger bowls containing a film of agar. Add one-fourth of the old culture to each freshly prepared finger bowl, and an equal volume of dilute solution A.

When the original source of amoebae is limited, as it may be when collected in the field, it may be necessary to start small cultures in Syracuse dishes rather than in the larger finger bowls. This apparently provides a better initial concentration of amoebae and makes the change of culture conditions less abrupt.

Prepare the Syracuse dishes with a thin layer of agar and embed two rice grains in each dish. Introduce the animals on hand with about 4 ml of the water in which they were collected; then add 4 ml of dilute solution A. In a successful culture amoebae rapidly proliferate. When some 200 organisms are present, add the culture to a rice-agar finger bowl, with 20 ml of dilute solution A. *Caution:* It is detrimental to have large ciliates such as *Stentor, Paramecium,* large hypotricha, *Philodina* (Fig. 2-10a), or *Stenostomum* in cultures of amoebae. Microscopic worms such as *Nais* (Fig. 4-3) or *Aeolosoma* (Fig. 1-18), or crustaceans such as *Cyclops* (Figs. 1-28, 12-15) or *Daphnia* (Fig. 1-26) also are harmful contaminants. Cultures containing such ciliates or worms may as well be discarded. Moderate populations of *Chilomonas* and *Colpidium* are beneficial as food organisms for amoebae, but these forms should not be present in such amounts that the medium is clouded by their presence. At times mold, such as *Dictyuchus,* may grow about the rice, but this does not seem to be detrimental; in fact, amoebae may be found congregated in the mycelia.

Two other solutions for synthetic pond water which may be used as medium for amoebae are as follows:

2. CHALKLEY'S SOLUTION. Combine the following with 1 liter of water:

NaCl	0.1 g
KCl	0.004 g
$CaCl_2$	0.006 g

---

[12] The agar, while not entirely necessary, helps to fix the rice.

3. HAHNERT'S SOLUTION. Combine the following with 1 liter of water:

KCl	0.004 g
$CaCl_2$	0.004 g
$CaH_4(PO_4)_2$	0.002 g
$Mg_3(PO_4)_2$	0.002 g
$Ca_3(PO_4)_2$	0.002 g

Allies of Amoeba. Many teachers have found that the large amoeba *Pelomyxa* (sometimes called *Chaos chaos*), which is 3 to 4 mm in length, is superior to *Amoeba proteus* (600 μ) or *A. dubia* (30 μ) (Fig. 12-2). The taxonomy of this giant form is still a matter of controversy.

PELOMYXA. It can be cultured in Brandwein's solution A, using the rice-agar method described earlier. In addition, add a pipetteful of *Paramecium, Blepharisma,* or *Stentor* (see Figs. 2-6, 2-8, 2-9), or all three, as food. If a rich supply of *Paramecium* is available, there should be good growth of *Pelomyxa* within a month. Since *Pelomyxa* is omnivorous, its food vacuoles will be colored red by engulfed *Blepharisma* or green by trapped *Stentor*. Various stages of rotifers and small worms may also be found in the vacuoles of *Pelomyxa*. *Pelomyxa* is characterized by having many nuclei (as many as a thousand), and it may contain up to 12 contractile vacuoles. When it divides it often divides into three parts instead of two.

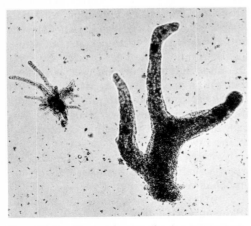

Fig. 12-2 Photomicrograph, showing a combined mount of *Amoeba proteus* and *Pelomyxa* (*Chaos chaos*). (Photo by Walter Dawn.)

ARCELLA. This shelled relative of amoeba may be cultured by method A or B.

DIFFLUGIA. Various species of this amoeba, such as *Difflugia oblonga, D. lobostoma,* and *D. constricta,* may be cultured by using a number of green algae as food organisms, using the following method, as described by Stump.[13] Place such algae as *Spirogyra, Zygnema,* or *Oedogonium* in Petri dishes or finger bowls. Cover the algae with spring or pond water. Thick cultures should develop in about 10 days. Add small quantities of fine sand for these organisms to build the intricate shells they carry about (see Fig. 12-7). In our experience, this method of cultivation is subject to the same criticism as method A. *Difflugia* can be readily cultured by method B, provided small amounts of fine sand are added to each culture.

ACTINOSPHAERIUM. Culture this form (Fig. 12-3) using method A or B; for best results, assure the presence of a moderate amount of *Paramecium*. To 30 ml of solution A in a rice-agar finger bowl, add 20 ml of a culture of *Paramecium,* and inoculate with five to ten specimens of *Actinosphaerium*. Prolific cultures are usually obtained in about 10 days to 2 weeks. At this time, it may be necessary to subculture.

KNOP'S SOLUTION. Both heliozoans, *Actinosphaerium* and *Actinophrys,* may be cultured in a modification of Knop's solution which is prepared by dissolving the following salts in distilled water to make 1 liter:

$MgSO_4$	0.25 g
$Ca(NO_3)_2$	1.00 g
KH_2PO_4	0.25 g
KCl	0.12 g
$FeCl_3$	trace

When these cultures are examined from week to week, a succession of living forms is found. In fact, the amoebae and related forms are the last to appear; that is, they begin to increase after the ciliates have reached a peak and the culture is "declining." In such a culture a frequent succession may be: small flagellates, *Colpoda,* hypotrichs, *Paramecium, Vorticella,* then

[13] A. B. Stump, "Method of Culturing Testaceae," described in Galtsoff *et al., op. cit.,* pp. 92–93.

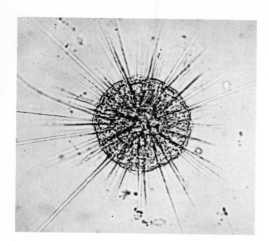

Fig. 12-3 *Actinosphaerium.* (Courtesy of Carolina Biological Supply Co.)

Amoeba. Paramecia often appear in a new culture during its second week, and amoebae after 2 to 6 weeks.

Slime molds. Many protozoologists consider slime molds to be colonial protozoans (amoeba-like). They are classified among the Sarcodina, Subclass Rhizopoda; sometimes slime molds are called mycetozoans. In addition, most botany texts classify slime molds as being similar to fungi—as myxomycophyta. However biologists wish to classify them, they offer a fascinating study of a complex life history of a plasmodium of fused myxamoebae. The plasmodium, when dried, divides into cysts, or sclerotia. (Students will want to read J. Bonner, *The Cellular Slime Molds* [Princeton Univ. Press, 1959] if they become interested in slime molds.) Methods of cultivation are given in Chapter 13. Refer also to aggregation of slime molds, Fig. 9-39.

Ciliates

Methods similar to those described for the subclass Rhizopoda may be used in collecting members of the class Ciliata. When only a small number of specimens have been collected, it may be necessary to use Syracuse dishes, as described under method B, in order to get concentrations for inoculation into larger finger bowls.

General methods. Several general methods may be used with success to culture most ciliates. Methods A and B have been described above for amoebae. When using method B for ciliates subculture every 2 weeks by dividing the liquid into two to four parts and adding fresh medium and rice grains.

METHOD C. Prepare a hay infusion in the following way. First, boil 1 liter of pond water, spring water, or tap water (listed in order of best results). As the water comes to a boil, add a handful of timothy hay and boil for an additional 10 min. Then cool the mixture, and allow it to stand for 2 days before inoculating it with the ciliated forms such as *Paramecium* or *Stentor*. If a culture of *Chilomonas* is available, add a few pipettefuls to the cooled culture and use immediately.

METHOD D. In this method add five grains of wheat to each 100 ml of boiled, then cooled, pond water. Let this mixture stand in exposed dishes for a day or so before inoculating with ciliates such as *Paramecium*. Add several pipettefuls of *Chilomonas* if it is available. A population peak should be attained within 2 weeks.

METHOD E. Prepare a thin, smooth paste by grinding 0.5 g of the yolk of a hard-boiled egg with a small amount of tap water or boiled pond water. Then add this paste to 500 ml of Brandwein's solution A, or boiled pond water, or tap water. Let this mixture stand for 2 days before inoculating it with the ciliated forms. Or inoculate the mixture immediately with *Chilomonas* and then add the ciliated forms that are to be cultivated without the 2 days of delay.

METHOD F. An excellent medium for most of the ciliates used in the classroom was described previously for amoebae as method B. The only modification is a variation in the number of uncooked rice grains. In culturing ciliates use about eight rice grains per finger bowl.

METHOD G. Add about ¼ package of dehydrated yeast (a package contains 7.5 g) to 250 ml of pond water, spring water, or tap water. Mix well and allow this culture medium to stand exposed to the air for several hours. Inoculate with the culture of protozoa you plan to maintain. At room temperature, rich cultures develop within a week. Keep the cultures covered after they have been inoculated with protozoa to prevent evaporation and contamination.

Paramecium grown in this medium often develop a darkened gray shade and have blackish food vacuoles.

METHOD H. In this method a pinch of skim-milk powder is added to 250 ml of spring water (or boiled, filtered pond water).[14] (For careful work with protozoa, experimentation with volumes and weights may be necessary, but most protozoa tolerate a wide range of concentration of this milk powder.)

Add a few pipettefuls of *Paramecium* to this culture medium, and maintain at about 22° C (72° F). In 2 to 3 days there should be a rapid proliferation with an abundance of forms undergoing fission. A population peak is reached in 5 days with this medium, and new cultures should be prepared within 2 weeks. At times, the addition of a pinch of milk powder to an old culture may be sufficient to renew the population peak for a short time.

Frings finds the protozoa reared in this medium to be large, with clear cytoplasm and clear food vacuoles; the macronucleus is usually visible. Other forms, such as *Tetrahymena* (Figs. 2-8c, 2-9f), *Colpoda, Oxytricha, Lacrymaria, Halteria* (Fig. 12-4), *Vorticella* (Fig. 2-8f), *Colpidium* (Fig. 12-5a), *Euplotes* (Fig. 12-5b), and *Stylonychia* (Fig. 2-9c) also have been successfully cultured with this medium. Amoebae do not maintain themselves in it. On the other hand, plant forms such as *Monas, Chilomonas* (Fig. 4-1a), *Pandorina* (Fig. 13-5b), *Euglena* (Fig. 2-11c), and *Peranema* (Fig. 2-12a) also thrive in this medium.

METHOD I. This method uses lettuce leaves rather than hay. Rub the outer dry

[14] H. Frings, "Dried Skim Milk Powder for Rearing Protozoa," *Turtox News* 26:1, January 1948.

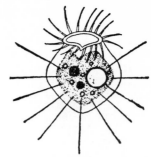

Fig. 12-4 *Halteria* (greatly magnified). (Reprinted with permission of The Macmillan Co., New York, from *Invertebrate Zoology* by R. W. Hegner. Copyright 1933 by The Macmillan Co., renewed 1961 by Jane Z. Hegner.)

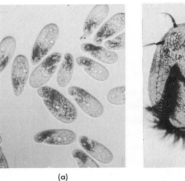

(a)   (b)

Fig. 12-5 (a) *Colpidium;* (b) *Euplotes* (stained). (Courtesy of General Biological Supply House, Inc., Chicago.)

leaves of lettuce through a fine-mesh wire strainer. Boil pond water, and when it cools, add 1 tsp of the lettuce to 1 liter of pond water. Boil this mixture for 1 min; let the jar stand covered, overnight. Later, divide the medium into small finger bowls or baby-food jars, and inoculate with a culture of *Paramecium*. Keep the containers covered.

In this method, DuShane and Regnery[15] suggest that a pinch of powdered milk be added to the culture after 12 days and then weekly. After a month's growth, subculture by dividing the old culture into

[15] G. DuShane and D. Regnery, *Experiments in General Biology,* Freeman, San Francisco, 1950.

four parts and inoculate into fresh material.

Some specific ciliates. Of the methods which have been described, these are especially recommended for specific ciliates.

PARAMECIUM AND COLPODA. Use methods C, D, E, F, and I. Methods C and E are superior.

PARAMECIUM BURSARIA. This green form (Fig. 2-8a) is easily cultivated by method F. Keep the culture in medium light; these organisms congregate near the source of light. The green alga *Chlorella* (see p. 103) lives in a symbiotic relationship with this species of *Paramecium*.

BLEPHARISMA. This pink ciliate (Fig. 2-8b) grows well when methods D and F are used. Method F seems to be superior. In old cultures look for giant carnivorous forms.

VORTICELLA. A modification of method E is the most satisfactory for this stalked form (Figs. 2-8f, 2-9d). Prepare the egg yolk–tap water medium as described (method E) and allow it to stand for 2 days. Do not add *Chilomonas*. Pour 40 ml of the supernatant fluid into a finger bowl and add *Vorticella*. The mature stalked forms adhere to the bottom of the finger bowls; young free-swimming forms often form a thick layer on the *surface* of the culture.

Subculture every 2 weeks. Scrape the bottom of the finger bowls to free the animals, and divide the culture into four parts. To each bowl add 30 ml of fresh medium. When a very heavy population is desired, pour off about 30 ml of the original liquid and add 30 ml of fresh liquid for replacement. In the same manner, remove contaminants by rinsing the bowls several times with Brandwein's solution A (since the *Vorticella* adhere to the bottom). Add fresh medium.

STENTOR. Both *Stentor polymorphus* and *S. coeruleus* may be cultured by methods D, E, and F. However, method F is superior. Excellent cultures of *Stentor* may be obtained when a pipetteful of *Paramecium* or *Blepharisma* (or both), as well as more *Chi-*

lomonas, is introduced into the culture periodically.

SPIROSTOMUM. This elongated ciliate (Fig. 12-6a) may be cultured by using methods D, E, and F. When the cultures become putrid, conditions seem to be favorable for this form.

DIDINIUM. Introduce several of these animals (Fig. 12-6b) into a rich culture of

(a)

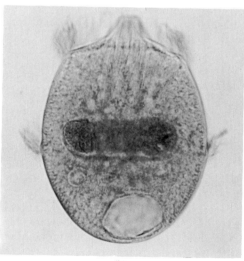

(b)

Fig. 12-6 (a) *Spirostomum* (magnified 33×); (b) *Didinium* (stained, magnified 525×). (Courtesy of General Biological Supply House, Inc., Chicago.)

Paramecium (prepared by method D or F). *Didinium* feed upon *Paramecium* and are found in a similar habitat. Within a week the *Paramecium* in the culture will probably have been consumed. Then inoculate the *Didinium* into fresh cultures of *Paramecium.*

You may preserve *Didinium* for later use by filtering an old culture through filter paper to retain the cysts. Dry the filter paper in air and store in envelopes. When you wish to start a culture of *Didinium,* add a dried sheet of this paper to a thriving culture of *Paramecium.* You may prefer to allow a culture of *Didinium* to dry out in its container. When you wish to revive the culture, add a fresh *Paramecium* culture to the dish.

COLPIDIUM. This ciliate (Figs. 2-8d, 2-9e) may be cultured in a medium made by boiling 100 ml of spring water or pond water to which 1.5 g of whole rye grains have been added. After 10 min of boiling, filter the fluid, then cool it. Expose to air for a day, and then inoculate with about 10 ml of an old culture of *Colpidium.* At a temperature of about 22° C (72° F) these cultures reach a peak in about a week. Also use method H or F.

Some interesting work has been done with bacteria-free cultures. Kudo[16] gives Kidder's formula for cultivating bacteria-free cultures of *Colpidium* (also axenic cultures, p. 587). In this method, add 10 g of brewer's yeast to 1 liter of distilled water. Then boil and filter through cotton. Again filter, this time through filter paper. To this solution, add 20 g of Difco dehydrated proteose-peptone medium. Sterilize the whole solution for 20 min at 15 lb pressure in an autoclave or pressure cooker. Inoculate with individual *Colpidium.*

STYLONYCHIA AND OXYTRICHA. Method F may be used for culturing *Stylonychia* (Fig. 2-9c) and *Oxytricha;* previous inoculation with *Chilomonas* is necessary.

EUPLOTES. For culturing this form (Fig. 12-5b), use method D or E; inoculate with *Chilomonas* as a preliminary step.

[16] R. Kudo, *Protozoology,* 4th ed., Thomas, Springfield, Ill., 1954, p. 884.

TETRAHYMENA. Method H is especially useful in culturing *Tetrahymena;* Method B is also effective. Prepare a 1 percent solution of Difco dehydrated proteose-peptone medium. Sterilize the solution in an autoclave or pressure cooker before inoculating with *Tetrahymena* (Figs. 2-8c, 2-9f).

One fact bears repetition. In raising certain carnivorous forms, the best results are produced when the food animal is raised separately and then added periodically to the culture medium. For example, we have said that *Stentor* grows best in media to which *Chilomonas* or *Colpidium* have been added. Similarly, *Didinium* cultures require *Paramecium* as a food organism, *Lionotus* feeds upon *Colpidium,* and *Actinobolus* consumes *Halteria.*

Flagellates

This class contains the protozoa with one or more flagella. Among the Mastigophora are some forms that may be classified as protozoa or as plant forms belonging to the Phytomastigina. An interesting feature of this group is its lengthwise fission in asexual reproduction.

The procedures described for collecting Rhizopoda and Ciliata should also be used for gathering flagellates. In addition, the green-surface "blooms" which may be found in ditches or ponds may often contain large numbers of *Euglena, Chlamydomonas,* and similar flagellates.

General methods. In general, three of the methods described previously are recommended for cultivating flagellates. Of course, the forms that contain chlorophyll require moderate light.

Method B (described for rhizopods) and methods D and E (described for ciliates) are also successful for flagellates.

METHOD J. An additional medium recommended for flagellates consists of a modification of method D (pond water and wheat grains). Boil four wheat grains in 80 ml of pond water. As soon as the medium is cool, add a few milliliters of pond water containing the flagellates that are to be cultured.

METHOD K. This method[17] uses a modification of Klebs' solution (below). To 100 ml of this solution in a glass battery jar, add 20 rice grains (which have been boiled for 5 min) and 900 ml of distilled water. Let this mixture stand for 2 days. Inoculate the mixture with an old culture of *Euglena,* and keep the jar in indirect sunlight. Direct rays of the sun should not strike this culture for more than 1 hr a day. Inoculate the culture with *Euglena* three times at 3-day intervals. If an old culture of *Euglena* with encysted forms is available (these may be found on the sides of the jar) inoculate the cysts along with the motile forms. After 2 to 3 weeks, add an additional 10 mg of tryptophane powder that has been dissolved in 25 ml of the modified Klebs' solution.

MODIFIED KLEBS' SOLUTION. To 1 liter of distilled water, add the following:

KNO_3	0.25 g
$MgSO_4$	0.25 g
KH_2PO_4	0.25 g
$Ca(NO_3)_2$	1.00 g
bacto-tryptophane broth powder (l-form)	0.01 g

Some specific flagellates

EUGLENA. For this flagellate use method K. The best methods for cultivating this autotroph are those of Jahn[18] and Hall.[19] These, however, require sterile conditions[20] and are tedious to prepare. Once they are established, the organisms may be maintained indefinitely in pure culture. Eventually, methods involving sterile conditions may replace those now employed. (For other culture methods, see green algae, p. 623.)

[17] Brandwein, *op. cit.*
[18] T. L. Jahn, "Studies on the Physiology of the Euglenoid Flagellates," Part III, "The Effect of Hydrogen Ion Concentration on the Growth of *Euglena gracilis* Klebs," *Biol. Bull.* 61:387, 1931.
[19] R. P. Hall, "On the Relation of Hydrogen Ion Concentration to the Growth of *Euglena anabaena* var. *minor* and *E. desos,*" *Arch. Protistol.* 79:239, 1933.
[20] A. K. Parpart, "The Bacteriological Sterilization of *Paramecium,*" *Biol. Bull.* 55:113–20, 1928, and G. W. Kidder, "The Technique and Significance of Control in Protozoan Culture," in G. N. Calkins and F. M. Summers, eds., *Protozoa in Biological Research,* Columbia Univ. Press, New York, 1941.

Euglena is classified in Family Euglenidae (Class Mastigophora) among the protozoans. When grown in light these flagellates photosynthesize food materials, as do typical autotrophs. However, since they sustain themselves and reproduce when grown in the dark and supplied with an organic source, they are often studied as examples of heterotrophs. Chloroplasts bleach out or fragment, and colorless forms persist as consumers of organic materials (see bleaching *Euglena* with streptomycin, p. 196).

CHLAMYDOMONAS. These small green flagellates, resembling algae, grow well under the conditions of methods K and B and the other methods described for growing green algae (Chapter 13).

PERANEMA. Method B yields excellent results; method J is also suitable.

CHILOMONAS. Raise this form with method D, E, or F, all of which yield excellent results.

ENTOSIPHON. Use method J or B, both of which are equally successful.

In this discussion we have omitted the Sporozoa and Suctoria since they are not commonly studied in high school biology classes.

Axenic cultures

Methods for establishing axenic cultures of protozoa (only one species of organism) were probably first devised by Parpart;[21] additional methods were developed by Claff,[22] Kidder,[23] and others.

An axenic culture of *Paramecium,* for example, can be established only after the organisms are washed free of bacteria, other protozoa, fungi, and multicellular microorganisms. After the desired forms are separated (by washing or use of antibiotics), the organisms must be grown in a suitable medium. Some of these synthetic

[21] *Ibid.*
[22] C. L. Claff, "A Migration-Dilution Apparatus for the Sterilization of Protozoa," *Physiol. Zool.* 13:334–40, 1940.
[23] G. W. Kidder, "The Technique and Significance of Control in Protozoan Culture," in Calkins and Summers, *op. cit.*

media contain an extensive list of ingredients. Manwell[24] lists a minimal, chemically defined medium for the ciliate *Tetrahymena* (as used by Elliott[25]) comprising over 11 amino acids, 2 carbon sources, 4 nucleic acids, 7 growth factors, and 5 inorganic salts (quantities in milligrams per liter).

Laboratory strains are available. See L. Provasoli, chairman, "A Catalog of Laboratory Strains of Free-living and Parasitic Protozoa" (*J. Protozool.* 5:1–38, 1958).

Uses of protozoa in the classroom

The organisms that can be cultured by the methods described above have many uses in the classroom during the year. When cultures are maintained routinely, living materials are available at any time for such studies as:

1. Microscopic examination (Chapter 2)
2. Behavior (Chapter 7)
3. Food-getting (Chapter 4)
4. Reproduction and heredity (Chapter 8)
5. Web of life (Chapter 11)

Students will no doubt find a considerable number of suggestions for long-term investigation in *Research Problems in Biology: Investigations for Students,* Series One–Four (BSCS, Anchor Books, Doubleday, 1963, 1965). Each of the investigations described is followed by suggestions for possible approaches and equipment and cautions concerning pitfalls. Some of the topics suggested for investigations in the field of protozoology are: "Responses of Algae or Protozoa to Ultraviolet Irradiation," "Water Balance in Protozoa," "Mutation in Bacteria," "Studies of Induction and Nature of Division Synchrony in Microbial Cultures," "Regeneration and Survival of Emicronucleated Ciliate Protozoa of the Genus *Euplotes*," "Biological Investigation

of *Herpetomonas*" (flagellated protozoan parasites of the digestive tract of some insects), and "Long-Time Effects of Dilute Lithium Chloride on the Form of the Ciliate Protozoan *Stentor coeruleus.*"

Classification of protozoa

The study of protozoa is a specialized area of biology. When an unfamiliar protozoan moves into view under the microscope, you may want students to identify it. Illustrations in a text on protozoology (see Bibliography) may be used as a guide for the identification of the unknown forms under the microscope.

In some circumstances, students may learn to use a key to identify organisms or as an introduction to a study of variations among living things (see Chapter 10). We shall not present detailed keys to classification of protozoa, but simply attempt, mainly with drawings, to focus observations on basic differences among the more common kinds of protozoans which might be found in quantity in pools of water, a ditch, or a container of rainwater.[26] For example, if students examined with a microscope a drop of pond water rich in protozoans they might find some of the forms shown in Fig. 12-7. Sketches of these representative forms could be drawn on the blackboard so that students could observe differences. What is one distinguishing characteristic of each type? Students easily distinguish that some forms are ciliated, some have one or more flagella, possibly some under view are amorphous masses of protoplasm with bulging false feet (pseudopodia), or some are highly organized cells with fine, spine-like projections.

Suppose we examine the flagellated forms closely. There may be single-celled types, such as *Euglena,* or colonial types, such as *Volvox.* Furthermore, some may have chlorophyll or other pigment (Subclass Phytomastigina); others may lack

[24] R. Manwell, *Introduction to Protozoology,* St. Martin's, New York, 1961, p. 562.
[25] A. M. Elliott, "Biology of *Tetrahymena,*" *Ann. Rev. Microbiol.* 13:76–96, 1959.

[26] We pass lightly over two major types of protozoa: Sporozoa (many of which are parasitic) and Suctoria (which are not widely abundant).

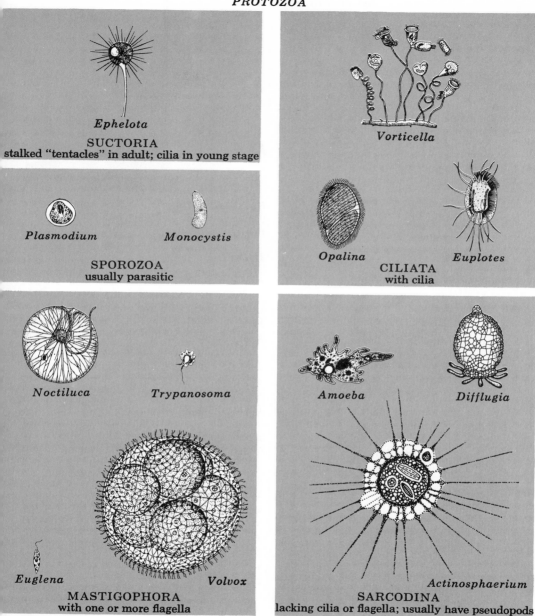

Fig. 12-7 A beginning classification of Protozoa.

color bodies (Subclass Zoomastigina) and be free living or, like trypanosomes, parasitic.

A more complex problem of variations exists among the ciliates. Have students look for differences among the members within Class Ciliata. The location of the "mouth," or peristome, and the arrangement of cilia or long, fused cilia (membranelles) in relation to the peristome are the basic distinguishing features of the four subclasses of Ciliata; see *Paramecium, Euplotes,* and *Vorticella.*

Microscopic examination of the contents of the intestine of an earthworm may reveal the sporozoan *Monocystis* (Fig. 11-12). Prepared slides of plasmodia that cause malaria may also be studied at this time. You may also want to have students examine the flagellates (Zoomastigina) in the intestine of termites (Fig. 11-16).

Students who want to go further in identifying genera of protozoans may consult the fine keys that are developed in textbooks of protozoology (see Bibliography).

Fresh-water invertebrates

Sponges

Only one family of sponges is found in fresh water, the Spongillidae. While marine sponges may be collected from considerable depths and dried for classroom use, they are difficult to keep alive without salt-water aquaria. In fact, fresh-water forms are also very difficult to keep in the laboratory. Specimens of fresh-water sponges may be found as crusty, brownish growths attached to submerged plants and rocks. Many of these sponges have the texture of raw liver.

Spongilla (Fig. 12-8) is difficult to keep alive more than a few weeks in an aquarium, and this is possible only when large specimens have been collected (as at the end of the summer and into fall). However, gemmules of these sponges may be cultivated in depression slides. Some will grow and attach themselves to the glass surfaces. Keep them in darkness at 20° C (68° F).

Because sponges as a group have limited usefulness in high school biology, they are given only brief mention here. A text in zoology, such as any one of those listed at the end of this chapter, will furnish a description of the phylum.

Fig. 12-8 *Spongilla,* a fresh-water sponge. (Courtesy of the American Museum of Natural History, New York.)

Hydra

While many jellyfish forms may be collected and preserved for class examination, we will discuss only one, *Hydra,* a form that may be maintained under fairly simple conditions in the laboratory.

This fresh-water coelenterate (Figs. 1-1, 1-2, 12-9) may be found in lakes or ponds, attached to submerged stems of water plants or on the underside of floating leaves of water lilies and water hyacinths.

In the laboratory, transfer the plants upon which hydras are found into finger bowls or small aquaria. Be certain to use the water in which the hydras have been living, if available, or water from a thriving aquarium. Add the aquarium water slowly, a glassful per day, to the original water in which the hydras were found, so that the organisms will become acclimated. Brandwein's solution A (p. 581) may be used in the absence of satisfactory aquarium water.

Keep the containers in medium light or semidarkness, at a temperature below 20° C (68° F). Within a day or so the hydras will be found on the surface of the water. They can be picked up with a pipette and transferred into new containers to start fresh cultures. (If the cultures remain contracted and fail to expand, the water that has been added is not

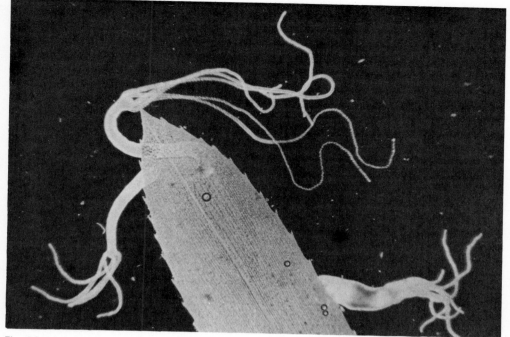

Fig. 12-9 Hydras attached to a submerged leaf. (Courtesy of General Biological Supply House, Inc., Chicago.)

suitable. Do not put more hydras in such a tank.)

Loomis and Lenhoff[27] recommend the following method for culturing hydras. Since they found that calcium ions are necessary for growth of hydras, they developed a chemically defined medium utilizing $CaCl_2$. To remove the toxic copper ions from tap water, they suggest using a chelating agent. In this solution the chelating agent is Versene (disodium ethylenediamine tetraacetate).

Prepare a stock solution of modified tap water:

$NaHCO_3$	20 g
Versene	10 g
$CaCl_2$	50 mg
tap water	1 liter

The sodium bicarbonate is used as a buffer

---

[27] W. F. Loomis and H. M. Lenhoff, "Growth and Sexual Differentiation of *Hydra* in Mass Culture," *J. Exptl. Zool.* 132:555–74, 1956. Other culturing methods are given in Galtsoff *et al., op. cit.*

to maintain the solution at pH 7.5 to 8.

Loomis and Lenhoff also found that as population density increased, sexual differentiation occurred (see also Fig. 8-15).

Keep the cultures in moderate light. Green hydras, *Chlorohydra viridissima*, in which *Chlorella* live symbiotically in the gastrodermis or endoderm, require more light. (Incidentally, these algae are passed to the next generation in the cytoplasm of the eggs of hydras.)

About twice a week feed the hydras a rich culture of *Dero, Tubifex, Daphnia*, or *Artemia* (culturing, pp. 594, 596, 598). Well-fed hydras grow rapidly and reproduce readily.

On the whole, the most successful hydras for cultivation in the laboratory aquarium belong to the group *Hydra oligactis*. Adequate green water plants are needed for a rich supply of oxygen.

At times periods of depression beset hydras; tentacles are contracted and the body becomes shortened. Depression may often be avoided by frequent changing of the water.[28]

Uses in the classroom. A student might design an investigation into the means of preventing depression of hydras. Students may also use hydras to investigate the following subjects.

1. Food-getting. Examine how the hydra uses nematocysts (p. 109) to capture *Daphnia* (Figs. 1-26a, 11-25a). Use starved hydras (forms that have not been fed for 24 hr) to study ingestion (p. 201).

2. Regeneration (p. 480). Hydras might be used in club work or project activities to study the nature of polarity. For example, can tentacles grow on the "wrong" end?

3. Sexual differentiation (p. 350).

4. Taxes (p. 300). Try to demonstrate the responses to touch, to food, to weak acids, and to light.

5. The glutathione feeding response (p. 202).

6. Rate of growth (p. 439) and grafting (p. 480).

7. Growth and differentiation at the tissue level. Fulton[29] suggests using a colonial hydroid, *Cordylophora lacustris*. The organisms may be cultured on slides slanted in a beaker of culture medium. Doubling time of hydranths is 3 days.

8. Preservation of organisms. Students might preserve some hydras to be used only when living hydras are not available. First, narcotize the forms by adding crystals of menthol or of Epsom salts (magnesium sulfate) to the finger bowl of water containing the hydras. When they are elongated and quiet, place them in 70 percent alcohol to fix and preserve them. (Refer also to Table 11-2.)

Planaria

Look for these small flatworms on the underside of submerged logs and under stones in ponds and lakes. Several varieties may be found in clear, running water, but the usual forms are the small, blackish *Planaria maculata* and the more frequent laboratory form, the brown *Dugesia tigrina* (Fig. 12-10). When you find some in a submerged log, wrap the whole log in wet newspaper and bring it into the laboratory. Submerge the log in a white enamel pan of water and peel off sections of wood. Usually the planarians float to the top. Planarians may also be baited by submerging a piece of raw beef liver or hard-boiled egg yolk (tied in cheesecloth), attached to a string, in a cold stream or lake. This method often attracts the larger form *Planaria dorotocephalia*. Brush off the gathered forms into collecting jars and submerge your bait in another part of the lake or stream.

Transfer the collected plant materials into larger glass jars and keep them in moderate light. Soon planarians may be found clustering on the surface of the water or adhering to the sides of the jars. Then pick them off with a pipette and isolate them in separate culture containers.

Because planarians are photonegative, they should be maintained in black or opaque containers; enameled pans are excellent. Change the water frequently, with fresh additions of aquarium water, spring water, or Brandwein's solution A (p. 581). Keep them at a temperature about 18° C (64° F). Once a week feed the planarians finely chopped raw beef liver; better still, since live food does not foul the water, feed bits of worms (*Tubifex* or *Enchytraeus* worms; culturing, p. 594). At other times, feed them bits of hard-boiled egg yolk.

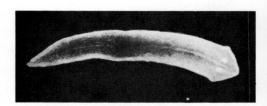

Fig. 12-10 *Dugesia tigrina,* the brown planarian. (Photo by Hugh Spencer.)

[28] Refer to W. F. Loomis, "Sex Gas in Hydra," *Sci. Am.,* April 1959, pp. 145–56.

[29] C. Fulton, "Culture of a Colonial Hydroid Under Controlled Conditions," *Science* 132:473–74, 1960.

Remove the excess food with a pipette after several hours to avoid fouling the water.

Uses in the classroom. Planarians are studied as a representative of the platyhelminthes (Chapter 1). These flatworms are classic material for studies in regeneration (Chapter 9). At times you may find planarians reproducing by fragmentation. They rarely reproduce sexually in the laboratory; orange cocoons may be found that occasionally hatch out some 4 to 6 small planarians in about 2 weeks (Fig. 12-11).

Use planarians to show taxes (pp. 298, 300). What happens when one is put on a slide and the glass is tapped? How do they respond to light? See also RNA and learning (p. 314) and ingestion (p. 27).

For gross examination under the microscope, you may find it necessary to narcotize the animals before putting them on a slide. To a small watch glass or Syracuse dish of pond water containing a few planarians, add a small amount of chloretone or a few crystals of Epsom salts or menthol (see p. 112). When the forms are quiet, lift them with a toothpick and arrange the animals on slides so that the proboscis is uppermost.

Rotifers

Half fill several jars with submerged plants from a pond, and then add pond water to fill the jars. In the laboratory, remove the covers and place the jars in moderate light. After a day or so rotifers (Fig. 2-10a) will be found congregated on the surface, where there is an abundant supply of oxygen.

Use a pipette to pick the rotifers out of the jar, and introduce them into finger bowls of pond water. Change the culture water frequently. Feed the rotifers *Chlorella, Euglena,* or *Chlamydomonas* (culturing, Chapter 13). Keep the finger bowls stacked to prevent evaporation of the medium.

In one simple procedure,[30] 1 g of nonfat dried milk is dissolved in 1 liter of tap water. Use a phosphate buffer (p. 172) to maintain the culture at pH 7 to 7.5. The condition of the local tap water may make it necessary to use a chelating agent, such as Versene (p. 625), to remove the copper ions; or use spring water or Brandwein's solution A (p. 581).

Vinegar eels

These nonparasitic roundworms (Fig. 1-13) feed upon the fungus "mother-of-vinegar." Because bottled vinegar has been pasteurized to inhibit the growth of these roundworms, bulk vinegar must be used as a source.

Add small quantities of bulk vinegar containing these worms to quart containers of pure, unadulterated cider vinegar. Then cover the cultures to prevent evaporation. Or add vinegar eels to small finger bowls containing two small cubes of raw apple in 150 ml of cider vinegar.

Wide fluctuations in temperature are tolerated by vinegar eels. Subculture the stock about four times during the year, adding a bit of the old culture to fresh cider vinegar. You may also purchase vinegar eels from supply houses (Appendix C).

Uses in the classroom. These small roundworms may be used in studies of

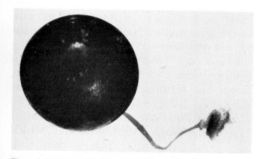

Fig. 12-11 Cocoon of black planarian. Planarian cocoons, about the size of radish seeds, consist of several fertilized eggs. (Note the short stalk on the cocoon.) (Courtesy of Carolina Biological Supply Co.)

[30] L. Lindner, H. Goldman, and P. Ruzicka, "Simple Methods for Rotifer Culture," *Turtox News* 39:74, 1961.

relationships among organisms in a "web of life" (Chapter 11). Where parasitic roundworms are not available for study, the vinegar eel may be substituted as a representative of the group (Chapter 1).

Earthworms

These annelids are readily collected at night or after a good rain, when they come to the surface.

Place several worms in wooden containers, such as cigar boxes. Place 2 to 6 in. of moist, rich soil or peat moss (sphagnum moss) in the boxes and lightly dampen the soil. Keep the animals covered and in a cool place (at a temperature about 15° C [59° F]). About twice a week feed them lettuce and bread soaked in milk. Bury the food.

Uses in the classroom. Earthworms are excellent organisms to use in studies of taxes. When the cover is removed so that the animals are exposed to light, watch the rapid burrowing movements. Earthworms may also be used to study chemotaxes (see tropisms and taxes, Chapter 7).

Earthworms are also favorable material for dissection. The reproductive, digestive, and circulatory systems, as well as the ventral nerve cord, may be studied as the worm is dissected (Fig. 1-24). Of course, earthworms may also be cultivated as food for frogs and some reptiles maintained in the laboratory. (See also *Monocystis,* p. 535.) Mature worms are also a source of living gametes (p. 356), and the nematode *Rhabditis* may be found in the nephridia (Fig. 11-13).

Enchytraeus

These white, semiaquatic segmented worms (Fig. 1-21) may be purchased from a supply house or collected from samples of damp, rich garden soil.

Cultivate these annelids in the same way as earthworms. Keep in covered boxes containing 2 to 4 in. of rich garden soil. Feed them lettuce and potatoes boiled in their skins. Alternate this food with cooked oatmeal. On occasion, feed them bread soaked in milk. Keep several small cultures going rather than one large culture. The cultures should be kept at temperatures about 20° C (68° F), and the food should be varied. Under these conditions, they multiply rapidly and many cocoons should be found among the masses of food.

Uses in the classroom. Enchytraeid worms are used mainly as food for other laboratory animals, such as fish, amphibia, and small reptiles. They may also be used to demonstrate taxes. Prepare wet mounts of *Enchytraeus* to show contraction of muscle, such as peristalsis along the length of the intestine.

Tubifex and other aquatic worms

Any of the oligochaete annelids described in the following paragraphs may be purchased from biological supply houses that distribute living materials (see Appendix C).

Tubifex (Fig. 1-22) may also be obtained readily from aquarium shops. In the field, *Tubifex* may be collected from the muddy bottom and decaying leaves of streams and ponds. They form tubes of mud held together by a secretion from epidermal cells. These forms usually have a reddish color due to dissolved erythrocruorin in the blood.

Members of the family Naididae, such as *Nais* (Fig. 4-3) and *Dero* (Fig. 1-19), carry on respiration through ciliated gills in the anal region. Naididae are abundant in old cultures of protozoa. Under the microscope, *Aulophorus* can be distinguished from *Dero* by its two microscopic, finger-like terminal processes, as well as gills (Fig. 1-20), which are not found in *Dero*. The Naididae are larger than members of the family Aeolosomatidae, which are found in similar places; they lack colored oil globules. *Nais* is about 3 mm long, comprising 15 to 37 segments. *Aeolosoma* (Fig. 1-18) consists of 8 to 10 segments and is only about 1 mm long. *Aeolosoma* reproduces asexually by transverse fission. The Naididae also lack dorsal bundles of setae in the anterior segments. *Aeolosoma* is readily identified by minute yellow, greenish, or red globules in the epithelium.

Introduce *Tubifex* and *Nais* into well-established aquarium tanks which contain an inch or so of muddy soil.

Culture *Nais,· Dero, Aeolosoma,* and *Aulophorus* by the methods described for maintaining protozoa, particularly method B or E.

Uses in the classroom. These aquatic worms serve as a food supply for other laboratory animals, such as fish and hydras. They may be used in laboratory and classroom investigations in the following areas.

1. Regeneration (p. 482).
2. Circulation of blood, which is visible in some forms, and peristaltic contractions, which are especially clear in wet mounts of these worms, under low and high power (pp. 95, 245).
3. Asexual reproduction (fragmentation and transverse fission); they are examples of hermaphroditic oligochaetes.

Snails and clams

Among mollusks there are wide variations in shape (Fig. 12-12). Such egg-laying forms as *Physa, Planorbis,* and *Lymnaea* may be found attached to water plants in ponds and lakes, or they may be purchased from aquarium stores. *Planorbis* is the form with a shell coiled in one plane like a watch spring. It lays eggs in clusters of jelly. A popular form for aquaria is the imported red variety, which lays pinkish masses of eggs. When both kinds are bred together, the common brown variety seems to be dominant. *Physa,* recognized by its sinistral spiral shell, lays eggs in long ribbons of jelly. *Lymnaea* is brownish-black in color with a dextrally coiled shell. It lays egg masses in jelly, usually found attached to stems of aquatic plants. A larger form, *Campeloma,* is a live-bearer and may be found in lakes or rivers attached to

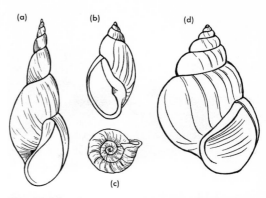

Fig. 12-12 Four fresh-water snails that thrive in a classroom aquarium: (a) *Lymnaea;* (b) *Physa;* (c) *Planorbis;* (d) *Campeloma.* (Courtesy of General Biological Supply House, Inc., Chicago.)

rocks or plants. Finally, *Helix* (Fig. 1-39), a land snail, may be found in moist, but not too acid, soil such as that in gardens or some wooded areas.

Raise *Helix* in a cool place in a moist terrarium with occasional feeding of lettuce. At times, also add whole oats rolled in calcium carbonate powder. Keep the aquatic snails such as *Planorbis* in an established aquarium. They normally feed upon algae, but when they increase in number they feed upon the aquatic plants (dissection, Chapter 1).

Uses in the classroom. When the snails reproduce, separate the developing eggs into covered Syracuse dishes. With a hand lens or binocular microscope, examine cleavage stages and ciliated veliger larvae. In the young embryos notice the beating heart. Snails furnish a ready source of living material for studies of developing eggs (p. 351).

Snails should be kept in aquaria in class to hold down the abundant growth of algae and to remove decaying materials in tanks.

Fresh-water mussels and clams are also good material to use for dissection (Fig. 12-13).

Land snails, which are available from some fish markets, can be distributed to students for many interesting studies of taxes (see Chapter 7).

Fig. 12-13 Fresh-water mussel. (Photo by Hugh Spencer.)

Some small crustaceans

Small crustaceans can be collected from their environment with the use of fine-mesh nets. Culture methods for most of these forms are similar to those described for *Daphnia*.

Daphnia

These small, laterally compressed water fleas (Order Cladocera) may be collected from ponds, lakes, or streams. They are characterized by a body enclosed in a transparent bivalve shell (Fig. 1-26a). A cleft marks off the head from the rest of the body. Large second antennae are modified as swimming appendages to assist the four to six pairs of swimming legs. During the spring and summer, females are usually found. Eggs generally develop parthenogenetically at these times. In the autumn, males appear and the "winter eggs" are fertilized. Female *Daphnia* may be recognized by the curved shape of the end of the intestine. In the male the intestine is a straight tube.

A great many successful methods have been described for maintaining *Daphnia*, using the fact that water fleas feed upon bacteria and nonfilamentous algae. Three methods which have proved successful are described below.

Using "green water." Fill large battery jars with tap water and let them stand overnight to permit evaporation of gases which may be harmful. Then put the battery jars in strong sunlight and inoculate them with nonfilamentous algae from a "soupy green" aquarium. After this "green water" has been standing for about 2 to 3 days, add *Daphnia* and several milliliters of hard-boiled egg yolk mashed into a paste with a bit of culture medium. You may also add a suspension of yeast to stimulate growth. This method produces a luxuriant growth of *Daphnia*. The temperature range may vary between 24° and 26° C (75° and 79° F). The sediment often contains viable eggs.

Using a modification of Knop's solution. In this method, a 6 percent stock solution is prepared (see below). For immediate use, add 5 liters of distilled water to 1 liter of the stock solution. This will yield a dilute 0.1 percent solution. When needed, this may be further diluted with an additional 4 liters of distilled water. Even this weak solution will maintain *Daphnia* adequately when the culture medium has been inoculated with nonfilamentous algae and allowed to stand in light until the water becomes tinged with a green color. About once a week, add a bit of hard-boiled egg yolk paste and a bit of yeast suspension.

MODIFIED KNOP'S SOLUTION. Combine the following materials with 1 liter of distilled water and pour into several battery jars:

KNO_3	1 g
$MgSO_4$	1 g
K_2HPO_4	1 g

Then add 3 g of calcium nitrate, $Ca(NO_3)_2$. As a result, a precipitate of calcium phosphate, $Ca_3(PO_4)_2$, is formed.

Using a culture of bacteria. Chipman[31] recommends the following method for culturing *Daphnia*. A rich growth of *Bacillus coli* is used as the food source. First, filter pond water through coarse filter paper. Then add about 90 g of garden soil and 17 g of cottonseed meal to 1 liter of this filtered pond water. Stir the mixture well and set it aside at room temperature

[31] W. Chipman, Jr., "Culture Medium for Cladocerans," *Science* 79:59–60, 1934.

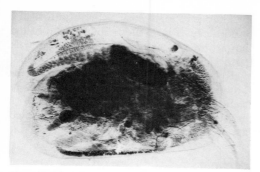

Fig. 12-14 *Cypris*, a small crustacean found in lakes and ponds (often mistaken for *Daphnia*). (Courtesy of General Biological Supply House, Inc., Chicago.)

for 5 days. Fermentation takes place and gases are formed. After the 5 days have elapsed, decant off the supernatant fluid and strain through muslin. An almost pure culture of *Bacillus coli* is produced. Correct the pH to 7.2 by adding sodium carbonate. Use Hydrion pH paper for testing the pH (or use a pH-meter if available).

Now dilute this fluid with pond water (1 part of strained fluid to 100 parts of pond water). Inoculate this culture medium with *Daphnia*. Keep the cultures in large battery jars. Each week prepare fresh stocks of cottonseed meal. Then add a small amount of the old culture each time a new medium is established. In this way inoculation with the original kind of bacteria is achieved.

Uses in the classroom. *Daphnia* serve as excellent food for small fish, tadpoles, and hydras.

Introduce a drop of a culture of these water fleas on a slide containing one or two hydras. Under a microscope watch ingestion. What is the role of the nematocysts (see p. 109)?

Use *Daphnia* to clear an aquarium that has become soupy green.

Demonstrate the rapidly beating heart of *Daphnia* under the low-power objective of the microscope (see p. 245). You may want to demonstrate the effect of narcotic drugs on the heartbeat as well. Use a hanging-drop preparation (Fig. 2-29), or

put bits of broken coverslips near the *Daphnia* as you prepare a wet mount, to avoid crushing the animals.

Small amounts of adrenalin and pituitrin cause a spontaneous shedding of the eggs from the dorsal brood sac.

Use wet mounts of *Daphnia* to study circulation, respiratory system, and peristalsis.

In some lake-dwelling *Daphnia*, students may study the change in head shape—from round to helmet shape—that occurs between spring and summer. What conditions cause this cyclomorphosis?

Cypris

At first glance this form (Figs. 1-26b, 12-14) is often mistaken for *Daphnia*. However, it has an opaque shell (this makes the study of internal anatomy difficult).

Cypris is laterally compressed and completely enclosed in a bivalve shell. It usually has seven pairs of appendages, and its antennae protrude from the shells and are used in swimming.

This form may be collected from ponds and streams.

Cyclops

This elongated crustacean (Subclass Copepoda) lacks a shell and has no abdominal appendages (Figs. 1-28, 12-15). It is characterized by the single compound eye located in the center of the head; it uses antennae for locomotion. During the summer months, females can be found carry-

Fig. 12-15 Female *Cyclops* with egg sacs. (Photo by Hugh Spencer.)

ing two brood pouches posterior to the body, as shown in Fig. 12-15. *Cyclops* may be found in brackish water as well as in fresh-water streams and lakes.

Culture methods are similar to those described for *Daphnia*. These forms are interesting for laboratory study and for introductory work with a microscope (see p. 95); they can be used as food for small invertebrates, fish, and amphibia.

Fairy shrimps

This small crustacean, *Eubranchipus* (Figs. 1-27, 12-16), Order Anostraca, is identified by its swimming motion: it swims with its ventral side uppermost. Its head bears stalked eyes, and its body is transparent. Fairy shrimps move by means of thoracic appendages, as shown.

They are found in shallow, stagnant ponds which may dry up during the summer months. Culture them like *Daphnia* and the other small crustaceans.

Gammarus

These fresh-water shrimps (Fig. 1-29) are found abundantly along fresh-water streams and along the seashore. They are members of Order Amphipoda, distinguished by laterally compressed bodies with gills borne on the legs. The first three pairs of legs are used as swimming legs, with the last three pairs modified as stiff processes used in jumping.

They are fairly easy to maintain in the laboratory. Place them in an established aquarium. A bit of hard-boiled egg yolk (pea size) may be added every 2 weeks.

Reproduction in the fresh-water forms occurs in the spring and summer. *Gammarus* has been used in biology laboratories in studies of heredity; eye color is studied most often.

Artemia

These brine shrimps (Fig. 12-17; larva, Fig. 8-19) are found in saline lakes, or they may be purchased as dried, resting eggs from aquarium shops or supply houses. They belong to Order Eubranchiopoda and are characterized by 10 to 30 pairs of leaf-like swimming limbs. In development they have a nauplius stage like *Cypris* and *Cyclops*. (On the other hand, *Daphnia* has direct development.)

Artemia are sensitive to light and orient themselves so that the ventral surface is placed toward light. Thus they often swim with the ventral region uppermost. In the female, lateral egg pouches are conspicuous. Rapid beating of the limbs is characteristic of this form (150 to 200 beats per minute). After each molting, the females are ready for mating. Batches of eggs may be laid as often as every 4 or 5 days when ample food is available. Eggs laid with abundant secretion usually remain dormant for some time, often for several months. A period of drying out seems to

Fig. 12-16 *Eubranchipus*, the fairy shrimp, a small crustacean that swims ventral side up. (Photo by L. C. Peltier, courtesy of *Nature Magazine*, Washington, D.C.)

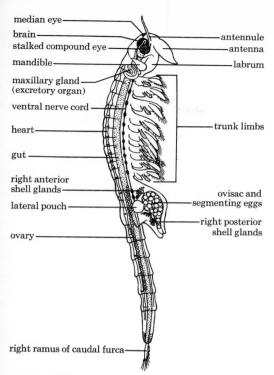

median eye
brain
stalked compound eye
mandible
maxillary gland (excretory organ)
ventral nerve cord
heart
gut
right anterior shell glands
lateral pouch
ovary
right ramus of caudal furca

antennule
antenna
labrum

trunk limbs

ovisac and segmenting eggs
right posterior shell glands

Fig. 12-17 Female *Artemia*, the brine shrimp. (Courtesy of General Biological Supply House, Inc., Chicago.)

shorten the time of hatching. Dried eggs retain their viability for several years, provided they are kept in a cool place. On the other hand, eggs that have a scant secretion when laid hatch out in 1 or 2 days as nauplius larvae.

HATCHING ARTEMIA EGGS. Dried brine shrimp eggs are available throughout the year and can readily be hatched to show a nauplius larva stage of crustaceans. Much effort has been given to develop standard conditions for hatching these eggs, since the larva are used commercially as live food for *Hydra*. You may wish to feed live larvae to *Hydra* or to planarian worms in studies of ingestion (pp. 109, 201). After considerable experimentation, Loomis and Lenhoff[32] have recommended the following procedure for hatching large quantities

[32] *Op. cit.*

of *Artemia* eggs (reduce quantities to fit your laboratory needs).

Prepare a stock solution of saturated sodium chloride (360 g per liter) by dissolving 5 lb of commercial table salt in 2 gal of hot tap water. Cool this solution, and dilute it with tap water (1:100). Seed wide, shallow, hatching dishes of the dilute salt solution with *Artemia* eggs (½ tsp of eggs per 500 ml of solution). Incubate for 48 hr at a constant temperature of 21° C (70° F). (At a temperature of 30° C [86° F] eggs hatch in 1 day; at 15° C [59° F] it takes 3 days.)

When the larvae are to be fed to hydras the larvae must be rinsed in aquarium water or conditioned water (see p. 606), since the salinity will kill the hydras. To collect the larvae, shine a light at one side of the container; then the phototactic larvae can be siphoned off with thin rubber tubing into a fine net (125-mesh).

Avoid overcrowding the culture of *Artemia* larvae, and supply them with non-filamentous algae and a yeast suspension. Algae scrapings from the sides of a tank have been found to develop as well in weak salt solutions and may be added to the culture of *Artemia*. However, there must be an adequate oxygen supply as well as food for the nauplius larvae to mature rapidly. Keep the containers in moderate light at a temperature that remains below 25° C (77° F).

Wood lice

These land isopods (*Oniscus*) and the related genera *Porcellio*, the sow bug, and *Armadillidium*, the pill bug (Fig. 12-18), are all well adapted for their way of life and are widespread. They may be found under stones, boards, and logs and in other dark, moist, undisturbed places.

The pill bug shows a characteristic response by rolling itself into a ball when it is disturbed. Wood lice and pill bugs are found in the same ecological conditions and are cultivated in the same way.

Their usual ecological conditions are best duplicated in the laboratory in a ter-

(a)

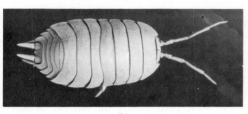

(b)

Fig. 12-18 (a) *Porcellio,* the sow bug; (b) *Armadillidium,* the pill bug. (a, photo by Hugh Spencer; b, courtesy of U.S. Dept. of Agriculture.)

rarium containing damp, rich humus with small rocks or logs under which the organisms can hide. Supply these isopods with bits of ripe fruit (apples), bits of lettuce, and at times pieces of raw potatoes. They will also accept small earthworms and insects.

Uses in the classroom. Isopods are interesting to have on hand for studies of adaptation of animals. Notice how their compressed bodies fit them for their environment. Use them in studies of taxes (p. 300); examine the contents of the intestine for flagellates (p. 538).

Marine aquaria

More and more biologists are keeping marine aquaria (Fig. 12-19).[33] These make especially fascinating studies for

[33] See I. L. Bayly, "Try a Salt-Water Aquarium," 2 parts: Part I, *Carolina Tips* 28:15–16, 1965, Part II, *Carolina Tips* 28:19–20, 1965.

classes in inland cities. Synthetic sea water can be purchased, along with small invertebrates and seaweeds, as a kit from some biological supply houses (see Appendix C). Use a hydrometer to establish the salt solution at a specific gravity of 1.017 to 1.022. Since some salts are absorbed by the organisms, the salts must be replenished. Each month add a level teaspoon of a mixture of 3 parts of rock salt and 1 part of Epsom salts to a 20-gal tank. Every 5 months add a small piece of plaster of Paris. (Avoid having brass, copper, or zinc in contact with sea water. Maintain in a cool place.)

If you live along the coast, you may collect live specimens (gather only small ones) together with a gallon of natural, unfiltered sea water. Maintain the specimens and sea water in a large, loosely covered battery jar. Within a week small crustaceans and diatoms should appear.

Since sea plants are not good oxygenators, attach an aerator and pump to the aquarium. Collect small starfish, small clams or oysters, sea anemones, and seaweeds. Light the tank with fluorescent lights to prevent a change in the color of the water. (Incandescent lights seem to increase the growth of microorganisms, turning the water yellow.) Many sea worms thrive in these tanks. To feed the starfish, clams, oysters, and anemones, remove them from the tank and place them in small containers so that the aquarium does not become contaminated by uneaten food. Feed the starfish bits of clams or oysters; brine shrimp eggs are suitable food for anemones.

Observe the means of locomotion of small starfish along the glass of the aquarium tank. Place a small oyster in a separate container, add a bit of lampblack to the water, and study the action of the incurrent and excurrent siphons as water moves in and out of the animal (anatomy, Chapter 1).

Marine algae may be cultured separately and added to the tank. Refer to L. Provasoli, J. McLaughlin, and M. Droop, "The Development of Artificial Media

sea urchin

starfish

sea cucumber

Fig. 12-19 A marine aquarium and some animals that might live in it. (Photos: top three, courtesy of Carolina Biological Supply Co.; right, courtesy of the American Museum of Natural History, New York, and U.S. Fish and Wildlife Service; bottom, courtesy of General Biological Supply House, Inc., Chicago.)

scallop

clam

for Marine Algae" (*Arch. Mikrobiol.* 25: 392–428, 1957).

Insects

Some insects may best be collected as pupae, others as adults or larvae. The planned classroom work will, no doubt, determine the kinds of insects that will be gathered and maintained. Following are some methods for keeping alive just a few kinds of insects; there are as many methods as there are kinds of insects (see Chapter 11).

Praying mantis

Collect egg masses in the fall or early spring. The egg cases are recognized as tan, foam-like masses attached to twigs (Fig. 4-4). Or purchase eggs from biological supply houses.

Keep the egg cases in covered terraria. With a gradual increase in temperature, hundreds of nymphs emerge. Supply the nymphs with dilute sugar solution or honey served in low, flat containers.

Uses in the classroom. In a study of reproduction among insects, the praying mantis is an illustration of a beneficial insect with incomplete metamorphosis: egg, nymph, adult. Some of the nymphs can be placed in corked vials and examined with a hand lens (pp. 202, 354).

Moths and butterflies

When pupal cases are collected in the fall, they should be stored in a cool place throughout the winter months. Place them in a box that can be left outside a window. In the spring put them into a small screened box or terrarium. Include twigs as supports for the emerging adults. Live pupae are generally heavier than dead or parasitized forms (Fig. 12-20). (For instance, *Cecropia* moth pupae, a common form that students bring to class, is often parasitized by the Ichneumon fly.) Some teachers on the West Coast have their students observe metamorphosis of the mourning cloak butterfly (*Vanessa*). The changing of larvae into the chrysalis stage is almost completed within an hour; in 7 days the butterfly emerges. Compare this development with that of the praying mantis or the grasshopper.

Uses in the classroom. In these living examples of insects, students may trace complete metamorphosis. At times, students may be fortunate enough to see a butterfly or moth emerge from a pupal case.

Drosophila

Directions have been given for raising fruit flies in class (see pp. 415–16).

Tribolium confusum

These beetles are often found in packaged flour and cereals; or cultures may be purchased. Keep both sexes in jars or finger bowls of slightly moistened whole-wheat flour, oatmeal, cornmeal, or bran. Where possible, adequate moisture may be supplied by attaching moist cotton to the cover of the container.

Metamorphosis is complete within 5

Fig. 12-20 The complete metamorphosis of *Samia cecropia*: (a) larva; (b) cocoon; (c) pupa removed from cocoon; (d) adult. (Courtesy of General Biological Supply House, Inc., Chicago.)

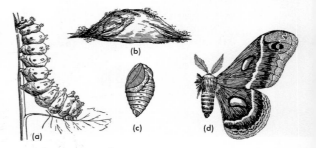

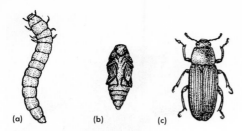

Fig. 12-21 The metamorphosis of *Tenebrio*: (a) larva; (b) pupa; (c) adult. (Courtesy of General Biological Supply House, Inc., Chicago.)

to 6 weeks when cultures are kept at temperatures between 28° and 30° C (82° to 86° F). Start new colonies by placing pupae in fresh food medium.

Uses in the classroom. *Tribolium* beetles show all the stages in complete metamorphosis: egg, larva, pupa, and adult. They are also fine organisms around which students may design experiments in physiology, genetics, and behavior.

Tenebrio beetles

Larvae, called mealworms, of these beetles (Fig. 12-21) may be purchased from aquarium shops.

Culture the mealworms in battery jars half filled with moist bran or oatmeal, covered with fine mesh so that adult beetles cannot fly off. When adult beetles develop, they should be fed bits of raw carrots or potatoes. At a temperature around 30° C (86° F), the complete life cycle may take 4 to 6 months.

Another method[34] describes the use of smooth, galvanized, iron boxes which are 2 × 1¼ × 1 ft deep. On the bottom of the box spread a mash used to feed chicks. Over this, place a layer of burlap, then another layer of mash. Alternate mash and burlap layers until some five or six layers of each are in place. This will support several hundred mealworms. In old cultures many eggs may be found. Start new cultures with a few organisms from the old cultures.

Uses in the classroom. While these bee-

[34] Galtsoff *et al., op. cit.,* p. 463.

tles also show complete metamorphosis in a reproductive cycle, they are especially bred as food for laboratory fish, amphibia, and some reptiles.

Aquatic insects

When collecting pond water, include submerged leaves and some bottom mud. Inspection of battery jars and finger bowls of pond water over a week or month will reveal larvae of many forms. Some of these forms were collected as immature aquatic larvae; others were collected as eggs laid on submerged leaves or in the mud. Possibly nymphs of stone flies, mayflies, or damsel flies or the predatory dragonfly nymphs will be found in the water (Figs. 11-25, 12-22). Separate the predatory

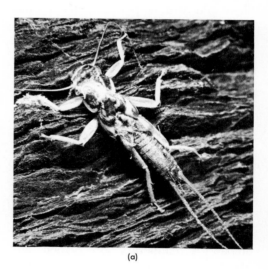

(a)

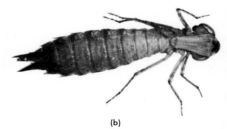

(b)

Fig. 12-22 Nymphs of two aquatic insects: (a) stone fly; (b) dragonfly. (a, photo by Walter Dawn; b, courtesy of Carolina Biological Supply Co.)

forms from the mayfly and damsel fly larvae, and maintain the larvae in gallon
battery jars. Avoid overcrowding the larvae, since a good oxygen supply is required
for their maintenance.

Adult insects collected by students can
be maintained in large tanks covered with
screening. Examine the adaptations of
such forms as the water boatman, upside-
down backswimmer, whirligig beetle, or
water bug (Fig. 12-23).

Uses in the classroom. Observing the
life cycle of several insects as part of a
study of reproduction is often desirable.
Some insects also furnish an excellent example of predator-prey relations in ecological communities, particularly of a
pond community (Chapter 11).

Circulation may be observed in some
small gill-bearing forms. It may be possible to add carmine powder and trace the
path of water. Use simple guides or keys
to identification such as those in J. Needham and P. Needham, *A Guide to the Study
of Fresh Water Biology,* 4th ed. (Comstock
[Cornell Univ. Press], 1941).

Activities of social insects

Directions for making a beehive may be
found in several books on bees (see Bibliography). Some teachers purchase wooden
ant houses and observation boxes (Fig.
12-24a) or an observation beehive (Fig.
12-24b) from a biological supply house
(Appendix C).

Making an ant colony. Students may
prepare a temporary observation ant
colony for class use. Partially fill a battery
jar or large box with slightly moistened
sandy soil. Collect an ant hill from the field
and place it in the container. Include in
the colony some workers and a queen.
Keep the container covered or in the dark.
Galleries made by the ants may be visible
through the glass sides of the container.
When possible place the jar in a basin of
water so that a moat is formed, preventing
the escape of ants.

Feed the ants on lettuce, carrots, and
potatoes (to provide moisture as well as

(a)

(b)

(c)

Fig. 12-23 Adult aquatic insects: (a) backswimmer; (b) whirligig beetles; (c) water bug
(male carrying eggs). (a, courtesy of U.S. Dept.
of Agriculture; b, c, photos by Lynwood M. Chace.)

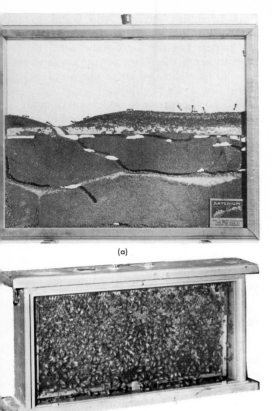

(a)

(b)

Fig. 12-24 Equipment for studying social insects: (a) observation ant house; (b) observation beehive. (a, courtesy of World of Adventure, Denver; b, courtesy of Univ. of Southern California.)

food) and on bread crumbs and dilute molasses or honey. On occasion add some dead insects. Remove all excess food to prevent the growth of molds.

Keeping a termite colony. Colonies of termites may be kept in the laboratory for a continuing study of social insects. A colony consists of a wingless large queen, winged males, wingless workers, and wingless soldiers. Inspect old tree stumps and wet logs for termite galleries. Gently strip off sections of the bark and wood to expose the insects and their eggs. Collect all stages with the wood in which they were thriving. In fact, move to the class-

room as much of the log as is practicable.

In the laboratory separate the termites from the wood with a camel's-hair brush to avoid injuring their soft bodies. Keep the insects, along with wood fragments, in covered finger bowls or Petri dishes. Add strips of moistened filter paper, then store in a dark place at room temperature. Keep moist by adding a few drops of water twice a week.

DeLong and Keagy[35] describe several excellent methods for making observation termite colonies in the laboratory. One simple method recommends the use of flat battery jars of the Delco type. Place a piece of balsa wood along the inside of each of the two wide sides of the jar. Then fill the jar about one-fourth full of earth. Place thin strips of balsa wood between the glass walls and the balsa wood sheets in order to leave a space for free movement of termites between the balsa wood layers and the glass walls. When termites are introduced into the jar, they establish themselves within a few hours. Tunneling may be observed in a short time.

USES IN THE CLASSROOM. These insects are splendid examples for a study of social life; they are also used in class as a source of the symbiotic flagellates that are found in their intestines. Prepare wet mounts of these flagellates as described on p. 537; see also p. 586.

Fish

There is scarcely a laboratory that does not have an aquarium. The aquarium is probably the best single device for maintaining many of the animals in the laboratory.

Types of aquaria

The healthiest fish are those that are placed in an aquarium of appropriate size. Overcrowding is usually detrimental. A pair of fish 1 in. long requires at least 1 gal of water. A 5-gal tank can house six pairs

[35] D. DeLong and R. Keagy, "Termite Cultures in the Laboratory," *Turtox News* 27:5, May 1949.

of fish, 1 in. or so long, together with the needed plants and other animals. In our experience, 4- to 5-gal tanks are the most suitable for aquaria in the laboratory or classroom. For demonstration or close study in the classroom the fish may be placed in smaller, more easily handled tanks.

For most purposes, rectangular tanks with slate or glass bottoms and chromium or iron frames in which thick glass is fitted are desirable. Cheap tanks usually end up costing more than good ones (damaged tables, time spent in repair).

Certain tropical fish require special heating and other arrangements. Their care is well treated in many texts (see the list of books at the end of this chapter).

How to prepare an aquarium

Proper, careful preparation and planting are necessary for success in maintaining animals. Before the animals are added, the tank should be prepared and planted.

Wash the tank with coarse sand and warm water. Avoid using very hot water; in many tanks the preparations used to cement the sides to the frame may soften. After several rinsings with cold water the aquarium should be two-thirds filled with cold water and allowed to stand for a day or so. During this time any leaks may be detected and any soluble matter in the tank will be dissolved. Discard this water.

When the aquarium has been thoroughly cleaned, it is ready for plants and animals. Cover the bottom of the tank with a ½-in. layer of coarse sand (gravel) that has been washed in boiling water. Embed a clean piece of clam shell at each end of the 5-gal aquarium to help neutralize acidity and to furnish calcium salts for the shells of the snails. Over this, put another ½-in. layer of clean sand. Excessive growth of certain algae may be avoided by embedding a 2-sq-in. strip of copper (or several copper coins) in the sand. Next, lay a large sheet of paper on top of the sand before pouring water into the tank; the paper will prevent the sand

from becoming stirred. Add water to a level of 1 or 2 in. from the top, and remove the paper. Let the tank stand for 1 or 2 days to bring the water to room temperature, to help dissolve air in the water, and to rid the water of chlorine.

It is good practice to add ½ gal of established aquarium water to a freshly prepared tank. A tank to which this "conditioned" water has been added develops more quickly than a completely new one. Better still, add 1 gal of a thriving *Daphnia* culture to the newly prepared aquarium. When neither of these two is available, the water should stand for 2 or 3 days. Then plants may be added along with the water in which they have been purchased or collected. Select rooted plants as well as floating plants for display and protection for young fish.

Plants for an aquarium

There is a variety of plants from which to choose. The common plants that grow well in a tank are *Lemna, Vallisneria,* corkscrew *Vallisneria, Sagittaria, Anacharis* (elodea), *Cabomba, Myriophyllum, Ludwigia, Potamogeton, Chrysosplenium, Herpestis, Utricularia, Lysimachia nummularia,* and *Cryptocoryne* (Fig. 12-25). It is recommended that the first plantings be of *Anacharis, Cabomba,* or *Vallisneria,* since these, in our experience, are the hardiest of the plants that may be purchased or collected. *Vallisneria* roots readily; it is pleasing to display and does not tangle (so that a fish net may be used easily). *Potamogeton, Chrysosplenium,* and *Lemna* are still hardier, but the first two must be collected and have a tendency to grow rapidly and crowd the tank. This may be beneficial to the fish but may not be satisfactory for display purposes.

The water ferns *Salvinia* and *Marsilia,* and algae such as *Nitella,* will do well after the water has been conditioned.

In any event, it may be desirable to use a variety of plants for display (Fig. 12-26). Place plants in the background of the tank where they will not interfere with examination and handling of the animals. Then

Fig. 12-25 Common aquarium plants: (a) *Vallisneria;* (b) corkscrew *Vallisneria;* (c) *Myriophyllum;* (d) *Anacharis;* (e) *Cabomba;* (f) *Lysimachia nummularia.* (a, c, d, e, photos by Hugh Spencer; b, f, courtesy of Carolina Biological Supply Co.)

Fig. 12-26 A thriving fresh-water aquarium. (Courtesy of General Biological Supply House, Inc., Chicago.)

add a rock or two for scenic effect and, more important, to afford a hiding place for the animals, especially for gravid females.

Snails

Place two snails in the aquarium for each gallon of water. The snails tend to keep the glass clean by removing encrusting algae. In addition, very young snails serve as food for some of the fish. *Lymnaea, Physa,* or *Planorbis* (Fig. 12-12) are suitable for this purpose. *Campeloma,* a very large snail, is excellent but requires more space then the others; three of them in a 5-gal tank are sufficient.

Light

Keep the aquarium in medium light; strong light favors the growth of algae,

which turn the water green. In general, northern or western exposure is most suitable. When a southern exposure is the only available location, the portion of the aquarium facing the direct rays of the sun may be covered with paper, aluminum foil, or glazed glass, or painted green. Should algae turn the water green, add quantities of *Daphnia.* After the water is cleared of the algae, the fish will feed upon the *Daphnia.* However, when there are many fish in the tank remove the fish before adding *Daphnia;* otherwise the fish will eat the water fleas before they have a chance at the algae. Add more snails if there is an excessive growth of filamentous algae.

Feeding

There are many fish foods on the market; most of them consist of dried, chopped shrimp, brine shrimp, ant pupae, or dried

Daphnia. Others have dried vegetables added. Any one of these is satisfactory for tropical and native fish, provided live food is added now and then (once a week is fine). Some fish, like the Bettas, do very well on dried food, while others, like the Japanese medaka, reproduce regularly only when fed living food daily.

Native fish prefer living food but will accept dried food. *Enchytraeus* (white worms)—either chopped or whole—*Tubifex* worms, *Daphnia,* and bits of fresh liver are also acceptable food.

Avoid overfeeding fish, for the excess food will foul the water, killing the fish. A pinch of dry food daily is enough for tropical fish about 3 or 4 in. long. During the first few days watch how much food the fish consume. Reduce the quantity the following day if food is not consumed. Remove the excess by siphoning. It is better to underfeed than to feed more than the animals take in a day.

Day-to-day care

Besides feeding the fish, keep the water level constant and remove dead plants or animals. When plants are growing rapidly some should be removed to prevent overcrowding. This is especially true of *Lemna* and *Salvinia,* two forms that grow profusely.

Also remove excess snails as they increase in number, for an excess will destroy plants. It is not necessary, especially for tropical fish, to change the water in the tank except when fouling or special care requires it. There is no cause for alarm if the water becomes yellowish or greenish; water that color is good, "conditioned" aquarium water. (A suitable pH range may be as wide as 6.8 to 7.2 for the average tank.)

Tropical fish

There are many tropical fish that add beauty and variety to a tank. Many hybrid forms have been developed as knowledge of the genetics of fish has accumulated. Only a few kinds of fish, typical and useful in the classroom, are described here. Although the life histories of the "fighting fish" *Betta* (Fig. 12-27e) and other bubble-nest builders are of unusual interest, they are not of general value for classroom use.

The guppy (*Lebistes reticulatus*) (Fig. 12-27f) is probably the most common of the tropical fish, and perhaps the most successful for the beginner to maintain. It is hardy and can withstand low room temperatures. It is a live-bearer (ovoviviparous) and reproduces readily.

Platy (*Platypoecilus*) (Fig. 12-27c) are not as hardy as the guppy but are interesting for display and study. They are also live-bearers. A temperature range from 20° to 25° C (68° to 77° F) is optimum for these fish.

Japanese medaka (*Oryzias latipis*) do not need as careful temperature control as do other tropical fish, but they do need live food for regular reproduction. When they are fed *Tubifex, Daphnia,* or *Enchytraeus,* the female produces about 8 to 20 eggs each morning. These eggs hang from the cloaca and can easily be removed if the fish is caught and transferred to a finger bowl of tank water. The eggs are clear and excellent for the study of embryological development. Students may examine these eggs under a binocular or stereoscopic microscope or magnifying glass.

Native fish

Killifish (*Fundulus heteroclitus*), the red-bellied dace (*Chrosomus erythrogaster*), the stickleback, the banded sunfish (*Mesogonistius chaetodon*), and the blue-gill sunfish (Fig. 12-27a) are desirable animals to have in the laboratory. These species should be kept separately in tanks well supplied with vegetation. In one sense they are easier to keep, since they can withstand a wide range in temperature.

In general, these fish will not accept the prepared foods although they may do so after some time. All living food and bits of raw meat, raw fish, or raw liver are readily accepted and eaten voraciously.

Fig. 12-27 Fish that thrive in a classroom aquarium: (a) blue-gill sunfish (bream); (b) goldfish; (c) platy; (d) special type of catfish for small tanks; (e) *Betta* (maintain in separate tank); (f) guppy. (Courtesy of *The Aquarium Magazine.*)

Goldfish (*Carassius auratus*) will thrive under the same conditions as tropical fish. Small goldfish (Fig. 12-27b) are preferable, since the larger ones require larger tanks. Goldfish do not need careful temperature control. Optimum temperature conditions range from 10° to 25° C (50° to 77° F). They are omnivorous and feed on plants in the tank and bits of boiled spinach in addition to other food.

Diseases of aquarium fish

Fungi, such as the water mold (*Saprolegnia*), and the protozoan called "water itch" (*Ichthyophthyrius*) are two of the many parasites that attack fish. When fish show gray patches on the fins or scales they should be isolated quickly, for these are symptoms of disease.

Fish that show these patches should be immersed in a 10 percent solution of table salt. After 1 hr remove the fish and wash in ordinary water. Usually the patches disappear after this heroic treatment. Nevertheless, put the fish into a 0.5 percent solution of potassium permanganate for 15 min. Then quarantine the affected fish in separate tanks and watch for the reappearance of the symptoms. (Avoid sharp changes in the temperature of the water during this treatment.)

Actually, so little is known about fish diseases that it seems most desirable to discard fish that have become diseased. This has proved to be the best and the cheapest procedure (even after experts have prescribed the "cure").

Amphibia

Salamanders and frogs are especially desirable to have on hand for studies in natural history and in the development of eggs (also in studies of induction of ovulation, p. 360). Students may prepare vivaria so that the tanks duplicate the natural habitat of these animals. (Stages in the development of frogs' eggs are shown in Fig. 9-22.)

Salamanders

Preparing vivaria. Long, low vivaria covered with glass are most desirable for salamanders. However, whatever type of container is used, it should provide a pool with a "beach" for most kinds of salamanders (Fig. 12-28).

The vivarium must be cleaned as thoroughly as if it were to house fish. At one end of the tank put a small glass dish or noncorrosive metal pan to serve as a pool. Cover the rest of the tank with coarse pebbles together with a few pieces of charcoal. This should ensure good drainage. Then cover the pebbles with a loam soil rich in humus, slanting the layers of soil away from the small pool to a height of 3 in. and keeping about 2 in. of water in the pool. Students can prepare an effective natural habitat by planting a "beach" around the pool, using layers of moss, such as *Sphagnum, Mnium, Dicranum,* or similar types. Place at least one rock in the water. Try planting the rest of the vivarium with small ferns, partridge berries, and a variety of other mosses that have different textures and shades of green.

Feeding. Almost all salamanders require living food such as *Daphnia, Tubifex, Lum-*

bricus (chopped), *Enchytraeus, Tenebrio* larvae, and *Drosophila.* On occasion, some forms such as *Triturus* (the red eft) may take fresh liver if it is dangled in front of them on a string. But dead animals and food that has not been eaten must be removed within an hour or so to avoid fouling the tank.

Temperature. The best temperature range seems to lie between 15° and 18° C (59° and 64° F). However, many salamanders can survive at temperatures as high as 25° C (77° F).

Red eft, or water newt (*Triturus viridescens*). This salamander (Fig. 12-29), now often called *Diemictylus viridescens,* is the one most easily reared. It can be easily handled, and its slow movements make it desirable for observation and study by students. Students may readily observe the two phases in the life cycle of the water

(a)

(b)

Fig. 12-29 *Diemictylus viridescens* (formerly called *Triturus*), the red-spotted newt: (a) aquatic phase; (b) terrestrial phase. (Photos by Hugh Spencer.)

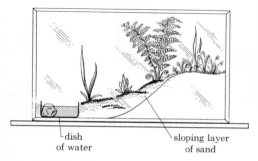

dish of water   sloping layer of sand

Fig. 12-28 Vivarium with a "beach" that is suitable for salamanders and frogs.

newt: the water and land phases. In the water phase, the animal is olive-green with carmine spots and yellow-speckled undersides. In this stage it may be kept with fish in an aquarium, although it is more desirable to keep it in a vivarium where it will remain in the pool of water. (Water plants should be included in the pool.) In this stage *Triturus* becomes sexually mature, and, if both sexes are present, fertilized eggs may be deposited on the water plants. Larvae hatch in water, and metamorphosis into the red phase—the land stage—occurs. The land stage is a beautiful creature which moves slowly over land and feeds on small insects and worms. After 2 to 3 years the red eft returns to the water and changes into the green phase.

In water, the newts feed on *Daphnia, Tubifex,* bits of *Enchytraeus,* and earthworms. On land, they consume larvae of *Drosophila* and *Tenebrio.*

The ambystomas (tiger, spotted, and Jefferson's salamanders). These animals are readily kept in vivaria. The young stage (axolotl) remains in water and feeds on *Daphnia, Tubifex,* earthworms, and *Enchytraeus.* Adults feed voraciously on earthworms, insects, and *Tenebrio* larvae.

A related form, the Mexican axolotl, may also be maintained in a laboratory. It may provide eggs for study. Mexican axolotls are of interest because they do not undergo metamorphosis but retain larval external gills, reproduce, and remain aquatic permanently. This characteristic seems a special adaptation existing among forms that live in arid regions. When fed thyroid glands of beef, these axolotls undergo metamorphosis.

Other salamanders that may be kept in a similar tank are the red-backed *Plethodon cinereus,* the slimy *Plethodon glutinosios,* the dusky *Desmognathus fuscus,* the two-lined *Eurycea bislineata,* the red *Pseudotriton ruber,* and the Pacific salamander. (Over the country, these have different common names.) These woodland salamanders feed upon small pieces of liver and lean beef, which should be offered to them on the end of a toothpick. There are a few forms such as *Amphiuma* and the red-bellied *Triturus pyrrhogaster* which thrive better in an aquarium. They may be fed the same diet as suggested previously.

Diseases. In the laboratory many salamanders may be affected by such fungi as *Saprolegnia.* The infection is symptomized by patches of fuzzy white thread on the tail or over the entire body. While most animals infected by fungi do not recover, some first-aid measures may be attempted. Isolate the infected animal and place it in 5 percent potassium permanganate for 10 min. After washing with cold water, place the animal in a jar of water to which a sheet of copper, copper wire, or copper filings have been added. Or add a drop of 1 percent copper sulfate solution to 200 ml of water. Keep the animal in this solution for about 2 days. It may be necessary to disinfect the vivarium with 5 percent formalin; then clean it thoroughly and replant it.

Frogs

Egg-laying time among frogs. Frogs' eggs and tadpoles, as well as adult frogs, may be collected by students and cultivated in the laboratory or classroom for long-range enriching experiences. The time of egg-laying is different for each species, and the surrounding temperature is a controlling factor. Therefore there are deviations from the time schedule given here; egg-laying occurs progressively later in the spring as one travels from the southern to the northern states.

As soon as the ice disappears, very early in March, *Rana sylvatica,* the wood frog, breeds, laying about 600 eggs. At this time the water temperature may be as low as 5° C (41° F). The eggs should be kept in water maintained at 5° to 15° C (41° to 59° F). The common leopard frog, *Rana pipiens* (Fig. 12-30a), lays eggs in March and April; about 2000 eggs are deposited at a time. The pond temperature at this time is about 15° C (59° F). Egg masses of *Rana palustris* are brownish in color and

(a)

(b)

Fig. 12-30 Frogs: (a) *Rana pipiens*, the leopard frog; (b) *R. catesbiana*, the bullfrog. (Photos by Hugh Spencer.)

larger than those of *R. pipiens*. This species breeds in April.

The green frog *Rana clamitans* and *R. catesbiana* (bullfrog, Fig. 12-30b) are both summer breeders. The males of both species can be distinguished from the females by the presence of a yellow throat and tympanic membranes larger than the eyes. In the females these membranes are the same size as the eyes. The eggs of *Rana clamitans* do not survive in temperatures below 12° C (54° F).

The spring peeper, *Hyla crucifer*, breeds in April, and its eggs are laid singly. Pairs found in amplexus will deposit eggs in the laboratory. This is the best way to collect these eggs.

Toads lay eggs in long strings of jelly. Those of *Bufo fowleri* and *B. americana* are found in early June, of *B. californica* in May. The young develop rapidly.

Vivaria. The eggs should be kept in finger bowls or in shallow aquaria containing 5 to 6 in. of water. Such a tank can be well stocked with plants to provide sufficient oxygen (Fig. 12-28).

It is best, according to our experience, to hatch the eggs in finger bowls. Jelly masses can be cut with scissors so that about 50 eggs are put in each bowl. Remove the unfertilized eggs. These can be spotted readily, for the fertilized eggs orient themselves so that the pigmented portion is uppermost. Thus, within the jelly masses, when the black surfaces are uppermost the eggs are fertile.

Cleavage stages may be studied with a hand lens, a stereoscopic microscope, or a dissecting microscope. The rate of development of frogs' eggs usually varies directly with the temperature. Viable temperatures vary from 15° to 24° C (59° to 75° F). As tadpoles hatch out they still have considerable yolk sac, and no feeding is needed. However, as they grow older, add scraps of boiled lettuce or spinach. At times, raw lettuce and spinach are accepted, as well as aquatic plants. Change the water twice weekly. After several weeks, pieces of hard-boiled egg yolk or small bits of raw liver may be added to the water. But quickly remove the excess food to prevent fouling of the water. When the hind legs have appeared and the forelimbs are just breaking through the operculum and skin, place the tadpoles in a combination water-woodland vivarium (as described for salamanders); they are ready to undergo metamorphosis into land forms.

Tadpoles of the larger species, *Rana clamitans* and *R. catesbiana,* as soon as they attain a length of ½ in., should be transferred to a larger aquarium having a 6-in. water level. Tadpoles of *Rana clamitans* do not complete metamorphosis until the following year. Bullfrog tadpoles take 2 to 4 years for complete metamorphosis. Tadpoles may be kept in a "balanced" aquarium. When they are kept separately, they can be fed scraps of lettuce, liver, or hard-boiled egg yolk.

When a few frogs are kept for display

purposes, a woodland vivarium is desirable. Such forms as bullfrogs are an exception; they fare better in 2 in. of water in a clean aquarium.

The problem of handling large quantities of frogs in the laboratory is a difficult one. The survival rate is high when they are kept at 10° C (50° F) in a granite sink containing 1 in. of water. Cover them with wire mesh; change the water daily. Better still, when facilities are available, keep the water running slowly in the sink. An unused sink in which the drain can be stoppered with a closed wire mesh tube about 6 in. high is ideal. A slanted board should be provided so the frogs may leave the water. Keep frogs away from zinc.

When possible, keep frogs in a drawer of a refrigerator in 1 in. of water or in large aquaria with water at a level of 1 in. Flush the frogs with a stream of water when the water in the tanks is changed each day; the water becomes fouled quickly. Dead frogs and frogs suffering from "red leg" should be isolated immediately, since the infection is highly contagious. To reduce infection, add a dilute solution of table salt (0.2 percent) to the water in which the frogs are kept.

Frogs that are most easily kept in beach vivaria as adults are *Rana pipiens, R. palustris,* and the green frog *R. clamitans.* They accept *Tenebrio* larvae, small earthworms, flies, and similar living materials as food. *Rana catesbiana* should be kept in water that is 2 to 3 in. deep. Provide a rock with a surface slightly above the water. These forms readily eat living things smaller than themselves, such as smaller frogs and earthworms.

The members of Hylidae, *Hyla crucifer* (spring peeper) and *H. versicolor* (tree frog, Fig. 12-31), may be collected by students and kept in a beach vivarium. *Hyla crucifer* feeds on *Drosophila* and small mealworms. Provide a few stout twigs as supports for the tree frogs in the terrarium.

Among the toads (Bufonidae), *Bufo fowleri* and *B. americana* both require terraria similar to that of the Hylidae. However, it must be remembered that toads

Fig. 12-31 *Hyla versicolor,* the common tree frog. (Photo by Hugh Spencer.)

are active burrowers, and they disrupt a well-managed terrarium.

Reptiles

Turtles

Vivaria. Most turtles, with the exceptions described below, should be kept in aquaria containing 2 to 4 in. of water. Cork floats can be added or a flat rock placed in one corner of the vivarium as a useful resting place. Students who care for the animals should change the water twice weekly to keep it clear.

Painted turtles, wood turtles, and box turtles (Fig. 12-32) may be kept in water (as described) or in a beach vivarium. However, box turtles seem to prefer a moist terrarium or vivarium rather than water. Segregate adult snapping turtles from other turtles, but small ones may be kept with other species.

Feeding. Most of the aquatic forms will accept bits of fish, ground raw meat, liver, earthworms, or dead frogs put in the water. In addition, most turtles will accept hard-boiled egg cut into slices, as well as lettuce and slices of apples. Box and wood turtles also take snails, slugs, and *Tenebrio* larvae.

When turtles become sluggish and show

Fig. 12-32 Turtles: (a) box turtle; (b) painted turtle. (Courtesy of U.S. Fish and Wildlife Service.)

a tendency to hibernate they should be placed in a cool place. Forced feeding, especially at this time, is often detrimental.

Sex differences. The most uniform guide for distinguishing sexes in turtles is the shape of the plastron, the under shell. The plastron in the female of many species is slightly convex, while in the male, it is slightly concave. During the breeding season, there is a swelling of the anal region in the male. A distinguishing characteristic among box turtles is eye color. Males usually have red eyes; females have yellow eyes. The males usually have longer claws than the females.

Lizards and alligators

Small specimens of the American alligator and several kinds of lizards can be reared in the laboratory. The horned "toad" (*Phrynosoma*), the skink (*Eumeces*), and the chameleon (*Anolis*) are the lizards most useful in school (Figs. 4-5, 12-33).

Vivaria. Chameleons and the larger skinks should be housed in a large terrarium. Include some twigs so the animals have room to climb. Spray the plants in the terrarium daily to supply water for

these lizards, since they seldom drink from a dish. They subsist mainly on live insects. It may be necessary to raise *Drosophila* for this purpose, especially during the winter months.

Young alligators survive in a vivarium of sand and rocks with a water trough embedded in the sand. Horned "toads" are maintained best in a similar desert vivarium containing about 5 in. of sand for burrowing, along with several rock piles for hiding. Embed a bowl of water up to the level of the sand. Students who care for the animals should provide several hours of direct sunlight but must also take care to ventilate the tank so that the temperature does not exceed 26° C (79° F).

Feeding. All these animals feed upon live insects, *Tenebrio* larvae, bits of earth-

Fig. 12-33 Lizards: (a) *Phrynosoma*, the horned "toad"; (b) *Eumeces fasciatus* (five-lined skink), female with eggs. (a, courtesy of the American Museum of Natural History, New York; b, Jack Dermid from National Audubon Society.)

worms, *Enchytraeus,* or similar living food. In addition, young alligators may take small frogs and fish. Chameleons and skinks may learn to accept small bits of raw liver or meat which are dangled before them. They do not feed regularly at low temperatures; 18° to 26° C (64° to 79° F) is the most suitable range.

Snakes (nonpoisonous)

Every biology laboratory or project room should have one or two snakes on display. It is one way to break down the inordinate fear many students have of these animals. The most desirable forms are those that are easy to keep and to handle. These are the garter, ribbon, hognosed, black, DeKay, and ring-necked snakes (Fig. 12-34). Many others such as the bull, milk, water, and green snakes may also be maintained by interested students, but in our experience the first-mentioned are the easiest to keep.

Vivaria. Mesh cages much like those used for mammals are best for housing snakes, although the mesh should be of smaller gauge so that the smallest snakes cannot escape. In addition, bottom pans of zinc are needed. Door openings at the top of the cage are the most convenient for handling the animals. Into such a vivarium place a pan or bowl of water and a few rocks. Keep the snakes at a temperature between 21° and 26° C (70° and 79° F).

Where all-mesh cages are not available, use ordinary aquarium tanks with tight-fitting zinc mesh tops. You may need a weight to hold down the cover. An aquarium completely enclosed by glass is undesirable since there is no provision for ventilation. When cages of mesh are used for DeKay and ring-necked snakes, check carefully that the size of the mesh is small enough to prevent their escape. Snakes may be kept in such terraria as described for *Hyla* or *Bufo* (p. 611). All snakes should be washed weekly by flushing with water. Their cages should be cleaned at the same time.

(a)

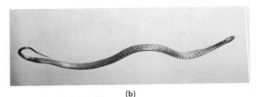

(b)

(c)

(d)

Fig. 12-34 Snakes: (a) garter; (b) DeKay; (c) ribbon; (d) black. (Courtesy of the American Museum of Natural History, New York.)

Feeding. Most snakes described here will feed upon readily available food. DeKay and ring-necked snakes feed on insects such as *Tenebrio* and on small earthworms; hog-nosed, garter, and ribbon snakes accept whole, large earthworms as well as insects, frogs or other amphibia, and lizards. Black snakes need live mammals; they may be fed a small rat every 2 weeks. On occasion, black snakes will accept a dead animal if it is waved in front of them.

General care. Some students are experts at handling snakes; they can do much to help other students overcome their fears. When snakes are handled gently each day, they will in turn become gentle. Large snakes should first be handled with thick gloves. This may not be necessary as time goes on as the snake shows signs of becoming accustomed to handling. Grasp a snake behind the head with one hand while the other hand is used to support its body.

Birds

Parakeets and canaries require little space and more or less routine care. Directions for maintaining these birds are given by the dealer at the time of purchase. In our experience, birds other than parakeets and canaries need more space and care then can be provided in the average classroom. However, where there are students who are expert in the handling of birds, this activity may become highly profitable.

On occasion, a crow may be housed in a large cage. Young ones seem to find the surroundings agreeable and can be conditioned by students studying behavior.

Mammals

White rats and hamsters (Fig. 12-35a) are the most satisfactory mammals to keep in the laboratory. Students become quite adept at handling them. Guinea pigs (Fig. 12-35b) and rabbits do not, ordinarily, demonstrate anything for which mice and rats are not suitable, and they require much more space and care. The gestation periods and breeding ages for all these mammals are given in Table 12-1.

The care of the rat will be described in

(a)

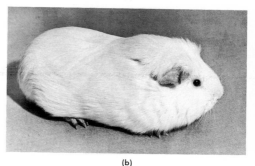

(b)

Fig. 12-35 Mammals useful in the laboratory: (a) golden hamster; (b) guinea pig. (Courtesy of Carolina Biological Supply Co.)

some detail; care of the mouse is similar. However, good ventilation is needed when mice are reared since they have an offensive odor—most noticeable in close quarters.

Cages. Most of the cages available commercially are satisfactory. The larger cages are best since they allow for exercise. The mesh should be large enough to allow droppings to fall through to a bottom pan (which should be made of a noncorrosive metal).

The door of the cage may be on the side or at the top. That on the top is most convenient since it permits easier handling. However, such cages cannot be stacked. Cages should be cleaned daily and fresh newspaper placed in the bottom pan. If cedar shavings are used instead of paper, the offensive odor of rats and mice is not as noticeable. Cleaning a cage the first or second day after delivery of a litter is not

TABLE 12-1
Gestation periods of some mammals

	gestation period (in days)	Breeding age
White rats	21–22	Females may be bred when 4 months old. Wean the young and separate the sexes after 21 days.
Mice	20–22	Breed females when 60 days old. Wean the young after 21 days.
Guinea pigs	63	Breed when 9 months old. Wean the young and separate the sexes at 4 to 5 weeks.
Golden hamsters	16–19	Breed the females when 60 days old.
Rabbits	30–32	Females are ready for mating when 10 months old. Wean the young and separate the sexes after 8 weeks.

advisable. In fact, care must be taken for several weeks in order not to disturb unnecessarily the mother and young. Disposable plastic cages are also useful.

Male and female rats should be segregated when 50 days old and should be kept segregated until mating is desirable. Remove the male as soon as the female is pregnant.

General care. Rats should be fed not more than once a day. They should be treated as pets (although many students will tend to overdo this). If they are, they will respond satisfactorily and reproduce readily. Rough treatment may result in viciousness and cause the mother to destroy her litter. From birth the rats should be handled gently and fondled. When this is done they do not bite, and, in fact, they are so conditioned that they will run forward to be handled. Such animals are especially desirable for the classroom.

Feeding. Rats should be fed a diet of bread, sometimes soaked in milk, in addition to lettuce, carrots, other vegetables, sunflower seeds, and similar foods. The bread should be broken and the carrots cut into portions equal to the number of rats in the cage. They will also accept hard-boiled eggs. Two or 3 drops of cod-liver oil on pieces of bread should be given twice a week. Provide a bowl of milk weekly.

Water must be supplied at all times. Water fountains, blown of one solid piece of glass, are available from biological supply houses (Appendix C). Substitute ones may be made in the laboratory by inserting a 6-in. length of ¼-in. glass tubing, straight or bent, through a rubber stopper in a suitable 300-ml bottle. Fire-polish the protruding end to a ⅛-in. opening and insert through the mesh of the cage (Fig. 12-36).

There are synthetic diets which may be used for rats.[36] When the young are to be weaned, feed them milk, bread soaked in milk, and lettuce. After they are 30 days old they can be fed the same diet as the adults. Pellet foods may then be added to the diet.

Care during the pregnancy and nursing periods. As soon as the female is pregnant, the male should be removed. A pregnant rat should be given strips of newspaper or paper toweling for nest building. Once a

[36] Refer to the *Turtox Service Leaflets* on *Care of White Rats and Mice* (No. 40) and *The Care and Breeding of the Golden Hamster* (No. 53), General Biological Supply House, Chicago. See also the fine booklet *Care and Feeding of Laboratory Animals*, Ralston Purina Co., St. Louis.

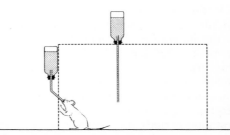

Fig. 12-36 Watering bottle for small mammals, showing two variations.

nest is built it is not necessary to change the paper, although this can be done after the young have hair and their eyes have opened (after 16 to 18 days).

The period of gestation is 21 to 22 days. The sex of the young rats may be distinguished by the fact that the distance between the anus and the genital papilla is greater in the male than in the female.

The young should be permitted to remain with the mother for 21 to 24 days, after which time they should be weaned. When the young are kept with the mother for a longer period there is a severe drain on the female.

A gentle female rat will respond favorably to handling during the nursing period and will not resent handling of the young, provided it is done by someone who has been responsible for her daily care. However, as mentioned above, special care should be taken during this period to avoid unnecessary disturbance of the mother and young, and they should be handled gently; otherwise the mother may turn vicious and destroy her litter. The mother should be fed whenever she is removed from her young. The female, especially during this nursing period, may react unfavorably to strangers. A bit of chocolate or carrot given to her as she is returned to her young will help ease her distress.

Readers who seek information about the reproductive patterns of wild animals of their own community may want to look into S. A. Asdell, *Patterns of Mammalian Reproduction* (Comstock [Cornell Univ. Press], 1946), which covers mammalian reproduction from the duckbills (Monotremes) to some of the higher hoofed forms.

Students interested in the average life span or maximum life span of invertebrates and vertebrates (and also trees and shrubs) will find a treasure in W. Spector, ed., *Handbook of Biological Data* (Saunders, 1956). Also read A. Comfort, "The Life Span of Animals" (*Sci. Am.,* August 1961).

CAPSULE LESSONS: Ways to get started in class

12-1. Students may form a "vivarium squad" (see Chapter 14), the function of which is to maintain invertebrates and aquaria of different fish (include live-bearers such as guppies, as well as egg-layers) in the laboratory. They may also collect amphibian eggs in season and make them available to other students and classes.

12-2. Encourage students to bring in all kinds of living materials that you want to maintain. Form a club of museum curators to care for the living materials. In this way, living forms will be available for study throughout the year, and in season. And your school may be in a position to help other schools in the vicinity if they call for living materials (student squads, p. 683).

12-3. There are times when you want to develop a "parade of the animal kingdom" in class. This may be a survey of the kinds of animals available in your community. Try to use fresh materials, examine them under the microscope, and keep them in vivaria or aquaria for observation. You may also want to survey animals by means of films, filmstrips, or slides.

In the catalogs of biological supply houses (Appendix C) you will find long lists of slides of invertebrates, as well as of vertebrates and plants. The catalogs of film distributors (Appendix B) offer a phenomenal list of films and filmstrips. Do you know *Beach and Sea Animals,* 2nd ed., and the series *Animal Life* (both EBF); *Tide Pool Life* (Simmel-Meservey); *Wonders of the Sea* (Teaching Films Custodians); *Butterflies, Pond Life,* and *Tiny Water Animals* (all three EBF)? A tremendous variety of animal life and functions is described in *Born to Die* (Skibo).

12-4. Are you familiar with the fine silent-film series produced by Rutgers Univ.: several reels on the *Parade of Invertebrates* and reels on *Annelids, Mollusks,* and *Arthropods?*

12-5. Students may be interested in collecting reptiles, and some students are experts on this subject. Have them as "guest speakers" in class programs or in assembly programs.

12-6. You may want to show films as vicarious field trips to survey animal forms and their adaptations.

12-7. Make available to students for browsing some of the BSCS laboratory blocks: E. A. Phillips, *Field Ecology* (Heath, 1964), H. Follansbee, *Animal Behavior* (Heath, 1965), F. Moog, *Animal Growth and Development* (Heath, 1963), and others.

12-8. Some students may want to plan research projects involving some aspect of the biology of some of these animals. Many ideas have been presented in passing in the previous chapters. Fish genetics may be one problem; a study of parasites of frogs from different localities may be another. Behavior among members of another group may hold the interest of certain students. Ideas for projects or long-range investigations are to be found in context throughout the book.

PEOPLE, PLACES, AND THINGS

Do you need help in identifying an insect a student came upon on his way to school? Find out which people in your community have made a hobby of knowing the insects; there may be some who gather organisms along the beaches at ebb tide, or in desert areas, or in the woods. Who knows the insect pests that attack crops or garden plants or the shade trees? Is there a local farmer who has a beehive?

You may want to look at L. Provasoli, chairman, "A Catalog of Laboratory Strains of Free-living and Parasitic Protozoa" (*J. Protozool.* 5:1–38, 1958).

There are many resources to tap: these people in the community, teachers at other schools and colleges. A museum or a zoological garden will help you identify many specimens. Biological supply houses will often assist you in identifying specimens; they also offer advice on how to maintain organisms.

Many students are experts in handling fish; many students or members of their family have tropical fish. We have had students at times who have exchanged snakes with other enthusiasts or with a nearby zoological park.

Departments of agriculture, a 4-H Club, or a local branch of the Audubon Society will help in identifying birds and suggesting ways to encourage birds to take up residence around school or in a nearby area. Directions for building birdhouses are readily available. You may have birdwatching groups in your community. Someone may speak to the class or club on a topic of general interest.

If you wish to keep rats or hamsters, you may purchase them from a supply house. Or sometimes parents dispose of these pets when their youngsters "tire" of them (possibly the parents tire first). Your school may fall heir to these animals. At times you may get animals from a hospital, college laboratory, or research laboratory.

BOOKS

Comprehensive references that encompass the broad fields of biology described in this chapter need special mention. The following are only a few of the broad references pertinent to this work. Refer also to the listings at the ends of Chapters 1, 2, 4, and 11; there is a comprehensive listing in Appendix A that includes fieldbooks and guides to identification of organisms.

Barnes, C. and L. G. Eltherington, *Drug Dosage in Laboratory Animals,* Univ. of California Press, Berkeley, 1964.

Beck, D. E., and L. F. Braithwaite, *Invertebrate Zoology: Laboratory Workbook,* 2nd ed., Burgess, Minneapolis, 1962.

Buchsbaum, R., *Animals Without Backbones: An Introduction to the Invertebrates,* 2nd ed., Univ. of Chicago Press, Chicago, 1948.

Bullough, W. S., *Practical Invertebrate Anatomy,* 2nd ed., Macmillan, London (St. Martin's, New York), 1958.

Edmondson, W. T., *see* Ward, H. B., and G. C. Whipple.

Galtsoff, P., *et al.,* eds. (J. Needham, chairman), *Culture Methods for Invertebrate Animals,* Comstock (Cornell Univ. Press), Ithaca, N.Y., 1937; Dover, New York, 1959.

Goin, C., and O. Goin, *Introduction to Herpetology,* Freeman, San Francisco, 1962.

Harris, R., ed., *The Problems of Laboratory Animal Disease,* Academic, New York, 1962.

Hegner, R. W., and K. A. Stiles, *College Zoology,* 7th ed., Macmillan, New York, 1959.

Hyman, L., *The Invertebrates,* 5 vols., McGraw-Hill, New York, 1940–59.

Kudo, R., *Protozoology,* 4th ed., Thomas, Springfield, Ill., 1954.

Manwell, R., *Introduction to Protozoology,* St. Martin's, New York, 1961.

Pennak, R., *Fresh-Water Invertebrates of the United States,* Ronald, New York, 1953.

Porter, G., and W. Lane-Petter, eds., *Notes for Breeders of Common Laboratory Animals,* Academic, New York, 1962.

Prosser, C. L., and F. A. Brown, Jr., *Comparative Animal Physiology,* 2nd ed., Saunders, Philadelphia, 1961.

Smith, R., *et al., Intertidal Invertebrates of Central California Coast,* Univ. of California Press, Berkeley, 1954.

Spector, W., ed., *Handbook of Biological Data,* Saunders, Philadelphia, 1956.

Ward, H. B., and G. C. Whipple, *Fresh-Water Biology,* 2nd ed., edited by W. T. Edmondson, Wiley, New York, 1959.

Catalogs of biological supply houses (Appendix C) and film distributors offer color slides of plants and animals that are grouped in sets focusing on some area of biology. A directory of distributors, with addresses, is given in Appendix B.

LOW-COST MATERIALS

The quality of free and low-cost materials available to the teacher in this area is excellent. A listing of distributors, with addresses, is given in Appendix D. Your state department of agriculture offers many aids. For example, we have seen booklets from Illinois on identification of mushrooms and books from Kansas on grasses, birds, and insects (including color plates). Many departments will put your name on their mailing list to receive announcements. Many agricultural colleges have a 4-H Club bulletin series such as that published by New York State College of Agriculture, Cornell Univ. This series includes such fine pamphlets as *Diseases and Insects in the Orchard, Pollination of Fruit Trees,* and *Insect Control.*

You may have your name placed on the mailing list of the U.S. Govt. Printing Office, Washington 25, D.C., to receive announcements of their new publications, including the Yearbooks of the U.S. Dept. of Agriculture. These publications range over all areas of biology.

Write to W. H. Freeman and Co., 660 Market St., San Francisco 4, Calif., for their catalog of available offprints of papers from *Scientific American* ($0.20 each).

Each of the biological supply houses offers aids for the teacher, as do many industrial concerns. We are familiar with only a few of the publications, but offer them as an example of the quality of materials which teachers may receive. Many are free; in some cases there are small handbooks available usually for about $1.00 or less. The following materials are available for the asking:

Care and Feeding of Laboratory Animals, Ralston Purina.

Carolina Tips, Carolina.

Living Specimens in the School Laboratory, General Biological.

Turtox News; Turtox Service Leaflets (set of some 60 leaflets), General Biological.

Ward's Bulletin; leaflets on techniques, Ward's.

Chapter 13

Growing Plants Useful
in the Classroom

Since there is no one best way to grow a specific plant, especially algae and bacteria, a number of selected methods which have proved useful for some forms are described in this chapter. The choice is offered to help meet the different needs of teachers and their students.

Algae

Culturing

In view of their ability to carry on photosynthesis, it might seem that algae should respond readily to culture. In fact, certain algae respond to specific methods, but some species do not. Others, such as *Oscillatoria, Chlorella,* and *Cladophora,* respond to a variety of methods.

Most of the algae useful in the biology classroom may be cultivated by one of the methods described in this section. A summary of the most suitable methods for each alga is given under the genus name of the alga, beginning on p. 626.

Special considerations. A successful culture medium for algae must provide the major nutrient salts, a usable source of nitrogen, a supply of carbon, some trace elements, and a suitable pH.

In addition, temperature and light conditions must be suitable for the successful culturing of algae. Most algae and plants grow well at 21° C (70° F). Care should be exercised to avoid temperatures higher than 27° C (80° F). Light from a north window is preferable, but, especially dur-

ing winter months, artificial light can be used to supplement daylight. A standard cool, white fluorescent light placed a few feet from the plants or cultures provides about 50 to 75 foot-candles of light intensity. For most successful growth of algae, a 16-hr light period should be alternated with an 8-hr period of darkness. Light intensity of some 200 foot-candles stimulates rapid growth, which is especially suitable for young cultures.

When artificial light is used exclusively, as in special culture rooms or greenhouses, a balance should be maintained between incandescent and fluorescent lamps in order to provide light in both the red and blue wavelengths. Fluorescent light—which is rich in blue wavelengths and deficient in red wavelengths—encourages the growth of low vegetation. Incandescent light—rich in red wavelengths and poor in blue wavelengths—stimulates the growth of tall, spindly plants with little supporting tissue.[1] Hence, a balanced combination of the two produces normal vegetative growth.

A source of carbon is required for the growth of algae, and it may be provided by adding 5 percent carbon dioxide to the air pumped into the medium. Or several drops of 0.5 percent bicarbonate solution

[1] Detailed discussions of wavelengths and the use of filters are given in two readily available sources: *Plant Culture with Artificial Light,* No. 60 *Turtox Service Leaflets,* General Biological Supply House, Chicago, and R. E. Barthelemy, J. R. Dawson, Jr., and A. E. Lee, *Innovations in Equipment and Techniques for the Biology Teaching Laboratory* (BSCS), Heath, Boston, 1964.

can be added to each 10 ml of culture. On a small scale, carbon dioxide can be added by delivering into the culture of such algae as *Chlorella* the carbon dioxide that is liberated by actively growing yeast cells in a fermenting medium (p. 642).

General method: A. One of the most successful methods for culturing a great number of different algae is simple: Introduce the algae into an established aquarium in which fresh-water animals and plants have been thriving for at least 2 months; avoid overcrowding.

Culture solutions: method B. Many solutions, including those that follow, have been used successfully for various algae.

1. MOLISCH'S SOLUTION. Dissolve the following materials in 1 liter of distilled water:

KNO_3	0.2 g
$MgSO_4$	0.2 g
K_2HPO_4	0.2 g
$CaSO_4$	0.2 g

Dissolve the calcium sulfate separately in 200 ml of water; then add this solution to the other salts dissolved in 800 ml of water.

Both this and Benecke's solution, which we have found to be more suitable, are similar to Klebs' solution (p. 587), which differs mainly in using both potassium nitrate and calcium nitrate.

2. BENECKE'S SOLUTION. Dissolve the following materials in 1 liter of distilled water:

$Ca(NO_3)_2$	0.5 g
$MgSO_4$	0.1 g
K_2HPO_4	0.2 g
$FeCl_3$ (1% solution)	trace (1 drop)

3. KNOP'S SOLUTION. Prepare a 0.6 percent Knop's solution for green algae by dissolving the following salts in 1 liter of distilled water, in this order:

KNO_3	1 g
$MgSO_4$	1 g
K_2HPO_4	1 g
$Ca(NO_3)_2$	3 g

In this solution, the calcium nitrate precipitates, forming a layer at the bottom of the bottle. Be sure to shake the bottle when using the solution. A solution of cane sugar (1 to 4 percent) is sometimes added to this dilution of Knop's solution to stimulate the formation of zoospores in some algae.

Use distilled water in preparing these solutions since tap water may contain chlorine or copper, both of which are highly toxic to algae.

The ability of these solutions to support the growth of various algae is greatly enhanced by adding to the culture jar 1 in. of soil taken from the pond where the algae have been growing. Good garden soil is a useful substitute. First, boil the soil in distilled water to destroy contaminating algae. Then you may have students add this, together with distilled water, to the prepared container. Let the medium stand for a day or so before adding algae to the culture medium. Such soil-solution cultures have been very satisfactory. You may need to subculture when the algae reach maximum growth (see specific descriptions of soil cultures in method E).

Method C. When filaments of coarse-filamented *Spirogyra*, such as *Spirogyra nitida,* are introduced into a thriving culture of *Daphnia* (preparation, p. 596), rapid growth takes place.[2] (*Note:* This method does not generally work with fine-filamented species.) In fact, the algae rapidly choke the container or aquarium in which the *Daphnia* are cultivated.

Keep the cultures in medium light, or in a north window. Each week add about 0.1 g of fresh, hard-boiled egg yolk smoothed between the fingers into a paste. The species of *Spirogyra* containing two or more spiral chloroplasts seem to respond more vigorously than the species with single chloroplasts.

In our experience, almost all the algae commonly used in elementary courses show vigorous growth when cultivated by this method. An attempt to discover improved methods of culturing algae is a very good project for an interested student.

A temperature range of 18° to 27° C

[2] P. F. Brandwein, "Preliminary Observations on the Culture of *Spirogyra,*" *Am. J. Botany* 27:195–98, 1940.

(65° to 80° F) is optimum in all the methods described. Higher temperatures are destructive, especially for methods A and C.

Method D: solid agar media. Algae may also be cultivated on solid medium. When completely sterile conditions can be obtained, these methods are successful. Here are two useful ones that include an agar base.

In preparing the media, mix the agar in the salt solution and gently bring to a boil. (See instructions for plating finger bowls or Petri dishes, p. 634.)

1. MODIFIED PFEFFER'S SOLUTION. Combine the following salts with 1 liter of distilled water. Add the $Ca(NO_3)_2$ last.

$Ca(NO_3)_2$	1.00 g
K_2HPO_4	0.25 g
$Fe_3(PO_4)_2$	0.05 g
$(NH_4)_2SO_4$	0.50 g

When this solution has been prepared, add 15 g of agar and 15 g of fructose.

2. MODIFIED KNOP'S SOLUTION. Combine the following salts with 1 liter of distilled water. Add the $Ca(NO_3)_2$ last.

$Ca(NO_3)_2$	1.00 g
KCl	0.25 g
$MgSO_4$	0.25 g
K_2HPO_4	0.25 g
$FeCl_3$	trace

To this solution add 20 g of agar and 20 g of glucose.

Method E: soil culture. Pringsheim[3] was one of the pioneers in introducing the use of soil-water cultures that closely duplicate a miniature artificial pond. He added either garden soil with seasonal compost or soil from an arable field (especially clay soils) to water to make a mud phase. This mixture seemed to supply organic matter, growth factors, and trace elements in amounts that apparently were optimum for the growth of algae.

To the bottom of test tubes or other containers, add dry soil that has been treated in a steam sterilizer for an hour or more on two consecutive days (to remove contaminants). Carefully add sterile, distilled water. If necessary, alter the pH of the culture: add calcium carbonate to raise the pH or add some peat to lower the pH.

Bold[4] describes a modified soil culture that uses a clear filtrate of soil, rather than soil itself. Dissolve 500 g of good field soil in 1 liter of distilled water; sterilize for 2 hr. When cool, filter the solution, and decant it several times until a clear filtrate is obtained. This is called a *stock* solution of soil extract, and it must be diluted for use. In actual testing situations, start with a dilute solution, and gradually add more extract until good growth results. We have found the following dilution useful. Combine 5 ml of stock soil solution with 94 ml of sterile distilled water. To this, add 1 ml of a 5 percent solution of KNO_3. Pour this solution into small finger bowls, baby-food jars, or similar containers. Inoculate with the pure type of alga desired, and cover with aluminum foil. Maintain in moderate light at 20° to 24° C (68° to 75° F). Use a phosphate buffer if necessary (p. 172).

Pringsheim also describes a technique in which the soil and water are sterilized as described above, but the mixture is allowed to settle so that the soil component remains in the bottom of the tubes (avoid shaking tubes).

Method E is recommended for pure culturing of unialgal cultures. If students study soil flora, use this method, omitting the sterilization of the soil. After a rich growth of algae results, algae can be isolated from the mixed cultures and subcultured as unialgal cultures (see p. 628).

Method F. Meyers[5] reports a method used equally successfully in large-scale cultivation of *Chlorella* and in sustaining vigorous growth of other algae. He recommends a freshly prepared Knop's solution, at pH 6.8, provided with necessary microelements by the addition of iron and

---

[3] E. G. Pringsheim, *Pure Cultures of Algae: Their Preparation and Maintenance,* Macmillan (Cambridge Univ. Press), New York, 1946.

[4] H. Bold, "The Cultivation of Algae," *Botan. Rev.* 8:69–138, 1942.

[5] J. Meyers, "Modified Knop's Solution for *Chlorella*," mimeo, personal communication, 1961.

Arnon's solution No. 5. A chelating agent is used to prevent the microelements from precipitating out of solution.

MEYERS' MODIFICATION OF KNOP'S SOLUTION. Dissolve the salts listed in A in 1 liter of distilled water. To this, add the chelating agent in B. Dissolve the microelements in C in a small amount of dilute sulfuric acid and add the mixture to the rest of the solution. Finally, adjust the pH with buffers.

A. Macroelements

$MgSO_4 \cdot 7H_2O$	2.50 g
KNO_3	1.25 g
KH_2PO_4	1.25 g

B. Chelating agent (ethylenediamine tetraacetic acid)

EDTA	0.50 g

C. Microelements (Arnon's solution)

		g/liter	ppm
Ca	(as $CaCl_2$)	0.084	30
B	(as H_3BO_3)	0.114	20
Fe	(as $FeSO_4 \cdot 7H_2O$)	0.050	10
Zn	(as $ZnSO_4 \cdot 7H_2O$)	0.088	20
Mn	(as $MnCl_2 \cdot H_2O$)	0.014	4
Mo	(as MoO_3)	0.007	4
Cu	(as $CuSO_4 \cdot 5H_2O$)	0.016	4
Co	(as $Co(NO_3)_2 \cdot 6H_2O$)	0.005	1

Cobalt, the last of the microelements listed above, is a specific requirement for the growth of some algae, such as *Euglena gracilis*. Cobalt is a constituent of vitamin B_{12}.

An excessively high initial nitrogen content is toxic to algae; as nitrogen is absorbed, renew it in the solution. Further, the uptake of nitrate by algae also causes a rise in pH of the medium; adjust pH to 6.3 to 6.8.

Method G. Among the other classic solutions for culturing algae is one prepared by Beijerinck. (He isolated pure cultures of algae in the 1890's in 1.5 percent agar cultures.)

BEIJERINCK'S SOLUTION. Add the following salts to 1 liter of distilled water:

NH_4NO_3	0.5 g
K_2PO_4	0.2 g
$MgSO_4 \cdot 7H_2O$	0.2 g
$CaCl_2 \cdot 2H_2O$	0.1 g

To this solution, Fogg[6] suggests that the following minerals be added:

Fe (as $FeCl_3$)	0.4 mg
Mn (as $MnSO_4$)	0.1 mg
Cu (as $CuSO_4$)	0.01 mg
Zn (as $ZnSO_4$)	0.01 mg

Method H. Bristol's solution is popular for algal cultures. Bold[7] recommends the following modification.

BOLD'S MODIFICATION OF BRISTOL'S SOLUTION. Prepare six stock solutions by dissolving each of the following salts in 400 ml of distilled water:

1. $NaNO_3$	10.0 g
2. $CaCl_2$	1.0 g
3. K_2HPO_4	3.0 g
4. KH_2PO_4	7.0 g
5. $MgSO_4$	3.0 g
6. NaCl	1.0 g

To prepare a dilute solution for culturing algae, add 10 ml each of all six stock solutions to 940 ml of distilled water. To this, add 1 drop of a 1.0 percent solution of $FeCl_3$ and 2 ml of Arnon's solution of microelements, described under method F.

In addition, Bold suggests that organic matter be added in the form of Difco yeast extract (1 g of yeast extract per liter of modified Bristol's solution); then sterilize.

Any of these solutions may be solidified for use in Petri dishes by adding 1.0 to 1.5 percent of agar.

Sources of algae

Algae of a given genetic homogeneity may be cultivated for biological investigations. Cultures may be obtained from Dr. R. C. Starr, Indiana Univ. You may also want to read Starr's paper "The Culture Collection of Algae at Indiana University" (*Am. J. Botany* 47:67–86, 1960). In it, he presents a list of cultures that are available for research work; he also gives directions for culturing and inducing sexual reproduction in specific algae.

[6] G. E. Fogg, "Famous Plants: 4, *Chlorella*," in M. L. Johnson, M. Abercrombie, and G. E. Fogg, eds., *New Biology*, Vol. XV, Penguin, Harmondsworth, Eng., 1953, p. 106.
[7] *Op. cit.*

When clonal cultures are not necessary for classroom examination, cultures may be purchased from biological supply houses (Appendix C).

For culturing marine algae, see p. 627.

Summary of methods for specific algae

The following are some successful methods for maintaining specific algae. Where two or more methods are cited they are listed in order of their effectiveness.

Oscillatoria (Fig. 11-26b). Use culture method B, C, A, E, or G. This blue-green alga is found as a thin scum or sheet, dark green or blackish in color, on the surface of stagnant water or damp soil, or on flowerpots. Culture *Oscillatoria* separately from green algae. Green algae are sometimes choked off by the surface growth of *Oscillatoria*.

Hydrodictyon (Fig. 11-26c). Use method C, A, G, or H. Collected from the surface of lakes, *Hydrodictyon* grows well in bright light.

Vaucheria (Fig. 8-41). Use method C. The form most commonly found is the vegetative stage. Klebs induced formation of antheridia and oogonia within 4 to 5 days by placing *Vaucheria* in a 2 to 4 percent cane sugar solution in *bright* light. *Vaucheria sessilis* is the common form found as green "felt" on flowerpots in greenhouses. In ponds and lakes in the spring the more common form is *Vaucheria geminata*.

Chara. Use method E, H, C, or B. *Chara* grows well when an inch of pond soil or sand is placed in the bottom of the culture jar.

Cladophora (Fig. 13-1b) and *Oedogonium* (Fig. 8-42). Use method A, C, or B.

Spirogyra (Fig. 2-11a). Use method C. Be sure the *Daphnia* are thriving in the culture before adding *Spirogyra*.

Nitella (Fig. 13-1a). Use method C.

Desmids (Fig. 13-2). Use method B or C. Prepare the cultures in finger bowls.

Diatoms (Fig. 13-3). Collect diatoms from submerged rocks and stems of plants,

(a)

(b)

Fig. 13-1 (a) *Nitella;* (b) *Cladophora.* (a, photo by Walter Dawn; b, courtesy of General Biological Supply House, Inc., Chicago.)

or from the surface of mud in ponds and ditches in the spring. Diatoms abound in the brownish coating on rocks and stems which are slippery to the touch. Culture them like desmids in finger bowls (see above).

Scenedesmus (Fig. 13-4). Use method E, H, or B. Like other desmids, culture in finger bowls.

Volvox (Fig. 11-27). Culture by method A, B, F, or G, in finger bowls.

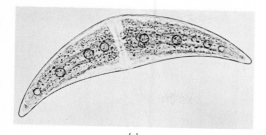

(a)

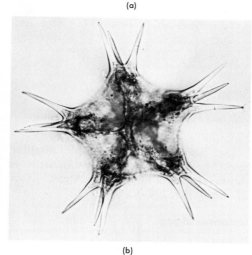

(b)

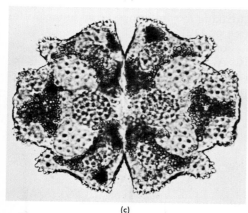

(c)

Fig. 13-2 Desmids: (a) magnified 325×; (b) magnified 600×; (c) magnified 785×. (a, b, c reduced by 23, 18, 26 percent, respectively, for reproduction.) (Winton Patnode from Photo Researchers, Inc.)

Chlamydomonas (Fig. 2-12b). Culture methods for this form are similar to those already suggested for flagellates, p. 586;

also use method E. Culture in flasks or finger bowls.

Eudorina, Pandorina (Fig. 13-5). Use method E, A, F, or G. Add 1 percent agar to solidify medium in Petri dishes.

Euglena (Fig. 2-11c). Cultivation of this flagellate was described in Chapter 12; also use method E. Culture in finger bowls or larger jars. To grow in the dark, add a carbon source (pp. 622–23).

Chlorella (Fig. 2-11d). Use method D (Knop's with agar), E, F, G, or H. This alga is a ubiquitous contaminant of culture media in the laboratory. A supply bottle of Knop's solution standing in the laboratory may turn green in 2 weeks. Examination of green water in an aquarium often reveals *Chlorella,* as well as *Euglena, Scenedesmus,* and *Chlamydomonas.*

With an organic carbon source, such as acetate or 10 percent glucose, or some organic acid, *Chlorella* flourishes in the dark and maintains its bright green chloroplasts. When *Euglena* is raised in the dark, chloroplasts fragment and non-green *Euglena* grows as a heterotroph (p. 493).

Marine algae. Methods for preparing and maintaining a marine aquarium are given on p. 600. Store ocean water in gallon glass containers in moderate light. Within a week to 10 days, marine algae become apparent—especially green, brown, and red algae. *Fucus* may be collected and maintained for limited periods of time. Male and female *Fucus* (Fig. 8-40), together with directions for a study of fertilization, may be purchased from supply houses.

SCHREIBER'S SOLUTION. This solution is suggested in the literature[8] as an enriched sea water culture medium. To 1 liter of sea water, add:

$NaNO_3$	0.10 g
Na_2HPO_4	0.02 g
Soil extract (commercial fertilizer 4-10-4)	50.00 ml

[8] From *Culture Leaflet,* No. 14, Ward's Natural Science Establishment, Rochester, N.Y.

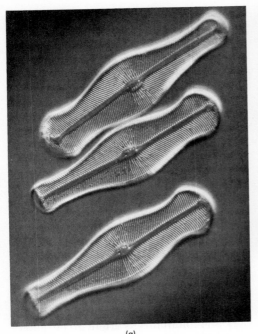

(a)

(b)

(See also, L. Provasoli, J. McLaughlin, and M. Droop, "The Development of Artificial Media for Marine Algae" [*Arch. Mikrobiol.* 25:392–428, 1957].)

Isolating algae

Algae may be separated from a mixed culture by using Beijerinck's device of streaking a transfer needle on an agar film (see p. 442).

For a more extensive discussion, see R. A. Lewin, "The Isolation of Algae" (*Rev. Algolog.* 5:181–97, 1959). See also the papers by Pringsheim[9] and Bold.[10]

[9] *Op. cit.*
[10] *Op. cit.*

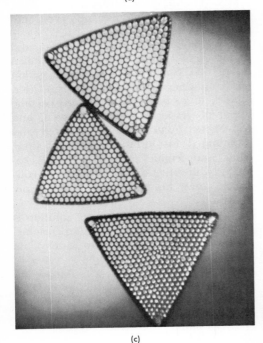

(c)

Fig. 13-3 Variety among diatoms: (a) *Gomphonema geminatum ag.;* (b) *Terpsinoë musica ehr.;* (c) *Triceratium.* (Carl Struewe from Monkmeyer.)

Fig. 13-4 *Scenedesmus.* (Courtesy of Carolina Biological Supply Co.)

Uses in the classroom

Algae of many kinds are used to show the structure of green plant cells (wet preparations, Chapter 2). *Spirogyra* (Fig. 2-11a) is the most commonly studied form in elementary work. This is unfortunate because *Spirogyra*—with its complex strands of cytoplasm, spiral chloroplasts, and pyrenoids—is one of the most difficult forms to examine, and it is difficult to keep in the laboratory.

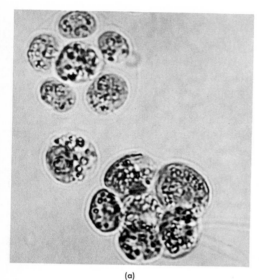

(a)

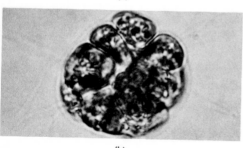

(b)

Fig. 13-5 (a) Living colonies of *Eudorina elegans* (magnified 3200 ×, reduced by 45 percent for reproduction). Flagella can be seen extending through the gelatinous sheath in the cluster at the right. (b) Living colony of *Pandorina morum* (magnified 2500 ×, reduced by 45 percent for reproduction). Moving flagella extending through the gelatinous sheath can be seen. (Photos by Walter Dawn.)

Protococcus (Fig. 11-28) and *Euglena* (Fig. 11-26a) show a conspicuous chloroplast which you may want students to see in their work in cell studies. *Euglena* is a motile flagellate with several chloroplasts. (A drop of Lugol's solution [p. 664] added to a slide of *Protococcus* will also bring out the nucleus.) At times you may prefer to substitute *Peranema* for *Euglena*, because its long flagellum is distinctly visible under low power. Although it has no chloroplasts (it is holozoic or saprozoic), it does undergo many changes in form and exhibits "euglenoid motion." One of the commonly occurring forms is *Peranema trichophorum* (Fig. 2-12a), which is some 60 to 70 μ long. *Euglena* may be used to show positive phototropic responses as well.

Euglena may also be studied as an example of an organism that reproduces by longitudinal fission. This form is sometimes used in assays of vitamins and other growth-promoting factors (p. 462). *Euglena* is a good choice for a test organism in studies of photosynthesis, since both green and non-green varieties of the same species are available (bleach with streptomycin); refer also to *Chlorella* (p. 627).

Study diatoms under the microscope for the beauty of the sculpturing in their siliceous shells. *Spirogyra* may be studied to show conjugation as well as plant structure, while such forms as *Vaucheria* and *Chlamydomonas* may be used to observe the evolution of sexuality in plants. Some suggestions for using *Chlorella* as a focus for developing key concepts in biology are described on p. 542, also p. 190.

Other uses for the study of algae will, of course, depend upon the nature of the course. In most courses in biology few algae are studied: mainly *Spirogyra* and some other forms casually studied to illustrate the phylum. Some teachers, however, use a variety of algae to illustrate different aspects of biology, for example, *Volvox* to illustrate heterosexuality and the prologue to metazoan structure, *Spirogyra* to illustrate pyrenoid activity as well as conjugation, *Ulothrix* to show zoospores, *Oedogonium* to show heterospory, *Chlorella* and *Euglena* to introduce modes of nutrition, and so on. These varied purposes, although acknowledged, are outside the scope of this book.

Phylum Schizomycophyta (bacteria)

Collecting

Because bacteria are found almost everywhere, they may be collected in many ways. When bacteria are to be studied as wet mounts, students may use sauerkraut juice or yoghurt as sources of harmless bacteria (see wet mounts and hanging-drop preparations, Chapter 2). The bacteria in sauerkraut juice and yoghurt are available for immediate use. For an alternate source of bacteria, soak some beans or peas for several days; as they decay, pipette off the fluid, which will be found to be teeming with large bacteria, probably *Bacillus subtilis* (Fig. 2-27a). Or, if you prefer, use the scum formed in the water of a vase of cut flowers. These kinds of bacteria may be examined adequately with the high power of a compound microscope when an oil immersion lens is not at hand.

Of course, special kinds of bacteria such as chromogenic, phosphorescent, or other specific genera of microorganisms may be obtained from biological supply houses (Appendix C). Obviously there is no use for pathogenic forms in a school laboratory. Besides, their use in high school laboratories is prohibited by law in many states. Furthermore, the literature in bacteriology indicates that under certain conditions some of the chromogenic forms are pathogenic and may cause serious infections.

In summary, the simplest means for collecting bacteria is to harvest them from decaying substances. Put some timothy hay, meat, a piece of potato, beans, or peas into distilled water and leave it exposed to the air. After a few days stopper the test tubes or small flasks lightly with cotton, and keep them in a comparatively warm place for the next few days.

Refer also to Chipman's method for obtaining almost pure cultures of *Bacillus coli* to culture *Daphnia,* p. 596.

Culturing

When bacteria are to be kept on hand *permanently* in the laboratory, a suitable medium for culture and subculture must be utilized. Or if you want to show how bacteria grow in colonies and that these are distinctive for each genus, use specific media for this purpose and keep under sterile conditions. Some ways to stain bacterial smears are described in Chapter 2. See methods for study of growth: turbidity and colony counts (pp. 443, 446).

A precaution. When bacteria are cultured in class or in the laboratory every precaution must be taken to avoid contamination with pathogenic forms. Treat each culture of bacteria as though it were pathogenic. Perhaps this needs a word of explanation. An old agar slant or Petri dish may have been exposed to the spray from a cough or someone's finger tips. The colonies that developed might be keyed out as nonpathogenic types. There may, however, be small numbers of pathogens that have failed to grow luxuriantly into conspicuous colonies because conditions were not optimum. (These bacteria may require a different pH, other salts, blood medium, and so on.) But they may be present in the culture nevertheless.

The methods below, therefore, include techniques for handling different kinds of bacteria as well as suggestions for avoiding contamination.

Preparation of culture medium for bacteria. For limited uses in the classroom, it is advantageous to purchase prepared sterilized test tubes of agar medium from supply houses. Or you may purchase dehydrated medium.[11] In the laboratory students can measure out accurately the necessary amount of dried medium, place it in a dry flask, and add distilled water. Then sterilize the solution (autoclave)

[11] Available from Difco Laboratories, Inc., Detroit 1, Mich., or Baltimore Biological Laboratory, 2201 Aisquith St., Baltimore 18, Md.

and pour it into plates or test tubes (see pp. 633-34).

In the following discussion some formulas for preparing a number of different kinds of culture media are described (for those that require special ingredients). The general procedures involved in handling bacteria are also described briefly. For more extensive work with bacteria the books in Appendix A may be helpful.

A suitable culture medium should contain a source of nitrogen such as peptone, plus inorganic salts, carbon, and possibly some growth-promoting substances. Certain bacteria require other substances for growth, such as serum or sugars or special salts. Meat infusions (where meat has been soaked in water for many hours) have given way to beef extracts, which may be added as the medium is prepared.

1. MEAT EXTRACT BROTH. Weigh out and combine the following materials with 1 liter of distilled water:

beef extract	3 g
peptone	10 g
NaCl	5 g

Heat this mixture slowly to 65° C (149° F), stirring until the materials are completely dissolved. Then filter through paper or cotton and adjust the pH to 7.2 to 7.6, by adding a bit of sodium bicarbonate. Pour the mixture through a funnel into test tubes, filling them one-third full; then stopper with cotton. Finally sterilize in an autoclave at 15 lb pressure for 15 min. This quantity should be sufficient to prepare 36 test tubes.

2. MEAT EXTRACT AGAR. A liquid broth may be solidified by adding agar or gelatin. For example, prepare the meat extract broth described above and, to 1 liter of the broth, add 20 to 30 g of agar. Heat slowly until the agar is dissolved. Then autoclave at 15 lb pressure for 15 min. Filter the solution through cotton and adjust the pH to about 7.5 by adding a few drops of normal NaOH solution. Then sterilize the medium again. (The melting point of agar is about 99° C [210° F], and it solidifies at about 39° C [102° F].)

3. NUTRIENT GELATIN. While gelatin was one of the first substances added to solidify media, it has limited usefulness today because it melts at room temperature in warm weather. It can be used, however, for detecting bacteria that produce a protein-splitting enzyme.

Prepare 1000 ml (1 liter) of meat extract broth. To this add 120 g of Bacto-gelatin, and heat in a double boiler. Restore the volume with distilled water, and adjust the pH to about 7.5 (by adding a bit of sodium bicarbonate).

Break a raw egg and mix a small amount of water with it. Add this to the solution to clarify it. Heat again very slowly until the egg becomes firmly solidified. Then filter the solution through cotton and pour it into test tubes (about 10 ml in each tube). Finally, sterilize the tubes in an Arnold sterilizer (Fig. 13-7) for 20 min on 3 successive days. Cool the solution rapidly after each sterilization. This quantity should be sufficient for about 36 test tubes.

4. POTATO MEDIUM. With a cork-borer, cut cylinders from large washed and peeled potatoes. Then cut the cylinders obliquely into wedge-shaped portions and leave them in running water overnight to reduce their acidity. On the next day put a wedge of potato into each of several test tubes. (Or use slices of potatoes in covered Petri dishes.) Add 3 ml of distilled water to each test tube, and stopper the tubes with non-absorbent cotton. Stand the tubes in a wire basket, but avoid packing them. Push down the cotton plugs so they won't pop out. Sterilize the tubes in an autoclave or pressure cooker for 20 min at 15 lb pressure. Be sure to allow air to escape from the pressure cooker before closing the valve. If a double boiler is used, heat to boiling for 1 hr.

Students may either expose these tubes to air, or inoculate them by touching the surface of the potato wedge with a platinum or Nichrome wire loop (p. 634) laden with bacteria. Finally, place the culture tubes or dishes in a warm, dark place for several days. Colonies of bacteria appear as spots of different texture in white, cream, buff, yellow, orange, and other colors.

5. BLOOD AGAR. Sometimes it is desirable to devise experiments to show that while bacteria do not develop colonies on nutrient agar, this is no proof that bacteria are not present. Several kinds of bacteria will not grow on nutrient agar unless blood is added. Many types of bacteria found in the throat may be transferred by coughing into a sterile nutrient agar Petri dish. Recall the danger of using pathogenic bacteria and the need for caution in handling and disposing of the used plates and test tubes (see pp. 633–34).

Cool 95 ml of nutrient agar to about 45° C (113° F), and add 5 ml of sterile, citrated blood prepared by adding 1 ml of sodium citrate to 10 ml of blood. Then pour into sterile Petri dishes; immediately cover the dishes.

6. ROBERTSON'S MEDIUM FOR ANAEROBES.[12] Mix 1 liter of distilled water, 500 g of ground beef heart (from which fat and blood vessels have been removed), and 10 g of peptone. Bring this mixture to a boil, simmer for 2 hr, and adjust the pH to about 8 (with sodium bicarbonate). Pour off the broth into flasks, and autoclave at 15 lb pressure for 15 min. Dry the meat portion by spreading it on clean filter paper and placing it in an oven at 56° C (133° F) for some 48 hr. Into each test tube put a small amount of the dried ground heart, and add about 10 ml of broth to each. Sterilize in an autoclave. When the broth is cool, adjust to a pH of about 7.5. Then resterilize the test tubes.

7. CARBOHYDRATE BROTH FOR FERMENTATION TESTS. This formula has been suggested for work in bacterial fermentation.[13] Make a 0.5 percent solution of the desired fermentable substance by first dissolving it in a small amount of hot water, and then adding this solution to 1 liter of meat extract broth. Then add to the mixture 1 ml of a 1.6 percent alcoholic solution of brom cresol purple (p. 664).

Fill fermentation tubes with about 6 ml of the solution, and then sterilize them in an autoclave at 7 lb pressure for 10 min. Or use the Arnold sterilizer (Fig. 13-7).

Sterilizing culture media. Naturally, when bacteria are to be cultured, conditions must be sterile; otherwise many kinds of bacteria and molds will contaminate the cultures. There are several ways to sterilize the materials you will need.

DRY HEAT. Glassware, including Petri dishes, test tubes, and flasks, may be sterilized by putting them in an oven at a temperature of 160° to 190° C (320° to 374° F) for at least 1 hr. At temperatures over 175° C (347° F) cotton and paper begin to char. When glassware is tightly packed in containers, it should be sterilized in an oven for at least 2 hr at 160° to 190° C.

STEAM UNDER PRESSURE (AUTOCLAVE). Sterilize culture media, cotton, rubber, and glassware in an autoclave (Fig. 13-6) set at 15 lb pressure for 15 min. For small quantities of medium, use a smaller pressure cooker. When large quantities of media are sterilized in flasks or bottles, keep them in the autoclave for ½ hr. Cover the cotton plugs on the flasks with paper to keep them dry.

When you use the autoclave or pressure cooker, follow this procedure: After the materials have been placed in the autoclave, open the escape or safety valve before turning on the steam. When the steam begins to flow, close the valve somewhat; when the steam escapes full force, close the valve so that only a small opening remains for steam to escape. Regulate the valve to maintain the desired pressure.

STEAM WITHOUT PRESSURE (ARNOLD STERILIZER). Certain kinds of media cannot be sterilized under pressure. Milk, nutrient gelatin, and any materials containing carbohydrates may be hydrolyzed or damaged by overheating. Sterilize these media in an Arnold sterilizer (Fig. 13-7). Bring the water in the pan to a boil, put

---

[12] U.S. War Dept., *Methods for Laboratory Technicians,* Tech. Man. 8-227, U.S. Govt. Printing Office, Washington, D.C., 1941, pp. 160–61.
[13] *Ibid.*

Fig. 13-6 Pressure cooker (which serves as an autoclave), for sterilizing with steam under pressure. (Courtesy of Central Scientific Co.)

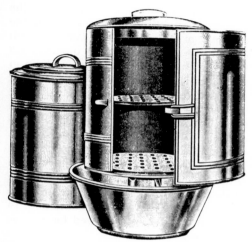

Fig. 13-7 Arnold sterilizer, for sterilizing with steam without pressure. (Courtesy of Central Scientific Co.)

the media in the inner compartment, and close it. Heat for some 30 min; then remove the media to room temperature. Repeat the boiling procedure on 2 successive days. (This is known as intermittent sterilization.)

STERILIZATION WITH CHEMICALS. Discarded cultures should be treated with steam under pressure or with a 2 to 5 percent solution of cresol; or the surface of the dishes should be covered with Lysol. Laboratory table tops may also be disin-

fected with cresol solution or Lysol. Wear rubber gloves when using a strong disinfectant.

BOILING. Some materials may simply be boiled for 30 min for effective sterilization. For example, small pieces of glassware, syringes, and pipettes may be sterilized (for all effective purposes) in this way.

Filling test tubes. After the medium has been prepared, it must be poured into sterilized test tubes in such a way that it does not streak the sides of the tubes (Fig. 13-8a). And when the medium contains agar it must be poured quickly so that the agar does not have time to solidify.

Set up a warmed glass funnel resting in a funnel warmer (Fig. 13-9) that permits

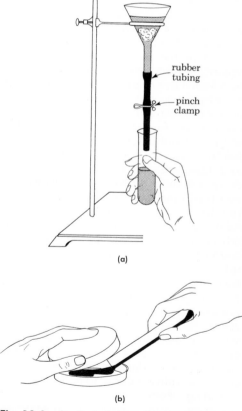

rubber
tubing

pinch
clamp

(a)

(b)

Fig. 13-8 Sterile techniques for handling culture media: (a) filling sterile test tube with warm culture medium (use pinch clamp to regulate flow); (b) plating a sterile Petri dish with warm medium.

Fig. 13-9 Funnel heater, with a heavy copper double wall jacket. (Courtesy of Carolina Biological Supply Co.)

circulation of warm water around the glass funnel. Place a cotton filter in the bottom of the glass funnel. Use a 4-in. square of absorbent cotton with the corners folded in toward the center to make a rounded filter. Wet the cotton, and press it into place. Attach a rubber tube and a clamp to the funnel. Line up the test tubes in racks. Hold one test tube at a time below the funnel, as shown in Fig. 13-8a, and insert the rubber tubing one-third down into the test tube. Pour the medium into the funnel and use the clamp to control the quantity of medium that flows into each test tube. Fill some tubes half full, others two-thirds full, and replace each cotton plug. Finally, sterilize all the test tubes, stand them in wire baskets, and put them in an autoclave for 20 min at 30 lb pressure. Remove the test tubes from the autoclave; set the half-full test tubes at an angle along a ledge so that they solidify with the medium slanted. (This gives a larger surface area on which to streak bacteria.) Use the test tubes that were filled two-thirds full to plate Petri dishes.

Plating Petri dishes. Sterilize Petri dishes by putting them in an autoclave for 15 min at 15 lb pressure, or wrap them individually in paper and put them in an oven. When the paper begins to char a bit,

the temperature is high enough. Leave the dishes in the oven for several hours. Petri dishes may be sterilized ahead of time in this way; if they are left wrapped in the paper, they may be stored for several weeks.

Now the prepared test tubes (two-thirds full of medium) are ready for use. If the agar medium has solidified, stand the test tubes in boiling water to melt the medium.

Wet down the table tops by wiping them with a damp sponge. Avoid air drafts while plating. Light a Bunsen burner, and work close to it. Remove one test tube of melted agar medium from the water bath, lift off the cotton plug, and bring the mouth of the test tube across the Bunsen flame to eliminate any contaminants around the mouth of the tube. Open the Petri dish slightly, and quickly pour the contents of one test tube into the dish (Fig. 13-8b). Cover the dish quickly, and tilt it around to distribute the medium in a uniformly thin film. Now set the dish aside to solidify. Proceed with the same technique until you have "plated" as many Petri dishes as you need. When they are all solidified, invert the dishes to prevent the condensing water vapor from falling back onto the agar surface. Then wrap each dish in paper until you are ready to inoculate it. Refrigerate the Petri dishes to help prevent contamination. In this way plates may be saved for a week before they are to be used.

Inoculating culture medium. Petri dishes or test tubes may be exposed to air (to permit them to become contaminated) in order to show that bacteria and mold spores are ever present in the air. But when you wish to establish a pure culture, inoculate a new culture medium in sterile Petri dishes or test tubes with a bit of the old culture. Transfer a bit of the bacterial culture (together with its agar) on an inoculating needle.

To prepare an inoculating needle, hold the end of a piece of glass rod in a Bunsen flame until the glass is softened. With a forceps, insert a 3-in. piece of Nichrome wire (No. 24) or platinum wire into the melted glass. When you want the needle for solid material, use a straight end of wire. When the medium is liquid, such as milk or broth, curve the wire into a small loop at the end. Fire-polish the other end of the glass handle.

To transfer or inoculate a fresh culture medium, the following procedure is useful. Heat the wire to red heat in a flame so that it is sterilized. Let the needle cool before touching it to the culture from which a transfer is to be made. Almost simultaneously remove the cotton plug from the sterile culture tube or flask and hold it in your hand. Place the mouth of the tube in a flame for a moment. Dip the flamed sterile needle tip into the old culture. Remove the needle from the old culture, and insert it into the sterile new tube, flask, or dish (Fig. 13-10). Streak the needle along the entire length of sterile agar surface. It isn't necessary to press the needle into the agar. Then flame the mouth of the test tube again and flame the plug (be careful that it does not catch fire). Replace the plug. Again flame the needle before putting it aside in a needle holder.

When you streak Petri dishes, raise the cover of the sterile Petri dish slightly and streak the surface in equally distant parallel lines (see Fig. 9-4). Quickly replace the cover and invert the dish. Tape the cover to the dish, and label the contents; place in an incubator at 37° C (99° F). Also put the test tubes in the incubator.

Broth-grown bacteria may be isolated in colonies by pouring a thin film of the broth over the surface of the sterile agar in a Petri dish. Do not invert these dishes. Be certain to keep the film thin. This technique is also used for isolating the kinds of bacteria in milk or tap water or other fluids.

Products of bacteria. The pH of a culture medium may be changed as a result of bacterial growth. You can determine what changes have occurred by adding a given quantity of an indicator to the medium when it is originally prepared. Table 13-1 lists three indicators which have been recommended.

Uses in the classroom

Often bacteria are collected from water, soil, air, milk, or other materials, if only to demonstrate that bacteria are found everywhere. You may also try to show that bacteria are visible as colonies and that bacteria may be identified by means of the color, shape, and texture of the colony as well as by their individual shape under the microscope (Fig. 2-27) and ability to hold specific stains. Furthermore, bacteria are also identified by their reaction with certain media.

In the classroom teachers often set up demonstrations to reveal the conditions needed for bacterial growth (turbidity, p. 445), or test the action of penicillin or other antibiotics on the growth of bacteria (see p. 450). There are other experiments which students may devise to test the effectiveness of many antiseptics (see p. 451). They may devise an experiment to test the effectiveness of preservatives: heavy sugar solution, salts, drying, spices, vinegar, and so forth. Some teachers grow bacteria in agar to which radioactive phosphorus has been added. Later, radioautographs are made (p. 233).

You may want to study carefully a special issue of *The American Biology Teacher*

Fig. 13-10 Method of subculturing: transfer from (A) into a new tube of medium (B).

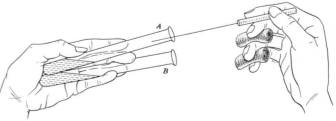

TABLE 13-1

Range of three indicators useful in testing culture media*

indicator	color change (acid to alkaline)	quantity per liter of medium (in ml)	pH range
Brom cresol purple (1.6% alcoholic solution)	yellow to purple	1	5.2–6.8
Brom thymol blue (0.04% aqueous solution)	yellow to blue	50	6.0–7.6
Phenol red (0.02% aqueous solution)	yellow to red	50	6.8–8.4

*U.S. War Dept., *Methods for Laboratory Technicians,* Tech. Man. 8-227, U.S. Govt. Printing Office, Washington, D.C., 1941.

(Vol. 22, No. 6, June 1960) devoted entirely to *Microbiology in Introductory Biology.* Prepared by the Soc. of Am. Bacteriologists, it includes detailed descriptions of a microbiology kit and preparation of transfer needles and pipettes. L. S. McClung offers suggestions on bacteriological techniques for the beginner. A second part of the same issue is a handbook in itself, listing reference books, films, papers from the literature, articles from *Scientific American,* and suggestions for projects and laboratory work.

Phylum Myxomycophyta (slime molds)

Collecting

The yellow plasmodia of slime molds may be found under the bark of fallen logs or under boards in swamps or moist woods. They may be found balled up or in the holes made by boring insects. A day or two after a rain is the best time to look for them. Whether slime molds should be considered animals (Mycetozoa) or plants (Myxomycophyta) does not concern us here; questions relating to controversies about taxonomy of forms are beyond the scope of this book (see Fig. 9-39).

When conditions of moisture and food are adequate, the plasmodia of slime molds will increase in size and their nuclei will divide repeatedly, so that ultimately the mycetozoan (or the myxomycete, if you are a botanist by inclination) may be several inches in diameter and contain some thousand nuclei. Cell walls are lacking, and slow, amoeboid movements can be observed under the microscope.

Small plasmodia may be obtained from the spores of the fruiting bodies of *Physarum* (Fig. 13-11a and b), *Didymium, Fuligo,* or *Stemonitis* (Fig. 13-11c). These fruiting bodies may be crumbled on filter paper or on solid agar.

Filter paper culturing

Camp[14] cultured slime molds on paper toweling and pulverized oatmeal, as follows. Wrap a finger bowl or similar container in filter paper to make a "drum." Then place this prepared finger bowl flat within a larger battery jar; the exposed filter paper should have a flat and smooth surface. It is better to have the open or upper side of the finger bowl on top. Then add 1 in. of water down the side of the battery jar. The filter paper will act as a wick and draw moisture to the top of the "drum" (Fig. 13-12).

[14] W. G. Camp, in *Bull. Torrey Botan. Club* 63:205–10, 1936.

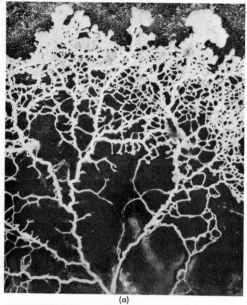

(a)

(b)

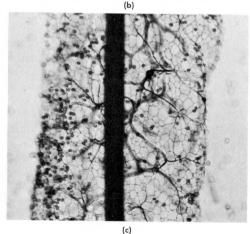

(c)

Fig. 13-11 Slime molds: (a) *Physarum poly-cephalum,* plasmodial stage (magnified 6 ×); (b) *Physarum polycephalum,* sporangial stage (magnified 15 ×, reduced by 62 percent for reproduction); (c) *Stemonitis,* whole mount. (a, b, photos by Hugh Spencer; c, courtesy of General Biological Supply House, Inc., Chicago.)

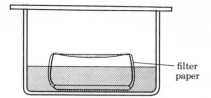

Fig. 13-12 Culturing slime molds on oatmeal on a wet filter paper "drum."

Place the plasmodia on the filter paper; then sprinkle about 0.1 g of pulverized oatmeal (rolled oats which have been ground in a mortar will suffice) onto the paper. Finally, cover the battery jar with a glass plate. Within 24 to 48 hr the oatmeal should be consumed by the spreading plasmodia. Avoid overfeeding, but as the plasmodia grow and spread over the paper, add larger quantities of oatmeal.

When the plasmodia have spread over the entire surface of the paper, place paper toweling on the inner surface of the battery jar so that, in effect, the inner surface is lined with paper. Then the toweling will also act as a wick to draw water. The plasmodia will move across the water moat between the filter paper and the toweling and begin to spread along the toweling. Place the oatmeal, however, on the central filter paper. Then after the plasmodia have spread over all the paper, remove the toweling and dry at room temperature. On the paper the yellow plasmodia quickly form sclerotia.

This dried paper may be cut into squares and stored in envelopes in a dry place. The plasmodia will remain viable for at least a year or two and may be used to prepare fresh cultures within this time.

To prepare a fresh culture, place a square of the dried culture paper on wet filter paper on the finger bowl arrangement already described. Allow 24 hr for the plasmodia to form, and place oatmeal powder in small quantities for food. Increase the amount of oatmeal as the plasmodia grow.

Students doing a project on plasmodia may prefer to prepare a culture of slime molds by crumbling fruiting bodies on a medium containing agar, such as the one that follows.

Agar medium for plasmodia

Soak 6 g of oatmeal in 200 ml of water for 1 hr. Then strain the liquid. To 100 ml of the clear filtrate add 1.5 g of plain agar, and heat until the agar is completely dissolved. Pour a thin layer of the agar into each of several Petri dishes. It is not necessary to sterilize the agar preparations if they are used immediately. Inoculate these dishes by crumbling fruiting bodies of plasmodia over them. Keep the cultures in a warm, dark place. Within about 3 days active plasmodia should be seen. Avoid high temperatures to control the growth of mold or bacteria contaminants.

Other methods of culturing myxomycetes have been described by Raper,[15] Howard,[16] and many other workers. Look into *Biological Abstracts* for current work.

Uses in the classroom

In the laboratory, slime molds may be used to illustrate myxamoebae and as an example of differentiation of structure (as described in Chapter 9).[17]

Phylum Eumycophyta (fungi)

Collecting molds and yeasts

For an unassorted culture of molds, scatter some dust on a piece of moist bread

Fig. 13-13 *Aspergillus* (magnified 680×, reduced by 41 percent for reproduction). (Courtesy of General Biological Supply House, Inc., Chicago.)

or fruit (apple or orange) and place it in Petri dishes or larger covered containers which have been lined with moist blotting paper. Store the containers in the dark. A good yield of sporangia of *Rhizopus* (Fig. 13-14b), the common white mold, should develop within a week. When the bread is left for a longer time, other fungi such as *Aspergillus* (Fig. 13-13) and *Penicillium* (Fig. 2-14) begin to appear.

Several other methods are described in the following pages. For example, dung media will yield many forms, such as *Mucor* (which is a close relative of *Rhizopus*), *Pilobolus,* certain Ascomycetes, and Basidiomycetes. Cheese, such as Roquefort, when covered with a bell jar or placed in a moist chamber, will yield *Penicillium.* This form may also be found growing on citrus fruits.

Yeasts may be found on the surface of dill pickle or sauerkraut juice or on the surface of bruised fruits, especially grapes. You may also purchase yeast in dried form and add it to culture media, as described on p. 642.

[15] K. Raper, "Isolation, Cultivation, and Conservation of Simple Slime Molds," *Quart. Rev. Biol.* 26:169–90, 1951.

[16] F. Howard, "Laboratory Cultivation of Myxomycetes Plasmodia," *Am. J. Botany* 18:624–28, 1931.
See also J. Ward, "The Enzymatic Oxidation of Ascorbic Acid in the Slime Mold *Physarum polycephalum,*" *Plant Physiol.* 40:58–69, 1955.

[17] You may want to read A. C. Lonert, "A Week-end with a Cellular Slime Mold," *Turtox News* 43:2, February 1965. The use of *Escherichia coli* as a food organism for *Dictyostelium discoideum* is described.
For a method of producing sclerotia rapidly—in 1½ to 4 hr, at 27° C (80° F)—refer to Lonert's paper "A High-Yield Method for Inducing Sclerotization in *Physarum polycephalum,*" *Turtox News* 43:4, April 1965.

Culturing molds

Most species of *Rhizopus* grow best at 30° C (86° F) in moderate light. *Rhizopus* and *Mucor* are often confused with each other, but they may be distinguished in this way. The sporangiophores in *Rhizopus* arise in a fascicle from a node on a stolon, while in *Mucor* there are single sporangiophores arising from mycelia (Fig. 13-14).

Asexual spores are abundant; you can also show a type of sexual reproduction called conjugation (Fig. 8-43b). Two different mating strains are needed for conjugation, a plus and a minus strain (purchase from a biological supply house, Appendix C). When a plus strain is streaked across a moistened slice of bread or agar dish and a minus strain streaked at the opposite end, zygospores (sexual spores) form along the center of the bread (Fig. 8-43).

Culture media with agar. The following are suitable solid media for culturing molds.

1. OATMEAL AGAR. Where pure cultures of one type of mold are to be prepared, sterile conditions must be observed. In this method, grind 5 g of oatmeal into 200 ml of water; boil this mixture for 20 min. Pour off the supernatant fluid through gauze to avoid including the oat residue. Add 4 g of agar to the fluid, and boil to dissolve. Finally, pour into sterile Petri dishes, cool, and later inoculate with spores of selected fungi.

2. POTATO-DEXTROSE AGAR. Peel and slice about 100 g of white potatoes and boil for 1 hr in 350 ml of distilled water. Strain through cheesecloth and restore the liquid to the original volume by adding more distilled water. To this 350 ml of potato filtrate add 1 g of dextrose and 10 g of agar; boil for ½ hr in a double boiler. Strain and sterilize; then pour the liquid into sterile Petri dishes. Later, inoculate with desired mold specimens in order to grow a pure culture.

A potato-agar medium may be prepared by the same procedure, omitting the dextrose.

3. PRUNE AGAR.[18] Boil about five prunes in water for 1 hr. Then pour off the supernatant liquid and make up to 200 ml. To this add 80 g of cane sugar and 10 g of agar. This quantity (that is, 200 ml of fluid) will be enough for about seven Petri dishes. Multiply all amounts for larger quantities of the medium.

4. PEA AGAR. Boil about 80 dried peas

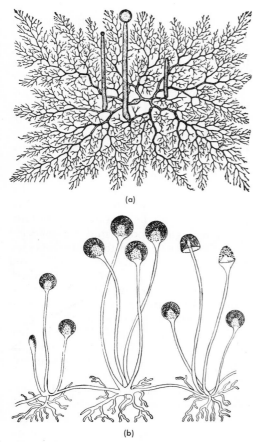

(a)

(b)

Fig. 13-14 Two common molds: (a) *Mucor*; (b) *Rhizopus*. (a, from W. H. Brown, *The Plant Kingdom*, 1935, reprinted through the courtesy of Blaisdell Publishing Co., a division of Ginn and Co.; b, from W. G. Whaley, O. P. Breland, et al., *Principles of Biology*, Harper & Row, 1954.)

[18] H. C. Gwynne-Vaughan and B. Barnes, *The Structure and Development of the Fungi*, Cambridge Univ. Press, New York, 1927, p. 333.

for 1 hr; then pour off the liquid and make up to 200 ml. To this add 5 g of agar. Boil in a double boiler as described previously, then strain and sterilize. Plate the medium in sterile Petri dishes.

5. DUNG AGAR. Soak some 200 g of horse, cow, or rabbit dung in water for 3 days. Then pour off the liquid and dilute with water until the liquid is straw colored. To this add about 2.5 g of agar for every 100 ml of diluted fluid. Boil the materials as described previously, and plate into sterile culture dishes.

Culture media without agar. Since it is often difficult to remove molds from solid media without damaging the rhizoids, several culture methods which omit agar have been developed.

6. KLEBS' SOLUTION. Klebs' solution (p. 587) can be considered one of these. (Dilute the stock solution 1:10 with distilled water.)

Another simple liquid culture medium can be prepared: 5 percent glucose solution.

7. JOHANSEN[19] recommends another medium. Dissolve the following in 1 liter of distilled water:

sucrose (cane sugar)	30.0 g
NH_4Cl	6.0 g
$MgSO_4$	0.5 g
K_2HPO_4	0.5 g

8. BARNES' MEDIUM.[20] Dissolve the following in 100 ml of distilled water:

K_3PO_4	0.1 g
NH_4NO_3	0.1 g
KNO_3	0.1 g
glucose	0.1 g

(If desired, 2.5 g of agar may be added to this glucose-salts solution.)

9. SABOURAUD'S MEDIUM. A useful medium for many kinds of fungi grown in the laboratory is Sabouraud's medium. Mix together and bring to a boil 10 g of peptone, 20 g of agar, and 1 liter of distilled water. To this add 40 g of maltose; filter if necessary. No adjustment for pH is required. Finally, sterilize for 30 min at 8 lb

[19] D. Johansen, *Plant Microtechnique*, McGraw-Hill, New York, 1940.
[20] Gwynne-Vaughan and Barnes, *op. cit.*, p. 333.

pressure and pour into Petri dishes, test tubes, or Syracuse dishes.

Summary of methods for specific molds. Specific molds thrive best on certain culture media. *Rhizopus,* for example, grows well with methods using bread and the potato-agar method (with or without sugar). For careful work the latter method is recommended.

The prune-agar method is recommended for culturing *Aspergillus.* This mold is also cultured readily on bread that has been exposed to the air for several hours and then soaked in a 10 percent solution of cane sugar, grape juice, or prune juice. Keep in a covered jar at room temperature.

Penicillium is best kept on its source, Roquefort cheese or citrus fruit. For pure cultures use the prune-agar method. Some teachers prefer to culture this mold in flasks.

Molds such as *Saprolegnia* (water mold) grow abundantly on decaying insects or small fish, even decaying radish seeds. When dead flies are put in pond water they soon develop a fuzz of growth of *Saprolegnia* and sporangia develop in 24 hr.

Small hymenomycetes among the Basidiomycetes may be kept in flasks containing dung agar. Spores may be collected from the hymenium of the desired species. Such forms as *Marasmius, Clitocybe,* and *Armillaria* have responded to this method.

Other forms, such as *Dictyuchus,* may be found and cultured in pond water to which wheat or rice grains have been added. The water must be changed when it becomes cloudy. *Dictyuchus* may also be found as a contaminant of protozoan cultures using rice.

Neurospora may be cultured easily in a commercially prepared medium such as Difco's Bacto-Neurospora Culture Agar (supply houses, Appendix C). Rehydrate this powdered medium, which contains yeast extract, peptone, and maltose, by adding 1 liter of distilled water to 65 g of the medium. Then heat it to boiling and sterilize it in an autoclave for 15 min at 15 lb pressure.

Culturing molds on slides

Vernon[21] recommends this method whereby slides containing molds may be prepared for individual students. Boil together 6 g of oatmeal, 1 g of agar, and 100 ml of distilled water. Then place a drop of the hot liquid on each of several glass slides. Cover with a coverslip until the medium has hardened. When cooled, remove the coverslip and cut the film of agar into several sections, separating each from the other to form little grooves. Inoculate with mold spores, of similar or different kinds, and ring the coverslips with petroleum jelly. Keep the slides in a moist chamber (preferably a Petri dish lined with damp filter paper). Aerial hyphae and sporangia or conidiophores will grow into the grooves, and they can be studied under the microscope.

Fenn[22] recommends a technique that provides a less dense growth of mold plants so that students can transfer entire plants more easily to a microscope slide (Fig. 13-15). Boil a sliced carrot in water for 20 min. After the liquid has cooled, soak strips of filter paper in the liquid. Sprinkle mold spores on the wet strips. We recommend that these strips be maintained in small stoppered vials or Petri dishes so that when they are distributed to students for their individual laboratory work the room will not become contaminated with mold spores. Furthermore, students may be allergic to spores. Maintain the dishes or vials for a few days at room temperature. Watch the pattern of growth and formation of spores. Transfer bits of mold to slides; mount in alcohol and examine under high power of the microscope.

Some parasitic fungi

Blackberry rust is one of the easiest of the parasitic fungi to keep in the labora-

(a)

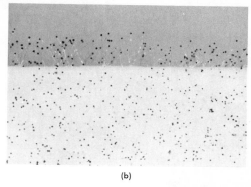

(b)

Fig. 13-15 Culturing the mold *Rhizopus:* (a) filter paper that has been saturated with carrot water is laid flat against a glass square which is slanted against the side of a crystallization dish; (b) magnification of the surface of the filter paper (magnified approximately 6.2×, reduced by 16 percent for reproduction). (Courtesy of Robert H. Fenn, Manchester Community College, Manchester, Conn.)

tory. Plant blackberries infected by rust in a window box; the rust, since it is autoecious, maintains itself on the plant.

Students also may collect several kinds of fungi, such as the powdery mildews which are parasitic on leaves of poplar, Virginia creeper, lilac, and willow (Fig. 13-16b). Others are found on cherry and apple leaves. Keep the leaves dried in envelopes until they are to be used; then soak the leaves in tap water for several hours and mount the scrapings carefully on clean slides in a drop of the fluid.

[21] T. R. Vernon, "An Improved Type of Moist Chamber for Studying Fungal Growth," *Ann. Botany* 45:733, 1931.

[22] R. H. Fenn, "A Simple Method for Growing *Rhizopus nigricans,*" *Turtox News* 40:226–27, 1962.

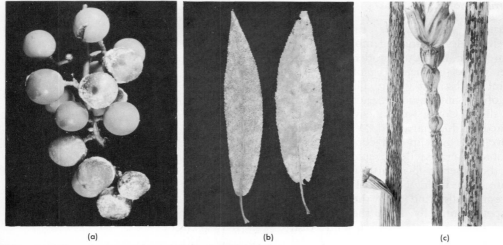

(a)

(b)

(c)

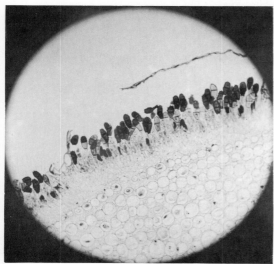

(d)

Fig. 13-16 Parasitic fungi: (a) downy mildew on grape; (b) powdery mildew of willow; (c) wheat rust; (d) microscopic view of wheat rust. (a, courtesy of Dr. T. Sproston, Univ. of Vermont; b, courtesy of Cornell Univ., Dept. of Plant Pathology; c, courtesy of U.S. Dept. of Agriculture; d, courtesy of General Biological Supply House, Inc., Chicago.)

peptone, or about 20 beans or peas, for rapid fermentation.

Insert a cotton plug into the container and set aside in a warm place (25° to 30° C [77° to 86° F]). Within 6 to 24 hr rapidly budding cells should be found (Fig. 8-6). Spore production is not typical in these species, although spores may be found on occasion on the surface of cakes of yeast (formerly used instead of dried yeast) which have been refrigerated for a week.

The method for staining cells with methylene blue, described in Chapter 2, may be used to dye these yeast cells. Students may notice that there is a very slow diffusion of methylene blue into the cells so that living cells are stained selectively. Also see growth (Chapter 9) and measuring liberation of CO_2 (Chapter 6).

Wheat rust (Fig. 13-16c and d) is also a parasitic fungus which may be studied.

Culturing yeast cells

To show budding as a means of asexual reproduction use *Saccharomyces ellipsoideus*, which is found on grapes, or *S. cerevisiae*, which is the commercial yeast. Prepare a 5 to 10 percent aqueous solution of molasses or diluted grape juice. Add one-half of a package of dried yeast to 500 ml of the medium. To this add 1 g of commercial

Lichens

Some lichens, such as the red-topped *Cladonia* or the gray, crusty *Parmelia* and

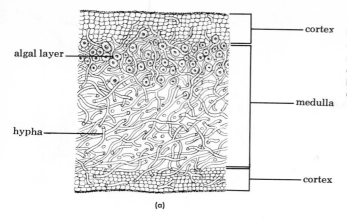

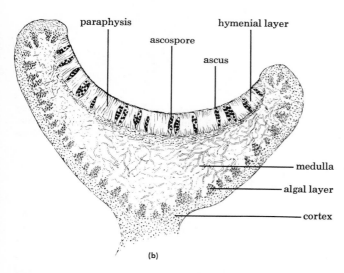

Fig. 13-17 *Physcia,* a lichen: (a) section showing association of algal and fungus elements; (b) section of fruiting body showing ascospores. (Courtesy of General Biological Supply House, Inc., Chicago.)

Physcia, are found on the bark of trees, especially fallen tree logs and tree stumps (Figs. 11-15, 13-17). Sections of bark covered with lichens may be put into a terrarium. Avoid excess moisture, as molds grow readily. Keep the terrarium uncovered.

Students may put dry, brittle specimens in water; after a few hours of soaking the lichens will probably become quite flexible and green.

Lichens are useful for study since they illustrate an interesting example of symbiosis (cooperative living of an alga and a fungus, see p. 537). Bits of the material may be teased apart in a drop of glycerin or water on a glass slide for examination under the microscope. More resistant forms can be crushed in a mortar. The algae carry on photosynthesis, while the fungi are better able to hold the water needed for food-making. You may see ascospores too (Fig. 13-17a).

Phylum Bryophyta

Collecting and culturing mosses and liverworts

When mosses and liverworts are collected, leave them affixed to a small amount of the soil on which they are grow-

Fig. 13-18 Three common mosses: (a) *Mnium affine;* (b) *Physcomitrium pyriforme;* (c) *Catharinea undulata.* (a, from G. F. Atkinson, *Botany,* Holt, Rinehart and Winston, 1905; c, from W. H. Brown, *The Plant Kingdom,* 1935, reprinted through the courtesy of Blaisdell Publishing Co., a division of Ginn and Co.)

ing. Transport the specimens to the laboratory in plastic bags, waxed paper, or newspaper. Place them in terraria in the laboratory or classroom (Fig. 13-21).

A terrarium for mosses and liverworts like *Marchantia* can be prepared by students in the following way. Place a 1-in. layer of coarse gravel or pebbles on the bottom of a tank. Over this spread a ½-in. layer of sand. Then add as a third layer a cover of garden loam about 1 in. deep. In this bed, sod the mosses which have been collected from the field. The water level within the terrarium should be halfway up the gravel layer, and the tank should be covered with a glass. Keep the tank in medium light. When molds show a tendency to grow in the tank, reduce the amount of water and remove the cover until they disappear.[23] Subsequently, keep the terrarium with the least amount of

water needed to keep the plants alive. During the winter months supply light from an electric light bulb.

Mosses such as *Mnium* (Fig. 13-18a), *Bryum, Fissidens, Dicranum,* and *Polytrichum* may be kept in this way. Liverworts such as *Pellia, Pallavacinia, Riccia, Marchantia* (Fig. 13-19), *Conocephalum,* and *Lunularia* have grown abundantly under identical conditions (Fig. 12-28). On the other hand, *Funaria* and *Polytrichum* can withstand a drier terrarium. Finally, the aquatic liverworts, *Ricciocarpus* and *Riccia,* grow abundantly in an ordinary aquarium.

Uses in the classroom

Mosses and ferns are good subjects for study of alternation of generations. The leafy moss gametophytes producing eggs and sperms are haploid, while the sporophyte is diploid. Meiosis occurs in the capsules, resulting in haploid spores; these

---

[23] Sometimes sprinkling the terrarium with a small amount of powdered sulfur will destroy the growth of molds.

(a)

(b)

(c)

Fig. 13-19 Life history of *Marchantia*: (a) thallus with gemmae; (b) thallus with antheridia; (c) thallus with sporophytes. (a, c, courtesy of Carolina Biological Supply Co.; b, courtesy of General Biological Supply House, Inc., Chicago.)

germinate into protonema. The cycle is repeated again (Fig. 8-45).

Also, students may cultivate the protonemata, which grow from spores, on sterile agar. See also the following methods of culturing.

Culturing moss protonemata

Several different nutrient solutions may be used for culturing moss protonemata. Four different methods utilizing several of these nutrient solutions are described below. Spores of mosses germinate well in a liquid solution; if desired, the medium may be solidified by the addition of agar to it.

Using Knop's solution. Crush a dry sporangium of *Funaria* or *Catherinea* (Fig. 13-19c) and liberate the spores onto the surface of some Knop's solution that has been diluted to one-third of its original strength (p. 623). The solution may be kept in Petri dishes or similar containers in moderate light. Spores germinate in 2 weeks, while branched protonemata are formed in 4 weeks. (Beijerinck's solutions, p. 625, can be substituted for Knop's solution.)

Using a solid medium. Some workers prefer this method. Prepare 98 ml of dilute Knop's solution (one-third its original strength) or Beijerinck's solution. Then add 2 g of agar. Boil this until the agar is dissolved and then restore the volume to 100 ml with distilled water. Filter through absorbent cotton into Petri dishes so that a ⅛-in.-thick film is formed. Cover the Petri dishes and allow the agar to cool and solidify. Crush a clean sporangium over the medium, replace the cover, and place the Petri dishes in subdued light.

Several cultures should be prepared, since some may become contaminated. Germination usually takes place in about 10 days. Buds and young gametophytes appear within 2 to 3 months. Students can prepare better cultures by using sterile medium and sterile dishes. And the unbroken sporangia may be sterilized in sodium hypochlorite solution (Clorox and oxolchlorite are commercial preparations). Rinse the sporangia in sterile water, transfer to the agar plates, and crush with sterile forceps. We find that spores of *Funaria* grow especially well in the laboratory.

Using Benecke's solution. Instead of Knop's agar nutrient, students may want to try a modification of Benecke's solution prepared in 100 ml of distilled water, as follows:

NH_4NO_3	0.2 g
$CaCl_2$	0.1 g
K_2HPO_4	0.1 g
$MgSO_4$	0.1 g
$FeCl_3$	trace

Dissolve 0.2 g of agar in 100 ml of the solution. The treatment of spores is similar to that described previously for the Knop's agar medium.

Using Shive's solution. Prepare several Petri dishes with a thin layer of Shive's solution to which agar has been added. Blow spores from crushed moss capsules over the surface. Then cover the Petri dishes and maintain them in moderate light.

1. SHIVE'S SOLUTION. Dissolve the following in 1 liter of distilled water:

$Ca(NO_3)_2 \cdot 4H_2O$	1.06 g
KH_2PO_4	0.31 g
$MgSO_4 \cdot 7H_2O$	0.55 g
$(NH_4)_2SO_4$	0.09 g
$FeSO_4 \cdot 7H_2O$	0.005 g

2. MODIFIED SHIVE'S SOLUTION. Bonner and Galston[24] recommend that minor elements be added to Shive's solution. For each liter of Shive's solution, 1 ml of the following solution should be added. Prepare the solution of minor elements by dissolving the following in 1 liter of distilled water:

H_3BO_3	0.6 g
$MnCl_2 \cdot 4H_2O$	0.4 g
$CuSO_4 \cdot 5H_2O$	0.05 g
$ZnSO_4$	0.05 g
$H_2MoO_4 \cdot 4H_2O$	0.02 g

[24] J. Bonner and A. Galston, *Principles of Plant Physiology*, Freeman, San Francisco, 1952, pp. 53–57.

Flowerpot culture method

Protonemata of mosses may be cultured using the method described for growing fern prothallia (p. 648). Gametophytes should appear within 3 weeks (life cycle, Fig. 8-45). If you also want to grow sporophytes, flood the gametophytes with water for about 1 hr so that fertilization may occur. Then let the excess water run off.

Students can examine the protonema stage under the microscope. They can mount it directly in a 10 percent glycerin solution on a clean slide. The color of chlorophyll remains well preserved, but for a permanent slide the coverslip should be sealed.

Mount the gametophyte of the moss *Funaria hygrometrica* in water. Examine the "typical" plant cells with many chloroplasts and prominent cell walls. Students may notice that the gametophyte is just one layer of cells in thickness. Or mount leaves of the moss *Mnium,* which is also a single layer of cells in thickness.

R. E. Anthony describes a careful development of a technique: "Greenhouse Culture of *Marchantia polymorpha* and Induction of Sexual Reproductive Structures" (*Turtox News* 40:2–5, 1962). Anthony starts new cultures in flats either by using gemmae-cups or by direct planting of the thallus. The results of using light to stimulate growth of antheridia and archegonia are shown in Fig. 13-20. Live sperms may be examined under the microscope.

Phylum Tracheophyta (vascular plants)

Ferns

Collecting. Terrestrial ferns may be kept in the laboratory if care is taken in transplanting them. Take plenty of soil along with the rhizomes and roots of ferns when they are uprooted. For large ferns, such as those of the Osmundaceae (for example, the cinnamon fern, *Osmunda cinnamomea*), dig in a circle the radius of which is 6 in., using the frond as the center of the circle.

(a)

(b)

Fig. 13-20 Effect of light on growth of antheridia and archegonia of *Marchantia polymorpha*: (a) flat showing a preponderance of antheridial structures approximately 25 days after start of light treatment; (b) flat with both antheridial and archegonial structures after 35 days of light treatment. (From Robert E. Anthony, "Greenhouse Culture of *Marchantia polymorpha* and Induction of Sexual Reproductive Structures," *Turtox News* 40:2–5, January 1962. Photos courtesy of Dr. Josiah Lowe, State Univ. College of Forestry, Syracuse, N.Y.)

(The rhizome extends about 6 in. into the soil.) For more delicate ferns, dig within a radius of 3 in. Pack the roots and the rhizomes together with the soil attached to them in moist newspaper, waxed paper, or plastic bags, then transport the entire plants in a plant-carrying case (vasculum) if possible.

(a)

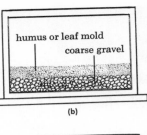

humus or leaf mold

coarse gravel

(b)

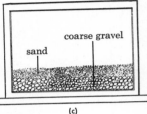

coarse gravel

sand

(c)

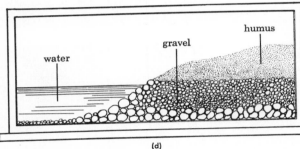

water

gravel

humus

(d)

Fig. 13-21 Habitat terraria for the classroom: (a) Wardian case, in which tall ferns thrive; (b)–(d) three habitat terraria that students can make: (b) woodland, (c) desert, (d) bog. (Courtesy of General Biological Supply House, Inc., Chicago.)

Culturing. In the laboratory, transplant the ferns into sufficiently large clay pots to avoid crowding. Before transplanting, prepare the pots with 1 in. of coarse gravel, broken tile, or broken clay pot on the bottom, and then about 1 in. of garden loam. Put the fern on top of this, and continue to add soil composed of equal parts of sand, peat, and garden loam. If the ferns were removed with enough soil around the roots and rhizomes, then the rhizomes probably have not been injured. Keep the plants in medium light and provide moisture by standing the flowerpots in water. Keep the soil moist but not wet. Such forms as the bracken fern (*Pteris aquilina*) and the hay-scented fern (*Dennstaedtia punctilobula*) can withstand fairly bright light but should not be put in direct or strong sunlight.

Water ferns such as *Salvinia, Azolla,* and *Marsilia* may be grown in aquaria which have been in use for a month or so. *Azolla* requires strong light, while *Salvinia* and *Marsilia* grow better in medium light.

Most of the small ferns need constant humidity, a condition which is best duplicated in the laboratory terrarium (Fig. 13-21). In this way students may grow Maidenhair and Royal Osmunda with success. A suitable terrarium soil consists of 1 part of coarse sand, 1 part of fine peat moss, and 2 parts of good garden loam.

Several common ferns, with their fruiting fronds, are shown in Fig. 13-22.

CULTURING FERN PROTHALLIA. Because it takes at least 6 weeks for fern gametophytes to grow, sow the spores well in advance of the time you will want gametophytes (life history, Fig. 8-46).

FLOWERPOT CULTURE OF FERNS. For this and the following method, use flowerpots and glass plates to maintain adequate moisture. Fill a large flowerpot with bro-

(a)

(b)

Fig. 13-22 Three common ferns, each with diagram of fruiting frond: (a) Christmas fern (*Polystichum acrosticoides*); (b) common polypody (*Polypodium vulgare*); (c) cinnamon fern (*Osmunda cinnamomea*). (Photographs: a, courtesy of U.S. Dept. of Agriculture; b, c, photo by Hugh Spencer; drawings: courtesy of General Biological Supply House, Inc., Chicago.)

(c)

ken pieces of tile or porous clay flowerpots to within 2 in. of the top. Then cover this with 1 in. of rich loam, and then a ½-in. layer of washed sand.

Sterilize the filled pot by placing it in boiling water for a few minutes or by pouring boiling water over the pot and its contents. However, keep the surface layer of fine sand intact. Next crush the sori found on the underside of ripe fern fronds (Fig. 13-22). Or spread ripe fern fronds on paper for a few days; the spores will fall on the paper. Sprinkle these spores over the sand surface of the flowerpot. Cover the pot with a glass plate to retain moisture and prevent contamination. Next stand the pot in a saucer of water. Add water to keep a level in the saucer and stand the pot in medium light. Should "damping off" by fungi (or mildewing) occur because the soil is too moist, students may

water the pot with 0.01 percent potassium permanganate solution.

Prothallia will develop from spores in about 3 to 4 weeks, producing a five- to ten-cell stage. Then the prothallia may be separated and subcultured in freshly prepared containers.

When gametophytes have been growing for about 5 months, they may be used to get sporophytes. Transfer them to the surface of sterile leaf mold packed into clay saucers. Water frequently and maintain a constant humidity. Within a few months sporophytes should begin to develop.

COSTELLO'S METHOD. In this method fern prothallia are grown over an inverted flowerpot. Clean a 4-in. clay flowerpot, and fill it with sphagnum moss (peat moss) or paper toweling; moisten the moss (or toweling) and pack it tightly. Immerse all the materials for 10 min in boiling water

Fig. 13-23 Fern prothallia growing in nutrient solution. (Courtesy of Carolina Biological Supply Co.)

(a)

to sterilize them. Then allow the pot to cool, and invert it into a flowerpot saucer filled with water or Knop's solution diluted to half its original strength. The contents of the pot will act as a wick to draw up water.

Dust fern spores over the moist surface of the inverted flowerpot. Cover with a battery jar or bell jar, and place the device in a cool place in medium light. Within 10 to 20 days, small prothallia should be growing on the outer surface of the flowerpot.

AGAR MEDIUM. The sterile agar method found suitable for moss protonemata may be used here. Use Benecke's, Knop's, or Beijerinck's solution. However, if small sporophytes are desired, wide-mouthed low jars may be used instead of Petri dishes. You may also grow fern prothallia on agar slants in test tubes or flasks.

OTHER METHODS. At other times, students may want to sow spores on Knop's agar solution on slides so that the growth of young prothallia may be studied under the microscope; seal the coverslips with petroleum jelly, and keep them in a moist chamber. Or prothallia may be grown in nutrient solution (Fig. 13-23).

The club mosses include the two common genera *Lycopodium* (Fig. 13-24a) and *Selaginella*. Horsetails, or scouring rushes, of which there is only one genus, *Equisetum* (Fig. 13-24b), may be collected and kept

(b)

Fig. 13-24 Relatives of the ferns: (a) *Lycopodium obscurum*, a club moss; (b) *Equisetum arvense*, a horsetail. (a, photo by Hugh Spencer; b, Roche photo.)

in a terrarium in which ferns and club mosses grow.

Students may want to try to grow *Equisetum* in the laboratory. Spores germinate quickly when they are shed. Shake the spores onto soil and cover with a pane of glass. For best results, sterilize the soil either by heating it in an oven or by watering it well with a dilute potassium permanganate solution (on the average some five small crystals to 1 liter of water). In about 3 to 5 weeks, antheridia should be visible; archegonia appear a little later.

Gymnospermae and Angiospermae

Culturing seed plants. While most seed plants grow readily in porous flowerpots of suitable size, several precautions must be followed (see also hydroponics, Chapter 9). There are many excellent books dealing with the care of specific house plants (see Bibliography). However, let us consider briefly a few of the requirements for raising seed plants.

1. Do not overwater plants. The soil should not be wet (nor should it be very dry). One of the best ways of watering plants is to immerse the entire flowerpot in water for 10 min. In most cases this treatment is sufficient to ensure moisture for the roots for from 4 to 5 days.

2. Do not place plants over a radiator. In many cases, this is the reason for poor growth of plants in the classroom and laboratory. Nor should a plant be kept too near the window, for in winter the side nearest the window is subjected to excessive cooling.

3. Examine the undersides of leaves for insect pests, such as aphids, mealy bugs, and other scale insects. When these are found they should be washed off with soap and warm water. Spray with a nicotine-soap solution.

4. Every month or so, water the plants with about 100 ml of Knop's solution (or add commercial mineral preparations). This will replenish the soil.

5. A good garden soil will bring about the greatest growth. But it is equally important that each flowerpot have broken tile or broken clay at the bottom to ensure drainage.

6. When the plant has outgrown the pot, it should be repotted. Carefully remove the roots and the soil. Tap the flowerpot to release the plant.

7. Good growth will be assured if the plants are kept in a Wardian case (Fig. 13-21a).

8. The number of plants may usually be increased by vegetative propagation. Chapter 9 describes several suitable methods.

What kinds of plants should you buy? Not all seed plants require equal amounts of light. In fact, teachers may be hard put to know the kinds of plants to grow in a northern exposure where there is little or no sunlight. A quick reference to the following lists[25] may help to establish a cheerful atmosphere in the classroom. Most of these plants are also useful in teaching biology.

PLANTS FOR FULL SUNLIGHT

Cactus	*Zygocactus truncatus*
Echeveria	*Echeveria*, sp.
Geranium	*Pelargonium*
Kalanchoë	*K. coccinea*
Oxalis	*O. rosea*
Patience plant	*Impatiens sultani* (or *Holstii*)
Primrose (fairy)	*Primula malacoides*
Spiraea	*Spiraea*

PLANTS FOR SUNLIGHT (2 TO 4 HR)

African violet	*Saintpaulia*
Asparagus "fern"	*Asparagus plumosus*
Begonias	Many varieties, *argenteo-guttata*
Coleus	*Coleus Blumei*
House iris (Apostle plant)	*Marica Northiana*
Nephthytis	*N. Afzelii*
Peperomia	*Peperomia maculosa* (and *arifolia*)
Pick-a-back	*Tolmiea Veitchii*

[25] Personal communication from M. Brooks, 1947, then Director of Gardening, New York City Board of Education.

Spider plant	*Anthericum*, sp.
Strawberry begonia (saxifrage)	*Saxifraga*

PLANTS FOR WINDOWS WITHOUT SUN (NORTHERN EXPOSURE)

Aloe	*Aloe arborescens*
Aspidistra	*Aspidistra lurida*
Begonia	*Begonia rex*
Boston ferns	*Nephrolepis exaltata bostoniensis*
Chinese evergreen	*Aglaonema modestum*
Chinese rubber plant	*Crassula arborescens*
Date palm	*Phoenix dactylifera*
Dumbcane	*Dieffenbachia seguine*
Gold-dust Dracaena	*Dracaena Sanderiana*
Grape ivy	*Vitis*
Hen and chickens	*Sempervivum*, sp.
Holly fern	*Cyrtomium falcatum*
India rubber tree	*Ficus elastica*
Ivy, English	*Hedera Helix*
Philodendron	*Philodendron cordatum*
Snake plant	*Sansevieria*
Stonecrop (live-forever)	*Sedum*, sp.
Tradescantia	*Tradescantia zebrina*

Germinating seeds. There are times when students may want to germinate seeds in the classroom either for class study or for individual projects. Seeds of gymnosperms germinate slowly, requiring 2 weeks to several months. Seeds of maples, oaks, or tulip trees grow more rapidly. When rapidly growing seeds are desired, use radish or mustard seeds, for they germinate within 24 hr. The larger seeds of beans, peas, corn grains, squash, and castor beans germinate in 1 to 2 days. When seeds are soaked overnight at room temperature (before sowing) germination is hastened. (Some students may be interested in tissue cultures made from germinating seeds, p. 468.)

All the stages in the germination of seeds can be observed by students. Line a small pan or cheese box with filter paper, or use sand. Sow soaked seeds and dampen the paper or sand. Cover with a glass aquarium and keep in moderate light. (For methods of disinfecting seed surfaces, see p. 457.)

Seeds can also be germinated on sphag-num moss to prevent "damping off." Dried sphagnum purchased from a florist can be shredded and is then ready for use. Plants may be grown directly in the shredded sphagnum.

Plants which are ordinarily difficult to root have been grown successfully in vermiculite, an insulating material which has the property of expanding into fluffy material when heated. It is also sterile when fresh and has extensive adsorbent surfaces that hold good quantities of water as well as air.

Youngsters can germinate seeds by placing them between blotting paper and the glass walls of a tumbler when they want to study the parts of a developing seed. In this way they may study the developing root system as well as the shoots. For other demonstrations using germinating seeds, refer to Chapter 9; see also rate of growth, p. 461.

Conditions for germination of some common seed plants are given in Table 13-2.

Seed plants: special methods

INSECTIVOROUS PLANTS. Insectivorous plants (Fig. 11-20) need special care. They should be placed in a glass-enclosed chamber (to maintain humidity). Feed sundew (*Drosera*) on fruit flies. For *Dionaea*, the Venus flytrap, drop *Tenebrio* larvae (mealworms) into the traps. Or bits of meat may be substituted. The pitcher plant (*Sarracenia*) will survive in such a terrarium without insect food. Water the plants once a month.

As the plants spread, and give rise to others, they may be repotted in a soil prepared by thoroughly mixing together 3 parts of garden loam, 1 part of peat moss, and 1 part of sand. The peat moss (sphagnum) will supply the needed acid condition.

CACTUS PLANTS. Cactus plants (Fig. 13-25) require still a different treatment. *Opuntia*, the pear cactus, may be planted in ordinary garden soil but should be watered only once every 2 weeks. Other cactus plants, such as *Cereus, Echinocactus,*

(a)

(b)

(c)

(d)

Fig. 13-25 Five cactus plants: (a) *Echinocereus*, lady-finger cactus; (b) *Opuntia linguiformis*, cow's-tongue cactus; (c) *Acistocactus*, fish-hook cactus; (d) *Selenicereus*, night-blooming cereus; (e) *Mammillaria*, pin-cushion cactus. (Courtesy of Carolina Biological Supply Co.)

(e)

Phyllocactus, and related forms, should be potted in soil made up of 4 parts of sand and pebbles and 1 part of garden loam. A good watering every 3 weeks is sufficient.

Seeds of cactus germinate slowly (2 weeks to months) in flats of light, sandy loam.

DODDER (CUSCUTA). This is a bright orange-colored seed plant which is often

TABLE 13-2
Conditions for germination of some plants*

plant	medium	temperature (degrees C)	length (in days)
Abies (fir)	F, S	20	10 to 28
Achillea (yarrow)	F, P	20	4 to 10
Allium (onion)	F, S, P	10–20	5 to 14
Alnus (alder)	F, S, P	20–30	6 to 21
Asparagus	F, S, P	20	10 to 28
Avena (oats)	S, P	20	4 to 10
Beta (beet)	S	20–30	7 to 14
Betula (birch)	S, P	20–30	30
Brassica (mustard)	F, P	20–30	3 to 10
Capsicum	F, P	20–30	14 to 28
Cucurbita	S	20–30	5 to 14
Daucus (wild carrot)	F, P	20–30	6 to 21
Fagus (beech)	S, P	20	6 to 28
Helianthus (sunflower)	F, S, P	20–30	4 to 10
Hordeum (barley)	S, P	20	3 to 10
Larix (larch)	P	20	10 to 28
Lathyrus (pea)	F, S	20	5 to 10
Linum (flax)	F, P	20–30	3 to 10
Lotus	F, S, P	20	6 to 14
Lupinus (lupine)	S	20	4 to 10
Lycopersicum (tomato)	F, P	20–30	3 to 10
Medicago (alfalfa)	F, P	20	3 to 10
Nasturtium	F, S, P	20	4 to 10
Nicotiana (tobacco)	F, P	20–30	5 to 14
Pastinaca (parsnip)	F, S	20–30	5 to 14
Phaseolus (bean)	S	20	4 to 10
Phleum (timothy)	S	20	4 to 10
Picea (spruce)	P	20	7 to 28
Pinus nigra (Austrian pine)	P	20	14 to 42
Pinus Strobus (white pine)	F, P	20–30	14 to 60
Pinus sylvestris (Scotch pine)	P	20	14 to 28
Pisum arvense (garden pea)	S	20	3 to 10
Quercus (oak)	S	20	28
Raphanus (radish)	F, P	20–30	3 to 10
Rumex Acetosella (sourgrass)	F, S, P	20	5 to 14
Secale (rye)	F, S	20	3 to 10
Spinacia (spinach)	S	20–30	4 to 14
Trifolium, sp. (clover)	F, S, P	20	3 to 10
Triticum (wheat)	S, P	20	3 to 10
Zea (corn)	S, P	20–30	4 to 10

Key for Table 13-2

F—Several pieces of wet filter or blotting paper placed below and above the seeds. Petri dishes are excellent.

S— Washed sand, thoroughly moistened but not wet. The seeds are embedded ½ to 1 in. below the surface.

P—Loam soil in a pot. The seeds are planted ½ to 1 in. below the surface. The soil should be loose.

*From H. C. Muller, "Methods for Establishing the Viability of Seeds of Various Plant Species," in E. Abderhalden, ed., *Handbuch der biologischen Arbeitsmethoden*, Urban & Schwarzenberg, Munich, 1924, Sec. XI, Vol. II, Part 121.

parasitic on clover, goldenrod, and other plants (see Fig. 11-8). There is enough food in the seed for the young plant to begin to develop a slender stem and roots by which it becomes attached to a host. Leaves are reduced to scale-like structures. Then it sends wedge-like haustoria into the host's stem, forming a junction with the vascular system of its host. Finally, its connection with the ground is severed and it depends upon the host. Dodder forms abundant white flowers which hang in clusters in late summer.

Collected specimens wilt quickly but may be preserved in 70 percent alcohol. Pressed specimens may be useful for temporary display. Hold the specimens on mounting paper with cellophane tape.

MISTLETOE (PHORADENDRON). This seed plant grows on branches of ash, elm, and hickory as well as many other trees. Mistletoe has green leaves and carries on photosynthesis, but it probably absorbs water and salts through its roots, which are embedded in the tree branches. Thus, it may be considered a partial parasite. Collect specimens and preserve them in the dry state on mounting paper.

INDIAN PIPE (MONOTROPA). This seed plant lacks chlorophyll and thus appears as a stark white stem with scale-like leaves (Fig. 11-14). It gets its food from humus and is a saprophyte, not a parasite.

When picked, the plant wilts quickly, developing a black-purple pigment. If the plants have been carefully transported undisturbed with their original soil and

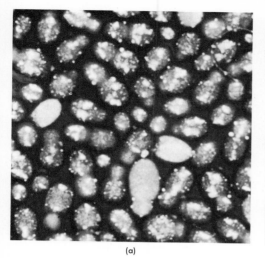 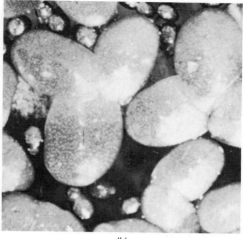

(a)   (b)

Fig. 13-26   (a) *Wolffia columbiana;* (b) *Lemna minor.* (Both are greatly magnified.) (Courtesy of General Biological Supply House, Inc., Chicago.)

put into a Wardian case (Fig. 13-21a), they may remain white for several weeks. Specimens may be preserved in alcohol or formalin.

WOLFFIA. In this smallest of flowering plants, the floating plant is reduced to an oval or rounded green body (the stem). There are no leaves or roots present (Fig. 13-26a). Absorption takes place through the underside of the stem. New fronds, which are modified branches, grow out from depressions on the underside of one end of the oval frond.

Wolffia is found floating on the surface of lakes and ponds, at times very abundantly. The plants multiply rapidly and lend a pleasing appearance to aquaria in the laboratory and classroom. Students may devise individual projects for studying the growth of these plants under a variety of conditions. It is relatively easy to measure the growth of these simple plants. What is the effect of growth-promoting chemicals (pp. 328–32)?

DUCKWEED (LEMNA). *Lemna* (Fig. 13-26b) and a closely related form, *Spirodela,* are found as small floating plants in lakes and ponds. The broad leaf-like expansions are not leaves but stems. A single rootlet lacking root hairs extends from the underside of the frond. They grow well on the surface of an aquarium and add an attractive touch. However, they are not good oxygenators.

MIMOSA. The sensitive plant *Mimosa pudica* grows readily from seeds planted in a light, sandy loam. Keep in sunlight, and invert a battery jar over the flowerpot in which the plants are contained. Young plants grow well at about 27° C (80° F).

CAPSULE LESSONS: Ways to get started in class

13-1. Have each student bring to class a plant specimen for identification. Develop the story of plants from one-celled forms to seed plants. As a club activity or a small group project teach students to use a key (Chapter 11) in identifying plants. Prepare a large bulletin board and mount dried specimens or drawings on it.

In season, maintain flowering twigs in containers of water in a hall cabinet. Other plants may be kept in a terrarium.

13-2. Have students plan a field trip to explore the school grounds. Examine many kinds of flowering plants such as magnolia, *Forsythia,* cherry, dandelion, and other plants in special biomes. (Suggestions for ways to develop the school grounds and for nature trails are given in Chapter 11.)

Have the class undertake a project such as this one: Develop a mimeographed sheet which outlines a specific trail and identifies plants along the way.

13-3. As a project, have students put name plates on the trees and shrubs in the school lawn.

13-4. When a field trip is planned, arrange to transplant ferns, mosses, possibly some spring flowers, small evergreens, and tree seedlings to a similar environment around the school grounds.

13-5. You may want to introduce students to plants that cannot be found on a field trip on the school grounds. Take a vicarious field trip by means of a film and study *Plant Traps: Insect Catchers of the Bog Jungle*, 2nd ed., or *Fungi* (both EBF). Or explore the series of nine filmstrips on *Classification of Plants* (EBF). Also check the catalogs of biological supply houses (Appendix C) for an abundant source of 2 × 2 in. color slides of all kinds of plants (and animals).

13-6. You may want to refer to the suggestions for "Capsule Lessons" given at the end of Chapter 11. Many of the ideas used plants and animals in illustrating the balance in nature. The materials needed for field trips are also indicated in Chapter 11.

13-7. Many suggestions for classwork with bacteria, algae, and higher plants are given in references listed throughout this chapter. Many contain ideas for extended student project work at home. Also refer to *Microbiology in Introductory Biology*, a special issue of *The American Biology Teacher* (Vol. 22, No. 6, June 1960) prepared by the Soc. of Am. Bacteriologists.

13-8. Explore the variety of laboratory work with plants in the laboratory manuals accompanying the three BSCS high school biology texts: the "Laboratory Investigations" section of *Biological Science: Molecules to Man* (Blue Version, Houghton Mifflin, 1963), *Student's Manual: Laboratory and Field Investigations* to *High School Biology* (Green Version, Rand McNally, 1963), and *Student Laboratory Guide* to *Biological Science: An Inquiry into Life* (Yellow Version, Harcourt, Brace & World, 1963). There are many activities relevant to the work of this chapter in such BSCS laboratory blocks as these (published by Heath): A. E. Lee, *Plant Growth and Development* (1963), D. Pramer, *Life in the Soil* (1965), A. S. Sussman, *Microbes: Their Growth, Nutrition, and Interaction* (1964), and W. P. Jacobs and C. E. LaMotte, *Regulation in Plants by Hormones* (1964).

PEOPLE, PLACES, AND THINGS

Is botany not your specialty? Suppose a student wants to work on a project in some specialized area in plant hormones or plant genetics. You may look into *Biological Abstracts*, which prints short summaries of work published by researchers in every area of biology. This is one way to discover who the active workers are in a special field of biology.

If there is a botanic garden or an agricultural college nearby, you may be able to plan a trip, borrow living materials or a film, or arrange to have someone sponsor a gifted student.

The many scholarly journals in all areas of biology contain a wealth of material that can be drawn upon for classroom activities. Among the many papers relevant to culturing of plants are R. A. Lewin, "The Isolation of Algae," *Rev. Algolog.* 5:181–97, 1959; L. Provasoli, J. McLaughlin, and M. Droop, "The Development of Artificial Media for Marine Algae," *Arch. Mikrobiol.* 25:392–428, 1957; and R. C. Starr, "The Culture Collection of Algae at Indiana University," *Am. J. Botany* 47:67–86, 1960.

BOOKS

The following are only a few of the broad references pertinent to work in culturing plants for the laboratory. Refer also to the listings at the ends of Chapters 1, 2, 4, and 11; there is a comprehensive listing of references in Appendix A, which includes field guides and guides to identification of plants.

Brunel, J., G. Prescott, and L. H. Tiffany, *The Culturing of Algae*, C. F. Kettering Foundation for the Study of Chlorophyll and Photosynthesis, Antioch College, Yellow Springs, O., 1950.

Burlew, J., ed., *Algal Culture: From Laboratory to Pilot Plant*, Carnegie Institution of Washington, Washington, D.C., 1953.

Lloyd, F., *The Carnivorous Plants*, Chronica Botanica, Waltham, Mass. (Ronald, New York), 1942.

Pringsheim, E. G., *Pure Cultures of Algae: Their Preparation and Maintenance*, Macmillan (Cambridge Univ. Press), New York, 1946.

Sinnott, E., *Plant Morphogenesis*, McGraw-Hill, New York, 1960.

Taylor, W., *Marine Algae of the Eastern Tropical and Subtropical Coasts of the Americas*, Univ. of Michigan Press, Ann Arbor, 1960.

We lack the space to indicate the vast array of 2 × 2 in. slides, in all areas of biology, that are available from biological supply houses (Appendix C). These companies will readily send their catalogs to you for your perusal.

LOW-COST MATERIALS

Your state department of agriculture has many bulletins which are available to teachers. Teachers may also order a limited number of free seedlings if they are doing useful work in that area. Have your name placed on the mailing list of the U.S. Govt. Printing Office, Washington 25, D.C., for notification of their publications. One of these that may interest you is C. M. Palmer, *Algae in Water Supplies* (U.S. Dept. of Health, Education, and Welfare, U.S. Govt. Printing Office). This manual, which costs $1.00, contains color plates that are useful guides for the identification of a number of common algae. Many state universities have an agricultural college which publishes materials available to teachers in that state, either free or at a small cost.

A tremendous number of fine books are available in paperback editions. For example, there are field guides to birds, to insects, and to flowers in editions which cost from $0.35 to $1.00. You may want to select titles from the list of science paperbacks, *An Inexpensive Science Library,* prepared by the Am. Assoc. for the Advancement of Science, 1515 Massachusetts Ave., N.W., Washington 5, D.C.

We have mentioned before, but it bears repetition, that the biological supply houses (Appendix C) publish directions for caring for plants and animals. Some of these are free leaflets; other directions are included with the specimens when they are purchased from the company. Several supply houses also publish small, inexpensive bulletins. Splendid booklets are also available from many of the large pharmaceutical firms.

Addresses of distributors of free and low-cost materials are given in Appendix D.

Chapter **14**

Stockroom and Facilities for Biology

The organization of the materials in this chapter is as follows:

1. Preparing solutions
2. Making visual "props"
3. Working with glass
4. Using student resources
5. Safety in the laboratory
6. Space for work
7. Storage facilities
8. Equipment and supplies

Preparing solutions

Students need practice in using the chemical balance: in finding true zero and in determining the sensitivity of the balance. Students working on a laboratory squad may well practice weighing in anticipation of preparing stock solutions for the laboratory and stockroom. (For specific directions, have students refer to one of the laboratory manuals in chemistry.)

It seemed useful to us to describe in detail the preparation of certain indicators or solutions at the point where they were first mentioned. If you do not find a solution here, use the index to locate it in context. A number of other general preparations useful in the laboratory have been included here, in alphabetical order. First, however, we shall discuss how to make up solutions of various strengths.

Strength of solutions

Percentage by weight. In common practice, when a dilute solution is prepared, such as a 1 percent solution of

sodium chloride, 1 g of the salt is added to 100 ml of water. Actually, this results in a solution the concentration of which is slightly less than 1 percent. When the concentration desired is greater than 10 percent, the error involved in the common practice becomes significant. Therefore, to make a 10 percent solution of sodium chloride that is accurate enough for most purposes, add 10 g of salt to a graduated cylinder; then add water up to the 100-ml mark.

Percentage by volume; making dilutions. To prepare a solution the concentration of which is measured by volume, begin by measuring out (in milliliters) a volume of the higher-percentage solution that is equal (in number of milliliters) to the percentage needed for the new solution. For example, when you have 70 percent alcohol on hand and want to prepare 50 percent alcohol, measure out 50 ml of the 70 percent alcohol. Then add enough distilled water to bring the volume to a number of milliliters equal to the percentage of the original solution (in this example, to 70 ml). As another example, suppose that you have 95 percent alcohol and want to prepare 70 percent alcohol. Measure out 70 ml of the 95 percent alcohol and add 25 ml of distilled water to make the total volume in milliliters equal to the percentage of the alcohol on hand.

Molar solutions. A *molar* solution is a solution containing one gram-molecule of the dissolved substance per liter of *solution* (not solvent). To prepare a molar solution, dissolve a number of grams equal to the molecular weight of the substance in water (or other solvent) and make up to 1 liter.

For example, sodium chloride has a molecular weight of 58.45. A molar solution of sodium chloride (written 1 M NaCl) contains 58.45 g of sodium chloride in 1 liter of solution. A molar solution of hydrochloric acid (1 M HCl) contains its molecular weight in grams (36.5) in 1 liter of solution.

We can make dilutions of molar solutions, such as 0.5 M, 0.1 M, and so forth. A 0.1 M solution of hydrochloric acid contains 36.5 × 0.1 = 3.65 g of HCl per liter of solution. A 0.4 M solution of sodium chloride contains 58.45 × 0.4 = 23.38 g of NaCl per liter of solution.

Normal solutions. One gram-equivalent of a substance in 1 liter of solution will result in a *normal* solution of that compound. A gram-equivalent is the amount of the substance equivalent to 1 gram-atom of hydrogen (1.008 g). Thus, a normal solution of an acid contains 1 gram-atom (1.008 g) of reacting hydrogen per liter of solution; any other normal solution can then replace or react quantitatively with an equal volume of such a solution.

To prepare normal solutions, study the formula of the acid, base, or salt to be dissolved. When there is one hydrogen atom or one hydroxyl group or one of any ion which will combine with one hydrogen or hydroxyl, a normal solution is the same as a molar solution.

When two hydrogen atoms are present, as in H_2SO_4, a normal solution contains half as much H_2SO_4 as a molar solution, because there are 2 gram-equivalents in every gram-mole. This is also true for $Ca(OH)_2$, and a normal solution is prepared by making a 0.5 M solution of calcium hydroxide.

A normal solution of $FeCl_3$ (which has three chlorines, each of which could react with one hydrogen) would be a 0.33 M $FeCl_3$ solution.

In general, a normal solution is prepared by dissolving in 1 liter of solution a quantity of the acid, base, or salt determined in the following way:

$$\frac{\text{number of grams needed for 1 } M \text{ solution}}{\substack{\text{number of equivalents to 1H} \\ \text{in each molecule}}}$$

or

$$\frac{\text{molecular weight of substance in grams}}{\text{valence}}$$

Ppm. The designation parts per million (abbreviated ppm) refers to parts by weight of a given substance, measured in milligrams of the substance in 1 liter of solution (that is, 1 liter of water weighs 1 million mg).

Rules of solubility. In preparing certain solutions, such as nutrient solutions for microorganisms or for studies in hydroponics, it is useful to know the relative solubilities of salts. The general rules of solubility are given in Table 14-1.

Some solutions for the stockroom

1. ACID-STARCH SOLUTION. Immediately before the acid-starch solution is needed, add 5 drops of yellowish nitric acid (containing nitrous acid) to 10 ml of a starch solution.

In an alternate method, add 1 ml of dilute $NaNO_2$ solution and 1 ml of dilute

TABLE 14-1
General rules of solubility

soluble in water

Compounds of sodium, potassium, and ammonium.

Sulfates (except lead and barium sulfate; calcium, strontium, and silver sulfate are slightly soluble).

Chlorates, nitrates, and acetates.

Chlorides (except silver and mercury chloride; lead chloride is slightly soluble).

insoluble in water

Phosphates, carbonates, oxides, sulfides, sulfites, and silicates (except those of sodium, potassium, and ammonium).

Hydroxides (except those of sodium, ammonium, potassium; calcium, barium, and strontium hydroxides are slightly soluble).

H_2SO_4 to 10 ml of the starch solution just before it is to be used.

2. ADHESIVE FOR JOINING GLASS TO METAL. A solution of sodium silicate has been recommended as an adhesive for joining glass and metal.[1] A paper gasket may be soaked in a solution of sodium silicate and inserted between the edges of glass and metal cutout.

3. ANTICOAGULANT 1 FOR BLOOD. Add 1 ml of this solution to 10 ml of blood. To prepare this solution, dissolve the following compounds in 100 ml of distilled water:

potassium oxalate	2 g
NaCl	6 g

4. ANTICOAGULANT 2 FOR BLOOD. Add 200 mg (0.2 g) of sodium citrate to 10 ml of blood.

5. AQUARIUM CEMENT. This type of cement has been recommended for repair of aquaria. It will stick to glass, metal, stone, or wood. The first four ingredients may be mixed together in the dry state. Then, just before using, add enough linseed oil to make a stiff putty. Allow 3 to 4 days for this cement to harden after it has been forced into crevices and smoothed over with a spatula. The following measurements are given as parts by weight.[2]

litharge	10 parts
plaster of Paris	10 parts
powdered rosin	1 part
dry white sand	10 parts
boiled linseed oil	

6. BENEDICT'S SOLUTION (QUALITATIVE). This is used in a test for the presence of simple sugars in foods, blood, and urine. In the presence of simple sugars, a yellow or reddish precipitate of cuprous oxide forms when the reagent is heated with the "unknown." The test will detect 0.15 to 0.20 percent dextrose.

This solution can be purchased ready-made, or it can be prepared as follows.

Dissolve the carbonate and the citrate in 800 ml of water; warm slightly to speed solution. Then filter. Dissolve the copper sulfate in 100 ml of water and slowly pour into the first solution. Stir constantly; let this cool and add distilled water to make 1 liter.

sodium or potassium citrate	173.0 g
Na_2CO_3 (crystalline)	200.0 g
(or 100 g anhydrous)	
$CuSO_4$ (crystalline)	17.3 g

7. BIURET REAGENT AND PAPER. Dissolve 2.5 g of copper sulfate in 1 liter of water to prepare a 0.01 M solution. Then prepare a second solution of 10 M sodium hydroxide by dissolving 440 g of sodium hydroxide in water and then making up to 1 liter. Before using, add about 25 ml of the copper sulfate solution to 1 liter of hydroxide solution. Or use Walker's modification of the Biuret reagent, prepared as follows.[3]

Add 1 percent copper sulfate solution, a drop at a time, with stirring, to a 40 percent (approximate) solution of sodium hydroxide until the mixture becomes a deep blue color.

Then immerse filter paper in the reagent, dry it, and cut it into small strips for use in tests for proteins.

8. BUFFER SOLUTION. The following salts should be weighed out and added to 1 liter of distilled water:

NaH_2PO_4	28.81 g
Na_2HPO_4	125.00 g

If a buffered solution of sodium chloride is needed, add 8.5 g of sodium chloride to 20 ml of this buffer solution, and make up to 1 liter with distilled water.

For other buffer solutions, see p. 172.

9. COBALT CHLORIDE PAPER. Immerse sheets of filter paper in a 5 percent aqueous cobalt chloride solution. Remove them and blot dry between other sheets of filter paper. Then dry them in an oven at 40° C (104° F). Cut the papers into strips. For immediate use, dry these strips quickly by putting them into a dry test tube and heating them over a flame until the paper turns from pink to blue.

[1] H. Bennett, ed., *Chemical Formulary,* Chemical Publishing Co., New York, 1951.
[2] *Ibid.*

[3] P. Hawk, B. Oser, and W. Summerson, *Practical Physiological Chemistry,* 13th ed., Blakiston (McGraw-Hill), New York, 1954, p. 171.

Store cobalt chloride paper in wide-mouthed, tightly stoppered bottles containing a layer of anhydrous calcium chloride covered with cotton. Should the blue papers change to pink, indicating the presence of moisture, heat them again in a dry test tube.

10. COPPER ACETATE PRESERVATIVE. This solution, which is especially effective for preserving the green coloring of plants, may be prepared as follows. Add enough copper acetate to 50 percent acetic acid to produce a saturated solution. To 250 ml of this saturated solution, add an equal volume of formalin. Then make up to 4 liters with water.

11. 2,6-DICHLOROPHENOL-INDOPHENOL. In 200 ml of water, dissolve 84 ml of sodium bicarbonate and 104 mg of the sodium salt of the indicator. Filter if necessary, and store in a refrigerator.

12. DIGITONIN SOLUTION. Dissolve 1 g of digitonin in 85 ml of 95 percent alcohol; add 15 ml of water, and mix thoroughly. This gives a 1 percent solution of digitonin in 80 percent alcohol.

13. FAAGO. This fixative is widely used, especially for fixing small specimens such as nematodes for permanent mounting on slides. Prepare the following solution in proportions to meet the needs of your class:

commercial formalin	5.0 ml
alcohol (50 percent)	90.0 ml
glycerin	1.5 ml
acetic acid	2.0 ml
osmic acid	trace

Let the specimens to be fixed stand in an evaporating dish containing hot FAAGO until the specimens are covered with a film of glycerin. Then add pure glycerin to the evaporating dish and let it stand for a few hours. Mount the worms or other small specimens in a drop of glycerin on a slide. Seal the coverslip with hot paraffin or with clear nailpolish.

Edmondson[4] suggests that small specimens be mounted (after fixing) in lacto-phenol. Transfer the specimens from the FAAGO into a mixture of formalin (5 percent) and lactophenol (5 percent). Allow the formalin to evaporate at 40° C (104° F) so that lactophenol remains. Finally, mount the specimen in pure lactophenol on a slide, and ring the coverslip with clear nailpolish.

14. FEHLING'S SOLUTION. Either purchase or prepare solutions 1 and 2 separately and store them separately in rubber-stoppered bottles. In testing for the presence of simple sugars add an equal amount of each solution to a test tube of the substance to be tested, and heat. A heavy yellow or reddish precipitate forms (cuprous oxide) if simple sugars are present.

solution 1
| $CuSO_4$ | 34.65 g | |
| distilled water | 500 | ml |

solution 2
KOH	125 g
potassium sodium tartrate	173 g
distilled water	500 ml

15. FERTILIZER FOR ACID-LOVING PLANTS.[5] Combine the following fertilizer ingredients. Then mix each pound of this with 5 cu ft of redwood or cypress sawdust.

$(NH_4)_2SO_4$	26 parts
superphosphate	31 parts
potash	190 parts

For a planting mixture, use half garden loam and half this mixture.

16. FORMALDEHYDE (0.5 PERCENT). Dilute 10 ml of 40 percent formaldehyde solution with 800 ml of water; adjust to pH 7. (See also p. 132.)

17. GRAFTING WAX NO. 1 (LIQUID).[6] Reduce the quantities given to suit your specific needs.

rosin	4	lb
tallow	0.5	lb
isopropyl alcohol	0.75	qt
turpentine	0.25	pt

Melt together the rosin and tallow, then cool a bit and dilute with alcohol and turpentine. Store in a tightly stoppered bottle.

[4] H. B. Ward and G. C. Whipple, *Fresh-Water Biology,* 2nd ed., edited by W. T. Edmondson, Wiley, New York, 1959.

[5] Bennett, *op. cit.*
[6] *Ibid.*

18. GRAFTING WAX NO. 2 (LIQUID)

rosin	4 lb
beeswax	1 lb
raw linseed oil	1 pt

Melt the rosin and add beeswax. Let this melt slowly; add linseed oil and mix thoroughly. After it has cooled, store it in a cool place.

19. GRAFTING WAX NO. 3 (MALLEABLE)

rosin	4 parts
beeswax	2 parts
tallow	1 part

Heat the ingredients together, and pour the melted material into water. When cool, shape into balls with the fingers.

20. HAYEM'S SOLUTION. This solution is used as the diluting fluid in preparing blood for red blood cell counts. It is often used as a stain for blood smears when 0.05 g of eosin is added to the solution indicated here. Before making the smear, mix 1 part of blood to 100 parts of this stain. Then make the blood smear on a clean slide (p. 118).

Weigh out these salts and add them to 100 ml of distilled water:

$HgCl_2$	0.25 g
Na_2SO_4	2.5 g
NaCl	0.5 g

21. INDICATORS. Indicators are dyes that are used to test the pH of a solution. As the hydrogen-ion content of the solution changes, rearrangement of the indicator molecule occurs. A color change results.

Figure 14-1 indicates the range of indicators from small variations within the acid or alkaline range and from one end of the scale to the other. Tables 14-2 and 14-3 give pH values of a variety of acids, bases, and common substances. Notice in Fig. 14-1 that some of the indicators commonly used in demonstrations are those that show a shift in pH around neutral (7). The strength and solvent recommended

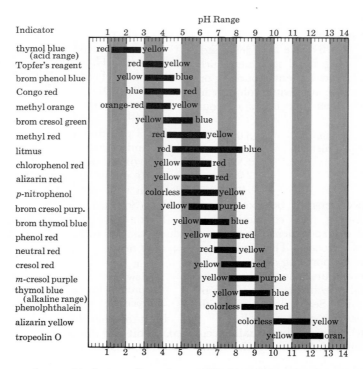

Fig. 14-1 Range of several indicators. (Data from P. Hawk, B. Oser, and W. Summerson, *Practical Physiological Chemistry*, 13th ed., Blakiston [McGraw-Hill], New York, 1954.)

TABLE 14-2
pH values of 0.1 N solutions of a variety of acids and bases*

acids (order of decreasing strength)	pH value	bases (order of increasing strength)	pH value
Hydrochloric acid	1.0	Sodium bicarbonate	8.4
Sulfuric acid	1.2	Borax	9.2
Phosphoric acid	1.5	Ammonia	11.1
Sulfurous acid	1.5	Sodium carbonate	11.36
Acetic acid	2.9	Trisodium phosphate	12.0
Alum	3.2	Sodium metasilicate	12.2
Carbonic acid	3.8	Lime (saturated)	12.3
Boric acid	5.2	Sodium hydroxide	13.0

*Data from *Handbook of LaMotte Chemical Control Units for Science and Industry,* 13th ed., LaMotte Chemical Products Co., Chestertown, Md., 1944.

are generally given on the bottle of indicator dye. The dilution and preparation for some of the common indicators follow.

Alizarin red. 1 percent aqueous solution.

Brom thymol blue. Add 0.04 g of brom thymol blue powder to 6.4 ml of $N/100$ NaOH; add 20 ml of absolute alcohol. Make up final volume to 100 ml with distilled water. In use, add 9 ml of liquid to be tested to 1 ml of stock solution. (*Note:* This is not the solution of brom thymol blue to be used in photosynthesis, p. 169.)

Congo red. 0.5 percent solution in 50 or 70 percent alcohol.

Methyl orange. 0.02 percent aqueous solution.

TABLE 14-3
Approximate pH of some common substances*

Apples	2.9–3.3	Human duodenal		Pickles, sour	3.0–3.5
Apricots (dried)	3.6–4.0	contents	4.8–8.2	Pimento	4.7–5.2
Asparagus	5.4–5.7	Human feces	4.6–8.4	Plums	2.8–3.0
Beans	5.0–6.0	Human gastric		Pumpkins	4.8–5.2
Beer	4.0–5.0	contents	1.0–3.0	Raspberries	3.2–3.7
Beets	4.9–5.6	Human milk	6.6–7.6	Rhubarb	3.1–3.2
Blackberries	3.2–3.6	Human saliva	6.0–7.6	Salmon	6.1–6.3
Bread, white	5.0–6.0	Human spinal fluid	7.3–7.5	Sauerkraut	3.4–3.6
Cabbage	5.2–5.4	Human urine	4.8–8.4	Shrimp	6.8–7.0
Carrots	4.9–5.2	Jams, fruit	3.5–4.0	Spinach	5.1–5.7
Cherries	3.2–4.1	Jellies, fruit	3.0–3.5	Squash	5.0–5.3
Cider	2.9–3.3	Lemons	2.2–2.4	Strawberries	3.1–3.5
Corn	6.0–6.5	Limes	1.8–2.0	Sweet potatoes	5.3–5.6
Crackers	7.0–8.5	Magnesia, milk of	10.5	Tomatoes	4.1–4.4
Dates	6.2–6.4	Milk, cow	6.4–6.8	Tuna	5.9–6.1
Flour, wheat	6.0–6.5	Molasses	5.0–5.4	Turnips	5.2–5.5
Ginger ale	2.0–4.0	Olives	3.6–3.8	Vinegar	2.4–3.4
Gooseberries	2.8–3.1	Oranges	3.0–4.0	Water, distilled	
Grapefruit	3.0–3.3	Peaches	3.4–3.6	(carbon-dioxide-free)	7.0
Grapes	3.5–4.5	Pears	3.6–4.0	Water, mineral	6.2–9.4
Hominy (lye)	6.9–7.9	Peas	5.8–6.4	Water, sea	8.0–8.4
Human blood plasma	7.3–7.5	Pickles, dill	3.2–3.5	Wines	2.8–3.8

*Data from *Handbook of LaMotte Chemical Control Units for Science and Industry,* 13th ed., LaMotte Chemical Products Co., Chestertown, Md., 1944.

Methyl red. 0.02 percent aqueous solution.

Neutral red. 1 percent solution in 50 percent alcohol.

Phenolphthalein. Prepare a 0.5 percent solution in alcohol by dissolving 0.5 g of phenolphthalein in 100 ml of 95 percent alcohol. For very sensitive tests, a 0.1 percent solution may be used.

Additional indicators may be prepared as follows.[7] Grind 0.05 g of the indicator (see list below) with the designated volume of 0.01 N sodium hydroxide solution in a mortar; add distilled water to make 125 ml of the indicator solution. (Increase proportions if larger quantities are required.)

Brom cresol green. 7.2 ml of 0.01 N NaOH.

Brom cresol purple. 9.3 ml of 0.01 N NaOH.

Brom phenol blue. 7.5 ml of 0.01 N NaOH.

Brom thymol blue. 8.0 ml of 0.01 N NaOH.

Chlorophenol red. 11.8 ml of 0.01 N NaOH.

Cresol red. 13.1 ml of 0.01 N NaOH.

Meta cresol purple. 13.1 ml of 0.01 N NaOH.

Phenol red. 14.1 ml of 0.01 N NaOH.

Thymol blue. 10.8 ml of 0.01 N NaOH.

22. IODINE (TINCTURE OF). Dissolve 70 g of iodine and 50 g of potassium iodide in 50 ml of distilled water. Then dilute to 1 liter with 95 percent alcohol.[8]

23. IODINE–POTASSIUM IODIDE SOLUTION. Dissolve 3 g of potassium iodide in 25 ml of water. Then add 0.6 g of iodine, and stir until dissolved. Make up to 200 ml with distilled water. Store in a dark bottle. This dilute solution may be used in determining amylase activity (see Lugol's solution).

24. LIGNIN TEST. A saturated solution of phloroglucin[9] (1,3,5-trihydroxybenzene) in alcohol is recommended as a test for the presence of lignin. First, mount plant tissue sections in this solution for a few minutes, then transfer to a drop of water (containing a minute trace of hydrochloric acid). If lignin is present it stains a bright reddish-violet color.

25. LIMEWATER. To distilled water add an excess of calcium hydroxide or calcium oxide. Cork the bottle, shake well, and let it stand for 24 hr. Then pour off the supernatant fluid (filter if necessary) and keep well stoppered.

The limewater should remain clear. When carbon dioxide is added, a milky precipitate of calcium carbonate is formed:

$$CO_2 + H_2O \longrightarrow H_2CO_3$$

$$Ca(OH)_2 + H_2CO_3 \longrightarrow$$
$$CaCO_3 + 2H_2O$$

26. LOCKE'S SOLUTION. This solution may be used to mount fresh mammalian blood samples or chick embryos. It is also a component of some special culture media for protozoa. Dissolve the following in 1 liter of distilled water:

NaCl	9.0 g
$CaCl_2$	0.2 g
$NaHCO_3$	0.2 g
KCl	0.4 g

This solution should be sterilized in an autoclave or an Arnold sterilizer.

27. LUBRICANT (STOPCOCK). Glycerin prevents sticking of ground-glass parts and is also useful in sealing ground-glass joints to prevent leaking of substances which are insoluble in it (such as ether).

28. LUGOL'S SOLUTION. This solution is used as a test for the presence of starch in food samples or in leaves and as a stain, especially for flagella, cilia, and the nuclei of cells.

Dissolve 10 g of potassium iodide in 100 ml of distilled water; then add 5 g of iodine. (See also I_2-KI solution.)

Gram's iodine stain solution may be made from this formula by adding 14 times its volume of water.

For very delicate work, this solution may be diluted 1 part to 10 parts of water.

29. MILLON'S REAGENT. This reagent is

---

[7] B. Meyer, D. Anderson, and C. Swanson, *Laboratory Plant Physiology,* 3rd ed., Van Nostrand, Princeton, N.J., 1955, p. 165.

[8] Iodine stains may be removed from clothing by washing the stain with a 10 percent solution of sodium thiosulfate in water. Then rinse in water.

[9] F. Emerson and L. Shields, *Laboratory and Field Exercises in Botany,* Blakiston (McGraw-Hill), New York, 1949.

used as a test for proteins. Dissolve 100 g of mercury in 200 ml of nitric acid (specific gravity 1.42). Dilute the solution with 2 volumes of distilled water.

30. MOTION-PICTURE FILM CEMENT. This type of cement has been recommended for splicing on direct positive safety film.[10] It evaporates very quickly; keep the solution in a tightly sealed bottle.

acetone	20 ml
Duco cement	10 ml
propylene oxide	10 ml

31. PHOTOGRAPHIC SOLUTION (NONCURL-ING).[11] Immerse prints after washing in a solution made by combining 12 parts of glycerin, 5 parts of alcohol, and 83 parts of water.

32. PHYSIOLOGICAL SALINE SOLUTION FOR COLD-BLOODED ANIMALS. Use as a mounting fluid in the preparation of temporary wet mounts. Prepare a 0.7 percent solution of sodium chloride in distilled water.

33. PHYSIOLOGICAL SALINE SOLUTION FOR WARM-BLOODED ANIMALS. Prepare a 0.9 percent solution of sodium chloride (dissolve 0.9 g in 100 ml of distilled water).

Plain distilled water is *hypotonic;* that is, cells will swell if placed in distilled water. Salts need to be added to produce an *isotonic* physiological solution. On the other hand, if the solution contains a greater concentration of salts than the contents of the cells, the cells will shrink. This is a *hypertonic* solution.

34. POTASSIUM PYROGALLATE SOLUTION. This solution is used to remove oxygen from air. Prepare the solution by combining 1 part by weight of pyrogallic acid, 5 parts of potassium hydroxide, and 30 parts of water (see also Chapters 3, 6).

35. PRESERVATIVE SOLUTION. Dissolve 1 g of thymol in 100 ml of toluene.

36. RENNIN SOLUTION. Prepare a 0.1 percent solution by grinding 1 g of rennin preparation with 50 ml of water to form a thin paste. Dilute with water to 1 liter.

37. RINGER'S SOLUTION NO. 1—FOR FROG TISSUE. This solution, isotonic for frog tissue, may be used as a mounting fluid for living frog tissue. The heart of a dissected frog will continue to beat for several hours if bathed in this buffered solution. Dissolve the following salts in 1 liter of distilled water:

KCl	0.14 g
NaCl	6.50 g
$CaCl_2$	0.12 g
$NaHCO_3$	0.20 g

38. RINGER'S SOLUTION NO. 2—FOR MAMMALIAN TISSUE. This solution, isotonic for mammalian tissue, may be used as mounting fluid for examination of living tissues. Prepare this solution in 1 liter of distilled water:

KCl	0.42 g
NaCl	9.0 g
$CaCl_2$	0.24 g
$NaHCO_3$	0.20 g

39. SEA WATER (ARTIFICIAL). Humason[12] describes Hale's procedure for preparing sea water with a salinity of 34.33 0/00 and a chlorinity of 19 0/00. Dissolve the following salts in distilled water, and make up to 1 liter. Increase the quantities two- or four-fold as the need demands.

NaCl	23.991 g
KCl	0.742 g
$CaCl_2$	1.135 g
or $CaCl_2 \cdot 6H_2O$	2.240 g
$MgCl_2$	5.102 g
or $MgCl_2 \cdot 6H_2O$	10.893 g
Na_2SO_4	4.012 g
or $Na_2SO_4 \cdot 10H_2O$	9.1 g
$NaHCO_3$	0.197 g
NaBr	0.085 g
or $NaBr \cdot 2H_2O$	0.115 g
$SrCl_2$	0.011 g
or $SrCl_2 \cdot 6H_2O$	0.018 g
H_3BO_3	0.027 g

(See also the formula for marine aquaria described in Chapter 12.)

40. SEED DISINFECTANT. In a description of seed germination (p. 457), we spoke of dipping seeds in Clorox (1 part to 6 parts of water) or dilute potassium permanganate solution (about five crystals to 1 liter of water). Dip seeds briefly into alcohol

[10] Bennett, *op. cit.*
[11] *Ibid.*

[12] G. Humason, *Animal Tissue Techniques,* Freeman, San Francisco, 1962.

(a wetting agent) and then into Clorox for 15 min. (See also p. 652.)

As an alternate technique, the following disinfectant solution has been suggested in the literature.[13] Dissolve 0.4 oz of mercuric chloride in 0.2 qt of boiling water; then add 3 gal of water to this solution. (*Caution:* Mercuric chloride is poisonous.)

Place the seeds in a cheesecloth bag and immerse in the liquid. For most seeds a 5-min immersion is sufficient. More hardy seeds can be immersed up to 10 min in this disinfectant. Then wash the seeds several times in water for 15 min; spread them out to dry.

41. SODIUM CHLORIDE SOLUTION (SATURATED). Add about 37 g of NaCl to 100 ml of water; if the resulting solution is not saturated, add a bit more of the salt.

42. SODIUM CHLORIDE SOLUTION (0.1 M). Dissolve 5.85 g of NaCl in water to make 1 liter of solution.

43. SODIUM HYDROXIDE SOLUTION (1 M). Dissolve 40 g of NaOH in 200 ml of water; dilute with water to 1 liter of solution.

Prepare a 2 M solution by adding 80 g to enough water to make 1 liter of solution.

44. SOIL ACIDIFIER.[14] Prepare the following and add it to the soil until the desired pH is obtained:

flowers of sulfur	1 part
$(NH_4)_2SO_4$	1 part
$Al_2(SO_4)_3$	1 part

45. STARCH PASTE (1 PERCENT). Add a small amount of cold water to 1 g of arrowroot starch and stir into a paste. Then add this to 100 ml of boiling water; stir constantly. Bring this to a boil, and then let it cool. This is a satisfactory strength for general use in demonstrations of salivary digestion.

46. SUCROSE SOLUTION (0.1 M). Dissolve 34.2 g of sucrose in water, and make up to 1 liter. Add toluene as a preservative.

47. SULFURIC ACID (0.1 M APPROXIMATELY). Slowly add 5.5 ml of concentrated sulfuric acid to water, allow to cool; mix well and dilute to 1 liter.

[13] Bennett, *op. cit.*
[14] *Ibid.*

48. TOLUENE-THYMOL PRESERVATIVE. See preservative solution.

49. WOOD STAIN (ACIDPROOF). Laboratory desks and tables may be protected against the action of strong acids. Commercial preparations are available, but the one described here is simple to prepare and inexpensive. First remove varnish, paint, grease, or other chemicals. Then with a paintbrush apply two coats of boiling hot solution 1. Allow the first coat to dry before applying the second layer. Then apply two coats of solution 2 in the same manner.

solution 1	
$CuSO_4$	125 g
$KClO_3$	125 g
water	1 liter
solution 2	
aniline oil (fresh)	150 g
HCl (concentrated)	180 ml
water	1 liter

When the wood surfaces are completely dry, wash off the excess chemicals with hot soapsuds. Then apply a finish of linseed oil, rubbed well into the wood. Later apply other coats of linseed oil and rub down well whenever the wood surface needs it.

Making visual "props"

An effective teacher in the classroom evokes images for children—with words, sometimes with gestures, other times with useful props. There are many kinds of props or aids. Where possible, use living materials—the real props. These may be brought to class, or youngsters may go to the living materials through field trips.

A three-dimensional model may be the next best experience. A slide, a photograph, a chart, a drawing on the board— all two-dimensional—also provide effective next-best experiences.

Only a brief introduction to the making of models, charts, and other props is presented here.

A visual aid should be simple and used sparingly, to clarify something confusing, to bring the attention of all students to the same thing, to provide a new view of material already studied, or as part of an evaluation of learning. Visual aids should not be "inserted" into a lesson because a good model of some sort is part of the equipment of the department. Students can become overwhelmed by seeing many charts or models, complex in design, in one period. (All of us know of some favorite piece of demonstration material we have which we think couldn't be simpler or more to the point. Yet we have found that all students do not have the same perception. They lack a common frame of reference. You may have found yourself explaining the model.)

Models

Real materials. A hen's egg may become a simple model. (Boil it first as a safety measure!) Use it to ask these questions: What kind of an animal arises from this egg? What does it contain that enables it to develop into a chicken, not an eagle? More than that—the egg develops into one kind of chicken—a Plymouth Rock, not a Leghorn.

Other times, students may bring in the realia for the lessons: a beef heart, a sheep's brain or eye, frogs, insects, a beef or hog kidney, beef lungs with trachea (haslet), arteries and veins, seeds, flowers and fruits, leaves showing variations, cocoons, water from a pond, guppies, starfish, mollusk shells, worms, and many other materials which are mentioned throughout this book. On field trips students may study the dynamic interrelationships among many living plants and animals.

Sometimes professionally made models supplement the real materials. Many times they simplify a concept by reducing the real thing to its essentials.

Models in clay. A ball of plasticine may represent a fertilized egg. What will it develop into? How does this one cell develop into many cells? Students can model several stages in cleavage. With a knife divide the ball in two, then each of these in two, to represent the four-celled stage.

The next cleavage is in a horizontal plane; each of the four cells divides and a ball of eight cells results. The clay ball can also be used to show the depression of the neural tube and how it is formed from the outer ectoderm of the embryo.

Clay in different colors can be utilized to make effective models. Each student in class may model pairs of chromosomes in clay. On a sheet of paper they can demonstrate mitotic divisions and the differences between mitosis and reduction division.

Students have modeled many tissue cells, the heart, and the entire respiratory and digestive systems in clay. However, the models fall apart fairly easily so that clay seems to be most useful in manipulating models in class before the eyes of students. However, some students have successfully created dioramas showing farming practices which favor conservation of land, life among the dinosaurs, and many other similar ideas.

Models in glass. You may draw an air sac of the lungs on the board, using colored chalk. A round-bottomed Florence flask is a useful three-dimensional model of an air sac. Have students draw capillaries on it with a glass-marking pencil as they explain the exchange of gases between the capillaries and the air sac. Or tie a meshwork of string around the bulb of the flask to represent a capillary network.

A "villus" can be made by inserting a small test tube (a "lacteal") into a larger test tube (the "outer membrane" of the villus).

Pieces of string, wire, and odds and ends. Tease apart some fibers in a square of gauze. Doesn't this resemble a capillary bed? Or use a length of hemp cord. Separate the fibers in the center section to represent capillaries. Students may dip one end of the cord in blue ink and the other end in red ink. This gives an idea of the conventional artery-capillary-vein representation.

Recall the rubber change mat in stationery stores? All the rubber projections may be considered villi in the lining of the small intestine.

Many kinds of plastic foam can be used for making models.[15]

We say that food materials pass along some 25 to 30 ft of digestive tract. Have you ever brought to class a length of clothesline and asked students to estimate 30 ft? Have students stretch out the clothesline along the length of the room to show 30 ft. Now, what happens to food as it passes along this long tube?

In Chapter 9 (see Fig. 9-26) a description was given for showing animated "chromosomes" in mitosis, using short pieces of electrical wiring, rubber bands, and string.

How are genes arranged on a pair of chromosomes? Are the same genes present in all the different pairs of chromosomes? Wooden beads of different colors may be threaded on wires. Make up pairs of different lengths with different-colored beads. Poppits are excellent props. These are plastic beads which fit into each other (and are not on thread) and can be pulled apart and remade into strings of beads of any length. Many chromosome aberrations can be demonstrated with these beads. With different-colored plastic beads students can show translocation of chromosome parts, fragmentation, crossover, and (with the help of some modeling clay or glue) nondisjunction of chromosome pairs. A dab of colorless nail polish will effectively hold the pairs together. A limitless variety of materials may be used in modeling a molecule of DNA (see also Chapters 8, 9).

Here is another simple device often used. Bring in a roll of the hard candies (Reed's) which are biconcave discs. These may represent red blood cells.

A cell model. Students may use agar to make a model of a cell. Prepare some agar in cold water, bring it to a boil, then let it cool. The agar may be packed in a cellophane "cell membrane." Embed colored marbles or other small particles in the cooling agar to represent nucleus, centrosome, and so forth. Or, if you wish to make a model of a plant cell, use green peas for chloroplasts.

Students can improvise in many ways. Suppose an indicator, such as phenolphthalein, is added to the agar medium as it is being prepared. Later on, if ammonium hydroxide is placed near the finished model of the "cell," diffusion of the alkaline into the "cell" will produce a pink color within the "cell." This offers a useful demonstration of diffusion through a cell membrane.

An artificial cell. To a test tube containing a small amount of dilute albumin from a raw egg, add a few drops of chloroform. Shake the materials together, mount a drop on a slide, and observe the artificial "cells" which form.

At times you may have reason to give the following demonstration. All these materials are, of course, nonliving, but this demonstration may help students visualize how a cell grows. Place a crystal of copper sulfate in a solution of potassium ferrocyanide. As the copper sulfate passes into solution, a membrane is formed around the crystal. This is a membrane of copper ferrocyanide which forms on contact with the surrounding potassium ferrocyanide. Watch the membrane expand and simulate the growth of a cell as water continues to pass through the membrane in one direction only. Note that the equal expansion in all directions is due to the entrance of water, not to the formation of new material (corresponding to new protoplasm which would occur in a living cell).

Models of plaster of Paris. Plaster of Paris may be poured into a mold shaped of plasticine, cardboard, or metal tins. Shape the plasticine to form a mold for a cell, such as an epithelial cell. If there are any parts which are to protrude in plaster, they should be indentations now in the plasticine. For example, indent the region where a nucleus should be in the cell. Build up a cuff—the outer edges of the cell mold—about 1¼ in. high. Apply

[15] Suggestions for urethane foam for individual student use are given in D. Naiman and A. Katz, "Production of Disposable Brain Models," *Turtox News* 42:8, August 1964.

petroleum jelly to the clay if needed. Then prepare the plaster of Paris. Add enough plaster of Paris (a good quality such as dental plaster) to water to form a consistency that is thick and creamy but thin enough to pour. Stir carefully to avoid introducing air bubbles to the plaster. Pour the plaster of Paris into the mold slowly so that air bubbles are not formed. Allow this to set and harden. Later peel away the clay mold. If desired, these plaster models may be smoothed down with fine sandpaper; they may be shellacked and painted different colors.

Professional-looking models may often be made (as a club activity).[16] These may become part of the stock of the science department. In the legend accompanying the photographs in Fig. 14-2 are the directions for especially fine models. Latex models may also be made; they have the advantages of being lighter in weight and less fragile. Or you may want to use the rubber liquid now available at arts and crafts suppliers; this liquid may be used to make molds of various sorts. The advantage of these rubber molds is that they can be used again and again.

Models in papier-mâché. In this method, strips of toweling or newspaper are dipped into a thin paste made of flour and water. Or strips of newspaper soaked in water may be worked over to make a malleable thick paste. Add 1 percent phenol to prevent spoilage. Then add this newspaper mash to flour and water to make an adhesive material. Construct the framework of the object to be modeled— such as a dinosaur, a cell, or cleavage stages—of cardboard and wire props. Then apply the mash of flour-soaked newspaper or apply overlapping strips of toweling soaked in flour paste. Apply successive layers after each preceding layer has dried thoroughly. About six or seven layers are needed. When the model is thoroughly dry, smooth it with fine sandpaper. These models may be painted in bright colors.

Models in wood. Some students are talented in working with wood. Fine lightweight models may be made of balsa wood. Students have made three-dimensional models of microorganisms, cells, and the heart. These can have the added advantage of hinged halves so that the model can be opened, revealing brightly painted interiors—a section through an amoeba or the chambers of the heart or the tubes of the kidneys.

With a jigsaw students can cut two-dimensional models—or, rather, plaques —of organs of the body or of tissue cells.

Models in clear plastic. Attractive, convenient models are available of plant and animal specimens mounted in clear plastic (see Figs. 14-3, 14-4). You may purchase models of representatives of the basic plant and animal phyla. Some students may want to try to duplicate these professionally made models. Most biological supply houses sell the materials required for construction of these plastic models and provide complete instructions. Clear plastic is a tricky medium to work in, but some students have, after practice, made better-than-average models which could be used in teaching. Following is a summary of the basic procedure to give you a notion of it.[17]

Specimens are fixed in Bouin's fixative or formaldehyde. Then the fixative is washed out with water. Next follows staining; acid carmine is a successful stain to use. Then the specimens are transferred through alcohols for dehydration: 70 percent, 85 percent, 90 percent, then absolute alcohol. Now the procedure begins to differ from regular microscope work. The alcohol is removed by transferring the specimens to anhydrous ether. Then the specimens are transferred to the uncatalyzed monomer and put into a vacuum dessicator to remove the ether. After the slow impregnation of plastic into the specimens, a catalyst is added to the plastic. Now additional plastic is poured

---

[16] General Biological Supply House offers directions and a kit of plastic materials (negative and positive modeling compounds).

[17] T. Romaniak, "The Use of Unsaturated Polyester Resins for Embedding Biological Material," *Ward's Natural Sci. Bull.* 20:39–42, January 1947.

1. The subject, an early embryo stage, is modeled in Plasteline (this one by Dr. J. Wilson, U. of Rochester Medical School) and then shellacked.

2. When dry, the model is covered with tissue paper (to protect it) and then completely covered with a ¾-inch layer of modeling clay.

3. A cardboard wall divides the clay layer into 2 equal parts. A coat of lard oil is brushed over all.

4. A ¾- to 1-inch layer of plaster is then built on one side up to the dividing wall. When it has set, the wall is removed and the edge of the plaster is smoothed.

5. Registration notches are cut in the edge, which is then brushed with lard oil, and plaster is applied to the other half in the same way.

6. The 2 halves of the jacket are separated, and the clay and tissue paper removed. A notch is cut in the top of each half. The inside of the jacket is shellacked, and, when dry, oiled and placed around the lightly oiled model.

7. The 2 halves of the jacket are tied together tightly, and molding glue is poured in, completely filling the space between the jacket and model.

8. One half of the jacket is removed, by prying with a screw driver. Then the glue is cut through along the plaster edge, so that the free half of the glue mold can be removed.

9. Then the Plasteline model is removed, and the inside of the glue mold is dusted with French chalk, then brushed with a saturated alum solution.

10. After it has cured for an hour, the inside of the mold is brushed with a mixture of stearine and kerosene, the halves are tied together, and plaster is poured in through the bottom.

11. It must be removed as soon as set, or the heat will melt the glue. The 2 halves of the mold are carefully separated, and the cast is removed.

12. The finished model. The glue mold can be used to make up to 20 casts, with care, by repeating the last two steps.

Fig. 14-2 Procedure for making professional-looking plaster-of-Paris models. These photographs and directions were supplied by Ward's Natural Science Establishment, Inc., Rochester, N.Y.; further instructions for sandpapering and applying lacquer, as well as formulas for the materials used, are given in *How Models Are Made, Ward's Natural Sci. Bull.*, No. 2.

into a tray until it gels sufficiently to support the specimens. After several hours a second layer is poured over the first. The specimens are then transferred into the plastic; this is placed in an oven for several hours. Then the blocks of plastic are left to cool and, finally, buffed and polished.

Preparation of skeletons

There may be students who want to prepare a skeleton of a small bird or mammal. A full description of the technique is beyond the scope of this book. However, students can get assistance from the biological supply houses.[18] In general, the technique involves these procedures: (1) Cut away as much of the flesh as possible from the bones and let the bones dry; (2) scrape the bones clean with special bone scrapers, and place the skeleton in an ammonia bath for several days; (3) bleach the skeleton in sodium hypochlorite, and then dry it; and (4) treat the skeleton with

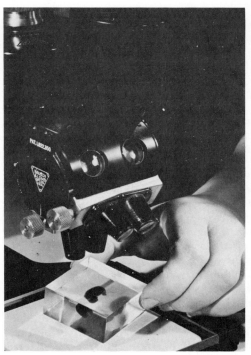

Fig. 14-3 Examining a plastic-mounted specimen under a binocular microscope. (Courtesy of Ward's Natural Science Establishment, Inc., Rochester, N.Y.)

[18] Ward's Natural Science Establishment, Inc., Rochester, N.Y., offers an excellent service bulletin: W. Kruse, *How to Make Skeletons*. There are careful descriptions with photographs showing skulls, ligamentary skeletons, and how to mount a skeleton.

(a)

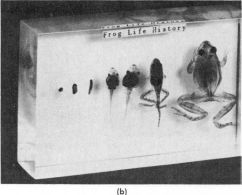

(b)

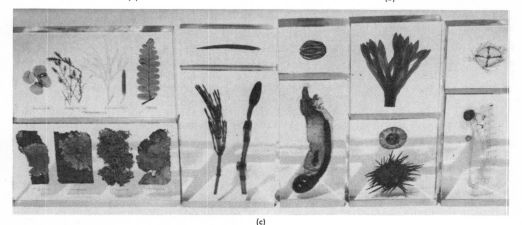

(c)

Fig. 14-4 A variety of specimens mounted in plastic. (a, courtesy of Ward's Natural Science Establishment, Inc., Rochester, N.Y.; b, courtesy of Carolina Biological Supply Co.; c, courtesy of General Biological Supply House, Inc., Chicago.)

carbon tetrachloride so that the oils are removed from the bones. Good ventilation, preferably a hood, is needed for each of these steps. All this is preliminary to the mounting of the bones to make a skeleton.

An acrylic spray may be used as a protective covering for skeletons.

Staining skeletons of chick embryos. Rugh[19] suggests a modification of the method of Spalteholz for staining skeletons of embryos of chicks that are more than 10 days old. This technique may also be used for adult amphibians or small fish.

Fix specimens for 2 weeks in 95 percent alcohol. Then transfer the hardened speci-

mens to 1 percent KOH. After 24 hr, transfer to tap water, and, with forceps, remove most of the fleshy tissue. Transfer the specimens to 95 percent alcohol for 3 hr; change into fresh alcohol for 3 hr more. Immerse the specimens in ether for 2 hr to dissolve fat tissues (or use acetone if there is little fat).

Again transfer to 95 percent alcohol for 6 hr, changing the alcohol once within this time. Then transfer to 1 percent KOH for 6 days. Immerse in alizarin red "S" for 12 hr, transfer to 1 percent KOH for 24 hr and into 10 percent NH_4OH for further clearing for the next 24 hr.[20] The NH_4OH

[19] R. Rugh, *Experimental Embryology: Techniques and Procedures,* 3rd ed., Burgess, Minneapolis, 1962, p. 417.

[20] R. Rugh, personal communication, October 20, 1964.

neutralizes the medium and tends to preserve the color longer. The resulting specimens should be transparent with stained skeletal elements. Store the specimens in 100 percent glycerin.

Humason[21] offers formulas for solutions and detailed procedures. For small fish, she suggests that the scales be removed and the specimens be fixed in 70 percent alcohol. Allow the specimens to remain in the fixative for several days, until hard. Small embryos of birds or mammals should be fixed in 95 percent alcohol after hair or feathers have been removed. Allow to remain in fixative for 3 days.

After the specimens have been in fixative, rinse in distilled water and then transfer to 2 percent KOH for 4 to 48 hr, depending on the size of the specimens and how long it takes for the skeleton to begin to show through the musculature. When the specimens are clear, transfer them to alizarin red working solution (see below for preparation). (It may take from 6 to 12 hr for the skeleton to be stained red; it may also be necessary to add fresh solution.)

Transfer the stained specimens to 1 or 2 percent KOH for 1 or 2 days to decolorize the soft tissues. The action can be speeded by using a lamp or direct sunlight. Next, clear the specimens in clearing solution No. 1 (described below) for 2 days; then transfer them to clearing solution No. 2 for 1 day. Immerse the specimens in pure glycerol (thymol may be added as a preservative); mount them on glass slides in museum jars or store them in bottles sealed to prevent evaporation.

Humason describes the improvement of Cumley in producing specimens with greater clarity. Gradually replace the glycerol with 95 percent alcohol, then absolute alcohol. Then transfer the specimens into toluene. Prepare a solution of naphthalene saturated with toluene and immerse the specimens in this for a few days; store specimens in anise oil saturated with naphthalene.

Commercial preparations are available from supply houses, mounted in either glycerin or clear blocks of plastic. The following solutions are needed for the procedures described above.

1. ALIZARIN RED "S" STOCK SOLUTION. Humason[22] recommends the method of Hollister. Prepare a saturated solution of alizarin red "S" (C.I. 58005) in 50 percent acetic acid. To 5 ml of this saturated solution, add 10 ml of glycerol and 60 ml of chloral hydrate (1 percent aqueous).

2. ALIZARIN WORKING SOLUTION. For use, dilute the alizarin stock solution, adding 1 ml of stock solution to 1000 ml of 1 or 2 percent KOH in distilled water.

3. CLEARING SOLUTION NO. 1 (OF HOOD AND NEILL). Combine the following:

KOH (2 percent)	150 ml
formalin (0.2 percent)	150 ml
glycerol	150 ml

4. CLEARING SOLUTION NO. 2 (OF HOOD AND NEILL). Combine the following:

KOH (2 percent)	100 ml
glycerol	400 ml

Blueprints of leaves

You may want to have some students prepare a file to show variety among leaves. Or, further, they may gather and file examples to serve for identification of trees and shrubs. This is one step in learning to identify plants, and it may result in an attractive display of materials.

One way is to prepare leaf skeletons which can be taped between two glass slides and projected on a screen for class study. In another method, which young children also enjoy, blueprints are made of leaves, as follows. Tape one side of a 6 × 6 in. square of cardboard to one side of a sheet of glass cut to the same size. Insert a sheet of blueprint paper of the same size or smaller between the glass and cardboard backing. Lay a leaf on the blueprint paper and hold it flat by pressing the glass over it. Expose this to sunlight. A few

[21] Op. cit., pp. 172–74.

[22] Ibid., p. 173.

trials will indicate the length of time needed. Remove the blueprint paper and rinse in water. Allow to dry, then flatten under a weight.

Outlines of leaves may be made on photographic paper, but the developing of the print is more time-consuming.

Skeletons of leaves

Students may wish to prepare skeletons of leaves that may be mounted between lantern-slide plates for projection. Such slides can be used for a series of discussions of plant identification.

It has been suggested[23] that leaves be immersed in a slow-boiling macerating solution for about 2 min, until the leaves turn a dark brown. The solution can be prepared as follows. Boil together 16 oz of water, 4 oz of sodium carbonate, and 2 oz of calcium oxide. Cool and then filter. If you find it necessary to convert ounces to another unit of measure, see the tables in Appendix F.

Transfer the leaves from the slow-boiling solution into a wide, shallow tray containing a small amount of water. Rub the leaves with a soft brush to remove the tissue from the supporting tissue or veins. If the tissue does not separate easily from the veins, return the leaves to the macerating fluid.

Bleach the resulting skeletons of leaves by immersing them for a few minutes in the following solution. To 1 qt of water, add 1 tbsp of chloride of lime and a few drops of acetic acid. When the skeletons have turned white, they can be pressed and mounted.

Dried specimens

From time to time, students may add properly dried algae, fungi, evergreens, and flowering plant specimens to the botanical collection of a biology department. Cambosco Scientific Co. (address in

[23] *Procedure for Preparing Leaf Skeletons,* General Biological Supply House, Chicago.

Appendix C) has distributed a fine booklet: I. M. Johnston, *The Preparation of Botanical Specimens for the Herbarium.* Most supply companies sell collecting cases, plant presses, driers, and mounting sheets.

Acrylic sprays form useful protective films to prevent fragmentation of delicate specimens of algae, leaves, and spore prints of fungi that may be in the school's herbarium collection. (*Caution:* Do not inhale the fumes of the spray.)

Freeze-drying of specimens

Many biological specimens may be frozen solid so that ice crystals replace the tissue fluids. They then can be dried at sub-zero temperatures, resulting in permanently rigid, dried specimens that exhibit little shrinkage.

The Smithsonian Institution (Washington 25, D.C.) offers an informative leaflet: R. Hower, *Freeze-drying Biological Specimens,* Information Leaflet 324. In it, the author describes how specimens can be placed in a deepfreeze so that tissue fluids are frozen and, as ice crystals, sublimed into a vapor, thus preventing shrinkage. This procedure requires temperatures of $-15°$ to $-20°$ C ($5°$ to $-4°$ F) during the drying process. A vacuum pump and cold trap, or condenser, are also required.

Taxidermy

There are many procedures for preserving skins of small animals. You may want to offer students an informative bulletin by L. McCain, *Brief Directions for Taxidermy Procedures and Animal Preparation* (Smithsonian Institution, Washington 25, D.C.). In it are lucid directions for the preparation and mounting of skeletons of small mammals, birds, snakes, and fish. The bulletin also offers an introduction to the tanning of hides and snake skins and the mounting and coloring of fish. Included are an excellent bibliography, a directory of courses in taxidermy, and names of taxidermists.

Radiobiology in the high school laboratory

Because there are so many fine references available to teachers in this specialized field, only a cursory attempt has been made in this book to describe methods of using radioactive tracers. Many of the supply houses offer suggestions. For example, General Biological offers a series of five investigations that may be done in high school, using bacteria, fish, and plants; the preparation of radioautographs of fish skeletons, leaves, and colonies of bacteria and molds is carefully described.

The U.S. Atomic Energy Comm. (Washington 25, D.C.) has prepared a number of guides for teachers, for example, *Laboratory Experiments with Radioisotopes for High School Demonstrations* and *Radioisotopes: Uses, Hazards, Controls*. Other references are suggested in footnotes in Chapter 3. Several fine papers in *Turtox News* (General Biological Supply House) describe procedures that may be repeated in class. Teachers may have in their files these papers: C. Levy, "Classroom Experiments in Radiobiology: The Effects of Radiation on Dormant Seeds" (*Turtox News* 40:14–15, 1962) and V. Carr and Brother T. Hennessy, "Use of Radioisotopes in High School Biology" (*Turtox News* 38:242–46, 1960).

Slides and transparencies

Complete libraries of slides and transparencies are available from many biological supply houses (Appendix C). Habitat studies, field work, parasitology, and all the other aspects of biology studied in high school and college are beautifully represented in either slides or transparencies.

At times, you may find it useful to flash slides, perhaps of tissues, on the blackboard. A student may outline the projected cells in chalk on the blackboard. When the projector is turned off and the lights are turned on, the image will still be on the blackboard, where it may be examined in detail. This may help to develop some point in a lesson.

Students may also be taught to accompany their talks with demonstrations. They will need to use a projector. Or it may be easier to use overlays and overhead projectors.

With a glass-marking pencil or with Higgins inks, students may make simple drawings on dry, clear, glass slides (2 × 2 in. or 3¼ × 4 in.). More successful slides may be made on ground glass using pencil, India ink, or crayons. A three-dimensional effect can be achieved with shading, or a blended color effect may be obtained by moistening crayons, very slightly, with water.

In addition, small, thin specimens, such as skeletons of leaves, pressed flowers, or butterflies may be inserted between two slides and bound with gummed tape (see also p. 674). General Biological offers a service leaflet on the making of lantern slides.

Overhead transparencies

Because of their ease of preparation and dynamic flexibility in use, overlays made by students add a dimension to a lesson.

Overhead transparencies covering many areas of biology are available in booklets. These are now listed in catalogs of supply houses. For example, Ward's offers an imprinted ditto master for running off unlabeled duplicates of its *Dynavue* series.

Tissue mounted on 35-mm film

As an alternative to glass coverslips and microscope slides, film can be used for mounting tissue. Tissue can be mounted directly on flexible plastic strips of perforated 35-mm film, without emulsion, cut into 3-in. lengths or in lengths to roll on reels. The resulting "slidestrips" can then be observed under a microscope or projected with a filmstrip projector.

This "slidestrip" technique is described in *Carolina Tips* (27:4, April 1964). Carolina Biological also supplies a "slidestrip kit," which is a technique developed by V. Bush (1955) and improved by several other workers. In brief, paraffin sections of tissue are placed in a drop of water on a length of 35-mm film, without emulsion, to which albumin fixative has been applied. The sections of tissue and film are placed on a warming table to spread the tissue. After drying, the tissue and film are transferred to xylol to remove the paraffin. After being passed through alcohols, the tissue is stained, dehydrated, and, finally, cleared in xylol. Several layers of an acrylic spray are applied to the tissue while the material is still wet with xylol.

Prepared slides of serial sections of early embryos of the pig and the chick, as well as series in physiology, are available from biological supply houses (see listing in Appendix C).

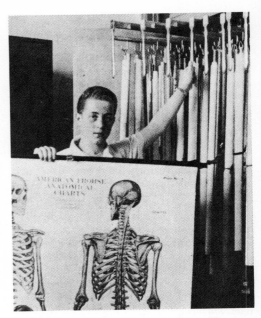

Fig. 14-5 Cabinet for storing charts. (Courtesy of Forest Hills High School.)

Making charts

Drawings made on the blackboard with chalk of different colors are effective when they are developed with the class as part of a lesson.

Or permanent charts can be prepared by drawing on window shades with ink and crayons or on muslin with wax crayons. When the wax-crayon drawing is finished, cover it with brown wrapping paper and iron the paper. The wax will melt and be absorbed into the muslin. (Protect the ironing board cover with paper too.) When they become soiled, these charts can be washed. The main advantage of these drawings on muslin is that they can be folded and kept in envelopes in a filing cabinet. Sew a 1-in. hem at the top of the chart and use metal eyelets so that the chart can be hung from hooks, or use thumbtacks. If you have a storage cabinet for other charts you may prefer to nail the muslin charts onto wooden rods like professional charts.

Supply houses provide professional charts of all kinds, and they also offer teachers leaflets and bulletins describing ways to make good-looking charts. All your charts may be stored in a cabinet such as the one in Fig. 14-5. Here a label is affixed to one end of the chart along with a loop of wire. Hang the charts according to your own filing system. Charts last longer in such a cabinet, since they are not handled or knocked about.

A picture file

Students readily accept the responsibility for maintaining a current, interesting bulletin board. You may want to train a squad of students to establish a picture file. Give them file folders and space. Excellent pictures (and reference reading materials) may be found in *Life, Scientific American, Natural History, National Geographic,* and similar magazines and bulletins. Valuable reprints are available from *Life.*

Students will soon have complete sets of bulletin board pictures for any topic in the year's work: behavior, physiology, heredity, disease prevention, conservation practices, reproduction, evolution, plant-animal relationships, mimicry, and camouflage.

Addenda

Making dissection pans. Rectangular metal cake pans of a uniform size make useful dissection pans. Melt together equal parts of paraffin and beeswax over a low flame. If a black wax is desired, add lampblack to the melted preparation. Then pour into metal pans to a depth of ½ in. and set aside to harden. Avoid stirring, for this creates air bubbles. Should air bubbles appear, they may be broken with a hot dissection needle while the wax preparation is still warm.

Labeling rock specimens. Sufficiently large rock specimens may be labeled by painting on a plaster of Paris surface, as follows.[24] Mix plaster of Paris and water to a creamy consistency. Apply a thin, smooth layer on the side of the specimen which is to be labeled. When the plaster has "set," apply a thin solution of glue as sizing over the surface. Labels may be printed with a permanent ink on the sized dry surface of the plaster. Then protect the label by brushing over the dry ink with melted paraffin. Very small specimens may be partly embedded in plaster.

Permanent labels for specimen bottles. Sometimes, to avoid smearing, printed labels can be placed inside a jar containing specimens preserved in alcohol.[25]

Print labels with India ink and set aside to dry until the ink no longer glistens; dip them into a jar containing 5 or 10 percent glacial acetic acid (use forceps). Then drain the labels on blotting paper and insert into the specimen jars containing 70 percent ethyl alcohol. Rather strong paper should be used for labels because glacial acetic acid tends to soften the paper.

24 Bennett, *op. cit.*
25 G. J. Spencer, "Rendering India Ink Labels Permanent in Alcohol," *Turtox News* 26:1, January 1948.

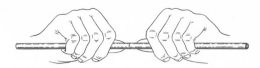

Fig. 14-6 Correct way to hold glass tubing to break it after it has first been scratched.

Working with glass

Cutting glass[26]

Students will be most successful when working with thick-walled glass tubing about 6 mm in diameter. With this tubing they may make capillary pipettes and glass bends (see also Chapter 9).

Cut glass by applying one firm scratch with a triangular file on a section of glass held securely on the top of a table. Then hold the tubing as shown in Fig. 14-6. (As a precaution, students should wrap each end with paper toweling or a soft cloth.) Hold the scratch away from the body; place the thumbs behind it. Pull on the tubing with the outside fingers and at the same time exert a forward push with the thumbs. An even break in the tubing should result. Fire-polish the cut ends in a flame. The bore of the opening of the tubing may be altered while the glass is heated. Overmelting of the glass will reduce the size; the opening can be increased by inserting the round end of the triangular file into the melted glass opening.

Making a capillary pipette

Cut glass tubing into sections about 7 to 9 in. in length; fire-polish both ends. Many sections of this tubing can be kept on hand until they are needed for making pipettes.

Between the thumb and first two fingers of both hands, rotate a center section of a length of glass in a flame until the glass is

26 Also refer to a companion volume, A. Joseph *et al., Teaching High School Science: A Sourcebook for the Physical Sciences,* Harcourt, Brace & World, New York, 1961, Chapter 25, for techniques in handling glass, metal, and plastics.

soft; remove the glass from the flame and pull it out. Hold it in position or place it on an asbestos sheet until it has hardened. Too rapid a pull will produce capillary sections with too narrow a bore. Cut off the lengths desired and carefully fire-polish.

When a long, tapered capillary tube is desired, rotate the glass tubing over a wing top or flame spreader.

The pulling out of soft glass test tubes is described in Chapter 6 in the section on making a microaquarium (see Fig. 6-2).

Bending glass

A good bend retains the same size bore throughout the bend. Rotate tubing or glass rod holding both ends lightly, and turn the tubing evenly so that all sides are heated equally. Use a wing top when a broad U-shaped bend is desired so that a larger area of glass may be heated. Remove the tubing from the flame when it starts to sag, and bend it to the desired shape, applying equal pressure with each hand. Hold the tubing in one plane until it hardens. Some workers remove the sagging tubing from the flame and take advantage of the natural sag by raising the ends upward so that the sag hangs down in a bend. Others lay the sagging tubing on asbestos and make the bend from a flat surface, thus ensuring that the bend is in one plane. A good right-angle bend can easily be made in this way by using the corner of the asbestos pad as a pattern for the right angle. When bends of a specific angle are needed, cut a pattern or draw one on asbestos. In any event, set the tubing aside on asbestos to cool. Students should learn to touch glassware cautiously before they grab it—hot glass looks exactly like cold glass!

Glass rods can be drawn out to varying thicknesses. Fine threads of glass are used for glass needles for careful work in embryology where bits of tissue are transplanted from a young embryo into an older one (see Chapter 9).

Glass stirring rods may be made easily:

Cut glass rods to desired lengths (depending on the size of the beakers used) and heat both ends (one at a time) to melting; while the rod is hot, press it against a sheet of asbestos to flatten the ends a bit.

A U-shaped tube: a simple manometer

Sections of glass tubing may be used as a manometer. Students may make one from 1-mm glass tubing. Heat the tubing over a flame spreader or wing top and shape into a symmetrical U-tube. At each end of the tube make a right-angle bend, add mercury or colored water, and mount the "manometer" on a board against a sheet of graph paper (Fig. 14-7).

Rubber tubing can be attached to one end in use.

Curved pipette

The curved pipette shown in Fig. 14-8 is useful in transferring eggs of frogs or snails, protozoa, or small crustaceans from one finger bowl to another. It fits into the curve of the finger bowl, so that the tip does not break off so readily.

Rotate a long section of glass over a fish tail or wing top. When the glass begins to sag, remove it from the flame and begin to make a U-shaped bend. However, quickly

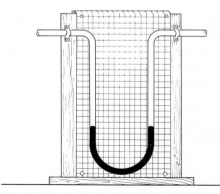

Fig. 14-7 A U-shaped tube mounted against graph paper to serve as a simple manometer. The tube may be filled with mercury, as shown, or with colored water.

Fig. 14-8 A curved pipette convenient for transferring delicate specimens, such as frogs' eggs, out of small finger bowls.

flick the wrists so that the palm of each hand is turned slightly inward forming a slight curve in another plane. After the glass cools, cut it in the center of the U-bend so that two pipettes result. Fire-polish the ends. The size of the bore should be made to fit the materials to be transferred; for example, it should be large if frogs' eggs are to be transferred.

Making a wash bottle

A wash bottle containing distilled water is a useful piece of equipment for careful work in the laboratory.

Students will need a flat-bottomed Florence or Erlenmeyer flask, a two-hole stopper, several lengths of glass tubing, and a 4-cm section of rubber tubing to which a small glass nozzle is attached.

Prepare the glass bends and assemble as shown in the completed wash bottle (Fig. 14-9). Both the glass tubing and the stopper should be wet while the glass tubing is being inserted into the stopper. Students should use toweling and hold the tubing next to the region inserted into the stopper. With a gentle twisting motion, the glass tubing can be inserted into the stopper. The glass tubing is safer to handle when the ends have been fire-polished beforehand.

Cleaning glassware

Most glassware can be washed with warm water and a detergent, but when experiments call for chemically clean glassware, extra precautions must be taken. After glassware has been washed by ordinary means, place it in a hot dichromate cleaning solution. This is prepared

by pouring 1 liter of concentrated sulfuric acid into 35 ml of a saturated aqueous solution of sodium dichromate. (*Caution:* Never pour the aqueous solution into the acid; avoid spilling on clothes or skin.)

After the glassware has been standing in this solution for several hours or overnight, rinse it continuously in warm tap water until all the chemical solution has been removed.

Glassware is clean when water completely wets the glass, forming a continuous film; on glassware that is not clean, water collects in drops on the surface.

Greasy and tarry materials on glassware may be removed by soaking the glassware in an alcohol–sodium hydroxide solution. This is prepared by dissolving 120 g of sodium hydroxide in 120 ml of water and diluting to 1 liter with 95 percent ethyl alcohol. Afterwards, rinse the glassware in tap or distilled water.

Repairing damaged glassware

At times, some sharp, uneven edges of pipettes, flasks, beakers, and test tubes may be repaired for subsequent use by filing the chipped edges or rims and then fire-polishing. It may be necessary to shape

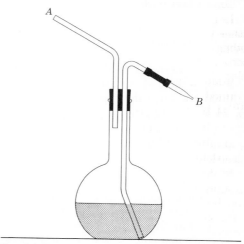

Fig. 14-9 A wash bottle can be used in two ways: Water can be simply poured out of A, or a fine stream of water can be obtained from B by blowing into A.

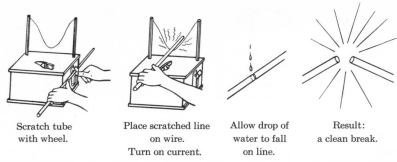

Scratch tube
with wheel.

Place scratched line
on wire.
Turn on current.

Allow drop of
water to fall
on line.

Result:
a clean break.

Fig. 14-10 Apparatus using electrically heated wire to cut glass. The unit shown operates on 115-volt AC, and a transformer controls the temperature of the cutting wire. (From W. M. Welch Manufacturing Co., Chicago.)

the opening with the blunt end of a glass file.

Large battery jars which have become cracked and useless for holding liquids may have a life as a terrarium if they are inserted into a block of plaster of Paris for support.

In the laboratory and classroom broken or irreparably cracked glassware should always be disposed of in crockery jars or in other special containers. Glassware should not be discarded haphazardly in wastepaper baskets.

Cutting glass with wire

Long-necked glassware or tubes which have been chipped can be cut off to serve other uses in the laboratory. Nokes[27] describes the following method for preparing electrically heated wire to cut glass.

Build (or buy) the wood stand as diagramed (Fig. 14-10). Connect a length of # 24 B & S gauge Nichrome wire between the two supports. Connect this with a suitable resistance to the main supply of electricity or to batteries.

Make a circular cut with a glass file around the circumference of the tube to be repaired. Then roll the tube along the hot wire so that the entire scratch or cut is heated. When the crack is completely made, the tube can be removed. Allow a drop of water to fall on the crack, and a

[27] M. O. Nokes, *Modern Glass Working and Laboratory Technique,* Chemical Publishing Co., New York, 1950.

clean break will result. Soft glass, up to a diameter of 2 cm, may be cut in this way. Larger glass tubes may need heat applied to the entire circle at the same time. Here a loop of Nichrome wire, attached to insulated handles, is more desirable.

Using student resources

Individual students have certain talents which a teacher can put to use in many ways. In fact, almost every teacher has a student following of a sort. In the beginning, a teacher may plan a program of activities with this unorganized group of students who probably have many common interests. In this way a squad of students may be born.

How can a squad, that is, a small group of students, help an individual teacher or several teachers who make up a department? And how, in turn, does this activity help these students to grow in competence?

Co-curricular and extracurricular activities in school provide opportunities for boys and girls to practice some of those so-called required learnings or developmental tasks that they all need to develop —skill in getting along with others, peers and adults; trying out one's abilities; getting practice in making judgments; self-evaluation concerning possible success in science as a vocation; manipulating tools of the scientist; gaining specific skills and a better understanding of what science is.

Many kinds of activities in which students may participate are listed here. They vary in their requirements for special skill or high ability. However, each kind of job has a status-giving quality for some child and may help him retain his individuality. Often through these informal relationships teachers come to know their students better; this rapport places the teacher in an effective role as friend and counselor.

Such activities benefit students in yet another way. Many teachers describe students' work in anecdotal accounts which are filed with the permanent cumulative records of students. These are useful later as recommendations for college or for future employers. Several schools have student governments which sponsor school service programs and allot service points for all kinds of work done in the school and community. The accumulation of 50 points may lead to a certificate, 75 points to a service pin presented at commencement.

What are some ways that teachers have used students as resources to increase the effectiveness of activities in the stockroom, the laboratory, and the science classroom? Let us begin by describing the work of students in squads in the laboratory and stockroom.

Laboratory squads

Probably the most pressing need for science teachers is assistance in the daily routine of getting materials from the stockroom, preparing solutions, setting up demonstrations or experiments, purchasing perishable materials, dismantling equipment, washing glassware, and storing materials in some organized way. Student assistants, as *organized* laboratory squads, can be trained in these routine tasks. The work may be done before school, during school hours, or after school. Students can be helpful in the bookkeeping job of keeping track of equipment, mainly expendable materials. This is a great help when it comes to ordering supplies and taking inventory.

In a small school the science teacher probably prepares his demonstrations on the run, for he carries a heavy teaching load. A squad of students can assist him at some assigned time that is mutually convenient. Students must, of course, work under a teacher's supervision.

In larger schools many procedures can be used. Where several science teachers share a common stock of materials and equipment, one teacher may be responsible for ordering supplies and taking inventory. In large school systems laboratory assistants have this special responsibility. Any one plan we describe cannot be feasible for all schools, large and small; teachers will select the possibilities which best meet their own needs. Specifically, what goes on in a supply laboratory and stockroom in a school?

Filling daily orders. Students may fill "orders" that teachers have submitted in advance of their classes. However, this presumes that teachers in the department have organized themselves and have developed a routine for making out orders for class teaching materials. This plan necessitates submitting orders a day in advance to a stockroom or laboratory center where the squad can have access to materials and can fill the orders: preparing solutions, obtaining models, and/or setting up demonstrations.

When teachers pool their own resources, this kind of procedure can be put into action. Under the guidance of a teacher (or a laboratory assistant), students can deliver this class order to the teacher's room when it is needed. When duplicate equipment is available, the teacher might get his materials in the morning and hold them for the day. Large pieces of equipment, such as microscopes, a manikin, dissection pans and kits, or a microprojector, may be delivered to the teacher at the beginning of the period. Then the material can be picked up again toward the end of that same period and returned to the laboratory or supply center. Students trained as a laboratory squad can make the deliveries and pick-ups.

How do teachers send in their daily orders for supplies to the supply center in the department? You and your colleagues may want to initiate a standard system in which each teacher submits his order on uniform-sized paper, a day in advance. You may want to use the pattern shown in Figs. 14-11 and 14-12. Teachers can check their orders against those of others; or at the end of the day one squad of students can examine the orders. When two teachers order the same materials for the next day, they are informed of this and they agree to share the materials or one accepts a substitute or possibly changes his lesson for the next day. The advantage is that the teacher knows beforehand whether he will get the materials he needs so that he has time to arrange another lesson.

That afternoon (or the next morning), students locate the materials that teachers want for the next day and stack small items on a tray, such as a lunchroom tray. Or they place large pieces or heavy equipment on a cart which can be wheeled through the halls to deliver to the teachers at the start of the day or the lesson for which the materials are needed. (Bottles

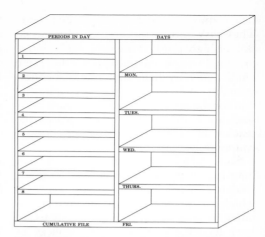

Fig. 14-12 Student-made wooden cabinet with shelves for teachers' daily order slips.

of strong acids and bases should be carried in wooden boxes as a safety precaution. Glassware, such as Petri dishes, should be stacked in boxes.)

For safety, avoid having the squad make deliveries when the student body is moving through the halls to their classes. For the same reason, when teachers are not going to keep their materials for the entire day, the pick-up of materials should occur a few minutes before the end of the class period. When this cannot be done, as occasionally happens, the teachers should assume the responsibility for getting the materials back to the common laboratory supply center. Otherwise the equipment will not be available for the next teacher who needs it.

When teachers need fresh materials such as elodea plants, frogs, protozoa cultures, a pig's uterus for dissection, blood-typing materials, germinating seedlings to show hereditary traits, or a thriving culture of fruit flies, a week's advance notice (in some cases, when material must be sent for, a longer time is needed) must be given to the teacher and the squad who are serving the science teachers. In practice, this works out easily when teachers confer with each other from time to time each month and plan to share perishable materials at the same time. When teachers outline

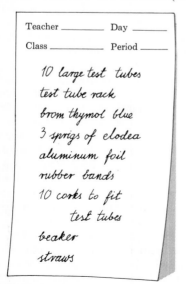

Teacher _____ Day _____

Class _____ Period _____

10 large test tubes
test tube rack
brom thymol blue
3 sprigs of elodea
aluminum foil
rubber bands
10 corks to fit
     test tubes
beaker
straws

Fig. 14-11 Replica of a teacher's order for materials.

their work at the beginning of the school year, fresh materials can be purchased then, or living materials can be collected and cultured in the laboratory or classroom (see Chapters 11–13) and made available when needed.

Cleaning up. Two or three students in the laboratory should be responsible for dismantling the equipment returned by teachers and for washing glassware used in class. Then all the materials are returned to labeled shelves, cabinets, or drawers.

Laboratory housekeeping. Plants must be watered, and aquaria and terraria need care. Someone must dust open shelves of models and glassware. Microscopes must be checked and lenses cleaned from time to time. New labels need to be put on cabinets and museum jars, and charts may need mending. Some fish tanks need patching and some broken glassware could be put to new use if the jagged edges were cut off and filed or fire-polished smooth. The wax lining some of the dissection trays becomes cracked and should be melted slowly over a Bunsen flame.

Possibly one student could add water colors to some of the black and white charts to freshen them and enliven students' interest. Who would prepare cotton-lined display boxes (Riker mounts) for the unassorted fossil specimens on hand? Some student on your squad, perhaps.

Safety in the laboratory. Every month over the school year, student squads in the lab should undertake a fire drill of their own design. Can the students on the squad use the fire extinguishers? Do they know when to use the sand in the sand buckets in the lab? Do they know how to use the fire blanket? All of these materials should be on hand. Also check the freshness of the supplies in the first-aid kit.

In the laboratory, students should not use a flame unless the teacher is present. Regulations should be designed concerning the storage of acids, bases, other dangerous chemicals such as chloroform or iodine, and so forth (safety procedures, pp. 689–93).

Squads for special services

Students with special interests and skills may receive training of many kinds which cannot be given in "class time." Either squads or special club activities may be developed in the following areas. In many cases, there are opportunities for enrichment here and also situations for students to discover where their interests and abilities lie.

Making models, charts, and slides. Many teachers initiate a routine for listing materials or equipment they would like to have for class use. Many teachers regularly peruse the catalogs of supply houses. Often they list their ideas on library cards at the moment they think of them. Possibly these cards are filed in a small metal box. At a given time each teacher submits these cards to one person in the department who has assumed this responsibility. These cards serve as a guide at the time for ordering supplies. But the main use of these teachers' ideas is to provide the stimulus for students who like doing things with their hands and brains. These students may build a piece of equipment such as an incubator or a device to show a tropism, or they may make a series of models to show the cleavage stages of a developing egg. Other students may make charts to show comparative anatomy among vertebrates or variations in some flowers, fish, or other forms not easily obtainable for class use as first-hand experiences. (See earlier sections in this chapter on model-making and chart-making.)

Students may supplement the supply of prepared slides in the department. For example, a club or squad may be trained to prepare blood smears (p. 118), whole mounts of small forms (p. 114), stained slides of protozoa (p. 123), and many other kinds of slides.

Curators of living things. When space is available, a museum of living things, plants and animals, can be maintained over the school year so that teachers may bring their classes to this area for many lessons in biology.

Where space is at a premium, a section of a classroom or a laboratory may be given over to three or four aquaria and at least one terrarium, and several shelves may be used to store cultures of protozoa and fruit flies. In another room (to avoid contamination of cultures of protozoa), different kinds of mold cultures may be grown. Perhaps there is also room for a cutting box—a box of sand or light soil—in which cuttings of plants and other forms of vegetative propagation may be grown.

Students may be trained to culture protozoa, subculture them every month or so (directions, Chapter 12), raise fruit flies (directions, Chapter 8), prepare mold cultures, grow fern prothallia (p. 648), and so forth. Guppies and other small tropical fish can be kept in one tank with snails and water plants. In this way, life cycles of fish and of snails' eggs can be studied in class. Fresh materials for study of plant cells are available at any time. In other aquaria students may maintain *Daphnia, Tubifex,* larvae of some insects, tadpoles, or other fresh-water forms they may obtain on field trips or from an aquarium shop (maintaining invertebrates and vertebrates, Chapter 12; plants, Chapter 13).

In each classroom there should be at least one aquarium and one terrarium. Battery jars of different sizes may be used also to accommodate living things that students bring to class (for example, land snails or cocoons). Small cages may be made of screening and wood frames by some students; observation ant jars may be prepared (see Fig. 12-24a). Directions have been given for collecting living things, keeping them in class, and also for preserving biological materials (see Chapter 11). If several classrooms house different living things, then space has been "stretched" and the science department has living culture centers in each classroom.

Soil-testing squads. Many schools have established excellent relations with the community by serving as a center for testing the kinds of soil people send in to the science department or science teacher.

There are many soil-testing kits on the market (see supply house catalogs) and students may be trained to give suggestions for improving the soil. This is a useful conservation practice. It may be possible to get help from the agricultural county agent in the community.

Landscaping the school grounds. A squad—a small core of interested students—may be trained to plan a nature trail around school (see p. 545), build birdhouses to attract birds, and plant trees, shrubs, and flower beds which would be useful to the teachers of biology. In this way, classes might be held on the school grounds for studies in reproduction, heredity, behavior, variations and evolution, interrelationships, and conservation. Students may label trees to help other students learn to identify some of the common trees and shrubs.

Book-room squad. A trained core of students can distribute books to classes at the beginning of the school year, or supplementary reference texts can be supplied to teachers whenever they are needed by a class. An up-to-the-minute inventory can be maintained if a tally sheet is posted on a bulletin board in the book room indicating books on shelves and books in use.

In fact, these students may take responsibility for an important enterprise in the department: encouraging students to build their own private science libraries. Some of the many excellent paperbound books might be sold (under a teacher's supervision) by a book-room squad before or after school one day a week as a special service.

Teachers will find ways to bring these readings into daily class use. Only a few books are listed below as a sampling for a student's library. Many teachers send for catalogs from publishers of paperbacks, or they use the comprehensive listing of paperbacks in science: H. Deason, ed., *A Guide to Science Reading* (Signet, New Am. Library, $0.60), an expanded edition of *An Inexpensive Science Library* (Am. Assoc. for the Advancement of Science).

New titles are being published constantly by many different publishers. A few of the many companies that publish paperback editions of science books and some of the available titles are listed below.

Doubleday & Co., Inc., 501 Franklin Ave., Garden City 31, N.Y.
> Asimov, I., *A Short History of Biology*, Am. Museum Science Books.
> Beck, W. S., *Modern Science and the Nature of Life*, Natural History Library Books.
> Eiseley, L., *Darwin's Century: Evolution and the Men Who Discovered It*, Anchor Books.
> Kalmus, H., *Genetics*, Natural History Library Books.
> Lanyon, W. E., *Biology of Birds*, Am. Museum Science Books.
> Simpson, G. G., *Horses*, Natural History Library Books.
> Wiener, N., *The Human Use of Human Beings: Cybernetics and Society*, rev. ed., Anchor Books.

Dover Publications, Inc., 180 Varick St., New York 14, N.Y.
> Beaumont, W., *Experiments and Observations on the Gastric Juice and the Physiology of Digestion*.
> Galtsoff, P., *et. al.*, eds. (J. Needham, chairman), *Culture Methods for Invertebrate Animals*.
> Leeuwenhoek, A. van, *Antony van Leeuwenhoek and His "Little Animals,"* coll., trans., and ed. by C. Dobell.
> Oparin, A., *The Origin of Life*, 2nd ed., trans. by S. Morgulis.
> Parsons, F. T., *How to Know the Ferns: A Guide to the Names, Haunts, and Habits of Our Common Ferns*, 2nd ed.
> Singer, C., *A Short History of Anatomy and Physiology: From the Greeks to Harvey*.
> Vallery-Radot, R., *The Life of Pasteur*, trans. by R. L. Devonshire.
> Winchester, A. M., *Heredity and Your Life: An Account of Everyday Human Inheritance*.

Golden Press, Inc., 630 Fifth Ave., New York 20, N.Y.

Harper & Row, Publishers, Inc., 49 E. 33 St., New York 16, N.Y. (Harper Torchbooks).

Holt, Rinehart and Winston, Inc., 383 Madison Ave., New York 17, N.Y. ("Modern Biology Series").

New Am. Library of World Literature, Inc., 501 Madison Ave., New York 22, N.Y.
> Asimov, I., *Chemicals of Life*, Signet Books.
> ———, *The Wellsprings of Life*, Signet Books.

Bates, M., *The Forest and the Sea*, Signet Books.

Carson, R., *The Sea Around Us*, Signet Books.

Malthus, T. K., J. Huxley, and F. Osborn, *On Population*, Mentor Books.

Peterson, R. T., *How to Know the Birds*, Signet Key Books.

Sherrington, C., *Man on His Nature*, Mentor Books.

Snow, C. P., *Two Cultures and a Second Look*, expanded ed. of *Two Cultures and the Scientific Revolution*, Mentor Books.

Storer, J. H., *The Web of Life*, Signet Books.

Penguin Books, Inc., 3300 Clipper Mill Rd., Baltimore 11, Md.
> Abercrombie, M., C. J. Hickman, and M. L. Johnson, *A Dictionary of Biology*.
> Fogg, G. E., *Growth of Plants*, Pelican Books.
> Romer, A., *Man and the Vertebrates*, 2 vols., Pelican Books.
> Vernon, M. D., *Psychology of Perception*, Pelican Books.
> Walker, K., *Human Physiology*, Pelican Books.

Pocket Books, Inc., 630 Fifth Ave., New York 20, N.Y.
> De Kruif, P., *Microbe Hunters*.
> Sproul, E. E., *The Science Book of the Human Body*.

Prentice-Hall, Inc., Englewood Cliffs, N.J. (Spectrum Books and "Foundations of Modern Biology Series").

Univ. of Michigan Press, 615 E. University, Ann Arbor, Mich. (Ann Arbor Science Library).

Research groups. Some of these special services squads regard themselves as clubs and develop a pride in membership. They hold meetings and discuss problems of importance to their operation.

Students with a common interest may approach a teacher with the plea that he sponsor their club, for they know he shares their interest. In this way a microscopy club is formed under a teacher's supervision, or a tropical fish club, or a walking club which goes on field trips on Saturdays or to a weekend school camp.

Each teacher may sponsor at least one "talented" student and give him scope to work on some investigation that holds long-range interest. Gradually in such individual work deep interest and enthusiasm are born and spread so that soon a large group of students select themselves for a consuming program of work. These students working on an original piece of work[28] gain skills in laboratory work, read more, and learn what the methods of science are by using them. Here, too, gifted students can often discover whether they have the aptitudes, abilities, and other personality factors needed for a career in science.

Some suggestions for investigations offered by biologists over the country are detailed in *Research Problems in Biology: Investigations for Students,* Series One–Four (BSCS, Anchor Books, Doubleday, 1963, 1965).

Students with some commitment to biology may participate in science fairs, science talent search examinations for scholarships, and seminars. They may also be guest speakers in classrooms, in school assembly programs, and at meetings of the PTA.[29]

Through these experiences students learn more about science, and do more; they also learn that leadership involves the ability to share what they know with a larger audience.

Science readings squad. This is a squad that can help to raise the interests of the entire student body if the members have a high level of initiative.

At the lowest level of operation, students might be assigned the task of writing a form letter to those industrial concerns (see Appendix D) known to supply low-cost teaching aids such as vitamin booklets, charts of vitamin deficiencies, pamphlets which describe how to make something, and so forth. At a higher level, students can handle magazine subscriptions (for example, *National Geographic, Life, Scientific American,* and *Science Newsletter*) and the publications of nearby museums and state and federal governments.

These students might have a bulletin board or hall case assigned to them. They could post articles, pictures, and copies of new paperbacks. Give them suggestions about sources of information on specific topics in biology such as heredity, health, foods and nutrition, safety and accident prevention, and identification of plants and animals in season (birds, insects, shells, and flowering plants). It may be possible for this squad of specialists to work with the librarians in preparing exhibits of books in science.

In many schools students have built attractive wall racks along which magazines may be inserted. This arrangement saves desk or table space in a classroom or laboratory.

Mimeographing squad. Some students like to put things into print. When a small squad of reliable students is trained to use a mimeograph or a rexograph machine they release "teacher time" for more creative work.

With a system for duplicating the written page, teachers may find it profitable to reproduce writings from original papers (described on pp. 193, 263, 493). You may devise a variety of pencil-and-paper tests which test reading ability, interpretation of data, synthesis, and application —all the activities that are necessary for an intelligent citizenry in a world of science.

Publishing squad. Classes in science might plan creative work. They might prepare a newsletter in biology or a small magazine for distribution in the department. Students who have special interests, are doing independent research, or enjoy writing for others might organize such a venture. The help of the mimeographing squad must be solicited and some typists must be put to work. This is a stimulating venture over the year. In fact, several small high schools might merge their efforts and widen their influence and "sales."

Projectionist squads. In many schools a teacher is assigned to supervise the audiovisual program for the entire school. This teacher is responsible for having slide and

[28] P. F. Brandwein, *The Gifted Student as Future Scientist,* Harcourt, Brace & World, New York, 1955.
[29] P. F. Brandwein *et al., Teaching High School Biology: A Guide to Working with Potential Biologists,* BSCS Bull. 2, AIBS, Washington, D.C., 1962.

film and overhead projectors, tape recorders, and microphones in good operating condition at all times. He must order films, plan a schedule of showings over the entire year, and train a squad of projectionists to operate the equipment.

Often one room is set apart and is equipped with heavy drapes; screens; storage closets for films, filmstrips, and slides; projectors for silent and sound films, filmstrips, and slides; and possibly a microprojector (Fig. 2-4) or other type of enlarger of the microscopic field of vision (Fig. 2-5). There may also be opaque projectors on hand to throw pictures from the printed page onto the screen. Equipment for splicing and rewinding film is also needed. Students must learn how to care for this expensive equipment and store it against drying out (storage facilities, p. 699).

It may be possible to assign one or two students to look through catalogs of free loan films, record these on library cards, and build a filing system of films in many areas of biology. Evaluation sheets might be used after showings so that teachers can indicate whether the film should be reordered.

Perpetuation of squads

We have discussed how students' skills can be put to use for a common purpose. How can squads already well trained be perpetuated? When young students are trained they can serve for several years. As they approach graduation, they should train younger students over at least a 3-month period to take over these tasks. In this way, although a teacher must supervise the students, the work is made routine. This saves time for the teacher and often brings to students in class activities in which they might not otherwise participate because of the extra time involved if teachers had to do them alone. Youngsters gain skill and self-confidence as they serve others in useful group work.

High grades in science should not be the decisive criterion for selecting students for these squads—each student brings his talent. Teachers agree that good will, fair play, and cooperation in sharing the load —many of the appreciations and attitudes we aim for in working with young people —often are more readily *caught* in this kind of work in groups than *taught* by teachers in classrooms.

Student squads in the classroom

Students readily accept a share in improving the atmosphere of the classroom. Working as small squads, they may turn an ordinary classroom into an attractive, distinctive place conducive to learning. These rooms can look like biology classrooms, places where interesting work is going on—different from the other classrooms students occupy during the day (Fig. 14-13).

Students may bring in cuttings or whole plants. These can be planted in window boxes and maintained by a few students. Another student may care for the fish tank. From nearby, students may bring in enough materials to set up plant and animal habitat studies—desert, woodland, or marsh (directions, p. 611; see also Figs. 13-21, 12-28).

Several students in each class may work as a squad or committee to gather pictures and articles on topics discussed in class. In this way, bulletin boards in class and in hallways may be kept current and be a source of stimulating discussion.

Students may exhibit their hobbies. Sometimes a student takes fine pictures or has a splendid collection of fossils, shells, or insects. These can be displayed in hall cases with appropriate legends.

A display of shiny glassware with labels draws many students. This might grow into an exhibit of "tools" of biology and how they are used. Simple ideas are dramatic; frequent changes are imperative.

Some teachers have arranged successful weekly exhibits through the spring and fall on flowering plants and mosses and ferns. Budding branches from trees and shrubs near the school can be placed in

Fig. 14-13 Classroom-laboratory facilities: (a) an attractive classroom-laboratory where students work on individual or small group projects; (b) a project room; (c) a student project area. (a, Phil Palmer from Monkmeyer; b, photograph by George Schutzer; c, courtesy of Bausch & Lomb, Inc., Rochester, N.Y.)

(a)

(b)

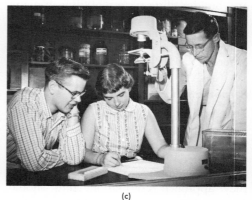

(c)

containers of uniform size. On small library cards, indicate for each plant the name, where it may be found, and an interesting fact about it.

Some students also plan contests in which a given number of plants, birds, or insects in the neighborhood must be identified correctly. Prizes may be small books or field guides.

We know, too, of some dramatic displays of shells, flowers, leaves, and fruit shapes which focused attention on variations among living things. Legends asked questions about the facts of heredity and variations and their role in effecting changes among living things.

Tutoring squads. In this category of squads we are not considering committee work that is part of classwork, but the long-range routines which should be established early in the year. Aside from maintaining growing, living things as already described, there is a worthy service in tutoring other students. There will be a student with a block of some sort or a blank area in past experience, or a student who takes a little longer to learn a point, or students who have been absent for long periods due to accidents or illness, and who thereby fall behind in their work. Many teachers arrange to have other students, often bright students who can communicate ideas, tutor the students who need remedial help or temporary "briefing" on new work.

Laboratory aides. What kind of class management is efficient and effective for distributing slides, medicine droppers, cultures for examination, and microscopes? How can dissection pans, toweling, kits, magnifying lenses, and specimens for study be distributed without confusion?

This is how some teachers organize activities for as many as 35 students in a group that may meet in a classroom if a biology laboratory is not available: Six students are assigned to act as laboratory aides in class. These students place lens paper and distribute slides and coverslips, culture bottles of material to be examined, and medicine droppers in six different "spots" in the room so that the class can divide itself and not crowd around one point of distribution. Microscopes are wheeled into class on a cart; when these are numbered it is possible to assign a specific microscope to each student or group of students. In this way the group of students shares responsibility for specific instruments.

Toward the end of the period, possibly 6 min before the end, students return microscopes to the cart, and the laboratory aides collect the materials which they distributed earlier in the period. Then the squad from the laboratory center picks up the materials (see laboratory squads, p. 681).

Safety in the laboratory

Basically, safety precautions for the classroom and laboratory involve good planning and common sense in the use of chemicals and equipment. Certain safety steps should be standard operating procedure for all the members of a science department. Students who work on squads should also be trained in their responsibility in establishing good safety practices.

A pamphlet called *For Greater Safety in Science Teaching*, rev. ed. (distributed by the New York City Board of Education) provides concise, yet comprehensive, general instructions concerning the responsibilities of the head of a science department, science teachers, and students in maintaining safety. A portion of that pamphlet is reproduced here.

DIRECTIONS FOR CHAIRMEN
OF SCIENCE DEPARTMENTS

Science Chairmen should:

1. Periodically, through departmental conferences at the beginning of each term, make all science teachers and laboratory assistants aware of the hazards in science instruction and stress the seriousness of accidents caused by carelessness.

2. File with the principal each term a signed statement that they have read the rules and that the provisions of this report have been complied with. A receipt from each teacher should be kept on file.

3. Notify the principal in writing each term and immediately in the case of emergencies of any hazards in either main building or annexes, such as:

   a. Defective gas fixtures, electrical outlets and connections.

   b. Seats so defective that they may cause injury.

   c. Inadequate storage cabinets.

   d. Lack of fire blankets, extinguishers, fire pails and sand.

4. See that a first aid cabinet is ordered and placed in each science laboratory and preparation room and elsewhere as needed, and that ample reserve stocks of tannic acid salve for possible burns, antidotes for poisons and first aid material are provided. A copy of the Red Cross booklet on First Aid should be kept with each first aid cabinet.

5. Inspect first aid cabinets and fire extinguishers at least once a term.

6. Make certain that combustible and dangerous materials such as poisons are kept securely locked in a metal cabinet. Acids should be

stored in an albarene closet, never in ordinary closets or wooden cabinets. Do not store chemicals which react with each other in close proximity such as glycerin and nitric acid, potassium chlorate and organic compounds, etc. No pupils should have access to such lockers and storerooms or closets in stockrooms.

7. Cyanides, except potassium ferrocyanide and ferricyanide, may not be stored in our schools.

8. See that the cabinet for the storing of sodium, potassium, calcium, and calcium carbide has printed on it boldly in white, the warning *"In case of fire, do not use water."*

9. See that pupils do not handle materials in cabinets reserved for dangerous substances.

10. Inspect chemical cabinets and record the dates of inspections.

11. Instruct laboratory assistants to keep tools and sharp-edged instruments under supervision and to exercise the greatest care when they are used by pupils.

12. In making electrical connections for the use of stereopticons, film projectors and the like, use heavy rubber-covered wall plugs (to avoid cracking if dropped) and good quality cable (double cords within heavy insulation) which will not kink or easily break. This will avoid short circuits and the danger of both electric shock and fire.

13. In order that squads may be properly supervised, limit the number of pupils to the maximum of three per period for each supervisor. In biology, such squads may be somewhat larger.

14. Be responsible for acquainting all squad members with safety rules and regulations at the beginning of their service.

15. Secure parents' consent in each case for the work of squad members.

16. Give a copy of these instructions to all new teachers and laboratory assistants.

17. Do not assign laboratory assistants to teach classes, except by special permission of the Superintendent.

18. Do not assign to science classes teachers not qualified to teach science.

19. Instruct the laboratory assistant to perform the actual procedures of each laboratory experiment or demonstration prior to the class session, to see that all materials and apparatus work properly.

20. Advise teachers to remove or regroup accumulated materials that are a safety hazard. This applies to preparation and storage rooms as well as to classrooms. Accumulations of glassware and chemicals on demonstration tables, of exhibits or projects that overload shelves, or miscellany kept on window sills are some examples.

21. See to it that a metal or earthenware waste jar is provided in every classroom where science experiments are performed. Such waste jar and *not* the wastepaper basket should be used for broken glassware, chemical residues, etc.

22. Discard unlabeled, contaminated, and undesirable reagents.

23. See that any demonstrations, experiments, or projects dealing with atomic energy or radioactivity are performed in accordance with safety practices in that field.

GENERAL DIRECTIONS FOR ALL SCIENCE TEACHERS AND LABORATORY ASSISTANTS

In reference to pupils

1. Pupils are to be under the direct supervision of a teacher at all times and in all places, as required by the By-Laws of the Board of Education.

2. Pupils are *not* to carry laboratory equipment or apparatus through the halls during the intervals when classes are passing.

3. Pupils are *not* to transport dangerous chemicals at any time except under the direct supervision of a teacher or laboratory assistant.

4. Pupils are *not* to handle materials on the demonstration desk except under the direction of the teacher.

5. Pupils are *not* to taste chemicals or other materials.

6. Before permitting pupils to work with sharp tools, the teacher must be assured that pupils are fully competent to use the tools.

7. At the beginning of each term, pupils should be instructed in general safety precautions.

8. Pupils should be specifically instructed regarding the dangers and the precautions required at the beginning of each laboratory period in which there is a special hazard.

9. Pupils should be cautioned about hazardous activities involving the use of chemicals outside the school—e.g., mixing chemicals to "see what happens," setting fire to gasoline cans, breaking open fluorescent tubes.

10. Pupils should be advised against experimenting with rocket fuels.

In reference to teachers

1. Teachers must
   a. Report immediately to the principal any injury or accident.

b. Arrange for the completion and filing of the proper forms to be signed by
   (1) the injured party
   (2) witnesses
   (3) teacher concerned, or laboratory assistant.

2. Teachers must be fully acquainted with first aid treatments.

3. Teachers should notify the chairman of the department of the existence or development of any hazard that comes to their attention.

4. Teachers are to perform classroom experiments only if they themselves have previously tried them out or have been properly instructed by the chairman or an experienced teacher.

5. When using volatile liquids which are flammable, such as alcohol, in a demonstration experiment, care must be taken that any flame in the room is at an absolutely safe distance from the volatile liquid.

6. Demonstrations involving explosive mixtures must be so arranged as to shield both pupils and teachers from the results of the explosion. Even when there is no likelihood of an explosion, pupils should be asked to evacuate seats directly in front of the demonstration table whenever there is any possibility of injury to them by the spattering of a chemical, an overturned burner, inhalation of fumes, etc.

7. Large storage bottles of dangerous chemicals such as acids and alkalies, if on shelves, are to be no more than two feet above the floor. If possible, they should be kept on the floor.

8. Never add water to concentrated sulphuric acid. If it is necessary to prepare diluted acid, the concentrated acid should always be added in small quantities to the water, stirring all the while.

9. Handle all corrosive substances with the greatest care. Special precautions should be taken with concentrated sulphuric acid, nitric acid, glacial acetic acid, and concentrated solutions of caustic alkalies and other corrosive chemicals as phenol, bromine and iodine, etc.

10. White phosphorus must be kept under water in a double container, one part of which is metal. This form of phosphorus must be cut only under water. If cut in the open air, the friction may be sufficient to ignite the material with very serious results. Use red phosphorus in place of white phosphorus whenever possible.

11. Residues of phosphorus should be completely burned in the hood before depositing in the waste jar.

12. Combustible materials of all types are to be kept in a metal cabinet or albarene closet provided with proper means for closing and locking.

13. Metallic sodium, potassium, and calcium and calcium carbide should not be stored above water solutions or vessels containing water. Metallic sodium and potassium, after the original container has been opened, must thereafter be kept under kerosene. These substances are corrosive and must not come in contact with the skin.

14. See precautions in the insertion and removal of glass tubing, thistle tubes, thermometers, etc., in reference to apparatus and rubber stoppers [under section for teachers of chemistry and physics, not reprinted here].

15. Do *not* demonstrate devices or equipment brought in by pupils before pre-testing.

16. When a motion picture machine or other projection apparatus is sent into non-science classrooms a carbon dioxide extinguisher should accompany the apparatus.

The same pamphlet offers specific provisions for the *biology* teacher.

SAFETY SUGGESTIONS FOR TEACHERS OF BIOLOGY AND GENERAL SCIENCE

Laboratory technique and procedure for pupils

1. See section under similar heading for Chemistry and Physics [not reprinted here].

2. Great care should be exercised by pupils in securing epithelial cells from the inside of the cheek for study under the microscope. Only smooth splints (wood) or the *blunt* edge of a flat toothpick should be used. Pointed instruments or any part of the scalpel should never be used for this purpose.

3. *Handling laboratory animals*

Rats, mice, guinea pigs and other laboratory animals should be handled gently by students so as not to unduly excite the animal into biting. Gloves made of thick rubber should be available and used whenever necessary, that is, when there is danger of biting (by excited animals, injured animals, new additions to cage, pregnant or feeding female, etc.). Students and visitors to the laboratory should be cautioned not to insert finger in wire mesh of cage. Appropriate signs should be displayed about cage such as "Keep Hands Off."

Only specially trained members of the laboratory squad should be permitted to handle laboratory animals. Poisonous snakes should not be kept in the laboratory.

4. *Use of formaldehyde*

Specimens preserved in formaldehyde should be thoroughly washed in running water for 24 hours before being handled by students. In taking specimens out of formaldehyde, students should either wear rubber gloves or else use tongs or forceps—depending on the size of the specimen. Adequate ventilation should be provided in any room where formaldehyde is used.

5. *Use of carbon tetrachloride*

Avoid unless absolutely necessary. Adequate ventilation should be provided in any room where carbon tetrachloride is used. The use of a hood is preferred.

6. *Precautions for field trips*

a. Pupils should be instructed about identification of poison ivy, poison sumac, copperhead snakes (found in Palisades), etc.

b. First aid kits should be taken along on all field trips.

c. Students should be instructed as to the proper clothing to take along on a field trip to avoid illnesses due to undue exposure, etc.

7. Pupils should not perform any experiments involving the heating of alcohol, strong acids or strong bases.

8. When pupils are to handle tools in constructing projects, etc., special instructions should be given where necessary.

9. Pupils should submit all electrical work to the instructor for approval before putting into use.

10. Pupils are not to handle steam pressure sterilizers, incubators, etc., without special instruction given beforehand.

General suggestions on technique and procedure for biology teachers and laboratory assistants

1. *Bread mold and pollen from flowers*

In handling flowers and bread mold, care should be taken that pollen or spores are not excessively distributed through the classroom. Some students may be allergic to pollen or spores.

2. *Blood experiments*

Only sterile needles or lancets should be used by the teacher for pricking his finger to draw blood. Blood should be drawn from the dorsal side of finger just in back of the cuticle. Rub finger with alcohol before pricking it and cover with bandaid afterward. A lancet should *not* be used more than *once*.

3. Biology teachers and laboratory assistants should observe the precautions listed for Chemistry and Physics teachers and laboratory assistants (previous section [not reprinted here]) before handling physical and chemical apparatus.

4. *Osmosis experiments*

In the osmosis experiment great care should be exercised in inserting the thistle tube through the rubber stopper. (A rubber stopper is commonly used to support the thistle tube in the clamp of the ring stand.) Always wet the tube and use a twisting motion while applying pressure. Do not grasp the thistle tube by the bowl; grasp the tubing of the thistle tube near the rubber stopper.

5. Only fresh materials should be used at all times. Do not use decayed or decaying material. If material is to be used for more than one day, keep it in a refrigerator or in formaldehyde.

6. *Extraction of chlorophyll*

a. Only pyrex or other hard glass test tubes should be used.

b. Use an electric heater of the immersion type or a water bath heated by an electric hot plate, instead of an open flame or a gas heated water bath for heating the alcohol.

c. Keep open flames away from alcohol or alcohol vapors. If alcohol ignites in beaker, cover beaker with a glass plate to extinguish. If burning alcohol runs over the table, use fire blanket. (These materials should be on hand at all times.)

7. *Operation of pressure cooker for sterilizing bacteria media*

a. Before operating the pressure cooker familiarize yourself thoroughly with the proper directions for operation.

b. Examine safety valve before use and make sure it is in working order.

c. Do not allow the pressure to go above 20 pounds.

d. Stop heating before removing the cover. Pressure should be down to normal before removing cover.

e. Be sure to open the stop cock before releasing the clamps.

8. Careful use of dissecting instruments and dissection material in laboratory work, through instruction and warning, will avoid cuts and possible infection. Especial care should be taken when cleaning scalpels and needles.

9. Pupils should not be required to prick fingers in order to supply blood.

10. Pathogenic bacteria should not be cultured. Exposed Petri dishes should be soaked in

a strong disinfectant (carbolic acid, cresol, lysol, etc.) before being washed. Rubber gloves should be worn during this operation. Dishes passed around for inspection should be bound together with Scotch tape.

11. Teachers and laboratory assistants should take special care to avoid the dangers indicated when pupils handle any of the following:

    a. Large paper cutter (cuts).

    b. Formaldehyde (fumes dangerous to the eyes and throat).

    c. Denatured alcohol and wood alcohol.

    d. Apparatus left hot by use, i.e., stereopticon, microprojector, etc. (burns). Do not put hot projectors into cases. Allow them to cool off first.

    e. Carpentry tools (wounds).

    f. Electrical apparatus and equipment (shocks).

12. Ether should not be used in a room where there is a flame or where flames may be shortly used. Containers in which ether has been used should be allowed to air out thoroughly before being washed. Put ether-soaked cotton or rags in a stone crock (not a waste paper basket). Set aside in a safe place to evaporate.

13. When testing for proteins, pains should be taken to avoid the possible spattering of the nitric acid when heating. This acid, like sulphuric acid and other acids, is corrosive and produces bad burns, even when cold. Pupils should never handle a test tube or bottle containing concentrated nitric acid. If the Biuret test for proteins is used, the same precautions should be exercised in handling the potassium hydroxide as in handling the nitric acid.

14. Wire loops used for transferring cultures of bacteria should be flamed after each transfer is made.[30]

Common accidents

Common accidents in the classroom or laboratory fall into three categories: (1) cuts from broken glassware, (2) burns from hot glass, and (3) chemical burns.

The number of each of these classes of accidents can be reduced by increased caution on the part of students. Throughout the year, periodically remind the students to exercise extreme caution when

working with glassware and chemicals. Most accidents caused by broken or hot glassware can be treated by the standard first-aid methods suggested in the latest *American Red Cross First Aid Textbook.*

Materials for treating chemical burns should also be in the first-aid cabinet.

Acid burns. These burns should be washed immediately with large quantities of water to remove the acid. Then neutralize any remaining acid by covering the burn with a saturated solution of sodium bicarbonate.

Alkaline burns. These burns should also be flushed immediately with large quantities of water. Then neutralize any remaining alkali with a saturated solution of boric acid.

First-aid kits. First-aid kits and cabinets should contain the materials listed by the Red Cross textbook, plus any special materials for treating chemical burns.

It is also convenient to prepare several tongue depressors covered with sterile gauze in a plastic bag or envelope. This may be useful in an emergency; for example, there may be an epileptic in class. In this case such a tongue depressor can be inserted between the teeth to prevent him from biting his tongue.

Ampules of ammonia are also useful in reviving a child who has fainted.

(See the end-of-chapter bibliography.)

Fireproofing fabrics. There may be occasion to fireproof materials in the classroom or laboratory. Bennett's compendium[31] suggests the following two methods. Small samples may be tested after they have soaked in the solution; then you may want to use it on a larger scale.

1. SOLUTION FOR FIREPROOFING HEAVY CANVAS MATERIALS. Impregnate the fabric with the following solution, squeeze out the excess, and allow the fabric to dry. (You may reduce the quantities given here to suit your needs.)

$(NH_4)_3PO_4$	1	lb
NH_4Cl	2	lb
water	0.5	gal

[30] From *For Greater Safety in Science Teaching,* rev. ed., New York City Board of Education, 110 Livingston St., Brooklyn 1, N.Y., 1961 ($0.25).

[31] *Op. cit.*

2. SOLUTION FOR FIREPROOFING LIGHT-
WEIGHT FABRICS. Impregnate the fabrics
with the following solution, squeeze out
the excess, and hang up to dry. Frequent
washing of the fabric will dissolve out the
fireproofing salts so that the fabric will
need to be treated again with the same
solution.

borax	10 oz
boric acid	8 oz
water	1 gal

Space for work

In older schools, teachers have used
their own good devices to find ways to
make over old things. Gradually they have
ordered new equipment. They have also
improvised. For example, in rooms which
are not equipped with gas, teachers use
cylinders of propane gas to which a burner
attachment is fitted (Fig. 14-14), provided
your Board of Education permits use of
propane burners. When a room lacks
running water, some teachers employ a
large battery jar of water, a water wash
bottle, or, for prolonged use, a carboy of
water with a rubber extension tube and
clamp.

Students willingly bring to class equipment to augment the facilities of the
school. In this way aluminum foil dishes,
baby-food jars, plastic boxes, and large
glass jars can be gathered in place of more
expensive glassware.

Some schools have solved their individual problems concerning older classrooms not equipped for science teaching
by purchasing a transportable demonstration table such as that in Fig. 14-15. These
units, along with the more elaborate kind
shown in Fig. 14-16, give teachers an
opportunity to provide more demonstrations and experiments in class.

Federal funds have been made available through the National Defense Education Act to aid in raising standards of classrooms, laboratories, and equipment.

Newer schools are equipped with outlets in classroom and laboratory for water,
gas, often a vacuum jet, and one for compressed air. The trend in newer schools

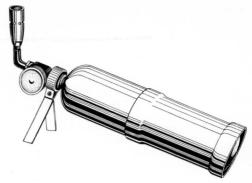

Fig. 14-14 One type of propane gas burner
that may be used in rooms not equipped with gas,
if a school system authorizes its use.

seems to be toward a combination classroom-laboratory so that the room serves
multiple purposes (Figs. 14-17, 14-18).

Science teachers are often asked to sug-

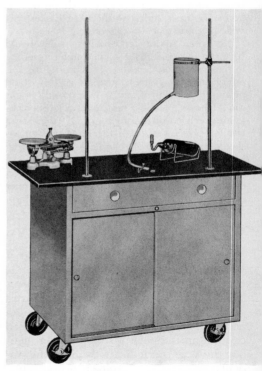

Fig. 14-15 A unit demonstration table, which
can be transported from one classroom to another. (Courtesy of W. M. Welch Manufacturing
Co., Chicago.)

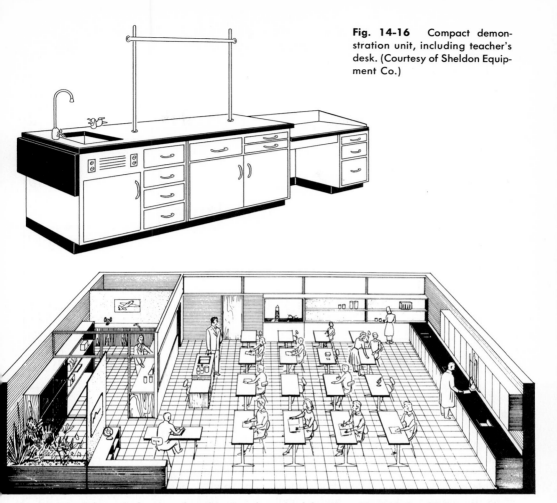

Fig. 14-16 Compact demonstration unit, including teacher's desk. (Courtesy of Sheldon Equipment Co.)

Fig. 14-17 Schematic drawing of combination classroom-laboratory. (From National Science Teachers Assoc., *School Facilities for Science Instruction*, 2nd ed., edited by J. S. Richardson, National Science Teachers Assoc., 1961.)

gest plans for classrooms (or lists of ideas they would want to see in practice) which may be incorporated in blueprints for a new school.

What facilities would you request for a new school? What kind of tables and chairs, and for what age group? Do you favor separate adjoining closets or closets and cabinets in a central supply laboratory? Do you prefer a combination classroom-laboratory? Is there space for a darkroom? What provisions would you have for a greenhouse or a nature room of some kind?

The kinds of activities you plan will decide the work space needed. In the following sections are suggestions for kinds of work space that teachers over the country find necessary in creative science teaching.

Classrooms and laboratories

If you are given the responsibility for ordering classroom furniture you will probably study supply house catalogs and those of companies which deal exclusively with school furniture. Specific recom-

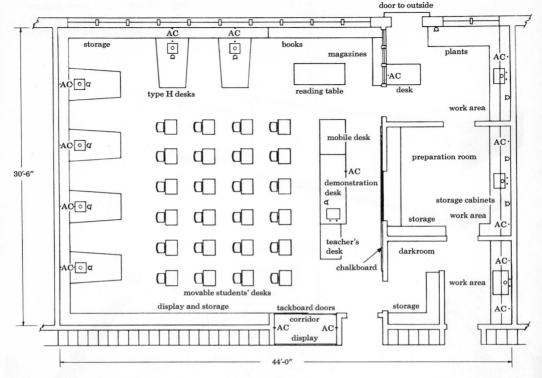

Fig. 14-18 Floor plans for science classroom-laboratories.

mendations are given for space per cubic centimeter for each student; chairs and tables of different heights are needed to provide for the many individual differences in maturation among high school students.

If space is limited, there is an advantage in having a laboratory arrangement at one side of a room and movable tables and chairs at the other side. In this way, students may be grouped at either end and the teacher and students do not have to speak and be heard across a large room (Fig. 14-19).

Much has been written on kinds of furniture and floor layouts for science rooms which include cabinets and closet space, provisions for bulletin boards, and shelves for aquaria. Perhaps the last word has been said for a long time to come in two well-conceived, comprehensive books: National Science Teachers Assoc., Com-

mittee on Science Facilities, *School Facilities for Science Instruction,* 2nd ed., edited by J. S. Richardson (National Science Teachers Assoc., 1961) and J. Schwab, Supervisor, *Biology Teachers' Handbook* (BSCS, Wiley, 1963). The former book, which is the source of the drawings reproduced in Figs. 14-17 and 14-18, contains chapters dealing with facilities for general science, biology, chemistry, physics, and special courses in science. There is also a chapter on planning for elementary science facilities, an excellent bibliography for extensive research in this area, and a splendid listing of suggested equipment and supplies for all the sciences. *Biology Teachers' Handbook* also offers a variety of floor plans, one of which is shown in Fig. 14-19. Also consult R. E. Barthelemy, J. R. Dawson, Jr., and A. E. Lee, *Innovations in Equipment and Techniques for the Biology Teaching Laboratory* (BSCS, Heath, 1964).

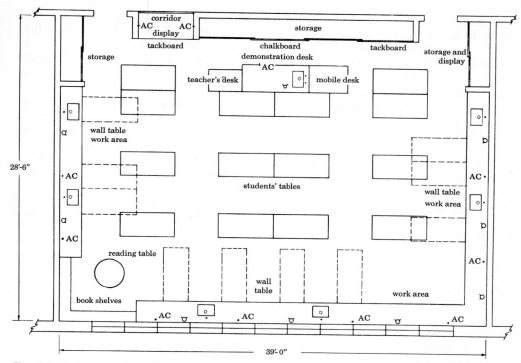

Fig. 14-18 (continued)   Floor plans for science classroom-laboratories. (Both floor plans from National Science Teachers Assoc., *School Facilities for Science Instruction*, 2nd ed., edited by J. S. Richardson, National Science Teachers Assoc., 1961.)

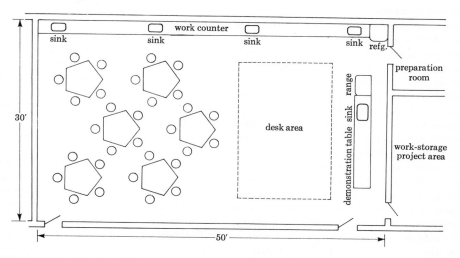

Fig. 14-19   Floor plan that utilizes space as a combination of laboratory work area and a desk area for students. This room is appropriate for group discussions. (After J. Schwab, Supervisor, *Biology Teachers' Handbook* [BSCS], Wiley, 1963.)

A Laboratory Planning Kit is available from Fisher Scientific Co. (branches in many cities) discussing the use of their "Unitized" laboratory furniture. Included in the kit are scaled cut-outs of fume hoods, open benches, wall cases, sinks, tables of many sizes, and laboratory stools. Graph paper is provided so that scale layouts of the laboratory furniture cut-outs can be made.

Other supply houses and furniture companies are listed at the end of this chapter.

Extending the classroom

The greenhouse. In many schools over the country a greenhouse is associated with student project work and field work activities on the school grounds or at school camps. Many students find work in a school greenhouse an enriching experience. This work may be an extension of class activities or a means for supplying living materials for classroom study.

Students may learn to grow algae, mosses, ferns, and seed plants. They can successfully raise plants by vegetative propagation. Some may go into soil-testing techniques while others begin studies in plant physiology.

School grounds. Very often teachers of biology are asked to contribute ideas for landscaping the school grounds. When flowering trees and shrubs are specially selected for planting around school, a teacher can conduct a short field trip (within a class period) to introduce the topic "How Flowers Function in Plant Reproduction." (This notion is developed further in Chapter 11 in the section on field trips.) At other seasons students may study seed formation, devices for seed dispersal, variations, and plant communities.

In fact, some teachers have laid out nature trails, have built artificial lakes which have later been stocked with plants and animals, and have, in general, stimulated the study of field biology. Clubs or classes in gardening are popular in many communities.

School camps. In some communities there is a cooperative bond between community and school to bring many first-hand outdoor nature experiences to students. Certainly students gain a greater appreciation of the biological world around them and the interacting factors in the conservation of living things. Often these are weekend excursions chaperoned by teachers and parents. When the school owns a camp, or rents one out of season, long-range activities can be planned.

School nature museum. Almost every teacher of biology grows some kinds of living things, and students have similar interests. When students and teachers pool their techniques, a small space can be found to exhibit living things for the whole school. Descriptions of relationships (classification) and interesting facts about the plants and animals can be typed as legends on library cards. This area can grow into a permanent, living nature museum. Students can be trained as curators (see p. 683).

Living-culture center. Many of the organisms for a school nature museum may be drawn from stocks that are maintained throughout the year in the biology classrooms or laboratories. In fact, a living-culture center which maintains protozoa cultures and representative members of the invertebrates and vertebrates may serve the elementary and junior high schools in the community. Students who work on individual projects may also get their supplies of organisms from the school living-culture center.

Just as space and facilities are needed for a greenhouse, there is need to plan for an animal room—a place for tanks and cages to house fish, amphibia, reptiles, hamsters, white rats, and, sometimes, as temporary guests, pigeons or parakeets. (Directions for collecting and maintaining invertebrates and vertebrates are given in Chapter 12.)

Project room. In some schools, teachers have set aside space for individual students to pursue an interest, a research problem, or a prolonged "original" experiment in science (Fig. 14-13b). Over a year or more these young people use the methods of scientists. They need shelves for storage of materials, a work desk, and locker space for storage of their equipment (see Figs. 14-18, 14-19). The facilities should include adequate electrical outlets and provisions for gas and water. A book shelf or cabinet for science reference books in the same room is also convenient.

Science library and committee room. When space is not the limiting factor a room should be provided for a small science library and a separate place for students to meet for conferences or committee work associated with their classwork. This library does not compete with the school library but contains more advanced texts and magazines for students engaged in individual research work. Suggestions for books to purchase may be found in Appendix A and on p. 684.

Film room. Many schools set aside a room for showing films that can be used by all the classes in the school. This room should be equipped with dark shades, drapes, and a large screen. Specially trained student squads (described on p. 687) can run the projectors (silent, sound, lantern, 2 × 2 in., opaque, microprojector, and so forth). These students also can keep the equipment in good condition in adequate storage facilities near the film room (see Fig. 14-24).

Catalogs of slides, films, and filmstrips may be obtained from supply houses (Appendix C) and from film distributors (Appendix B). You may also want to check on the availability of free loan films from industry (consult end-of-chapter listings of films and Appendix D, sources of low-cost materials). Also refer to the extensive bibliography at the end of this chapter.

Darkroom. Many young people are camera enthusiasts and develop their own pictures. Thoughtful planning of space in a department or school may produce a spare closet which can be converted into a darkroom.[32]

Teacher's corner. Provisions for a small,

snug, confined area for the teacher reaps high returns in teacher efficiency. While an "office" sounds formidable to some, a small nook to which a teacher may repair to meet a student or colleague in semi-privacy, to work over records, or to read relieves the teacher of the strains which mount from the continued buzz and activity of many busy people in a school day.

Storage facilities

Closed storage space, that is, compact closets and cabinets, are in greater demand than open shelves. When stored in closed cabinets equipment and supplies do not accumulate dust, and they are not exposed to fumes or fluctuations in humidity and temperature.

Demonstration materials such as models, skeletons, glassware, and projectors may be stored in cabinets such as the ones shown in Figs. 14-20 and 14-21. The cabinet shown in Fig. 14-20b has many drawers in which rocks, fossil specimens, or small demonstration objects may be stored. When stored in a hanging position (see Fig. 14-5), many charts can be stored conveniently, and they are not exposed to wear. Each chart can be wired at one end and hooked onto the sliding arms of the cabinet. If each arm holds the charts dealing with a given topic in biology, student aides in the laboratory (p. 681) can find the charts quickly when they are preparing a teacher's order for the day.

The small wooden slide box shown in Fig. 14-22a is convenient for storing prepared slides for microscopic examination. However, when large numbers of sets of slides are stored flat, as in the cabinet shown in Fig. 14-22b, they take less space, the specimens do not slip in the mounting fluid, and all the slides are visible at a glance, so that a student aide can quickly select the desired slides for a day's lesson. Broken slides or coverslips can be detected at the end of a class period (see repair of broken coverslips, Chapter 2).

Microscopes can be stored and easily counted in a cabinet made up of indi-

[32] Improvised darkrooms and other photographic devices are described in Joseph *et al., op. cit.,* pp. 487–92.

(a)

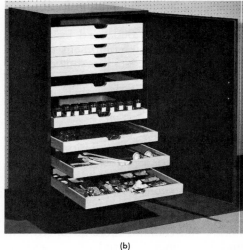

(b)

Fig. 14-20 Storage cabinets: (a) cabinet for general storage of equipment, including microscopes; (b) smaller cabinet of drawers for general storage of instruments, jar mounts, small specimens, and so forth. (b, Courtesy of Ward's Natural Science Establishment, Inc., Rochester, N.Y.)

vidual sections (Fig. 14-23). There is less chance for tumbling or knocking of the microscopes in this type of storage cabinet. Stereoscopic microscopes can be stored in individual cases.

Films may be stored in a cabinet such as that in Fig. 14-24. Lantern slides or 2 × 2 in. slides may be stored in grooved boxes or drawers.

Open-shelf storage space may be covered with transparent plastic sheeting to reduce the accumulation of dust.

Chemicals—especially acids, strong bases, and such substances as chloroform, ether, iodine, and hormones—should be stored in facilities that can be locked. A part of a storeroom may have a special soapstone section which is locked by doors of metal wiring.

Equipment and supplies

What kinds of equipment and supplies does a teacher of biology need? And what is the cost? Many guides and check lists are available. When funds became available to schools as a result of the National Defense Education Act, the Council of Chief State School Officers, with the assistance of Educational Facilities Laboratory, Inc., and others, prepared a *Purchase Guide for Programs in Science, Mathematics, and Modern Foreign Languages* (Ginn, 1959). Master lists are offered of equipment and supplies, categorized as basic, standard, and advanced. The following excerpts from the revised edition of this guide (Ginn, 1965) indicate the kind of specifications suggested.

FIRST-AID CABINET WITH SUPPLIES

Biology—basic, one for each classroom . . .

To provide essential supplies required for immediate treatment of personnel due to accidents or emergencies occurring in the laboratories. This cabinet should be checked at frequent intervals to keep its contents replenished.

Specifications: This cabinet shall be sturdily constructed of steel and be not less than 7 in. ×

8 in. × 2 in. The door shall operate easily. Inside of door shall provide for a first aid hand-book, a listing of cabinet supplies, and a pair of scissors. The cabinet shall be stocked with the following items which should be available in refill size and form when needed:

adhesive bandages, assorted sizes	1 box
adhesive compresses, 10 per pkg.	2 pkg.
adhesive plaster, water-proof, ½ in. × 2½ yd.	1 roll
adhesive plaster, water-proof, 1 in. × 2⅓ yd.	1 roll
absorbent cotton, 1-oz. pkg.	1 pkg.
gauze bandage, 1 in. × 10 yd.	3 rolls
gauze bandage, 2 in. × 10 yd.	1 roll
gauze pads, sterile, 12 in pkg.	1 pkg.
safety pins, 10 on card	1 card
scissors	1
tourniquet	1
hand soap	1 cake
boric acid ointment	1 tube
petroleum jelly	1 tube
any skin antiseptic approved in "New and Non-Official Remedies" published by the American Medical Association	1 oz.
aromatic ammonia inhalant capsules	1 pkg.
first-aid instruction book	1
tincture of Merthiolate, 1 oz.	1 bottle
boric acid solution, 4 per cent, 1 oz.	1 bottle
triangle bandage, not less than 40 in.	1
applicators, cotton	36
tweezers	1
medicine droppers	4

Fig. 14-21 A storage cabinet that also serves as a display case. Many combinations can be made from this type of unitized cabinet. (Courtesy of Ward's Natural Science Establishment, Inc., Rochester, N.Y.)

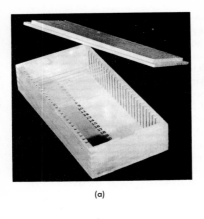

(a)

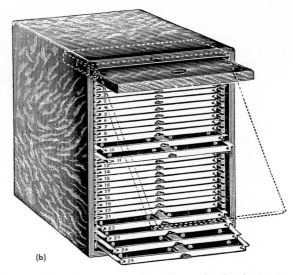

(b)

Fig. 14-22 Storage of microscope slides: (a) small box for temporary handling or for distribution to students; (b) in this cabinet of drawers, slides are held flat and visible. (a, courtesy of General Biological Supply House, Inc., Chicago; b, from *Turtox* catalog.)

Fig. 14-23 Microscope storage cabinet with adjustable shelves. (Courtesy of Hamilton Manufacturing Co., Two Rivers, Wis.)

Fig. 14-24 Cabinet for compact filing of reels of film.

KYMOGRAPH

Biology—standard, one for each classroom . . .

A device for recording, in graph form, observable biological and physical phenomena. Such phenomena as sine waves, propagation frequency and period of pendulums, and springs, characteristics of vibrating forks, normal muscle reaction times, as well as those showing effects of chemical, electrical or other stimulation may be recorded.

Accessories for biological experiments shall include such items as an induction stimulator, muscle lever, scale pan, weights, tambour, electrodes, plethymograph, clamps, signal magnet, heart lever, frog board, frog clips, pneumograph, and tuning fork. . . .

Specifications: This apparatus shall consist of a variable-speed, electrically-driven drum, adjustable in height, to which recording paper can be fastened.

Purchaser shall specify type of accessories desired and whether ink writing or smoke writing style recording is desired. Instructions shall be included with all equipment and accessories.

LACTOMETER

Biology—advanced, one for each school

To determine the temperature of milk and the specific gravity and percentage of water in whole and skim milk.

Specifications: The lactometer shall consist of a glass hydrometer, approximately 12 in. long, with both specific gravity and temperature scales enclosed in the stem. The instrument shall be calibrated at 60° F and shall show specific gravity on a central scale, graduated from 14°–42° Quevenne in 1° divisions, and temperature on a scale above the Quevenne scale, graduated from 20°–110° F in 2° divisions with corrections to be made when the temperature is above or below 60° F indicated. On each side of the Quevenne scale shall be 2 additional scales, with different colored backings, one showing the percentage of water in whole milk to 0.1% and the other the percentage of water in skim milk to 0.1%. Instructions shall be furnished.

LAMP, ALCOHOL

Biology—basic, one for each 2 students . . .

Provides portable source of heat; useful only when hot source is not required.

Specifications: This equipment shall consist of a glass, flat-bottom container of not less than 4 oz. capacity with a threaded neck opening in the top. The top shall be of the metal-screw type which holds an approximately ¼ in. wick. A metal cap which fits snugly over the top shall be provided for extinguishing the flame.[33]

The Biological Sciences Curriculum Study has prepared check lists of equipment, living and preserved materials, and chemicals to be used as guides in ordering supplies for its own courses. (See Fig. 14-25 for an example of such a check list.) Use

[33] Reprinted from *1965 Purchase Guide for Programs in Science and Mathematics,* prepared by The Council of Chief State School Officers, with the assistance of the National Science Foundation and others, Ginn and Co., Boston, 1965.

the catalogs of the various suppliers to determine the cost of the items. Also see *BSCS Newsletter* 18 (June 1963) and *BSCS Newsletter* 21 (April 1964) for complete listings. (You will want to refer to the other publications of BSCS, listed in the bibliography at the end of this chapter.)

Biological supply houses also have prepared basic materials that follow new courses, such as the BSCS versions (see end-of-chapter listings). Furthermore, several supply houses offer to compile for a teacher a cost list of the materials needed to accompany a specific laboratory manual–workbook that the teacher may plan to adopt. (See Fig. 14-26 for Ward's, Fig. 14-27 for Welch, and Fig. 14-28 for Cenco. General Biological and Carolina also have check lists.)

The quantity of material to be ordered depends upon which activities are to be carried on and how they are to be done—as a classroom experiment by all students, by groups of four students working together, by an individual as a project, or as a class demonstration. The amount of material also depends on the annual budget for equipment and supplies.

Some teachers may find the check lists from biological supply houses useful for taking inventory as a first step in planning a science program. Each teacher, of course, will select those activities which fit his own school situation.

Many of the small materials may be purchased in local hardware stores and aquarium shops; some may be brought to class by students and teachers from home supplies. Many of the fresh materials may be collected in season by students on field trips. Other materials and basic equipment you will need to order from supply houses (Appendix C).

PEOPLE, PLACES, AND THINGS

We find that teachers are always looking for new ways to develop a concept in the science classroom. Do you belong to a professional science teachers' organization? Many groups of teachers plan techniques meetings in which they

Chap. 1   The Web of Life
  1.1   Basic   Observing Living Things
  1.2   Basic   An Experiment: The Germination of Seeds
  1.3   Basic   Interrelationships of Producers and Con-
                     sumers
  1.4   Basic   Use of the Microscope: Introduction
  1.5   Basic   Use of the Microscope: Biological Material
  1.6   Optional   Use of the Stereoscopic Microscope

Chap. 2   Individuals and Populations
  2.1   Basic   Population Growth: A Model
  2.2   Basic   Study of a Yeast Population
  2.3   Basic   Factors Limiting Populations
  2.4   Optional   Effect of an Abiotic Environmental Factor
                     on a Population

Chap. 3   Communities and Ecosystems
  3.1   Basic   The Study of a Biotic Community
  3.2   Optional   Competition Between Two Species of Plants
  3.3   Highly Rec. A Comparative Study of Habitats

(a)

Equipment and Supplies†
(quantities for 1 class of 28 students)

The following list does not include items students can supply or items to be purchased locally from petty cash (e.g., toothpicks, newspapers, aluminum foil, plastic bags, oatmeal, fresh vegetables, etc.).

	Amt	Equipment	Suppliers*
		LARGE	
BASIC	1	Autoclave or pressure cooker—22 qt.	A,F,G,H,M,N,Q,S,V,W,Z,a,b,h,i,k,m,n
	2	Aquaria	A,F,G,H,M,N,P,Q,S,U,W,Y,Z,a,p,d,h,i,k,m,n
	2	Balances—triple beam (.01 gm)	A,F,G,H,M,N,Q,S,V,W,Y,Z,a,b,h,i,k,m,n
	2	Hotplates—electric, or stove	A,F,G,H,M,N,Q,S,V,W,X,Y,Z,a,b,d,h,i,k,m,n
	1	Incubator	A,F,G,H,J,M,N,Q,S,V,W,X,Z,a,b,d,h,i,j,k,m,n
	7	Lamps—gooseneck	F,G,H,M,N,Q,S,V,W,a,b,d,h,i,k,n
	14	Microscopes—compound monocular	A,B,D,F,G,H,K,L,M,N,Q,S,U,V,W,Y,Z,a,b,f,h,i,k,m,n
	7	—stereo binocular	A,B,D,F,G,H,K,L,M,N,Q,S,U,V,W,Y,Z,a,b,f,h,i,k,m,n
	1	Refrigerator	A,F,H,N,Q,S,W,Z,a,b,h,i,k,m,n
ALT. BASIC	1	Temperature gradient box	H,Q,d,h,i
HIGH. REC.	1	Aquarium—5 gal	A,F,G,H,M,N,Q,S,U,W,Y,Z,a,b,d,h,i,k,m,n
	2	Balances—triple beam (.01 gm)	A,F,G,H,M,N,Q,S,V,W,Y,Z,a,b,h,i,k,m,n
	7	Lenses—oil immersion	A,B,F,G,H,M,N,Q,S,W,Z,a,f,h,i,k,m,n
OPTIONAL	1	Drying oven	A,F,G,H,M,,N,Q,S,W,Z,a,b,d,h,i,k,m,n
	3	Temperature gradient boxes	H,Q,b,d,h,i
		SMALL	
BASIC	28	Blood lancets—sterile, disposable	A,F,G,H,I,M,N,Q,S,U,V,W,X,Z,a,h,i,k,m,n
	1	Borer—cork	A,F,G,H,M,N,Q,S,V,W,Z,a,d,h,i,k,m,n
	14	Brushes—camel's hair	A,F,G,H,I,M,N,Q,S,V,W,Z,a,h,i,k,m,n

(b)

Fig. 14-25   A sampling of how equipment and supplies are suggested for laboratory activities in one of the BSCS courses (Green Version): (a) the laboratory exercises are classified (second column) as "Basic," "Optional," and "Highly Recommended"; (b) suppliers of equipment for these classifications are indicated by letters (right-hand column), which refer to a key in the *BSCS Newsletter*. (Excerpts from pages in *BSCS Newsletter* 21, April 1964; courtesy of Biological Sciences Curriculum Study.)

share successful demonstrations and ask help in improving less successful techniques. Along with the professional journals of such organizations as the National Science Teachers Assoc. and the National Assoc. of Biology Teachers, local groups of teachers often publish a small newsletter in which they exchange techniques and ideas for project work with students.

We know of a school where the science department worked closely with the shop depart-

ment in creating a science-shop course in which practice was given in the use of tools—wood lathes, jigsaws, and the like. Such a course might be planned for teachers in a community, and also opened to students who plan to major in science. Students who plan to specialize in science need practice in manipulating skills.

Many comunities are concerned with improving science facilities, developing plans for building a new wing or an entire school, or remodeling

ITEM NAME AND DESCRIPTION	CAT. NO.	Blue Version	Green Version	Yellow Version	Animal Growth	Plant Growth	Microbes	Interdependence	Field Ecology	Plant Hormones	Unit of Quantity	Price per Unit of Quantity	NUMBER OF UNITS YOU DESIRE	Price Extension
		Used In Versions			Used In Lab Blocks									
1. EQUIPMENT														
Abrasive powder, 400 grit	C 4060.40			✓							1 lb.	$1.70		
Aerator, aquarium	C 656.5	✓	✓	✓	✓	✓					1	6.00		
Aerosol applicator unit	C 1790			✓		✓					1	2.95		
Air pump	C 656.1	✓	✓								1	21.25		
Anemometer	C 5770								✓		1	59.95		
Aquarium, 3 gallon	C 655.2	✓	✓	✓							1	5.50		
Aquarium, 5 gallon	C 655.3	✓	✓	✓							1	6.50		
Aquarium, 10 gallon	C 655.5	✓	✓	✓							1	11.75		
Aquarium, 15 gallon	C 655.6	✓									1	16.50		
Aquarium gravel	C 651	✓	✓	✓							10 lbs.	.60		

Fig. 14-26 Part of a page from Ward's order booklet keying items in the BSCS check list against the company's supply catalog. (Courtesy of Ward's Natural Science Establishment, Inc., Rochester, N.Y.)

Order	Quan. Rec.	Welch Cat. No.	Items	Unit Quan.	Unit Price	Total Amount
_____	X	22211	Mushroom Growth Kit, for 1 class (Exp. 12.3)	ea	$ 4.00	_____
_____	21	50506	Planaria (Exp. 29.1)	Culture	3.50	_____
_____	14	51011	Tadpoles (Exp. 29.1)	ea	.20	_____
			CULTURES - FOR 1 CLASS OF 28			
_____	X	50716	Daphnia (Exp. 22.1)	Culture	3.50	_____
_____	X	51951	Drosophila, brown vestigial (Exp. 31.1)	Culture	3.50	_____
_____	X	51851	Drosophila, brown (Exp. 31.1)	Culture	3.50	_____

Fig. 14-27 Part of a page from a list of equipment available from Welch for use with the BSCS Yellow Version. (Courtesy of The Welch Scientific Co., Skokie, Ill.)

CENCO EQUIPMENT LIST FOR

BSCS LABORATORY BLOCK

Plant Growth and Development

1963 COMMERCIAL EDITION

DIRECTIONS: Quantities "recommended" are for a class of 28 students. The Unit Package indicates the minimum quantity that can be supplied. For the definitions of other markings (dns, *, etc.) see the footnotes at the end of this section.

Cenco No.	Description of Item	Rec. Qty.	Unit Pkg.	Unit Price	Order Here Units	Ext
	LABORATORY APPARATUS AND SPECIAL SUPPLIES					
88039-2	Cellophane, dark blue, (20" x 25' roll)	15 ft	1 rl	1.40	....	....
88039-1	Cellophane, dark green, (20" x 25' roll)	5 ft	1 rl	1.40	....	....
88039-4	Cellophane, dark red, (20" x 25' roll)	15 ft	1 rl	1.40	....	....
12465-2	Cork Borer, Set of 6 ($^3/_{16}$" to $^1/_2$")	1	1 set	2.75	....	....
12525	Cotton, Non-absorbent	1 lb	1 lb	1.20	....	....
68205	Cutter, Tissue	7	1	2.40	....	....
12750-11 or	Distillation Unit, Electric	1 or	1	135.00	....	....
12820	Demineralizer (requires No. 12821 filters)	1	1	39.50	....	....
12821-1	Resin Filter (for No. 12820)	–	1	2.25	....	....
12821-2	Resin Filter (for No. 12820) (pkg of 6)	–	1 pkg	11.70	....	....
65310	Eyepiece Micrometer	14	1	7.50	....	....
50245	Finger Print Ink, 2 oz. btl.	1	1	1.50	....	....

Fig. 14-28 Part of a page from a list of equipment available from Cenco for use with a BSCS laboratory block. (Courtesy of Central Scientific Co.)

standard classrooms into science rooms. Where can you go for help? Visit other schools and colleges or write to schools which have specialized facilities. You may also wish to consult professional education journals; in them you will find the names of manufacturers of equipment and supplies who will send you their catalogs.

JOURNALS

The following list is only a sampling of specialized journals for teachers of biology and the many professional journals in education. Sources of audio-visual materials are listed in Appendix B; supply houses are listed in Appendix C.

The American Biology Teacher, National Assoc. of Biology Teachers.

American Scientist, Soc. of Sigma Xi, 51 Prospect St., New Haven, Conn.

Canadian Nature, Audubon Soc. of Canada, 181 Jarvis St., Toronto 2, Ont., Can.

Cornell Rural School Leaflets, New York State College of Agriculture, Cornell Univ., Ithaca, N.Y.

Journal of Research in Science Teaching, National Assoc. for Research in Science Teaching.

Look and Listen (Britain's Audio-visual Aids Journal), 62 Doughty St., London, W.C. 1, Eng.

Metropolitan Detroit Science Review, Metropolitan Detroit Science Club.

National Geographic, National Geographic Soc., 16 and M Sts., N.W., Washington 6, D.C.

Natural History, Am. Museum of Natural History, Central Park W. and 79 St., New York 24, N.Y.

Nature, Macmillan & Co., St. Martin's St., London, W.C. 2, Eng.

Review of Educational Research, Am. Educational Research Assoc., 1201 16 St., N.W., Washington 6, D.C.

School Science and Mathematics, Central Assoc. of Science and Mathematics Teachers, Menasha, Wis.

Science, Am. Assoc. for the Advancement of Science, 1515 Massachusetts Ave., N.W., Washington 5, D.C.

Science and Children, National Science Teachers Assoc., 1201 16 St., N.W., Washington 6, D.C.

Science Books, Am. Assoc. for the Advancement of Science, 1515 Massachusetts Ave., N. W., Washington 5, D.C.

Science Education, National Assoc. for Research in Science Teaching, C. M. Pruitt, ed., Univ. of Tampa, Tampa, Fla.

Science News Letter, Science Service, 1719 N St., N.W., Washington 6, D.C.

The Science Teacher, National Science Teachers Assoc., 1201 16 St., N.W., Washington 6, D.C.

Scientific American, Scientific American, 415 Madison Ave., New York 17, N.Y.

The Times Science Review, The Times Publishing Co., Printing House Sq., London, E.C. 4, Eng.; 25 E. 54 St., New York 22, N.Y.

World Science Review, 11 Eaton Place, London, S.W. 1, Eng.

SCIENCE EDUCATION: PROFESSIONAL RESOURCES

Am. Assoc. for the Advancement of Science, *Guidelines for Preparation Programs of Teachers of Secondary School Science and Mathematics,* 1515 Massachusetts Ave., N.W., Washington 5, D.C., 1961.

————, *Science Teaching Materials for Elementary and Junior High Schools,* J. Mayor, Director, 1963.

Am. Institute of Biological Sciences (AIBS), *Equipment and Literature Recommended for Use in Teaching High School Biology,* 2000 P St., N.W., Washington 6, D.C., 1964.

————, Speakers Bureau, *Visiting Biologists Program for High Schools: Grades 9–12.*

Am. Medical Assoc., *Mental Retardation,* 535 N. Dearborn St., Chicago, Ill., 1965.

Am. Physiological Soc., *Laboratory Experiments in Elementary Human Physiology,* 9650 Wisconsin Ave., Washington 14, D.C., 1961.

————, *Laboratory Experiments in General Physiology,* S. Tipton, ed.

Am. Phytopathological Soc., Committee on Teaching, *Sourcebook of Laboratory Exercises in Plant Pathology,* A. Kelman, Director, North Carolina State College, Raleigh, 1964.

Ashbaugh, B., and M. Beuschlein, *Things to Do in Science and Conservation,* Interstate Printers and Publishers, Danville, Ill., 1960.

Biological Sciences Curriculum Study (BSCS), A. B. Grobman, Director.

MATERIALS FOR THE STUDENT

1. Textbooks
   Biological Science: An Inquiry into Life (Yellow Version), Harcourt, Brace & World, New York, 1963.
   Biological Science: Molecules to Man (Blue Version), Houghton Mifflin, Boston, 1963.
   High School Biology (Green Version), Rand McNally, Chicago, 1963.
   High School Biology: Special Materials, Special Student Committee; experimental use, 1963–64.
   Biological Science: Interaction of Experi-

ments and Ideas (a BSCS Second Course), Prentice-Hall, Englewood Cliffs, N.J., 1965.

2. Laboratory blocks (Heath, Boston; others in preparation; each accompanied by a *Teacher's Supplement*)
   Follansbee, H., *Animal Behavior,* 1965.
   Glass, B., *Genetic Continuity,* 1965.
   Jacobs, W. P., and C. E. LaMotte, *Regulation in Plants by Hormones,* 1964.
   Lee, A. E., *Plant Growth and Development,* 1963.
   Moog, F., *Animal Growth and Development,* 1963.
   Phillips, E. A., *Field Ecology,* 1964.
   Pramer, D., *Life in the Soil,* 1965.
   Richards, A. G., *The Complementarity of Structure and Function,* 1963.
   Sussman, A. S., *Microbes: Their Growth, Nutrition, and Interaction,* 1964.

3. *Research Problems in Biology: Investigations for Students,* Series One–Four, Anchor Books, Doubleday, Garden City, N.Y., 1963, 1965.

MATERIALS FOR THE TEACHER

1. Books
   Barthelemy, R. E., J. R. Dawson, Jr., and A. E. Lee, *Innovations in Equipment and Techniques for the Biology Teaching Laboratory,* Heath, Boston, 1964.
   Schwab, J., Supervisor, *Biology Teachers' Handbook,* Wiley, New York, 1963.

2. Bulletins (AIBS, Washington, D.C.)
   Brandwein, P. F., *et al., Teaching High School Biology: A Guide to Working with Potential Biologists,* BSCS Bull. 2, 1962.
   Grobman, A. B., *et al., BSCS Biology—Implementation in the Schools,* BSCS Bull. 3, 1964.
   Hurd, P. DeH., *Biological Education in American Secondary Schools: 1890–1960,* BSCS Bull. 1, 1961.

3. Newsletter (BSCS, Univ. of Colorado, Boulder)

4. Pamphlet Series (Heath, Boston)

5. Techniques films (Thorne Films, Boulder, Colo.)

Brandwein, P. F., "Elements in a Strategy for Teaching Science in Elementary School," The Burton Lecture, in J. Schwab and P. F. Brandwein, *The Teaching of Science* (Inglis and Burton lectures), Harvard Univ. Press, Cambridge, Mass., 1962.

———, *The Gifted Student as Future Scientist,* Harcourt, Brace & World, New York, 1955.

———, F. G. Watson, and P. E. Blackwood, *Teaching High School Science: A Book of Methods,* Harcourt, Brace & World, New York, 1958.

Burnett, R. W., *Teaching Science in the Secondary School,* Rinehart, New York, 1957.

Buros, O., ed., *Mental Measurements Yearbook,* Gryphon Press, Highland Park, N.J., 1959.

Coler, M., ed., *Essays on Creativity in the Sciences,* New York Univ. Press, New York, 1963.

College Entrance Examination Board, *Advanced Placement Program Syllabus,* Princeton, N.J., 1958.

Conservation Education Assoc., *Directory of Workshop Courses and Adult Training Programs in Conservation of Natural Resources,* Conservation Education Assoc., c/o W. Clark, Eastern Montana College, Billings, 1965.

Council of Chief State School Officers *et al., 1965 Purchase Guide for Programs in Science and Mathematics,* Ginn, Boston, 1965.

Educational Press Assoc., *America's Education Press,* Yearbook 28, Glassboro State College, Glassboro, N.J., 1963.

Eysenck, H., *Experiments in Motivation,* Macmillan, New York, 1964.

Feldman, D., *Electron Micrographs in the Teaching of Biology,* Bur. of Publications, Teachers College, Columbia Univ., New York, 1964.

Goldstein, P., *How to Do an Experiment,* Harcourt, Brace & World, New York, 1957.

Jersild, A. T., and R. Tasch, *Children's Interests and What They Suggest for Education,* Bur. of Publications, Teachers College, Columbia Univ., New York, 1951.

Klopfer, L., *The Cells of Life,* "History of Science Cases," Science Research Associates, Chicago, 1964.

Koppelman, R., *Continental Classroom: The New Biology,* Learning Resources Institute, New York, 1961.

Lawson, C., and R. Paulson, eds., *Laboratory and Field Studies in Biology: A Sourcebook for Secondary Schools,* Holt, Rinehart and Winston, New York, 1960.

Lumsdaine, A., and R. Glaser, eds., *Teaching Machines and Programmed Learning,* National Education Assoc., Washington, D.C., 1960.

McClelland, D., *The Achieving Society,* Van Nostrand, Princeton, N.J., 1961.

McCollum, H., *Let's Look at Our Facilities for Teaching Science,* State Education Dept., Baton Rouge, La., 1960.

Madsen, K., *Theories of Motivation,* 2nd ed., Howard Allen, Inc., Cleveland, 1961.

Martin, W. E., *Facilities and Equipment for Science and Mathematics: Requirements and Recommenda-*

tions of State Departments of Education, U.S. Govt. Printing Office, Washington, D.C., 1960.

Miller, D., and G. Blaydes, *Methods and Materials for Teaching the Biological Sciences*, 2nd ed., McGraw-Hill, New York, 1962.

Montgomery, D. H., ed., *Directory of Marine Sciences Programs and Facilities on the Pacific Coast* (annual), D. H. Montgomery, Biological Sciences Dept., California State Polytechnic College, San Luis Obispo, Calif.

Moore, S., and J. Viorst, eds., *Wonderful World of Science*, Bantam, New York, 1960.

Munzer, M., and P. F. Brandwein, *Teaching Science Through Conservation*, McGraw-Hill, New York, 1960.

National Aeronautics and Space Administration, Scientific and Technical Information Division, *Bioastronautics Data Book*, P. Webb, ed., Washington, D.C., 1964.

National Education Assoc., *Quality Science for Secondary Schools*, Bull. of National Assoc. of Secondary School Principals, Vol. 44, No. 260, 1201 16 St., N.W., Washington 6, D.C., 1960.

National Science Foundation Course Center, *Science Course Improvement Projects*, National Science Foundation, Washington, D.C., 1962.

National Science Teachers Assoc., *New Developments in High School Science Teaching*, 1201 16 St., N.W., Washington 6, D.C., 1960.

————, *Planning for Excellence in High School Science*, 1961.

————, Committee on Science Facilities, *School Facilities for Science Instruction*, 2nd ed., edited by J. S. Richardson, 1961.

————, *Science for the Academically Talented Student in the Secondary School*, 1959.

National Soc. for the Study of Education, *Rethinking Science Education*, Yearbook 59, Part I, Univ. of Chicago Press, Chicago, 1960.

————, *Science Education in American Schools*, Yearbook 46, Part I, Univ. of Chicago Press, Chicago, 1947.

Obourn, E., and C. Koelsche, *Analysis of Research in the Teaching of Science*, U.S. Govt. Printing Office, Washington, D.C., 1959.

George Peabody College for Teachers, *Free and Inexpensive Learning Materials* (annual), Nashville, Tenn.

Pepe, T., *Free and Inexpensive Educational Aids*, Dover, New York, 1960.

Perlman, P., *General Laboratory Techniques*, Franklin, Englewood, N.J., 1964.

Richardson, J. S., *Science Teaching in Secondary Schools*, Prentice-Hall, Englewood Cliffs, N.J., 1957.

————, *et al.*, eds., *Resource Literature for Science Teachers*, 2nd ed., Ohio State Univ. Press, Columbus, 1961.

Schwab, J., "The Teaching of Science as Enquiry," The Inglis Lecture, in J. Schwab and P. F. Brandwein, *The Teaching of Science* (Inglis and Burton lectures), Harvard Univ. Press, Cambridge, Mass., 1962.

Science Clubs of America, *Sponsors' Handbook* (annual), Science Service, 1719 N St., N.W., Washington 6, D.C.

Science Manpower Project Committee, *Policies for Science Education*, Bur. of Publications, Teachers College, Columbia Univ., New York, 1960.

Soc. of Wood Science and Technology, *Study: Education in Wood Science and Technology*, G. Garratt, ed., Yale Univ., New Haven, Conn., 1962.

Stone, D., *Modern High School Biology*, Science Manpower Project, Bur. of Publications, Teachers College, Columbia Univ., New York, 1959.

Sutton, R., *Demonstration Experiments in Physics*, McGraw-Hill, New York, 1938.

Taylor, C., ed., *Creativity: Progress and Potential*, McGraw-Hill, New York, 1964.

————, and F. Barron, *Scientific Creativity: Its Recognition and Development*, Wiley, New York, 1963.

Textbooks in Print (annual), Bowker, New York.

Thomas, R., and S. Swartout, *Integrated Teaching Materials*, Longmans, Green, New York, 1960.

UNESCO, *Sourcebook for Science Teaching*, rev. ed., UNESCO Publications Center, 801 Third Ave., New York 22, N.Y., 1964.

U.S. Dept. of Health, Education, and Welfare, *Aids for Teaching Science: A Suggested Checklist for Assessing a Science Program*, U.S. Govt. Printing Office, Washington, D.C., 1964.

————, *Science Publications: An Annotated Guide to Selected Listings*, U.S. Govt. Printing Office, Washington, D.C., 1963.

U.S. Govt. Printing Office, *Selected United States Government Publications* (biweekly), Washington, D.C.

Welte, A., J. Dimond, and A. Friedl, *Your Science Fair*, 2nd ed., Burgess, Minneapolis, 1962.

Section *FOUR*

Appendixes:
References for
the teacher

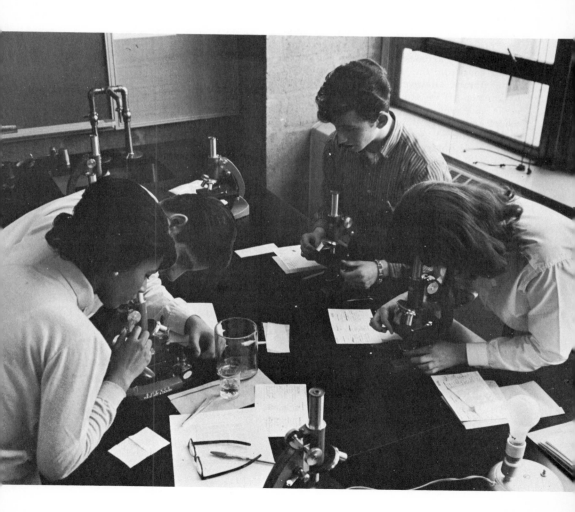

Appendix **A**

Bibliography: Books for a Library
of Biological Sciences

Some of these books are recommended for the professional library of the teacher of biology. Many others are suggested as a starting point in building a school library of books in biological sciences for the use of both students and teachers. In general, these are technical references, not popularized accounts of biology. There are many fine popularized accounts and paperbound books, a selection of which might constitute the hard core of a student's home library. However, our listing would be without end if we attempted to list them all.

We have seen many of these books in high school and college libraries, and in this light they are offered as suggestions in building a professional library.

You will want to refer to catalogs of paperbound publications, the most nearly comprehensive of which is the monthly publication *Paperbound Books in Print* (R. R. Bowker Co., 1180 Ave. of the Americas, New York 36, N.Y.). Also look into the cumulative listing of titles in science, medicine, and technology published in the United States: P. Steckler, ed., *American Scientific Books: 1960–1962* (Bowker, 1962). Many students and teachers refer to H. Deason, ed., *A Guide to Science Reading* (Signet, New Am. Library, 1963, $0.60), an annotated listing of science paperbacks.

Directories of book publishers are included in such reference works as *American Book Trade Directory, Literary Market Place, Books in Print,* and *Cumulative Book Index.* The first three are annuals, published by Bowker (New York); the fourth is a monthly, published by Wilson (New York).

General science

Abercrombie, M., C. J. Hickman, and M. L. Johnson, *A Dictionary of Biology,* Aldine, Chicago, 1962.

Asimov, I., *A Short History of Biology,* Natural History Press, Doubleday, Garden City, N.Y., 1964; Am. Museum Science Books, Natural History Press, Doubleday, 1964.

——, *The Wellsprings of Life,* Abelard-Schuman, New York, 1961; Signet, New Am. Library, New York, 1962.

Bacq, Z. M., and P. Alexander, *Fundamentals of Radiobiology,* rev. ed., Pergamon (Macmillan), New York, 1961.

Balabukha, V. S., ed., *Chemical Protection of the Body Against Ionizing Radiation,* Pergamon, New York, 1964.

Barthelemy, R. E., J. R. Dawson, Jr., and A. E. Lee, *Innovations in Equipment and Techniques for the Biology Teaching Laboratory* (BSCS), Heath, Boston, 1964.

Beck, W. S., *Modern Science and the Nature of Life,* Harcourt, Brace & World, New York, 1957.

Beveridge, W., *The Art of Scientific Investigation,* rev. ed., Norton, New York, 1957.

Bier, W. C., ed., *Problems in Addiction: Alcohol and Drug Addiction,* Fordham Univ. Press, New York, 1962.

Biological Sciences Curriculum Study (BSCS), *Research Problems in Biology: Investigations for Students,* Series One–Four, Anchor Books, Doubleday, Garden City, N.Y., 1963, 1965.

Bodenheimer, F. S., *History of Biology,* Banner, New York, 1958.

Bonner, J., *The Ideas of Biology,* Harper & Row, New York, 1962; Harper Torchbook, 1962.

Brandwein, P. F., *et al., Teaching High School Biology: A Guide to Working with Potential Biologists,* BSCS Bull. 2, AIBS, Washington, D.C., 1962.

Broda, E., *Radioactive Isotopes in Biochemistry,* Van Nostrand, Princeton, N.J., 1960.

Bronowski, J., *The Common Sense of Science,* Harvard Univ. Press, Cambridge, Mass., 1953.

Brown, M. J., *Scientific Vocabulary for Beginning Zoology Students and Non-scientific Students,* Pageant, New York, 1957.

Bugelski, B., *The Psychology of Learning,* Holt, Rinehart and Winston, New York, 1956.

Butterfield, H., *The Origins of Modern Science: 1300–1800,* rev. ed., Macmillan, New York, 1957.

Christensen, B., and B. Buchmann, eds., *Progress in Photobiology* (The Finsen Memorial Congress, Copenhagen, 1960), Elsevier, Amsterdam, 1961.

Clagett, M., ed., *Critical Problems in the History of Science,* Univ. of Wisconsin Press, Madison, 1959.

Claus, W., ed., *Radiation Biology and Medicine,* Addison-Wesley, Reading, Mass., 1958.

Clifton, C. E., ed., *Annual Review of Microbiology,* Annual Reviews, Palo Alto, Calif., 1962.

Conant, J., *On Understanding Science,* Yale Univ. Press, New Haven, Conn., 1947; Mentor, New Am. Library, New York, 1951.

———, ed., *Harvard Case Histories in Experimental Science,* reprint ed., 2 vols., Harvard Univ. Press, Cambridge, Mass., 1957.

Crombie, A. C., *Medieval and Early Modern Science,* 2 vols., Doubleday, Garden City, N.Y., 1959.

Crow, W. B., *A Synopsis of Biology,* Williams & Wilkins, Baltimore, 1960.

Dampier, W. C., ed., *Readings in the Literature of Science,* Harper & Row, New York, 1959.

Dauvillier, A., *The Photochemical Origin of Life,* Academic, New York, 1965.

Davidsohn, I., *see* Todd, J., and A. Sanford.

Dawes, B., *A Hundred Years of Biology,* Macmillan, New York, 1952.

Dillon, L., *The Science of Life,* Macmillan, New York, 1964.

Dobell, C., *see* Leeuwenhoek, A. van.

Duquesne, M., *Matter and Antimatter,* Harper & Row, New York, 1960.

Edmondson, W. T., *see* Ward, H. B., and G. C. Whipple.

Frank, P., *Modern Science and Its Philosophy,* Harvard Univ. Press, Cambridge, Mass., 1949; Braziller, New York, 1955.

Gabriel, M., and S. Fogel, eds., *Great Experiments in Biology,* Prentice-Hall, Englewood Cliffs, N.J., 1955.

Gerard, R. W., *Unresting Cells,* new ed., Harper & Row, New York, 1949; Harper Torchbook, 1961.

Goldstein, P., *How to Do an Experiment,* Harcourt, Brace & World, New York, 1957.

Gray, L. H., *see* Lea, D. E.

Guilford, J. P., *Personality,* McGraw-Hill, New York, 1959.

Hall, T. S., *A Sourcebook in Animal Biology,* Hafner, New York, 1951.

Hall, V. E., ed., *Annual Review of Physiology,* Annual Reviews, Palo Alto, Calif., 1963.

Handbook of Toxicology, 5 vols., prepared under the direction of the Committee on Handbook of Biological Data, Division of Foods and Nutrition, National Research Council, National Academy of Sciences, Saunders, Philadelphia, 1957–59.

Harvey, O. J., D. E. Hunt, and H. M. Schroder, *Conceptual Systems and Personality Organization,* Wiley, New York, 1961.

Harvey, W., *The Circulation of Blood and Other Writings of William Harvey,* trans. by K. Franklin, Everyman's Library, Dutton, New York, 1963.

Heftmann, E., ed., *Chromatography,* Reinhold, New York, 1961.

Henderson, I., and W. Henderson, *A Dictionary of Biological Terms,* 8th ed., edited by J. Kenneth, Van Nostrand, Princeton, N.J., 1963; previously published under the title *A Dictionary of Scientific Terms.*

Hepler, O., *Manual of Clinical Laboratory Methods,* 4th ed., Thomas, Springfield, Ill., 1960.

Heyerdahl, T., *Kon-Tiki: Across the Pacific by Raft,* trans. by F. H. Lyon, Rand McNally, Chicago, 1950; Pocket Books, New York, 1957.

Hilgard, E. R., *Introduction to Psychology,* 3rd ed., Harcourt, Brace & World, New York, 1962.

———, *Theories of Learning,* 2nd ed., Appleton-Century-Crofts, New York, 1956.

Hollaender, A., ed., *Radiation Biology,* 3 vols., McGraw-Hill, New York, 1954–56.

Hooke, R., *Micrographia, or Some Physiological Descriptions of Minute Bodies—Observations by Robert Hooke,* orig. ed., 1665; Dover, New York, 1961.

Hull, L., *History and Philosophy of Science,* Longmans, Green, New York, 1959.

Huntsman, A., *Life and the Universe,* Univ. of Toronto Press, Toronto, 1959.

Jackson, F., and P. Moore, *Life in the Universe,* · Norton, New York, 1962.

Jaeger, E., *A Sourcebook of Biological Names and Terms,* 3rd ed. rev., Thomas, Springfield, Ill., 1962.

Johnson, W., and W. Steere, eds., *This Is Life,* Holt, Rinehart and Winston, New York, 1962.

Kenneth, J., *see* Henderson, I., and W. Henderson.

Kluyver, A., and C. Van Niel, *The Microbe's Contribution to Biology,* Harvard Univ. Press, Cambridge, Mass., 1956.

Kurtz, S., ed., *Electron Microscopic Anatomy,* Academic, New York, 1964.

Lawson, C., and R. Paulson, eds., *Laboratory and Field Studies in Biology: A Sourcebook for Secondary Schools,* Holt, Rinehart and Winston, New York, 1960.

Lea, D. E., *Actions of Radiations on Living Cells,* 2nd ed., revised by L. H. Gray, Cambridge Univ. Press, New York, 1955.

Leeuwenhoek, A. van, *Antony van Leeuwenhoek and His "Little Animals,"* coll., trans., and ed. by C. Dobell, Harcourt, Brace & World, New York, 1932; Dover, New York, 1960.

Leicester, H., *The Historical Background of Chemistry,* Wiley, New York, 1956.

Luck, J. M., ed., *Annual Review of Biochemistry,* Annual Reviews, Palo Alto, Calif., 1963.

MacArthur, R., and J. Connell, *The Biology of Populations,* Wiley, New York, 1965.

McCarthy, R., ed., *Drinking and Intoxication: Selected Readings in Social Attitudes and Controls,* College and Univ. Press, New Haven, Conn., 1963.

Machlis, L., ed., *Annual Review of Plant Physiology,* Annual Reviews, Palo Alto, Calif., 1963.

Marsland, D., *Principles of Modern Biology,* 4th ed., Holt, Rinehart and Winston, New York, 1964.

Mavor, J. W., *General Biology,* 5th ed., Macmillan, New York, 1959.

Medawar, P., *The Future of Man,* Basic Books, New York, 1960.

Merrill, P., *Space Chemistry,* Univ. of Michigan Press, Ann Arbor, 1963.

Milne, L., and M. Milne, *The Biotic World and Man,* 3rd ed., Prentice-Hall, Englewood Cliffs, N.J., 1965.

Moment, G., ed., *Frontiers of Modern Biology,* Houghton Mifflin, Boston, 1962.

Morison, R., *Scientist,* Macmillan, New York, 1964.

Morrow, C. A., *Biochemical Laboratory Methods for Students of the Biological Sciences,* Wiley, New York, 1927.

Mowrer, O. H., *Learning Theory and Behavior,* Wiley, New York, 1960.

Nagel, E., *The Structure of Science: Problems in the Logic of Scientific Explanation,* Harcourt, Brace & World, New York, 1961.

National Research Council, Institute of Laboratory Animal Resources, *Laboratory Animals Resources,* Part II, *Animals for Research,* rev. ed., National Research Council, Washington, D.C., 1963.

Nigrelli, R., *Metabolites of the Sea,* Heath, Boston, 1963.

Nordenskiöld, E., *The History of Biology,* Tudor, New York, 1960.

Northrop, F., *The Logic of the Sciences and the Humanities,* Macmillan, New York, 1947; Meridian, World Publishing, Cleveland.

Organization for Economic Cooperation and Development, *New Thinking in School Biology,* McGraw-Hill, New York, 1963.

Pennak, R., *Collegiate Dictionary of Zoology,* Ronald, New York, 1964.

Perlman, P., *General Laboratory Techniques,* Franklin, Englewood, N.J., 1964.

Piltz, A., ed., *Science Publications,* U.S. Govt. Printing Office, Washington, D.C., 1963.

Putnam, W., *Geology,* Oxford Univ. Press, New York, 1964.

Racker, E., *Mechanisms in Bioenergetics,* "Advanced Biochemistry Series," Academic, New York, 1965.

Riedman, S., and E. Gustafson, *Portraits of Nobel Laureates in Medicine and Physiology,* Abelard-Schuman, New York, 1963.

Royal College of Physicians of London, *Smoking and Health: A Summary on Smoking in Relation to Cancer of the Lung and Other Diseases,* Pitman, New York, 1962.

Sanderson, R., *Teaching Chemistry with Models,* Van Nostrand, Princeton, N.J., 1962.

Schwab, J., and P. F. Brandwein, *The Teaching of Science* (Inglis and Burton lectures), Harvard Univ. Press, Cambridge, Mass., 1962.

Schwartz, G., and P. Bishop, eds., *Moments of Discovery,* 2 vols., Basic Books, New York, 1958.

Seliger, H. H., and W. D. McElroy, *Light: Physical and Biological Action,* "A.E.C. Monograph Series on Radiation Biology and Industrial Hygiene," Academic, New York, 1965.

Shelford, V., *The Ecology of North America,* Univ. of Illinois Press, Urbana, 1963.

Shilling, C., ed., *Atomic Energy Encyclopedia in the Life Sciences,* Saunders, Philadelphia, 1964.

Simpson, G. G., *Principles of Animal Taxonomy,* Columbia Univ. Press, New York, 1961.

———, *This View of Life,* Harcourt, Brace & World, New York, 1964.

———, and W. S. Beck, *Life: An Introduction to Biology,* 2nd ed., Harcourt, Brace & World, New York, 1965.

Singer, C., *From Magic to Science,* Dover, New York, 1959.

———, *A History of Biology,* 3rd rev. ed., Abelard-Schuman, New York, 1959.

Smith, R. F., *Guide to the Literature of the Zoological Sciences,* 6th ed., Burgess, Minneapolis, 1962.

———, ed., *Annual Review of Entomology,* Annual Reviews, Palo Alto, Calif., 1963.

Spector, W., ed., *Handbook of Biological Data,* Saunders, Philadelphia, 1956.

Stent, G., *Molecular Biology of Bacterial Viruses,* Freeman, San Francisco, 1963.

Sutton, O. G., *et al., The World Around Us,* Macmillan, New York, 1960.

Talalay, P., ed., *Drugs in Our Society,* Johns Hopkins Press, Baltimore, 1964.

Telfer, W., and D. Kennedy, *The Biology of Organisms,* Wiley, New York, 1965.

Thompson, Sir D'A., *On Growth and Form,* 2 vols., Cambridge Univ. Press, New York, 1952; abridged ed. by J. Bonner, 1961.

Thomson, G., *The Inspiration of Science,* Oxford Univ. Press, New York, 1961.

Thorpe, W., *Biology and the Nature of Man,* Oxford Univ. Press, New York, 1962.

Todd, J., and A. Sanford, *Clinical Diagnosis by Laboratory Methods,* 13th ed., edited by I. Davidsohn and B. Wells, Saunders, Philadelphia, 1962.

Toulmin, S., *The Philosophy of Science,* Harper & Row, New York, 1960.

Vallery-Radot, R., *The Life of Pasteur,* trans. by R. L. Devonshire, Doubleday, Garden City, N.Y., 1923; Dover, New York.

Villee, C., *Biology,* 4th ed., Saunders, Philadelphia, 1962.

Waddington, C., *The Ethical Animal,* Atheneum, New York, 1961.

———, *The Nature of Life,* Atheneum, New York, 1962.

Ward, H. B., and G. C. Whipple, *Fresh-Water Biology,* 2nd ed., edited by W. T. Edmondson, Wiley, New York, 1959.

Weber, R., *Physics for Teachers: A Modern Review,* McGraw-Hill, New York, 1964.

Weisz, P., *Elements of Biology,* McGraw-Hill, New York, 1961.

———, *The Science of Biology,* 2nd ed., McGraw-Hill, New York, 1963.

Wightman, W. P. D., *The Growth of Scientific Ideas,* Yale Univ. Press, New Haven, Conn., 1951.

Witherspoon, J., and R. Witherspoon, *The Living Laboratory,* Doubleday, Garden City, N.Y., 1960.

Wolf, G., *Isotopes in Biology,* Academic, New York, 1964.

Zweifel, F. W., *A Handbook of Biological Illustration,* Univ. of Chicago Press, Chicago, 1961.

Physiology of plants and animals

Adolph, E. F., *The Development of Homeostasis,* Academic, New York, 1962.

Allen, M. B., ed., *Comparative Biochemistry of Photoreactive Systems,* Academic, New York, 1960.

Asimov, I., *The Human Brain,* Houghton Mifflin, Boston, 1964.

———, *The Wellsprings of Life,* Abelard-Schuman, New York, 1961; Signet, New Am. Library, New York, 1962.

Audus, L. J., *Plant Growth Substances,* 2nd ed., Interscience (Wiley), New York, 1960.

———, ed., *The Physiology and Biochemistry of Herbicides,* Academic, New York, 1964.

Barrington, E., *An Introduction to General and Comparative Endocrinology,* Oxford Univ. Press, New York, 1963.

Beaumont, W., *Experiments and Observations on the Gastric Juice and the Physiology of Digestion,* 1833, 1902; Dover, New York, 1960.

Berman, W., *Experimental Biology,* Sentinel, New York, 1963.

Best, C., and N. Taylor, *The Physiological Basis of Medical Practice,* 7th ed., Williams & Wilkins, Baltimore, 1961.

Biochemical Soc., *Biochemistry of Fish,* Symposium 6, ed. by R. T. Williams, Cambridge Univ. Press, New York, 1951.

Blanchard, J. R., and H. Ostvold, *Literature of Agricultural Research,* Univ. of California Press, Berkeley, 1958.

Bonner, J., and A. Galston, *Principles of Plant Physiology,* Freeman, San Francisco, 1952.

Borek, E., *Atoms Within Us,* Columbia Univ. Press, New York, 1961.

Bourne, G., ed., *Medical and Biological Problems of Space Flight,* Academic, New York, 1963.

Brown, J. H., ed., *Physiology of Man in Space,* Academic, New York, 1963.

Brown, M. E., ed., *The Physiology of Fishes,* Vol. II, Academic, New York, 1957.

Bullough, W. S., *Vertebrate Reproductive Cycles,* 2nd ed., Wiley, New York, 1961.

Bünning, E., *The Physiological Clock,* 2nd ed., Academic, New York, 1964.

Butler, E., ed., *Biological Specificity and Growth,* Princeton Univ. Press, Princeton, N.J., 1955.

Calvin, M., and J. Bassham, *The Photosynthesis of Carbon Compounds,* Benjamin, New York, 1962.

Cannon, W., *Bodily Changes in Pain, Hunger, Fear, and Rage,* 2nd ed., Branford, Newton Centre, Mass., 1953; Harper Torchbook, Harper & Row, New York, 1963.

——, *The Wisdom of the Body,* rev. ed., Norton, New York, 1939; Norton Library, 1963.

Cantarow, A., and B. Schepartz, *Textbook of Biochemistry,* 3rd ed., Saunders, Philadelphia, 1962.

Carlson, A., V. Johnson, and H. M. Cavert, *The Machinery of the Body,* 5th ed., Univ. of Chicago Press, Chicago, 1961.

Carthy, J. D., *An Introduction to the Behavior of Invertebrates,* Macmillan, New York, 1958.

Cassidy, H., *Fundamentals of Chromatography,* Interscience (Wiley), New York, 1957.

Cloudsley-Thompson, J. L., *Rhythmic Activity in Animal Physiology and Behaviour,* Vol. I of J. F. Danielli, ed., "Theoretical and Experimental Biology," Academic, New York, 1961.

Cochrane, V. W., *Physiology of Fungi,* Wiley, New York, 1958.

Cornelius, C. E., and J. J. Kaneko, eds., *Clinical Biochemistry of Domestic Animals,* Academic, New York, 1963.

Corner, G. W., *The Hormones in Human Reproduction,* rev. ed., Princeton Univ. Press, Princeton, N.J., 1947; Atheneum, New York, 1963.

Crocker, W., and L. V. Barton, *Physiology of Seeds,* Chronica Botanica, Waltham, Mass. (Ronald, New York), 1953.

D'Amour, F. E., *Basic Physiology,* Univ. of Chicago Press, Chicago, 1961.

Davidsohn, I., *see* Todd, J., and A. Sanford.

Davson, H., *A Textbook of General Physiology,* 3rd ed., Little, Brown, Boston, 1964.

Dethier, V. G., *The Physiology of Insect Senses,* Methuen (Wiley), New York, 1963.

——, and E. Stellar, *Animal Behavior,* 2nd ed., Prentice-Hall, Englewood Cliffs, N.J., 1964.

Dukes, H. H., *et al., The Physiology of Domestic Animals,* 7th ed., Comstock (Cornell Univ. Press), Ithaca, N.Y., 1955.

Eccles, J. C., *The Physiology of Synapses,* Academic, New York, 1963.

Eiduson, S., *et al., Biochemistry and Behavior,* Van Nostrand, Princeton, N.J., 1964.

Eigsti, O. J., and P. Dustin, *Colchicine: In Agriculture, Medicine, Biology, and Chemistry,* Iowa State Univ. Press, Ames, 1955.

Eisenstein, A., ed., *The Biochemical Aspects of Hormone Action,* Little, Brown, Boston, 1964.

Elsasser, W. M., *Physical Basis of Biology,* Pergamon (Macmillan), New York, 1958.

Evans, L. T., ed., *Environmental Control of Plant Growth,* Academic, New York, 1963.

Faires, R., and B. Parks, *Radioisotope Laboratory Techniques,* 2nd ed., Pitman, New York, 1960.

Feldman, S., *Techniques and Investigations in the Life Sciences,* Holt, Rinehart and Winston, New York, 1962.

Fishman, A. P., and D. W. Richards, eds., *Circulation of Blood: Men and Ideas,* Oxford Univ. Press, New York, 1964.

Florkin, M., *Unity and Diversity in Biochemistry,* Pergamon (Macmillan), New York, 1960.

——, and H. Mason, eds., *Comparative Biochemistry,* 7 vols., Academic, New York, 1960–64.

Fogg, G. E., *Growth of Plants,* Pelican, Penguin, Baltimore, 1962.

——, *The Metabolism of Algae,* Wiley, New York, 1953.

Follansbee, H., *Animal Behavior* (BSCS laboratory block), Heath, Boston, 1965.

Frere, M., *et al., The Behavior of Radioactive Fallout in Soils and Plants,* National Academy of Sciences, Washington, D.C., 1963.

Fruton, J. S., and S. Simmonds, *General Biochemistry,* 2nd ed., Wiley, New York, 1958.

Galston, A., *The Life of the Green Plant,* 2nd ed., "The Foundations of Modern Biology Series," Prentice-Hall, Englewood Cliffs, N.J., 1964.

Giese, A., ed., *Photophysiology,* 2 vols.: Vol. I, *General Principles; Action of Light on Plants,* 1964, Vol. II, *Action of Light on Animals and Microorganisms; Photobiochemical Mechanisms; Bioluminescence,* 1964; Academic, New York.

Gilmour, D., *Biochemistry of Insects,* Academic, New York, 1961.

Gorbman, A., and H. Bern, *A Textbook of Comparative Endocrinology,* Wiley, New York, 1962.

Gray, J., *How Animals Move,* Cambridge Univ. Press, New York, 1959; Pelican, Penguin, Baltimore, 1964.

Guyton, A. C., *Function of the Human Body,* 2nd ed., Saunders, Philadelphia, 1964.

Hafez, E. S. E., ed., *The Behavior of Domestic Animals,* Williams & Wilkins, Baltimore, 1962.

Hanrahan, J. S., and D. Bushnell, *Space Biology,* Basic Books, New York, 1960.

Harrison, K., *A Guidebook to Biochemistry,* Cambridge Univ. Press, New York, 1959.

Harrow, B., and A. Mazur, *Textbook of Bio-*

chemistry, 8th ed., Saunders, Philadelphia, 1962.

Hawk, P., B. Oser, and W. Summerson, *Practical Physiological Chemistry*, 13th ed., Blakiston (McGraw-Hill), New York, 1954.

Heftmann, E., ed., *Chromatography*, Reinhold, New York, 1961.

Heilbrunn, L. V., *An Outline of General Physiology*, 3rd ed., Saunders, Philadelphia, 1952.

Hillman, W., *The Physiology of Flowering*, Holt, Rinehart and Winston, New York, 1962.

Hutner, S., and A. Lwoff, eds., *Biochemistry and Physiology of the Protozoa*, 3 vols.: Vol. I ed. by A. Lwoff, 1951, Vol. II ed. by S. Hutner, 1955, Vol. III ed. by S. Hutner, 1964; Academic, New York.

Jacobs, W. P., and C. E. LaMotte, *Regulation in Plants by Hormones* (BSCS laboratory block), Heath, Boston, 1964.

James, W., *An Introduction to Plant Physiology*, 6th ed., Oxford Univ. Press, New York, 1963.

Jenkins, J., and D. Paterson, eds., *Studies in Individual Differences*, Appleton-Century-Crofts, New York, 1961.

Jenkins, P., *Animal Hormones: A Comparative Survey*, Part I, *Kinetic and Metabolic Hormones*, Pergamon, New York, 1961.

Kabat, E., *Blood Group Substances: Their Chemistry and Immunochemistry*, Academic, New York, 1956.

Kamen, M., *Primary Processes in Photosynthesis*, "Advanced Biochemistry Series," Academic, New York, 1963.

King, B., and M. Showers, *Human Anatomy and Physiology*, 5th ed., Saunders, Philadelphia, 1963.

LaMotte Chemical Controls Catalog, LaMotte Chemical Products Co., Chestertown, Md.

Lanyon, W. E., *Biology of Birds*, Natural History Press, Doubleday, Garden City, N.Y., 1964; Am. Museum Science Books, Natural History Press, Doubleday, 1964.

Laverack, M., *The Physiology of Earthworms*, Pergamon, New York, 1963.

Lee, A. E., *Plant Growth and Development* (BSCS laboratory block), Heath, Boston, 1963.

Leopold, A. C., *Auxins and Plant Growth*, Univ. of California Press, Berkeley, 1955.

Lewin, R., ed., *Physiology and Biochemistry of Algae*, Academic, New York, 1962.

Lindauer, M., *Communication Among Social Bees*, Harvard Univ. Press, Cambridge, Mass., 1961.

McElroy, W., *Cell Physiology and Bio-Chemistry*, 2nd ed., "The Foundations of Modern Biol-

ogy Series," Prentice-Hall, Englewood Cliffs, N.J., 1964.

Machlis, L., and J. Torrey, *Plants in Action*, Freeman, San Francisco, 1956.

Mackinnon, D., and R. Hawes, *An Introduction to the Study of Protozoa*, Oxford Univ. Press, New York, 1961.

Mayer, E., *Introduction to Dynamic Morphology*, Academic, New York, 1963.

Mazia, D., and A. Tyler, *General Physiology of Cell Specialization*, McGraw-Hill, New York, 1963.

Meyer, B., D. Anderson, and R. Böhning, *Introduction to Plant Physiology*, Van Nostrand, Princeton, N.J., 1960.

———, D. Anderson, and C. Swanson, *Laboratory Plant Physiology*, 3rd ed., Van Nostrand, Princeton, N.J., 1955.

Mitchell, J., G. Livingston, and P. Marth, *Test Methods with Plant-regulating Chemicals*, U.S. Dept. of Agriculture Handbook 126, U.S. Govt. Printing Office, Washington, D.C., 1958.

Moog, F., *Animal Growth and Development* (BSCS laboratory block), Heath, Boston, 1963.

Moulder, J., *Biochemistry of Intracellular Parasitism*, Univ. of Chicago Press, Chicago, 1962.

National Research Council, Committee on Photobiology, *Photosynthetic Mechanisms of Green Plants*, National Research Council, Washington, D.C., 1963.

Parker, G. H., *Animal Colour Changes*, Cambridge Univ. Press, New York, 1948.

Patton, R., *Introduction to Insect Physiology*, Saunders, Philadelphia, 1963.

Pauling, L., *College Chemistry*, 3rd ed., Freeman, San Francisco, 1964.

Penfield, W., and L. Roberts, *Speech and Brain Mechanisms*, Princeton Univ. Press, Princeton, N.J., 1959.

The Photochemical Apparatus: Its Structure and Function, Brookhaven Symposia in Biology, 11, Brookhaven National Laboratory, Upton, N.Y., 1959.

Pierce, J., and E. David, Jr., *Man's World of Sound*, Doubleday, Garden City, N.Y., 1958.

Prescott, D., ed., *Methods in Cell Physiology*, Vol. I, Academic, New York, 1964.

Prosser, C. L., and F. A. Brown, Jr., *Comparative Animal Physiology*, 2nd ed., Saunders, Philadelphia, 1961.

Ramsey, J., *Physiological Approach to the Lower Animals*, Cambridge Univ. Press, New York, 1952.

Roeder, K. D., *Nerve Cells and Insect Behavior*,

Harvard Univ. Press, Cambridge, Mass., 1963.

Rogers, W. P., *The Nature of Parasitism,* Vol. II of J. F. Danielli, ed., "Theoretical and Experimental Biology," Academic, New York, 1961.

Ross, H., *A Textbook of Entomology,* 3rd ed., Wiley, New York, 1965.

Salisbury, F., *The Flowering Process,* Pergamon, New York, 1963.

Savory, T., *Instinctive Living: A Study of Invertebrate Behaviour,* Pergamon, New York, 1959.

Scheer, B., *Animal Physiology,* Wiley, New York, 1963.

Schmidt-Nielsen, K., *Animal Physiology,* 2nd ed., "The Foundations of Modern Biology Series," Prentice-Hall, Englewood Cliffs, N.J., 1964.

———, *Desert Animals: Physiological Problems of Heat and Water,* Oxford Univ. Press, New York, 1963.

Schmitt, F. O., ed., *Macromolecular Specificity and Biological Memory,* M.I.T. Press, Cambridge, Mass., 1962.

Scott, J., *Animal Behavior,* Univ. of Chicago Press, Chicago, 1958.

Selye, H., *The Stress of Life,* McGraw-Hill, New York, 1956.

Smythe, R. H., *Animal Vision: What Animals See,* Thomas, Springfield, Ill., 1961.

Steward, F. C., ed., *Plant Physiology: A Treatise* Vol. IB, *Photosynthesis and Chemosynthesis,* Academic, New York, 1960.

Strafford, G., *Plant Metabolism,* Harvard Univ. Press, Cambridge, Mass., 1963.

Thorpe, W., *Learning and Instinct in Animals,* Harvard Univ. Press, Cambridge, Mass., 1956.

———, and O. Zangwill, *Current Problems in Animal Behaviour,* Cambridge Univ. Press, New York, 1961.

Todd, J., and A. Sanford, *Clinical Diagnosis by Laboratory Methods,* 13th ed., edited by I. Davidsohn and B. Wells, Saunders, Philadelphia, 1962.

Turner, C. D., *General Endocrinology,* 3rd ed., Saunders, Philadelphia, 1960.

Umbreit, W., R. Burris, and J. Stauffer, *Manometric Techniques,* Burgess, Minneapolis, 1957.

Van Slyke, D., and J. Plazin, *Micromanometric Analyses,* Williams & Wilkins, Baltimore, 1961.

Wain, R., and F. Wightman, eds., *The Chemistry and Mode of Action of Plant Growth Substances,* Academic, New York, 1956.

Wald, G., *et al., Twenty-six Afternoons of Biology,* Addison-Wesley, Reading, Mass., 1962.

Walker, B., W. Boyd, and I. Asimov, *Biochemistry and Human Metabolism,* 3rd ed., Williams & Wilkins, Baltimore, 1957.

Walsh, E. O., *An Introduction to Biochemistry,* Macmillan, New York, 1961.

Waring, H., *Color Change Mechanisms of Cold-blooded Vertebrates,* Academic, New York, 1963.

Waterman, T. H., ed., *The Physiology of Crustacea,* Vol. I, *Metabolism and Growth,* Academic, New York, 1960.

Webster, G., *Nitrogen Metabolism in Plants,* Harper & Row, New York, 1959.

Welsh, J., and R. Smith, *Laboratory Exercises in Invertebrate Physiology,* Burgess, Minneapolis, 1960.

Went, F. W., *The Experimental Control of Plant Growth,* Ronald, New York, 1957.

Wigglesworth, V. B., *The Physiology of Insect Metamorphosis,* Cambridge Univ. Press, New York, 1954.

Williams, R. J., *Biochemical Individuality,* Wiley, New York, 1956.

Williams, R. T., *see* Biochemical Soc.

Withrow, R., ed., *Photoperiodism and Related Phenomena in Plants and Animals,* Am. Assoc. for the Advancement of Science, Washington, D.C., 1959.

Wolken, J., *Euglena: An Experimental Organism for Biochemical and Biophysical Studies,* Rutgers Univ. Press, New Brunswick, N.J., 1961.

Wooldridge, D., *The Machinery of the Brain,* McGraw-Hill, New York, 1963.

Youmans, W., *Human Physiology,* rev. ed., Macmillan, New York, 1962.

Young, W. C., ed., *Sex and Internal Secretions,* 3rd ed., 2 vols., Williams & Wilkins, Baltimore, 1961.

Cell physiology

Ariens, E. J., ed., *Molecular Pharmacology,* 2 vols., Academic, New York, 1964.

Baldwin, E., *Nature of Biochemistry,* Cambridge Univ. Press, New York, 1962.

Billingham, R., and W. Silvers, *Transplantation of Tissues and Cells,* Wistar Institute Press, Philadelphia, 1961.

Bourne, G., *Cytology and Cell Physiology,* 3rd ed., Academic, New York, 1964.

———, *Division of Labor in Cells,* Academic, New York, 1962.

Brachet, J., and A. Mirsky, eds., *The Cell: Bio-*

chemistry, Physiology, Morphology, 6 vols.: Vol. I, 1959, Vols. II, III, 1961; Academic, New York.

Brieger, E. M., *Structure and Ultrastructure of Microorganisms,* Academic, New York, 1963.

Butler, J. A., *Inside the Living Cell,* Basic Books, New York, 1959.

Cantarow, A., and B. Schepartz, *Textbook of Biochemistry,* 3rd ed., Saunders, Philadelphia, 1962.

Cellular Regulatory Mechanisms, Cold Spring Harbor Symposia on Quantitative Biology, 26, Cold Spring Harbor Laboratory of Quantitative Biology, Cold Spring Harbor, N.Y., 1961.

Cheldelin, V., *Metabolic Pathways in Microorganisms,* Wiley, New York, 1961.

Danielli, J. F., ed., *General Cytochemical Methods,* 2 vols., Academic, New York, 1958, 1961.

DeRobertis, E., W. Nowinski, and F. Saez, *Cell Biology,* 4th ed., Saunders, Philadelphia, 1965.

Dixon, M., and E. C. Webb, *Enzymes,* 2nd ed., Academic, New York, 1964.

Fogg, G. E., ed., *Cell Differentiation,* Symposia of the Soc. for Experimental Biology, 17, Academic, New York, 1963.

Gerard, R. W., *Unresting Cells,* new ed., Harper & Row, New York, 1949; Harper Torchbook, 1961.

Giese, A., *Cell Physiology,* 2nd ed., Saunders, Philadelphia, 1962.

Gray, L. H., *see* Lea, D. E.

Hawker, L. E., *et al., An Introduction to the Biology of Microorganisms,* St. Martin's, New York, 1961.

Hayashi, T., *Subcellular Particles,* Ronald, New York, 1959.

Kuyper, C., *The Organization of Cellular Activity,* Am. Elsevier, New York, 1962.

Lea, D. E., *Actions of Radiations on Living Cells,* 2nd ed., revised by L. H. Gray, Cambridge Univ. Press, New York, 1955.

Lehninger, A. L., *The Mitochondrion,* Benjamin, New York, 1964.

Levine, L., ed., *Symposium on the Cell in Mitosis,* Academic, New York, 1963.

Mackinnon, D., and R. Hawes, *An Introduction to the Study of Protozoa,* Oxford Univ. Press, New York, 1961.

Paul, J., *Cell and Tissue Culture,* 2nd ed., Williams & Wilkins, Baltimore, 1960.

Picken, L., *The Organization of Cells and Other Organisms,* Oxford Univ. Press, New York, 1960.

Porter, E., and M. Bonneville, *An Introduction to the Fine Structure of Cells and Tissues,* 2nd ed., Lea & Febiger, Philadelphia, 1964.

Prescott, D., ed., *Methods in Cell Physiology,* Vol. I, Academic, New York, 1964.

Richards, A. G., *The Complementarity of Structure and Function* (BSCS laboratory block), Heath, Boston, 1963.

Stace, C., *A Guide to Subcellular Botany,* Longmans, London, 1963.

Stern, H., and D. Nanney, *The Biology of Cells,* Wiley, New York, 1965.

Strehler, B., *Time, Cells, and Aging,* Academic, New York, 1962.

Swanson, C., *The Cell,* 2nd ed., "The Foundations of Modern Biology Series," Prentice-Hall, Englewood Cliffs, N.J., 1964.

Umbreit, W., R. Burris, and J. Stauffer, *Manometric Techniques,* Burgess, Minneapolis, 1957.

Van Slyke, D., and J. Plazin, *Micromanometric Analyses,* Williams & Wilkins, Baltimore, 1961.

White, M., *Animal Cytology and Evolution,* 2nd ed., Cambridge Univ. Press, New York, 1954.

Microtechnique

Bloom, W., and D. Fawcett, *A Textbook of Histology,* 8th ed., Saunders, Philadelphia, 1962.

Clark, G. L., ed., *Encyclopedia of Microscopy,* Reinhold, New York, 1961.

Conn, H. J., *Biological Stains,* 7th ed., Williams & Wilkins, Baltimore, 1961.

———, M. Darrow, and V. Emmel, *Staining Procedures Used by the Biological Stain Commission,* 2nd ed., Williams & Wilkins, Baltimore, 1960.

DeRobertis, E., W. Nowinski, and F. Saez, *Cell Biology,* 4th ed., Saunders, Philadelphia, 1965.

Gatenby, J., *see* Lee, A. B.

Gray, P., *Handbook of Basic Microtechnique,* 3rd ed., McGraw-Hill, New York, 1964.

———, *The Microtomist's Formulary and Guide,* Blakiston (McGraw-Hill), New York, 1954.

Gurr, E., *Encyclopedia of Microscopic Stains,* Williams & Wilkins, Baltimore, 1960.

Guyer, M. F., *Animal Micrology,* 5th ed., Univ. of Chicago Press, Chicago, 1953.

Hooke, R., *Micrographia, or Some Physiological Descriptions of Minute Bodies—Observations by Robert Hooke,* orig. ed., 1665; Dover, New York, 1961.

Humason, G., *Animal Tissue Techniques,* Freeman, San Francisco, 1962.

Jensen, W., *Botanical Histochemistry,* Freeman, San Francisco, 1962.

Jones, R. M., *see* McClung, C. E.

Lee, A. B., *Microtomist's Vade-Mecum,* 11th ed., edited by J. Gatenby and H. Beams, Blakiston (McGraw-Hill), New York, 1950.

McClung, C. E., ed., *Handbook of Microscopical Technique for Workers in Animal and Plant Tissues,* 3rd ed., revised by R. M. Jones, Harper & Row, New York, 1950.

Pease, D., *Histological Techniques for Electron Microscopy,* 2nd ed., Academic, New York, 1964.

Weesner, H., *General Zoological Microtechnique,* Williams & Wilkins, Baltimore, 1960.

Botany

Alexopoulos, C., *Introductory Mycology,* 2nd ed., Wiley, New York, 1962.

Bailey, L. H., *Manual of Cultivated Plants,* rev. ed., Macmillan, New York, 1949.

Barnett, H. L., *Illustrated Genera of Imperfect Fungi,* Burgess, Minneapolis, 1960.

Benson, L., *Plant Classification,* Heath, Boston, 1957.

Billings, W., *Plants and the Ecosystem,* Wadsworth, Belmont, Calif., 1964.

Bold, H., *The Plant Kingdom,* 2nd ed., "The Foundations of Modern Biology Series," Prentice-Hall, Englewood Cliffs, N.J., 1964.

Bonner, J., *The Cellular Slime Molds,* Princeton Univ. Press, Princeton, N.J., 1959.

Brunel, J., G. Prescott, and L. H. Tiffany, *The Culturing of Algae,* C. F. Kettering Foundation for the Study of Chlorophyll and Photosynthesis, Antioch College, Yellow Springs, Ohio, 1950.

Burlew, J., ed., *Algal Culture: From Laboratory to Pilot Plant,* Carnegie Institution of Washington, Washington, D.C., 1953.

Carlquist, S., *Comparative Plant Anatomy,* Holt, Rinehart and Winston, New York, 1961.

Coulter, M., *The Story of the Plant Kingdom,* 3rd ed., revised by H. Dittmer, Univ. of Chicago Press, Chicago, 1964.

Cronquist, A., *Introductory Botany,* Harper & Row, New York, 1961.

Dittmer, H., *see* Coulter, M.

Erdtman, G., *An Introduction to Palynology,* 2 vols.: Vol. I, *Pollen Morphology and Plant Taxonomy—Angiosperms,* Ronald, New York, 1952.

Esau, K., *Anatomy of Seed Plants,* Wiley, New York, 1960.

————, *Plant Anatomy,* 2nd ed., Wiley, New York, 1965.

Fernald, M., *see* Gray, A.

Fisk, E., and W. Millington, *Atlas of Plant Morphology,* 2 portfolios: Portfolio I, *Photomicrographs of Root, Stem, and Leaf,* 1959, Portfolio II, *Photomicrographs of Flower, Fruit, and Seed,* 1962; Burgess, Minneapolis.

Foster, A., and E. Gifford, *Comparative Morphology of Vascular Plants,* Freeman, San Francisco, 1959.

Fritsch, F., and E. Salisbury, *Plant Form and Function,* new and rev. ed., Bell & Sons, London, 1953.

Fuller, H., and A. Carothers, *The Plant World,* 4th ed., Holt, Rinehart and Winston, New York, 1963.

Garrett, S., *Biology of Root-infecting Fungi,* Cambridge Univ. Press, New York, 1956.

Godwin, H., *Plant Biology,* 4th ed., Cambridge Univ. Press, New York, 1945.

Gray, A., *Manual of Botany,* 8th Centennial ed., rewritten and expanded by M. Fernald, American Book, New York, 1950.

Greulach, V., and J. E. Adams, *Plants: An Introduction to Modern Botany,* Wiley, New York, 1962.

Heslop-Harrison, J., *New Concepts in Flowering Plant Taxonomy,* Harvard Univ. Press, Cambridge, Mass., 1956.

Hylander, C., *The World of Plant Life,* 2nd ed., Macmillan, New York, 1956.

International Bureau for Plant Taxonomy and Nomenclature, "International Directory of Botanical Gardens," compiled by R. Howard, *Regnum Vegetable,* Vol. 28, Utrecht, Netherlands, 1963.

Knobloch, I., *Selected Botanical Papers,* Prentice-Hall, Englewood Cliffs, N.J., 1963.

McCracken, E., M. Newcomb, and S. Pady, *General Botany Laboratory Manual,* Burgess, Minneapolis, 1962.

Mayer, A., and A. Poljakoff-Mayber, *Germination of Seeds,* Macmillan, New York, 1963.

Miller, R., *Morphology and Anatomy of Roots,* Scholar's Library, New York, 1961.

Muller, W., *Botany: A Functional Approach,* Macmillan, New York, 1963.

Porter, C., *Taxonomy of Flowering Plants,* Freeman, San Francisco, 1959.

Pringsheim, E. G., *Pure Cultures of Algae: Their Preparation and Maintenance,* Macmillan (Cambridge Univ. Press), New York, 1946.

Sinnott, E., *Plant Morphogenesis,* McGraw-Hill, New York, 1960.

————, and K. Wilson, *Botany: Principles and*

Problems, 6th ed., McGraw-Hill, New York, 1963.

Steward, F. C., *Plants at Work,* Addison-Wesley, Reading, Mass., 1964.

Uphof, J., *Dictionary of Economic Plants,* Hafner, New York, 1959.

Vorontsova, N., and L. Liosner, *Asexual Propagation and Regeneration,* Pergamon, New York, 1960.

Weisz, P., and M. Fuller, *Science of Botany,* McGraw-Hill, New York, 1962.

Withner, C., *The Orchids,* Ronald, New York, 1959.

Zoology

Barnard, F., R. Menzies, and M. Băcescu, *Abyssal Crustacea,* Columbia Univ. Press, New York, 1962.

Barnes, R. D., *Invertebrate Zoology,* Saunders, Philadelphia, 1963.

Beck, D. E., and L. F. Braithwaite, *Invertebrate Zoology: Laboratory Workbook,* 2nd ed., Burgess, Minneapolis, 1962.

Berman, W., *How to Dissect,* Sentinel, New York, 1961.

Borradaile, L., *Manual of Elementary Zoology,* 14th ed., edited by W. Yapp, Oxford Univ. Press, New York, 1963.

Brown, F., *Selected Invertebrate Types,* Wiley, New York, 1950.

Buchsbaum, R., *Animals Without Backbones: An Introduction to the Invertebrates,* 2nd ed., Univ. of Chicago Press, Chicago, 1948.

———, and L. Milne, in collaboration with M. Buchsbaum and M. Milne, *The Lower Animals: Living Invertebrates of the World,* Doubleday, Garden City, N.Y., 1960.

Bullough, W. S., *Practical Invertebrate Anatomy,* 2nd ed., Macmillan, London (St. Martin's, New York), 1958.

Carthy, J. D., *An Introduction to the Behavior of Invertebrates,* Macmillan, New York, 1958.

Cheng, T., *Biology of Animal Parasites,* Saunders, Philadelphia, 1964.

Corliss, J., *The Ciliated Protozoa,* Pergamon, New York, 1961.

Dugdale, C., *Manual for Dissection of the Cat,* Burgess, Minneapolis, 1949.

Du Porte, E. M., *Manual of Insect Morphology,* Reinhold, New York, 1959.

Eaton, T. H., Jr., *Comparative Anatomy of the Vertebrates,* 2nd ed., Harper & Row, New York, 1960.

Eddy, S., C. Oliver, and J. Turner, *Guide to the Study of the Shark, Necturus, and the Cat,* 3rd ed., Wiley, New York, 1960.

Francis, E., *The Anatomy of the Salamander,* Oxford Univ. Press, New York, 1934.

Galtsoff, P., *et al.,* eds. (J. Needham, chairman), *Culture Methods for Invertebrate Animals,* Comstock (Cornell Univ. Press), Ithaca, N.Y., 1937; Dover, New York, 1959.

Gans, C., and J. F. Storr, *Comparative Anatomy Atlas,* Academic, New York, 1962.

Gardner, E., D. Gray, and R. O'Rahilly, *Anatomy,* Saunders, Philadelphia, 1963.

Goin, C., and O. Goin, *Introduction to Herpetology,* Freeman, San Francisco, 1962.

Goodnight, C., M. Goodnight, and P. Gray, *General Zoology,* Reinhold, New York, 1964.

Guthrie, M., and J. Anderson, *General Zoology,* Wiley, New York, 1957.

Harrison, B., *Dissection of the Shark,* Burgess, Minneapolis, 1949.

Hausman, L., *Essentials of Zoology,* Doubleday, Garden City, N.Y., 1963.

Hegner, R. W., *Parade of the Animal Kingdom,* Macmillan, New York, 1942.

———, and K. A. Stiles, *College Zoology,* 7th ed., Macmillan, New York, 1959.

Hyman, L., *The Invertebrates,* 5 vols., McGraw-Hill, New York, 1940–59.

Roscoe B. Jackson Memorial Laboratory, staff, *The Biology of the Laboratory Mouse,* ed. by G. D. Snell, Blakiston, Philadelphia, 1941; Dover, New York, 1956.

Kudo, R., *Protozoology,* 4th ed., Thomas, Springfield, Ill., 1954.

Lanham, U., *The Fishes,* Columbia Univ. Press, New York, 1962.

———, *The Insects,* Columbia Univ. Press, New York, 1964.

Lenhoff, H. M., and F. Loomis, eds., *The Biology of Hydra and of Some Other Coelenterates,* Univ. of Miami Press, Coral Gables, Fla., 1961.

Lockhart, R., G. Hamilton, and F. Fyfe, *Anatomy of the Human Body,* Lippincott, Philadelphia, 1959.

Manwell, R., *Introduction to Protozoology,* St. Martin's, New York, 1961.

Marshall, A., ed., *Biology and Comparative Physiology of Birds,* 2 vols., Academic, New York, 1960, 1961.

Moore, J., *Principles of Zoology,* Oxford Univ. Press, New York, 1957.

Needham, J., *see* Galtsoff, P., *et al.*

Nicol, J., *The Biology of Marine Animals,* Interscience (Wiley), New York, 1960.

Patton, R., *Introduction to Insect Physiology,* Saunders, Philadelphia, 1963.

Pettingill, O., *A Laboratory and Field Manual of Ornithology,* 3rd ed., Burgess, Minneapolis, 1956.

Porter, G., and W. Lane-Petter, eds., *Notes for Breeders of Common Laboratory Animals,* Academic, New York, 1962.

Romer, A., *The Vertebrate Body,* 3rd ed., Saunders, Philadelphia, 1962.

Snedigar, R., *Our Small Native Animals: Their Habits and Care,* rev. ed., Dover, New York, 1963.

Tartar, V., *Biology of Stentor,* Pergamon, New York, 1961.

Torrey, T., *Morphogenesis of Vertebrates,* Wiley, New York, 1962.

Villee, C., W. Walker, and F. Smith, *General Zoology,* 2nd ed., Saunders, Philadelphia, 1963.

Walker, W., *Vertebrate Dissection,* 2nd ed., Saunders, Philadelphia, 1960.

Wallace, G., *Introduction to Ornithology,* 2nd ed., Macmillan, New York, 1963.

Welty, C. J., *The Life of Birds,* Saunders, Philadelphia, 1962.

Winchester, A., and H. Lovell, *Zoology,* 3rd ed., Van Nostrand, Princeton, N.J., 1961.

Yapp, W., *see* Borradaile, L.

Young, J., *The Life of Vertebrates,* 2nd ed., Oxford Univ. Press, New York, 1962.

Microbiology and health

Alexander, M., *Introduction to Soil Microbiology,* Wiley, New York, 1961.

Bedson, S., *et al., Virus and Rickettsial Diseases of Man,* 3rd ed., Williams & Wilkins, Baltimore, 1961.

Best, C., and N. Taylor, *The Human Body,* 4th ed., Holt, Rinehart and Winston, New York, 1963.

Bodansky, M., and O. Bodansky, *Biochemistry of Disease,* 2nd ed., Macmillan, New York, 1952.

Brock, T., ed., *Milestones in Microbiology,* Prentice-Hall, Englewood Cliffs, N.J., 1961.

Burnet, F. M., *The Integrity of the Body: A Discussion of Modern Immunological Ideas,* Harvard Univ. Press, Cambridge, Mass., 1962.

———, *Principles of Animal Virology,* 2nd ed., Academic, New York, 1960.

Burrows, W., *Textbook of Microbiology,* 18th ed., Saunders, Philadelphia, 1963.

Campbell, D., *et al., Methods in Immunology,* Benjamin, New York, 1963.

Carpenter, P., *Immunology and Serology,* 2nd ed., Saunders, Philadelphia, 1964.

———, *Microbiology,* Saunders, Philadelphia, 1961.

Chandler, A., and C. Read, *Introduction to Parasitology,* 10th ed., Wiley, New York, 1961.

Cheldelin, V., *Metabolic Pathways in Microorganisms,* Wiley, New York, 1961.

Clendening, L., ed., *Sourcebook of Medical History,* Harper & Row, New York, 1942; Dover, New York, 1960.

Cochrane, V. W., *Physiology of Fungi,* Wiley, New York, 1958.

Culture Media: Materials and Apparatus for the Microbiological Laboratory, 4th ed., Baltimore Biological Laboratory, Baltimore, 1956.

Dagley, S., and D. Wild, *Bacterial Metabolism,* Academic, New York, in preparation.

De Kruif, P., *Microbe Hunters,* Harcourt, Brace & World, New York, 1932; Pocket Books, New York.

The Difco Manual of Dehydrated Culture Media and Reagents, 9th ed., "Difco Supplementary Literature," Difco Laboratories, Detroit, 1953.

Dubos, R., *Louis Pasteur: Free Lance of Science,* Little, Brown, Boston, 1950.

Duddington, C. L., *Micro-Organisms as Allies,* Macmillan, New York, 1961.

Esau, K., *Plants, Viruses, and Insects,* Harvard Univ. Press, Cambridge, Mass., 1961.

Faust, E., C. Beaver, and C. Jung, *Animal Agents and Vectors of Human Disease,* 2nd ed., Lea & Febiger, Philadelphia, 1962.

Foster, J., *Chemical Activities of Fungi,* Academic, New York, 1949.

Fraenkel-Conrat, H., *Design and Function at the Threshold of Life: The Viruses,* Academic, New York, 1962.

Frobisher, M., *Fundamentals of Microbiology,* 7th ed., Saunders, Philadelphia, 1962.

Galigher, A., and E. Kozloff, *Essentials of Practical Microtechnique,* Lea & Febiger, Philadelphia, 1964.

Garrett, S., *Soil Fungi and Soil Fertility,* Pergamon, New York, 1963.

Harris, R., ed., *The Problems of Laboratory Animal Disease,* Academic, New York, 1962.

Hawker, L. E., *et al., An Introduction to the Biology of Microorganisms,* St. Martin's, New York, 1961.

Heukelekian, H., and N. Dondero, eds., *Principles and Applications in Aquatic Microbiology,* Wiley, New York, 1964.

Jacob, F., and E. Wollman, *Sexuality and the Genetics of Bacteria,* rev. ed., Academic, New York, 1961.

Jacobs, M., and M. Gerstein, *Handbook of Microbiology,* Van Nostrand, Princeton, N.J., 1960.

Kavanagh, K., ed., *Analytical Microbiology,* Academic, New York, 1963.

Lapage, G., *Animals Parasitic in Man,* Penguin, Baltimore, 1957; Dover, New York, 1963.

May, J., *Ecology of Human Disease,* MD Publications, New York, 1958.

Miller, M. W., *The Pfizer Handbook of Microbial Metabolites,* McGraw-Hill, New York, 1961.

Nickerson, W., *Biology of Pathogenic Fungi,* Ronald, New York, 1947.

Oginsky, E., and W. Umbreit, *An Introduction to Bacterial Physiology,* 2nd ed., Freeman, San Francisco, 1959.

Pelczar, M., and R. Reid, *Microbiology,* 2nd ed., McGraw-Hill, New York, 1965.

Rainbow, C., and A. Rose, eds., *Biochemistry of Industrial Microorganisms,* Academic, New York, 1963.

Seeley, H. W., Jr., and P. J. VanDemark, *Microbes in Action,* Freeman, San Francisco, 1962.

Sistrom, W. R., *Microbial Life,* Holt, Rinehart and Winston, New York, 1962.

Smith, K., *Viruses,* Cambridge Univ. Press, New York, 1962.

Smith, W., ed., *Mechanisms of Virus Infection,* Academic, New York, 1963.

Soc. of Am. Bacteriologists, Committee on Bacteriological Technic, *Manual of Microbiological Methods,* ed. by H. J. Conn, McGraw-Hill, New York, 1957.

Spector, W., ed., *Antibiotics,* Vol. II of *Handbook of Toxicology* (5 vols.), Saunders, Philadelphia, 1957.

Stanier, R., M. Doudoroff, and E. Adelberg, *The Microbial World,* 2nd ed., Prentice-Hall, Englewood Cliffs, N.J., 1963.

Sussman, A. S., *Microbes: Their Growth, Nutrition, and Interaction* (BSCS laboratory block), Heath, Boston, 1964.

Taliaferro, W. H., L. G. Taliaferro, and B. N. Jaroslow, *Radiation and Immune Mechanisms,* "A.E.C. Monograph Series on Radiation Biology and Industrial Hygiene," Academic, New York, 1964.

Thimann, K. V., *The Life of Bacteria,* 2nd ed., Macmillan, New York, 1963.

Umbreit, W., *Modern Microbiology,* Freeman, San Francisco, 1962.

U.S. Dept. of Agriculture, *Yearbook of Agriculture, 1953: Plant Diseases; Yearbook of Agriculture, 1956: Animal Diseases,* U.S. Dept. of Agriculture, Washington 25, D.C.

Westcott, C., *Plant Disease Handbook,* 2nd ed., Van Nostrand, Princeton, N.J., 1960.

Williams, G., *Virus Hunters,* Knopf (Random), New York, 1959.

Wynder, E., *The Biologic Effects of Tobacco,* Little, Brown, Boston, 1955.

Zinsser, H., *Rats, Lice, and History,* Little, Brown, Boston, 1935.

Heredity and development

Allard, R., *Principles of Plant Breeding,* Wiley, New York, 1960.

Allen, J., ed., *The Molecular Control of Cellular Activity,* McGraw-Hill, New York, 1962.

Anfinsen, C., *The Molecular Basis of Evolution,* Wiley, New York, 1959.

Arey, L. B., *Developmental Anatomy: A Textbook and Laboratory Manual of Embryology,* 7th ed., Saunders, Philadelphia, 1965.

Asimov, I., *The Genetic Code,* Signet, New Am. Library, New York, 1962.

————, and W. Boyd, *Races and People,* Abelard-Schuman, New York, 1955.

Auerbach, C., *The Science of Genetics,* Harper & Row, New York, 1961.

Balinsky, B. I., *An Introduction to Embryology,* 2nd ed., Saunders, Philadelphia, 1965.

Barnett, A., *The Human Species,* rev. ed., Penguin, Baltimore, 1961.

Beale, G., *Genetics of Paramecium aurelia,* Cambridge Univ. Press, New York, 1954.

Berrill, N. J., *Growth, Development, and Pattern,* Freeman, San Francisco, 1961.

Bonner, D., and S. Mills, *Heredity,* 2nd ed., "The Foundations of Modern Biology Series," Prentice-Hall, Englewood Cliffs, N.J., 1964.

Bonner, J., *The Cellular Slime Molds,* Princeton Univ. Press, Princeton, N.J., 1959.

————, *The Evolution of Development,* Cambridge Univ. Press, New York, 1958.

Borek, E., *The Code of Life,* Columbia Univ. Press, New York, 1965.

Boyer, S., *Papers on Human Genetics,* Prentice-Hall, Englewood Cliffs, N.J., 1963.

Brachet, J., *The Biochemistry of Development,* Pergamon, New York, 1960.

Burdette, W., ed., *Methodology in Mammalian Genetics,* 2 vols.: Vol. I, *Methodology in Human Genetics,* 1962, Vol. II, *Methodology in Basic Genetics,* 1963; Holden-Day, San Francisco.

Carson, H., *Heredity and Human Life,* Columbia Univ. Press, New York, 1963.

Carter, C., *Human Heredity,* Pelican, Penguin, Baltimore, 1962.

Child, C. M., *Patterns and Problems of Development,* Univ. of Chicago Press, Chicago, 1941.

Cole, H., *Gonadotropins,* Freeman, San Francisco, 1964.

Coon, C., *The Origin of Races,* Knopf (Random), New York, 1962.

Crane, M., and W. Lawrence, *The Genetics of Garden Plants,* 4th ed., Macmillan, New York, 1952.

Darlington, C. D., *Chromosome Botany and Origins of Cultivated Plants,* new ed., Hafner, New York, 1963.

Demerec, M., and B. P. Kaufmann, *Drosophila Guide: Introduction to the Genetics and Cytology of Drosophila melanogaster,* 7th ed., Carnegie Institution of Washington, Washington, D.C., 1961.

———, ed., *Biology of Drosophila,* Wiley, New York, 1950.

Dunn, L. C., ed., *Genetics in the Twentieth Century,* Macmillan, New York, 1951.

Ebert, J., *Development,* Holt, Rinehart and Winston, New York, 1965.

Elliott, F., *Plant Breeding and Cytogenetics,* McGraw-Hill, New York, 1958.

Ford, E., *Ecological Genetics,* Methuen (Wiley), New York, 1964.

Fraenkel-Conrat, H., *Design and Function at the Threshold of Life: The Viruses,* Academic, New York, 1962.

Fuller, J., and W. Thompson, *Behavior Genetics,* Wiley, New York, 1960.

Gardner, E., *Principles of Genetics,* 2nd ed., Wiley, New York, 1964.

Gardner, L., ed., *Molecular Genetics and Human Disease,* Thomas, Springfield, Ill., 1961.

Garner, R., *The Grafter's Handbook,* new ed., Oxford Univ. Press, New York, 1959.

Glass, B., *Genetic Continuity* (BSCS laboratory block), Heath, Boston, 1965.

Goldschmidt, R., *Theoretical Genetics,* Univ. of California Press, Berkeley, 1955.

Gowen, J., ed., *Heterosis,* Hafner, New York, 1952.

Grant, V., *The Origin of Adaptations,* Columbia Univ. Press, New York, 1963.

Hadorn, E., *Developmental Genetics and Lethal Factors,* Wiley, New York, 1961.

Hamburger, V., *A Manual of Experimental Embryology,* rev. ed., Univ. of Chicago Press, Chicago, 1960.

Hanson, E., *Animal Diversity,* 2nd ed., "The Foundations of Modern Biology Series," Prentice-Hall, Englewood Cliffs, N.J., 1964.

Harris, H., *Human Biochemical Genetics,* Cambridge Univ. Press, New York, 1959.

Hartman, P., and S. Suskind, *Gene Action,* Prentice-Hall, Englewood Cliffs, N.J., 1964.

Hayes, W., *The Genetics of Bacteria and Their Viruses,* Wiley, New York, 1964.

Ingram, V., *The Hemoglobins in Genetics and Evolution,* Columbia Univ. Press, New York, 1963.

Jacob, F., and E. Wollman, *Sexuality and the Genetics of Bacteria,* rev. ed., Academic, New York, 1961.

Kalmus, H., *Variation and Heredity,* Humanities, New York, 1958.

King, R., *Genetics,* Oxford Univ. Press, New York, 1962.

Lawler, S., and L. Lawler, *Human Blood Groups and Inheritance,* Harvard Univ. Press, Cambridge, Mass., 1957.

Lenz, W., *Medical Genetics,* trans. by E. Lanzl, Univ. of Chicago Press, Chicago, 1963.

Levine, R., *Genetics,* Holt, Rinehart and Winston, New York, 1962.

Li, C. C., *Human Genetics: Principles and Methods,* McGraw-Hill, New York, 1961.

Lwoff, A., *Biological Order,* M.I.T. Press, Cambridge, Mass., 1962.

MacArthur, R., and J. Connell, *The Biology of Populations,* Wiley, New York, 1965.

McKusick, V., *Human Genetics,* Prentice-Hall, Englewood Cliffs, N.J., 1964.

Meeuse, B. J., *The Story of Pollination,* Ronald, New York, 1961.

Moore, J., *Heredity and Development,* Oxford Univ. Press, New York, 1963.

Needham, J., *A History of Embryology,* 2nd ed., revised with A. Hughes, Abelard-Schuman, New York, 1959.

Neel, J. V., and W. J. Schull, *Human Heredity,* Univ. of Chicago Press, Chicago, 1954.

Newby, W., *A Guide to the Study of Development,* Saunders, Philadelphia, 1960.

Patten, B., *The Early Embryology of the Chick,* 4th ed., McGraw-Hill, New York, 1951.

———, *Foundations of Embryology,* 2nd ed., McGraw-Hill, New York, 1964.

Patterson, J., and W. Stone, *Evolution in Genus Drosophila,* Macmillan, New York, 1952.

Penrose, L., *Outline of Human Genetics,* 2nd ed., Wiley, New York, 1963.

Peters, J., ed., *Classic Papers in Genetics,* Prentice-Hall, Englewood Cliffs, N.J., 1959.

Purdom, C. E., *Genetic Effects of Radiation,* Academic, New York, 1963.

Raven, C. P., *Morphogenesis: The Analysis of Molluscan Development,* Pergamon, New York, 1958.

Ravin, A. W., *The Evolution of Genetics,* Academic, New York, 1965.

Rothschild, N., *Fertilization,* Wiley, New York, 1956.

Rugh, R., *Experimental Embryology: Techniques and Procedures,* 3rd ed., Burgess, Minneapolis, 1962.

————, *Vertebrate Embryology: The Dynamics of Development,* Harcourt, Brace & World, New York, 1964.

Sager, R. A., and F. Ryan, *Cell Heredity,* Wiley, New York, 1961.

Salisbury, G., and N. VanDemark, *Physiology of Reproduction and Artificial Insemination of Cattle,* Freeman, San Francisco, 1961.

Schmitt, F. O., ed., *Macromolecular Specificity and Biological Memory,* M.I.T. Press, Cambridge, Mass., 1962.

Sinnott, E., L. C. Dunn, and T. Dobzhansky, *Principles of Genetics,* 5th ed., McGraw-Hill, New York, 1958.

Snyder, L. H., and P. R. David, *The Principles of Heredity,* 5th ed., Heath, Boston, 1957.

Spemann, H., *Embryonic Development and Induction,* Yale Univ. Press, New Haven, Conn., 1938.

Spiess, E., *Papers on Animal Population Genetics,* Little, Brown, Boston, 1962.

Srb, A., R. Owen, and R. Edgar, *General Genetics,* 2nd ed., Freeman, San Francisco, 1965.

Stern, C., *Principles of Human Genetics,* 2nd ed., Freeman, San Francisco, 1960.

Strauss, B., *Chemical Genetics,* Saunders, Philadelphia, 1960.

Sutton, H., *Genes, Enzymes, and Inherited Diseases,* Holt, Rinehart and Winston, New York, 1961.

Thompson, Sir D'A., *On Growth and Form,* 2 vols., Cambridge Univ. Press, New York, 1952; abridged ed. by J. Bonner, 1961.

Timakov, V., *Microbial Variation,* Pergamon, New York, 1959.

Torrey, T., *Morphogenesis of Vertebrates,* Wiley, New York, 1962.

Waddington, C., *How Animals Develop,* rev. ed., Harper Torchbook, Harper & Row, New York, 1962.

————, *New Patterns in Genetics and Development,* Columbia Univ. Press, New York, 1962.

————, ed., *Biological Organisation, Cellular and Sub-cellular,* Pergamon, New York, 1959.

Wagner, R. P., and H. K. Mitchell, *Genetics and Metabolism,* 2nd ed., Wiley, New York, 1964.

Wallace, B., and A. Srb, *Adaptation,* 2nd ed., "The Foundations of Modern Biology Series," Prentice-Hall, Englewood Cliffs, N.J., 1964.

Wardlaw, C., *Embryogenesis in Plants,* Methuen (Wiley), New York, 1955.

Williams, R. J., *Biochemical Individuality,* Wiley, New York, 1956.

Willier, B., and J. Oppenheimer, *Foundations of Experimental Embryology,* Prentice-Hall, Englewood Cliffs, N.J., 1964.

————, P. Weiss, and V. Hamburger, eds., *Analysis of Development,* Saunders, Philadelphia, 1955.

Winge, O., *Inheritance in Dogs,* Comstock (Cornell Univ. Press), Ithaca, N.Y., 1950.

Evolution

Andrews, H. N., *Studies in Paleobotany,* Wiley, New York, 1961.

Asimov, I., *Chemicals of Life: Enzymes, Chemicals, Hormones,* Abelard-Schuman, New York, 1954; Signet, New Am. Library, New York, 1962.

Barrington, E., *Hormones and Evolution,* Van Nostrand, Princeton, N.J., 1964.

Bernal, J., *The Physical Basis of Life,* Routledge & Kegan Paul, London, 1951.

Berrill, N. J., *The Origin of the Vertebrates,* Oxford Univ. Press, New York, 1955.

Blum, H., *Time's Arrow and Evolution,* 2nd ed., Harper Torchbook, Harper & Row, New York, 1962.

Bondi, H., *et al., Rival Theories of Cosmology: A Symposium and Discussion of Modern Theories of the Structure of the Universe,* Oxford Univ. Press, New York, 1960.

Boschke, F., *Creation Still Goes On,* McGraw-Hill, New York, 1964.

Calvin, M., *Chemical Evolution,* 2 vols.: Vol. I, *From Molecule to Microbe,* 1961, Vol. II, *The Origin of Life on Earth and Elsewhere,* 1961; Oregon State System of Higher Education, Eugene.

Carson, H., *Heredity and Human Life,* Columbia Univ. Press, New York, 1963.

Colbert, E., *Evolution of the Vertebrates,* Wiley, New York, 1955; Science Editions Paperbacks, Wiley, 1961.

Darlington, C. D., *Evolution of Genetic Systems,* rev. and enl. ed., Basic Books, New York, 1958.

Dauvillier, A., *The Photochemical Origin of Life,* Academic, New York, 1965.

Delevoryas, T., *Morphology and Evolution of Fossil Plants,* Holt, Rinehart and Winston, New York, 1962.

Dobzhansky, T., *Mankind Evolving: The Evolution of the Human Species,* Yale Univ. Press, New Haven, Conn., 1962.

Dodson, E., *Evolution: Process and Product,* Reinhold, New York, 1960.

Dowdeswell, W. H., *The Mechanism of Evolution,* 2nd ed., Heinemann, London, 1958; Harper Torchbook, Harper & Row, New York, 1960.

Ehrensvärd, G., *Life: Origin and Development,* Univ. of Chicago Press, Chicago, 1962; Phoenix, Univ. of Chicago Press, 1962.

Ehrlich, R., and R. Holm, *The Process of Evolution,* McGraw-Hill, New York, 1963.

Fisher, R. A., *The Genetical Theory of Natural Selection,* 2nd ed., Dover, New York, 1959.

Florkin, M., *Biochemical Evolution,* Academic, New York, 1949.

———, *Unity and Diversity in Biochemistry,* Pergamon (Macmillan), New York, 1960.

———, ed., *Aspects of the Origin of Life,* Pergamon, New York, 1960.

Ford, E., *Ecological Genetics,* Methuen (Wiley), New York, 1964.

Gregory, W., *Evolution Emerging,* 2 vols., Macmillan, New York, 1951.

Haber, F., *The Age of the World: Moses to Darwin,* Johns Hopkins Press, Baltimore, 1959.

Heusser, C., *Late-Pleistocene Environments of North Pacific North America,* Am. Geographical Soc. of New York, New York, 1961.

Hoagland, M., and R. Burhoe, eds., *Evolution and Man's Progress,* Columbia Univ. Press, New York, 1962.

Huxley, J., *The Wonderful World of Life: The Story of Evolution,* Doubleday, Garden City, N.Y., 1958.

———, *et al.,* eds., *Evolution as a Process,* 2nd ed., Macmillan, New York, 1963.

Lack, D., *Darwin's Finches,* Harper Torchbook, Harper & Row, New York, 1961.

Lasker, G., *The Evolution of Man: A Brief Introduction to Physical Anthropology,* Holt, Rinehart and Winston, New York, 1961.

Life, editors of, and L. Barnett, *The Wonders of Life on Earth,* Time, New York, 1960.

Martin, P., *The Last 10,000 Years: A Fossil Pollen Record of the American Southwest,* Univ. of Arizona Press, Tucson, 1964.

Mayr, E., *Animal Species and Evolution,* Harvard Univ. Press, Cambridge, Mass., 1963.

Moody, P., *Introduction to Evolution,* 2nd ed., Harper & Row, New York, 1962.

Nairn, A., ed., *Descriptive Palaeoclimatology,* Interscience (Wiley), New York, 1961.

Oparin, A., *Life: Its Nature, Origin, and Development,* trans. by A. Synge, Academic, New York, 1962.

———, *The Origin of Life on Earth,* 3rd ed., rev. and enl., trans. by A. Synge, Academic, New York, 1957.

Portmann, A., *Animal Camouflage,* Univ. of Michigan Press, Ann Arbor, 1959.

Romer, A., *The Vertebrate Story,* rev. ed. of *Man and the Vertebrates,* Univ. of Chicago Press, Chicago, 1959.

Ross, H., *A Synthesis of Evolutionary Theory,* Prentice-Hall, Englewood Cliffs, N.J., 1962.

Simpson, G. G., *Life of the Past: An Introduction to Paleontology,* Yale Univ. Press, New Haven, Conn., 1953.

———, *The Meaning of Evolution,* Yale Univ. Press, New Haven, Conn., 1949; Yale Paperbounds, 1960.

Thiel, R., *And There Was Light,* Knopf (Random), New York, 1957; Mentor, New Am. Library, New York, 1960.

Waddington, C., *The Strategy of the Genes,* Macmillan, New York, 1957.

Wallace, A., *The Malay Archipelago,* rev. ed., Macmillan, New York, 1869; Dover, New York, 1962.

Wallace, B., and A. Srb, *Adaptation,* 2nd ed., "The Foundations of Modern Biology Series," Prentice-Hall, Englewood Cliffs, N.J., 1964.

Ecology and conservation

Andrewartha, H., *Introduction to the Study of Animal Populations,* Univ. of Chicago Press, Chicago, 1961.

Barclay, G., *Techniques of Population Analysis,* Wiley, New York, 1958.

Barnes, H., *Oceanography and Marine Biology,* Macmillan, New York, 1959.

Bates, M., *The Forest and the Sea,* Random, New York, 1960; Signet, New Am. Library, New York, 1961.

———, *Man in Nature,* 2nd ed., "The Foundations of Modern Biology Series," Prentice-Hall, Englewood Cliffs, N.J., 1964.

Bennett, H., *Elements of Soil Conservation,* 2nd ed., McGraw-Hill, New York, 1955.

Billings, W., *Plants and the Ecosystem,* Wadsworth, Belmont, Calif., 1964.

Buchsbaum, R., and M. Buchsbaum, *Basic Ecology,* Boxwood Press, Pittsburgh, 1957.

Cain, S., and G. de Oliveira Castro, *Manual of Vegetation Analysis,* Harper & Row, New York, 1959.

Cameron, T., *Parasites and Parasitism,* Wiley, New York, 1956.

Carson, R., *The Sea Around Us,* rev. ed., Oxford Univ. Press, New York, 1961; Signet, New Am. Library, New York, 1964.

Chapman, V., *Salt Marshes and Salt Deserts of the World,* Wiley, New York, 1960.

Darlington, P., *Zoogeography: The Geographical Distribution of Animals,* Wiley, New York, 1957.

Daubenmire, R. F., *Plants and Environment,* 2nd ed., Wiley, New York, 1959.

Dice, L., *Natural Communities,* Univ. of Michigan Press, Ann Arbor, 1952.

Dietrich, G., *General Oceanography,* trans. by F. Ostapoff, Wiley, New York, 1963.

Edmondson, W. T., *see* Ward, H. B., and G. C. Whipple.

Elton, C. S., *The Ecology of Invasions by Animals and Plants,* Wiley, New York, 1958.

Forbes, R., and A. Meyer, *Forestry Handbook,* Ronald, New York, 1955.

Frey, D., ed., *Limnology in North America,* Univ. of Wisconsin Press, Madison, 1963.

Friedlander, C., *Heathland Ecology,* Harvard Univ. Press, Cambridge, Mass., 1961.

Hanson, H., and E. Churchill, *The Plant Community,* Reinhold, New York, 1961.

Hardy, A., *The Open Sea, Its Natural History: The World of Plankton,* Houghton Mifflin, Boston, 1956.

Hazen, W., *Readings in Population and Community Ecology,* Saunders, Philadelphia, 1964.

Hubbs, C. L., ed., *Zoogeography,* Am. Assoc. for the Advancement of Science, Washington, D.C., 1958.

Hutchinson, G. E., *A Treatise on Limnology,* Vol. I, Wiley, New York, 1957.

Johannesson, B., *The Soils of Iceland,* University Research Institute, Reykjavík, Ice., 1960.

Klopfer, P., *Behavioral Aspects of Ecology,* Prentice-Hall, Englewood Cliffs, N.J., 1962.

Lack, D., *The Natural Regulation of Animal Numbers,* Oxford Univ. Press, New York, 1954.

Macan, T., *Fresh Water Biology,* Wiley, New York, 1963.

MacArthur, R., and J. Connell, *The Biology of Populations,* Wiley, New York, 1965.

Macfadyen, A., *Animal Ecology: Aims and Methods,* 2nd ed., Pitman, New York, 1963.

Metcalf, C., and W. Flint, *Destructive and Useful Insects,* 4th ed., revised by R. Metcalf, McGraw-Hill, New York, 1962.

Moore, H., *Marine Ecology,* Wiley, New York, 1958.

Munzer, M., and P. F. Brandwein, *Teaching Science Through Conservation,* McGraw-Hill, New York, 1960.

Neal, E., *Woodland Ecology,* Harvard Univ. Press, Cambridge, Mass., 1958.

Nicol, J., *The Biology of Marine Animals,* Interscience (Wiley), New York, 1960.

Nikolsky, G., *Ecology of Fishes,* trans. by L. Birkett, Academic, New York, 1963.

Odum, E., *Ecology,* Holt, Rinehart and Winston, New York, 1963.

————, with H. Odum, *Fundamentals of Ecology,* 2nd ed., Saunders, Philadelphia, 1959.

Oosting, H., *The Study of Plant Communities: An Introduction to Plant Ecology,* 2nd ed., Freeman, San Francisco, 1956.

Phillips, E. A., *Field Ecology* (BSCS laboratory block), Heath, Boston, 1964.

Popham, E. J., *Some Aspects of Life in Fresh Water,* 2nd ed., Harvard Univ. Press, Cambridge, Mass., 1961.

Pramer, D., *Life in the Soil* (BSCS laboratory block), Heath, Boston, 1965.

Reid, G., *Ecology of Inland Waters and Estuaries,* Reinhold, New York, 1961.

Reitz, L. P., ed., *Biological and Chemical Control of Plant and Animal Pests,* Am. Assoc. for the Advancement of Science, Washington, D.C., 1960.

Sears, M., ed., *Oceanography,* Am. Assoc. for the Advancement of Science, Washington, D.C., 1961.

Shelford, V., *The Ecology of North America,* Univ. of Illinois Press, Urbana, 1963.

Storer, J. H., *The Web of Life,* Devin-Adair, New York, 1953; Signet, New Am. Library, New York, 1956.

Telfer, W., and D. Kennedy, *The Biology of Organisms,* Wiley, New York, 1965.

U.S. Dept. of Agriculture, *Yearbook of Agriculture, 1957: Soil; Yearbook of Agriculture, 1958: Land; Yearbook of Agriculture, 1961: Seeds; Yearbook of Agriculture, 1963: A Place to Live,* U.S. Dept. of Agriculture, Washington 25, D.C.

Ward, H. B., and G. C. Whipple, *Fresh-Water Biology,* 2nd ed., edited by W. T. Edmondson, Wiley, New York, 1959.

Welch, P., *Limnological Methods,* Blakiston (McGraw-Hill), New York, 1948.

Wiens, H., *Atoll Environment and Ecology,* Yale Univ. Press, New Haven, Conn., 1962.

Williams, C. B., *Patterns in the Balance of Nature*

and Related Problems in Quantitative Ecology, Vol. III of J. F. Danielli, ed., "Theoretical and Experimental Biology," Academic, New York, 1964.

Natural history and identification of plants and animals

General

Bor, N., *The Grasses of Burma, Ceylon, India, and Pakistan, Excluding Bambusae,* Pergamon, New York, 1960.

Borradaile, L., *Manual of Elementary Zoology,* 14th ed., edited by W. Yapp, Oxford Univ. Press, New York, 1963.

Buchsbaum, R., *Animals Without Backbones: An Introduction to the Invertebrates,* 2nd ed., Univ. of Chicago Press, Chicago, 1948.

Bullough, W. S., *Practical Invertebrate Anatomy,* 2nd ed., Macmillan, London (St. Martin's, New York), 1958.

Cameron, T., *Parasites and Parasitism,* Wiley, New York, 1956.

Campbell, C., *et al., Great Smoky Mountains Wildflowers,* 2nd enl. ed., Univ. of Tennessee Press, Knoxville, 1964.

Carson, R., *The Edge of the Sea,* Houghton Mifflin, Boston, 1955; Signet, New Am. Library, New York, 1959.

Cushman, J., *Foraminifera: Their Classification and Economic Use,* 4th ed., Harvard Univ. Press, Cambridge, Mass., 1948.

Davis, C., *The Marine and Fresh Water Plankton,* Michigan State Univ. Press, East Lansing, 1955.

Edmondson, W. T., *see* Ward, H. B., and G. C. Whipple.

Farmer, W., *One Hundred Common Marine Animals of San Diego,* San Diego Soc. of Natural History, San Diego, Calif., 1964.

Fox, H., and H. Vevers, *The Nature of Animal Colors,* Macmillan, New York, 1960.

Galtsoff, P., *et al.,* eds. (J. Needham, chairman), *Culture Methods for Invertebrate Animals,* Comstock (Cornell Univ. Press), Ithaca, N.Y., 1937; Dover, New York, 1959.

Hardy, A., *The Open Sea, Its Natural History: The World of Plankton,* Houghton Mifflin, Boston, 1956.

Hausman, L., *Beginner's Guide to Fresh Water Life,* Putnam's, New York, 1950.

———, *Beginner's Guide to Seashore Life,* Putnam's, New York, 1949.

Hedgpeth, J., *see* Ricketts, E., and J. Calvin.

Jaeger, E., *Desert Wildlife,* Stanford Univ. Press, Stanford, Calif., 1961.

Jaques, H., ed., "How to Know" nature series, Wm. C. Brown, Dubuque, Iowa, 1946–63.

Jordan, E., *Hammond's Illustrated Nature Guide,* Hammond, New York, 1955.

Kingsbury, J., *Poisonous Plants of the United States and Canada,* Prentice-Hall, Englewood Cliffs, N.J., 1964.

Light, S. F., *Intertidal Invertebrates of the Central California Coast,* rev. ed. of *Laboratory and Field Text in Invertebrate Zoology,* rev. by R. I. Smith, *et al.,* Univ. of California Press, Berkeley, 1954.

Little, V., *General and Applied Entomology,* 2nd ed., Harper & Row, New York, 1963.

MacGinitie, G., and N. MacGinitie, *Natural History of Marine Animals,* McGraw-Hill, New York, 1949.

Maheshwari, J., *The Flora of Delhi,* Counsel of Scientific and Industrial Research, New Delhi, 1963.

Miner, R. W., *Field Book of Seashore Life,* Putnam's, New York, 1950.

Morgan, A., *Field Book of Animals in Winter,* Putnam's, New York, 1939.

———, *Field Book of Ponds and Streams,* Putnam's, New York, 1930.

Muenscher, W., *Aquatic Plants of the United States,* Comstock (Cornell Univ. Press), Ithaca, N.Y., 1944.

Murie, O., *Field Guide to Animal Tracks,* Houghton Mifflin, Boston, 1954.

Needham, J., and P. Needham, *A Guide to the Study of Fresh-Water Biology,* 4th ed., Comstock (Cornell Univ. Press), Ithaca, N.Y., 1938.

———, *see* Galtsoff, P., *et al.*

Pennak, R., *Fresh-Water Invertebrates of the United States,* Ronald, New York, 1953.

Ray, C., and E. Ciampi, *The Underwater Guide to Marine Life,* Barnes & Noble, New York, 1956.

Ricketts, E., and J. Calvin, *Between Pacific Tides,* 3rd ed., revised by J. Hedgpeth, Stanford Univ. Press, Stanford, Calif., 1962.

Shelford, V., *The Ecology of North America,* Univ. of Illinois Press, Urbana, 1963.

Smith, R. I., *et al., see* Light, S. F.

Summer, L., and J. Dixon, *Birds and Mammals of the Sierra Nevada* (with records from Sequoia and Kings Canyon National Parks), Univ. of California Press, Berkeley, 1953.

Swain, T., ed., *Chemical Plant Taxonomy,* Academic, New York, 1963.

Thorpe, W., *Learning and Instinct in Animals,*

Harvard Univ. Press, Cambridge, Mass., 1956.

Ward, H. B., and G. C. Whipple, *Fresh-Water Biology*, 2nd ed., edited by W. T. Edmondson, Wiley, New York, 1959.

Welch, P., *Limnology*, 2nd ed., McGraw-Hill, New York, 1952.

Wiggins, I., and J. Thomas, *Flora of the Alaskan Arctic Slope*, Univ. of Toronto Press, Toronto, 1961.

Yapp, W., *see* Borradaile, L.

Algae

Forest, H., *Handbook of Algae, with Special Reference to Tennessee and the Southeastern United States*, Univ. of Tennessee Press, Knoxville, 1954.

Prescott, G., *Algae of the Western Great Lakes Area*, Bull. 30, Cranbrook Institute of Science, Bloomfield Hills, Mich., 1951.

Smith, G., *The Fresh-Water Algae of the United States*, 2nd ed., McGraw-Hill, New York, 1950.

Taylor, W., *Marine Algae of the Northeastern Coast of North America*, 2nd rev. ed., Univ. of Michigan Press, Ann Arbor, 1957.

Tiffany, L. H., and M. Britton, *The Algae of Illinois*, Univ. of Chicago Press, Chicago, 1952.

Mushrooms

Christensen, C., *Common Fleshy Fungi*, Burgess, Minneapolis, 1955.

Hesler, L., *Mushrooms of the Great Smokies: A Field Guide to Some Mushrooms and Their Relatives*, Univ. of Tennessee Press, Knoxville, 1960.

Johnson, T. W., and F. K. Sparrow, *Fungi in Oceans and Estuaries*, Hafner, New York, 1962.

Smith, A., *The Mushroom Hunter's Field Guide*, rev. and enl. ed., Univ. of Michigan Press, Ann Arbor, 1963.

Thomas, W., *Field Book of Common Mushrooms*, Putnam's, New York, 1948.

Lichens

Hale, M., *Lichen Handbook*, Smithsonian Institution, Washington, D.C., 1961.

Mosses

Bodenberg, E., *Mosses: A New Approach to the Identification of Common Species*, Burgess, Minneapolis, 1954.

Conard, H. S., *How to Know the Mosses and Liverworts*, ed. by H. E. Jaques, Wm. C. Brown, Dubuque, Iowa, 1956.

Ferns

Cobb, B., *Field Guide to the Ferns and Their Related Families of Northeastern and Central North America*, Houghton Mifflin, Boston, 1956.

Durand, H., *Field Book of Common Ferns*, rev. ed., Putnam's, New York, 1949.

Parsons, F., *How to Know the Ferns: A Guide to the Names, Haunts, and Habits of Our Common Ferns*, Scribner's, New York, 1899; 2nd ed., Dover, New York, 1961.

Shaver, J., *Ferns of Tennessee*, George Peabody College for Teachers, Nashville, Tenn., 1954.

Wild flowers

Abrams, L., *Illustrated Flora of the Pacific States*, 5 vols., Stanford Univ. Press, Stanford, Calif., 1923–60.

Benson, L., *Plant Classification*, Heath, Boston, 1957.

Dana, W., *How to Know the Wild Flowers*, Scribner's, New York, 1895; rev. by C. Hylander, Dover, New York, 1963.

Gilbert-Carter, H., *Glossary of the British Flora*, 3rd ed., Cambridge Univ. Press, New York, 1964.

Gleason, H., *The New Britton & Brown's Illustrated Flora of N.E. United States and Adjacent Canada*, 3 vols., New York Botanical Garden, New York, 1952.

Gottscho, S., *A Pocket Guide to Wild Flowers*, Dodd, Mead, New York, 1951; Washington Square Press, New York, 1960.

Greene, W. F., and H. L. Blomquist, *Flowers of the South: Native and Exotic*, Univ. of North Carolina Press, Chapel Hill, 1953.

Hausman, L., *Beginner's Guide to Wild Flowers*, Putnam's, New York, 1948.

House, H. D., *Wild Flowers*, Macmillan, New York, 1961.

Hutchinson, J., *The Families of Flowering Plants*, 2nd ed., 2 vols.: Vol. I, *Dicots*, 1959, Vol. II, *Monocots*, 1959; Oxford Univ. Press, New York.

Hylander, C., *Wild Flower Book*, Macmillan, New York, 1954.

Jepson, W. L., *A Manual of the Flowering Plants of California*, Sather Gate Bookshop (Univ. of California Press), Berkeley, 1926.

Kearney, T., *et al.*, *Arizona Flora*, 2nd rev. ed., Univ. of California Press, Berkeley, 1960.

Martin, A., and W. Barkley, *Seed Identification Manual*, Univ. of California Press, Berkeley, 1961.

Mathews, F., *Field Book of American Wild Flow-*

ers, rev. and ed. by N. Taylor, Putnam's, New York, 1955.

Moldenke, H., *American Wild Flowers,* Van Nostrand, Princeton, N.J., 1949.

Porter, C., *Taxonomy of Flowering Plants,* Freeman, San Francisco, 1959.

Stefferud, A., *How to Know the Wild Flowers,* Holt, Rinehart and Winston, New York, 1950; Signet, New Am. Library, New York, 1950.

Taylor, N., *see* Mathews, F.

Wherry, E., *Wild Flower Guide: Northeastern and Midland United States,* Doubleday, Garden City, N.Y., 1948.

Trees and shrubs

Graves, A., *Illustrated Guide to Trees and Shrubs,* rev. ed., Harper & Row, New York, 1956.

Harlow, W., *Fruit Key and Twig Key* (two books bound as one), self-copyright, 1941, 1946; Dover, New York, 1959.

————, *Trees of the Eastern United States and Canada,* McGraw-Hill, New York, 1942; Dover, New York, 1957.

Hottes, A., *The Book of Shrubs,* 6th ed., De La Mare (Dodd, Mead), New York, 1952.

————, *The Book of Trees,* 3rd ed., De La Mare (Dodd, Mead), New York, 1952.

Little, E., Jr., and F. Wadsworth, *Common Trees of Puerto Rico and the Virgin Islands,* U.S. Dept. of Agriculture, Washington, D.C., 1964.

Longyear, B., *Trees and Shrubs of the Rocky Mountain Region,* Putnam's, New York, 1927.

Muenscher, W., *Keys to Woody Plants,* 6th rev. ed., Comstock (Cornell Univ. Press), Ithaca, N.Y., 1950.

Petrides, G., *A Field Guide to the Trees and Shrubs,* Houghton Mifflin, Boston, 1958.

Platt, R., *American Trees,* Dodd, Mead, New York, 1952.

Shells

Abbott, R., *How to Know the American Marine Shells,* Signet Key, New Am. Library, New York, 1961.

Keen, A., *Marine Molluscan Genera of Western North America: An Illustrated Key,* Stanford Univ. Press, Stanford, Calif., 1963.

Morris, P., *A Field Guide to the Shells of Our Atlantic and Gulf Coasts,* new ed., Houghton Mifflin, Boston, 1951.

————, *A Field Guide to the Shells of the Pacific Coast and Hawaii,* Houghton Mifflin, Boston, 1952.

Perry, L., and J. Schwengel, *Marine Shells of the Western Coast of Florida,* new and rev. ed.

of L. Perry, *Marine Shells of the Southwest Coast of Florida,* Paleontological Research Institution, Ithaca, N.Y., 1955.

Verrill, A., *Shell Collector's Handbook,* Putnam's, New York, 1950.

Insects and spiders

Barker, W., *Familiar Insects of America,* Harper & Row, New York, 1960.

Borror, D. J., and D. M. DeLong, *An Introduction to the Study of Insects,* rev. ed., Holt, Rinehart and Winston, New York, 1964.

Crompton, J., *The Life of the Spider,* Houghton Mifflin, Boston, 1951; Mentor, New Am. Library, New York, 1954.

Dillon, E., and L. Dillon, *A Manual of Common Beetles of Eastern North America,* Harper & Row, New York, 1961.

Emerton, J., *The Common Spiders,* Ginn, London, 1902; Dover, New York, 1961.

Farb, P., and the editors of *Life, The Insects,* Time, New York, 1962.

Ford, E., *Moths,* Macmillan, New York, 1955.

Frisch, K. von, *The Dancing Bees,* Harcourt, Brace & World, New York, 1955; Harvest, Harcourt, Brace & World, 1961.

Frost, S., *Insect Life and Insect Natural History,* 2nd ed., Dover, New York, 1959.

Kalmus, H., *101 Simple Experiments with Insects,* Doubleday, Garden City, N.Y., 1960.

Klots, A., *Field Guide to the Butterflies of North America, East of the Great Plains,* Houghton Mifflin, Boston, 1951.

Needham, J., and M. Westfall, *A Manual of Dragonflies of North America,* Univ. of California Press, Berkeley, 1954.

Ross, H., *A Textbook of Entomology,* 3rd ed., Wiley, New York, 1965.

Swain, R., *The Insect Guide,* Doubleday, Garden City, N.Y., 1948.

U.S. Dept. of Agriculture, *Yearbook of Agriculture, 1952: Insects,* U.S. Dept. of Agriculture, Washington 25, D.C.

Zim, H., and C. Cottam, *Insects,* Simon and Schuster, New York, 1951.

RECORDING

The Songs of Insects, 12 in., 33⅓ rpm, Cornell Univ. Records, Ithaca, N.Y., 1958.

Fish

Axelrod, H., and L. Schultz, *Handbook of Tropical Aquarium Fishes,* McGraw-Hill, New York, 1955.

————, and W. Vorderwinkler, *Encyclopedia of Tropical Fishes,* Sterling, New York, 1958.

————, *Salt-Water Aquarium Fish,* Sterling, New York, 1956.

Herald, E., *Living Fishes of the World,* Doubleday, Garden City, N.Y., 1961.

Innes, W., *Goldfish Varieties and Water Gardens,* 3rd ed., Innes, Philadelphia, 1960.

Lanham, U., *The Fishes,* Columbia Univ. Press, New York, 1962.

Perlmutter, A., *Guide to Marine Fishes,* New York Univ. Press, New York, 1961.

Walden, H., *Familiar Freshwater Fishes of America,* Harper & Row, New York, 1964.

Amphibians and reptiles

Barker, W., *Familiar Reptiles and Amphibians of America,* Harper & Row, New York, 1964.

Boys, F., and H. Smith, *Poisonous Amphibians and Reptiles: Recognition and Bite Treatment,* Thomas, Springfield, Ill., 1959.

Carr, A., *Handbook of Turtles: The Turtles of the United States, Canada, and Baja California,* Comstock (Cornell Univ. Press), Ithaca, N.Y., 1952.

Cochran, D. M., *Living Amphibians of the World,* Doubleday, Garden City, N.Y., 1961.

Conant, R., *A Field Guide to Reptiles and Amphibians,* Houghton Mifflin, Boston, 1958.

Ditmars, R., *Field Book of North American Snakes,* Doubleday, Garden City, N.Y., 1939.

————, *Reptiles of the World,* Macmillan, New York, 1936.

Goin, C., and O. Goin, *Introduction to Herpetology,* Freeman, San Francisco, 1962.

Klauber, L., *Rattlesnakes: Their Habits, Life Histories, and Influence on Mankind,* 2 vols., Univ. of California Press, Berkeley, 1956.

Oliver, J., *The Natural History of North American Amphibians and Reptiles,* Van Nostrand, Princeton, N.J., 1955.

Smith, H., *Handbook of Lizards,* Comstock (Cornell Univ. Press), Ithaca, N.Y., 1946.

Wright, A. H., and A. Wright, *Handbook of Frogs and Toads of the United States and Canada,* 3rd ed., Comstock (Cornell Univ. Press), Ithaca, N.Y., 1949.

RECORDING

Voices of the Night, rev. and expanded, recorded by P. Kellogg and A. Allen, 12 in., 33⅓ rpm, Cornell Univ. Records, Ithaca, N.Y., 1953.

Birds

Allen, A., *Book of Bird Life,* 2nd ed., Van Nostrand, Princeton, N.J., 1961.

Audubon, J., *Birds of America,* Introduction and descriptive captions by L. Griscom, Macmillan, New York, 1950.

Barruel, P., *Birds of the World: Their Life and Habits,* trans. by P. Barclay-Smith, Oxford Univ. Press, New York, 1954.

Darling, L., and L. Darling, *Bird,* Houghton Mifflin, Boston, 1962.

Headstrom, B. R., *Birds' Nests: A Field Guide,* new ed., Ives Washburn (McKay), New York, 1961.

Heinroth, O., and K. Heinroth, *The Birds,* Univ. of Michigan Press, Ann Arbor, 1958.

Mason, C., *Picture Primer of Attracting Birds,* Houghton Mifflin, Boston, 1952.

Nice, M., *Studies in the Life History of the Song Sparrow,* Am. Museum of Natural History, New York, 1937, 1943; 2 vols., Dover, New York, 1964.

Palmer, R., ed., *Handbook of North American Birds,* Vol. I, *Loons Through Flamingos,* Yale Univ. Press, New Haven, Conn., 1962.

Peterson, R. T., *A Field Guide to the Birds,* 2nd rev. and enl. ed., Houghton Mifflin, Boston, 1947.

————, *How to Know the Birds,* Houghton Mifflin, Boston, 1949; Signet Key, New Am. Library, New York, 1949.

————, G. Mountfort, and P. Hollom, *A Field Guide to the Birds of Britain and Europe,* Houghton Mifflin, Boston, 1954.

Pettingill, O., *A Guide to Bird Finding East of the Mississippi,* Oxford Univ. Press, New York, 1951.

————, *A Guide to Bird Finding West of the Mississippi,* Oxford Univ. Press, New York, 1953.

————, *A Laboratory and Field Manual of Ornithology,* 3rd ed., Burgess, Minneapolis, 1956.

Phillips, A., and G. Monson, *A Checklist of Arizona Birds,* Univ. of Arizona Press, Tucson, 1964.

Pough, R., *Audubon Guides: All the Birds of Eastern and Central North America,* Doubleday, Garden City, N.Y., 1953.

Saunders, A., *A Guide to Bird Songs,* rev. and enl. ed., Doubleday, Garden City, N.Y., 1951.

Thorpe, W., *Bird Song: The Biology of Vocal Communication and Expression in Birds,* Cambridge Univ. Press, New York, 1961.

Welty, C. J., *The Life of Birds,* Saunders, Philadelphia, 1962.

RECORDINGS

American Bird Songs, 2 vols., recorded by P. Kellogg and A. Allen, Cornell Univ. Records, Ithaca, N.Y., 1954, 1955.

Bird Songs of Dooryard, Field, and Forest, 3 vols., recorded by J. Stillwell and N. Stillwell, $33\frac{1}{3}$ rpm, Ficker Recording Service, 27 Arcadia Rd., Old Greenwich, Conn., 1961.

Birds on a May Morning, 12 in., $33\frac{1}{3}$ rpm, Droll Yankwes, Providence, R.I., 1964.

Fassett, J., *Music and Bird Songs,* 10 in., $33\frac{1}{3}$ rpm, Cornell Univ. Records, Ithaca, N.Y., 1952.

A Field Guide to Bird Songs of Eastern and Central North America (to accompany R. T. Peterson, *A Field Guide to the Birds,* 2nd rev. and enl. ed., Houghton Mifflin, Boston, 1947), 2 vols., 12 in., $33\frac{1}{3}$ rpm, recorded by P. Kellogg and A. Allen, in collaboration with R. T. Peterson, Houghton Mifflin, Boston, 1959.

A Field Guide to Western Bird Songs (to accompany R. T. Peterson, *A Field Guide to Western Birds,* 2nd rev. and enl. ed., Houghton Mifflin, Boston, 1961), 3 vols., 12 in., $33\frac{1}{3}$ rpm, Houghton Mifflin, Boston, 1962.

Florida Bird Songs, 10 in., 78 rpm, Cornell Univ. Records, Ithaca, N.Y., 1959.

The Sounds of a South American Rain Forest, 12 in., $33\frac{1}{3}$ rpm, Folkways Records, New York, 1952.

The Swamp in June, 12 in., $33\frac{1}{3}$ rpm, Droll Yankwes, Providence, R.I., 1965.

Mammals

Allen, G., *Bats,* Harvard Univ. Press, Cambridge, Mass., 1939; Dover, New York, 1962.

Bourlière, F., *Mammals of the World,* Knopf (Random), New York, 1955.

Burt, W., and R. Grossenheider, *A Field Guide to the Mammals,* Houghton Mifflin, Boston, 1952.

Harrop, A., *Reproduction in the Dog,* Williams & Wilkins, Baltimore, 1960.

Lilly, J., *Man and Dolphin,* Doubleday, Garden City, N.Y., 1961.

Palmer, E., *Fieldbook of Mammals,* Dutton, New York, 1957.

Palmer, R., *Mammal Guide,* Doubleday, Garden City, N.Y., 1954.

Sanderson, I., *How to Know the American Mammals,* Little, Brown, Boston, 1951; Signet, New Am. Library, New York, 1951.

———, *Living Mammals of the World,* Doubleday, Garden City, N.Y., 1955.

Walker, E., *Mammals of the World,* 3 vols., Johns Hopkins Press, Baltimore, 1964.

Appendix B

Distributors of
Audio-Visual Materials

Source materials:
educational film media

You will want to consult these resource materials to keep informed about methods and materials—including new films, tapes, and other audio-visual devices—in the field of educational communication.

Also refer to the directory of film distributors (p. 732) for sources of films and other aids that can be purchased, rented, or borrowed. Check the film libraries of colleges, state departments of education, and health agencies in your own area for lists of available free films and filmstrips.

Allison, M., E. Jones, and E. Schofield, *A Manual for Evaluators of Films and Filmstrips,* UNESCO, Paris, 1956.

Am. Institute of Biological Sciences, *Sourcelist for Biology Films and Filmstrips for Secondary and College Level,* 2000 P St., N.W., Washington 6, D.C.

Am. Psychological Assoc., *Film Series on Scientific Developments in Psychology,* 1333 16 St., N.W., Washington 6, D.C.

Brown, J. W., R. B. Lewis, and F. F. Harcleroad, *A-V Instruction: Materials and Methods,* 2nd ed., "Curriculum and Methods in Education Series," McGraw-Hill, New York, 1964.

Dale, E., *Audio-Visual Methods in Teaching,* Dryden (Holt, Rinehart and Winston), New York, 1957.

Educational Film Library Assoc., *Evaluation of Current Films* (monthly), 250 W. 57 St., New York 19, N.Y.

Educational Media Council, *EMI: The Educational Media Index,* 14 vols., McGraw-Hill, New York, 1964.

Educational Screen, Inc., *Blue Book of Audio-Visual Materials,* Chicago.

Educators' Progress Service, *Educators' Guide to Free Films; Educators' Guide to Free Materials; Educators' Guide to Free Science Materials; Educators' Guide to Free Scripts and Transcriptions; Educators' Guide to Free Slidefilms; Educators' Guide to Free Tapes,* Randolph, Wis.

Field Enterprises Educational Corp., *Sources of Free and Inexpensive Educational Materials,* Merchandise Mart Plaza, Chicago 54, Ill.

General Electric Film Library, *Motion Pictures and Slide Films,* P.O. Box 5970A, 840 S. Canal St., Chicago, Ill.; offices in many other cities.

General Motors, *General Motors Motion Picture Catalog,* Detroit 2, Mich.

Hulfish, J., ed., *The Audio-Visual Equipment Directory,* National Audio-Visual Assoc., Fairfax, Va., 1965.

Landers Film Reviews, *Source Directory,* Landers Associates, Coliseum St., Los Angeles, Calif., 1965.

McGraw-Hill Book Co., *Filmstrips: A Descriptive Index and User's Guide,* New York.

Modern Talking Picture Service, Inc., *Modern Index and Guide to Free Educational Films from Industry,* 3 E. 54 St., New York 22, N.Y.

Molstad, J., *Sources of Information Educational Media,* U.S. Office of Education, Washington 25, D.C., 1963.

National Education Assoc., *National Tape Recording Catalog: 1962-63,* 1201 16 St., N.W., Washington 6, D.C., 1962.

National Soc. for the Study of Education, *Audio-Visual Materials of Instruction,* Yearbook 48, Part I, Univ. of Chicago Press, Chicago, 1949.

New York State College of Agriculture, *Films, Recordings, and Slides,* Cornell Univ., Ithaca, N.Y.

New York State Dept. of Commerce, *Film Library Catalog,* 112 State St., Albany 7, N.Y.

Reid, S., A. Carpenter, and A. Daugherty, *A Directory of 3660 16-mm Film Libraries,* U.S.

Dept. of Health, Education, and Welfare, Washington 25, D.C., 1960.

Rufsvold, M., and C. Guss, *Guides to Newer Educational Media,* Am. Library Assoc., Chicago, 1961.

Shell Film Library, *Shell Motion Picture Catalog,* Shell Oil Co., 50 W. 50 St., New York 22, N.Y.; offices in other cities.

Teaching Film Custodians. Inc., *Films for Classroom Use,* 25 W. 43 St., New York 36, N.Y.

U.S. Dept. of Health, Education, and Welfare, *Mental Health Motion Pictures,* U.S. Govt. Printing Office, Washington 25, D.C.

———, *United States Government Films for Public Educational Use,* U.S. Govt. Printing Office, Washington 25, D.C.

United World Films, Inc., *United States Government Films for Schools and Colleges,* 221 Park Ave. South, New York 3, N.Y.; depository agency of films of U.S. Office of Education and those of many other government agencies.

Univ. of California, Audio-Visual Center, *Short Films in Microbiology,* D. Reynolds, Univ. of California, Davis.

———, Audio-Visual Service, *Lifelong Learning; Motion Picture Films on Predatory-Prey Relationships,* Univ. of California, Berkeley.

Univ. of Colorado, Audio-Visual Service, *Cinephotographic Techniques in Morphology,* Univ. of Colorado, Boulder.

H. W. Wilson Co., *Educational Film Guide; Filmstrip Guide,* 950 University Ave., Bronx 52, N.Y.

Films and filmstrips

We hope that teachers will find this directory useful in ordering films and filmstrips, recordings, tapes, and slides. Listings always seem to be incomplete or out of date; new materials in audio-visual aids are produced regularly, and quotations of price vary. We recommend that teachers obtain catalogs from some of the distributors listed, and also refer to the suggested films and other materials listed at the end of each chapter.

Many "free" films are available from industry, and from state and government agencies (conservation, wildlife, forestry, agriculture, and so forth). In many cases, these films are distributed through film libraries in your city. The only charge may be for postage.

In the Directory we have listed one address for each company; some of the larger ones maintain branch offices or are affiliated with film libraries. In ordering materials for your school try to locate a nearby film library or branch office, which may be much nearer to you than the address we give, and will thereby speed your orders.

Films, especially free films in popular demand, need to be ordered weeks, even months, in advance. When possible, plan large units of work for the school year early in September, and mark off several alternate days for each area of work for showing films. Then order early and submit alternate choices of days whenever feasible. In this way films reach the school a day or two before classes so that they augment the day's lesson, rather than detract from it because they do not fit into the sequence of work.

Films should be previewed for relevance to the day's work, suitability for the age level of students, and appropriateness in your community. Many teachers develop a file of their own approved films in this way, using library cards indicating the source of the film, rental or free loan, and a brief summary of the film or filmstrip (or attach the guide).

There are many techniques in showing films. A film or filmstrip need not be shown in its entirety; start the film at the place you want to use in your lesson to illustrate a point which cannot be demonstrated firsthand. In using a filmstrip, a teacher may ask questions, direct the observation of students, reverse and repeat a section for closer study and interpretation, or give students the time to ask questions while the filmstrip is still on hand. Some of the very short films illustrating a technique, such as that of handling fruit flies, may be shown a second time to present the fine details after a broad overview. For some of these special techniques, filmloops are also available (see Ealing filmloops of the BSCS techniques films).

At the end of the film or filmstrip, students may be asked to summarize the main ideas; to develop in words the concepts that were progressively developed. (Also read Section V in the accompanying volume: P. F. Brandwein, F. G. Watson, and P. E. Blackwood, *Teaching High School Science: A Book of Methods* [Harcourt, Brace & World, 1958].)

Suggestions for facilities for storing films and other audio-visual materials are described in Chapter 14.

Film distributors

While only the central office is given for most of these film distributors, many of them have branch offices in a number of the larger cities.

Before you order films, check the office nearest you or inquire of your local film library.

United States

Abbott Laboratories, Film Service Dept., North Chicago, Ill.

Allis-Chalmers Manufacturing Co., Film Section, Box 512, Milwaukee 1, Wis.

Almanac Films Inc., 29 E. Tenth St., New York 3, N.Y.

Am. Can Co., 100 Park Ave., New York 17, N.Y.

Am. Cancer Soc., 219 E. 42 St., New York 17, N.Y.

Am. Cyanamid Co., Lederle Laboratories, Film Library, Pearl River, N.Y.

Am. Film Registry, 831 S. Wabash, Chicago, Ill.

Am. Guernsey Cattle Club, 70 Main St., Peterborough, N.H.

Am. Heart Assoc., 44 E. 23 St., New York 10, N.Y.

Am. Hereford Assoc., Hereford Dr., Kansas City 5, Mo.

Am. Institute of Biological Sciences, 2000 P St., N.W., Washington 6, D.C.

Am. Museum of Natural History, 79 St. & Central Park West, New York 24, N.Y.

Am. National Cattlemen's Assoc., 801 E. 17 St., Denver 18, Colo.

Am. Osteopathic Assoc., Public and Professional Service, 212 E. Ohio St., Chicago 11, Ill.

Am. Petroleum Institute, 1271 Ave. of the Americas, New York 20, N.Y.

Am. Pharmaceutical Assoc., 2215 Constitution Ave., N.W., Washington 7, D.C.

Am. Potash Institute, 1102 16 St., N.W., Washington 6, D.C.

Am. Waterways Operators, Inc., 1319 F St., N.W., Washington 4, D.C.

Association Films Inc., 347 Madison Ave., New York 17, N.Y.; many branch offices.

Audio Productions Inc., 630 Ninth Ave., New York 36, N.Y.

Australian News & Information Bur., 636 Fifth Ave., New York 20, N.Y.

Bausch & Lomb Inc., Film Service, 635 St. Paul St., Rochester 2, N.Y.

Beet Sugar Development Foundation, Robertson Bldg., Fort Collins, Colo.

Bell System; contact your local office, or American Telephone and Telegraph Co., Film Library, 195 Broadway, New York 7, N.Y.

Brandon Films Inc., 200 W. 57 St., New York 19, N.Y.

Bray Studios, 729 Seventh Ave., New York 19, N.Y.

Brice, *see* Phase Films.

Bur. of Communication Research, 267 W. 25 St., New York 1, N.Y.

Bur. of Mines, Graphic Services, 4800 Forbes St., Pittsburgh 13, Pa.

California Chemical Co., Advertising and Public Relations, 200 Bush St., San Francisco 20, Calif.

California Spray Chemical Corp., Lucos & Ortho Way, Richmond, Calif.

Canadian Consulate General, Film Library, 111 N. Wabash Ave., Chicago 2, Ill.

Carolina Biological Supply Co., Burlington, N.C.

Carousel Films Inc., 1501 Broadway, New York 36, N.Y.

J. I. Case Co., 700 State St., Racine, Wis.

Cenco Educational Films, 1700 Irving Park Rd., Chicago 13, Ill.

Cereal Institute Inc., 135 S. LaSalle St., Chicago 3, Ill.

Chemagro Corp., P.O. Box 4913, Hawthorn Rd., Kansas City 20, Mo.

Churchill Films, Educational Film Sales Dept., 6671 Sunset Blvd., Los Angeles 28, Calif.

CIBA Chemical and Dye Co., Division of CIBA Corp., Fair Lawn, N.J.

Colburn Film Services Inc., Film Library, 164 N. Wacker Dr., Chicago 6, Ill.

Colonial Film and Equipment Co., 71 Walton St., N.W., Atlanta 6, Ga.

Connecticut Light and Power Co., P.O. Box 2010, Hartford 1, Conn.

Conservation Foundation, 30 E. 40 St., New York 16, N.Y.

Contemporary Films, 267 W. 25 St., New York 1, N.Y.

Coronet Instructional Films, 65 E. South Water St., Chicago, Ill.; or nearest film library.

Cox Enterprises, 2900 S. Sawtelle Blvd., Los Angeles 24, Calif.

Davey Tree Expert Co., City Bank Bldg., Kent, Ohio.

De Kalb Agricultural Assoc., Inc., Educational Dept., 310 N. Fifth St., De Kalb, Ill.

Denoyer-Geppert Co., 5235 Ravenswood Ave., Chicago 40, Ill.

Walt Disney Productions, Educational Film Division, 500 S. Buena Vista Ave., Burbank, Calif.; branch offices in Illinois and New York.

DoAll Co., Film Librarian, 254 N. Laurel Ave., Des Plaines, Ill.

Dow Chemical Co., Film Library, Public Relations Dept., Midland, Mich.

Dowling Productions, 1056 S. Robertson Blvd., Los Angeles 35, Calif.

E. I. du Pont de Nemours & Co., Inc., Advertising Dept., Motion Picture Distribution, Wilmington 98, Del.

Ealing Corp., 2225 Massachusetts Ave., Cambridge, Mass.

Eastman Kodak Co., Audio-Visual Service, 348 State St., Rochester 4, N.Y.

Educational Film Library Assoc., 250 W. 57 St., New York 19, N.Y.

Educational Testing Service, Rosedale Rd., Princeton, N.J.

Encyclopaedia Britannica Films Inc. (EBF), 1150 Wilmette Ave., Wilmette, Ill.

Ethyl Corp., 100 Park Ave., New York 17, N.Y.

Farm Film Foundation, Southern Bldg., Washington 5, D.C.

Ferry-Morse Seed Co., 111 Whisman Ave., Mountain View, Calif.

Film Assoc. of California, 11014 Santa Monica Blvd., Los Angeles, Calif.

Films of the Nations Distributors Inc., 305 E. 86 St., New York 28, N.Y.

Filmscope Inc., Box 397, Sierra Madre, Calif.

Fish and Wildlife Service, U.S. Dept. of the Interior, P.O. Box 128, College Park, Md.

Gateway Productions Inc., 1859 Powell St., San Francisco 11, Calif.

General Electric Corp., Advertising and Sales Promotion Dept., 1 River Rd., Schenectady 5, N.Y.; or nearest G.E. film library.

General Mills, Inc., Film Library, 9200 Wayzata Blvd., Minneapolis 26, Minn.

General Motors Corp., Dept. of Public Relations, General Motors Bldg., Detroit 2, Mich.

Handel Film Corp., 6926 Melrose Ave., Los Angeles 38, Calif.

Harcourt, Brace & World, Inc., 757 Third Ave., New York 17, N.Y.

Hayes Spray Gun Co., Film Division, 98 N. San Gabriel Blvd., Pasadena, Calif.

Heidenkamp Nature Pictures, 538 Glen Arden Dr., Pittsburgh 8, Pa.

H. J. Heinz Co., 1062 Progress St., Pittsburgh 30, Pa.

Hy-Line Poultry Farms, 1206 Mulberry St., Des Moines 9, Iowa.

Ideal Pictures, Inc., 58 E. South Water St., Chicago 1, Ill.

Indiana Univ. Films, Audio-Visual Center, Bloomington, Ind.; or Educational Film Library Assoc. See also NET.

Institute of Visual Communication Inc., 420 Lexington Ave., New York, N.Y.

Institutional Cinema Service Inc., 29 E. Tenth St., New York 3, N.Y.

Instructional Films Inc., 1150 Wilmette Ave., Wilmette, Ill.

International Film Foundation Inc., 475 Fifth Ave., New York 17, N.Y.

International Paper Co., 220 E. 42 St., New York 17, N.Y.

Iowa State Univ., Bur. of Visual Instruction, Ames, Iowa.

Jam Handy Organization, Inc., 2821 E. Grand Blvd., Detroit 11, Mich.; many branch offices.

Kansas State College, Dept. of Poultry Husbandry, Manhattan, Kan.

Knowledge Builders, Visual Education Center Bldg., Floral Park, L.I. 1, N.Y.

Library Films Inc., 79 Fifth Ave., New York 11, N.Y.

McGraw-Hill Publishing Co., Text-film Division, 330 W. 42 St., New York 36, N.Y.; also distributors of AIBS Films.

Mallinckrodt Chemical Works, Advertising Dept., 3600 N. Second St., St. Louis 7, Mo.

Metropolitan Life Insurance Co., 1 Madison Ave., New York 10, N.Y.

Michigan Dept. of Conservation, Film Loan Service, Lansing 26, Mich.

Milk Industry Foundation, Chrysler Bldg., New York 17, N.Y.

Minneapolis-Moline, Inc., Hopkins, Minn.

Modern Talking Picture Service, Inc., 1212 Ave. of the Americas, New York 36, N.Y.; branch offices in many cities.

Monsanto Chemical Co., 800 N. Lindbergh Blvd., St. Louis 66, Mo.

Moody Institute of Science, 11428 Santa Monica Blvd., Los Angeles 25, Calif.

National Agricultural Chemicals Assoc., 1145 19 St., N.W., Washington 6, D.C.

National Audubon Soc., 1130 Fifth Ave., New York 28, N.Y.

National Cancer Institute, Bethesda 14, Md.

National Cotton Council of America, 1918 N. Parkway, Memphis 12, Tenn.

National Education Assoc., Publications Division, 1201 16 St., N.W., Washington 6, D.C.

National Film Board of Canada, 680 Fifth Ave., New York 19, N.Y.; for rental, inquire of Contemporary Films.

National Kidney Disease Foundation, Inc., 143 E. 35 St., New York 16, N.Y.

National Safety Council, 425 N. Michigan Ave., Chicago 11, Ill.

National Tuberculosis Assoc., 1790 Broadway, New York 19, N.Y.

National Vitamin Foundation, 149 E. 78 St., New York 21, N.Y.

NET Film Service, Audio-Visual Center, Indiana Univ., Bloomington, Ind.; also 10 Columbus Circle, New York 19, N.Y.

Netherlands Flower-Bulb Institute, Inc., 29 Broadway, New York 6, N.Y.

New Holland Machine Co., Film Dept., New Holland, Pa.

New York Botanical Garden, Bronx Park, N.Y.

New York State College of Agriculture, Film Library, Cornell Univ., Ithaca, N.Y.

New York State Dept. of Commerce, Film Library, Albany, N.Y.

New York State Soc. for Medical Research, 2 E. 63 St., New York 21, N.Y.

New York State Univ., College of Forestry, Syracuse Univ., Syracuse 10, N.Y.

New York Times, Education Office, 229 W. 43 St., New York 36, N.Y.

New York Univ., Film Library, Washington Square, New York 3, N.Y.

New Zealand Embassy, 19 Observatory Circle, N.W., Washington 8, D.C.

North Carolina State College, Dept. of Visual Aids, Raleigh, N.C.

Northern Films, Box 98, Main Office Station, Seattle 11, Wash.

Ohio State Univ., Teaching Aids Laboratory, 1988 N. College Rd., Columbus 10, Ohio.

Olin Mathieson Chemical Corp., % K. M. Baker, 460 Park Ave., New York 22, N.Y.

Sam Orleans Film Productions, Inc., 211 W. Cumberland Ave., Knoxville 15, Tenn.

Chas. Pfizer & Co., Inc., Educational Services Dept., 235 E. 42 St., New York 17, N.Y.

Pfizer Medical Film Library, 267 W. 25 St., New York 17, N.Y.

Pharmaceutical Manufacturers Assoc., 1411 K St., N.W., Washington 5, D.C.

Phase Films, 656 Austin Ave., Sonoma, Calif.

Photo Lab, Inc., 3825 Georgia Ave., N.W., Washington 11, D.C.

Popular Science Monthly, 355 Lexington Ave., New York 17, N.Y.

Post Pictures Corp., 445 E. 86 St., New York 28, N.Y.

Potomac Films, 1536 Connecticut Ave., Washington 6, D.C.

Prudential Insurance Co. of America, Prudential Plaza, Newark 1, N.J.; branch offices in many cities.

Richfield Oil Corp., Film Library, 555 S. Flower St., Los Angeles 17, Calif.

Rohm & Haas Co., Washington Square, Philadelphia 5, Pa.

Rothschild Film Corp., 1046 E. 18 St., Brooklyn 30, N.Y.

Roberts Rugh, % Columbia Univ. Medical Center, 630 W. 168 St., New York 32, N.Y.

Science Slides Co., 22 Oak Dr., New Hyde Park, N.Y.

Scientific Supplies Co., 600 S. Spokane St., Seattle 4, Wash.; color slides.

Shell Oil Co., Film Library, 50 W. 50 St., New York 22, N.Y.

Sinclair Refining Co., 600 Fifth Ave., New York 20, N.Y.; branch offices in many cities.

Smith, Kline & French Laboratories, Medical Film Center, 1500 Spring Garden St., Philadelphia 1, Pa.

Soc. for Visual Education, Inc., 1345 W. Diversey Parkway, Chicago 14, Ill.

Soc. of Am. Bacteriologists, Committee on Visual Aids, School of Medicine, Univ. of Pennsylvania, Philadelphia 4, Pa.

Socony-Mobil Oil Co., Inc., 150 E. 42 St., New York 17, N.Y.

E. R. Squibb & Sons, Division of Olin Mathieson Chemical Corp., 745 Fifth Ave., New York 22, N.Y.

Standard Oil Co. of California, 225 Bush St., San Francisco 20, Calif.

Standard Oil of New Jersey, Public Relations Dept., 30 Rockefeller Plaza, New York 20, N.Y.

State Univ. of Iowa, Audio-Visual Center, Ames, Iowa; plant-science films.

Stauffer Chemical Co., Public Relations Dept., 380 Madison Ave., New York 17, N.Y.

Sterling Movies, U.S.A., Inc., 43 W. 61 St., New York, N.Y.

Swedish Film Center, Dept. of Creativism Inc., 1780 Broadway, New York, N.Y.

Swift & Co., 115 W. Jackson Blvd., Chicago 4, Ill.

Syracuse Univ., College of Forestry, Syracuse 10, N.Y.

Teaching Film Custodians, Inc., 25 W. 43 St., New York 36, N.Y.; for long-term loans. For short loans, inquire of regional film library.

Tennessee Valley Authority, Division of Agricultural Relations, Knoxville, Tenn.

Texaco Inc., 135 E. 42 St., New York 17, N.Y.

Thorne Films Inc., 1220 University Ave., Boulder, Colo.; BSCS techniques shorts.

Time, Inc., *Life* Filmstrips, Time & Life Bldg., New York 20, N.Y.

Union Pacific Railroad, Livestock and Agricultural Dept., 1416 Dodge St., Omaha 2, Nebr.

U.S. Atomic Energy Comm., Division of Public

Information, Washington 45, D.C.; films also available from field offices of AEC.

U.S. Dept. of Agriculture, Motion Picture Service, Washington 45, D.C.; also refer to films of the Graduate School, Dept. of Agriculture.

U.S. Forest Service, Dept. of Agriculture, Washington 25, D.C.; branch offices in many cities.

U.S. Public Health Service, Communicable Disease Center, Atlanta 22, Ga.

U.S. Soil Conservation Service, Dept. of the Interior, Washington, D.C.; for films, write to your state conservation office.

United World Films, Inc., 221 Park Ave. South, New York 3, N.Y.

Univ. of California, Educational Film Sales Dept., Univ. Extension, Los Angeles 24, Calif.

Univ. of California, Univ. Extension, Visual Dept., 2272 Union St., Berkeley, Calif.

Univ. of Iowa, Bur. of Visual Instruction, Iowa City, Iowa.

Univ. of Nebraska, Bur. of Audio-Visual Instruction, Extension Division, Lincoln 8, Nebr.

Utica Duxbak Corp., 815 Noyes St., Utica, N.Y.

Velsicol Chemical Corp., Advertising Dept., 341 E. Ohio St., Chicago 11, Ill.

Venard Organization, 113 N.E. Madison Ave., Peoria, Ill.

R. Vishniac, Yeshiva Univ., 526 W. 187 St., New York 33, N.Y.

Ward's Natural Science Establishment, Inc., P.O. Box 1712, Rochester, N.Y.

West Coast Lumbermen's Assoc., Yeon Bldg., Portland, Ore.

Western Phosphates, Inc., Film Dept., P.O. Box 893, Salt Lake City 10, Utah.

Westinghouse Electric Corp., 3 Gateway Center, Pittsburgh 30, Pa.

Roy Wilcox Productions, Allen Hill, Meriden, Conn.

Wilner Films and Slides, P.O. Box 231, Cathedral Station, New York 25, N.Y.

Wisconsin Alumni Research Foundation, P.O. Box 2059, Madison, Wis.

Wisconsin Conservation Dept., Film Library, Madison, Wis.

Wistar Institute, 36 St. and Woodland Ave., Philadelphia, Pa.

Wool Bureau, Inc., 360 Lexington Ave., New York 17, N.Y.

Yeshiva Univ., Audio-Visual Center, 526 W. 187 St., New York 33, N.Y.

Zurich-American Life Insurance Co., 111 W. Jackson St., Chicago 3, Ill.

United Kingdom

Science and Film, % Scientific Film Assoc., 164 Shaftesbury Ave., London, W.C. 2.

Canada

Audio-Visual Supply Co., Toronto General Trusts Bldg., Winnipeg, Man.

Canadian Film Institute, 142 Sparks St., Ottawa, Ont.

Canadian Industries Ltd., P.O. Box 10, 1253 McGill College Ave., Montreal, P.Q.

General Films Ltd., Head Office, 1534 13 Ave., Regina, Sask.; inquire of regional office in your area.

Radio-Cinema, 5011 Verdun Ave., Montreal, P.Q.

Shell Oil Co. of Canada, Ltd., Chemical Division, 505 University Ave., Toronto, Ont.

Appendix C

Supply Houses, Manufacturers, and Distributors

Teachers will also want to refer to the "Annual Buyer's Guide to Advertised Products" in the "Instrument" issue (October) of *Science*.

United States

Alconox Inc., 853 Broadway, New York 3, N.Y.

Allied Chemical Corp., 61 Broadway, New York 6, N.Y.

Aloe Scientific, 1831 Olive St., St. Louis 3, Mo.

Am. Hospital Supply Corp., 40-05 168 St., Flushing, N.Y.

Am. Optical Co., Instrument Division, Box 3, Buffalo 15, N.Y.

Am. Sterilizer Co., 2424 W. 23 St., Erie, Pa.

Am. Type Culture Collection, 2029 M St., N.W., Washington 6, D.C.; bacteria.

Ames Co., Inc., 819 McNaughton Ave., Elkhart, Ind.

Baltimore Biological Laboratory, Division of B-D Laboratories, Inc., 2201 Aisquith St., Baltimore 18, Md.

Bausch & Lomb Inc., 635 St. Paul St., Rochester 2, N.Y.

Becton, Dickinson & Co., Stanley St., East Rutherford, N.J.; medical instruments.

Biological Research Products, 243 W. Root St., Stockyards Station, Chicago 9, Ill.

California Biological Service, 1612 W. Glen Oaks Blvd., Glendale, Calif.

California Corp. Biochemical Research, 3625 Medford St., Los Angeles 63, Calif.

Cambosco Scientific Co., Inc., 342 Western Ave., Brighton Station, Boston, Mass.

Cambridge Instrument Co., 420 Lexington Ave., New York 17, N.Y.

Carolina Biological Supply Co., Burlington, N.C.

Central Scientific Co., 1700 W. Irving Park Rd., Chicago, Ill.

Certified Blood Donor Service, 146-16 Hillside Ave., Jamaica 35, N.Y.

Chemical Rubber Co., 2310 Superior Ave., Cleveland 14, Ohio.

CIBA Pharmaceutical Co., Division of CIBA Corp., 556 Morris Ave., Summit, N.J.

Clay-Adams Co., 141 E. 25 St., New York 10, N.Y.

Coleman Instruments Corp., 42 Madison St., Maywood, Ill.

Coors Porcelain Co., Golden, Colo.

Corning Glass Works, 75 Crystal St., Corning, N.Y.

Denoyer-Geppert Co., 5235 Ravenswood Ave., Chicago 40, Ill.

Difco Laboratories Inc., 920 Henry St., Detroit 1, Mich.; prepared media.

Disposable Laboratory Cages Inc., 15th & Bloomingdale, Melrose Park, Chicago, Ill.

Dow Chemical Co., Midland, Mich.

Duralab Equipment Corp., 980 Linwood St., Brooklyn 8, N.Y.

Ealing Corp., 2225 Massachusetts Ave., Cambridge, Mass.

Eastman Kodak Co., Organic Chemicals Division, Rochester 3, N.Y.

Eimer and Amend, 633 Greenwich St., New York 14, N.Y.

Elgeet Optical Co., 838 Smith St., Rochester, N.Y.; microscopes.

Equipto, 618 Prairie Ave., Aurora, Ill.; storage, shelving, carts.

Falcon Plastics, Division of Becton, Dickinson & Co., 6016 W. Washington Blvd., Culver City, Colo.; see also Becton, Dickinson.

Fisher Scientific Co., 711 Forbes Ave., Pittsburgh, Pa.; branches in many cities, also makers of Fisher Unitized Laboratory furniture.

General Biochemicals Inc., 9 Laboratory Park, Chagrin Falls, Ohio.

General Biological Supply House, Inc. (Turtox), 8200 S. Hoyne Ave., Chicago 20, Ill.

Gradwohl Laboratories, 3514 Lucas Ave., St. Louis 3, Mo.

Graf-Apsco Co., 5868 N. Broadway, Chicago 40, Ill.

Greiner Scientific Corp. and Palo Laboratory Suppliers Inc., 20–26 N. Moore St., New York 12, N.Y.

Harshaw Chemical Co., 1945 E. 97 St., Cleveland 6, Ohio.

Harvard Apparatus Co., Millis, Mass.

Hyland Laboratories, 4501 Colorado Blvd., Los Angeles 39, Calif.; tissue cultures, blood-typing materials.

International Chemical and Nuclear Corp., 13332 E. Amar Rd., City of Industry, Calif.

Kelly-Koett Manufacturing Co., 24 E. Sixth St., Covington, Ky.

Kewaunee Manufacturing Co., Adrian, Mich.

Kimble Glass, Owens-Illinois Bldg., Toledo, Ohio.

Knickerbocker Blood Bank, 251 W. 42 St., New York, N.Y.

Laboratory Construction Co., 8811 Prospect Ave., Kansas City 32, Mo.

LaMotte Chemical Products Co., Chestertown, Md.; indicators, soil testing.

Lane Corp., 156 W. 86 St., New York 7, N.Y.; cabinets, storage.

La Pine Scientific Co., 6001 S. Knox Ave., Chicago 29, Ill.

Lederle Laboratories, Midtown Rd., Pearl River, N.Y.

E. Leitz Inc., 468 Park Ave. South, New York 16, N.Y.

Lemberger Co., P.O. Box 482, Oshkosh, Wis.; living specimens.

Los Angeles Biological Laboratories, 2977 W. 14 St., Los Angeles 6, Calif.

Macalaster Scientific Co., Inc., 80 Arsenal Rd., Watertown, Mass.

Mallinckrodt Chemical Works, St. Louis, Mo.; also New York.

Marine Biological Laboratory, Woods Hole, Mass.

Matheson Co., East Rutherford, N.J.; chemicals, cylinders of compressed gases.

Merck and Co., Rahway, N.J.

Michigan Scientific Co., 6780 Jackson Rd., Ann Arbor, Mich.

Monsanto Chemical Co., 800 N. Lindbergh Blvd., St. Louis 66, Mo.

Nalge Co., Inc., Rochester 2, N.Y.; plastics.

Nutritional Biochemicals Corp., 21010 Miles Ave., Cleveland 2, Ohio.

Nystrom and Co., 3333 N. Elston Ave., Chicago 18, Ill.

Ohaus Scale Corp., 1050 Commerce Ave., Union, N.J.

Pacific Biological Laboratories, Box 63, Albany Station, Berkeley, Calif.

Pacific Bio-Marine Supply Co., P.O. Box 285, Venice, Calif.; specimens.

Penn-Chem Corp., 232–238 N. Marshall St., Lancaster, Pa.

Perkin-Elmer Corp., Instrument Division, Norwalk, Conn.

Chas. Pfizer & Co., Inc., 235 E. 42 St., New York 17, N.Y.

Polaroid Corp., Cambridge 39, Mass.

Schwarz BioResearch, Inc., Mountain View Ave., Orangeburg, N.J.

Science Education Products (B-D Laboratories, Inc.), 1660 Laurel St., San Carlos, Calif.

Science Teaching Aids Co., Box 386, Pell Lake, Wis.

Scientific Glass Apparatus Co., Bloomfield, N.J.

Scientific Products Division, Am. Hospital Supply Corp., 1210 Leon Place, Evanston, Ill.

Sheldon Equipment Co., 149 Thomas St., Muskegon, Mich.

Sjöström Co. Inc., 1717 N. Tenth St., Philadelphia 20, Pa.; science circle laboratory furniture.

Southwestern Biological Supply Co., P.O. Box 4084, Dallas, Tex.

Sprague-Dawley Inc., P.O. Box 2071, Madison 5, Wis.; laboratory rats.

Standard Scientific Supply Corp., 808 Broadway, New York 3, N.Y.

Stansi Scientific Co., 1237 N. Honore St., Chicago 22, Ill.

Swift Instruments, Inc., 952 Dorchester Ave., Boston 25, Mass.; microscopes.

Arthur H. Thomas Co., Vine & Third Sts., Philadelphia 5, Pa.; reagents, apparatus.

Torsion Balance Co., 35 Monhegan, Clifton, N.J.

Tracerlab, 1601 Trapelo Rd., Waltham, Mass.

U.S. Hospital Supply Corp., 838 Broadway, New York 3, N.Y.

U.S. Stoneware, 1515 N. Harlem, Oak Park, Ill.; plastics and synthetics.

Ward's Natural Science Establishment, Inc., P.O. Box 1712, Rochester, N.Y.

Waring Products Corp., Winsted, Conn.; blendors.

Wilkens-Anderson Co., 4525 W. Division St., Chicago 51, Ill.

Will Scientific Inc., Box 1050, Rochester 3, N.Y.

Windsor Biology Gardens, Moore's Creek Rd., Bloomington, Ind.

Winthrop Laboratories, 90 Park Ave., New York 16, N.Y.

Worthington Biochemical Corp., Freehold, N.J.

Carl Zeiss, Inc., 444 Fifth Ave., New York 17, N.Y.

United Kingdom

Aldis Brothers, Ltd., Sarehole Rd., Hall Green, Birmingham 28; optical.

Cambridge Instrument Co., Ltd., 13 Grosvenor Place, London, S.W. 1.

C. Hearson and Co., Ltd., 68 Willow Walk, Bermondsey, London, S.E. 1.

Laboratory Glassware Manufacturers, 200 Ravenscroft Rd., Beckenham, Kent.

National Collection of Type Cultures, Chelsea Gardens, London, S.W.; bacteria.

Scientific Instrument Manufacturers' Assoc. of Great Britain, Ltd., 17 Princess Gate, London, S.W. 7.

Further information concerning British scientific instrument manufacturers may be found by consulting the *Directory and Buyer's Guide of British Instruments* obtainable from Scientific Instrument Manufacturers' Assoc. of Great Britain.

Canada

Beaconing Optical and Precision Materials Co., Ltd., 455 Craig W., Montreal, P.Q.

Canadian Laboratory Supplies, Ltd., 403 St. Paul W., Montreal, P.Q.

Fine Chemicals of Canada, Ltd., Toronto, Ont.

General Optical Co., Ltd., Montreal, P.Q.

Otto Klein, P.O. Box 74, Outremont, P.Q.

Richards Glass Co., Ltd., Toronto, Ont.

Appendix **D**

Suppliers of Free
and Low-Cost Materials

Most of the companies and institutions listed here distribute useful free material on request; some charge a small sum, well under a dollar. We cannot, of course, list all of the distributors of free and inexpensive materials, for we do not know them all. We have not attempted to list the specific materials available, since these change fairly often whereas the companies offering these materials do not usually change their public relations policies.

We recommend that the teacher send for a list of booklets, charts, films, or filmstrips available from the companies, and then select the materials useful for classwork. These can be filed in manila envelopes or file folders, properly labeled, ready for teacher or student use. We hope that individual students in class will not send for materials, for this creates a strain on any company's policy of good will.

There is much abuse as well as use of free teaching aids. When you have found a successful way to use some aid in the classroom, why not submit the method to a teachers' publication so that other teachers may gain from your experience?

Directory of distributors

Aetna Life Insurance Co., Public Education Dept., 151 Farmington Ave., Hartford 15, Conn.

Am. Angus Breeders' Assoc., 3201 Frederick Blvd., St. Joseph, Mo.

Am. Can Co., 100 Park Ave., New York 17, N.Y.

Am. Cancer Soc., 219 E. 42 St., New York 19, N.Y.

Am. Dental Assoc., Order Dept., 222 E. Superior St., Chicago 11, Ill.

Am. Diabetes Assoc., 18 E. 48 St., New York 17, N.Y.

Am. Educational Publishers, 400 S. Front St., Columbus 15, Ohio.

Am. Forest Products Industries Inc., 1816 N St., N.W., Washington 6, D.C.

Am. Gas Assoc., 605 Third Ave., New York, N.Y.

Am. Genetic Assoc., 1507 M St., N.W., Washington 5, D.C.

Am. Guernsey Cattle Club, Peterborough, N.H.

Am. Hampshire Sheep Assoc., Stuart, Iowa.

Am. Heart Assoc., Inc., 44 E. 23 St., New York 10, N.Y.

Am. Hereford Assoc., 715 Hereford Dr., Kansas City 5, Mo.

Am. Institute of Baking, Consumer Service Dept., 400 E. Ontario St., Chicago 11, Ill.

Am. Museum of Natural History, 79 St. & Central Park West, New York 24, N.Y.

Am. National Red Cross, Washington 6, D.C.

Am. Optical Co., Instrument Division, Buffalo 15, N.Y.

Am. Osteopathic Assoc., Division of Public and Prof. Services, 212 E. Ohio St., Chicago 11, Ill.

Am. Petroleum Institute, Committee on Public Affairs, 1271 Ave. of the Americas, New York, 20, N.Y.

Am. Potash Institute, 1102 16 St., N.W., Washington 6, D.C.

Am. Sheep Producers Council, Film Division, Railway Exchange Bldg., Denver, Colo.

Am. Soybean Assoc., Hudson, Iowa.

Aquarium Publishing Co., 51 E. Main St., Norristown, Pa.

Armour and Co., Audio-Visual Director, 401 N. Wabash Ave., Chicago 90, Ill.

Babson Bros. Dairy Research Service, Film Dept., 2843 W. 19 St., Chicago 23, Ill.

Bausch & Lomb Inc., 635 St. Paul St., Rochester 2, N.Y.

Becton, Dickinson & Co., Advertising and Sales Promotion, Stanley St., East Rutherford, N.J.

Bell & Howell Co., Audio-Visual Dept., 7100 McCormick Rd., Chicago 45, Ill.

Better Light, Better Sight Bur., 750 Third Ave., New York 17, N.Y.

Better Vision Institute, 230 Park Ave., New York 17, N.Y.

Borden Co., Educational Services, 350 Madison Ave., New York 17, N.Y.

Bristol-Meyers Co., 630 Fifth Ave., New York 20, N.Y.

Bur. of Land Management, U.S. Dept. of the Interior, Washington 25, D.C.

California Dept. of Mental Hygiene, Sacramento 14, Calif.

California Fruit Growers Exchange, P.O. Box 5030, Metropolitan Station, Los Angeles 54, Calif.

California Redwood Assoc., Service Library, 576 Sacramento St., San Francisco 11, Calif.

California State Dept. of Natural Resources, Conservation Education, State Office Bldg., Sacramento 14, Calif.

Cambosco Scientific Co., Inc., 342 Western Ave., Brighton, Mass.

Campbell Soup Co., Camden 1, N.J.

Carnegie Institution of Washington, Cold Spring Harbor, L.I., N.Y.

Carolina Biological Supply Co., Burlington, N.C.

J. I. Case Co., 700 State St., Racine, Wis.; conservation pamphlets and filmstrips.

Cereal Institute Inc., Educational Service, 135 S. LaSalle St., Chicago 3, Ill.

Chicago Natural History Museum, Educational Dept., Roosevelt Rd., Chicago 5, Ill.

Church & Dwight Inc., 70 Pine St., New York 5, N.Y.; pictures of birds.

Conservation Foundation, 30 E. 40 St., New York 16, N.Y.

Continental Baking Co., Inc., 3362 Park Ave., New York 20, N.Y.

Corning Glass Works, Public Relation Dept., 717 Fifth Ave., New York 22, N.Y.

Cream of Wheat Corp., 730 Stinson Blvd., Minneapolis 13, Minn.

Crown Zellerbach Corp., 1 Bush St., San Francisco 19, Calif.

Denoyer-Geppert Co., 5235 Ravenswood Ave., Chicago 40, Ill.

Diamond Crystal Salt Co., 916 S. Riverside Ave., St. Clair, Mich.

DoAll Co., Film Librarian, 254 N. Laurel Ave., Des Plaines, Ill.

Dow Chemical Co., Midland, Mich.

E. I. du Pont de Nemours & Co., Inc., Motion Picture Section, 1007 Market St., Wilmington 98, Del.

Eastman Kodak Co., Audio-Visual Service, 343 State St., Rochester 4, N.Y.

Evaporated Milk Assoc., 228 N. LaSalle St., Chicago 1, Ill.

Farm Film Foundation, Southern Bldg., Washington 5, D.C.

Farmers' and Manufacturers' Beet Sugar Assoc., Education Dept., Saginaw 5, Mich.

Fels and Co., 73 St. & Woodland Ave., Philadelphia 42, Pa.

Fisher Scientific Co., 711 Forbes St., Pittsburgh 19, Pa.

Ford Motor Co., Educational Affairs Dept., Dearborn, Mich.

General Biological Supply House, Inc. (Turtox), 8200 S. Hoyne Ave., Chicago 20, Ill.

General Dynamics Corp., San Diego, Calif.

General Mills, Inc., 9200 Wayzata Blvd., Minneapolis 26, Minn.

John Hancock Mutual Life Insurance Co., Public Relations Dept., 200 Berkeley St., Boston 17, Mass.

Harper & Row, Publishers, Elementary–High School Division, 2500 Crawford Ave., Evanston, Ill.

Health Publications Institute, Inc., Raleigh, N.C.

H. J. Heinz Co., 1062 Progress St., Pittsburgh 12, Pa.

Hershey Chocolate Corp., Hershey, Pa.

Hoffman-LaRoche Inc., Kingsland St., Nutley 10, N.J.

Holstein-Friesian Assoc. of America, Extension Dept., 1 Main St., Brattleboro, Vt.

Illinois Natural History Survey, 189 Natural Resources Bldg., Urbana, Ill.

Illinois State Dept. of Health, Springfield, Ill.

Illuminating Engineering Soc., 345 E. 47 St., New York 23, N.Y.

Indiana Conservation Dept., Division of Parks, Lands, and Waters, Indianapolis 9, Ind.

International Assoc. of Ice Cream Manufacturers, 910 17 St., N.W., Washington, D.C.

International Harvester Co., 180 N. Michigan Ave., Chicago 1, Ill.

International Paper Co., 220 E. 42 St., New York 17, N.Y.

Johnson and Johnson, Education Dept., New Brunswick, N.J.

Kellogg Co., Home Economics Services, Battle Creek, Mich.

Knox Gelatine Inc., Knox Ave., Johnstown, N.Y.

Kraft Food Co., 500 N. Peshtigo Court, Chicago 90, Ill.

Laboratory of Applied Psychology, Box 2162, Yale Station, New Haven, Conn.

LaMotte Chemical Products Co., Chestertown, Md.

Lederle Laboratories, Division of Am. Cyanamid Co., Public Services Division, Pearl River, N.Y.

E. Leitz, Inc., 468 Park Ave. South, New York 16, N.Y.

Louisiana Dept. of Wildlife & Fisheries, Visual Education Dept., 126 Civil Courts Bldg., New Orleans 16, La.

Maine Dept. of Agriculture, State Office Bldg., Augusta, Maine.

Manufacturing Chemists Assoc., 1825 Connecticut Ave., N.W., Washington 9, D.C.

Marineland of Florida, St. Augustine, Fla.

Maryland Soc. for Medical Research, 522 W. Lombard St., Baltimore, Md.

Mentor Books (New Am. Library of World Literature), 1301 Ave. of the Americas, New York, N.Y.

Merck and Co., Rahway, N.J.

Charles E. Merrill Books, Inc., 1300 Alum Creek Dr., Columbus 15, Ohio.

Metropolitan Life Insurance Co., School Service Dept., 1 Madison Ave., New York 16, N.Y.

Minneapolis-Moline Co., Minneapolis 1, Minn.; conservation materials.

Mobile Azalea Trail, Film Distributing Committee, P.O. Box 172, Mobile, Ala.

National Academy of Sciences, 2101 Constitution Ave., Washington 25, D.C.

National Aeronautics and Space Administration, Educational Audio-Visual Branch AFEE3, Washington 25, D.C.

National Assoc. of Manufacturers, 277 Park Ave., New York 20, N.Y.

National Audubon Soc., 1130 Fifth Ave., New York 28, N.Y.

National Cancer Assoc., 1739 H St., N.W., Washington 6, D.C.

National Cancer Institute, 9000 Wisconsin Ave., Bethesda, Md.

National Canners Assoc., Information Division, 1133 20 St., N.W., Washington 6, D.C.

National Chinchilla Breeders Assoc. Inc., P.O. Box 1806, Salt Lake City 10, Utah.

National Committee for Careers in Medical Technology, 1785 Massachusetts Ave., N.W., Washington 6, D.C.

National Confectioners Assoc., 36 S. Wabash Ave., Chicago 3, Ill.

National Cotton Council of America, Audio-Visual Service, 1918 N. Parkway Ave., Memphis 1, Tenn.

National Dairy Council, 111 N. Canal St., Chicago 6, Ill.

National Epilepsy League, Inc., 203 N. Wabash, Chicago, Ill.

National Foundation, 800 Second Ave., New York 17, N.Y.

National Institute of Mental Health, Public Health Service, 9000 Wisconsin Ave., Bethesda, Md.

National Kidney Disease Foundation, Film Dept., 342 Madison Ave., New York, N.Y.

National Live Stock and Meat Board, 36 S. Wabash, Chicago, Ill.

National Lumber Manufacturers, 1619 Massachusetts Ave., N.W., Washington 6, D.C.

National Safety Council, 425 N. Michigan Ave., Chicago 11, Ill.

National Soc. for Medical Research, 111 Fourth St., S.E., Rochester, Minn.

National Soc. for the Prevention of Blindness, 16 E. 40 St., New York 16, N.Y.

National Tuberculosis Assoc., 1790 Broadway, New York 19, N.Y.

National Vitamin Foundation, Inc., 250 W. 57 St., New York 21, N.Y.

National Wildlife Federation, Education Dept., 1412 16 St., N.W., Washington, D.C.

New Hampshire Dept. of Health, Office of Health Education, 66 South St., Concord, N.H.

New York City Cancer Committee, 7 E. 52 St., New York 21, N.Y.

New York City Dept. of Health, 125 Worth St., New York 13, N.Y.

New York Heart Assoc. Inc., 10 Columbus Circle, New York 19, N.Y.

New York State College of Agriculture, Cornell Univ., Ithaca, N.Y.

New York State Dept. of Health, Office of Health Education, 84 Holland Ave., Albany, N.Y.

New York Zoological Society, Bronx Park, New York 60, N.Y.

North American Aviation Inc., Atomics Division, Public Relations Dept., P.O. Box 309, Canoga Park, Calif.

Northwest Nut Growers, 1601 N. Columbia Blvd., Portland 17, Ore.

Nuclear-Chicago Corp., 333 E. Howard Ave., Des Plaines, Ill.; also branch offices.

Nutrition Foundation Inc., 99 Park Ave., New York 16, N.Y.

Pet Milk Co., Arcade Bldg., St. Louis 66, Mo.

Chas. Pfizer & Co., Inc., 267 W. 25 St., New York 1, N.Y.

Poultry and Egg National Board, Film Dept., 8 S. Michigan Ave., Chicago 3, Ill.

Quaker Oats Co., Advertising Dept., Merchandise Mart Plaza, Chicago 54, Ill.

Ralston Purina Co., Nutrition Service, St. Louis 2, Mo.

Richfield Oil Corp., 555 S. Flower St., Los Angeles 54, Calif.

A. I. Root Co., Medina, Ohio; pamphlets on bees.

Science Clubs of America, Science Service, 1719 N St., N.W., Washington 6, D.C.

Science Research Associates Inc., 259 E. Erie St., Chicago, Ill.

Science Service, 1719 N St., N.W., Washington 6, D.C.

Shell Oil Co., 50 W. 50 St., New York, N.Y.

Smith, Kline & French Laboratories, Medical Film Center, 1500 Spring Garden St., Philadelphia 1, Pa.

Soc. for Visual Education Inc., 1345 W. Diversey Parkway, Chicago 14, Ill.

Soil Conservation Service, U.S. Dept. of Agriculture, Washington 25, D.C.

Southern Pine Assoc., National Bank of Commerce Bldg., New Orleans, La.

Standard Brands Inc., 625 Madison Ave., New York 22, N.Y.

Standard Oil of New Jersey, Education Dept., 30 Rockefeller Plaza, New York 20, N.Y.

Sugar Information Inc., 52 Wall St., New York 5, N.Y.

Sunsweet Growers Inc., Box 670, San Jose, Calif.

Swift & Co., Agricultural Research Dept., Union Stockyards, Chicago 9, Ill.

Tea Council of the USA, Film Library, 16 E. 56 St., New York 22, N.Y.

Tennessee Valley Authority, Health & Safety Dept., Chattanooga 1, Tenn.

United Fruit Co., Education Dept., Pier 3, North River, New York 6, N.Y.

U.S. Atomic Energy Comm., Washington 25, D.C.

U.S. Dept. of Agriculture, Experimental Fur Station, Petersburg, Alaska.

U.S. Dept. of Health, Education, and Welfare, Education Division, Washington 25, D.C.

U.S. Fish & Wildlife Service, Dept. of the Interior, Washington 25, D.C.

U.S. Forest Service, San Francisco 11; Milwaukee 3; Missoula, Mont.; Portland 8, Ore.

U.S. Govt. Printing Office, Washington 25, D.C.

U.S. Plywood Corp., 777 Third Ave., New York 16, N.Y.

U.S. Public Health Service, Communicable Disease Center, Atlanta 22, Ga.

U.S. Steel, Public Relations Dept., 71 Broadway, New York 6, N.Y.

Univ. of Illinois, Dept. of Horticulture, College of Agriculture, Urbana, Ill.

Upjohn Co., 1001 E. 87 St., Chicago 19, Ill.; many branch offices; vitamin booklets.

Vermiculite Institute, 208 S. LaSalle St., Chicago 4, Ill.

Ward's Natural Science Establishment, Inc., P.O. Box 1712, Rochester, N.Y.

Washington State Apple Comm., P.O. Box 18, Wenatchee, Wash.

Westinghouse Electric Corp., 3 Gateway Centers, Pittsburgh 30, Pa.

Weyerhaeuser Co., Tacoma 1, Wash.

Wheat Flour Institute, 309 W. Jackson Blvd., Chicago 6, Ill.

Zurich-American Life Insurance Co., 111 W. Jackson, Chicago, Ill.

Additional reference aids

There are many compilations of free and inexpensive materials which have been published to meet the constant search of teachers for aids for the classroom. For those teachers who will want to go beyond this directory we have added a short bibliography of some publications that we have encountered.

Catalog of Free Educational Material on the Banana and Related Subjects, United Fruit Co., Educational Service, Pier 3, North River, New York 6, N.Y.

Catalog of Man and Nature Publications, Am. Museum of Natural History, 79 St. & Central Park West, New York 24, N.Y.

Choosing Free Materials for Use in the Schools, Am. Assoc. of School Administrators, National Education Assoc., 1201 16 St., N.W., Washington 6, D.C.; $0.50.

Conservation Teaching Aids, Michigan Dept. of Conservation, Education Division, Lansing 26, Mich.

Elementary Teachers Guide to Free Curriculum Materials, Educators' Progress Service, Randolph, Wis.

Free and Inexpensive Educational Aids, rev. ed., T. Pepe, Dover, New York, 1962.

Free and Inexpensive Learning Materials, George Peabody College for Teachers, Division of Surveys and Field Services, Nashville 5, Tenn.

Free and Inexpensive Teaching Materials for Science Education, M. Beuschlein and J. Sanders, *Chicago Schools Journal,* Vol. 34, Nos. 5, 6, 1953; available as reprints.

General Motors Aids to Educators, General Motors Corp., Detroit 2, Mich.

Health Materials and Resources for Oregon Teachers, State Dept. of Education, Salem, Ore.

Hobby Publications, U.S. Govt. Printing Office, Washington 25, D.C.

1001 Things Free, 2nd ed., M. Weisinger, Bantam Books, New York, 1957.

Sources of Free and Inexpensive Educational Materials, Field Enterprises Inc., Education Division, Merchandise Mart Plaza, Chicago 54, Ill.

Sources of Free and Inexpensive Materials in Health Education, Teachers College, Curriculum Laboratory, Temple Univ., Philadelphia, Pa.

Sources of Free and Inexpensive Pictures for the Classroom, B. Miller, Box 369, Riverside, Calif.

Sources of Teaching Materials, C. Williams, Bur. of Educational Research, Ohio State Univ., Columbus, Ohio.

Sponsors' Handbook; Thousands of Science Projects, Science Service, 1719 N St., N.W., Washington 6, D.C.

Teaching Aids, School Service, Westinghouse Electrical Corp., 306 Fourth Ave., Pittsburgh 30, Pa.

Using Free Materials in the Classroom, Assoc. of Supervision and Curriculum Development, National Education Assoc., 1201 16 St., N.W., Washington 6, D.C.; $0.75.

A Wonderful World for Children, P. Cardozo, Bantam Books, New York, 1962.

Appendix E

Biology Field Stations

We think of field stations and nature camps as a kind of bibliography for the professional growth of teachers. Many universities offer courses at camps and field stations, as well as summer institutes and special graduate programs. Many well-known biological research stations attract research biologists and teachers of biology. Some are open the year around, others for a 6- to 8-week summer session, for research work only.

Many teachers will be interested in knowing about the many scientific associations over the country. Refer to *Scientific and Technical Societies of the United States and Canada,* Public. 900 of the National Academy of Sciences, National Research Council, 2101 Constitution Ave., Washington 25, D.C.

A directory of Pacific Coast field laboratory stations is available: D. H. Montgomery, ed., *Directory of Marine Sciences Programs and Facilities on the Pacific Coast* (annual), D. H. Montgomery, Biological Sciences Dept., California State Polytechnic College, San Luis Obispo, Calif.

Some teachers may also be interested in a directory of summer programs in conservation: *Directory of Summer 1965 Workshops, Courses, and Adult Training Programs in Conservation of Natural Resources in the United States and Canada,* Conservation Education Assoc., February 1, 1965; distributed through the office of Dr. Wilson F. Clark, Eastern Montana College, Billings, Mont.

For further information concerning summer institutes, consult the National Science Foundation, 1951 Constitution Ave., N.W., Washington, D.C. This listing of only a few of the field stations in biology is offered by way of suggestion; further information may be obtained by writing to the director of the specific field station.

Gulf Coast Research Laboratory
Ocean Springs, Miss.
Year-round and summer session:
Marine geology, zoology.

Institute of Marine Biology
Univ. of Oregon, Eugene, Ore.
Summer session:
Marine biology, algology, invertebrate zoology.

Marine Biological Laboratory
Woods Hole, Mass.
Summer sessions:
Research and courses in embryology, physiology, marine botany, invertebrate zoology, marine ecology.

Marine Laboratory
1 Rickenbacker Causeway, Miami 54, Fla.
Year-round and summer session:
Marine biology, geology, ichthyology, oceanography.

Mount Desert Island Biological Laboratory
Salisbury Cove, Maine
Summer work:
Research, no courses.

Purdue Univ. Conservation Education Camp
Lafayette, Ind.
Summer session:
Conservation education.

Science Lodge, Institute of Arctic and Alpine Research
Univ. of Colorado, Boulder, Colo.
Year round:
No classes, accommodations and cooperation on field research.

Scripps Institution of Oceanography
Univ. of California (at La Jolla)
Year round:
No classes, research and graduate study in marine geophysics, physical oceanography, submarine geology, marine biology, marine chemistry, geochemistry.

Trout Lake Biological Laboratory
Univ. of Wisconsin, Madison, Wis.
Summer session:
Research station; no courses (courses in hydro-biology at University campus).

Univ. of Michigan Biological Station
Ann Arbor, Mich.
Summer session:
Marine biology, botany, fresh-water biology, zoology.

Univ. of New Hampshire
Durham, N.H.
Summer session:
Marine invertebrate zoology.

Virginia Fisheries Laboratory
Gloucester Point, Va.
Year round:
Marine biology, fisheries science.

Walla Walla College Biological Station
Anacortes, Wash.
Summer session:
Marine invertebrates, marine botany, ichthyology, ornithology, oceanography, fisheries biology, parasitology, invertebrate physiology, entomology.

West Coast Nature School
San Jose State College, San Jose 14, Calif.
Summer session:
Field biology.

Appendix **F**

Reference Tables

Metric system*

Length

The unit of length is the meter (m).

1 micron (μ) = 10^{-6} meter
$\quad\quad\quad\quad\; = 10^{-4}$ centimeter (cm)
$\quad\quad\quad\quad\; = 10^{-3}$ millimeter (mm)
$\quad\quad\quad\quad\; = 1000$ millimicrons (mμ)
$\quad\quad\quad\quad\; = 10,000$ Ångström units (Å)
$\quad\quad\quad\quad\; = \frac{1}{25,000}$ inch (in.)
1 millimicron = 10^{-6} millimeter
1 millimeter = 1000 microns
1 Ångström unit = 10^{-10} meter
$\quad\quad\quad\quad\quad\;\; = 10^{-8}$ centimeter
$\quad\quad\quad\quad\quad\;\; = 10^{-7}$ millimeter
1 centimeter = 0.01 meter
$\quad\quad\quad\quad\;\; = 10$ millimeters
1 kilometer (km) = 1000 meters

Weight

The unit of weight is the gram (g).

1 centigram (cg) = 0.01 gram
1 milligram (mg) = 0.001 gram
1 microgram (μg) = 0.000001 gram
1 kilogram (kg) = 1000 grams

Volume

The unit of volume is the liter; this is the volume of 1 kilogram of water at 4°C at standard atmospheric pressure.

1 liter = 1000.028 cubic centimeters (cc)
$\quad\quad\; = 1000$ milliliters (ml)
1 milliliter = 0.001 liter

*Adapted from P. Perlman, *General Laboratory Techniques*, Franklin Publishing Co., Englewood, N.J., 1964, pp. 380–83.

United States system

Length

1 foot (ft) = 12 inches (in.)
1 yard (yd) = 3 feet = 36 inches
1 rod = 5½ yards = 16½ feet
1 mile (statute) = 1760 yards = 5280 feet
1 league = 3 miles
1 nautical mile
$\quad$(international) = 6076.1 feet
$\quad\quad\quad\quad\quad\quad\; = 7\frac{1}{3}$ cablelengths
1 fathom (mariners') = 6 feet
1 cablelength = 120 fathoms

Weight

TROY WEIGHT

1 pennyweight (dwt) = 24 grains (gr)
1 ounce (oz) = 20 pennyweights = 480 grains
1 pound (lb) = 12 ounces = 5760 grains

APOTHECARIES' WEIGHT

1 scruple (sc) = 20 grains (gr)
1 dram (dr) = 3 scruples = 60 grains
1 ounce (oz) = 8 drams = 480 grains
1 pound (lb) = 12 ounces = 5760 grains

AVOIRDUPOIS WEIGHT

1 dram (dr) = $27\frac{11}{32}$ grains (gr)
1 ounce (oz) = 16 drams = 437½ grains
1 pound (lb) = 16 ounces = 7000 grains

Volume

APOTHECARIES' FLUID MEASURE

1 fluid ounce (fl oz) = 8 fluid drams (fl dr)
$\quad\quad\quad\quad\quad\quad\quad\quad = 480$ minims
1 pint (pt) = 16 fluid ounces
1 quart (qt) = 2 pints = 32 fluid ounces

1 gallon (gal) = 4 quarts
                = 231 cubic inches (cu in.)

DRY MEASURE

1 quart (qt) = 2 pints (pt)
             = 67.2 cubic inches (cu in.)
1 peck = 8 quarts = 537.6 cubic inches

Equivalents of metric and U.S. systems

Length

1 millimeter (mm) = 0.03937 inch (in.)
1 centimeter (cm) = 0.3937 inch
1 meter (m) = 39.37 inches = 3.2808 feet (ft)
1 kilometer (km) = 0.6214 mile
1 inch = 2.54 centimeters
1 foot = 30.48 centimeters
1 yard (yd) = 91.44 centimeters
            = 0.9144 meter
1 mile = 1.6093 kilometers

Area

1 square centimeter (sq cm) = 0.155 square
                              inch (sq in.)
1 square meter (m²) = 1550.0 square inches
                    = 10.764 square feet
                      (sq ft)
                    = 1.196 square yards
                      (sq yd)
1 square inch = 6.4516 square centimeters
1 square yard = 0.8361 square meter

Volume

1 cubic centimeter (cc) = 0.0610 cubic inches
                          (cu in.)
1 cubic meter (m³) = 35.3145 cubic feet (cu ft)
                   = 1.3079 cubic yards (cu yd)
1 cubic inch = 16.3872 cubic centimeters
1 cubic yard = 0.7646 cubic meter

Capacity

1 milliliter (ml) = 0.2705 fluid dram (fl dr)
                  = 0.0338 fluid ounce (fl oz)
1 liter = 33.8148 fluid ounces
        = 2.1134 pints (pt)
        = 1.0567 quarts (qt)
        = 0.2642 gallon (gal)
1 fluid dram = 3.697 milliliters
1 fluid ounce = 29.573 milliliters
1 quart = 946.332 milliliters

1 gallon = 3.785 liters
1 cubic inch (cu in.) = 16.387 milliliters
1 cubic foot (cu ft) = 28.316 liters

Weight

1 gram (g) = 15.432 grains (gr)
           = 0.03527 avoirdupois ounce
             (avdp oz)
           = 0.03215 apothecaries' or troy
             ounce
1 kilogram (kg) = 35.274 avoirdupois ounces
                = 32.151 apothecaries' or troy
                  ounces
                = 2.2046 avoirdupois pounds
                  (avdp lb)
                = 2.6792 apothecaries' or troy
                  pounds
1 grain = 64.7989 milligrams (mg)
1 avoirdupois ounce = 28.3495 grams
1 apothecaries' or troy ounce = 31.1035 grams
1 avoirdupois pound = 453.5924 grams
1 apothecaries' or troy pound = 373.2418
                              grams

Kitchen measure and other approximations

1 small test tube ≅ 30 milliliters (ml)
1 large test tube ≅ 70 milliliters
1 tumbler or 1 cup = 16 tablespoons (tbsp)
                   = 8 fluid ounces (fl oz)
                   = ½ pint (pt)
                   ≅ 240 milliliters
1 teacup ≅ 4 fluid ounces = 120 milliliters
1 tablespoon ≅ ½ fluid ounce ≅ 16 milliliters
1 teaspoon (tsp) ≅ 4 milliliters
                 ≅ 60 drops (depending on
                   bore of medicine dropper)
1 milliliter ≅ 25 drops (depending on size of
               bore)

Using coins as substitutes for weights

coin	approximate weight (grams)
Dime	2.5
Penny	3.25
Nickel	5.0
Quarter	6.5
Half dollar	13.0
Silver dollar	26.0

(A dime is about 1 mm thick.)

Useful formulas

The following formulas are useful for calculating the volume or surface area of a cell:

$\pi = 3.1416$
Circumference of a circle $= \pi d = 2\pi r$
Area of a circle $= \pi r^2$
Surface area of a cylinder $= \pi dh = 2\pi rh$
Volume of a cylinder $= \pi r^2 h$
Surface area of a sphere $= 4\pi r^2$
Volume of a sphere $= \frac{1}{3}\pi r^3$

A typical cell, such as the alga *Chlorella*, is a sphere. Therefore, the volume of the cell may be determined as follows. Using an eyepiece micrometer (or other methods described on p. 92), measure the diameter of the cell. Divide the diameter by 2 to determine the radius r. Calculate the volume from the formula $\frac{1}{3}\pi r^3$. In a similar manner, the surface area of the nucleus of this cell may be computed from the formula $4\pi r^2$.

Wavelengths of various radiations*

radiation	Ångström units (Å) †
Cosmic rays	0.0005 and under
Gamma rays	0.005–1.40
X-rays	0.1–100
Ultraviolet	2920–4000
Visible spectrum	4000–7000
Violet	4000–4240
Blue	4240–4912
Green	4912–5750
Yellow	5750–5850
Orange	5850–6470
Red	6470–7000
Maximum visibility	5560
Infrared	over 7000

pH notation

A pH value is designated as the logarithm of the number of liters of a solution needed to get 1 gram-ion of hydrogen (1.008 g of H^+). Pure water is only slightly ionized; there is 1 gram-ion of hydrogen in 10,000,000 liters of pure water, that is, its H-ion concentration is 10^{-7}. We say

*From *Teaching High School Science: A Sourcebook for the Physical Sciences* by Joseph, Brandwein, Morholt, Pollack, and Castka, © 1961, by Harcourt, Brace & World, Inc., and reproduced with their permission.
† To convert to inches, multiply by 3.937×10^{-9}; to convert to centimeters, multiply by 1×10^{-8}.

the pH value of pure water is 7, since 7 is the log of 10,000,000. A neutral point, then, is pH 7.

Since these are logarithmic values, a solution with a pH 6.0 contains 10 times as many H-ions as a solution having a pH value of 7.0.

Buffered solutions are mixtures of weak acids and their salts, or of weak bases and their salts. For information concerning the preparation of buffered solutions, see pp. 172–73, 660.

Powers of ten

$10^0 = 1$	$10^0 = 1$
$10^1 = 10$	$10^{-1} = 0.1$
$10^2 = 100$	$10^{-2} = 0.01$
$10^3 = 1000$	$10^{-3} = 0.001$
$10^4 = 10,000$	$10^{-4} = 0.0001$
$10^6 = 1$ million	$10^{-6} =$ one millionth
$10^9 = 1$ billion	$10^{-9} =$ one billionth

EXAMPLES

$$(4.5 \times 10^8)(5 \times 10^4) = (4.5 \times 5) \times 10^{8+4}$$
$$= 22.5 \times 10^{12}$$

$$\frac{4.5 \times 10^{-8}}{5 \times 10^4} = \frac{4.5}{5} \times 10^{-8-4}$$
$$= 0.9 \times 10^{-12}$$

Greek alphabet

A	α	alpha
B	β	beta
Γ	γ	gamma
Δ	δ	delta
E	ϵ	epsilon
Z	ζ	zeta
H	η	eta
Θ	θ	theta
I	ι	iota
K	κ	kappa
Λ	λ	lambda
M	μ	mu
N	ν	nu
Ξ	ξ	xi
O	o	omicron
Π	π	pi
P	ρ	rho
Σ	σ	sigma
T	τ	tau
Υ	υ	upsilon
Φ	ϕ	phi
X	χ	chi
Ψ	ψ	psi
Ω	ω	omega

Temperature conversion

It is simple to convert Fahrenheit into centigrade temperatures (and vice versa) without memorizing any formulas. All that need be remembered is that $0°\,C = 32°\,F$ (the freezing point of water) and that each Fahrenheit degree is only ⅝ as large as a centigrade degree. Thus if you knew the temperature in Fahrenheit degrees, you would subtract 32 and take ⅝ of the result to find the temperature in centigrade degrees. And if you knew the centigrade temperature, you would multiply it by ⅝ and add 32 to the result to find the Fahrenheit temperature.

The illustration on this page is useful in understanding the relationship between these two temperature scales; or it can be used for the conversion itself: Simply read directly across from one thermometer scale to the other.

$$°F = °C \times ⅝ + 32$$
$$°C = °F - 32 \times ⅝$$

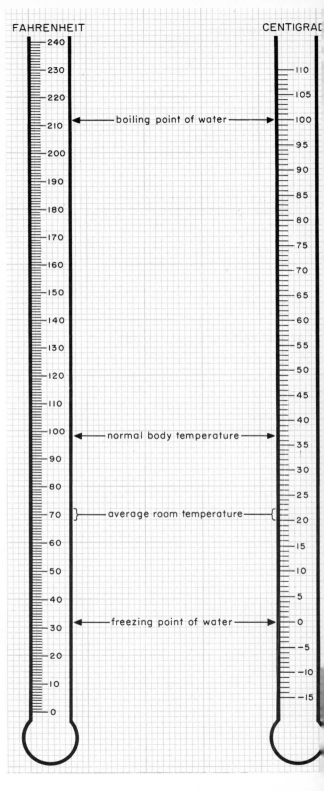

FAHRENHEIT CENTIGRADE

boiling point of water

normal body temperature

average room temperature

freezing point of water

Table of atomic weights based on carbon 12*

	symbol	atomic number	atomic weight		symbol	atomic number	atomic weight
Actinium	Ac	89	(227)	Mercury	Hg	80	200.59
Aluminum	Al	13	26.9815	Molybdenum	Mo	42	95.94
Americium	Am	95	(243)	Neodymium	Nd	60	144.24
Antimony	Sb	51	121.75	Neon	Ne	10	20.183
Argon	Ar	18	39.948	Neptunium	Np	93	(237)
Arsenic	As	33	74.9216	Nickel	Ni	28	58.71
Astatine	At	85	(210)	Niobium	Nb	41	92.906
Barium	Ba	56	137.34	Nitrogen	N	7	14.0067
Berkelium	Bk	97	(247)	Nobelium	No	102	(253)
Beryllium	Be	4	9.0122	Osmium	Os	76	190.2
Bismuth	Bi	83	208.980	Oxygen	O	8	15.9994
Boron	B	5	10.811	Palladium	Pd	46	106.4
Bromine	Br	35	79.909	Phosphorus	P	15	30.9738
Cadmium	Cd	48	112.40	Platinum	Pt	78	195.09
Calcium	Ca	20	40.08	Plutonium	Pu	94	(242)
Californium	Cf	98	(249)	Polonium	Po	84	(210)
Carbon	C	6	12.01115	Potassium	K	19	39.102
Cerium	Ce	58	140.12	Praseodymium	Pr	59	140.907
Cesium	Cs	55	132.905	Promethium	Pm	61	(147)
Chlorine	Cl	17	35.453	Protactinium	Pa	91	(231)
Chromium	Cr	24	51.996	Radium	Ra	88	(226)
Cobalt	Co	27	58.9332	Radon	Rn	86	(222)
Copper	Cu	29	63.54	Rhenium	Re	75	186.2
Curium	Cm	96	(247)	Rhodium	Rh	45	102.905
Dysprosium	Dy	66	162.50	Rubidium	Rb	37	85.47
Einsteinium	Es	99	(254)	Ruthenium	Ru	44	101.07
Erbium	Er	68	167.26	Samarium	Sm	62	150.35
Europium	Eu	63	151.96	Scandium	Sc	21	44.956
Fermium	Fm	100	(253)	Selenium	Se	34	78.96
Fluorine	F	9	18.9984	Silicon	Si	14	28.086
Francium	Fr	87	(223)	Silver	Ag	47	107.870
Gadolinium	Gd	64	157.25	Sodium	Na	11	22.9898
Gallium	Ga	31	69.72	Strontium	Sr	38	87.62
Germanium	Ge	32	72.59	Sulfur	S	16	32.064
Gold	Au	79	196.967	Tantalum	Ta	73	180.948
Hafnium	Hf	72	178.49	Technetium	Tc	43	(99)
Helium	He	2	4.0026	Tellurium	Te	52	127.60
Holmium	Ho	67	164.930	Terbium	Tb	65	158.924
Hydrogen	H	1	1.00797	Thallium	Tl	81	204.37
Indium	In	49	114.82	Thorium	Th	90	232.038
Iodine	I	53	126.9044	Thulium	Tm	69	168.934
Iridium	Ir	77	192.2	Tin	Sn	50	118.69
Iron	Fe	26	55.847	Titanium	Ti	22	47.90
Krypton	Kr	36	83.80	Tungsten	W	74	183.85
Lanthanum	La	57	138.91	Uranium	U	92	238.03
Lawrencium	Lw	103	(257)	Vanadium	V	23	50.942
Lead	Pb	82	207.19	Xenon	Xe	54	131.30
Lithium	Li	3	6.939	Ytterbium	Yb	70	173.04
Lutetium	Lu	71	174.97	Yttrium	Y	39	88.905
Magnesium	Mg	12	24.312	Zinc	Zn	30	65.37
Manganese	Mn	25	54.9380	Zirconium	Zr	40	91.22
Mendelevium	Md	101	(256)				

*( ) Numbers in parentheses are the mass numbers of the most stable or best-known isotope.

Periodic chart of the elements*

The periodic table:

IA ... **VIIA** / **INERT GASES**

| 1 H 1.00797 | IIA | | | | | | | | | | | IIIA | IVA | VA | VIA | 1 H 1.00797 | 2 He 4.0026 |

Row 2: 3 Li 6.939 | 4 Be 9.0122 | | | | | | | | | | | 5 B 10.811 | 6 C 12.01115 | 7 N 14.0067 | 8 O 15.9994 | 9 F 18.9984 | 10 Ne 20.183

Row 3: 11 Na 22.9898 | 12 Mg 24.312 | IIIB | IVB | VB | VIB | VIIB | ——VIII—— | | | IB | IIB | 13 Al 26.9815 | 14 Si 28.086 | 15 P 30.9738 | 16 S 32.064 | 17 Cl 35.453 | 18 Ar 39.948

Row 4: 19 K 39.102 | 20 Ca 40.08 | 21 Sc 44.956 | 22 Ti 47.90 | 23 V 50.942 | 24 Cr 51.996 | 25 Mn 54.9380 | 26 Fe 55.847 | 27 Co 58.9332 | 28 Ni 58.71 | 29 Cu 63.54 | 30 Zn 65.37 | 31 Ga 69.72 | 32 Ge 72.59 | 33 As 74.9216 | 34 Se 78.96 | 35 Br 79.909 | 36 Kr 83.80

Row 5: 37 Rb 85.47 | 38 Sr 87.62 | 39 Y 88.905 | 40 Zr 91.22 | 41 Nb 92.906 | 42 Mo 95.94 | 43 Tc (99) | 44 Ru 101.07 | 45 Rh 102.905 | 46 Pd 106.4 | 47 Ag 107.870 | 48 Cd 112.40 | 49 In 114.82 | 50 Sn 118.69 | 51 Sb 121.75 | 52 Te 127.60 | 53 I 126.9044 | 54 Xe 131.30

Row 6: 55 Cs 132.905 | 56 Ba 137.34 | 57 †La 138.91 | 72 Hf 178.49 | 73 Ta 180.948 | 74 W 183.85 | 75 Re 186.2 | 76 Os 190.2 | 77 Ir 192.2 | 78 Pt 195.09 | 79 Au 196.967 | 80 Hg 200.59 | 81 Tl 204.37 | 82 Pb 207.19 | 83 Bi 208.980 | 84 Po (210) | 85 At (210) | 86 Rn (222)

Row 7: 87 Fr (223) | 88 Ra (226) | 89 ‡Ac (227)

† Lanthanum series:
58 Ce 140.12 | 59 Pr 140.907 | 60 Nd 144.24 | 61 Pm (147) | 62 Sm 150.35 | 63 Eu 151.96 | 64 Gd 157.25 | 65 Tb 158.924 | 66 Dy 162.50 | 67 Ho 164.930 | 68 Er 167.26 | 69 Tm 168.934 | 70 Yb 173.04 | 71 Lu 174.97

‡ Actinium series:
90 Th 232.038 | 91 Pa (231) | 92 U 238.03 | 93 Np (237) | 94 Pu (242) | 95 Am (243) | 96 Cm (247) | 97 Bk (247) | 98 Cf (249) | 99 Es (254) | 100 Fm (253) | 101 Md (256) | 102 No (253) | 103 Lw (257)

*( ) Numbers in parentheses are the mass numbers of the most stable or best-known isotope. Atomic weights are based on carbon 12, conforming to the conventions adopted in 1961 by the International Union of Pure and Applied Chemistry.

Some radioactive isotopes

isotope	chemical formula	half life	beta energy (Mev)	gamma energy (Mev)
Iodine-131	NaI	8.05 days	0.61	0.364
Phosphorus-32	NaH_2PO_4	14.3 days	1.71	none
Iron-59	$FeCl_3$	45 days	0.46	1.102, 1.290
Strontium-89	$SrCl_2$	50.4 days	1.46	none
Sulfur-35	H_2SO_4	86.7 days	0.167	none
Calcium-45	$CaCl_2$	165 days	0.25	none
Zinc-65	$ZnCl_2$	245 days	0.33	0.511, 1.114
Sodium-22	$NaCl$	2.58 years	0.54	1.277
Carbon-14	Na_2CO_3	5770 years	0.156	none
Chlorine-36	KCl	300,000 years	0.71	none

Index

Aquarium (*cont.*)
    interdependence of living things
        in, 533
    maintenance of, 548, 609, 619
    marine, 600, *601*
    plants for, 606–08, *607*
    preparation of, 606
    snails in, 595
    student curators of, 684
    types of, 605–06
Aquatic animals. *See* Animals,
    aquatic
Aquatic plants. *See* Plants,
    aquatic
Aquatic worms. *See* Worms,
    aquatic
Arachnids, 47
Arbacia. See Sea urchins
Arcella, 582
Armadillidium (pill bug), 599–*600*
Armillaria, 640
Arnold sterilizer, 632–*33*
Arnon, D., 157, 160–61, 190
Artemia (brine shrimp), *599*
    anatomy of, *599*
    characteristics of, 598
    culturing of, 599
    development of, 598
    eggs of
        development of, 354
        hatching of, 599
        obtaining, 598
        preservation of, 598–99
    as food for *Hydra,* 591
    larvae of, 354, *356,* 599
    mounting of in syrup, 115
    obtaining, 598
    sensitivity of to light, 354, 598
    where to find, 42, 598
Arteries
    in frog's tongue, 247
    oxygen supply of blood in, 249
Arterioles
    effect of contractors and dilators
        on; 247–48
    in frog, 246–48
    in goldfish's tail, 246
    use of epinephrine and histamine
        to show vasocontractor ef-
        fect on, 248
Artesian wells, 527, 528
Artificial pollination, 380–81, 428,
    548
Ascaris
    anatomy of, 28, 29–*31, 32*
    collecting of, 560
    dissection of, 29–31
    mitosis in eggs of, *471*
    preservation of, 560
    use of to show antipepsin action,
        214

Aschelminths, microscopic exami-
    nation of, 27–33, 29, 30, 31,
    32
Ascorbic acid. *See* Vitamin C
Ascospores, of lichens, 643
Asexual reproduction. *See* Repro-
    duction, asexual
Aspergillus, 638
    collecting of, 638
    culturing of, 640
    spore formation in, 344
Aspirator, use of to measure respi-
    ration, 270
Assay methods. *See* Test(s)
Assimilate stream, in plants, 231
Asterias. See Starfish
Atmometer, use of to demonstrate
    lifting power of leaves, 238–
    39
Atomic weights, table of, 751
ATP (adenosine triphosphate), *430*
    formation of, 160, 262, 291
    in photosynthesis, 156, 157, *159,*
        161
    role of in active transport, 291
    as source of energy, 291, 312, 313
    See also Mitochondria
Auditory nerve, 320
Aulophorus (segmented aquatic
        worm), 34, *35*
    culturing of, 595
    identification of, 594
Aurelia, 22, *25*
Auricles, 245
Autoclave, use of to sterilize cul-
    ture media, 632, *633*
Autogamy, 424–25
Autonomic nervous system, 320,
    321
Autosomal linkage, in *Drosophila,*
    420–21
Autosomal traits, 389–93, *391*
Autotrophs
    chemosynthetic, 161–62, 533
    metabolism of, 493–94
    nutrition of, 161, 231, 533
    photosynthetic, 533
Auxanometer, *453*
"Auxin," as general term, 328
Auxins
    comparison of with gibberellic
        acid, 331
    effect of on phototropism, 297
    interaction of with kinetin, 330
    investigation of, 327–28
    and polarity, 337, 479–*80*
    production of, 297
    role of in growth, 326, 327, *328,*
        489
    source of, 191
Avogadro's law, 281

Axenic cultures of protozoa, 587–88
Azolla, 648

Bacillus coli, use of for culturing
        Daphnia, 596–97
Bacillus subtilis, 120
    as inhibitor of *Sarcina subflava,*
        450–51
Backcross, to determine gene pu-
    rity, *417, 436*
Backswimmer, *604*
Bacteria, *104, 120*
    antibiotic-resistant forms of, 452–
        53
    bioluminescence of, 148–*49*
    collecting of, 630
    colonies of, 448–49
    counting of, 444–45, 451–52
    cultures free of, 587–88
    culturing of
        in Petri dishes, 192, 448–52,
            630–*35, 633*
        in prescription bottles, 447–*48*
    enzymes in, 217
    fission in, 119
    genetic transformation of, 452
    growth of
        conditions required for, 449–
            53, 490
        effect of antiseptics on, 451
        effect of penicillin on, *450,* 490
        effect of temperature on, 450,
            490
        inhibition of by other bacteria,
            450–51
        prevention of, 457–58, 490
    identification of, 635
        with Gram's stain, *120*
    investigations using, 452
    irradiation of, 452
    isolation of, 635
    making smears of, 119–21
    microscopic examination of, *104*
    motility in, 119
    mounting of, *104*–05, 119–20
    nitrogen-fixing, *538*
        use of for culturing *Daphnia,*
            596–97
        use of on school grounds, 547
    pathogenic, avoidance of, 630,
        692
    pure, purchasing cultures of, 105
    sexuality in, 404, 406
    soil, utilization of starch by, 451
    staining smears of, 119–21, *120,*
        132
    subculturing of, *635*
    uses of in classroom, 635–36
    where to find, 449, 489–90
Bacteria of decay
    action of, 547, 569

Collodion, use of to locate stomates, 234
Collodion membrane, diffusion through, 206, *207*
Colony counter, 446–*47*
Color blindness
  inheritance of, 434
  test for, 388–89
Colorimetry, 171–72
Colpidium, 584
  culturing of, 586
  as food for amoebae, 581
  microscopic observation of, *100, 101*
Colpoda
  culturing of, 585
  silver impregnation of, 126
Column adsorption chromatography. See Chromatography
Combustible materials, storage of, 691
Compensating flask, 179–80
Competitive inhibitor, 146
Competitor, as heterotroph, 533
Compost, use of to test soil acidity, 569
Compounds, chemical, 495
Compounds, testing of by chromatography, 137
Concentrating cultures, 111, 123–24
Concepts, development of through "block laboratory approach," 188–96, 542–44
Condensation of water, *527*
Conditioning. See Learning
Congo red
  dilution of, 663
  pH range of, *662*
  as vital stain, 121
Conjugation
  versus autogamy, 424–25
  definition of, 349
  in molds, 639
  in *Paramecium, 352,* 424, 426
  in protozoa, 349–50, *352*
  in *Spirogyra,* 345, *370,* 629
Conklin's hematoxylin, as stain for chick embryos, 131, 133
Conocephalum, 644
Conservation, 228, 524–26, 569–70
Constant-pressure respirometer, 284
Consumers, in food chain, 569
Contact, response to, 300
Contour line, 526
Contour map
  of pond, 531–32
  of stream, 532
Contour plowing
  demonstration of, *525*
  use of on school grounds, 547

usefulness of, 569
Controls, student planning of, 8–9
Cooley's disease, 390
Coordination of responses
  in animals, 323–26
  through hormones, 323–26
  in plants, 326–32
  role of ciliated epithelial tissue in, 323
Coplin jar, *119*
Copper acetate solution, for preserving green color of plants, 661
Copper sulfate
  anhydrous, use of to remove water from absolute alcohol, 129
  use of to demonstrate cell model, 668
Cork, dyes for, 130
Corms, regeneration by, 475–76
Corn
  chloroplasts in, *183*
  crossing of, 435
  heredity in, 394–97, *395, 396*
  pollen grains of, 395–96
  staining kernels of, 396–97
  stamens on tassel of, *84*
Cornmeal medium, for *Drosophila,* 416
Cornmeal–molasses–rolled oats medium, for *Drosophila,* 416
Corolla, of flowers, 79, 81
Corpuscles. See Blood cells
Corrosive substances, handling of, 691
Costello's method, for culturing fern prothallia, 649–50
Counterstaining, 130
Counting chamber, for cells, 443–44
Cover crops
  role of, 522, 524, 526, 569
  use of on school grounds, 547
Coverslip, repairing of, 116
Coverslip bridges, 488
Crabs
  preservation of, 565
    See also Hermit crab; King crab
Crayfish
  anatomy of, 43, *44*
  collecting of, 561
  dissection of, 43, 44
  preservation of, 561, 565
Cream, hydrolysis of by pancreatic lipase, 216
Cream of wheat medium, for *Drosophila,* 416
Cresol red
  pH range of, *662*
  preparation of, 664

Crocus, corm of, *475–76*
Crop rotation, 538, 569
Crustaceans, nauplius larvae stage of, 599
Crustaceans, small fresh-water
  collecting of, 596
  culturing of, 596–600
  examination of, 39–43, *40–44*
  mounting of in glycerin jelly, 115
Crystal violet
  for staining bacteria, 119–20
  use of for Gram's stain technique, 120
Culture media
  inoculation of, 634–35
  range of indicators for testing, 636
  sterilization of, 632–*33*
  transfer of, 635
  See also Agar medium; Culturing
Cultures
  concentrating of, 111
  study of in depression slides, 122–23
  succession of living forms in, 582–83
Culturing
  of *Actinosphaerium,* 582
  of *Aeolosoma,* 595
  of algae, 622–28
  of amoebae, 579–82
  of *Arcella,* 582
  of *Armillaria,* 640
  of *Artemia* (brine shrimp), 599
  of *Aspergillus,* 640
  of bacteria, 192, 447–52, *448, 630–35, 633*
  of *Blepharisma,* 585
  of bread molds, 639–*41*
  of cactus plants, 652–53
  of *Cereus,* 652–53
  of *Chara,* 626
  of *Chilomonas,* 587
  of *Chlamydomonas,* 587, 627
  of *Chlorella,* 627
  of ciliates, 583–86
  of *Cladophora,* 626
  of *Clitocybe,* 640
  of *Colpidium,* 586
  of *Colpoda,* 585
  of crustaceans, small, 596–600
  of *Cyclops,* 598
  of *Daphnia,* 596–97
  of *Dero,* 595
  of desmids, 626, 627
  of diatoms, 626, 628
  of *Dictyuchus,* 640
  of *Didinium,* 585–86
  of *Difflugia,* 582
  of *Drosophila,* 415–*16*
  of duckweed (*Lemna*), 655
  of earthworms, 594

Dero (cont.)
　as food for *Hydra,* 591
　ingestion by, 201
　microscopic observation of, 102
　regeneration in, 482
　respiration of, 594
　use of in cell study, 95
　use of in microaquarium, 266
　where to find, 594
Desert terrarium, *648*
Desmids, *627*
　culturing of, 626, 627
　where to find, 554
Deviation, chance as factor in, 407–08
De Vries, H., theory of, 516–17
Dextrins
　formation of in starch digestion, 210, 211
　test for, 218
Dialysis, for separating molecules, 134–35
Diaphragm, action of, 287–88
Diastase
　commercial, 208–09
　digestion of starch by, 208, 227
　　in starch-agar medium, 208
　effect of dilution on, 209
　effect of temperature on activity of, 209
　experiment using commercial, 208–09
　extraction of from germinating barley seeds, 207–08
　production of during germination of seeds, 208
Diatoms, *628*
　culturing of, 626, 628
　study of, 629
　where to find, 626
Dichromate cleaning solution for glassware, 679
Dicotyledons
　root of, cross section of, *80*
　seeds of, 378
　similarity of, 500
　stem of, cross section of, *78*
Dicranum, 644
Dictyuchus, 640
Didinium, 585–86
Diemictylus viridescens, 611–12
Diet
　balanced, 227–28
　calories in, 222, 224
　in different countries, 227
　for mice, 463
　recommended, 222, 224, 228
　for testing Vitamin B$_1$ deficiency, 220
　for testing Vitamin D deficiency, 219

Differentiation
　in *Acetabularia,* 346, 404, *406*
　in frog's embryos, 463–67, *464–66,* 483–*85, 484*
　in plant tissue, 479
　and regeneration, 474–85, *475–82*
　in slime molds, 479
　time as factor in, 483
Difflugia, 582
Diffusion
　and digestive system, 204–06, 205, 207, 226
　through frog's intestine, 206, *207*
　of gases, 204, 258
　of liquids, 204
　through a membrane, 204–*06, 205, 207,* 227, 240–41, 258, 668
　nature of, 150, 151
　versus osmosis, 206*n,* 240
Digestion, chemical
　and absorption, 206
　in alimentary canal, 210–12, 214–17
　in animals, 209–17
　　of starch, *210–11*
　in bacteria, 217
　and diffusion, 204–*06, 205, 207*
　and dilution of enzymes, 209, 212
　enzyme action in, in animals, 209–11
　　of amylase in pancreatic juice, 215
　　of amylase in saliva, 211–12
　　of antipepsin, in *Ascaris,* 214
　　of bile salts, 215
　　of invertase, in hydrolysis of sugar, 212
　　of lipase, 215–16
　　of pepsin, 213–14
　　of ptyalin, effect of inhibitor on, 210
　　of ptyalin, effect of temperature on, 211–12
　　of ptyalin, in test tubes, 205, *210–11*
　　of ptyalin, on slides, 211
　　of rennin in, 214
　　of salivary enzymes on starch, 205, 209–12, *210*
　　of trypsin in, 214–15
　enzyme action in, in plants, 207–09
　　of amylase in germinating seedlings, 148, 207
　　of bromelin, as protein-splitter, 209
　　of diastase, 207–09, 227
　　effect of dilution of an enzyme (diastase) on, 209
　　of invertase, 148, 209, 215

　　protein-splitting, 209
　and extraction of digestive juices, 217
　of fats and oils, 215–16
　in insectivorous plants, 217
　in molds, 217
　in mouth, 210–12
　of nutrients, 205, 207–09, 215–16
　and pH, 203, 210
　in plants, 207–09
　　of proteins, 209
　　of starch, 207–09
　of proteins, 209
　salivary, 13, 205, *210,* 211, 227, 307, 310
　　digestion of starch by, 205, 209, *210*
　in small intestine, 214–17
　of starch, 8, 13, 205, 207–09, *210–11,* 227
　in stomach, 213–14, 215, 216
　and temperature, 209, 210, 211
　in termites, 537
Digestive systems
　in *Ascaris,* 29, *32*
　in cockroach, *46,* 47
　in crayfish, 43, *44*
　in earthworm, 37, *38*
　in fetal pig, 69, *70*
　in fish, 60, *62*
　food tube in, 203–04
　in fresh-water clam, 51–52, *53*
　in frog, 62, *63,* 203–04, 226
　in grasshopper, *45,* 47
　in many-celled microorganisms, 203
　model of, 667, 668
　in pigeon, *68,* 69
　in planarians, 27, *28*
　in protozoa, 200–01
　in rats, *72*
　in rotifers, 31–32, *33*
　in sandworm, 38–*39*
　in starfish, 56, *57*
　in turtle, *66,* 67
　See also Anatomy, internal
Digitonin solution, 661
Dileptus, 100
Dilution
　of alcohol, 658
　effect of on enzyme activity, 209, 212
Dimetrodon, 497
Dinitrophenol, interference of with ATP formation, 291
Dinosaurs, *497*
Dionaea. See Venus flytrap
Dioxane, use of to mount small forms, 116
Dip nets, *552*

Fetus, nourishment of, 433
Feulgen nuclear reagent, for staining protozoa, 124
Fibrin, 252–53
Fibrovascular bundles
  examination of, 77, 232, 257
  preservation of, 77
  transport of organic substances in, 231
Ficus elastica, 236, 237
Fiddler crabs, 564
Field stations, listing of, 745–46
Field trips
  equipment for, 11–12
  indoor, 548
  planning of, 11–12, 545
  precautions for, 692
  and school grounds, 547–48, 570, 698
  varieties of, 545, 546–47
Field work
  as essential part of biology, 545, 572
  values of, 545
Films
  cement for, 665
  directory of distributors of, 733–36
  making of by students, 10
  room for, 699
  storage of, 700, 702
Filter paper
  use of to culture bread molds, 641
  use of to culture slime molds, 636–38, 637
  use of to germinate seeds, 652
Filter paper chromatography. See Chromatography
Finches, adaptive radiation among, 509, 512
Fire
  ecology of, 541
  equipment for dealing with, 689, 692
Fireproofing of fabrics, 693–94
First-aid cabinet, 700–01
First-aid kit, 693
Fish
  adaptations of, 508
  anatomy of, 59–60, 62
  anesthetization of, 112
  for aquarium, 610
  blood system of, 60, 62
  brain of, 60, 62
  collecting of, 562
  conditioning in, 315
  digestion in, 60
  diseased, treatment for, 610
  dissection of, 59–60, 62
  eggs of, mitosis in, 471
  experimental embryology of, 467

feeding of, 608–09
genetics of, 434
  readings on, 411
gills of, 289
heart of, 503
maintenance of, 605–10
native, 609–10
ovaries of, 357
preservation of, 562, 565
reproductive system of, 60, 62
respiratory system of, 60, 62
scales of, mounting of, 115
staining skeletons of, 673
testes of, 357
tropical, 609, 610
types of useful in classrooms, 609–10
See also Goldfish
Fissidens, 644
Fission
  in bacteria, 119
  in Euglena, 629
  mitotic stages in, 470
  in Paramecium, 343
  in protozoa, 343, 433
  rate of, 442
  as type of asexual reproduction, 343–44
  use of aquatic worms to study, 595
Fixation
  of protozoa, 124
  of small specimens, use of FAAGO for, 661
  of tissue, 126–28
Fixative(s)
  formalin as, 132
  preparation of, 126–28, 473
  for tissue cells, 126–28
  uses of for unicellular organisms, 126
Flagella, staining of, 121–22
Flagellates
  algal, 104
  collecting of, 586
  concentrating a culture of, 111
  culturing of, 586–87
  microscopic observation of, 103, 104, 588
  response of to light, 298
  symbiotic, 537–38, 605
  in termite intestines, 537–38
Flatworms, as parasites, 534–35
  See also Planarians
Fleas, mounting of, 115
Flemming's fixative, 127
Flies
  as food for frogs, 614
  nymphs of, 553, 603
  obtaining mitochondria from, 287
  pinning of, 567

use of in experiment disproving spontaneous generation, 341, 342
  See also Damsel fly; Dragonfly; Drosophila (fruit flies); Mayfly
Floor plans, for classrooms and laboratories, 695, 696–98, 697
Florence flask, use of to represent air sac, 249, 293, 667
Florigens, 330
Flowering, effect of light and dark periods on, 333
Flowerpot culture method
  for culturing fern prothallia, 648–49
  for culturing moss protonemata, 646
Flowers
  adaptations of, 79, 81
  artificial pollination of, 380–81
  classification of, using key, 376, 377
  complete, 79
  composite, 376, 377
  dissection of, 80, 83, 378
  generalized, 80
  identification of, 80, 433
  incomplete, 79–80
  inflorescence in, 377, 378
  leguminous, 376, 377
  ovaries of, 81, 378, 379
  parts of, 79–80, 81, 83, 84, 376–78, 377, 433
  pistils of, and germination of pollen grains, 300
  pollen tubes of, 433
  rose family of, 376, 377
  stamens of, 84
  student allergies to, 692
  types of, 80, 81, 82, 377
Fluorescence, of chloroplast pigments, 183, 185
Fluorescent light, for plant growth, 622
Food
  absorption of in small intestine, 216–17
  caloric values of, 225
  as limiting factor in survival, 499–500
  making of by autotrophs, 533
  oxidation of, 266–67, 293
  preservation of from bacterial growth, 490
  storage of, growth of seedlings in relation to, 458
  synthesis of by Chlorella, 544
  testing nutrients in, 217–21
  use of by consumers. See Food-getting

Fungi (*cont.*)
evolution of sexuality among, 346–47
in lichens, 537
parasitic, *641–42*
preservative for, 128
production of carbon dioxide by, *277*
sporulation in, *105, 433*
See also Molds
Funnel heater, *634*
Furniture, for classrooms and laboratories, 695–98

Galton apparatus, for demonstrating laws of probability, 408–09
Galvani, L., discovery of, 311–12
"Galvanic" forceps, 311
Gametangia, in *Oedogonium*, 370
Gametogenesis, occurrence of meiosis during, 347
Gammarus
anatomy of, *43*
characteristics of, 598
culturing of, 598
mounting of, 115
use of in classroom, 598
where to find, 598
Garden pea
flower of, *81*
leaves of, *76, 77*
use of in Mendel's work on heredity, 409–10
Garter snake, *616–17*
Gases
diffusion of, 204
measuring exchange of, 178–80
See also Carbon dioxide; Oxygen
Gastric digestion, 213–17
Gastric juice, artificial, 213
Gastrocnemius muscle of frog, dissection of, *311*
Gastropods, 49–50
Gastrotrichs, *102*
anatomy of, 33, *34*
Gates' fluid, as fixative for plant tissues, 127
Geiger counter, use of to trace uranium nitrate in stem, 233
Gelatin, use of to simulate fossil formation, 498
Gelatin broth, for bacteria, 631
Gelatin solution, use of to retard motion of protozoa, 112
Generations, alternation of. *See* Alternation of generations
Genes. *See* Chromosomes; Heredity; Inheritance, human
Gentian violet, as basic dye, 130
Geotaxis, 301

Geotropism, *301–04, 302*
versus hydrotropism, 304
in roots, *302–04*
in stems, *301–02*
Geranium
consumption of carbon dioxide by, 172–73, 174–75
production of carbon dioxide by, *273*
production of oxygen by, 181–82
starch-making by, 173–74, 197
stem cuttings of, 477
Germination
of pollen grains, 382
of seed plants, 654
of seedlings
consumption of oxygen during, 281–82
conversion of starch to sugar during, 227
effect of mineral deficiency on, 460
production of carbon dioxide during, 273, *274–75*
production of heat during, *286–87, 293*
of seeds, 433, 450, 456–58, *457*
consumption of oxygen during, 279–*81, 280,* 293
disinfectants for, 457–58
light requirements for, 333–34
materials required for, 652
nutrient solutions for, 459–60
production of carbon dioxide during, *274–77, 275, 276,* 293
production of diastase during, 208
rate of, 501, 516
test of viability in, 457
time required for, 652
use of to show growth of root hairs, 232–*33*
Gestation period, in some mammals, 368, 618
Gibberellic acid, 330–32, *331*
Gibberellin, effect of on plants, 454, 455
Giemsa stain, 119
Gilson's fluid, as fixative, 127–28
Glass
bending of, 678
cutting of, *677,* 680
joining of to metal, 660
working with, 677–80
Glass models, use of in classroom, 667
Glass slides. *See* Slides
Glass stirring rods, 678
Glassware
cleaning of, 679
repair of, 679–80
storage of, 699

Glucose
change of sucrose to, 162–63
diffusion of through membrane, 205
oxidation of, 262–65
storage of in plants, 162
tests for, 135, 162, 163, 205, 218 in urine, 292
transformation of to alcohol in yeast extracts, 279*n*
Glutathione
extraction of from yeast cells, 202
role of in feeding response of *Hydra,* 202, 592
Glycerin
use of to mount small specimens, 115, 661
use of to seal ground-glass joints, 664
Glycogen, staining of, 121–22
Glycollic acid, 191
Goldbeater's membrane, 8, 204, 205
Goldberger, J., discovery of cause of pellagra by, 228
Goldfish
anesthetization of, 112
circulation in tail of, 245–*46,* 257
conditioning of, 315
embryology of, *358,* 359
kidney of, 241
maintenance of, 610
use of in aquarium, *610*
Gonadotropic activity, effect of light on, 332–33
Gonionemus, 22–23
Gorgodera, as parasite in frog, *534*
Grafting
of *Acetabularia, 406*
of plant stems, *479*
regeneration by, 404, *406, 479*
of tissues of embryo frog, 362
use of *Hydra* to study, 592
waxes for, 661–62
Gram's stain technique
materials needed for, 120–21
preparation of stain for, 664
use of to identify bacteria, 120
Grassfrogs, 563
Grasshopper
anatomy of, *45, 47*
digestive system of, 45, 47
dissection of, 45, 47
mounting of, 565, *567*
pinning of, 567
preservation of, 564
reproductive system of, 45, 47
spermatogenesis in, *348*
Gravity
response of land snails (*Helix*) to, 301

Gravity (*cont.*)
response of *Paramecium* to, 301
response of roots to, 303–04
response of seedlings to, *302*
response of stems to, *301*
Green plants. See Plants, green
"Green water," use of for culturing *Daphnia*, 596
Greenhouse, 698
Ground water
action of, 524, 528
as source of well water, 527
spreading of, 569
Growth
of algae. See *Chlorella*
of animals, *443*, 462–67, *464–66*
of excised leaves, 469
of excised plant embryos, 468–69
of excised roots, 468
exponential
formula for, 440
limits to, *442–43*
exponential phase of, 440–41
of *Chlorella*, 544
of duckweed fronds, 456
graphs of, *439–41*, *440*, *443*
of human population, *443*
at intracellular level, 440
of leaves, 453–*54*
measurement of, *439–41*, *440*
in animals, 439–*40*
by gain in height, *453*
by gain in volume, 448
by gain in weight, 439, 448, 453
by increase in clones, 446–48
in leaves, 453–*54*
methods for, 443
in microorganisms, *441*–49, *444*, *445*, *447*, *448*
in rats, *440*
in roots, 453
in stems, 453
in total plant, 453–56
by turbidity, 445–46
of microorganism populations, 442–48
of motile forms, 449
pattern of, 489
peculiarities in, 454
phases of, 440–41
of plants. See Plants, growth of
rate of, 439–40
of duckweed fronds, 456
of tadpoles, 463
use of *Hydra* to study, 592
readings on, 454
in roots, *453*
of seedlings, 458, 461–62, 490
of stems, 453
vegetative, light requirements for, 622

of yeast cells, 449
Growth hormones
effect of on animals, 324–26
effect of on plants, 326–32, *328*, *330*, *331*
effect of on regeneration, 479, 548
See also Auxins; Ductless glands
Growth-regulating substances, 327–330, 454–*55*
See also Auxins; Gibberellic acid
Guard cells
of *Sedum*, *175*
study of, 75, 77
turgor in, 175–76
Gudernatsch, F., experiment of, 338
Guinea pigs, *617*
anesthetization of, 112, 113
breeding age of, 618
gestation period of, 618
Gum arabic, for mounting small forms, 115
Gum tragacanth solution, for retarding motion of protozoa, 112
Guppy (*Lebistes reticulatus*), 609, *610*
Guttation, 240
Gymnosperms
classification of, *514*
cross section of stem of, *78*
life cycle of, 378, *380*
similarity of, 500
See also Pines
Gyri, of brain, 319

Haas and Reed's "A to Z" solution, for plants, 460
Habits
formation of, 336, 337
in human beings, 316
value of, 337
Habrobracon, 421–*23*, *422*, 427
Hahnert's solution, for amoebae, 582
Hair loops, for transplanting embryonic tissue, 486–*87*
Haldane, J. B. S., view of on formation of chemical compounds, 495
Hales, S., experiments of, 242
Haliotis, *50*
Halteria, *584*
Hämmerling, J. S., experiments of, 346, 404
Hamsters, *617*
breeding age of, 618
gestation period of, 618
heredity in, 411
Hanging drop preparation, 122, *123*

Hanging garden, use of to show hydrotropism, 304
Hardy–Weinberg law, 407, 431, 508, 513
Harris' hematoxylin, as nuclear stain, 129–30, 131
Hartig's method of collecting sieve-tube exudate, 231
Harvestfish, as example of mutualism, *539*
Harvey, E. B., work of on *Hydra*, 480
Harvey, W., study of blood circulation by, 257
Haslet, use of to study mammalian respiration, 289, 293
Hay infusion, 580, 583
Hay-scented fern (*Dennstaedtia punctilobula*), 648
Hayem's solution, as diluting fluid for blood counts, 662
Heart, 244–45
in aquatic worm, 245
in clam, 53, 244
in *Daphnia*, 40, 245, 258, 597
dissection of, 245, 257
in earthworm, 248
effect of adrenalin on, *244*, 326
effect of minerals on, *244*
effect of temperature on, *244*
in embryo snails, 244–45
in frog, 64, 244, 325–26
of mammal, *245*
model of, *259*
muscles in. See Muscle, cardiac
role of in circulation, 258
of sheep, *85*
study of, 259
use of kymograph to study effect of drugs on, 326
vertebrate, evolution of, *503*
Heat
dry, use of to sterilize culture media, 632
production of by plants, 286
removal of by evaporation, 291, 294
Height retardation of plants, 454–55
Helix
anatomy of, *52*
culturing of, 595
feeding of, 595
geotaxis in, 301
responses of, 301
use of to measure oxygen consumption, 284
where to find, 595
See also Land snails
Heller's nitric acid ring test, 292
Helmont, Jan van, experiment of, 193, 197

Model(s) (*cont.*)
in clay, 667
of digestive tract, 668
of DNA, *428–29, 432*
of ear, *320*
of egg, 667
of cleavage stages, 432
of eye, *319*
of flower, *84*
of fossils, 497–98
of frog, *84*
of genes, 410, 434, 668
in glass, 667
of heart, *259*
of human birth, *369*
of human nervous system, *321*
of intestines, 226, 668
of leaf, *84*
of nucleotides, 432
in papier-mâché, 669
in plaster of Paris, 668–69, *670–71*
in plastic, 669, *671, 672*
of prehistoric animals, *597*
real materials for, 667
of sporangia, 345
storage of, 699
of valves in veins, *249*
of villus, 249, 667
in wood, 669
Molar solutions, 658–59
Molds
bread
handling of, 692
See also Mucor; Rhizopus
collecting of, 638
conjugation in, 639
culturing of, 639–*41*
enzymes in, 217
extracellular digestion by, 217
food-getting by, 227
inhibitor of, 415
microscopic observation of, *105*
mounting of, 105, 116
removal of from cultures of *Drosophila*, 416
rhizoids of, 217
sexual reproduction in, 370, 372, *373, 374*
spores of. *See* Sporangia (spores)
use of sulfur to destroy, 644*n*
Molds, slime. *See* Slime molds
Molecules
active transport of across kidney membrane, 291
diffusion of
in air, 204, 258
in liquid, 204
through membrane, 204–06, 207, 227, 240–41, 258, 668
movement of
in chromatography, 137
in electrophoresis, 141

separation of by dialysis, 134–35
Molisch's reaction, as test for sugars, 218
Molisch's reagent, 218
Molisch's solution, for culturing algae, 623
Mollusks
anatomy of, 48–*55, 50–54*
characteristics of, 48–49
dissection of, 49
shells of, *566*
preserving, 565, 566
soft-bodied, preservation of, 564
variations among, 503–04
Monera, use of term, 89*n*
Monocotyledons
root of, cross section of, *79*
seeds of, 378
similarity of, 500
stem of, cross section of, *71*
Monocystis, as parasite of earthworm, *535,* 590
Moonsnail shells, variations among, *427*
Mormoniella
breeding of, 435
Sarcophaga as host for, 423, *424*
use of to study genetics, 423, *424*
Mosquitoes
mounting larvae of, 115
pinning of, *568*
Mosses, *644, 645*
alternation of generations in, 374, *375*
collecting of, 555, 643–44
culturing protonemata of, 645–46
effect of light on growth of sexual structures of, 646, *647*
life history of, *645*
maintenance of, 644
preservative for, 128
use of to germinate seeds, 652
uses of in classroom, 644–45
Moths
cocoon of, *602*
larvae of, *602*
as host for *Habrobracon,* 421–*22*
metamorphosis of, *602*
pinning of, 567
pupae of, 354–55, *602*
collecting, 602
maintaining, 602
Motion-picture film cement, 665
Mounting
of bacteria, 119–20
in balsam, 115, 131–32
of chick embryos, 133
in dioxane, 116
in glycerin jelly, 115
in gum arabic, 115
of insects, *567–68*
of permanent slides, 126–32

in plastic, 669, *671, 672*
of small specimens, 114–16, 661
in syrup, 115
Mounting fluid
preparation of, 132
Ringer's solution as, 665
Mourning cloak butterfly (*Vanessa*), metamorphosis of, 602
Mucor
collecting of, 638, 639
culturing of, 639–40, 641
cyclosis in, 106
as distinguished from *Rhizopus,* 639
See also Molds
Mungo bean, cross section of root of, *80*
Muscle
cardiac
contractions of, 313
maceration of, 114
mounting of, 108, 114
contractions of, 310–14
by electric current, 311–13
readings on, 321
tracing pattern of with kymograph, 312–13
of frog
dissection of, *311*
response of to electric shock, *312*
smooth, 313
microscopic observation of, *108*
striated
beef, 107, 108
contractions of, 313
maceration of, 114
microscopic observation of, 107, *108*
mounting of, 107, 108, 114
transmission of stimulus to, 310–12
Muscle fibrils, 313
Muscle tissue
of frog, staining, 107
maceration of, 113–14
preparing mount of, 114
staining of, 114
use of to show oxidation in living cells, 267–68
Museum, establishment of in school, 548
Museum jars, 548, 564
Mushrooms
respiration in, 277
spore formation in, 344
Mussels, fresh-water, *596*
dissection of, 50–54
use of for dissection, 595
Mustard family, 500
Mutants, biochemical
detection of in *Neurospora,* 400–02, *401*

Photolysis of water, in presence of chloroplasts, 170
Photometer
making of, 210, *445*
use of to measure growth of microorganisms, 228, *445–46*, 462
Photoperiodism, 332–35, *334*
effect of on reproduction in fish, 357*n*
investigation of, 548
region in spectrum affecting, 332, 333
Photophosphorylation, 157–*59*
Photosensitivity, in *Blepharisma*, 10, 298, 336
Photosynthesis
absorption of carbon dioxide in, 166–76
action spectrum of, 164–*66, 165*
of *Chlorella*, 167–68, 169, 176, 187, 188, 190, 191, 196, 544
in chloroplasts, 157–*59*
designing investigations to study, 180
equation for, 156
evidence of, 162–88, *173, 174, 177, 180, 181*
growth of organisms as evidence of, 188
liberation of gas as evidence of, 176–*77, 180*
measurement of, 178–*80, 178*
and microscopic examination of chloroplasts of leaves, 106, *183*, 187
versus oxidation, 293
phases of, 156
as process, 155–62
products of, first, 162–63
rate of, 190
in algae, 176
in *Chlorella*, 272
effect of carbon dioxide on, *173–74*
in elodea (*Anacharis*), 176–*77*
in green plants, 166–67
and light wavelength, 187
measurement of, 163–66, 167, *178*, 271–72
region in spectrum affecting, 164–66, *165*, 186–87, 332
release of oxygen in, 156, 176–82, *177, 180, 181*
measurement of, *178*, 179, *180*, 187
role of chlorophyll in, 164–66, 182–88
role of chloroplasts in, 186–87
role of guard cells in, *175–76*
role of light in, 163–66, 191
use of *Euglena* to study, 629

use of radioactive tracers to study, 157, *158, 159*
Photosynthesizing systems, evolution of, 160–61
Photosynthetic phosphorylation, 157, *159*, 160–61
Photosynthetic quotient (*PQ*), 167
Phototaxis, 296, 297–99
Phototropism, 296–*97*
region in spectrum affecting, 332
Phycomyces blakesleeanus, 221
Phyllocactus, 653
Physa, 595
ciliated epithelial tissue in, 54
egg-laying by, 595
as food for fish, 608
obtaining, 595
use of to study eggs and sperms of snails, 351
See also Fresh-water snails; Snails
Physalia (Portuguese man-of-war), 22, *24*
Physarum polycephalum, 637
Physcia
as example of symbiosis, *537*
section of, *643*
where to find, 643
See also Slime molds
Physcomitrium pyriforme, 644
Physiological saline solution, for animals, 665
Phytochrome, as light-sensitive pigment, 333
Phytohormones, 326
See also Auxins
Picture file, 676–77
Pig
embryo of, *367*
fetus of. *See* Fetal pig
tapeworm in, 535
uterus of, dissection of, *366–67*, 433
Pigeon, *68–69*
use of to show Vitamin B_1 deficiency, 220
Pigmentation, inheritance of, 390
Pigments, chloroplast. *See* Chloroplast pigments
Pill bug (*Armadillidium*), 599–*600*
Pines
classification of, *515*
cross section of leaf of, *75*
seeds of, *571*
See also Gymnosperms
Pinning, of insects, *567–68*
Pipefish, breeding pouch of male, *358*
Pipettes, for transferring specimens. *See* Transfer pipettes
Pistils, 79, 80, 81, 376
Pitcher plant (*Sarracenia*)
culturing of, 652

as insectivorous plant, 539, *540*
Pithing of frog, 307, *308–09*
Pituitary glands
and coordination, 324
stimulation of by light in animals, 332–33
use of to induce ovulation in frog, 360–*61*
Pituitrin, effect of on shedding eggs by *Daphnia,* 597
Planarians, *592*
anatomy of, 25, 27, *28*
anesthetization of, 112, 593
chemotaxis in, 300
cocoons of, *593*
collecting of, 592
conditioning in, 27, 314
culturing of, 592
cutting patterns for, *14, 482*
as experimental animal, 482
feeding of, 592–93
feeding of *Artemia* larvae to, 599
fresh-water, 560
ingestion by, 201
killing of, 112
long-range investigation on, 14–16
movement of, 27
organ systems of, 27, *28*
pharynx of, 27
phototaxis in, 27, 298
pigmentation of, 27
preservation of, 564
regeneration in, 14–16, *482*, 489
responses of, 27, 298, 300
uses of in classroom, 25, 27, 593
Plankton, 532–33
Planorbis, 595
as food for fish, 608
Planting mixture, 661
Plants
absorption of carbon dioxide by, during photosynthesis in, 167–70, 172–*74, 173*
absorption of oxygen by, 282–83
alternation of generations in, 347
versus animals, 89*n*
for aquarium, 606–08, *607*
aquatic, green
absorption of carbon dioxide by, 168–70
absorption of oxygen by, 283
collecting of, 552
production of carbon dioxide by, 272–73
production of oxygen by, 176–81, *177, 180*
behavior in, 337–38
breeding of, 428, 435
cells of, comparison of with animal cells, 96
chemotropism in, 300

Reproduction (*cont.*)
  in plants, *370–82, 371–74,* 433,
    434
    See also Meiosis; Mitosis
  use of protozoa to study, 588
Reptiles
  collecting of, 563, 619
  heart of, 503
  maintenance of, 614–17
  preservation of, 563
Research by students, 224, 226,
    685–86
Resins, for mounting tissue on
    slides, 114
Resources
  natural
    careless use of, 541
    conservation of, 570
  of students. *See* Student squads
Respiration
  in animals, 265–72
  equation for, 262
  materials exchanged in, 293
  measuring carbon dioxide pro-
    duced during, *178–80,* 275–
    77, 293
  measuring oxygen consumed dur-
    ing, 280, *282*
  in plants, 265–72
  production of heat by seedlings
    during, 286, 287, 293
  rate of, 271–72
    factors affecting, 282–83
    in plants, 166–67
    variation of with age, 288
    volumeter for measuring, *178,*
      180
  See also Oxidation
Respiration chamber, use of in
    measuring volume of carbon
    dioxide produced, *270,* 277
Respiratory quotient (R.Q.), 262
Respiratory systems, comparison of
    in animals, 289
Respirometer, for measuring con-
    sumption of oxygen, 283–*84*
Responses
  of animals
    to chemicals, 299–300
    to contact, 300
    coordination of, 323–26
    "educated," 314
    to electricity, *299*
    to gravity, 301
    to light, 297–99, 332, *598*
    See also Learning
  of plants
    to chemicals, 300
    to contact, 304–*05*
    coordination of, 326–32
    to gravity, *301*–04, *302*
    to heat, 304–05

to light, 296–*97,* 332, 333–35,
    334
  to water, *304*
reflex
  in frog, 306–07, *309, 310*
  in man, 307, 310
  summary demonstration of, 305–
    06
Retina, 319
Reversible shift, 333–34
Rh blood factor
  distribution of, *255*
  inheritance of, 385–86, 387,
    434–35
  typing of, 255
Rhabditis, as parasite of earthworm,
    535, *536,* 594
Rhizobium leguminosarum, 538
Rhizoids, of molds, 217
Rhizomes
  of Bermuda grass, *475*
  cross section of, *78*
  plants suitable for regeneration
    by, 474
  reproductive function of, 79
  of Solomon's-seal, *474*
Rhizopods, 579–83
Rhizopus
  collecting of, 638, 639
  conjugation in, 370, 372, *373*
  culturing of, 640, *641*
  as distinguished from *Mucor, 639*
  spore formation in, 344
  See also Molds
Rhymes, learning of, 317
Ribbon snake, *616*–17
Riboflavin
  daily requirements for, 223
  test for, 221
Ribonuclease, 314
Riccia, 644
Ricciocarpus, 644
Rice, starch grains in, *165*
Riker mounts, *568*
Ring-necked snake, 616–17
Ringer's solution
  composition of, 244
  for frog's tissue, 665
  for mammalian tissue, 665
  use of as mounting fluid, 665
RNA (ribonucleic acid)
  extraction of, 140, 142
  hormonal role in production of,
    347
  as messenger, *430*
  as nucleic acid, 134
  use of planarians to study, 27
Robertson's medium, for anaerobic
    bacteria, 632
Rock specimens, labeling of, 677
Rocks, factors causing change of
    into soil, 524, 528

Root hairs
  examination of, 123, 232, 257
  growth of, 232, *233,* 456
Root stocks, regeneration by, 474
Root tips
  excision of, 468
  mitosis in, *117*–18, 470
  See also Onion root tips
Rooting preparations, 328–30
Roots
  binding force of, 14, 522
  cross sections of, *79, 80*
  effect of auxins on, 327
  excised, growing of, 468, 490
  exudation pressure of, 239–*40*
  formation of, 327
  functions of, 79
  geotropism in, 302–04
  growth of, 327
    effect of tropisms on, 337
    measuring, 453
    region of greatest, *453*
  hydrotropism in, *304*
  modifications of, 79
  phototropism in, 297
  production of carbon dioxide by,
    273–74
  regeneration by, 476
  region of sensitivity in, 303–04
  reproductive function of, 79
  responses of, 297, 302–*04*
  stimulation of by growth hor-
    mones, 329
Rotifers, *102*
  anatomy of, 31–*33,* 102
  anesthetization of, 112
  collecting of, 560, 593
  culturing of, 593
  feeding of, 593
  organ systems of, 31–*33*
  preservation of, 560
  use of in cell study, 96
Roundworms
  anatomy of, 27–31, *29, 30, 31, 32*
  mounting of, 115
  nonparasitic, 28, *31*
  preservation of, 564
Roux, W., work of, 467, 483
Royal Osmunda fern, 648
Runners (stolons)
  regeneration by, *475*
  reproductive function of, *79*
Runoff water, and soil erosion,
    526

Sabouraud's medium, for culturing
    molds, 640
Saccharomyces cerevisiae, use of to
    show budding, 642
Saccharomyces ellipsoideus, use of to
    show budding, 642
Sachs' solution, for plants, 459

Starch
breakdown of by diastase, 208,
227
digestion of by saliva, 8, 13, *210*,
211, 227
hydrolysis of
enzyme activity in, 209–12
in plants, 207–09
insolubility of, *8*, 205
making of. *See* Starch-making
tests for, 135, 218
utilization of by soil bacteria, 451
Starch-forming enzyme, test for in
peas, 297
Starch grains
identification of plants by, 164
in oats, *165*
in peas, 397
in potato, 164, *165*
in rice, *165*
in sweet potato, *165*
Starch-making
in algae, 164
by green plants, *163–64*, 173–74,
197
absorption of carbon dioxide
during, 273
role of light in, *9*, *163–64*
See also Photosynthesis
Starch paste, 666
Starfish, *601*
anatomy of, 55–58, *57*
anesthetization of, 55, 112
cleavage stages of, *353*
digestive system of, 56, 57
dissection of, 55–58
feeding of, 600
fertilization in, 351, 353
locomotion of, 600
preservation of, 565
sperms of, 351, 353, 354
water-vascular system of, 56, 57
Starling, E. H., work of, 324
Stationary growth phase, 441
of duckweed fronds, 456
Steam, use of to sterilize culture
media, 632–33
Steam pressure sterilizers, 692
Stegosaurus, *497*
Stem cuttings, 476–77
effect of leaves and buds on
growth of, 327
and gibberellins, 330–31
and growth hormones, 328–30
and kinetin, 330
See also Cuttings; Regeneration;
Soilless cultures; Vegetative
propagation
Stemonitis, *637*
Stems
adaptations of, 78–79
cross sections of, *77, 78*

cuttings of. *See* Stem cuttings
dicotyledonous, 77, *78*
exudation of water from, 239
fibrovascular bundles in. *See*
Fibrovascular bundles
function of, 78
geotropism in, *301–02*
grafting of, *479*
and growth hormones, 326–32,
479
identifying plant by growth of,
77–78
measuring growth of, 453
monocotyledonous, *77*
movement of water up, *240, 241,*
257
tracing of with uranium nitrate, 233
phototropism in, 296–*97*
polarity in, 479, *480*
responses of, 296, *301–02*
study of, 77–79
woody, mounting of, 113
Stentor, *201*
culturing of, 585
feeding of, 585
as food for *Pelomyxa*, 582
ingestion by, 201
microscopic observation of, 95,
96, 104
response of to contact, 300
use of in cell study, 95, 96
Sterilization
of culture media, 632–33
of glassware, 679
of moss sporangia, 645–46
of Petri dishes, 634, 692–93
of seeds, 457–58, 468
of wire loops, 693
Sterilizers, for culture media, 632–
33
Stigma, 80, 81
Stimulus, substitute
response to, 315, 316
Stomates
entry of carbon dioxide through,
75, 174–75
evaporation of water through,
232, 234, *235, 237*
location of
with cobalt chloride paper, 257
with collodion, 234
number of, 175, 257
study of, 75, 77, 174, 175
turgor in guard cells surrounding,
175–76
Stone fly, nymph of, *603*
Storage cabinets, *676, 699–700,*
701, 702
Storage facilities, need for, 699–700
Strawberry, stolon of, *475*
Streaking, *442,* 635

Stream, 532
Streptococcus lacti, found in yoghurt,
104, 105
Strip cropping, 569
Strobe disc, use of to slow down
Daphnia heartbeat, 245
Student squads, 680–81
book-room, 684–85
in classroom, 687–88
laboratory, 681–83
for landscaping school grounds,
545, 547, 684
mimeographing, 686
perpetuation of, 687
projectionist, 686–87
publishing, 686
safety rules for, 690
science readings, 686
soil-testing, 684
for special services, 683–87
tutoring, 688–89
Study habits
discussion of, 336, 337
improvement of, 316–18
Sturgeon, as example of adaptation, *508*
Stylonychia
culturing of, 586
microscopic observation of, 101
Stylonychia mytilus, 101
Subculturing
of amoebae, 581
of bacteria, *635*
of ciliates, 583, 584–85
of *Drosophila,* 415
See also Culturing; Replica plating
Subsoil, location of, 524
Succession
ecological, 528, 531–32
of microorganisms in depression
slides, 122
of protozoa in a culture, 580
Succinic acid, molecular structure
of, 146
Succinic dehydrogenase, 145–46
Sucrose
action of invertase on, 148, 209,
212, 215
change of to glucose, 162–63
Sucrose solution, 666
Sudan III, as vital stain, 121
Sudan III test, for fats and oils, 136
Sugars
in green leaves, 162
simple, as products of photosynthesis, 162–63
tests for, 135, 136, 148, 162–63,
205, 218
Sulci of brain, 319
Sulfate, need of for plant growth,
522

Sulfonamides, in destroying bacteria, 146–47
Sulfur, use of to destroy molds, 644n
Sulfuric acid, handling of, 691
Sulfuric acid solution, 666
Sundew (Drosera), 539, 540
 culturing of, 652
Sunfish, 609, 610
Suppliers, directories of, 731–44
Supplies. See Equipment and supplies
Surface tension
 demonstration of in model of phagocytosis, 251
 reduction of by bile salts, 215
Survival
 among animals, 540
 conditions for, 433
 and overcrowding. See Overcrowding
 among plants, 499, 540
 of seeds, 13, 540, 572
 See also Adaptation
Suspension
 counting bacteria in, 451–52
 counting cells in, 443–44
Sweep net, 555
Sweet potato
 starch grains in, 165
 vegetative propagation of, 476
Symbiosis (mutualism)
 examples of, 537–39, 538
 of flagellates in termites, 537–38
 food-getting by, 533, 537–39
 lichens as examples of, 537, 643
 of nitrogen-fixing bacteria and legumes, 538
 of Paramecium bursaria and Chlorella, 585
Syracuse dishes, beginning cultures of protozoa in, 579, 581
Syrup, use of for mounting small forms, 115

T₄ bacteriophage, 452
Tadpoles
 chromosomes in tail fin of, 473–74
 circulation in gills and tail of, 246
 collecting of, 612
 embryological development of, 360, 464–66, 465, 483–88, 484, 485, 505
 growth rate of, 463
 metamorphosis of, 613
  effect of iodine on, 325
  effect of thyroxin on, 324–25
 regeneration of tails of, 432, 483, 489

Taenia solum (tapeworm), as parasite, 535
Talks, by students, 11
Tallquist scale, to determine hemoglobin content of blood, 252
Tannic acid content of ink, 299n
Tannic acid salve, 689
Tapeworms, 535, 560
Taste, sense of, 322–23
Tatum, E. L., experimental work of on Neurospora, 221, 400, 401, 402, 435
Taxes
 use of earthworms to study, 594
 use of Enchytraeus to study, 594
 use of Hydra to study, 592
 use of isopods to study, 600
 use of land snails to study, 595
 use of planarians to study, 593
 See also Responses
Taxidermy, 674
Teachers. See Science teachers
Tegosept M (preservative), 415n
Temperature
 for algae, 622
 centigrade-Fahrenheit conversion of, 750
 effect of on activity of amylase, 211–12
 effect of on activity of diastase, 209
 effect of on activity of ptyalin, 210
 effect of on aerobic respiration rate in germinating seeds, 274
 effect of on bacterial growth, 450, 490
 effect of on capillaries in frog's webbed foot, 247
 effect of on development of eggs, 463
  of frog, 359–60
 effect of on fermentation, 278
 effect of on growth of Chlorella, 544
 effect of on heartbeat of frog, 244
 effect of on pancreatic digestion, 215
 effect of on pepsin digestion, 114
 effect of on photosynthesis of Chlorella, 544
 effect of on respiration rate, 284
 for protozoa cultures, 579
 response of skin receptors to, 322
 of soil, 526–27
Temporary slides. See Slides
Tenebrio
 culturing of, 603

as food for fish, amphibia, and reptiles, 603
as food for snakes, 616
larvae of, 603
 as food for Dionaea, 652
 as food for frogs, 614
 as food for red eft, 612
 as food for salamanders, 611
 as food for turtles, 614
 mouth parts of, 202
 obtaining, 603
 metamorphosis of, 603
 pupae of, 603
 use of in classroom, 603
Tent caterpillar
 egg masses of, 354
 nest of, 356
Termites
 colony of, 605
 digestion in, 227, 537, 538
 symbiotic flagellates of, 537–38
Terracing, soil
 use of blackboard to show, 525
 use of on school grounds, 547
 usefulness of, 569
Terrarium
 bog, 548, 648
 desert, 648
 for ferns, 648
 for mosses and liverworts, 644
 preparation of, 548
 student curators of, 684
 woodland, 648
Test cross, to determine genotype, 411, 417, 436
Test tubes
 filling of with culture media, 633–34
 making microaquarium in, 265, 266
 sealing of, 265, 266
Testes
 of birds, 332
 of chickens, 363
 of fish, 357
 of frogs, 62–63, 64, 358
  extraction of DNA from, 143
 of grasshoppers, spermatogenesis in, 348
 of mammals, 366
Test(s)
 for chloride, 217, 219
 for fats, 219
 for food nutrients, 217–21, 227
 for mineral requirements, 222
  of rats, 462–63
 for minerals, 219
 for proteins, 218–19
 for riboflavin, 221
 for starch, 218
 for sugars, 218

Test(s) (cont.)
for vitamin B$_{12}$, 462
for vitamin C, 220–21, 227
for vitamin requirements
of pigeons, 220
of plants, 221–22
of rats, 219–20
for water, 219
Tetrahymena
culturing of, 586
microscopic observation of, 100, 101
silver impregnation of, 126
Tetrazolium salts, use of to identify viable seeds, 457
Thalassemia factor, inheritance of, 390
Thermobarometer, 178, 179, 281–82
Thiamine
daily requirements for, 223
need of by Phycomyces blakesleeanus, 221
testing for, 222
See also Vitamin B$_1$
Thigmotaxis, 300
Thigmotropism
in climbing vines, 304
in Mimosa, 305
Thistle tube
handling of, 692
use of in experiment showing osmosis, 205–06
use of to measure volume of carbon dioxide produced, 276
Thymol blue
pH range of, 662
preparation of, 664
Thymus glands, extraction of DNA from, 143
Thyrodopteryx efemoriformis, 354–55, 357
Thyroid gland, 324
Thyroidectomy, of rat, 324
Thyroxin, feeding of to frog tadpoles, 324–25
Thyroxin solution, 325
Thyroxin–wheat powder, 325
Ticks
collecting of, 561
preservation of, 561, 565
Time, in hydrolysis of starch, 212
Tissue cells, study of, 105–11, 106, 108, 109
Tissues
clearing of, 128–29
dehydration of, 128–29
embedding of, 128–29, 130
fixing of, 126–28
freehand sectioning of, 113

of Hydra, 109, 592
maceration of, 113–14
mounting of on 35-mm film, 675–76
staining of, 129–32
Titration
to measure amount of carbon dioxide produced, 268–69, 270, 271, 274–75
to measure amount of oxygen dissolved in water, 284–85
to measure enzyme reactions, 144, 148
Toads
anatomy of. See Frogs, anatomy of
eggs of, 613
ingestion by, 203
Tobacco, heredity in, 397–98
Toluene-thymol solution, 666
Tongue, sense receptors in, 322–23
Tools, handling of by students, 690, 692
Topfer's reagent, pH range of, 662
Topsoil, examination of, 14, 521, 524
Touch
localization of, 321–22
response to, 304–05
Touch-me-not (Impatiens), use of to study fibrovascular bundles, 77
TPN (triphosphopyridine nucleotide), role of in photosynthesis, 158, 159
TPNH (reduced TPN), production of, 156, 262
Tracers, radioactive, 233–34
and plant nutrition, 228
use of to study photosynthesis, 157–59
use of to trace iodine intake by thyroid gland, 256
Tradescantia
cyclosis in cells of, 107
flower of, 84
geotropism in, 301–02, 337
stomates in, 175
Traits
autosomal, 389–93
in Drosophila, 420–21
in human beings, 389–94
inheritance of, 12–13, 434
table of, in human beings, 392
sex-linked
in Drosophila, 412, 418–20, 419
in human beings, 388–89
Transfer pipettes, 487, 679
making of, 488, 678–79
Transparencies, overhead, 675

Transpiration
demonstrations showing, 232–41, 233, 235–41 257–58
lifting power of, 238–40, 239
measurement of, 236–38, 237
weight loss in leaf due to, 236, 237
Transpiration-cohesion-tension theory, 231–32
Transplanting
in Acetabularia, 404, 406
of embryonic tissue, 483, 484–88, 486
in frogs, 484–85, 486
Trees
loss of water from twigs of, 236, 240, 241
succession of, 570
uses of, 526
Tribolium confusum, 602–03
Tricaine methanesulfonate solution, for anesthetization, 112
Trichinella (roundworm), use of to study parasite-host relationship, 28, 30
Triturus viridescens. See Diemictylus viridescens
Tropeolin O, pH range of, 662
Tropical fish, 609
Tropisms, 337, 454
See also Responses
Trypsin, digestion of proteins by, 214–15
Tryptophan reaction, 136
Tubers
growing of, 476
production of carbon dioxide by, 277
regeneration by, 476
See also Vegetative propagation
Tubifex (segmented worm)
anatomy of, 35, 36
anesthetization of, 35
collecting of, 594
cutting of, 482
as food for fish, 609
as food for Hydra, 22, 591
as food for salamanders, 611
as food for water newts, 612
heartbeat in, 245, 258
maintaining of, 595
mounting specimens of, 35
obtaining, 594
phototaxis in, 299
regeneration in, 482
responses of, 299
study of, 35, 95, 595
Tulip bulbs, 476
Turbatrix aceti (vinegar eel), 28, 31
Turbidity, use of in counting cells, 445–46
error in, 447

Turgor, in guard cells, 175–76
Turgor movement, in *Mimosa*, 304–05
Turtles, *615*
  anatomy of, *66, 67*
  dissection of, 66, 67
  feeding of, 614–15
  maintenance of, 614–16
  organ systems of, 66, 67
  sex identification of, 615
  vivarium for, 614
Tutoring squads, 688–89
2,4-D (2,4-dichlorophenoxyacetic acid), as herbicide, 328
2,6-dichlorophenol-indophenol
  preparation of, 661
  reduction of by illuminated chloroplasts, 187
Tyrannosaurus rex, 497

U-tube
  use of as manometer, *678*
  use of to measure water loss from leaves, *238*
  use of to show lifting power of leaves, *239*
  use of to show oxygen consumed by germinating seeds, 281
  use of to show response of protozoa to electricity, 299
Ulothrix
  microscopic observation of, *103,* 104
  study of, 629
  where to find, 554–55
Ulva, alternation of generations in, 345, *346*
Unialgal cultures, 624
Uranium nitrate, use of to trace circulation in stem, 233
Urea
  action of urease on, 294
  as source of nitrogen for *Chlorella*, 443*n*
  transformation of into ammonia, equation for, 290
Urease
  action of, 290–91, 294
  and transformation of urea into ammonia, equation for, 290
Urethane foam, 668*n*
Urethane solution, for anesthetization, 112, 113
Urine, 292
Urodeles
  gastrulae of, 483
  larvae of, 473–74
  operating solution for, 486
Urosalpinx cinerea (oyster drill), *566*
U.S. Atomic Energy Commission, as source of radioactive isotopes, 234

Uterus
  of pig, dissection of, *366–67*
  types of, *367–69*
Utricularia, as insectivorous plant, 539, *541*

Vacuoles, microscopic observation of, 123
Vacuum bottles, substitution for, 286
Vallisneria
  cyclosis in, 106
  use of in aquarium, 606, *607*
Valves, in heart, 248, *249*
Variations
  in adaptation, *509*
  in animals, 435, 570, 572
  among birds, *509*
  demonstrations of, 427–28
  among frogs, 504, 507
  among human beings, 504
  among mollusks, 503–04
  among moonsnail shells, *427*
  in plants, *500*–02, 516, 570, 572
  within species, 503–04, 507
  in structure, 509
  types of, 427–28
  See also Mutations
Vasodilator effects, of histamine, 248
Vaucheria
  culturing of, 626
  reproduction in, 370, *372*
  study of, 629
  where to find, 554–55, 626
Vegetable juice, testing of for vitamin C, 220–21
Vegetative propagation
  methods of, *474–79, 475–78, 480* 548
  preparing examples of, 13
  value of, 434
  of woody plants, 489
  See also Regeneration
Veins
  circulation in, in frog's tongue, 247
  valves in, 248, *249*
Venae cavae, 245
Ventricles, 245
Venules
  in frog's foot, 246
  in goldfish's tail, 246
Venus flytrap, 539–*40*
  culturing of, 652
  leaves of, 77
  responses of, 305
  thigmotropism in, 305
Venus mercenaria (marine clam), *566*
Vermiculite, use of to germinate seeds, 652
Vessels, blood, 245, 248–*49*

Vestigial structures, 503, *505,* 516
Villus, model of, 249, 667
Vinegar eels
  anatomy of, *31*
  collecting of, 593
  culturing of, 593
  feeding of, 593
  as nonparasitic roundworm, 28
  uses of in classroom, 593–94
Viosterol, use of in experimental diet, 219
Visual aids, 666–67
Visual receptors, absence of in blind spot, 322
Vital stains, 121–22
Vitamin A, daily requirements for, 223
Vitamin B_1
  effect of on pigeons, 220
  See also Thiamine
Vitamin B_{12}, assay of using *Euglena,* 462
Vitamin C
  daily requirements for, 223
  destruction of by bicarbonate of soda, 221
  test for, 220–21, 227
Vitamin D
  daily requirements for, 223
  effect of on rats, 219–20
Vitamins
  assay of, use of *Euglena* to study, 462, 629
  in foods, 219–22
  requirements for, of plants, 221–22
Vitreous humor, 319
Vivarium
  for frogs, *611,* 613–14
  for lizards and small alligators, 615
  for nonpoisonous snakes, 616
  principle for construction of, 578
  for salamanders, *611*–12
  for turtles, 614
Vöchting, H., experiments of on polarity, 479, 502
Vogt, W., experiments of on fate maps in frog's eggs, 362, 483
Volatile liquids, handling of, 691
Volume, gain in as measure of growth, 448
Volumeter
  principle of, 179
  use of to measure carbon dioxide production, 277
  use of to measure oxygen consumption, *281*–82, *283*
  use of to measure oxygen produced during photosynthesis, 178, 179, 180